D1376939

# Progress in Inorganic Chemistry

## Volume 55

## Advisory Board

# PROGRESS IN INORGANIC CHEMISTRY

*Edited by*

KENNETH D. KARLIN

DEPARTMENT OF CHEMISTRY
JOHNS HOPKINS UNIVERSITY
BALTIMORE, MARYLAND

VOLUME 55

WILEY-INTERSCIENCE
A John Wiley & Sons, Inc., Publication

Published by John Wiley & Sons, Inc., Hoboken, New Jersey
Published simultaneously in Canada



For general information on our other products and services or for technical support, please contact our Customer Care Department within the United States at (800) 762-2974, outside the United States at (317) 572-3993 or fax (317) 572-4002.

Wiley also publishes its books in a variety of electronic formats. Some content that appears in print may not be available in electronic formats. For more information about Wiley products, visit our web site at www.wiley.com.

Wiley Bicentennial Logo: Richard J. Pacifico

Library of Congress Catalog Card Number 59-13035
ISBN 978-0-471-68242-4

Printed in the United States of America

10 9 8 7 6 5 4 3 2 1

# Contents

# CHAPTER 1

# Elucidation of Electron- Transfer Pathways in Copper and Iron Proteins by Pulse Radiolysis Experiments

**OLE FARVER**

*Institute of Analytical Chemistry, University of Copenhagen, 2100 Copenhagen, Denmark*

**ISRAEL PECHT**

*Department of Immunology, The Weizmann Institute of Science, 76100 Rehovot, Israel*

CONTENTS



*Progress in Inorganic Chemistry, Vol. 55* Edited by Kenneth D. Karlin

# I. INTRODUCTION

## A. Biological Electron Transfer

Electron-transfer (ET) reactions play a central role in all biological systems ranging from energy conversion processes (e.g., photosynthesis and respiration) to the wide diversity of chemical transformations catalyzed by different enzymes (1). In the former, cascades of electron transport take place in the cells where multicentered macromolecules are found, often residing in membranes. The active centers of these proteins often contain transition metal ions [e.g., iron, molybdenum, manganese, and copper ions] or cofactors as nicotinamide adenine dinucleotide (NAD) and flavins. The question of evolutionary selection of specific structural elements in proteins performing ET processes is still a topic of considerable interest and discussion. Moreover, one key question is whether such structural elements are simply of physical nature (e.g., separation distance between redox partners) or of chemical nature (i.e., providing ET pathways that may enhance or reduce reaction rates).

Biological ET is characterized by the use of redox centers that are spatially fixed in macromolecules and thus separated by the protein matrix and usually prevented from coming in direct contact with solvent. Therefore intramolecular ET becomes a central part of the biological function of the redox proteins, and the ET rates are expected to decrease exponentially with the separation distance of the redox centers that is generally quite large ($>1.0$ nm). Further, usually only very modest structural changes of the active centers accompany the redox changes, thus minimizing the activation energy required for ET.

During the last decades there has been a remarkable progress in determination of three-dimensional (3D) structures of proteins, including many involved in ET, which has opened the way for a detailed study of the relationship between structure and reactivity of this large and diverse group of molecules (2).

A key challenge in studies of biological redox processes is trying to define and understand the parameters that control the rates of ET. These parameters include (a) driving force (i.e., change in free energy of the reaction); (b) the reorganization energy (i.e., the energy of the reactants at the equilibrium nuclear configuration of the products); (c) the distance separating electron donor and acceptor; and finally (d) the nature of the medium separating the two redox centers.

This chapter reviews results and current insights emerging primarily from pulse radiolysis (PR) studies of intramolecular ET in multisite proteins, mainly iron- and copper-containing redox enzymes, with emphasis on interactions between the different redox centers.

## B. Electron-Transfer Theory

Rates of ET are expected to depend on the energy required for bond-length and bond-angle changes of the reactants, as well as solvent reorganization accompanying the ET process. In proteins, however, these processes involve the polypeptide matrix as part of the medium that is far less homogeneous than solvent molecules surrounding small molecule ET partners. Further, conformational changes preceding or following ET in macromolecules may affect the free energy changes of the reaction. Moreover, while small molecules exchange electrons in solutions where they are in close, sometimes direct contact, in proteins the redox reaction partners are held in fixed positions by and within the polypeptide matrix. Hence, they are prevented from coming into direct inner-sphere contact. Therefore, the distance between electron donor and acceptor is one decisive parameter affecting long-range ET (LRET) rates. Considerable efforts have been devoted to studies of these processes, and the comprehension of intra- and interprotein mediated ET reactions has advanced significantly, largely due to the determination of 3D structures of an ever-increasing number of redox proteins. In addition, the theoretical models for analyzing LRET have also advanced to a stage where they can more readily be employed and tested experimentally. Still, some very interesting questions remain to be answered: (1) How does the ET rate depend on the nature of the medium separating the two redox partners in a protein system? (2) To what extent did the structures of redox proteins undergo evolutionary selection in order to optimize their function for specificity and control of biological ET, and if so, what are the structural corollaries of this selection? These are some of the issues that will be addressed in this chapter.

Several excellent reviews on ET theory are available (3–5). Here only an outline, necessary for the discussion of this topic, is presented. Long-range ET in proteins is characterized by a weak interaction between electron donor, D, and acceptor, A, and in the nonadiabatic limit, where the D–A distance is large (>1.0 nm), the rate constant is proportional to the square of the electronic coupling between the electronic states of reactant and product, represented in the form of a tunneling matrix element, $H_{DA}$. For intramolecular ET the rate constant is given by Fermi's golden rule (3):

$$k = \frac{2\pi}{\hbar} H_{DA}^2 \cdot (FC) \tag{1}$$

The Franck–Condon factor (FC) for nuclear movements can for relatively small vibrational frequencies, where $k_B T > h\nu$, be treated classically, a condition that often applies to biological ET. In polar solvents like water, the reorientation of solvent molecules contributes considerably to the total reorganization energy, $\lambda_{tot}$, in response to changes in charge distribution of the reaction partners. The reorganization energy can be defined as the energy of the reactants at the equilibrium nuclear configuration of the products (3, 4). If D and A are viewed as conducting spheres, the dielectric model illustrates an important feature, namely, that the more polar the medium, the larger becomes $\lambda_{tot}$ (3). Therefore, reorganization energy requirements are expected to decrease dramatically when the redox centers reside in a low-dielectric medium, (e.g., the hydrophobic interior of a protein). In a nonpolar environment, the reorganization energy requirement for the surrounding medium vanishes. Another contribution to the reorganization energy comes from changes in bond lengths and angles of the coordination sphere that accompany ET. The nuclear factor also expresses the relationship between reorganization energy and driving force that further influences the ET rate (cf. Eq. 2). The other factor included in the FC term is the driving force of the ET reaction that is given by the difference in reduction potentials of electron acceptor and donor. These potentials are rather sensitive to the structure and environment of the metal site and may thus be tuned by subtle conformational changes.

The electronic motion, however, requires a quantum mechanical approach, and the semiclassical Marcus equation may be expressed as (3):

$$k = \frac{2\pi}{\hbar} \frac{H_{DA}^2}{(4\pi\lambda_{tot}RT)^{\frac{1}{2}}} e^{-(\Delta G^\circ + \lambda_{tot})^2/(4\lambda_{tot}RT)} \tag{2}$$

where $\Delta G^\circ$ is the reaction free energy and $\lambda$ the nuclear reorganization energy. Since wave functions decay exponentially with distance, the tunneling matrix element, $H_{DA}$ will decrease with the distance, $(r - r_0)$, as:

$$H_{DA} = H_{DA}^0 \cdot e^{-\beta(r - r_0)/2} \tag{3}$$

where $H_{DA}^0$ is the electronic coupling at direct (van der Waals) contact between electron donor and acceptor (where $r = r_0$) and the decay rate of electronic coupling with distance is determined by the coefficient, β. In LRET, there is no direct electronic coupling between D and A. Instead, the coupling is mediated by the electronic states of the intervening atoms via superexchange. Both theoretical and experimental studies demonstrate that the chemical nature and structure of the protein medium separating electron donor and acceptor must be included in a theoretical analysis of the electronic coupling, $H_{DA}$, of the redox centers. In one theoretical treatment that has proven to be very useful (6, 7), the protein medium is divided into small elements linked by covalent bonds, hydrogen bonds and through-space contacts, and each type of link is then assigned a coupling decay factor (with distances in nm):

$$\begin{aligned} \varepsilon_C &= 0.6 \\ \varepsilon_H &= \varepsilon_C^2 \cdot \exp[-17(r - 0.28)] \\ \varepsilon_S &= 1/2\varepsilon_C \cdot \exp[-17(r - 0.14)] \end{aligned} \tag{4}$$

for covalent (*C*), hydrogen bonded (*H*), and van der Waals interaction (*S*). The ET pathways may now be identified by analyzing the bonding interactions that maximize $H_{DA}$, which now can be expressed as:

$$H_{DA} = P \cdot \prod_i \varepsilon_{C(i)} \prod_j \varepsilon_{H(j)} \prod_k \varepsilon_{S(k)} \tag{5}$$

The prefactor, *P*, depends on the electronic coupling of donor and acceptor with the bridging orbitals. A correlation between β (cf. Eq. 3) and $\varepsilon_C$ (Eq. 4) is easily demonstrated: An individual strand of a β-sheet protein defines a linear tunneling pathway along the peptide, spanning a distance $(r - r_0)$ of 0.34 nm per residue (three covalent bonds). Thus, inserting a β value of 10 $nm^{-1}$ in Eq. 3, the decay factor, $\varepsilon_C$ of Eq. 4 becomes 0.6 per covalent bond.

The activation enthalpy is related to the thermodynamic parameters (3):

$$\Delta H^{\neq} = \frac{\lambda}{4} + \frac{\Delta H^\circ}{2}\left(1 + \frac{\Delta G^\circ}{\lambda}\right) - \frac{(\Delta G^\circ)^2}{4\lambda} \tag{6}$$

The entropy of activation includes a contribution from the distance dependence of the electronic coupling (3) (cf. Eq. 3):

$$\Delta S^{\neq} = \Delta S^* - R\beta(r - r_0) \tag{7}$$

where β is the electronic coupling decay factor, and $\Delta S^*$ is related to the standard entropy change, $\Delta S^\circ$ (3):

$$\Delta S^* = \frac{1}{2}\Delta S^\circ(1 + \Delta G^\circ/\lambda) \tag{8}$$

Particularly β-sheet proteins, being composed of extended polypeptide chains interconnected by hydrogen bonds, give rise to coupling pathways along the peptide backbone. Many studies have demonstrated that the distance decay constant, β, is $\sim 10\,\text{nm}^{-1}$ (8). Using edge-to-edge separation distances between ligands involves some ambiguity since it is often difficult to define the atoms that comprise the edges of a donor and an acceptor. Experimental evidence now supports the notion that metal–metal distances are more appropriate (4) and this is the length scale employed here. In conclusion, the folded polypeptide, which constitutes the scaffold for the metal ion coordination sphere and provides the path for electron tunneling, plays the major role in determining the thermodynamic and electronic properties, and hence ET reactivity of a given pair of redox centers.

## II. PULSE RADIOLYSIS

The main method employed in our studies described here is PR (9). Introduced and developed in the early 1960s, PR has found a broad range of important applications in both chemistry and biochemistry. Of considerable significance and interest are PR studies of ET processes in proteins. The method is based upon excitation and decomposition of solvent molecules by short pulses (typically 0.1–1 μs) of high energy (2–10 MeV) accelerated electrons yielding several primary products that can be employed for induction of additional reactions. Thus, though PR is essentially a higher energy analogue of flash photolysis, the latter method uses photoexcitation of specific solutes rather than bulk solvent, which distinguishes the two methods and provides the former with some clear advantages. For example, as the solvent is the source of the reactive species and no chromophore is required, essentially any reaction partner for the radicals formed can be chosen.

Different types of electron accelerator systems have been adopted for pulse radiolysis (10). In biochemical studies, pulsed electron accelerators have most often been used; these instruments produce short pulses of accelerated electrons with adequate energy to ensure uniform irradiation of the solution. The most common and versatile detection system is optical absorption, although other techniques have also been applied, including electrical conductivity, resonance Raman spectroscopy, or electron paramagnetic resonance (EPR) (11). When using optical detection, the analytical light beam is directed through the sample in a quartz cuvette perpendicular to the electron beam. After passing through the irradiated solution, the light beam is guided through a system of mirrors and lenses to a monochromator isolated from the radiation zone by a protective wall. Light of a selected wavelength then reaches the photomultiplier and the signal is finally transferred to a computer system via an analog–digital converter for further processing.

Introducing the electron pulses into dilute aqueous solutions under anaerobic conditions causes, as stated above, primary changes in the solvent (12). In such experiments, water molecules undergo conversion mainly into OH radicals and hydrated electrons [$e^-(aq)$] and, to a lesser extent, H atoms, $H_2$ and $H_2O_2$ molecules.

Yields of the reaction products are usually presented as *G* values giving the number of chemical species produced per 100 eV of absorbed energy: $G(e^-(aq)) = 2.9$; $G(OH) = 2.8$; $G(H) = 0.55$; $G(H_2) = 0.45$; $G(H_2O_2) = 0.75$ (12).

The hydrated electron and the hydroxyl radical are exceptionally reactive and present thermodynamic extremes of reducing and oxidizing potentials, respectively. These primary products, though having their own applications, are usually converted into less reactive and more selective agents using protocols devised by radiation chemists (12). Hence, they provide the possibility of inducing a wide range of ET processes, as illustrated by one useful procedure transforming $e^-(aq)$ [with a reduction potential, $E° = -2.8$ V vs. standard hydrogen electrode (SHE)] into a milder reductant, the $CO_2^-$ radical ($E° = -1.8$ V vs. SHE) (13). First, the hydrated electrons are converted into an additional equivalent of OH radicals by the following reaction in $N_2O$ saturated solutions:

$$e^-(aq) + N_2O + H_2O \rightarrow N_2 + OH + OH^- \quad (9)$$

Second, the 2 equiv of OH radicals, as well as the hydrogen atoms, then react with formate anions to produce the $CO_2^-$ radical in a diffusion-controlled process:

$$HCO_2^- + OH/H \rightarrow H_2O/H_2 + CO_2^- \quad (10)$$

By analogy, other reducing and oxidizing radicals can be produced by similar protocols. Thus, uncharged 1-methylnicotinamide radicals (1-MNA*), which are formed by the following reaction sequence, have also been employed: The OH radicals are scavenged by *tert*-butanol to produce a relatively inert radical species that hardly reacts with copper- or heme containing proteins:

$$OH + Me_3OH \rightarrow H_2O + \cdot CH_2Me_2OH \quad (11)$$

1-methylnicotinamide chloride ($1\text{-}MNA^+ Cl^-$) reacts with solvated electrons to produce 1-MNA* with a reduction potential, $E^0 = -1.0$ V vs. SHE (13):

$$1\text{-}MNA^+ + e^-(aq) \rightarrow 1\text{-}MNA^* \quad (12)$$

The choice of radicals to be employed for reaction with a certain protein is based on its reactivity and specificity that in turn is determined by reduction potentials and physical properties (e.g., accessibility and electrical charge).

Nonfunctional redox centers in exposed areas of proteins present an interesting target; cystine disulfide residues in particular efficiently compete for reducing $CO_2^-$ radicals to produce disulfide radical anions, a reaction that has been utilized in several cases, yet may also be a problem as it may disfavor designed reactions with functional redox centers. An additional advantage of the PR method over flash photolysis is noteworthy: Usually, the whole spectral range is available for monitoring induced ET reactions, as stated above because the reactive species are solvent derived rather than from a chromophore of the solute. The combination of a wide range of reactivity of the produced reagents with time resolution that extends from nanoseconds to minutes along with convenient spectroscopic monitoring has made the PR technique highly useful in studies of a wide range of chemical and biochemical ET processes.

Finally, the PR method also enables to perform systematic titrations of a given protein by e.g. sequential introduction of reduction equivalents. Adding a series of pulses was found to be of considerable importance in the study of multisite enzymes, where the distribution and rates of ET depend on their degree of reduction (see below).

The potential of pulse radiolysis for studies of biological redox processes was recognized many years ago (14). However, it was initially employed for studies of radiation damage and only later on was it shown to be an effective tool for investigating ET processes to and within proteins. A great advantage in studies of biochemical redox processes is the capability of the PR method to produce the reactive (reducing or oxidizing) species *in situ* and almost instantaneously (i.e., on time scales relevant to those of biological processes). Depending on pH, concentration, and choice of scavengers, the primary aqueous radicals produced by the radiation pulse can be converted into particular inorganic or organic reagents that are appropriate for the reaction to be investigated. Two main interests have guided PR studies of redox proteins: First, elucidation of reaction mechanisms of these proteins, and, second, resolving the parameters that determine the rates of ET within proteins. Obviously, these complement each other. The fast progress attained during the last two decades in resolving 3D structures of a large number of redox active proteins has provided insights that are essential for a meaningful analysis and interpretation of the kinetic results derived from PR studies.

## III. COPPER PROTEINS

### A. Azurins, a Model System

Azurins are single copper proteins that function as electron mediators in the energy conversion systems of many bacteria (15). While azurins isolated from distinct bacteria are highly homologous, subtle sequence differences do exist, conferring upon these proteins some variation in reactivity and redox potentials. Azurins have a characteristic β-sandwich structure (16) and contain a single disulfide bridge [Cys3–Cys26] at one end of the molecule separated from the blue (or type1, T1) copper ion by a distance of 2.6 nm (Fig. 1).

$CO_2^{-}$ radicals reduce azurin either at the Cu(II) site (followed at 625 nm; $\varepsilon = 5,700\,M^{-1}\,\mathrm{cm}^{-1}$) or the disulfide center (followed at 410 nm; $\varepsilon = 10,000\,M^{-1}\,\mathrm{cm}^{-1}$) (Eqs. 13 and 14) with similar, nearly diffusion controlled rates (17). In those molecules where disulfide radicals were formed, they were found to decay by reducing the Cu(II) ion via intramolecular ET (Eq. 15).

$$\mathrm{RSSR{-}Az(Cu^{II})} + \mathrm{CO_2^{-}} \rightarrow \mathrm{RSSR{-}Az(Cu^{I})} + \mathrm{CO_2} \quad (13)$$

$$\mathrm{RSSR{-}Az(Cu^{II})} + \mathrm{CO_2^{-}} \rightarrow \mathrm{RSSR^{-}{-}Az(Cu^{II})} + \mathrm{CO_2} \quad (14)$$

$$\mathrm{RSSR^{-} - Az(Cu^{II})} \rightarrow \mathrm{RSSR{-}Az(Cu^{I})} \quad (15)$$

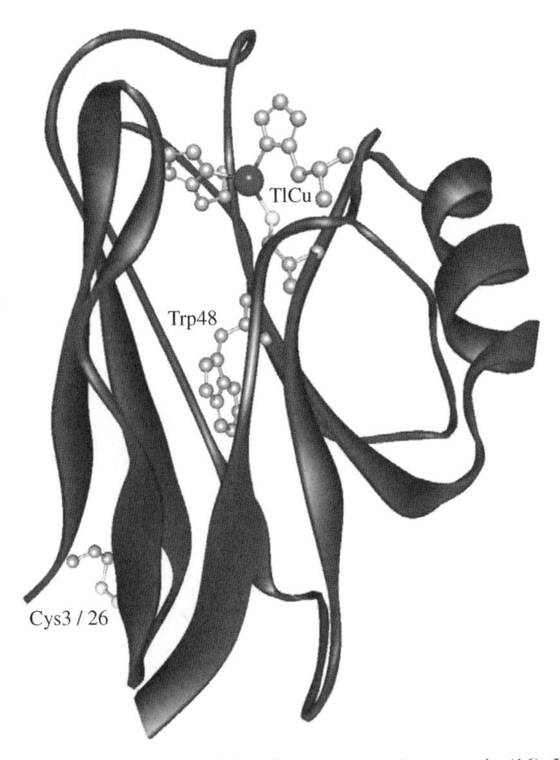


Figure 1. Three-dimensional structure of *Pseudomonas aeruginosa* azurin (16). In addition to the protein backbone, the side chains of three copper ligating residues, His46, His117, and Cys112 are shown near the top together with the disulfide bridge (bottom) and Trp48 (center). Coordinates were taken from the Protein Data Bank (PDB), code 4AZU. (See color insert.)

Examples of time-resolved absorption changes occurring upon reaction of azurin with $CO_2^-$ radicals are illustrated in Fig. 2. Following a fast, direct bimolecular reduction of the two redox active centers at either 625 nm (Eq. 13) or 410 nm (Eq. 14), a slower, concentration independent, unimolecular process takes place

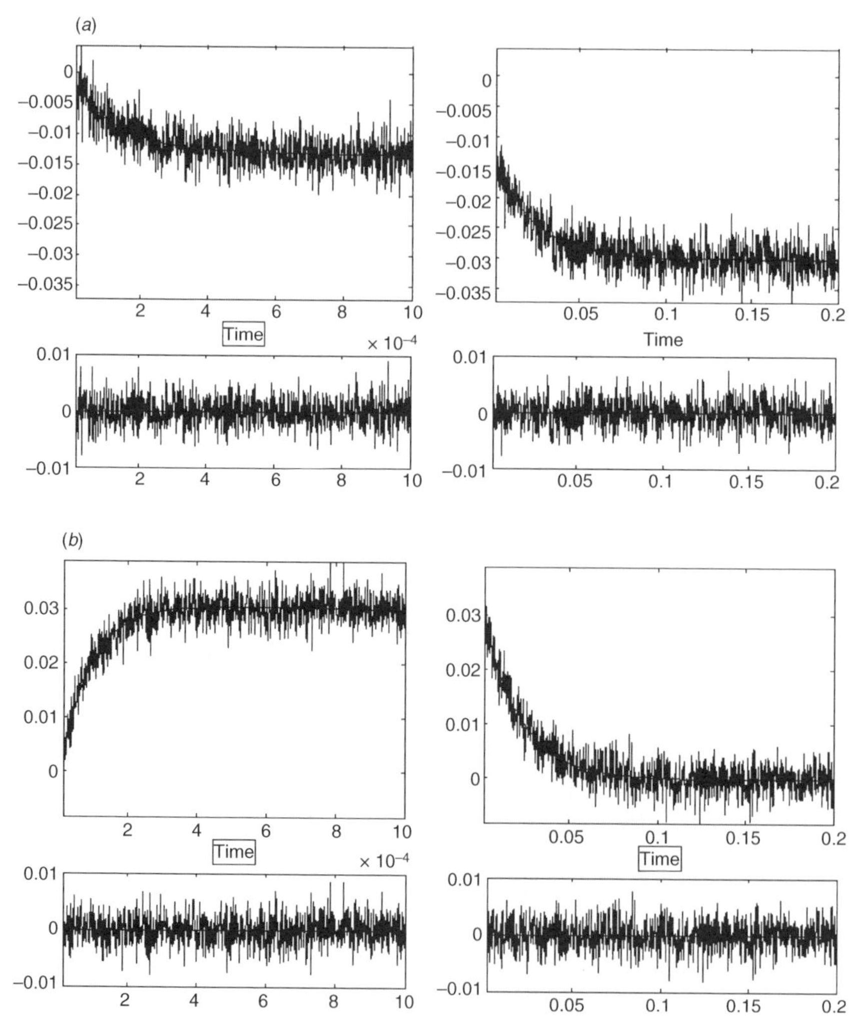


Figure 2. Time-resolved absorption changes induced by reaction of pulse radiolytically produced $CO_2^-$ radicals with *P. aeruginosa* azurin. (*a*) Reduction of Cu(II) followed at 625 nm. (*b*). Formation and decay of the disulfide radical anion measured at 410 nm. Protein concentration is 10μ*M*, where $T = 298$ K; pH 7.0; 0.1 *M* formate; 10 m*M* phosphate; $N_2O$ saturated; pulse width 0.4 μs optical path 12.3 cm. Time is in seconds; the left panel shows the faster phase, while the right one shows the reaction taking place at the slower phase. The lower panels show the residuals of the calculated fits to the data.

attributable to electron transfer from $RSSR^-$ to T1Cu(II) (Eq. 15). For wild-type *P. aeruginosa* azurin the rate of intramolecular ET is $k_{ET} = 44 \pm 7\,s^{-1}$ at pH 7.0 and 25°C at a driving force, $-\Delta G^\circ = 68.9\,kJ\,mol^{-1}$ (17). From temperature-dependence studies of the internal ET reaction the activation enthalpy and activation entropy were determined to be, $\Delta H^{\neq} = 47.5 \pm 4.0\,kJ\,mol^{-1}$; $\Delta S^{\neq} = 56.5 \pm 7.0\,J\,K^{-1}\,mol^{-1}$, respectively. Intramolecular ET between the Cys3/Cys26 disulfide radical anion and the blue Cu(II) center was further studied in a large number of both wild type and single site azurin mutants (17–24). Rates and activation parameters derived from these studies are given in Table I. A linear relationship between activation enthalpy and entropy for a series of homologous reactions is often found (25, 26); and, a plot of the activation enthalpy against the activation entropy for a range of both wild-type and single-site azurin mutants (Fig. 3) is linear with a slope of $T_C = 258 \pm 6\,K$ (and a correlation coefficient of 0.99). Such *enthalpy–entropy compensation* is commonly found for closely related reactions, with observed slopes in the 250–315 K range for reactions in aqueous solution (26, 27). Hypotheses explaining enthalpy–entropy compensation include changes in solvent reorganization, particularly in hydrogen bonding solvents (26, 27). While such linear behavior may indeed be related to the properties of solvent water, deviation of certain data from the linear relationship is usually rationalized by a divergent property of that particular reaction. In the following sections, some of the divergent cases will be addressed. As a consequence of the compensation temperature, $T_C$, being close to the experimental temperature range, the observed free energy of activation, $\Delta G^{\neq}$, at 298 K is virtually the same ($63 \pm 4\,kJ\,mol^{-1}$) over the full range of azurins investigated, although a small systematic decrease in $\Delta G^{\neq}$ with decreasing $\Delta S^{\neq}$ is observed and is reflected in a small but steady increase in the ET rate.

The possibility of introducing single-site mutations in azurins enabled a detailed analysis of structure–reactivity relationships where, for example, the impact of specific amino acid substitutions on the rate of intramolecular ET could be investigated. In order to understand better the role of the polypeptide matrix separating electron donor and acceptor on ET reactivity, the structure-dependent theoretical model developed by Beratan et al. (6, 7) was employed to identify relevant ET pathways (cf. Section I.B). In this model, the total electronic coupling of a pathway is calculated as a repeated product of the couplings of the individual links. The optimal pathway connecting the two redox sites, $\prod\varepsilon$, is thus identified (cf. Eq. 5).

Pathway calculations for the aforementioned intramolecular ET reaction were performed using the high-resolution 3D structures of *P. aeruginosa* azurin and its mutants, where available. For other mutants, structures based on two-dimensional nuclear magnetic resonance (2D NMR) studies and energy minimization calculations were employed. The pathway calculations predict similar ET routes in all the azurins shown in Fig. 4: One longer path through the peptide

TABLE I
Kinetic and Thermodynamic Data for the Intramolecular reduction of Cu(II) by $RSSR^-$ in AZURIN; pH 7.0

| Azurin | $k_{298}$ ($s^{-1}$) | $E'$ (mV) | $-\Delta G^\circ$ (kJ $mol^{-1}$) | $\Delta H^{\neq}$ (kJ $mol^{-1}$) | $\Delta S^{\neq}$ ($JK^-$ $mol^{-1}$) |
|---|---|---|---|---|---|
| *Wild Type* | | | | | |
| *P. aeruginosa*[a] | 44 ± 7 | 304 | 68.9 | 47.5 ± 2.2 | −56.5 ± 3.5 |
| *P. fluorescense*[b] | 22 ± 3 | 347 | 73.0 | 36.3 ± 1.2 | −97.7 ± 5.0 |
| *Alc. spp.*[a] | 28 ± 1.5 | 260 | 64.6 | 16.7 ± 1.5 | −171 ± 18 |
| *Alcalegenes faecalis*[b] | 11 ± 2 | 266 | 65.2 | 54.5 ± 1.4 | −43.9 ± 9.5 |
| *Alc. denitrificans* | 42 ± 4 | 305 | 69.0 | 43.5 ± 2.5 | −67 ± 9 |
| *Mutant* | | | | | |
| *D23A*[c] | 15 ± 3 | 311 | 69.6 | 47.8 ± 1.4 | −61.4 ± 6.3 |
| *F110S*[d] | 38 ± 10 | 314 | 69.9 | 55.5 ± 5.0 | −28.7 ± 4.5 |
| *F114A*[e] | 72 ± 14 | 358 | 74.1 | 52.1 ± 1.3 | −36.1 ± 8.2 |
| *H35G*.aq[f] | 15 ± 2 | <300 | <68.5 | 42.1 ± 3.5 | −81 ± 5 |
| *H35Q*[g] | 53 ± 11 | 268 | 65.4 | 37.3 ± 1.3 | −86.5 ± 5.8 |
| *H117G*.aq[g] | 7 ± 3 | <300 | <68.5 | 22.0 ± 3.2 | −155 ± 11 |
| *H117G.im*[h] | 149 ± 17 | 240 ± 20 | 62.7 | 54.5 ± 3.9 | −22 ± 1 |
| *17S*[d] | 42 ± 8 | 301 | 68.6 | 56.6 ± 4.1 | −21.5 ± 4.2 |
| *M44K*[g] | 134 ± 12 | 370 | 75.3 | 47.2 ± 0.7 | −46.4 ± 4.4 |
| *M64E*[d] | 55 ± 8 | 278 | 66.4 | 46.3 ± 6.2 | −56.2 ± 7.2 |
| *M121H*[f] | 21 ± 47 | 215 | 60.3 | 28.0 ± 2.1 | −127 ± 8 |
| *M121L*[d] | 38 ± 7 | 412 | 79.3 | 45.2 ± 1.3 | −61.5 ± 7.2 |
| *V31W*[h] | 285 ± 18 | 301 | 68.6 | 47.2 ± 2.4 | −39.7 ± 2.5 |
| *W48F*[h] | 35 ± 7 | 301 | 68.6 | 46.3 ± 5.9 | −58.3 ± 6.0 |
| *W48F*[h] | 80 ± 5 | 304 | 68.9 | 43.7 ± 6.7 | −61.9 ± 9.7 |
| *W48*[h] | 50 ± 5 | 314 | 69.9 | 49.8 ± 4.9 | −44.0 ± 3.5 |
| *W48Y*[h] | 85 ± 5 | 323 | 70.7 | 52.6 ± 6.9 | −30.2 ± 3.6 |
| *W48L*[e] | 40 ± 4 | 323 | 70.7 | 48.3 ± 0.9 | −51.5 ± 5.7 |
| *W48M*[e] | 33 ± 5 | 312 | 69.7 | 48.4 ± 1.3 | −50.9 ± 7.4 |

[a]Ref. (17).
[b]Ref. (18).
[c]Ref. (21).
[d]Ref. (22).
[e]Ref. (19).
[f]Ref. (24).
[g]Ref. (20).
[h]Ref. (23).

chain to the copper-ligating imidazole of His46, and one shorter path through the buried indole ring of Trp48, necessitating a through-space jump to this residue. The electronic coupling factors were found to be $2.5 \times 10^{-7}$ and $3.0 \times 10^{-8}$, respectively. However, in this analysis, the electronic interaction between the Cu(II) ion and its ligands was not included. It has been demonstrated

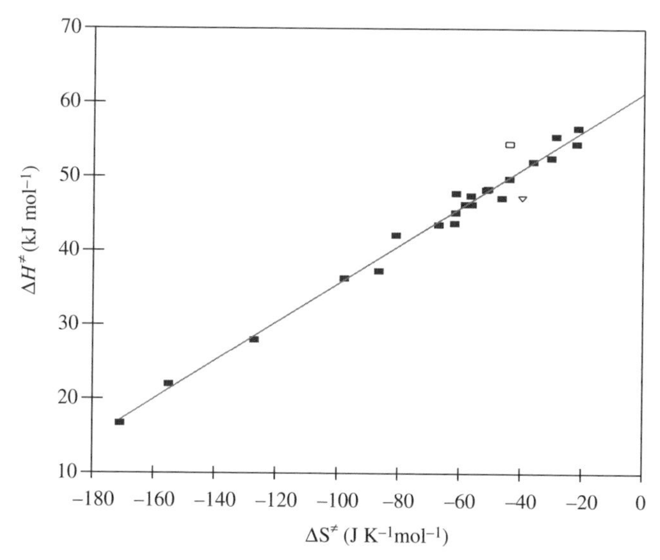


Figure 3. Activation enthalpy–activation entropy compensation plot. Activation enthalpy values determined at 298 K plotted as a function of activation entropy. The straight line is the result of a linear least-squares fit to all data points and has a slope of $258 \pm 6$ K (24). Two points, representing WT *A. faecalis* azurin (□) and the V31W mutant (∇), respectively, are discussed in the text.

that the high degree of anisotropic covalency in the copper coordination site would enhance ET through the Cys112 thiolate ligand (28). By similar arguments, from the ligand coefficients of $\Psi_{\mathrm{HOMO}}$ (HOMO = highest occupied molecular orbital) in azurin calculated by Larsson et al. (29), it can be estimated that ET through the Cys thiolate would be enhanced by a factor of ~150 over ET via one of the His imidazole ligands. The pathway calculations combined with the notion of anisotropic covalency would therefore suggest that the "Trp48" pathway would offer better coupling than one going through His46. Since the same LRET pathway from $RSSR^-$ to Cu(II) applies to all azurins studied so far, it was possible, from the kinetic data and activation parameters to calculate the reorganization energy, $\lambda_{\mathrm{tot}} = 1.0 \pm 0.05$ eV, and the experimental decay factor, $\beta = 10.0 \pm 0.5\ \mathrm{nm}^{-1}$ (22).

In order to probe the possible influence of aromatic residues on internal ET, an investigation was initiated on single-site azurin mutants in which Trp48 had been substituted by other amino acids, with both aromatic and nonaromatic side chains. In the experiments, the rate constants for intramolecular ET were determined as a function of temperature (23). The results are set out in Table I together with the standard free energies of reaction ($\Delta G^{\circ}$), the activation enthalpy ($\Delta H^{\neq}$) and activation entropy ($\Delta S^{\neq}$). It is clear that substitution of

Trp48 by other amino acids only has a small effect on the kinetic parameters after correcting for changes in driving force.

In further studies of this aspect, another mutant was constructed in which Val31 was substituted by Trp, thus producing a double-Trp mutant (V31W azurin) where the two indole rings are placed in neighboring positions (23). The spatial relationship between the two indole rings, in the V31W mutant, was investigated by two-dimensional nuclear overhauser enhancement spectroscopy (2D NOESY) and total correlated spectroscopy (TOCSY) experiments. Two spin systems consisting of four peaks (tryptophans) could immediately be identified from the TOCSY spectra; these systems were assigned to residues 31 and 48 (23). A large number of residues exhibited chemical shift values identical with those of the corresponding residue in the wild-type protein (30). The chemical shifts of the four protons of the Trp48 side chain are within 0.1 ppm of those of the wild-type protein, indicating similar orientation. Thus, the side chain of Trp31 is probably positioned above the plane of the Trp48 indole, since the signals of the Trp31 side chain are upshifted. Both tryptophans have nuclear overhauser effects (NOEs) between their side chains and methyl groups of an isoleucine and a valine, probably Ile7 and Val95. These NOEs put further constraints on the orientation of the Trp31 side chain. The two ring systems are not stacked in a parallel fashion, but they form an oblique angle relative to each other. Thus, the NMR data show that the regions in the mutant located behind Trp 48 (relative to Trp31) have the same structure as the equivalent regions in the wild-type (WT) protein. Energy minimization calculations have also been performed on this mutant and show a close (van der Waals) contact of the two indole rings consistent with the observation of NOEs between the ring protons (23).

The $RSSR^-$ to Cu(II) LRET in the V31W azurin mutant was found to take place with a rate constant of 285 $s^{-1}$ (298 K, pH 7.0, and similar driving force as in WT azurin), which is a considerably faster reaction than for any other azurin studied so far (cf. Table I). The high rate strongly suggests that the main ET route is the Trp48 pathway, since the one through His46 should not be affected directly by this mutation. The activation enthalpy and entropy of this LRET were also examined. The dependence of the activation enthalpy on reorganization energy is given by Eq.6. In azurins, where Trp48 has been exchanged by other amino acid residues, $\Delta H^{\neq}$ is constant within experimental error, consistent with the previous assumption that the reorganization energies do not change significantly in this series.

The entropy of activation, which includes a contribution from the separation distance dependence of the electronic coupling is given by Eq.7. It is seen that the increase in rate in V31W azurin follows from a more favorable entropy of activation (Table I), which is larger by 16.8 $J\,K^{-1}mol^{-1}$ compared with WT azurin. Since $\Delta S^{\circ}$ can safely be assumed to be the same for intramolecular ET in

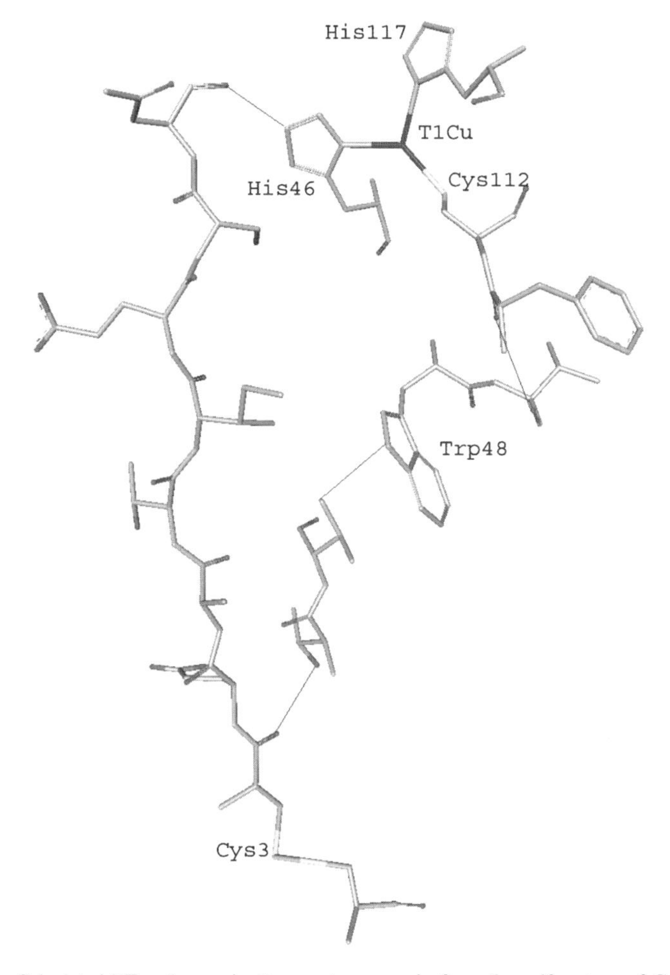


Figure 4. Calculated ET pathways in *P. aeruginosa* azurin from the sulfur atom of Cys3 to the copper ligands His46 and Cys112 applying the Beratan and Onuchic model (6, 7). The left path consists of 27 covalent bonds and one hydrogen bond, while the right pathway includes 21 covalent bonds, two hydrogen bonds, and one van der Waals contact. Noncovalent connections are also shown. Coordinates were taken from the PDB, code 4AZU.

WT and V31W azurins, the increase in entropy would according to Eq.7 correspond to a decrease in $\beta(r - r_0)$ from the previously determined value of 24.6 in WT to 22.6 in V31W azurin. A smaller exponential decay factor, $\beta$, for the mutant is also reflected in the electronic coupling matrix element, $H_{DA}$, between electron donor and acceptor, which was found to be $2.1 \times 10^{-7}$ eV (23), or an improvement of a factor 2.6 relative to WT azurin ($H_{DA} = 0.8 \times 10^{-7}$ eV). In contrast, a calculation of the electronic coupling factor that treats all covalent bonds equally yielded $\Pi\varepsilon = 0.9 \times 10^{-8}$ for V31W azurin as compared with $3.0 \times 10^{-8}$ for WT azurin (19). The relative positions of Trp31 and Trp48 in Val31Trp azurin may enhance the interaction between D and A, since the aromatic ring systems are in van der Waals contact; this close contact (with a large electronic overlap) may give rise to resonance-type tunneling through the two indole rings. Aromatic residues placed in appropriate positions may enhance ET through proteins by providing more effective coupling through their extended $\pi^*$ orbitals, since the energy gap between the tunneling electron and the aromatic $\pi$ system is significantly smaller than that involving $\sigma$ orbitals. A single aromatic residue placed midway between D and A in a predominantly $\sigma$-ET pathway is not advantageous by itself, however, since $\sigma \rightarrow \pi \rightarrow \sigma$ ET will be energetically unfavorable.

However, several aromatic residues placed in successive positions or aromatic molecules in direct contact with either D or A would act as an extended relay that could enhance the electronic coupling (24).

The relevance of the Trp48 route is further confirmed by the pronounced deviation of two data sets from the linear relationship between the activation entropy and enthalpy (cf. Fig. 3), namely for WT *A. faelcalis* azurin (□) and for the *P. aeruginosa* Val31Trp azurin mutant (∇). The WT azurin from *A. f.* is unique among azurins since it has Val at position 48 (instead of a Trp). The absence of the aromatic residue here is noteworthy, since Fig. 3 demonstrates that the point for this particular azurin clearly falls above the line (an unusually slow ET rate), whereas the point for the V31W mutant with two stacked aromatic residues falls below the line (reflecting the fast ET rate).

Aromatic residues have been found in proteins at positions that probably enhance the electronic coupling in systems that have been selected by evolution for efficient ET. Examples are the tryptophan mediated reduction of quinone in the photosynthetic reaction center (31), the methylamine dehydrogenase (MADH): amicyanin system, where a Trp residue is placed at the interface between the two proteins (32), as well as the [cytochrome *c* peroxidase–cytochrome *c*] complex, where a Trp seems to have a similar function (33).

In order to further distinguish between the possible pathways for ET in azurins, several azurin mutants were produced where two of the copper ligating amino acids, His46 and His117, were systematically replaced by a noncoordinating glycine, while the third T1 ligand, Met121, was replaced by a histidine (24). Of course these modifications of the copper-binding site also change other properties of the site (e.g., its redox potential and the charge distribution on the metal ion). An influence of electron delocalization on the electronic coupling between donor and acceptor will appear as an apparent change in the activation entropy (cf. Eq. 7). Indeed, a decrease in electron density on the cysteine was evident in the observed larger EPR hyperfine splitting ($A_z$) of the different mutants, Met121His, His46Gly, and His117Gly (34).

The relevance of the His46 pathway relative to the Trp48 route was investigated in experiments using the aforementioned mutants, where the His ligands had been replaced by glycine (24). The former pathway would be blocked when the linkage to the copper ion via the hydrogen bond from Asn10 to His46 is lost (cf. Fig. 3). However, this mutation does considerably modify the coordination sphere of the copper ion (35), a change that could affect the ET rate. Therefore, in order to investigate the influence of the change in the electronic properties caused by removal of a His ligand and its effect on the ET rate, the kinetics of ET in a similar mutant, *P. aeruginosa* His117Gly azurin, were also studied (24). The His117 provides the other imidazole residue coordinated to the copper ion in azurin. The Cu(II) ion in both *P. aeruginosa* His46Gly and *P. aeruginosa* His117Gly mutants was found to be accessible to external ligands, which upon coordination to the metal ion, obviously perturb its spectroscopic features (35–37).

When imidazole (im) is added in sufficient concentration to coordinate the copper ion in *P. aeruginosa* azurin mutants His46Gly or His117Gly, the green mutants turn blue with absorption bands at 628 nm (His117Gly · im) and 621 nm (His46Gly · im), which are close to that of WT-azurin (626 nm) (35, 36). Other spectroscopic features, including those from EPR, electron–nuclear double resonance (ENDOR), and resonance Raman measurements, are restored, implying that the structures of the mutants, specifically metal-binding site geometries, are maintained despite the replacement of histidine by imidazole.

Intramolecular ET in His117Gly · im azurin is considerably faster than in WT *P. aeruginosa* azurin in spite of the lower driving force (Table I). Since their activation enthalpies are the same within experimental error, the reorganization energy is probably not changed significantly (cf. Eq. 6). In spite of the large difference in ET rates (cf. Table I), the points for His117Gly · aq, His46Gly · aq, His117Gly · im, and WT *P.a.* azurin fit perfectly on the free energy ($\Delta H^{\neq}$ vs. $\Delta S^{\neq}$ compensation) plot shown in Fig. 3. In His46Gly · aq and His117Gly · aq, water molecules can enter the copper coordination sphere and give rise to rather large changes in the solvation sphere. Inserting an imidazole, however, which is perfectly accommodated in the pocket of H117G azurin, will prevent water from approaching the redox site. Nevertheless, all three data points extending over the full range of the plot (Fig. 3) lie on the straight line as expected, consistent with the operation of one and the same ET mechanism, thereby providing further support for the Trp48 ET pathway from the disulfide radical to Cu(II).

A strong pH dependence is observed for the rate constant of the intramolecular $RSSR^{-} \rightarrow Cu(II)$ LRET in all the different wild-type and single-site azurin mutants studied so far, with the rate constant increasing by an order of magnitude upon decreasing the pH from 8 to 4 (21). In order to rationalize the influence of pH on ET reactivity, the different parameters that determine the LRET rates were considered: Driving force, reorganization energy, distance between electron donor and acceptor, and the nature of the protein medium separating the redox couple.

The reduction potential of *P. aeruginosa* azurin increases by 60–70 mV upon changing pH from 10 to 5 (38–40). The NMR studies addressed the question of how much protonation of the two conserved titratable histidines in azurin, His35 and His83, may increase the Cu(II)/Cu(I). It was found that the contributions from these two residues are 50 and 13 mV, respectively (40). There is a problem here, however, since the reduction potential of the H35K azurin mutant displays essentially the same pH dependence as WT (39), which seems to speak against this hypothesis, unless the protonated lysyl ε-amine group has an unusually low p*K* value.

All other wild-type and mutated azurins studied so far exhibit a similar change of ~60 mV in reduction potential upon going from neutral solutions to lower pH (38–40). For the WT *P. aeruginosa* azurin, this would correspond to an increase in driving force ($-\Delta G^{\circ}$) from 68.9 to 74.7 kJ mol$^{-1}$. Equation 2 predicts an increase in intramolecular ET rate constant from 44 to 61 s$^{-1}$, which is far less than observed experimentally ($k = 285\,s^{-1}$ at pH 4.0 and 25°C) (21). Moreover, the same pH dependence is also observed for LRET in H35Q azurin, where no protonation of residue 35 is possible (20); this would exclude any effect of His35 protonation on either reduction potential or intramolecular ET kinetics. Another candidate for modifying the LRET driving force is Asp23. According to the 3D structure of *P. aeruginosa* azurin (16), this residue is proximal to the electron donor, the $RSSR^-$ radical, with its peptide carbonyl hydrogen bonded to the amide N of Cys26 and with one of its carboxyl oxygens within hydrogen-bond distance to two neighboring residues that could increase the p*K* of Asp23 from the regular value to 6.2. However, any key role of this residue can now also be excluded, since the same LRET pH dependence is observed for the D23A mutant as with all the other azurins studied so far. Finally, the $RSSR^-$ radical anion has a p*K* of ~6 (41), but protonation is expected to lower the reduction potential, owing to the elimination of electrostatic charge, and thus we tend also to exclude this residue as a cause of the rate acceleration.

Most importantly, as shown in the following, it is impossible to reconcile any increase in driving force with the observed pH dependence of the rate for WT azurin; with $\lambda = 1.0\,eV$, $\beta(r - r_0) = 24.6$, and $-\Delta G^{\circ} = 0.71\,eV$, Eq. 2 gives a maximum rate constant, $k_{max} = 210\,s^{-1}$, which is still smaller than the experimentally observed value at low pH, 285 s$^{-1}$. Further increase in the driving force above the value of the reorganization energy would only bring the system into the inverted region where the rate constant will decrease again (cf. Eq. 2). For the same reason, a rate acceleration caused by changes in reorganization energy, λ, due to protonation of a specific protein site can be excluded. It should also be emphasized that the rate increase at low pH is due to an entropy effect rather than to a more favorable enthalpy term. The exponential term in Eq. 3 is also included in the calculated entropy of activation (cf. Eq. 7). Hence, the rate increase logically is the result of slightly better electronic coupling between electron donor and acceptor. Indeed, the increase in specific rate of LRET in WT *P. aeruginosa* azurin from 44 to 285 $s^{-1}$ can be accounted for by reducing the exponential term, $\beta(r - r_0)$ from 24.6 calculated for WT azurin at pH 7 (22) to 23.6. Careful examination of the 3D structures of azurin determined at both pH 5.5 and 9.0 (16) shows no structural changes in the region of the calculated electron tunneling pathways. The only major conformational change observed involves a Pro36–Gly37 main chain peptide bond flip, and these residues are not involved in the pathways. Still, a decrease in ET distance of 0.3 nm would be sufficient for rationalizing the rate increase observed at low pH. An alternative explanation would be a slight decrease in the distance decay factor, $\beta$, from 10.0 to 9.6 $m^{-1}$. Obviously, more experimental work is required combined with a more detailed examination of possible hydrogen bonds present in azurin, in order to unequivocally identify the cause(s) for the observed marked pH induced acceleration of the LRET rate constant.

In order to gain further insight into the possible impact of the solvent on LRET in *P. aeruginosa.* azurin, rates of intramolecular ET in water have been compared with those in deuterium oxide (42). Unexpectedly, the kinetic isotope effect, $k_H/k_D$, was found to be *smaller* than unity (0.7 at 298 K), primarily as a result of differences in activation entropies in $H_2O$ ($-56.5\,J\,K^{-1}\,mol^{-1}$) and in $D_2O$ ($-35.7\,J\,K^{-1}\,mol^{-1}$), which in turn suggests a distinct role for protein solvation in the two media. This notion is further supported by results of voltammetric measurements where the reduction potential of Cu(II)/Cu(I) was found to be 10 mV more positive in $D_2O$ at 298 K. The standard entropy changes also differ ($-57\,J\,K^{-1}\,mol^{-1}$ in water and $-84\,J\,K^{-1}\,mol^{-1}$ in deuterium oxide) (42) and thus make different contributions to the activation entropies (cf. Eq. 8). Isotope effects are also inherent in the nuclear term of the Gibbs free energy, as well as in the tunneling factor. A slightly larger thermal protein expansion in $H_2O$ than in $D_2O$ ($0.001\,nm\,K^{-1}$) is sufficient to account for both activation and standard entropy differences. Thus, differences in driving force and thermal expansion seem to be the simplest rationales for the observed isotope effect (42). These observations once more underscore the important role of solvent in affecting the rates of internal ET in proteins.

A very different approach to studies of internal ET in azurin employed an Asn42Cys mutant which under oxidizing conditions forms a dimer where the two azurin monomers are covalently linked via the Cys42–Cys42 disulfide bridge (43). The 3D structure of the dimer has been determined and the short intermolecular disulfide link was found to cause a strong steric constraint (44). This new type of engineered azurin was employed in order to investigate ET between the pulse radiolytically produced disulfide radical ion and Cu(II) over a considerably shorter distance than in the monomer (1.28 nm for Cys42 to Cu in the dimer as compared with 2.59 nm for Cys3–Cys26 to Cu in the previous studies). In order to avoid possible interference from reduction of the native Cys3–Cys26 disulfide bond, a triply mutated azurin was constructed and expressed, where the latter two cysteines were substituted by alanines, Cys3Ala–Cys26Ala–Asn42Cys. Earlier structural studies of an azurin mutant, where this disulfide bridge has been eliminated (Cys3Ala–Cys26Ala) established that the overall structure of the protein is not changed and the only difference is in the immediate proximity of the mutated residues (45).

Reacting the azurin dimer with pulse radiolytically produced $CO_2^-$ radical anions, the intermolecular disulfide bridge becomes reduced forming the $RSSR^-$ radical in an essentially diffusion controlled reaction ($k_1 \simeq 10^9\,M^{-1}\,s^{-1}$). In contrast to the behavior of monomeric azurins, no competing bimolecular reduction of the blue Cu(II) center by $CO_2^-$ was observed. Disulfide reduction was followed by concentration independent, intramolecular $RSSR^- \rightarrow Cu^{(II)}$ ET (Fig. 5):

$$RSSR^-\text{—}Az(Cu^{II}) \xrightarrow{k_2} RSSR\text{—}Az(Cu^{I}) \tag{16}$$

The process was studied at pH 7.0 over a dimer concentration range from 5 to 54 $\mu M$ and monitored at both 410 nm ($RSSR^-$ absorption, $\varepsilon_{410} = 10,000\,M^{-1}\,cm^{-1}$) and at 625 nm [Cu(II) absorption, $\varepsilon_{625} = 5,000\,M^{-1}\,cm^{-1}$) (43).

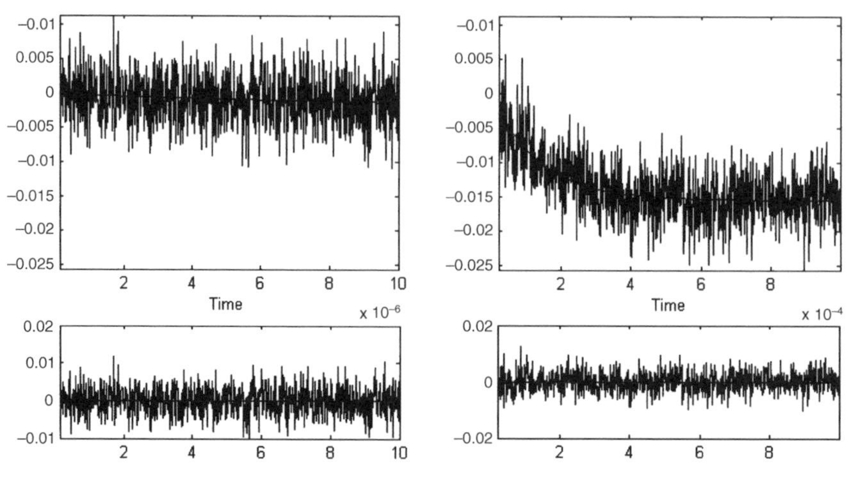


Figure 5. Time-resolved absorption changes, induced by reaction of $CO_2^-$ radicals, due to intramolecular ET from the internal disulfide radical anion to Cu(II) in the C3/C26A–N42C azurin dimer measured at 625 nm. Protein concentration was 20$\mu M$, where $T = 299$ K; pH 7.0; 0.1 $M$ formate; 10 m$M$ phosphate; $N_2O$ saturated; pulse width 1.5 μs; optical path 3 cm. Time is in seconds; the left panel shows the faster phase, while the right one shows the reaction taking place at the slower phase. The lower panels show residuals of the fits to the data.

The intramolecular ET rate constant, $k_2$, was found to be $7200 \pm 100\,s^{-1}$ at 25°C and pH 7.0 (43). The employed protocol exposed each solution to only few pulses causing < 10% reduction of the molecules present. Hence, the probability of reducing more than one of the Cu(II) ions in a dimer is negligible. From the temperature dependence of the internal ET rate (studied from 3.2 to 40.0°C) the activation parameters were derived (Table II). Finally, results of intramolecular ET measurements in the triply mutated azurin dimer showed that its rate constant is in good agreement with the Beratan and Onuchic (6) tunneling pathway model. The polypeptide chain in the azurin dimer links $S_\gamma$ of Cys42 with $N_\delta$ of His46, which provides one of the copper ligands (Fig. 6). This pathway consists of 17 covalent bonds and the distance between the copper ion and the sulfur atom of the same monomer is 1.29 nm. Driving force optimized rate constants for ET in a β-sheet protein can be described by an average coupling decay constant of $10.0\,nm^{-1}$ (46), which leads to an activationless $k_{max} = 10^5\,s^{-1}$ (i.e., when the driving force, $-\Delta G^\circ$ equals the reorganization energy, $\lambda$). Assuming that the previously determined reorganization energy and driving force for intramolecular ET between the Cys3–Cys26 $RSSR^-$ and the copper center in WT *P. aeruginosa* azurin ($\lambda_{tot} = 1.0\,eV$ and $-\Delta G^\circ = 0.71\,eV$) (22) are also applicable to the present mutant dimer, a rate constant of $4 \times 10^4\,s^{-1}$ at 298 K is calculated that is still fivefold larger than the experimentally observed rate, $k_{298} = 7200\,s^{-1}$. Though not an unreasonably large discrepancy, a rationale may be considered for the divergence: In native *P. aeruginosa* azurin the ET pathway includes $S_\lambda$ of Cys112, while tunneling from the external C42/C42 disulfide bridge to the copper center proceeds via the $N_\delta$ of His46. It has already been pointed out that there is a high degree of anisotropic covalency in the blue Cu(II) center (28, 29): While 50% of the electron density is concentrated on the Cu–sulfur bond only 4% is found on

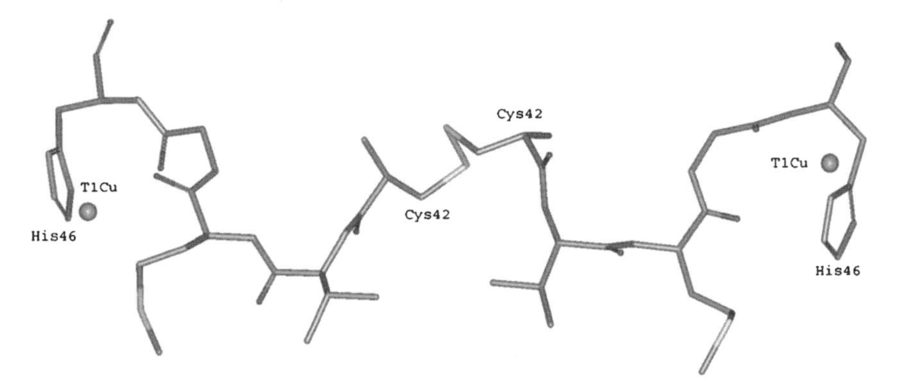


Figure 6. Electron-transfer pathway in the azurin dimer mutant (43). The path connects $S_\gamma$ of Cys42 with $N_\delta$ of His46, which is one of the copper ligands, and consists of 17 covalent bonds resulting in a very effective electronic coupling of the two redox centers. Calculations were based on the Beratan and Onuchic model (6, 7). Coordinates were taken from the PDB, code 1JVO.

each of the ligating imidazoles, which would drastically diminish the electronic coupling in the dimer compared with the former one and cause the observed lower rate.

An additional interesting observation emerged from a comparison of the activation parameters calculated here for the azurin dimer with those obtained earlier for azurins as well as for related blue copper containing enzymes (cf. later sections): In all (single copper) azurins, the $RSSR^-$ to copper(II) LRET is controlled by a relatively large activation enthalpy, while in multicopper proteins, including the present azurin dimer as well as systems like cytochrome *c* oxidases, the activation enthalpies are relatively small and the large negative activation entropies were found to be rate determining, although in these proteins the connecting ET pathways are considerably shorter. One rationale could be that major solvent reorganization takes place in all monomeric azurins upon ET where the intramolecular Cys3–Cys26 disulfide bridge is considerably more solvent exposed than is the intermolecular Cys42–Cys42 cystine of the dimer. This hypothesis is corroborated by the observed excellent linear correlation between the activation enthalpy and entropy data presented in Table II that point to an overriding influence of solvent effects.

Finally, the ET reactivity of the binuclear $Cu_A$ site present in cytochrome *c* oxidase and nitrous oxide reductase illustrates an additional interesting

TABLE II
Rate Constants and Activation Parameters for Internal ET in Different Copper-Containing Proteins

| Protein | ET Process | $k_{298}$ ($s^{-1}$) | $\Delta H^{\neq}$ (kJ $mol^{-1}$) | $\Delta S^{\neq}$ (J $K^{-1}mol^{-1}$) | ET Distance (nm) |
|---|---|---|---|---|---|
| *P. aeruginosa* azurin[a] | $RSSR^- \rightarrow Cu^{2+}$ | $44 \pm 7$ | $47.5 \pm 4$ | $-56.5 \pm 7.0$ | 2.56 |
| C3/26A-N42C dimer[b] | $RSSR^- \rightarrow Cu^{2+}$ | $7200 \pm 100$ | $17.7 \pm 2.0$ | $-112 \pm 6$ | 1.28 |
| CuNiR[c] | $T1Cu^I \rightarrow T2Cu^{II}$ | $185 \pm 12$ | $22.7 \pm 3.4$ | $-126 \pm 11$ | 1.27 |
| Ascorbate oxidase[d] | $T1Cu^I \rightarrow T2/T3Cu^{II}$ | $201 \pm 8$ | $9.1 \pm 1.1$ | $-170 \pm 9$ | 1.22 |
| Cytchrome *c* oxidase (bovine)[e] | ${Cu_A}^I \rightarrow$ heme-$c^{III}$ | $13{,}000 \pm 1{,}200$ | $11.4 \pm 0.9$ | $-128 \pm 11$ | 1.96 |
| Cytchrome *c* oxidase (bacterial)[f] | ${Cu_A}^I \rightarrow$ heme-$c^{III}$ | $20{,}400 \pm 1{,}500$ | $22.2 \pm 1.2$ | $-88 \pm 2$ | 1.96 |

[a]Ref. 17.
[b]Ref. 43.
[c]Ref. 47.
[d]Ref. 48.
[e]Ref. 49.
[f]Ref. 50.

application of PR in azurin. The $Cu_A$ serves as the electron uptake center in the above enzymes, and its discovery raised a considerable debate regarding the causes for the evolution of multiple forms of copper electron mediation centers (i.e., T1 and $Cu_A$). An azurin mutant, the so-called "purple" azurin, has been engineered by Lu and co-workers (51). In this mutant, the amino acids forming the blue type 1 (T1) copper site have been replaced by residues constituting the $Cu_A$ site of *Paracoccus denitrificans* cytochrome *c* oxidase. The close overall structural similarity between the native blue copper azurin and the engineered purple $Cu_A$ azurin has been demonstrated by X-ray crystallography (51). Since the T1 and the $Cu_A$ centers are placed in the same location in azurin, while all other structural elements remain the same, the purple mutant is an optimal system for examining the unique electron mediation properties of the binuclear $Cu_A$ center. Analysis of the ET path from the disulfide radical anion to $Cu_A$ shows that the same number of covalent bonds, the same two hydrogen bonds, and the through-space jump are all also found in the purple $Cu_A$ azurin structure. Therefore, the same pathway is most likely operative in the purple azurin. The rate constant of the intramolecular process, $k_{ET} = 650 \pm 60\,s^{-1}$ at 298 K and pH 5.1 (52), is almost threefold faster than for the same process in the wild-type single blue copper azurin from *P. aeruginosa* ($250 \pm 20s^{-1}$ at this pH), in spite of a smaller driving force (0.69 eV for purple $Cu_A$ azurin vs. 0.77 eV for the blue copper center). Apparently, the presence of a binuclear $Cu_A$ center as the primary electron uptake site in cytochrome *c* oxidase, the terminal enzyme of the respiratory system, rather than the more prevalent T1 copper site, constitutes an illustration of evolutionary selection of a structure better suited for specific requirements of the biological function.

Di Bilio et al. (53) analyzed the intramolecular LRET in Ru(II) modified *P. aeruginosa* azurin and calculated the reorganization energies of both electron donor and acceptor centers from the temperature dependence of the observed rate constants. For the blue, T1 copper center $\lambda_{T1} = 0.82\,eV$ was reported. Now, from the observed rate of intramolecular ET in purple azurin, $\lambda_{CuA} = 0.4\,eV$ was calculated for the $Cu_A$ center (52) (i.e., only 50% of the reorganization energy required for the blue copper site, supporting the notion that $Cu_A$ is indeed a redox center with improved ET properties). In conclusion, the purple $Cu_A$ center transfers electrons more efficiently than the blue copper center. This is mainly a result of the low reorganization energy of the mixed-valence [Cu(1.5)–Cu(1.5)] site.

Summarizing the results of studies of intramolecular ET in azurins, it is important to stress that in all probability, the induced ET between the disulfide radical and the blue copper(II) center is not part of any physiological function, and the role of the disulfide is structural, solely. Still, azurin has turned out to be a very useful model system for examination of different parameters controlling LRET rates in proteins. The impact of specific structural changes introduced by single-site mutations has been studied in order to obtain a better understanding of these parameters, namely, driving force, reorganization energy, as well as distance and structure of the medium separating the redox centers. The ET reactions are well described within the framework of Marcus theory (3), and the electronic coupling model developed by Beratan and Onuchic (6) works well for ET pathway determination. Experience obtained through studies of ET in azurins has been exploited in further studies on multicentered metalloenzymes, which is the subject of the following sections.

### B. Copper-Containing Oxidases and Reductases

Intramolecular ET between distinct copper centers is part of the catalytic cycles of many copper-containing redox enzymes, such as the multicopper oxidases, ascorbate oxidase, and ceruloplasmin, as well as the copper-containing nitrite reductases. Examination of internal LRET in these proteins is of considerable interest as it may also provide insights into the evolution of selected ET pathways; in particular, whether and how the enzymes have evolved in order to optimize catalytic functions. With the increase in the number of known high-resolution 3D structures of transition metal containing redox enzymes, studies of structure–reactivity relationships have become feasible and indeed many have been carried out during the last two decades.

The blue multicopper oxidases catalyze the four-electron reduction of dioxygen to water by four sequential one-electron oxidations of their respective substrates (54). These enzymes are widely distributed in Nature, from bacteria, fungi, and plants to mammals. All contain at least four copper ions bound to sites of the following types: (a) The blue type 1 site (T1), also found in azurin (Section III.A), characterized by an intense charge-transfer band in the 600-nm region ($\varepsilon \sim 5,000\,M^{-1}\,cm^{-1}$) and a narrow hyperfine coupling constant ($A_{\|} < 10^{-3}\,cm^{-1}$) extracted from EPR spectra (2). (b) A type 2 (T2) copper center (lacking intense absorption bands) with a *normal* EPR spectrum. (c) A copper ion pair, called type 3 (T3), which in the oxidized state is characterized by an intense absorption in the near ultraviolet (UV) region ($\varepsilon \sim 4,000\,M^{-1}\,cm^{-1}$) and by strong antiferromagnetic coupling. The 3D structural studies have shown that T2 and T3 are proximal and together form a trinuclear cluster (2). The physiological function of T1 is sequential uptake and delivery of single electrons from substrate molecules to the trinuclear center where dioxygen binds is reduced to water. Thus, the enzymatic cycle of the oxidases feature a ping–pong mechanism:

$$2AH_2 + O_2 \xrightarrow{\text{oxidase}} 2A + 2H_2O \qquad (17)$$

Intramolecular ET between T1 and the trinuclear T2/T3 center is therefore expected to play a crucial role in the molecular mechanism of this class of enzymes. High-resolution 3D-crystal structures are now available for ascorbate oxidase (55–58), human ceruloplasmin (59), and its yeast homolog, Fet3p (60), as well as fungal laccase (61). These structures have proven to be very helpful in analysis and interpretation of kinetic results.

Pulse radiolysis was applied to studying ET within multicentred blue oxidases three decades ago (62). However, at that early time, without knowledge of the 3D structures of any of the multicopper proteins, only general conclusions about possible intramolecular ET processes could be made. The PR studies of human ceruloplasmin, hCp, showed that first a diffusion controlled reaction of the enzyme with hydrated electrons takes place, forming transient optical absorptions that are primarily due to surface exposed disulfide radicals. The radicals were found to decay concomitantly with the reduction of the solvent less-accessible T1Cu(II) site by intramolecular ET (62). Thus, even at that stage some insights were gained that were greatly extended later using structural information as it became available.

### 1. Ascorbate Oxidase

The earliest high-resolution 3D structure of a copper oxidase was reported in 1989, namely, for ascorbate oxidase from *Cucurbita pepo medullosa* (AO cf. Fig. 7), first in a fully oxidized state and eventually also in the reduced state as well as with different exogenous ligands (55–58). Ascorbate oxidase is a 140-kDa dimer, consisting of two identical subunits; each monomer includes three domains with the T1 copper site residing within one domain, while three copper ions form a trinuclear center, consisting of both T2 and T3 in a structural arrangement formed by four histidine imidazoles from one domain and four from another: Of the three copper ions, two are coordinated to six histidines whose nitrogen atoms (five $N_\varepsilon$ and one $N_\delta$ atom) are arranged in a trigonal prism, reminiscent of the hemocyanin copper site structure, and were, by analogy, called T3 The remaining copper ion, coordinated to two histidines, would then be a T2 site. Importantly, none of the imidazoles is involved in bridging copper ions. The two copper ions constituting the T3 site are bridged by an oxygen atom (either as $OH^-$ or $O^{2-}$) and at the T2 site a hydroxide ion or water molecule is coordinated to the copper ion. The trinuclear site is accessible to exogenous ligands and may bind anions such as $F^-$, $N_3^-$, and $CN^-$ (55–58).

Like all other blue copper oxidases, AO catalyzes the four-electron reduction of dioxygen, $O_2$, to water. Electrons are taken up sequentially by the T1 copper(II) center, while dioxygen coordinates to and is reduced at the T2/T3 copper cluster. Thus, intramolecular ET is central to the function of the enzyme (54). Although no knowledge of the 3D structure was available, early results obtained by pulse radiolysis using hydrated electrons as well as the $CO_2^-$ and reduced nitroaromatic radicals as reductants established that the T1 copper is the

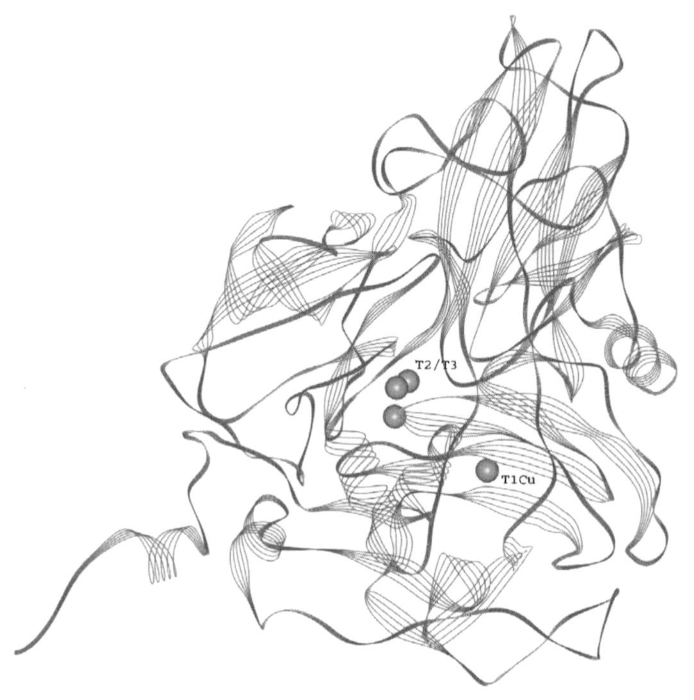


Figure 7. Three-dimensional structure of *C. pepo medullosa* ascorbate oxidase. The protein backbone is shown together with the four copper ions. Coordinates were taken from the PDB, code 1AOZ.

acceptor site (63, 64). At that time no clear evidence for intramolecular ET was presented (63). In a later study by the same group, both native AO and a modified enzyme where the T2 copper ion had been removed (T2D) were reduced pulse radiolytically using $ArNO_2^-$, $CO_2^-$, or $O_2^-$ radicals (64). The reduction stoichiometry of native AO by the former agents required 6–7 equiv while for T2D only 3 equiv were necessary (all based on 610-nm absorption). Reduction by $O_2^-$ radicals, performed in the presence of excess $O_2$, yielded rather limited net reduction (as expected from this experimental protocol). In contrast, T2D AO underwent full reduction of the $T1Cu^{II}$, clearly showing the requirement of an intact trinuclear center for a functional catalytic cycle (64).

Later pulse radiolysis experiments demonstrated that, following a bimolecular reduction of $T1Cu^{II}$ by the reducing radicals, $T1Cu^{I}$ reoxidation at 605 nm and concomitant $T3Cu^{II}$ reduction at 330 nm could be monitored independently and directly (48) (cf. Fig. 8). The rate of the latter reactions was independent of protein concentration, consistent with being a unimolecular process. An unprecedented feature in AO was that at least two parallel phases of intramolecular ET were resolved. For the first phase, a rate constant of $200\ s^{-1}$ was determined

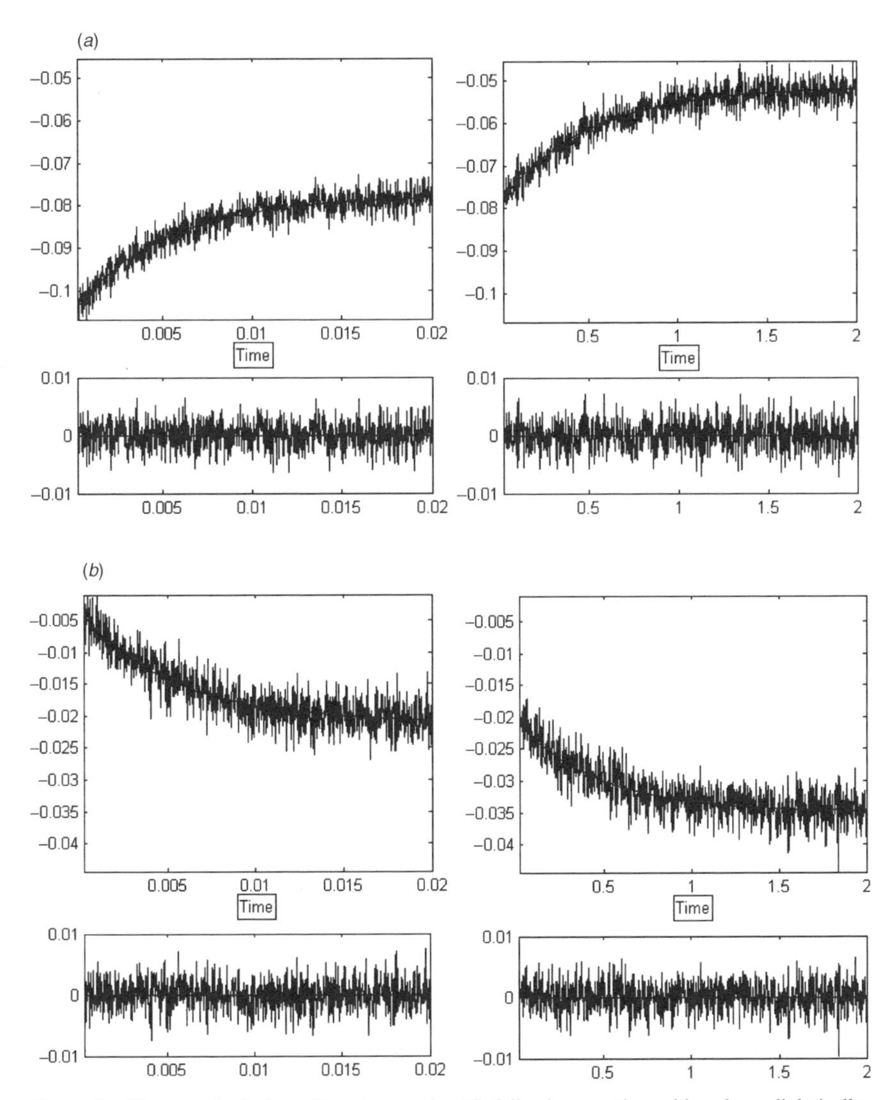


Figure 8. Time resolved absorption changes in AO following reaction with pulse-radiolytically produced $CO_2^-$ radicals (48). (*a*) Absorption changes at 610-nm monitoring reoxidation of a 5.5-$\mu M$ solution of the protein by $CO_2^-$ radicals. The initial, fast bimolecular reduction of $T1Cu^{II}$ has a half-life of $< 100\,\mu s$ and is thus not resolved on the time scales shown here. (*b*) Absorption changes at 330 nm following intramolecular reduction of the T3 copper by $T1Cu^{I}$. $T = 286\,K$; pH 5.5; 0.1 *M* formate; 10 m*M* phosphate; $N_2O$ saturated; pulse width 1.0 μs; optical path 12.3 cm. Time is in seconds; the left panel shows the faster phase, while the right one shows the reaction taking place at the slower phase. The lower panels show residuals of the fits to the data.

at 25°C, pH 5.5 under anaerobic conditions, while a slower reaction was monitored with a rate constant of $2\,s^{-1}$ (48). The activation parameters for the faster phase are presented in Table II. Two reduction phases with similar rate constants were later reported by another group using different organic radicals as reductants, but monitoring only the 610-nm chromophore (65). The observation of two different ET processes is rather puzzling and could be rationalized by the existence of distinct conformers since the two rates can be related to differences in activation entropies ($-170 \pm 9$ and $-215 \pm 16\,J\,K^{-1}\,mol^{-1}$, respectively), which are attributed to changes in electronic coupling between electron donor and acceptor in the two AO monomers. The observed internal ET reactions are rather slow considering that the centers are connected by a short covalent pathway with a direct distance between the T1 and the nearest T3

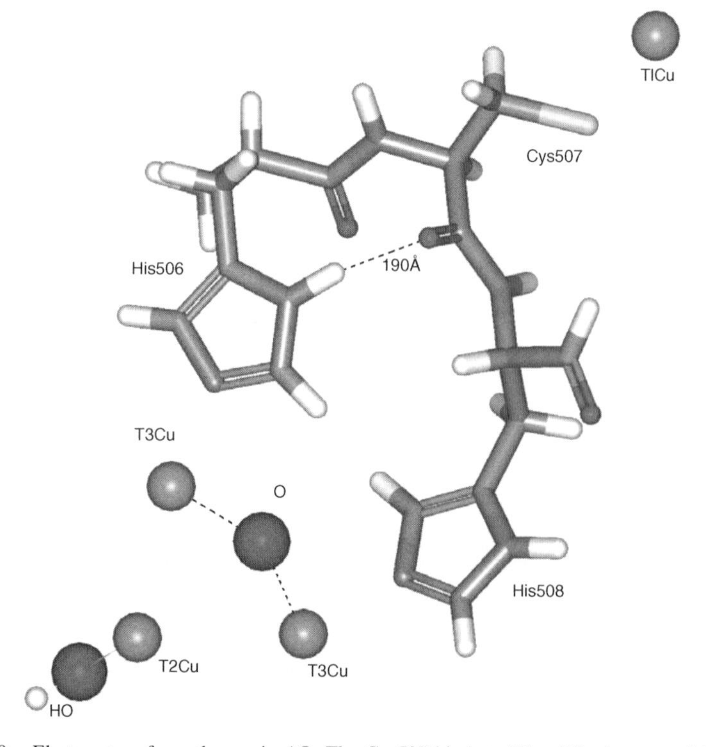


Figure 9. Electron-transfer pathways in AO. The Cys507 binds to T1, while the two neighboring residues, His506 and His508 bind to the two T3 copper ions. This provides a very short electronic coupling between Cys507 and either His506 or His508, both consisting of nine covalent bonds. An alternative pathway may be envisaged through the carbonyl oxygen atom of Cys507 and $N_\delta$ of His506. Calculations, showing that all three pathways have essentially the same electronic coupling (48), were based on the Beratan and Onuchic model (6, 7). Coordinates were taken from the PDB, code 1AOZ. (See color insert.)

copper ion of 1.2 nm (Fig. 9). One of the T1 copper ligands is a thiolate from Cys507, while imidazoles of the two neighboring histidines, His506 and His508, coordinate to the two T3 copper ions. Both pathways consist of only 11 covalent bonds. Thus, the shortest ET pathway linking the two redox centers comprise Cys507 and either His506 or His508 giving a decay factor of $\varepsilon = 3.6 \times 10^{-3}$. An alternative ET route is provided by a hydrogen bond between the carbonyl oxygen of Cys507 and His506 with an even better electronic coupling, $\varepsilon = 5.2 \times 10^{-3}$ (48). It is furhter noteworthy that similar to the azurin system discussed above the cysteine thiolate ligand takes part in ET to and from the copper ion.

A possible reason for the surprisingly slow ET processes is that the driving force is close to zero and the reorganization energy is quite large ($1.5 \pm 0.1$ eV), suggesting that the reaction is gated either by substrate ($O_2$) binding or by an even larger reorganization energy of the trinuclear center due to conformational changes (48). In this context, it is interesting to compare the above rates with the rate of the initial internal $Cu_A$(I) to heme-*a*[Fe(III)] ET step in cytochrome *c* oxidase (COX) reduction (cf. Section V.A). While in AO the internal ET is slowed down by properties of the acceptor, this is most likely not the case in either mammalian or bacterial COX. The trinuclear copper center in AO is solvent exposed, and redox changes are expected to be accompanied by large reorganization energy requirements. In COX the electron acceptor, heme-*a*, is buried in the solvent inaccessible interior of the protein and, as shown below, the reorganization energy is quite small. This result may explain why the rate of intramolecular ET is 17,000–30,000 $s^{-1}$, that is several orders of magnitude faster than in the blue oxidase, although the shortest distance separating electron donor [$Cu_A$(I)] and heme-*a*(III) in COX is 2.0 nm through a hydrogen-bonded system, contrasting the covalent pathway distance of 1.2 nm found in AO.

Assuming that the linkage between the copper sites in the blue oxidases has been optimized by evolution, the following pertinent questions arise: (1) Is the intramolecular ET rate in AO controlled during the multielectron reduction and oxidation? (2) Does the rate of intramolecular ET depend on the number of electron equivalents taken up by the molecules? (3) Does the rate of ET relate to the conformational changes that were resolved by the structural studies? (4) Finally, how can the turnover number of the enzyme, determined under steady state conditions (12,000–14,000 $s^{-1}$), be considerably faster than the above rate of intramolecular ET? Since AO molecules reduced to different degrees (from 0–95%) yielded identical ET rates, the notion that the rate depends on the number of electron equivalents taken up can be rejected (48). Also, enzymes that have been activated or pulsed by going through a turnover of 1 m*M* ascorbate and 0.25-nm $O_2$ prior to the PR experiment did not show different rates of internal ET. However, in a study where a controlled small amount of dioxygen (15–65 μ*M*) was introduced into the solution, a conspicuous difference in reactivity was observed (66). A new and faster phase of intramolecular ET was discovered with a rate constant of $1100\,s^{-1}$ (18°C, pH 5.8), which was maintained as long as $O_2$ remained in the solution. At the same time, large spectral changes took place, most probably due to interaction between $O_2$ and the partially reduced T2/T3 center. Dioxygen or its reduction intermediates coordinated to the trinuclear center would most probably increase the driving force of ET considerably and cause the observed increase in rate of intramolecular ET. Still, it is unclear whether the first species interacting with $O_2$ are partially reduced AO molecules or whether the fully oxidized enzyme is capable of such binding (66). Calculations show that an increase in driving force of 0.1 eV would suffice to produce the observed fivefold increase in rate. However, as this increase does not account for the experimentally observed turnover, an interaction between AO and $O_2$ is not adequate for attaining maximal enzymatic activity. Under optimal physiological conditions, the concentration of the reducing substrate, ascorbate, is sufficient for maintaining a steady state of fully reduced copper centers, an observation that was exploited in a flash photolysis study where reoxidation of fully reduced AO was followed using a laser generated triplet state of 5-deazariboflavin as electron acceptor (67). Subsequent to the assumed one-electron oxidation of the T2/T3 center, rapid electron transfer from $T1Cu^{I}$ and (presumably) to the trinuclear center was monitored with a rate constant of $\sim 10,000\,s^{-1}$, which is comparable to the turnover rate, so it is possible that this is the rate-limiting step in catalysis. Why this fast internal ET process could not be resolved in the PR studies of the oxidized enzyme remains unclear.

### 2. *Ceruloplasmin*

Mammalian ceruloplasmin (Cp) is an exceptional case among the blue copper oxidase family members for several reasons: First and foremost, its function as an oxidase has been a subject of considerable debate for a long time; though its capacity to catalyze ferrous ion oxidation by dioxygen has been well established and its involvement in iron metabolism strongly supported by considerable physiological and genetic evidence, only recently has its physiological function as a ferroxidase been fully accepted (68). Another conspicuous characteristic is the copper content, and hence the number and nature of possible active sites, topics that have been debated for quite some time. The 3D structure determination of human Cp (hCp) clearly resolved the latter issue by revealing the presence of three distinct T1 sites in addition to a single trinuclear cluster (59) (cf. Fig. 10). Finally, Cp is the only known mammalian member of the blue copper protein family.

The more recent advances in understanding copper and iron homeostasis are based to a large extent on studies of ferroxidase in lower organisms. In yeast, a correlation between the ferroxidase activity and a high-affinity iron uptake complex has been demonstrated. Further, the ferroxidase present in *Saccharomyces cerevisiae*, Fet3p, has been extensively studied and its 3D structure was recently determined (60).

Human ceruloplasmin consists of a single polypeptide chain with a MW of 132 kDa folded in six cupredoxin domains arranged in a triangular array. Each domain comprises a β-barrel, constructed in a Greek key motif, typical for the cupredoxins. Three of the six copper ions are bound to T1 sites present in domains 2, 4, and 6, whereas the other three copper ions form a trinuclear cluster, bound at the interface between domains 1 and 6 (Fig. 10). The spatial relation between the trinuclear center and the nearest T1 site (A, in domain 6) closely resembles that found in AO and was taken to further support the proposal that hCp has an oxidase function. The three T1 sites are separated from each other by a distance of 1.8 nm, a distance that might still allow for internal ET at reasonable rates and could also increase the probability for electron uptake. The coordination sphere of the T1 site in domain 4 (T1B) is identical with that of domain 6 (T1A). The third type 1 center (T1C), however,

Figure 10. The 3D structure of human Ceruloplasmin. The protein backbone is shown together with the three T1 copper ions and the trinuclear T2/T3 site. Coordinates were taken from the PDB, code 1KCW.

contains a nonligating Leu residue instead of the "usual" Met. Redox titrations along with EPR and UV–VIS (visible) spectroscopy demonstrated that hCp contains three paramagnetic copper(II) ions, two in T1 sites and one in a T2 center (69). In analogy with ascorbate oxidase, two of the copper ions in the trinuclear site are bound to six histidines and assigned as the T3 site, while the third copper (most distant from T1) is coordinated to two histidines only, and is designated as the T2 site. By further analogy to the ascorbate oxidase structure, an oxygen atom connects the two T3 coppers, while another is bound to the T2 copper ion. An additional relevant structural feature is the domain 6 cysteine (Cys1021) that provides the thiolate ligand to the T1 site; it is placed between the neighboring histidine residues (His1020 and His1022) coordinated to the T3 copper pair (59). As in ascorbate oxidase, this sequence motif most likely provides the ET path between T1 and the trinuclear center (cf. Fig. 9).

The 3D structure of hCp further resolved five disulfide bridges distributed evenly throughout the protein in domains 1–5. All five disulfides are near the bottom of a β-barrel, and in two domains the T1 copper centers (T1B and T1C) are placed at the opposite end of the barrel. The only domain lacking a disulfide is domain 6, which contains T1A and the trinuclear copper centers.

The reduction potentials of two of the T1 copper centers were determined to be 580 and 490 mV, respectively (69). However, the experimental accuracy did not allow for a precise determination of whether hCp contains three or four nonparamagnetic copper ions. For azurin, where the copper ligating Met was substituted by Leu or Val, the T1 copper reduction potentials were found to increase from 307 to 412 to 445 mV (WT $<$Leu $<$Val) at pH 7.0 (39). Moreover, X-ray absorption spectroscopy suggested that the resting oxidized enzyme contains one permanently reduced T1 center, and that this site cannot be involved in the catalytic process, since its reduction potential is at least 1.0 V (70). Thus, based on the above electrochemical results and the 3D structure of hCp, this high-potential T1 site can be assigned to T1C of domain 2 (Fig. 10). Recently, the 3D structure determination of fungal laccase from *Trametes versicolor* has shown that a phenylalanine residue replaces methionine in the strongly oxidizing T1 copper(II) center (61).

A considerable body of results accumulated during earlier decades from activity studies of hCp now awaits a more meaningful analysis using the available 3D structure. Catalysis of amine oxidation by hCp, in particular biogenic ones present in plasma, cerebral, spinal, and intestinal fluids as well as of ferrous ions, which is probably physiologically relevant, has been studied extensively (68, 71). The mechanism of dioxygen reduction by hCp at the trinuclear center is of particular interest, as the presence of three distinct T1 sites raises the question of which centers are involved in internal ET to the single $O_2$ reduction site. This mechanistic question prompted us to initiate ET studies by PR.

The $CO_2^-$ radicals react with disulfide groups in hCp at diffusion controlled rates to produce the $RSSR^-$ radicals monitored at 410 nm (where $RSSR^-$ radicals exhibit an absorption maximum) while no direct reduction of $T1Cu^{II}$ was observed (72). Instead, the electrons were further transferred to a $T1(Cu^{II})$ center in an intramolecular process with a rate constant of $28 \pm 2\,s^{-1}$ at 279 K. An *intra*molecular electron equilibration with the T2/T3 center was then observed with a rate constant of $2.9 \pm 0.6\,s^{-1}$ determined independently at both 610 nm ($T1Cu^{II}$ absorption maximum) and at 330 nm, where the oxidation state of the trinuclear center can be monitored independently [Fig. 11(*a* and *b*)]. The internal ET process

$$T1(Cu^{I})T2/T3(Cu^{II}) \rightleftarrows T1(Cu^{II})T2/T3(Cu^{I}) \tag{18}$$

has an equilibrium constant of 0.17 at 279 K (72). After introduction of one-electron equivalent into this particular T1 center, intramolecular ET to the trinuclear site ceases to take place. Nevertheless, a new intramolecular ET from disulfide radicals to another $T1(Cu^{II})$ site is observed, but slowed by almost a factor of 10 to $3.9 \pm 0.8\,s^{-1}$; in addition, further reduction was monitored at 610 nm on a 4-s time scale (Fig. 11*c*). The dependence of the observed rate constants on the degree of reduction of the enzyme is shown in Fig. 12. The first T1 center to be reduced is most probably the one closest to the trinuclear site (T1A), since its reoxidation only takes place as long as less than one reducing equivalent is added to the hCp molecules. The $T1(Cu^{II})$ with the highest reduction potential (580 mV) is the obvious candidate here, as in reductive titrations, the first 50% of the total absorption at 610-nm decays within 3 min, while the further reduction proceeds much more slowly (69). From the equilibrium constant of the internal ET equilibration process, a difference in reduction potential between T1A and T3 Cu(II)/Cu(I) of 43 mV was calculated, giving a reduction potential for $T3(Cu^{II}/Cu^{I})$ of 537 mV. The potential for the second (lower potential) site, T1B, was reported to be 490 mV (69), which places the reduction potential of the trinuclear site midway between those of the two redox active T1 centers. Identifying T1A as the primary electron acceptor among the T1 centers is further substantiated by examination of the relative solvent exposure of the disulfide groups (72) and their distance from the T1 centers (see below); this assignment agrees well with the proposed Fe(II) binding site at Asp 975, which is in hydrogen-bond contact with the T1Cu ligating His1026 (60).

All three T1 centers are well protected from the solvent by the protein matrix. Thus, it is not surprising that only indirect copper reduction by radicals via the exposed disulfide groups is observed (72). In fact, this behavior is reminiscent of results of our studies of intramolecular ET in azurins where, in addition to the direct, bimolecular reduction of both the disulfide group and the T1 site, LRET takes place from the disulfide radical to $T1(Cu^{II})$ over a distance of 2.6 nm

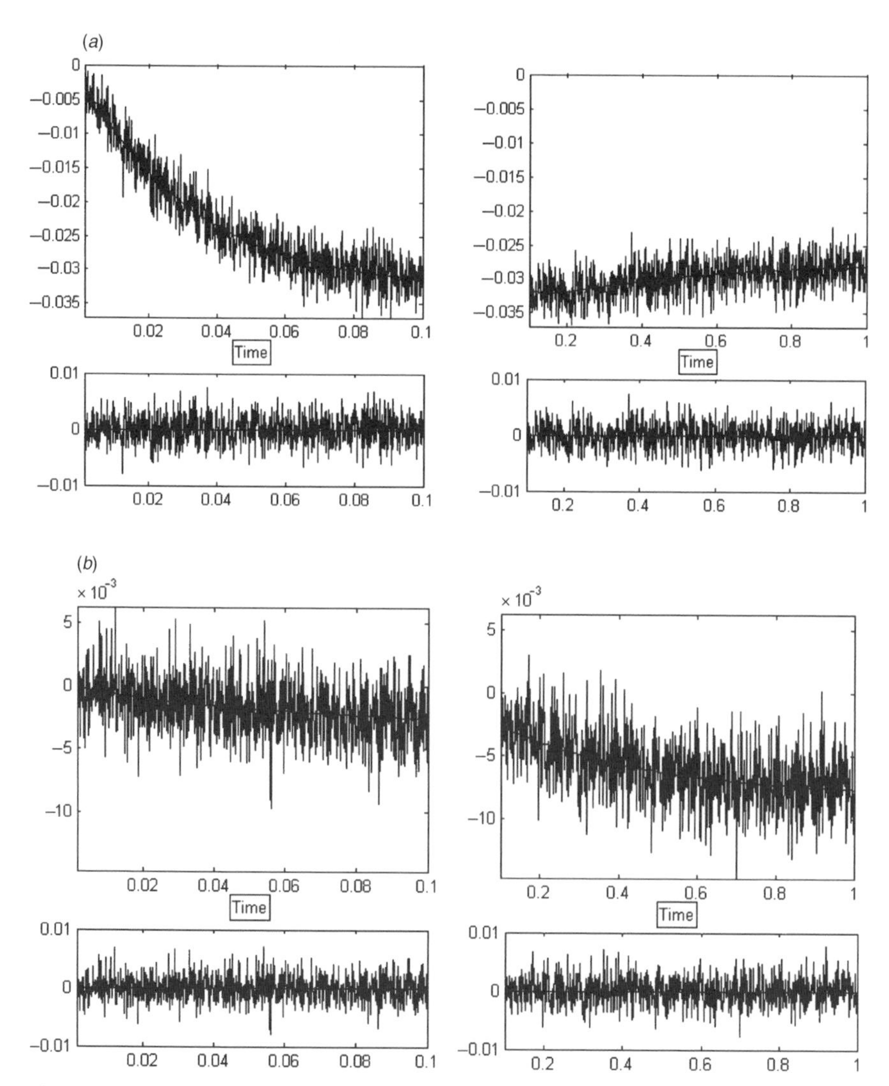


Figure 11. Time-resolved absorption changes in human Ceruloplasmin (72). (*a*) Reduction of $T1(Cu^{II})$ by $RSSR^-$ followed by $T1(Cu^I)$ reoxidation; monitored at 610 nm after 0.5 reduction equivalents have been accumulated by the enzyme (cf. Fig. 12). (*b*) T3[Cu(II)] reduction followed at 330 nm in a protein solution after accumulation of 0.6 reduction equivalents (cf. Fig. 12). (*c*) Monitoring $T1Cu^{II}$ reduction at 610 nm after 1.5 reduction equivalents were accumulated by the enzyme. Notice the absence of $T1Cu^I$ reoxidation here. $T = 278$ K; pH 7.0; 0.1 *M* formate; 10 m*M* phosphate; $N_2O$ saturated; pulse width 0.5 μs; optical path 12.3 cm. Time is in seconds; the left panel shows the faster phase, while the right one shows the reaction taking place at the slower phase. The lower panels show residuals of the calculated fits to the data.

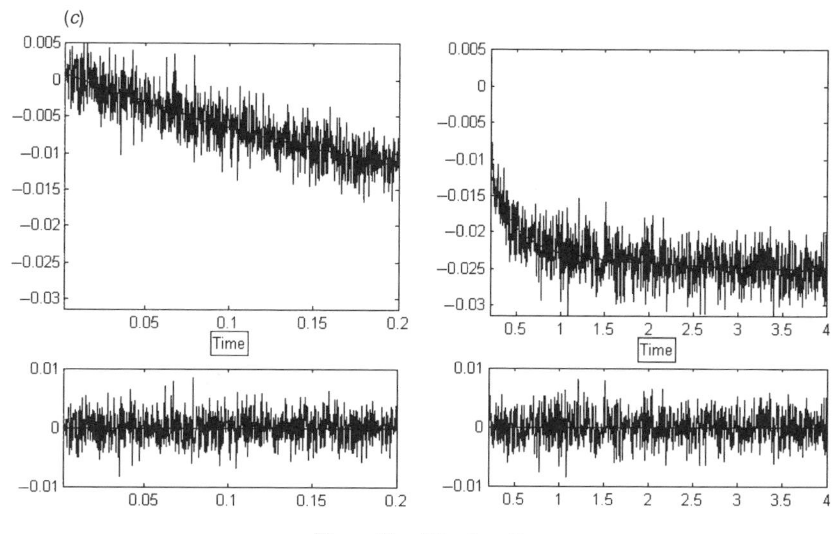


Figure 11. *(Continued)*

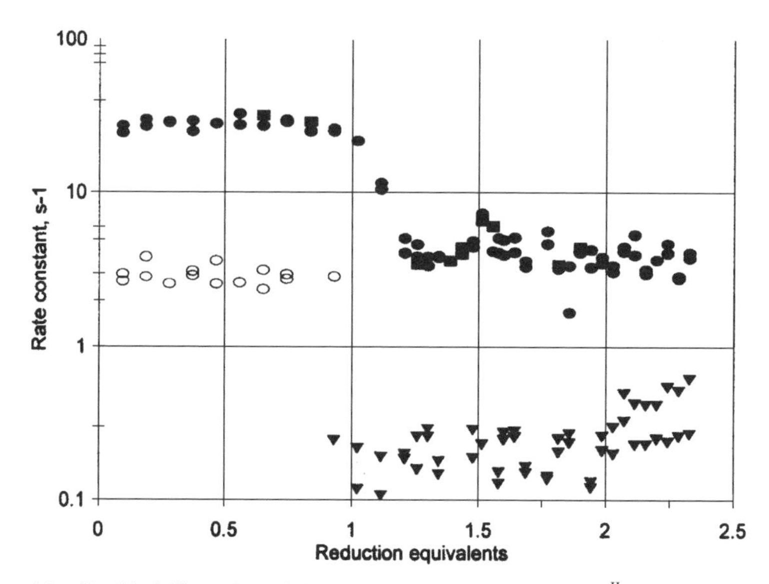


Figure 12. Graphical illustration of the observed rate constants [T1($Cu^{II}$)] reduction, T1($Cu^{I}$) reoxidation as well as $RSSR^-$ decay] in human Ceruloplasmin as a function of the number of reduction equivalents introduced into the solution by $CO_2^-$ radicals. Symbols: (●) reduction of T1($Cu^{II}$); (○) reoxidation of T1($Cu^{I}$); (■) reoxidation of $RSSR^-$; (▼) slow reduction phase of T1($Cu^{II}$).

(Section III.A). Thus the main difference between Az and hCp is that no direct Cu(II) reduction is observed in hCp, in line with the more limited exposure of the T1 sites compared with that of some disulfide bonds.

For hCp, driving forces of 0.99 and 0.90 eV were determined for intramolecular ET from disulfide radicals to T1A and T1B, respectively. Further, the electronic couplings between the three T1 centers and all disulfide sites have been calculated. The optimal ET pathway connects the highly solvent exposed disulfide Cys858/Cys881 of domain 5 and the T1A center of domain 6, which is also the T1 site adjacent to the trinuclear T2/T3 copper center (Fig. 13) (72). Furthermore, the domain 5 disulfide bridge is surrounded by a number of arginines, whose positive charges make this site even more attractive for negatively charged $CO_2^-$ radicals. The pathway connecting the latter disulfide group with the T1A ligating His975, shown in Fig. 13, consists of 20 covalent bonds, 2 hydrogen bonds, and 1 van der Waals contact, giving a coupling decay factor of $\prod \varepsilon = 1.0 \times 10^{-8}$. Assuming similar reorganization energies for electron donor and acceptor as found in azurin, a theoretically calculated rate constant of $10\,s^{-1}$ is obtained, in good agreement with the experimentally observed rate constant of the first phase of intramolecular T1 reduction.

Electronic couplings between pairs of the three different T1 centers have also been calculated. These centers are connected pairwise by 13 covalent and 2 hydrogen bonds, resulting in an electronic decay coupling factor of $1.7 \times 10^{-4}$. However, in spite of this relatively short distance, intramolecular ET between these sites is not expected. The high-potential T1C is, as mentioned above, most likely in the reduced state even in the resting (as isolated) enzyme and thus not involved in ET at all. The driving force for T1A to T1B ET is very low or even

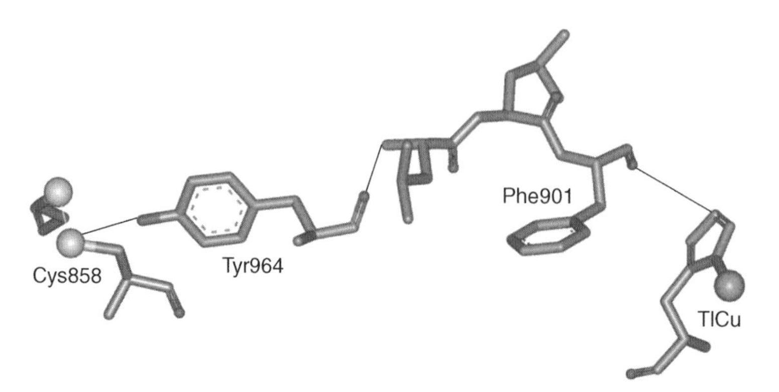


Figure 13. Result of ET pathway calculations in human Ceruloplasmin of the connection between the Cys858–Cys881 disulfide group in domain 5 and $T1(Cu^{II})$ of domain 6 (72). The pathway consists of 20 covalent bonds, one hydrogen bond, and two van der Waals contacts. The distance between the T1 copper ion and $S_\gamma$ of Cys858 is 2.59 nm. Calculations were based on the Beratan and Onuchic model (6, 7). Coordinates were taken from the PDB, code 1KCW. (See color insert.)

negative. Indeed, no kinetic evidence for intramolecular T1 copper-to-copper ET was observed in PR experiments. As mentioned above, the T1A center is in direct covalent connection with the T3 site: The Cys thiolate coordinating T1[Cu] is placed between two His residues that are both coordinated to the T3 copper pair. These two pathways consist of nine covalent bonds, with a direct separation distance between the two copper ions of 1.2 nm. An additional pathway is provided via the carbonyl oxygen of Cys, which is hydrogen bonded to the $N_\delta$ of the His1020 imidazole. Such an arrangement of alternate ET pathways is homologous to the one in AO shown in Fig. 9. Similarly, a calculation of the relative electronic coupling between electron donor and acceptor was found to be $\varepsilon = 0.01$ [i.e., a very effective coupling (72)]. Therefore, the T1A copper ion is in all probability the one involved in ET equilibration with the trinuclear site.

The 3D structure of yeast ferroxidase, Fet3p, has been determined to 2.8-Å resolution in crystals prepared from an *endo*-H deglycosylated protein (60). It consists of three cupredoxin domains with a single T1 site residing in domain 3. The trinuclear center is positioned at the interface of domains 2 and 3 in an arrangement similar to the one observed in other multicopper oxidases, including hCp. The specificity of Fet3p for Fe(II) as substrate has also been inferred from the presence of an array of negatively charged residues, serving as potential iron ligands adjacent to the T1 site. Significantly, both histidine imidazoles coordinated to the T1 copper appear to be hydrogen bonded to two of the latter negative residues, forming the suggested Fe(II) binding site that features a short, effective ET path between these two metal ions. Interestingly, this arrangement is missing in laccases. However, it appears to be homologous to the iron binding site in hCp with the putative ligands Glu272, Glu935, His940, and Asp975 (60). The latter is in hydrogen bond contact with His1026, which is one of the ligands of the T1 center in hCp. Thus, there is a very short effective ET pathway from Fe(II) to T1Cu(II) (60) in hCp, consistent with the proposed physiological role of hCp as a ferroxidase.

Finally, it is of interest to compare intramolecular ET in hCp with the corresponding processes in laccase and ascorbate oxidase. As in hCp, pulse radiolytically produced $RSSR^-$ radicals are also the primary reaction products in tree laccase (73), and the reduction equivalents are further transferred to the $T1(Cu^{II})$ center in an intramolecular process. The rate of $T1(Cu^{I})$ reoxidation by intramolecular ET to T2/T3 takes place unimolecularly with a rate constant of $2\,s^{-1}$ at room temperature, similar to that observed in hCp ($2.9\,s^{-1}$), which is hardly surprising, since the structural arrangements of the T1-T2/3 sites in these two proteins are quite similar (the driving forces also are comparable). The situation in ascorbate oxidase, however, was found to be more complex (Section III.B.2.), since two intramolecular ET phases were observed between $T1(Cu^{I})$ and $T3(Cu^{II})$ with rate constants of 200 and $2\,s^{-1}$, respectively. The observation of two different rates in the AO dimer were explained by differences in activation entropy, which were attributed to differences in electronic coupling in the two monomers (48). Intramolecular ET in hCp is thus reminiscent of the similar process in laccase and to a limited extent in AO. It is noteworthy that steady-state kinetic measurements of hCp activity with Fe(II) as reducing substrate yield a turnover rate of $2.2\,s^{-1}$ (71), which is similar to the rate constant observed for intramolecular T1 to T3 ET. Intramolecular ET thus seems to be the rate-limiting step in the catalytic cycle of this enzyme.

One important point worth stressing is that practically all PR experiments, including those of hCp, were performed under strictly anaerobic conditions (i.e., in absence of an oxidizing substrate). In addition, the reduction potential of T1A is higher than that of T3, which is probably the reason why only slightly more than two electrons were taken up by hCp (72). Coordination of dioxygen to the trinuclear site, which occurs under physiological conditions, will undoubtedly increase the reduction potential of this site and thus the driving force for intramolecular T1($Cu^{I}$) to T2/T3($Cu^{II}$) ET. Under such conditions, further ET from reduced T1 copper to the oxygen coordinated trinuclear center is expected to take place in order to fulfill the requirement for four-electron reduction of dioxygen to water. Finally, the question remains why hCp contains two additional T1 centers that apparently play a minor (if any) role in enzymatic processes. The more recent finding showing that they are absent in the yeast analog of this enzyme (70) makes this puzzle even more intriguing. Thus, it had earlier been suggested that besides being a ferroxidase, hCp may also serve as a copper reservoir in human copper metabolism (67), which then assigns two important roles to hCp in human metal ion metabolism.

### 3. *Copper Nitrite Reductase*

The bacterial dissimilatory copper-containing nitrite reductases (CuNiR), which have been isolated from different bacterial sources, constitute another interesting family of copper containing enzymes, as they partake in the biological denitrification process of reducing nitrate to dinitrogen (74). The interest in these enzymes is amplified by the existence of another family of bacterial nitrite reductases containing heme centers at their active sites. The copper NiRs are homotrimers (109-kDa molecular mass) with two copper ions per monomer in the catalytic unit. There is one T1 copper and one T2 center. The enzyme catalyzes the one-electron reduction of nitrite to nitric oxide (75):

$$NO_2^- + e^- + 2H^+ \rightarrow NO + H_2O \qquad (19)$$

The T1 center serves as the electron uptake site from azurin or pseudoazurin (74). Binding and reduction of substrate nitrite takes place at the T2 site. Hence, as in the blue oxidases discussed above, the internal T1 $\rightarrow$ T2 ET in this copper enzyme is an essential part of the catalytic cycle. In fact, it has been suggested that the rate of ET from T1 to T2 is controlled by changes in the T2 reduction potential induced upon nitrite binding (76, 77). Thus, perturbations of the potentials of the two copper centers (and in turn the driving force of the reaction) are expected to control both the rates and direction of internal ET.

Three-dimensional structures have been determined for NiRs isolated from several different bacteria; the overall structures appear to be very similar (cf. Fig. 14) (75). Each monomer contains a T1 copper (bound to two histidines, a cysteine, and a methionine) and a T2 copper coordinated to two histidines from one monomer and one from another. A water molecule is the fourth ligand to this copper site. Crystallographic studies support the notion that T2 is the site at which nitrite is reduced. The T1Cu(II) site exhibits strong absorption in the visible region where the T2 center is not observed. Small differences in the ligand geometry of the T1 site, evident from the 3D structures of *Alcaligenes xylosoxidans* nitrite reductase (AxNiR) and *Achromobacter cycloclastes* nitrite reductase (AcNiR), determine whether the protein appears blue (as in AxNiR, $\varepsilon_{595} \sim 6300\,M^{-1}\,\text{cm}^{-1}$) (78) or green due to the presence of an additional strong band at lower wavelength (as in AcNiR, $\varepsilon_{458} \sim 4800\,M^{-1}\,\text{cm}^{-1}$) (79). Each of the blue NiRs exhibits an axial EPR signal like those seen in plastocyanin and azurin, in contrast to the rhombic EPR signals of the green NiRs.

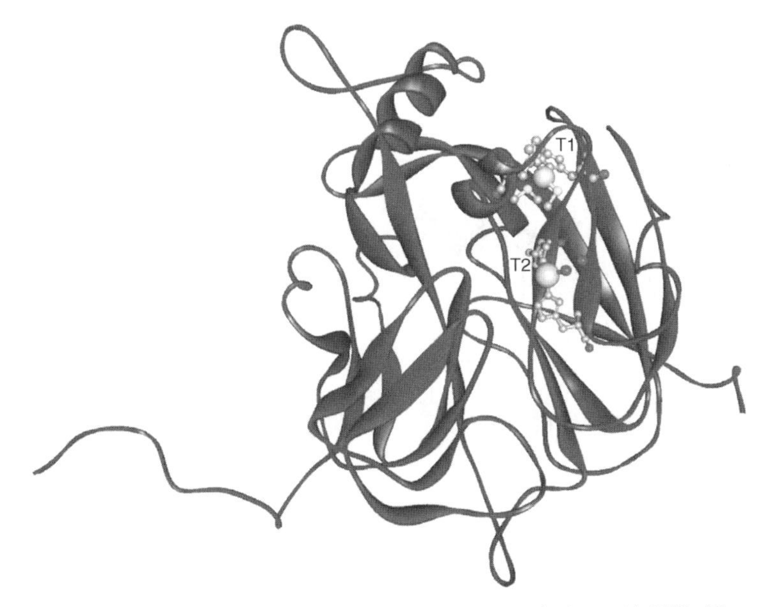


Figure 14. The 3D structure of CuNiR from *Achromobacter cycloclastes* (AcNiR). The protein backbone is shown together with the type1 and type 2 copper ions and their ligands. Coordinates were taken from the PDB, code 1NDT.

Pulse radiolysis studies have been performed on both AxNiR and AcNiR using either $CO_2^-$ or 1-MNA* radicals (80–84). In the first PR experiments of Suzuki et al. (80, 81), direct bimolecular reduction of the $T1Cu^{II}$ site in both the blue AxNiR and green AcNiR established this copper center as the primary electron uptake site using either of the above reducing radicals. Following fast $T1Cu^{II}$ reduction, a slow unimolecular reoxidation process has been observed that was assigned to intramolecular equilibration between $T1Cu^{I}$ and $T2Cu^{II}$. For AcNiR the rate of internal ET from $T1Cu^{I}$ to $T2Cu^{II}$ was found to be $1100\,s^{-1}$, while for AxNiR, ET was significantly slower, with a rate constant of $360\,s^{-1}$ at pH 7.0 (80, 81). Temperatures at which the experiments were performed were not reported. In later PR studies using either 1-MNA* or $CO_2^-$ as reducing radicals, the internal ET kinetics were examined over an extended temperature range (82–84). A typical example of a time-resolved spectrum reflecting reduction of AxNiR is shown in Fig. 15. Rate constants and driving forces for the internal ET in this enzyme are given in Table III. The observed rate constant is the sum of the forward and the backward rate constants: $k_{obs} = k_F + k_B$:

$$T1Cu^{I}T2Cu^{II} \underset{k_B}{\overset{k_F}{\rightleftarrows}} T1Cu^{II}T2Cu^{I} \tag{20}$$

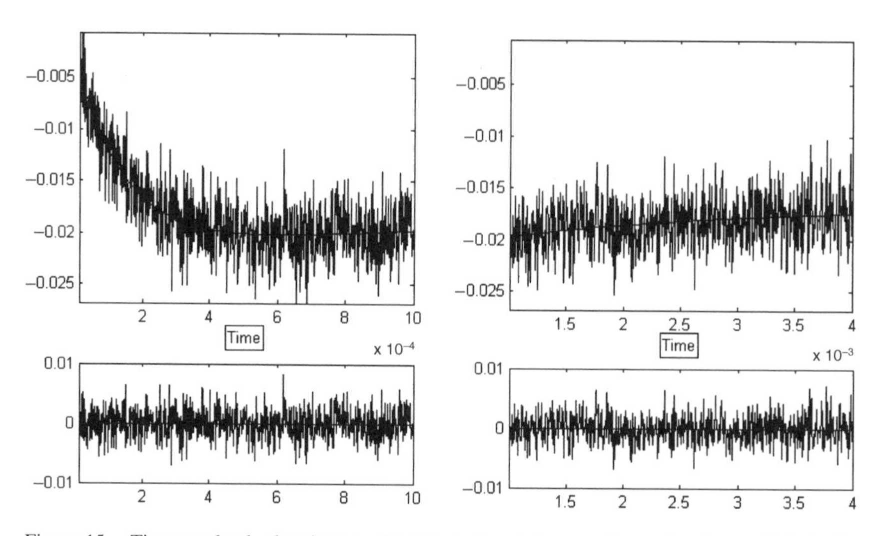


Figure 15. Time-resolved absorbance changes induced by reaction of pulse radiolytically produced 1-MNA* radicals with CuNiR from *Alcaligenes xylosoxidans* (AxNiR). $T = 298$ K; pH 7.0; 0.1 *M* *tert*-BuOH; 1 m*M* 1-MNA; 10 m*M* phosphate; Ar saturated; pulse width 1.0 µs; optical path 12.3 cm. Time scale is in seconds; the left panel shows the faster phase, while the right one shows the reaction taking place at the slower phase. The lower panels show residuals of the theoretical fits to the data.

TABLE III
Kinetic and Equilibrium Data for Internal ET in CuNiRs[a]

| Protein | $k_{298}$ ($s^{-1}$) | $-\Delta G^\circ$ (eV) | $\Delta G^{\neq}$ (eV) | $\Delta S^{\neq}$ (meV/T) | $\lambda_{tot}$ (eV) | $\lambda_{T1}$ (eV) |
|---|---|---|---|---|---|---|
| Blue AxNiR[b] | 450 ± 30<br>185[e]/265[e] | −0.009 | 0.32 | −1.31 ± 0.11 | 1.26 ± 0.08 | 0.77 ± 0.05[c] |
| Green AcNiR[d] | 1030 ± 80<br>334[e]/696[e] | −0.019 | 0.30 | −1.09 ± 0.09 | 1.16 ± 0.07 | 0.57 ± 0.07 |

[a]298 K; pH 7.0; 5 m*M* phosphate
[b]Ref. 82.
[c]Refs. 53 and 85.
[d]Ref. 84.
[e]Rate constants for forward and backward ET process, respectively;

Comparative EPR, extended X-ray absorption fine structure (EXAFS), and UV–Vis spectroscopic studies of reduced and oxidized AxNiR demonstrate that reduction of the enzyme by ascorbate–phenazine methosulfate in the absence of nitrite results in the loss of the water molecule bound to T2 Cu in the oxidized protein (86). The change in coordination geometry of the T2 Cu ion from a near tetrahedral $(His)_3$–$H_2O$ site to a trigonal $(His)_3$ site induced by the reduction would make a major contribution to both the reduction potential and difference in reorganization energy compared with T1Cu, where only minor changes in geometry are associated with a change in redox state (87,88).

The EXAFS measurements of AxNiR showed further that binding of the competitive inhibitor azide is not observed when the T2 Cu ion is reduced, consistent with the observation that the enzyme is inactivated by reduction in the absence of substrate (86). The decreased affinity of the reduced T2 site for nitrite is also apparent in an X-ray crystallographic study of CuNiR from *A. faecalis* that showed a low occupancy of nitrite in crystals of reduced enzyme soaked with this substrate (89). Thus, nitrite must bind to the $T2Cu^{II}$ ion during NiR turnover *before* ET occurs from the reduced T1 center. A nitrite-induced gating of ET during catalysis has been suggested to involve the longer linkage between the two Cu centers formed by amino acid residues His89 (a T1Cu ligand) and the T2 Cu ligand His94 (AxNiR numbering) (86).

The subtle structural differences in the T1 copper sites of enzymes isolated from different bacteria provide a good opportunity to resolve their contribution to the reorganization energies of the copper centers. It is noteworthy that the green enzyme exhibits a higher rate of intramolecular ET than the blue CuNiR in spite of a lower driving force ($-\Delta G^\circ$) for the process (cf. Table III). It is also interesting that the activation entropy is larger for the green CuNiR as compared with the blue enzyme. Since the activation entropy includes an electronic term, electron tunneling from/to the green T1 copper center could be slightly more advantageous.

The ET pathways in both blue and green NiRs are very short, consisting of the T1Cu ligand Cys130 and the neighboring His129 ligand of T2Cu (AcNiR numbering); altogether 11 covalent bonds corresponding to a 1.26-nm distance between the two Cu ions (Fig. 16), which leads to an electronic coupling factor of $\varepsilon = 3.6 \times 10^{-3}$. The pathway is reminiscent of the link between T1 and the trinuclear site in blue oxidases, thus being a structural feature shared by both families of multicopper reductases and oxidases. Gray et al. (90) determined rates of bond-mediated electron tunneling in modified iron–sulfur proteins where electron donor and acceptor are separated by a similar Cys-His bridge as found in the CuNiRs. The experimentally determined rate contants are essentially coupling limited (i.e., $k_{max} \sim 2 \times 10^7 - 2 \times 10^8\ s^{-1}$).

With identical ET pathways in the two WT CuNiRs, the observed higher rate constant in the green enzyme despite a smaller driving force must be due to a difference in reorganization energies. For green AcNiR a $\lambda_{tot} = 1.16 \pm 0.07$ eV was calculated, and in blue AxNiR $\lambda_{tot} = 1.26 \pm 0.08$ eV (Table III) (84). The reorganization energy of the blue T1 copper center was found to vary from 0.72 eV in plastocyanin (92) to 0.82 eV in azurin (53). Comparison of the 3D structures of the blue sites of azurin, plastocyanin, and (blue) AxNiR shows similar geometries and metal ligand distances (75). Thus, it is safe to assume a reorganization energy of the T1 center in the blue WT AxNiR to be in the same range ($\lambda_{T1} = 0.77 \pm 0.05$ eV). The reorganization energy of the T2 copper center may now be calculated from the relation $\lambda_{tot} = \lambda_{T1}/2 + \lambda_{T2}/2$ (3) giving $\lambda_{T2} = 1.75$ eV which, as expected is much larger than that calculated for T1.

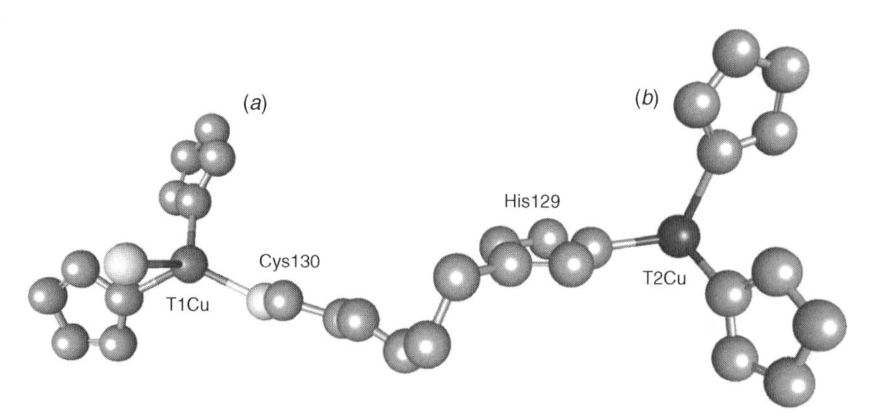


Figure 16. Calculated ET pathway between T1Cu (*a*) and T2Cu (*b*) in CuNiR from *A. cycloclastes* (AcNiR). The short path consists of the T1 ligand, Cys130 and the neighboring His129, which coordinates to T2Cu. The connection consists of 11 covalent bonds, and the distance between the copper centers is 1.26 nm. Calculations were based on the Beratan and Onuchic model (6, 7). Coordinates were taken from the PDB, code 1NDT. (See color insert.)

The T2 center is solvent accessible as it is involved in nitrite binding, reduction, protonation, and product release via a 1.3-nm deep hydrophobic channel (84). In contrast, the T1 center is buried inside the protein, 0.6 nm below the Connolly surface of the molecule and isolated from solvent. Nonetheless, the reorganization energy calculated for the T2 copper center is still below values quoted in the literature for low molecular weight copper complexes: For $Cu(phen)_2^{2+/+}$ (phen = 1,10-phenanthroline), for example, the reorganization energy has been determined to be 2.4 eV (91).

Although the T1 center of green AcNiR exhibits the "classic" coordination sphere of a T1 site with an axial S(Met), the distortion of the site weakens the Cu—S(Cys) bond, as demonstrated by spectral features with a decrease in the dominant S(Cys)$\pi$ Cu(II) charge-transfer intensity together with a more significant S(Cys)$\sigma$ Cu(II) intensity that causes increased absorption around 450 nm changing the color from blue to green (92).

Since the structures of the T2 carrying domains of AcNiR and AxNiR are essentially identical, the respective reorganization energies, $\lambda_{T2}$ are expected to be equal in the two proteins, and from $\lambda_{tot}$ values determined for the distorted green copper center in AcNiR a $\lambda_{T1} = 0.57 \pm 0.07$ eV can be calculated (84). Thus, the more asymmetric flattened tetrahedral T1 site of the green CuNiR gives rise to a lower reorganization energy than the distorted tetrahedral geometry of the blue T1 ($\lambda_{T1} = 0.77 \pm 0.05$ eV) (cf. Table III). It is noteworthy that $\lambda_{T1}$ here is still larger than that of the binuclear (purple) $Cu_A$ center found to be 0.4 eV when inserted in a mutated azurin (52) (cf. Section III.A).

In conclusion, the tetragonal distortion, possibly arising from small shifts in the loop carrying the Met ligand in the blue and green enzymes, emphasizes the subtle, yet important, role of the geometric and electronic structure of the T1 site in determining its reactivity (84). Further studies of the contribution of structure to driving force and reorganization energy in proteins are clearly needed, and single site CuNiR mutants are expected to be useful in this respect.

Several mutants of the blue AxNiR have already been produced, and in one of these the weaker T1 ligand, Met144, was substituted by a nonligating alanine, and the 3D structure of this mutant has been determined at 2.2-Å resolution (93). The mutant still maintains 30% activity relative to the WT enzyme, yet the reduction potential of T1 was found to have increased from 240 to 314 mV. The T2 potential (230 mV) is expected to remain constant, since the T2 structure is the same as that of the WT. The change in driving force caused by the above mutation turned out to have a dramatic influence on the kinetics of intramolecular ET: No direct bimolecular reduction of $T1Cu^{II}$ is observed in the M144A mutant. Rather, $T2Cu^{II}$ is reduced directly by $CO_2^-$ or 1-MNA* radicals, after which reverse ET takes place, that is $T2Cu^{I}$ to $T1Cu^{II}$ with a rate constant of 425 $s^{-1}$ (83). The 3D model of M144A shows that the T1 site is more protected from solvent compared to WT CuNiR, since the T1 copper ion is significantly disordered and has moved 0.3 Å away from the cavity, together with the ligating His139 residue (93). The His139 is essential for bimolecular reduction of $T1Cu^{II}$ by physiological electron donors, since it has been shown that a His139Ala mutant is unreactive with natural reducing substrates (azurin or pseudoazurin) (94). Notably, reduction of the *Alc. xylosoxidans* His139Ala mutant by dithionite/methyl viologen was also shown to proceed via reaction at the $T2Cu^{II}$ site (94).

The PR studies of CuNiR demonstrate that internal ET can be the rate-determining step for catalytic activity. Control of the internal ET rates seems to be attained through ligand changes at T2, that is, the substrate reduction site (including binding of the oxidizing substrate here) and the reorganization energy of the electron acceptor site, that is, $T1Cu^{II}$. The ligand changes modify both driving force and reorganization energy, while electronic coupling is maintained by an ET pathway consisting of a short, direct covalent peptide stretch connecting electron donor and acceptor.

## IV. IRON-CONTAINING PROTEINS

### A. $cd_1$ Nitrite Reductase

The bacterial, iron-containing $cd_1$ nitrite reductases constitute another family of enzymes catalyzing the one-electron reduction of nitrite to nitric oxide (74). These enzymes are homodimers of 60-kDa subunits, each containing one heme-*c* and one heme-$d_1$. Extensive studies have established heme-*c* as the electron entry site, whereas heme-$d_1$ is the catalytic center where nitrite is reduced (95). Three-dimensional structures of two different cytochromes $cd_1$ have been determined in oxidized and reduced states: *P. pantotrophus*, Pp-NiR (96, 97) and *P. aeruginosa*, Pa-NiR (98, 99). In both enzymes, heme-*c* is covalently linked to the *N*-terminal α-helical domain and heme-$d_1$ is bound noncovalently to the *C*-terminal ß-propeller domain (Fig. 17). Intramolecular ET between *c* and $d_1$ hemes is an essential step in the catalytic cycle; this reaction has been studied by several groups using different methods (95, 100–106). The rate constants for Pa-NiR are on the order of $1\ s^{-1}$ (95, 100–102, 104), while intramolecular ET in Pp-NiR is significantly faster (rate constant of $1.4 \times 10^3\ s^{-1}$) (103). Such a pronounced difference in rates is particularly interesting, as the distances separating the two heme centers within a subunit, are similar (1.1-nm edge-to-edge or 2.0-nm Fe-to-Fe) in the two proteins (96, 98). Likewise, differences in reaction driving forces do not explain such a difference in rates. Interestingly, it was found that the rate of intramolecular electron transfer in Pa-NiR was accelerated four orders of magnitude by cyanide coordination to the oxidized heme-$d_1$ (104); it was suggested that, in the absence of cyanide, a coordination change takes place upon reduction of this heme,

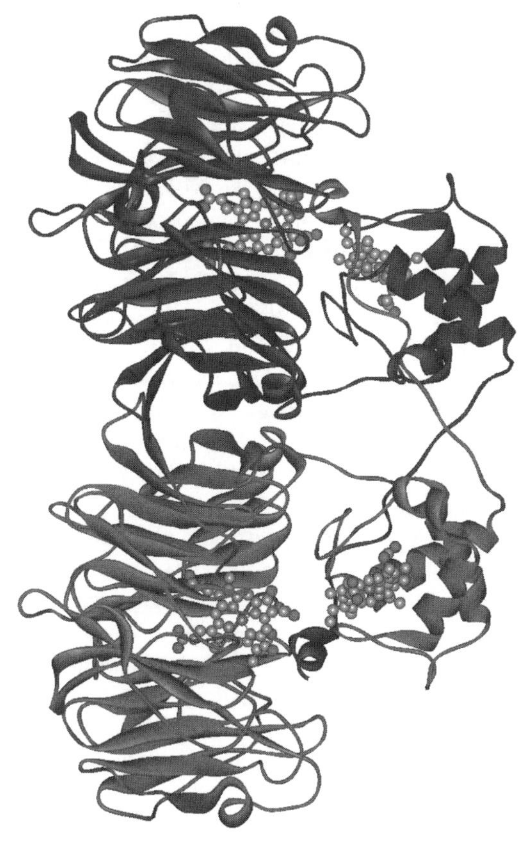

Figure 17. The 3D structure of *P. aeruginosa* $cd_1$NiR. The two subunits containing the heme-*c* centers are seen on the right-hand side while the heme-$d_1$ centers are buried in the subunits shown to the left in the figure. The coordinates were taken from the PDB, code 1NNE. (See color insert.)

which might be a major rate-determining factor in Pa-NiR ET. Thus, structural differences among the $cd_1$-NiRs may be decisive in modulating the different intramolecular ET reaction rates. Indeed, unexpectedly for such closely related enzymes, the structural details of the proteins differ significantly: The "as isolated" Pp-NiR heme-*c* iron(III) has His–His axial ligands, whereas the heme-$d_1$ iron(III) axial ligands are Tyr–His (96). Upon reduction, the heme-*c* iron(II) ligands switch to His–Met concomitantly with dissociation of the tyrosine ligand, leaving the heme-$d_1$ iron(II) five coordinate (107). These differences are an apparent feature of the "as isolated" Pp enzyme, which causes it to be essentially inactive. A transition to a catalytically active form can be induced by reoxidation of the fully reduced enzyme by hydroxylamine (108). Thus, His–His heme-*c* and Tyr-OH ligation are most likely inert conformers formed during the isolation procedure, and their functional role is unclear.

In contrast to Pp-NiR, heme-*c* in Pa-NiR is His–Met coordinated in both oxidation states; the axial heme-$d_1$ ligands are hydroxide and His in the oxidized state, whereas only one His is bound in the five-coordinated reduced state (vacant/His) (99). A remarkable feature of Pa-NiR is the so-called arm exchange or domain swapping of the *N*-terminal region, which places Tyr10 of one monomer close to the heme-$d_1$ of the other one (cf. Fig. 17). In the oxidized enzyme, Tyr10 is hydrogen bonded by its OH group to the hydroxide ligand of heme-$d_1$, thereby preventing access of the substrate to the catalytic site (98). No such "domain swapping" was observed in Pp-NiR. Instead, a segment of the heme-*c* *N*-terminus makes an excursion into the $d_1$ domain and the OH of Tyr25 coordinates directly to the heme-$d_1$ iron of the same monomer (96).

Redox related structural changes are apparently less dramatic in the Pa enzyme; they take place mainly in the neighborhood of heme-$d_1$ (109), which upon nitrite reduction is expected to exchange at least one of its ligands. Still, as the reduced form has been prepared by diffusion of the reductant into crystals of the enzyme, some changes in geometry could have been missed. Indeed, marked flexibility has been established in the recently determined 3D structures of two single-site mutants of the Pa enzyme where the two invariant active site histidines, His327 and His369 were substituted with Ala, which caused a 60° rotation of the heme-*c* domain relative to that of heme-$d_1$ (110). Since enzymatic activities of these mutants were also significantly reduced, the results illustrate the functional importance of these residues and their long-range structural impact.

Sequence comparison of enzymes isolated from the different bacteria shows a striking difference between Pp and Pa $cd_1$NiRs and that isolated from *P. stutzeri*, (Ps-NiR), where the *N*-terminal arm, including the tyrosine residues mentioned above, is missing (98). The absence of this peptide stretch in the Ps enzyme suggests that there may be differences in the intramolecular ET mechanism. In view of these very interesting differences in primary structure of $cd_1$-NIR from Pa, Pp, and Ps, as well as the observed major differences in internal ET rates of the two former proteins, PR investigation of ET between the *c*- and $d_1$-hemes was extended to the Pa- and Ps- NiRs.

1-Methylnicotinamide radicals (1-MNA*), produced pulse radiolytically, react with Ps-NiR concomitantly with the appearance of an absorption increase at $\sim$ 554 nm, a characteristic wavelength for heme-*c* reduction [Fig. 18(*a*)] (105, 106). At 670 nm, where heme-$d_1$ reduction can be independently monitored, no absorption changes were observed in the fast time domain [Fig. 18(*b*)], indicating that no parallel bimolecular reduction of the heme-$d_1$ center by 1-MNA* is taking place. Therefore it was concluded that only heme-*c* is reduced directly by 1-MNA* and in an essentially diffusion controlled process, with a second-order

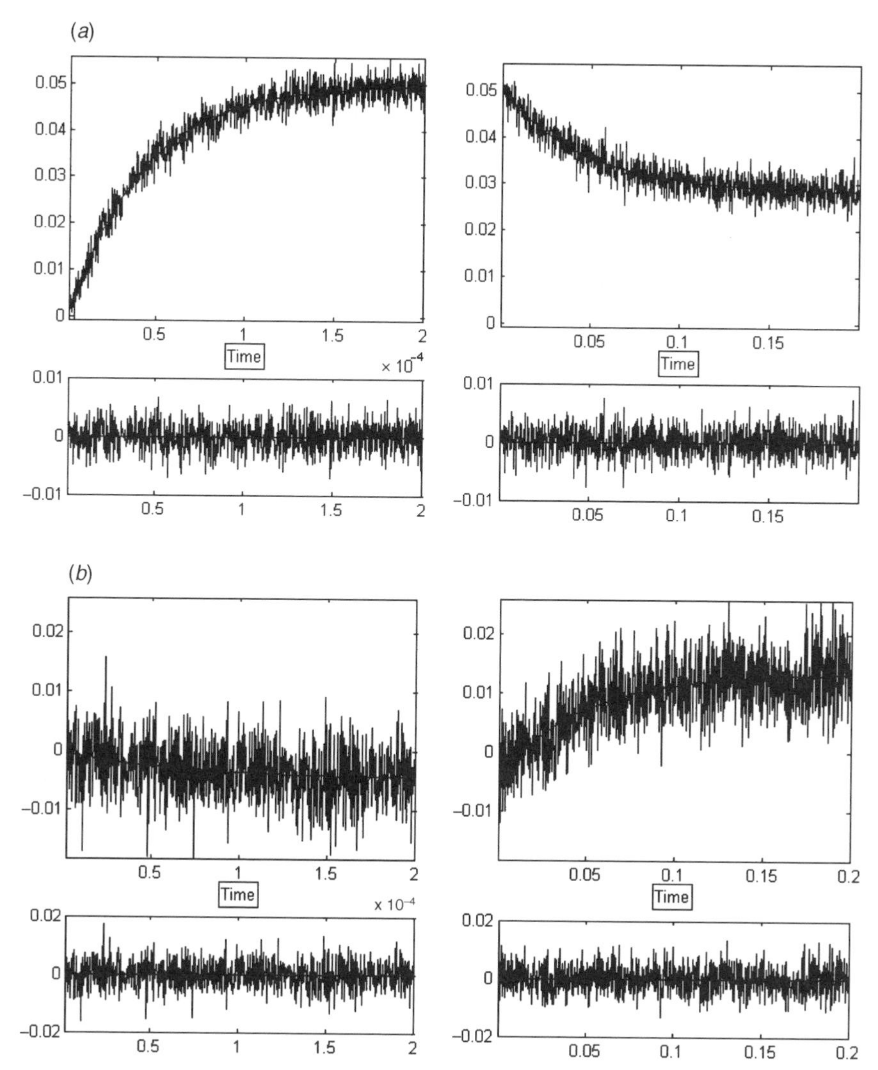


Figure 18. Time resolved absorption changes in $cd_1$NiR from *P. stutzeri* induced upon reaction with PR produced 1-MNA* radicals. (*a*) Monitoring redox changes in heme-*c* at 554 nm. (*b*) Measurement at 670 nm, showing heme-$d_1$ reduction on the slower time scale (left panel). The conditions were: 24-μ*M* enzyme; 5 m*M* 1-MNA; 5 m*M* phosphate; 0.1 *M tert*-BuOH; $T = 298$ K; pH 7.0; Ar saturated; 1-cm optical path; pulse width 1.0 μs. Time is in seconds; the left panel shows the faster phase, while the right one shows the reaction taking place at the slower phase. The lower panels show residuals of the fits to the data.

rate constant of $10^9\,M^{-1}\,s^{-1}$ at pH 7.0 and 298 K. Following the initial direct heme-*c* reduction, a decrease in the 554-nm absorption was observed on a slower time scale [Fig. 18(*a*); right panel], indicating reoxidation of the heme-*c* Fe(II) center. Indeed, concomitant absorption increases could be monitored at both 460 nm (not shown) and 670 nm [Fig. 18(*b*); right panel] where reduced heme-$d_1$ has a strong absorption band. Analyzing these spectral changes characteristic for both heme-*c* and heme-$d_1$ yielded a rate constant $k_{ET} = 23\,s^{-1}$ at 298 K and pH 7.0, independent of 1-MNA* and initial protein concentrations, implying intramolecular ET between heme-*c* and heme-$d_1$ (105, 106). Analysis of the reaction amplitudes, however, made it clear that the internal electron-transfer processes do not proceed to completion: At 298 K, only 50% of the reduction equivalents initially taken up by oxidized heme-*c* are transferred to the heme-$d_1$ Fe(III), consistent with an equilibrium constant of ~1.0:

$$\text{heme-}c\,\text{Fe}^{\text{II}}\ \text{heme-}d_1\,\text{Fe}^{\text{III}} \underset{k_{21}}{\overset{k_{12}}{\rightleftarrows}} \text{heme-}c\,\text{Fe}^{\text{III}}\ \text{heme-}d_1\,\text{Fe}^{\text{II}} \tag{21}$$

The observed rate constant, $k_{ET}$, is the sum of rate constants for the forward and back ET reactions: $k_{ET} = k_{12} + k_{21}$. Combining the observed rate and equilibrium constants, $k_{12}$ and $k_{21}$ can be calculated. Similar PR studies have been performed on $cd_1$, nitrite reductase from Pa, and qualitatively the same results

TABLE IV
Kinetic and Thermodynamic Results for Internal ET Equilibration between Heme-*c* and Heme-$d_1$ at 298 K and pH 7.0

| | ***Pseudomonas aeruginosa***[a] | | |
|---|---|---|---|
| Rate Constants ($s^{-1}$) | $k_{obs} = 2.9$ | $k_{12} = 0.7$ | $k_{21} = 2.2$ |
| $\Delta H^{\neq}$ ($kJ\,mol^{-1}$) | | $+49.0 \pm 3.8$ | $+84.7 \pm 8.2$ |
| $\Delta S^{\neq}$ ($J\,K^{-1}\,mol^{-1}$) | | $-83 \pm 7$ | $+46.1 \pm 4.5$ |
| Equilibrium data | $K = 0.3$ | | |
| $\Delta H^\circ$ ($kJ\,mol^{-1}$) | $-34.5 \pm 5.2$ | | |
| $\Delta S^\circ$ ($J\,K^{-1}\,mol^{-1}$) | $-125 \pm 18$ | | |
| | ***Pseudomonas stutzeri***[b] | | |
| Rate constants ($s^{-1}$) | $k_{obs} = 23$ | $k_{12} = 11.7$ | $k_{21} = 11.3$ |
| $\Delta H^{\neq}$ ($kJ\,mol^{-1}$) | | $+46.2 \pm 4.8$ | $+71.6 \pm 4.8$ |
| $\Delta S^{\neq}$ ($J\,K^{-1}\,mol^{-1}$) | | $-71.0 \pm 1.5$ | $+13.6 \pm 1.5$ |
| Equilibrium data | $K = 1.0$ | | |
| $\Delta H^\circ$ ($kJ\,mol^{-1}$) | $-24.9 \pm 2.5$ | | |
| $\Delta S^\circ$ ($J\,K^{-1}\,mol^{-1}$) | $-83 \pm 8$ | | |

[a]Ref. 111.
[b]Ref. 105.

were obtained although the reactions are considerably slower. Table IV summarizes the observed rate and equilibrium data for Ps-NiR (105) together with preliminary results of PR studies on Pa-NiR (111), all obtained under conditions of *limited* reduction (i.e., up to one-electron equivalent per mole enzyme).

The electron-transfer pathway between the $c$ and $d_1$ sites (Fig. 19) for the reduced Pa protein involves Cys50 that covalently connects heme-$c$ to the protein and two water molecules, one of which forms a hydrogen bond to heme-$d_1$. The pathway includes 12 covalent bonds and 3 less efficient hydrogen bonds, over a 2.0-nm distance between the Fe atoms. The electronic coupling is calculated to be $\varepsilon = 1.0 \times 10^{-4}$. Thus, this pathway has a significantly lower electronic coupling than that linking the T1 and T2 sites in CuNiR discussed in Section III.B.3. Everything else being equal, the rate of intramolecular between electron donor and acceptor is expected to be three orders of magnitude lower in heme NiR compared to the same process in CuNiR.

It is of interest to compare the rate constants determined for intramolecular ET in Ps-NiR (23 $s^{-1}$ at 298 K, pH 7.0) with the corresponding ones reported for Pa- and Pp-NiRs (cf. Table V). An early stopped-flow study of the reduction kinetics of Pa-NiR by an excess of reduced azurin yielded a rate constant of 0.25 $s^{-1}$ at pH 7.0 and 298 K (100). In another stopped-flow study of Pa-NiR, Silvestrini et al. (95) determined a rate constant of 1 $s^{-1}$ at 293 K and pH 8 for the internal ET from heme-$c$ to oxidized NO-bound heme-$d_1$. Schichman and Gray (102) have performed stopped-flow measurements on Pa-NiR using $Fe(edta)^{2-}$, where edta = ethylenediaminetetraacetic acid, as reductant and obtained a rate constant of 0.3 $s^{-1}$ at pH 7.0 and 298 K. Interestingly, PR studies

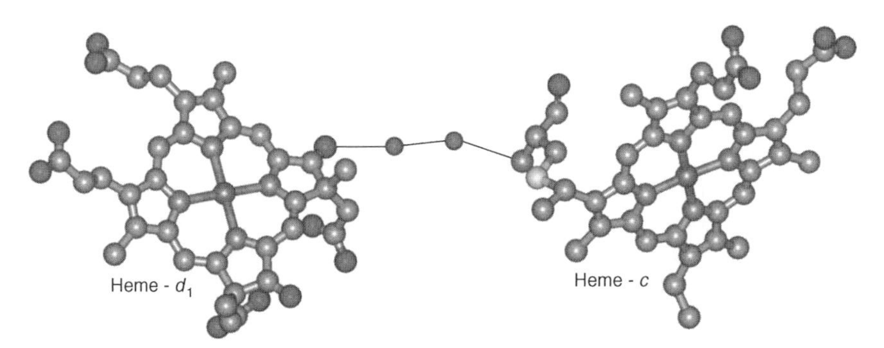


Figure 19. Calculated ET pathway between heme-$c$ and heme-$d_1$ in *P. aeruginosa* $cd_1$NiR. The pathway includes Cys50, covalently connected to heme-$c$, a van der Waals contact to a water molecule, and two hydrogen bonds, the latter of which connects to a carbonyl oxygen on the $d_1$ heme ring. The distance between the iron atoms is 2.0 nm. Calculations were based on the Beratan and Onuchic model (6, 7). Coordinates were taken from the PDB, code 1NNE. (See color insert.)

TABLE V

Rate Constants for Intramolecular ET in Nitrile Reductases from Different Sources

| Source | $k_{ET}$ ($s^{-1}$) | Conditions | References |
|---|---|---|---|
| *Paracoccus pantotrophus* | 1400 | pH 7.0 | 103 |
| *P. aeruginosa* | 0.25 | pH 7.0; 298 K | 100 |
| *P. aeruginosa* | 0.3 | pH 7.0; 298 K | 102 |
| *P. aeruginosa* | 3 | pH 7.0 | 104 |
| *P. aeruginosa* | 2.9 | pH 7.0; 298 K | 111 |
| *Pseudomonas stutzeri* | 23 | pH 7.0; 298 K | 105, 106 |

on Pa-NiR revealed slightly faster rates of intramolecular ET (cf. Table V) (104, 111). As already mentioned, Kobayashi et al. (103), using pulse radiolysis to study the intramolecular heme-$c$ to heme-$d_1$ ET in Pp-NiR, reported a considerably larger rate constant, $1.4 \times 10^3\ s^{-1}$ at pH 7.0. However, employing rather larger radiation pulses, they observed an essentially quantitative transfer of the reduction equivalents from the reduced heme-$c$ to oxidized heme-$d_1$, implying an equilibrium constant in Pp-NiR, which is far from unity. Thus, the rates determined in studies of Ps-NiR (105) fall between those observed for the two other nitrite reductases, raising the question of what causes such a pronounced difference in reactivity among such structurally and functionally similar enzymes.

Driving force is first considered as the cause for this difference in ET rates. In the absence of external ligands, the reduction potentials of heme-$c$ and heme-$d_1$ in Pa-NiR are similar, ~280 mV (102). In contrast, the reported irreversibility of the intramolecular $c \rightarrow d_1$ ET in Pp-NiR indicates that the difference in reduction potentials between these sites must be > 100 mV (103). The internal ET equilibrium constant in Ps-NiR is similar to the one found in Pa-NiR, with a value of ~1.0 at 298 K and pH 7.0 (105), an observation that excludes differences in driving force as being responsible for the observed ten to hundred fold faster rate of intramolecular ET in Ps- compared with Pa-NiR. Hence, the causes could be subtle differences in the nature of the intervening media (i.e., the ET pathway) and/or differences in reorganization energies.

Relevant structural features are therefore considered: The sixth ligand in Pa-NiR, a hydroxide ion, is hydrogen bonded to Tyr10 belonging to the $N$-terminal stretch of the second monomer (98), but no homologous tyrosine is present in the Ps-NiR sequence. The oxidized heme-$c$ domain of Pa-NiR exhibits the classical $c$-type cytochrome coordination by His and Met, and it has been shown by magnetic circular dichroism (MCD) and EPR spectroscopy that, in the oxidized state, heme-$c$ of Ps-NiR also has His-Met axial ligation, while studies of heme-$d_1$ suggest that the axial ligands are His-Tyr or possibly His-hydroxide (112). Major conformational changes were observed upon reduction of Pp-NiR: The Tyr ligand dissociates from the heme-$d_1$ iron coordination sphere, leaving the site five coordinate (107). However, as indicated above the oxidized "as isolated" Pp-enzyme is catalytically inactive and further discussion is not warranted, owing to a lack of information on the structures of species on which kinetics data are available. Much smaller conformational changes were found to take place in Pa-NiR: Reduction of the *c*-heme is not accompanied by ligand exchange, while reduction and hydroxide ion dissociation are closely linked at heme-$d_1$, thereby providing access to external ligands, such as oxidizing substrates (99). Thus, the much higher rate of intramolecular ET in Pp-NiR compared with Pa-NiR could be due to the lack of reactivity of heme-$d_1$ in the latter enzyme with hydroxide coordinated to the iron(III) center, which makes the $OH^-$ ligand dissociation rate limiting for ET. Indeed, this notion is supported by 3D structural studies on Pa-NiR with a reduced heme-*c* and an oxidized heme-$d_1$, where the latter still binds to the hydroxide ligand (113). However, the faster intramolecular ET observed in Pp-NiR (103) could also be due to the experimental conditions of the PR studies: With His–His axial heme-*c* ligation the reduction potential will be considerably lower than with Met–His coordination after a redox turnover. In the enzymatically active state, the driving force for ET from reduced heme-*c* to oxidized heme-$d_1$ is expected to be considerably smaller. It would therefore be interesting to perform ET measurements on Pp-NiR that has undergone turnover and ligand switching that transforms it into the catalytically active state.

Unfortunately, no 3D structure is available for Ps-NiR. Still, as mentioned above, sequence comparison demonstrates a high degree of homology with the two other $cd_1$ nitrite reductases (98), although the two above-mentioned major differences between Pa- and Ps-NiR sequences deserve attention. The first is the lack in Ps-NiR of the *N*-terminal stretch containing Tyr10 found in the Pa enzyme, since it may be involved in the reaction of the proteins. The second is related to the domain-swapping observed in Pa-NiR. Obviously, both these structural differences may substantially affect ET pathways in the two proteins. Interestingly, kinetic studies of the photodissociation of CO from fully reduced Pa- and Ps-NiR yielded evidence for fast global structural changes, probably excluding a major role for the extended *N*-terminal peptide in this process (114). A second transient decay was observed in Ps-NiR on a much slower time scale, with a rate constant of $8\,s^{-1}$. It is not clear whether this rate depends on CO concentration, but it is noteworthy that it is quite similar to that of the ET rate observed for Ps-NiR. A better understanding will obviously depend on acquiring the 3D structure of this protein.

It is of interest to compare rate constants and activation parameters of the intramolecular ET in Ps-NiR with those determined for analogous intramolecular ET processes in other multicentered redox enzymes. Reversible intramolecular ET reaction between type 1 and 2 sites in the copper-containing nitrite reductases (CuNiR isolated from *A. xylosoxidans* and *A. cycloclastes*) have already been discussed in Section III.B.3. Rate constants for $T1Cu^{I}$ to T2 $Cu^{II}$ ET are 185 and 334 $s^{-1}$, respectively, that is, more than 10-fold faster than the forward ET process in Ps-NiR (11 $s^{-1}$) under similar conditions (105, 106). Comparing the activation parameters of the two processes provides a rationale for this difference: The ET activation enthalpy in AxCuNiR is much smaller (22.7 kJ $mol^{-1}$) (82) than values 46 – 48 kJ $mol^{-1}$ determined for Ps- and Pa-NiR (Table IV), suggesting that structural reorganization plays a much smaller role in the copper enzymes, as their redox sites are constrained by the more rigid ß-sheet structure (Fig. 14). Further, it seems plausible that the markedly slower intramolecular ET rate constants in Ps-NiR compared with those determined for bovine or bacterial cytochrome *c* oxidases (cf. Section V.A below) are due to functionally related structural changes, such as ligand substitution in the coordination spheres of the hemes: Reduction of Pa-NiR, heme-$d_1$ requires, as suggested above, dissociation of the hydroxide ion ligand, and the rate of ET from heme-*c* to heme-$d_1$ is possibly governed by dissociation of this $OH^-$ ligand; this event could indeed be the built-in gating mechanism controlling ET rate, and a similar gating mechanism could apply to Ps-NiR. The activation parameters determined for intramolecular ET processes demonstrate some noteworthy features: Particularly the difference between activation entropies for heme-$d_1$ reduction (process 12 in Eq. 22) and reoxidation (process 21 in Eq. 22) deserves attention: A channel leading to the heme-$d_1$ pocket in Pa-NiR is composed of polar side chains and main chain atoms. It has been shown that the Tyr10 side chain in Pa-NiR rotates upon heme-$d_1$ reduction, leaving the catalytic site open for binding of either substrates or solvent molecules (98), which would lead to an expected decrease in activation entropy, as observed for Ps-NiR. Upon reoxidation and release of exogenous ligands, such as reaction products or water molecules, this site closes again, thereby causing increase in activation entropy, in accordance with experimental observations (105).

The above results provide some insights into the intricate control that may operate in internal ET processes in multicenter redox enzymes. Specifically, it is of interest to compare control mechanisms exerted on sites that function only in mediating ET (e.g., T1 or $Cu_A$ sites) with those that function as catalytic centers, that is, sites that interact with substrates ($d_1$-heme, T2 sites). Indeed, while the former centers are characterized by relatively low reorganization energies, the latter have relatively high ones.

A different control mechanism results from allosteric interactions between distinct sites; this is a central regulatory mechanism employed by proteins. While there are numerous examples of allosteric control of the activity of enzymes, receptors, and transport proteins, regulation of ET in redox enzymes has rarely been documented, and no kinetic analysis of such processes has been performed. Interestingly, site–site interactions involved in control of electron distribution and transfer rates have recently been reported between the heme sites in both Pa- and Ps-NiR (106, 111).

As discussed above, earlier studies on Pa- and Pp-NiR were performed under conditions where the number of reduction equivalents (i.e., pulses) introduced into the protein solution was limited; as a result, enzyme molecules were reduced by only one-electron equivalent (103–105). Also, the equilibrium and activation parameters given in Table IV were determined for this first part of the reductive titration of the enzyme, that is, where only up to *one* reduction equivalent was taken up (105). However, adding sequential pulses into solutions of either Pa- or Ps-NiR under anaerobic conditions resulted in an accumulation of reduction equivalents in the heme sites, eventually leading to a fully reduced enzyme (106). Surprisingly, upon introducing more than one-electron equivalent to the Ps enzyme the internal $c$ to $d_1$ ET rates decreased by more than two orders of magnitude as illustrated in Fig. 20(*a*), showing that the intramolecular ET rate depends on the degree of enzyme reduction. Similarly, the internal electron distribution between $c$ and $d_1$ heme sites in each monomeric subunit depends on the number of reduction equivalents already taken up by the enzyme [Fig. 20(*b*)]. The same pattern has been observed over the whole temperature range examined (3–40°C) first for Ps and later also for Pa-NiR (106, 111).

The dependence of heme–heme ET rates and equilibria on the degree of enzyme reduction has been analyzed using a model that involves electron uptake by the $c$-hemes followed by equilibration between heme-$c$ and -$d_1$ within the same subunit (106). Intersubunit ET can be excluded, because the heme–heme separation distances in the dimer are too large to allow the reaction to occur in the time domain investigated. Intermolecular ET between enzyme dimers is also unlikely in the examined time window. Reaction schemes of all the major intramolecular ET reactions are shown in Fig. 21, while Table VI gives results of the equilibrium analysis for Pa- and Ps-NiR based on a model (Fig. 21) that encompasses all four equilibration steps in which intrasubunit ET can take place. Standard enthalpy and entropy changes for the intraprotein ET equilibria are also presented in Table VI. The exceptionally large changes in both enthalpy and entropy indicate that distinct mechanisms likely are operative in the different reaction steps.

The detailed internal electron distribution equilibria calculations were based on observed absorbance values (as described in the legend to Fig. 20(*b*)). Large changes accompany the transformation from one reduction state to the other in both Pa- and Ps-NiR (Figs. 20 and 21). The following discussion will focus on equilibrium distributions in the latter enzyme, although it is obvious from the data presented in Table VI that essentially similar equilibrium distributions are found in both enzymes.

While there is equidistribution with a single electron per enzyme dimer, a second equivalent leads to an asymmetric distribution (species 5 dominating),

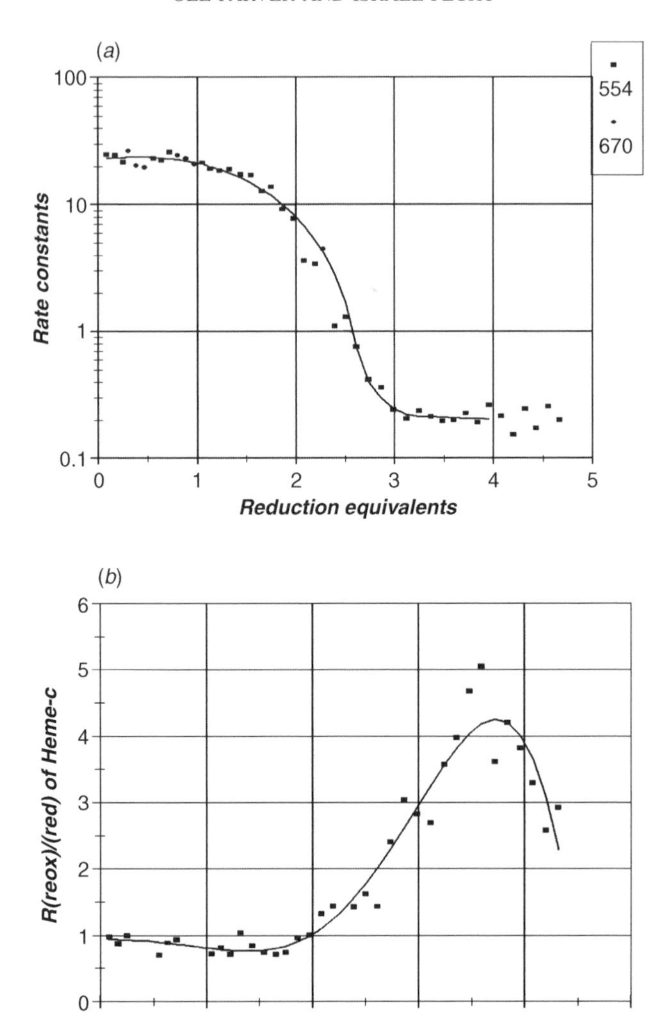


Figure 20. ($a$). Rate constants of intramolecular ET between heme-$c$ and heme-$d_1$ in the *P. stutzeri* enzyme as a function of reduction equivalents taken up. Symbols: (■) indicates heme-$c$[Fe(II)] reoxidation (554 nm) while (•) represents heme-$d_1$($Fe^{III}$) reduction (670 nm). ($b$) Ratios between observed amplitudes of heme-$c$ reoxidation and heme-$c$ reduction, respectively, following each pulse. The first (fast, bimolecular) heme-$c$(III) reduction is a measure of the number of reduction equivalents taken up by the enzyme ($A_{tot}$). Part of heme-$c$(II) is then being reoxidized by intramolecular ET to heme-$d_1$(III), and is called $A_{reox}$. The remaining part of the reduced heme-$c$(II) is $A_{red}$. Thus, for each pulse $A_{tot} = A_{reox} + A_{red}$, and $R(\text{reox}/\text{red}) = A_{reox}/A_{red}$. The points have been plotted against total number of reduction equivalents taken up by the enzyme during the pulse radiolytic reduction titration. The extended line was calculated from the model using equilibrium constants given in Table VI.

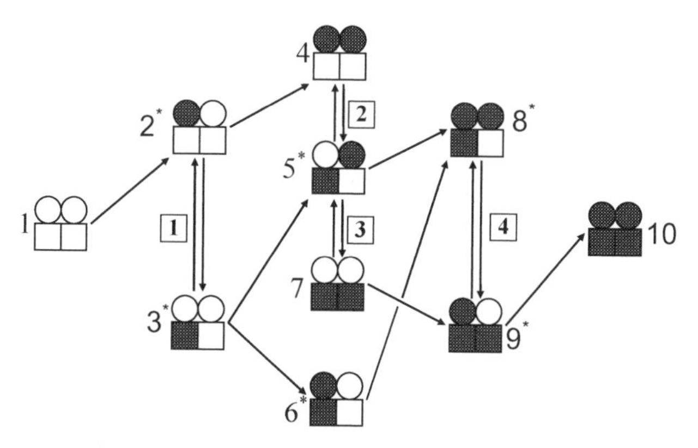


Figure 21. Reaction scheme describing the stepwise reduction of *P. aeruginosa* and *P. stutzeri* NiR. Symbols: Circle: Heme-*c*; Square: Heme-$d_1$. Empty symbols represent an oxidized site, filled ones a reduced site. Each protein subunit includes one heme-*c* and the heme-$d_1$ below it. A single arrow represents intermolecular reduction of heme-*c* by the external reductant. These reactions are all assumed to be irreversible and to occur with the same, near diffusion controlled, rate constant. The double arrows represent reversible intramolecular electron transfer within a subunit. It is assumed that heme-$d_1$ is not reduced by *inter*molecular ET, and that there is no intramolecular electron transfer between the two subunits. The asterisk indicates that there are two forms of the species. They differ by the subunit on which the electron resides. For example, and are two forms of species 5.

since $K_2$ and $K_3$ clearly show that the equilibrium distribution among species 4, 7, and 5 is markedly shifted toward the latter, that is, cases where one heme-*c* or -$d_1$ site is reduced. Introducing the third electron equivalent shifts the equilibrium in the opposite direction causing, as $K_4$ shows, predominance of species

TABLE VI
Equilibrium Results During Full Pulse Radiolytic Reduction of $cd_1$ NiR[a] at 298 K

| | | | | |
|---|---|---|---|---|
| | | ***P. aeruginosa***[b] | | |
| $K_i$ (298 K) | $K_1 = 0.3$ | $K_2 = 90$ | $K_3 = 0.09$ | $K_4 = 80$ |
| $\Delta H^\circ$ (kJ mol$^{-1}$) | $-34.5 \pm 5.2$ | $+80.0 \pm 17.6$ | $-78.6 \pm 25.1$ | $+102 \pm 31$ |
| $\Delta S^\circ$ (J K$^{-1}$ mol$^{-1}$) | $-125 \pm 18$ | $+306 \pm 58$ | $-284 \pm 88$ | $+377 \pm 106$ |
| | | ***P. stutzeri***[c] | | |
| $K_i$ (298 K) | $K_1 = 1$ | $K_2 = 30$ | $K_3 = 0.05$ | $K_4 = 25$ |
| $\Delta H^\circ$ (kJ mol$^{-1}$) | $-24.9 \pm 2.5$ | $+124 \pm 20$ | $-113 \pm 25$ | $-43 \pm 13$ |
| $\Delta S^\circ$ (J K$^{-1}$ mol$^{-1}$) | $-83 \pm 8$ | $+436 \pm 65$ | $-400 \pm 68$ | $-122 \pm 45$ |

[a]The equilibrium constants were determined from the amplitude changes following each pulse and fitted as demonstrated in Fig. 16(*b*). The four equilibria are defined in Fig. 20.
[b]Ref. 111.
[c]Ref. 105.

where both $d_1$ hemes (species 9) are reduced. The parallel marked changes in the kinetics parameters are also interesting: After adding a second electron equivalent to the enzyme [Fig. 20(*a*)], a conspicuous drop is seen in the rate of internal heme $c(Fe^{II}) \rightarrow$ heme $d_1(Fe^{III})$ ET. The observed intramolecular rate constant, which is an average value weighted by the amplitudes of the different species present follows from analysis of the intramolecular electron distribution where the observed change corresponds to the first two forward rates being much faster than the latter two. The decline in the internal ET rate provides clear kinetic evidence for negative cooperativity between the two heme-$d_1$ sites. Moreover, the calculated electron distributions and rate constants obtained using this negative cooperativity model accord closely with the experimental data.

The specific rates observed for the individual internal ET steps [Fig. 20(*a*) in Ps-NiR deserve attention]: Assuming that the distance between the iron centers of the two hemes in one subunit is the same as found in *Pa* and *Pp* NiR (2.06 nm), and using a procedure outlined by Gray and Winkler (4), we calculate that the activationless intramolecular ET rate for species 2 $\rightleftarrows$ species 3 is $k_{max} = 10^4\ s^{-1}$. The experimentally observed rate constant is $11\ s^{-1}$ at *zero* driving force ($K = 1.0$ at 298 K), which corresponds to a reorganization energy for heme-*c* to heme-$d_1$ ET of $\lambda_{tot} = 0.7$ eV ($67\ kJ\ mol^{-1}$), a value in the range expected for heme reorganization (0.8 eV). From Table VI it is obvious that formation of species 5 from half-reduced species 4 and 7, $K_2$ and $K_3^{-1}$, respectively (where either both hemes-*c* or both hemes-$d_1$ are fully oxidized), proceeds with essentially the same driving force (~0.08 eV). Thus, the observed 50-fold decline in rate constant for formation of species 5 in ET equilibration steps 3 and 2, respectively (cf. Fig. 21), suggests that changes in the driving force are not of major influence. Neither are changes in reorganization energy since this would require an increase in $\lambda_{tot}$ of no less than 0.3 eV. Hence, the decrease in rate is likely due to changes, probably in structure. Redox-induced conformational changes (e.g., reported for both *Pa*- and *Pp*-$cd_1$ NiRs) in *Ps*-$cd_1$ NiR may provide a rationale for the steep decline in rates. Specifically, a structural change reducing the electronic coupling between donor and acceptor would cause a marked decrease in ET rate constant: Breaking one hydrogen bond, forcing a through-space jump across a 0.28-nm distance, could account for a ~50-fold drop in rate constant. Such rate modulation caused by intrinsic, site–site interactions within the protein provides an interesting, though rare, illustration of intersubunit "gating" of ET reactions. Since it has so far been observed in both the Pa- and the Ps-enzymes, it makes an allosteric regulation mechanism of $cd_1$-nitrite reductases a more general one.

The above results raise an intriguing question: Why has evolution selected such a quite elaborate control mechanism for an enzyme that catalyzes a relatively simple, one-electron transfer process? This question becomes even more puzzling in view of the fact that some bacteria employ a copper-containing enzyme, CuNiR (cf. Section III.B.3), to catalyze the very same process without the aforementioned allosteric control.

One possible rationale for these observations is related to the chemical properties of the low-spin heme-$d_1$ active site, to which the enzymatic reaction product, NO, is assumed to bind with high affinity, thereby inhibiting its activity. A way to minimize product inhibition is by limiting nitrite oxidation of [$Fe^{II}$]heme-$d_1$ only in those species with an oxidized heme-$c$ in the same subunit that will prevent rapid, intramolecular reduction of the heme $d_1$ Fe(III) that has just been formed. The species that fulfill this requirement are 3, 5, and 7 (9 can react in two ways, only one of which fits the requirements). The observed pattern of intramolecular rate constants supports this reactivity by favoring species 5, but complete control would also have to include maintaining the enzyme at a low steady-state level of reduction so that species 8, 9, and 10 are not produced.

An alternative explanation that has been considered is based on the fact that the bacteria from which $cd_1$ NiRs have been isolated have evolved under intermittently anaerobic and facultatively aerobic conditions (74). The catalytic dioxygen reduction activity of $cd_1$ NiR is well established, although its physiological significance is still not fully understood. Several of these enzymes are usually present in a bacterium and the systematic knockout mutagenesis of oxygen reductase activities, including cytochrome $cd_1$ to reveal its contribution to cellular oxygen metabolism, has not been carried out. Still, one might consider the aforementioned site–site interactions as being the result of an evolutionary adjustment process (105).

In conclusion, the PR studies of the $cd_1$ NiRs provide a clear example of an allosteric control mechanism of intraprotein ET reactivity; this system is an attractive model for internal control of charge migration and distribution in proteins. The immediate candidate for observation of similar regulation is in one of nature's key players in biological energy conversion, namely, cytochrome $c$ oxidase, which will be discussed later (cf. Section VI.A).

## V. COPPER VERSUS IRON NITRITE REDUCTASES: FINAL COMMENTS

Copper- and heme-containing NiRs are both key enzymes in denitrification. They are both homooligomers and their subunits contain two distinct redox-active metal centers, an electron accepting site and a catalytic electron delivery center where the single electron reduction of nitrite to NO takes place. Thus, PR studies providing comparison of the two enzyme families are helping to resolve the different mechanisms of control of intramolecular ET reactivity. Internal electron transfer could be a rate-determining step in the catalytic cycle of both enzymes.

Intramolecular ET reactivity depends on the reduction potentials of the donor and acceptor, that is, the driving force as well as reorganization energies and the distance and electronic coupling efficiency of electron tunneling. Internal ET in the CuNiRs could be promoted through ligand changes at the substrate reduction site (including binding of the oxidizing substrate) and the reorganization energy of the electron acceptor site, that is, $T1Cu^{II}$. Ligand changes at the T2 site may modify the driving force and reorganization energy while keeping a constant, strong electronic coupling. The ET pathway was proposed to consist of a short, direct covalent peptide patch connecting electron donor and acceptor. Changes in ligation at the substrate reduction site of the $cd_1$NiRs are also assumed to be utilized, but as detailed above, allosteric interactions exert control of the internal ET rates, probably by changes in pathways that affect coupling efficiency. The proposed path (in addition to the covalent bonds) involves three hydrogen bonds to and from water molecules; this network could have direct bearing on the observed mechanism of allostery. This is probably another interesting illustration of an evolutionary impact on development and selection of ET pathways, in order to adapt to the specific requirements of a catalytic process. Further progress in understanding the ET reactivity of these proteins will require additional structural, thermodynamic, and kinetics investigations of both wild-type and single-site mutants of these enzymes.

## VI. PROTEINS WITH MIXED-METAL ION CONTENT

### A. Cytochrome *c* Oxidase

Cytochrome *c* oxidase (COX) is the terminal enzyme in the respiratory system of most aerobic organisms and catalyzes the four electron transfer from *c*-type cytochromes to dioxygen (115, 116). The A-type COX enzyme has three different redox-active metal centers: A mixed-valence copper pair forming the so-called $Cu_A$ center, a low-spin heme-*a* site, and a binuclear center formed by heme-$a_3$ and $Cu_B$. The $Cu_A$ functions as the primary electron acceptor, from which electrons are transferred via heme-*a* to the heme-$a_3$/$Cu_B$ center, where $O_2$ is reduced to water. In the B-type COX heme-*a* is replaced by a heme-*b* center. The intramolecular electron-transfer reactions are coupled to proton translocation across the membrane in which the enzyme resides (117–123) by a mechanism that is under active investigation (119, 124–126). The resulting electrochemical proton gradient is used by ATP synthase to generate ATP.

The availability of high resolution 3D structures (cf. Fig. 22) of both mammalian (bovine) and bacterial (*P. denitrificans*) (A-type) and *Thermus thermophilus* (B-type) cytochrome *c* oxidases in both oxidized and reduced states (127–134) stimulated the interest in studying the intramolecular ET

Figure 22. The 3D structure of *P. denitrificans* COX. Positions of the four redox centers are shown. Coordinates were taken from the PDB, code 1QLE.

thermodynamics and kinetics of both the $Cu_A$-heme-*a* equilibration and the further ET to the heme-$a_3$/$Cu_B$ binuclear center. A wide range of methods has been used for investigation of electron transfer between cytochrome *c* and COX, as well as ET within COX (135–143). The limited time resolution of stopped-flow techniques led to the application of alternative approaches that include: Flow-flash experiments of the reaction of the electrostatic COX–cytochrome *c* complex with $O_2$ (144, 145), time-resolved measurements of the reverse electron transfer from the binuclear center to the oxidized heme-*a* and $Cu_A$ upon photolysis of a three-electron reduced CO-inhibited enzyme (146), and light-induced electron injection into COX or the electrostatic cytochrome *c*–cytochrome oxidase complex (122, 147–150). Results of all these experiments support $Cu_A$ being the initial electron acceptor, followed by electron transfer to heme-*a* with an apparent rate constant at room temperature on the order of $2 \times 10^4\ s^{-1}$.

The ET kinetics for both mammalian (bovine) (49, 151) and bacterial [*P. denitrificans* (50); *T. thermophilus* (152)] enzymes have been investigated by pulse radiolysis using 1-MNA* radicals along with attempts to correlate the results with the known 3D structures. The oxidized $Cu_A$ center in bovine COX was reduced by 1-MNA* (Eq. 23), with a rate constant, $k = 3 \times 10^9\ M^{-1}\ s^{-1}$ at 25°C, that is, an essentially diffusion-controlled process. Following its fast reduction, $Cu_A$ undergoes partial reoxidation (Eq. 24) as revealed by an increase in absorption at 830 nm [Fig. 23(*a*)]. The rate constant of this step, which is $(1.67 \pm 0.15) \times 10^4\ s^{-1}$ 25°C, pH 7.4 for bovine COX, was found to be independent of enzyme concentration, and therefore assigned as an *intra*molecular process (49). The value is in good agreement with the reciprocal τ-value determined by Kobayashi et al. (151) in PR studies of this enzyme.

The reaction occurs simultaneously with an absorption increase at 605 nm [Fig. 23(*b*)], where the reduced hemes absorb strongly [80% contribution from heme-*a* (153)], further supporting its assignment to *intra*molecular ET between reduced $Cu_A$ ($Cu_A{}^I$) and oxidized heme-*a* ($Fe_a{}^{III}$) (49). The process does not go to completion, however, and from the amplitudes of the $Cu_A$ reduction and reoxidation at 830 nm [cf. Fig. 23(*a*)], the equilibrium constant could be determined.

The reaction scheme for the ET processes can be depicted as follows:

$$e^-(aq) + MNA^+ \rightarrow MNA^* \tag{22}$$

$$MNA^* + Cu_A^{II}Fe_a^{III} \rightarrow Cu_A^{I}Fe_a^{III} + MNA^+ \tag{23}$$

$$Cu_A^{I}Fe_a^{III} \rightleftarrows Cu_A^{II}Fe_a^{II} \tag{24}$$

The kinetic and thermodynamic parameters for electron transfer from $Cu_A$ to heme-*a* in both mammalian and bacterial COX are presented in Table VII.

The equilibrium constant of $3.4 \pm 0.5$ at 25°C, pH 7.0 for the rapid electron equilibration step (Eq. 24) in bovine COX (49) corresponds to a difference in reduction potentials between the heme-*a* [$Fe^{III}/Fe^{II}$] and $Cu_A$ [$Cu^{II}/Cu^{I}$] couples of $+31 \pm 4$ mV, which differs from earlier values, where an $E^0 = 276$ mV was reported for heme-*a* and 288 mV for $Cu_A$, that is, a difference of $-12$ mV (154). The discrepancy is not surprising, however, considering the disparate experimental conditions employed in the earlier studies: 0.1 *M* phosphate buffer saturated with 1 atm CO, which maintains heme $a_3$-$Cu_B$ in the reduced state. The observed equilibrium constant of 3.4 is in good agreement with results obtained by Kobayashi et al. (151) ($K \sim 1-4$) and by Einarsdóttir and co-workers ($K = 2$) (155). In experiments where a binuclear polypyridine ruthenium(II) complex (bound electrostatically to cytochrome oxidase) was

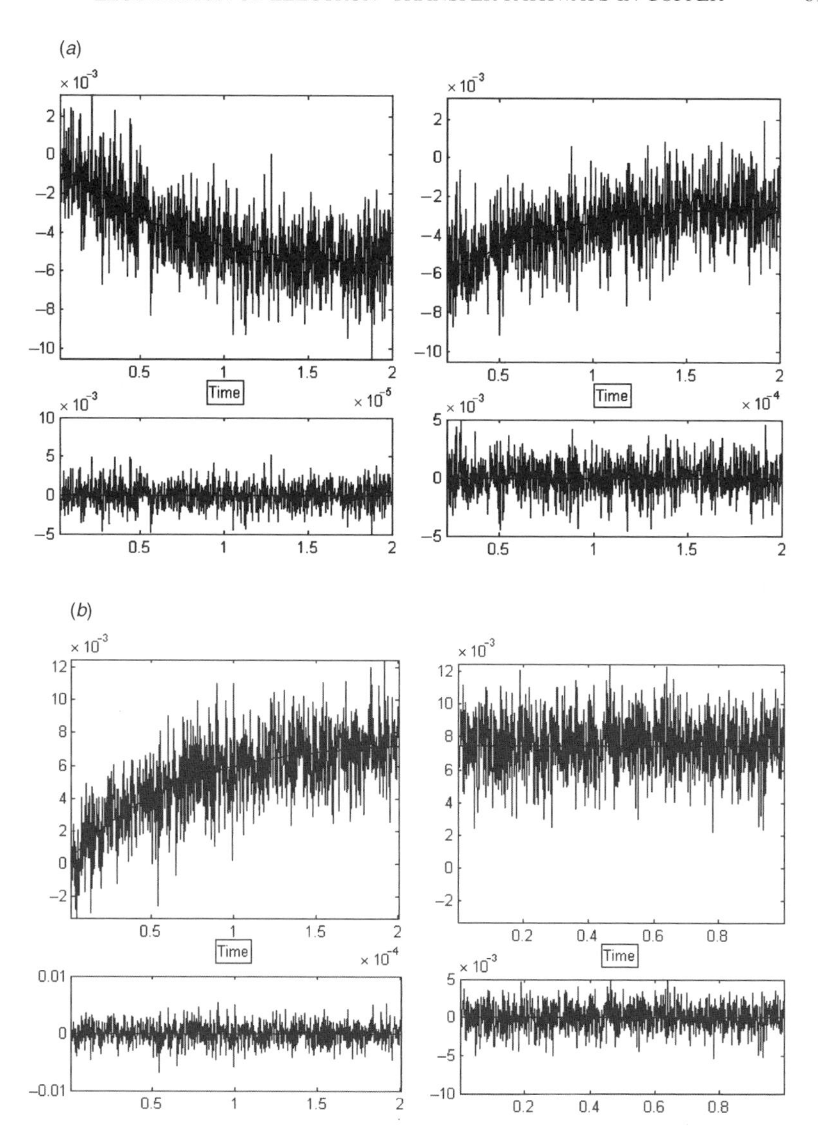


Figure 23. Time-resolved absorption changes in bovine COX induced by reaction with pulse-radiolytically produced 1-MNA radicals. (*a*) Bimolecular $Cu_A$ reduction (fast time scale) and intramolecular reoxidation (slow time scale) monitored at 830 nm. Enzyme concentration: 41 μ*M*; 1-MNA concentration: 5 m*M*. Buffer: 5 m*M* phosphate; 0.25% Tween20; Argon saturated. pH 7.4. Temperature: 298 K. Optical path length: 12.3 cm. Pulse width: 0.75 μs. (*b*) Heme-*a* reduction (both time scales) at 605 nm. Enzyme concentration: 15 μ*M*; 1-MNA concentration: 5 m*M*. Buffer: 5 m*M* phosphate; 0.25% Tween20; Argon saturated. pH 7.4. Temperature: 298 K. Optical path length: 3.1 cm. Pulse width: 0.2 μs. Time is in seconds; the left panel shows the faster phase, while the right one shows the reaction taking place at the slower time scale. The lower panels show residuals of the fits to the data.

TABLE VII

Kinetics and Thermodynamic Parameters for $Cu_A$–heme *a* ET Equilibration at 298 K and pH 7.5

| Kinetics Parameters | $k$ ($s^{-1}$) | $\Delta H^{\neq}$ (kJ $mol^{-1}$) | $\Delta S^{\neq}$ (J $K^{-1}$ $mol^{-1}$) |
|---|---|---|---|
| Forward (bacterial)[a] | $20{,}400 \pm 1{,}500$ | $22.2 \pm 1.2$ | $-88 \pm 2$ |
| (bacterial)[b] | $3{,}380 \pm 420$ | | |
| (bovine)[c] | $13{,}000 \pm 1{,}200$ | $11.4 \pm 0.9$ | $-128 \pm 11$ |
| Reverse (bacterial)[a] | $10{,}030 \pm 800$ | $24.6 \pm 1.3$ | $-86 \pm 2$ |
| (bacterial)[b] | $7{,}320 \pm 800$ | | |
| (bovine)[c] | $3{,}700 \pm 300$ | $13.4 \pm 1.0$ | $-131 \pm 11$ |

| Equilibrium Data | $K$ | $\Delta H^{\circ}$ (kJ $mol^{-1}$) | $\Delta S^{\circ}$ (J $K^{-1}$ $mol^{-1}$) |
|---|---|---|---|
| (bacterial)[a] | $2.0 \pm 0.1$ | $-2.4 \pm 0.7$ | $-2.6 \pm 2.4$ |
| (bacterial)[b] | $0.53 \pm 0.04$ | | |
| (bovine)[c] | $3.4 \pm 0.5$ | $-2.0 \pm 0.3$ | $3 \pm 5$ |

[a]*Paracoccus denitrificans* Ref. 50.
[b]*Thermus thermophilus* Ref. 152.
[c]Ref. 49.

used as a photoactivated reducing agent, an equilibrium constant for electron transfer between $Cu_A$ and heme-*a* was determined to be 8 (150). These constants were determined while the heme $a_3$-$Cu_B$ site was in the oxidized state, as in the pulse radiolysis studies, which suggests that the heme-*a* potential may be influenced by the oxidation state of the binuclear Fe–Cu center. In fact, it has already been observed that heme-*a* is a stronger oxidant in the fully oxidized enzyme: Moody and Rich (156) reported a reduction potential for heme-*a* of 310 mV, while the $Cu_A$ potential is significantly lower, 260 mV (pH 7.4, 25°C). The difference in potentials of +50 mV is close to the value noted above (Eq. 24) for the rapid electron equilibration step (49). Thus, the driving force of the $Cu_A$ to heme-*a* ET will be larger in molecules where the terminal electron acceptor site is oxidized, which is expected and suggests allosteric interactions among the sites in the enzyme that modulate the potentials as a result of the reduction.

Similar PR experiments performed on *P. denitrificans* COX (50) have also revealed a rapid $Cu_A$–heme-*a* electron equilibration step with an observed rate constant of $30{,}430 \pm 2{,}300\ s^{-1}$ and with an equilibrium constant of $2.0 \pm 0.1$ at 25°C, pH 7.0 (cf. Table VII) corresponding to a difference in reduction potentials between the heme-*a* [$Fe^{III}/Fe^{II}$)] and $Cu_A$ [$Cu^{II}/Cu^{I}$] couples of $+18 \pm 1$ mV. For $Cu_A$ in this enzyme, a midpoint potential of 213 mV versus SHE was found under the aforementioned conditions (157), while the potential of heme-*a* was reported to be 428 mV versus SHE (158). The observed equilibrium constant thus disagrees considerably with an equilibrium constant calculated from these potential differences ($K = 4300$). This is not surprising, however, since in spectroelectrochemical measurements, electrons are equilibrating both *intra-* and *inter*molecularly, while in the kinetic measurements, the observed ET equilibration between $Cu_A$ and heme-*a* is occurring *within* single molecules. The equilibrium electron distribution between heme-*a* and $Cu_A$ exhibits exceptionally small temperature dependence, in good agreement with results of bovine COX (Table VII). The values are also in good agreement with earlier results of Morgan et al. (146), suggesting that the ET reaction requires rather limited sites reorganization.

Importantly, no direct reduction of the heme-*a* site by the radicals was observed in the PR experiments of A-type COX. Also, there was no indication in these experiments for the heme-$a_3$ site being reduced, even on a 1-s time scale. Intramolecular ET between heme-*a* and heme-$a_3$ would result in an absorbance decrease at 605 nm, where heme-*a* contributes fourfold more to the absorption than heme-$a_3$ (153). Thus, our PR studies of bovine and *P. denitrificans* COX have so far failed to resolve ET to the oxidized heme-$a_3$–$Cu_B$ binuclear center,in full agreement with earlier work where reoxidation of heme-*a* by the binuclear center was not observed (147, 151). In addition, estimation of the equilibrium electron distribution from the difference in reduction potentials of heme-*a* and heme-$a_3$ [80 mV at pH 7.7 and 10°C for bovine COX (159) and 223 mV in COX from *P. denitrificans* (158)] shows that reduction of heme-*a* would be favored over that of heme-$a_3$ by a factor of 22 in bovine COX (153) and 5900 in COX from the bacterial source (158). Therefore, ET from reduced heme-*a* to oxidized heme-$a_3$ is not expected to occur under the anaerobic conditions employed.

However, recently a PR study has been performed on *T. thermophilus (Tt)* cytochrome $ba_3$ (152). This enzyme typifies the B-type COX having high functional and structural similarity to the above discribed type A cytochrome *c* oxidases. A special feature of cytochrome $ba_3$ is that the low-spin heme-*b*, which substitutes for the low-spin heme-*a* of A-type oxidases, is spectrally distinct, and therefore provides an additional tool for investigating ET in COX. As with the other oxidases, the fastest resolved event was the bimolecular reduction of $Cu_A$. This reaction was also followed by reoxidation of $Cu_A$, with a rate constant of $k_{obs} = 11,200 \pm 1300\,s^{-1}$ at 24.8°C, independent of initial [1-MNA*] and [$ba_3$] concentration (Table VII) (152). Concomitantly, cytochrome *b* was found to undergo reduction, which suggests an intramolecular ET equilibration between $Cu_A$ and cytochrome *b*. At a longer time domain, however, reduced cytochrome *b* undergoes partial reoxidation. This process was also found to be independent of reactant concentration with $k_{obs} \sim 770 \pm 85\,s^{-1}$ at 25°C (152). At 445 nm, a wavelength characteristic for reduced cytochrome $a_3$, a large absorption increase was observed at a similar time scale, suggesting that the reaction monitored is indeed an ET equilibration between cytochromes *b* and $a_3$.

The rate of intramolecular ET from $Cu_A$ to heme-*b* in *Tt* COX is slower compared with the same electron equilibration taking place in the A-type COX discussed above, which is to be expected from a smaller driving force of this reaction. Also noteworthy is the different behavior with respect to the high-spin cytochrome $a_3$. No electron transfer to $a_3$ was observed in those studies even when both the $Cu_A$ and cytochrome *a* centers were reduced. These observations suggest rather different, relative reduction potentials among the three redox sites in the A-type oxidases as compared to the B-type oxidase. A great number of studies have been performed in measuring and interpreting the redox potentials of A-type oxidases, but no coherent explanation has emerged. However, several studies suggest that the low-spin heme may indeed have a higher reduction potential than the $a_3$-$Cu_B$ site (158), which would be consistent with this site having a higher electron affinity in A-type oxidases.

Activation parameters for both forward and reverse ET reactions between $Cu_A$ and heme-*a* were calculated from the equilibrium and kinetic PR studies (cf. Table VII). It should be added that Kobayashi et al. (151) earlier reported an activation enthalpy for the bovine enzyme of $11.7\,kJ\,mol^{-1}$ for the overall reaction ($k_f + k_r$), in excellent agreement with the results presented in Table VII.

An electron-tunneling pathway from $Cu_A$ to heme-*a* has been suggested to proceed via a hydrogen bond between the His204 ligand of subunit II (bovine enzyme numbering) and a carbonyl group on loop XI–XII (Arg439 of subunit I) (49, 160–162). Another hydrogen bond from the amide on the same arginine residue to one of the propionyl side chains on heme-*a* provides a continuation of the pathway. Altogether the path includes 14 covalent bonds and 2 hydrogen bonds. The overall metal to metal distance is 1.98 nm, corresponding to an electronic coupling decay factor $\prod \varepsilon = 1.0 \times 10^{-4}$. Essentially the same electron-tunneling pathway was proposed for the *P. dentrificans* enzyme (cf. Fig. 24) (50).

The electronic coupling energy between $Cu_A$ and heme-*a* in bovine and bacterial A-type COX was determined to be $2.9 \times 10^{-6}$ eV (49, 50). Brzezinski (163) has estimated the electronic coupling energy between $Cu_A$ and heme-*a* in bovine COX to be $H_{DA} = 3.8 \times 10^{-6}$ eV, which is in excellent agreement with the experimentally determined value. Applying Eq. (2) above gives $\lambda_{tot} = 0.40$ eV for the electron equilibration in bovine COX (49) and 0.32 eV in the bacterial COX (50), in good agreement with the value estimated by Brzezinski (0.3 eV) (163). Based on a driving force of 0.05 eV Ramirez et al. (160) calculated that in order to account for the maximum (activationless) rate constant of $4 \times 10^4\,s^{-1}$ to $8 \times 10^5\,s^{-1}$ the value of the reorganization energy must be between 0.15 and 0.5 eV.

Paula et al. (120) estimated the lower limit of the electron-transfer rate constant for ET between $Cu_A$ and heme-*a* to be $\sim 2 \times 10^5\,s^{-1}$ from

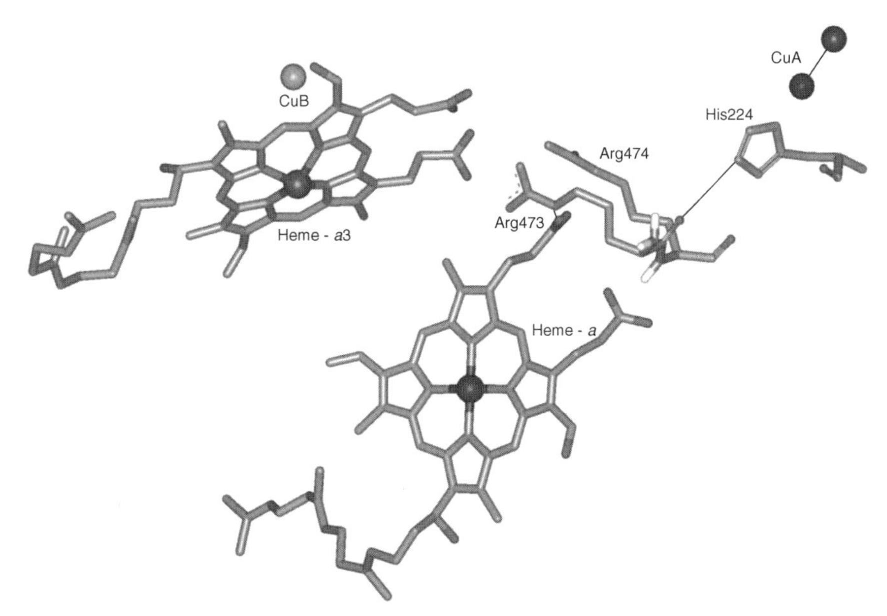


Figure 24. Electron-transfer pathway between $Cu_A$ and heme-*a* in *P. denitrificans* COX. The path consists of 14 covalent bonds and two hydrogen bonds. The direct distance between to two metal ion centers is 2.0 nm. The binuclear heme-$a_3$/$Cu_B$ site is also shown. Calculations were based on the Beratan and Onuchic model (6, 7). Coordinates were taken from the PDB, code 1QLE. (See color insert.)

measurements where reduction of $O_2$ to water by fully reduced cytochrome *c* oxidase using the flow–flash method was employed, which corresponds to a reorganization energy close to 0.3 eV, in good agreement with the value reported above. In this system, electron transfer between $Cu_A$ and heme-*a* is limited by proton uptake (120, 164).

The reason for the higher rate of internal ET in bacterial cytochrome *c* oxidase is the smaller energy requirement for reorganization of the redox centers in *P. dentrificans* COX ($\lambda_{tot} = 0.32\,eV$) (50) than in bovine COX ($\lambda_{tot} = 0.40\,eV$) (49), according with the observation that in the bovine enzyme, the subunits containing the heme regions are surrounded by additional water molecules that are not present in the bacterial enzyme: The medium surrounding a metalloprotein active site affects the reorganization energy associated with the ET reaction. Thus, a hydrophobic active site will lead to smaller reorganization energies than a hydrophilic one, and consequently the kinetics of intraprotein ET will be very sensitive to the active site environment.

As discussed above, the rate constant for intramolecular ET is $k_{ET} = 650\,s^{-1}$ at 298 K in purple azurin (52) (Section III.A). The electron donor in the latter protein is a disulfide radical anion ($RSSR^-$) situated at the opposite end of the azurin molecule, and a pathway consisting of 19 covalent bonds, 2 hydrogen bonds, and 1 0.38-nm van der Waals contact was identified in this protein, yielding $\Pi\varepsilon = 4.7 \times 10^{-8}$. Assuming that the reorganization energy previously calculated for the $Cu_A$ center in purple azurin ($\lambda_{CuA} = 0.4\,eV$) remains the same in both enzymes, we can calculate that the reorganization energies of heme-*a* are $\lambda_{hemea} = 0.40\,eV$ in the bovine (49) and 0.24 eV in the bacterial (50) enzyme. Thus, the reorganization energies of the electron donor and acceptor in COX are unusually low compared with typical values observed for ET in many other proteins, which are often in the range 0.7–1.3 eV (160).

In conclusion, quantitative analysis of both the time courses and amplitudes associated with intramolecular electron transport between $Cu_A$ and heme-*a* in type A (bovine and bacterial) cytochrome *c* oxidase has yielded microscopic rate constants, activation parameters, and equilibrium constants as presented in Table VII. The results are in excellent agreement with those of earlier studies and provide a rigorous framework for understanding the electron transfer interactions between $Cu_A$ and heme-*a*. The experimentally determined rate constants fit well with those theoretically calculated using results of previous work on intramolecular ET in copper proteins. By using the value determined for the reorganization energy of the $Cu_A$ center in purple azurin, similarly small reorganization energy has been found for the heme-*a* center. The challenge ahead is understanding the next step in reduction of this enzyme (i.e., the heme-*a* to $a_3/Cu_B$ site).

## B. Xanthine Dehydrogenase and Oxidase

Xanthine dehydrogenase (XDH) is a multicentered enzyme that catalyzes oxidation of xanthine to uric acid and concomitantly reduces $NAD^+$ to NADH. This enzyme form is the one isolated from mammalian cells (165). However, it is converted into xanthine oxidase (XO) by oxidation of sulfhydryl residues or by proteolysis. Thus, while XDH uses $NAD^+$ as oxidizing substrate, XO utilizes dioxygen and catalyzes further the hydroxylation of a wide variety of purine, pyrimidine, pterin, and aldehyde substrates. The enzymes isolated from a wide range of organisms have comparable molecular weights, as well as identical redox centers. The enzymes are homodimers of molecular weight 290 kDa, and the monomers act independently in catalysis. Each subunit contains a molybdopterin cofactor together with two spectroscopically distinct [2Fe–2S] centers (Fe/S I and II) and an flavin adenine dinucleotide(FAD) cofactor (165).

The 3D structures have been reported for bovine milk xanthine oxidoreductase in both the XDH form at 2.1 Å and in the XO form at 2.5-Å resolution (165). The structure of XDH has been further refined (1.98-Å resolution); it features a bound inhibitor that interacts with the channel leading to the molybdenum–pterin active site, but without directly coordinating to the molybdenum ion (166).

The internal electron-transfer processes in milk XO have been the subject of an elegant application of the pulse radiolysis method by Hille and co-workers (167, 168) with the goal of understanding the catalytic mechanism of this enzyme. Indeed, these workers have identified the individual internal electron-transfer steps between the redox centers that constitute the reductive half of the catalytic cycle. Oxidation of xanthine to uric acid catalyzed by XO is performed by the molybdopterin center (Mo-pt), followed by rapid intramolecular electron transfer via the iron–sulfur centers to FAD. The reduction of the natural oxidant, dioxygen, occurs through FAD, leading to the formation of superoxide anions or hydrogen peroxide, depending on the reduction state of the protein (167). In the first study (168), several different pulse radiolytically produced radicals were employed in order to probe the above reaction sequence. The reaction of bovine milk XO with $CO_2^-$ radicals was found to yield two distinct species at diffusion controlled rates. One species was identified as an $RSSR^-$ radical, decaying in a first-order process with a rate constant of $60\,s^{-1}$. Concomitantly, a second species was observed, and spectral changes monitored at 550 nm were consistent with formation of the blue flavosemiquinone of XO. These processes were followed by ET from the flavosemiquinone to one of the two iron–sulfur clusters, the Fe–S II center, with a rate constant of $290\,s^{-1}$. The reactivity of fully reduced XO was also investigated by PR, employing the oxidizing azide ($N_3$) free radical, which led to fast transient formation of the flavosemiquinone followed by subsequent oxidation of the Fe–S centers with a rate constant of $170\,s^{-1}$, independent of the protein concentration (167).

In a later pulse radiolysis study, 1-MNA* or deazalumiflavin radicals were used as reductants (168). Both radicals selectively introduced reduction equivalents into the enzyme molecules, enabling the monitoring of the internal ET cascade. The molybdenum center was found to be the preferred electron uptake site of the enzyme. The ensuing process was a fast electron equilibration ($k_{obs} = 8500\,s^{-1}$) between the molybdenum ion and one of the two iron–sulfur centers (tentatively assigned as Fe–S I). Subsequent equilibration between this center and the flavin ($k_{obs} = 125\,s^{-1}$) completes the catalytic half-cycle, and links the catalytic, substrate oxidizing site with the dioxygen reduction center. The unambiguous demonstration of the involvement of the Fe–S centers in the ET chain of the reductive half-cycle is noteworthy. All these assignments were primarily based on the wavelength dependence of the observed kinetics phases (168) (see Fig. 25).

Kobayashi et al. (169) studied the reduction of the parental enzyme, XDH, by several different free radicals produced by PR: Salicylate anion radicals ($SL^-$), nicotinamide adenine dinucleotide radicals (NAD*), and 1-methylnicotinamide radicals (1-MNA*). These free radicals are potential substrates that interact with

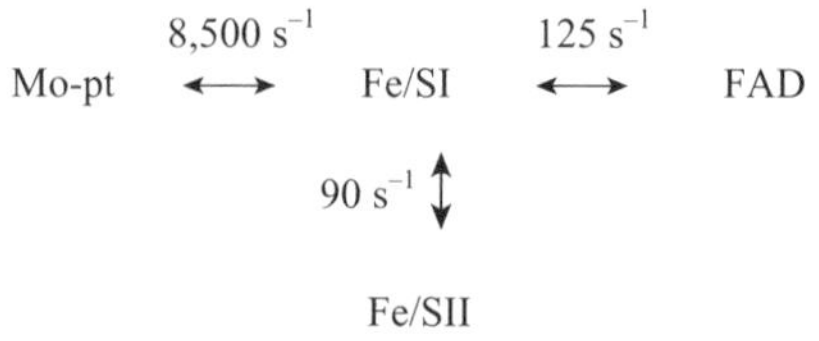

$$\text{Mo-pt} \overset{8{,}500\ s^{-1}}{\longleftrightarrow} \text{Fe/SI} \overset{125\ s^{-1}}{\longleftrightarrow} \text{FAD}$$

$$\text{Fe/SI} \ \updownarrow\ 90\ s^{-1} \ \text{Fe/SII}$$

Figure 25. Reaction scheme for the proposed internal electron equilibration between the different redox centers in xanthine oxidase. [Adapted from Ref. (167)].

either the molybdenum electron uptake site ($SL^{-}$) or the electron delivery site, FAD (NAD* and 1-MNA*). Two phases of reduction were observed when XDH was reduced by $SL^{-}$. The faster one is a bimolecular reduction of one of the Fe–S centers with a rate constant of $2.9 \times 10^7\,M^{-1}\,s^{-1}$, while the slower phase represents unimolecular ET to FAD with a rate constant of $510\,s^{-1}$ at pH 7.5. The possible involvement of the molybdenum center in the above reaction was probed using a modified enzyme having oxidized FAD and Fe–S centers together with a redox-inert Mo(IV) site. While the faster reaction phase was unaffected, the slower one could no longer be detected, suggesting that the latter reaction is due to intramolecular electron transfer from the molybdenum center to FAD (169). The NAD* radicals were found to react primarily with FAD in a bimolecular process with a second-order rate constant of $1.4 \times 10^7\,M^{-1}\,s^{-1}$, yielding the blue semiquinone radical, which is not surprising, since FAD is the catalytic site for NAD reduction in the enzyme. In contrast, it is unexpected that 1-MNA* radicals react with a Fe–S center *also* in a bimolecular ET process with a rate constant of $6.5 \times 10^7\,M^{-1}\,s^{-1}$ (169). Notably, the reduction of the Fe–S center was not affected in the modified enzyme having a redox-inert Mo(IV) center, which is in contrast to the findings by Hille et al. (167) in a similar experiment using XO.

The 3D structure of XO (165, 166) offers a possibility of correlating structure with the kinetics observations. The distance between the Mo ion and the closest iron atom in the Fe–S I cluster is 1.49 nm, while the two nearest Fe atoms of the two Fe–S clusters are separated by a distance of 1.26 nm. Finally, the distance from C7 of FAD to the closer of the two sulfur atoms in the Fe–S II cluster is only 0.66 nm (166). The total distance between Mo and FAD is 2.9 nm. Pathway calculations using the HARLEM program (170) demonstrate that the redox centers are relatively strongly coupled through a series of covalent bonds, three short van der Waals contacts, and a single hydrogen bond (cf. Fig. 26). The reported reduction potentials of the four redox centers (171, 172) demonstrate that the arrangement of the redox centers is in accordance with the notion that electrons are transferred from Mo via the two Fe–S centers to FAD in a thermodynamically favorable process. From the relative reduction potentials (172) we estimate a total driving force $-\Delta G^{\circ} = 0.18\,\text{eV}$.

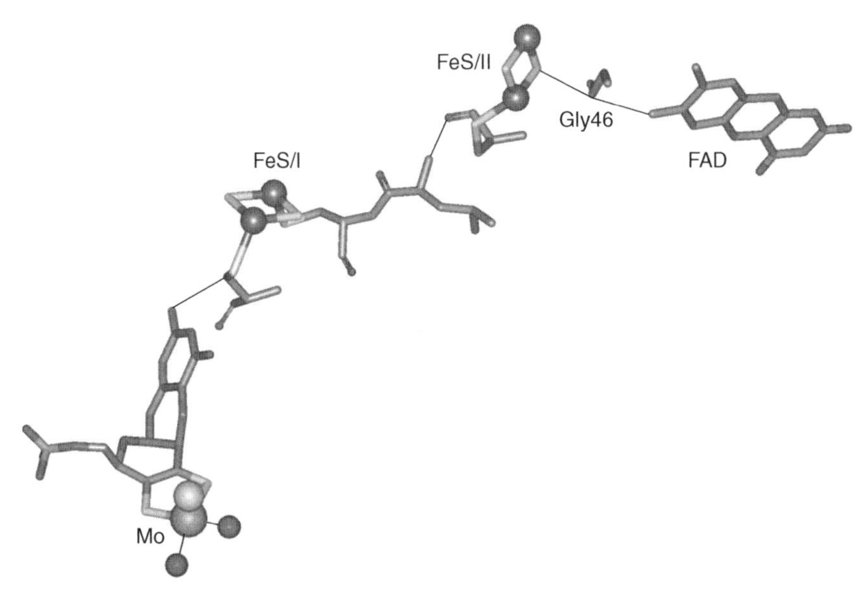


Figure 26. Electron-transfer pathway from the molybdenum–pterin center to FAD in xanthine oxidase. Besides the two mercapto groups, Mo is shown coordinated to inorganic sulfur and two oxygen atoms. The total distance between Mo and FAD is 2.9 nm. The redox centers are coupled through a series of covalent bonds, three short van der Waals contacts, and a single hydrogen bond. Calculations were based on the Beratan and Onuchic model (6, 7). Coordinates were taken from the PDB, code 1FIQ. (See color insert.)

The PR studies on XO and XDH provide qualitatively similar sequences of internal ET reactions and illustrate the usefulness of this method in delineating enzymatic redox reaction mechanisms (167–169, 172). No one to date has attempted to obtain more quantitative information regarding the ET reactivity of the individual sites. Better characterization of the individual ET steps in XO will surely be required for this purpose.

## VII. CONCLUSIONS

As detailed in this chapter, pulse radiolysis has been useful in investigating electron-transfer processes in redox proteins, primarily those containing transition metal ions in the active sites. There were two main, complementary objectives guiding these studies. One was resolving the detailed mechanism of the function(s) performed by the protein and the second was attaining an understanding of the electron-transfer process within the polypeptide matrix separating the redox centers and defining the parameters controlling their rates.

Azurin has been extensively investigated using pulse radiolysis; it is found to be a very convenient and informative model system. Triggering internal ET from the $RSSR^-$ radical ion to the Cu(II) site over a 2.7-nm separation, though being an artificial process, turned out to be useful for examining the impact of specific structural differences on ET rates. Furthermore, the production of "purple" azurin with a $Cu_A$ site substituting the T1 made it possible to examine the reorganization energy of the former site in an independent framework.

As a next step we have discussed attempts to determine the parameters of intraprotein ET steps that are part of functional processes, namely, catalytic cycles, assumed to be a result of evolutionary selection.

There have been several interesting studies, where the application of pulse radiolysis yielded a detailed insight into the operative mechanisms of redox–active enzymes. Notably, investigations of the copper-containing nitrite reductases have clearly established the distinct roles of the two different copper sites, T1 as the electron acceptor and T2 as the active site that reduces the nitrite ion substrate to NO. Studies of the analogous, heme containing $cd_1$ nitrite reductases have provided insight into the catalytic reaction mechanism, also revealing a rare case of allosteric control of electron-transfer rates and distribution among the sites of these enzymes.

Still, mechanisms of certain enzymes investigated by pulse radiolysis are not fully understood. Our limited knowledge of the elementary steps of both internal electron and proton-transfer processes in the cytochrome *c* oxidases is a case in point. Use of pulse radiolysis has contributed to the establishment of the $Cu_A$ site being the initial electron acceptor, as well as to the determination of electron-transfer rates and the equilibrium between $Cu_A$ and the heme-*a* site. The driving force apparently controls further transfer of the reduction equivalents from heme-*a* to the binuclear heme-$a_3$–$Cu_B$ center. Characterizing the latter process is of considerable importance for understanding how the binuclear center undergoes reduction (whether by a sequential two electron transfer or a concerted one-step process), which may have implications for the second half of the catalytic cycle, that is, dioxygen reduction and probably also for the coupled proton transfer that takes place.

The elegant studies of xanthine oxidase using pulse radiolysis illustrate the power of this method in resolving the complex catalytic reaction scheme of an enzyme that involves four distinct redox centers. Still, in these studies the individual ET steps were not characterized in a more quantitative manner. Investigation of the internal ET steps leading to disulfide bond formation catalyzed by the disulfide bond-forming enzyme Erv2p from *A. xylanus* is another formidable challenge (173).

We have discussed structure–function relations of a diverse group of electron mediating proteins with both natural (i.e., wild type) and artificial structural features. The review thus provides an interesting illustration of evolution's profound impact on development and selection of ET sites and pathways, thereby adjusting the rates to requirements of specific reactions. Clearly, parameters like driving force, distance, and the nature of the medium separating electron donor and acceptor all contribute to this elaborate control.

## ACKNOWLEDGMENTS

Financial support of several of the studies described in this review has been extended to IP by the Minerva Foundation, Munich, FRG. IP also wishes to acknowledge the kind hospitality of the Beckman Institute at The California Institute of Technology during writing this review. The authors are most grateful to Harry B. Gray for his many comments and suggestions leading to great improvement of the text.

This review is dedicated to the memory of a dear friend, Dr. Michael Goldberg, with whom our research on protein ET reported in this chapter was initiated.

## ABBREVIATIONS

| | |
|---|---|
| 1-MNA | 1-Methylnicotinamide |
| 2D NMR | Two-dimensional NMR |
| 3D | Three dimensional |
| AcNiR | *Achromobacter cycloclastes* nitrite reductase |
| AO | Ascorbate oxidase |
| ATP | Adenosine triphosphate |
| aq | aqueous |
| AxNiR | *Alcaligenes xylosoxidans* nitrite reductase |
| COX | Cytochrome *c* oxidase |
| Cp | Ceruloplasmin |
| edta | Ethylenediaminetetraacetic acid |
| ENDOR | Electron nuclear double resonance |
| EPR | Electron paramagnetic resonance |
| ET | Electron transfer |
| EXAFS | Extended X-ray absorption fine structure |
| FAD | Flavin adenin dinucleotide |
| FC | Franck–Condon |
| hCp | Human ceruloplasmin |
| HOMO | Highest occupied molecular orbital |
| im | imidazole |
| LRET | Long-range ET |
| MADH | Methylamine dehydrogenase |
| MCD | Magnetic circular dichroism |

| | |
|---|---|
| NAD | Nicotinamide adenine dinucleotide |
| NiR | Nitrite reductase |
| NOE | Nuclear Overhauser effect |
| NOESY | Nuclear Overhauser enhancement spectroscopy |
| PDB | Protein Data Bank |
| phen | Phenanthroline |
| PR | Pulse radiolysis |
| SHE | Standard hydrogen electrode |
| TOCSY | Total correlated spectroscopy |
| tot | Total |
| Tt | *Thermus thermophilus* |
| WT | Wild type |
| XDH | Xanthine dehydrogenase |
| XO | Xanthine oxidase |

## REFERENCES


1. M. K. Johnson et al. Eds., *Adv. Chem. Ser., 226* (1990).
2. A. Messerschmidt, R. Huber, K. Wieghardt, and T. Poulos, *Handbook of Metalloproteins*, John Wiley & Sons, Inc., New York, 2001.
3. R. A. Marcus and N. Sutin, *Biochim. Biophys. Acta, 811*, 265 (1985).
4. H. B. Gray and J. R. Winkler, *Q. Rev. Biophys.*, *36*, 341 (2003).
5. J. Jortner and M. Bixon, Eds., *Electron Transfer; From Isolated Molecules to Biomolecules, Adv. Chem. Phys.*, *107* (1999).
6. D. N. Beratan, J. N. Betts, and J. N. Onuchic, *Science*, *252*, 1285 (1991).
7. D. N. Beratan, J. N. Onuchic, J. R. Winkler, and H. B. Gray, *Science*, *258*, 1740 (1992).
8. R. Langen, I. J. Chang, J. P. Germanas, J. H. Richards, J. R. Winkler, and H. B. Gray, *Science, 268*, 1733 (1995).
9. G. E. Adams, E. M. Fielden, and B. D. Michael, Eds., *Fast processes in Radiation Chemistry and Biology*, John Wiley & Sons, Inc., New York, 1975.
10. M. Ebert, J. P. Keene, A. J. Swallow, and J. H. Baxendale, Eds., *Pulse Radiolysis*, Academic Press, New York, 1965.
11. M. S. Matheson and L. M. Dorfman, *J. Chem. Phys.*, *32*, 1870 (1960).
12. E. J. Hart, *Ann. Rev. Nuclear Sci.*, *15*, 125 (1965).
13. M. H. Klapper and M. Faraggi, *Quart. Rev. Biophys. 12*, 465 (1979).
14. J. W. Boag and G. E. Adams, *Cellular Radiation Biology*, The Williams and Wilkins Company, Baltimore, 1965.
15. E. T. Adman, *Topics in Molecular and Structural Biology: Metalloproteins*, P. M. Harrison, Ed., Chemie Verlag, Weinheim, Germany, 1985, pp. 1–42.
16. H. Nar, A. Messerschmidt, R. Huber, M. van de Kamp, and G. W. Canters, *J. Mol. Biol.*, *221*, 765 (1991).
17. O. Farver and I. Pecht, *Proc. Natl. Acad. Sci, U.S.A.*, *86*, 6968 (1989).

18. O. Farver and I. Pecht, *J. Am. Chem. Soc.*, *114*, 5764 (1992).
19. O. Farver, L. K. Skov, T. Pascher, B. G. Karlsson, M. Nordling, L. G. Lundberg, T. Vänngård, and I. Pecht, *Biochemistry*, *32*, 7317 (1993).
20. O. Farver, L. K. Skov, M. van de Kamp, G. W: Canters, and I. Pecht, *Eur. J. Biochem.*, *210*, 399 (1992).
21. O. Farver, N. Bonander, L. K. Skov, and I. Pecht, *Inorg. Chim. Acta, 243*, 127 (1996).
22. O. Farver, L. K. Skov, G. Gilardi, G. van Pouderoyen, G. W. Canters, S. Wherland, and I. Pecht, *Chem. Phys.*, *204*, 271 (1996).
23. O. Farver, L. K. Skov, S. Young, N. Bonander, B. G. Karlsson, T. Vänngård, and I. Pecht, *J. Am. Chem. Soc.*, *119*, 5453 (1997).
24. O. Farver, L. J. Jeuken, G. W. Canters, and I. Pecht, *Eur. J. Biochem.*, *267*, 3123 (2000).
25. S. Wherland and H. B. Gray, in *Biological Aspects of Inorganic Chemistry*, D. Dolphin, Ed., John Wiley & Sons, Inc., New York, 1977, pp. 289–368.
26. E. Grünwald and C. Steel, *J. Am. Chem. Soc.*, *117*, 5687 (1995).
27. J. E Leffler, *J. Org. Chem.*, *20*, 1202 (1955).
28. E. I. Solomon, R. K. Szilagyi, S. D. George, and L. Basumallick, *Chem. Rev.*, *104*, 419 (2004).
29. S. Larsson, A. Broo, and L. Sjölin, *J. Phys. Chem.*, *99*, 4860 (1995).
30. M. van de Kamp, G. W. Canters, S. S. Wijmenga, A. Lommen, C. W. Hilbers, H. Nar, A. Messerschmidt, and R. Huber, *Biochemistry*, *31*, 10194 (1992).
31. M. Plato, M. E. Michel-Beyerle, M. Bixon, and J. Jortner, *FEBS Lett.*, *249*, 70 (1989).
32. L. Chen, R. C. Durley, F. S. Mathews, and V. L. Davidson, *Science*, *258*, 1748 (1992).
33. H. Pelletier and J. Kraut, *Science*, *258*, 1748 (1992).
34. C. M. Groeneveld, R. Aasa, B. Reinhammar, and G. W. Canters, *J. Inorg. Biochem.*, *31*, 143 (1987).
35. G. van Pouderoyen, C. R. Andrew, T. M. Loehr, J. Sanders-Loehr, S. Mazumdar, H. Allen, H. A. O. Hill, and G. W. Canters, *Biochemistry*, *35*, 1397 (1996).
36. T. den Blaauwen and G.W. Canters, *J. Am. Chem. Soc.*, *115*, 1121 (1993).
37. S. J. Kroes, J. Salgado, G. Parige, C. Luchinat, and G. W. Canters, *J. Biol. Inorg. Chem.*, *1*, 551 (1996).
38. C. S. St. Clair, W. R. Ellis, and H. B. Gray, *Inorg. Chim. Acta, 191*, 149 (1992).
39. T. Pascher, B. G. Karlsson, M. Nordling, B. G. Malmström, and T. Vänngård, *Eur. J. Biochem.*, *212*, 289 (1993).
40. M. van de Kamp, G. W. Canters, C. R. Andrew, J. Sanders-Loehr, C. J. Bender, and J. Peisach, *Eur. J. Biochem.*, 218, 229 (1993).
41. C. von Sonntag, in *Sulfur-Centered Reactive Intermediates in Chemistry and Biology*, Eds., C. Chatgilialoglu and K.-D. Asmus, Plenum Press, New York, 1990, p. 359.
42. O. Farver, J. Zhang, Q. Chi, I. Pecht, and J. Ulstrup, *Proc. Natl. Acad. Sci. U.S.A.*, *98*, 4426 (2001).
43. O. Farver, G. W. Canters, I. van Amsterdam, and I. Pecht, *J. Phys. Chem. A.*, *107*, 6757 (2003).
44. I. M. C. van Amsterdam, M. Ubbink, L. J. C. Jeuken, M. P. Verbeet, O. Einsle, A. Messerschmidt, and G. W. Canters, *Chem.-Eur. J.*, *7*, 2398 (2001).
45. N. Bonander, J. Leckner, H. Guo, B. G. Karlsson, and L. Sjölin, *Eur. J. Biochem.*, *267*, 4511 (2000).
46. J. R. Winkler and H. B. Gray, *J. Bioinorg. Chem.*, *2*, 399 (1997).

47. O. Farver, R. R. Eady, Z. H. L. Abraham, and I. Pecht, *FEBS Lett., 436*, 239 (1998).
48. O. Farver and I. Pecht, *Proc. Natl. Acad. Sci. U.S.A.*, *89*, 8283 (1992).
49. O. Farver, Ó. Einarsdóttir, and I. Pecht, *Eur. J. Biochem., 267*, 950 (2000).
50. O. Farver, E. Grell, B. Ludwig, H. Michel, and I. Pecht, *Biophys. J.*, *90*, 2131 (2006).
51. M. Hay, J. H. Richards, and Y. Lu, *Proc. Natl. Acad. Sci. U.S.A.*, *93*, 461 (1996).
52. O. Farver, Y. Lu, M. C. Ang, and I. Pecht, *Proc. Natl. Acad. Sci. U.S.A.*, *96*, 899 (1999).
53. A. J. DiBilio, M. G. Hill, N. Bonander, B. G: Karlsson, R. M. Villahermosa, B. G. Malmström, J. R. Winkler, and H. B. Gray, *J. Am. Chem. Soc.*, *119*, 9921 (1997).
54. O. Farver and I. Pecht, *Electron Transfer Reactions in Multi-Copper Oxidases*, A. Messerschmidt, Ed., World Scientific Publications, Singapore 1997, pp. 355–389.
55. A. Messerschmidt, A. Rossi, R. Ladenstein, R. Huber, M. Bolognesi, G. Gatti, A. Marchesini, R. Petruzzelli, and A. Finazzi-Agró, *J. Mol. Biol.*, *206*, 513 (1989).
56. A. Messerschmidt, R. Ladenstein, R. Huber, M. Bolognesi, L. Avigliano, A. Marchesini, R. Petruzzelli, A. Rossi, and A. Finazzi-Agró, *J. Mol. Biol.*, *224*, 179 (1992).
57. A. Messerschmidt, W. Steigemann, R. Huber, G. Lang, and P. M. H. Kroneck, *Eur. J. Biochem. 209*, 597 (1992).
58. A. Messerschmidt, H. Luecke, and R. Huber, *J. Mol. Biol.*, *230*, 997 (1993).
59. I. Saitzeva, V. Saitzev, G. Card, K. Moshkov, B. Bax, A. Ralph, and P. Lindley, *J. Biol. Inorg. Chem. 1*, 15 (1996).
60. A. B. Taylor, C. S. Stoj, L. Ziegler, D. J. Kosman, and P. J. Hart, *Proc. Natl. Acad. Sci. U.S.A., 102*, 15459 (2005).
61. K. Piontek, M. Antorini, and T. Choinowski, *J. Biol. Chem., 277*, 37663 (2002).
62. M. Faraggi and I. Pecht, *J. Biol. Chem. 248*, 3146 (1973).
63. P. O'Neill, E. M. Fielden, A. Finazzi-Agró, and L. Avigliano, *Biochem. J., 209*, 167 (1983).
64. P. O'Neill, E. M. Fielden, L. Avigliano, G. Marcozzi, A. Ballini, and A. Finazzi-Agró, *Biochem. J.*, 222, 65 (1984).
65. P. Kyristis, A. Messerschmidt, R. Huber, G. A. Salmon, and A. G. Sykes, *J. Chem. Soc.*, *Dalton Trans. 1993*, 731 (1993).
66. O. Farver, S. Wherland, and I. Pecht, *J. Biol. Chem., 269*, 22933 (1994).
67. J. T. Hazzard, A. Marchesini, P. Curir, and G. Tollin, *Biochim. Biophys. Acta, 1208*, 166 (1994).
68. E. D. Harris, *Nutr. Rev. 53*, 170 (1996).
69. J. Deinum and T. Vänngård, *Biochim. Biophys. Acta, 310*, 321 (1973).
70. T. E. Machonkin, H. H. Zhang, B. Hedman, K. O. Hodgson, and E. I. Solomon, *Biochemistry, 37, 9570* (1998).
71. L. Avigliano and A. Finazzi-Agrò, in *Electron Transfer Reactions in Multi-Copper Oxidases*, A. Messerschmidt, Ed., World Scientific Publications, Singapore, pp. 251–284.
72. O. Farver, L. Bendahl, L. K. Skov, and I. Pecht, *J. Biol. Chem.*, *274*, 26135 (1999).
73. O. Farver and I. Pecht, *Mol. Cryst. Liq. Cryst. Sci. Technol., 194*, 215 (1991).
74. W. G. Zumft, *Microbiol. Mol. Biol. Rev., 61*, 533 (1997).
75. E. T. Adman and M. E. P. Murphy, in *Handbook of Metalloproteins*, Vol. 2, A. Messerschmidt, R. Huber, K. Wieghardt, and T. Poulos, Eds., John Wiley & Sons, Inc., New York, 2001, pp. 1381–1390.
76. F. K. Yousafzai and R. R. Eady, *J. Biol. Chem.*, *277*, 34067 (2002).
77. K. Kobayashi, S. Tagawa, Deligeer, and S. Suzuki, *J. Biochem.*, *126*, 408 (1999).

78. Z. H. L. Abraham, D. J. Lowe, and B. E. Smith, *Biochemical J.*, *295*, 587 (1993).
79. C. L Hulse, J. M. Tiedje, and B. A. Averill, *Anal. Biochem.*, *172*, 420 (1988).
80. S. Suzuki, T. Kohzuma, Deligeer, K. Yamaguchi, N. Nakamura, S. Shidara, K. Kobayashi, and S. Tawaga, *J. Am. Chem. Soc. 116*, 11145 (1994).
81. S. Suzuki, Deligeer, K. Yamaguchi, K. Kataoka, K. Kobayashi, S. Tawaga, T. Kohzuma, S. Shidara, and H. Iwasaki, *J. Biol. Inorg. Chem.*, *2*, 265 (1997).
82. O. Farver, R. R. Eady, Z. H. L. Abraham, and I. Pecht, *FEBS Lett.*, *436*, 239 (1998).
83. O. Farver, R. R. Eady, G. Sawers, M. Prudêncio, and I. Pecht, *FEBS Lett.*, *561*, 173 (2004).
84. O. Farver, R. R. Eady, and I. Pecht, *J. Phys. Chem. A*, *108*, 9005 (2004).
85. A. J. Di Bilio, C. Dennison, H. B. Gray, B. E. Ramirez, and A. G. Sykes, *J. Am. Chem. Soc.*, *120*, 1998.
86. R. W. Strange, L. M. Murphy, F. E. Dodd, Z. H. L. Abraham, R. R. Eady, B. E. Smith, and S. S. Hasnain, *J. Mol. Biol.*, *287*, 1001 (1999).
87. E. N. Baker, *J. Am. Chem. Soc.*, *112*, 7817 (1990).
88. H. B. Gray, B. G. Malmstrom, and R. J. P. Williams, *J. Biol. Inorg. Chem.*, *5*, 551 (2000).
89. M. E. P. Murphy, S. Turley, and E. T. Adman, *J. Biol. Chem.*, *270*, 27458 (1995).
90. E. Babini, I. Bertini, M. Borsari, F. Capozzi, C. Luchinat, X. Zhang, G. L. C. Moura, I. V. Kurnikov, D. N. Beratan, A. Ponce, A. J. Di Bilio, J. R. Winkler, and H. B. Gray, *J. Am. Chem. Soc.*, *122*, 4532 (2000).
91. J. R. Winkler, P. Wittung-Stafshede, J. Leckner, B. G. Malmström, and H. B. Gray, *Proc. Natl. Acad. Sci. U.S.A.*, *94*, 4246 (1997).
92. E. I. Solomon, R. K. Szilagyi, S. DeBeer George, and L. Basulallick, *Chem. Rev.*, *104*, 419 (2004).
93. M. J. Ellis, M. Prudêncio, F. E. Dodd, R. W. Strange, G. Sawers, R. R. Eady, and S. S. Hasnain, *J. Mol. Biol. 316*, 51 (2002).
94. M. Prudêncio, G. Sawers, F. K. Yousafzai, and R. R. Eady, *Biochemistry*, *41*, 3430 (2002).
95. M. C. Silvestrini, M. G. Tordi, G. Musci, and M. Brunori, *J. Biol. Chem. 265*, 11783 (1990).
96. V. Fülöp, J. W. B. Moir, S. J. Ferguson, and J. Hajdu, *Cell*, *81*, 369 (1995).
97. A. Jafferji, J. W. A. Allen, S. J. Ferguson, and V. Fülöp, *J. Biol. Chem.*, *275*, 25089 (2000).
98. D. Nurizzo, M.-C. Silvestrini, M. Mathieu, F. Cutruzzola, D. Bourgeois, V. Fülöp, J. Hajdu, M. Brunori, M. Tegoni, and C. Cambillau, *Structure, 5*, 1157 (1997).
99. D. Nurizzo, F. Cutruzzola, M. Arese, D. Bourgeois, M. Brunori, C. Cambillau, and M. Tegoni, *Biochemistry*, *37*, 13987 (1998).
100. S. R. Parr, D. Barber, and C. Greenwood, *Biochem. J.*, *167*, 447 (1977).
101. Y. Blatt and I. Pecht, *Biochemistry*, *18*, 2917 (1979).
102. S. A. Schichman and H. B. Gray, *J. Am. Chem. Soc.*, *103*, 7794 (1981).
103. K. Kobayashi, A. Koppenhöfer, S. J. Ferguson, N. J. Watmough, and S. Tagawa, *Biochemistry*, *36*, 13611 (1997).
104. K. Kobayashi, A. Koppenhöfer, S. J. Ferguson, and S. Tagawa, *Biochemistry*, *40*, 8542 (2001).
105. O. Farver, P. H. M. Kroneck, W. G. Zumft, and I. Pecht, *Biophys. Chem.*, *98*, 27 (2002).
106. O. Farver, P. H. M. Kroneck, W. G. Zumft, and I. Pecht, *Proc. Natl. Acad. Sci. U.S.A.*, *100*, 7622 (2003).

107. P. A. Williams, V. Fülöp, E. F. Garman, N. F. W. Saunders, S. J. Ferguson, and J. Hajdu, *Nature (London)*, *389*, 406 (1997).
108. J. W. A. Allen, N. J. Watmough, and S. J. Ferguson, *Nat. Struct. Biol.*, *7*, 885 (2000).
109. K. Brown, V. Roig-Zamboni, F. Cutrozzola, M. Arese, W. Sun, M. Brunori, C. Cambillau, and M. Tegoni, *J. Mol. Biol.*, *312*, 541 (2001).
110. F. Cutruzzola, K. Brown, E. K. Wilson, A. Bellelli, M. Arese, M. Tegoni, C. Cambillau and M. Brunori, *Proc. Natl. Acad. Sci. U.S.A.*, *98*, 2232 (2001).
111. O. Farver, M. Brunori, F. Cutruzzola, and I. Pecht, to be published.
112. M. R. Cheesman, S. J. Ferguson, J. W. B. Moir, D. R. Richardson, W. G. Zumft, and A. J. Thomson, *Biochemistry*, *36*, 16267 (1997).
113. A. Jüngst, S. Wakabayashi, H. Matsubara, and W. G. Zumft, *FEBS Lett.*, *279*, 205 (1991).
114. E. K. Wilson, A. Bellelli, F. Cutruzzola, W. G. Zumft, A. Gautierrez, and N. S. Scrutton, *Biochem.*, *J. 355*, 39 (2001).
115. O.-M. H. Richter and B. Ludwig, *Rev. Physiol. Biochem. Pharmacol.*, *147*, 47 (2003).
116. M. Wikström, *Biochem. Biophys. Acta*, *1655*, 241 (2004).
117. M. K. F. Wikström, *Nature (London)*, *266*, 271 (1977).
118. G. T. Babcock and M. Wikström. *Nature (London)*, *356*, 301 (1992).
119. H., Michel, J. Behr, A. Harrenga, and A. Kannt, *Annu. Rev. Biophys. Biomol. Struct.*, *27*, 329 (1998).
120. S. Paula, A. Sucheta, I. Szundi, and Ó. Einarsdóttir, *Biochemistry*, *38*, 3025 (1999).
121. M. Wikström, M., *Biochim. Biophys. Acta*, *1458*, 188 (2000).
122. M. Ruitenberg, A. Kannt, E. Bamberg, B. Ludwig, H. Michel, and K. Fendler, *Proc. Natl. Acad. Sci. U.S.A.*, *97*, 4632 (2000).
123. M. Wikström and M. I. Verkhovski, *Biochim. Biophys. Acta.*, *1555*, 128 (2002).
124. H. Michel, *Proc. Natl. Acad. Sci. U.S.A.*, *95*, 12819 (1998).
125. C. Backgren, G. Hummer, M. Wikström, and A. Puustinen, *Biochemistry*, *39*, 7863 (2000).
126. D. Bloch, I. Belevich, A. Jasaitis, C. Ribacka, A. Puustinen, M. I. Verkhovsky, and M. Wikström, *Proc. Natl. Acad. Sci. U.S.A.*, *101*, 529 (2004).
127. S. Iwata, C. Ostermeier, B. Ludwig, and H. Michel, *Nature (London)*, *376*, 660 (1995).
128. T. Tsukihara, H. Aoyama, E. Yamashita, T. Tomizaki, H. Yamaguchi, K. Shinzawa-Itoh, R. Nakashima, R. Yaono, and S. Yoshikawa, *Science*, *269*, 1069 (1995).
129. T. Tsukihara, H. Aoyama, E. Yamashita, T. Tomizaki, H. Yamaguchi, K. Shinzawa-Itoh, R. Nakashima, R. Yaono, and S. Yoshikawa, *Science*, *272*, 1136 (1996).
130. C. Ostermeier, A. Harrenga, U. Ermler, and H. Michel, *Proc. Natl. Acad. Sci. U.S.A.*, *94*, 10547 (1997).
131. S. Yoshikawa, K. Shinzawa-Itoh, R. Nakashima, R. Yaono, E. Yamashita, N. Inoue, M. Yao, M. J. Fei, C. Peters-Libeu, T. Mizushima, H. Yamaguchi, T. Tomizaki, and T. Tsukihara, *Science*, *280*, 1723 (1998).
132. A. Harrenga and H. Michel, *J. Biol. Chem.*, *274*, 33296 (1999).
133. T. Soulimane, G. Buse, G. P. Bourenkov, H. D. Bartunik, R. Huber, and M. E. Than, *EMBO J.*, *19*, 1766 (2000).
134. M. Svensson-Ek, J. Abramson, G. Larsson, S. Törnroth, P. Brzezinski, and S. Iwata, *J. Mol. Biol.*, *321*, 329 (2002).
135. C. Greenwood and Q. H. Gibson, *J. Biol. Chem.*, *242*, 1782 (1967).

136. M. T. Wilson, C. Greenwood, M. Brunori, and E. Antonini, *Biochem. J.*, *147*, 145 (1975).
137. S. Ferguson-Miller, D. Brautigam, and E. Margoliash, *J. Biol. Chem.*, *251*, 1104 (1976).
138. R. Rieder and H.R. Bosshard, *J. Biol. Chem.*, *255*, 4732 (1980).
139. B. Ludwig Q. H. Gibson, *J. Biol. Chem.*, *256*, 10092 (1981).
140. T. M. Antalis and G. Palmer, *J. Biol. Chem.*, *257*, 6194 (1982).
141. R. Bisson, G. C. M. Steffens, R. A. Capaldi, and G. Buse, *FEBS Lett.*, *144*, 359 (1982).
142. Ó. Einarsdóttir, M. G. Choc, S. Weldon, and W. S. Caughey, *J. Biol. Chem.*, *263*, 13641 (1988).
143. V. Drosou, F. Malatesta, and B. Ludwig, *Eur. J. Biochem.*, *269*, 2980 (2002).
144. B. C. Hill, *J. Biol. Chem.*, *266*, 2219 (1991).
145. B. C. Hill, *J. Biol. Chem.*, *269*, 2419 (1994).
146. J. E. Morgan, P. M. Li, D.-J. Jang, M. Q. A. El-Sayed, and S. I. Chan, *Biochemistry*, *28*, 6975 (1989).
147. T. Nilsson, *Proc. Natl. Acad. Sci. U.S.A.*, *89*, 6497 (1992).
148. P. Brzezinski and M. T. Wilson, *Proc. Natl. Acad. Sci. U.S.A.*, *94*, 6176 (1997).
149. A. A. Konstantinov, S. Siletsky, D. Mitchell, A. Kaulen, and R. B. Gennis, *Proc. Natl. Acad. Sci. U.S.A.*, *94*, 9085 (1997).
150. D. Zaslavsky, R. C. Sadoski, K. Wang, B. Durham, R. B. Gennis, and F. Millett, *Biochemistry*, *37*, 14910 (1998).
151. K. Kobayashi, H. Une, and K. Hayashi, *J. Biol. Chem.*, *264*, 7976 (1989).
152. O. Farver, Y. Chen, J. A. Fee, and I. Pecht, *FEBS Lett. 580*, 3417 (2006).
153. G.-L. Liao, and G. Palmer, *Biochim. Biophys. Acta, 1274*, 109 (1996).
154. H. Wang, D. F. Blair, W. R. Ellis, H. B. Gray, and S. I. Chan, *Biochemistry*, *25*, 167 (1986).
155. A. Sucheta, I. Szundi, and Ó, Einarsdóttir, *Biochemistry*, *37*, 17905 (1998).
156. A. J. Moody and P. R. Rich, *Biochim. Biophys. Acta*, *1015*, 205 (1990).
157. P. Hellwig, personal communication.
158. P. Hellwig, S. Grzybek, J. Behr, B. Ludwig, H. Michel, and W. Mäntele, *Biochemistry*, *38*, 1685 (1999).
159. D. F. Blair, C. T. Martin, J. Gelles, H. Wang, G. W. Brudvig, T. H. Stevens, and S. I. Chan, *Chem. Scr.*, *21*, 43 (1983).
160. B. E. Ramirez, B. G. Malmström, J. R. Winkler, and H. B. Gray, H. B. *Proc. Natl. Acad. Sci. U.S.A.*, *92*, 11949 (1995).
161. J. R. Winkler, B. G. Malmström, and H. B. Gray, *Biophys. Chem.*, *54*, 199 (1995).
162. J. J. Regan, B. E. Ramirez, J. R. Winkler, H. B. Gray, and B. G. Malmström, *J. Bioenerg. Biomembr.*, *30*, 35 (1998).
163. P. Brzezinski, *Biochemistry*, *35*, 5611 (1996).
164. P. Ädelroth, M. Svensson-Ek, D. M. Mitchell, R. B. Gennis, and P. Brzezinski, *Biochemistry*, *36*, 13824 (1997).
165. C. Enroth, B. T. Eger, K. Okamoto, T. Nishino, T. Nishino, and E. F. Pai, *Proc. Natl. Acad. Sci. U.S.A.*, *97*, 10723 (2000).
166. A. Fukunari, K. Okamoto, T. Nishino, B. T. Eger, E. F. Pai, M. Kamezawa, I. Yamada, and N. Kato, *J. Pharmacol. Exp. Ther.*, *311*, 519 (2004).
167. R. F. Anderson, R. Hille, and V. Massey, *J. Biol. Chem.*, *261*, 15870 (1986).
168. R. Hille and R. F. Anderson, *J. Biol. Chem.*, *266*, 5608 (1991).

169. K. Kobayashi, M. Miki, K. Okamoto, and T. Nishino, *J. Biol. Chem.*, *268*, 24642 (1993).
170. I. V. Kurnikov and D. N. Beratan, *J. Chem. Phys.*, *105*, 9561 (1996).
171. J. Hunt, V. Massey, W. R. Dunham, and R. H. Sands, *J. Biol. Chem.*, *268*, 18685 (1993).
172. R. Hille, *Biochemistry*, *30*, 8522 (1991).
173. E. Gross, C. S. Sevier, N. Heldman, E. Vitu, M. Bentzur, C. A. Kaiser, C. Thorpe, and D. Fass, *Proc. Natl. Acad. Sci. U.S.A.*, *103*, 299 (2006).

# CHAPTER 2

# Peptide- or Protein-Cleaving Agents Based on Metal Complexes

**WOO SUK CHEI and JUNGHUN SUH**

*Department of Chemistry*
*Seoul National University*
*Seoul 151-747, Korea*

CONTENTS



*Progress in Inorganic Chemistry, Vol. 55* Edited by Kenneth D. Karlin

# I. INTRODUCTION

Catalysis by metal ions in organic reactions is classified into three categories: Lewis acid catalysis, catalysis via redox reactions, and catalysis via organometallic compounds (1). Lewis acid catalysis by metal ions also plays an important role in the action of many metalloenzymes. Virtually all types of organic reactions are catalyzed by these metalloenzymes. For example, Zn(II) metalloenzymes are the most typical metalloenzymes involving metal ions as the Lewis acid catalysts. The Zn(II) metalloenzyme is known to be an essential component of a large number of enzymes isolated from different species (2, 3). They participate in biological reactions encompassing synthesis and degradation of all major metabolites, that is, carbohydrates, lipids, proteins, and nucleic acids. The Zn(II) enzymes are found in each class of the six categories designated by the International Union of Biochemistry on the basis of the types of organic reactions (3).

Many enzymes contain two or more metal ions in the active site, exploiting collaboration among the metal centers in the catalytic action. Examples of multinuclear metalloenzymes catalyzing hydrolysis of acyl derivatives and related compounds are methionine aminopeptidase (4), metallo-β-lactamase (5), proline dipeptidase (prolidase) (6), urease (7), and agmatinase (8). In addition, there are a large number of multinuclear metalloenzymes that catalyze several other types of reactions (e.g., nucleic acid hydrolysis, synthetic transformations, or oxidation–reduction).

Design of catalysts mimicking the catalytic principles of enzymes is among the great challenges of modern chemistry (9, 10). Catalytic antibodies are examples of semisynthetic artificial enzymes (11–14). Fully synthetic molecules also have been designed as enzyme mimics by using either peptidic (15, 16) or nonpeptidic (17–24) molecules.

Peptide-cleaving metal-complexes are mimics of metalloenzymes, that have shown great promise for use in a variety of applications (25), including protein sequencing and proteomics. When the metal-assisted protein cleavage is accomplished under nondenaturing conditions of temperature and pH, their use can be extended to include the study of protein function and solution structure, the analysis of protein folding, the mapping of enzyme active sites, metal- and ligand-binding sites, ligand induced conformational changes in protein structure, the generation of semisynthetic proteins, the proteolytic cleavage of bioengineered fusion proteins, and therapeutics.

In the area of artificial enzymes, emphasis is given to biomimetic agents that are regenerated after chemical conversion of the substrates, and consequently lead to catalytic turnover. In the early stage of designing artificial enzymes, the main objective was to reproduce the major characteristics of enzymatic action (e.g., complex formation with substrates, large degrees of rate acceleration, and high selectivity). To design more elaborated artificial enzymes, attempts were later made to overcome limitations of enzymes (e.g., instability to heat, incompatibility with organic solvents, inapplicability to abiotic reactions, and too narrow selectivity).

If an artificial enzyme can solve an important biological problem that cannot be treated with natural enzymes, it may be regarded as an artificial dream enzyme. For example, artificial proteases removing protein aggregates allegedly are responsible for Alzheimer's, Parkinson's, or mad cow disease, are among the artificial dream enzymes.

We have been involved in the synthesis of metal-based enzyme-like catalysts for hydrolysis of peptide bonds of proteins. We have been interested in hydrolysis as the target reaction since only one molecule is involved as the substrate, except for the water molecule. We selected peptide bonds as the targets since protein hydrolysis is important in the era of genomics. Proteomics and peptide bonds are quite stable with a half-life (26, 27) of spontaneous hydrolysis at (pH 7 at 25°C) being ~500 years. We chose complexes of transition metal ions as the catalytic centers because a single metal ion can hydrolyze peptide bonds effectively as described in Section II. On the other hand, collaboration among multiple catalytic groups is necessary to hydrolyze peptide bonds when organic functionalities are used to build the catalytic center (28–30).

Artificial metallopeptidases require knowledge of inorganic, organic, biochemistry, and polymer chemistry for their design and have potential practical applications in the pharmaceutical and food industries. Our studies on artificial peptidases have been carried out toward two goals. The first goal is to design synthetic peptidases hydrolyzing essentially all proteins (broad substrate selectivity) at selected sites of the substrates (high cleavage-site selectivity). Synthetic peptidases of this type have been prepared on insoluble supports so that they can have practical applications in the protein industry. The second is to prepare synthetic peptidases cleaving only the target protein or oligopeptide (high substrate selectivity) in the presence of many other biomolecules. Such a synthetic peptidase may be used as a new medicine.

This chapter describes our efforts made toward immobile artificial metallopeptidases with broad substrate selectivity and high cleavage-site selectivity as well as the soluble artificial metallopeptidases highly selective for disease-related proteins or oligopeptides. As a background of our studies, this chapter describes results of some related studies carried out by other investigators (25, 31) in efforts to design reagents based on metal complexes useful for various applications.

## II. ROLES OF METAL IONS ACTING AS LEWIS ACID CATALYSTS IN PEPTIDE HYDROLYSIS AND RELATED REACTIONS

### A. Mechanistic Studies on Small Molecules

For designing enzyme-mimicking catalysts exploiting metal ions as catalytic centers, it is necessary to understand catalytic repertories (32, 33) of metal ions acting as Lewis acid catalysts in the hydrolysis of peptide bonds and related carboxyl derivatives (e.g., esters). Although esters are much easier to hydrolyze than peptides, important mechanistic information for catalysis in peptide hydrolysis can be obtained from that in ester hydrolysis.

#### *1. Catalysis by a Metal Ion Itself*

Unlike protons, another Lewis acid, transition metal ions, or rare-earth metal ions can be present in relatively high concentrations even at neutral or basic pH values and can possess multiple positive charges. In addition, transition metal ions or rare-earth metal ions can strongly bind specific parts of organic molecules through complex formation.

**1.1. Enhancement of Electrophilicity of Substrates.** As oxygen atoms are coordinated to metal ions, carbonyl groups become more electrophilic. In addition, negative charges developed on carbonyl oxygen atoms during the reactions of peptides, or esters can be stabilized by interaction with metal ions.

Coordination of the carbonyl oxygen to the metal ion and attack of hydroxide ion at the metal-bound carbonyl group (**A**) are assumed to occur in the hydrolysis of a number of esters or peptides, as exemplified by hydrolysis of **B** (34) or **C** (XR = OEt or $NH_2$) (35, 36). For the reaction path indicated by **C**, the half-life was 1.5 min at pH 6 at 25°C, when XR = OEt and 80 min at pH 9 at 25°C, when XR = $NH_2$.

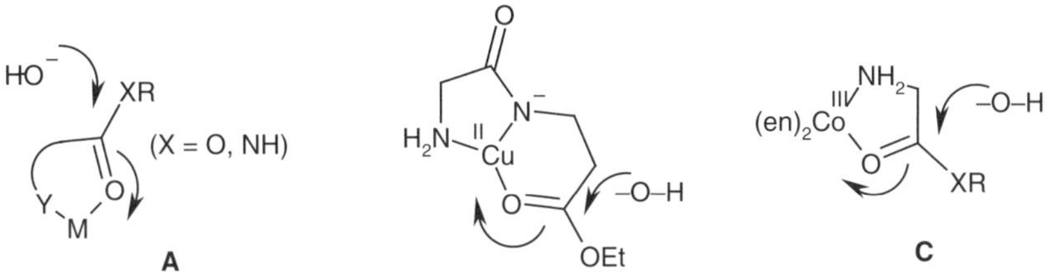

HO⁻ XR (X = O, NH) Y M O

**A**

O N⁻ $H_2N$ II Cu O –O–H OEt

**B**

$NH_2$ III $(en)_2Co$ O –O–H XR

**C**

**1.2. Activation of Leaving Groups.** When the alkoxy oxygen atom is bound to a metal ion, its leaving ability from electrophilic carbon centers is enhanced, leading to catalysis in ester hydrolysis as exemplified by **D** (37).

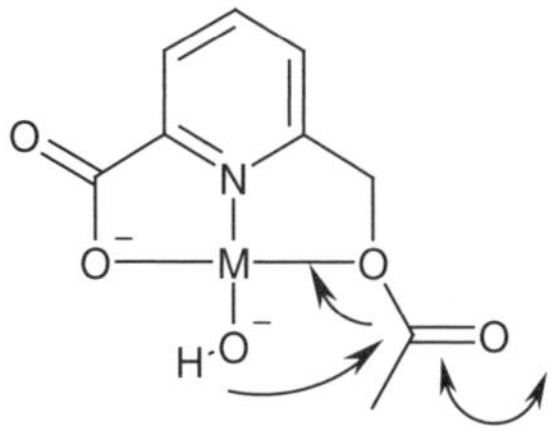

O N O⁻ M O H-O⁻ O

M: $Ni^{II}$, $Co^{II}$

**D**

Even the leaving ability of oxide ion from acyl centers is raised upon coordination to metal ions. For example, metal-bound carboxylate ion can be hydrolyzed through the attack of hydroxide ion at the acyl carbon of the bound carboxylate ion (**E**) accompanied by C–O bond cleavage instead of metal–O cleavage (38–40). The oxide ion is much more basic than the hydroxide ion, with the leaving group ability being extremely low. It is remarkable, therefore, to note that even the oxide ion can be activated upon coordination to a metal ion leading to effective expulsion from a procarbonyl carbon.

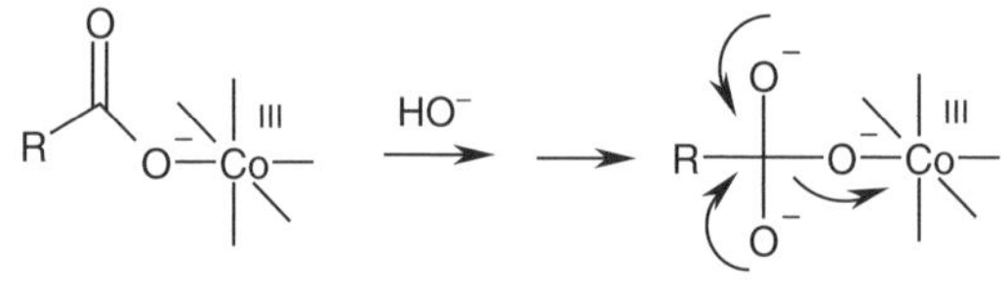

E

Amide nitrogen atoms of ordinary peptide bonds do not coordinate to metal ions since the nitrogen atom contains a partial positive charge due to resonance in the amide bond (**F**). In a β-lactam, the strained ring prohibits resonance and the nitrogen atom is basic enough to coordinate to a metal ion (**G**) leading to catalysis in amide hydrolysis (41, 42).

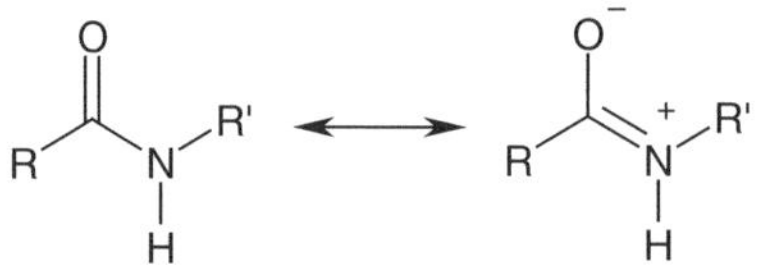

M: $Cu^{II}$, $Ni^{II}$

F 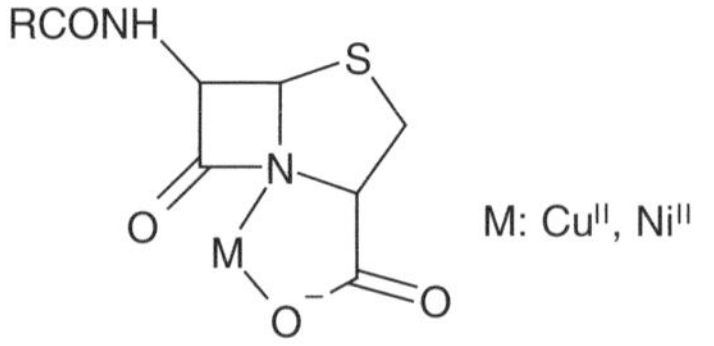 G

## 2. *Catalysis by Metal-Bound Water*

When metal ions are dissolved in water, metal-bound water molecules and metal-bound hydroxide ions are also produced and play important catalytic roles.

**2.1. Attack as a Nucleophile.** In the Cu(II)-catalyzed hydrolysis of the acetyl ester of 2-pyridinecarboxaldoxime, the metal-bound water molecule makes nucleophilic attack at the carbonyl carbon of the bound ester (**H**), instead of the kinetically equivalent attack by an external water molecule at the ester linkage bound by the metal ion (43). The basicity and nucleophilicity of water would decrease upon coordination to metal ions. Efficient nucleophilic attack by the Cu(II)-bound water in **H** is attributable to the general base assistance from another water molecule and to the efficient intramolecular reaction between the nucleophile and the ester.

In the hydrolysis of Co(III)-bound amide, such as **I** (R = H), the rate was independent of pH at pH $< 5$, with a half-life of 2 min at 20°C. This reaction

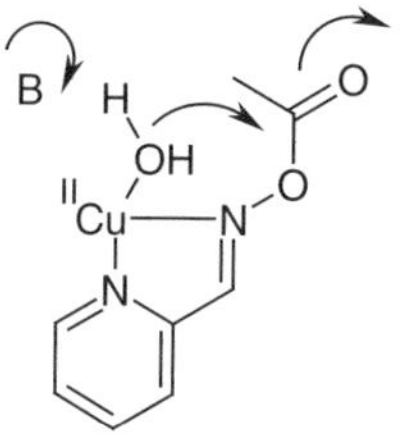

H

path may be ascribed to the intramolecular nucleophilic attack of metal-bound water molecules at the amide group (**I**) (36), where en = ethylenediamine. This path, however, is better explained in terms of nucleophilic attack by the metal-bound hydroxide ion and participation of the specific acid in the rate-controlling expulsion of amine leaving groups (**J**) (44), by analogy with the mechanism of hydrolysis of maleamic acid derivatives (**K**) (45).

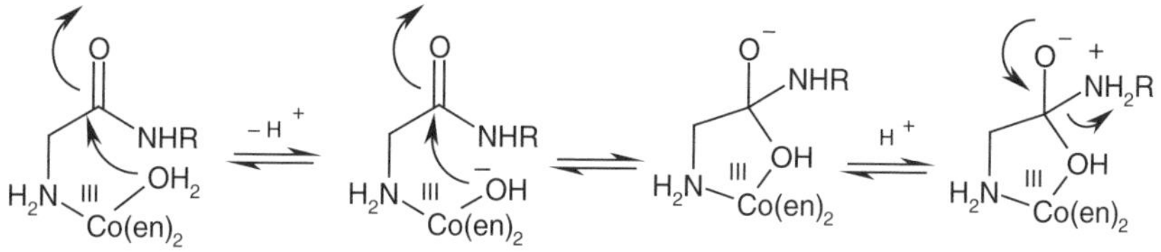

I

J

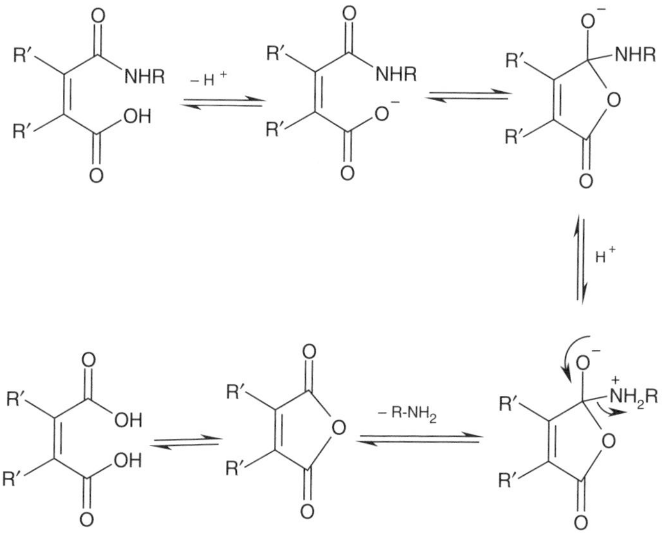

K

**2.2. Assistance as a General Acid.** Upon coordination to a metal ion, water becomes a weak acid and may act as a general acid catalyst, donating a proton in the rate-determining transition state. This catalytic role has been demonstrated in the hydrolysis of an alkyl amide cocatalyzed by a metal ion and carboxylate anion in dimethyl sulfoxide (DMSO) containing 5%(v/v) water (**L**) (46, 47). Here, the alkoxide anion formed by the nucleophilic attack of the carboxylate anion at the carbonyl carbon is stabilized by the metal ion. In the rate-determining expulsion of the amine leaving group from the tetrahedral intermediate, the metal-bound water acts as a general acid to protonate the leaving nitrogen anion (**M**). The resulting anhydride undergoes hydrolysis to complete the reaction.

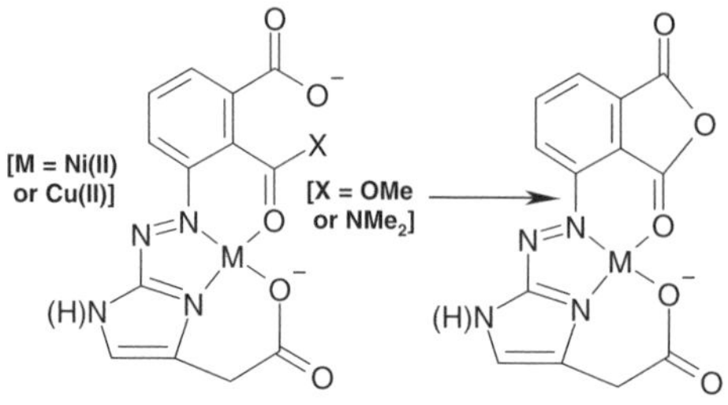

## *3. Catalysis by Metal-Bound Hydroxide Ion*

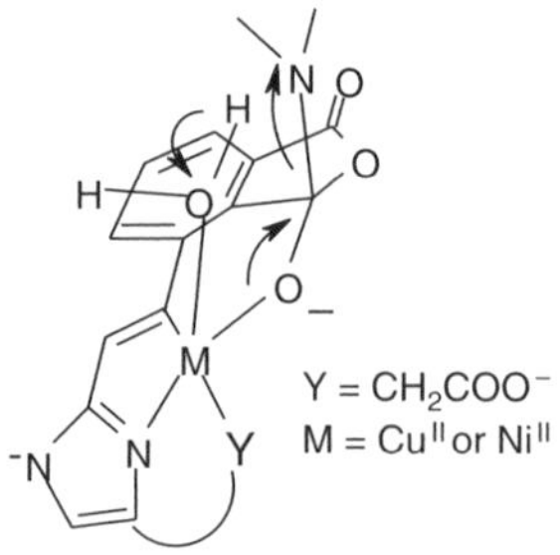

Depending on the nature of metal complexes, metal-bound hydroxide ions can be present in considerable concentrations at neutral or even acidic pH values (48).

**3.1. Attack as a Nucleophile.** Hydrolysis of esters or amides can occur through the nucleophilic attack of metal-bound hydroxide ions, as exemplified by **N** (49, 50). In most cases, however, this mechanism is not easily differentiated from the kinetically equivalent attack by hydroxide ion at the metal-bound carbonyl carbon (**A**) (38). In the case of substitutionally inert complexes of Co(III), $^{18}O$-tracer experiments revealed that both of the two mechanisms occur in the hydrolysis of the bound amino acid esters and amides (36, 49, 51, 52). At a pH 7–8, ionization of Co(III)-bound amide **I** (R = H) produced **N** in >90% of the total concentration, and **N** was hydrolyzed with a half-life of 100 min at 20°C (36).

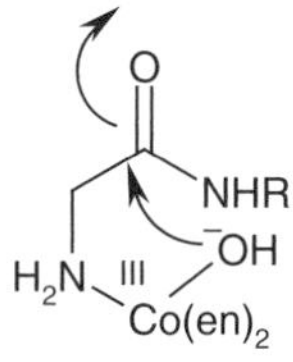

**N**

Efficient amide hydrolysis of an amide bond by a Co(III)-bound hydroxide ion has been demonstrated (**O**) (53, 54). In this intramolecular catalytic system, the metal-bound hydroxide ion is positioned in proximity to the carbonyl carbon of the scissile amide bond contained in a strained ring.

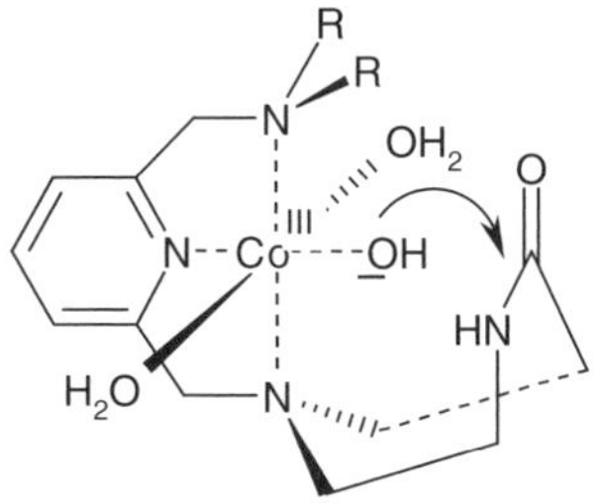

**O**

For most of the hydrolysis of esters or amides catalyzed by exchange-labile metal ions, whether the reaction involves attack by hydroxide ion on the metal-bound carbonyl group or attack by the metal-hydroxo ion at the unactivated carbonyl group is not clearly differentiated. In rare instances, the two mechanistic possibilities were resolved by kinetic methods as exemplified by the hydrolysis of an aryl ester catalyzed by a Zn(II) complex, which proceeds through the attack of (**P**) by an hydroxide ion at the metal-bound carbonyl group rather than attack of (**Q**) by the metal-bound hydroxide ion (55).

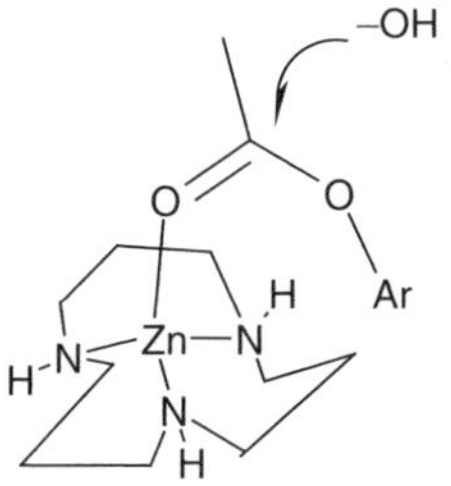

**P** 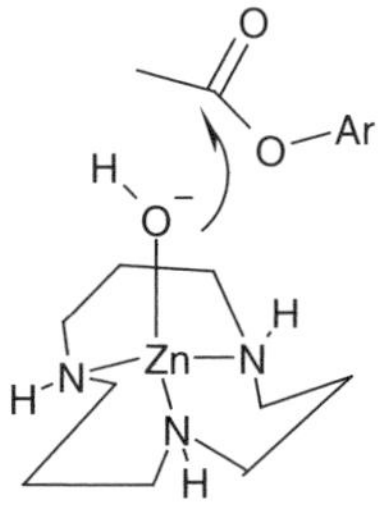 **Q**

### B. Mechanistic Studies on a Metallopeptidase

Information on catalytic roles of metal ions acting as Lewis acid catalysts in metallopeptidases provides insights into designing artificial metallopeptidases. Here, the mechanistic studies on carboxypeptidase A is described as an example. Since it is very difficult to determine the catalytic roles of such metal ions directly by using the metalloenzymes, several lines of circumstantial evidence have been collected to obtain clues for the catalytic roles.

Carboxypeptidase A, a Zn(II)-exopeptidase, is the most intensively investigated among metallopeptidases. X-ray crystallographic studies on carboxypeptidase A revealed that the active-site Zn(II) ion, Glu(270), Tyr(248), and Arg(145) are among the catalytic groups located in the vicinity of the active site (56), and mechanistic studies have been focused on elucidation of the roles of the Zn(II) ion and Glu(270).

In one of the most often proposed mechanisms, the Glu(270) carboxylate makes nucleophilic attack at the carbonyl carbon of the substrate (**R**). In other widely proposed mechanisms, the Glu(270) carboxylate acts as a general base to assist the attack of water molecule (**S**) or the Zn(II)-bound water molecule (**T**) at the carbonyl group. The mechanisms of **S** and **T** do not involve the covalent intermediate, whereas that of **R** produces an anhydride intermediate formed between the acyl portion of the substrate and Glu(270).

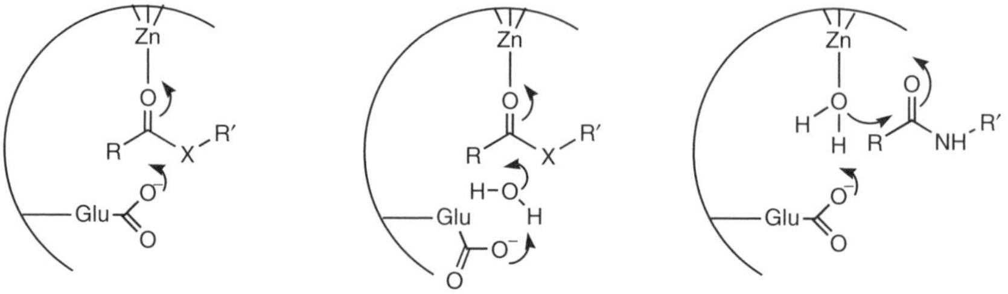

Zn O R X R′ O⁻ Glu O

Zn O R X R′ H–O H Glu O⁻ O

Zn H O H R O NH R′ O⁻ Glu O

(X = O, NH)

**R** **S** **T**

In support of the mechanism of **R**, accumulation of intermediates during the carboxypeptidase A-catalyzed ester hydrolysis has been reported (57, 58), and several lines of evidence were consistent with the nucleophilic attack of Glu(270) leading to the anhydride intermediate (59, 60). Other pieces of evidence have been presented for the esterase action of carboxypeptidase A in support of the anhydride mechanism, such as resonance Raman spectroscopic measurements or trapping of the anhydride intermediate with a radioactive borohydride derivative (61–63). Structures of the intermediates observed during the esterase action of the enzyme are not, however, unambiguously determined. In addition, whether the same mechanism is operative during the peptidase action is not clear. In this regard, it is noteworthy that the model study involving **L** revealed that the peptidase action of carboxypeptidase A (CPA) would proceed through the mechanism of **R** if the esterase action involves the same mechanism (47).

The mechanism of **S** was proposed on the basis of the results of an $^{18}$O-exchange experiment (64). Results of the $^{18}$O-exchange experiment were later shown to be compatible with the mechanism of **R** on the basis of a model study (44) as well.

The mechanism of **T** was proposed on the basis of X-ray crystallographic studies (65) on complexes (**U**) of carboxypeptidase A formed with phosphonate inhibitors and the linear correlation between log $K_i$ for the phosphonate inhibitors and $\log k_{cat}/K_m$ for analogous amide substrates (66). The linear correlation was taken to suggest that the rate-determining transition state for the peptidase action would resemble **U** and that the enzymatic action proceeds through **V**.

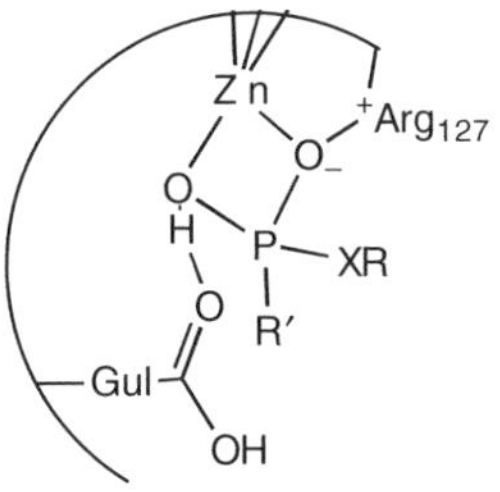


**U** (X = NH or O)

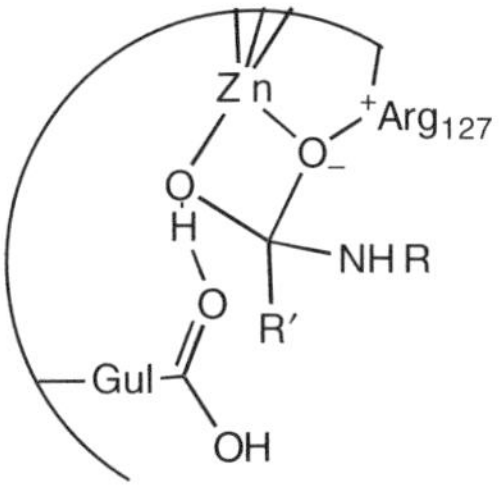


**V**

The linear correlation between $\log K_i$ and $\log k_{cat}/K_m$ is an example of linear correlation between $\Delta G^\circ$ and $\Delta G^\ddagger$ of two different reactions. Huge amounts of such linear correlation data (e.g., Hammett plots or Brønsted plots) are reported in the literature for various organic reactions (67). The correlation, however, does not mean that the transition state corresponding to $\Delta G^\ddagger$ resembles the species corresponding to $\Delta G^\circ$. The linear correlation observed with carboxypeptidase A, therefore, cannot be regarded as evidence for the mechanism of **T** (33).

Although a vast amount of mechanistic studies have been carried out for carboxypeptidase A, conclusive evidence is not available for the catalytic roles of the active-site Zn(II) ion.

## III. PROTEIN-CLEAVING AGENTS REQUIRING OXIDO-REDUCTIVE ADDITIVES

### A. Agents for Oxidative Cleavage of Proteins

Early studies in the area of metal-promoted peptide cleavage employed metal complexes in combination with oxidoreductive additives to achieve either oxidative (68–82) or hydrolytic (83) cleavage of the target proteins. For the oxidative cleavage, various redox active metal ions, such as Cu(I)–Cu(II), Cr(III), Ce(V), Fe(II)–Fe(III), Ni(II), and V(V) were employed.

Metal complexes [e.g., a Cu(I) complex of 1,10-phenanthroline (phen) or and Fe(II) complex of ethylenediaminetetraacetato (edta, ligand)] conjugated to deoxyribonucleic acid (DNA) binding ligands oxidatively cleave nucleic acids in the presence of redox additives (84, 85). The redox chemistry of those coordination complexes is related to the oxidative cleavage of polypeptide backbones of proteins. Oxidative cleavage of peptide bonds usually involve reactive oxygen species yielding multiple fragments (76, 86, 87).

Target-selective (i.e., reacting with the target molecule selectively) oxidative cleavage of proteins by metal complexes has been achieved by attaching the metal complexes to organic moieties that recognized the target proteins. For example, calmodulin, a multifunctional calcium receptor protein, was chosen as the target protein and edta was covalently tethered to the calmodulin antagonist trifuoperazine to obtain **W** (68). Owing to the trifuoperazine moiety, **W** recognized calmodulin, leading to oxidative cleavage of the protein in the presence of Fe(II), $O_2$, and dithiothreitol to produce six major fragments. Similarly, edta was attached to biotin to obtain **X** (69). Upon reaction with the Cu(II) or Fe(III) complex of **X** in the presence of $O_2$ and mercaptoethanol, streptavidin was cleaved. Since biotin has a very high affinity for streptavidin, the metal complex of **X** selectively delivered the redox active metal center in close proximity to the polypeptide backbone at the biotin binding site of streptavidin, resulting in the oxidative protein cleavage.

Analysis of the solution structure of a protein can be carried out by using redox-active metal centers (25). For example, $[Cu(phen)_2]^+$ was exploited in the structural analysis of *Escherichia coli* lactose permease, a paradigm for membrane transport proteins (73). By the reaction of **Y** with a functional lactose permease mutant, phen was covalently linked to a single cysteine residue situated within one of the 12 membrane-spanning α-helices of the protein. After Cu(II) ion was bound to the phen moiety of the modified protein and ascorbate was added in the presence of $O_2$, efficient, localized scission (up to 30%) of the protein backbone was observed in adjacent helices, providing information on the structure of the protein.

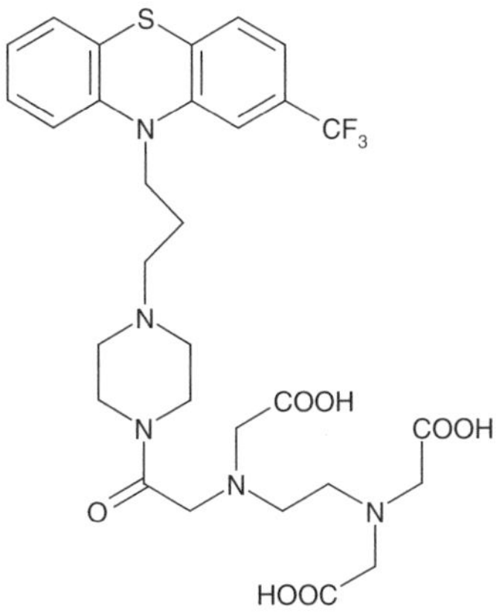

X = NH or $HN(CH_2)_5CONH$

**W**

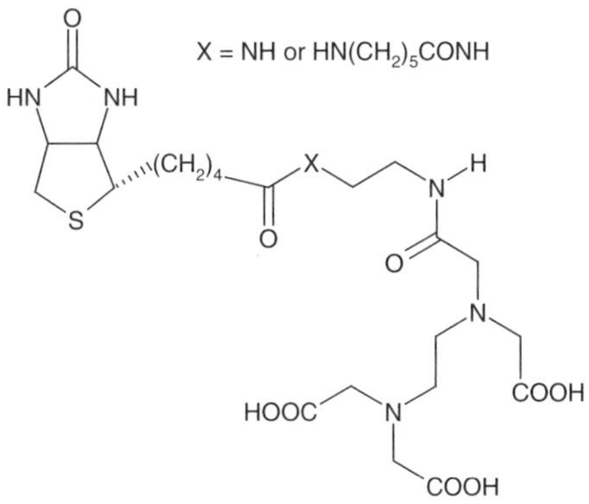

**X**

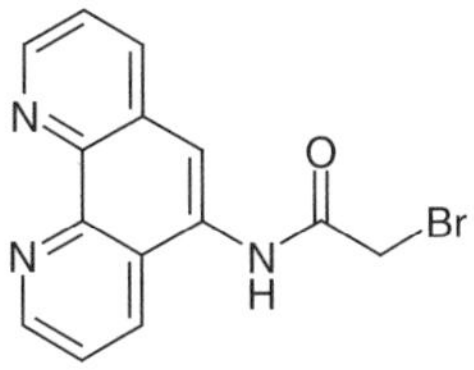

**Y**

Another example of the use of an affinity ligand in the design of a protein-cleavage reagent selective for the target protein is the Cu(II) complex of **Z**. By taking advantage of the tight binding of sulfonamide inhibitors to carbonic anhydrase, the Cu(II) complex of **Z** was mixed with carbonic anhydrase (70). Upon addition of ascorbate, the Cu(II) complex bound to the active site of carbonic anhydrase was reduced to the corresponding Cu(I) complex, which efficiently cleaved the enzyme in the presence of $O_2$ to yield a discrete set of

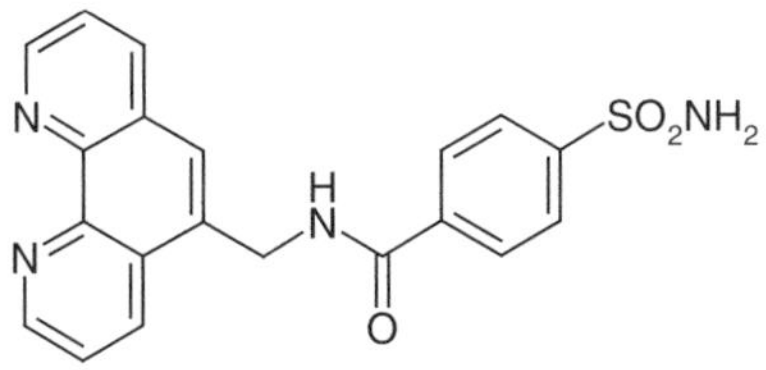

**Z**

cleavage fragments. Upon oxidative cleavage of peptide bonds, the amine moieties were oxidized, whereas the carboxyl moieties were generated intact.

The peptide cleavage by an Fe complex of edta involves production of diffusible hydroxyl radicals that presumably lead to multiple products (70). On the other hand, the $[Cu(phen)_2]^+$ reagent appears to react via a nondiffusible copper-oxo intermediate. Nevertheless, those reagents require coreactants (e.g., mercaptans or ascorbate). In addition, the cleavage reagents should be added in excess of the substrate proteins. Oxidative cleavage of peptide bonds produces substances with unknown safety risks. Therefore oxidative cleavage of peptide bonds by various metal complexes are not useful for therapeutic or industrial purposes.

### B. Agents for Hydrolytic Cleavage of Proteins

The Fe(III) complex of edta cleaves peptide bonds leading to the formation of hydrolysis products when $H_2O_2$ and ascorbic acid are added as coreactants. For example, the reaction of **AA** with carbonic anhydrase resulted in covalent attachment of the Fe(III)–edta complex to Cys(212) to form **AB** (83). Upon addition of $H_2O_2$ and ascorbic acid to **AB**, the peptide bond of Leu(189)–Asp(190) was quickly cleaved to form the hydrolysis products. The mechanism proposed for the hydrolytic cleavage of the peptide bond is indicated by **AC**. Although peptide bonds are cleaved by hydrolysis through the catalytic action of the metal complex, this method requires addition of equivalent amounts of $H_2O_2$ and ascorbic acid. This reagent is not, therefore, suitable as the catalytic center of artificial proteases applicable for protein industries or for peptide-cleaving catalytic drugs.

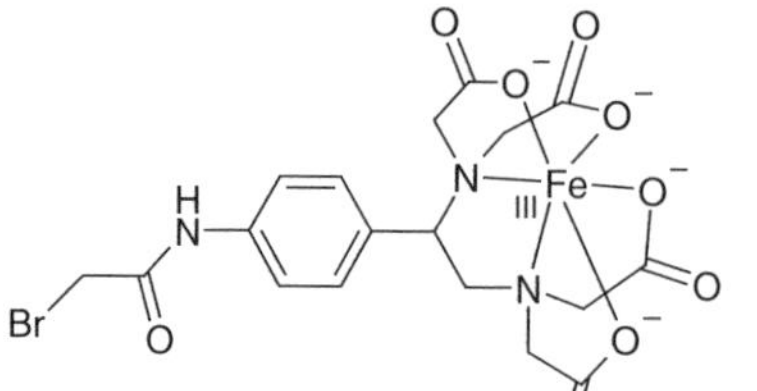

**AA**

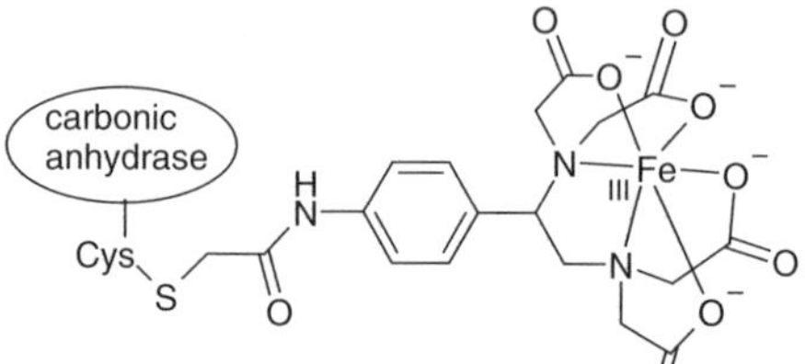

**AB**

## IV. PEPTIDE- OR PROTEIN-HYDROLYZING AGENTS LACKING SUBSTRATE SELECTIVITY

### A. Soluble Agents

Hydrolysis of oligopeptides or proteins is promoted or catalyzed by complexes of various transition metal ions or lanthanide ions. Often, the hydrolysis reactions proceed very slowly even in the presence of the metal ions, requiring high temperatures and/or extended incubation. In many cases, the metal ion is more strongly bound by the hydrolysis product than by the substrate, and thus fails to achieve catalytic turnover.

Several research groups have been involved in designing soluble metal-based agents for peptide hydrolysis, often reporting important results (25, 31). A very recent review article (25) described those studies comprehensively. Instead of repeating the comprehensive review on those results, only some of the soluble metal complexes manifesting fairly high peptide-hydrolyzing activities are described in this chapter.

### 1. *Agents Based on Co(III) Complexes*

A peptide group is activated by attaching electron-withdrawing groups to the acyl portion or an aryl ring to the amine nitrogen atom or by introducing conformational strain around the amide group. Unactivated petides, which are more difficult to hydrolyze than activated peptides, are the main targets of biomimetic catalysts. Hydrolysis of unactivated peptides is sometimes remarkably accelerated when metal ions are tethered to the substrates so that the reaction between the metal center and the peptide group becomes an intramolecular process. For example, **C** (XR = $NH_2$) and **N** (R = H) are hydrolyzed with half-lives of 80 min at pH 9 and 25°C and 100 min at pH 7–8 and 20°C, respectively. As mentioned above, the half-life for peptide bonds is ~ 500 years at 25°C and pH 7 in the absence of any catalysts. Despite the remarkable rate acceleration, the Co(III) ion of **C**, **I**, and **N** acts as a promoter instead of a catalyst. The exchange-inert Co(III) is not released from the product after peptide hydrolysis and, thus, catalytic turnover is prevented.

The Co(III) ion is first tethered to the substrate and subjected to kinetic studies after separation of the resulting Co(III) complex in the case of **C**, **I**, and **N**. In those Co(III) complexes, two cis-oriented coordination sites are exploited for peptide hydrolysis; one for binding of substrate and the other for performing the catalytic roles.

Intermolecular catalysis in hydrolysis of carboxyl amides, carboxyl esters, or phosphodiesters by Co(III) ion untethered to the substrate was accomplished when 3,3′,3″-triaminotripropylamine (trpn: **AD**) or 1,4,7,10-tetraazacyclododecane (cyclen: **AE**) was employed as the chelating ligand of the Co(III) ion (88–90). In particular, parameter values for the hydrolysis of formylmorpholine catalyzed by $[Co(cyclen)(OH_2)]^{3+}$ were measured at 60°C in $D_2O$ (**AF**, **AG**). Inspite of the exchange-inertness (91) of Co(III), the carboxylate or phosphate anion coordinated to the Co(III) ion complexed by cyclen or trpn dissociated readily without affecting the overall rate of the catalytic process.

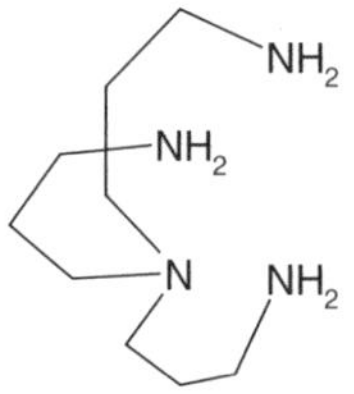

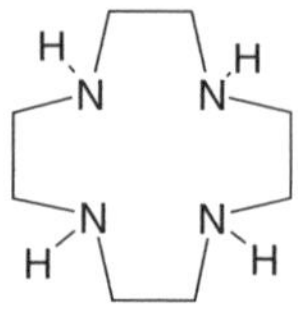

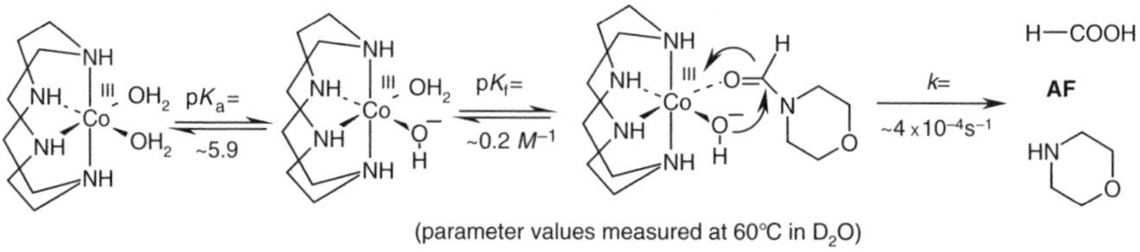

In a recent study, incubation of **AH-a** with β-Ala-His in a 1:2 molar ratio at pH 7.5 and 45°C for 6 h produced the crystals containing **AH-c** (92). Here, the reaction proceeded through complex formation as indicated by **AH-b**. Whether the product dissociates readily from **AH-c** leading to catalytic turnover, however, was not examined.

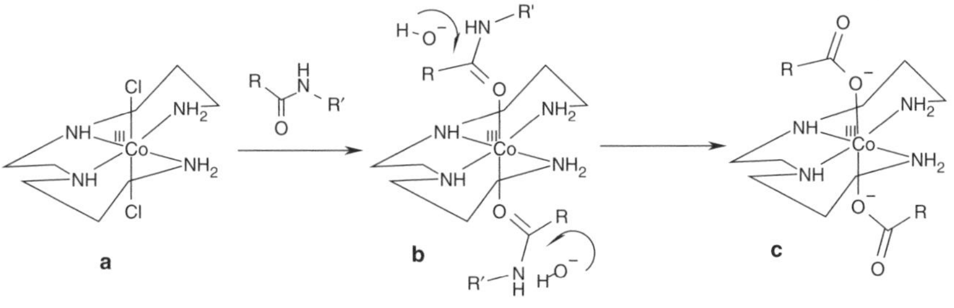

**AH**

Intermolecular reaction between Co(III) complexes and a protein leading to the cleavage of the protein was reported (93). By using $[Co(H_2O)(NH_3)_5]^{3+}$ or $[Co(H_2O)_2(NH_3)_4]^{3+}$ (1 m*M*), hen egg lysozyme (15 μ*M*) was hydrolyzed at pH 7 and 37°C (35–45% after 26 h). The protein was selectively cleaved at Ala(110)–Trp(111). It was suggested that the Co(III) center was selectively anchored in the proximity of the scissile peptide bond.

### 2. *Agents Based on Pt(II) or Pd(II) Complexes*

On the reaction of $[PtCl_4]^{2-}$ with sulfur-containing amino acid derivatives, Pt was tethered to the sulfur atom. The Pt(II) center promoted hydrolysis of the adjacent peptide bond as indicated by **AI** (94). At the optimum pH ($\leq 2$), the half-life for peptide cleavage was as short as 3–4 h at 40°C.

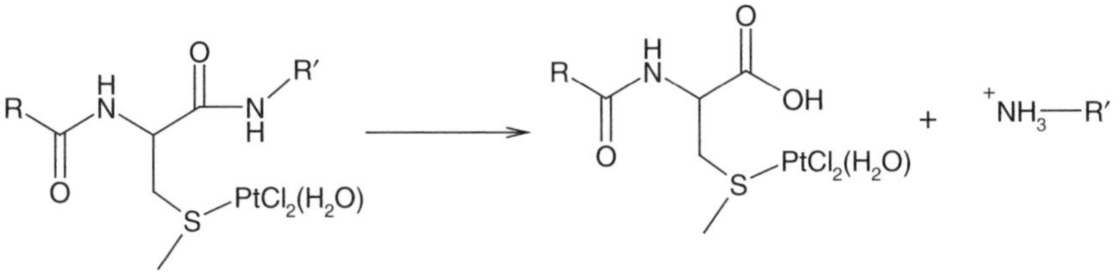

**AI**

The study on metal ions anchored at the sulfur atom was extended to Pd(II) complexes in view of much faster ligand substitution of Pd(II) compared with Pt(II) (95, 96). Under optimal conditions, the Pd(II) complexes promoted the peptide hydrolysis of sulfur-containing oligopeptides with the half-lives as short as 13 min at 40°C. Both Pt(II) and Pd(II) complexes were, however, active only under acidic conditions (pH 1–2). This results because of the high Lewis acidity of Pt(II) and Pd(II) ions that facilitates deprotonation of an adjacent amide N–H group and the subsequent coordination of the resulting anion to the metal center as indicated by structure **b** of **AJ**. In addition, anchorage of the metal ion by the sulfur atom provides a very high effective concentration of the metal ion toward the amide nitrogen atom. As the pH is raised from 1–2, the metal–substrate complex is transformed into the inactive species **b**. At pH $\leq 2$, **a** is the dominant species that undergoes peptide hydrolysis through the mechanism indicated by either **c** or **d**.

**AJ**

As indicated by **AI** and **AJ**, the carboxylic side of the metal-bearing residue is cleaved with short peptide substrates. With longer peptides and proteins, the Pd(II) reagents cleaved peptide bonds located on the amino side of the Pd(II)-anchoring residue as indicated by **AK** (97).

By taking advantage of easy ligand substitution of Pd(II), site-selective (i.e., reacting with a specific site of the substrate selectively) cleavage of cytochrome *c* with untethered promoter was achieved with Pd(II) complexes (98). After

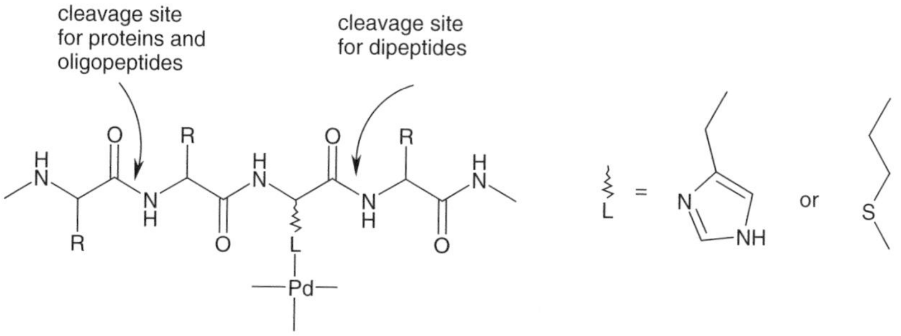

AK

incubation with Pd(II) complexes (1.3 m*M*) for 2 days at pH 1.7 and 40°C, cytochrome *c* (1.3 m*M*) was cleaved selectively at a peptide linkage in proximity to a sulfur-containing residue in 80% yield.

In acetone solutions containing 1 m*M* $HClO_4$ and up to 0.3 *M* $H_2O$, Pd(II) complexes are bound to the indole ring of tryptophan leading to hydrolysis of the adjacent peptide bond as indicated by **AL** (99).

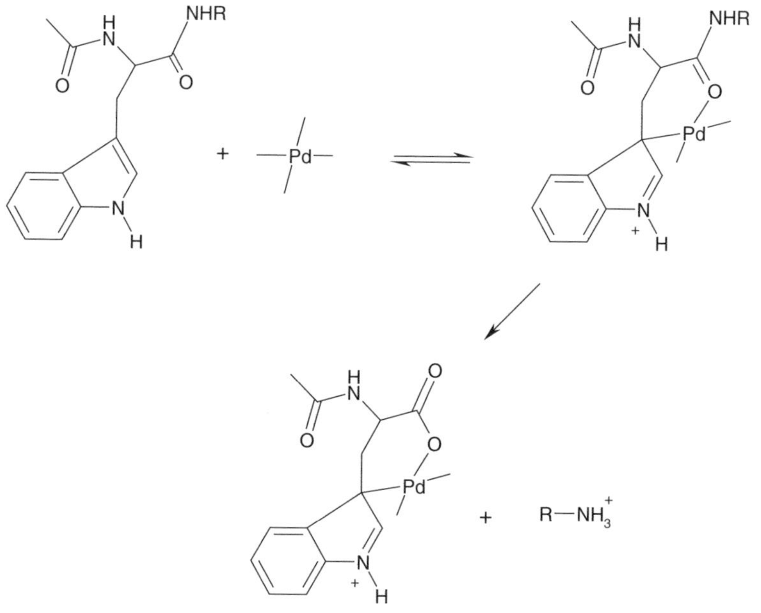

AL

Although site-selective cleavage of a protein or amino acid derivatives with intermolecular catalysts is achieved by using Pd(II) complexes, the Pd(II) complexes require highly acidic conditions and, in some cases, acetone medium. The Pd(II) complexes may be useful for practical applications that allow such conditions.

In a recent study, hydrolysis of peptide bonds at neutral pH values with Pd(II) complexes was achieved (100). Since the amide nitrogen of an *N*-acyl-L-proline derivative lacks ionizable hydrogen, formation of hydrolytically inactive Pd(II)–amidate complex indicated by **AJ-b** is not possible for the peptide bond of the acyl-proline. When AcAla-Ala-Pro-Ala-naphthyl amide (1 m*M*) was treated with $[Pd(H_2O)_4]^{2+}$ (10 m*M*) at pH 7.0 and 60°C, the Ala-Pro peptide bond was hydrolyzed exclusively with the half-life of 3 h.

In order to provide the peptide-cleaving agent with a recognition site, a Pd(II) complex was linked to β-cyclodextrin (100). When the Pd(II)–cyclodextrin conjugate was treated with an oligopeptide containing three proline residues (Arg-Pro-Pro-Gly-Phe-Ser-Pro-Phe-Arg), the Pd(II) reagent recognized the proline adjacent to phenylalanine resulting in sequence specific peptide cleavage. As illustrated by **AM** where the bucket symbolizes cyclodextrin, cyclodextrin forms an inclusion complex with the phenyl ring of phenylalanine and the Pd(II) ion polarizes the carbonyl group of the scissile peptide linkage. Although sequence-selective peptide-cleavage is achieved, there are many obstacles to apply this agent to proteins.

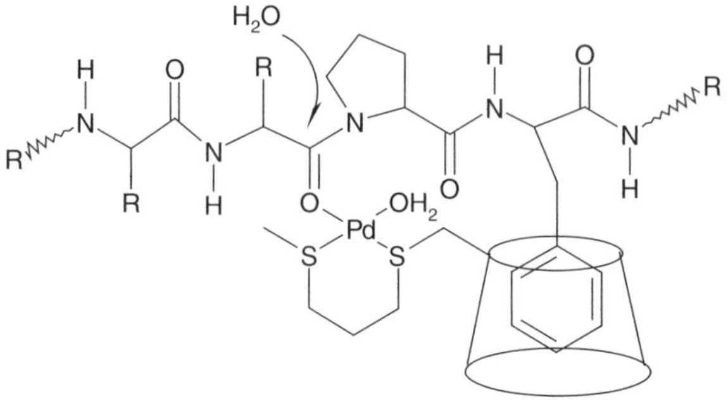

**AM**

### 3. *Agents Based on Cu(II) Complexes*

Application of Cu(II) complexes to catalytic peptide hydrolysis at neutral pH was also attempted. When **AN** (5–10 m*M*) and a formamide derivative

(10–30 m*M*) were heated in $D_2O$ at pD 8.0 and 100°C, the amide was hydrolyzed with a half-life of 3–150 h (101). On the other hand, **AO** failed to manifest any catalytic activity toward the formamides. To explain the different activity of **AN** and **AO**, a mechanism of **AP** was proposed that required two coordination sites for binding of the carbonyl carbon of the amide bond and the nucleophilic hydroxo ligand. Here, the metal ion plays several catalytic roles; polarization of the carbonyl group, delivery of the nucleophilic hydroxo ligand, and provision of an acid needed for protonation of the leaving nitrogen atom. Although catalytic turnover was achieved with **AN** in the hydrolysis of peptide bonds, the catalytic activity was very low since the kinetic measurements were performed at 100°C.

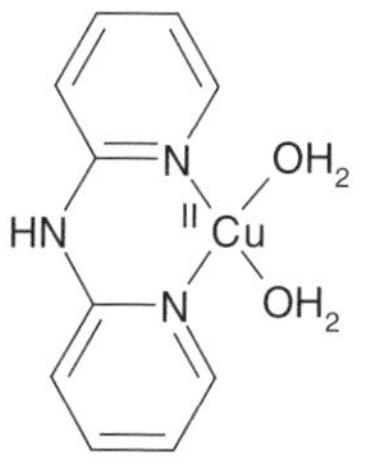

**AN** 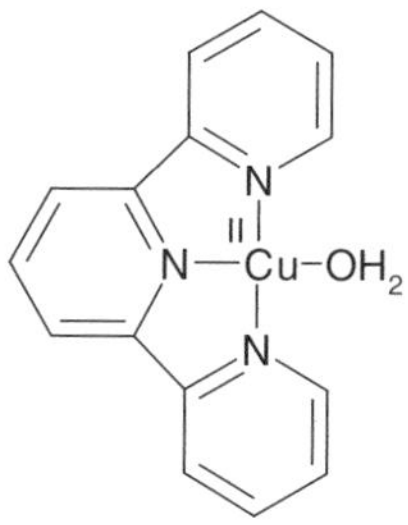 **AO**

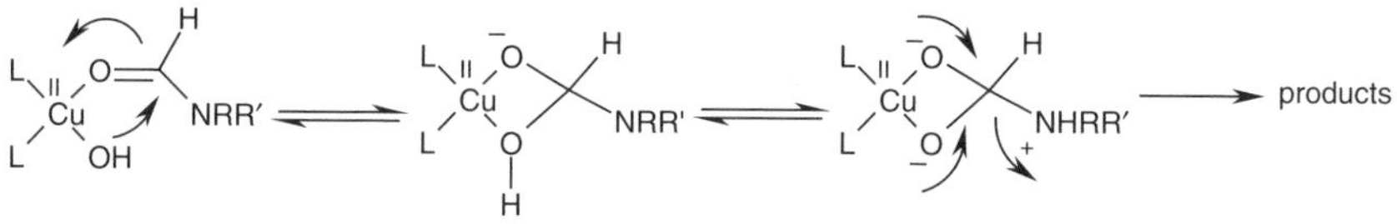

**AP**

Dinuclear Cu(II) complex **AQ** hydrolyzed dimethylformamide (DMF) with a half-life of ~ 2 h at room temperature, although the reaction was stoichiometric rather than catalytic (102). In the mechanism of **AR**, either the polarization of the scissile carbonyl group or delivery of the hydroxo ligand is proposed for each of the two Cu(II) ions.

The activity of the Cu(II) center for peptide hydrolysis was examined by employing a chelating ligand ([9]ane$N_3$: **AS**) (103). When Gly-Gly (1.3 m*M*) was incubated with $Cu^{(II)}$([9]ane$N_3$)$Cl_2$ (1.0 m*M*) at pH 8.1 and 50°C, 15% of the peptide bond was hydrolyzed after 7 days. When bovine serum albumin (BSA) (0.21 m*M*) was incubated with $Cu^{(II)}$([9]ane$N_3$)$Cl_2$ (1.5 m*M*) at pH 7.8 and 50°C, 15% of the protein was degraded after 13 days. In a later study (104), up

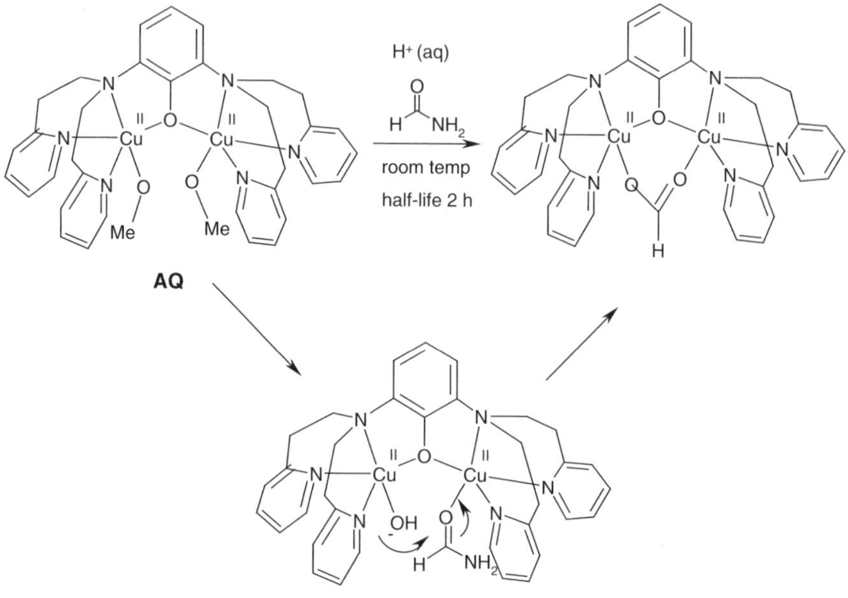

$H^+$ (aq)

room temp

half-life 2 h

**AQ**

**AR**

to 17% of myoglobin (1 m*M*) was degraded after incubation with Cu(II) compounds (5 m*M*), such as $CuCl_2$ for 3 days at pH 8.3 and 50°C.

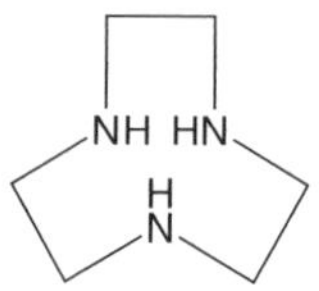

**AS**

The protein-cleaving activity of Cu(II) was considerably improved when **AT** was used (105). With **AT** (150 μ*M*), BSA (1.1 μ*M*) was degraded with a half-life of 30 min at pH 7.2 and 50°C. The proteolytic activity of the Cu(II) complex is enhanced apparently due to the imine formation between the aldehyde group of **AT** and the ammonium groups of the protein substrate, which raises the effective molarity of the Cu(II) center toward the substrate. The *N*-termini of the cleavage products were, however, all blocked by **AT** due to the imine formation between the *N*-terminal amino groups and the aldehyde group of **AT**.

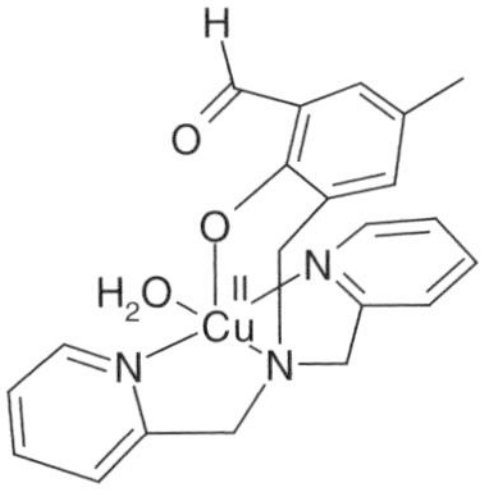

**AT**

### 4. *Agents Based on Other Metal Complexes*

Peptide-hydrolyzing activity of Ce(IV) has been examined. In the presence of 10 m*M* $Ce(NH_4)_2(NO_3)_6$, small peptides (10 m*M*) were hydrolyzed with half-lives of ~ 2 h at pH 7.0 and 50°C (106). The reaction was, however, accompanied by formation of hydroxide gels. When the gelation was avoided by solubilizing Ce(IV) with γ-cyclodextrin and Gly-Gly (10 m*M*) was incubated with the Ce(IV)–cyclodextrin conjugate (10 m*M*) at pH 8.0 and 60°C, 39% of the peptide was hydrolyzed after 24 h (107). Coordination of cyclodextrin to Ce(IV) apparently lowered the catalytic activity of the metal center considerably. The mechanism of **AU** or **AV** was proposed for the peptide hydrolysis of the Ce(IV) complexes.

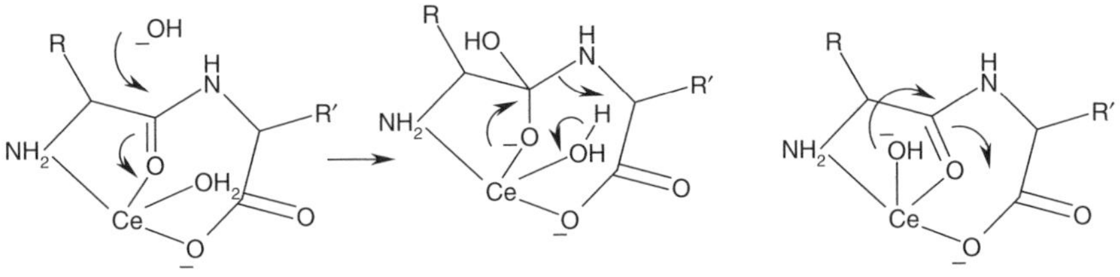

**AU** **AV**

In a recent study, the peptide-hydrolyzing activity of Zr(IV) was greatly enhanced by addition of crown ether **AW** (108). When various dipeptides (2 m*M*) were incubated with $ZrCl_4$ (10 m*M*) and **AW** (19–22 m*M*) at pH 7.0–7.3 and 60°C for 20 h, peptide hydrolysis proceeded in 17–97% yields. This reaction was complicated by precipitation of Zr(IV) complexes.

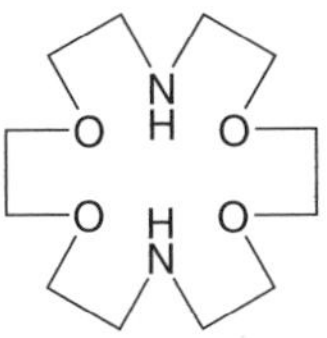

**AW**

When an organometallic Mo(IV) complex (**AX**) was mixed with cysteine-containing oligopeptides, **AX** was bound to the thiolate group of the cysteine residue (109). In the resulting complex, 25% of the adjacent amide group was hydrolyzed within 20 h at pH 7.4 and 60°C through the mechanism of **AY**. The cleavage product was strongly bound by the Mo(IV) complex (**AZ**) and the action of **AX** was stoichiometric rather than catalytic.

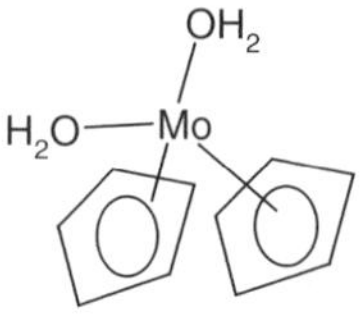

**AX** 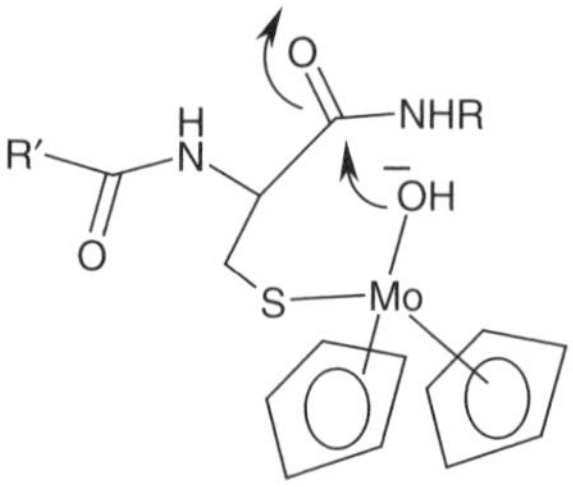 **AY** 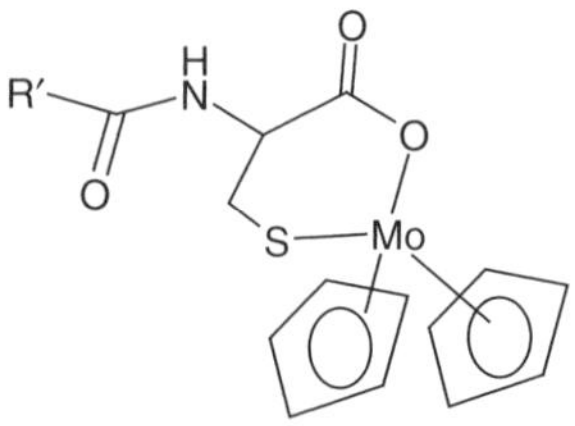 **AZ**

## B. Immobilized Agents

High reaction rates at ambient temperatures and near neutral pH values are necessary to design artificial proteases applicable to food industries, catalytic turnover in the peptide hydrolysis, and hydrolysis of a broad range of protein substrates at selected sites. In addition, easy separation of the catalysts from protein hydrolysates is required. Construction of catalytic centers directly on immobile supports is, therefore, advantageous to designing artificial proteases applicable to protein industries.

### *1. Activation of Agents Based on Cu(II) Complexes*

As discussed in Section IV.A.4, hydrolysis of peptide bonds of oligopeptides or proteins by various soluble metal complexes untethered to the substrate is usually very slow under physiological conditions (pH 7.5 and 37°C). In some cases, fairly fast peptide hydrolysis was achieved, but the product was not released from the metal complex, preventing catalytic turnover. To design an effective synthetic metalloproteases manifesting high activity and catalytic turnover at near physiological pH values, it is necessary to raise the activity of metal complexes substantially and to facilitate release of hydrolysis products from the metal complexes.

There was an attempt to raise the proteolytic activity of metal complexes either by immobilizing the catalytic center on hydrophobic resins or by raising the effective molarity of the catalytic center. We tried to achieve catalytic turnover by using the metal center solely for the peptide hydrolysis and by introducing an auxiliary binding site to the chelating ligand of the catalytic metal center. As described in Section IV.A.4, several investigators have used a single metal center both for recognition of a functional group of the substrate and for promotion of peptide hydrolysis. The resulting hydrolysis product forms a chelate ring with the metal center, often hampering catalytic turnover.

The intrinsic reactivity of the metal complexes in peptide hydrolysis may be enhanced by optimizing the structure of the ligand of the complex or by changing the microenvironment. In enzyme–substrate complexes, polar interactions, such as hydrogen-bonding and dipole–dipole interactions, as well as electrostatic interactions between the substrate and the enzyme, contribute significantly to the stabilization of the transition states (110). Those polar interactions are enhanced in the hydrophobic microenvironments provided by the enzyme. Thus, Nature enhanced the intrinsic reactivity of some catalytic groups of enzymes by optimizing the polarity of microenvironments. If the microenvironment of the catalytic center of a protein-cleaving agent based on a metal complex is modified, the proteolytic activity may be greatly enhanced. One of the simple ways to obtain hydrophobic microenvironment in water is to attach the catalytic center to a hydrophobic synthetic polymer.

When the Cu(II) complex of cyclen (**AE**) was attached to cross-linked polystyrene, the proteolytic activity of the Cu(II) complex of cyclen was enhanced remarkably (111). Poly(chloromethylstyrene-*co*-divinylbenzene) (PCD) was prepared as a derivative of polystyrene in which all of the styryl residues contained chloromethyl groups. By the substitution of chloro groups of PCD with various nucleophiles, PCD derivative **BA** was prepared. γ-Globulin was hydrolyzed upon incubation with **BA**, whose half-life was as short as 1 h at 25°C and pH 7 in the presence of excess **BA**. The rates for hydrolysis of γ-globulin by the Cu(II) complex of cyclen itself dissolved in water were also measured. Comparison of rate data collected at the same catalyst concentrations revealed that the proteolytic activity of the Cu(II) complex of cyclen was

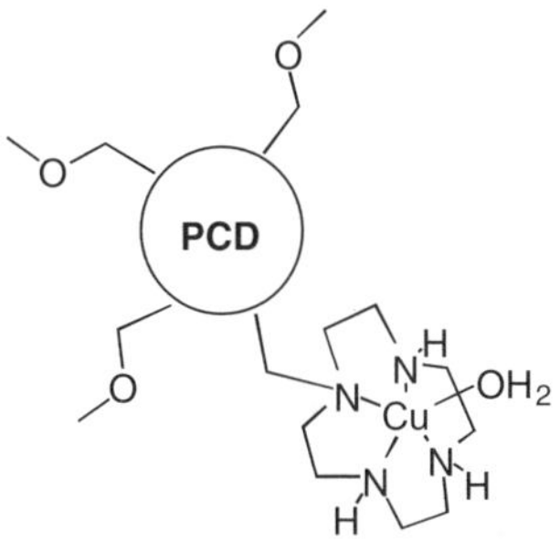

**BA**

enhanced by $\sim 10^3$-times upon attachment to the polystyrene. Considering that only a small fraction of the Cu(II) complexes of cyclen are present on the open surface of PCD and participate in the hydrolysis of γ-globulin, the degree of activation should be substantially $>10^3$-fold. Since the catalytic center was immobilized on the insoluble polystyrene, release of the hydrolysis products from the metal center was assured by analyzing the proteins or peptides dissolved in the buffer solution.

The hydrophobic microenvironments created on the cross-linked polystyrene may raise the intrinsic peptide-hydrolyzing activity of the metal center, as noted above. The microdomains created on the synthetic polymer may facilitate complexation of the protein substrate with the catalytic center. Possible inactivation of the metal center by formation of hydroxo- or oxo-bridged dimers or oligomers can be prevented upon immobilization of the metal center to a solid support. Higher pH values inaccessible by a soluble metal complex can be attained by the corresponding immobilized metal complex. The enhancement in the proteolytic activity of the Cu(II) complex of cyclen upon attachment to the polystyrene may be attributed to some of these effects.

### 2. *Site-Selective Amide Hydrolysis by Cu(II) Complex Combined With the Guanidinium Group*

Since metal complexes with high activity for peptide hydrolysis were secured, it was subsequently attempted to achieve substrate selectivity in the hydrolysis of peptide bonds by the metal complexes. The active site of **BB** was constructed on the surface of partially chloromethylated cross-linked polystyrene (PCPS) (112). Here, the active site was chiral since L-arginine was used to introduce the guanidinium portion.

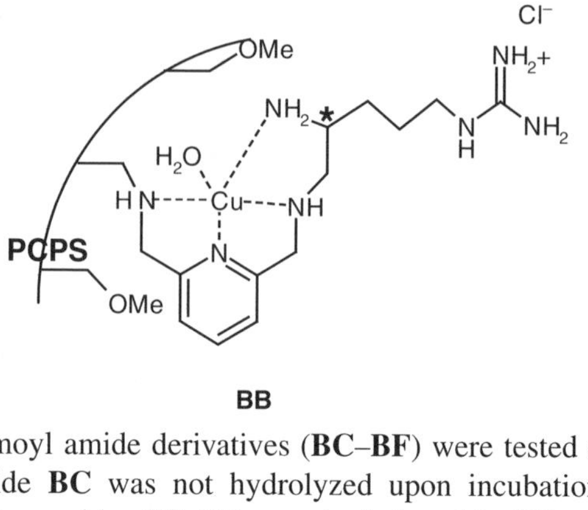

**BB**

Several cinnamoyl amide derivatives (**BC–BF**) were tested as substrates for **BB**. Neutral amide **BC** was not hydrolyzed upon incubation with **BB**, but carboxyl-containing amides **BD–BF** were hydrolyzed by **BB** with the optimum activity observed at pH 9 [half-life at 50°C and 4 m*M* total Cu(II) concentration: 5–10 h depending on the substrate]. Both acetyl L-phenylalanine and acetyl D-phenylalanine were also hydrolyzed by **BB**. The L-isomer was hydrolyzed faster than the D-isomer by 50% as the catalyst was chiral.

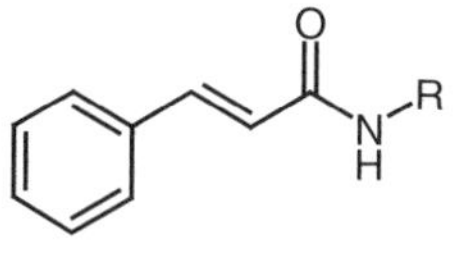

**BC:** R = $CH_3$
**BD:** R = $CH_2COOH$
**BE:** R = $CH_2CH_2COOH$
**BF:** R = $CH_2CH_2CH_2COOH$

Failure to observe any hydrolysis with **BC** lacking the carboxylate moiety strongly suggests that the guanidinium group of **BB** plays an essential role in recognizing the carboxyl-containing amides, as indicated in the proposed mechanism of **BG**. In carboxypeptidase A (56, 57), the active-site Zn(II) ion plays essential catalytic roles, whereas the guanidinium of Arg(145) recognizes the carboxylate anion of the substrates, thus making the enzyme an exopeptidase. In this regard, **BB** may be considered as a model of carboxypeptidase A.

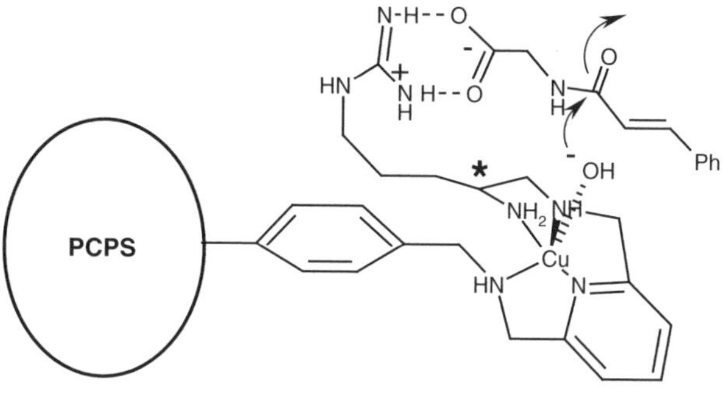

**BG**

### 3. *Site-Selective Amide Hydrolysis by Three Convergent Cu(II) Complexes*

An effective multinuclear artificial metalloenzyme would be obtained if an artificial active site comprising of two or more proximal metal centers is designed. A trinuclear artificial metallopeptidase was prepared by using **BH** (113). The bowl-shaped complex **BH** was originally synthesized by the one-pot template condensation using $CuCl_2 \cdot 2H_2O$, tris(2-aminoethyl)amine, and paraformaldehyde (114). Upon treatment of **BH** with excess NaH, at least three of the six N—H protons were neutralized producing the anion of **BH**. As indicated by Scheme 1, **BH** was attached to PCD by mixing the anion of **BH** with PCD. In

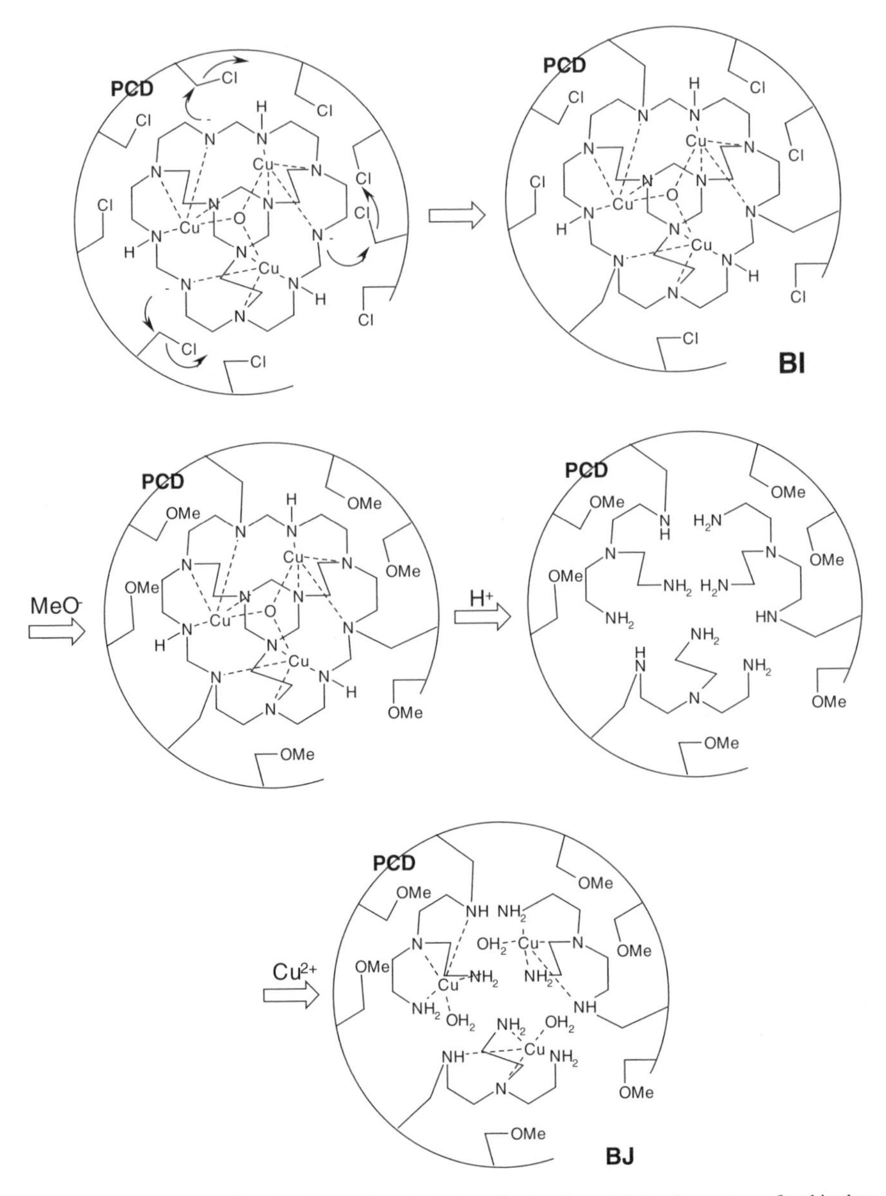


Scheme 1. Synthesis of a trinuclear active site by transferring the catalytic elements confined in the molecular bowl to a polystyrene derivative. [Adapted from Ref. (113).]

PCD, each styryl residue contains a chloromethyl group that reacts readily with one of the nitrogen anions of **BH**. Once a nitrogen atom of **BH** is attached to PCD, the reaction between the remaining nitrogen anions with the chloromethyl groups of PCD becomes an intramolecular process. In **BI**, **BH** is connected to PCD by triple attachment. After the chloro groups of **BI** were reacted with methoxide anion, the resulting resin was treated with acid, and then with $CuCl_2$ to produce **BJ**. The artificial active site in **BJ** contains three convergent Cu(II) complexes of tris(2-aminoethyl)amine.

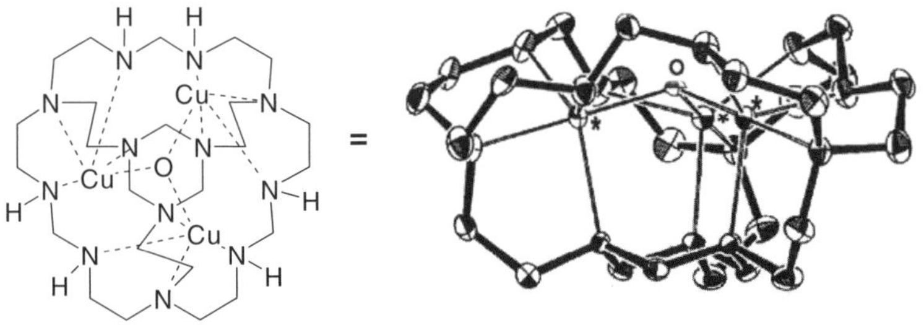

**BH**

The neutral amide **BC** was not affected, on incubation with **BJ**, whereas carboxyl-containing amides **BD–BF** were hydrolyzed with optimum activity observed at pH 8 [half-life at 50°C and 3 m*M* total Cu(II) concentration: 10–15 h, depending on the substrate]. The mechanism of **BK** was proposed to account for the substrate selectivity manifested for the carboxyl-containing amides. Here, one Cu(II) acts as the binding site to recognize the substrate and the other two Cu(II) as catalytic groups.

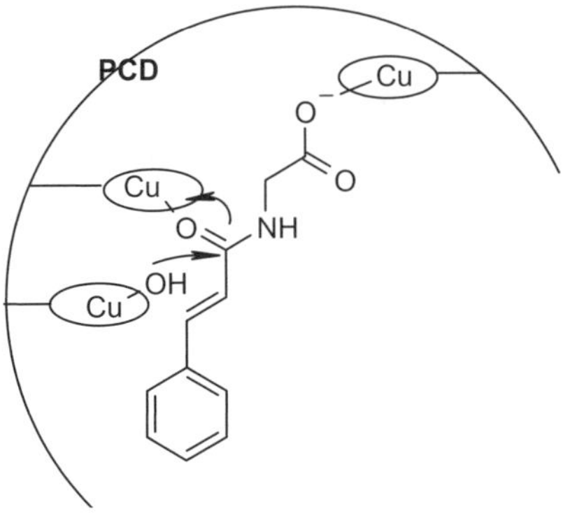

**BK**

The kinetic data were analyzed in terms of the Michaelis–Menten scheme (Eq. 1). Parameter $k_{cat}/K_m$ stands for the reactivity of the catalyst (C) toward the substrate (S) and $k_{cat}$ represents that of the complex formed between the catalyst and the substrate (CS). The $k_{cat}$ value estimated for the amide hydrolysis by **BB** or **BJ** was $\gg 0.1\,h^{-1}$ at pH 8.5 and 50°C or $\gg 0.2\,h^{-1}$ at pH 8 and 50°C, respectively. This may be compared with the $k_{cat}$ of $0.18\,h^{-1}$ (at the optimum pH of 9 and 25°C) measured with a catalytic antibody (115) with peptidase activity elicited by a joint hybridoma and combinatorial antibody library approach in the hydrolysis of an amide substrate.

$$C + S \underset{k_{-1}}{\overset{k_1}{\rightleftharpoons}} CS \xrightarrow{k_2} C + P \tag{1}$$

$$k_{cat} = k_2 \tag{2}$$

$$K_m = (k_{-1} + k_2)/k_1 \tag{3}$$

### 4. *Site-Selective Protein Cleavage by Multinuclear Cu(II) Complexes Contained in a Catalytic Module*

Multinuclear metallocatalyst **BJ** manifested both catalytic activity and substrate selectivity in the hydrolysis of small peptides. The metal centers of the artificial active site of **BJ** were utilized both in substrate recognition and in catalytic conversion. The structure of the active site obtained by using the bowl-shaped molecule, however, is unknown. In addition, it is not possible to synthesize a variety of artificial multinuclear metalloenzymes by the method of transferring catalytic elements confined in a prebuilt cage to a synthetic polymer.

To develop a methodology applicable to designing a wide range of multinuclear polymeric metallocatalysts, active sites were prepared by attaching a molecular entity comprising various catalytic elements with known structure (a catalytic module) to a polymeric backbone. Thus, new multinuclear polymeric artificial metalloproteases (**BL**–**BN**) were synthesized by preparation of catalytic modules containing one, two, or four metal-chelating sites followed by attachment of the modules to a polystyrene and addition of metal ions to the chelating sites (116).

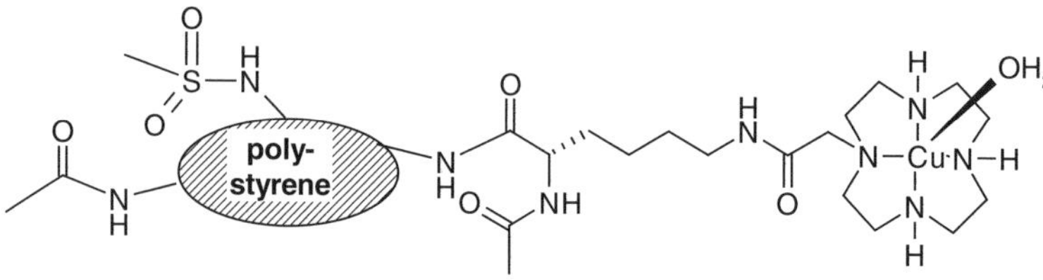

**BL**

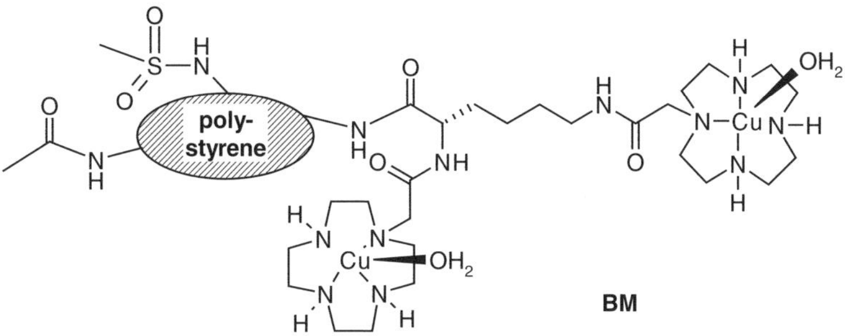

Proteolytic activity of **BL–BN** was examined by using horse heart myoglobin, bovine serum γ-globulin, or BSA as the substrate. γ-Globulin and albumin were not cleaved appreciably over the period of 24 h at pH 7–10 and 37 or 50°C. On the other hand, considerable proteolytic activity was observed with myoglobin.

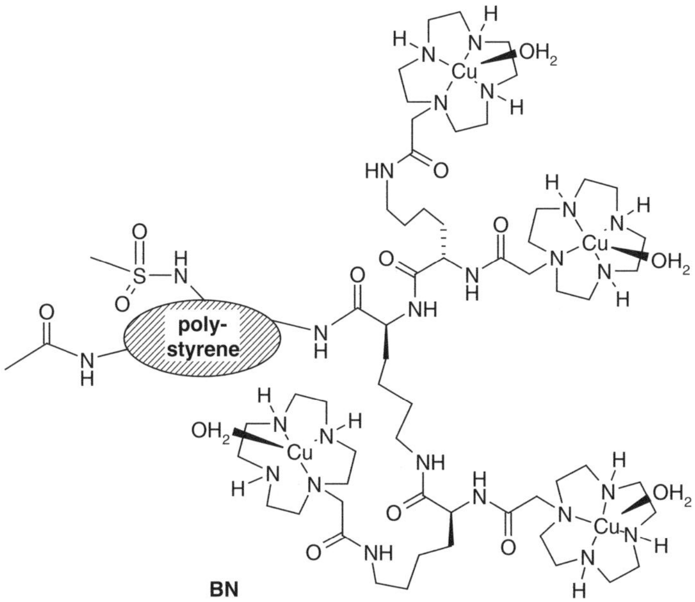

Figure 1 illustrates kinetic data obtained for **BL–BN** at their optimum pH values. The ratio of $k_{cat}/K_m$ values was estimated as 1:13:100 for **BL–BN** and

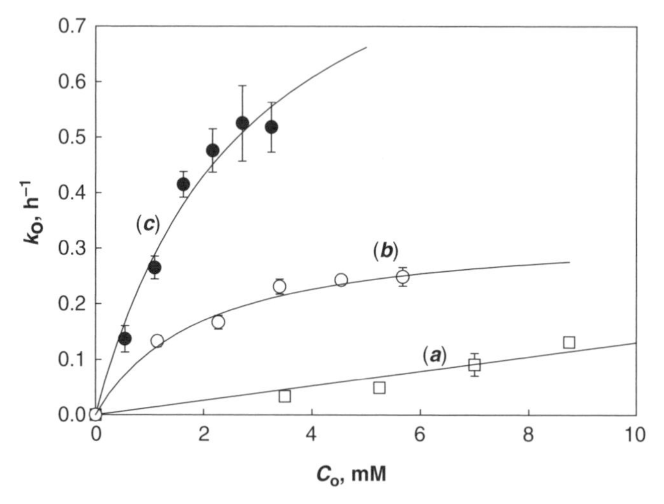


Figure 1. The plot of $k_0$ versus $C_0$ for the hydrolysis of myoglobin catalyzed by **BL** at pH 9.0 and 50°C (*a*), **BM** at pH 9.0 and 50°C (*b*), or **BN** at pH 9.0 and 37°C (*c*). (~ Adapted from Ref. (116).]

the ratio of $k_{cat}$ as 1:10 for **BM** and **BN**. As the catalytic group density of the module is increased in the polystyrene-based catalysts, up to 100-fold enhancement is achieved in the reactivity of the catalyst. In the polystyrene-based catalysts, catalytic activity was considerably improved simply by raising the catalytic group density without deliberate positioning of the catalytic groups.

Intermediate proteins accumulated in amounts detectable by sodium dodecyl sulfate polyacrylamide gel electrophoresis (SDS–PAGE) during the cleavage of myoglobin by **BM**. On the other hand, accumulation of intermediate proteins was not detected by SDS–PAGE when myoglobin was incubated with **BL** or **BN**. Matrix-assisted laser desorption–ionization time-of-flight mass spectrum (MALDI–TOF MS) taken for the intermediate protein mixture obtained by incubation of myoglobin with **BM**, indicated the presence of protein fragments obtained by cleavage of myoglobin at two different positions. Analysis employing *N*-terminal sequencing by Edman degradation (117), *C*-terminal sequencing with carboxypeptidase A coupled with MALDI–TOF MS identified Gln(91)–Ser(92) and Ala(94)–Thr(95) as the initial cleavage site and demonstrated the hydrolytic nature of the protein cleavage by **BM**.

### 5. *Site-Selective Cleavage of a Broad Range of Proteins by Cu(II) Complex Combined With an Aldehyde Group*

Although **BM** achieved high site selectivity for the initial cleavage of myoglobin, **BM** failed to cleave other common proteins (e.g., γ-globulin or albumin). One of the next challenges in the area of synthetic artificial proteases was to design catalysts that hydrolyze a wide range of protein substrates (broad substrate selectivity) by cleaving the substrate at selected positions on the polypeptide backbone (high site selectivity). In order to possess both the broad substrate selectivity and the high site selectivity as digestive proteases (118) or proteasomes (119), the artificial proteases should be able to form complexes with a variety of protein substrates effectively and to cleave peptide bonds at selected positions in the resulting complexes.

For this purpose, the aldehyde group was employed as the binding site of the artificial proteases because the aldehyde group can form imine bonds with the ε-amino groups of lysine residues exposed on the surface of proteins. Since the imine bonds are readily hydrolyzed, the artificial protease equipped with an active site containing the aldehyde group may be able to form complexes with a variety of proteins reversibly. If a catalytic group with high proteolytic activity is positioned in proximity to the aldehyde group, the catalytic group may cleave a peptide bond in the vicinity of the lysine residue of the substrate complexed to the artificial protease. The cyclen–aldehyde conjugates built on the backbone of polystyrene are indicated by **BO** and **BP** (120). As the control catalysts lacking the aldehyde groups, **BO**$^{cont}$ and **BP**$^{cont}$ were also prepared.

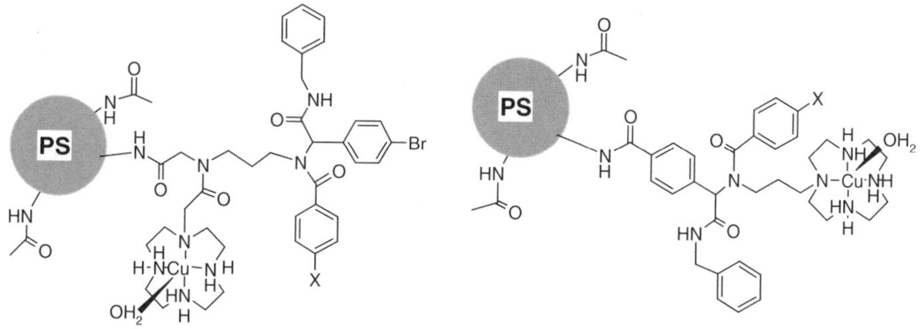

**BO** (X = CHO)

**BO**$^{cont}$ (X = H)

**BP** (X = CHO)

**BP**$^{cont}$ (X = H)

Proteolytic activity of **BO**, **BO**$^{cont}$, **BP**, and **BP** $^{cont}$ was examined by using horse heart myoglobin, bovine serum γ-globulin, BSA, human serum albumin, chicken egg white lysozyme, and chicken egg ovalbumin as the substrates. Examples of the kinetic data are illustrated in Fig. 2. The values of $k_{cat}$, $K_m$, and $k_{cat}/K_m$ for the cleavage of the protein substrates by **BO**, **BO**$^{cont}$, **BP**, and **BP** $^{cont}$ are summarized in Table I. As demonstrated by the kinetic data summarized in Table I, **BO** and **BP** effectively degraded all of the six protein substrates. The

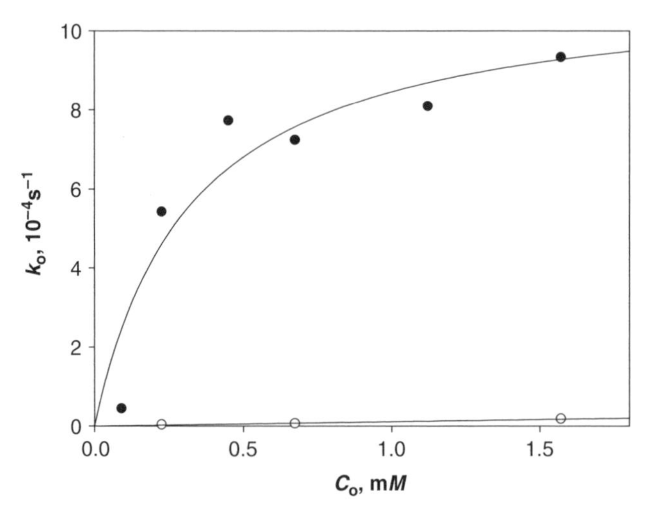


Figure 2. The plot of $k_0$ versus $C_0$ for the hydrolysis of myoglobin catalyzed by **BO** (•) and **BO**$^{cont}$ (○) at pH 9.5 and 50°C. [Adapted from Ref. (120).]

values of $k_{cat}$ represent the maximum values of $k_0$ achievable under the conditions of $C_0 \ggg S_0$. The highest value of $k_{cat}$ listed in Table I corresponds to the half-life of 4.6 min. The values of $k_{cat}/K_m$ for myoglobin degradation by **BO** and **BP** are 20–80 times greater than that by **BM**. Catalysts **BO** and **BP** are even more effective for γ-globulin and BSA, since these proteins were not cleaved by **BM**. On the other hand, **BO**$^{cont}$ and **BP**$^{cont}$ lacking the aldehyde groups in the active sites exhibited negligible proteolytic activity. By positioning the aldehyde group in proximity to the Cu(II) complex of cyclen attached to a cross-linked polystyrene, both broad substrate selectivity and high proteolytic rate are achieved.

The broad substrate selectivity and high proteolytic rate can be attributed to the imine formation between the aldehyde group of the artificial protease and the amino groups of the protein substrate as schematically indicated by **BQ**. Essentially all globular proteins possess lysine residues that provide ammonium cations on the protein surfaces. Imine formation between the lysine-amino group of the protein substrate and the aldehyde group of the polystyrene-based artificial protease would anchor the protein on polystyrene. The broad substrate selectivity manifested by **BO** and **BP** is attributable to the anchorage of the protein substrates through imine bonds. Upon formation of the covalent complex, cleavage of the polypeptide backbone of the bound protein substrate could occur more readily since the reaction between the catalytic Cu(II) center and the scissile peptide bond becomes entropically more favorable.

Attempts were made to trap the imine intermediates with $NaB(OAc)_3H$, which is known to reduce imines much faster than aldehydes. Since reductive amination

TABLE I
Values of Kinetic Parameters for the Cleavage of Various Protein Substrates by **BO** and **BP** at 50°C and at the Optimum pH[a]

| Substrate | Catalyst | pH | $k_{cat}$ ($10^{-4}$ $s^{-1}$) | $K_m$ (m$M$) | $k_{cat}/K_m$ ($s^{-1}M^{-1}$)[b] |
|---|---|---|---|---|---|
| Myoglobin | **BO** | 9.5 | 11 | 0.32 | 3.5 (0.005) |
| Myoglobin | **BP** | 9.5 | 6.7 | 0.77 | 8.7 (0.003) |
| γ-Globulin | **BO** | 9.5 | 22 | 1.3 | 1.7 (0.008) |
| γ-Globulin | **BP** | 9.5 | 7.8 | 0.57 | 1.4 (0.005) |
| BSA | **BO** | 9.5 | 8.0 | 0.92 | 0.87 (0.005) |
| BSA | **BP** | 9.5 | 8.7 | 1.2 | 0.73 (0.005) |
| Human serum albumin | **BO** | 9.5 | 18 | 0.79 | 2.3 (0.008) |
| Human serum albumin | **BP** | 9.0 | 6.7 | 1.1 | 0.61 (0.006) |
| Lysozyme | **BO** | 9.5 | 6.2 | 0.13 | 4.8 (0.002) |
| Lysozyme | **BP** | 9.5 | 6.0 | 1.3 | 0.46 (0.002) |
| Ovalbumin | **BO** | 9.5 | 25 | 1.2 | 2.1 (0.003) |
| Ovalbumin | **BP** | 9.5 | 15 | 2.2 | 0.68 (0.002) |

[a]Adapted from Ref. 120.
[b]Values in parentheses are $k_{cat}/K_m$ measured with the respective control polymer (**BO**$^{cont}$ or **BP**$^{cont}$) under the same conditions.

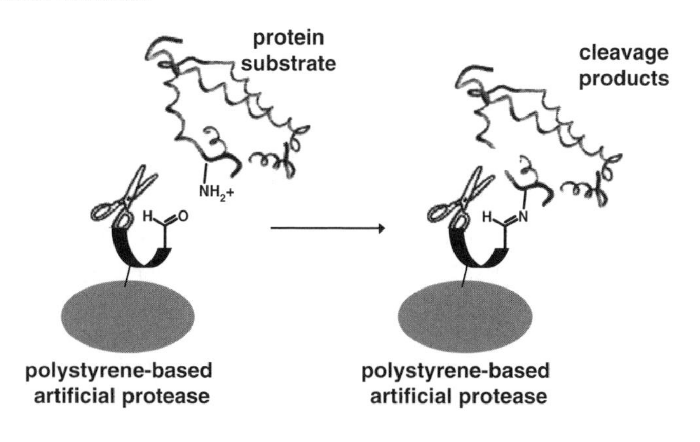


**BQ**

with $NaB(OAc)_3H$ is usually carried out under weakly acidic conditions (121) and **BO** and **BP** are deactivated under acidic conditions, the trapping experiments were carried out at pH 6.0. When myoglobin ($S_0 = 4.9$ μ$M$) was degraded by **BO** ($C_0 = 0.45$ m$M$) at pH 6.0 and 50°C, $k_0$ was $6.8 \times 10^{-6}$ $s^{-1}$. After disappearance of myoglobin was complete, 88% of the total amino acid residues initially introduced by myoglobin were recovered in the buffer solution. When 5 m$M$ $NaB(OAc)_3H$ was initially introduced to the reaction mixture, $k_0$ was $1.3 \times 10^{-4}$ $s^{-1}$ and only 15% of the total amino acid residues were recovered in the buffer solution. When the same reaction was carried out first without $NaB(OAc)_3H$ to complete the degradation of myoglobin, $NaB(OAc)_3H$ was added to the degradation mixture, and the mixture was incubated for 3 days, 72% of the total amino acid residues were recovered in the buffer solution. The enhanced rate for disappearance of the protein substrate is consistent with trapping of the imine intermediate with the borohydride reagent. Reduced amount of amino acid residues recovered from the reaction mixture incubated with the borohydride reagent also supports trapping of the imine intermediate.

After **BO** or **BP** was incubated with BSA at various pH values until the protein was almost completely degraded, the polymer was recovered and used again to cleave BSA. The activity of the recovered polymer was almost identical with that of the fresh polymer. The active sites of the artificial proteases, therefore, were not appreciably damaged during incubation with the proteins under the conditions adopted for collection of kinetic data.

During incubation of a protein substrate with **BO** or **BP**, aliquots of the buffer solution were taken to measure MALDI–TOF MS. An example is illustrated in Fig. 3. Disappearance of myoglobin, formation and decay of some intermediate proteins, and accumulation of product proteins are shown in the MALDI–TOF MS. Structural analysis of protein fragments was performed by treatment with carboxypeptidase A or trypsin followed by MALDI–TOF MS measurement. This allowed identification of several cleavage sites for each protein substrate: Phe(33)–Thr(34), Gln(91)–Ser(92), Ala(94)–Thr(95), and His(116)–Ser(117) for horse heart myoglobin; Gln(33)–Cys(34), Gln(140)–Ile(141), Phe(205)–Gly(206), Trp(213)–Ser(214), and Val(230)–Thr(231) for BSA; Ile(25)–Ala(26), Gln(33)–Cys(34), Glu(297)–Met(298), and Met(329)–Phe(330) for human serum albumin; Ala(90)–Ser(91), Asn(103)–Gly(104), Arg(114)–Cys(115), and Arg(125)–Gly(126) for chicken egg lysozyme. Myoglobin showed preference for cleavage at the amine side of hydroxyl-containing amino acids (e.g., serine or threonine). When all of the protein substrates are considered, the cleavage sites are not simply explained in terms of the amino acid sequences.

*In vivo*, digestive proteases and proteasomes mostly hydrolyze unfolded or precleaved proteins. For example, pepsin cleaves proteins unfolded in the acidic medium of the stomach producing protein fragments, which are cleaved further by other proteases in the intestine. Proteasomes, the main enzymes of the non-lysosomal pathway of protein degradation in cells of higher organisms, cleave cellular proteins unfolded with adenosine triphosphate (ATP) (119). In the industrial application of artificial proteases (122, 123), however, it is desirable to hydrolyze a variety of proteins without addition of denaturing agents. Hydrophilic residues (e.g., ammonium or carboxylate anions) are exposed on the surface of undenatured globular proteins. To synthesize artificial proteases recognizing undenatured protein molecules, we chose to design recognition sites targeting the ammonium groups exposed on the surface of protein

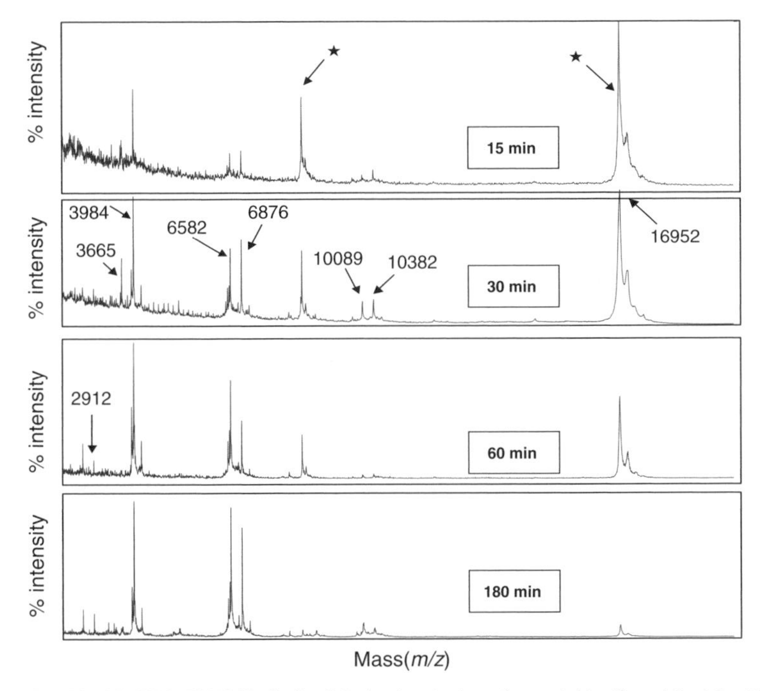


Figure 3. The MALDI–TOF MS obtained during incubation of myoglobin ($S_0 = 4.9\ \mu M$) with **BO** ($C_0 = 0.45\ mM$) at pH 9.5 and 50°C. The peaks labeled with asterisks ($m/z = 16952$ and 8476) are those of myoglobin. The peaks with $m/z$ of 10382, 10089, 6876, 6582, 3984, 3663, and 2912 are identified as Gly(1)–Ala(94), Gly(1)–Gln(91), Ser(92)–Gly(153), Thr(95)–Gly(153), Ser(117)–Gly(153), Gly(1)–Phe(33), and Ser(92)–His(116), respectively. [Adapted from Ref. (120).]

substrates. As we expected, artificial proteases **BO** and **BP** cleaved all of the protein substrates tested in this study by using both the aldehyde and the Cu(II) complex of cyclen in the active site. Furthermore, the artificial proteases manifested high cleavage-site selectivity.

## V. PEPTIDE- OR PROTEIN-CLEAVING CATALYSTS SELECTIVE FOR TARGET PEPTIDE OR PROTEIN: A NEW PARADIGM IN DRUG DESIGN

The reactivities of metal complexes in peptide hydrolysis under physiological conditions are usually very low. As described in Section IV.B.5, we increased the intrinsic proteolytic activity of Cu(II) complexes by attaching them to cross-linked polystyrene derivatives.

Attempts were also made to raise the protease-like activity of metal complexes by increasing their effective molarities toward the scissile peptide bonds. The concept of effective molarity has been proposed to estimate the effectiveness of an intramolecular catalytic group (45). In the hydrolysis of maleamic acid (**K**), carboxylic acids act as nucleophilic catalysts for hydrolysis of peptide bonds. The effectiveness of the intramolecular catalysis by the carboxyl group was expressed as the molarity of the corresponding intermolecular catalysts. As summarized in Table II, remarkably high effective molarities were achieved for hydrolysis of various maleamic acid derivatives. In addition, very high variations in the effective molarities were observed by small structural variations. By changing the alkyl substituents attached to the olefinic carbons, up to $10^9$-fold rate differences were achieved. This implies that enzymes may achieve high catalytic rates by forming productive complexes with the substrates, leading to high effective molarities of the catalytic groups toward the substrate. This also suggests that metal complexes with low intrinsic peptide-cleaving activity may become effective catalysts if they can form highly productive complexes with

TABLE II
Effective Molarities of Carboxyl Groups in Hydrolysis of Maleamic Acids

| Compound | Effective Molarity | Compound | Effective Molarity |
|---|---|---|---|
| 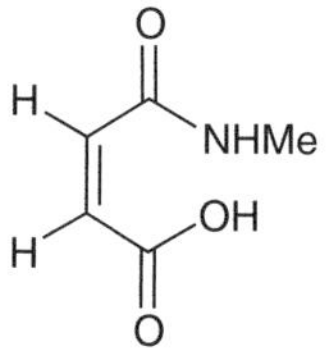 H, H, O, NHMe, OH, O | $2 \times 10^9$ | 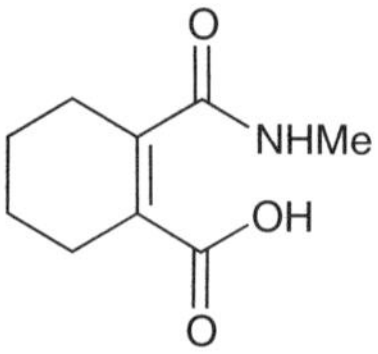 O, NHMe, OH, O | $1 \times 10^{12}$ |
| 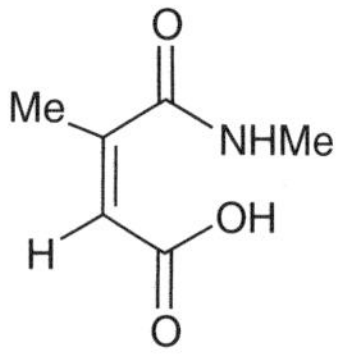 Me, H, O, NHMe, OH, O | $6 \times 10^{10}$ | 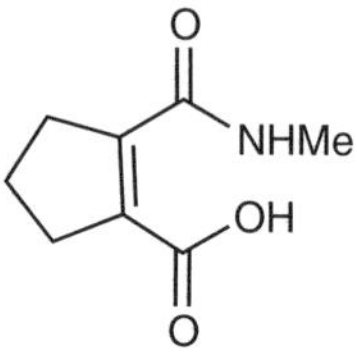 O, NHMe, OH, O | $8 \times 10^4$ |
| 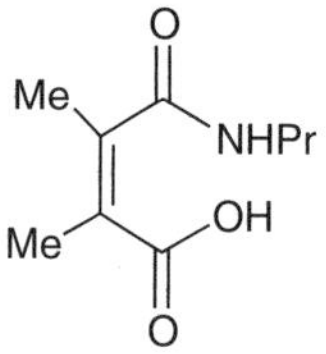 Me, Me, O, NHPr, OH, O | $3 \times 10^{13}$ | 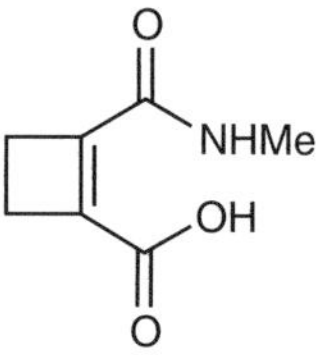 O, NHMe, OH, O | $8 \times 10^3$ |

their substrates and can possess sufficiently high effective molarities toward the scissile peptide bond.

Designing derivatives of metal complexes that can form productive complexes with substrates and can manifest high effective molarities toward peptide bonds is not easy. Nature has adopted macromolecular polypeptides as the backbones of enzymes to tune the positions of catalytic elements in enzyme–substrate complexes, and thus, to achieve high effective molarities of the catalytic groups. The idea of macromolecular systems for the catalyst–substrate complexes may be applied not only to a macromolecular catalyst and a small substrate as in enzymatic systems, but also to a small catalyst and a macromolecular substrate. If a protein is used as the substrate, even a small catalyst may form a productive complex by utilizing the three dimensional (3D) structure of the protein substrate (22).

If a protein-cleaving catalyst based on a metal complex contains a recognition site selective for the protein substrate, the agent can manifest substrate selectivity. In the complex formed between the catalyst and the protein, the metal complex should take a highly productive position in order to possess a high effective molarity and, consequently, cleave the peptide bond effectively.

If the target protein is an enzyme, a receptor, an ion channel, or a toxin related to a disease, the catalyst may be used as a protein-cleaving catalytic drug. As illustrated by the cartoon of **BR**, one molecule of the protein-cleaving catalytic drug can inactivate a multiple number of the target protein (22). On the other hand, the corresponding conventional drug can deactivate only up to an equivalent amount of the target protein. Thus, the protein-cleaving catalytic drug can reduce the drug dosage and the drug toxicity substantially. The activities of proteins related to cancers or of those originating from bacteria or viruses may be modulated by cleavage of the protein backbones.

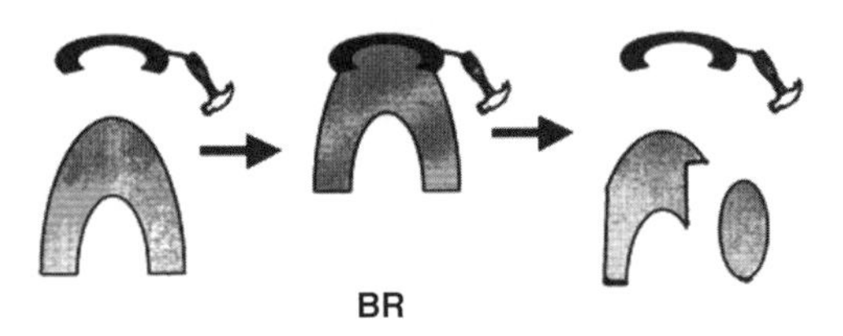

**BR**

Conventional drugs for modulation of various proteins recognize the active sites of the target proteins. No conventional drugs can be, therefore, designed for proteins lacking active sites. Toxic proteins, such as β-amyloid aggregates related to Alzheimer's disease (124) or prion related to mad cow disease (125), do not have active sites. Protein-cleaving catalysts of **BR** do not necessarily bind to the active site of the protein. The target-selective protein-cleaving catalysts, therefore, can be used as medicines to cleave the toxic proteins lacking active sites.

### A. Protein-Cleaving Catalysts Selective for Myoglobin

The first protein-cleaving catalyst selective for a target protein was synthesized by using myoglobin as the target (126, 127). Metal complexes were used as the catalytic center for the protein-cleaving agents. The recognition site of the catalyst was selected from a combinatorial library of cyclen-containing peptide nucleic acid (PNA) oligomers. The PNA oligomers are analogues of DNA oligomers where the backbone is synthetic peptide oligomers instead of deoxyriboses connected by phosphodiester bonds (128, 129). Proteins can recognize the nucleobases of DNAs or ribonucleic acids (RNAs). Conversely, nucleobases may be utilized as the components of recognition sites targeting protein substrates. In addition, automated synthesizers are available for PNA oligomers.

By using an automated synthesizer, combinatorial chemical libraries containing derivatives of PNA oligomers were constructed with the general structure of **BS**. The library can be presented as cyclenAc(Q)$_n$LysNH$_2$, where Q is a PNA monomer A*, G, T*, or C, indicated in **BT**. The PNA contains nucleobases that

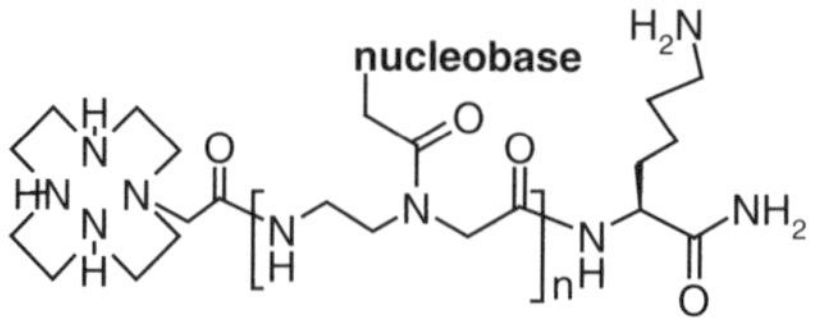

**BS**

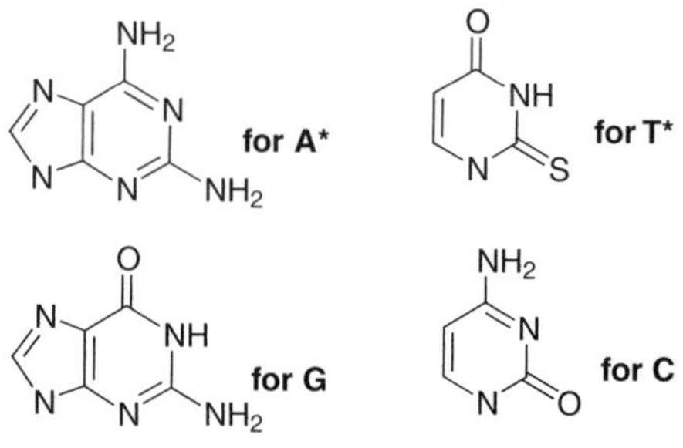

**BT**

can be used for base pairing with nucleobases of DNA. Modified nucleobases A* and T* were used instead of A and T. Both A* and T* recognize T and A, respectively but however, they do not recognize each other (130). Base pairing between two PNAs with complementary sequences present in the library, therefore, can be suppressed by using A* and T*.

The library of cyclenAc(Q)$_n$LysNH$_2$ (total concentration: ~70 $\mu M$) was mixed with an aqueous solution of $Cu^{(II)}Cl_2$ (350 $\mu M$) to generate the library of $Cu^{(II)}$cyclenAc(Q)$_n$LysNH$_2$, where Cu(II) is bound to the cyclen moiety. The Cu(II) ion forms a strong complex with cyclen ($\log K_f = 16.8$ at pH 7) (131). With the library containing the Cu(II) complex of cyclen attached to 7- or 8-mer PNAs, no evidence was obtained for cleavage of proteins (10 $\mu M$), such as BSA, $\gamma$-globulin, elongation factor P, gelatin A, gelatin B, and horse heart myoglobin at 37°C and pH 7, as checked by electrophoresis (SDS–PAGE). The library containing the Cu(II) complex of cyclen attached to 9-mer PNAs clearly showed activity for cleavage of myoglobin. The number of catalyst candidates contained in the chemical library of 9-mer PNAs is $2.6 \times 10^5$. Four groups of library with the known PNA monomers positioned next to the Cu(II) complex of cyclen were subsequently synthesized, tested for their activity, and A* was identified as the best monomer for that position. By repeating the search for the other nine positions occupied by PNA monomers, **BU** was chosen as the ligand of the best catalyst. The binding site for myoglobin was, therefore, obtained by using the pyrimidine and purine bases of **BU**.

CyclenAc-A*-A*-T*-T*-C-G-A*-A*-C-LysNH$_2$

**BU**

To check whether other metal ions complexed to **BU** cleave myoglobin, metal ions, such as Co(III), Fe(III), Hf(IV), Pt(IV), Zr(IV), Pd(II), and Ce(IV), were tested. Among the metal ions tested, only Co(III) manifested the protein-cleaving activity upon complexation to **BU**. An example of kinetic data obtained with $Co^{(III)}$**BU** is illustrated in Fig. 4. Up to 2.5 or 6 molecules of myoglobin were cleaved by each molecule of $Cu^{(II)}$**BU** or $Co^{(III)}$**BU**, respectively, for the data of Fig. 4, indicating the catalytic nature of the action of $Cu^{(II)}$**BU** and $Co^{(III)}$**BU**.

Compared with Cu(II), Co(III) manifested the higher protein-cleaving activity upon complexation to **BU**. It is noteworthy that Co(III) complexes may be more suitable for medical uses compared with Cu(II) complexes since metal transfer to metal-abstracting materials in the living body should be substantially slower for the Co(III) complexes due to the exchange inertness of Co(III). The dependence of $k_0$ on $C_0$ measured at pH 7.5 is illustrated in Fig. 5. Although the kinetic data are somewhat scattered, the two straight lines drawn in Fig. 5 intersect at $C_0 = [\text{myoglobin}]_0$. This intersection agrees with

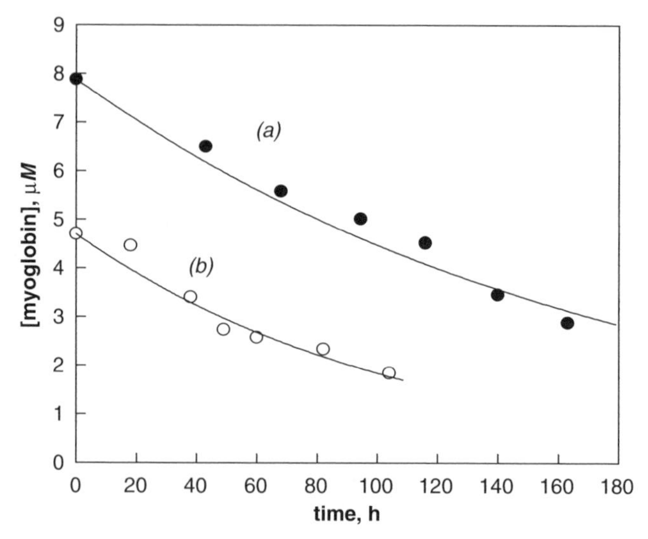


Figure 4. Decrease in [myoglobin] during incubation of myoglobin with $Cu^{(II)}$**BU** (•; curve a, $[\text{myoglobin}]_0 = 7.9\ \mu M$, $[Cu^{(II)}\mathbf{BU}]_0 = 2.0\ \mu M$) or $Co^{(III)}$**BU** (o; curve b, $[\text{myoglobin}]_0 = 4.7\ \mu M$, $[Co^{(III)}\mathbf{BU}]_0 = 0.47\ \mu M$) at pH 7.5 and 37°C. The curves were obtained as indicated in the text: $k_0 = 5.7 \times 10^{-3}\ h^{-1}$ for curve a and $9.4 \times 10^{-3}\ h^{-1}$ for curve b. [Adapted from Ref. (126).]

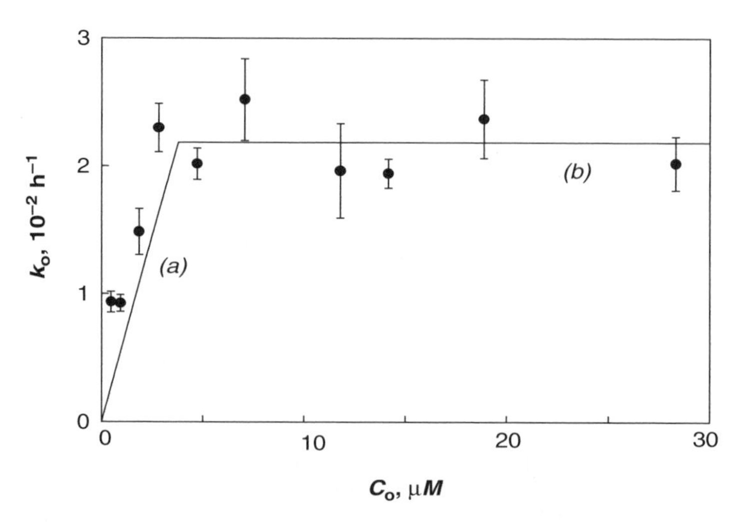


Figure 5. The plot of $k_0$ versus $C_0$ for cleavage of myoglobin ($S_0 = 4.7\ \mu M$) by $Co^{(III)}$**BU** at pH 7.5 and 37°C (0.05 $M$ buffer; addition of 0.5 $M$ NaCl did not affect the rate data appreciably). Straight lines **a** ($C_0 < S_0$) and **b** ($C_0 > S_0$) stand for $v_0/S_0$ ($v_0$: initial velocity) and $k_0$, respectively, predicted by Michaelis–Menten scheme under the condition of $C_0 \gg K_M$. [Adapted from Ref. (126).]

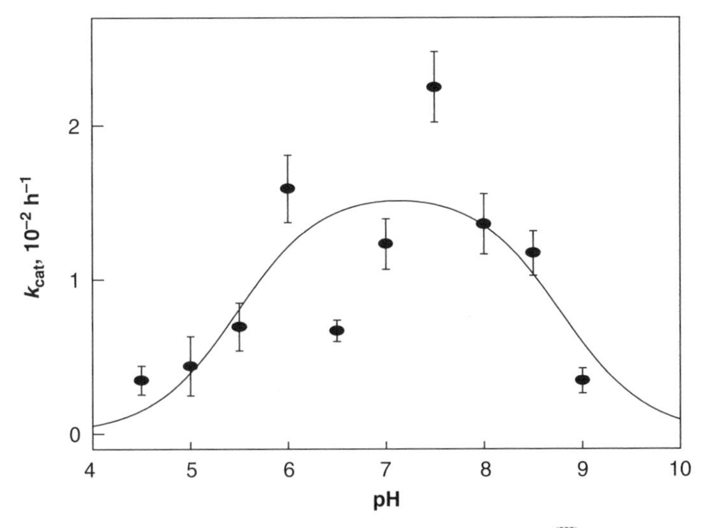


Figure 6. The pH dependence of $k_{cat}$ for cleavage of myoglobin by $Co^{(III)}$**BU** at 37°C. The bell-shaped curve was obtained by analyzing the data by treating the $Co^{(III)}$**BU**–myoglobin complex as a diprotonic acid ($pK_{a1} = 5.50$, $pK_{a2} = 8.68$) and by assuming that the monoprotonated species is reactive. [Adapted from Ref. (126).]

strong binding of myoglobin to $Co^{(III)}$**BU**: In terms of Michaelis–Menten parameters, $K_m \ll C_0$ and, thus, $K_m \ll 5\ \mu M$. Furthermore, $k_0$ measured with $C_0 >$ [myoglobin]$_0$ corresponds to $k_{cat}$.

The $k_{cat}$ values thus measured at various pH values are illustrated in Fig. 6. If ionization of myoglobin or **BU** is disregarded, the $pK_a$ values (5.50, 8.68) estimated from analysis of the bell-shaped pH profile may be assigned to the ionization of water molecules coordinated to the Co(III) ion of the $Co^{(III)}$**BU**–myoglobin complex. Note that the catalyst is most active at the physiological pH. The $k_{cat}$ measured at pH 7.5 and 37°C corresponds to a half-life of 30 h.

The effectiveness of protein-cleaving catalytic drugs exploiting metal complexes as the catalytic centers would depend on the structure of the chelating ligands of the catalytic centers. In this regard, various additional chelating ligands (**BV–CD**: Q = A*-A*-T*-T*-C-G-A*-A*-C-Lys) were tested.

**BV** **BW** **BX**

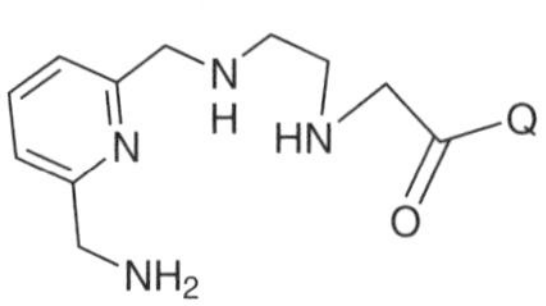

**BY** 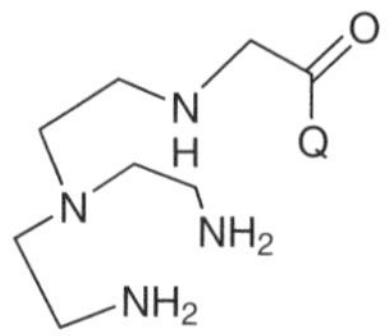 **BZ** 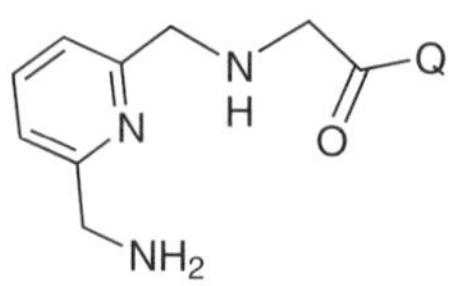 **CA**

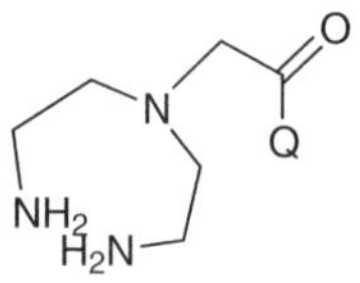

**CB** 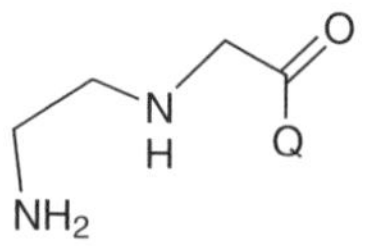 **CC** 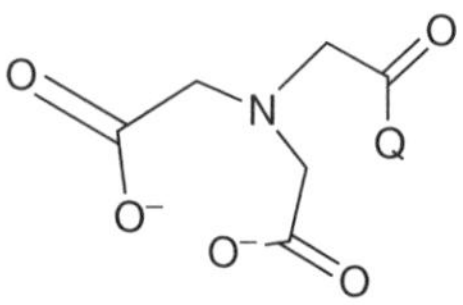 **CD**

Myoglobin-cleaving activity was observed with the Co(III) complex of **BV** or **BW**, for which the dependence of $k_0$ on $C_0$ was the same as that observed with that of **BU** with $K_m \ll C_0$ and $K_m \ll 10\ \mu M$. For Co$^{(III)}$**BV** and Co$^{(III)}$**BW**, the half-lives were 70 and 250 h at 37°C, respectively, at the optimum pH. With **BX**–**CD** in combination with Co(III) ion, myoglobin-cleaving activity was not observed.

The MALDI–TOF MS of reaction mixtures obtained by incubation of myoglobin with the Co(III) complexes of **BU**–**BW** revealed that myoglobin was dissected into either one (**BV**, **BW**) or two (**BU**) pairs of proteins. Characterization of protein fragments by *N*-terminal sequencing and MALDI–TOF MS identified the cleavage sites as Leu(89)–Ala(90) and Leu(72)–Gly(73) for Co$^{(III)}$**BU** and Leu(89)–Ala(90) for Co$^{(III)}$**BV** and Co$^{(III)}$**BW**.

When other proteins (e.g., albumin, γ-globulin, elongation factor P, gelatin A, and gelatin B) were incubated with Cu$^{(II)}$**BU** or Co$^{(III)}$**BU**, protein cleavage was not observed. When treated with $CuCl_2$, the Cu(II) complex of cyclen, or the Co(III) complex of cyclen, myoglobin was not degraded. An analogue of Co$^{(III)}$**BU** was prepared where the PNA residue next to the cyclenAc unit was C instead of A* as indicated by **CE**. No caıalytic activity was observed for Co$^{(III)}$**CE** and Cu$^{(II)}$**CE** in the cleavage of myoglobin. Thus, interaction of Cu$^{(II)}$**BU** or Co$^{(III)}$**BU** with myoglobin is highly selective.

CyclenAc-C-A*-T*-T*-C-G-A*-A*-C-LysNH$_2$

**CE**

### B. Protein-Cleaving Catalyst Selective for Peptide Deformylase

As described in Section V.A, the first artificial protease selective for a target protein has been discovered by designing synthetic homogeneous catalysts that recognized myoglobin and cleaved the polypeptide backbone of myoglobin by using metal complexes as the catalytic center. Although the myoglobin-cleaving catalysts were the first target-selective artificial proteases, their molecular weights (~3000) were too large to use the catalysts as drugs and to analyze the mechanism of the catalytic action. In addition, myoglobin is not related to a disease. To establish drug discovery exploiting peptide-cleaving catalysts, it is desirable to design artificial proteases possessing considerably smaller molecular weights as well as high selectivity for proteins directly related to diseases. The first peptide-cleaving catalyst meeting those criteria has been designed by using peptide deformylase as the target (132).

Peptide deformylase is involved in deformylation of the formyl-methionyl derivatives of proteins formed in the prokaryotic translational systems, and thus its inhibitors are searched as candidates for new antibiotic drugs (133). The active peptide deformylase has Fe(II) ion in the active site, which reacts readily with oxygen. To obtain a stable variant, the Fe(II) ion is often substituted with Zn(II), although Zn(II)-peptide deformylase has reduced activity by two to three orders of magnitude. *Escherichia coli* Zn(II)-peptide deformylase was used as the target enzyme in this study.

The catalyst for cleavage of peptide deformylase was searched with a library of catalyst candidates synthesized by the Ugi reaction (Scheme 2) (134). In this multicomponent condensation reaction, the mixture of a carboxylic acid, an amine, an aldehyde, and an isocyanide produces an *N*-acyl amino acid amide. The catalyst candidates, therefore, are *N*-acylamino acid amides containing various polar and nonpolar pendants as well as the Co(III) complex of cyclen. The Co(III) complex of cyclen (135) was chosen as the proteolytic center in view of the results described in Section V.A. Cyclen with three secondary amines protected with *tert*-butyloxycarbonyl (*t*-boc) groups was incorporated in either the carboxyl or the amine component of the Ugi reaction. Later, the *t*-boc groups were removed and Co(III) ion was inserted to the cyclen portion.

For 4 cyclen-containing carboxylic acids, 18 amines, 17 aldehydes, and 5 isocyanides were used to build the library. For 3 cyclen-containing amines, 18 carboxylic acids, 17 aldehydes, and 5 isocyanides were utilized to construct the library. In total, therefore, ~7500 combinations of the four components were employed for the Ugi reaction. As each Ugi condensation reaction produces a stereogenic carbon atom, the total number of catalyst candidates prepared in this study was ~15,000. The number of compounds to be placed in one reaction vessel was adjusted to produce 9 different combinations and, consequently, 18 catalyst candidates in each vessel.

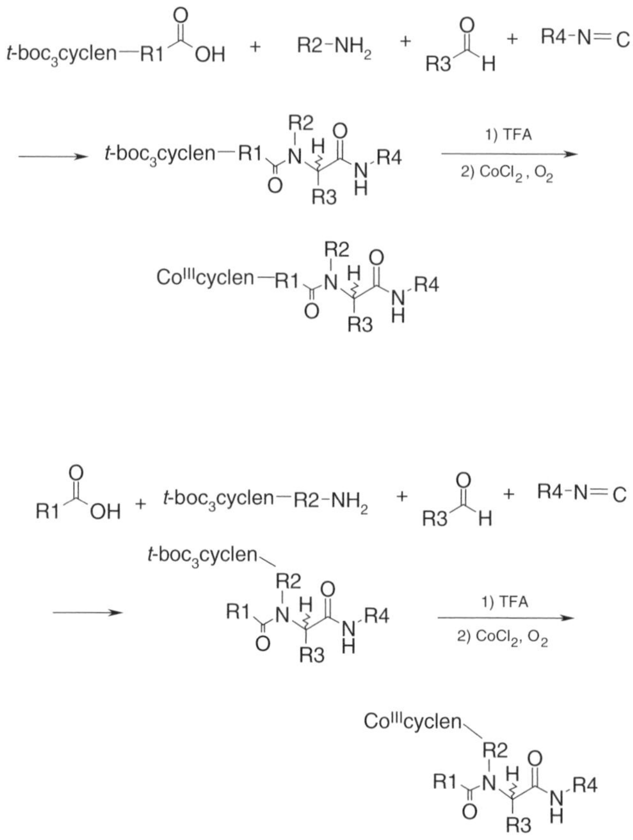

Scheme 2. Synthesis of members of a chemical library of catalyst candidates for cleavage of peptide deformylase. (TFA = Trifluoroacetic acetic acid; *t*-boc = *tert*-butyloxycarbonyl) [Adapted from Ref. (132).]

A mixture of 18 catalyst candidates generated in one vessel was screened together for the peptide-cleaving activity toward peptide deformylase. In a typical screening condition, the concentration of peptide deformylase was 5 $\mu M$, whereas that of each catalyst candidate was 1.5–3 $\mu M$ assuming that the overall yield for the synthetic steps was 100%. After the solution was incubated overnight at pH 7.5 and 37°C, whether new protein fragments was formed by the cleavage of peptide deformylase was examined by MALDI–TOF MS. When the formation of some protein fragments was indicated by a batch of the catalyst candidates, each of the Co(III) complexes of cyclen derivatives contained in the batch was individually generated, and then incubated with peptide deformylase to identify the active compound. The only compound that manifested peptide

**CF**

deformylase-cleaving activity positively was Co(III) complex **CF**, which was subsequently synthesized on a large scale.

The MALDI–TOF MS (Fig. 7) of a reaction mixture obtained by incubation of peptide deformylase (peak a; $m/z = 19{,}198$) with **CF**, disclosed that peptide deformylase was dissected to produce a new peak (peak b) with an $m/z$ value of 17,236. Peak b may be formed by cleavage of peptide deformylase at either Gln(152)–Arg(153) or Ala(17)–Lys(18), which produces a protein fragment with $m/z$ of 17,236 or 17,243, respectively.

The cleavage site was identified as Gln(152)–Arg(153) by treating the cleavage product with carboxypeptidase A. When treated with carboxypeptidase A, peak b produced new MALDI–TOF MS peaks with $m/z$ reduced by 128, 256, and 385. These are consistent with the theoretical values of 128, 256, and 384 for the *C*-terminal amino acid residues of Lys(150)–Gln(151)–Gln(152). The successful *C*-terminal sequencing of the cleavage product with carboxypeptidase A demonstrates that the carboxylic group is generated from cleavage of the peptide backbone with the Co(III) complex of cyclen of the catalyst. In Section

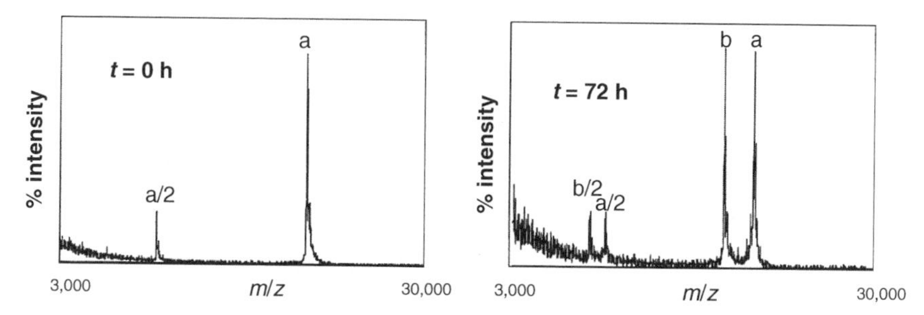


Figure 7. The MALDI–TOF MS of peptide deformylase (5.2 $\mu M$; peaks a and a/2) before and 72 h after incubation with **CF** (1.0 $\mu M$; calculated by assuming that only one enantiomer of **CF** is active) at pH 7.5 (0.05 *M* hepes) (*N*-(2-hydroxyethyl)piperazine-*N'*-ethanesulfonic acid = hepes) and 37°C. [Adapted from Ref. (132).]

V.A, generation of the amine moiety from the cleavage of myoglobin by the Co(III) complex of cyclen of $Co^{(III)}$**BV** was confirmed by *N*-terminal sequencing of the cleavage product. Those *C*- and *N*-terminal sequencing experiments provide evidence for the hydrolytic nature of the cleavage of the polypeptide backbones of protein substrates by the Co(III) complex of cyclen.

The MALDI–TOF MS data indicated that the optimum pH was 7.5. The amount of proteins, such as peak a of Fig. 7, can be estimated fairly accurately by taking the MALDI–TOF MS several times (136). Based on 10 different MALDI–TOF MS measurements, it appeared that about one-half of peptide deformylase was cleaved on incubation with 0.19 equiv of the catalyst for 72 h (Fig. 7), which corresponds to $k_0$ of $0.010\,h^{-1}$. Here, the initial fraction of peptide deformylase complexed with the catalyst cannot exceed 20%. The $k_0$ observed when peptide deformylase is fully bound to the catalyst is $k_{cat}$. The lower limit of $k_{cat}$ is, therefore, estimated as $0.050\,h^{-1}$. For the myoglobin-cleaving catalyst $Co^{(III)}$**BU**, $k_{cat}$ was $0.022\,h^{-1}$ under the same conditions (126, 127).

To gain insights into the mechanism of the cleavage of peptide deformylase by **CF**, docking experiments were performed for the complex formed between peptide deformylase and **CF**. The docking simulations indicated that the (*S*) isomer of **CF** produces more stable complex with peptide deformylase than the (*R*) isomer. Figure 8(*a*) shows the lowest–energy conformation of (*S*)-**CF** in the peptide deformylase surface thus predicted. In the complex, the catalytic head of the Co(III) complex of cyclen and the central acyclic chain of the catalyst interact with the *C*-terminal α-helix, while the three aromatic tails make contact

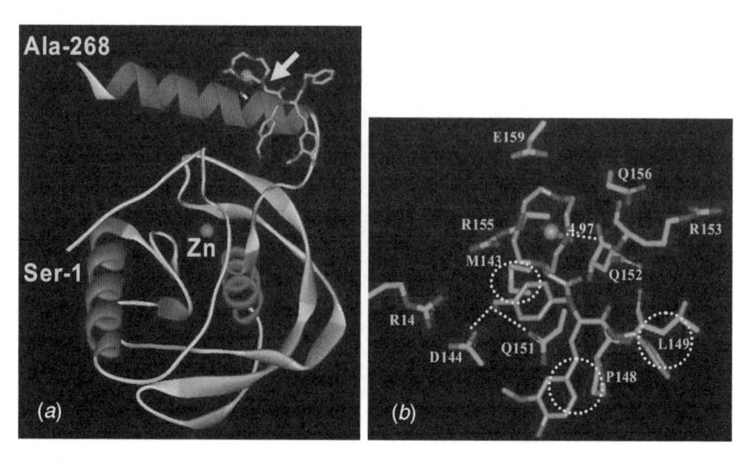


Figure 8. The lowest energy conformation of **CF**–peptide deformylase complex predicted by the docking simulations: Full view [(*a*) arrow indicates the cleavage site] and expanded view [(*b*) P148–E159 come from the helix and M143 and D144 from the loop; the catalyst is shown in pink]. [Adapted from Ref. (132).]

with the helical and the loop structures residing above the active site. Thus, the catalyst would not recognize other proteins that do not have these helical and loop structures. An expanded view [Fig. 8(*b*)] disclosed several modes of interactions between the catalyst and the side chains of peptide deformylase. The hydroxo ligand of the Co(III) ion, the putative nucleophile attacking the peptide group of Gln(152)–Arg(153), is situated in proximity of Gln(152) and Gln(156). Each of the three phenyl rings of the catalyst forms independent van der Waals contact with the side chain of Met(143), Pro(148), or Leu(149). Hydrogen bonds are formed between cyclen N—H of **CF** and the carboxylate group of Glu(159) and between the ammonium group of the catalyst and the side chains of both Asp(144) and Gln(151).

The multiple number of the interactions between **CF** and peptide deformylase suggested by the simulations experiments accounts for the high selectivity manifested by **CF** and peptide deformylase. When tested with 15 other proteins, **CF** did not cleave the proteins. The organic moiety of **CF** selected from the chemical library containing ~15,000 compounds forms a productive complex with peptide deformylase. The Co(III) center of **CF** appears to be located in a highly productive position in the peptide deformylase-catalyst complex. The high effective molarity of the Co(III) center thus achieved apparently led to effective peptide cleavage.

This study reported the first peptide-cleaving catalyst selective for a protein related to a disease. Moreover, mechanistic analysis of the protein cleavage was performed for the first time owing to the moderate size (MW of ligand of **CF** = 644 where MW = molecular weight) of the catalyst.

### C. Oligopeptide-Cleaving Catalysts Selective for Angiotensins or Melanin-Concentrating Hormone

Successful design of peptide-cleaving catalysts selective for myoglobin or peptide deformylase demonstrated that target-selective proteolytic catalysts can be obtained in principle for any soluble proteins. Attempts to provide a firmer basis for the new paradigm of drug discovery based on peptide-cleaving catalysts include designing catalysts for oligopeptides related to diseases, so that the peptide-cleaving catalysts cover not only proteins, but also oligopeptides. As the substrates for the first oligopeptide-cleaving catalysts, we have chosen human angiotensin I and angiotensin II (137), in view of their physiological importance (138–140).

Angiotensin I (**CG**) is a decapeptide formed by the proteolytic cleavage of angiotensinogen by rennin. Angiotensin I is further cleaved by angiotensin converting enzyme to form the octapeptide, angiotensin II (**CH**). Therapeutic manipulation of the pathway involving angiotensin I and angiotensin II has become very important in treating hypertension and heart failure. Angiotensin converting enzyme inhibitors and angiotensin II receptor antagonists, for example, can be used to decrease arterial pressure, ventricular afterload, blood volume, and hence ventricular preload, as well as inhibit and reverse cardiac and vascular hypertrophy.

Asp-Arg-Val-Tyr-Ile-His-Pro-Phe-His-Leu

**CG**

Asp-Arg-Val-Tyr-Ile-His-Pro-Phe

**CH**

By exploiting the Co(III) complex of cyclen as the essential part of the catalyst candidates for cleavage of angiotensin I or II, two kinds of chemical libraries of the catalyst candidates were constructed by using multicomponent condensation reactions. Members of the first chemical library were synthesized according to Scheme 3 by the Ugi condensation reaction (134). Members of the second chemical library were synthesized according to Scheme 4 by a variation (141) of the Ugi condensation.

Members of the first library were synthesized by using 11 cyclen-containing carboxylic acids, 2 cyclen-containing amines, 9 aldehydes, and 5 isocyanides, from which 990 combinations are possible for the condensation reaction. Members of the second library were generated by using 40 carboxylic acids, 2 cyclen-containing amines, and 10 isocyanides. Thus, 800 combinations are available for the condensation reaction. Each condensation reaction of Schemes

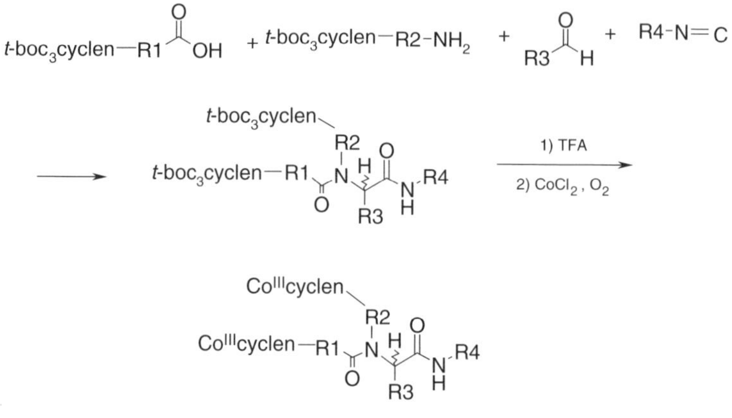

*t*-boc$_3$cyclen—R1—C(=O)OH + *t*-boc$_3$cyclen—R2-NH$_2$ + R3—CHO + R4-N=C

⟶ *t*-boc$_3$cyclen—R1—C(=O)—N(R2—*t*-boc$_3$cyclen)—CH(R3)—C(=O)—NH—R4

1) TFA
2) CoCl$_2$, O$_2$

⟶ Co$^{III}$cyclen—R1—C(=O)—N(R2—Co$^{III}$cyclen)—CH(R3)—C(=O)—NH—R4

Scheme 3. Synthesis of members of the first chemical library of catalyst candidates for cleavage of angiotensins. [Adapted from Ref. (137).]

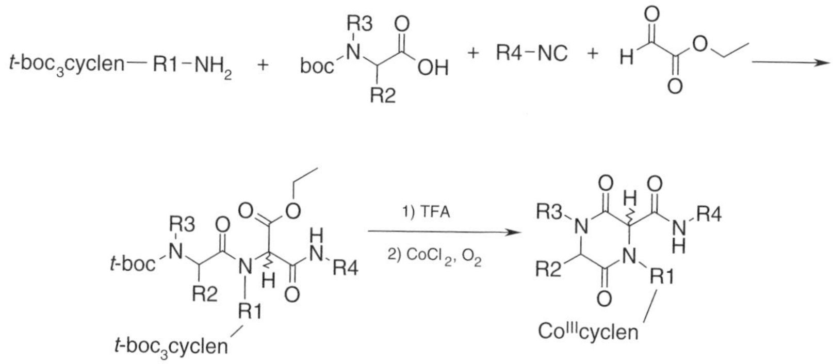

Scheme 4. Synthesis of members of the second chemical library of catalyst candidates for cleavage of angiotensins. [Adapted from Ref. (137).]

4 and 5 creates a stereogenic center, and, thus, a total of 1980 and 1600 condensation products can be obtained for the first and the second libraries, respectively. One combination of the reactants was subject to the condensation reaction in one reaction vessel.

The activity of the catalyst candidates for angiotensin cleavage was initially screened by measuring MALDI–TOF MS after incubation of angiotensin I ($S_0 = 0.02$ m*M*) with the catalyst candidates ($C_0 = 0.2$ m*M* for each catalyst candidate assuming 100% overall synthetic yield) at 37°C and pH 7.5 for 24 h. Whether smaller fragments were formed was checked by the MALDI–TOF MS measurement.

About 70 catalyst candidates showed positive responses to the screening for the activity to cleave angiotensin I. By comparing the activity at reduced concentrations of the seemingly active compounds, the Co(III) complexes of two cyclen derivatives, **CI** and **CJ**, were selected as the most active for the two libraries. To carry out detailed investigation, **CI** and **CJ** were synthesized and purified on a large scale by the condensation reactions of Schemes 4 and 5. Since **CI** contained (*S*)-proline, two diastereomers [(*SS*)-**CI** and (*SR*)-**CI**)] were obtained by the route of Scheme 3. By column chromatography, the diastereomeric mixture was partially separated to produce a 4:1 mixture of (*SS*)-**CI** and (*SR*)-**CI** in addition to pure (*SR*)-**CI**. For **CJ**, an enantiomeric mixture of (*S*)-**CJ** and (*R*)-**CJ** was produced by the reactions of Scheme 4.

Both angiotensin I and II formed products with molecular weights smaller than the parent molecules by 45 upon reaction with the angiotensin-cleaving agents as checked by MALDI–TOF MS. The MALDI–TOF MS taken after treatment of the cleavage products with carboxypeptidase A indicated that the *C*-terminal amino acid residues of angiotensin I and II were not affected by the angiotensin-cleaving reagents. The MALDI–TOF MS taken after treatment of

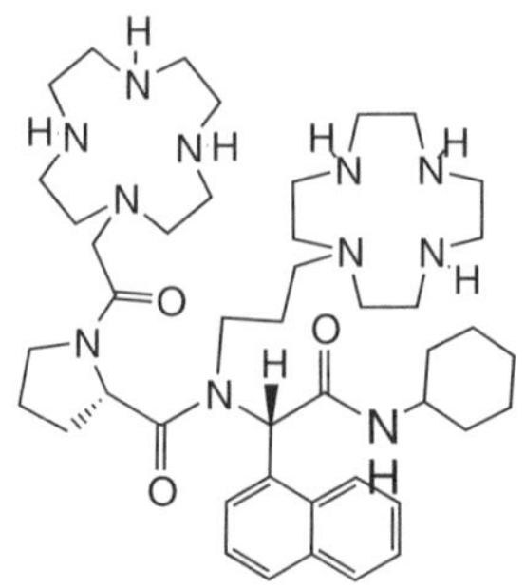

(SS)-**CI**

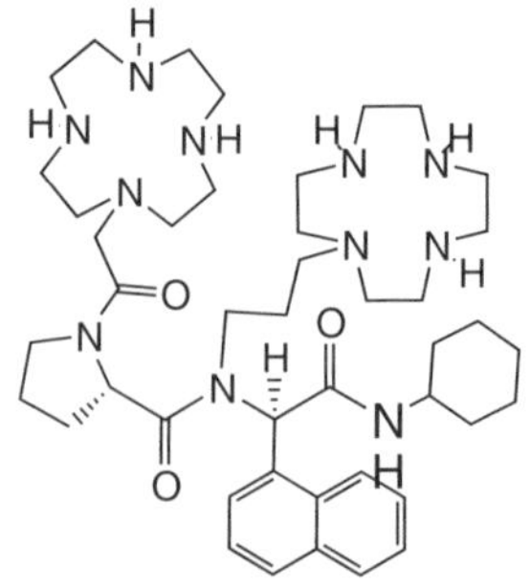

(SR)-**CI**

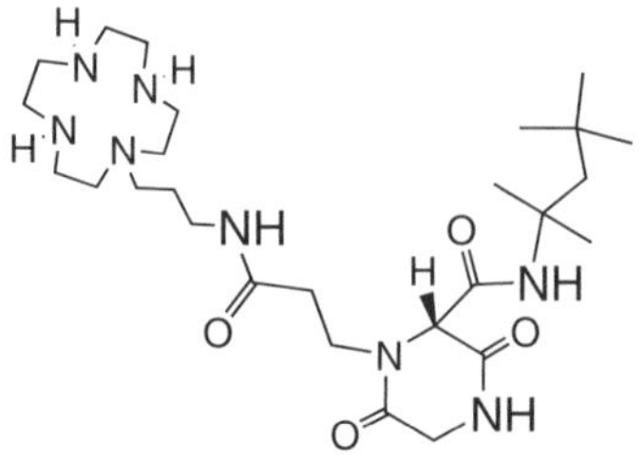

(S)-**CJ**

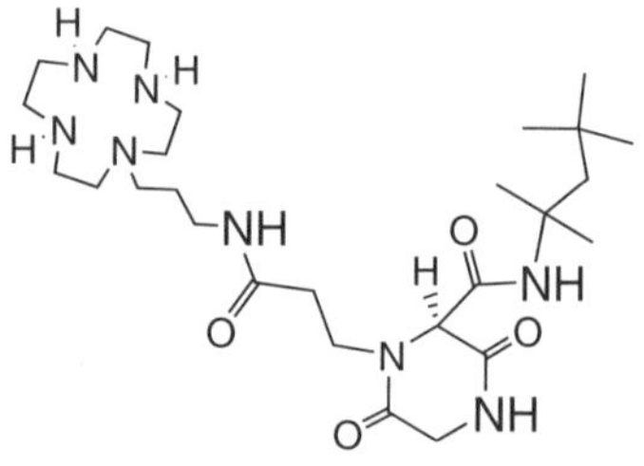

(R)-**CJ**

the cleavage products with phenyl isothiocyanate, the amine-labeling reagent used in Edman degradation (117) for *N*-terminal sequencing of peptides, revealed that the cleavage products did not have the *N*-terminal amino group. On reaction with *p*-toluenesulfonylhydrazide, a reagent readily reacting with aldehydes or ketones to form the corresponding hydrazones (142), MALDI–TOF MS revealed that the cleavage products produced adducts whose molecular weights agreed with formation of the hydrazones. This indicated that the *N*-terminal amino group of angiotensin I or II was converted to a carbonyl group by the action of the angiotensin-cleaving agents.

The results are consistent with conversion of the *N*-terminal aspartate of angiotensin I or II to pyruvate residue leading to the formation of **CK** or **CL**, respectively. In order to confirm the structure of the cleavage products, **CK** and **CL** were synthesized. The authentic samples of **CK** or **CL** and the cleavage products obtained by the action of the angiotensin-cleaving agents produced identical quadrupole TOF MS, positively identifying **CK** and **CL** as the cleavage products.

Arg-Val-Tyr-Ile-His-Pro-Phe-His-Leu

**CK**

Arg-Val-Tyr-Ile-His-Pro-Phe

**CL**

In previous studies, the Co(III) and the Cu(II) complexes of cyclen were used as catalytic centers for hydrolysis of peptide bonds as described above and for hydrolysis of phosphodiester bonds of DNA (143, 144). It was surprising, therefore, that the compounds containing the Co(III) complex of cyclen cleaved angiotensins I and II by converting the *N*-terminal aspartate residue to pyruvate residue instead of hydrolyzing the peptide backbone.

In order to determine the stereochemistry of the active catalysts, (*SS*)-**CI**, (*SR*)-**CI**, and (*R*)-Y were separately prepared by stepwise synthetic routes. Kinetic measurement with (*SS*)-$Co^{(III)}$**CI**, *SR*-$Co^{(III)}$**CI**, and (*R*)-$Co^{(III)}$**CJ** confirmed that (*SS*)-$Co^{(III)}$**CI** and (*S*)-$Co^{(III)}$**CJ** were the catalytically active species. The kinetic data also revealed the catalytic nature of the reaction.

Pseudo-first-order rate constants ($k_0$) for the cleavage of angiotensin I or II catalyzed by (*SS*)-$Co^{(III)}$**CI** or (*S*)-$Co^{(III)}$**CJ** were collected under the conditions of $C_0 > S_0$. The pH dependence of $k_0$ measured at a fixed $C_0$ showed that the optimum activity was manifested at pH 8.0. The $k_0$ values of 0.04–0.6 $h^{-1}$ were observed for the cleavage of angiotensins I and II, when $C_0$ was 1.5–9 m*M*. The values of Michaelis–Menten parameters obtained for the cleavage of angiotensin I or II catalyzed by (*SS*)-$Co^{(III)}$**CI** or (*S*)-$Co^{(III)}$**CJ** are summarized in Table III.

When the Co(III) complex of cyclen was incubated with angiotensin I or II, no reaction was detected. When the racemic **CJ** was mixed with an equivalent

TABLE III

Values of Kinetic Parameters Estimated for Cleavage of Angiotensins I and II by (*SS*)-$Co^{(III)}$**CI** or (*S*)-$Co^{(III)}$**CJ** at pH 8.0 and 37°C

| Substrate | Catalyst | $k_{cat}/K_m$, $h^{-1}M^{-1}$ | $k_{cat}$, $h^{-1}$ | $K_m$, m*M* |
|---|---|---|---|---|
| Angiotensin I | (*SS*)-$Co^{(III)}$**CI** | $67 \pm 1$ | $\gg 0.63$ | $\gg 9.3$ |
| Angiotensin I | (*S*)-$Co^{(III)}$**CJ** | $57 \pm 2$ | $\gg 0.12$ | $\gg 2.0$ |
| Angiotensin II | (*SS*)-$Co^{(III)}$**CI** | $47 \pm 4$ | $\gg 0.10$ | $\gg 1.8$ |
| Angiotensin II | (*S*)-$Co^{(III)}$**CJ** | $25 \pm 1$ | $\gg 0.04$ | $\gg 1.5$ |

[a]Adapted from Ref. 137.

amount of Fe(III), Ni(II), Cu(II), or Zn(II) ion, cleavage of angiotensin I or II was not observed. When the racemic $Co^{(III)}$**CJ** was mixed with various oligopeptides lacking *N*-aspartate residues, no peptide-cleavage was detected.

Catalysts for oxidative decarboxylation of the *N*-terminal aspartate residue contained in an oligopeptide are unprecedented in both biological and chemical systems. In biological systems, malic enzyme catalyzes conversion of aspartate to pyruvate (145). The substrate of malic enzyme is the free amino acid, whereas those of the angiotensin-cleaving catalysts discovered in this study are oligopeptides containing *N*-terminal aspartate residue.

The mechanism (145) generally proposed for the enzymatic action summarized in Scheme 5 may give insights into the mechanism for the action of the angiotensin-cleaving catalysts. The mechanism of Scheme 5 involves oxidation of the amino group of aspartate to the iminium group by the action of $NAD^+$ (nicotinamine adenine dinucleotide) or $NADP^+$ (nicotinamide dinucleotide phosphate, oxidized form). The iminium group facilitates the decarboxylation reaction leading to the enamine, which undergoes tautomerization to form the imine and the subsequent hydrolysis to produce the ketone. Mechanism for the oxidative decarboxylation of the *N*-terminal aspartate residue contained in angiotensin I or II by the catalytic action of (*SS*)-$Co^{(III)}$**CI** or (*S*)-$Co^{(III)}$**CJ** may be tentatively proposed by analogy with the mechanism proposed for malic enzyme. In the action of (*SS*)-$Co^{(III)}$**CI** or (*S*)-$Co^{(III)}$**CJ**, $O_2$ is the most evident oxidant available for the oxidative decarboxylation. The Co(III) center most likely acts as a catalyst for the oxidation reaction in the action of the angiotensin-cleaving catalysts. After oxidation of the amino group, the substrate could undergo reaction pathways essentially identical with those of the malic enzyme-catalyzed reaction.

In view of potential pharmaceutical applications of the target-selective oligopeptide-cleaving catalysts, it was attempted to discover peptide-cleaving catalysts selective for other disease-related peptide hormones containing *N*-terminal aspartate. As the next target, human melanin-concentrating hormone (**CM**) was selected. Melanin-concentrating hormone is a cyclic oligopeptide containing 19 amino acid residues with the *N*-terminal residue being Asp. Melanin-concentrating hormone, a neuropeptide highly expressed in the lateral hypothalamus, has an important role in the regulation of energy balance and body weight in rodents and is a target of designing drugs for obesity (146–149)

From the three kinds of libraries of catalyst candidates used in the studies of catalysts for cleavage of peptide deformylase and angiotensins, the diastereomeric mixture of (*SS*)-**CN** and (*RS*)-**CN** was found to have the activity to cleave melanin-concentrating hormone (150). Examination of the activity of (*SS*)-**CN** synthesized separately by a stepwise synthetic route revealed that (*SS*)-**CN** was the active isomer. Upon incubation with (*SS*)-**CN**, the *N*-terminal aspartate residue of melanin-concentrating hormone was converted to pyruvate residue by

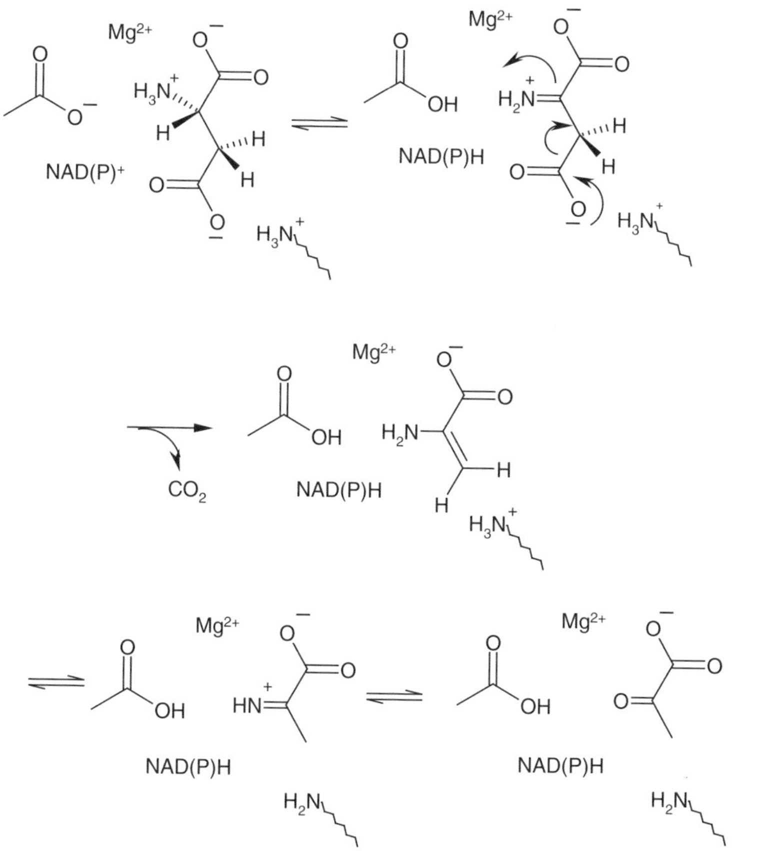

Scheme 5. Mechanism proposed for malic enzyme, where NAD(P) = nicotinamide adenine dinucleotide phosphate. (NADPH = reduced NAD(P) Adapted from Ref. (137).]

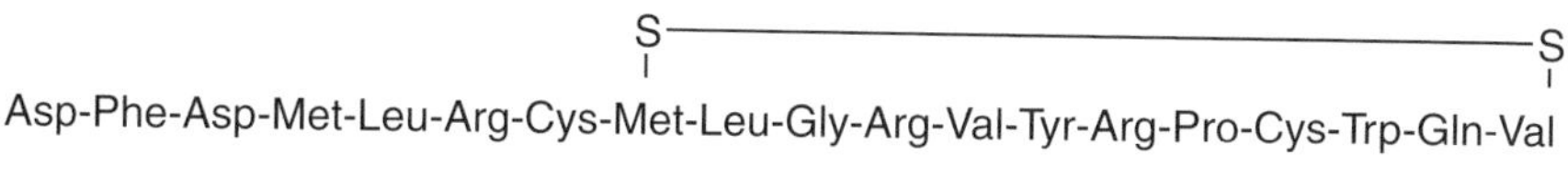

Asp-Phe-Asp-Met-Leu-Arg-Cys-Met-Leu-Gly-Arg-Val-Tyr-Arg-Pro-Cys-Trp-Gln-Val

(Cys–S—S–Cys disulfide bridge)

**CM**

oxidative decarboxylation. This was confirmed by MALDI–TOF MS of the reaction product with or without treatment with *p*-toluenesulfonylhydrazide that had been employed in the study with angiotensins.

In Fig. 9, the dependence of $k_0$ on $C_0$ for the reaction measured at the optimum pH of 8.0 is illustrated. The highest value of $k_0$ indicated in this figure

(*SS*)-**CN** (*RS*)-**CN**

corresponds to the half-life of 3 h. Such a short half-life may be achieved with a much lower catalyst concentration when the structure of the catalyst is improved. Overexpression of melanin-concentrating hormone produces obesity in mice. Conversely, melanin-concentrating hormone receptor-deficient mice are

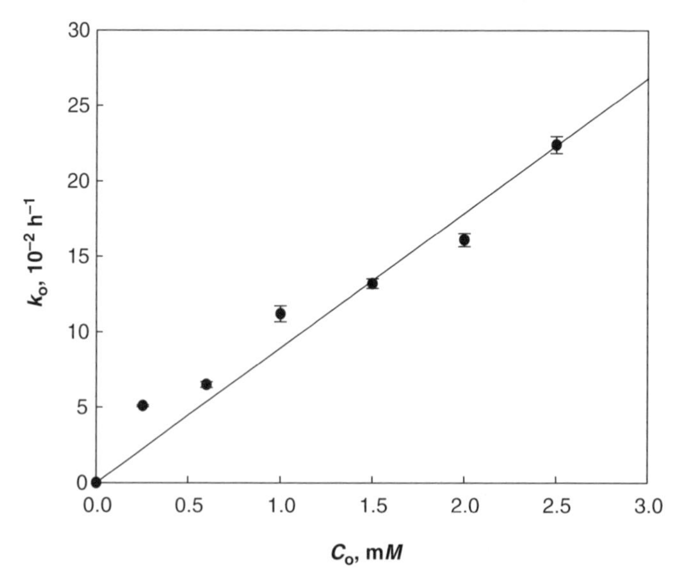


Figure 9. Plot of $k_0$ versus $C_0$ ($S_0 = 0.2$ m*M*) for the disappearance of melanin-concentrating hormone by (*SS*)–**CN** at pH 8.0 and 37°C. [Adapted from Ref. (150).]

resistant to the orexigenic action of melanin-concentrating hormone and maintain leanness (146). Catalysts that can cleave melanin-concentrating hormone with a half-life of 3 h or less at submicromolar concentrations might be, therefore, used as drugs for obesity.

Results of this study indicate that the concept of the peptide-cleaving catalysts can be expanded to include oligopeptides as the targets and nonhydrolytic reactions as the means for cleavage.

## VI. PROSPECTS

By attaching Cu(II) complexes to polystyrene backbones, we succeeded to enhance the intrinsic reactivity of the metal center in hydrolyzing peptide bonds. In addition, the immobilization of the catalytic groups on polystyrene produces catalysts suitable for industrial applications. The immobile artificial proteases can overcome thermal, chemical, and mechanical instabilities of natural proteases. Broad substrate selectivity, high proteolytic rate, and high cleavage-site selectivity are the three major objectives in designing artificial proteases applicable to protein industry. By introducing an auxiliary binding group in proximity to the Cu(II) complex of cyclen attached to polystyrene, remarkable improvement has been achieved in those three major goals.

Future studies in the area of artificial metalloproteases with broad substrate selectivity would be directed toward higher reaction rates, higher amino acid specificity, and longer catalyst lives. To achieve these goals, it may be attempted to search for other solid supports for the artificial metalloproteases to enhance the catalytic activity of the metal centers further. Nonspecific binding of protein substrates or cleavage products to the solid supports should be minimized. The structure of artificial active sites built on the solid support can be improved by incorporating more catalytic elements. Catalytic principles employed by natural proteases may be mimicked in designing more elaborated active sites. When the artificial metalloproteases are sufficiently improved, they might be exploited in industrial applications.

Artificial metalloproteases selective for target proteins have been obtained by attaching the Co(III) complex of cyclen to binding sites. Thus, it has been demonstrated that target-selective artificial metalloproteases can be designed for any soluble proteins. It the target protein is related to a disease, the cleavage agent based on a metal complex can be used as a catalytic drug. High selectivity for the target protein is important for the catalytic drug, since interaction with nontarget biomolecules can cause undesirable side effects. To reduce the drug dosage, high catalytic rate for the peptide cleavage is necessary. Both the substrate selectivity and the catalytic rate can be improved through designing better binding sites by using chemical libraries.

Future studies in the area of target-selective artificial metalloproteases would include designing lead compounds for catalytic drugs targeting soluble proteins related to various diseases. Proteins related to cancers or those originating from bacteria or viruses would be the primary targets since destruction of those proteins would induce little side effects. Many soluble protein toxins are known (151), and protein-cleaving agents based on metal complex would provide a new method for detoxification of the toxins. Another challenging area of the protein-cleaving agents is the cleavage of protein aggregates causing diseases (e.g., Alzheimer's, Parkinson's, and mad cow disease).

Targets for peptide-cleaving agents have been expanded to oligopeptides. Future studies in the area of cleavage of oligopeptides would include improvement of binding constants and catalytic rates. Alternatively, inhibitors for the oligopeptide-cleaving catalyst may be searched by using chemical libraries. Such inhibitors with high affinity toward the oligopeptide can be used as substitutes of receptor antagonists for the target oligopeptides. Receptor antagonists for many oligopeptide hormones are used as drugs. Since the antagonist blocks the receptor, the antagonist can interfere with other functions of the receptor, causing undesirable side effects. Such side effects can be avoided by using a molecule that forms a tight complex with the hormone, thus prohibiting its interaction with the receptor.

Over a period of $\sim 10^9$ years (1 gigaennium) natural selection has led to enzymes with remarkable catalytic properties. Perhaps in the early decades of the third millennium, directed selection will produce equally or more effective and versatile synthetic biocatalysts (23).

## ACKNOWLEDGMENT

This work was supported by the National Research Laboratory Program (305-20050012) of Korea Science and Engineering Foundation.

## ABBREVIATIONS

| | |
|---|---|
| ATP | Adeniosine triphosphate |
| *t*-boc | *tert*-Butyloxycarbonyl |
| BSA | Bovine serum albumin |
| CPA | carboxypeptidase A |
| cyclen | 1,4,7,10-tetraazacyclododecane |
| 3D | Three dimensional |
| DMF | Dimethylformamide |
| DMSO | Dimethyl sulfoxide |

| | |
|---|---|
| DNA | Deoxyribonucleic acid |
| edta | Ethylenediaminetetraacetato (ligand) |
| en | Ethylenediamine |
| MALDI–TOF | Matrix-assisted laser desorption–ionization time of flight |
| MW | Molecular weight |
| $NAD^+$ | Nicotinamide adenosine dinucleotide |
| $NADP^+$ | Nicotinamide adenosine nucleotide phosphate |
| $NAD(P)^+$ | $NAD^+$ or $NADP^+$ |
| NAD(P)H | Reduced $NAD^+$ or $NADP^+$ |
| phen | 1,10-Phenanthroline |
| PCD | Poly(chloromethylstyrene-*co*-divinylbenzene) |
| PCPS | Partially chloromethylated cross-linked polystyrene |
| PNA | Peptide nucleic acid |
| RNA | Ribonucleic acid |
| SDS–PAGE | Sodium dodecyl sulfate polyacrylamide gel electrophoresis |
| TFA | Trifluoroacetic acid |
| trpn | 3,3′,3″-Triaminotripropylamine |

## REFERENCES


1. M. L. Bender, *Mechanisms of Homogeneous Catalysis from Protons to Proteins*, Wiley-Interscience, New York, 1971, p. 212.
2. B. L. Vallee and W. E. C Wacker, *Handbook of Biochemistry and Molecular Biology*, 3rd ed., Vol. 2, G. D. Fasman, Ed., CRC Press, Cleveland, OH, 1976, pp. 276–292.
3. B. L. Vallee, *Zinc Enzymes*, Vol. 1, Chapter 1, I. Bertini, C. Luchinat, W. Maret, and M. Zeppezauer, Eds., Birkhäuser, Boston, 1986.
4. T. H. Tahirov, H. Oki, T. Tsukihara, K. Ogasahara, K. Yutani, K. Ogata, Y. Izu, S. Tsunasawa, and I. Kato, *J. Mol. Biol.*, *284*, 101 (1998).
5. R. Paul-Soto, R. Bauer, J.-M. Frère, M. Galleni, W. Meyer-Klaucke, H. Nolting, G. M. Rossolini, D. de Seny, M. Hernandez-Valladares, M. Zeppezauer, and H.-W. Adolph, *J. Biol. Chem.*, *274*, 13242 (1999).
6. W. L. Mock and Y. Liu, *J. Biol. Chem.*, *270*, 18437 (1995).
7. S. J. Lippard, *Science*, *268*, 996 (1995).
8. N. Carvajal, V. López, M. Salas, E. Uribe, P. Herrera, and J. Cerpa, *Biochem. Biophys. Res. Commun.*, *258*, 808 (1999).
9. R. Breslow, *Acc. Chem. Res.*, *28*, 146 (1995).
10. R. Breslow, Ed., *Artificial Enzymes*, John Wiley & Sons, Inc., Weinheim, 2005.
11. R. A. Lerner, S. J. Benkovic, and P. G. Schultz, *Science*, *252*, 659 (1991).
12. P. G. Schultz and R. A. Lerner, *Acc. Chem. Res.*, *26*, 391 (1993).
13. P.G. Schultz and R. A. Lerner, *Science*, *269*, 1835 (1995).

14. D. Hilvert, *Artificial Enzymes*, R. Breslow, Ed., John Wiley & Sons, Inc., Weinheim, 2005, pp. 89–108.
15. K. S. Broo, H. Nilsson, J. Nilsson, A. Flodberg, and L. Baltzer, *J. Am. Chem. Soc.*, *120*, 4063 (1998).
16. B. Baumeister, N. Sakai, and S. Matile, *Org. Lett.*, *3*, 4229 (2001).
17. I. M. Klotz, *Enzyme Mechanisms*, M. I. Page and A. Williams, Eds., Royal Society of Chemistry, London, 1987, pp. 14–34.
18. J. Suh, *Polymeric Materials Encyclopedia*, J. C. Salamone, Ed., CRC Press, Boca Raton, FL, 1996, pp. 8230–8237.
19. J. Suh, *Advances in Supramolecular Chemistry*, Vol. 6, G. W. Gokel, Ed., JAI Press, London, 2000, pp. 245–286.
20. J. Suh, *Synlett*, *9*, 1343 (2001).
21. P. Hodge, *Chem. Soc. Rev.*, *26*, 417 (1997).
22. J. Suh, *Acc. Chem. Res.*, *36*, 562 (2003).
23. I. M. Klotz and J. Suh, *Artificial Enzymes*, Chapter 3, R. Breslow, Ed., John Wiley & Sons, Inc., Weinheim, 2005.
24. M.-s. Kim and J. Suh, *Bull. Korean Chem. Soc.*, *26*, 1911 (2005).
25. K. B. Grant and M. Kassai, *Curr. Org. Chem.*, *10*, 1021 (2006).
26. A. Radzicka and R. Wolfenden, *J. Am. Chem. Soc.*, *118*, 6105 (1996).
27. R. M. Smith and D. E. Hansen, *J. Am. Chem. Soc.*, *120*, 8910 (1998).
28. J. Suh and S. S. Hah, *J. Am. Chem. Soc.*, *120*, 10088 (1998).
29. J. Suh and S. Oh, *J. Org. Chem.*, *65*, 7534 (2000).
30. S. Oh, W. Chang, and J. Suh, *Bioorg. Med. Chem. Lett.*, *11*, 1469 (2001).
31. A. Sigel and H. Sigel, Eds., *Metal Ions in Biological Systems*, Vol. 38, Dekker, New York, 2001.
32. J. Suh, *Acc. Chem. Res.*, *25*, 273 (1992).
33. J. Suh, *Perspectives on Bioinorganic Chemistry*, Vol. 3, R. W. Hay, J. R. Dilworth, and K. B. Nolan, Eds., JAI Press, London, 1996, pp. 115–149.
34. R. W. Hay and K. B. Nolan, *J. Chem. Soc., Dalton Trans.*, 2542 (1974).
35. D. A. Buckingham, J. Dekkers, A. M. Sargeson, and M. Wein, *J. Am. Chem. Soc.*, *94*, 4032 (1972).
36. C. J. Boreham, D. A. Buckingham, and F. R. Keene, *J. Am. Chem. Soc.*, *101*, 1409 (1979).
37. T. H. Fife and T. J. Przystas, *J. Am. Chem. Soc.*, *104*, 2251 (1982).
38. M. A. Wells and T.C. Bruice, *J. Am. Chem. Soc.*, *99*, 5341 (1977).
39. C. Andrade and H. Taube, *J. Am. Chem. Soc.*, *86*, 1328 (1964).
40. R. B. Jordan and H. Taube, *J. Am. Chem. Soc.*, *88*, 4406 (1966).
41. N. P. Gensmantel, P. Proctor, and M. I. Page, *J. Chem. Soc. Perkin Trans. 2*, 1725 (1980).
42. M. A. Schwartz, *Bioorg. Chem.*, *11*, 4 (1982).
43. J. Suh, M. Cheong, and M. P. Suh, *J. Am. Chem. Soc.*, *104*, 1654 (1982).
44. J. Suh and H. Han, *Bioorg. Chem.*, *12*, 177 (1984).
45. A. J. Kirby, *Adv. Phys. Org. Chem.*, *17*, 183 (1980).
46. J. Suh, B. K. Hwang, and Y. H. Koh, *Bioorg. Chem.*, *18*, 207 (1990).
47. J. Suh, T. H. Park, and B. K. Hwang, *J. Am. Chem. Soc.*, *114*, 5141 (1992).

48. F. Basolo and R. G. Pearson, *Mechanisms of Inorganic Reactions*, 2nd ed., John Wiley & Sons, Inc., New York, 1968, p. 32.
49. D. A. Buckingham, J. M. Harrowfield, and A. M. Sargeson, *J. Am. Chem. Soc.*, *96*, 1726 (1974).
50. C. J. Boreham, D. A. Buckingham, D. J. Francis, A. M. Sargeson, and L. G. Warner, *J. Am. Chem. Soc.*, *103*, 1975 (1981).
51. D. A. Buckingham, F. R. Keene, and A. M. Sargeson, *J. Am. Chem. Soc.*, *96*, 4981 (1974).
52. P. A. Sutton and D. A. Buckingham, *Acc. Chem. Res.*, *20*, 357 (1987).
53. J. T. Groves and R. R. Chambers, Jr., *J. Am. Chem. Soc.*, *106*, 630 (1984).
54. J. T. Groves and L. A. Baron, *J. Am. Chem. Soc.*, *111*, 5442 (1989).
55. J. Suh, S. J. Son, and M. P. Suh, *Inorg. Chem.*, *37*, 4872 (1998).
56. W. N. Lipscomb, *Acc. Chem. Res.*, *3*, 81 (1970).
57. J. Suh, W. Cho, and S. Chung, *J. Am. Chem. Soc.*, *107*, 4530 (1985).
58. J. Suh, B. K. Hwang, I. Jang, and E. Oh, *J. Biochem. Biophys. Methods*, *22*, 167 (1991).
59. J. Suh, S.-B. Hong, and S. Chung, *J. Biol. Chem.*, *261*, 7112 (1986).
60. J. Suh, S. Chung, and G. B. Choi, *Bioorg. Chem.*, *17*, 64 (1989).
61. M. W. Makinen, K. Yamamura, and E. T. Kaiser, *Proc. Natl. Acad. Sci. U.S.A*, *73*, 3882 (1976).
62. M. E. Sander and H. Witzel, *Biochem. Biophys. Res. Commun.*, *132*, 681 (1985).
63. B. M. Britt and W. L. Peticolas, *J. Am. Chem. Soc.*, *114*, 5295 (1992).
64. R. Breslow and D. Wernick, *J. Am. Chem. Soc.*, *98*, 259 (1976).
65. H. Kim and W. N. Lipscomb, *Biochemistry*, *30*, 8171 (1991).
66. J. E. Hanson, A. P. Kaplan, and P. A. Bartlett, *Biochemistry*, *28*, 6294 (1989).
67. R. A. Y. Jones, *Physical and Mechanistic Organic Chemistry*, 2nd ed., Chapter 3, Cambridge University Press, Cambridge, U.K., 1983.
68. A. Schepartz and B. Cuenoud, *J. Am. Chem. Soc.*, *112*, 3247 (1990).
69. D. Hoyer, H. Cho, and P. G. Schultz, *J. Am. Chem. Soc.*, *112*, 3249 (1990).
70. J. Gallagher, O. Zelenko, A. D. Walts, and D. S. Sigman, *Biochemistry*, *37*, 2096 (1998).
71. J. Wu, D. M. Perrin, D. S. Sigman, and H. R. Kaback, *Proc. Natl. Acad. Sci. U.S.A.*, *92*, 9186 (1995).
72. M. B. Shimon, R. Goldshleger, and S. J. D. Karlish, *J. Biol. Chem.*, *273*, 34190 (1998).
73. A. Buranaprapuk, S. P. Leach, C. V. Kumar, and J. R. Bocarsly, *Biochim. Biophys. Acta*, *1387*, 309 (1998).
74. H. Y. Shrivastava and B. U. Nair, *Biochem. Biophys. Res. Commun.*, *279*, 980 (2000).
75. H. Y. Shrivastava and B. U. Nair, *Biochem. Biophys. Res. Commun.*, *285*, 915 (2001).
76. R. Goldshleger and S. J. D. Karlish, *Proc. Natl. Acad. Sci. U.S.A.*, *94*, 9596 (1997).
77. G. Mocz and I. R. Gibbons, *J. Biol. Chem.*, *265*, 2917 (1990).
78. S. Soundar and R. F. Colman, *J. Biol. Chem.*, *268*, 5264 (1993).
79. E. Zaychikov, E. Martin, L. Denissova, M. Kozlov, V. Markovtsov, M. Kashlev, H. Heumann, V. Nikiforov, A. Goldrarb, and A. Mustaev, *Science*, *273*, 107 (1996).
80. B. Cuenoud, T. M. Tarasow, and A. Schepartz, *Tetrahedron Lett.*, *33*, 895 (1992).
81. G. Mocz, *Eur. J. Biochem.*, *179*, 373 (1989).
82. C. R. Cremo, J. A. Loo, C. G. Edmonds, and K. M. Hatlelid, *Biochemistry*, *31*, 491 (1992).
83. T. M. Rana and C. F. Meares, *Proc. Natl. Acad. Sci. U.S.A*, *88*, 10578 (1991).
84. P. B. Dervan, *Science*, *232*, 464 (1986).

85. D. S. Sigman, T. W. Bruice, A. Mazumber, and C. L. Sutton, *Acc. Chem. Res.*, *26*, 98 (1993).
86. R. C. Bateman, Jr., W. W. Youngblood, W. H. Busby, Jr., and J. S. Kizer, *J. Biol. Chem.*, *260*, 9088 (1985).
87. I. E. Platis, M. R. Ermacora, and R. O. Fox, *Biochemistry*, *32*, 12761 (1993).
88. J. Chin and M. Banaszczyk, *J. Am. Chem. Soc.*, *111*, 2724 (1989).
89. J. H. Kim and J. Chin, *J. Am. Chem. Soc.*, *114*, 9792 (1992).
90. M. K. Takasaki, J. H. Kim, E. Rubin, and J. Chin, *J. Am. Chem. Soc.*, *115*, 1157 (1992).
91. N. N. Greenwood and A. Earnshaw, *Chemistry of the Elements*, Reed, Oxford, 1997, p. 1123.
92. M. K. Saha and I. Bernal, *Chem. Commun.*, 612 (2003).
93. C. V. Kumar, A. Buranaprapuk, A. Cho, and A. Chaudhari, *Chem. Commun.*, 597 (2000).
94. I. E. Burgeson and N. M. Kostić, *Inorg. Chem.*, *30*, 4299 (1991).
95. L. Zhu and N. M. Kostić, *Inorg. Chem.*, *31*, 3994 (1992).
96. L. Zhu and N. M. Kostić, *J. Am. Chem. Soc.*, *115*, 4566 (1993).
97. N. M. Milović and N. M. Kostić, *J. Am. Chem. Soc.*, *124*, 4759 (2002).
98. L. Zhu, L. Qin, T. N. Parac, and N. M. Kostić, *J. Am. Chem. Soc.*, *116*, 5218 (1994).
99. N. V. Kaminskaia, T. W. Johnson, and N. M. Kostić, *J. Am. Chem. Soc.*, *121*, 8663 (1999).
100. N. M. Milović, J. D. Badjić, and N. M. Kostić, *J. Am. Chem. Soc.*, *126*, 696 (2004).
101. J. Chin, V. Jubian, and K. Mreje, *J. Chem. Soc. Chem. Commun.*, 1326 (1990).
102. N. N. Murthy, M. Mahroof-Tahir, and K. D. Karlin *J. Am. Chem. Soc.*, *115*, 10404 (1993).
103. E. L. Hegg and J. N. Burstyn, *J. Am. Chem. Soc.*, *117*, 7015 (1995).
104. L. Zhang, Y. Mei, Y. Zhang, S. Li, X. Sun, and L. Zhu, *Inorg. Chem.*, *42*, 492 (2003).
105. M. C. B. de Oliveira, M. Scarpellini, A. Neves, H. Terenzi, A. J. Bortoluzzi, B. Szpoganics, A. Greatti, A. S. Mangrich, E. M. de Souza, P. M. Fernandez, and M. R. Soares, *Inorg. Chem.*, *44*, 921 (2005).
106. T. Takarada, M. Yashiro, and M. Komiyama, *Chem. Eur. J.*, *6*, 3906 (2000).
107. M. Yashiro, T. Takarada, S. Miyama, and M. Komiyama, *J. Chem. Soc., Chem. Commun.*, 1757 (1994).
108. M. Kassai, R. G. Ravi, S. J. Shealy, and K. B. Grant, *Inorg. Chem.*, *43*, 6130 (2004).
109. A. Erxleben, *Inorg. Chem.*, *44*, 1082 (2005).
110. R. B. Silverman, *The Organic Chemistry of Drug Design and Drug Action*, Academic Press, San Diego, CA, 1992, pp. 98–145.
111. B.-B. Jang, K.-P. Lee, D.-H. Min, and J. Suh, *J. Am. Chem. Soc.*, *120*, 12008 (1998).
112. J. Suh and S.-J. Moon, *Inorg. Chem.*, *40*, 4890 (2001).
113. S.-J. Moon, J. W. Jeon, H. Kim, M. P. Suh, and J. Suh, *J. Am. Chem. Soc.*, *122*, 7742 (2000).
114. M. P. Suh, M. Y. Han, J. H. Lee, K. S. Min, and C. Hyeon, *J. Am. Chem. Soc.*, *120*, 3819 (1998).
115. C. Gao, B. J. Lavey, C.-H. L. Lo, A. Datta, P. Wentworth, Jr., and K. D. Janda, *J. Am. Chem. Soc.*, *120*, 2211 (1998).
116. C. E. Yoo, P. S. Chae, J. E. Kim, E. J. Jeong, and J. Suh, *J. Am. Chem. Soc.*, *125*, 14580 (2003).
117. P. F. Nielsen, B. Landis, M. Svoboda, K. Schneider, and M. Przybylski, *Anal. Biochem.*, *191*, 302 (1990).
118. L. Polgár, *Mechanisms of Protease Action*, CRC Press, Boca Raton, FL, 1989.
119. A. F. Kisselev and A. L. Goldberg, *Chem. Biol.*, *8*, 739 (2001).
120. S. H. Yoo, B. J. Lee, H. Kim, and J. Suh, *J. Am. Chem. Soc.*, *127*, 9593 (2005).

121. A. F. Abdel-Magid, K. G. Carson, B. D. Harris, C. A. Maryanoff, and R. D. Shah, *J. Org. Chem.*, *61*, 3849 (1996).

122. O. P. Ward, *Comprehensive Biotechnology*, Vol. 3, M. Moo-Young, Ed., Pergamon, Oxford, 1985, pp. 789–818.

123. J. Tramper, *Applied Biocatalysis*, J. M. S. Cabral, D. Best, L. Boross, and J. Tramper, Eds., Harwood Academic Publishers, Chur, 1994, pp. 1–46.

124. D. Schubert, Ed., *The Structure and Function of Alzheimer's Amyloid Beta Proteins*, R. G. Landes Co., Austin, TX, 1994.

125. S. B. Pruisiner, Ed., *Prion Biology and Diseases*, Cold Spring Harbor Laboratory, Cold Spring Harbor, 1999.

126. J. W. Jeon, S. J. Son, C. E. Yoo, I. S. Hong, J. B. Song, and J. Suh, *Org. Lett.*, *4*, 4155 (2002).

127. J. W. Jeon, S. J. Son, C. E. Yoo, I. S. Hong, and J. Suh, *Bioorg. Med. Chem.*, *11*, 2901 (2003).

128. E. Uhlmann, A. Peyman, G. Breipohl, and D. W. Will, *Angew. Chem. Int. Ed. Engl.*, *37*, 2796 (1998).

129. P. E. Nielsen, *Curr. Opin. Struct. Biol.*, *9*, 353 (1999).

130. J. Lohse, O. Dahl, and P. E. Nielsen, *Proc. Natl. Acad. Sci. U.S.A*, *96*, 10804 (1999).

131. R. M. Izatt, K. Pawlak, J. S. Bradshaw, and R. L. Bruening, *Chem. Rev.*, *91*, 1721 (1991).

132. P. S. Chae, M.-s. Kim, C.-S. Jeung, S. D. Lee, H. Park, S. Lee, and J. Suh, *J. Am. Chem. Soc.*, *127*, 2396 (2005).

133. P. T. R. Rajagopalan, S. Grimme, and D. Pei, *Biochemistry*, *39*, 779 (2000).

134. A. Dömling and I. Ugi, *Angew. Chem. Int. Ed. Engl.*, *39*, 3168 (2000).

135. A. J. Clarkson, D. A. Buckingham, A. J. Rogers, A. G. Blackman, and C. R. Clark, *Inorg. Chem.*, *39*, 4769 (2000).

136. K. Hollemeyer, W. Altmeyer, and E. Heinzle, *Anal. Chem.*, *74*, 5960 (2002).

137. M.-s. Kim, J. W. Jeon, and J. Suh, *J. Biol. Inorg. Chem.*, *10*, 364 (2005).

138. L. H. Opie, *Angiotensin-Converting Enzyme Inhibitors: Scientific Basis for Clinical Use*, 2nd ed., John Wiley & Sons, Inc., New York, 1994.

139. R. M. Touyz and E. L. Schiffrin, *Pharmacol. Rev.*, *52*, 639 (2000).

140. W. C. De Mello, Ed., *Renin Angiotensin System and the Heart*, John Wiley & Sons, New York, 2004.

141. C. Hulme and M.-P. Cherrier, *Tetrahedron Lett.*, *40*, 5295 (1999).

142. M. Gobbini, P. Barassi, A. Cerri, S. De Munari, G. Fedrizzi, M. Santagostino, A. Schiavone, M. Torri, and P. Melloni, *J. Med. Chem.*, *44*, 3821 (2001).

143. C.-S. Jeung, C. H. Kim, K. Min, S. W. Suh, and J. Suh, *Bioorg. Med. Chem. Lett.*, *11*, 2401 (2001).

144. C.-S. Jeung, J. B. Song, Y.-H. Kim, and J. Suh, *Bioorg. Med. Chem. Lett.*, *11*, 3061 (2001).

145. D. Liu, C.-C. Hwang, and P. F. Cook, *Biochemistry*, *41*, 12200 (2002).

146. D. J. Marsh, D. T. Weingarth, D. E. Novi, H. Y. Chen, M. E. Trumbauer, A. S. Chen, X. M. Guan, M. M. Jiang, Y. Feng, R. E. Camacho, Z. Shen, E. G. Frazier, H. Yu, J. M. Metzger, S. J. Kuca, L. P. Shearman, S. Gopal-Truter, D. J. MacNeil, A. M. Strack, D. E. MacIntyre, L. H. T. Van der Ploeg, and S. Qian, *Proc. Natl. Acad. Sci. U.S.A*, *99*, 3240 (2002).

147. G. Segal-Lieberman, R. L. Bradley, E. Kokkotou, M. Carlson, D. J. Trombly, X. Wang, S. Bates, M. G. Myers Jr., J. S. Flier, and E. Maratos-Flier, *Proc. Natl. Acad. Sci. U.S.A*, *100*,10085 (2003).

148. W. T. Gibson, P. Pissios, D. J. Trombly, J. Luan, J. Keogh, N. J. Wareham, E. Maratos-Flier, S. O'Rahilly, and I. S. Farooqu, *Obes. Res.*, *12*, 743 (2004).

149. C. G. Bell, D. Meyre, C. Samson, C. Boyle, C. Lecoeur, M. Tauber, B. Jouret, D. Jaquet, C. Levy-Marchal, M. A. Charles, J. Weill, F. Gibson, C. A. Mein, P. Froguel, and A. J. Walley, *Diabetes*, *54*,3049 (2005).

150. M. G. Kim, M.-s. Kim, S. D. Lee, and J. Suh, *J. Biol. Inorg. Chem.*, *11*, 867 (2006).

151. M. W. Parker, Ed., *Protein Toxin Structure*, R. G. Landes Co., Austin, TX, 1996.

# CHAPTER 3

# Coordination Polymers of the Lanthanide Elements: A Structural Survey

**DANIEL T. DE LILL**

*Department of Chemistry*
*The George Washington University*
*Washington, DC 20052*

**CHRISTOPHER L. CAHILL**

*Department of Chemistry*
*The George Washington University*
*Washington, DC 20052*
and
*Geophysical Laboratory*
*Carnegie Institution of Washington*
*Washington, DC 20015*

CONTENTS



*Progress in Inorganic Chemistry, Vol. 55* Edited by Kenneth D. Karlin

# I. INTRODUCTION

The study of porous and framework materials has seen a tremendous rise in popularity over the past 10–15 years. Arguably, this increase has stemmed from the introduction of a new class of compounds, namely, metal–organic frameworks (MOFs), or more generally speaking, coordination polymers (CPs). A more thorough distinction will be made between these terms later, but for the moment, let us define these materials as nonmolecular assemblies of metal centers (or polynuclear clusters) polymerized through multifunctional organic linker molecules. The resulting topologies of such compounds make them particularly attractive for applications, such as enantioselective molecular recognition (1), catalysis (2), gas storage and separations (3), ion exchange (4), and magnetic materials (5).

From a synthetic point of view, much of the appeal of MOFs and CPs lies in the potential for these materials to be viewed as "targets" of reactions (6). The use of metal centers with appropriately paired organic molecules wherein the local coordination geometry preferences of the metal center are exploited lets one think in terms of *assembling* higher dimensional solids. This finding is in stark contrast to earlier, high-temperature approaches to extended solid-state structures. Indeed, the vast majority of MOF and CP materials have been synthesized at room temperature, or at most under solvothermal conditions.

By considering the recent surge in research activity in this area, it is not surprising that a number of review articles have appeared in the literature. Key articles by (among others) Férey (7), Rao and co-workers (8), Yaghi and co-workers (9–11), Moulton and Zaworotko (12, 13), Janiak (14), Kepert (15),

James (16), Kitigawa and co-workers (17–19), Wuest (20), and Chen and co-workers (21) explore the chemistry, structures, and properties of MOFs and CPs from a variety of perspectives. Indeed, a recent special issue of the *Journal of Solid State Chemistry* was dedicated to these very topics (22). One will notice almost immediately that this field is dominated by materials based on *d*-block transition metal compositions. Lanthanide (Ln)-containing materials have been much scarcer, perhaps for reasons to be discussed herein (e.g., a tendency to exhibit higher coordination numbers) (23). With this in mind, however, recent advances in polymeric Ln-containing materials suggest that these compounds are as structurally diverse and that the unique luminescence behavior of the *f*-elements may be harnessed for applications, such as sensing and molecular recognition (23–30). Such inherent properties may extend the applications of framework materials beyond those introduced above.

The purpose of this chapter is to highlight materials consisting of extended topologies built from Ln element metal centers. We hope to demonstrate the richness of these materials, as well as inspire others to utilize these elements in their own work in extended solids. Further, a somewhat unusual approach to discussing MOFs and CPs is taken in that materials that predate this terminology are included. Extended solids have clearly existed for quite some time, yet their labeling as MOFs or CPs is a relatively recent (~1990s) phenomenon. Lastly, when one considers the rate at which new materials are appearing in the literature (a testament to the richness of these systems!), it is simply not possible to imagine a completely comprehensive review. Instead, an attempt has been made to include representative compounds that exemplify some of the key structural features of MOFs and CPs.

## II. TERMINOLOGY

Before beginning a systematic treatment of specific structure types, it is perhaps important to develop some terminology that is useful in describing MOFs and CPs. Generally speaking, a *CP* will be any hybrid material (i.e., one that contains metal centers linked through multifunctional organic components), with an overall structural topology that is at least one dimensional (1D). This definition will exclude all molecular materials and many salts. The dimensionality must arise from direct connectivity between the metal centers and linkers and contain ionic and/or covalent bonds. Hydrogen bonding will not be included as a contributor to our dimensionality designation. A *MOF* is actually a subspecies of a CP, in that their structures must be three dimensional (3D) in overall topology through ionic and/or covalent bonding only. In a recent review by James (16), this definition is extended somewhat to encompass materials that exhibit some degree of porosity.

A generic representation of these concepts is shown in Fig. 1. Generic, implies that these structural definitions are, at this point, independent of the composition

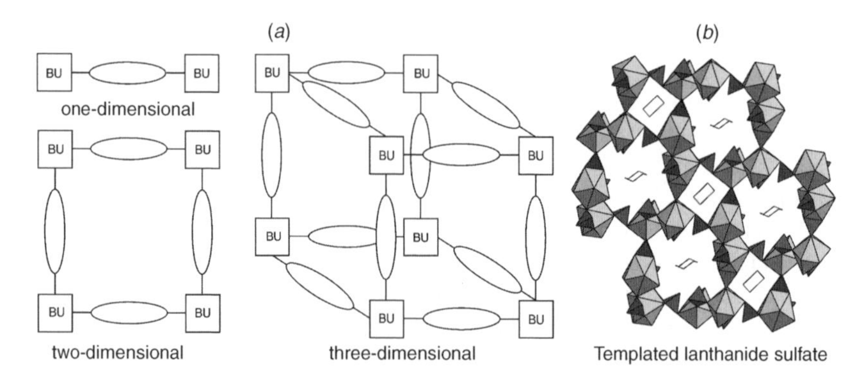


Figure 1. (*a*) A skeletal representation of coordination polymers and a MOF. The inorganic building units (BU) are connected through organic linker molecules (tethered ovals) to form structures of varying dimensionality. (*b*) The templated material, $[Nd_2(SO_4)_4(H_2O)_2][C_4N_2H_{12}]$ (31), is not a CP or MOF by our classification scheme. In this figure, lighter polyhedra are $NdO_9$ and the dark tetrahedra are $SO_4^{2-}$ units. The guest piperazine ($C_4N_2H_{12}$) is shown as a line drawing.

metal center or inorganic component. The first image is a 1D CP, as evidenced by connectivity in one direction. The inorganic component is labeled as BU as is defined for the Ln elements below. Examples of 2D and 3D structures follow, as does a hybrid, yet non-CP material for comparison, where the organic component is a guest or template and not coordinated to the inorganic host. The term MOF is reserved for the 3D example only. These designations are similar to those found in recent reviews by Férey (7), Janiak (14), and as described in a recent review of actinide containing CPs (32).

A material exemplifying many of the MOF concepts is shown in Fig. 2. Briefly, this structure, GWMOF-2 (33) consists of chains of edge-shared $NdO_9$ polyhedra (along [010]) that are tethered to one another through bridging tridentate and bridging bidentate adipate groups.

The term building unit is used extensively in the literature of extended and framework hybrid solids. In general, this term, whether modified by *primary* or *secondary*, refers to the basic inorganic repeat unit that when polymerized, forms the foundation of the inorganic component of the structure. In aluminosilicate

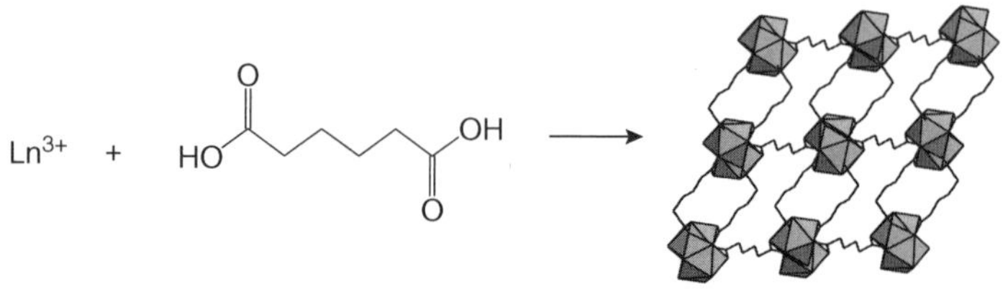

Figure 2. A schematic of MOF assembly using GWMOF-2 as an example (33) $Nd^{3+}$ ions are coordinated to the bifunctional adipic acid molecules to form an overall 3D structure (right). This representation is consistent with our presentation throughout this chapter: Ln coordination is shown as polyhedra and organic components are shown as black lines.

zeolites, for example, the term primary BU refers to a single $TO_4$ tetrahedron. Secondary building units (or SBUs) in this system then refer to theoretical topological entities that when repeated define the entire framework structure. In MOFs and CPs, the primary building unit is referred to as the metal center and its first shell of coordination (34). As in the zeolite system, this results in metal-centered polyhedra, yet by virtue of the variety of metal species encountered, is not restricted to simply tetrahedral environments. For the lanthanides, the primary building unit may be thought of as the $Ln^{3+}$ cation and its first shell of coordination (e.g., the familiar dodecahedron, tricapped trigonal prism, or bicapped square antiprism). The introduction of an additional SBU definition to describe both the coordination geometry as well as any oligomerization of primary building units into larger species, such as dimers, chains, or slabs, may also be of use. These definitions and the evolution of the characterization schemes are presented in Fig. 3. Shown first [Fig. 3(*a*)] is a silicate example of a primary $SiO_4$ BU (ball-and-stick and polyhedral representations are on the left). On the right is a nodal representation (each dot = one $SiO_4$ tetrahedron) of BUs polymerized to form commonly observed SBUs. Figure 3(*b*) shows a similar evolution of a $ZnO_4$ tetrahedron to a tetranuclear SBU as observed in MOF-5, a now famous example of a transition metal MOF with attractive sorption properties (35). Figure 3(*c*) contains frequently observed BUs of the Ln cations

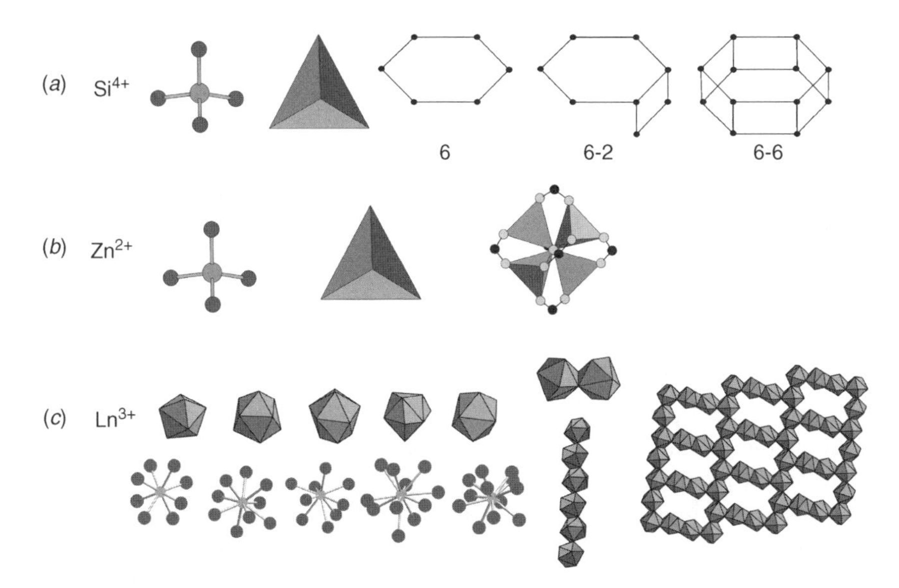


Figure 3. (*a*) Primary and secondary building units in the aluminosilicate zeolite system. (*b*) Primary and secondary building units as found in transition metal MOFs. (*c*) Primary (CN = 8–12) and secondary building units (dimers, chains, and slabs) present in Ln containing CPs and MOFs. These polyhedral representations of various coordination geometries will be used throughout this chapter. For clarity, coordinated anions are typically shown as vertices in polyhedral representations.

$$D^{a,b}_{c,d}$$

Scheme 1. A classification scheme to describe the overall dimensionality and local Ln geometry in CPs and MOFs. The superscripts describe the lanthanide coordination geometry, whereas the subscripts refer to any additional metal centers present, as in heterometallic compounds. $D$ = overall dimensionality of compound; $a$ = building unit dimensionality of *f*-block metal; $b$ = number of centers in zero-dimensional building unit of first metal; $c$ = building unit dimensionality of *d*-block metal (if present); $d$ = number of centers in zero-dimensional building unit of second metal center. This labeling scheme is represented in the tables for all referenced compounds.

followed by SBUs encountered in coordination polymer materials. Although BUs and SBUs are present in solid-state structures, such entities are not necessarily reflective of solution speciation prior to crystallization.

Our review of Ln CPs will demonstrate that these elements lend themselves to polymerization of primary BUs to form secondary BUs with some frequency. As such, we introduce a classification scheme that in addition to reporting the overall dimensionality, highlights the local structure of the Ln metal centers and any subsequent polymerization. Note that this scheme is analogous to that already utilized for actinide CPs in a previous review (32) (see Scheme 1).

Classification of MOFs has been studied extensively, yet this complementary approach is offered in an effort to focus on local geometry and explore the identity of any SBUs present. Topological studies tend to treat the BUs (secondary or otherwise) as nodes and focus less on the composition of the nodes themselves. This scheme is thus an effort to remain cognizant of the inorganic components of these structures as SBU variety may ultimately be considered a desirable or even tunable characteristic of this family of materials.

## III. LANTHANIDE-COORDINATION CHEMISTRY: A BRIEF REVIEW

### A. Molecular Materials

A brief review of Ln coordination chemistry is warranted as a precursor to appreciating the structural systematics of extended solids comprised thereof. In addition, this section aims to exemplify a number of Ln-specific properties and coordination preferences with the goal of distinguishing these materials from transition metal MOFs and CPs. It is felt that there are a number of attractive features of the *f*-elements (specifically their luminescent behavior) that make Ln-MOFs potentially more appropriate for certain applications (e.g., sensing).

As mentioned, the field of MOFs and CPs has been thus far dominated by *d*-block transition metal compositions with Ln compounds following behind. A line of reasoning that may explain this is a diminished capacity for prediction of

Ln coordination geometry. Lanthanide elements possess large ionic radii and an almost complete absence of crystal-field effects. In addition, the $4f$ orbitals are highly penetrating and are thus almost completely shielded by the filled $5s$ and $5p$ orbitals, leading to highly ionic character in the bonds formed. As a result, the coordination environment tends to be more spherical in nature, as opposed to the familiar octahedral, square planar, and so on found in the transition metals. Lanthanides do exhibit a range of coordination geometries (some of which were shown in Fig. 3), yet these tend to have higher values, often ~8–12. This lack of directional bonding at the metal center itself suggests that ligand geometry is thus essential for influencing both local and extended structure (36, 37).

The most common oxidation state of the Ln ions is +3, although many divalent and tetravalent species are known. The predominance of trivalent ions can be seen from the electron configurations of these elements: $[Xe]4f^n5d^16s^2$ ($n = 0$ for La, $n = 14$ for Lu), where successive ionizations of $5d$ and $6s$ electrons leave the $4f$ electrons as the valence. This also explains the well-known lanthanide contraction, wherein a small (~16%), yet steady reduction in ionic radii is observed across this series with increasing nuclear charge (36, 37).

Considering the larger size of Ln ions (as opposed to *d*-block metals), another challenge is *completing* their coordination spheres. Multidentate ligands may display a tendency to chelate rather than polymerize. Thus, judicious choice of organic linker molecules is essential for construction of Ln centered materials. Further, smaller molecules (e.g., solvents) with little steric hindrance will often coordinate to open sites and can lead to some difficulty in predicting the final coordination geometry of the Ln sphere (38). Polymerization of individual Ln polyhedra (BUs) to form oligomeric species (SBUs) is another way in which coordination spheres are satisfied. This tendency to form chains and sheets containing direct Ln—O—Ln linkages through point, edge, or face sharing is a distinguishing feature of Ln-CPs and will be highlighted throughout this chapter.

The coordination chemistry of molecular complexes of lanthanide elements may be regarded as mature and indeed rather extensive. This is not surprising considering the range of applications, both realized and potential, of these materials. Many of these have been summarized quite nicely in recent reviews by Bunzli, Piguet, and Parker and include: phosphors, Lewis acid catalysts, electroluminescent devices, fluoroimmunoassays, tags for luminescence microscopy, magnetically addressable liquid crystals, superconducting materials, therapeutics, nuclear magnetic resonance (NMR) shift reagents, and magnetic resonance imaging (MRI) contrast agents (38–43). Precise control of coordination geometry around Ln ions is of course critical for harnessing and tuning the properties of these materials yet for reasons introduced above, such control imparts unique challenges when considering the spherical coordination environments of the Ln elements. The research groups of Bunzli and Piguet (38–41) have been particularly interested in the photophysical behavior of these materials and as a result have made significant contributions to our understanding of the relationship between coordination chemistry and luminescent behavior. Indeed, this area has particular relevance to this chapter as sensitized luminescence via the antenna effect is an area of growing interest within Ln-CPs. Review articles and a text highlighting other components of Ln coordination chemistry (including behavior in aqueous solution) have appeared from Parker and co-workers (42,43) and Aspinall (36) and are excellent resources to consult for thorough overviews of both recent advances in these areas, as well as fundamental properties of the lanthanides.

In light of these resources, we hesitate to present an in-depth treatment of Ln coordination chemistry here, and instead simply call attention to a few characteristics of particular importance to polymeric materials, specifically, the predominance of N- and O- donor ligands in the first coordination sphere of Ln ions. Carboxylate functional groups as organic linkers have been used extensively in CP construction and chelating N-donor ligands have been shown to influence polymerization (and ultimately, topology), as well as luminescent properties.

Interesting potentials exist for polymeric materials containing heteropolymetallic architectures. In the molecular regime, such mixed Ln/Ln' or Ln/An materials are being explored as (e.g.) new light harvesting compounds. Ideally, these could serve as both luminescent probes and contrast agents if two different Ln elements with complementary properties could be assembled in close proximity to one another. Further, separation of Ln and actinide elements is of particular interest with respect to the nuclear fuel cycle. An extensive amount of work in these areas and a recent review by Bunzli and Piguet (41) highlights the inherent challenge to the synthesis of these types of materials: Differentiating between Ln ions is quite difficult when one considers that the only real difference between these species is a slight difference in ionic radii (41). Interestingly, this challenge in the molecular regime can actually be an asset to inorganic–organic hybrid materials. The similarity in ionic size gives rise to isomorphous substitution as is well known from solid-state chemistry and mineralogy. Since, as will be seen upon inspection of compounds herein, Ln-CPs and MOFs often contain a secondary building unit, wherein Ln polyhedra are polymerized, the potential to place specific Ln ions in close proximity to one another exists by default.

### B. Solid-State Materials

Several classes of solid-state materials have particular relevance to polymeric hybrid materials as well. A very through review by Wickleder (44) describes a wide range of compounds wherein lanthanide ions are coordinated to complex anions (e.g., phosphates, sulfates, silicates, and molybdates). Typically, these materials tend to form extended structures through edge, face, or point sharing of the complex anion polyhedra with the Ln ion as shown in Fig. 4. This leads to

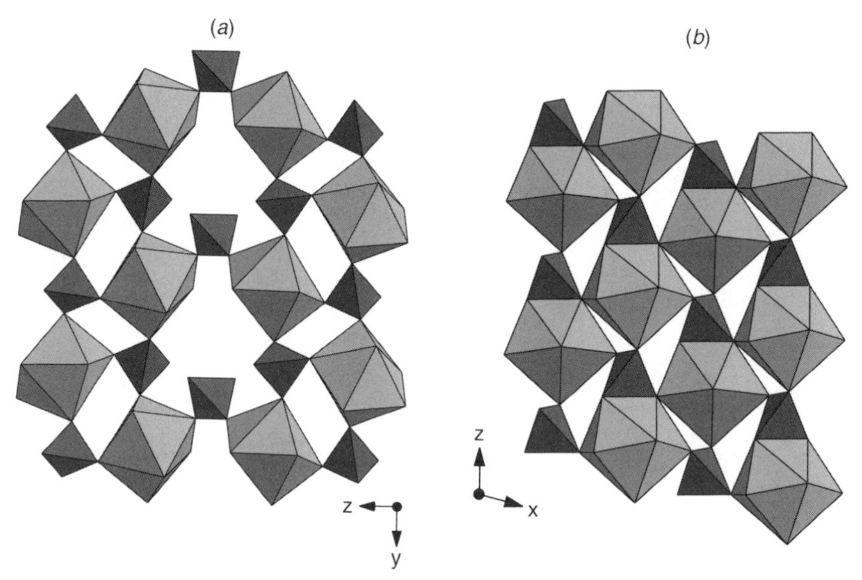


Figure 4. (*a*) The structure of $Eu_2(SO_4)_3 \cdot 8H_2O$ (45). Gray polyhedra are Ln sites whereas the bridging, point shared tetrahedra are sulfate anions. Extra-framework water molecules have been omitted for clarity. (*b*) The structure of the mineral monazite, $REEPO_4$ (46). In this case, the tetrahedra represent phosphate anions. Both of these solid-state structures feature the ways in which tetrahedral anions may coordinate to $Ln^{3+}$ centers, in particular point and edge sharing. This connectivity will be revisited in the sulfonate and phosphonate sections (Sections V. F and G).

a variety of interesting topologies ranging from clusters to rings to chains to slabs. These materials also have the ability to accommodate charge-balancing cations due to the negative charges introduced by the coordinated complex anion. Often, alkali metals or alkaline earths will be incorporated into the structure to serve this charge-balancing role. It is also seen that topology may be influenced through appropriate choice of charge-balancing cation (31,47), and a representative structure of this type was shown in Fig. 1 to highlight the difference between CPs and other inorganic–organic hybrids. Such templating of structures where extraframework cations can range from monatomic species to complex organic molecules, is a popular route to polymeric hybrid materials in many systems (48). Templating is just beginning to appear in the MOF literature and will be explored as part of this chapter.

Ultimately, the chemistry of lanthanide-containing MOFs and CPs can be regarded as a blend between the coordination chemistry of molecular and solid-state materials. The structural themes present in solid-state materials (e.g., edge sharing of polyhedra) and the coordination preferences seen in molecular compounds (e.g., N-, O- donors) are essentially what form the structural basis for extended topologies.

A final comment to bear in mind as we explore the polymeric materials presented in this chapter is that new or unique coordination geometries as compared to the molecular regime will not be seen. Instead, manifestations of bonding themes well known from molecular materials (e.g., multidentate ligands and spherical coordination geometries will be seen). Further, whereas chelation is often a goal of supramolecular assemblies for applications described above, CP and MOF chemistry in some regard strives to avoid chelation (or at least do so judiciously) in an effort to promote polymerization into higher dimensional topologies.

## IV. INCLUSION CRITERIA AND RESOURCES

As a final component of introduction, here we discuss our criteria for what should be included in this chapter as well as our sources of information. The primary resource for materials discussed herein has been the Cambridge Crystallographic Data Centre Database (V5.27, November 2005) (49). The initial searches were conducted with ConQuest (V1.8) using the following criteria: Ln–O–C ("any" bond selected between O and C) yielded 5673 hits. A subset of 1136 polymeric structures was extracted using appropriate filters within ConQuest. This sizable subset was reduced to 795 entries by visual inspection by excluding polyoxometallates and related hybrid materials. From this list, ~150 structures were selected and presented herein. These are, in our opinion, reflective of not only the current state of the science with respect to MOF and CP synthesis, but also draws on much earlier reported compounds whose chemistry was explored prior to the current definitions as MOFs or CPs. Several liberties were taken with respect to the inclusion of some materials that were found particularly interesting and exclusion of some that were perhaps not representative of the structural themes trying to be developed. Note that this chapter is meant to be representative and not necessarily comprehensive, as the latter would be too extensive. Further, it is hoped that present day researchers may benefit from this revisiting of older literature in terms of stimulating some new ideas to couple to the more recently developed MOF–CP vocabulary.

The material to be discussed in this chapter has been arranged into a series of tables based on the local coordination geometry of the Ln metal center. This includes a variety of O-donors, N-donors, and the use of multiple linker molecules and/or transition metals. Each category will be taken in turn and structural features from representative compounds will be highlighted. In addition, each table is arranged according to the overall dimensionality of the material. Recall that a coordination polymer has connectivity in at least 1D, thus distinguishing such compounds from molecular structures. Each section therefore attempts to display an evolution of dimensionality from 1D to 3D (i.e., true

MOFs). Finally, the CCDC structure code for each material (where available) has been included in addition to the primary literature reference. All figures were produced using CrystalMaker [V5.2, from the published crystallographic information files (CIFs)] or ChemWindow (V3.1).

## V. EXAMPLES

### A. The Ln—O Coordination: Aliphatic Linkers

The first family of materials to be highlighted is arguably one of the richest and among the more popular systems studied in recent years. Table I contains a number of representative compounds in this area. Note that the most common form of linker molecule are those with carboxylate functional groups. This finding is to be expected considering the hard acid behavior of the $Ln^{3+}$ ions, as well as the hard base characteristics of $-COO^-$ groups. The local coordination geometry of the individual $Ln^{3+}$ centers will be highlighted in each figure and range from monodentate to bridging tridentate. Further, Fig. 5 contains a separate depiction of linker molecules.

The oxalates may be regarded as the simplest carboxylate linker molecules in Ln-MOF and CP systems. The $C_2O_4^{2-}$ anion is found almost exclusively in bridging bidentate coordination and in turn promotes a range of topologies. Figure 6 shows AOXNDH (CCDC code, used throughout), or $[Nd_2(C_2O_4)_3(H_2O)_6]\cdot 4H_2O$, wherein monomeric $NdO_9$ polyhedra are polymerized through oxalate linkages to form six-membered rings and ultimately neutral porous sheets. The local coordination consists of three oxalate linkers for every two $Nd^{3+}$ centers with the remaining three positions on each occupied by water molecules. This example is representative of some of the themes developed in the coordination chemistry, Section III. In this case, solvent molecules remain bound to the $Ln^{3+}$ center to complete the Ln-coordination sphere. Similar local and extended geometry may be seen in ERHOXL $[Er_2(C_2O_4)_4(H_2O)_2]\cdot 4H_2O$ (63) and in a neutral 3D framework in OJONOQ $[Er_2(C_2O_4)_3(H_2O)_6]\cdot 12H_2O$ (79). The EHAXIU $[La_2(C_2O_4)_4(H_2O)_2]\cdot 2(CH_6N_3)$ (Fig. 7) compound is an interesting example of a structure containing anionic sheets that formed when there are two oxalates for each $Ln^{3+}$ center (68). This results in four-membered rings and requires protonated charge balancing guanidinium ions in the interlayer regions.

Moving toward longer aliphatic carboxylates is a succinate FIJYAZ $[Pr_2(C_4H_4O_4)_3(H_2O)_2]\cdot H_2O$ (Fig. 8). A 3D framework is observed in this structure, the neutrality of which is anticipated from the 2:3 Ln/linker ratio. Bridging bidentate and bridging tridentate coordination is observed in this compound, with each displayed by the single, unique succinate linker. Glutarate

TABLE I
Ln–O Aliphatics

| Code | Compound | Linker | Dim | a | b | c | α | β | γ | Space Group | Reference |
|---|---|---|---|---|---|---|---|---|---|---|---|
| **DIQZAE** | $[Nd(C_5H_6O_4)(H_2O)_4] \bullet Cl \bullet 2H_2O$ | Glutarate | $1^{0,2}$ | 7.877 | 8.781 | 10.444 | 110.69 | 95.24 | 103.48 | *P1* | 50 |
| ELOFUG | $[Pr_2(C_8H_{20}O_8)(H_2O)_8(Cl)_2] \bullet 4(Cl) \bullet 4(H_2O)$ | Erythritol | $1^{0,2}$ | 7.5099 | 10.186 | 10.4213 | 85.394 | 84.2916 | 78.6868 | *P-1* | 51 |
| FUTVUL | $[Ho(C_4H_9NO_3)_2(H_2O)_5] \bullet 6(Cl)$ | L-Threonine | $1^{0,1}$ | 10.508 | 7.395 | 18.708 | 90 | 90 | 90 | $P2_12_12_1$ | 52 |
| HANJEL | $[Nd_2(C_7H_{15}O)_6(C_4H_{10}O_2)]$ | 1,2-Dimethoxyethane | $1^{0,2}$ | 10.797 | 10.984 | 12.112 | 109.08 | 100.32 | 90.8 | P-1 | 53 |
| JAZLEB | $[Ho(C_4H_6NO_4)(H_2O)_5]Cl_2 \bullet H_2O$ | Aspartate | $1^{0,1}$ | 10.0708 | 10.0708 | 11.845 | 90 | 90 | 120 | $P3_1$ | 54 |
| MIWWEU | $[La(C_6H_{12}NO_4)_2] \bullet (Cl) \bullet 3(H_2O)$ | *N,N*-bis(2-Hydroxyethyl) glycine | $1^{0,1}$ | 21.287 | 6.792 | 14.344 | 90 | 90 | 90 | $Pca2_1$ | 55 |
| AFUQUN | $[Er(C_6H_6NO_6)(H_2O)]$ | Nitrilotriacetate | $2^{0,1}$ | 6.7262 | 6.5427 | 19.8 | 90 | 93.444 | 90 | $P2_1/n$ | 56 |
| **AOXNDH01** | $[Nd_2(C_2O_4)_3(H_2O)_6] \bullet 4H_2O$ | bb-Oxalate | $2^{0,1}$ | 11.1919 | 9.612 | 10.257 | 90 | 114.42 | 90 | $P2_1/c$ | 57 |
| **BALXAO** | $[(Eu_2(C_4H_2O_4)(C_4HO_4Cl)_2(H_2O)_4] \bullet 4(H_2O)$ | Fumarate; 2-chlorofumarate | $2^{0,2}$ | 8.872 | 9.422 | 9.545 | 103.52 | 101.355 | 94.076 | *P-1* | 58 |
| BUHVOP | $[Er(C_4H_5O_6)_2(H_2O)_2]$ | Tartrate | $2^{0,1}$ | 5.995 | 5.995 | 36.433 | 90 | 90 | 90 | $P4_12_12$ | 59 |
| DIFQOY | $[Pr(C_4H_6NO_4)(H_2O)_4] \bullet 2(Cl) \bullet (H_2O)$ | Iminiodiacetate | $2^{1}$ | 9.5728 | 16.3086 | 9.0007 | 90 | 112.114 | 90 | $P2_1/c$ | 60 |
| EFIYOH | $[Sm(C_4H_4O_6)_2(H_2O)_3]$ | Tartrate | $2^{0,1}$ | 6.103 | 6.103 | 36.826 | 90 | 90 | 90 | $P4_12_12$ | 61 |
| EJIDOQ | $[Pr_2(C_6O_4H_8)_3] \bullet (H_2O)_{10}$ | Adipate | $2^{0,2}$ | 9.242 | 9.764 | 10.682 | 67.75 | 84.58 | 61.79 | *P-1* | 62 |
| ERHOXL | $[Er_2(C_2O_4)_2(HC_2O_4)_2(H_2O)_2] \bullet 4(H_2O)$ | bb-Oxalate | $2^{0,1}$ | 8.6664 | 8.6664 | 6.4209 | 90 | 90 | 90 | *P4/n* | 63 |

| | | | | | | | | | | | |
|---|---|---|---|---|---|---|---|---|---|---|---|
| FIYQEK | $[Eu(C_6H_5O_7)(H_2O)]$ | Citrate | $2^{0,2}$ | 19.243 | 6.1288 | 15.645 | 90 | 103.616 | 90 | *C2/c* | 64 |
| ISOJOP | $[Tb_2(C_6H_6O_4)_3(H_2O)_5]$ | 1,1-Cyclobutanedicarboxylate | $2^{0,3}$ | 15.885 | 8.489 | 19.189 | 90 | 106.02 | 90 | *P2₁/n* | 65 |
| KEZHIG | $[Tb(C_2O_4)_2(H_2O)] \bullet (H_5O_2)$ | bb-Oxalate | $2^{0,1}$ | 8.7649 | 8.7649 | 12.7907 | 90 | 90 | 90 | *P4₂/n* | 66 |
| AROVAE | $(La_2(C_5H_6O_4)_3(H_2O)_3] \bullet 2(H_2O)$ | Glutarate | $3^1$ | 11.438 | 13.869 | 15.635 | 90 | 109.75 | 90 | *P2₁/n* | 67 |
| AZIYIR | $[Nd_2(C_6H_8O_4)_3(H_2O)_2]$ | Adipate | $3^1$ | 8.0752 | 11.637 | 13.209 | 106.76 | 93.419 | 107.225 | *P-1* | 33 |
| **EHAXIU** | $[La(C_2O_4)_4(H_2O)_4] \bullet 2(CH_6N_3)$ | Oxalate | $2^{0,1}$ | 6.618 | 12.636 | 13.19 | 90 | 93.007 | 90 | *P2₁/c* | 68 |
| EJIFAE | $[Nd_2(C_6O_4H_8)_3(H_2O)_5]$ | Adipate | $3^{0,2}$ | 23.657 | 14.184 | 8.881 | 90 | 105.81 | 90 | *C2/c* | 62 |
| FAQYUR | $[Nd_2(C_5H_6O_4)_3(H_2O)_2] \bullet 4(H_2O)$ | Glutarate | $3^1$ | 8.1174 | 15.1841 | 19.8803 | 90 | 93.762 | 90 | *C2/c* | 69 |
| FEYQIK | $[Pr_2(C_6H_8O_4)_3(H_2O)_4] \bullet (C_6H_{10}O_4) \bullet 4(H_2O)$ | Adipate | $3^1$ | 8.655 | 9.9366 | 11.717 | 74.71 | 69.795 | 85.959 | *P-1* | 25 |
| **FEYQOQ** | $[Pr_2(C_6H_8O_4)_3(H_2O)_2] \bullet (H_2O)$ | Adipate | $3^2$ | 8.6786 | 10.3831 | 13.6693 | 101.35 | 106.1204 | 93.9593 | *P-1* | 25 |
| **FEYQUW** | $[Pr_2(C_6H_8O_4)_3(H_2O)_2] \bullet (C_{10}H_8N_2)$ | Adipate | $3^1$ | 21.9415 | 7.7878 | 19.6649 | 90 | 90 | 90 | *Pbcn* | 25 |
| **FIJYAZ** | $[Pr_2(C_4H_4O_4)_3(H_2O)_2)] \bullet (H_2O)$ | succinate | $3^1$ | 20.2586 | 7.9489 | 13.9716 | 90 | 121.641 | 90 | *C2/c* | 70 |
| **GOPLUS** | $[Pr_2(C_{12}H_{20}N_2O_4)_3(H_2O)_4](ClO_4)_6 \bullet 2(H_2O)$ | 1,4-Diazoniabicyclo(2.2.2)octane-1,4-dipropionate | $3^1$ | 8.085 | 14.316 | 29.775 | 90 | 103.04 | 90 | *P2₁/n* | 71 |
| GUXCOR | $[Nd_2(C_{12}H_{22}N_2O_2)_4(NO_3)_6] \bullet (C_6H_{12})$ | Bis(*N,N′*-ethylenebis(acetylacetoneimine) | $3^{0,1}$ | 11.5549 | 15.386 | 20.339 | 90 | 102.792 | 90 | *P2₁/n* | 72 |
| ITAHEQ | $[Eu_2(C_{12}H_{14}O_4)_3]$ | 1,3-Adamantanedicarboxylate | $3^{0,1}$ | 16.466 | 16.466 | 7.7375 | 90 | 90 | 120 | *P6₃cm* | 73 |
| JAHXUM | $[Gd_4(C_8H_4O_4)_7(H_2O)_2] \bullet 2(Cu(C_{10}H_8N_2)_2)$ | 1,3-Benzenedicarboxylate | $3^{0,2}$ | 26.359 | 14.3982 | 25.3106 | 90 | 113.016 | 90 | *C2/c* | 74 |

(*Continued*)

TABLE I
(*Continued*)

| Code | Compound | Linker | Dim | a | b | c | α | β | γ | Space Group | Reference |
|---|---|---|---|---|---|---|---|---|---|---|---|
| **JOSNAG** | $[La_2(C_6H_8O_4)_3(H_2O)_4] \bullet 6(H_2O)$ | Adipate | $3^{0,2}$ | 9.239 | 9.79 | 10.709 | 68.04 | 84.49 | 62.01 | *P-1* | 75 |
| MAGXOI | $[Nd(C_6H_6NO_6)]$ | Nitrilotriacetate | $3^{1}$ | 6.622 | 7.0334 | 17.426 | 90 | 90 | 90 | $P2_12_12_1$ | 76 |
| **MASXIN** | $[La(C_5H_6O_4)(C_5H_7O_4)(H_2O)] \bullet H_2O$ | Glutarate | $3^{1}$ | 16.4027 | 8.76805 | 19.5576 | 90 | 90 | 90 | *Pbca* | 77 |
| MAZJAY | $[Eu_2(C_4H_4O_5)_3(H_2O)_2] \bullet 5(H_2O)$ | Oxydiacetate | $3^{0,1}$ | 16.986 | 15.374 | 9.538 | 90 | 90 | 90 | *Ama2* | 78 |
| OJONOQ | $[Er_2(C_2O_4)_3(H_2O)_6] \bullet 12(H_2O)$ | bb-Oxalate | $3^{0,1}$ | 30.869 | 30.869 | 7.2307 | 90 | 90 | 120 | *R-3* | 79 |

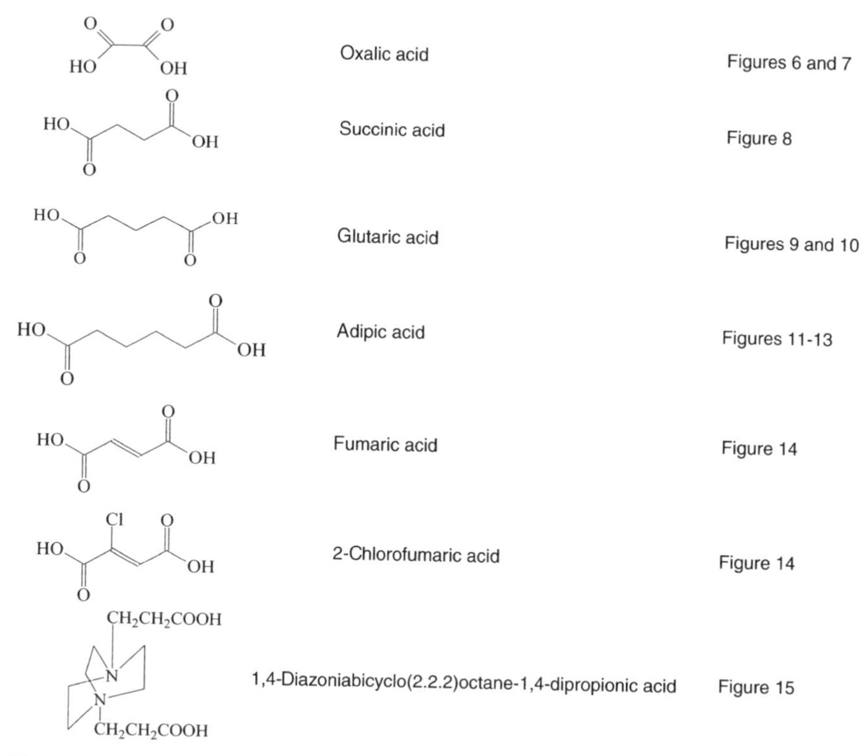


Figure 5. Pictorial representation of the linkers used in the aliphatic-based CPs described in the corresponding figure.

linkages can be found in 1D DIQZAE $[Nd_2(C_5H_6O_4)_2(H_2O)_8]Cl_2 \cdot 4H_2O$ (50) (Fig. 9) and 3D MASXIN $[La(C_5H_6O_4)(C_5H_7O_4)(H_2O)] \cdot H_2O$ (77) (Fig. 10). Both of these materials serve as the first examples (for this chapter) of polymerization of primary BUs to form SBUs: dimers in DIQZAE and infinite chains in MASXIN. The compound DIQZAE consists of $Nd_2O_8(H_2O)_8$ dimers linked into cationic chains that run along [100] and are charge balanced by uncoordinated $Cl^-$ anions. Both bridging tridentate and bridging bidentate carboxylate geometries are observed. The compound MASXIN contains chains of $LaO_9(H_2O)$ polyhedra that run along [010] and are linked to one another in two directions through glutarate molecules. This overall 3D topology is similar to that seen in FEYQUW $[Pr_2(C_6H_8O_4)_3(H_2O)_2] \cdot (C_{10}H_8N_2)$ (25), a structure that also contains chains of hydrated polyhedra as SBUs that are tethered through adipate molecules in bridging tridentate and bridging bidentate coordination (Fig. 11) (25). This material is an example of a templated MOF in

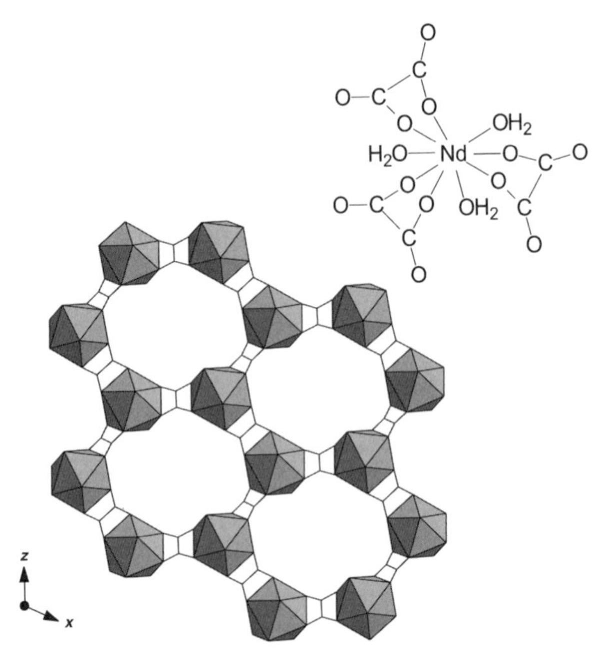


Figure 6. The structure of $[Nd_2(C_2O_4)_3(H_2O)_5]\cdot 4H_2O$ (AOXNDH) shown down [010]. Interlayer water molecules have been omitted for clarity. The gray polyhedra represent the $Ln^{3+}$ centers and black lines are the organic linker species. The inset is a skeletal representation of the local geometry of the $Ln^{3+}$ center. This representation format will be used throughout this chapter.

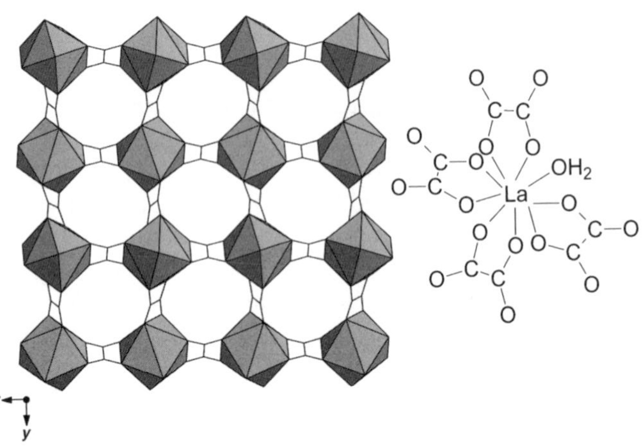


Figure 7. The structure of $[La_2(C_2O_4)_4(H_2O)_2]\cdot 2(CH_6N_3)$ (EHAXIU) shown down [100]. Charge balancing guanidinium cations have been omitted.

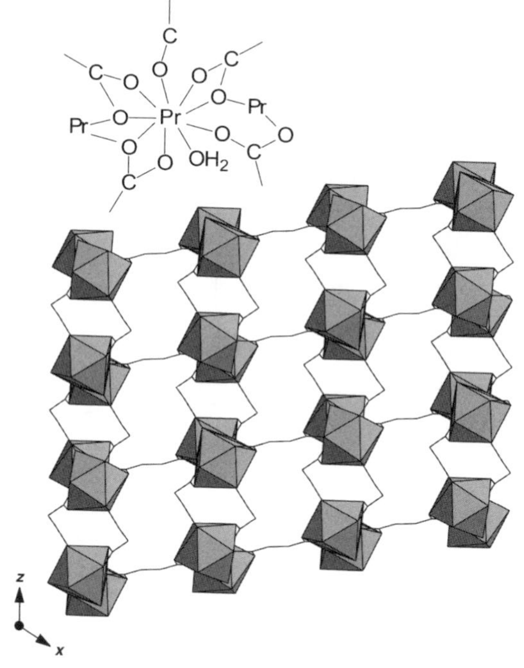


Figure 8. The 3D structure of FIJYAZ $[Pr_2(C_4H_4O_4)_3(H_2O)_2]\cdot H_2O$, shown down [010]. Extra-framework water molecules are not shown.

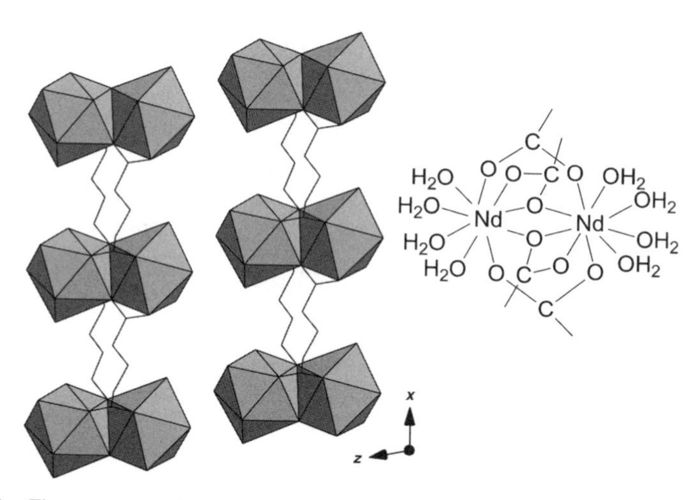


Figure 9. The structure of DIQZAE $[Nd_2(C_5H_6O_4)_2(H_2O)_8]Cl_2\cdot 4H_2O$. Cationic chains of $Nd_2O_8(H_2O)_8$ dimers propagate along [100] and are charge balanced by uncoordinated $Cl^-$ anions (not shown).

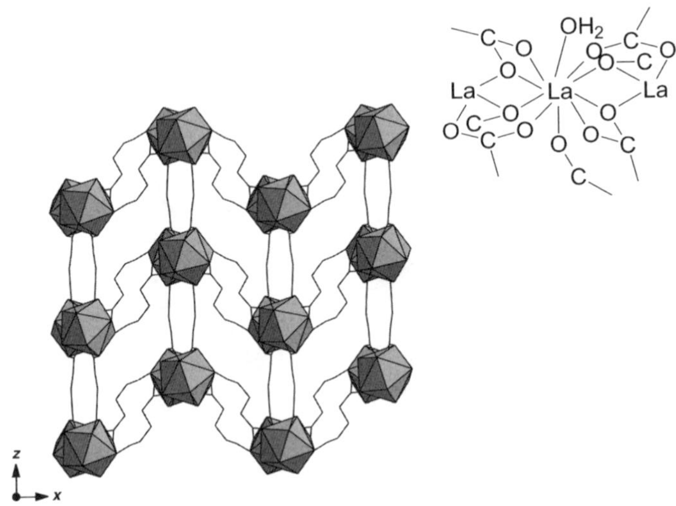


Figure 10. The 3D structure of MASXIN $[La(C_5H_6O_4)(C_5H_7O_4)(H_2O)]\cdot H_2O$. Chains of $LaO_9(H_2O)$ polyhedra propagate into the page and are linked through glutarate groups.

that neutral 4,4′-bipyridine (bpy) molecules reside in the extra-framework region and have conceivably influenced the resulting topology. Further, these guest molecules serve as chromophores in that their excitation in the ultraviolet (UV) can transfer energy to the Ln host and promote luminescence in the visible

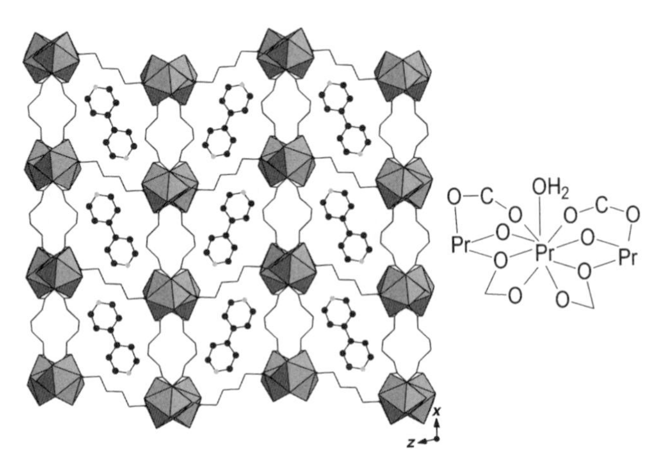


Figure 11. The 3D structure of FEYQUW $[Pr_2(C_6H_8O_4)_3(H_2O)_2]\cdot(C_{10}H_8N_2)$. Chains of hydrated polyhedra propagate into the page and are linked through adipate groups. Neutral bpy molecules reside inside the channels of this compound.

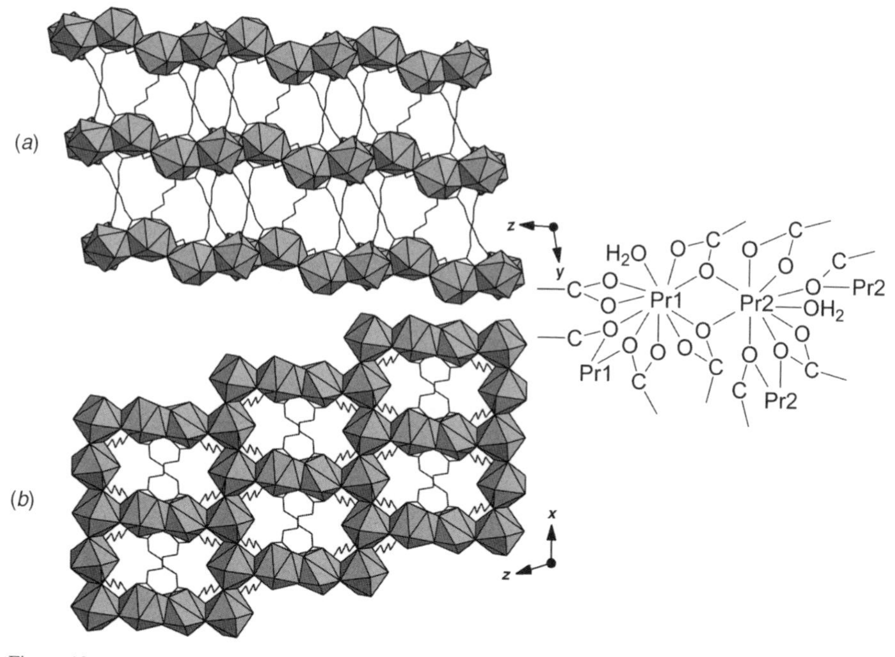


Figure 12. Two views of FEYQOQ $[Pr_2(C_6H_8O_4)_3(H_2O)_2]\cdot 4H_2O$. (*a*) Side view showing the stacking of layers (or slabs) of hydrated $PrO_9$ and $PrO_{10}$ polyhedra. (*b*) A single slab wherein intraslab coordination of adipate molecules is observed and shown.

region. Such indirect excitation, also known as the antenna effect, was introduced above and is currently an area of significant interest within polymeric Ln materials.

Polymerization of $LnO_9$ and $LnO_{10}$ polyhedra to form a slab-like SBU is seen in FEYQOQ $[Pr_2(C_6H_8O_4)_3(H_2O)_2]\cdot 4H_2O$ (Fig. 12) (25). The structure of this material is 3D in that each slab is tethered to the other through adipate groups in bidentate or bridging tridentate coordination. Additional adipate coordination is noted to span the intraslab pores (Fig. 12), yet these species do not appear to contribute structurally to the overall dimensionality, and instead merely fill space and complete the $Pr^{3+}$ coordination spheres.

The Ln adipates are indeed a rather rich system and several other examples of polymeric materials are noted (Table I). One final adipate on which to comment is JOSNAG $[La_2(C_6H_8O_4)_3(H_2O)_4]\cdot 6H_2O$ (Fig. 13) (75). Local La-carboxylate coordination displays bridging bidentate, bridging tridentate, and bidentate geometries to form $La_2O_{16}$ dimer units. These SBUs are in turn polymerized along [100] through bridging bidentate carboxylate groups. The resulting chains

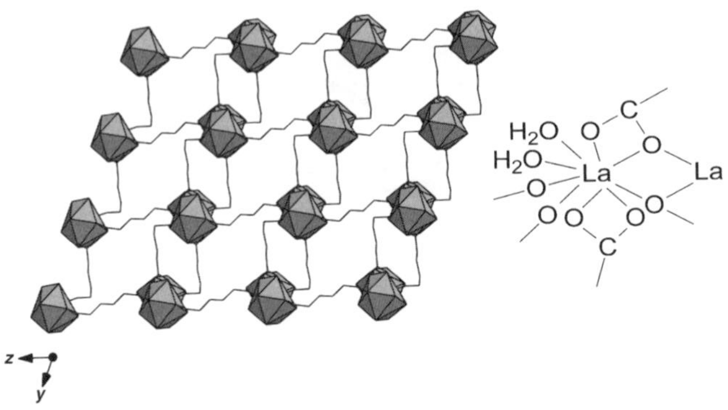


Figure 13. The 3D structure of JOSNAG $[La_2(C_6H_8O_4)_3(H_2O)_4]\cdot 6H_2O$ shown down [100].

of dimeric SBUs are tethered through adipate groups in manners analogous to the structures described above. Thus we see a range of BUs and SBUs in the Ln-adipate system. Arguably the flexibility of the carboxylate linker allows for subtle variation in packing and ultimately gives rise to a rich structural diversity. Further, the syntheses of these materials, although all aqueous, have been performed from a range of starting compositions and at room temperature to ~180°C. Some topological variation has been described as a function of synthesis temperature (33), yet discerning any type of synthesis–structure relationships beyond that is difficult given the number of variables across this family of materials.

Two final Ln-aliphatic carboxylate materials to be described are GOPLUS $[Pr_2(C_{12}H_{20}N_2O_4)_3(H_2O)_4](ClO_4)_6\cdot 2H_2O$ (71) and BALXAO $[Eu_2(C_4H_2O_4)(C_4HO_4Cl)_2(H_2O)_4]\cdot 4H_2O$ (58), both of which make use of modification of the backbone of the linker molecule (and not the functional group) to direct formation of unique topologies. The compound BALXAO actually utilizes two linker molecules: fumarate and 2-chlorofumarate (Fig. 14). Again we see edge-shared polymerization of Ln polyhedra to form dimeric SBUs that are tethered through bidentate and bridging tridentate carboxylate groups. The chloro group on the substituted fumarate linker remains uncoordinated to the Eu centers and points toward the center of the structure's void space. In some respects, this may be regarded as a functionalization of the CP wall, an area of interest that has been explored more extensively in the *d*-block containing CPs and MOFs (18).

GOPLUS consists of $PrO_7(H_2O)$ polyhedra polymerized through edge sharing into infinite chains along [100] (Fig. 15). The chains are linked through 1,4-diazoniobicyclo[2.2.2]octane-1,4-dipropionate groups that, like BALXAO, is essentially a carboxylate linker with a modified backbone. The bulky diazonibicylco[2.2.2]octane groups may contribute to the overall assembly of this 3D

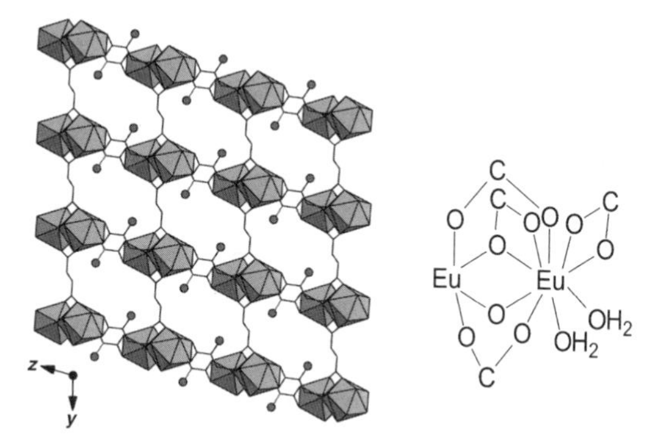


Figure 14. The structure of BALXAO $[Eu_2(C_4H_2O_4)(C_4HO_4Cl)_2(H_2O)_4]\cdot 4H_2O$. Gray spheres off of fumarate linkers are chlorine atoms.

topology as sterics from these species may play a role. Bridging tridentate and monodentate coordination is observed in this structure, the latter being somewhat less common for polymeric materials.

## B. The Ln—O Coordination: Aromatic Linkers

Table II includes a representative sampling of a number of Ln-MOFs and CPs built from aromatic linker molecules. Note that this class of materials is dominated by the benzene-carboxylate (and related) species as shown in Fig. 16. This is an attractive linker for a number of reasons, including its planarity and rigidity that to a first assumption, may suggest a more restricted structural diversity when compared to (e.g.) the aliphatics. Despite these features, however, no less rich coordination chemistry or resultant topologies in this family are seen.

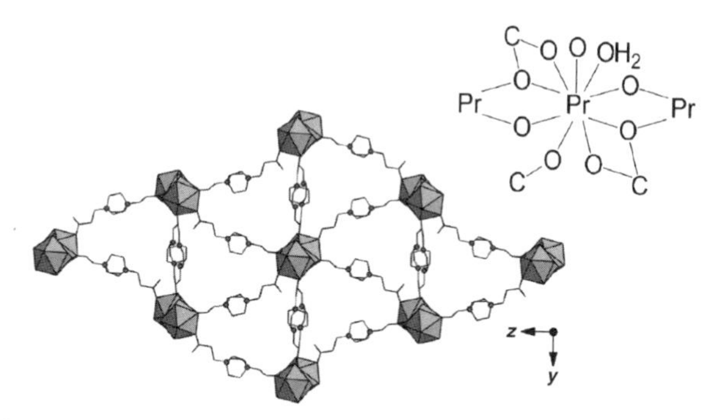


Figure 15. The structure of GOPLUS $[Pr_2(C_{12}H_{20}N_2O_4)_3(H_2O)_4](ClO_4)_6\cdot 2H_2O$. Linker molecules are 1,4-diazoniobicyclo[2,2,2]octane-1,4-dipropionate, and hence gray spheres are N atoms. Uncoordinated $H_2O$ molecules have been omitted for clarity.

TABLE II
Ln–O Aromatics

| Code | Compound | Linker | Dim | a | b | c | α | β | γ | Space Group | Reference |
|---|---|---|---|---|---|---|---|---|---|---|---|
| AMIYUQ | $[Nd_2(C_{14}H_8O_4)_3(H_2O)_2]$ | 2,2′-Biphenyldicarboxylate | $1^1$ | 20.9868 | 21.5521 | 8.3162 | 90 | 103.95 | 90 | *C2/c* | 80 |
| **ARUKIH** | $[Pr(C_7H_3NO_4)_2(H_2O)_2] \bullet 4(H_2O)$ | 2,6-Pyridinedicarboxylate | $1^{0,1}$ | 14.127 | 11.308 | 13.02 | 90 | 101.972 | 90 | $P2_1/c$ | 81 |
| AXECUB | $[Gd_2(C_{14}H_8O_4)_3]$ | 2,6-Pyridinedicarboxylate | $1^1$ | 20.9526 | 21.4295 | 8.2384 | 90 | 104.006 | 90 | *C2/c* | 82 |
| BEKQIS | $[Eu(C_5H_3SO_2)_3(C_5H_4SO_2)_2]$ | a-Thiophene carboxylate | $1^{0,1}$ | 20.5629 | 14.1736 | 9.7736 | 90 | 91.998 | 90 | *Cc* | 83 |
| KIHYIJ | $[Nd_2(C_6H_4NO_2)_8] \bullet (C_2H_{10}N_2) \bullet 6(H_2O)$ | 2-Pyridinecarboxylate | $1^{0,1}$ | 13.654 | 21.813 | 20.136 | 90 | 99.36 | 90 | *Cc* | 84 |
| **AKOVUR** | $[(Gd(C_9H_3O_6)(H_2O)_3] \bullet 1.5(H_2O)$ | 1,3,5-Benzenetricarboxylate | $2^{0,1}$ | 20.454 | 9.973 | 15.251 | 90 | 125.68 | 90 | *C2/c* | 85 |
| BABNAT | $[Nd(C_7H_4NO_2)_3(H_2O)_2]$ | *p*-Aminobenzoate | $2^{0,1}$ | 9.882 | 22.81 | 9.851 | 90 | 100.02 | 90 | $P2_1/n$ | 86 |
| DANKAE | $[La_2(C_{17}H_{18}N_2O_4)_2(H_2O)_9] \bullet 6(Cl) \bullet 2(H_2O)$ | bis(4,4′-Trimethylenedipyridine-*N,N*′-diacetate) | $2^{0,1}$ | 9.681 | 9.651 | 27.593 | 90 | 98.71 | 90 | *Pn* | 87 |
| **EHEYUL** | $[Nd_2(C_8H_4O_4)_4(H_2O)] \bullet (H_2O)$ | 1,3-Benzenedicarboxylate | $2^1$ | 10.408 | 13.954 | 21.347 | 90 | 95.557 | 90 | $P2_1/c$ | 88 |
| FIPPIE | $[Pr_3(C_7H_3NO_4)_5(H_2O)_8] \bullet 8(H_2O)$ | 2,6-Pyridinedicarboxylate | $2^{0,1}$ | 22.3971 | 13.9013 | 17.865 | 90 | 114.487 | 90 | *C2/c* | 89 |
| GAXBEM | $[Ce_2(C_4O_4)_3(H_2O)_4]$ | Squarate | $2^1$ | 7.143 | 16.94 | 6.994 | 90 | 101.24 | 90 | $P2_1/c$ | 90 |

| | | | | | | | | | | | |
|---|---|---|---|---|---|---|---|---|---|---|---|
| OBEMAJ | $[Nd_3(C_7H_3NO_4)_5(H_2O)_8](NH_4) \bullet 5(H_2O)$ | 2,6-Pyridinedicarboxylate | $2^{0,1}$ | 22.205 | 13.774 | 17.783 | 90 | 114.55 | 90 | *C2/c* | 91 |
| **WAKZUE** | $[Gd_2(C_{12}H_7SO_4)_3(H_2O)_4] \bullet 8(H_2O)$ | Thiophenylisophthalate | $2^{0,1}$ | 14.1195 | 10.6954 | 15.1149 | 90 | 102.529 | 90 | *P2/n* | 24 |
| XATCOK | $[Eu_2(C_{15}H_{14}N_2O_4)_2(H_2O)_3(NO_3)_2](NO_3)_4 \bullet 7(H_2O)$ | 1,3-Bis(pyridine-4-carboxylate)propane | $2^{0,1}$ | 10.509 | 11.438 | 12.628 | 69.77 | 71.27 | 89.8 | *P-1* | 92 |
| AHONEQ | $[Eu(C_8H_4O_4)(H_2O)_4] \bullet (H_2O)$ | 1,3-Benzenedicarboxylate | $3^{0,2}$ | 10.36699 | 12.43 | 10.1742 | 90 | 94.023 | 90 | *P2<sub>1</sub>/c* | 93 |
| ATIJOC | $[La_4(C_8H_4O_4)_6(H_2O)_8]$ | 1,4-Benzenedicarboxylate | $3^{0,2}$ | 6.2428 | 8.9794 | 25.3905 | 88.954 | 86.896 | 70.92 | *P-1* | 94 |
| **BIDHIG** | $[Nd_2(C_{12}H_6N_2O_4)_3(H_2O)_4]$ | 2,2′-Bipyridine-4,4′-dicarboxylate | $3^{0,1}$ | 6.0294 | 10.9061 | 13.8392 | 101.486 | 93.799 | 100.933 | *P-1* | 95 |
| FATSAV | $[Tm_2(C_7H_3NO_4)_3(H_2O)_3] \bullet (H_2O)$ | 3,5-Pyridinedicarboxylate | $3^{0,2}$ | 14.579 | 11.193 | 14.839 | 90 | 94.009 | 90 | $P2_1/n$ | 96 |
| FIJFUA | $[Tb_2(C_{18}H_8O_4)_3C_3H_7NO)_2(H_2O)_4] \bullet 2(C_3H_7NO)$ | 2,7-Pyrenedicarboxylate | $3^{0,1}$ | 9.8294 | 12.1098 | 14.6279 | 109.75 | 103.613 | 100.14 | *P-1* | 97 |
| **IBEXUJ** | $[Pr(C_{14}H_8O_4)_2(H_2O)_2]$ | 4,4′-Biphenyldicarboxylate | $3^{0,1}$ | 27.698 | 8.673 | 9.939 | 90 | 90 | 90 | *Pbcn* | 98 |
| MEWNAD | $[La_2(C_{12}O_{12})(H_2O)_6]$ | Benzenehexacarboxylate | $3^{0,1}$ | 13.642 | 6.769 | 10.321 | 90 | 90 | 90 | *Pnnm* | 99 |
| NACXIZ | $[La(C_5H_2N_2O4)(H_2O)_2]$ | 2,6-Dioxo-1,2,3,6-tetrahydropyrimidine-4-carboxylate | $3^1$ | 10.221 | 16.976 | 4.6996 | 90 | 90 | 90 | $Pna2_1$ | 100 |

*(Continued)*

TABLE II
*(Continued)*

| Code | Compound | Linker | Dim | a | b | c | α | β | γ | Space Group | Reference |
|---|---|---|---|---|---|---|---|---|---|---|---|
| **PIQSOX** | $[Ce_2(C_{10}H_4O_8)_2(H_2O)_2]$ | 1,2,4,5-Benzenetetracarboxylate | $3^{0,2}$ | 6.419 | 9.414 | 9.604 | 88.49 | 74.68 | 76.68 | *P-1* | 101 |
| UJEFOE | $[Yb_2(C_{12}H_6O_4)_3H_2O)] \bullet (H_2O)$ | 2,6-Naphthalenedicarboxylate | $3^{0,1}$ | 23.2842 | 8.2568 | 17.501 | 90 | 97.836 | 90 | *C2/c* | 102 |
| **QACTUJ** | $[Tb_2(C_8H_4O_4)_3(H_2O)_4]$ | 1,4-Benzenedicarboxylate | $3^{0,1}$ | 6.142 | 10.0694 | 10.0956 | 102.247 | 91.118 | 101.518 | *P-1* | 103 |
| XULQAW | $[Tb_2(C_{12}H_6O_4)_3(H_2O)_2] \bullet (H_2O)$ | 2,6-Naphthalenedicarboxylate | $3^{0,3}$ | 17.1005 | 15.245 | 24.969 | 90 | 106.093 | 90 | $P2_1/n$ | 104 |
| **XUWVUG** | $[Eu_3(C_9H_7O_6)] \bullet 3(H_2O)$ | 1,3,5-Benzenetricarboxylate | $3^2$ | 17.64 | 3.689 | 12.385 | 90 | 91.247 | 90 | $P2_1$ | 105 |
| IYEYOA01 | $[Nd(C_8H_6NO_2)_3H_2O]$ | *trans*-3-(3-Pyridyl)acrylate | $1^1$ | 6.2543 | 12.7228 | 15.635 | 112.093 | 90.608 | 94.739 | *P-1* | 106 |

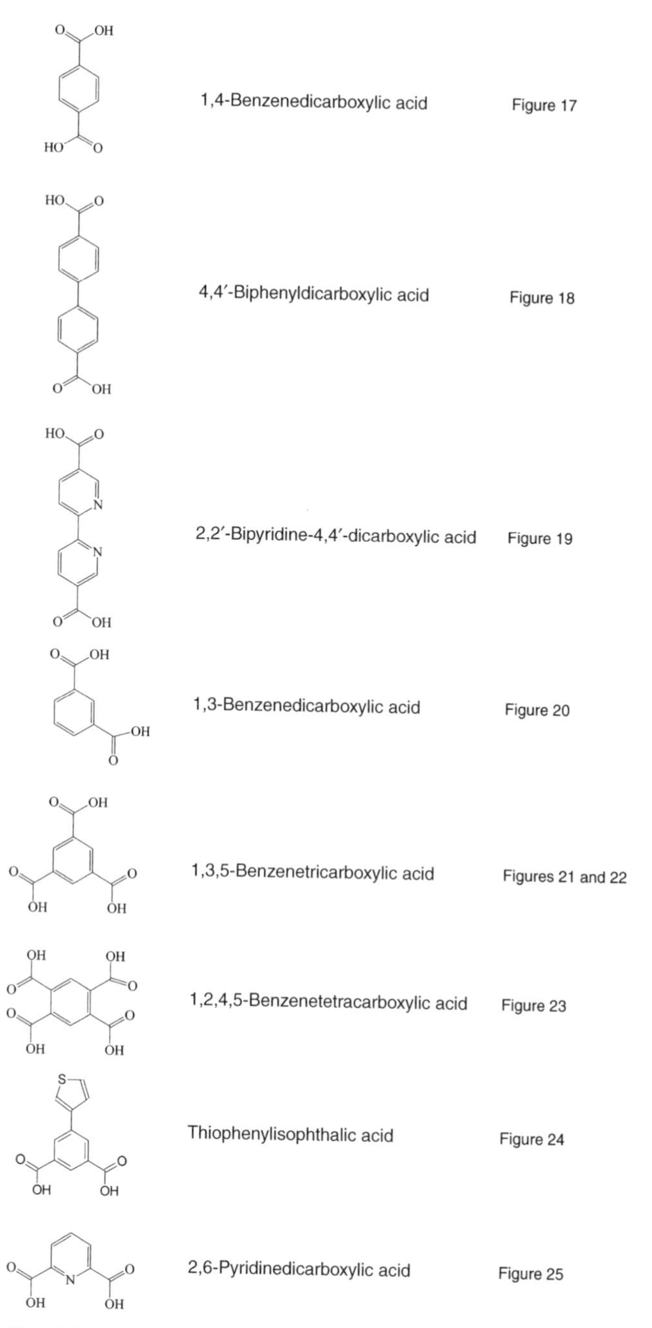


Figure 16. Pictorial representation of the linkers used in the aromatic-based CPs described in the corresponding figure.

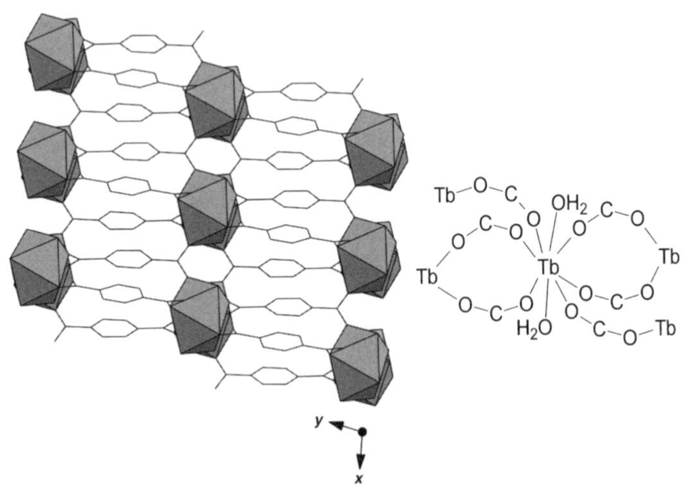


Figure 17. The structure of QACTUJ $[Tb_2(C_8H_4O_4)_3(H_2O)_4]$.

An introductory example of a Ln-aromatic MOF constructed from 1,4-benzene dicarboxylate (BDC) is QACTUJ $[Tb_2(C_8H_4O_4)_3(H_2O)_4]$ (103) (Fig. 17). In this material, each $Tb^{3+}$ center is coordinated to six BDC monodentate oxygen anions and two water molecules. The overall coordination of the BDC is bridging bidentate and thus each $Tb^{3+}$ center is polymerized into pseudo-chains (no edge sharing) along [010]. This compound demonstrates a potential for applications in sorption and catalysis in that the framework remains intact after thermal removal of the coordinated water molecules. Further, the $Tb^{3+}$ centers are luminescent upon excitation of the BDC linkers, the lifetimes of which are shown to be a function of bound and/or sorbed species ($H_2O$ vs. $NH_3$). Such behavior may be compared to FEYQUW (above), where a guest molecule served as the sensitizing species.

Tuning the distance between carboxylate functional groups in 1,4-BDC to utilize 4,4′-biphenyldicarboxylate (BPDC) results in IBEXUJ $[Pr(C_{14}H_8O_4)(C_{14}H_9O_4)(H_2O)_2]$ (Fig. 18) (98). Each 4,4′-BPDC linker is in bridging bidentate coordination at one end and monodentate coordination at the other, and displays a less common terminal carboxyl oxygen. Like QACTUJ, this structure displays pseudo-chains along [100], wherein the monomeric $Pr^{3+}$ centers are bridged by carboxylate groups. Cross-linking of the BPDC groups to the $Pr^{3+}$ centers results in an overall 3D topology. The $Eu^{3+}$ analogue of IBEXUJ displays characteristic emission features.

These structural features continue with BIDHIG $[Nd_2(C_{12}H_6O_4N_2)_3(H_2O)_4]$ (95) (Fig. 19), wherein each $Nd^{3+}$ center is coordinated to eight oxygen atoms, two of which are water molecules. The linker is deprotonated 2,2′-bipyridine-4,4′-

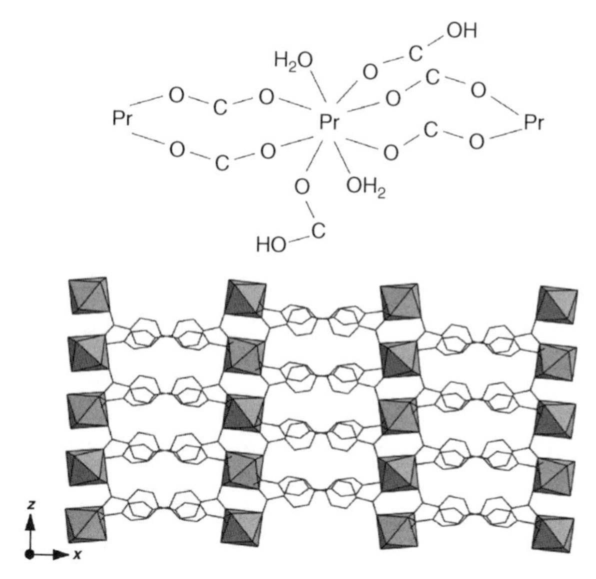


Figure 18. The structure of IBEXUJ $[Pr(C_{14}H_8O_4)(C_{14}H_9O_4)(H_2O)_2]$.

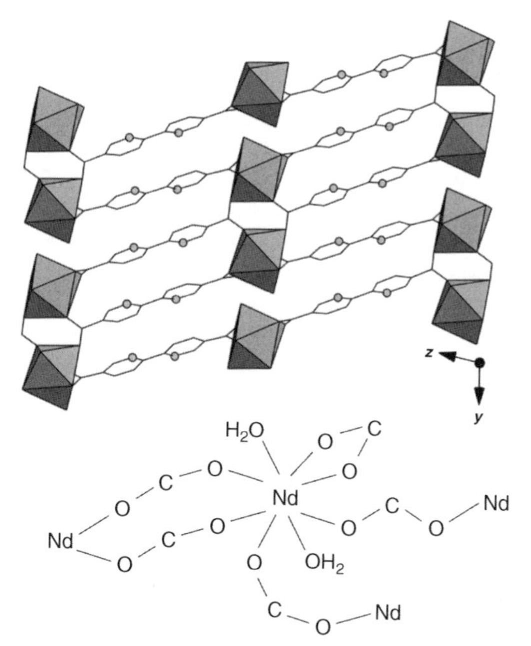


Figure 19. The structure of BIDHIG $[Nd_2(C_{12}H_6O_4N_2)_3(H_2O)_4]$. The linker molecules are 2,2′-bipyridine-4,4′-dicarboxylate groups, and thus gray spheres are N atoms.

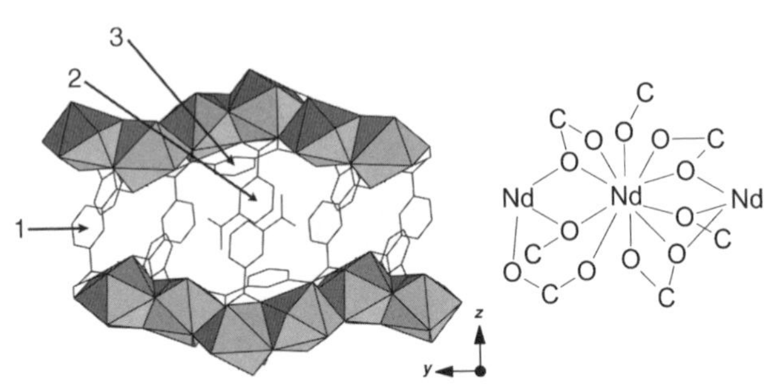


Figure 20. The structure of EHEYUL $[Nd(C_8H_4O_4)_3(C_8H_5O_4)(H_2O)]$. Three distinct coordination modes of 1,3-BDC groups are shown. Type 1 connects the chains of $NdO_{10}$ polyhedra; Type 2 decorates the chains as one group remains unbound; Type 3 is an intrachain coordination.

dicarboxylic acid and is found in both bidentate and bridging bidentate coordination. The $Nd^{3+}$ centers remain monomeric, yet are bridged into pseudo-dimers at the bridging bidentate ends of the linker. Like many materials in this family, structural rearrangement, yet retention of crystallinity, is observed upon dehydration.

In addition to changing the distance between carboxylate functional groups as in the above two examples, much work has been done where the positions of the carboxylate moieties are altered around a BDC ring. The compound EHEYUL $[Nd(C_8H_4O_4)_3(C_8H_5O_4)(H_2O)]$ (88) (Fig. 20) is a 2D CP that contains chains of alternating edge- and face-shared $NdO_{10}$ polyhedra linked through 1,3-BDC groups.

Increasing the number of carboxylate groups, as with 1,3,5-benzenetricarb-oxlic acid (BTC) as a linker, results in AKOVUR $[Gd(C_9H_3O_6)(H_2O)_3]\cdot 1.5H_2O$ (Fig. 21) (85). This structure contains monomeric $Gd^{3+}$ polyhedra, each bound to three bidentate 1,3,5-BTC groups and three $H_2O$ molecules to give an overall ninefold coordination. This gives rise to a 2D structure that exhibits paramagnetic behavior over the range 2–300 K.

The compound XUWVUG $[Eu_3(C_9H_3O_6)(OH)(H_2O)]\cdot 3H_2O$ (Fig. 22) (105) is another compound constructed from 1,3,5-BTC linkers. In this material, each $EuO_9$ polyhedron is edge and face shared to form a 2D slab SBU. The layers are in turn linked through either bidentate or monodentate geometry on the BTC linkers to form an overall 3D topology. When comparing this material to AKOVUR, both of which contain 1,3,5-BTC, it is interesting to contrast the degree of polymerization in each structure. The compound AKOVUR was synthesized from a gel preparation, whereas XUWVUG was prepared at 140°C. Increased polymerization (i.e., monomers to slabs), may be related to the synthesis temperature.

Another example of benzene carboxylate-containing materials is PIQSOX $[Ce_2(C_{10}H_2O_8)(C_{10}H_4O_8)(H_2O)_2]$ (Fig. 23) (101). This compound uses 1,2,4,5-benzenetetracarboxylate linkers to assemble a 3D structure consisting of $Ce_2O_{16}$ dimers that are connected through bridging bidentate carboxylate groups. Other

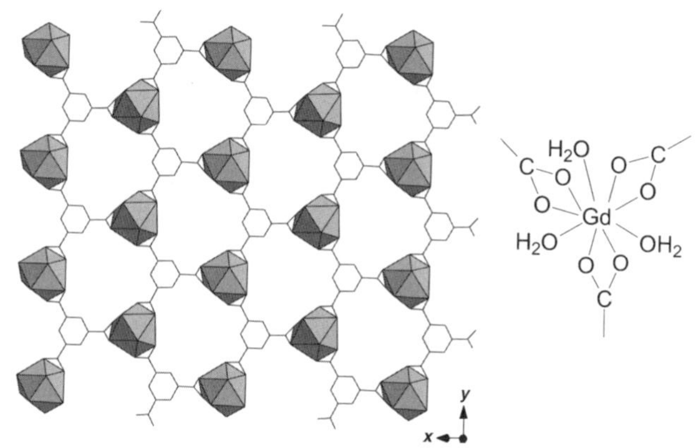


Figure 21. The structure of AKOVUR $[Gd(C_9H_3O_6)(H_2O)_3] \cdot 1.5H_2O$. Layers are formed when monomeric $GdO_6(H_2O)_3$ polyhedra are polymerized through 1,3,5-BTC groups. Interlayer water molecules are not shown.

coordination modes in this structure include bidentate, monodentate, and bridging tridentate.

Thiophene substituted isophthalic acid has been used as a linker with the goal of enhancing Ln emission, as well as introducing a polymerizable moiety for additional coordination and functionalization. The compound WAKZUE $[Gd_2(C_{12}H_6O_4S)_3(H_2O)_6] \cdot 6H_2O$ (Fig. 24) (24) contains monomeric $GdO_8$

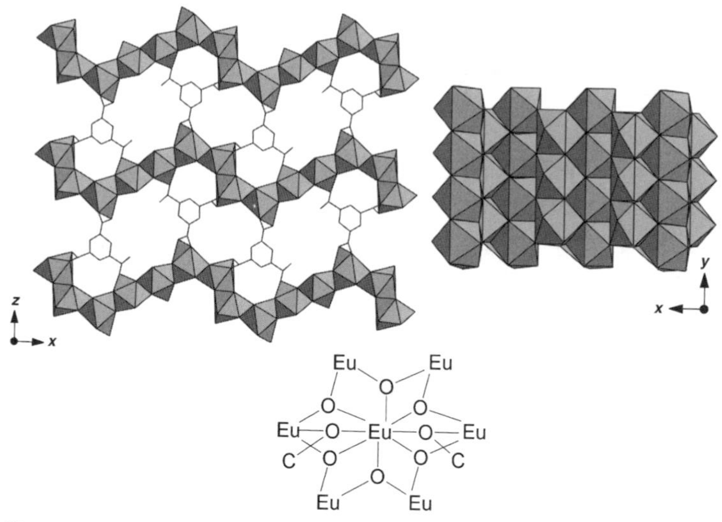


Figure 22. The structure of XUWVUG $[Eu_3(C_9H_3O_6)(OH)(H_2O)] \cdot 3H_2O$. The $EuO_9$ polyhedra are edge or face shared to form slabs that are in turn bridged through 1,3,5-BTC linkers. In this figure, neither the coordinated nor interlayer water molecules are shown.

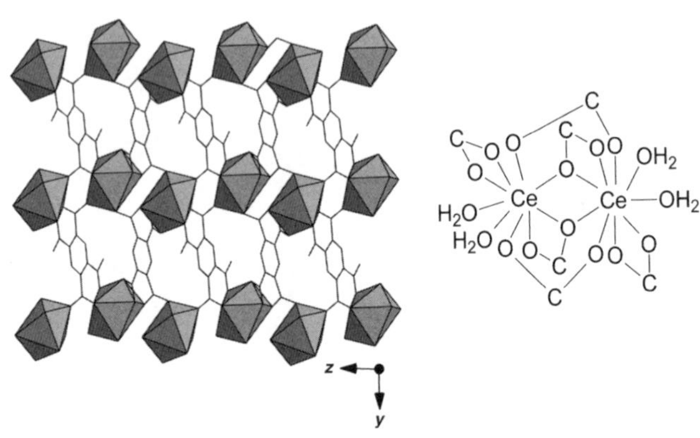


Figure 23. The structure of PIQSOX $[Ce_2(C_{10}H_2O_8)(C_{10}H_4O_8)(H_2O)_2]$. 1,2,4,5-BTC linkers polymerize $CeO_7(H_2O)_2$ monomers into a 3D structure.

polyhedra polymerized through one monodentate, one bidentate, and two bridging bidentate carboxylate groups from four unique isophthalate molecules. There are three water molecules on each $Gd^{3+}$ center as well and no coordination to the sulfur on the thiophene is observed.

Pyridinedicarboxylic acids have been effective linker molecules in the construction of CPs and MOFs as well. There are many examples in *d*-block

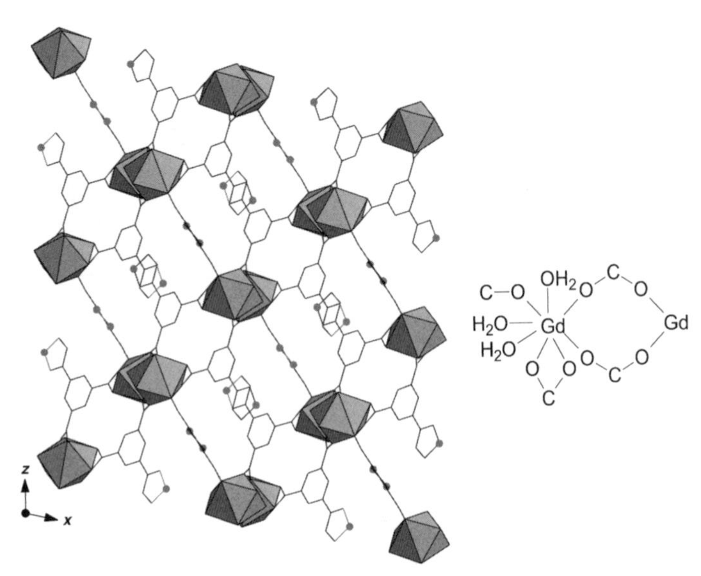


Figure 24. The structure of WAKZUE $[Gd_2(C_{12}H_6O_4S)_3(H_2O)_6]\cdot 6H_2O$. Uncoordinated water molecules are not shown. Gray spheres are S atoms on the isophthalate linkers.

and actinide systems and several from the lanthanides, which are given in Table II. Part of the appeal of this linker is its heterofunctional nature (e.g., the presence of two distinct functional groups: carboxylate and pyridyl). The positions of the carboxylate groups may be altered as a design variable and this combination of features leads to a range of topologies, including those containing *d-/f-* metal pairings (below). A strictly Ln example is ARUKIH: $[Pr(C_7H_3O_4N)(C_7H_4O_4N)(H_2O)_2]\cdot4H_2O$ (Fig. 25) (81). This material consists of $PrO_5(H_2O)_2N_2$ monomers that are polymerized into chains (along [001]) through bridging bidentate carboxylate groups. The linker molecule is 2,6-pyridine dicarboxylic acid and is in tridentate coordination with each $Pr^{3+}$ center.

Figure 26 is a summary of observed Ln-carboxylate coordination geometries observed in the materials discussed above. Note the variety of connectivity and this skeletal presentation is common in the literature of Ln-MOFs and CPs. Given the complexity of many structures, as well as the spherical nature of the Ln coordination sphere, many publications use these representations to distinguish between compounds in a series.

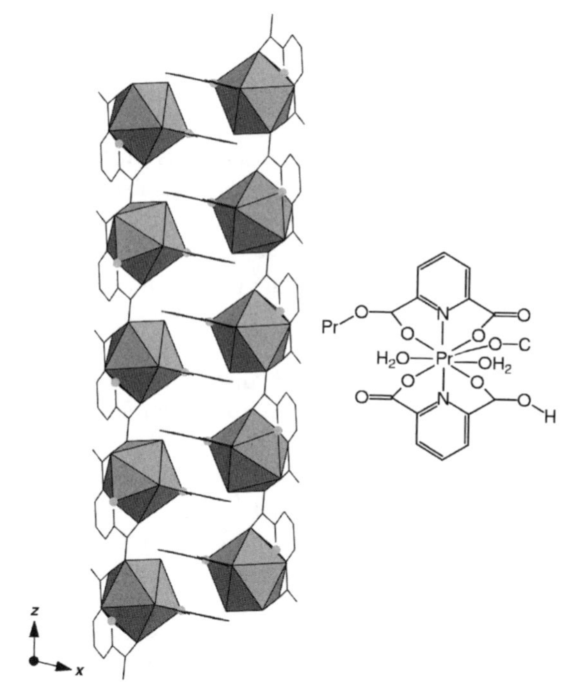


Figure 25. The structure of ARUKIH $[Pr(C_7H_3O_4N)(C_7H_4O_4N)(H_2O)_2]\cdot4H_2O$. 2,6-Pyridine dicarboxylate linkers are in tridentate coordination with the $Pr^{3+}$ centers and polymerization to form chains is observed through bridging carboxylate groups.

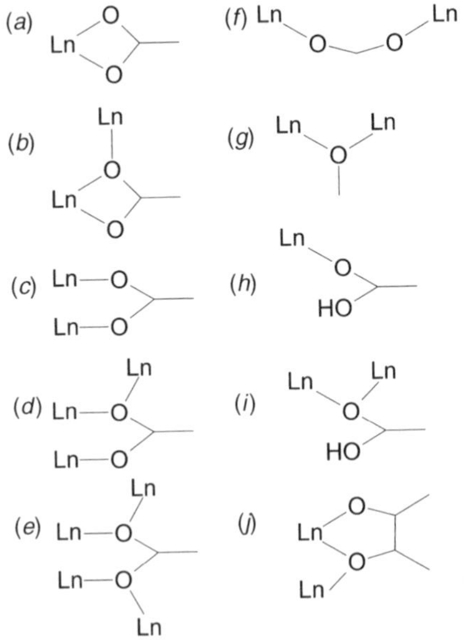

Figure 26. A summary of Ln-carboxylate coordination geometries seen in the structures discussed herein.

## C. Heterometallic (*d*-/*f*-) Systems

The potential to produce heterometallic materials, in particular those consisting of *d*- and *f*- metal pairings, is attractive for a number of reasons. Coupling of these metals within controlled architectures has the potential to influence magnetic and conductivity properties as assembly can impart steric constraints using various ligands that effectively isolate metal centers from one another at fixed and ideally, controlled distances (107–113). From a purely structural point of view, however, the appeal can be though of in terms of introducing another building unit (the *d*-metal) with a geometry distinct from that of the lanthanide. Further, we have made the argument thus far that control over Ln geometry is a challenge, yet *d*-block transition metals often display relatively predictable coordination preferences, thus imparting an additional degree of design potential. Lastly, these ions may be distinguished from one another somewhat easily based on their hard–soft acid–base properties. The Ln ions may be thought of as harder than their *d*-block counterparts, and thus appropriate choice of a heterofunctional linker molecule can be utilized to assemble both species into extended architectures that contain characteristics of each metal center. Table III includes a number of materials utilizing this approach, and Fig. 27 contains several linker molecule examples.

The first compound to be described, AJAYIT $[Cu_3(C_4H_5O_4N)_6Er_2]\cdot 7H_2O$ (121), utilizes iminodiacetate as its linker molecule and displays $Er^{3+}$–$Cu^{2+}$

TABLE III
Heterometallic (*d-/f-*) Systems

| Code | Compound | Linker | Dim | a | b | c | α | β | γ | Space Group | Reference |
|---|---|---|---|---|---|---|---|---|---|---|---|
| GEGWOE | $[EuHg_2(C_4H_6NO)_4(NO_3)_3]$ | 2-Pyrrolidone | $1^{0,1}{}_{0,1}$ | 9.879 | 14.594 | 18.211 | 90 | 93.58 | 90 | *P2₁/c* | 114 |
| **QUPYOP** | $[Er_2Cu_3(C_7H_3NO_4)_6(H_2O)_{12}] \bullet 4(H_2O)$ | 2,5-Pyridinedicarboxylate | $1^{0,1}{}_{0,1}$ | 7.3601 | 13.5424 | 15.1256 | 72.582 | 76.263 | 80.152 | *P-1* | 108 |
| BAJQAF | $[Ce_2Fe_4(C_7H_4O_3)_8(C_{10}H_8N_2)_4(H_2O)_{11}] \bullet 2(C_7H_5O_3) \bullet (C_2H_6O) \bullet 3(H_2O)$ | Salicylate, 2,2′-bipyridyl | $2^{0,1}{}_{0,1}$ | 13.8916 | 19.8401 | 22.2319 | 86.592 | 78.871 | 83.81 | *P-1* | 115 |
| BEGKAA | $[GdCu_2(C_{10}H_{18}N_4O_2)_2(H_2O)_3Co(CN)_6] \bullet 7(H_2O)$ | 1,4,8,11-Tetraazacyclotetradecane-2,3-dionate | $2^{0,1}{}_{0,1}$ | 27.413 | 10.13 | 19.067 | 90 | 129.81 | 90 | *C2/c* | 116 |
| HAGWIV | $[Dy_2Cu_3(C_7H_6N_2O_6)_3(H_2O)_7] \bullet 16(H_2O)$ | 1,3-Propylenebis(oxamate) | $2^{0,1}{}_{0,1}$ | 18.637 | 16.025 | 19.685 | 90 | 112.37 | 90 | *P2₁/n* | 117 |
| PUTZUZ | $[Gd_2Co_3(C_{10}H_{12}N_2O_8)_3(H_2O)_{11}] \bullet 12(H_2O)$ | Ethylenediamine-*N,N,N′, N′*-tetraacetate | $2^{1}{}_{1}$ | 15.799 | 14.603 | 14.69 | 90 | 116.64 | 90 | *C2* | 118 |
| VANPAC | $[La_2(C_{12}H_8FeO_4)_3(CH_3OH)_4]$ | Ferrocene-1,1′-dicarboxylate | $2^{0,1}$ | 9.5149 | 10.3318 | 11.0945 | 92.139 | 109.367 | 97.289 | *P-1* | 119 |
| VUCLOU | $[Sm_2Cu_3(C_7N_2O_6)_3(H_2O)_9] \bullet 14(H_2O)$ | 1,3-Propylenebis(oxamate) | $2^{0,1}{}_{0,1}$ | 16.076 | 16.079 | 21.256 | 90 | 90 | 90 | *Pcc2* | 120 |
| **AJAYIT** | $[Er_2Cu_3(C_4H_5NO_4)_6] \bullet 7(H_2O)$ | Iminodiacetate | $3^{0,1}{}_{0,1}$ | 13.3169 | 13.3169 | 14.297 | 90 | 90 | 120 | *P3c1* | 121 |
| FAHTUE | $[Eu_2Mn_3(C_7H_3NO_4)_6(H_2O)_6] \bullet 6.5(H_2O)$ | 2,6-Pyridinedicarboxylate | $3^{0,1}{}_{0,1}$ | 15.288 | 15.288 | 15.663 | 90 | 90 | 120 | *P6/mcc* | 122 |

*(Continued)*

TABLE III
*(Continued)*

| Code | Compound | Linker | Dim | a | b | c | α | β | γ | Space Group | Reference |
|---|---|---|---|---|---|---|---|---|---|---|---|
| FIHZOM | $[LaCu_2(C_7H_3NO_4)_4(H_2O)_4] \bullet 9(H_2O)$ | 2,6-Pyridinedicarboxylate | $3^{0,1}{}_{0,1}$ | 21.06 | 17.305 | 15.15 | 90 | 131.239 | 90 | *C2/c* | 123 |
| ILEBOQ | $[Tb_2(C_{12}H_8FeO_4)_3(H_2O)_4] \bullet 4(H_2O)$ | Ferrocene-1,1′-dicarboxylate | $3^{0,1}$ | 11.302 | 10.32 | 17.133 | 90 | 91.48 | 90 | *P2/n* | 124 |
| MOBMUL | $[Gd_2Cu(C_7H_3NO_4)_4(H_2O)_6]$ | 2,4-Pyridinedicarboxylate | $3^{0,1}{}_{0,1}$ | 6.1066 | 17.7753 | 16.1658 | 90 | 99.813 | 90 | $P2_1/c$ | 125 |
| NERNIH | $[Dy_2Co_2(C_{10}H_8N_2O)_8(NO_3)_4(H_2O)_2(SCN)_4] \bullet (CoC_4N_4S_4) \bullet 3(CH_4O) \bullet 4(H_2O)$ | 1-(4′-Pyridyl)pyridin-4-one-isothiocyanate | $3^{0,1}{}_{0,1}$ | 34.565 | 17.06 | 12.646 | 90 | 99.88 | 90 | *C2* | 126 |
| NIJZIP | $[Nd_2Cu_3(C_4H_4O_5)_6(H_2O)_6] \bullet 3(H_2O)$ | Diglycolate | $3^{0,1}{}_{0,1}$ | 15.091 | 15.091 | 14.71 | 90 | 90 | 120 | *P6/mcc* | 127 |
| OBUBIW | $[La_2Cu_3(C_4H_4O_5)_6(H_2O)_4] \bullet 3(H_2O)$ | Diglycolate | $3^{0,2}{}_{0,1}$ | 15.522 | 15.522 | 13.712 | 90 | 90 | 120 | $P\text{-}6_2c$ | 128 |
| PAKLAO | $[La_2Cu(C_4O_4)_4(H_2O)_{16}] \bullet 2(H_2O)$ | Squarate | $3^{0,1}{}_{0,1}$ | 6.78 | 32.311 | 8.162 | 90 | 111.55 | 90 | $P2_1/c$ | 129 |
| POPXIB | $[NdNi(C_{10}H_{12}N_2O)_4(NO_3)_3(SCN)_2] \bullet 0.25(H_2O)$ | 4-Picolylpyrrolidin-2-one | $3^{0,1}{}_{0,1}$ | 9.722 | 16.828 | 17.332 | 69 | 83.05 | 81 | *P-1* | 130 |
| **YATRER** | $[Cu_3(C_8H_6NO_2)_6Nd_2(NO_3)_6]$ | *trans*-3-(3-Pyridyl) acrylate | $3^{0,1}{}_{0,1}$ | 14.130 | 14.130 | 49.480 | 90 | 90 | 120 | *R-3c* | 106 |

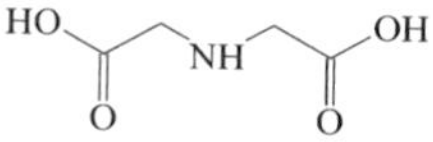

Iminodiacetic acid Figure 28

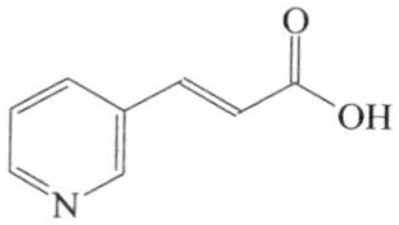

*trans*-3-(3-pyridyl)acrylic acid Figure 29

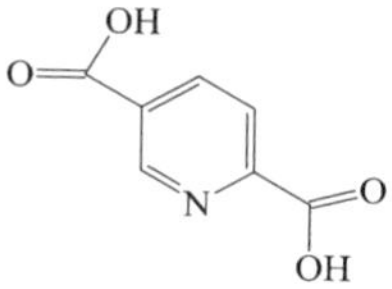

2,5-Pyridinedicarboxylic acid Figure 30

Figure 27. Pictorial representation of the linkers used in the mixed LN-TM-based CPs described in the corresponding figure.

segregation based on coordination preferences (Fig. 28). The $Er^{3+}$ centers are in ninefold coordination with three monodentate and three bridging tridentate carboxylate groups. In contrast, the $Cu^{2+}$ site is in octahedral coordination where the equatorial plane is occupied by two imino groups and two oxygen atoms of the iminodiacetate that effectively chelate the metal center. These

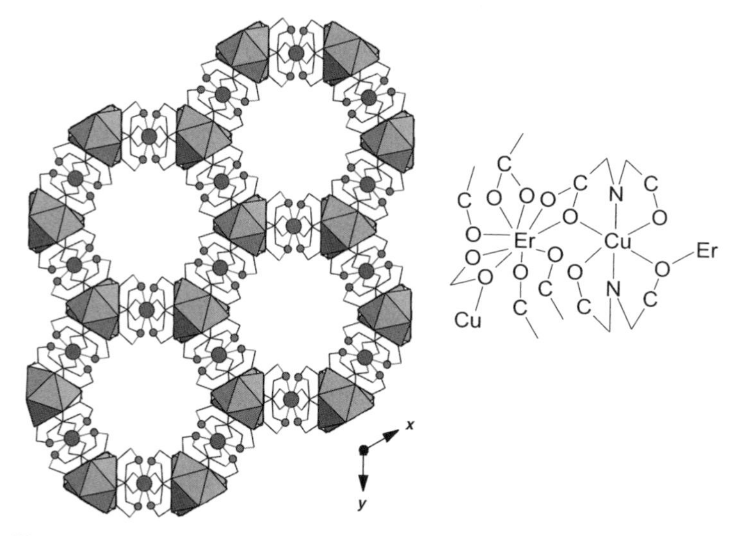


Figure 28. The structure of AJAYIT $[Cu_3(C_4H_5O_4N)_6Er_2] \cdot 7H_2O$. This overall 3D topology results from polymerization of two distinct coordination sites: $Er^{3+}$ and $Cu^{2+}$. These are segregated based on coordination preferences. Large gray spheres are $Cu^{2+}$, whereas small gray spheres are N atoms.

oxygen atoms are in bridging bidentate coordination with the $Cu^{2+}$ and neighboring $Er^{3+}$ sites. The axial positions are occupied by oxygen atoms that are in bridging tridentate coordination with $Er^{3+}$. Iminodiacetate and indeed nitrilotriacetate are common ligand anions in Ln chemistry (see, e.g., DIFQOY and MAGXOI in Table I, respectively). For the latter, N is often coordinated to the Ln center by default considering the proximity to the carboxylate groups and is an effective chelating ligand as shown in ARUKIH (above). For structures containing transition metals, however, this geometry is not as common, as the *d* metals will tend to bind to the N-donor sites, either exclusively or in addition to the oxygen groups, whereas the Ln are primarily coordinated to O-donor sites.

Another material to demonstrate this hard–soft acid–base preferences to assemble *d-/f-* metal architectures is YATRER $[Cu_3(C_8H_6NO_2)_6Nd_2(NO_3)_6]$ (Fig. 29) (106). This compound uses *trans*-3-(3-pyridyl)acrylic acid (3-HPYA)

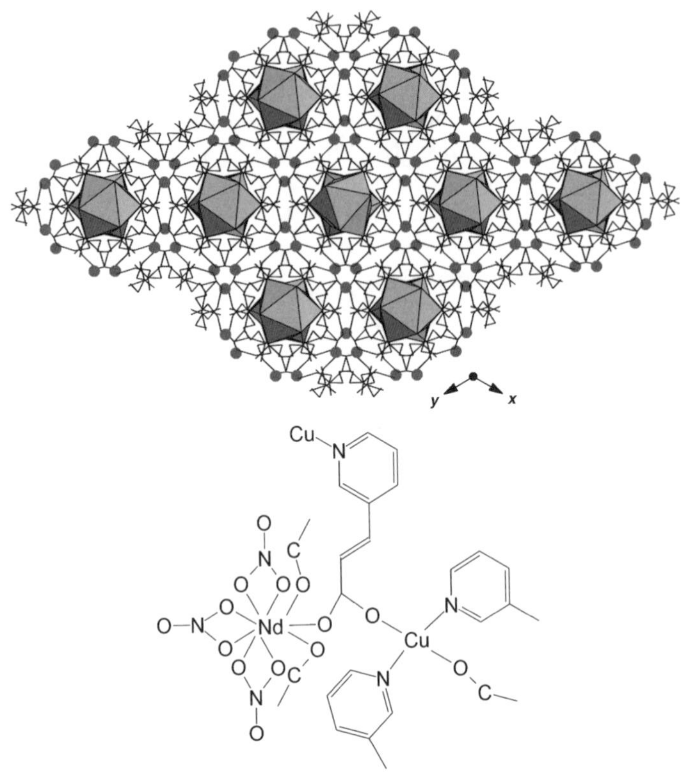


Figure 29. The structure of YATRER $[Cu_3(C_8H_6NO_2)_6Nd_2(NO_3)_6]$. Monomeric $NdO_9$ polyhedra are connected to $CuN_2O_2$ monomers through *trans*-3-(3-pyridyl)acrylate groups. An overall 3D topology results from this connectivity. Gray spheres are $Cu^{2+}$ sites.

to assemble a 3D structure, where $Nd^{3+}$ is in ninefold coordination with three monodentate 3-HPYA linkers and three bidentate nitrate groups. The $Cu^{2+}$ center is in square-planar geometry with two cis nitrogen atoms from two different 3-HPYA linkers and two monodentate oxygen atoms from two additional linkers. The homometallic end member of this system, $[Nd(C_8H_6NO_2)_3H_2O]$ (IYEYOA01), was listed in Table I and contains no $Nd^{3+}$–N coordination. The $Eu^{3+}$ analogue of this material demonstrates sensitized luminescence, whereas the heterometallic Eu–Cu material does not, suggesting that the $Cu^{2+}$ may influence or effectively tune the properties of this material.

A final example of these coordination preferences can be found in QUPYOP $[Cu_3(C_7H_3O_4N)_6Er_2(H_2O)_{12}]\cdot 4H_2O$ (Fig. 30) (108). The linkers used here are pyridine-2,5-dicarboxylic acid ($H_2$pydc) and demonstrate the same hard–soft coordination seen above. Each $Er^{3+}$ center is coordinated to eight oxygen atoms, three of which are three distinct monodentate pydc linkers and five from $H_2O$ molecules. The two unique $Cu^{2+}$ sites are square planar and bound to two bidentate N-/O- pairings from two pydc linkers in trans configuration. One of these sites has an additional $H_2O$ molecule coordinated axially. This connectivity gives rise to an overall 1D topology and the magnetic behavior for the $Gd^{3+}$ analogue has been investigated (108).

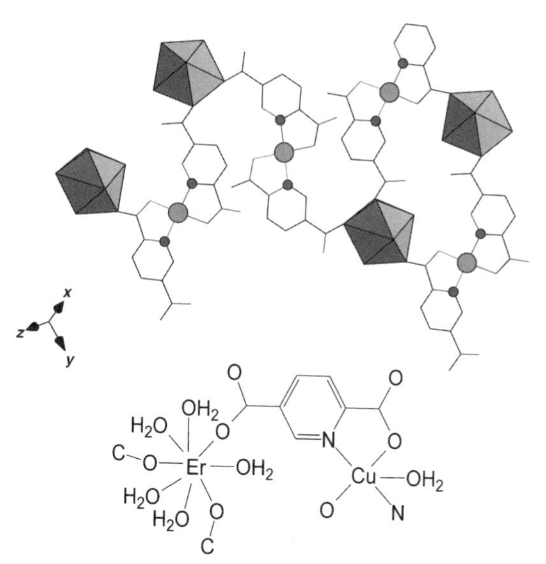


Figure 30. The structure of QUPYOP $[Cu_3(C_7H_3O_4N)_6Er_2(H_2O)_{12}]\cdot 4H_2O$. Large gray spheres are $Cu^{2+}$ sites, whereas smaller spheres are N atoms. Segregation of metal centers is again seen based on coordination preferences.

### D. Structures With Ln–O–N Linkages

The use of bipyridine-*N,N′*-dioxide ligands (Table IV) has been explored extensively by Champness and co-workers (23,131). The rationale for use of these types of ligands is similar to what we have been seeing so far throughout this chapter: A complementarity of the O-donor group for the hard Lewis acid Ln ion sites. Further, the O-donor sites are somewhat extended off of the body of the linker molecule (Fig. 31) and thus can minimize steric crowding. This ultimately results in highly connected frameworks derived from (e.g.) eight-coordinated lanthanide nodes. Other groups have explored these linkers as well and a representative example is AXINAW $[La_2(C_{12}H_{12}N_2O_2)_4(H_2O)_2(CF_3SO_3)_2]\cdot[2(CF_3O_3S)\ 6(H_2O)]$ shown in Fig. 32 (132). This material actually contains two linkers: 1,2-bis(4-pyridyl)ethane-*N,N′*-dioxide and isonicotinc acid *N*-oxide. Each $La^{3+}$ ion is coordinated to nine oxygen atoms from one $CF_3SO_3^-$ anion, two aqua ligands, two isonicotinic *N*-oxide and three *N,N′*-dioxide linkers. The isonicotinic groups are interesting in that they exhibit coordination through both the *N*-oxide and carboxylate functional groups. The $La^{3+}$ centers are bridged into dimers through a less commonly observed μ2 briding of the *N*-oxide groups of the *N,N′*-dioxide ligand. Ultimately, a 2D structure is realized with π stacking interactions between the layers.

### E. Structures With Multiple Linkers

#### 1. *Oxalate and Second Linker*

Table V presents a number of compounds that utilize two linker molecules (Fig. 31) to promote extended topologies, one of which in each being the oxalate anion. A component of the rational of this approach is to make use of small and somewhat predictable linker molecules (the oxalate) to direct local geometry, and then pair this entity with a separate linker molecule. For example, many Ln-oxalates display honeycomb layers like that seen in AOXNDH (Fig. 6) and the challenge is to then connect the layers (146). Also note, however, that many of these materials that contain oxalate linkers do so somewhat serendipitously. Of the structures listed in Table V, ACOXEV, EBAMUQ, JOVDON, JOVDUT, XUPBUF, and XUPCAM were all synthesized without oxalate precursors and instead display *in situ* oxalate formation and subsequent complexation. *In situ* ligand synthesis will be visited again in Section V.F (see sulfonates) and is not uncommon in hydrothermal Ln systems.

A representative mixed oxalate–second linker compound is ACOXEV $[Tb_4(C_2O_4)_2(C_5H_6O_4)_4(H_2O)_2]$ (Fig. 33) (140). The structure may be thought of as containing chains of two distinct $Tb_2O_{15}$ dimers that alternatively share edges and faces to propagate along [010]. These chains are then tethered through

TABLE IV
Ln–O–N

| Code | Compound | Linker | Dim | a | b | c | α | β | γ | Space Group | Reference |
|---|---|---|---|---|---|---|---|---|---|---|---|
| BENCED | $[Yb_2(C_{12}H_{10}N_2O_2)_2(CH_3O)_2(NO_3)_6]$ | 1,2-Bis(4′-pyridyl) ethane-*N,N*′-dioxide | $1^{0,1}$ | 11.089 | 7.8615 | 21.8828 | 90 | 101.783 | 90 | *P*2[1]/*c* | 133 |
| OBIFUB | $[Eu(C_{10}H_8N_2O_2)(C_8F_3H_4S)_3] \bullet (CH_4O)$ | 4,4′-Bipyridine-*N,N*′-dioxide | $1^{0,1}$ | 10.813 | 10.8971 | 18.628 | 95.161 | 94.963 | 111.134 | *P-1* | 134 |
| **AXINAW** | $[La_2(C_{12}H_{12}N_2O_2)_4(H_2O)_2(CF_3SO_3)_2] \bullet 2(CF_3O_3S) \bullet 6(H_2O)$ | 1,2-Bis(4′-pyridyl) ethane-*N,N*′-dioxide; isonicotinate-N-oxide | $2^{0,2}$ | 10.5878 | 12.6587 | 14.5687 | 98.778 | 96.343 | 103.102 | *P-1* | 132 |
| EFOKUF | $[Tb(C_{10}H_8N_2O_2)(CH_3OH)(NO_3)_3]$ | 4,4′-Bipyridine-*N,N*′-dioxide | $2^{0,1}$ | 15.228 | 8.3001 | 13.996 | 90 | 92.705 | 90 | *C2/c* | 135 |
| IXEDIY | $[La_4(C_{10}H_8N_2O_2)_{15}(CH_3O)_2] \bullet 12(CF_3O_3S) \bullet 4(CH_4O) \bullet 7.2(H_2O)$ | 4,4′-Bipyridine-N,N′-dioxide | $2^{0,1}$ | 16.9322 | 18.5089 | 18.7708 | 90.056 | 111.528 | 102.21 | *P-1* | 136 |
| WUHJOY | $[La_2(C_{12}H_{12}N_2O_2)_2(NO_3)_6(CH_3OH)_2] \bullet 2(CH_4O) \bullet 2(CH_2Cl_2)$ | 1,2-Bis(4-pyridyl) ethane-*N,N*′-dioxide | $2^{0,1}$ | 9.609 | 10.801 | 13.566 | 78.38 | 82.011 | 89.895 | *P-1* | 137 |
| QIBTAW | $[La_4(C_{10}H_8N_2O_2)_{10}(CH_3OH)_{10}(Cl)_3] \bullet Cl \bullet 8(C_{24}H_{20}B) \bullet 22(CH_4O)$ | 4,4′-Bipyridine-*N,N*′-dioxide | $3^{0,1}$ | 15.134 | 47.621 | 22.431 | 90 | 91.947 | 90 | *Pc* | 138 |

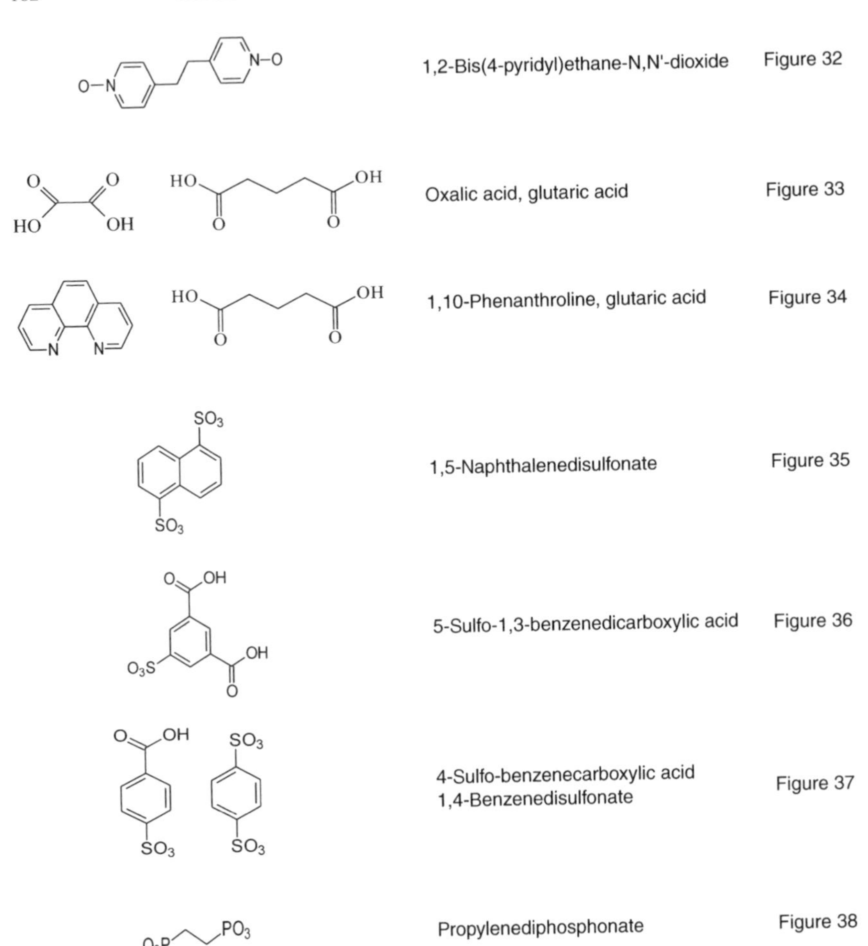


Figure 31. Pictorial representation of the linker molecules used in Ln–O–N containing materials, as well as those using multiple linkers, sulfonates, and phosphonates. Figures where examples can be found are listed.

bridging tridentate and bridging bidentate glutarate groups (approximately along [001]) to form an overall 3D structure. The oxalate groups are in bidentate coordination with two distinct $Tb^{3+}$ ions, yet do not contribute to the overall dimensionality as their connectivity is strictly *intra*chain.

## 2. *The Ln–O Linkages With Chelating Ln–N Coordination*

As discussed in the introduction, a challenge to directing the formation of Ln-CPs and MOFs is the high coordination number and spherical coordination

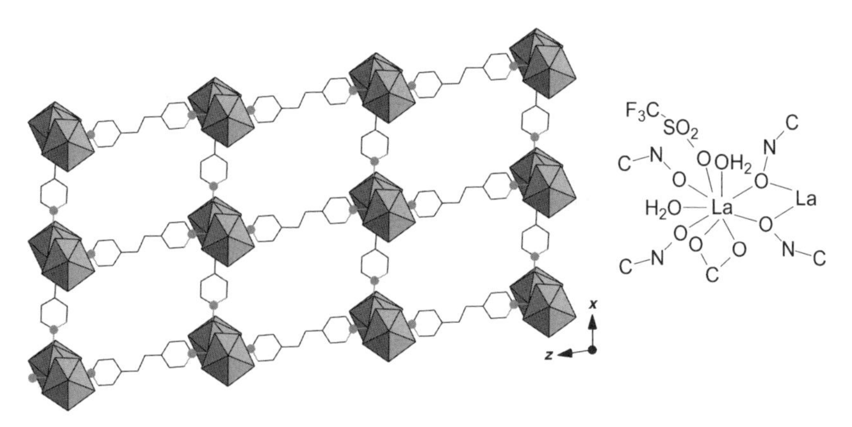


Figure 32. The structure of AXINAW $(La_2(C_{12}H_{12}N_2O_2)_4(H_2O)_2(CF_3SO_3)_2]\cdot[2(CF_3O_3S)$ $6(H_2O)]$. For clarity, uncoordinated water molecules and $CF_3O_3S$ groups are not shown. Gray spheres are N atoms.

geometry displayed by Ln metal centers. A dominant theme thus far has been for solvent molecules (primarily $H_2O$ molecules in our examples) to fill out the coordination sphere of a Ln ion after coordination to linker molecules has been satisfied. A way in which to subvert this is to use chelating ligands in addition to the multifunctional linker molecules (Fig. 31). Table VI contains a number of examples of materials exploring this approach. One example, QAFGIO [Ho $(C_5H_6O_4)$(phen)Cl] (151) (Fig. 34) uses a combination of phen and glutaric acid to produce an overall 1D structure consisting of chains of edge shared $(Ho_2O_8N_2Cl_2)$ dimers. Each $Ho^{3+}$ is coordinated to one Cl anion, five oxygen atoms from four distinct glutarate groups in bidentate, bridging bidentate, and bridging tridentate coordination, and two nitrogen atoms from a chelating phen group. The chains propagate along [100], yet experience interchain $\pi$–$\pi$ interactions among the phen groups. Thus these chelating groups force SBU propagation in one direction and are available for secondary interactions themselves. Further, the use of these conjugated ligands in particular may be explored as sensitizing species (157).

## F. Lanthanide-Sulfonates

Linkers with sulfonate functional groups are another class of O-donor species that have been successful in polymerizing Ln ions into extended architectures. This is not surprising considering the similar geometry between a sulfate $(SO_4^{2-})$ anion and the functional groups on sulfonate linkers. Further, the $SO_4^{2-}$ anion has been utilized extensively as a complex anion for the assembly of templated materials and extended, noncoordination polymeric materials (44).

TABLE V
Ln-Oxalate and Second Linker

| Code | Compound | Second Linker | Dim | a | b | c | α | β | γ | Space Group | Reference |
|---|---|---|---|---|---|---|---|---|---|---|---|
| OBENOZ | $[Eu_2(C_8H_3NO_6)_2(C_2O_4)(H_2O)_6] \bullet 4(H_2O)$ | 5-Nitro-1,3-benzenedicarboxylate | $1^{0,1}$ | 7.451 | 9.234 | 11.585 | 73.989 | 71.656 | 79.707 | *P-1* | 139 |
| **ACOXEV** | $[Tb_4(C_5H_6O_4)_4(C_2O_4)_2(H_2O)_2]$ | Glutarate | $3^{1}$ | 9.514 | 9.066 | 19.706 | 90 | 97.896 | 90 | $P2_1$ | 140 |
| EBAMUQ | $[Nd_2(C_{24}H_{25}NO_9)_2(C_2O_4)_4(H_2O)_2] \bullet 12(H_2O)$ | 2-(4-Carboxyphenyloxyethyl) amine | $3^{0,1}$ | 10.888 | 11.638 | 13.793 | 95.367 | 103.259 | 108.532 | *P-1* | 141 |
| FITHUM | $[Gd_2(C_4H_9NO_5P)_2(C_2O_4)_2] \bullet (H_2O)$ | Hydrogen ((carboxylatemethyl) (methyl)ammonium) methylphosphonate | $3^{0,1}$ | 14.004 | 10.7125 | 15.6716 | 90 | 98.036 | 90 | *C2/c* | 142 |
| IJASER | $[Sm_2Co(C_5H_2O_4)_2(C_2O_4)_2(H_2O)_6] \bullet 2(H_2O)$ | 2,6-Dioxo-1,2,3,6-tetrahydropyrimidine-4-carboxylate | $3^{0,1}{}_{0,1}$ | 8.2266 | 9.2895 | 9.7934 | 75.347 | 69.708 | 72.665 | *P-1* | 143 |
| JOVDON | $[Ce_2(C_4O_4)_2(C_2O_4)(H_2O)_8] \bullet 3(H_2O)$ | Squarate | $3^{0,1}$ | 23.825 | 14.044 | 6.673 | 90 | 90 | 90 | *Pbcn* | 144 |
| JOVDUT | $[Eu_2(C_4O_4)_2(C_2O_4)(H_2O)_4]$ | Squarate | $3^{0,2}$ | 9.543 | 9.828 | 9.926 | 90 | 118.63 | 90 | $P2_1/c$ | 144 |
| OBENUF | $[Gd_2(C_8H_4O_4)_2(C_2O_4)(H_2O)] \bullet 2(H_2O)$ | 1,3-Benzenedicarboxylate | $3^{0,1}$ | 13.483 | 12.814 | 7.076 | 90 | 104.188 | 90 | $P2_1/c$ | 139 |

| | | | | | | | | | | | |
|---|---|---|---|---|---|---|---|---|---|---|---|
| OBEPER | $[Tm_2(C_8H_3NO_6)_2(C_2O_4)(H_2O)_4] \bullet 2(H_2O)$ | 5-Nitro-1,3-benzenedicarboxylate | $3^{0,1}$ | 11.211 | 12.851 | 9.689 | 90 | 111.726 | 90 | *P2₁/c* | 139 |
| RIFQIG | $[Ce(CHO_2)(C_2O_4)]$ | Formate | $3^{1}$ | 7.322 | 10.825 | 6.738 | 90 | 90 | 90 | *Pnma* | 145 |
| XUGJEO | $[Nd_4(CO_3)_2(C_2O_4)_6(H_2O)_2] \bullet 4(NH_4) \bullet 2(H_2O)$ | Carbonate | $3^{1}$ | 8.7065 | 9.53 | 10.3274 | 73.35 | 86.9 | 80.5 | *P-1* | 146 |
| XUPBUF | $[Tb_2(C_7H_3NO_4)_2(C_2O_4)(H_2O)_2]$ | 2,4-Pyridinedicarboxylate | $3^{0,1}$ | 17.693 | 23.712 | 9.7052 | 90 | 90 | 90 | *Fdd2* | 147 |
| XUPCAM | $[Tb_2(C_7H_3NO_4)_2(C_2O_4)(H_2O)_4] \bullet 2(H_2O)$ | 3,5-Pyridinedicarboxylate | $3^{0,1}$ | 7.6118 | 9.8253 | 14.748 | 90 | 98.14 | 90 | *P2₁* | 147 |

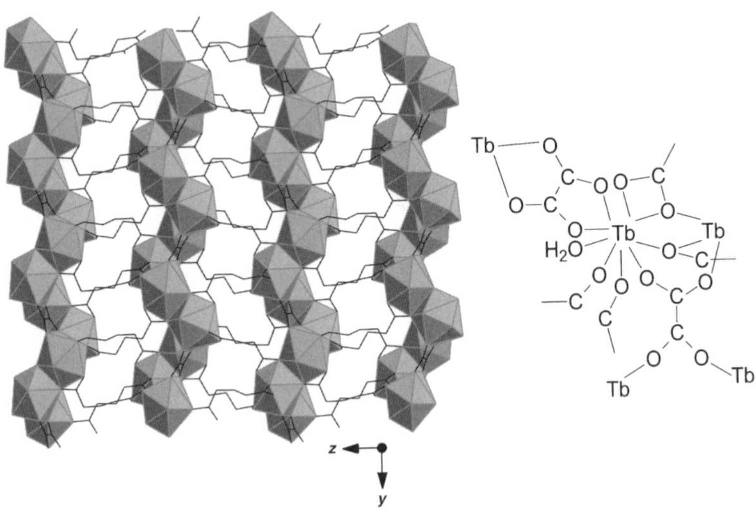


Figure 33. The structure of ACOXEV $[Tb_4(C_2O_4)_2(C_5H_6O_4)_4(H_2O)_2]$, shown down [100].

Table VII contains a number of materials utilizing sulfonate functionalized organic linkers (Fig. 31). One such example, UFUQUH $[La_2(OH)_2(C_{10}H_6S_2O_6)(H_2O)_2]$ (Fig. 35) (166) consists of 1,5-naphthalenedisulfonate linkers (1,5-NDS) coordinated to $La_2O_{14}$ dimers. Each $La^{3+}$ ion is in eightfold coordination with two $OH^-$ groups, one water molecule, and five oxygen atoms (each in bridging bidentate coordination) from five distinct 1,5-NDS molecules. Edge sharing through the $OH^-$ groups gives rise to the dimers that are then polymerized through the sulfonate groups. One oxygen atom in each sulfonate group is uncoordinated to any La ion and is thus terminal, yet likely participates in hydrogen bonding with neighboring (coordinated) $H_2O$ groups. An overall 3D topology is realized when one considers the resulting topology to be chains of dimers that propagate along [100] that are cross-linked. The $Nd^{3+}$ analogue of this material exhibits photoluminescent behavior when excited at 799 nm and the $La^{3+}$, $Nd^{3+}$, and $Pr^{3+}$ analogues each exhibit catalytic activity and selectivity for oxidation and epoxide ring opening.

The compound FAGVUF $[Ln(C_8H_3SO_7)(H_2O)_4]$ (Fig. 36) (164), where Ln = Eu, Gd or Ce, makes use of a mixed carboxylate–sulfonate linker, 5-isophthalic acid. In this 2D structure, each $Ln^{3+}$ ion is coordinated to four water molecules, four oxygen atoms from two bidentate carboxylate groups, and one oxygen atom from a sulfonate group. Polymerization gives rise to a layered topology, wherein the Ln centers remain monomeric and the planar linkers promote coordination in two dimensions. In contrast to UFUQUH (above), two oxygen atoms on each sulfonate group remain terminal.

TABLE VI
Ln—O Plus Ln—N Chelating

| Code | Compound | Linkers | Dim | a | b | c | α | β | γ | Space Group | Reference |
|---|---|---|---|---|---|---|---|---|---|---|---|
| EMADOL | $[La_2(C_{14}H_8O_4)_3(C_{12}H_8N_2)(H_2O)] \bullet 2(H_2O)$ | 2,2′-Biphenyldicarboxylate, 1,10-phenanthroline | $1^1$ | 12.39 | 15.419 | 15.951 | 117.53 | 95.18 | 107.28 | *P-1* | 148 |
| EMAFAZ | $[Tb_2(C_{14}H_8O_4)_3(C_{12}H_8N_2)_2(H_2O)_2] \bullet 4(H_2O)$ | 2,2′-Biphenyldicarboxylate, 1,10-phenanthroline | $1^{0,1}$ | 24.48 | 12.104 | 22.94 | 90 | 115.571 | 90 | *C2/c* | 148 |
| FORPEU | $[Eu(CHO_2)_3(C_{12}H_8N_2)(H_2O)]$ | Formate; 1,10-phenanthroline | $1^{0,1}$ | 13.68 | 10.71 | 6.714 | 90 | 90 | 118.9 | $P2_1$ | 149 |
| HUVWUQ | $[Nd_2(C_8H_4O_4)_3(C_{12}H_8N_2)(H_2O)] \bullet (H_2O)$ | 1,2-Benzenedicarboxylate, 1,10-phenanthroline | $1^2$ | 7.605 | 12.972 | 18.773 | 109.778 | 91.657 | 103.951 | *P-1* | 150 |
| **QAFGIO** | $[Ho(C_5H_6O_4)(C_{12}H_8N_2)(Cl)]$ | Glutarate; 1,10-phenanthroline | $1^{0,2}$ | 8.189 | 12.522 | 16.168 | 90 | 93.96 | 90 | $P2_1/n$ | 151 |
| AVUWOD | $[Er(C_{14}H_9O_5)_2(C_{12}H_8N_2)]$ | 4,4′-oxybis(benzoate); 1,10-phenanthroline | $2^{0,1}$ | 13.897 | 17.5455 | 28.6986 | 90 | 91.0013 | 90 | *C2/c* | 152 |
| AVUXEU | $[Pr_2(C_7H_{10}O_4)_3(C_{12}H_8N_2)_2(H_2O)] \bullet 3(H_2O)$ | Pimelate; 1,10-phenanthroline | $2^{0,2}$ | 12.267 | 13.903 | 15.348 | 89.008 | 86.468 | 64.754 | *P-1* | 153 |
| BEPDIK | $[La_2(C_4H_2O_4)_3(C_{12}H_8N_2)_2(H_2O)_2]$ | Fumerate; 1,10-phenanthroline | $2^{0,2}$ | 9.0567 | 10.072 | 10.61 | 72.54 | 77.83 | 70.02 | *P-1* | 154 |

(*Continued*)

TABLE VI
(*Continued*)

| Code | Compound | Linkers | Dim | a | b | c | α | β | γ | Space Group | Reference |
|---|---|---|---|---|---|---|---|---|---|---|---|
| FENBAC | $[La_2(C_{12}H_6O_4)_3(C_{12}H_8N_2)_2(H_2O)_2]$ | 1,4-Naphthalenedicarboxylate; 1,10-phenanthroline | $2^{0,2}$ | 11.842 | 14.525 | 16.168 | 84.198 | 69.245 | 89.964 | *P-1* | 155 |
| GAFCEW | $[Eu_2(C_8H_5NO_4)_3(C_{12}H_8N_2)_2(H_2O)_2]$ | 2-Aminoterephthalate; 1,10-phenanthroline | $2^{0,1}$ | 10.497 | 10.87 | 11.355 | 70.836 | 87.853 | 65.793 | *P-1* | 156 |
| IKAXUN | $[Eu(C_8H_5O_4)_2(C_{12}H_8N_2)(H_2O)]$ | 1,2-Benzenedicarboxylate, 1,10-phenanthroline | $2^{0,2}$ | 12.565 | 16.005 | 12.891 | 90 | 102.173 | 90 | $P2_1/c$ | 157 |
| ISULOX | $[Eu(C_6H_2N_2O_4)(C_{12}H_8N_2)(NO_3)(H_2O)] \bullet (H_2O)$ | 2,3-Pyrazinedicarboxylate; 1,10-phenanthroline | $2^{0,1}$ | 12.3427 | 10.4806 | 15.7966 | 90 | 101.947 | 90 | $P2_1/c$ | 158 |
| OKUQEQ | $[Pr_2(C_8H_4O_4)_3(C_{12}H_8N_2)(H_2O)] \bullet 0.5(H_2O)$ | 1,3-Benzenedicarboxylate; 1,10-phenanthroline | $2^{0,2}$ | 14.505 | 20.576 | 23.388 | 90 | 90 | 90 | *Pbca* | 159 |
| QAFGEK | $[Nd_2(C_5H_6O_4)_3(C_{12}H_8N_2)_2]$ | Glutarate; 1,10-phenanthroline | $2^{0,2}$ | 8.81 | 13.4 | 16.31 | 84.09 | 83.81 | 75 | *P-1* | 151 |
| DACXOV | $[Tb_2(C_{12}H_6O_4)_3(C_{12}H_8N_2)_2] \bullet (C_{12}H_8O_4)$ | 2,6-Naphthalenedicarboxylate; 1,10-phenanthroline | $3^{0,1}$ | 10.7822 | 12.1806 | 12.3369 | 94.2028 | 106.3049 | 100.9193 | *P-1* | 160 |
| IKAYUO | $[Yb_2(C_8H_4O_4)_3(C_{12}H_8N_2)_2(H_2O)]$ | 1,4-Benzenedicarboxylate, 1,10-phenanthroline | $3^{0,1}$ | 10.349 | 11.052 | 19.431 | 105.464 | 91.3 | 93.655 | *P-1* | 157 |
| OHENIY | $[Tb_2(C_{14}H_9O_5)_3(C_{10}H_8N_2)_2]$ | 4,4′-Oxybis(benzoate); 2,2′-bipyridyl | $3^{0,2}$ | 30.896 | 10.183 | 20.432 | 90 | 118.75 | 90 | *C2/c* | 161 |
| OKUQAM | $[Er_4(C_8H_4O_4)_6$ [illegible] | 1,3-Benzenedicarboxylate; 1,10-phenanthroline | $3^1$ | 8.196 | 24.75 | 29.54 | 90 | 90 | 90 | $C222_1$ | 159 |

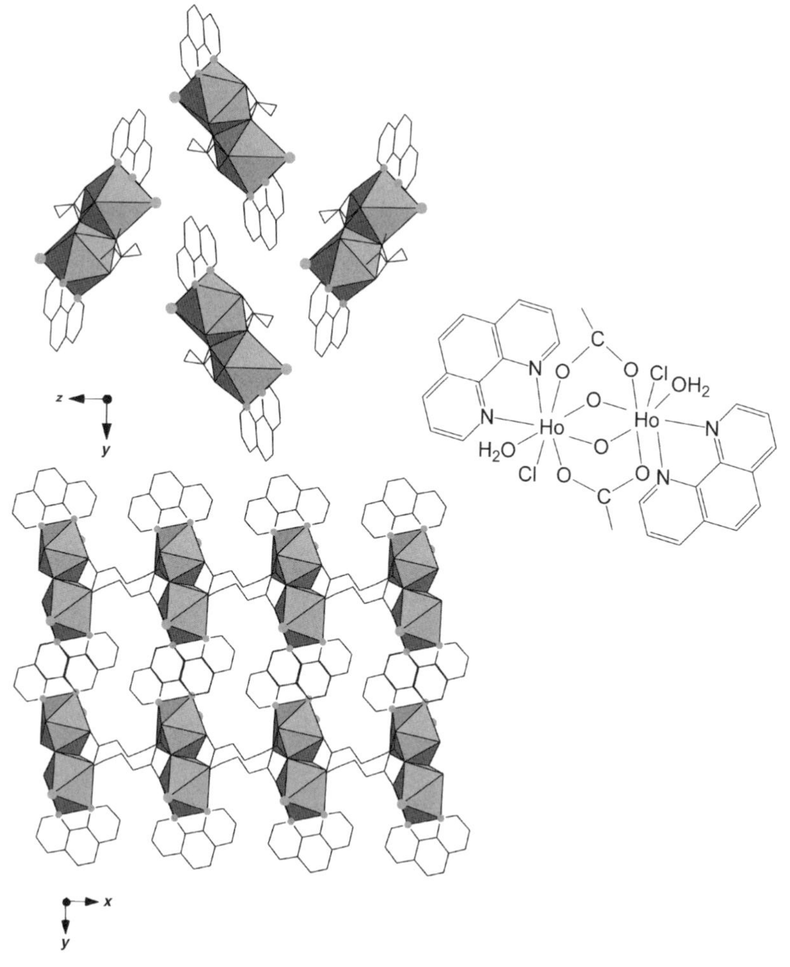


Figure 34. The structure of QAFGIO $[Ho(C_5H_6O_4)(phen)Cl]$, where 1,10-phenanthroline = phen.1,10-Phenanthroline chelates Ho–O dimers that are then connected through glutarate groups to form chains along [100]. The large phenanthroline groups also participate in π stacking.

A final example of sulfonate linkages is found in XENWUI $[Eu_2(C_6H_4S_2O_6)(C_7H_4SO_5)_2(H_2O)_2]$ (Fig. 37). This material contains two different sulfonate linkers, 4-sulfobenzoic acid and benzene-1,4-disulfonate, the latter of which formed *in situ* by reaction of 1,4-benzenedicarboxylic acid with $H_2SO_4$ under hydrothermal conditions. Such *in situ* ligand syntheses are becoming increasingly common in MOF and CP chemistry and the reader is encouraged to see an excellent review by Zhang for more details (168). As for the structure of

TABLE VII
Ln-Sulfonates

| Code | Compound | Linkers | Dim | a | b | c | α | β | γ | Space Group | Reference |
|---|---|---|---|---|---|---|---|---|---|---|---|
| IZAVOU | $[La_2(C_8H_3SO_7)_2(C_{12}H_8N_2)_2(H_2O)_3] \bullet H_2O$ | 5-Sulfo-1,3-benzenedicarboxylate | $1^{2}$ | 10.1655 | 15.156 | 13.8914 | 90 | 108.302 | 90 | *P2/n* | 162 |
| YOYSOU | $[Tb_2(C_6H_2S_2O_8)_2(C_3H_8NO)_4] \bullet 2(C_3H_7NO)$ | 1,2-Benzenediol-3,5-disulfonate | $1^{0,2}$ | 25.696 | 9.651 | 9.464 | 90 | 96.82 | 90 | *P2₁/n* | 163 |
| **FAGVUF** | $[Eu(C_8H_3SO_7)(H_2O)_4]$ | 5-Sulfo-1,3-benzenedicarboxylate | $2^{0,1}$ | 7.267 | 16.534 | 10.373 | 90 | 90 | 90 | *Pna2₁* | 164 |
| WEMYAO | $[Sm(C_7H_3SO_6)_2(H_2O)_4] \bullet 2(H_2O)$ | 5-Sulfonyloxysalicylate | $2^{0,1}$ | 22.807 | 9.282 | 10.853 | 90 | 90 | 90.07 | *P2₁/a* | 165 |
| **UFUQUH** | $[La_2(OH)_2(C_{10}H_6S_2O_6)(H_2O)_2]$ | 1,5-Naphthalenedisulfonate | $3^{0,2}$ | 5.7263 | 10.6104 | 11.5533 | 76.615 | 77.619 | 87.937 | *P-1* | 166 |
| **XENWUI** | $[Eu_2(C_6H_4S_2O_6)(C_7H_4SO_5)_2(H_2O)_2]$ | 1,4-Benzenedisulfonate | $3^{0,1}$ | 15.9019 | 20.5627 | 8.1489 | 90 | 90 | 90 | *Pnma* | 167 |

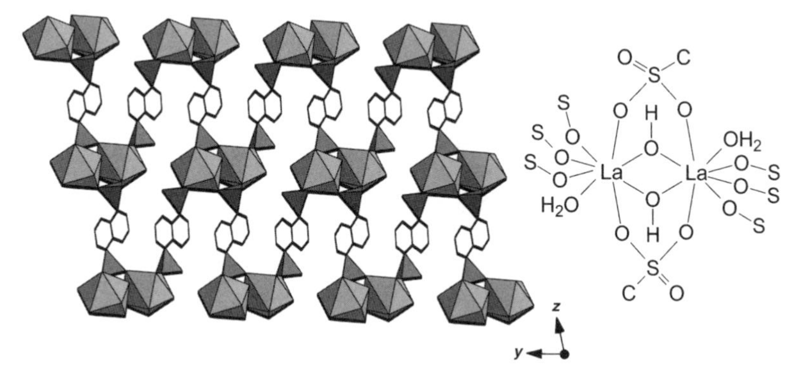


Figure 35. The structure of UFUQUH $[La_2(OH)_2(C_{10}H_6S_2O_6)(H_2O)_2]$. Tetrahedra are bridging (point-shared) sulfonate groups of the 1,5-naphthalenedisulfonate linkers.

XENWUI, there are two crystallographically unique $Eu^{3+}$ sites: Eu1 is in ninefold coordination with one terminal and two bridging aqua ligands, four sulfonate oxygens, and two carboxylate oxygens. The Eu2 atom is coordinated to four sulfonate oxygens, two bridging water molecules, and two carboxylate oxygens. The sulfonate groups each leave one oxygen atom terminal and are all bridging bidentate, which is also the coordination geometry of each carboxylate group.

## G. Lanthanide-Phosphonates

Phosphonate linkages (Fig. 31) may be thought of as an extension to the sulfonate examples just mentioned. Indeed the tetrahedral geometry of the phosphonate functional groups are analogous to $PO_4^{3-}$ tetrahedra as found in

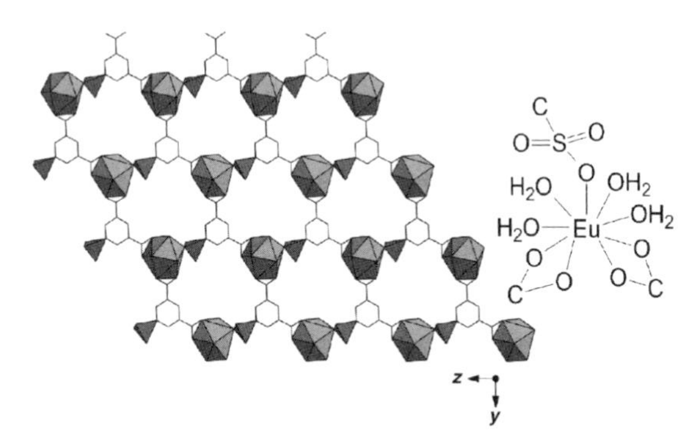


Figure 36. The structure of FAGVUF $[Ln(C_8H_3SO_7)(H_2O)_4]$. The linker is 5-isophthalic acid, a mixed sulfonate–carboxylate species. Gray tetrahedra are point-shared sulfonate groups.

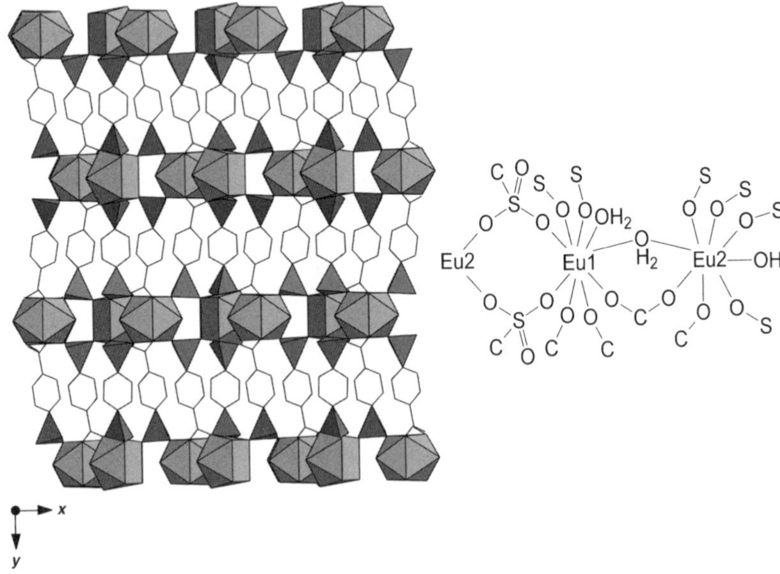


Figure 37. The structure of XENWUI [$Eu_2(C_6H_4S_2O_6)(C_7H_4SO_5)_2(H_2O)_2$]. Gray tetrahedra are bridging (point-shared) sulfonate groups on benzene-1,4-disulfonate and 4-sulfobenzoate linkers.

solid-state structures and minerals, such as monazite, xenotime, and rhabdophane. Table VIII lists a number of MOFs and CPs containing phosphonate linkers. Two representative examples are BIVWAE and BIVVOR. These compounds, $GdH[O_3P(CH_2)_3PO_3]$ and $GdH[O_3P(CH_2)_2PO_3]$, respectively, are part of the same study where the length of the aliphatic chain between phosphonate functional groups was varied (173). The inorganic component of these two compounds, that is, the Gd-phosphate slab, is identical for both materials. The only difference is thus the interlayer spacing imparted by the varying organic chain length (Fig. 38). The slabs consist of $Gd^{3+}$ ions coordinated to eight oxygen atoms from the phosphonate groups, the oxygens of which are coordinated to at least one $Gd^{3+}$ center. Individual $Gd^{3+}$ ions are edge shared to form chains along [100] that are then stitched by alternating edge point-shared phosphonate tetrahedra.

This structural investigation also highlights a challenge that we have neglected to mention thus far: The structures of both of these compounds were determined from X-ray powder diffraction data. These materials resisted forming single crystals of suitable size and quality for single-crystal analysis. This problem is often encountered with polymeric materials and our own experience suggests that the challenge is even greater when the inorganic component of these materials is increased, that is, when the structures contain

TABLE VIII
Ln-Phosphonates

| Code | Compound | Linkers | Dim | a | b | c | α | β | γ | Space Group | Reference |
|---|---|---|---|---|---|---|---|---|---|---|---|
| BUQXIU | $[Gd(C_{24}H_{18}P_2O_8)_2(H_2O)_4] \bullet 12(H_2O)$ | (*R*)-2,2′-Diethoxy-1,1′-binaphthalene-6,6′-diphosphonate | $2^{0,1}$ | 7.771 | 23.8998 | 32.6762 | 90 | 90 | 90 | $P2_12_12_1$ | 169 |
| HUNCOI | $[Gd(C_2H_5P_2O_6)(H_2O)_2] \bullet (H_2O)$ | 1-Hydroxyethylidene-bis(phosphonate) | $2^2$ | 9.69 | 9.744 | 20.585 | 90 | 90 | 90 | *Pbca* | 170 |
| LONPAF | $[Eu_2(C_3H_4NPO_5)_2(H_2O)_7(ClO_4)] \bullet 3(ClO_4) \bullet (H_2O)$ | *N*-Phosphonomethyl-glycinate | $2^{0,1}$ | 17.788 | 10.706 | 18.56 | 90 | 113.37 | 90 | $P2_1/c$ | 171 |
| AZILEA | $[Pr_4(C_7H_9NPO_5)_4(H_2O)_7] \bullet 5(H_2O)$ | 4-Carboxypiperidyl phosphonate | $3^1$ | 23.4805 | 10.1588 | 23.0055 | 90 | 105.63 | 90 | *Cc* | 172 |
| **BIVVOR** | $[Gd(C_3H_7P_2O_6)]$ | Propylenediphosphonate | $3^2$ | 8.2141 | 18.9644 | 5.2622 | 90 | 111.999 | 90 | *C2/m* | 173 |
| **BIVWAE** | $[Gd_2(C_2H_5P_2O_6)_2]$ | Ethylenediphosphonate | $3^2$ | 5.2918 | 15.975 | 8.338 | 90 | 111.491 | 90 | $P2_1/c$ | 173 |
| QAHTOI | $[Pr(C_3H_4PO_5)]$ | Carboxyethylphosphonate | $3^2$ | 8.3617 | 7.1899 | 5.4586 | 90 | 103.784 | 90 | $P2_1/m$ | 174 |

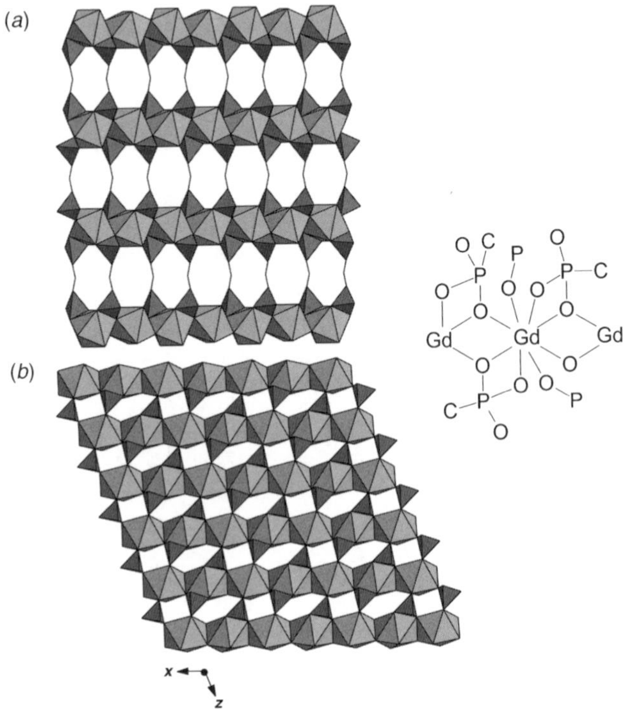


Figure 38. (*a*) The structure of BIVVOR $GdH[O_3P(CH_2)_2PO_3]$ shown down [001]. (*b*) The Ln-phosphonate slab common to both BIVVOR and BIVWAE.

significant degrees of *inorganic* polymerization (i.e., complex SBUs), such as Ln-phosphonate sheets.

## VI. CONCLUSIONS

The field of polymeric hybrid materials is currently in a period of tremendous growth and development. This chapter has made every effort to present representative compounds and provide a snapshot of the current rationale and structural design principles for the synthesis of extended Ln-containing materials. Given the frequency with which new materials are being reported, more regular contributions similar to this chapter will be needed. In terms of future directions within the field of Ln-MOFs (and arguably *d*-metal materials as well), it is our opinion that much promise lies in tailoring the organic components of these systems. Indeed, most of the linker molecules featured in this chapter are off-the-shelf compounds. This is understandable considering the vast library of multifunctional organic molecules available for purchase. The synthesis of designed linker molecules, however, is only beginning to be exploited. Considering the extensive coordination chemistry of *molecular* Ln materials, especially those with ligands designed for many of the applications mentioned in the introduction (MRI contrast agents, sensitization, *d-/f-* functional assemblies, etc.), a natural extension would be to move beyond the primary Ln coordination sphere and introduce additional functional groups with the goal of producing higher dimensional topologies. Along this line is *in situ* ligand synthesis, wherein linker molecules are formed from decomposition of other starting materials during hydro(solvo)thermal reaction. Although reports of this phenomenon are increasingly common, one could argue that most instances are serendipitous (yet indeed explicable and reproducible). A challenge is thus to harness this activity and perhaps couple the Lewis acid catalytic potential of Ln elements to design *in situ* reactions of interest.

This chapter has been admittedly structural in nature, with little mention of the applications of these materials. Various recent reviews (175–177) have concentrated on harnessing coordination polymers materials for a range of uses, but have in general been limited in scope to those that contain transition metal elements. This result may not be entirely surprising being that there are significantly more transition metal based materials in the literature than those containing $f$-block elements. Additionally, the success of Yaghi and co-workers with MOF-5 (35,178) has highlighted the concept of reticular design to engineer topologies for specific functions, directing attention toward this compound and others similar to it. It is arguable that the synthesis of lanthanide-containing coordination polymers and metal–organic frameworks has trailed that of *d*-metal compounds, perhaps due to some of the inherent challenges associated with lanthanide elements (see Section III). Nonetheless, there are many examples in the literature where potential uses of lanthanide-containing coordination polymers are explored. Further, it is our belief that the advantages of using lanthanides in the syntheses of these materials are just now being realized.

As mentioned earlier, one distinct advantage of using the Ln elements is to harness their unique optical features, namely, luminescence. Indeed, there are countless examples of this very phenomenon within the literature, with feasible applications in display technologies or sensors. Most of the luminescence studies done on these materials are usually based on $Eu^{3+}$ or $Tb^{3+}$ emission (25,30,161,179). The reason for concentrating on these two ions is due to the ease with which they can be sensitized by an aromatic organic linker. Just the same, other lanthanides have been studied in these systems, including $Dy^{3+}$ (180), $Ho^{3+}$ (181), $Er^{3+}$ (181), $Pr^{3+}$ (182), and $Gd^{3+}$ (24). Additionally, certain crystal structures of lanthanide-containing CPs have the potential to produce second-harmonic generation, which could lead to applications in nonlinear optics. One $Eu^{3+}$ sample shows a SHG that is 16.8 times that of urea (183) and another using

$Nd^{3+}$ shows a signal that is 1.1 times greater than urea (184). Since the lanthanide ions are predominately in the +3 oxidation state, examinations into the magnetic properties of these materials have also been conducted. As expected, these materials tend to follow the Curie–Weiss law and exhibit paramagnetic behavior at moderate-to-high temperatures (157,164). As the temperature is lowered, it seems that within polynuclear systems there is weak ferromagnetic and/or antiferromagnetic coupling exhibited by a variety of compounds (157,185,186).

Aside from the specific properties induced through the lanthanide metal center, structurally derived characteristics, such as porosity, have been explored as well. Preliminary studies conducted on newly synthesized compounds usually include porosity measurements, as traditionally done with gas sorption isotherms (105,187). Quite recently, Daiguebonne and Guillou (188) introduced a promising mathematical approach for calculating porosity based solely on single-crystal X-ray data, which is suitable for compounds with low thermal stability or are highly likely to decompose upon desorption. Additionally, the sorption of ammonia on a terbium-benzenedicarboxlyate has been noted (26), as well as the sorption of a variety of organic solvents by other Tb-containing CPs (97). Of course, CPs are being studied quite extensively as a potential candidate for hydrogen storage materials. Very inspiring results have been obtained in this regard (100,189,190). Finally, any application that these materials may be used for will be limited primarily by how stable these compounds are. Thermal stability for these materials usually lies within the 150–300°C range (191–196), but some materials do possess uncharacteristically higher stability of ~500°C (197–199). It is also not uncommon to observe phase transitions of these materials at various temperatures, which could have potential uses as molecular storage materials (200,201). Thus, CPs in general possess the capability of being very useful materials in a variety of applications, and even more so with the introduction of a lanthanide as the metal center of choice.

## ACKNOWLEDGMENTS

Research efforts in the Cahill group are supported by the National Science Foundation (NSF) and the United States Department of Energy (DOE). DTD is grateful to the ARCS Foundation of Metro Washington for a tuition scholarship.

## ABBREVIATIONS

| | |
|---|---|
| 1,5-NDS | 1,5-Naphthalenedisulfonate |
| 1D | One dimensional |
| 2D | Two dimensional |

| | |
|---|---|
| 3D | Three dimensional |
| 3-HPYA | *trans*-3-(3-Pyridyl)acrylic acid |
| bb | Bridging bidentate |
| BDC | 1,4-Benzenedicarboxylic acid |
| BPDC | 1,4-Biphenyldicarboxylic acid |
| bpy | 4,4′-Dipyridyl |
| BTC | 1,3,5-Benzenetricarboxylic acid |
| BU | Building unit |
| CCDC | Cambridge Crystallographic Data Centre |
| CP | Coordination polymer |
| $H_2$pydc | Pyridine-2,5-dicarboxylic acid |
| Ln | Lanthanide |
| MOF | Metal–organic framework |
| MRI | Magnetic resonance imaging |
| NMR | Nuclear magnetic resonance |
| phen | 1,10-Phenanthroline |
| SBU | Secondary building unit |
| SHG | Second harmonic generation |
| TM | Transition metal |
| UV | Ultraviolet |

## REFERENCES


1. W. Lin, *J. Solid State Chem.*, *178*, 2486 (2005).
2. T. Uemura, K. Kitagawa, S. Horike, T. Kawamura, S. Kitagawa, M. Mizuno, and K. Endo, *Chem. Commun. (Cambridge, U K)*, 5968 (2005).
3. A. J. Fletcher, K. M. Thomas, and M. J. Rosseinsky, *J. Solid State Chem.*, *178*, 2491 (2005).
4. C. Thompson, N. R. Champness, A. N. Khlobystov, C. J. Roberts, M. Schroder, S. J. B. Tendler, and M. J. Wilkinson, *J. Micros. (Oxford, U K)*, *214*, 261 (2004).
5. B. O. Patrick, C. L. Stevens, A. Storr, and R. C. Thompson, *Polyhedron*, *24*, 2242 (2005).
6. T. J. Barton, L. M. Bull, W. G. Klemperer, D. A. Loy, B. McEnaney, M. Misono, P. A. Monson, G. Pez, G. W. Scherer, J. C. Vartuli, and O. M. Yaghi, *Chem. Mater.*, *11*, 2633 (1999).
7. G. Férey, *Chem. Mater.*, *13*, 3084 (2001).
8. C. N. R. Rao, S. Natarajan, and R. Vaidhyanathan, *Angew. Chem., Inter. Ed.*, *43*, 1466 (2004).
9. N. W. Ockwig, O. Delgado-Friedrichs, M. O'Keeffe, and O. M. Yaghi, *Acc. Chem. Res.*, *38*, 176 (2005).
10. M. Eddaoudi, D. B. Moler, H. Li, B. Chen, T. M. Reineke, M. O'Keeffe, and O. M. Yaghi, *Acc. Chem. Res.*, *34*, 319 (2001).
11. J. Kim, B. Chen, T. M. Reineke, H. Li, M. Eddaoudi, D. B. Moler, M. O'Keeffe, and O. M. Yaghi, *J. Am. Chem. Soc.*, *123*, 8239 (2001).
12. B. Moulton and M. J. Zaworotko, *Chem. Rev.*, *101*, 1629 (2001).
13. B. Moulton and M. J. Zaworotko, *Curr. Opin. Solid State Mater. Sci.*, *6*, 117 (2002).

14. C. Janiak, *Dalton Trans.*, 2781 (2003).
15. C. J. Kepert, *Chem. Commun. (Cambridge, U K)*, 695 (2006).
16. S. L. James, *Chem. Soc. Rev.*, *32*, 276 (2003).
17. K. Uemura, R. Matsuda, and S. Kitagawa, *J. Solid State Chem.*, *178*, 2420 (2005).
18. S. Kitagawa, S.-i. Noro, and T. Nakamura, *Chem. Commun. (Cambridge, U K)*, 701 (2006).
19. S. Kitagawa and K. Uemura, *Chem. Soc. Rev.*, *34*, 109 (2005).
20. J. D. Wuest, *Chem. Commun. (Cambridge, U K)*, 5830 (2005).
21. B.-H. Ye, M.-L. Tong, and X.-M. Chen, *Coord. Chem. Rev.*, *249*, 545 (2005).
22. JSSC *J. Solid State Chem.*, *178*, 2409 (2005).
23. R. J. Hill, D.-L. Long, P. Hubberstey, M. Schroder, and N. R. Champness, *J. Solid State Chem.*, *178*, 2414 (2005).
24. A. de Bettencourt-Dias, *Inorg. Chem.*, *44*, 2734 (2005).
25. D. T. de Lill, N. S. Gunning, and C. L. Cahill, *Inorg. Chem.*, *44*, 258 (2005).
26. T. M. Reineke, M. Eddaoudi, M. Fehr, D. Kelley, and O. M. Yaghi, *J. Am. Chem. Soc.*, *121*, 1651 (1999).
27. T. M. Reineke, M. Eddaoudi, D. Moler, M. O'Keeffe, and O. M. Yaghi, *J. Am. Chem. Soc.*, *122*, 4843 (2000).
28. T. M. Reineke, M. Eddaoudi, M. O'Keeffe, and O. M. Yaghi, *Angew. Chem., Inter. Ed.*, *38*, 2590 (1999).
29. R. Cao, D. F. Sun, Y. C. Liang, M. C. Hong, K. Tatsumi, and Q. Shi, *Inorg. Chem.*, *41*, 2087 (2002).
30. J.-Y. Wu, T.-T. Yeh, Y.-S. Wen, J. Twu, and K.-L. Lu, *Crystal Growth Design*, *6*, 467 (2006).
31. M. Dan, J. N. Behera, and C. N. R. Rao, *J. Mater. Chem.*, *14*, 1257 (2004).
32. C. L. Cahill and L. A. Borkowski, *Structural Chemistry of Inorganic Actinide Compounds*, I. G. Tananaev, Ed., Elsevier, Amsterdam, The Netherlands, 2007, in press.
33. L. A. Borkowski and C. L. Cahill, *Inorganic Chemistry Communications*, *7*, 725 (2004).
34. C. Baerlocher, W. M. Meier, and D. H. Olson, *Atlas of Zeolite Framework Types*; 5th Rev. ed., Elsevier Science B.V., Amsterdam, The Netherlands, 2001.
35. H. Li, M. Eddaoudi, M. O'Keeffe, and M. Yaghi, *Nature (London)*, *402*, 276 (1999).
36. H. C. Aspinall, *Chemistry of the f-Block Elements*, 2001.
37. N. Kaltsoyannis and P. Scott, *The f elements*, Vol. 76, Oxford University Press, New York, 1999.
38. C. Piguet and J.-C. G. Bunzli, *Chem. Soc. Rev.*, *28*, 347 (1999).
39. J.-C. G. Bunzli, *Acc. Chem. Res.*, *39*, 53 (2006).
40. J.-C. G. Bunzli and C. Piguet, *Chem. Soc. Rev.*, *34*, 1048 (2005).
41. J.-C. G. Bunzli and C. Piguet, *Chem. Rev.*, *102*, 1897 (2002).
42. D. Parker, *Chem. Soc. Rev.*, *33*, 156 (2004).
43. D. Parker, R. S. Dickins, H. Puschmann, C. Crossland, and J. A. K. Howard, *Chem. Rev.*, *102*, 1977 (2002).
44. M. S. Wickleder, *Chem. Rev.*, *102*, 2011 (2002).
45. J. N. Low, T. R. Sarangarajan, K. Panchanatheswaran, and C. Glidewell, *Acta Crystallogr., Sect. E: Struct. Rep. Online*, *E60*, i142 (2004).
46. Y. Ni, J. M. Hughes, and A. N. Mariano, *Ame. Mineral.*, *80*, 21 (1995).
47. Y. Xing, Y. Liu, Z. Shi, H. Meng, and W. Pang, *J. Solid State Chem.*, *174*, 381 (2003).

48. M. E. Davis and R. F. Lobo, *Chem. Mater.*, *4*, 756 (1992).
49. F. H. Allen, *Acta Crystallogr., Sect. B: Structural Sci.*, *B58*, 380 (2002). Cambridge Structure Database, Version 5,7, November 2005.
50. J. Legendziewicz, B. Keller, I. Turowska-Tyrk, and W. Wojciechowski, *New J. Chem.*, *23*, 1097 (1999).
51. L. Yang, Y. Su, Y. Xu, Z. Wang, Z. Guo, S. Weng, C. Yan, S. Zhang, and J. Wu, *Inorg.Chem.*, *42*, 5844 (2003).
52. T. Glowiak and D. C. Ngoan, *Acta Crystallogr., Sect. C: Cryst. Struct. Commun.*, *44*, 41 (1988).
53. D. M. Barnhart, D. L. Clark, J. C. Huffman, R. L. Vincent, and J. G. Watkin, *Inorg. Chem.*, *32*, 4077 (1993).
54. I. Csoregh, P. Kierkegaard, J. Legendziewicz, and E. Huskowska, *Acta Chem. Scand.*, *43*, 636 (1989).
55. Y. Inomata, T. Takei, and F. S. Howell, *Inorg. Chim. Acta*, *318*, 201 (2001).
56. C.-D. Wu, C.-Z. Lu, H.-H. Zhuang, and J.-S. Huang, *Acta Crystallogr., Sect. C: Cryst. Struct. Commun.*, *58*, m283 (2002).
57. W. Ollendorf and F. Weigel, *Inorg. Nuclear Chem. Lett.*, *5*, 263 (1969).
58. S. C. Zhu, J. M. Shi, Q. Y. Liu, and C. J. Wu, *Pol. J. Chem.*, *76*, 1747 (2002).
59. F. C. Hawthorne, I. Borys, and R. B. Ferguson, *Acta Crystallogr., Sect. C: Cryst. Struct. Commun.*, *39*, 540 (1983).
60. J.-R. Li, Z.-M. Wang, C.-H. Yan, L.-P. Zhou, and T.-Z. Jin, *Acta Crystallogr., Sect. C: Cryst. Struct. Commun.*, *55*, 2073 (1999).
61. C.-D. Wu, X.-P. Zhan, C.-Z. Lu, H.-H. Zhuang, and J.-S. Huang, *Acta Crystallogr., Sect. E: Struct. Rep. Online*, *58*, m228 (2002).
62. A. Dimos, D. Tsaousis, A. Michaelides, S. Skoulika, S. Golhen, L. Ouahab, C. Didierjean, and A. Aubry, *Chem. Mater.*, *14*, 2616 (2002).
63. H. Steinfink and G. D. Brunton, *Inorg. Chem.*, *9*, 2112 (1970).
64. S.-G. Liu, W. Liu, J.-L. Zuo, Y.-Z. Li, and X.-Z. You, *Inorg. Chem. Commun.*, *8*, 328 (2005).
65. Z. Rzaczynska, A. Bartyzel, and T. Glowiak, *J. Coord. Chem.*, *56*, 1525 (2003).
66. A. E. Prozorovskii, A. B. Yaroslavtsev, and Z. N. Prozorovskaya, *Zh. Neorg. Khim. (Russ.) (Russ. J. Inorg. Chem.)*, *34*, 2622 (1989).
67. B. Benmerad, A. Guehria-Laidoudi, S. Dahaoui, and C. Lecomte, *Acta Crystallogr., Sect. C: Cryst. Struct. Commun.*, *60*, m119 (2004).
68. F. Fourcade-Cavillou, and J.-C. Trombe *Solid State Sci.*, *4*, 1199 (2002).
69. F. Serpaggi and G. Ferey *J. Mater. Chem.*, *8*, 2737 (1998).
70. F. Serpaggi and G. Ferey, *Microporous Mesoporous Mater.*, *32*, 311 (1999).
71. J.-G. Mao, H.-T. Wu, J.-Z. Ni, H.-J. Zhang, and T. C. W. Mak, *J. Chem. Crystallogr.*, *28*, 177 (1998).
72. P. C. Junk and M. K. Smith, *Polyhedron*, *22*, 331 (2003).
73. F. Millange, C. Serre, J. Marrot, N. Gardant, F. Pelle, and G. Ferey, *J. Mater. Chem.*, *14*, 642 (2004).
74. Y.-F. Zhou, F.-L. Jiang, D.-Q. Yuan, B.-L. Wu, Rui-HuWang, Z.-Z. Lin, and M.-C. Hong, *Angew. Chem., Int. Ed.*, *43*, 5665 (2004).
75. V. Kiritsis, A. Michaelides, S. Skoulika, S. Golhen, and L. Ouahab, *Inorg. Chem.*, *37*, 3407 (1998).
76. Q.-Z. Zhang, C.-Z. Lu, and W.-B. Yang, *Z. Anorg. Allg. Chem.*, *630*, 1550 (2004).

77. B. Benmerad, A. Guehria-Laidoudi, F. Balegroune, H. Birkedal, and G. Chapuis, *Acta Crystallogr., Sect. C: Cryst. Struct. Commun., 56*, 789 (2000).
78. P. F. Aramendia, R. Baggio, M. T. Garland, and M. Perec, *Inorg. Chim. Acta, 303*, 306 (2000).
79. M. Camara, C. Daiguebonne, K. Boubekeur, T. Roisnel, Y. Gerault, C. Baux, F. L. Dret, and O. Guillou, *C. R. Chim., 6*, 405 (2003).
80. A. Thirumurugan, S. K. Pati, M. A. Green, and S. Natarajan, *J. Mater. Chem., 13*, 2937 (2003).
81. S. K. Ghosh, and P. K. Bharadwaj, *Inorg. Chem., 42*, 8250 (2003).
82. A. Thirumurugan, S. K. Pati, M. A. Green, and S. Natarajan, *Z. Anorg. Allg. Chem., 630*, 579 (2004).
83. L. Yuan, M. Yin, E. Yuan, J. Sun, and K. Zhang, *Inorg. Chim. Acta, 357*, 89 (2004).
84. P. Starynowicz, *Acta Crystallogr., Sect. C: Cryst. Struct. Commun., 47*, 294 (1991).
85. C. Daiguebonne, Y. Gerault, O. Guillou, A. Lecerf, K. Boubekeur, P. Batail, M. Kahn, and O. Kahn, *J. Alloys Comp., 275*, 50 (1998).
86. M. S. Khiyalov, I. R. Amiraslanov, K. S. Mamedov, and E. M. Movsumov, *Zh. Strukt. Khim. (Russ.)(J. Struct. Chem.), 22*, 113 (1981).
87. J.-G. Mao, H.-J. Zhang, J.-Z. Ni, S.-B. Wang, and T. C. W. Mak, *Polyhedron, 18*, 1519 (1999).
88. L.-P. Zhang, Y.-H. Wan, and L.-P. Jin, *Polyhedron, 22*, 981 (2003).
89. B. Zhao, L. Yi, Y. Dai, X.-Y. Chen, P. Cheng, D.-Z. Liao, S.-P. Yan, and Z.-H. Jiang, *Inorg. Chem., 44*, 911 (2005).
90. J.-C. Trombe, J.-F. Petit, and A. Gleizes, *New J. Chem. (Nouv. J. Chim.), 12*, 197 (1988).
91. Z. Rzaczynska, W. Brzyska, R. Mrozek, W. Ozga, and T. Glowiak, *J. Coord. Chem., 43*, 321 (1998).
92. J.-G. Mao, T. C. W. Mak, H.-J. Zhang, J.-Z. Ni, and S.-B. Wang, *J. Coord. Chem., 47*, 145 (1999).
93. C.-G. Zhang, J. Zhang, Z.-F. Chen, Z.-J. Guo, R.-G. Xiong, and X.-Z. You, *J. Coord. Chem., 55*, 835 (2002).
94. A. Thirumurugan and S. Natarajan, *Eur. J. Inorg. Chem.*, 762 (2004).
95. B. Schoknecht and R. Kempe, *Z. Anorg. Allg. Chem., 630*, 1377 (2004).
96. S. Si, C. Li, R. Wang, and Y. Li, *J. Coord. Chem., 57*, 1545 (2004).
97. N. L. Rosi, J. Kim, M. Eddaoudi, B. Chen, M. O'Keeffe, and O. M. Yaghi, *J. Am. Chem. Soc., 127*, 1504 (2005).
98. Y.-B. Wang, W.-J. Zhuang, L.-P. Jin, and S.-Z. Lu, *J. Mol. Struct., 705*, 21 (2004).
99. S. S.-Y. Chui, A. Siu, X. Feng, Z. Y. Zhang, T. C. W. Mak, and I. D. Williams, *Inorg. Chem. Commun., 4*, 467 (2001).
100. X. Li, Q. Shi, D. Sun, W. Bi, and R. Cao, *Eur. J. Inorg. Chem.*, 2747 (2004).
101. O. M. Yaghi, H. Li, and T. L. Groy, *Z. Kristallogr.-New Cryst. Struct., 212*, 457 (1997).
102. F. A. Almeida Paz, and J. Klinowski, *Chem. Commun. (Cambridge, U K)*, 1484 (2003).
103. T. M. Reineke, M. Eddaoudi, M. Fehr, D. Kelley, and O. M. Yaghi, *J. Am. Chem. Soc., 121*, 1651 (1999).
104. D. Min and S. W. Lee, *Bull. Korean Chem. Soc., 23*, 948 (2002).
105. C. Serre and G. Ferey, *J. Mater. Chem., 12*, 3053 (2002).
106. N. S. Gunning and C. L. Cahill, *Dalton Trans.*, 2788 (2005).
107. A. C. Rizzi, R. Calvo, R. Baggio, M. T. Garland, O. Pena, and M. Perec, *Inorg. Chem., 41*, 5609 (2002).
108. Y. Liang, M. Hong, W. Su, R. Cao, and W. Zhang, *Inorg. Chem., 40*, 4574 (2001).

109. Y. C. Liang, R. Cao, M. C. Hong, D. F. Sun, Y. J. Zhao, J. B. Weng, and R. H. Wang, *Inorg. Chem. Commun.*, *5*, 366 (2002).
110. Y. Liang, R. Cao, W. Su, M. Hong, and W. Zhang, *Angew. Chem., Inter. Ed.*, *39*, 3304 (2000).
111. R. E. P. Winpenny, *Chem. Soc. Rev.*, *27*, 447 (1998).
112. B.-Q. Ma, S. Gao, G. Su, and G.-X. Xu, *Angew. Chem., Inter. Ed.*, *40*, 434 (2001).
113. X.-M. Chen, Y.-L. Wu, Y.-Y. Yang, S. M. J. Aubin, and D. N. Hendrickson, *Inorg. Chem.*, *37*, 6186 (1998).
114. D. M. L. Goodgame, D. J. Williams, and R. E. P. Winpenny, *Chem. Commun.*, 437 (1988).
115. G. B. Deacon, C. M. Forsyth, T. Behrsing, K. Konstas, and M. Forsyth, *Chem. Commun.*, 2820 (2002).
116. H.-Z. Kou, B. C. Zhou, and R.-J. Wang, *Inorg. Chem.*, *42*, 7658 (2003).
117. O. Guillou, O. Kahn, R. L. Oushoorn, K. Boubekeur, and P. Batail, *Inorg. Chim. Acta*, *198*, 119 (1992).
118. T. Yi, S. Gao, and B. Li, *Polyhedron*, *17*, 2243 (1998).
119. G. Dong, L. Yu-ting, D. Chun-ying, M. Hong, and M. Qing-jin, *Inorg. Chem.*, *42*, 2519 (2003).
120. O. Guillou, R. L. Oushoorn, O. Kahn, K. Boubekeur, and P. Batail, *Angew. Chem., Int. Ed.*, *31*, 626 (1992).
121. M.-Y. Xu, F.-H. Liao, J.-R. Li, H.-L. Sun, Z.-S. Li, X.-Y. Wang, and S. Gao, *Inorg. Chem. Commun.*, *6*, 841 (2003).
122. B. Zhao, X.-Y. Chen, P. Cheng, D.-Z. Liao, S.-P. Yan, and Z.-H. Jiang, *J. Am. Chem. Soc.*, *126*, 15394 (2004).
123. Y.-P. Cai, C.-Y. Su, G.-B. Li, Z.-W. Mao, C. Zhang, A.-W. Xu, and B.-S. Kang, *Inorg. Chim. Acta*, *358*, 1298 (2005).
124. X. Meng, G. Li, H. Hou, H. Han, Y. Fan, Y. Zhu, and C. Du, *J. Organomet. Chem.*, *679*, 153 (2003).
125. Y. Liang, R. Cao, M. Hong, D. Sun, Y. Zhao, J. Weng, and R. Wang, *Inorg. Chem. Commun.*, *5*, 366 (2002).
126. D. M. L. Goodgame, D. A. Grachvogel, A. J. P. White, and D. J. Williams, *Inorg. Chem.*, *40*, 6180 (2001).
127. J.-G. Mao, J.-S. Huang, J.-F. Ma, and J.-Z. Ni, *Transition Met. Chem.*, *22*, 277 (1997).
128. Q.-D. Liu, J.-R. Li, S. Gao, B.-Q. Ma, F.-H. Liao, Q.-Z. Zhou, and K.-B. Yu, *Inorg. Chem. Commun.*, *4*, 301 (2001).
129. A. Bouayad, C. Brouca-Cabarrecq, J.-C. Trombe, and A. Gleizes, *Inorg. Chim. Acta*, *195*, 193 (1992).
130. D. M. L. Goodgame, S. Menzer, A. T. Ross, and D. J. Williams, *Chem. Commun.*, 2605 (1994).
131. N. R. Champness, *Dalton Trans.*, 877 (2006).
132. L.-P. Zhang, M. Du, W.-J. Lu, and T. C. W. Mak, *Polyhedron*, *23*, 857 (2004).
133. L.-P. Zhang, W.-J. Lu, and T. C. W. Mak, *Polyhedron*, *23*, 169 (2004).
134. C. Seward and S. Wang, *Can. J. Chem.*, *79*, 1187 (2001).
135. D.-L. Long, A. J. Blake, N. R. Champness, C. Wilson, and M. Schroder, *Chem.-Eur. J.*, *8*, 2026 (2002).
136. D.-L. Long, R. J. Hill, A. J. Blake, N. R. Champness, P. Hubberstey, D. M. Proserpio, C. Wilson, and M. Schroder, *Angew. Chem., Int. Ednge.*, *43*, 1851 (2004).
137. W.-J. Lu, L.-P. Zhang, H.-B. Song, Q.-M. Wang, and T. C. W. Mak, *New J. Chem. (Nouv. J. Chim.)*, *26*, 775 (2002).

138. D.-L. Long, A. J. Blake, N. R. Champness, C. Wilson, and M. Schroder, *J. Am. Chem. Soc.*, *123*, 3401 (2001).
139. S. Si, C. Li, R. Wang, and Y. Li, *J. Mol. Struct.*, *703*, 11 (2004).
140. P. Thomas and J. C. Trombe, *J. Chem. Crystallogr.*, *30*, 633 (2000).
141. S. Neogi, G. Savitha, and P. K. Bharadwaj, *Inorg. Chem.*, *43*, 3771 (2004).
142. J.-L. Song and J.-G. Mao, *Chem.-Eur. J.*, *11*, 1417 (2005).
143. X. Li, R. Cao, D. Sun, Q. Shi, W. Bi, and M. Hong, *Inorg. Chem. Commun.*, *6*, 815 (2003).
144. J. C. Trombe, J. F. Petit, and A. Gleizes, *Eur. J. Solid State Inorg. Chem.*, *28*, 669 (1991).
145. S. Romero, A. Mosset, and J. C. Trombe, *J. Solid State Chem.*, *127*, 256 (1996).
146. J. C. Trombe, J. Galy, and R. Enjalbert, *Acta Crystallogr., Sect. C: Cryst. Struct. Commun.*, *C58*, m517 (2002).
147. D. Min and S. W. Lee, *Inorg. Chem. Commun.*, *5*, 978 (2002).
148. Y. Wang, X. Zheng, W. Zhuang, and L. Jin, *Eur. J. Inorg. Chem.*, 3572 (2003).
149. L. A. Aslanov, V. M. Ionov, V. B. Ribakov, and I. D. Kiekbaer, *Koord. Khim. (Russ. )(Coord. Chem.)*, *4*, 1598 (1978).
150. Y.-H. Wan, L.-P. Jin, and K.-Z. Wang, *J. Mol. Struct.*, *649*, 85 (2003).
151. L.-P. Zhang, Y.-H. Wan, and L.-P. Jin, *J. Mol. Struct.*, *646*, 169 (2003).
152. Y. Wang, Z. Wang, C. Yan, and L. Jin, *J. Mol. Struct.*, *692*, 177 (2004).
153. L. Huang, L.-P. Zhang, and L.-P. Jin, *J. Mol. Struct.*, *692*, 169 (2004).
154. L. Huang and L.-P. Zhang, *J. Mol. Struct.*, *692*, 249 (2004).
155. X.-J. Zheng, L.-P. Jin, S. Gao, and S.-Z. Lu, *Inorg. Chem. Commun.*, *8*, 72 (2005).
156. C.-B. Liu, C.-Y. Sun, L.-P. Jin, and S.-Z. Lu, *New J. Chem. (Nouv. J. Chim.)*, *28*, 1019 (2004).
157. Y. Wan, L. Zhang, L. Jin, S. Gao, and S. Lu, *Inorg. Chem.*, *42*, 4985 (2003).
158. M.-L. Hu, J.-X. Yuan, F. Chen, and Q. Shi, *Acta Crystallogr., Sect. C: Cryst. Struct. Commun.*, *60*, m186 (2004).
159. Y.-H. Wan, L.-P. Zhang, and L.-P. Jin, *J. Mol. Struct.*, *658*, 253 (2003).
160. X.-J. Zheng, Z.-M. Wang, S. Gao, F.-H. Liao, C.-H. Yan, and L.-P. Jin, *Eur. J. Inorg. Chem.*, 2968 (2004).
161. G.-F. Liu, Z.-P. Qiao, H.-Z. Wang, X.-M. Chen, and G. Yang, *New J. Chem.*, *26*, 791 (2002).
162. M.-L. Hu, Q. Miao, Q. Shi, and M.-D. Ye, *Acta Crystallogr., Sect. C: Cryst. Struct. Commun.*, *60*, m460 (2004).
163. H.-Y. Sun, C.-H. Huang, G.-X. Xu, Z.-S. Ma, and N.-C. Shi, *Polyhedron*, *14*, 947 (1995).
164. Z. Wang, M. Strobele, K.-L. Zhang, H.-J. Meyer, X.-Z. You, and Z. Yu, *Inorg. Chem. Commun.*, *5*, 230 (2002).
165. Z. G. Aliev, T. A. Baranova, L. O. Atovmyan, and S. B. Pirkes, *Koord. Khim. (Russ. )(Coord. Chem.)*, *20*, 150 (1994).
166. N. Snejko, C. Cascales, B. Gomez-Lor, E. Gutierrez-Puebla, M. Iglesias, C. Ruiz-Valero, and M. A. Monge, *Chem. Commun.*, 1366 (2002).
167. R.-G. Xiong, J. Zhang, Z.-F. Chen, X.-Z. You, C.-M. Che, and H.-K. Fun, *J. Chem. Soc., Dalton Trans.*, 780 (2001).
168. X.-M. Zhang, *Coord. Chem. Rev.*, *249*, 1201 (2005).
169. O. R. Evans, H. L. Ngo, and W. Lin, *J. Am. Chem. Soc.*, *123*, 10395 (2001).
170. J.-P. Silvestre, N. Q. Dao, and M.-R. Lee, *Phosphorus, Sulfur, Silicon, Relat. Elem.*, *176*, 173 (2001).

171. E. Galdecka, Z. Galdecki, P. Gawryszewska, and J. Legendziewicz, *New J. Chem.*, *24*, 387 (2000).
172. C. Serre, N. Stock, T. Bein, and G. Ferey, *Inorg. Chem.*, *43*, 3159 (2004).
173. F. Serpaggi and G. Ferey, *J. Mater. Chem.*, *8*, 2749 (1998).
174. F. Serpaggi and G. Ferey, *Inorg. Chem.*, *38*, 4741 (1999).
175. C. J. Kepert, *Chem. Comm.*, *7*, 695 (2006).
176. U. Mueller, M. Schubert, F. Teich, H. Puetter, K. Schierle-Arndt, and J. Pastre, *J. Mater. Chem.*, *16*, 626 (2006).
177. C. J. Janiak, *Chem. Soc., Dalton Trans.*, 2781 (2003).
178. O. M. Yaghi and M. Eddaoudi, *Abstracts of Papers, 223rd ACS National Meeting*, Orlando, FL, INOR-194, (2002).
179. Y. Kim, M. Suh, and D.-Y. Jung, *Inorg. Chem.*, *43*, 245 (2004).
180. Y.-Q. Sun, J. Zhang, and G.-Y. Yang, *Chem. Commun.*, 1947 (2006).
181. X. Guo, G. Zhu, Q. Fang, M. Xue, G. Tian, J. Sun, X. Li, and S. Qiu, *Inorg. Chem.*, *44*, 3850 (2005).
182. S. Mondal, M. Mukherjee, S. Chakraborty, and A. K. Mukherjee, *Cryst. Growth Des.*, *6*, 940 (2006).
183. J.-m. Shi, W. Xu, Q.-y. Liu, F.-l. Liu, Z.-l. Huang, H. Lei, W.-t. Yu, and Q. Fang, *Chem. Commun.*, 756 (2002).
184. R. Vaidhyanathan, N. Srinivasan, and C. N. R. Rao, *J. Solid State Chem.*, *177*, 1444 (2004).
185. A. Michaelides, S. Skoulika, E. G. Bakalbassis, and J. Mrozinski, *Cryst. Growth Design*, *3*, 487 (2003).
186. J. Legendziewicz, B. Keller, I. Turowska-Tyrk, and W. Wojciechowski, *New J. Chem.*, *23*, 1097 (1999).
187. L. Pan, E. B. Woodlock, X. Wang, and C. Zheng, *Inorg. Chem.*, *39*, 4174 (2000).
188. C. Daiguebonne, N. Kerbellec, K. Bernot, Y. Gerault, A. Deluzet, and O. Guillou, *Inorg. Chem.*, *45*, 5399 (2006).
189. L. Pan, K. M. Adams, H. E. Hernandez, X. Wang, C. Zheng, Y. Hattori, and K. Kaneko, *J. Am. Chem. Soc.*, *125*, 3062 (2003).
190. N. L. Rosi, J. Eckert, M. Eddaoudi, D. T. Vodak, J. Kim, M. O'Keeffe, and O. M. Yaghi, *Science (Washington, DC, U S)*, *300*, 1127 (2003).
191. L. Pan, B. Parker, X. Huang, D. H. Olson, J. Lee, and J. Li, *J. Am. Chem. Soc.*, *128*, 4180 (2006).
192. A.-Q. Wu, Y. Li, F.-K. Zheng, G.-C. Guo, and J.-S. Huang, *Cryst. Growth Des.*, *6*, 444 (2006).
193. M. Du, X.-J. Jiang, and X.-J. Zhao, *Chem. Commun.*, *44*, 5521 (2005).
194. B.-B. Ding, Y.-Q. Weng, Z.-W. Mao, C.-K. Lam, X.-M. Chen, and B.-H. Ye, *Inorg. Chem.*, *44*, 8836 (2005).
195. M. Du, C.-P. Li, and X.-J. Zhao, *Cryst. Growth Des.*, *6*, 335 (2006).
196. S. K. Ghosh, J. Ribas, M. S. El Fallah, and P. K. Bharadwaj, *Inorg. Chem.*, *44*, 3856 (2005).
197. S. Y. Yang, L. S. Long, Y. B. Jiang, R. B. Huang, and L. S. Zheng, *Chem. Mater.*, *14*, 3229 (2002).
198. D. Sun, S. Ma, Y. Ke, T. M. Petersen, and H.-C. Zhou, *Chem. Commun.*, *21*, 2663 (2005).
199. J. Perles, M. Iglesias, M.-A. Martin-Luengo, M. A. Monge, C. Ruiz-Valero, and N. Snejko, *Chem. Mater.*, *17*, 5837 (2005).
200. N. Mahe and T. Bataille, *Inorg. Chem.*, *43*, 8379 (2004).
201. X.-L. Hong, Y.-Z. Li, H. Hu, Y. Pan, J. Bai, and X.-Z. You, *Cryst. Growth Des.*, *6*, 1221 (2006).

## CHAPTER 4

# Supramolecular Chemistry of Gases

**DMITRY M. RUDKEVICH**

*Department of Chemistry and Biochemistry*
*The University of Texas at Arlington*
*Arlington, TX 76019*

CONTENTS



*Progress in Inorganic Chemistry, Vol. 55* Edited by Kenneth D. Karlin

# I. INTRODUCTION

Molecular chemistry is based on the covalent bond, while weaker, noncovalent forces define supramolecular chemistry. Supramolecular chemistry studies intermolecular interactions and assemblies of molecules (1). Over the last four decades, revolutionary advances of supramolecular chemistry have resulted in successful complexation, sensing, and separation of ionic species and neutral organic molecules (2). Molecular recognition of gases is still at the early stage (3). This is surprising. Gases compose the atmosphere and play important roles in medicine, science, technology, and agriculture. Hydrogen is extremely promising in the design of energy-rich fuel-cell devices. Oxygen is of great importance for medicine and also for steel making. Nitrogen is utilized in space technology and in ammonia production. Chemical, biomedical, and food industries widely use $O_2$, CO, $CO_2$, $N_2$, $NH_3$, $Cl_2$, and $C_2H_4$. Both $CO_2$ and $N_2O$ are major greenhouse gases (4). Upon burning fuels, oil, coal, wood, and natural gas, huge amounts of $CO_2$ are released into the air. As a result, $CO_2$ in the atmosphere is accumulating faster than the Earth's natural processes (plants and aqueous resources) can absorb it. Excessive emission of $N_2O$ to the atmosphere is caused by the widespread use of nitrogen fertilizers and industrial manufacturing of nylon (4). Both $CO_2$ and $N_2O$ are also involved in a number of important biochemical processes; they are called blood gases and $N_2O$ is heavily used in anesthesia. Another crucial group of gases is $NO_X$: The sum of NO, $NO_2$, $N_2O_3$, $N_2O_4$, and $N_2O_5$. Nitric oxide (NO) holds multiple roles in the human organism (5). It serves as an important messenger in signal transduction processes in smooth muscle cells and neurons and is also the key intermediate in global denitrification. Other $NO_X$ gases are extremely toxic pollutants derived from fossil fuel combustion, power plants, and large-scale industrial processes. The $NO_X$ gases participate in the formation of ground-level ozone, global warming, and they also form toxic chemicals, nitrate particles, and acid rain–aerosols. The $NO_X$ gases are aggressively involved in damaging nitrosation processes in biological tissues (6). Also important are $SO_2$ and $H_2S$ gases, which contribute significantly to the production of smog and acid rain (7).

Extensive circulation of gases necessitates the development of novel methods of their detection and monitoring under a variety of conditions. Another unresolved issue is their chemical fixation in environmentally benign processes.

Applying principles and techniques of supramolecular chemistry and molecular recognition for these purposes is just at the very beginning. However, there are a lot of exciting, unique opportunities and promises. In this chapter, general principles and rules toward design and synthesis of receptor molecules for gases will be overviewed. These are responsible for the binding selectivity and the biochemical action, and usually precede the covalent fixation. Gases surrounding us are not ideal gases and they interact with other molecules. Yet, their recognition and complexation remain a challenge. Experimentally, gases are more difficult to handle than liquids and solids. Their molecules are neutral, and electrostatic-binding interactions are much less effective than they would be for the similar size cations and anions. Furthermore, gases are small, of a 2–4-Å range. To achieve their binding, very close contacts with receptor sites are needed. A higher degree of design is required to achieve complementarity. Often, a combination of different binding forces is necessary. Solvation may well compete with the recognition event.

The possibilities of hydrogen bonding and Lewis acid–base interactions, dipole–dipole attractions, as well as encapsulation strategy will be discussed. Reversible covalent fixation of gases will also be introduced. Experimental accomplishments in all these directions will be analyzed. Further, existing approaches toward studying supramolecular complexes with gases will be reviewed. Thermodynamics and kinetics of supramolecular interactions with gases will be discussed, as well as the geometry and stereochemistry of gas complexes. This knowledge should allow for the development of more specific and selective receptors and molecular containers for gases; potentially useful chemical processes and catalysts for their utilization; novel gas sensors and chemically modified materials for gases and from gases.

Nature effectively employs molecular recognition for discriminating between blood gases. The differences in the $O_2$ and CO binding by hemes are the most spectacular example (8). In addition to the iron–gas interaction, the histidine residue on the distal porphyrin face of hemoglobin and myoglobin is involved in hydrogen bonding with $O_2$ (**1**, Fig. 1). Such hydrogen bonding not only affects

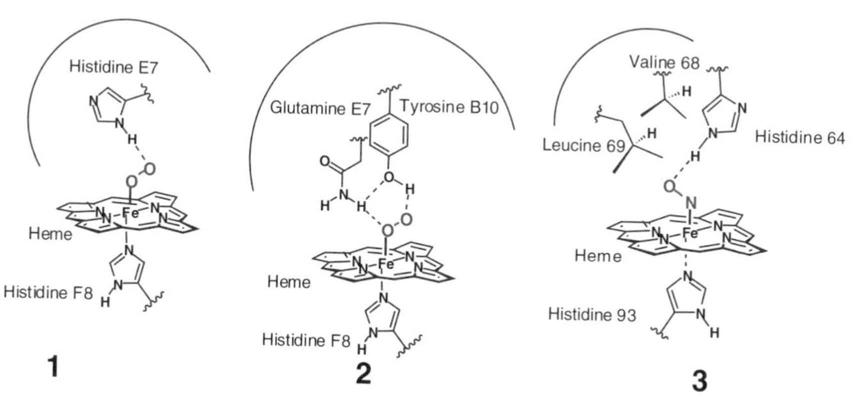


Figure 1. Molecular recognition of gases in natural systems.

the affinity, but also may stabilize the oxy form and prevent autooxidation. The distal cavity is also important. The walls polarity, functional composition, and the three-dimensional (3D) arrangement of amino acids all play their roles.

In oxygen-avid *Ascaris* hemoglobin, glutamine and tyrosine residues actively participate in hydrogen bonding with $O_2$ (**2**, Fig. 1) (9). In the crystal structure of complex **2**, the tyrosine hydroxyl is perfectly positioned to form a strong hydrogen bond with the distal atom of the complexed $O_2$, and the glutamine forms a somewhat weaker hydrogen bond to the oxygen atom coordinated with the iron. There is also a hydrogen bond between the tyrosine and glutamine fragments. Such hydrogen-bonding network is believed to be responsible for the four orders of magnitude stronger binding for $O_2$ compared to human hemoglobin.

The crystal structures of heme protein complexes with NO suggest that the distal cavity is important in gas binding. Being rather hydrophobic, such cavities possibly help to exclude the noncoordinated $H_2O$ molecule prior to the NO complexation. In the X-ray structure of ferrous nitric oxide form of native sperm whale myoglobin (**3**, Fig. 1), along with the Fe–NO interaction, the histidine 64 residue on the distal porphyrin face of myoglobin forms a hydrogen bond with the gas molecule (10). Overall, the NO binding event takes place in a tight cavity formed by lipophilic leu69 and val68 fragments and his64.

In medicine, Noble (Xe, Ar, Kr, Ne, He) and some other lipophilic gases (e.g., $N_2$, $H_2$, and $N_2O$) produce anesthesia, but the exact chemical mechanism of this process is still not fully understood. Binding energies between anesthetic sites and the gases, most probably include London dispersion and charge-induced dipole attractions. These are sufficient to overcome the concurrent unfavorable decrease in entropy, which occurs when a gas molecule occupies the protein site. Recent infrared (IR) spectroscopic studies of the interactions between the anesthetic $N_2O$ and lysozyme, cytochrome *c*, myoglobin, Hb (hemoglobin), serum albumin, and cytochrome *c* oxidase showed that $N_2O$ interacts with these proteins and occupies hydrophobic sites in their interiors (11).

These examples from Nature directly point out at the binding forces to employ and explore: hydrogen bonding, metal–gas interactions (Lewis acid–base interactions), and cavity effects.

## II. HYDROGEN BONDING WITH GASES

Hydrogen bonding is one of the most important interactions utilized by Nature, which is responsible for biotic self-assembly and enzyme selectivity. Besides the natural examples, more information is available on hydrogen bonds with gases. Gas-phase adducts of such acids (e.g., HF, HCl, HBr, HCN with $N_2$, CO, $CO_2$, OCS) have been described (12). The geometry of the B•••H–X (B = N, O, S; X = halogen) complexes was deduced from the *ab initio* calculations and also from the advanced spectroscopic measurements, such as molecular beam Fourier transform (FT) microwave spectroscopy, vibration spectroscopy, and photoelectron spectroscopy. Zero-electron kinetic energy (ZEKE) photoelectron spectroscopy and resonance-enhanced multiphoton ionization (REMPI) spectroscopy proved to be very useful to characterize the ArOH•••gas (gas = Ar, $N_2$, CO) molecular clusters (13). At the same time, much less is known about hydrogen bonding with gas molecules in solution.

A wide variety of heme models have been prepared, which are based on sterically hindered picket fence, picnic basket, capped, strapped, hanging base, and other superstructured metalloporphyrins (14). In some cases, the increased affinity toward $O_2$ for these porphyrins was interpreted by favorable hydrogen bonding between the complexed $O_2$ molecule and the hydrogen-donating functional fragment within the porphyrin structure. As a recent example, dendritic iron porphyrins (**4**, Fig. 2) were introduced possessing a highly

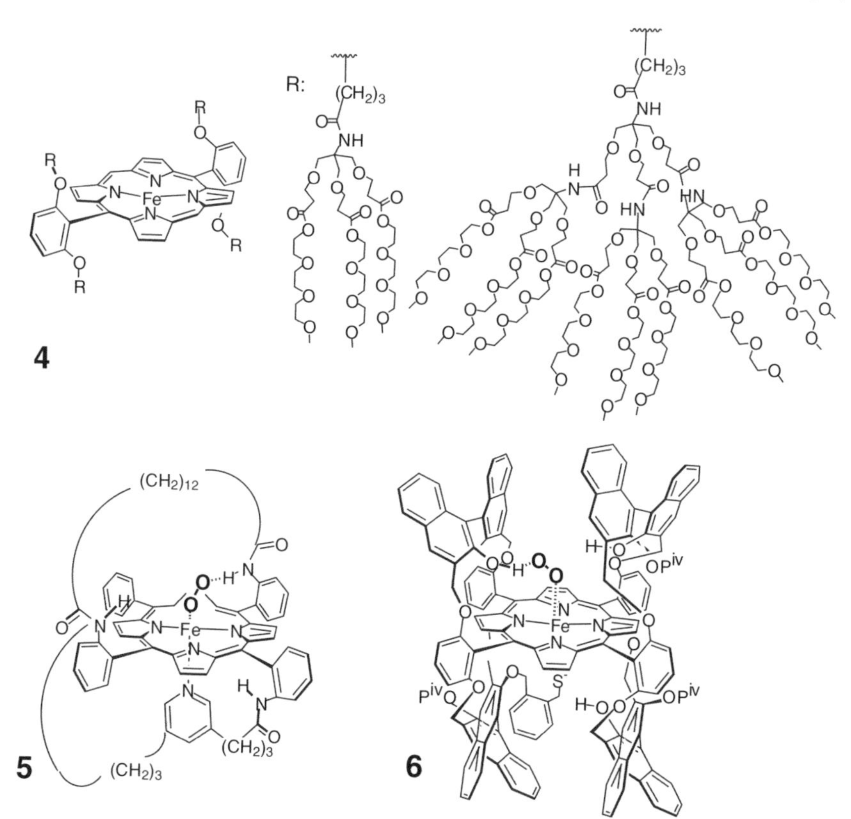

Figure 2. Superstructured porphyrins for complexation of $O_2$ employ metal–ligand interactions and hydrogen bonding.

protected gas-binding pocket. These porphyrins exhibited increased affinity toward $O_2$. To explain this, hydrogen bonding between the coordinated $O_2$ and the C(O)—NH function of the dendritic surrounding was proposed (15).

In the early proton nuclear magnetic resonance ($^1$H NMR) studies with the $O_2$ complexes of amide functionalized, basket-handle porphyrins (**5**, Fig. 2), a pronounced downfield shift of the amide C(O)—NH was observed in apolar solvent, indicating the C(O)—NH•••$O_2$ hydrogen bonding (14a,16). This hydrogen bond, however, was weak, since the estimated nitrogen–$O_2$ distance is $\sim$ 4 Å.

Later, Naruta and co-workers (17) presented Raman spectroscopic evidence for hydrogen bonding in the complex of twin-coronet porphyrin (**6**) and $O_2$ (Fig. 2). Specifically, the hydrogen bond between the bound $O_2$ and the OH groups of the binaphthyl moieties was studied through H/D exchange experiments. When the OH groups in **6** were deuterated, the frequencies of the ν(O—O) bands were shifted $\sim$2 $cm^{-1}$ to a higher field. The estimated distances between the binaphthyl oxygens and the bound $O_2$ are short and in the range of 3 Å.

Direct evidence of hydrogen bonding in oxyhemoglobin was obtained by echo–anti–echo ($^1$H, $^{15}$N) heteronuclear multiple quantum coherence (HMQC) nuclear magnetic resonance (NMR) experiments (18). Through analysis of the pattern of cross-peaks for the distal histidines in the HMQC spectrum, it was found that the histidine NH hydrogen in the $O_2$ complex is stabilized against solvent exchange by a hydrogen bond with $O_2$.

Other avenues have also been explored. One of them is designing $CO_2$-philic molecules and structural fragments, whose attachment to chelating agents, surfactants, catalysts, and other functional molecules would enhance the solubility of such compounds in *sc*$CO_2$ (*sc* = supercritical) (19). It was found that carbonyl derivatives are very useful for these purposes. Spectroscopic techniques and computational studies identified interesting complexes between $CO_2$ and carbonyl-containing compounds (20). For example, two-point interactions between a C(O)Me group and $CO_2$ were identified for complexes of peracetylated carbohydrates (**7–9**) and $CO_2$ (21, 22). This involves *both* the acetate C=O•••$CO_2$ electrostatics and C—H•••$O_2$C hydrogen bonding (Fig. 3). As a result, derivatives **7–9** have an enhanced solubility in supercritical $CO_2$. *Ab initio* calculations of $CO_2$ complexes with simple carbonyl compounds were performed, giving binding energy values of 2.69 kcal $mol^{-1}$ per complex. In the Raman spectra of the complexes between acetaldehyde and $CO_2$, the acetaldehyde carbonyl band maximum red shifted from 1746.0 to 1743.5 $cm^{-1}$, and the aldehyde CH proton band maximum blue shifted from 2717.0 to 2718.3 $cm^{-1}$ (22). Both observations were consistent with the computer predicted structures. These studies may also help to understand interactions between $CO_2$ and carbonyl-containing membrane polymers.

It was also found that secondary cis-amides interact with $N_2O$ and $CO_2$ in apolar solution (Fig. 3) (23). For example, upon saturation of the solution of

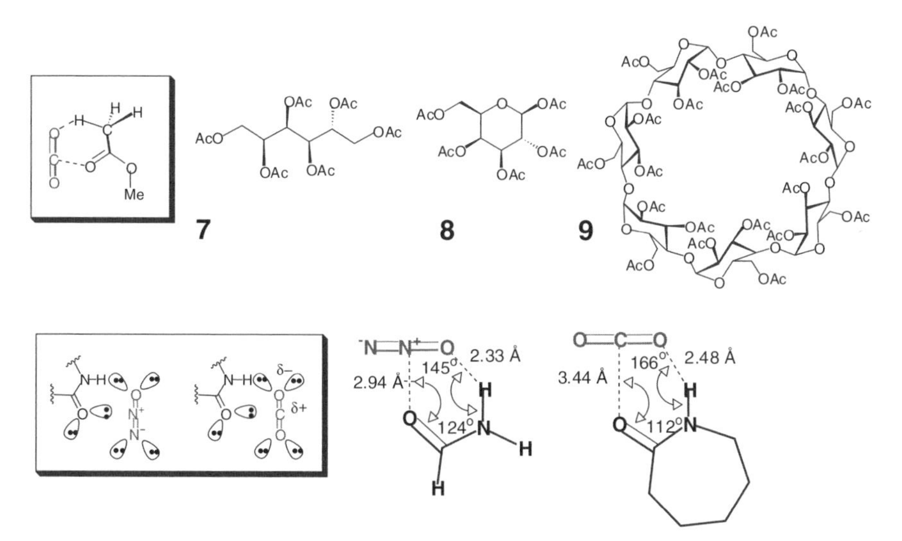


Figure 3. Both $N_2O$ and $CO_2$ are involved in hydrogen bonding and electrostatic interactions with carbonyl-containing compounds. Complexes with acetates and cis-amides have been observed experimentally. Bottom right: calculated geometrical parameters for complexes between cis-amides and $N_2O$ and $CO_2$ (23).

caprolactam with $N_2O$ at room temperature, the NH singlet shifted upfield $\Delta\delta \sim 0.25$ in $CDCl_3$ and benzene-$d_6$. *Ab initio* calculations of the complex between formamide (H—C(O)—$NH_2$), also possessing a cis-amide arrangement, and $N_2O$ confirm that a hydrogen bond is possible between the amide N—H hydrogen and partially negatively charged oxygen of the gas molecule. The N—H•••O=$N^+$=$N:^-$ distance of 2.33 Å was computed. This places the amide basic C=O oxygen right in front of the central, electron deficient nitrogen of the $N_2O$ molecule. There is an electrostatic attraction between the former lone pair and the latter partial positive charge. The C=O•••$N^+$ distance of 2.94 Å was calculated. Taken together, two-point noncovalent interaction occurs, which is not possible for trans-amides. The lower limit of $\Delta G^{295} \sim 0.9\,\text{kcal mol}^{-1}$ was estimated for the $N_2O$ complexation. This is typical for weak interactions in apolar solution. Analogous conclusions were obtained for $CO_2$, which is isoelectronic to $N_2O$ (23). These studies point out at possible interactions between anesthetic gases and proteins.

## III. LEWIS ACID–BASE INTERACTIONS WITH GASES

Lewis acid–base interactions with gases are closely related to hydrogen bonding. However, they are stronger, and therefore bring more applications.

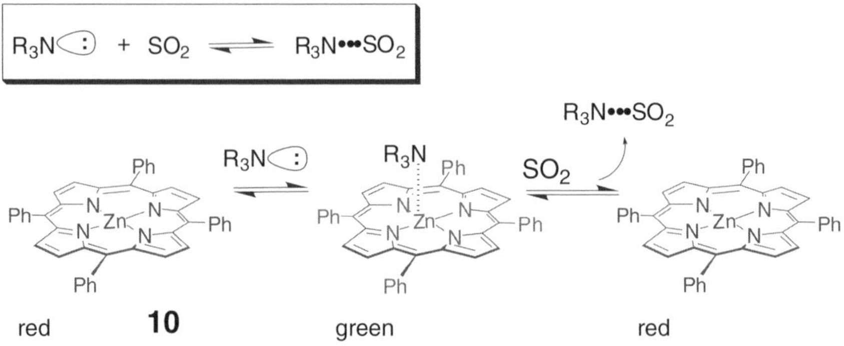

Figure 4. An indicator-displacement assay for colorimetric sensing of $SO_2$ gas (25).

That $SO_2$ and secondary or tertiary amines form stable charge-transfer (CT) complexes, both in solution and in the solid state, has been known for decades (24). A supramolecular approach for potential detection of $SO_2$ was recently introduced, which includes an indicator-displacement assay (25). When amines were added to Zn-tetraphenylporphyrin (**10**) in $CHCl_3$, the solution changed from red to dark green (Fig. 4). A bathochromic shift of $\Delta\lambda \sim 10$ nm was observed for the Soret band, indicating the formation of **10**•amine complexes. After this, $SO_2$ gas was introduced, and the original red color of the solution was restored. The Soret band returned to its position for free porphyrin (**10**). The **10**•amine complexes dissociated, and new $SO_2$•amine adducts formed. Porphyrin **10** thus served as an indirect colorimetric indicator for $SO_2$. The system discriminates between $SO_2$ and such typical exhaust gases as $CO_X$, $NO_X$, and $H_2O$.

While there are obvious ultraviolet (UV) changes simply upon addition of $SO_2$ to an amine, incorporating the porphyrin in the assay brings the response into the visible region of the spectrum. It should be possible to modify both amines and porphyrins to achieve more colorful responses, and also to detect $SO_2$ in aqueous solutions and at the gas–solid interface.

Inorganic and organometallic compounds are widely known to react with $N_2$, NO, $N_2O$, CO, $CO_2$, $O_2$ and others through metal–gas bond formation. Some of these processes are reversible and related to supramolecular chemistry. Collman et al. (26) discovered controlled gas exchange within the cavity of ruthenium picnic-basket porphyrins (**11**, Fig. 5). The Ru(II)–gas interactions were the driving force for these processes, but the cavity also played a crucial role, especially in the kinetics and regioselectivity. First, so-called bis(solvent) complex **11**•$(thf)_2$ where thf = tetrahydrofuran (ligand) was prepared by photolysis of the corresponding carbonyl derivative **11**•CO•thf. Treatment of this with bulky 1,5-dicyclohexylimidazole (1,5-DCI) resulted in formation of

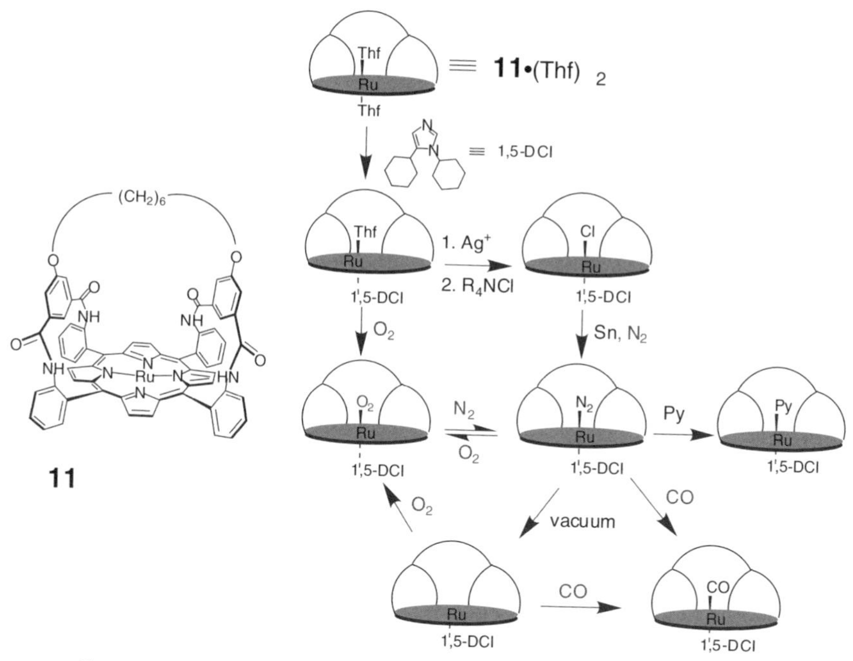


Figure 5. Exchange experiments with gases in picnic-basket porphyrin **11** (26).

**11**•$(thf)_{in}$•$(1,5\text{-DCI})_{out}$ complex, with the thf molecule coordinated inside the cavity. Dioxygen quantitatively replaced the internally coordinated thf molecule in toluene solution, and this process appeared to be reversible. Further, by introducing $N_2$ reversible exchange between the corresponding dioxygen **11**•$(O_2)_{in}$•$(1,5\text{-DCI})_{out}$ and dinitrogen **11**•$(N_2)_{in}$•$(1,5\text{-DCI})_{out}$ complexes was achieved. Dinitrogen complex **11**•$(N_2)_{in}$•$(1,5\text{-DCI})_{out}$ was also converted to the carbon monoxide relative **11**•$(CO)_{in}$•$(1,5\text{-DCI})_{out}$, simply by exposure to CO gas, and also to the gas-free derivative **11**•$(-)_{in}$•$(1,5\text{-DCI})_{out}$. Upon exposure to $O_2$, this was readily converted to **11**•$(O_2)_{in}$•$(1,5\text{-DCI})_{out}$. The gas exchange processes were effectively monitored by visible (vis), IR, and $^1H$ NMR spectroscopy (26).

van Koten and co-workers (27) proposed square-planar *N,C,N*-terdentate pincer Pt(II) complexes (**12**) for reversible sensing of $SO_2$. In these, $SO_2$ is directly bound to the metal center. In the presence of $SO_2$, complexes **12** spontaneously adsorbed the gas and formed pentacoordinated, square-pyramidal adducts **12**•$SO_2$ (Fig. 6). The process is fast and reversible, both in apolar solution ($CH_2Cl_2$, $CHCl_3$, $C_2F_4Br_2$, and toluene) and in the solid state. Importantly, a characteristic color change of the complex from colorless-to-bright orange occurred upon $SO_2$ entrapment. The exchange of $SO_2$ is fast at room temperature,

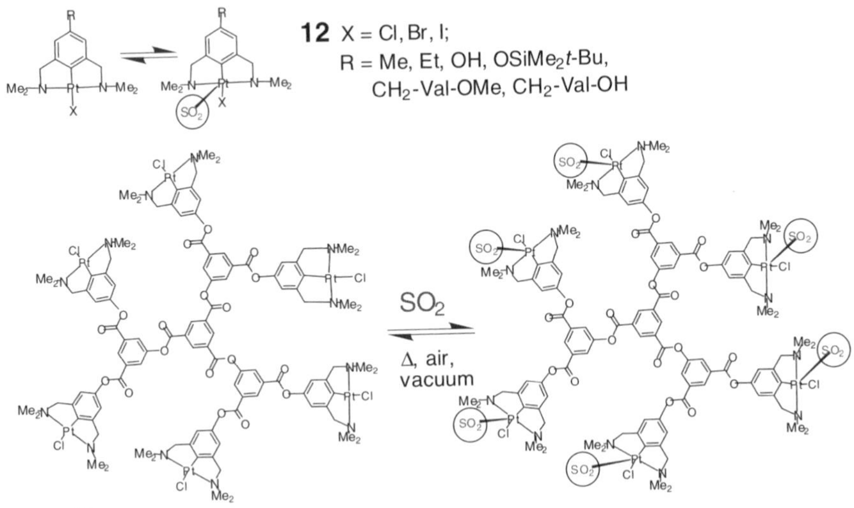

Figure 6. Reversible binding of $SO_2$ by pincer Pt(II) complexes **12** and **13** (27). (Adapted from (3a).)

but can be slowed down and studied by conventional $^1$H NMR spectroscopy at 167–188 K (in $C_2F_4Br_2$). For example, the exchange rate constants are $k = (1.5 \pm 0.5) \times 10^8\ s^{-1}$ at 298 K and $k = (2.34 \pm 0.08) \times 10^3\ s^{-1}$ at 174 K. The VT NMR (VT = variable temperature) analysis of the equilibrium constant between complex **12** and adduct **12•**$SO_2$ (X = I) gave $\Delta H^\circ = -36.6 \pm 0.8$ kJ mol$^{-1}$, $\Delta S^\circ = -104 \pm 3$ J K$^{-1}$ mol$^{-1}$, and $K_{298} = 9 \pm 4\ M^{-1}$.

Crystalline-state reactions between **12** and $SO_2$ produced the orange-colored, reversible adducts, which may be used for gas storage and optoelectronic switchable devices. Molecules **12** were also covalently linked to amino acid valine (Val) through the *N*-terminus. This may provide a peptide-labeling tool for diagnostic applications. In biological systems, such organoplatinum-labeled peptides may be traced and located specifically by highlighting the platinum site with $SO_2$.

In addition, dendritic organoplatinum(II) complexes (**13**) were prepared as reusable $SO_2$ sensing and storing materials (Fig. 6). Such dendrimers offer monodisperse, globular nanostructures, with defined size and shape. Through multiple, multivalent interactions, macromolecular receptors (**13**) display an increased affinity toward the gas.

While searching for materials capable of $O_2$ gas storage, the dendritic hemocyanin model was prepared (28). Hemocyanin proteins are known to bind $O_2$ at a site containing two copper atoms, directly ligated by protein side chains. From the synthesized Cu(I) dendrimer, a multioxygen complex was obtained and 10–11 molecules of $O_2$ per dendritic molecule were found. It was shown that 60–70% of the Cu(I) centers were involved in oxygen binding. The high local concentration of oxygenating equivalents in such a hot dendrimer may be of interest for synthetic chemistry.

Borovik and co-workers (29–31) developed porous organic materials for reversible binding of CO, $O_2$, and NO by means of gas chemical coordination to the metal centers. For immobilization of metal centers, templated copolymerization was employed (Fig. 7). Material **14**, for example, contained immobilized four-coordinate Co(II) centers, and the cobalt concentration ranged from 180 to 230 mol $g^{-1}$, with an average pore diameter of 25 Å (31). Polymer **14** bound NO in toluene solution and even on the air–solid interphase, but was relatively inert toward other biologically important gases $O_2$, $CO_2$, and CO. Nitric oxide could be slowly released from **14** under ambient conditions. For example, after 30 days ~80% of NO was lost. Heating the sample accelerated the gas release.

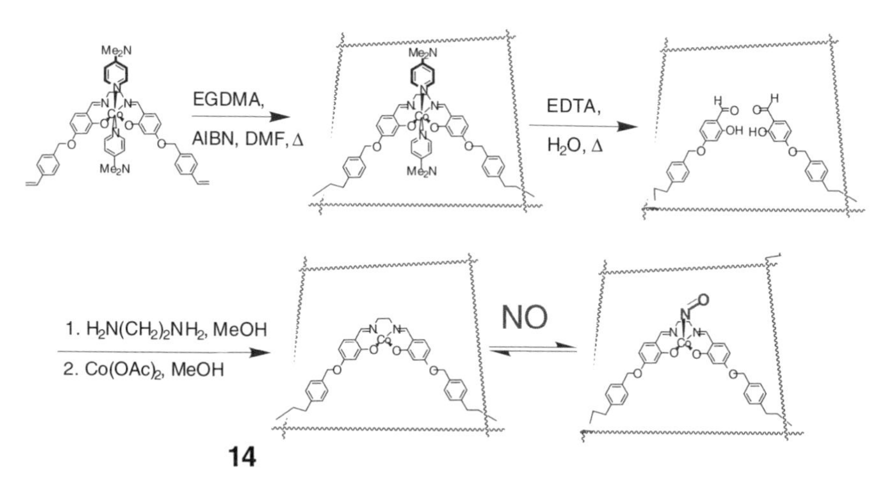


Figure 7. Preparation of porous organic material (**14**) for reversible binding of NO gas (31). The template Shiff-base complex was used, along with the cross-linking agent EGDMA (EGDMA = ethylene glycol dimethylacrylate), and solvent, which served as the porogen during the radical polymerization process. The pre-formed kinetically inert $Co^{III}$–DMAP (DMAP = *N*, *N*-dimethylaminopyridine) bis(salicyledene) ethylenediamine = salen) complex contains the styrene-modified salen ligand, which is covalently attached to the porous methacrylate host. The Co(III) ions and DMAP were removed under acidic conditions to afford the bis(aldehyde) material. This further reacted with ethylenediamine (en) to reform the immobilized tetradentate salen, which readily binds Co(II) ions to form **14**.

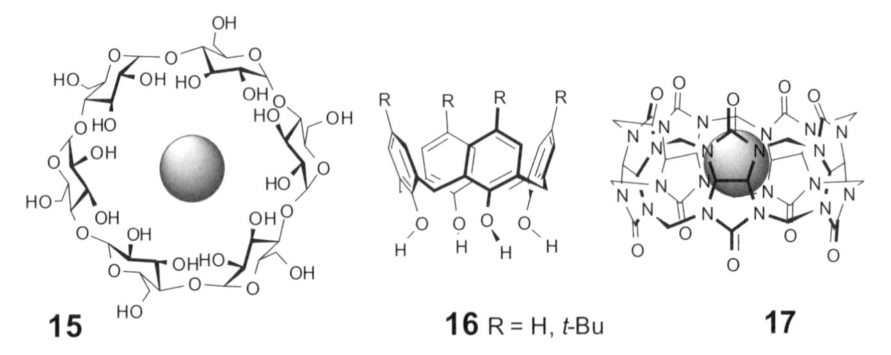


Figure 8. Open-ended molecular containers, or cavitands, for gases: cyclodextrins, calixarenes, and cucurbiturils.

Such controlled release of NO may be important for medical applications. Finally, the pronounced color change resulting from the conversion of **14**–**14**•NO can be useful for sensing.

## IV. ENCAPSULATION OF GASES

### A. Synthetic Molecular Containers: Cavitands and Carcerands

Early gas encapsulation experiments were performed with naturally occurring cyclodextrins in the 1950s (32). α-Cyclodextrin (**15**, Fig. 8), for example, interacted with $Cl_2$, Kr, Xe, $O_2$, $CO_2$, $C_2H_4$, $CH_4$, $C_2H_6$, $C_3H_8$, and $n$-$C_4H_{10}$ in water, under an elevated pressure of these gasses, with the formation of stable clathrates. The **15**•$CO_2$ complex was isolated simply by precipitation, and this was used for the purification of the cyclodextrin itself (33).

To obtain quantitative information about guest encapsulation processes with cyclodextrins, $^{129}$Xe NMR spectroscopy was later employed (34). This gas is highly polarizable, but inert and hydrophobic. Two isotopes are easily accessible to NMR spectroscopy: $^{129}$Xe ($I = \frac{1}{2}$, natural abundance of 26%) and $^{131}$Xe ($I = \frac{3}{2}$, abundance 21%). The NMR parameters of these isotopes are very sensitive to the environment, in which Xe is located. This feature has been used to probe structural and dynamic properties of host–guest complexes with Xe, both in the solid state and in solution. For example, in protein NMR studies, weak Xe–protein interactions influence the $^{129}$Xe chemical shift depending on the accessible protein surface and allow the monitoring of the protein void spaces and conformations.

In the early 1980s, Cram (35a) designed synthetic molecular containers for encapsulation. Cavitands were the most popular among them. By definition, cavitands are open-ended molecular containers. A number of gases, including $O_2$, were initially tested, however, the wide-open inner cavities of cavitands did not allow the tight encapsulation (36). More recently, Atwood et al. (37) showed that $CH_4$, $CF_4$, $C_2F_6$, $CF_3Br$, and other low-boiling halogenated alkanes could be reversibly entrapped and retained within the lattice voids of a crystalline calix[4]arene framework (see **16**, Fig. 8). Ripmeester discovered that the calix[4]arene (**16**) (R = *t*-Bu) cavities in such crystals are directly involved in the gas complexation (38). The Atwood's team further demonstrated that calix[4]arene (**16**) (R = *t*-Bu) dimerizes in a crystalline phase into a hourglass-shaped cavity, capable of gas entrapment (39). These crystals soak up gases when stored in air. Absorption of $CO_2$ was particularly rapid, but CO, $N_2$, and $O_2$ were also trapped. Of special importance, the calixarene crystals selectively absorbed $CO_2$ from a $CO_2$—$H_2$ mixture, leaving the $H_2$ behind. This phenomenon can be used for purification of $H_2$. Very recently, Atwood showed that calix[4]arene crystals can also absorb $H_2$ at higher pressures (40). Cavity-containing solid materials for gas entrapment, storage, and release have thus emerged.

Decamethylcucurbit[5]uril (**17**) (a smaller relative of cucurbituril) was found to entrap a variety of lipophilic gases such as $H_2$, He, Ne, Ar, $N_2$, $O_2$, $N_2O$, NO, CO, $CO_2$, $CH_4$, and acetylene in water (Fig. 8) (41). The gas encapsulation was studied by conventional NMR spectroscopy in $D_2O$ and also single-crystal X-ray analysis. In the case of $CH_4$, the encapsulated $CH_4$ singlet appeared at −0.87 ppm in the $^1$H NMR spectrum. Weak $^{13}$C NMR resonance at 127.4 ppm for the dissolved $CO_2$ transformed into a strong signal at 127.1 ppm upon encapsulation. In the presence of **17**, typically insoluble acetylene exhibited signals at 1.02 ppm ($^1$H NMR) and 73.6 ppm ($^{13}$C NMR).

Most of the gases were encapsulated by simply bubbling through the $D_2O$ solution of decamethylcucurbit[5]uril (**17**) at room temperature. On the other hand, larger gases (e.g., $CH_4$, Kr, and Xe) required heating up to 80°C. Most probably, competition with earlier encapsulated water molecules was responsible for such behavior. By circulating the air through the powder of **17**, a 5% level of $CO_2$ was decreased to 0.01%. Although with some lower capacity compared to 5-Å molecular sieves, decamethylcucurbit[5]uril (**17**) offers much lower regeneration temperature (110 vs. 350°C).

Carcerands are another popular class of Cram's molecular containers. They are close-surface host-molecules with enforced inner cavities, which are large enough to incarcerate smaller organic molecules, and portals of which are too narrow to allow the guest to escape without breaking covalent bonds (35). Complexes of carcerands (carceplexes) represent an extreme form of host–guest complex stabilities. Carcerands hold guests permanently. In fact, most carceplexes include guests during their synthesis, upon covalent shell closure. When concave cavitand derivatives **18** and **19** were coupled in the presence of $Cs_2CO_3$

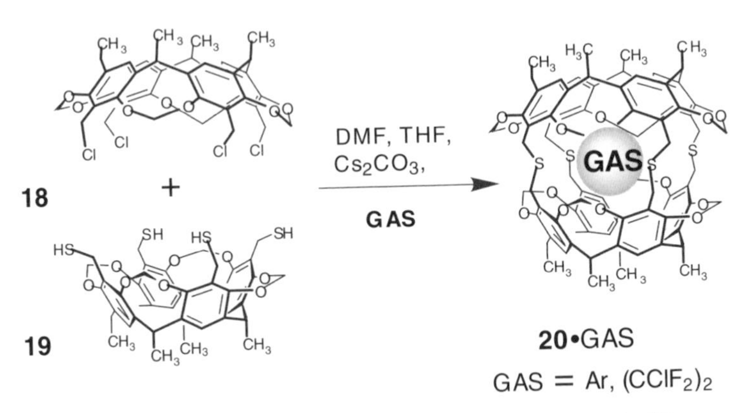


Figure 9. Cram's carcerand (**20**) encapsulates gases permanently (42).

in DMF–THF (demetylform amide = DMF) solution under an Ar atmosphere, carcerand (**20**) formed, which contained encapsulated Ar (Fig. 9) (42). The gas presence was confirmed by elemental analysis and fast atom bombardment mass spectrometry (FABMS). The analytical data indicated that only 1 among 150 shell-closures encapsulates the gas. The diameter of Ar is ~3.1 Å, which is greater than the portal diameter (~2.6 Å), and the gas apparently could not escape. When the reaction between **18** and **19** was performed under the atmosphere of Freon 114, this gas was also trapped (42).

Hemicarcerands are carcerands with larger portals, which allow for the entrapped guest to escape at high temperature, but to remain incarcerated under conditions of their isolation and characterization at ordinary temperatures (35). Hemicarcerand (**21**) reversibly encapsulated $O_2$, $N_2$, $CO_2$, and Xe (Fig. 10) (43). The exchange between the free and occupied **21** was slow on the NMR time scale. The $K_{assoc}$ constants values of 180 $M^{-1}$ ($N_2$), 44 $M^{-1}$ ($O_2$), and 200 $M^{-1}$ (Xe) were obtained in $CDCl_3$ at 22°C, assuming a 1:1 stoichiometry. The inner-cavity volume in **21** is relatively large (~100 $Å^3$) and the volume of gas molecules are within the 40-$Å^3$ range. Recently, He, $H_2$, $N_2$, $N_2O$, and $CO_2$ were encapsulated by hemicarcerand **21** in the solid state, so the scope of encapsulation was expanded from solution to the gas–solid interface (44). These gases were shown to replace each other in the solid (**21**). For example, upon flashing the powder containing **21** and **21**•$N_2$ with $CO_2$ or $N_2O$, hemicarceplexes, **21**•$CO_2$ or **21**•$N_2O$ were obtained. Being identical to the exchange in solution, these data somewhat diminish the role of solvent in gas encapsulation.

Very recently, Naruta and co-workers (45) demonstrated reversible encapsulation of hydrocarbons into the hemicarcerand container **22** (Fig. 10). Methane, ethane, propane, cyclopropane, ethylene, and acetylene were trapped in $CDCl_3$, showing the $K_{assoc}$ values between 10 (ethane, cyclopropane) and 130 $M^{-1}$

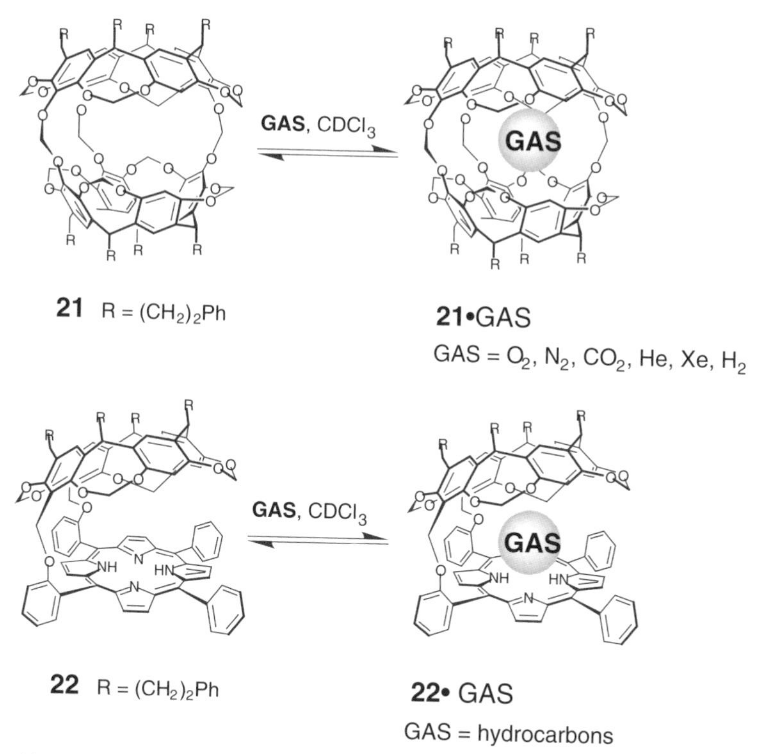


Figure 10. Encapsulation of gases within hemicarcerand (**21**) (43, 44) and cavitand–porphyrin (**22**) (45).

(acetylene). The presence of a porphyrin fragment in **22** should allow for the metal insertion and thus creation of catalytic vessels for hydrocarbons in the future.

So-called cryptophane-A (**23**) and its derivatives represent another class of hemicarcerands (Fig. 11). With the inner cavity of 95 $Å^3$, host **23** readily encapsulated $CH_4$ with $K_{assoc} = 130\,M^{-1}$ and Xe with much higher $K_{assoc}$ of $3000\,M^{-1}$ in $(CDCl_2)_2$ (46). Likewise for **21**, the gas molecules were better guests for cavity **23** than the solvent molecules, which were simply too big (>70 $Å^3$) to enter and/or occupy the interior. This provided us with a strong driving force for the gas entrapment.

Pines and co-workers (47) attached cryptophane-A to biotin. Cage **24** was used as a $^{129}$Xe NMR biosensor to monitor biotin–avidin binding. Free Xe in water generated a signal at 193 ppm, while the **24**•Xe peak was located at 70 ppm. Upon addition of avidin, a third peak appeared 2.3 ppm downfield of the **24**•Xe peak, which was assigned to the protein-bound **24**•Xe complex. When the concentration of avidin was increased, the intensity of this signal increased. The proposed methodology offers the capability of attaching different

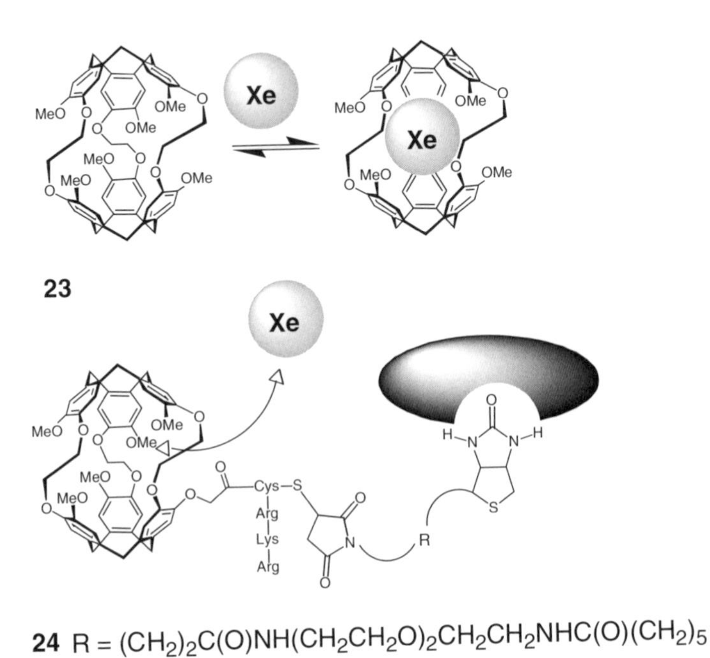


Figure 11. Cryptophane-A (**23**) and its complexes with Xe (46). Biotinylated cryptophan-A (**24**) for binding to proteins (47). [Adapted from (3a).]

ligands to different cages, thus forming Xe-based sensors associated with distinct resolved chemical shifts.

## B. Self-Assembling Capsules

Self-assembly is another way to build molecular containers (48). These are called self-assembling capsules and form through reversible noncovalent interactions between two or more concave subunits. Capsules possess enclosed cavities. Similar to carcerands and hemicarcerands, capsules completely surround their guests, but may release them without breaking covalent bonds. Rebek's tennis balls (**25**) assembled through hydrogen bonding around a molecule of $CH_4$, ethylene, or Xe in $CDCl_3$ solution (Fig. 12) (49). The gas molecules volumes are 28, 40, and 42 Å$^3$, respectively, and the inner cavity of capsule **25** is 50 Å$^3$. The exchange between free and trapped guest species was slow on the NMR time scale. For example, uncomplexed $CH_4$ exhibited a singlet at 0.24 ppm, while the encapsulated species gave a peak at −0.91 ppm. Both signals could be seen

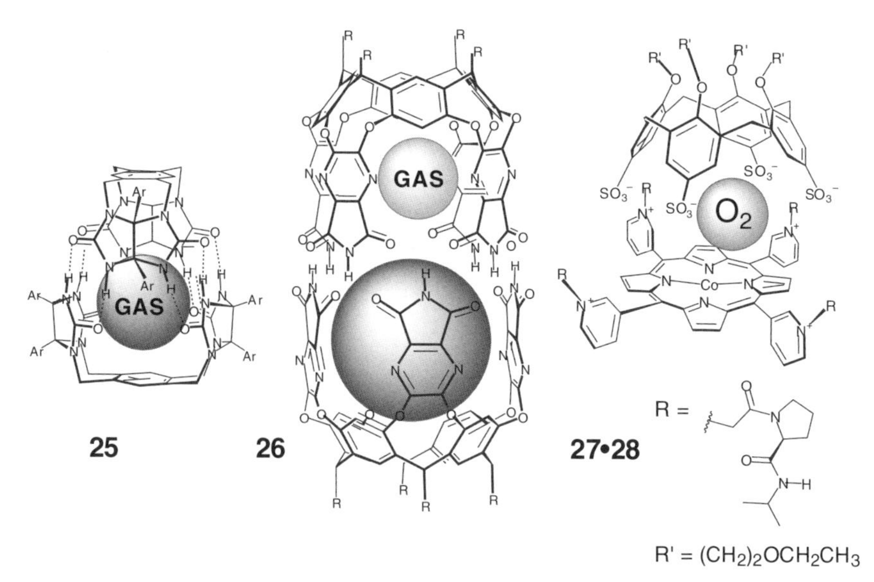


Figure 12. Self-assembilng capsules for gas complexation.

simultaneously. For this gas, $\Delta H = -9\,\text{kcal mol}^{-1}$, $\Delta S = -20$ e.u., and $K^{273} = 300\,M^{-1}$ were estimated in $CDCl_3$.

Binary coencapsulation of ethane, cyclopropane, and isomeric butanes together with larger organic guest (e.g., *p*-xylene, naphthalene, and anthracene) was observed for nanoscale cylindrical capsule **26** (50). Note that these gases alone could not be trapped (Fig. 12).

Reinhoudt and co-workers (51) targeted $O_2$ and prepared water-soluble models of heme–protein active sites. These were self-assembling, heterodimeric capsules of cationic Co(II) porphyrin (**27**) and tetrasulfonato calix[4]arene (**28**) (Fig. 12). A millipore VMWP membrane, loaded with an aqueous solution of **27** (2 m*M*), **28** (6 m*M*), and 1-methyl imidazole (20 m*M*), showed a facilitated transport of $O_2$. The facilitation factor of 1.15 and the $O_2/N_2$ selectivity of 2.3–2.4 were obtained, while the membrane loaded only with $H_2O$ showed the facilitation factor of 1.0 and the $O_2/N_2$ selectivity of 2.0.

## C. Fullerenes

With current interests in fabricating nanoscale structures for biomedical and materials applications, much attention is focused on fullerenes and nanotubes. Fullerene ($C_{60}$, **29**) possesses an inner cavity of ~4-Å diameter and is known to permanently trap Noble gases (Fig. 13) (52). The process is rather extreme and

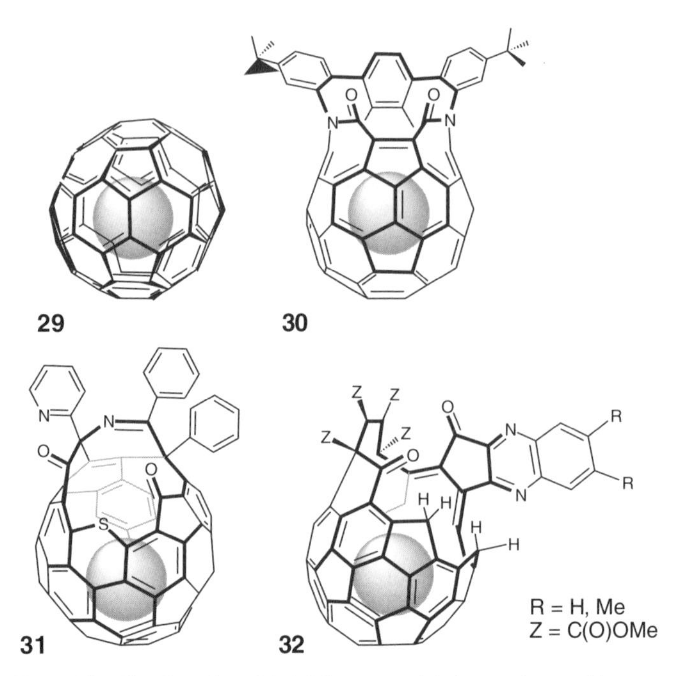


Figure 13. Simple and modified fullerenes and their complexes with gases.

requires heating at 650°C and 2000–3000-atm pressures. Under these conditions, Noble gases He, Ne, Ar, Kr, and Xe were incorporated into the $C_{60}$ and $C_{70}$ interiors. Even so, the fractional occupational levels of only 0.1% were achieved. Mechanistically, it was proposed that the gas molecule enters through a window formed by the reversible breaking of one or more carbon–carbon bonds. Theoretical barriers, for example, He to penetrate the benzene rings of fullerene were calculated to be $\geq 200$ kcal $mol^{-1}$. The encapsulated gases can be released by breaking of one or more carbon–carbon bonds upon heating under vacuum. Other methods to entrap gases in $C_{60}$ are also extreme and based on the bombardment by the corresponding gas ions–fast atoms or generating high kinetic energy atoms in a nuclear reaction hot-atom chemistry.

Alternatively, fullerenes can be opened through so-called molecular surgery (53). This requires heavy synthetic efforts, but the results are exciting. Open-neck $C_{60}$-based bis(lactam) (**30**) encapsulated gases much more easily (Fig. 13) (53). With the distance between parallel rings in the orifice of ~4.1 Å, calculated insertion barriers for He and $H_2$ are 24.5 and 41.4 kcal $mol^{-1}$, respectively. This corresponds to the insertion temperatures of 124 and

397°C, respectively. Experimentally, $^3He$ gas was inserted into **30** at 475 atm, ~300°C over 7.5 h (1.5% maximal yield). For $H_2$, it took 100 atm, 400°C and 48 h (5% yield). Both complexes were studied by NMR spectroscopy. Complex **30**•He showed a $^3He$ NMR signal at −10.1 ppm, relative to the free $^3He$ gas at 0 ppm. The encapsulated inside **30** $H_2$ was seen at −5.43 ppm.

With a larger orifice (3.75 × 5.64 Å) on the surface of the $C_{60}$ derivative **31**, quantitative encapsulation of $H_2$ molecule was accomplished (at 200°C, 800 atm) (Fig. 13) (54). The encapsulated $H_2$ molecule exhibited a sharp singlet at −7.25 ppm in the $^1H$ NMR spectrum (*o*-dichlorobenzene-$d_4$). It was also detected by matrix-assisted laser desorption ionization–time of flight (MALDI–TOF) mass spectrometry (MS). The activation energy for the $H_2$ escape of ~34 kcal $mol^{-1}$ was experimentally obtained. Spectacularly, a single $H_2$ molecule encapsulated inside **31** was recently observed by a single-crystal X-ray diffraction analysis (55).

The 13-membered orifice in complex **31**•$H_2$ can be closed via a four-step synthetic procedure, resulting in completely sealed **29**•$H_2$ (56). This complex is as stable as $C_{60}$ itself. For example, the encapsulated $H_2$ does not escape when heated at 500°C for 10 min. Using the same strategy, it could be possible to prepare **29**•$D_2$, **29**•HD, and other endohedral fullerenes.

Much faster guest exchanges were observed for fullerenes with larger openings. Derivative **32** possesses a 20-membered orifice of 4.2 × 6.5 Å and was found to trap one $H_2O$ molecule (Fig. 13) (57). In the $^1H$ NMR spectrum of **32**•$H_2O$, a sharp singlet at −11.4 ppm in tetrachloroethane-$d_2$ was detected for the encapsulated guest. This, however, disappeared upon addition of $D_2O$, indicating rather frequent guest mobility.

## D. Carbon Nanotubes

Recent studies have revealed that some industrially and biologically important gases can be adsorbed and stored inside single-walled carbon nanotubes (SWNTs) (58). Initial synthesis of SWNTs typically produces sealed structures, which prevents the gas adsorption within the interiors. Chemical treatments (e.g., oxidation) lead to open-ended SWNTs. In addition to the interiors, gases may also be adsorbed in the outside surfaces, groove sites, and in the interstitial sites of nanotubes. Calculations showed that the SWNTs' interiors are energetically more favorable than their surfaces (59). Entrapping isotopes of Noble gases by SWNTs may improve their use in medical imaging, where it would be desirable to confine the gas physically before injection (60). Single-walled carbon nanotubes also trap $N_2$, $O_2$, NO, $CO_2$, and $CF_4$ (61). Storage of $H_2$ in SWNTs is extremely promising in the design of energy-rich fuel-cell electric devices. This area is currently a subject of extensive studies (62).

### E. Solid Networks and Materials

That certain porous solids form inclusion complexes with smaller molecules has been known for years, but with recent developments in materials chemistry and nanotechnology, this field is booming.

Among clathrates, gas hydrates are probably known the best. Hydrogen-bonded inclusion complexes of water are called clathrate hydrates (63). Clathrate hydrates with gases have a long history, dated back to Faraday, who in 1820s prepared crystalline hydrates of $Cl_2$ and $Br_2$. Later, also in the nineteenth century, hydrates of most major gases were isolated, but it took another century before first X-ray crystallographic studies were published. In gas clathrate hydrates, voids are much larger than in ice, and the nets are unstable unless occupied. Cages formed by water crystals under ambient temperatures ($<300$ K) and moderate pressures ($>6$ atm) may accommodate $CH_4$ and other gaseous hydrocarbons, fluorinated and chlorinated hydrocarbons and their hybrids, $CO_2$, $N_2$, $N_2O$, $SO_2$, $H_2S$, $H_2Se$, $O_2$, Xe, Ar, and Kr. Recently, interest in hydrates of $CH_4$ and $CO_2$ grew. Carbon dioxide can be buried as a hydrate in the deep ocean in order to reduce the release of this greenhouse component to atmosphere. Hydrates of $CH_4$ can serve as a source of hydrocarbons and, in general, organic carbon. Literature on other various clathrates with gases is also available (64).

Extensive studies are being undertaken in the search for stable nano- or microporous networks utilizing coordination polymers. Porosity of these polymers can be programmed depending on the applications. A variety of metal–organic microporous materials (MOMs) and frameworks (MOFs) have been designed for gas storage and transport (65). Both MOMs and MOFs were found to adsorb $N_2$, $O_2$, Ar, $CO_2$, $N_2O$, $H_2$, $CH_4$, and acetylene. The remarkable ability of certain MOFs in the sorption of $H_2$ makes them very attractive candidates for a vehicular hydrogen-based economy. The U.S. Department of Energy's ambitious 2010 target for storage of $H_2$ is 6 wt%, and should be 9 wt% by 2015. To date, however, the storage capacities of MOFs are not sufficiently high for practical applications. The best results have been achieved so far are within 1–3 wt% (66).

Much less effort has been devoted to assessment of pure organic solids as gas sorbents, in which building blocks are linked by covalent bonds. Organic molecules typically adhere to close-packing principles and do not afford porous structures. Calix[4]arenes offer a remarkable exception. Atwood et al. (37) showed that $CH_4$, $CF_4$, $C_2F_6$, $CF_3Br$, and other low-boiling halogenated alkanes could be reversibly entrapped and retained within the lattice voids of a crystalline calix[4]arene framework. Such gas-storing crystals appeared to be extremely stable and release their guests only at elevated temperatures, several hundreds Celsuis above their boiling points. As mentioned earlier, Ripmeester

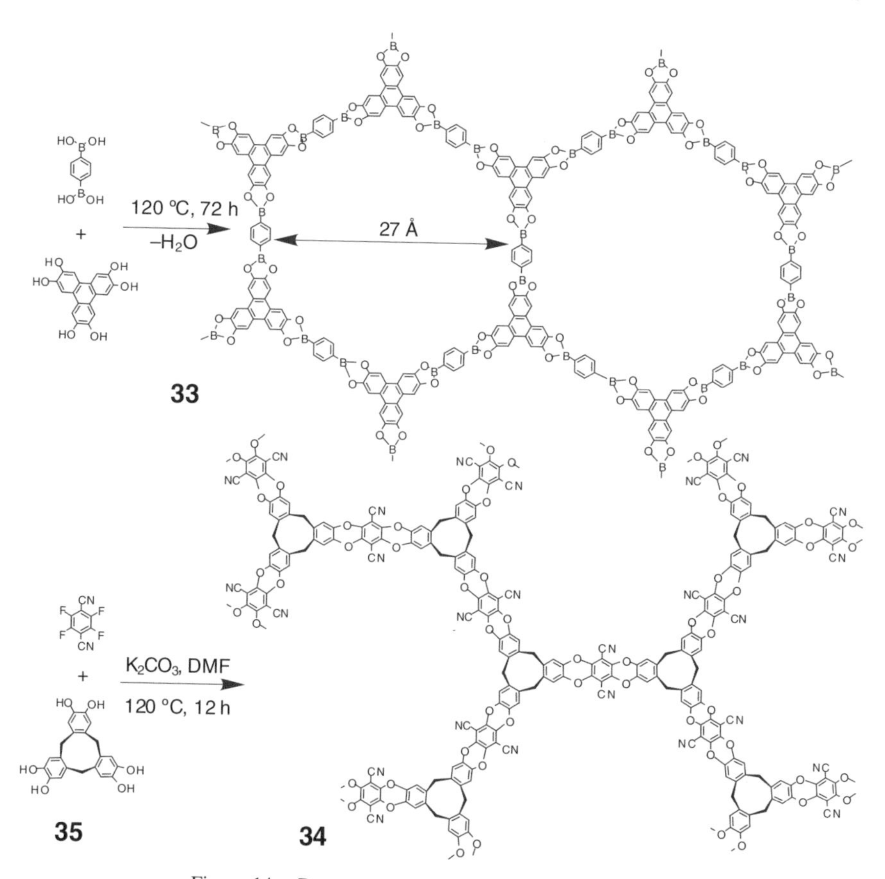


Figure 14. Porous organic materials for gas entrapment.

and co-workers (38) further discovered that calix[4]arene cavities in such crystals are directly involved in the gas complexation.

Very recently covalent organic frameworks (COFs) **33** have been designed and synthesized by simply condensation of phenyl diboronic acid with hexahydroxytriphenylene (67) (Fig. 14). Material **33** is composed of expanded porous graphitic layers, with pore sizes up to 27 Å. Covalent organic framework (**33**) is thermally stable till 500–600°C and have a surface area of $1590\,m^2\,g^{-1}$. This exceeds the highest reported surface area of $1300\,m^2\,g^{-1}$ for macroporous ordered silica.

Polymers with intrinsic microporosity (PIMs) have recently been introduced (68). Polymer **34**, which contains preformed, bowl-shaped cyclotricatechylene (**35**) cavities, has been prepared (Fig. 14) (69). This material has a surface area of $830\,m^2\,g^{-1}$ and adsorbs 1.43 wt% $H_2$ (at 1 bar, 77 K). In the future, it will be important to construct PIMs with much larger accessible surface areas (> $2000\,m^2\,g^{-1}$) and maintain an ultramicroporous structure to retain maximal interactions between the walls and $H_2$. For this, calixarene cavities should be of great interest.

Ripmeester and co-workers (70) showed that $H_2$ storing capacities of clathrate hydrates could be increased to 4 wt% by simply adding THF. Prior to this study, the pressure required to maintain the stability of clathrate hydrates with $H_2$ was very high (~2 kbar). The pressure can now be decreased to 100 bar by simply cooccluding THF molecules in the clathrate. While the system's hydrogen capacity still falls short of the Department of Energy requirement, it offers some advantages. For example, the storage materials ($H_2O$ and THF) are cheap, and the environmental and health hazards are relatively low.

## V. REVERSIBLE CHEMICAL FIXATION OF GASES

### A. Short Introduction

In many applications, the major drawback of noncovalent complexes with gases lies in their relatively low thermodynamic stability. An alternative approach is based on the reversible chemical transformation of gases upon complexation. In this case, they produce reactive intermediates with higher affinities for the receptor molecules. Higher stabilities of such host–guest complexes result in better sensors and also offer attractive opportunities in the design of chemical reagents from gases, as well as conceptually novel materials for gases and from gases. Below, this approach is demonstrated for $NO_X$ and $CO_2$ gases.

### B. Calixarenes and $NO_X$ Gases

Kochi and co-workers (71) showed that when converted to the cation radical, calix[4]arene (**36**) can strongly complex NO gas with the formation of cationic calix-nitrosonium species **37**. In these, the NO molecule is transformed into the nitrosonium cation ($NO^+$) (Fig. 15). Strong charge-transfer interactions between $NO^+$ and the π-surface of **36** places the guest molecule between the cofacial aromatic rings at a distance 2.4 Å, which is shorter than the typical van der Waals contacts. A value for the association constant $K_{assoc} > 5 \times 10^8\,M^{-1}$ was determined (in $CH_2Cl_2$). Charge-transfer complex **37** is deeply colored and can be used for colorimetric sensing of NO gas. Similar complexes were also obtained for calix[4]arenes in their *cone* and *partial cone* conformations.

Host–guest complexes formed upon reversible reaction between $NO_2$ and simple calix[4]arenes were also studied (72). Nitrogen dioxide ($NO_2$) is a paramagnetic gas of an intense brown-orange color. It exists in equilibrium

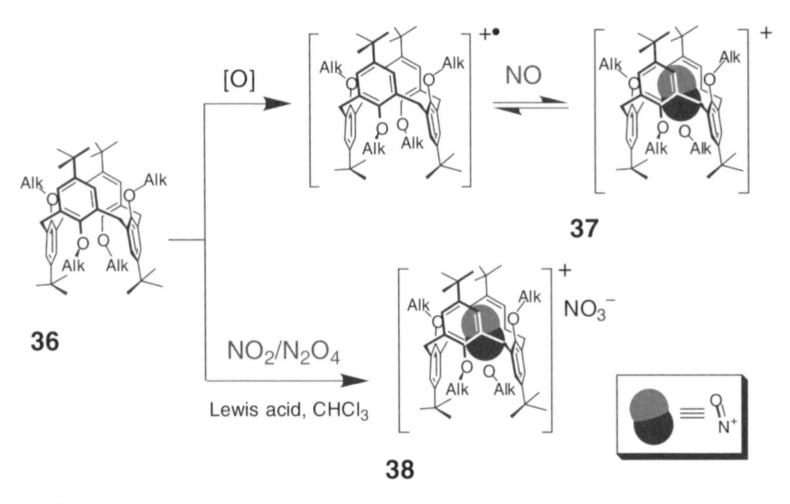


Figure 15. Encapsulation complexes of $NO_X$ gases with calix[4]arenes. Formation of calixarene–$NO^+$ complexes (71–73). [Adapted from (72).]

with its dimer $N_2O_4$, which is colorless. The dynamic interconversion between $NO_2$ and $N_2O_4$ makes it impossible to study either of these species alone. The $N_2O_4$ may disproportionate to ionic $NO^+NO_3^-$ while interacting with simple aromatic derivatives. It was found that tetrakis-*O*-alkylated calix[4]arenes in their *cone* and *1,3-alternate* conformations react with $NO_2/N_2O_4$ to form stable nitrosonium complexes, for example, **38** (Fig. 15) (73). These complexes are deeply colored as well. They are strong ($K_{assoc} \gg 10^6 M^{-1}$, $\Delta G^{295} \gg 8$ kcal $mol^{-1}$), but dissociate upon addition of water or alcohols. More stable calixarene–nitrosonium complexes were isolated upon addition of Lewis acids (e.g., $SnCl_4$ and $BF_3$—$Et_2O$).

The visible spectrum of complex **38** showed a broad charge-transfer (CT) band at $\lambda_{max} \sim 560$ nm ($\varepsilon = 8 \times 10^3 M^{-1} cm^{-1}$). While neither calixarenes nor $NO_2$ absorb in this region, addition of as little as ~1 equiv $NO_2$ to the solution of **36** in chlorinated solvent results in appearance of the CT band. Its absorbance grows upon addition of larger quantities of $NO_2$ and reaches saturation when ~10 equiv $NO_2$ is added (74). Accordingly, calix[4]arenes can detect $NO_2$ even at micromolar–ppm concentrations. Interestingly, wider calix[5]-, calix[6]-, and calix[8]arenes do not form encapsulation complexes with $NO^+$.

Based on these findings, calixarene materials have been synthesized for entrapment of $NO_X$ gases. Specifically, calix-silica gel (**39**) (73) and calix[4]-arene-based periodic mesoporous silica (**40**) (75) were prepared (Fig. 16). In the $NO_2$ entrapment experiments, a stream of the gas was passed through columns

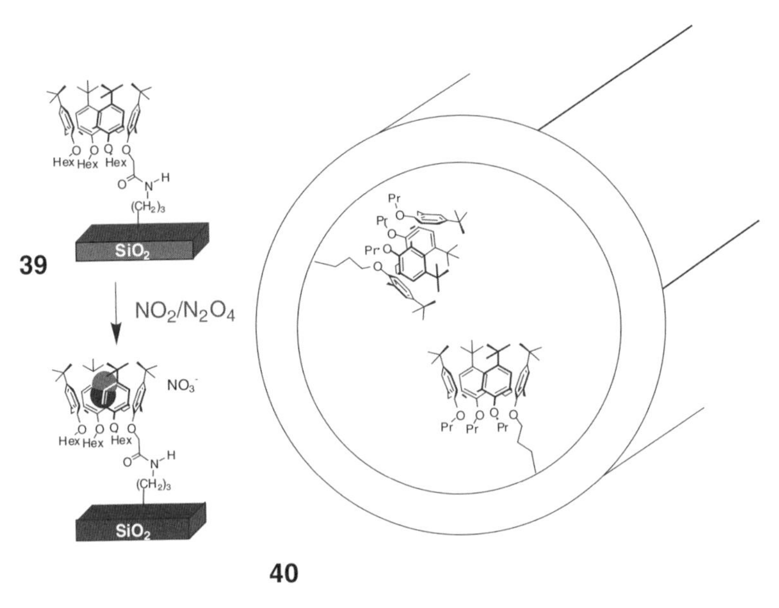


Figure 16. Calixarene-based materials for $NO_X$ entrapment and storage.

loaded with the materials **39** and **40**, instantly producing a dark purple color, indicative of $NO^+$ complexation. The complexation was also confirmed by IR spectroscopy, where the characteristic calixarene–$NO^+$ stretch was observed at $\nu \sim 1920\,cm^{-1}$. The solid-supported complexes were stable for hours, especially for mesoporous silica material **40**. In this case, the $NO^+$ entrapment was also detected by elemental analysis. The materials can be regenerated by simply washing with alcohol and reused.

Calixarene–nitrosonium complex **38** and its analogues can release $NO^+$ and serve as nitrosating agents (76). Nitrosating reagents are important in synthetic organic chemistry, where nitrosation holds a special place. Alkyl nitrites, nitrosamines, nitrosamides, and nitrosothiols are widely used in medicine as NO-releasing drugs. In total synthesis and methodology, –N=O serves as an important activating group, allowing easy transformations of amides to carboxylic acids and their derivatives. In addition, nitrosation mimics interactions between biological tissues and $NO_X$ gases.

Secondary amide substrates (**41**) reacted with complex **38** with remarkable selectivity. Chemical properties of the encapsulated $NO^+$ are different from those in bulk solution and controlled by the cavity. The cavity in **38** protects highly reactive $NO^+$ species from the bulk environment. The complex is quite stable toward moisture and oxygen, and can be handled, for at least several days, without a drybox and/or an $N_2$ atmosphere. On the other hand, it can be decomposed within a few minutes by addition of larger quantities of $H_2O$ or alcohols, regenerating the free calixarene. Accordingly, complex **38** represents an encapsulated reagent.

When mixed with the equimolar solution of amides R'C(O)NHR (**41**) in $CHCl_3$, complex **38** reacted quickly at room temperature, yielding up to 95% of *N*-nitrosamides (**42**, Fig. 17) (76). Dark-blue solutions of **38** lost their color upon addition of the substrates, which is a reasonable visual test for the reaction. Among the variety of amides (**41**), only those possessing *N*-Me substituents were transformed to the corresponding *N*-nitrosamides (**42**). No reaction occurred for substrates with bulkier groups. Consequently, no color discharge

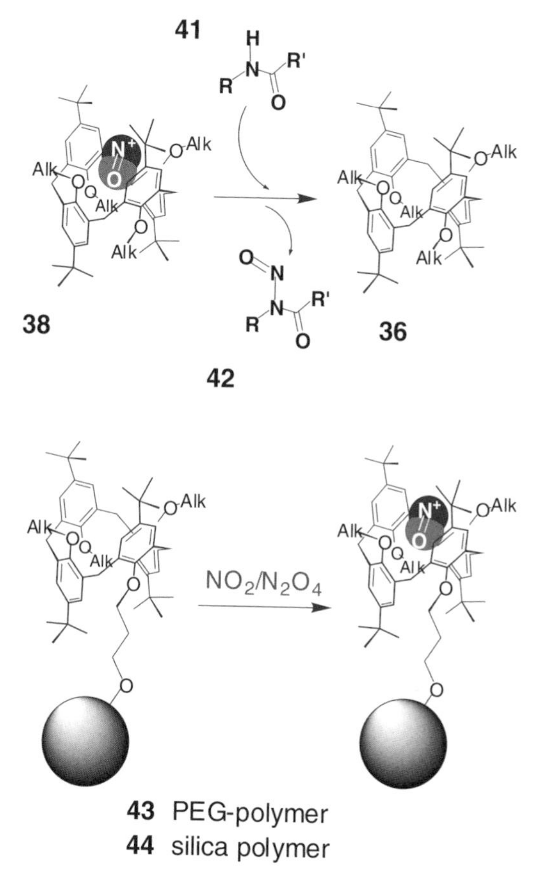


Figure 17. Nitrosating reagents and nitrosating materials, based on calix[4]arenes. [PEG = poly(ethylene glycol.]

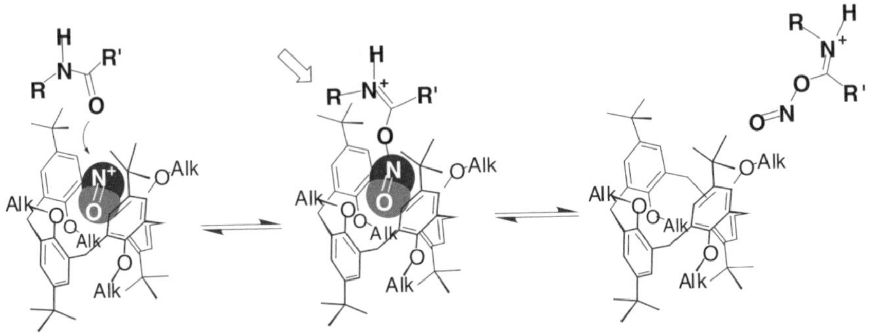

Figure 18. The proposed mechanism of nitrosation of secondary amides by encapsulated calixarene-based reagents. The dimensions of R groups of substrate are critical and control the formation of the corresponding *O*-nitroso intermediate.

was observed. The calixarene cavity was obviously responsible for such unique size–shape selectivity. Larger substrates simply cannot reach the encapsulated $NO^+$ (see the proposed mechanism in Fig. 18).

Polymer-supported nitrosating reagents have been prepared (74, 76). Among the advantages of such reagents are the ease of their separation from the reaction mixture, their recyclability, and the simplification of handling toxic and odorous $NO_X$ gases. Particularly useful are soluble polymers, as they overcome problems associated with the heterogeneous nature of the reaction conditions.

A variety of polymers are commercially available, but $NO_2/N_2O_4$ (and also other $NO_X$) react with many of them, causing destruction and aging (77). As a free radical, $NO_2$ aggressively attacks double bonds in polybutadiene, polyisoprene and their copolymers, ester groups in poly(methyl methacrylate) (PM MA), and also amide fragments in polyamides and polyurethanes. Furthermore, $NO^+$, generated from various $NO_X$, reacts with alkenes and other double-bond containing structures (78). Considering this, PEG and silica gels were employed as solid supports as these are robust and stable (74, 76).

Deep purple $NO^+$-storing, PEG-supported material, **43**•$(NO^+)_n$, was obtained upon simply bubbling $NO_2/N_2O_4$ through the solution of polymer **43** in $CH_2Cl_2$ for 2–3 min, followed by brief flushing with $N_2$ to remove the remaining $NO_2/N_2O_4$ gases (Fig. 17). Material **43**•$(NO^+)_n$ effectively nitrosated amides (**41**) in $CH_2Cl_2$ and preference for the less bulky *N*-Me amide was observed as well. In the preparation of insoluble nitrosating materials, commercial silica gel of relatively high porosity (150 Å) was activated in 18% HCl at reflux, and then reacted with the corresponding calixarene siloxane in $CH_2Cl_2$ to give material **44** (Fig. 17). The presence of calix[4]arene units in **44** was confirmed by the appearance of characteristic absorption bands in the IR spectrum. From the thermogravimetric analysis (TGA) and elemental analyses, a calixarene loading of ~10% was estimated. This rather modest level appeared to be reproducible, even when larger quantities of the calixarene were employed, and may be due to the steric bulkiness of the calixarene fragment.

The dark-blue nitrosonium-storing silica gel (**44**•$(NO^+)_n$) was prepared by bubbling $NO_2/N_2O_4$ through the suspension of **44** in $CH_2Cl_2$ for 5–10 s, followed by filtration and washing with $CH_2Cl_2$. Material **44**•$(NO^+)_n$ is quite robust and does not change the color for several days. For nitrosation, it was suspended in dry $CH_2Cl_2$, an equimolar amount of amides **41** was then added, and the reaction mixture was stirred at room temperature for 24 h. The reactant's color disappeared, thus visually indicating the reaction progress. Material **44** was separated by simple filtration. Yields of nitrosamides **42** were determined by $^1$H NMR spectroscopy, integrating signals of the product versus the starting compounds. The size–shape selectivity trend, observed for the solution experiments with complexes **38**, was clearly seen in this case as well. After at least three independent runs, the averaged yields of nitrosamides (**42**) with less hindered *N*-Me fragments were established as up to 30%, while bulkier *N*-Et and *N*-Pr derivatives (**42**) formed in much smaller quantities ($\leq 8\%$). In control experiments, involving starting silica gel, no visible amounts of **42** were seen, again emphasizing the role of calixarene cavity in the described reactions.

Encapsulation of $NO^+$ species by calix[4]arene and calix[4]arene- based materials offer size–shape selectivities, which were previously unknown for existing, conventional nitrosating agents. The $NO^+$ species, generated from $NO^+$-salts, $N_2O_3$, $NO_2/N_2O_4$, $NO/O_2$, NO/air, and $NaNO_2/H_2SO_4$, are typically not selective.

### C. Synthetic Nanotubes and $NO_X$ Gases

Filling SWNTs with gases is a rapidly emerging research area (58). Potential applications include using SWNTs as gas-storing cylinders and, further, as reaction vessels. Among the problems, however, are far from trivial protocols for chemical opening of SWNTs, still not fully understood mechanisms of their filling, as well as identification of the encapsulated material. Synthetic analogues of SWNTs have recently been introduced, both covalently built (79) and self-assembling (80). Organic synthesis offers a variety of sizes and shapes. However, the stability of the encapsulation complexes here is generally weak, which diminishes their capabilities as storing materials. Moreover, self-assembling tubes are stable only under specific conditions (80).

As discussed above, calix[4]arenes reversibly interact with $NO_2/N_2O_4$ and entrap reactive $NO^+$ cations within their π-electron-rich interiors, one per cavity (72, 73). Very high $K_{assoc} \gg 10^6\, M^{-1}$ values ($\Delta G^{295} \gg 8$ kcal mol$^{-1}$) for these processes were determined, and the complexes were also kinetically stable. These features were used in the design of calixarene-based nanotubes

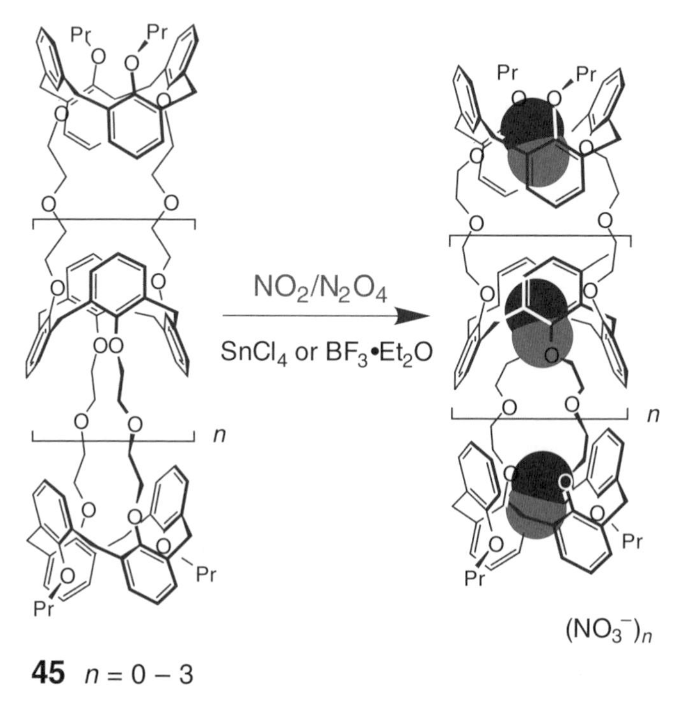


Figure 19. Synthetic nanotubes based on calixarenes and their reactions with $NO_2/N_2O_4$ gases.

(**45**, Fig. 19) (81–83). In the nanotubes, *1,3-alternate* calix[4]arenes are rigidly connected from both sides of their rims with pairs of diethylene glycol linkers. In this calixarene conformation, two pairs of phenolic oxygens are oriented in opposite directions, providing a diverse means to modularly enhance the tube length. The bridge length is critical since it not only provides relatively high conformational rigidity of the tubular structure, but also seals the walls, minimizing the gaps between the calixarene modules. The calixarene tubes possess defined inner tunnels of 6 Å diameter and may entrap multiple $NO^+$, one per cavity. Finally, they can be emptied at will, in a nondestructive manner.

Exposure of **45** to $NO_2/N_2O_4$ in chlorinated solvents results in the rapid encapsulation of $NO^+$ cations within its interior (Fig. 19). The complexes were characterized by UV–vis, Fourier transferinfrared (FTIR) and $^1H$ NMR spectroscopies. The $NO^+$ entrapment process is reversible, and addition of water quickly regenerated starting tube **45**.

In the solid state, the nanotube units **45** ($n = 1$, Fig. 19) pack head to tail, in straight rows, resulting in infinitely long cylinders (82). The neighboring nanocylinders aligned parallel to each other (Fig. 20). In each nanocylinder, molecules **45** are twisted by 90° relative to each other, and the Ar–O–Pr

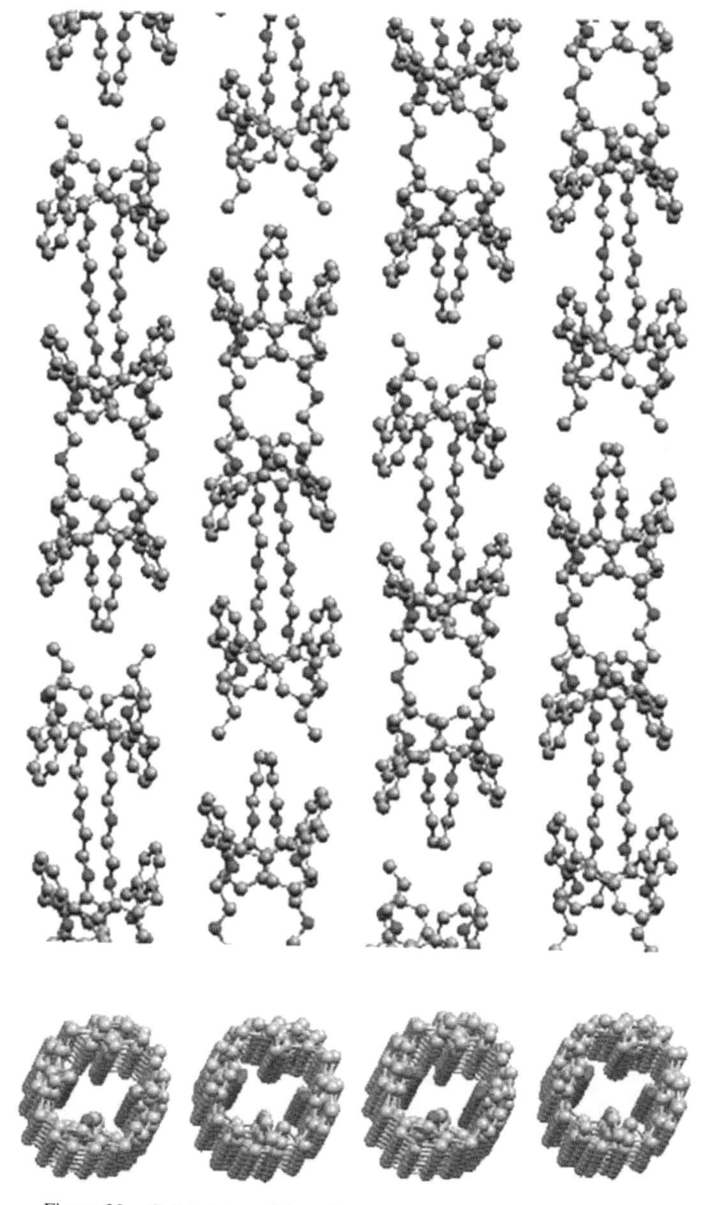

Figure 20. Solid-state packing of nanotube **45** ($n = 1$) (X-ray structure).

propyl groups effectively occupy the voids between the adjacent molecules. In such an arrangement, the intermolecular distance between two neighboring tubes in the nanocylinder is ~6 Å. The nanocylinders are separated from each other by ~9 Å. This supramolecular order comes with the tube length and is without precedent for conventional, shorter calixarenes. The unique linear nanostructures maximize their intermolecular van der Waals interactions in the crystal through the overall shape simplification. Such a unique arrangement resembles that of SWNTs bundling.

Taken together, in addition to SWNTs, synthetic nanotubes are now available for filling purposes. They pack in tubular bundles and can be reversibly filled with foreign guests upon reacting with $NO_X$ gases. Thus far, only certain guests can fill the interiors, but they are charged, which is important for the design of nanowires. Chemical (redox) reactions with entrapped $NO^+$ inside the nanotube can, in principle, be achieved. Given the ability of calix[4]arenes to react with $NO_X$ gases even in the solid state (73), it should be possible to achieve the nitrosonium flow along the infinite nanocylinders in the solid-state bundles of **45**.

## D. Nanostructures and Supramolecular Materials from Carbon Dioxide

It has been known for decades that $CO_2$ smoothly reacts with amines at ordinary conditions to yield carbamates (84, 85). Carbamates are thermally unstable and release $CO_2$ upon heating. Polymer-bound amines have been employed in industry as reusable $CO_2$ scrubbers, removing $CO_2$ from industrial exhaust streams (86). Imprinted polymers have been introduced, in which a template can be attached, and then removed through a carbamate linker (87). Reactions between $CO_2$ and immobilized amines have been employed for gas sensing (88). It has been shown, that exposure of long-chain primary $RNH_2$ and secondary RR′NH alkyl amines to $CO_2$ results in the formation of thermally responsive alkylammonium alkylcarbamate organogels (89). In another example, the amine-containing ionic liquid 1-(3′-aminopropyl)-3-butylimidazolium tetrafluoroborate reacted with $CO_2$ with the formation of the corresponding carbamate dimers (90). This process can be used to capture $CO_2$ in the purification of industrial gas mixtures.

It was recently suggested that carbamate bonds could be employed for a wider variety of dynamic covalent chemistry (DCC) experiments (88) and DCC is quickly emerging as a promising alternative to noncovalent self-assembly (91). This experiment offers an elegant opportunity of performing supramolecular chemistry with covalent bonds. One of the most important advantages here is the robustness of covalently organized structures, which on the other hand can be reversibly broken, at will. Of particular interest are supramolecular polymers and supramolecular materials.

Reversibly formed polymers are usually called supramolecular polymers (92). They represent a novel class of macromolecules, in which monomeric units are held together by reversible bonds–forces. These are self-assembling polymers, and thus far hydrogen bonds, metal–ligand interactions, and van der Waals forces have been employed to construct them. Supramolecular polymers combine features of conventional polymers with properties resulting from the bonding reversibility. Structural parameters of supramolecular polymeric materials, in particular their two (2D)- and three-dimensional (3D) architectures, can be switched on–off through the main chain association–dissociation processes. On the other hand, their strength and degree of polymerization depend on how tightly the monomeric units are aggregated. A strategy was introduced to build supramolecular polymeric chains, which takes advantage of dynamic chemistry between $CO_2$ and amines (93) and, in addition, utilizes hydrogen bonds (85).

Monomeric units were designed, which strongly aggregate in apolar solution and possess $CO_2$-philic primary amino groups on the periphery. Calixarenes were employed as self-assembling units. Calix[4]arene tetraurea dimers are probably the most studied class of strong hydrogen-bonding aggregates (94, 95). These dimers form in apolar solution with $K_D \geq 10^6\ M^{-1}$ and are held together by a seam of 16 intermolecular C=O•••H–N hydrogen bonds. This results in a rigid inner cavity of ~200 Å$^3$ that reversibly encapsulates a solvent molecule or a benzene-sized guest.

Supramolecular polymer **46** is based on carbamate chemistry (Fig. 21). This is a 3D molecular network, which employs $CO_2$ as a cross-linking agent (96). In monomer **47**, two calixarene tetraurea moieties are linked with a dipeptide, di-*l*-lysine chain. Calixarenes were attached to the ε-$NH_2$ ends, so the dilysine module orients them away from each other, in roughly opposite directions. Such arrangement also prevents intramolecular assembly. A hexamethyleneamine chain was then attached to the carboxylic side of the dipeptide. Its amino group and the α-$NH_2$ group of **47** react with $CO_2$, providing cross-linking.

Viscosity studies of solutions **47** in apolar solvents confirmed the formation of polymer **48**. Thus, concentrating $CHCl_3$ solutions of **47** from 5 to just 40 m*M* leads to a greater than or equal to fivefold increase in viscosity. This was not observed for the model, nonpolymeric calixarene tetraurea precursors. From specific viscosities measurements, the degree of polymerization for linear chains **48** of $\sim 2.8 \times 10^2$ was estimated at 20 m*M*, which corresponds to the average molar mass of $\sim 7.6 \times 10^5$ g mol$^{-1}$. Bubbling $CO_2$ through a solution of **48** in $CHCl_3$ or benzene yields cross-linked material **46**, which is a gel.

The main chains in **46** are held together by a hydrogen-bonding assembly of capsules, and multiple carbamate $-N^+H_3$•••$O^-C(O)NH-$ bridges cross-link these chains. This is a 3D network, since the side amine groups are oriented in all three directions. The carbamate bridges were detected by $^{13}C$ NMR spectroscopy.

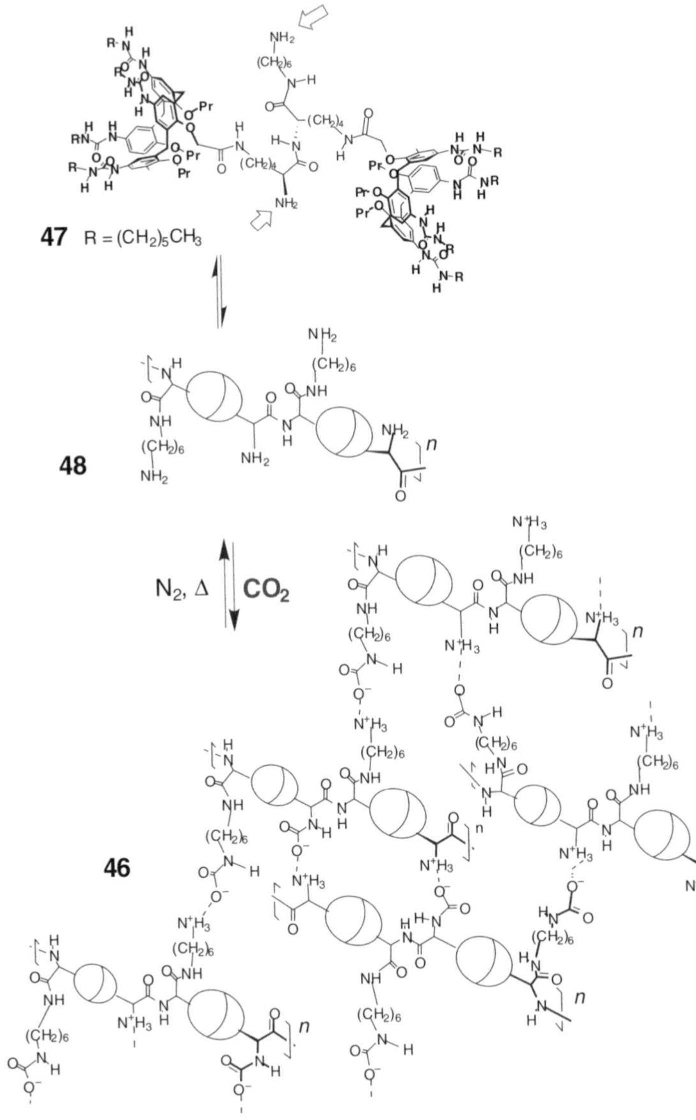

Figure 21. Supramolecular calixarene-based polymers can be reversibly cross-linked by $CO_2$.

Finally, visual insight into the aggregation mode and morphology in **46** was obtained by scanning electron microscopy (SEM) of dry samples, or xerogels. While the precursors show only negligible fiber formation, a 3D network was obvious for **46**.

Reaction of calixarene (**47**) and $CO_2$ is special, because it converts linear supramolecular polymeric chains (**48**) into supramolecular, 3D polymeric networks (**46**). These are also switchable and can be transformed back to the linear chains (**48**) without breaking them. While supramolecular cross-linked polymers are known (92), they break upon dissociation of the noncovalent aggregates that compose them. Material **46** is different, as it only releases $CO_2$ and keeps hydrogen-bonding intact.

Supramolecular polymer **46** serves for entrapment and controlled release of organic guests (Fig. 22) (97). On a molecular level, multiple voids are generated between the carbamate-lysine fragments in **46**, which are of 15–20-Å dimensions. Gel **46** was used to trap commercial dyes (e.g., coumarins and porphyrins). Absorption spectrophotometry was employed to monitor their release. In a typical experiment, peptide **47** was dissolved in a small volume of $CHCl_3$, and then coumarin 314 or tetraphenylporphyrin were added. Carbon dioxide was bubbled through the solution for 5 min. Colored gels, **46•**(guest)$_n$, were formed. The guests can be stored in dried gels indefinitely and released only upon the gel dissipation. The guests release was accomplished through changing solvent polarity (hydrogen-bond breaking) and/or temperature ($CO_2$ release/carbamate breaking) (97). It was also found that lowering the pH also facilitates the $CO_2$ release due to the carbamate hydrolysis.

Gel **49** was obtained from benzene and benzene–$CHCl_3$ solutions of bis(calixarene) (**50**) and $CO_2$ (Fig. 23) (98). The polymeric chains here possess multiple fluorophore units (pyrene moieties) brought together through hydrogen bonding and carbamate bridges. Accordingly, material **49** is fluorescent and may act as a vehicle for energy migration. The aggregation degree, and therefore the fluorophore local concentrations, can be controlled and switched on–off, as described earlier. Formation of the carbamate bridges in **49** was routinely confirmed by $^{13}C$ NMR spectroscopy. As previously described, they can be broken after heating solution **49** for a few minutes at ~100°C and bubbling $N_2$ through it. The SEM pictures of the corresponding xerogels revealed, in particular, well-defined pores of ~1–3 mm diameter, which can be used for guest–solvent entrapment.

In photophysical experiments, a striking contrast in fluorescent behavior was noticed of xerogels (**49**) obtained from benzene and from 95:5 benzene–nitrobenzene solutions (Fig. 23) (98). The former is strongly fluorescent ($\lambda_{ex} = 347$ nm), but the latter is not. Nitrobenzene is known to quench fluorescence of pyrene. Incorporated within the gel's pores, molecules of nitrobenzene appear to be in close proximity to the multiple pyrene donors, and energy transfer is effective. In another experiment, dropwise addition of nitrobenzene (up to 10% v/v) to the benzene suspension of fluorescent xerogel **49**, initially obtained from benzene, resulted in the fluorescence disappearing within a few seconds. These observations could be useful in the design of switchable light harvesting materials.

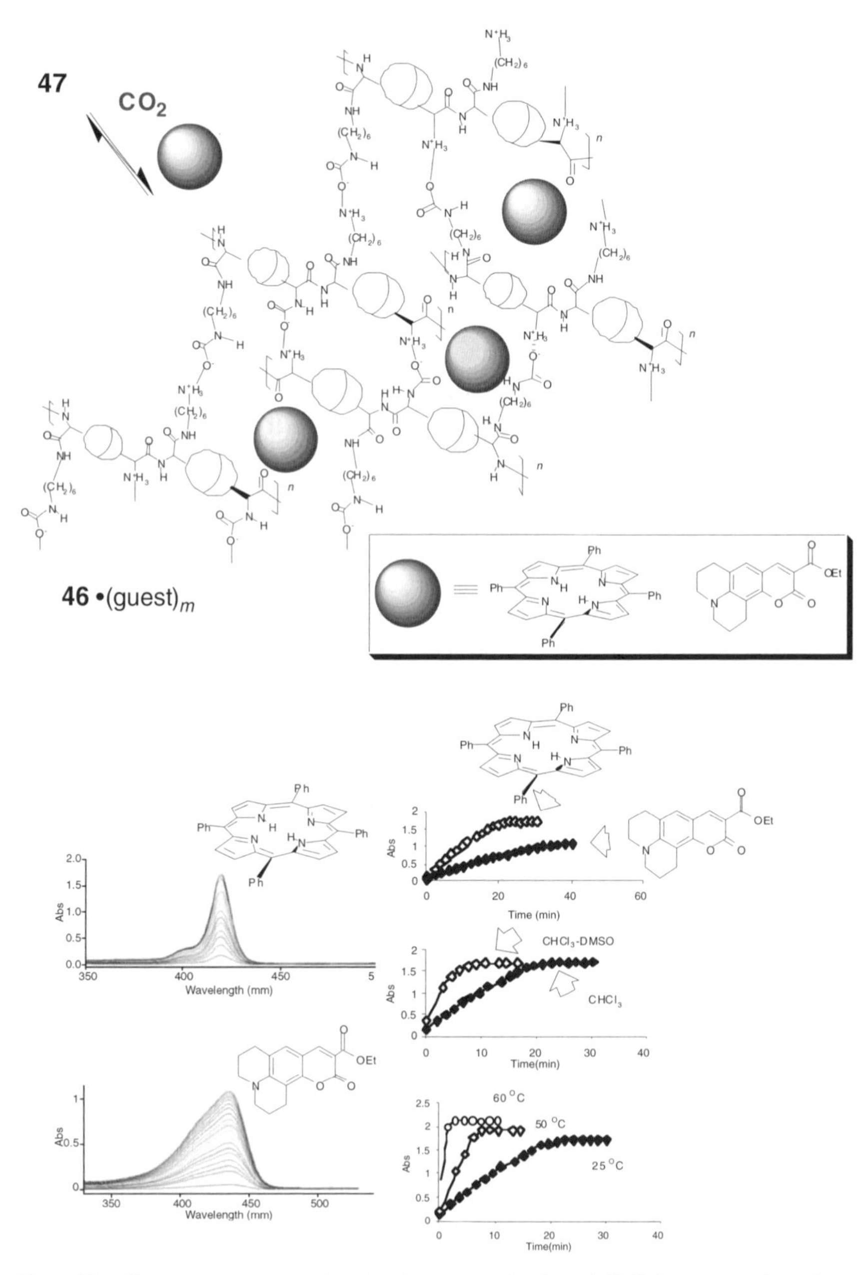


Figure 22. Guest entrapment experiments using supramolecular gel **46**. Below: guest release from material **46** can be followed by absorption spectroscopy. The release rates are dependent on temperature, solvent polarity, and other factors.

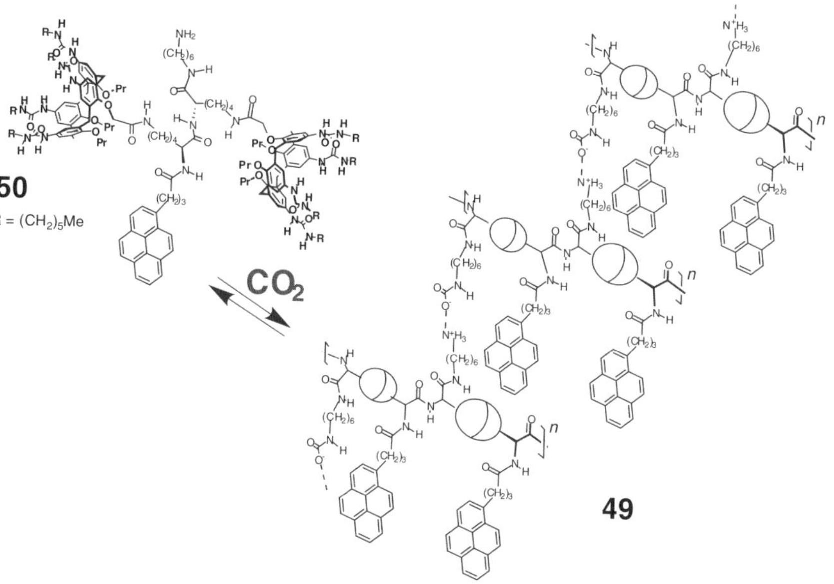

Figure 23. Fluorescent supramolecular material **49** is made using $CO_2$ as a cross-linking agent.

## VI. CONCLUSIONS

It is now clear that chemists have reached a molecular-level understanding of forces, which bind gases, trap them, fix in time and in space, and further chemically transform them. This understanding, of course, requires an inspiration from Nature, which indeed offers superb examples of gas binding and storing. Lewis acid–base, dipole–dipole interactions, hydrogen bonding, van der Waals forces, and encapsulation within enclosed spaces, and smart combination of all these apply in molecular recognition and complexation of various gases. Reversible covalent chemistry with gases has also emerged. A number of interesting receptors, sensors, and materials for gases, based on supramolecular approaches, already have been constructed.

Despite the tiny dimensions and simple atomic composition, each gas is different in its physical and chemical properties and requires an individual approach. However, it is possible to rationalize kinetics, thermodynamics, and selectivity of molecular hosts for gases on the basis of receptor–substrate electronic and geometrical complementary. It is also clear that widely accepted supramolecular concepts (e.g., the preorganization, multiplication of binding sites, and encapsulation) are applicable for gases. As for all other areas of supramolecular chemistry, spectroscopic techniques and molecular modeling are crucial to elucidate the specific contributions of various binding forces, as well as solvent effects. Finally, supramolecular chemistry has found ways to chemically transform gases into synthetically useful and selective reagents, smart nanostructures and materials. Most certainly, this is just a beginning of much bigger developments.

## ACKNOWLEDGMENTS

The American Chemical Society Petroleum Research Fund, the US National Science Foundation, the Texas Higher Education Coordinating Board—Advanced Technology Program, and the Alfred P. Sloan Foundation are acknowledged for partial support of this work.

## ABBREVIATIONS

| | |
|---|---|
| 2D | Two dimensional |
| 3D | Three dimensional |
| AIBN | Azobisisobutyronitrile |
| $^{13}$CNMR | Carbon-13 nuclear magnetic resonance |
| COF | Covalent organic framework |
| CT | Charge transfer |
| DCC | Dynamic covalent chemistry |
| DMAP | *N*,*N*-Dimethylaminopyridine |
| DMF | Dimethylformamide |
| EDTA | ethylenediaminetetraacetic acid (solvent) |
| EGDMA | Ethylene glycol dimethylacrylate |
| FAB | Fast atom bombardment |
| FT | Fourier transformation |
| FTIR | Fourier transform infrared |
| $^{1}$H NMR | Proton nuclear magnetic resonance |
| HMQC | Heteronuclear multiple quantum coherence |
| IR | Infrared |
| MALDI–TOF | Matrix-assisted laser desorption ionization–time of flight |
| MOF | Metal–organic microporous framework |
| MOM | Metal–organic microporous material |
| MS | Mass spectrometry |
| NMR | Nuclear magnetic resonance |
| NO | Nitric oxide |
| $NO_2$ | Nitrogen dioxide |
| $N_2O$ | Nitrous oxide |

| | |
|---|---|
| $N_2O_4$ | Dinitrogen teroxide |
| PEG | Polyethylene glycol |
| PIM | Polymer with intrinsic microporosity |
| PMMA | Poly(methyl methacrylate) |
| REMPI | Resonance-enhanced multiproton ionization |
| salen | Bis(salicylidene)ethylenediamine |
| SEM | Scanning electron microscopy |
| SWNT | Single-walled carbon nanotubes |
| TGA | Thermogravimetric analysis |
| Thf | Tetrahydrofuran (ligand) |
| UV | Ultraviolet |
| vis | Visible |
| VT NMR | Variable temperature NMR |
| ZEKE | Zero-electron kinetic energy |

## REFERENCES

1. (a) J.-M. Lehn, *Supramolecular Chemistry. Concepts and Perspectives.* VCH, Weinheim-New York-Basel-Cambridge-Tokyo, 1995. (b) J. W. Steed and J. L. Atwood, *Supramolecular Chemistry*, John Wiley & Sons, Inc., Chichester-New York-Weinheim-Brisbane-Singapore-Toronto, 2000. (c) P. D. Beer, P. A. Gale, and D. K. Smith, *Supramolecular Chemistry*, Oxford University Press, UK, 1999. (d) H. Dodziuk, *Introduction to Supramolecular Chemistry*, Kluwer Academic Publishers, Dordrecht-Boston-London, 2002.
2. *Comprehensive Supramolecular Chemistry, Vol. 1. Molecular Recognition: Receptors for Cationic Guests*, G. W. Gokel, Ed.; *Molecular Recognition: Receptors for Molecular Guests, Vol. 2*, F. Vögtle, Ed.; Pergamon, New York, 1996.
3. (a) D. M. Rudkevich, *Angew. Chem. Int. Ed. Engl.*, *43*, 558 (2004). (b) D. M. Rudkevich and A. V. Leontiev, *Aust. J. Chem.*, *57*, 713 (2004).
4. (a) D. S. Schimel, J. I. House, K. A. Hibbard, P. Bousquet, P. Ciais, P. Peylin, B. H. Braswell, M. J. Apps, D. Baker, A. Bondeau, J. Canadell, G. Churkina, W. Cramer, A. S. Denning, C. B. Field, P. Friedlingstein, C. Goodale, M. Heimann, R. A. Houghton, J. M. Melillo, B. Moore, III, D. Murdiyarso, I. Noble, S. W. Pacala, I. C. Prentice, M. R. Raupach, P. J. Rayner, R. J. Scholes, W. L. Steffen, and C. Wirth, *Nature (London)*, *414*, 169 (2001). (b) X. Xiaoding, and J. A. Moulijn, *Energy Fuels*, *10*, 305 (1996). (c) N. H. Batjes, *Biol. Fertil. Soils*, *27*, 230 (1998). (d) A. V. Leontiev, O. A. Fomicheva, M. V. Proskurnina, and N. S. Zefirov, *Russ. Chem. Rev.*, *70*, 91 (2001). (e) W. C. Trogler, *Coord. Chem. Rev.*, *187*, 303 (1999).
5. (a) L. J. Ignarro, *Angew. Chem. Int. Ed. Engl.*, *38*, 1882 (1999). (b) S. Pfeiffer, B. Mayer, and B. Hemmens, *Angew. Chem. Int. Ed. (Engl.)*, *38*, 1714 (1999). (c) A. R. Butler and D. L. H. Williams, *Chem. Soc. Rev.*, 233 (1993).
6. (a) M. T. Lerdau, J. W. Munger, and D. J. Jacob, *Science*, *5488*, 2291 (2000). (b) M. Kirsch, H.-G. Korth, R. Sustmann, and H. de Groot, *Biol. Chem.*, *383*, 389 (2002).
7. (a) L. A. Komarnisky, R. J. Christopherson, and T. K. Basu, *Nutrition*, *19*, 54 (2003). (b) O. Herbarth, G. Fritz, P. Krumbiegel, U. Diez, U. Franck, and M. Richter, *Environ. Tox.*, *16*, 269 (2001). (c) C. N. Hewitt, *Atm. Environ.*, *35*, 1155 (2001).

8. B. A. Springer and S. G. Sligar, *Chem. Rev.*, *94*, 699 (1994).
9. D. E. Goldberg, *Chem. Rev.*, *99*, 3371 (1999).
10. E. A. Brucker, J. S. Olson, M. Ikeda-Saito, and G. N. Phillips, Jr., *Proteins*, *30*, 352 (1998).
11. V. Sampath, X. J. Zhao, and W. S. Caughey, *J. Biol. Chem.*, *276*, 13635 (2001).
12. A. C. Legon, *Angew. Chem. Int. Ed. (Engl.)*, *38*, 2686 (1999).
13. C. E. H. Dessent and K. Müller-Dethlefs, *Chem. Rev.*, *100*, 3999 (2000).
14. (a) M. Momenteau and C. A. Reed, *Chem. Rev.*, *94*, 659 (1994). (b) G. E. Wuenschell, C. Tetreau, D. Lavalette, and C. A. Reed, *J. Am. Chem. Soc.*, *114*, 3346 (1992). (c) J. P. Collman, X. Zhang, K. Wong, and J. I. Brauman, *J. Am. Chem. Soc.*, *116*, 6245 (1994). (d) C. K. Chang, Y. Liang, and G. Aviles, *J. Am. Chem. Soc.*, *117*, 4191 (1995).
15. (a) A. Zingg, B. Felber, V. Gramlich, L. Fu, J. P. Collman, and F. Diederich, *Helv. Chim. Acta*, *85*, 333 (2002). (b) B. Felber, C. Calle, P. Seiler, A. Schweiger, and F. Diederich, *Org. Biomol. Chem.*, 1090 (2003).
16. J. Mispelter, M. Momenteau, D. Lavalette, and J.-M. Lhoste, *J. Am. Chem. Soc.*, *105*, 5165 (1983).
17. (a) F. Tani, M. Matsu-ura, S. Nakayama, M. Ichimura, N. Nakamura, and Y. Naruta, *J. Am. Chem. Soc.*, *123*, 1133 (2001). (b) S. Nakayama, F. Tani, M. Matsu-ura, and Y. Naruta, *Chem. Lett.*, 496 (2002).
18. J. A. Lukin, V. Simplaceanu, M. Zou, N. T. Ho, and C. Ho, *Proc. Natl. Acad. Sci. U.S.A.*, *97*, 10354 (2000).
19. E. J. Beckman, *Chem. Commun.*, 1885 (2004).
20. P. Raveendran, Y. Ikushima, and S. L. Wallen, *Acc. Chem. Res.*, *38*, 478 (2005).
21. V. K. Potluri, J. Xu, R. Enick, E. Beckman, and A. D. Hamilton, *Org. Lett.*, *4*, 2333 (2002).
22. (a) P. Raveendran and S. L. Wallen, *J. Am. Chem. Soc.*, *124*, 7274 (2002). (b) P. Raveendran and S. L. Wallen, *J. Am. Chem. Soc.*, *124*, 12590 (2002). (c) M. A. Blatchford, P. Raveendran, and S. L. Wallen, *J. Am. Chem. Soc.*, *124*, 14818 (2002).
23. G. V. Zyryanov, E. M. Hampe, and D. M. Rudkevich, *Angew. Chem. Int. Ed. (Engl.)*, *41*, 3854 (2002).
24. (a) Z. Florjanczyk and D. Raducha, *Pol. J. Chem.*, *69*, 459 (1995). (b) N. Maier, J. Schiewe, H. Matschiner, C.-P. Maschmeier, and R. Boese, *Phosphues Sulfur Silica*, *91*, 179 (1994). (c) M. W. Wong and K. B. Wiberg, *J. Am. Chem. Soc.*, *114*, 7527 (1992). (d) J. E. Douglas and P. A. Kollman, *J. Am. Chem. Soc.*, *100*, 5226 (1978). (e) J. Grundnes and S. D. Christian, *Acta Chem. Scand.*, *23*, 3583 (1969). (f) D. L. A. de Faria and P. S. Santos, *Magn. Res. Chem.*, *25*, 592 (1987).
25. A. V. Leontiev and D. M. Rudkevich, *J. Am. Chem. Soc.*, *127*, 14126 (2005).
26. J. P. Collman, J. I. Brauman, J. P. Fitzgerald, J. W. Sparapani, and J. A. Ibers, *J. Am. Chem. Soc.*, *110*, 3486 (1988).
27. (a) M. Albrecht, R. A. Gossage, M. Lutz, A. L. Spek, and G. van Koten, *Chem. Eur. J.*, *6*, 1431 (2000). (b) M. Albrecht, R. A. Gossage, U. Frey, A. W. Ehlers, E. J. Baerends, A. E. Merbach, and G. van Koten, *Inorg. Chem.*, *40*, 850 (2001). (c) M. Albrecht, G. Rodriguez, J. Schoenmaker, and G. van Koten, *Org. Lett.*, *2*, 3461 (2000). (d) M. Albrecht and G. van Koten, *Adv. Mater.*, *11*, 171 (1999). (e) M. Albrecht, M. Lutz, A. L. Spek, and G. van Koten, *Nature (London)*, *406*, 970 (2000).
28. R. J. M. K. Gebbink, A. W. Bosman, M. C. Feiters, E. W. Meijer, and R. J. M. Nolte, *Chem. Eur. J.*, *5*, 65 (1999).
29. J. F. Krebs and A. S. Borovik, *J. Am. Chem. Soc.*, *117*, 10593 (1995).
30. A. C. Sharma and A. S. Borovik, *J. Am. Chem. Soc.* *122*, 8946 (2000).

31. K. M. Padden, J. F. Krebs, C. E. MacBeth, R. C. Scarrow, and A. S. Borovik, *J. Am. Chem. Soc.*, *123*, 1072 (2001).

32. (a) F. Cramer and F. M. Henglein, *Chem. Ber.*, *90*, 2572 (1957). (b) F. D. Cramer, *Rev. Pure Appl. Chem.*, *5*, 143 (1955). (c) F. Cramer, *Angew. Chem.*, *64*, 437 (1952).

33. E. Fenyvesi, L. Szente, N. R. Russel, and McNamara, *Comprehensive Supramolecular Chemistry, Vol 3. Cyclodextrins*, J. Szejtli, and T. Osa, Eds.; Pergamon, New York 1996, p. 305.

34. (a) K. Bartik, M. Luhmer, S. J. Heyes, R. Ottinger, and J. Reisse, *J. Magn. Reson. B*, *109*, 164 (1995). (b) L. Dubois, S. Parres, J. G. Huber, P. Berthault, and H. Desvaux, *J. Phys. Chem. B*, *108*, 767 (2004). (c) S. M. Rubin, M. M. Spence, A. Pines, and D. E. Wemmer, *J. Magn. Reson.*, *152*, 79 (2001).

35. (a) D. J. Cram and J. M. Cram, *Container Molecules and their Guests*, Royal Society of Chemistry, Cambridge, 1994. (b) D. M. Rudkevich, *Bull. Chem. Soc. Jpn.*, *75*, 393 (2002). (c) A. Jasat and J. C. Sherman, *Chem. Rev.*, *99*, 931 (1999).

36. D. J. Cram, K. D. Stewart, I. Goldberg, and K. N. Trueblood, *J. Am. Chem. Soc.*, *107*, 2574 (1985).

37. J. L. Atwood, L. J. Barbour, and A. Jerga, *Science*, *296*, 2367 (2002).

38. G. D. Enright, K. A. Udachin, I. L. Moudrakovski, and J. A. Ripmeester, *J. Am. Chem. Soc.*, *125*, 9896 (2003).

39. J. L. Atwood, L. J. Barbour, and A. Jerga, *Angew. Chem. Int. Ed. Engl.*, *43*, 2948 (2004).

40. P. K. Thallapally, G. O. Lloyd, T. B. Wirsig, M. W. Bredenkamp, J. L. Atwood, and L. J. Barbour, *Chem. Commun.*, 5272 (2005).

41. Y. Miyahara, K. Abe, and T. Inazu, *Angew. Chem. Int. Ed. Engl.*, *41*, 3020 (2002).

42. (a) D. J. Cram, S. Karbach, Y. H. Kim, L. Baczynskyj, and G. W. Kalleymeyn, *J. Am. Chem. Soc.*, *107*, 2575 (1985). (b) D. J. Cram, S. Karbach, Y. H. Kim, L. Baczynskyj, K. Marti, R. M. Sampson, and G. W. Kalleymeyn, *J. Am. Chem. Soc.*, *110*, 2554 (1988).

43. D. J. Cram, M. E. Tanner, and C. B. Knobler, *J. Am. Chem. Soc.*, *113*, 7717 (1991).

44. A. V. Leontiev and D. M. Rudkevich, *Chem. Commun.*, 1468 (2004).

45. J. Nakazawa, J. Hagiwara, M. Mizuki, Y. Shimazaki, F. Tani, and Y. Naruta, *Angew. Chem. Int. Ed. Engl.*, *44*, 3744 (2005).

46. (a) K. Bartik, M. Luhmer, J.-P. Dutasta, A. Collet, Reisse, *J. Am. Chem. Soc.*, *120*, 784 (1998). (b) T. Brotin, A. Lesage, L. Emsley, and A. Collet, *J. Am. Chem. Soc.*, *122*, 1171 (2000). (c) T. Brotin and J.-P. Dutasta, *Eur. J. Org. Chem.*, 973 (2003).

47. (a) M. M. Spence, S. M. Rubin, I. E. Dimitrov, E. J. Ruiz, D. E. Wemmer, A. Pines, S. Q. Yao, F. Tian, and P. G. Schultz, *Proc. Natl. Acad. Sci. U.S.A.*, *98*, 10654 (2001). (b) C. Hilty, T. J. Lowery, D. E. Wemmer, and A. Pines, *Angew. Chem. Int. Ed. Engl.*, *45*, 70 (2006).

48. (a) D. M. Rudkevich, *Functional Artificial Receptors*, T. Shrader and A. D. Hamilton, Eds., Wiley-VCH, New York, 2005, p. 257. (b) D. M. Rudkevich, *Calixarene 2001*; Z. Asfari, V. Böhmer, J. Harrowfield, and J. Vicens, Eds., Kluwer Academic Publishers, Dordrecht The Netherlands, 2001, p. 155. (c) F. Hof, S. L. Craig, C. Nuckolls, and J. Rebek, Jr., *Angew. Chem. Int. Ed. Engl.*, *41*, 1488 (2002).

49. (a) N. Branda, R. Wyler, and J. Rebek, Jr., *Science*, *263*, 1267 (1994). (b) N. Branda, R. M. Grotzfeld, C. Valdes, and J. Rebek, Jr., *J. Am. Chem. Soc.*, *117*, 85 (1995).

50. A. Shivanyuk, A. Scarso, and J. Rebek, Jr., *Chem. Commun.*, 1230 (2003).

51. R. Fiammengo, K. Wojciechowski, M. Crego-Calama, P. Timmerman, A. Figoli, M. Wessling, and D. N. Reinhoudt, *Org. Lett.*, *5*, 3367 (2003).

52. (a) M. Saunders, H. A. Jimenez-Vazquez, J. R. Cross, S. Mroczkowski, M. L. Gross, D. E. Giblin, and R. J. Poreda, *J. Am. Chem. Soc.*, *116*, 2193 (1994). (b) M. Saunders, R. J. Cross, H. A. Jimenez-Vazquez, R. Shimshi, and A. Khong, *Science*, *271*, 1693 (1996).
53. Y. Rubin, T. Jarrosson, G.-W. Wang, M. D. Bartberger, K. N. Houk, G. Schick, M. Saunders, and R. J. Cross, *Angew. Chem. Int. Ed. Engl.*, *40*, 1543 (2001).
54. Y. Murata, M. Murata, and K. Komatsu, *J. Am. Chem. Soc.*, *125*, 7152 (2003).
55. H. Sawa, Y. Wakabayashi, Y. Murata, M. Murata, and K. Komatsu, *Angew. Chem. Int. Ed. Engl.*, *44*, 1981 (2005).
56. K. Komatsu, M. Murata, and Y. Murata, *Science*, *307*, 238 (2005).
57. S. Iwamatsu, T. Uozaki, K. Kobayashi, S. Re, S. Nagase, and S. Murata, *J. Am. Chem. Soc.*, *126*, 2668 (2004).
58. (a) M. Monthioux, *Carbon*, *40*, 1809 (2002). (b) D. A. Britz and A. N. Khlobystov, *Chem. Soc. Rev.*, 637 (2006).
59. J. Zhao, A. Buldum, J. Han, and J. P. Lu, *Nanotechnology*, *13*, 195 (2002).
60. G. E. Gadd, M. Blackford, S. Moricca, N. Webb, P. J. Evans, A. M. Smith, G. Jacobsen, S. Leung, A. Day, and Q. Hua, *Science*, *277*, 933 (1997).
61. (a) A. I. Kolesnikov, J.-M. Zanotti, C.-K. Loong, P. Thiyagarajan, A. P. Moravsky, R. O. Loutfy, and C. J. Burnham, *Phys. Rev. Lett.*, *93*, 035503 (2004). (b) C. Matranga and B. Bockrath, *J. Phys. Chem. B*, *108*, 6170 (2004). (c) O. Byl, P. Kondratyuk, S. T. Forth, S. A. FitzGerald, L. Chen, J. K. Johnson, and J. T. Yates, Jr., *J. Am. Chem. Soc.*, *125*, 5889 (2003). (d) O. Byl, P. Kondratyuk, and J. T. Yates, Jr., *J. Phys. Chem. B*, *107*, 4277 (2003). (e) A. Fujiwara, K. Ishii, H. Suematsu, H. Kataura, Y. Maniwa, S. Suzuki, and Y. Achiba, *Chem. Phys. Lett.*, *336*, 205 (2001).
62. (a) A. C. Dillon, K. M. Jones, T. A. Bekkedahl, C. H. Kiang, D. S. Bethune, and M. J. Heben, *Nature (London)*, *386*, 377 (1997). (b) C. Liu, Y. Y. Fan, M. Liu, H. T. Cong, H. M. Cheng, and M. S. Dresselhaus, *Science*, *286*, 1127 (1999). (c) H. Cheng, G. P. Pez, and A. C. Cooper, *J. Am. Chem. Soc.*, *123*, 5845 (2001). (d) W.-F. Du, L. Wilson, J. Ripmeester, R. Dutrisac, B. Simard, and S. Denommee, *Nano Lett.*, *2*, 343 (2002). (e) M. Volpe and F. Cleri, *Chem. Phys. Lett.*, *371*, 476 (2003).
63. G. A. Jeffrey, *An Introduction to Hydrogen Bonding*, Oxford University Press, Oxford-New York, 1997.
64. (a) T. C. W. Mak and B. R. F. Bracke, *Comprehensive Supramolecular Chemistry, Vol. 6*, J. Szejtli and T. Osa, Eds., Pergamon, New York, 1996, p. 23. (b) H. G. McAdie, *Can. J. Chem.*, *44*, 1373 (1966). (c) S. Takamizawa, E. Nakata, H. Yokoyama, K. Mochizuki, and W. Mori, *Angew. Chem. Int. Ed. Engl.*, *42*, 4331 (2003). (d) C. S. Clarke, D. A. Haynes, J. M. Rawson, and A. D. Bond, *Chem. Commun.*, 2774 (2003).
65. (a) S. L. James, *Chem. Soc. Rev.*, *32*, 276 (2003). (b) C. Janiak, *J. Chem. Soc., Dalton Trans.*, 2781 (2003). (c) N. L. Rosi, J. Eckert, M. Eddaoudi, D. T. Vodak, J. Kim, M. O'Keeffe, and O. M. Yaghi, *Science*, *300*, 1127 (2003). (d) L. Pan, M. B. Sander, X. Huang, J. Li, M. Smith, E. Bittner, B. Bockrath, and J. K. Johnson, *J. Am. Chem. Soc.*, *126*, 1308 (2004). (e) T. Düren, L. Sarkisov, O. M. Yaghi, and R. Q. Snurr, *Langmuir*, *20*, 2683 (2004). (f) M. Eddaoudi, J. Kim, N. Rosi, D. Vodak, J. Wachter, M. O'Keeffe, and O. M. Yaghi, *Science*, 295, 469 (2002). (g) S. Kitagawa, R. Kitaura, and S. Noro, *Angew. Chem. Int. Ed. Engl.*, *43*, 2334 (2004).
66. J. L. C. Rowsell and O. M. Yaghi, *Angew. Chem. Int. Ed. Engl.*, *44*, 4670 (2005).
67. A. P. Cote, A. I. Benin, N. W. Ockwig, M. O'Keeffe, A. J. Matzger, and O. M. Yaghi, *Science*, *310*, 1166 (2005).
68. N. B. McKeown, P. M. Budd, K. J. Msayib, B. S. Ghanem, H. J. Kingston, C. E. Tattershall, S. Makhseed, K. J. Reynolds, and D. Fritsch, *Chem. Eur. J.*, *11*, 2610 (2005).

69. N. B. McKeown, B. S. Ghanem, K. J. Msayib, P. M. Budd, C. E. Tattershall, K. Mahmood, S. Tan, D. Book, H. W. Langmi, and A. Walton, *Angew. Chem. Int. Ed. Engl.*, *45*, 1804 (2006).

70. H. Lee, J. Lee, D. Y. Kim, J. Park, Y.-T. Seo, H. Zeng, I. L. Moudrakovski, C. I. Ratcliffe, and J. A. Ripmeester, *Nature (London)*, *434*, 743 (2005).

71. (a) R. Rathore, S. V. Lindeman, K. S. S. Rao, D. Sun, and J. K. Kochi, *Angew. Chem. Int. Ed. Engl.*, *39*, 2123 (2000). (b) S. V. Rosokha and J. K. Kochi, *J. Am. Chem. Soc.*, *124*, 5620 (2002). (c) S. V. Rosokha, S. V. Lindeman, R. Rathore, and J. K. Kochi, *J. Org. Chem.*, *68*, 3947 (2003).

72. D. M. Rudkevich, Y. Kang, A. V. Leontiev, V. G. Organo, and G. V. Zyryanov, *Supramol. Chem.*, *17*, 93 (2005).

73. (a) G. V. Zyryanov, Y. Kang, S. P. Stampp, and D. M. Rudkevich, *Chem. Commun.*, 2792 (2002). (b) G. V. Zyryanov, Y. Kang, and D. M. Rudkevich, *J. Am. Chem. Soc.*, *125*, 2997 (2003).

74. Y. Kang and D. M. Rudkevich, *Tetrahedron*, *60*, 11219 (2004).

75. C. Liu and L. Fu, J. Economy, *Macromol. Rapid Commun.*, *25*, 804 (2004).

76. (a) Y. Kang, G. V. Zyryanov, and D. M. Rudkevich, *Chem. Eur. J.*, *11*, 1924 (2005). (b) G. V. Zyryanov and D. M. Rudkevich, *Org. Lett.*, *5*, 1253 (2003).

77. G. B. Pariiskii, I. S. Gaponova, and E. Y. Davydov, *Russ. Chem. Rev.*, *69*, 985 (2000).

78. A. V. Stepanov and V. V. Veselovsky, *Russ. Chem. Rev.*, *72*, 327 (2003).

79. (a) A. Harada, J. Li, M. Kamachi, *Nature (London)*, *364*, 516 (1993). (b) A. Ikeda, M. Kawaguchi, and S. Shinkai, *Anal. Quim. Int. Ed. Engl.*, *93*, 408 (1997). (c) J.-A. Perez-Adelmar, H. Abraham, C. Sanchez, K. Rissanen, P. Prados, and J. de Mendoza, *Angew. Chem. Int. Ed. Engl.*, *35*, 1009 (1996). (d) S. K. Kim, W. Sim, J. Vicens, and J. S. Kim, *Tetrahedron Lett.*, *44*, 805 (2003). (e) S. K. Kim, J. Vicens, K.-M. Park, S. S. Lee, and J. S. Kim, *Tetrahedron Lett.*, *44*, 993 (2003). (f) Y. Kim, M. F. Mayer, and S. C. Zimmerman, *Angew. Chem. Int. Ed. Engl.*, *42*, 1121 (2003).

80. (a) D. T. Bong, T. D. Clark, J. R. Granja, and M. R. Ghadiri, *Angew. Chem. Int. Ed. Engl.*, *40*, 988 (2001). (b) S. Matile, A. Som, and N. Sorde, *Tetrahedron*, *60*, 6405 (2004). (c) T. Yamaguchi, S. Tashiro, M. Tominaga, M. Kawano, T. Ozeki, and M. Fujita, *J. Am. Chem. Soc.*, *126*, 10818 (2004). (d) S. Tashiro, M. Tominaga, T. Kusukawa, M. Kawano, S. Sakamoto, K. Yamaguchi, and M. Fujita, *Angew. Chem. Int. Ed. Engl.*, *42*, 3267 (2003). (e) M. Tominaga, S. Tashiro, M. Aoyagi, and M. Fujita, *Chem. Commun.*, 2038 (2002).

81. G. V. Zyryanov and D. M. Rudkevich, *J. Am. Chem. Soc.*, *126*, 4264 (2004).

82. V. G. Organo, A. V. Leontiev, V. Sgarlata, H. V. R. Dias, and D. M. Rudkevich, *Angew. Chem. Int. Ed. Engl.*, *44*, 3043 (2005).

83. V. Sgarlata, V. G. Organo, and D. M. Rudkevich, *Chem. Commun.*, 5630 (2005).

84. D. B. Dell'Amico, F. Calderazzo, L. Labella, F. Marchetti, and G. Pampaloni, *Chem. Rev.*, *103*, 3857 (2003).

85. D. M. Rudkevich and H. Xu, *Chem. Commun.*, 2651 (2005).

86. (a) T. Yamaguchi, C. A. Koval, R. D. Nobel, and C. Bowman, *Chem. Eng. Sci.*, *51*, 4781 (1996). (b) T. Yamaguchi, L. M. Boetje, C. A. Koval, R. D. Noble, and C. N. Bowman, *Ind. Eng. Chem. Res.*, *34*, 4071 (1995). (c) P. Kosaraju, A. S. Kovvali, A. Korikov, and K. K. Sirkar, *Ind. Eng. Chem. Res.*, *44*, 1250 (2005).

87. (a) C. D. Ki, C. Oh, S.-G. Oh, and J. Y. Chang, *J. Am. Chem. Soc.*, *124*, 14838 (2002). (b) J. Alauzun, A. Mehdi, C. Reye, and R. J. P. Corriu, *J. Am. Chem. Soc.*, *127*, 11204 (2005).

88. (a) E. M. Hampe and D. M. Rudkevich, *Chem. Commun.*, 1450 (2002). (b) E. M. Hampe and D. M. Rudkevich, *Tetrahedron*, *59*, 9619 (2003). (c) L. C. Brousseau, III, D. J. Aurentz, A. J.

Benesi, and T. E. Mallouk, *Anal. Chem.*, *69*, 688 (1997). (d) P. Herman, Z. Murtaza, and J. Lakowicz, *Anal. Biochem.*, *272*, 87 (1999).

89. (a) M. George and R. G. Weiss, *J. Am. Chem. Soc.*, *123*, 10393 (2001). (b) M. George, and R. G. Weiss, *Langmuir*, *18*, 7124 (2002). (c) M. George and R. G. Weiss, *Langmuir*, *19*, 1017 (2003). (d) M. George and R. G. Weiss, *Langmuir*, *19*, 8168 (2003).
90. E. D. Bates, R. D. Mayton, I. Ntai, and J. H. Davis, Jr., *J. Am. Chem. Soc.*, *124*, 926 (2002).
91. (a) J.-M. Lehn, *Chem. Eur. J.*, *5*, 2455 (1999). (b) S. J. Rowan, S. J. Cantrill, G. R. L. Cousins, J. K. M. Sanders, and J. F. Stoddart, *Angew. Chem. Int. Ed. Engl.*, *41*, 898 (2002).
92. (a) A. W. Bosman, L. Brunsveld, B. J. B. Folmer, B. J. B.; Sijbesma, R. P.; Meijer, and E. W. *Macromol. Symp.*, *201*, 143 (2003). (b) J.-M. Lehn, *Polym. Int.*, *51*, 825 (2002). (c) U. S. Schubert and C. Eschbaumer, C. *Angew. Chem. Int. Ed. Engl.*, *41*, 2892 (2002). (d) A. T. ten Cate and R. P. Sijbesma, *Macromol. Rapid Commun.*, *23*, 1094 (2002). (e) L. Brunsveld, B. J. B. Folmer, E. W. Meijer, and R. P. Sijbesma, *Chem. Rev.*, *101*, 4071 (2001). (f) R. F. M. Lange, M. van Gurp, and E. W. Meijer, *J. Polym. Sci. A*, *37*, 3657 (1999). (h) R. K. Castellano, D. M. Rudkevich, and J. Rebek, Jr., *Proc. Natl. Acad. Sci. U.S.A.*, *94*, 7132 (1997).
93. D. M. Rudkevich and G. A. Woldemariam, H. Xu, *Polym. Prepr. (Am. Chem. Soc., Div. Polym. Chem.)*, *46*, 1162 (2005).
94. (a) K. D. Shimizu and J. Rebek, Jr., *Proc. Natl. Acad. Sci. U.S.A.*, *92*, 12403 (1995). (b) J. Rebek, Jr., *Chem. Commun.*, 637 (2000).
95. (a) O. Mogck and V. Böhmer, W. Vogt, *Tetrahedron*, *52*, 8489 (1996). (b) O. Mogck, E. F. Paulus, V. Böhmer, I. Thondorf, and W. Vogt, *Chem. Commun.*, 2533 (1996).
96. H. Xu and D. M. Rudkevich, *Chem. Eur. J.*, *10*, 5432 (2004).
97. H. Xu and D. M. Rudkevich, *Org. Lett.*, *7*, 3223 (2005).
98. H. Xu and D. M. Rudkevich, *J. Org. Chem.*, *69*, 8609 (2004).

# CHAPTER 5

# The Organometallic Chemistry of Rh-, Ir-, Pd-, and Pt-Based Radicals: Higher Valent Species

**BAS DE BRUIN**

*University of Amsterdam*
*Van 't Hoff Institute for Molecular Chemistry*
*Department of Homogeneous and Supramolecular Catalysis*
*1018 WV, Amsterdam, The Netherlands*

**DENNIS G. H. HETTERSCHEID and ARJAN J. J. KOEKKOEK**

*Radboud University Nijmegen*
*Institute for Molecules and Materials*
*Department of Molecular Materials*
*6525 ED Nijmegen, The Netherlands*

**HANSJÖRG GRÜTZMACHER**

*ETH-Hönggerberg*
*Department of Chemistry and Applied Biosciences*
*8093 Zürich Switzerland*

CONTENTS



*Progress in Inorganic Chemistry, Vol. 55* Edited by Kenneth D. Karlin

# I. INTRODUCTION

A tremendous amount of information is available concerning the structure and reactivity of organometallic species of the second- and third-row groups 9 (VIIIB) and 10 (VIII) platinum metals (Rh, Ir, Pd, and Pt). This is not surprising considering the fact that these species are frequently used as homogeneous and heterogeneous catalysts for a variety of important transformations, both in industry as well as in academic research laboratories (hydrogenations, hydroformylations, cross-couplings, allylic couplings, methatesis reactions etc.). Even the automobile-exhaust catalysts (meanwhile present in almost everyone's car) are largely based on the catalytic properties of groups 9 (VIIIB) and 10 (VIII) transition metals. Apart from catalysis, these metal complexes find various other applications, for example, in supramolecular assemblies, hydrogen storage, and photovoltaic systems [Ir-based triplet emitters in organic light-emitting diode (OLEDs)]. Partly, the widespread use of these transition metal complexes may simply rely on the fact that they form rather stable metal–ligand bonds, which helps to develop synthetic techniques.

Our understanding of the reactivity of these metals is based on studies of diamagnetic complexes in homogeneous solutions. Much less is known of the structure, reactivity, and physical properties of their *paramagnetic* counterparts. There are several reasons for this. First, paramagnetic complexes are often highly reactive (especially to $O_2$) and difficult to handle, thus often requiring low-temperature syntheses, inert atmospheres, and nonreactive solvents. Also the characterization of these species can be a problem. Characterisation problems not only arise due to the short lifetime of some of these species. More generally, paramagnetic species can be difficult to characterize and are often overseen, especially when nuclear magnetic resonance (NMR) is used as the spectroscopic tool for characterization of products. Nevertheless, over the past years interest in the properties and reactivity of such paramagnetic organometallic species is rising. This is triggered by the idea that they might play an active role in the reactivity of these metals, perhaps even as crucial intermediates in some catalytic reactions. This holds for homogeneous catalysts, but perhaps chances to stabilize and utilize these species as catalysts are even higher for heterogenized systems. Special interest in open-shell organometallic species also comes from their expected higher and different reactivity compared to their closed-shell counterparts.

In this chapter and in an upcoming review (Part II), the chemistry of mononuclear paramagnetic organometallic complexes of the second- and third-row groups 9(VIIIB) and 10(VIII) platinum metals (Rh, Ir, Pd, Pt) are summarized. In this chapter, the higher valent species (Rh(II), Ir(II), Rh(IV), Ir(IV), Pd(III), Pt(III) and complexes with oxidized redox noninnocent ligands) are described. In the next chapter, the lower valent species (Rh(0), Ir(0), Pd(I), Pt(I), and complexes with reduced redox noninnocent ligands will be discussed.

The number of publications on this subject is growing, especially over the last few years. The excellent reviews of Pandey (1) and DeWit (2) were used in this chapter as a basis for much of the older work on this subject, and we tried to give a complete overview to date.

### A. Ligand Radicals versus Metal Radicals: The Concepts of "Ligand Noninnocence"

When considering the properties and reactivity of paramagnetic species, an important question to answer is the location of the unpaired electrons. This chapter considers second- and third-row transition metal complexes that usually adopt a low-spin configuration. Thus, in general, these species contain only one unpaired electron ($S = \frac{1}{2}$ systems). This does not mean that their electronic structure is always as simple as it seems, because the unpaired electron can be located at the metal, at the ligand, or inbetween. Some species should be regarded as distinct metal or ligand radicals (redox isomerism), whereas in other species the ligand and metal radical descriptions are limiting resonance structures of the real electronic structure.

In classic Werner-type coordination chemistry, the amount of (half) filled *d* orbitals is usually quite clear from the formal oxidation state of the metal. The latter is easily obtained by completely assigning the electron density of metal–ligand bonds to the more electronegative ligand atoms (ionic model), then assigning formal charges of the ligand atoms on the basis of the ligand Lewis structures, and subsequently counting back the formal oxidation state being the difference between the total charge of the complex and the sum of all ligand charges. Together with crystal-field (ligand-field) theory, such formal oxidation states usually give a good approximation of the so-called *d*-electron configuration of the metal. The essential feature of using this model is the assumption that all metal–ligand interactions are mainly ionic with only small covalent contributions. This means that in using molecular orbital (MO) descriptions the assumption is made that the (filled) valence ligand orbitals used for ligand–metal bonding (either $\sigma$ or $\pi$ bonding) are always substantially lower in energy than the metal *d* orbitals (see Fig. 1).

In many cases, organometallic complexes also contain $\pi$-acceptor ligands. These ligand types contain relatively low-lying empty orbitals capable of accepting electron density from (the appropriate $\pi$-type) metal *d* orbitals. Textbook examples of such ligands are $CN^-$, CO, olefins, and phosphine-type ligands. The essential assumption made when regarding binding of $\pi$-acceptor ligands is that the valence ligand orbitals for M—L $\pi$ back-bonding are always substantially higher in energy than the metal *d* orbitals.

A formal oxidation state becomes less meaningful when the covalency increases. For example, when the $\sigma$-donating ligand orbitals and the metal *d* orbitals lie very close in energy, substantial mixing of the ligand and metal *d* orbitals occurs

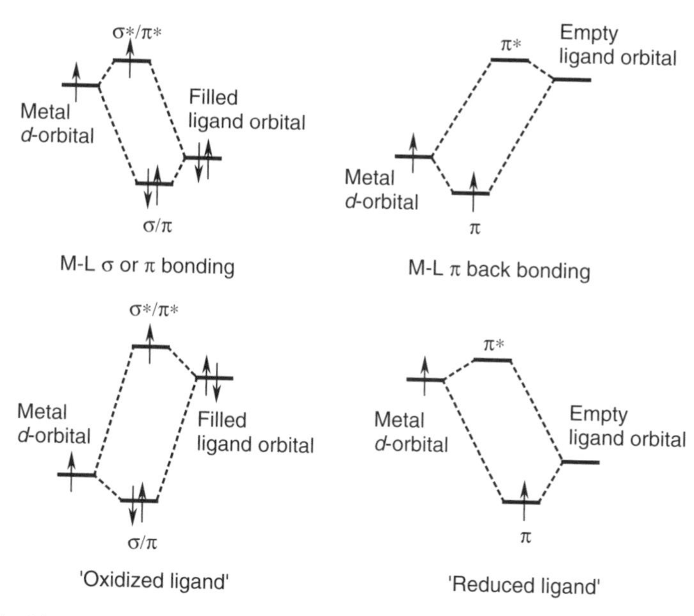


Figure 1. Ligand–metal donation and metal–ligand back-donation models and their extremes, where ligands become 'redox noninnocent'.

and the bonding and antibonding combinations are no longer of mainly ligand or metal character, respectively. In such cases, the formal oxidation state becomes misleading and no longer predicts the *d*-electron configuration accurately.

*"Redox Noninnocent" Ligands.* The situation becomes even less clear with unusually high- or low-metal oxidation states. For complexes in high oxidation states, the ligand orbitals of some specific ligand types can actually end-up higher in energy than the metal *d* orbitals. In such cases, we speak of a "redox noninnocent" ligand, because this situation leads to oxidation of the ligand, whereas oxidation of the metal was expected. This has important consequences, as the ligand(s) now contain(s) unpaired spin-density (corresponding to *x* electrons), and the filling of the metal '*d* orbitals' $(d^{n+x})$ is in reality higher than predicted by the formal oxidation state $(d^n)$. Obviously, "redox noninnocent" behavior of ligands is also possible when π-acceptor-type orbitals lie lower in energy than the metal *d* orbitals. This situation typically involves complexes in low oxidation states and leads to reduction of the ligands. Also in these cases, the ligand(s) end-up containing unpaired spin-density, and the actual filling of the metal '*d* orbitals' $(d^{n-x})$ is now lower than predicted by the formal oxidation state

of the metal. The distribution of the spin density over the metal and 'redox noninnocent' ligands can be subtle and quite difficult to assign.

*"Chemically Noninnocent" Ligands.* Ligands that play an active role in chemical reactions are sometimes referred to as 'chemically noninnocent' ligands. Chemical noninnocence is often synonymous with 'redox non-innocence'. Spin delocalization over the metal and a 'redox noninnocent' ligand usually stabilizes the complex, but unpaired spin-density located at the ligand can also render the ligand susceptible to chemical reactions.

*Innocent Radical Ligands.* Common ligands are sometimes substituted with noncoordinating persistent organic radical fragments, for example, nitronyl nitroxide radical fragments (3). For metal complexes containing these ligands, there is no real confusion about the oxidation state of the metal, and these radical ligands are thus not referred to as 'redox noninnocent'. Likewise, for paramagnetic species obtained by radical additions of organic radicals to organometalic complexes with large extended $\pi$-systems, such as $C_{60}$ fullerenes (4, 5), it is usually quite clear that the unpaired electron will be most concentrated on the ligand rather than the metal. Although a number of complexes containing 'innocent radical' ligands have been described, the paramagnetic nature of these ligands in most cases does not influence the intrinsic organometallic chemistry of the complex. We consider such examples not relevant to the chemistry described in this chapter.

### B. Basic Interpretation of EPR Spectra Concerning $S = \frac{1}{2}$ Systems

In most cases, electron paramagnetic resonance (EPR) spectra of the $S = \frac{1}{2}$ system species described here are relatively noncomplicated (compared to $S = n/2$ systems), and should be rather diagnostic for the location of the unpaired electron. The absolute $g$ values and hyperfine couplings (if resolved) are informative for the relative spin–orbit couplings and metal–ligand interactions. Most of the reported EPR spectra concern measurements in frozen solutions, and thus reveal $g$ anisotropy, which contains valuable information about the electronic structure of the species. For a thorough understanding of EPR spectroscopy of transition metal complexes, including spectral interpretations, we refer to the excellent overview of Goodman and Raynor (6). However, because EPR spectroscopy is not a very common technique among many organometallic chemists, a very restricted overview of some of the important issues is given below.

Electron paramagnetic resonance spectroscopy relies on microwave radiation induced resonance of electron spins in an external magnetic field, $B$ (Fig. 2). The effective magnetic field experienced by the electron is, however, influenced by

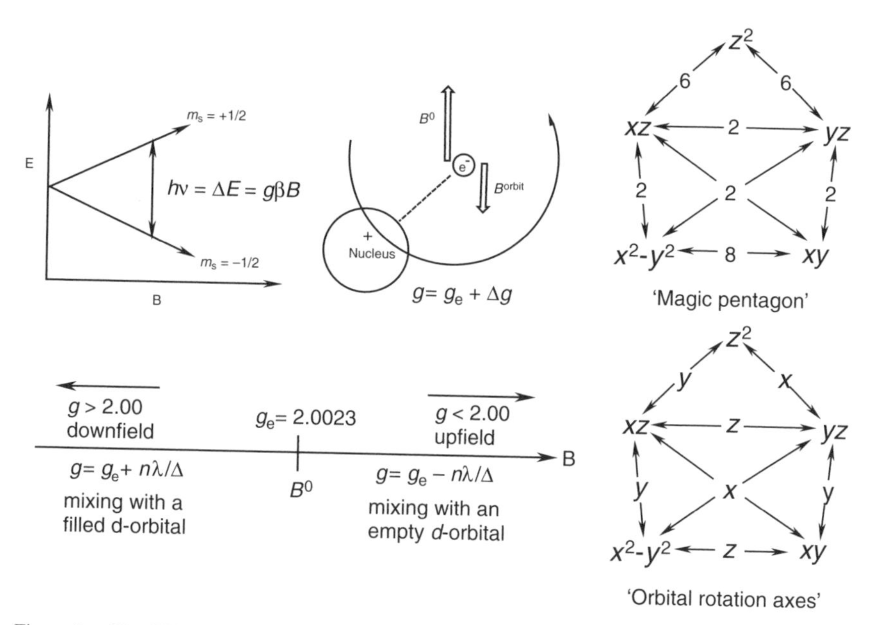


Figure 2. The EPR resonance conditions, the influence of orbital motion on $g$ values and the so-called "magic pentagon" to determine the "orbital mixing number" $n$.

orbital motions of the electron, which change the resonance conditions by inducing a local magnetic field. This leads to a shift of the observed $g$ values in an EPR spectrum. In addition, the electron spin couples with nuclear spins, giving rise to splitting patterns of the signal, referred to as hyperfine couplings (or superhyperfine couplings in case of coupling with ligand atoms).

A free electron, or an electron in a nondegenerate orbital with no mixing with exited states, has only a spin contribution to its angular momentum ($S$). In the latter case, $g$ values close to $g_e$ (2.00321), the free-electron value, are expected. However, the ground state is quantum mechanically allowed (to a first approximation) to mix to a certain extend with exited states via the spin–orbit operator $\lambda \cdot L \cdot S$. For reasons of simplicity, this can be most conveniently regarded as a "mixing" of the singly occupied MO (SOMO) with other filled or empty orbitals. This results in orbital contributions to the total angular momentum of the electron, even if the SOMO is nondegenerate. The orbital angular momentum amplifies or reduces the effective magnetic field experienced by the electron, depending on whether the orbital contributions arise from "mixing" with a filled or an empty orbital, respectively. Therefore, resonance conditions occur at lower or higher external magnetic fields using a constant microwave frequency. This is expressed in EPR terminology in observing $g$ values above and below $g_e$, respectively (Fig. 2). For a transition metal centered radical, the amount of orbital mixing depends on both the spin–orbit coupling constant ($\lambda$) of the metal atom at which the electron mainly resides and the energy difference ($\Delta$) between the SOMO and the "mixed-in" empty or filled orbitals. This gives rise to the approximate $g$ values: $g = g_e + n \cdot \lambda/\Delta$ in case of "mixing" with a filled $d$ orbital and $g = g_e - n \cdot \lambda/\Delta$ for mixing with an empty $d$ orbital. The "orbital mixing number" $n$ depends on which of the $d$-orbitals "mix" according to the so-called "magic pentagon" (6) shown in Fig. 2.

Because the mixing of the SOMO with other orbitals by means of spin–orbit coupling depends on the orientation of the molecule with respect to the applied magnetic field, $B$, different $g$ values are observed in different directions in solid-state spectra (e.g., frozen solutions). Optimal "mixing" of the SOMO with a filled or empty $d$-orbital occurs when the magnetic field lies along the "orbital rotation axis" between these orbitals, as shown in Fig. 2. The above so-called qualitative orbital rotation method is not highly accurate, and deviations from the formula $g = g_e +/- n \cdot \lambda/\Delta$ become increasingly significant with increasing spin–orbit coupling constants $\lambda$, on going from the first-row, via the second-row, to the third-row transition metals. Nevertheless, for a qualitative interpretation of EPR spectra, the above method remains very useful.

The amount of different $g$ values (actually the main components) observable in solid-state EPR spectra thus depends on the symmetry of the molecule. Molecules with low symmetry are likely to reveal three different $g$ values, associated with the main axes of the molecule (rhombic $g$ values: $g_x \neq g_y \neq g_z$). Axially symmetric molecules are likely to reveal two different $g$ values (axial g-values: $g_x \neq g_y \neq g_z$), and highly symmetric species (e.g., tetrahedral complexes) give rise to isotropic $g$ values ($g_x \neq g_y \neq g_z$). Examples of such spectra are shown in Fig. 3. For EPR spectra recorded in solution, rapid tumbling of the molecule also results in isotropic $g$ values, which are actually the weight average of the anisotropic $g$ values of the species observed in frozen solution (Fig. 3). For transition metal complexes of concern to this chapter, containing only one unpaired electron ($S = \frac{1}{2}$ systems), species with exact isotropic $g$ values ($g_x = g_y = g_z$) should not exist due to expected Jahn–Teller distortions. Therefore, any isotropic $g$ values observed in frozen solutions are likely the result of poor resolution: nonresolved $g$ anisotropy due to broad lines.

Hyperfine couplings are also anisotropic in solid-state spectra, but are not always resolved. Although all atoms containing spin density give rise to hyperfine couplings along all $g$ values, the coupling constants are not the same along the different directions and *resolved* hyperfine couplings are often only observable along a single $g$ value and for a restricted number of atoms (see Fig. 3). But, even a restricted number of resolved hyperfine couplings gives valuable information about the (electronic) structure of the molecule, especially in comparison with values obtained from density functional theory (DFT) calculations.

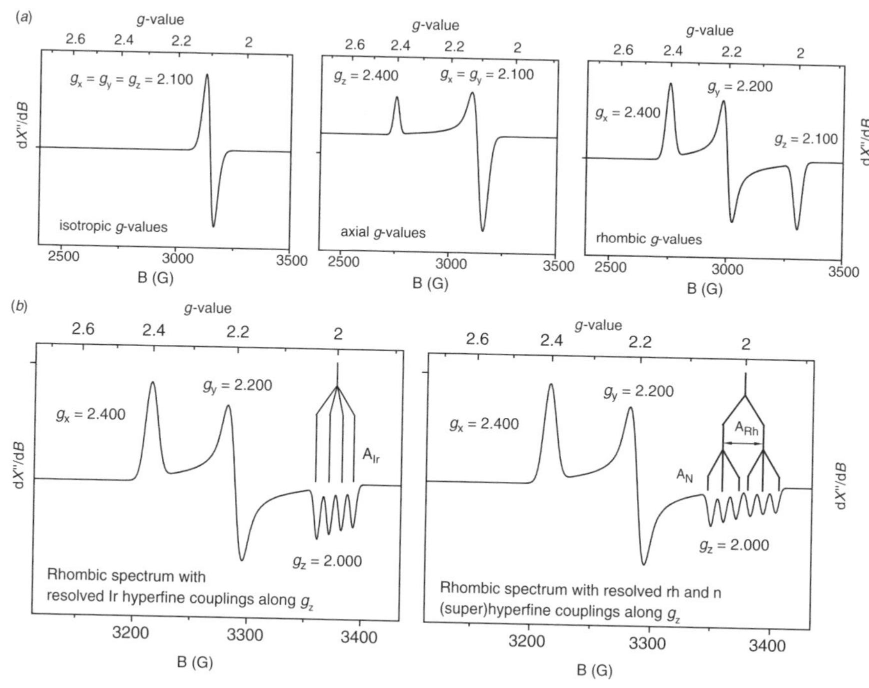


Figure 3. Some illustrative examples of anisotropic EPR spectra, such as those observed in frozen solutions (spectral simulations). (*a*) Typical spectra of species with isotropic, axial, and rhombic $g$ values (from left to right). (*b*) Rhombic spectra revealing resolved (super)hyperfine couplings along a single direction. The spectrum in the bottom-left reveals iridium hyperfine coupling (1 nucleus, $I = 3/2$) along $g_z$. The spectrum at the bottom-right reveals both rhodium hyperfine coupling (1 nucleus, $I = 1/2$) and nitrogen superhyperfine coupling (1 nucleus, $I = 1$) along $g_z$.

*EPR Characteristics of Metal-Centered Radicals.* Generally, electron-spin relaxation processes are much faster for metal-centered radicals compared to organic radicals (or ligand-centered radicals). Therefore, EPR spectra of metal-centered radicals are usually recorded at low temperatures ($<100$ K) to prevent substantial line broadening.

For systems where the unpaired spin primarily resides in a metal *d* orbital, one obviously expects large metal hyperfine interactions. But even when the spin density is primarily located in a metal-based orbital, some small mixing with ligand orbitals often leads to substantial ligand hyperfine couplings of the same order of magnitude. Therefore, one cannot entirely rely on the hyperfine couplings.

Metal-centered radicals usually have large deviations of the measured $g$ values from the free electron value $g_e = 2.0023$. The reason for the large deviations from $g = g_e$ stems from the larger spin–orbit coupling constants ($\lambda$) of the heavier metal atoms compared to the lighter ligand atoms, facilitating spin–orbit coupling. In addition, the energy differences ($\Delta$) between the half-filled (SOMO), filled occupied MO closest in energy to the HOMO–SOMO (HOMO-1), and empty lowest unoccupied MO (LUMO) *d* orbitals (or those substantially mixed with ligand orbitals) are usually small. In most cases, metal-centered radicals reveal a relatively large *g* anisotropy (with often one of the anisotropic *g* values close to $g \sim 2$). Typical EPR spectra indicative for metal-centered radicals usually contain *g* values $g > 2.10$ or $g < 1.95$, but a definitive assignment of a species as metal or ligand radical cannot be made on the basis of the *g* values alone.

To assist the reader in interpreting the EPR data of metal centered radicals presented in this chapter (and the upcoming article, Part II), we refer to Fig. 4 showing the expected *g* anisotropy of $d^7$ and $d^9$ complexes in their most common geometries, based on the above magic pentagon qualitative orbital rotation method. For the octahedral complexes, Jahn–Teller compressed and Jahn–Teller elongations are considered. For square-planar complexes, the energy of the $d_{z^2}$, $d_{xz}$, $d_{yx}$, and $d_{xy}$ orbitals depends on the nature of the ligands ($\pi$ donor, $\pi$ acceptor, pure $\sigma$ donors) and the three different *d*-orbital configurations shown in Fig. 4 need to be considered.

*EPR Characteristics of Ligand-Centered Radicals.* The light atoms of ligands have small spin–orbit coupling constants, $\lambda$. Very often organic radicals also have larger energy differences between the half-filled (SOMO), filled (HOMO-1), and empty (LUMO) orbitals compared to metal-centered radicals. Therefore, *g* anisotropies and absolute deviations from $g = g_e$ are often small for systems where the unpaired spin primarily resides in a ligand-based orbital (even when the molecule has a low symmetry). Metal hyperfine interactions are also usually small in these situations. Observation of sharp and strong EPR spectra of metal complexes in solutions at room temperature is usually a strong indication that the unpaired electron is located in a ligand-based orbital with only weak mixing with metal orbitals. Sometimes such spectra even reveal a well-resolved hyperfine pattern from ligand atoms (e.g., see Fig. 5), but this is not a general rule (small spin densities at the metal, e.g., can lead to substantial line broadening at higher temperatures).

*Exceptions.* The above characteristics are useful, but one should keep in mind that these are not much more than a rule-of-thumb. There is a sound theoretical basis for many possible exceptions. For example, a primarily metal-centered radical can reveal only small deviations from $g = g_e$ when

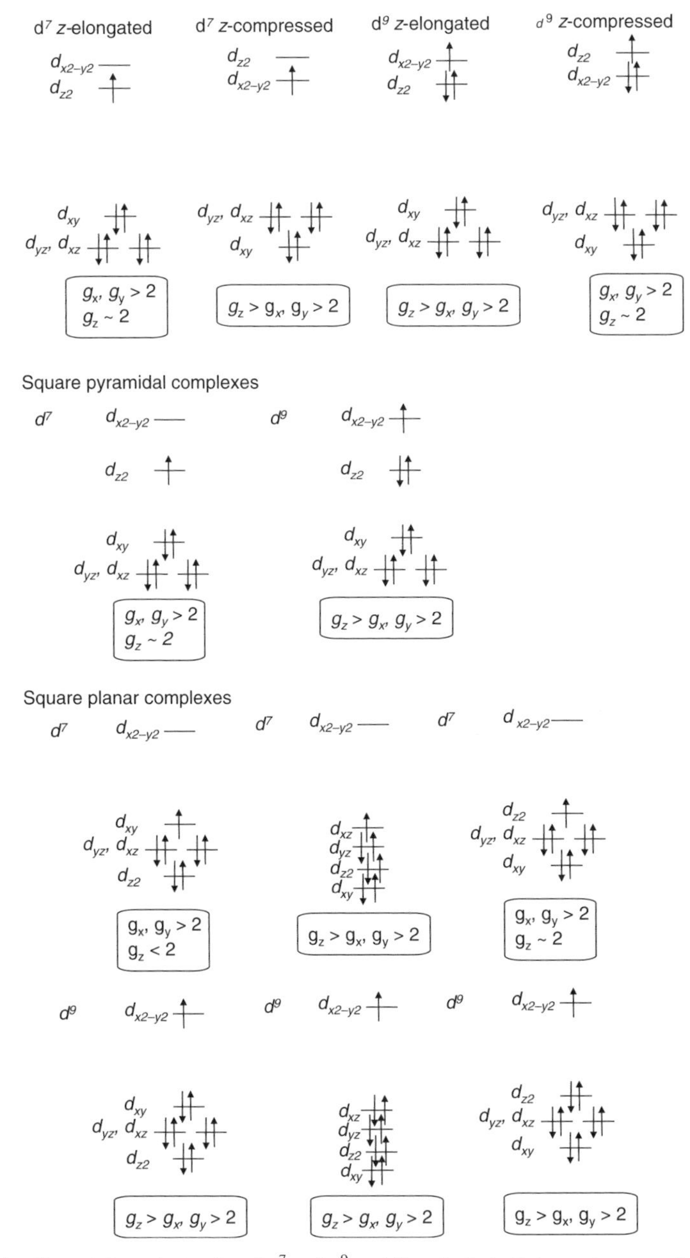


Figure 4. Expected $g$ anisotropies of $d^7$ and $d^9$ metalloradicals in their most commonly observed geometries (based on the 'magic pentagon' orbital rotation method).

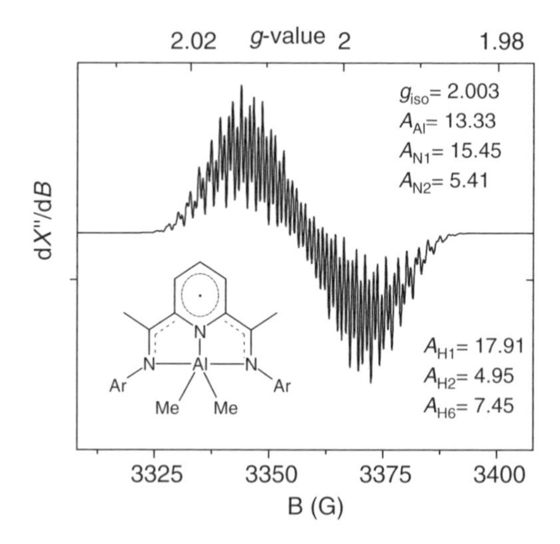


Figure 5. Illustrative example of an EPR spectrum of a ligand-centred radical species in isotropic solution at room temperature. The spectrum reveals hyperfine couplings with all nitrogen and hydrogen atoms of the 1e-reduced pyridine-2,6-diimine ligand.

the energy differences (Δ) between the SOMO and the HOMO-1 and LUMO orbitals are large.

## II. PERSISTENT METALLORADICALS

This section describes stable metalloradicals Rh(II), Ir(II), Pd(III), and Pt(III) species for which not much reactivity has been reported. Many other metalloradicals described in Section IV (*reactivity of transient metalloradicals*) are also persistent under certain conditions, but because of their high reactivity under other conditions, they are classified as transient metalloradicals nonetheless.

### A. Carbonyl–Phosphine Rh(II) and Ir(II) Complexes

Although its iridium equivalents were already studied in the 1980s, (7–10), and the electrochemistry of $[Rh^{II}(H)(CO)(PPh_3)_3]^+$ was already reported in the 1970s, (9,10) the spectroscopical analysis of $[Rh^{II}(H)(CO)(PPh_3)_3]^+$ was described only recently by Bond and co-workers (11). The unusually stable 17 electron $[Rh^{II}(H)(CO)(PPh_3)_3]^+$ complex is easily obtained by oxidation [$1e^-$-process at $-0.48$ V vs. ferrocene/ferrocenium ($Fc/Fc^+$)] of the neutral diamagnetic rhodium(I) precursor (Fig. 6) (9–11). Formation of $[Rh^{II}(H)(CO)(PPh_3)_3]^+$

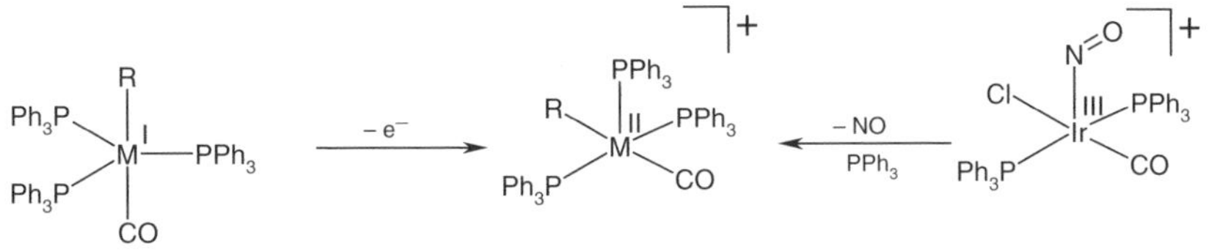

Figure 6. One electron oxidation of $[M^{I}(R)(CO)(PPh_3)_3]$ or release of nitrogen monoxide from $[Ir^{III}(Cl)(NO^{-})(CO)(PPh_3)]^{+}$ yields paramagnetic rhodium or iridium complexes of the type $[M^{II}(R)(CO)(PPh_3)_3]^{+}$.

is so facile that it may need to be considered as an active species in a variety of processes catalyzed by the often used homogeneous catalyst $[Rh^{I}(H)(CO)(PPh_3)_3]$. Unlike its Rh(II) precursor, the $Rh^{III}$—H species obtained by oxidation of $[Rh^{II}(H)(CO)(PPh_3)_3]^{+}$ is not stable, and loses a proton as witnessed by its electrochemical behavior (EC mechanism).

The rhombic EPR spectrum of $[Rh^{II}(H)(CO)(PPh_3)_3]^{+}$ reveals hyperfine coupling with Rh and a strong superhyperfine coupling with the axial phosphorus atom (see Table I) (11). Apparently, oxidation of the trigonal-bipyramidal Rh(I) complex yields a square-pyramidal Rh(II) species, which is an expected Jahn–Teller distortion. Somewhat smaller, but nevertheless quite large hyperfine couplings were observed for the two equatorial phosphorus nuclei. The large P hyperfine interactions suggest a significant delocalization of the unpaired electron over the metal and the P donors. Hyperfine coupling to the equatorial hydride is poorly resolved for $[Rh^{II}(H)(CO)(PPh_3)_3]^{+}$. The spectrum changes slightly when the hydrogen is replaced by a deuterium atom [$A_x(D) = 6$ MHz, $A_y(D) = 4$ MHz for $[Rh^{II}(D)(CO)(PPh_3)_3]^{+}$). Quite similar structures and EPR spectra were obtained for the iridium analogue $[Ir^{II}(H)(CO)(PPh_3)_3]^{+}$ (12) and similar iridium(II) alkyls $[Ir^{II}(R)(CO)(PPh_3)_3]^{+}$ (R = Me, Ph, $CH_2CN$) (7) (Table I).

The $[Ir^{II}(Cl)(CO)(PPh_3)_3]^{+}$ complex was obtained by thermal–photochemical decomposition of $[Ir^{III}(Cl)(CO)(PPh_3)_2(NO)]^{+}$ in the presence of triphenylphosphine. The +II oxidation state of iridium was confirmed by a strong absorption band at 2007 $cm^{-1}$, in between the CO stretching frequencies of iridium(III) carbonyls in the region of 2040–2080 $cm^{-1}$ and iridium(I) carbonyls in the region of 1930–1970 $cm^{-1}$. The CO frequencies and EPR data of $[Ir^{II}(Cl)(CO)(PPh_3)_3]^{+}$ are quite similar to the other iridium(II) species shown in Table I.

The $[Rh^{II}(\eta^3\text{-TMPP})_2]^{2+}$ [TMPP = tris(2,4,6-trimethoxyphenyl)phosphine] complex reacts reversibly with CO and isonitriles to form stable square-planar diamagnetic carbonyl and isonitrile complexes (13, 14). In the octahedral $[Rh^{II}(\eta^3\text{-TMPP})_2]^{2+}$ complex, the two TMPP ligands are coordinated to rhodium(II) with the two P donors in mutual cis positions. Four methoxy

TABLE I
The EPR and Infrared (IR) Parameters for $[M^{II}(R)CO(PPh_3)_3]^{+(a)}$

| Complex | $\upsilon$(CO) $cm^{-1}$ | $g_x$ $g_y$ $g_z$ | $A_x(^{31}P_{ax})$ $A_y(^{31}P_{ax})$ $A_z(^{31}P_{ax})$ | $A_x(M)$ $A_y(M)$ $A_z(M)$ | References |
|---|---|---|---|---|---|
| $[Ir^{II}(Me)(CO)(PPh_3)_3]^+$ | 2008 | 2.22 | 506 | | 7 |
| | | 2.19 | 500 | | |
| | | 1.97 | 623 | 85 | |
| $[Ir^{II}(Ph)(CO)(PPh_3)_3]^+$ | 2010 | 2.25 | 538 | | 7 |
| | | 2.16 | 517 | | |
| | | 1.97 | 626 | 85 | |
| $[Ir^{II}(CH_2CN)(CO)(PPh_3)_3]^+$ | 2015 | 2.24 | 542 | | 7 |
| | | 2.15 | 520 | | |
| | | 1.97 | 640 | 105 | |
| $[Ir^{II}(H)(CO)(PPh_3)_3]^+$ | 1990 | 2.22 | 457 | | 7 |
| | | 2.15 | 442 | | |
| | | 1.98 | 554 | 105 | |
| $[Ir^{II}(D)(CO)(PPh_3)_3]^+$ | 2017 | 2.22 | 435 | 25 | 7, 12 |
| | | 2.15 | 442 | 21 | |
| | | 1.98 | 554 | 108 | |
| $[Ir^{II}(Cl)(CO)(PPh_3)_3]^+$ | 2007 | 2.33 | 587 | | 8, 12 |
| | | 2.24 | 570 | | |
| | | 2.06 | 677 | | |
| $[Rh^{II}(H)(CO)(PPh_3)_3]^+$ | | 2.12 | 520 | 36 | 11 |
| | | 2.07 | 510 | 12 | |
| | | 2.00 | 644 | 48 | |
| $[Rh^{II}(H)(CO)(PPh_3)_3]^+$ | | 2.12 | 534 | 24 | 11 |
| | | 2.07 | 527 | 9 | |
| | | 2.00 | 658 | 11 | |

[a]Hyperfine couplings in megahertz (MHz).

groups, two of each TMPP ligand, complete the octahedral geometry. Magnetic susceptibility measurements (1.80 $\mu_B$ in the solid state; 2.10 $\mu_B$ by the Evans method) and EPR measurements (axial symmetry with $g_\perp = 2.250$ and $g_\parallel = 2.004$) indicate an $S = \frac{1}{2}$ system with the unpaired electron occupying the $d_{z^2}$ orbital. The methoxy groups opposite to each other are somewhat elongated due to a Jahn–Teller distortion, as expected for a $d^7$ ion in an octahedral ligand environment (14).

Treatment of $[Rh^{II}(\eta^3\text{-TMPP})_2]^{2+}$ with CO generates a new species, which is probably the bis-CO complex $[Rh^{II}(CO)_2(\eta^2\text{-TMPP})_2]^{2+}$ (Fig. 7). Because formation of this reactive species is slow, there is still a large excess of $[Rh^{II}(\eta^3\text{-TMPP})_2]^{2+}$ present in the early stages of the reaction. As a result, electron transfer takes place between the starting material and $[Rh^{II}(CO)_2(\eta^2\text{-TMPP})_2]^{2+}$, yielding $[Rh^{III}(\eta^3\text{-TMPP})_2]^{3+}$ and $[Rh^{I}(CO)_2(\eta^1\text{-TMPP})_2]^+$ (14, 15).

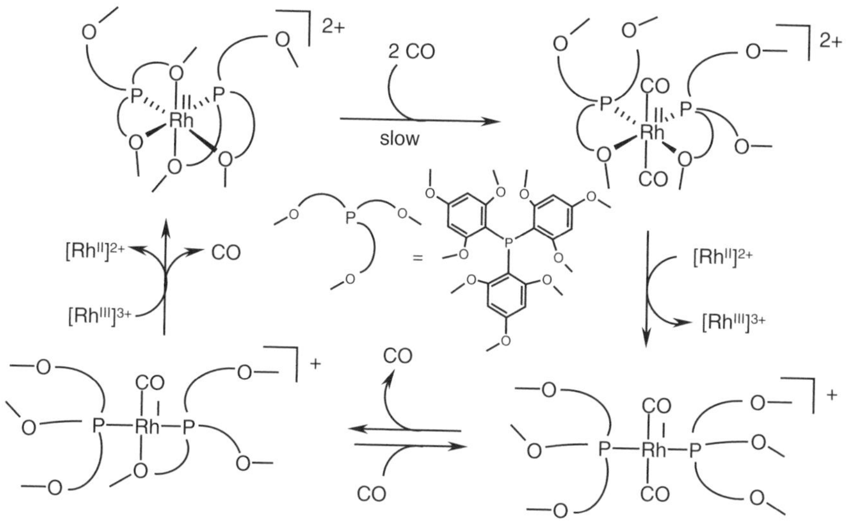

Figure 7. Reversible CO induced disproportionation of $[Rh^{II}(\eta^3\text{-TMPP})_2]^{2+}$.

Under conditions of pumping or purging with inert gas, $[Rh^{I}(CO)_2(\eta^1\text{-TMPP})_2]^+$ loses a CO ligand to form $[Rh^{I}(CO)(\eta^1\text{-TMPP})(\eta^2\text{-TMPP})]^+$. This species is oxidized at a much lower potential than the bis-CO complex, and is therefore oxidized by $[Rh^{III}(\eta^3\text{-TMPP})_2]^{3+}$ causing loss of another CO ligand and regeneration of $[Rh^{II}(\eta^3\text{-TMPP})_2]^{2+}$ (14, 15).

Treatment of $[Rh^{II}(\eta^3\text{-TMPP})_2]^{2+}$ with *tert*-butylisonitrile yields a new paramagnetic species: $[Rh^{II}(\eta^1\text{-TMPP})_2(CN\textit{t}\text{-Bu})_2]^{2+}$ (16). X-ray diffraction revealed that $[Rh^{II}(\eta^1\text{-TMPP})_2(CN\textit{t}\text{-Bu})_2]^{2+}$ adopts a distorted square-planar structure with mutually trans phosphine and isocyanide ligands. Somehow a cis–trans isomerization has taken place. The trans Rh—P bonds are slightly longer than the cis Rh—P bonds in the starting material, as a result of a stronger trans influence of a phosphine donor compared to an ether.

The EPR spectra of polycrystalline $[Rh^{II}(\eta^1\text{-TMPP})_2(CN\textit{t}\text{-Bu})_2]^{2+}$ have axial symmetry with *g*-values $g_\perp = 2.45$ and $g_\parallel = 1.96$, of which $g_\parallel$ reveals hyperfine coupling with $^{103}Rh$ ($A_\parallel = 62\,G$). In frozen solution, $g_\perp$ shows Rh hyperfine coupling as well. Magnetic susceptibilities of $\mu_{eff} = 2.04\mu_B$ (solid phase) and $2.20\,\mu_B$ (Evans method) were measured. In contrast to $[Rh^{II}(CO)_2(\eta^2\text{-TMPP})_2]^{2+}$, $[Rh^{II}(\eta^1\text{-TMPP})_2(CN\textit{t}\text{-Bu})_2]^{2+}$ is not reduced by $[Rh^{II}(\eta^3\text{-TMPP})_2]^{2+}$ (16).

Somewhat related to the above isocyanide complexes, paramagnetic isocyanide–$AsPh_3$ complexes of iridium have also been reported (Fig. 8). The EPR parameters of the complexes $[Ir(L)_n(AsPh_3)(CNPhMe)]$ largely depends on the nature of the anionic *O,O*-ligand/*N,O*-ligand L (17, 18).

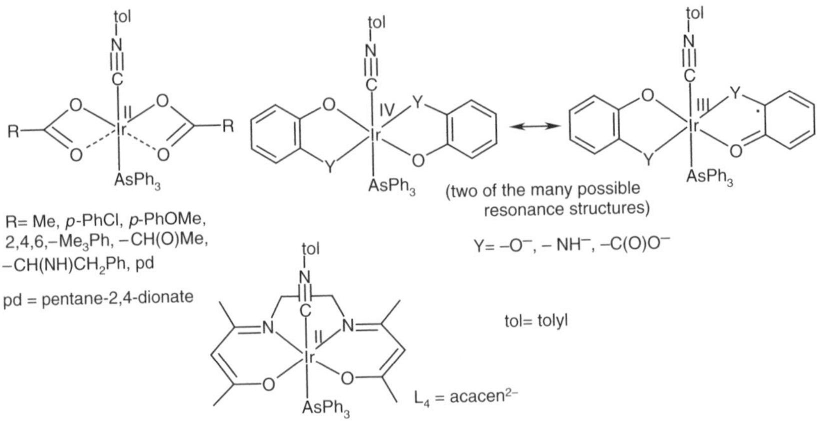

R= Me, p-PhCl, p-PhOMe, 2,4,6,–$Me_3Ph$, –CH(O)Me, –CH(NH)$CH_2Ph$, pd

pd = pentane-2,4-dionate

(two of the many possible resonance structures)

Y= $-O^-$, $-NH^-$, $-C(O)O^-$

tol= tolyl

$L_4$ = $acacen^{2-}$

Figure 8. Paramagnetic [Ir$(L)_n$(p-tolyl-isocyanide)($AsPh_3$)] complexes.

With the ligand *N,N′*-ethylenebis(acetylacetoneiminate) ($acacen^{2-}$), quite large deviations from $g = g_e$ were observed. For the other ligands, $g$ values are much closer to $g_e$ (Table II). For the doubly deprotonated *o*-dihydroxybenzene, *o*-aminohydroxybenzene, and *o*-carboxylic acid-hydroxybenzene type ligands, complexes with a formal Ir(IV) oxidation state are formed, which at that time were believed to be real Ir(IV) species with a $d^5$ electron configuration. Meanwhile, it is well known that these type of ligands are actually "redox non-innocent" (also see Section III.B), and Ir(III)(ligand radical) or Ir(II)(ligand radical)$_2$ descriptions are more likely electronic configurations for these species. The small deviations of the $g$-values from $g_e$ for the complexes with innocent carboxylato complexes are, however, not so easily understood. A detailed reinvestigation, including DFT

TABLE II

The $\mu_{eff}$ and EPR parameters reported for [Ir$(L)_n$($AsPh_3$)(CNPhMe)] Complexes[a]

| Ligand (L) | $\mu_{eff}$ (RT) | $g_{\parallel}$ | $g_{\perp}$ | $A_{\parallel}^{As}$ | $A_{\perp}^{As}$ |
|---|---|---|---|---|---|
| acacen ( = $L_4$) | 1.76 | 3.19 (broad) | 1.9 (broad) | nr | nr |
| *o*-OPhO–[b] | | 1.997 | 1.906 | nr | nr |
| *o*-OPhNH–[b] | 1.74 | 1.986 | 1.940 | nr | nr |
| *o*-OPh$CO_2$–[b] | 2.30 | 1.996 | 1.912 | nr | nr |
| *o*-NHPh$CO_2$–[b] | 1.96 | 1.988 | 1.930 | nr | nr |
| –$O_2$CCH(O)Me | 2.04 | 2.008 | 1.940 | nr | nr |
| –$O_2$CCH(NH)$CH_2$Ph | 1.82 | 1.991 | 1.894 | nr | nr |
| –$O_2$C–Ar | 1.67 | 2.039 | 2.015 | 94 | 37 |
| –$O_2$C–pd | | 2.06 | 2.011 | 95 | 12 |

[a]Hyperfine couplings in MHz; nr = not resolved, Ar = p-PhCl, p-PhOMe, 2,4,6-$Me_3$Ph, pd = pentane-2,4-dionate, Ar = aromat.

[b]Most likely ligand radical complexes.

calculations, would be required to obtain a better understanding of the electronic structure of these early examples.

### B. The π-Arene–Phosphine Rh(II) Complexes

A series of rhodium(II) bis(phosphine), $\eta^6$-arene, piano stool complexes of type **I**, **II**, **III**, and **IV** (Fig. 9) has been described in detail (19–22). The relative kinetic and thermodynamic stabilities of these complexes were investigated by electrochemical methods. The kinetic stability proved to be dependent on the electron-donating character of the arene ligand. Whereas the Rh(I) precursor of the benzene complex (type **II**, $n = 2$) reveals irreversible waves in cyclic voltammetry at every scan rate, the toluene and *m*-xylene analogues give reversible oxidation waves at 200 and 20 $mV\,s^{-1}$, respectively. Even higher substituted arenes give reversible oxidation waves at all scan rates. This trend is probably due to the fact that more electron-donating arenes coordinate stronger to the rhodium(II) center, thus preventing their substitution. Furthermore, the more substituted arene ligands protect the metal center from external attack through the increased steric bulk of the ligand (19). The Rh(I)/Rh(II) couple occurs at lower potentials upon increasing the arene substitution, which reflects the corresponding increase of electron density at the rhodium center.

Whereas the toluene analogue of **I** is stable, the toluene analogue of **II** is only stable for a propyl linker between both phosphines ($n = 3$). In the case of ethyl or butyl linkers, irreversible oxidation waves are observed upon oxidation (Rh(I) / Rh(II) couple) at low scan rates. This is explained in terms of accommodation of the rhodium(II) state (19). The DFT calculations suggest that widening of the P—Rh—P angle and lengthening of the Rh—P bonds occurs upon oxidation of rhodium(I) to rhodium(II). This was confirmed by the elucidation of the X-ray structure of **IV** ($X = CH_2$) and its rhodium(I) precursor (20). A much lower $E_{1/2}$ is found for the propyl spacer, compared to the ethyl and butyl spacers. This

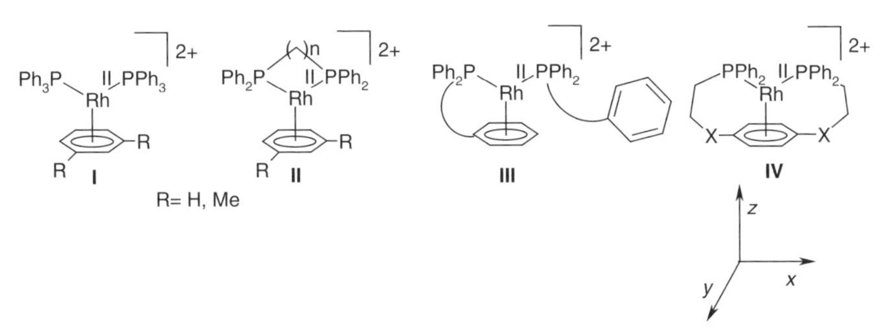


Figure 9. Structure of rhodium(II) complexes of type **I–IV**.

indicates that the ring strain in the case of $n = 3$ is much lower. Even lower $E_{1/2}$ values are found for the rhodium(II) species of type **I** with nonchelating ligands. The relative high stability of **III** and **IV** is caused by kinetic stabilization of the rhodium(II) from loss of the arene ligand by the chelated P–arene ligand arrangement.

Complexes of type **IV** were prepared by chemical oxidation of their rhodium(I) precursor upon addition of $Ag^+$ (20). Both $^1H$ and $^{31}P\{^1H\}$ NMR spectra exhibit broad, nearly featureless spectra, indicative of a paramagnetic species. The frozen solution EPR spectrum of **IV** (X = O) displays a rhombic spectrum with no resolved hyperfine couplings, with the following $g$ values and line widths: $g_1 = 2.390$ (85 MHz), $g_2 = 2.038$ (140 MHz), and $g_3 = 1.997$ (90 MHz). In the case of $X = CH_2$ the EPR spectrum exhibits an axial pattern with no resolved hyperfine splittings. Computer simulation reveals a slightly rhombic $g$-tensor with $g_1 = 2.363$ (82 MHz), $g_2 = 2.0245$ (77 MHz), $g_3 = 2.0045$ (77 MHz). The large $g$-anisotropy is clearly indicative for a metal centered radical. This was confirmed with DFT calculations. The nearly axial $g$ tensor ($g_{\parallel} = 2.36$ and $g_{\perp} \sim 2.015$) is consistent with a mainly $d_{yz}$ SOMO (with the $z$ axis along the Rh–arene bond and the X-axis along the P—Rh—P plane, see Fig. 9) (20).

Upon treatment of **IV** with small molecules (e.g., CO or *tert*-butyl isocyanide), reduction to a square-planar Rh(I) species is observed with two (or more) CO or isocyanide ligands and two phosphine ligands coordinated and with the arene no longer in the coordination sphere of the metal (Fig. 10). The same reaction products are obtained upon treatment of the rhodium(I) precursor of **IV** with CO or isocyanide. This reaction, however, is much slower (20, 22). This finding can be explained. In the case of the 17 VE (VE = valence electron) rhodium(II) species, addition of an external ligand proceeds via an associative step, which is fast. In the case of the 18 VE rhodium(I) analogue, however, first one of the ligands has to dissociate before CO or isocyanide is able to coordinate. Such a reaction sequence is much slower. Although traces of $CO_2$ were found, no explicit explanation was given for the observed reduction of rhodium(II) to rhodium(I).

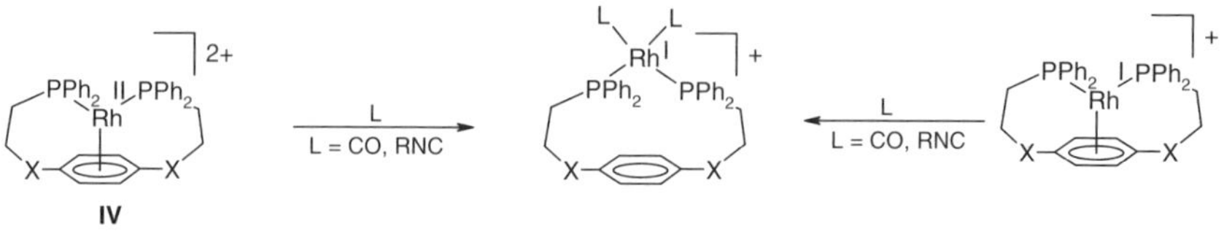

Figure 10. Arene dissociation upon addition of CO or isocyanides for **IV** and its rhodium(I) precursor.

### C. [(σ-Aryl)$_4$M•] and [(σ-Aryl)$_2$(L)$_2$M•] Type Complexes (M = Rh, Ir, Pt)

Over the past years quite a number of paramagnetic $[M(Aryl)_2L_2]^{n-}$ and $[M(Aryl)_4]^{n-}$ type complexes have been reported (Fig. 11). The complex $[NBu_4]^+[Pt^{III}(C_6Cl_5)_4]^-$ has been obtained electrochemically and by treatment of the platinum(II) precursor $[NBu_4]_2[Pt^{II}(C_6Cl_5)_4]$ with either one of the oxidants chlorine, bromine, iodine, $TlCl_3$, or $[p\text{-}CH_3C_6H_4N_2]^+[BF_4]^-$ (23, 24). Magnetic susceptibility values of $\mu_{eff} = 2.57$ and $2.42\ \mu_B$ at 260 and 80 K were obtained, which is in agreement with an $S = \frac{1}{2}$ spin system with a substantial spin–orbit contribution from mixing with lower lying states. The isoelectronic $[Rh^{II}(C_6Cl_5)_4]^{2-}$ was obtained from the reaction of $[Rh^{III}Cl_3(tht)_3]$ (tht = tetrahydrothiophene) with an excess of $LiC_6Cl_5$, which in this case not only acts as an aryllating reagent, but also as a one-electron reducing agent (25). A similar procedure was followed for the synthesis of *trans*-$[Ir^{II}(mes)_2(SEt_2)_2]$ (mes = mesetyl) (from *mer*-$[IrCl_3(SEt_2)_3]$ and excess of $Mg(mes)_2$•$(thf)_2$) [thf = tetrahydrofuran (ligand)] (26) and *trans*-$[Rh^{II}(2,4,6\text{-}i\text{-}PrC_6H_2)_2(tht)_2]$ (from $[Rh^{III}Cl_3(tht)_3]$ and lithium aryl etherate) (27). The diethyl sulfide ligand is quite labile and can be substituted by $PMe_3$, $PMe_2Ph$, $PEt_2Ph$, pyridine and 4-*tert*-butylpyridine.

The complexes $[Ir^{II}(C_6Cl_5)_4]^{2-}$ (28), $[Rh^{II}(C_6Cl_5)_2(cod)]$ (29), $[Rh^{II}(C_6Cl_5)_2(P(OPh_3)_3)_2]$ (29), and $[Ir^{II}(C_6Cl_5)_2(cod)]$ (30) (cod = 1,5-cyclooctadiene) were obtained by oxidation of the *in situ* prepared metal(I) precursors, either electrochemically or chemically with the oxidants $I_2$ or $Ag^+$. The $[Rh^{II}(C_6Cl_5)_2(L)_2]$ complexes [L = $P(OPh_3)_3$, $P(OMe_3)_3$, $P(Ph_3)_3$, Py (pyridene), 1,2-bis(diphenylphosphino)ethane (dppe), bis(diphenylphosphino)methane (dppm)) have been obtained by displacement of cod from $[Rh^{II}(C_6Cl_5)_2(cod)]$ (cod = 1,5-cycloctadiene (29).

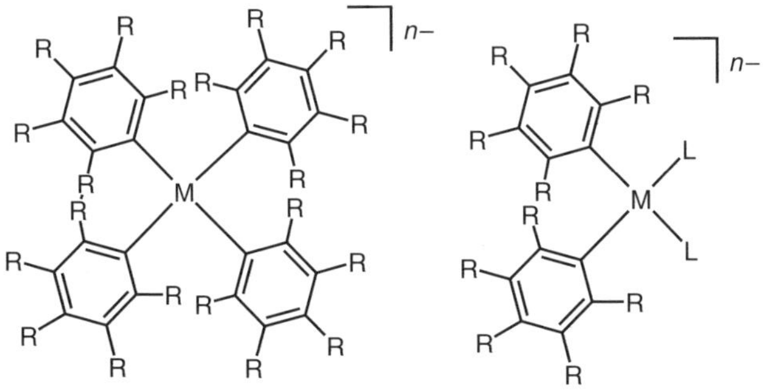

M= Rh$^{II}$, Ir$^{II}$, Pt$^{III}$

L= PR$_3$, P(OR)$_3$, py, SR$_2$, tht, dpe, dpm

Figure 11. $[(\sigma\text{-aryl})_4M\bullet]^{n-}$ and $[(\sigma\text{-aryl})_2(L)_2M\bullet]^{n-}$ type complexes (M = Rh$^{II}$, Ir$^{II}$, Pt$^{III}$).

TABLE III
The EPR Data for $[M(aryl)_2L_2]$ and $[M(aryl)_4]^{2-}$ species

| Complex | $g_x$ | $g_y$ | $g_z$ | References |
|---|---|---|---|---|
| $[Rh^{II}(C_6Cl_5)_4]^{2-}$ | 2.74 | 2.60 | 1.94 | 25 |
| $[Ir^{II}(C_6Cl_5)_4]^{2-}$ | 2.131 | 2.126 | 2.110 | 28 |
| $[Ir^{II}(C_6Cl_5)_2(cod)]$ | 3.00 | 2.79 | 1.85 | 30 |
| $[Pt^{III}(C_6Cl_5)_4]^-$ | 3.18 | 3.01 | 1.60 | 32 |
| $[Pt^{III}(C_6Cl_4H\text{-}p)_4]^-$ | 3.09 | 3.09 | 1.62 | 32 |
| $[Rh^{II}(C_6Cl_5)_2(cod)]$ | 2.52 | 2.45 | 1.99 | 29 |
| $[Rh^{II}(C_6Cl_5)_2\{P(OPh)_3\}_2]$ | 2.79 | 2.29 | 1.96 | 29 |
| $[Rh^{II}(C_6Cl_5)_2\{P(OMe)_3\}_2]$ | 2.84 | 2.32 | 1.95 | 29 |
| $[Rh^{II}(C_6Cl_5)_2\{P(Ph)_3\}_2]$ | 2.83 | 2.39 | 1.95 | 29 |
| $[Rh^{II}(C_6Cl_5)_2(py)_2]$ | 2.66 | 2.59 | 1.94 | 29 |
| $[Rh^{II}(C_6Cl_5)_2(dppe)]$ | 2.58 | 2.48 | 1.99 | 29 |
| $[Rh^{II}(C_6Cl_5)_2(dppm)]$ | 2.80 | 2.41 | 1.96 | 29 |
| $[Ir^{II}(mes)_2(SEt_2)_2]$ | 2.29 | 2.14 | 1.90 | 26 |
| $[Rh^{II}(2,4,6\text{-}i\text{-}PrC_6H_2)(tht)_2]$ | 2.96 | 2.58 | 1.85 | 27 |
| $[Rh^{II}(C_6Cl_5)_2(tht)_2]$ | 2.73 | 2.60 | 1.95 | 31 |

Rhombic EPR spectra have been observed for these Rh and Ir species. The large $g$ anisotropies (Table III) are consistent with the proposed electronic structures, with the unpaired electron mainly residing in a $d_{xy}$ or $d_{z^2}$ orbital (with the $z$ axis perpendicular to the coordination plane of rhodium). The EPR spectrum of $[Rh^{II}(2,4,6\text{-}i\text{-}PrC_6H_2)(tht)_2]$ shows coupling of the unpaired electron to the $^{103}Rh$ ($I = ½$) nucleus at all three $g$ values (Table III) (27). For all other complexes, no resolved (super)hyperfine structures were observed. Substitution of diethyl sulfide in $[Ir^{II}(mes)_2(SEt_2)_2]$ by phosphines or N-donor ligands produces EPR silent paramagnetic products. It seems that mixing of the ground-state SOMO with multiple exited states that are very close in energy to the SOMO results in very fast relaxation processes, thus causing substantial broadening (EPR silence). Fast relaxation processes also cause $[Ir^{II}(porphyrin)]$ species to be EPR silent (33).

Substantially larger spin–orbit couplings are observed for $[Ir^{II}(C_6Cl_5)_2(cod)]$ ($g_{av} = 2.55$) compared to $[Rh^{II}(C_6Cl_5)_2(cod)]$ ($g_{av} = 2.32$), which is usual for third-row compared to second-row transition metals. For $[Ir^{II}(C_6Cl_5)_4]^{2-}$ ($g_{av} = 2.12$), however, *smaller* spin-orbit couplings were observed than for $[Rh^{II}(C_6Cl_5)_4]^{2-}$ ($g_{av} = 2.43$). Although this was not stated by the authors, this might be an indication that in $[Ir^{II}(C_6Cl_5)_4]^{2-}$ the unpaired electron resides in another metal $d$ orbital (most likely a $d_{xz}$ or $d_{yz}$ orbital, see Fig. 5) than in $[Rh^{II}(C_6Cl_5)_4]^{2-}$. Similar differences in electronic structures of other isostructural Rh and Ir compounds have been observed (33). In the case of $[Pt^{III}(C_6Cl_5)_4]^-$ and $[Pt^{III}(C_6Cl_4H\text{-}p)_4]^-$ ($g_{av} = 2.60$) larger orbital contributions are found compared to their rhodium analogues, as one can expect from the large spin–orbit coupling constant ($\lambda$) of the platinum centre (32).

The geometry of $[Pt^{III}(C_6Cl_5)_4]^-$ was elucidated by X-ray diffraction. Like its $Pt^{II}$ precursor, $[Pt^{III}(C_6Cl_5)_4]^-$ adopts a square planar geometry. The bond lengths of the $Pt^{III}$ complex (2.094 Å) and its $Pt^{II}$ precursor (2.086 Å) are very similar (24). Similar square-planar geometries were observed for $[Ir^{II}(C_6Cl_5)_4]^{2-}$ (28), $[Ir^{II}(C_6Cl_5)_2(cod)]$ (30), $[Ir^{II}(mes)_2(PMe_3)_2]$ (26), $[Ir^{II}(mes)_2(SEt_2)_2]$ (26), $[Rh^{II}(2,4,6\text{-}i\text{-}PrC_6H_2)_2(tht)_2]$(27), and $[Rh^{II}(C_6Cl_5)_2\{P(OMe)_3\}_2]$ (29).

All these complexes proved to be (moderately) air and moisture stable, both as a solid and in solution. Upon binding of a fifth ligand, however, the complexes become quite sensitive. In some cases, pentacoordinated structures were obtained, but frequently this resulted in dissociation of one of the aryl groups as a radical with net reduction of the metal center (see Section IV.B for examples of related reactions). For example, pentacoordinated $[Ir^{II}(C_6Cl_5)_2(cod)(PPh_3)]$, prepared by treatment of $[Ir^{I}(C_6Cl_5)_2(cod)]^-$ with $[(PPh_3)Ag(ClO_4)]$, is very unstable and decomposes to diamagnetic $[Ir^{I}(C_6Cl_5)(cod)\ (PPh_3)]$ and a $C_6Cl_5^\bullet$ radical (30). Treatment of $[M^{II}(C_6Cl_5)_4]^{2-}$ ($M = Rh, Ir$) with CO yields $[Rh^{I}(C_6Cl_5)_2(CO)_2]^-$ (25) or $[Ir^{I}(C_6Cl_5)_3(CO)]^{2-}$ (28) and treatment of $[Ir^{II}(mes)_2(SEt_2)_2]$ with CO yields $[Ir^{I}(mes)(CO)_2(SEt_2)_2]$ (26). In the case of $[Rh^{II}(2,4,6\text{-}i\text{-}PrC_6H_2)_2(tht)_2]$ a new EPR spectrum was obtained after addition of CO. At 77 K, it consists of several signals with $g \sim 2$, which suggest that the unpaired electron is located at the (coordinated) CO ligand. Removal of all volatiles resulted in regeneration of $[Rh^{II}(2,4,6\text{-}i\text{-}PrC_6H_2)(tht)_2]$, which indicates that coordination of CO is reversible (27). In contrast to the previous examples, in this case no decomposition to a Rh(I) species upon release of an aryl radical takes place. This, on the other hand was achieved by treatment of $[Rh^{II}(2,4,6\text{-}i\text{-}PrC_6H_2)_2(tht)_2]$ with isonitriles.

As mentioned before, $[Pt^{III}(C_6Cl_5)_4]^-$ is quite stable toward a variety of reagents, even at reflux, but at room temperature it quickly reacts with NO to form $[Pt^{II}(C_6Cl_5)_4(NO)]^-$ (24). X-ray diffraction measurements reveal a linear coordination of the NO ligand, suggestive for a $NO^+$ ligand (formally). Upon heating, or prolongued stirring in dichloromethane or acetone, NO is expelled resulting in regeneration of $[Pt^{III}(C_6Cl_5)_4]^-$. The $[Pt^{II}(C_6Cl_5)_4(NO)]^-$ complex can also be obtained by treatment of $([NBu_4]^+)_2[Pt^{II}(C_6Cl_5)_4]^{2-}$ with $[NO^+][ClO_4]^-$. Dissociation of gaseous NO can be prevented when the reaction is carried out in an NO atmosphere (24). Treatment of $[Ir^{II}(mes)_2(SEt_2)_2]$ with NO, however, yields $[Ir(mes)_2(SEt_2)_2(NO)]$, containing a bent, $NO^-$ ligand (formally) (26). This was confirmed with IR spectroscopy, showing a band at 1519 $cm^{-1}$ typical for $NO^-$, and therefore the iridium center was assigned to be in the oxidation state +III.

One-electron oxidation of $[Rh^{II}(C_6Cl_5)_4]^{2-}$ to $[Rh^{III}(C_6Cl_5)_4]^-$ was observed upon addition of chlorine, bromine, or iodine (25). Further, treatment of

TABLE IV
Selected $Pt^{II}/Pt^{III}$ Redox Couples Relevant to Formation of (Transient) Paramagnetic Pt(III) Species[a]

| Compound[b] | $E_{1/2}$ | $E_{peak}$ (irrev.) | Solvent | References |
|---|---|---|---|---|
| $[Pt^{II}(C_6Cl_5)_4]^{2-}/[Pt^{III}(C_6Cl_5)_4]^-$ | −0.007 | | $CH_2Cl_2$ | 24 |
| $[Pt^{II}(B_9C_2H_{11})_2]^{2-}/[Pt^{III}(B_9C_2H_{11})_2]^-$ | −1.170 | | MeCN | 37 |
| $[Pt^{II}(mes)_2(bpy)]/[Pt^{III}(mes)_2(bpy)]^+$ | 0.45 | | THF | 34, 35 |
| $[Pt^{II}(mes)_2(bpm)]/[Pt^{III}(mes)_2(bpm)]^+$ | 0.53 | | THF | 34 |
| $[Pt^{II}(mes)_2(bpz)]/[Pt^{III}(mes)_2(bpz)]^+$ | 0.59 | | THF | 34 |
| $[Pt^{II}(mes)_2(bpym)]/[Pt^{III}(mes)_2(bpym)]^+$ | 0.54 | | THF | 34 |
| $[Pt^{II}(mes)_2(tap)]/[Pt^{III}(mes)_2(tap)]^+$ | 0.60 | | THF | 34 |
| $[Pt^{II}(mes)_2(dppz)]/[Pt^{III}(mes)_2(dppz)]^+$ | 0.49 | | THF | 34 |
| $[Pt^{II}(mes)(Cl)(bpy)]/[Pt^{III}(mes)(Cl)(bpy)]^+$ | | 0.83 | MeCN | 34 |
| $[Pt^{II}(o\text{-}CF_3Ph)_2(bpy)]/[Pt^{III}(o\text{-}CF_3Ph)_2(bpy)]^+$ | | 0.91 | MeCN | 34 |
| $[Pt^{II}(Ph)_2(bpy)]/[Pt^{III}(Ph)_2(bpy)]^+$ | | 0.52 | MeCN | 34 |
| $[Pt^{II}(mes)_2(dmso)_2]/[Pt^{III}(mes)_2(dmso)_2]^+$ | | 0.89 | THF | 34 |
| $[Pt^{II}(mes)_2(phen)]/[Pt^{III}(mes)_2(phen)]^+$ | 0.45 | | THF | 36 |
| $[Pt^{II}(mes)_2(5,6\text{-dmphen})]/[Pt^{III}(mes)_2(5,6\text{-dmphen})]^+$ | 0.45 | | THF | 36 |
| $[Pt^{II}(mes)_2(4,7\text{-dmphen})]/[Pt^{III}(mes)_2(4,7\text{-dmphen})]^+$ | 0.43 | | THF | 36 |
| $[Pt^{II}(mes)_2(3,8\text{-dmphen})]/[Pt^{III}(mes)_2(3,8\text{-dmphen})]^+$ | 0.44 | | THF | 36 |
| $[Pt^{II}(mes)_2(2,9\text{-dmphen})]/[Pt^{III}(me)_2(2,9\text{-dmphen})]^+$ | 0.43 | | THF | 36 |
| $[Pt^{II}(Ph)_2(2,9\text{-dmphen})]/[Pt^{III}(Ph)_2(2,9\text{-dmphen})]^+$ | | 1.02 | THF | 36 |
| $[Pt^{II}(mes)_2(2,9\text{-dmphen})]/[Pt^{III}(mes)_2(2,9\text{-dmphen})]^+$ | | 1.01 | THF | 36 |
| $[Pt^{II}(mes)_2(tmphen)]/[Pt^{III}(mes)_2(tmphen)]^+$ | 0.42 | | THF | 36 |
| $[Pt^{II}(Me_2)(\alpha\text{-diimine})]/[Pt^{III}(Me_2)(\alpha\text{-diimine})]^+$ | | +0.6–0.7 | MeCN | 38 |

[a]Redox potentials versus $Fc/Fc^+$.
[b]Demethyl sulfoxide = dmse, phen = 1,10-phenanthroline.

$[Ir^{II}(mes)_2(PMe_3)_2]$ with hydrogen yields $[Ir^{V}H_5(PMe_3)_2]$. Mechanistic details of these reactions are not known.

A variety of related $[Pt^{III}(Ar)_2(L)_2]^+$ complexes (Ar = mes, Ph; L = DMSO, $L_2$ = bpy (2,2′-bipyridine), phen, and analogues) have a measurable lifetime, some of which might even be stable according to electrochemical measurements (Table IV) (34–36).

### D. Organometallic Ir(IV) Complexes

The Rh and Ir species in oxidation state +IV are usually very reactive (see Section IV.B), and only a few stable, unequivocal organometallic Ir(IV) radicals are known to date. The first reported stable example is the neutral tetrakis-mesityl Ir(IV) compound $[Ir^{IV}(mes)_4)]$ (39). This unusual species was obtained in ~20% yield by reaction of partially dehydrated $IrCl_3.nH_2O$ with mesityllithium at room temperature (*RT*). The route to this species is rather unclear, but a disproportionation reaction, $2Ir^{III} \rightarrow Ir^{II} + Ir^{IV}$, was suggested as one of the

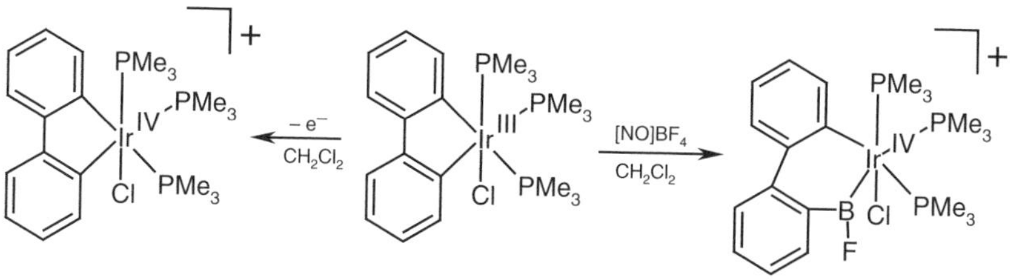

Figure 12. Formation of stable organometallic aryl-$Ir^{IV}$ species upon oxidation of an $Ir^{III}$(biphenyl-2,2′-diyl) precursor.

possible mechanisms. The X-ray structure of [$Ir^{IV}$(mes)$_4$)] reveals a slightly distorted tetrahedral geometry without any agostic interactions. The EPR data ($g_{\parallel} = 2.487$, $g_{\perp} = 2.005$ with some nonresolved Ir hyperfine features) are consistent with a tetrahedral, low-spin $d^5$ molecule having a single unpaired electron.

Organometallic Ir(IV) species were also obtained by 1e$^-$ oxidation of diamagnetic [$Ir^{III}$(biphenyl-2,2′-diyl)] complexes (see Fig. 12) (40). Oxidation of [$Ir^{III}$(biphenyl-2,2′-diyl)(Cl)($PMe_3$)$_3$] (+319 mV vs Fc/Fc$^+$) leads to the stable [$Ir^{IV}$(biphenyl-2,2′-diyl)(Cl)($PMe_3$)$_3$]$^+$ cation (Table V). The EPR data of this complex confirm the formation of a metal centered radical ($g_x = 2.37$, $g_y = 2.28$, $g_z = 1.84$, no resolved hyperfine couplings). Extended Hückel calculations indicated that the easy oxidation of this complex is due to the presence of a high energy HOMO in the [$Ir^{III}$(biphenyl-2,2′-diyl)(Cl)($PMe_3$)$_3$] species. The highest filled π level of the biphenyl-2,2′-diyl ligand combines in an antibonding way with the filled Ir $d_{xy}$ orbital, which results in a high-lying HOMO with large metal character. Removal of 1e$^-$ from this orbital yields the SOMO of the [$Ir^{IV}$(biphenyl-2,2′-diyl)(Cl)($PMe_3$)$_3$]$^+$ cation, in good agreement with the EPR data.

Interestingly, oxidation of [$Ir^{III}$(biphenyl-2,2′-diyl)(Cl)($PMe_3$)$_3$] with [NO]$BF_4$ yields a different paramagnetic species in which a BF unit has been inserted into one of the Ir—C bonds. This species was characterized by X-ray diffraction, and high-resolution mass spectrometry confirmed the formulation.

TABLE V
Selected M(IV)/M(III) Redox Couples Relevant to Formation of Paramagnetic M(III) Species (M = Pd, Pt) and M(IV) Species (M = Ir)[a]

| Compound | $E_{1/2}$ | $E_{peak}$ (irrev.) | Solvent | References |
|---|---|---|---|---|
| [$Pt^{IV}$($B_9C_2H_{11}$)$_2$]/[$Pt^{III}$($B_9C_2H_{11}$)$_2$]$^-$ | −0.657 | | MeCN | 37 |
| [$Ir^{III}$(biphenyl-2,2′-diyl)(Cl)($PMe_3$)$_3$]/ [$Ir^{IV}$(biphenyl-2,2′-diyl)(Cl)($PMe_3$)$_3$]$^+$ | +0.319 | | | 40 |

[a]Redox potentials versus Fc/Fc$^+$.

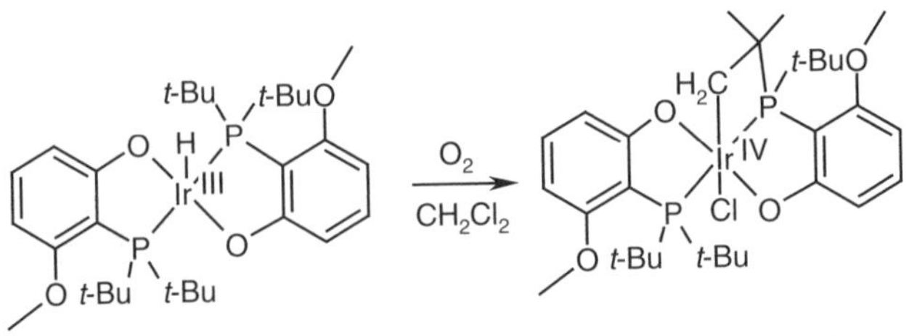

Figure 13. Formation of a stable alkyl-$Ir^{IV}$ species upon oxidation of an $Ir^{III}$–H precursor.

The EPR spectrum of this species reveals an axial $g$ tensor with $g$ values indicative of a metal centered radical ($g_x = g_y = 2.15$, $g_z = 1.96$, no resolved hyperfine couplings).

Another claimed organometallic $Ir^{IV}$ species also contains a potentially noninnocent ligand. This species was obtained as one of the products by oxidation of a coordinatively unsaturated (—OArP($t$-Bu)$_2$)$_2$$Ir^{III}$—H species with $O_2$ in $CH_2Cl_2$. This results in a stable alkyl–$Ir^{IV}$ species containing a four-membered $\kappa^2$-$C,P$–$CH_2C(Me_2)P$– moiety obtained by some sort of C—H activation of one of the $t$-Bu methyl groups (Fig. 13) (41). The peculiar structure of this paramagnetic species was confirmed by X-ray diffraction. The EPR spectrum of this species reveals a rhombic spectrum with a large $g$ anisotropy, with $g$ values $g_{11} = 2.49$, $g_{22} = 2.38$, $g_{33} = 2.23$. Regarding the large deviations of the $g$ values from $g = g_e$, this might be a real (i.e., metal-centered radical) organometallic Ir(IV) species, but we do have some doubts about the oxidation state in this case. A structurally similar Ir(II) species containing two coordinated HOAr (i.e., protonated at the phenolate oxygen) fragments cannot be excluded on the basis of the provided data. It was mentioned that the species reveals observable hyperfine interactions with only one $^{31}P$ nucleus along one direction, but no further information was given about the coupling constants or any eventual Ir hyperfine interactions.

## E. Pyrazolyl-Type Rh(II) and Ir(II) Carbonyl Complexes

Chemical and electrochemical oxidation of $[M^I(HBPz'_3)(CO)(L)]$ ($P_z$ = pyrazolyl) and $[Rh^I(HCPz'_3)(CO)(L)]^+$ complexes has recently been described [L = CO, $PPh_3$, $PCy_3$, $P(NMe_2)_3$, P($p$-tolyl)$_3$, P($m$-tolyl)$_3$, $P(OPh)_3$, $AsPh_3$], in which $HBPz'_3$ represents a tris-pyrazolylborate anion and $HCPz'_3$ represents the neutral carbon equivalent (Fig. 14) (42–46).

Whereas a coordination geometry with the ligand in a $\kappa^2$-coordination mode is preferred for the M(I) species, a $\kappa^3$ square-pyramidal structure is preferred for the M(II) analogues (M = Rh, Ir). As a result, reorganization takes place upon oxidation. For small ligands, reorganization is fast on the cyclic voltammogram

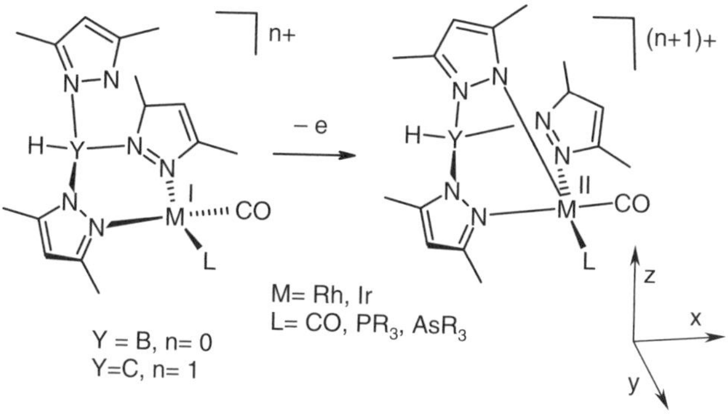

Figure 14. Formation of $[M^{II}(HBPz'_3)(CO)(L)]$ and $[M^{II}(HCPz'_3)(CO)(L)]$.

time scale and reversible redox waves are observed (44). The reorganization of the coordination geometry can be followed by means of infrared (IR) spectroscopy. The C—O stretching frequency ν(CO) is ∼ 60–70 $cm^{-1}$ higher in energy in the case of a square-pyramidal M(II) complex, compared to a square-planar M(I) species.

With bulky ligands (e.g., $PCy_3$) (Cy = cyclohexyl), the reorganization is slow resulting in irreversible oxidation waves at high oxidation potentials, with characteristics of an EC type mechanism (45). Quite a number of rhodium(II) analogues were successfully isolated, by treatment of their rhodium(I) precursors $[Rh^I(HBPz'_3)(CO)_2]$ with ferrocenium salts in the presence of additional monodentate phosphine ligands. These species were analyzed by EPR spectroscopy in frozen solution (Table VI). Except for the $P(OPh)_3$ ligand, all frozen solution spectra revealed resolved N and Rh hyperfine patterns for the $g_z$ component. This, together with the $g$ values $g_x$, $g_y > 2$ and $g_z \sim 2$ (see Fig. 4), points to a square-pyramidal structure with one of the pyrazolyl groups at the apical position (with the $z$ axis perpendicular to the square plane). The largely Rh-based SOMO consists of a Rh $d_{z^2}$ orbital in antibonding combination with a lone pair of the apical N-donor (Rh—N σ* interaction). The spectroscopically derived geometry was confirmed by elucidation of the solid-state structure of $[Rh^{II}(HBPz'_3)(CO)(PPh_3)][PF_6]$ by X-ray diffraction.

Although one-electron oxidation of $[Rh^I(CO)_2\{(HB(Pz')_3\}]$ initially yields the EPR spectrum of its rhodium(II) analogue, the final product of this reaction is the $\kappa^2$-$N$-protonated $HB(HPz)(Pz)_2$ rhodium(I) complex $[Rh^I(CO)_2\{(HB(HPz)(Pz)_2\}]^+$ (43). The same product is obtained upon protonation of $[Rh^I(CO)_2\{(HB(Pz')_3\}]$. Although the authors did not give an explanation for this reaction, the intermediate rhodium(II) species probably picked up an hydrogen atom from the solvent. Electrochemical reduction of $Rh^I(CO)_2\{(HB(HPz)(Pz)_2\}]^+$ leads to deprotonation of the complex to form $[Rh^I(CO)_2\{(HB(Pz')_3\}]$, and reduction $H^+$ to $H_2$ (43).

TABLE VI
EPR Data for $[M^{II}(HBPz'_3)(CO)(L)]$ and $[M^{II}(HCPz'_3)(CO)(L)]$ Complexes[a]

| Complex | $g_x$ <br> $g_y$ <br> $g_z$ | $A_x(M)^b$ <br> $A_y(M)$ <br> $A_z(M)$ | $A_x(^{14}N)$ <br> $A_y(^{14}N)$ <br> $A_z(^{14}N)$ | $A_x(^{31}P)$ <br> $A_y(^{31}P)$ <br> $A_z(^{31}P)$ | References |
|---|---|---|---|---|---|
| $[Rh^{II}(HBPz'_3)(CO)_2]^+$ | 2.318 | 66 | 78 | nr | 43 |
| | 2.147 | nr | 81.5 | nr | |
| | 1.989 | 73.5 | 93.8 | nr | |
| $[Rh^{II}(HBPz'_3)(CO)(PPh_3)]^+$ | 2.256 | 51 | 75 | nr | 42, 43 |
| | 2.277 | | 77.3 | nr | |
| | 1.993 | 86.6 | 83 | 14.4 | |
| $[Rh^{II}(HBPz'_3)(CO)(PCy_3)]^+$ | 2.277 | nr | nr | nr | 43 |
| | 2.167 | nr | nr | nr | |
| | 1.993 | 77.3 | 86.3 | 14.7 | |
| $[Rh^{II}(HBPz'_3)(CO)(P(NMe_2)_3)]^+$ | 2.234 | 63.3 | 63.3 | nr | 43 |
| | 2.157 | 30.3 | 73.2 | nr | |
| | 1.995 | 84.8 | 84.8 | nr | |
| $[Rh^{II}(HBPz'_3)(CO)P(p\text{-tolyl})_3]^+$ | 2.260 | 63 | 63 | nr | 43 |
| | 2.164 | nr | 75 | nr | |
| | 1.993 | 76.7 | 87.8 | 21.3 | |
| $[Rh^{II}(HBPz'_3)(CO)(P(m\text{-tolyl})_3)]^+$ | 2.256 | 54 | 75 | nr | 43 |
| | 2.161 | 31.2 | 74.6 | nr | |
| | 1.994 | 76.7 | 88 | 18.6 | |
| $[Rh^{II}(HBPz'_3)(CO)(P(OPh)_3)]^+$ | 2.233 | nr | nr | nr | 43 |
| | 2.220 | nr | nr | nr | |
| | 1.993 | 81.2 | 83.3 | nr | |
| $[Rh^{II}(HCPz'_3)(CO)(PPh_3)]^{2+}$ | 2.272 | 66 | 66 | nr | 45 |
| | 2.180 | 42 | 72 | nr | |
| | 1.993 | 81.8 | 114 | 19.2 | |
| $[Rh^{II}(HCPz'_3)(CO)(AsPh_3)]^{2+}$ | 2.287 | nr | nr | | 45 |
| | 2.170 | nr | nr | | |
| | 1.995 | 64.8 | 92.6 | | |
| $[Ir^{II}(HBPz'_3)(CO)(PPh_3)]^+$ | 2.544 | 60[c] | ~240 | nr | 46 |
| | 2.238 | 45 | ~120 | nr | |
| | 1.927 | 187.4 | 88.4 | nr | |

[a]Hyperfine couplings in megahertz (MHz). nr = not resolved.
[b]$^{103}$Rh: I = 1/2 (nat. ab. 100%); $^{191}$Ir: I = 3/2 (37.3%); $^{193}$Ir: I = 3/2 (62.7%).
[c]Derived from fitting the complicated Ir hyperfine pattern due to large Ir quadrupole couplings.

### F. The Rh(II)-, Ir(II)-, and Pd(III)-Cyanide Complexes Trapped in KCl/NaCl Host Lattices

A few reports describe the formation of $Rh^{II}$—CN, $Ir^{II}$—CN, and $Pd^{II}$—CN species upon electron irradiation of $K_3[M^{III}(CN)_6]$ (M = Rh, Ir, Pd) in KCl or NaCl host lattices (47–52). For the rhodium species, this leads to formation of three different species: $[Rh^{II}(CN)_6]^{4-}$, $[Rh^{II}(Cl)_2(CN)_4]^{4-}$, and a third $Rh^{II}$ species which, to the best of our knowledge, remains yet to be identified. All

TABLE VII
The EPR Data of $Rh^{II}$– and $Ir^{II}$–CN Complexes in KCl and NaCl Host Lattices[a]

| Complex | $g_x$ $g_y$ $g_z$ | $A_x(M)$ $A_y(M)$ $A_z(M)$ | $A_x(^{14}N)$ $A_y(^{14}N)$ $A_z(^{14}N)$ | $A_x(Cl)$ $A_y(Cl)$ $A_z(Cl)$ | $\rho_M$ $(s + d_{z^2})$ | $\rho_{Cl}$ $(s + p)$ | $\rho_N$ $(s + p)$ | References |
|---|---|---|---|---|---|---|---|---|
| $[Rh^{II}(CN)_6]^{4-}$ | 2.128 | +24 | +12 | | 44% | | 1.6% (x2) | 47, 49 |
| | 2.128 | +24 | +12 | | | | | |
| | 1.997 | −34 | +13.5 | | | | | |
| $[Rh^{II}(CN)_4(Cl)_2]^{4-}$ | 2.297 | −111 | nr | +37 | 22% | 11% | | 47 |
| | 2.297 | −111 | nr | +37 | | (x2) | | |
| | 1.995 | −122 | nr | +83 | | | | |
| Third Rh(II) species | 2.160 | nr | nr | nr | | | | 47 |
| | 2.160 | nr | nr | nr | | | | |
| | 1.990 | nr | nr | nr | | | | |
| $[Ir^{II}(CN)_5]^{3-}$ | 2.203 | +31.7 | ax: +11.4 | | 50% | | ax: 5.9% | 48, 51 |
| | 2.211 | +35.1 | eq: −0.6 | | | | eq: 1.28% | |
| | 1.967 | −66.9 | ax: +11.4 | | | | (x4) | |
| | | | eq: 0.6 | | | | | |
| | | | ax: +18.6 | | | | | |
| | | | eq: +6.9 | | | | | |
| $[Pd^{III}(CN)_4(Cl)_2]^{3-}$ | $g_\perp = 2.116$ $-2.130$[b] $g_\parallel = 2.00$ | | | | | | | 52 |

[a]Hyperfine couplings in MHz, nr = not resolved, ax = axial, eq = equatorial, ρ = spin population in %.
[b]Temperature dependent.

three $Rh^{II}$ species reveal axial EPR spectra (Table VII). The complex $[Rh^{II}(CN)_6]^{4-}$ reveals (super)hyperfine couplings with $^{103}Rh$ and 2 equiv N atoms. $[Rh^{II}(Cl)_2(CN)_4]^{4-}$ reveals (super)hyperfine couplings with $^{103}Rh$ and 2 equiv Cl atoms (albeit in a complex patterns due from the presence of $^{35}Cl$—$^{35}Cl$, $^{35}Cl$—$^{37}Cl$, and $^{37}Cl$—$^{37}Cl$ species according to the natural abundances of the Cl isotopes). These data suggest that in both cases the SOMO mainly consists of an Rh $d_{z^2}$ orbital with antibonding contributions from orbitals of the axial $CN^-$ and $Cl^-$ ligands (Fig. 15). The reported spin populations estimated from the EPR data and semiempirical calculations are listed in Table VII.

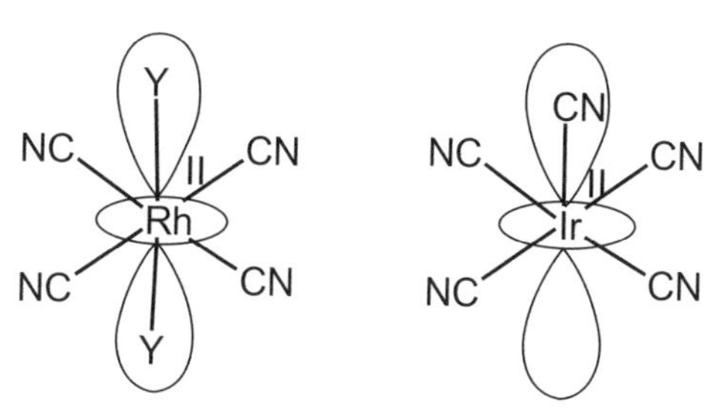


Figure 15. Structure and SOMO of Rh(II)– and Ir(II)–cyanide complexes trapped in KCl and NaCl host lattices.

Similar experiments with $K_3[Ir^{III}(CN)_6]$ resulted in formation of two paramagnetic species, one being $[Ir^{II}(CN)_5]^{3-}$. The second species is most likely an organic dihydrogen cyanide ($H_2CN$) radical (53). In contrast to the results reported for Rh, the $Ir^{II}$ species now reveals a rhombic EPR spectrum indicative for an $\sim C_{4v}$ symmetry as a result of the loss of one of the $CN^-$ ligands. Ligand hyperfine coupling with 4 equiv nitrogens (equatorial, weak couplings) and 1 nonequiv nitrogen (axial, strong coupling) was observed, indicative of a square-pyramidal geometry with the unique CN ligand pointing to the primarily Ir $d_{z^2}$ based SOMO (Fig. 15).

## G. Palladium Carbollides

Organometallic Pd(III) species are extremely rare. Among the few known examples is a palladium–carbollyl complex. The synthesis of $[Pd^{III}\{\pi\text{-}(3)\text{-}1,2\text{-}B_9C_2H_{11}\}_2]^-$ by treatment of an equimolar ratio of $[Pd^{II}\{\pi\text{-}(3)\text{-}1,2\text{-}B_9C_2H_{11}\}_2]^{2-}$ and $[Pd^{IV}\{\pi\text{-}(3)\text{-}1,2\text{-}B_9C_2H_{11}\}_2]$ has been reported (Fig. 16) (37, 54). The deep red product is probably isostructural with its Ni analogue $[Ni^{III}\{\pi\text{-}(3)\text{-}1,2\text{-}B_9C_2H_{11}\}_2]^-$ and has a magnetic susceptibility of $\mu_{eff} = 1.68\,\mu_B$, which is in agreement with a $d^7$ ground state. The EPR data were not reported. Recently, however, the related (non-organometallic) coordination compound 1,4-$Br_2$-1,2,5-$(PMe_2Ph)_3$-*closo*-1-$PdB_{11}H_8$ was characterized (55). The EPR spectrum of this species shows a near isotropic signal with $g_{av} = 2.02$. Despite the small deviations from $g = g_e$, the species was assigned as a metal centered radical.

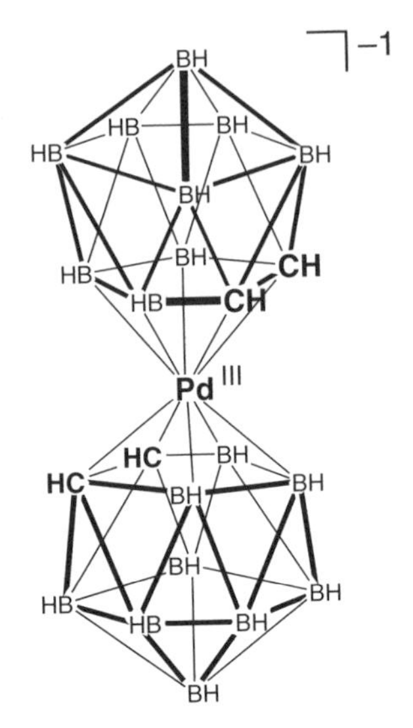


Figure 16. Structure of a $Pd^{III}$-bis-($\pi$-(3)-1,2-dicarbollyl) complex.

### H. Dinuclear Species, Multinuclear Wires, and Clusters

Several paramagnetic organometallic clusters, multinuclear wires, and dinuclear species involving platinum, palladium, rhodium, and iridium have been reported. It is not our intention to give a comprehensive overview of all paramagnetic multinuclear species here, because we intend to focus on the chemistry of the mononuclear paramagnetic complexes in this chapter. However, some of these species bear some similarity to the mononuclear species described elsewhere in this chapter, and for that reason it is good to present a short overview of the various types of multinuclear paramagnetic species.

Most of the multinuclear paramagnetic clusters and wires are connected by metal–metal bonds, often supported by bridging ligands. Examples are the Krogmann salts ($M_3[Pt\{CN\}_4]X_{0.3}$) (56), and the wire-like tri- and tetranuclear Rh and Ir compounds of the type $[Rh_3(\mu_3\text{-}L)_2(CO)_2(\eta_4\text{-cod})_2]^{2+}$ (57), $[Rh_2Ir(\mu_3\text{-}OMe_2napy)_2(CO)_2(cod)_2]^{2+}$ (58), ($OMe_2$ napy = 5,7-dimethyl-1,8-naphyridine-2-onace) and $[M_4(\mu\text{-}PyS_2)_2(\text{diolefin})_4]^+$ (M = Rh, Ir) (59). Interestingly, these "molecular wires" have the unpaired electron delocalized over multiple metal centers, suggestive of potential conducting properties of such wires.

Several mixed-valent dinuclear rhodium and iridium complexes of the type $[M_2]^{5+}$ have been reported (M = Rh, Ir). These species can formally be regarded as a paramagnetic M(II) species bound to a closed-shell M(III) ion via metal–metal interactions. However, the actual electronic configuration of the $[Rh_2]^{5+}$ species is highly dependent on the type and number of ligands bound to the axial sites. For symmetrical complexes, the unpaired electron is delocalized over both metal atoms and occupies a metal–metal bonding or antibonding orbital with $\pi^*$ $\{[Rh_2(\mu\text{-acetate})_4L_2]^+$ (L = $H_2O$, MeCN, acetone, etc.)$\}$, $\sigma$ $\{[Rh_2(\mu\text{-}O_2CR)_4L_2]^+$ (R = Et, $CF_3$; L = phosphine, phosphate)$\}$, or $\delta^*$ $\{[Rh_2(\mu\text{-acetate})_{4-m}\{\mu\text{-MeC-(O)NH}\}_n(NCMe)_2]^+$ (n = 1–4)$\}$ symmetry (60). The SOMO of the $[Ir_2]^{5+}$ species $[Ir_2(\mu\text{-}O_2CH)_2Cl_2(CO)_2(L)_2]^+$ (L = P, As ligand) is a $\sigma_{Ir\text{–}Ir}$ orbital with some $\sigma^*_{Ir\text{–}ligand}$ character (61). Binding of an additional ligand can have a large effect on the electronic structure. For example, symmetrical $[Rh_2(\mu\text{-}L{\cap}L)_4]^{5+}$ species have the unpaired electron delocalized over the two Rh sites with an equal distribution over both metal ions, but coordination of an additional ligand (CO, Cl, C≡CH, or C≡C– C≡C-$SiMe_3$) to one of both rhodium atoms yields asymmetric species with the unpaired electron located on only one of the metal ions (62, 63). The electronic configuration of these species (based on EPR spectroscopy and DFT calculations) is thus quite close to that of a mononuclear metalloradical.

Various mixed-valent $[Rh_2]^{3+}$ species, which formally attain $Rh^I$–$Rh^{II}$ oxidation states, have also been reported (60). The metal contribution to the

SOMO is mainly a $d_{z^2}$ orbital, consistent with characterization of the SOMO as the Rh—Rh σ* MO. When the metal coordination planes are tilted, the metal hybrid contribution to the SOMO changes with the angle and $d_{xz}$ or $d_{yz}$ character is incorporated. Similar to the $[M_2]^{5+}$ clusters, the electronic structure of $[M_2]^{3+}$ clusters changes when the cluster becomes asymmetric and the unpaired electron becomes more localized on one metal center. For example, the Rh and Ir sites containing the cod fragment in the complexes $[Rh(\eta_4\text{-cod})(\mu\text{-}RNNNR)_2Ir(CO)_2]^+$ and $[Ir(\eta_4\text{-cod})(\mu\text{-}RNNNR)_2Rh(CO)_2]^+$ bear most of the unpaired electron density (80 and 75%, respectively) (64). This is even more drastic in the case of $[I_2Rh([Rh(\eta_4\text{-cod})(\mu\text{-}RNNNR)_2Pd(\eta_3\text{-}C_3H_5)]^-$, which has 90% of the unpaired electron spin density located at the rhodium atom, with only 10% at palladium (65). The DFT calculations further suggest that π-donor ligands, such as iodide, raise the energy of the rhodium $d_{z^2}$ orbital, resulting in an almost entirely rhodium $d_{z^2}$ based SOMO, essentially without any metal–metal σ* character left.

## III. PERSISTENT LIGAND RADICAL COMPLEXES OBTAINED BY LIGAND OXIDATION

A number of paramagnetic Rh, Ir, Pd, and Pt complexes containing redox noninnocent radical ligands have been reported. These species are not always easily distinguishable from their redox tautomeric metal centered radical descriptions, as described in the introduction.

Ligand radical complexes based on both oxidized ligands and reduced ligands have been described for groups 9(VIIIB) and 10(VIII) organometallic species. Complexes containing *reduced* redox noninnocent ligands will be described in a forthcoming review. This section focuses on complexes with *oxidized* ligands.

### A. Aminyl Radicals and Related Species

Recently, it was demonstrated that aminyl radicals are substantially stabilized by coordination to Rh(I). Deprotonation of cationic $[Rh^I(trop_2N\mathit{H})(bpy)]^+$ (5-*H*-dibenzo-[*a,b*]cycloheptene-5-yl-trop) species generates a stable (under inert atmosphere) neutral amido complex $[Rh^I(trop_2N^-)(bpy)]$. Remarkably, one-electron oxidation of the latter with $[Fc]^+$ does not result in a Rh(II) species, but yields a stable aminyl radical complex $[Rh^I(trop_2N\bullet)(bpy)]^+$. This is the first example of a stable aminyl radical coordinated to a transition metal ever reported (66). Both detailed EPR spectroscopic investigations ($g_{11} = 2.0822$, $g_{22} = 2.0467$, $g_{33} = 2.0247$, $A^{N}_{iso} = 45$ MHz), and DFT calculations revealed that

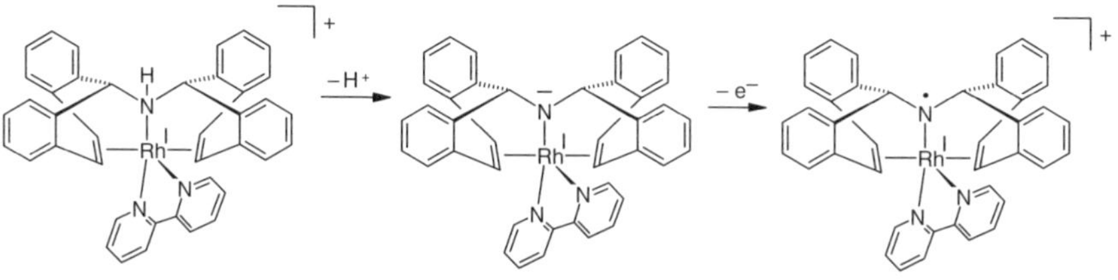

Figure 17. Formation of a stable $Rh^{I}$–aminyl radical complex by one-electron oxidation of a $Rh^{I}$–amido complex.

the unpaired electron is delocalized over the $R_2N$ (~57%) fragment and Rh (~ 30%), but with a clear preference for the N atom (Fig. 17).

The N-centered radical is stable in organic solvents for days, and even when samples are brought in contact with air or water the EPR signals persist for hours. The aminyl radical fragment nevertheless behaves as a nucleophilic radical, capable of hydrogen atom abstraction reactions from $Bu_3Sn$–H or RSH (R = Ph-, *t*-Bu-, $MeOOCCH_2$–), thus regenerating the $[Rh^{I}(trop_2N\mathit{H})(bpy)]^+$ starting compound and $Bu_3Sn$-$SnBu_3$ or RS–SR coupled products.

Quite similar experiments were performed with a rhodium complex of the methyl-substituted $trop_2N$–Me ligand (Fig. 18). Deprotonation of $[Rh^{I}(trop_2N{-}Me)(PPh_3)]^+$ occurs at the methyl moiety of the $trop_2N$–Me ligand, resulting in formation of a rhodaaza-cyclopropane ring structure (67). The $[Rh^{I}(trop_2N{-}CH_2{-})(PPh_3)]$ complex reveals a reversible oxidation wave in the cyclic voltammogram at a remarkably low potential (−0.51 vs. $Fc/Fc^+$) ($Fc/Fc^+$ = ferrocene/ferrocenium). Preparative one-electron oxidation of this species yields a stable cationic radical analogue. The EPR data ($g_{11} = 2.121$, $g_{22} = 2.014$, $g_{33} = 2.016$) reveal hyperfine couplings to Rh, N, the aza-cyclopropane methylene carbon and hydrogen atoms, and the trop olefinic carbon and hydrogen atoms. Accordingly, DFT calculations reveal that the unpaired electron is strongly delocalized over rhodium (47%), the methylene (14%), and the four olefinic carbons (15, 15, 3, and 3%). Assignment of an

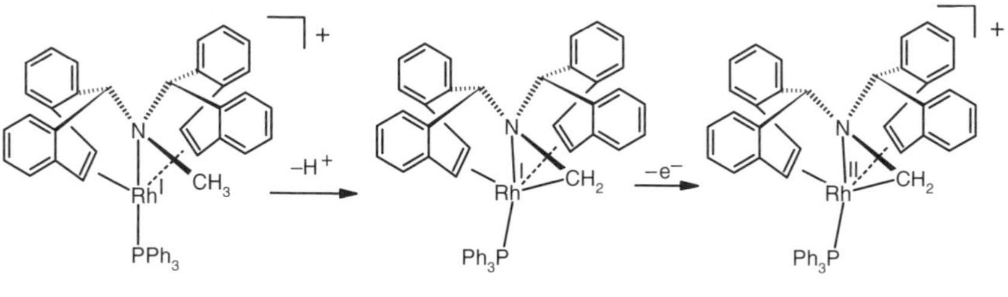

Figure 18. Formation of highly delocalized rhoda-aza-cyclopropane radicals from $trop_2N$–Me ligands.

oxidation state to the metal is not very meaningful in this case. This species is better described as a highly delocalized radical.

## B. Complexes With "Redox Noninnocent" Phenolate-Type Ligands

In 1992, the synthesis of a paramagnetic $Ir^{IV}$(cod) species containing two 3,6-di-*tert*-butylcatecholate (3,6-DBCat) ligands was claimed (68). From many studies in later years, however, it has become most obvious that catecholate-quinone-type ligands are redox noninnocent ligands (69–74). The metal quinone, $M^{I}$(Q), metal semiquinone, $M^{II}(SQ^{\cdot -})$, and metal catecholate, $M^{III}(Cat^{2-})$, are description of either one electronic ground state (mesomeric form) or are distinct electronic states that are in equilibrium (redox isomerism) (Fig. 19).

The original assignment as $[Ir^{IV}(3,6\text{-DBCat})_2(cod)]$ was mainly based on X-ray diffraction results. The C–C lengths reflect an aromatic ring of a catecholate fragment and the C–O distance is longer than that of a semiquinone structure. However, a description of $[Ir^{III}$(3,6-DBCat)(3,6-DBSQ)(cod)] in which the DBSQ and the DBCat fragments are randomly ordered in the X-ray structure, cannot be ruled out. In this case, one would find very similar average bond lengths. In addition, the system might even be highly delocalized.

In a toluene solution at room temperature, a broad EPR resonance (200 G) at a $g$ value of 1.986 was observed. In frozen solution, the signal splits to give a slightly rhombic spectrum with components $g_1 = 1.948$, $g_2 = 1.942$, and $g_3 = 2.006$. Since an EPR spectrum was obtained at room temperature and the $g$ values are close to $g_e$, we suggest that the unpaired electron is mainly located at the organic fragment, rather than the metal. Therefore, we prefer the $[Ir^{III}$(3,6-DBCat)(3,6-DBSQ)(cod)] description. The lack of resolved ligand superhyperfine couplings could point to some influence of the metal though, because a small spin population at the metal can cause substantial line broadening. Also, the X-ray structure of $[Rh^{I}(3,6\text{-DBSQ})(CO)_2]$ was determined (75). The square-planar molecules in the crystal are oriented with the rhodium centers on top of each other, with a Rh–Rh distance of 3.3 Å, similar to other stacked rhodium(I) dimers and oligomers. The absorption spectrum of $[Rh^{I}(3,6\text{-DBSQ})(CO)_2]$ in pentane at room temperature shows a

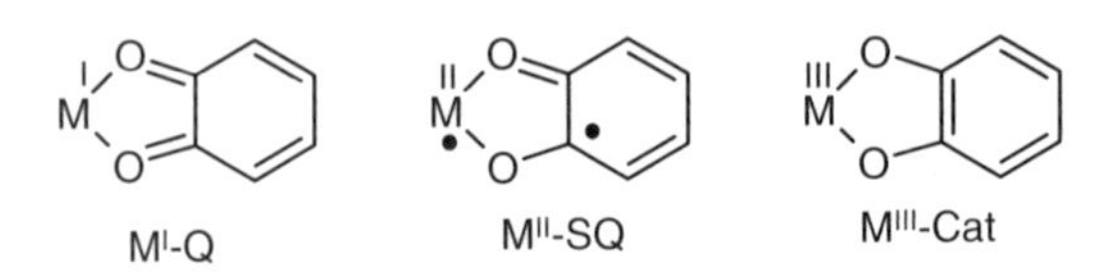


Figure 19. Redox noninnocence of catecholate–quinone-type ligands.

MLCT band at 430 nm, which was assigned as the $2a_1(d_{z_2}) \rightarrow 2b_1[\pi * (SQ)]$ transition. Spectra at lower temperatures and in the solid phase give rise to a new broad signal at 1500 nm. This temperature and concentration dependence of the absorption spectra, indicate that stacking of monomer complexes via interaction of the $d_{z^2}$ orbitals also takes place in solution. The EPR spectrum showing three lines due to coupling with 2 equiv hydrogens of 3,6-DBSQ (3,6-DBSQ = 3,6-di-*tert*-butyl semiquinonate) is in agreement with the $[Rh^{I}(3,6\text{-DBSQ})(CO)_2]$ description.

### C. Complexes With "Redox Noninnocent" Orthometallated Indole-Based Ligands

Oxidation of the [Pd(tbu-iepp-c)Cl] and [Pd(p-iepp-c)Cl] complexes shown in Fig. 20 with 1 equiv of Ce(IV) afforded the corresponding one-electron oxidized species (76). (Htb-iepp = 3-[*N*-2-pyridylmethyl-*N*-2-hydroxy-3,5-di(-*tert*-butyl)benzylamina] ethylindole; *H* p-iepp = 3-(*N*-2-pyridylmethyl-*N*-4-hydroxybenzylamine] ethylindole; *H* denotes a dissociable proton and tbu-iepp-c p-iepp-c and tbu-iepp-c denote p-iepp and tbu-iepp bound to Pd(II) through a carbon atom respectively (76). The EPR spectroscopy (one sharp signal at $g = 2.004$) and absorption spectroscopy (peak at ~550 nm) are consistent with an indole π radical rather than a Pd(III) or a phenoxyl radical.

### D. Complexes With "Redox Noninnocent" Dithiole-2-Thionate-Type Ligands

Complexes of the type $[Rh^{III}(Cp)(C_3S_5)]$ and $[M^{III}(Cp^*)(C_3S_5)]$ [M = Rh, Ir, $C_3S_5^{2-}$ = 4,5-disulfanyl-1,3-dithiole-2-dithionate(2-)] have been oxidized

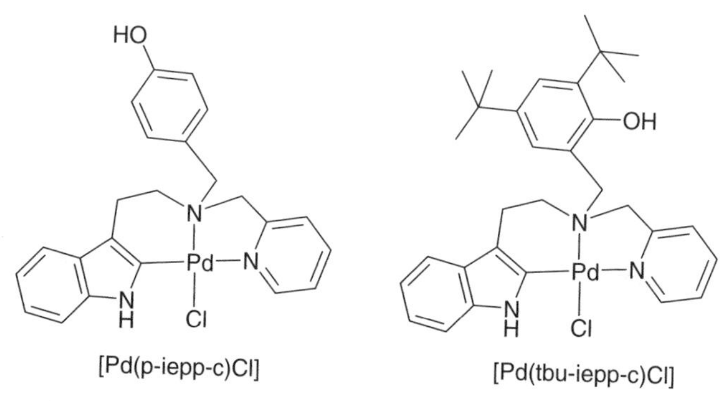


Figure 20. The structure of the precursors [Pd(tbu-iepp)Cl] and [Pd(p-iepp-c)Cl] prior to one-electron oxidation.

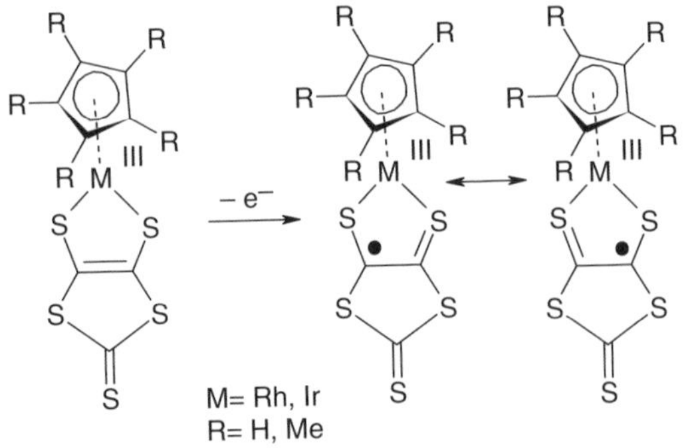

Figure 21. Formation of ligand radicals upon oxidation of rhodium and iridium dithiole-2-dithionate complexes.

chemically and electrochemically. Reversible waves observed with cyclic voltammetry measurements reveal that the oxidized species are rather stable. Chemical oxidation with $Br_2$ affords the coordinatively saturated, paramagnetic bromo adducts $[Rh(Cp)(C_3S_5)(Br)]$ and $[M(Cp^*)(C_3S_5)(Br)]$ (77, 78). The *formally* M(IV) species are in reality ligand centered radical complexes containing a singly oxidized $C_3S_5^-$ radical ligand, as indicated by the isotropic EPR spectra reported for $[M^{III}(Cp^*)(C_3S_5)]$ in dichloromethane at room temperature (M = Rh: $g = 2.010$; $A^{Rh} = 14$ MHz, $A^{Br} = 9$ MHz; M = Ir: $g = 2.000$; $A^{Ir}$ and $A^{Br}$ not resolved). The observation of EPR spectra at room temperature, the small deviations from $g = g_e$ and the weak metal and Br hyperfine couplings all point to the presence of ligand centred radicals. For the Rh complex, this was confirmed with DFT calculations revealing 87% of the Mulliken spin population to be present at the $C_3S_5$ ligand (Fig. 21).

## IV. REACTIVITY OF TRANSIENT (METALLO) RADICALS

Although some of the species desribed in this section are stable in certain solvents or under certain conditions, they reveal a remarkably high reactivity toward a variety of substrates. This section focuses on the diverse radical-type reactivity of open-shell organometallic species of the groups 9(VIIIB) and 10(VIII) transition metals.

### A. Reactivity of Paramagnetic Cyclopentadienyl Complexes

Paramagnetic rhodocene and iridocene derivatives ($[M^{II}(CpR_n)_2]$), as well as the half-sandwich ($[M^{II}(Cp)(L)_n]$) complexes of rhodium and iridium (Fig. 22) are generally very reactive. The most commonly observed reaction types involve Cp–Cp ligand coupling, and metal–metal bond formation.

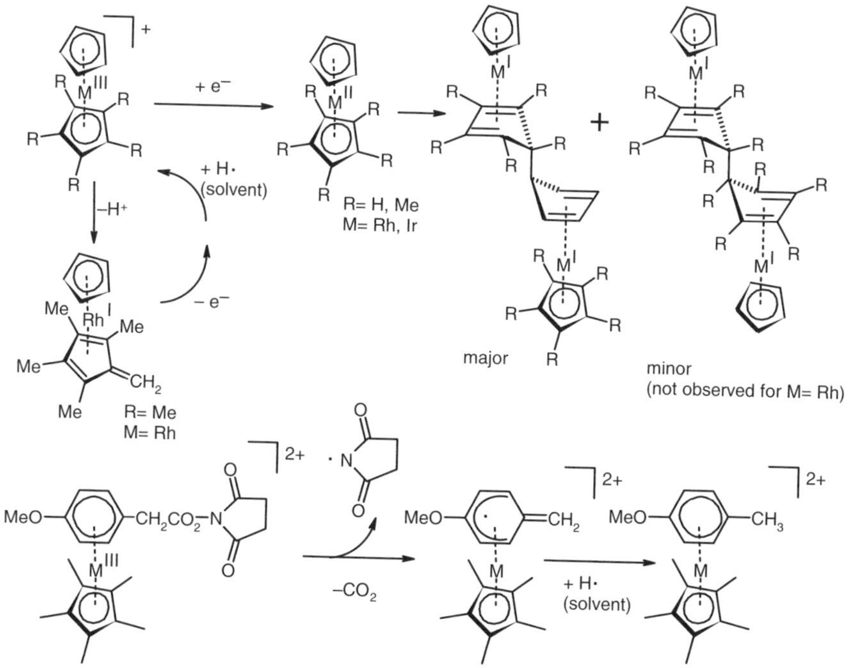

Figure 22. Redox chemistry of $[(CpR_5)_2M^{III}]^+$ species (M = Rh, Ir, R = H, Me).

Many reports describe the irreversible (electro)chemical reduction of rhodocenium and iridocenium cations $[M^{III}(Cp)_2]^+$ and related species containing Cp* or indenyl ligands (Cp = $\eta^5$-$C_5H_5$, Cp* = $\eta^5$-$C_5Me_5$, indenyl = $\eta^5$-$C_9H_7$) (79–83).

Reduction of rhodocenium or iridocenium salts $[M^{III}(Cp)_2]^+$ leads to neutral rhodocene and iridocene complexes (formally $[M^{II}(Cp)_2]$; M = Rh, Ir) with a short lifetime of ~2 s at room temperature (79, 80). This leads to irreversible waves in cyclic voltammetry measurements, indicative of an EC mechanism. The chemical reaction following the reduction involves a dimerization reaction by intermolecular C—C coupling between two Cp rings, thus yielding diamagnetic, binuclear $[M^I(Cp)(\mu$-$\eta^4,\eta^4$-$C_5H_5$—$C_5H_5)M^I(Cp)]$ species (Fig. 22). At lower temperatures, electrochemically reversible waves are observed, thus indicating somewhat longer lifetimes (80). Reduction of $[M^{III}(Cp)_2]^+$ in a melt with Na/K, followed by sublimation on a cold finger cooled with liquid $N_2$ gave polycrystalline samples (black for Rh, colorless for Ir), which upon warming to room temperature convert to yellow $[M^I(Cp)(\mu$-$\eta^4,\eta^4$-$C_5H_5$—$C_5H_5)M^I(Cp)]$ species. These black and colorless samples were reported to contain the mononuclear $[M^{II}(Cp)_2]$ species. Thus obtained polycrystalline samples gave axial EPR spectra, revealing: $g_{\parallel} = 2.033$, $g_{\perp} = 2.003$ (no resolved Rh hyperfine couplings) for $[Rh^{II}(Cp)_2]$, and $g_{\parallel} = 2.033$, $g_{\perp} = 2.001$ [1.997 in ref. (79)], $A^{191/193}Ir_{\parallel} < 1 \times 10^{-4}\ cm^{-1}$, $A^{191/193}Ir_{\perp} = 5.1 \times 10^{-4}\ cm^{-1}$ (15 MHz) for $[Ir^{II}(Cp)_2]$ (79, 84). Given the small deviations from $g = g_e$, and the very small metal hyperfine couplings, it is rather doubtful whether these spectra really correspond to $[M^{II}(Cp)_2]$ species with a "sandwich" structure (also see the discussion in relation to the tetra-phenyl substituted analogue $[M^{II}(C_5Ph_4H)_2]$).

Reduction of the nonsymmetrically substituted iridium complex $[Ir^{III}(Cp^*)(Cp)]^+$ leads mostly to Cp*—Cp coupling, with minor Cp*—Cp* and no Cp—Cp coupling. The result is a 2:1 mixture of the dinuclear complexes $[Ir^{I}(Cp^*)(\mu\text{-}\eta^4,\eta^4\text{-}C_5H_5\text{—}C_5Me_5)Ir^{I}(Cp)]$ and $[Ir^{I}(Cp)(\mu\text{-}\eta^4,\eta^4\text{-}C_5Me_5\text{—}C_5Me_5)Ir^{I}(Cp)]$ (Fig. 22) (83). A similar reduction of the rhodium analogue $[Rh^{III}(Cp^*)(Cp)]^+$ gave selective formation of only the asymmetrical Cp—Cp* bridged product $[Rh^{I}(Cp^*)(\mu\text{-}\eta^4,\eta^4\text{-}C_5H_5\text{—}C_5Me_5)Rh^{I}(Cp)]$ (81).

The instability of bis(cyclopentadienyl) Rh(II) species contrasts the reported synthesis of the reasonably stable bis-π-arene Rh(II) sandwich species $[Rh^{II}(C_6Me_6)_2](X)_2$ ($X^- = AlCl_4^-$, $AlBr_4^-$; $X^-$ can be replaced by $PF_6^-$) from $RhCl_3$, $AlBr_2$, Al, and hexamethylbenzene. The species can be reduced to the corresponding Rh(I) species. No EPR data have been reported for $[Rh^{II}(C_6Me_6)_2]^{2+}$ (85).

The formation of these type of dinuclear complexes could be taken as evidence that at least some of the spin density of the neutral $[M^{II}(\eta^5\text{-}C_5R_5)_2]$-type species is actually located at the Cp-type $\eta^5\text{-}C_5R_5$ ligands. If the spin density would be equally distributed over the Cp and Cp* ligands in the nonsymmetrically substituted neutral $[M^{II}(Cp^*)(Cp)]$ complexes, one would expect formation of three types of dinuclear species; the ones depicted in Fig. 22, and in addition the nonobserved products $[M^{I}(Cp^*)(\mu\text{-}\eta^4,\eta^4\text{-}C_5H_5\text{—}C_5H_5)M^{I}(Cp^*)]$ arising from intermolecular C—C coupling of two nonsubstituted Cp rings. In that case, one would even expect preferential formation of the latter for steric reasons. If, on the other hand, the spin density is located preferentially on either the Cp or Cp* ring, only one symmetric product would be expected. A tentative explanation for the preferential formation of the nonsymmetrical species $[M^{I}(Cp^*)(\mu\text{-}\eta^4,\eta^4\text{-}C_5H_5\text{—}C_5Me_5)M^{I}(Cp)]$ was suggested by the authors (81–83). Reduction of the diamagnetic cationic species will result in rather nucleophilic 19-VE radicals $[M^{II}(Cp^*)(Cp)]$, which could attack any nonreacted cationic starting material with a steric preference for attack at the nonsubstituted Cp ring of $[M^{III}(Cp)(Cp^*)]^+$. With the assumption that the Cp* ligand contains a larger amount of the total spin density of the radical species $[M^{II}(Cp)(Cp^*)]$, this mechanism would explain the preferred C—C coupling between Cp* and Cp. Reduction of thus obtained dinuclear monocationic intermediates yields the final products.

Deprotonation of $[Rh^{III}(Cp)(Cp^*)]^+$ with *tert*-BuOK at a Cp* methyl group gives the $[Rh^{I}(\eta^4\text{-tetramethylfulvene})(Cp)]$ species (Fig. 22). One-electron oxidation of this species with ferrocenium salts regenerates $[Rh^{III}(Cp)(Cp^{*})]^{+}$, most likely involving hydrogen abstraction from the solvent by the reactive cation radical intermediate $[Rh(Cp)(Me_4C_5{=}CH_2)]^{+}$ (82). The analogous Ir chemistry proved to be less straightforward. However, thermal C—N bond homolysis of the related species $[Ir^{III}(Cp^{*})(\eta^6$-*p*-MeO—$C_6H_4$—$CH_2$—$CO_2$—succinimidyl)] produces a $[Ir(Cp^{*})(\eta^6$-*p*-MeO—$C_6H_4{=}CH_2)$ radical intermediate, which subsequently abstracts a hydrogen atom from the solvent to yield $[Ir(Cp^{*})(\eta^6$-*p*-MeO—$C_6H_4$—Me) (86), (see Fig. 22).

Reduction of the Cp*-indenyl rhodium complex $[Rh^{III}(Cp^{*})(\eta^5\text{-}C_9H_7)]^{+}$ has been reported to yield dark blue solutions of the persistent Rh(II) sandwich radical complex $[Rh^{II}(Cp^{*})(\eta^5\text{-}C_9H_7)]$ (81), (Fig. 23), revealing an axial EPR spectrum ($g_{\perp} = 2.0118$, $g_{\parallel} = 1.8846$, $g_{iso} = 1.96$, no reported $^{103}$Rh hyperfine couplings).

The enhanced stability was ascribed to the electron-withdrawing character of the indenyl ligand. Reduction of the analogous iridium complex $[Ir^{III}(Cp^{*})(\eta^5\text{-}C_9H_7)]^{+}$ did not result in a stable Ir(II) species. Formation of a tetranuclear complex was observed instead (Fig. 23). This reaction proceeds via loss of an

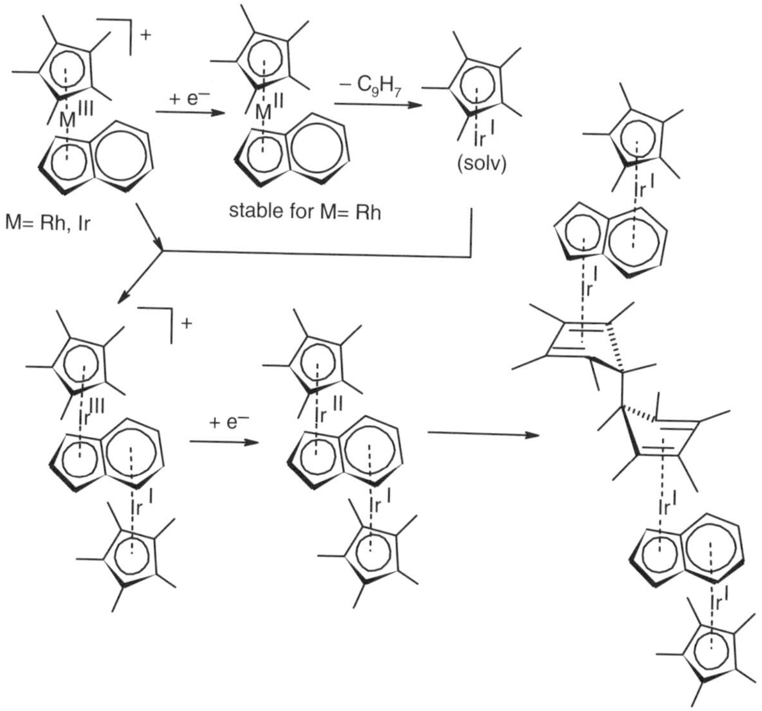

Figure 23. Redox chemistry of indenyl $[M^{III}(Cp^{*})(\eta^5\text{-}C_9H_7)]^{+}$ complexes (M = Rh, Ir).

indenyl radical from the electrochemically generated neutral Ir(II) species $[Ir^{II}(Cp^*)(\eta^5\text{-}C_9H_7)]$. The remaining half-sandwich complex then forms an indenyl bridged dinuclear species with the nonreduced starting material. Subsequent further reduction yields the tetranuclear species via intermolecular C—C coupling of two Cp* ligands in a similar reaction as observed for $[M^{II}(Cp^*)(Cp)]$ species (see Figs. 22 and 23).

Half-sandwich complexes $[Rh^I(Cp)(CO)(L)]$ with L = CO, $PPh_3$, $PMe_3$, and $P(OPh)_3$ all reveal irreversible oxidation waves in cyclic voltammograms. In all cases, the intermediate 17 VE electron monocationic species couple to give dinuclear species. For L = CO and $PPh_3$, the oxidation leads to formation of fulvalene complexes (87–89). This reaction could proceed via C—C coupling of two Cp ligands with a net loss of $H_2$, followed by Rh—Rh bond formation [path (*a*) in Fig. 24].

Remarkably, with the ligands L = $PMe_3$ or L = $P(OPh_3)$, no Cp—Cp ligand coupling nor formation of a fulvalene complex was observed. These complexes only dimerize under formation of a Rh—Rh bond (90). Based on studies of anthryl–methylene substituted analogues (see below), it has been suggested that the nature of the ligands L (Fig. 24) affect the spin distribution over the metal and the Cp ring, with stronger σ-donors promoting metalloradical behavior and stronger π-acids promoting Cp-ligand radical character (91). The π-acids like CO are thus suggested to promote dimerization via ligand–ligand coupling, as in route (*a*), Fig. 24, whereas σ donors like $PMe_3$ should promote metal–metal coupling as in route (*b*). However, this still does not give a satisfactory explanation for the fact that $PPh_3$ promotes fulvalene formation, whereas the comparably stronger π-acid $P(OPh)_3$ does not. Reduction cleaves the metal–metal bonds to form Rh(I) species in all cases.

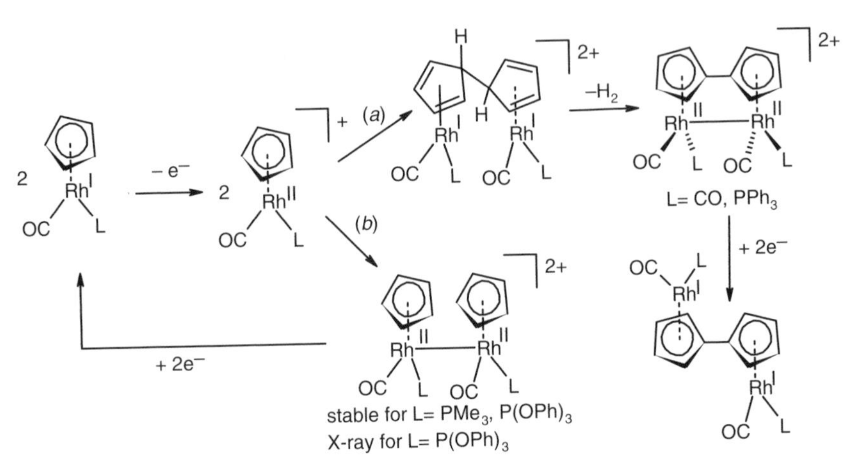


Figure 24. Redox chemistry of $[Rh^I(Cp)(CO)(L)]$ complexes.

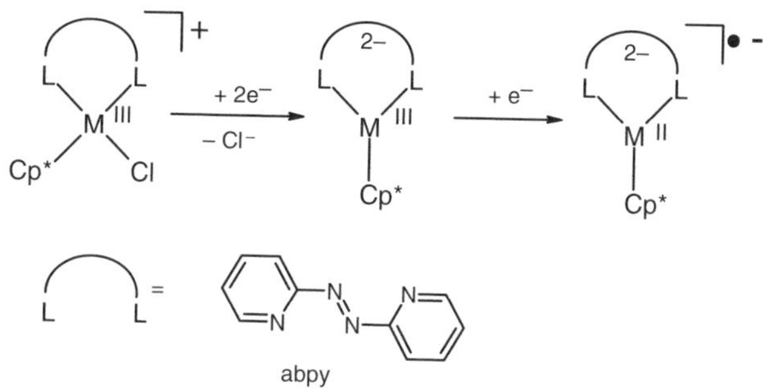

Figure 25. Reduction of $[Rh^{III}(Cp^*)Cl(abpy)]^+$ to $[Rh^{II}(Cp^*)(abpy^{2-})]^-$ (abpy 2,2-azobipyrydyl) via the diamagnetic species $[Rh^{III}(Cp^*)(abpy^{2-})]$.

Electrochemically reversible oxidation has been reported for the related complex $[Rh^I(Cp)(ethene)_2]$. Coulometrically generated samples revealed a broad spectrum around $g = 1.9855$ without a resolved hyperfine structure in dimethoxyethane solution at $-70°C$ (frozen solutions at 20–100 K would have probably resulted in more informative spectra) (92, 93). A detailed characterization of the radical cation is lacking so far. Similar reports concern the electrochemically reversible reduction of $[Rh^{III}(Cp)(S_2C_4F_6)_2]$ (of which the interpretation is even less clear due to the presence of the potentially noninnocent $S_2C_4F_6$ ligand) (92, 94).

Reduction of $[Rh^{III}(Cp^*)Cl(abpy)]^+$ initially occurs at the "redox noninnocent" abpy ligand (Fig. 25). Two-electron reduction thus yield the diamagnetic neutral complex $[Rh^{III}(Cp^*)(abpy^{2-})]$. Further reduction of this species yields the anionic radical complex $[Rh(Cp^*)(abpy)]^-$ (95, 96). The redox noninnocence of the abpy ligand allow the electronic structures $[Rh^{II}(Cp^*)(L^{2-})]^-$, $[Rh^I(Cp^*)(L^-)]^-$, and $[Rh^0(Cp^*)(L^0)]^-$. *In situ* EPR measurements reveal an isotropic $g$ value of 2.042 with the anisotropic components: $g_1 = 2.161$, $g_2 = 2.002$, and $g_3 = 1.945$. These data are indicative for a species with significant spin population at the metal center. Therefore, both $[Rh^{II}(Cp^*)(abpy^{2-})]^-$ and $[Rh^0(Cp^*)(abpy^0)]^-$ seem reasonable descriptions. Since abpy is a strong $\pi$ acceptor, the authors favor the Rh(II) description.

In order to understand fluorescence quenching of anthyl units in Rh(I) and Ir(I) complexes of the type shown in Fig. 26, the redox chemistry of these complexes has been investigated (91). There are two main redox processes observed with cyclic voltammetry in THF; a reversible antryl centered reduction and an irreversible metal centered oxidation in all cases. The observation of irreversible oxidation waves in THF indicates that the electrode generated cationic species are not stable at room temperature. Apparently, however, the use of the solvent 1,1,1,3,3,3-hexafluoropropan-2-ol (HFP) somewhat stabilizes these species. Oxidation of the M(I) species with thalium(III) trifluoroacetate in

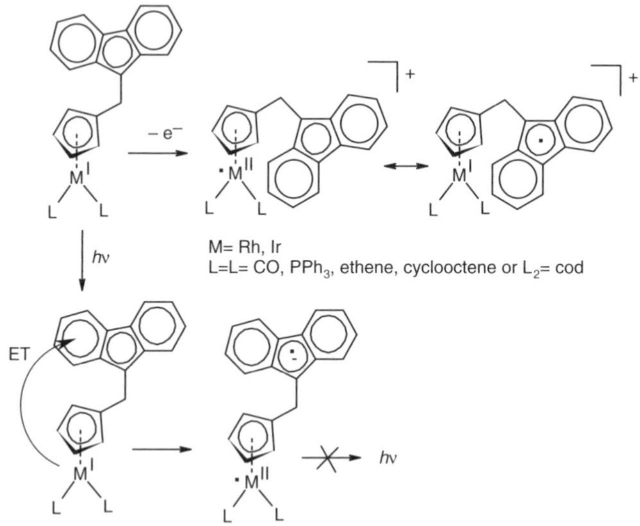

Figure 26. Redox chemistry of $[M^I(Cp–CH_2–antryl)(L)_2]$ complexes (M = Rh, Ir).

HFP or HFP–$CH_2Cl_2$ mixtures allowed EPR detection in solutions between $-10$ and $+10$°C. A detailed evaluation of the observed isotropic hyperfine couplings with the aid of spectral simulations and DFT calculations revealed that a strong electronic communication exists between the antryl moiety and the M(Cp) subunit. Oxidation causes a change in the position of the antryl unit with respect to M(Cp), orienting it in a sort of cisoid conformation, thus allowing a strong communication between the antryl, metal, and Cp units. The *d* electrons of the metal are in $\pi$-symmetry orbitals and in this way effectively overlap with $\pi$ orbitals of the antryl fragment. The DFT calculations confirm that the odd electron is strongly delocalized over the whole molecular skeleton (as illustrated by the resonance structures in Fig. 26). This explains that these species are somewhat stabilized compared to [Rh(Cp)(CO)(L)] species with nonsubstituted Cp rings. The probability of finding the unpaired electron at the metal center was found to correlate with the electronics of the ligands L, with the $\sigma$-donor $PPh_3$ promoting metalloradical behavior and $\pi$-acids CO and olefins promoting Cp–$CH_2$-antryl ligand radical character. The fate of the radical cations has not been investigated. An intramolecular photon initiated electron transfer from the (Cp)M unit to the antryl unit was proposed to explain the quenching of antryl-based fluorescence.

Oxidation of $[Rh^I(Cp)(cot)]$ (cot = 1,3,5,7-cyclooctatetraene also results in dimerization, but does not result in Cp–Cp coupling, neither in Rh–Rh bond formation. Instead intermolecular C–C coupling involving two cot ligands is observed (Fig. 27) (97).

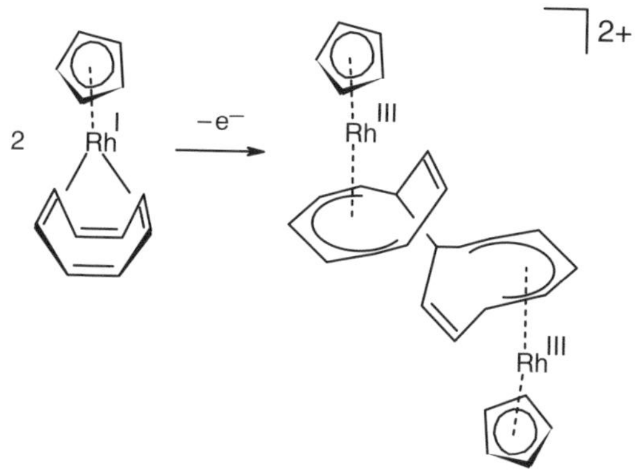

Figure 27. Oxidation of [Rh(Cp)(cot)] resulting in cot–cot coupling.

Quite remarkably, paramagnetic Cp*Pt radicals tend to escape from undesirable $d^7$ and $d^9$ electron configurations via a net loss or gain of hydrogen atoms, instead of the ligand–ligand or metal–metal coupling reactions observed for related Rh species (98).

Oxidation of the Cp*–diolefin complex $[Pt^{II}(Cp^*)(pentamethylcyclopentadiene)]^+$ results in formation of $[Pt^{IV}(Cp^*)_2]^{2+}$ via a net loss of a hydrogen atom from the proposed $[Pt^{III}(Cp^*)(pentamethylcyclopentadiene)]^+$ intermediate. Reduction of $[Pt^{IV}(Cp^*)_2]^{2+}$ restores the original $[Pt^{II}(Cp^*)(pentamethylcyclopenta\text{-}diene)]^+$ complex, which was proposed to proceed via hydrogen atom abstraction from the solvent by the $[Pt^{III}(Cp^*)_2]^+$ intermediate (Fig. 28).

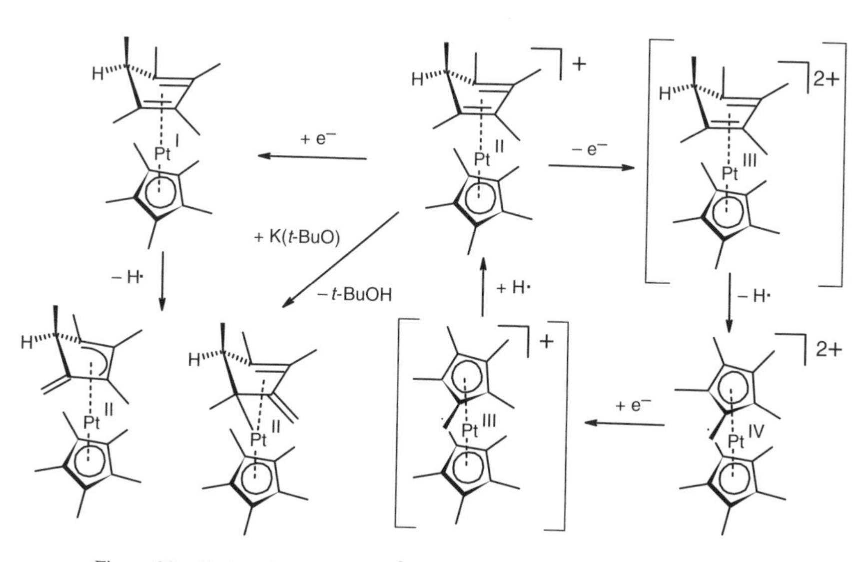


Figure 28. Redox chemistry of $[Pt^I(Cp^*)(penta\text{-}methylcyclopentene)]$ species.

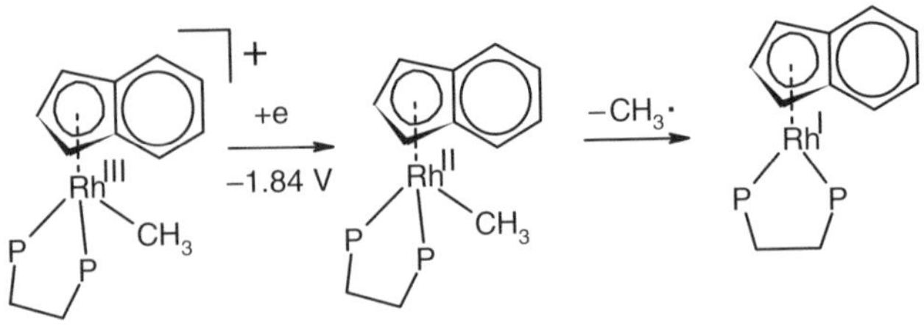

Figure 29. The Rh—C bond homolysis following reduction of a $[Rh^{III}Me(indenyl)(bis\text{-}phosphine)]$.

Reduction of $[Pt^{II}(Cp^*)(pentamethylcyclopentadiene)]^+$ also leads to a net loss of a hydrogen atom, but not from the ring carbon in this case, but from an allylic methyl group of the pentamethylcyclopentadiene moiety. This occurs at both possible positions, that is from the methyl group next to the CH fragment and its neigbouring methyl group. This leads to the $\eta^3$-allylic species and a $\sigma,\pi$-olefin complexes shown in Fig. 28. Deprotonation of $[Pt^{II}(Cp^*)(pentamethylcyclopentadiene)]^+$ with K($t$-BuO) yielded the same mixture of species.

Reduction of the indenyl complex $[Rh^{III}Me(indenyl)(Ph_2PCHMeCH_2PPh_2)]^+$ has been reported to result in formation of $[Rh^I(indenyl)(Ph_2PCHMeCH_2PPh_2)]$ via Rh—C bond homolysis from the Rh(II) intermediate (Fig. 29) (99).

Only a few stable paramagnetic cyclopentadienyl-type complexes of Rh and Ir are known. As mentioned above, $[Rh^{II}(Cp^*)(\eta^5\text{-}C_9H_7)]$ appears to be stable. All other stable complexes also contain electron-withdrawing substituents at the cyclopentadienyl ligands. The complex $[Rh^{II}(\eta^5\text{-}C_5Me(CO_2Me)_4)_2]$ has been reported, but a definitive proof of its identity was not provided (100). Definitive stable paramagnetic Rh(II) complexes containing the tetra- and penta-phenyl substituted cyclopentadienyl ligands $C_5HPh_4^-$ ($CpPh_4$) and $C_5Ph_5^-$ ($CpPh_5$) allowed unequivocal EPR characterization of cationic $[Rh^{II}(\eta^5\text{-}C_5Ph_5)(cod)]^+$ complexes (1,3 and 1,5 isomers) (101)] and even the X-ray structure determination of the neutral radical $[Rh^{II}(C_5HPh_4)_2]$ (102). These phenyl substituted Cp ligands allow efficient delocalization of the unpaired electron over the metal, the Cp ligand and its phenyl substituents. At the same time, the Ph substituents provide the required steric bulk to prevent dimerization reactions (see Fig. 30).

The $[Rh^{III}(C_5HPh_4)_2]$ complex reveals a reversible reduction wave in the cyclic voltammogram. One-electron reduction of this species yields the stable neutral radical species $[Rh^{II}(C_5HPh_4)_2]$. X-ray diffraction reveals the expected sandwich structure of a rhodocene derivative for $[Rh^{II}(C_5HPh_4)_2]$. The EPR data of this complex in a toluene glass at 77 K reveal a rhombic $g$ tensor ($g_x = 1.952$, $g_y = 2.030$, $g_z = 1.771$). Hyperfine coupling with the $^{103}Rh$ nucleus is resolved for the low-field signal ($A_y = 74$ MHz). The SOMO was interpreted to consist of a mixture of the ligand orbitals and the nearly degenerate $d_{xy}$ and $d_{yx}$ orbitals of the rhodium nucleus. Note that for this well-characterized species the $g$

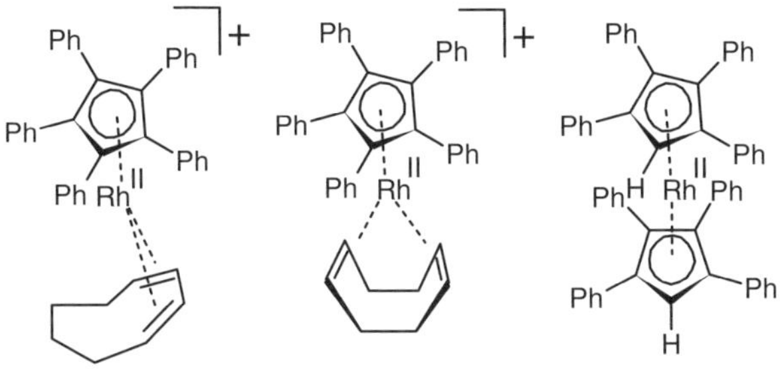

Figure 30. Persistent Rh(II) radicals from phenyl substituted cyclopentadienyl ligands.

anisotropy and absolute deviations from $g = g_e$ are much larger than reported for the nonsubstituted rhodocene $[Rh^{II}(Cp)_2]$ (79, 84). This is taken as an indication that the reported EPR spectra of $[Rh^{II}(Cp)_2]$, with $g$ values very close to $g = g_e$, do not likely belong to the sandwhich rhodocene species. We suspect, strengthened by some preliminary DFT EPR property calculations (Table VIII), that the reported signals for $[Rh^{II}(Cp)_2]$ actually belong to some kind of an organic $\pi$ radical (e.g., the Cp radical), possibly complicated by weak interactions with (diamagnetic) Rh or Ir species.

The $CpPh_5^-$ ligand also allows stabilization of paramagnetic half-sandwich complexes. Both the 1,3- and the 1,5-isomers of $[Rh^{II}(\eta^5\text{-}C_5Ph_5)(cod)]^+$ are persistent radicals at room temperature (Fig. 30) (101). Isomerisation of the 1,3-isomer into the 1,5-isomer does not occur, unless the species is oxidized beyond

TABLE VIII

Comparison of Experimental and DFT Calculated EPR Parameters of $[(Cp)_2Rh^{II}]$ and $[(C_5HPh_4)_2Rh^{II}]$

| Compound | | $g_{iso}$ | $g_1$ | $g_2$ | $g_3$ | $A_1^{(a)}$ | $A_2^{(a)}$ | $A_3^{(a)}$ | $\rho^{Rh(b)}$ | References |
|---|---|---|---|---|---|---|---|---|---|---|
| $[(Cp)_2Rh^{II}]$ | Exp | | 2.033 | 2.001 | 2.001 | nr | nr | nr | | 79, 84 |
| | DFT$^{(c)}$ | | *1.85* | *1.78* | *1.29* | *82* | *24* | *22* | *0.55* | 103 |
| $[(C_5HPh_4)_2Rh^{II}]$ | Exp | | 2.030 | 1.952 | 1.771 | 74 | nr | nr | | 102 |
| | DFT$^{(c)}$ | | *1.910* | *1.835* | *1.443* | *74* | *21* | *20* | *0.48* | 103 |
| Cp Radical | Exp | 2.004 | | | | | | | | 104, 105 |
| | DFT$^{(c)}$ | *2.003* | *2.003* | *2.003* | *2.002* | | | | | 103 |

$^a$The $^{103}$Rh hyperfine couplings in MHz; nr = not resolved.
$^b$The Mulliken spin populations at Rh calculated with ADF2000.02, for details see Ref. (106).
$^c$The EPR property calculations using ADF (BP86 functional, ZORA/V basis sets) using Turbomole [BP86 functional, SV(P) basis set] optimized geometries as input for ADF. Methods described in more detail in the supplementary material of Ref. (106).

the Rh(II) oxidation state (slow isomerization of the dication). As expected, the $[Rh^{II}(\eta^5\text{-}C_5Ph_5)(1,3\text{-cod})]^+$ radical gives rise to a rhombic EPR spectrum ($g_{11} = 2.329$, $g_{22} = 2.120$, $g_{33} = 2.003$). The 1,5-isomer $[Rh^{II}(\eta^5\text{-}C_5Ph_5)(\mathit{1,5}\text{-cod})]^+$ reveals an axial EPR spectrum ($g_{\parallel} = 2.181$, $g_{\perp} = 1.997$) indicative of an approximate $C_{2v}$ symmetry. Neither of the species reveals resolved hyperfine couplings with the $^{103}Rh$ nucleus, preventing assignments of the orbital composition of the SOMO. Remarkably, the observed $g$-anisotropies of these complexes is smaller than their cobalt analogues. Given larger spin–orbit constants for the heavier Rh nucleus, the opposite is expected. The authors interpret this phenomenon as an indication for less metal character to the SOMO, which in turn is the result of increased metal–ligand covalency on going from cobalt to rhodium.

Cyclic voltammetry measurements revealed that the 1,3-isomer $[Rh^I(\eta^5\text{-}C_5Ph_5)(1,3\text{-cod})]$ ($E_{1/2} = -0.01$ V vs. $Fc/Fc^+$) is easier to oxidize than the 1,5-isomer $[Rh^I(\eta^5\text{-}C_5Ph_5)(1,5\text{-cod})]$ ($E_{1/2} = 0.09$ V vs. $Fc/Fc^+$), presumably because the Rh(I) complex of 1,3-isomer is less stable than the 1,5-isomer and the oxidized Rh(II) complex profits from an additional agostic interaction between Rh(II) and an allylic hydrogen atom (a geometry not accesible for the 1,5-isomer).

The CO ligand substitutions at diamagnetic $[Rh^I(\eta^5\text{-}C_5Ph_5)(CO)_2]$ complexes do not easily proceed under thermal conditions. In the presence of ferrocenium, oxidatively induced substitutions proceed much more rapidly. Formation of $[Rh^I(\eta^5\text{-}C_5Ph_5)(CO)(P(OPh)_3)]$ from $[Rh^I(\eta^5\text{-}C_5Ph_5)(CO)_2]$ and $P(OPh)_3$ is even reported to be catalytic in the oxidant (107). In the same paper, the authors report the characterization of paramagnetic $[Rh^{II}(\eta^5\text{-}C_5Ph_5)(CO)(L)]^+$ species [$L = PPh_3$, $P(OPh)_3$, $AsPh_3$] with IR and EPR spectroscopy in $CH_2Cl_2$ solutions at room temperature (RT). The succesfull characterization of these $C_5Ph_5$-species further prompted the authors to study the formation and EPR characterization of the instable Cp* analogue $[Rh^{II}(Cp^*)CO)(PPh_3)]^+$ at low temperatures (Table IX).

TABLE IX
Spectroscopic Data of $[Rh^{II}(\eta^5\text{-}C_5Ph_5)(CO)(L)]^+$ [$L = PPh_3$, $P(OPh)_3$, $AsPh_3$] and $[Rh^{II}(Cp^*)CO)(PPh_3)]^+$ Species

| Compound | ν (CO) ($cm^{-1}$) | $g$ (−196°C) | $g_{iso}$ (RT) | $A_{iso}$ (RT) (MHz) |
|---|---|---|---|---|
| $[Rh^{II}(C_5Ph_5)(CO)(PPh_3)]^+$ | 2041 | $g_{\parallel} = 2.142$<br>$g_{\perp} = 2.019$ | 2.059 | 43 ($^{31}P$) |
| $[Rh^{II}(C_5Ph_5)(CO)(AsPh_3)]^+$ | 2039 | | 2.060 | 49 ($^{75}As$) |
| $[Rh^{II}(C_5Ph_5)(CO)(P(OPh)_3)]^+$ | 2060 | | 2.049 | 49 ($^{31}P$) |
| $[Rh^{II}(Cp^*)(CO)(PPh_3)]^+$ | | $g_{\parallel} = 2.115$<br>$g_{\perp} = 2.025$ | | |

## B. Paramagnetic Intermediates in Oxidatively Induced Reductive Eliminations

One-electron oxidation of stable closed-shell dialkyl-metal and hydrido-alkyl-metal species frequently results in reductive elimination reactions. These reactions are reported to proceed via a homolytic pathway (1), involving homolytic bond splitting of a M—C bond, thus generating alkyl radicals and closed-shell metal complexes. In addition, heterolytic pathways (2) have been reported, involving formation of closed-shell organic fragments via concerted reductive eliminations generating C—C or C—H bonds, thus leaving the metal in an open-shell state (Fig. 31). For Pd and Pt, the homolytic pathway seems to prevail, although bimolecular alkyl radical transfer has also been observed. The oxidized Rh and Ir species mainly decompose via concerted reductive eliminations, but also other decomposition pathways have been observed.

### *1. Rhodium and Iridium Systems*

In 1993, Tilset described the intriguing redox behavior of $[Rh^{III}(Cp^*)Me_2(PPh_3)]$ (108). Voltammetric experiments revealed that this compound is oxidized at a rather low redox potential (0.04 V vs. $Fc/Fc^+$) in $MeCN/CH_2Cl_2$ mixtures (Table X). At low scan rates, cyclic voltammetry revealed an irreversible $2e^-$ oxidation process, whereas at higher scan rates a quasi reversible $1e^-$ oxidation was observed. The reaction yielded mainly $[Rh^{III}(Cp^*)(PPh_3)(NCMe)_2]^{2+}$. Chemical oxidation of $[Rh^{III}(Cp^*)(PPh_3)Me_2]$ with $Fc^+$ in MeCN yielded the same product. The overall $2e^-$ oxidation was accompanied by formation of ethane, thus revealing C—C bond formation via a net oxidatively induced reductive elimination. The oxidation of a mixture of $[Rh^{III}(Cp^*)Me_2(PPh_3)]$ and $[Rh^{III}(Cp^*)(CD_3)_2(PPh_3)]$ yielded MeMe and $CD_3CD_3$, with only minor (<10%) formation of the $CH_3CD_3$ cross-over

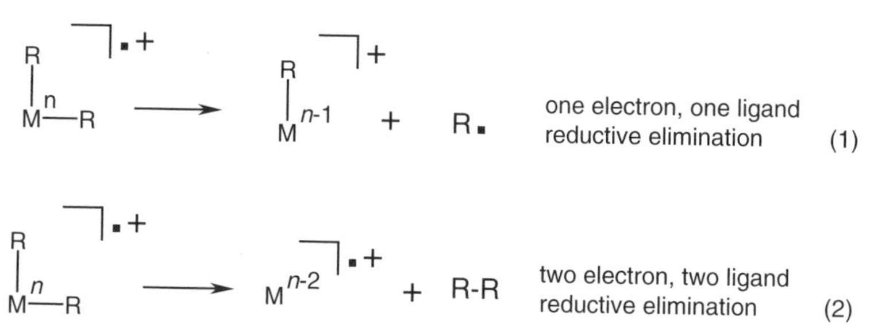


Figure 31. Homolytic and concerted pathways in oxidatively induced reductive elimination reactions via open-shell intermediates.

TABLE X
Selected M(III)/M(IV) Redox Couples Relevant to Formation of (Transient) Paramagnetic M(IV) Species (M = Rh, Ir)[a]

| Compound | $E_{1/2}$ | $E_{peak}$ (irrev.) | Solvent | References |
|---|---|---|---|---|
| $[Rh^{III}(\eta^5\text{-}C_9H_7)(Me)(Ph_2PCH(Me)CH_2PPh_2)]^+$/ $[Rh^{IV}(\eta^5\text{-}C_9H_7)(Me)(Ph_2PCH(Me)CH_2PPh_2)]^{2+}$ | | −1.845 | MeCN | 99 |
| $[Rh^{III}(Me_3)(tacn)]/[Rh^{III}(Me_3)(tacn)]^+$ | −0.15[b] | | MeCN | 109 |
| $[Rh^{III}(Cp^*)(Me_3)(PPh_3)]/[Rh^{IV}(Cp^*)(Me_3)(PPh_3)]^+$ | 0.04[b] | | MeCN/ $CH_2Cl_2$(1:1) | 108 |
| $[Ir^{III}(Cp^*)(Me_2)(PPh_3)]/[Ir^{IV}(Cp^*)(Me_2)(PPh_3)]^+$ | 0.04[b] | | MeCN | 110, 111 |
| $[Ir^{III}(Cp^*)(Me)(H)(PPh_3)]/[Ir^{IV}(Cp^*)(Me)(H)(PPh_3)]^+$ | −0.01[b] | | MeCN | 110 |
| $[Ir^{III}(Cp^*)(Me_2)(PMe_3)]/[Ir^{IV}(Cp^*)(Me_2)(PMe_3)]^+$ | | −0.13 | $CH_2Cl_2$ | 111 |
| $[Ir^{III}(Cp^*)(Me_2)(dmso)]/[Ir^{IV}(Cp^*)(Me_2)(dmso)]^+$ | +0.18[b] | | DMSO | 112 |

[a] Redox potentials versus $Fc/Fc^+$.
[b] Quasireversible at high cyclic voltammetry scan rates.

product. This indicated that the reductive elimination of ethane took place intramolecularly. Kinetic measurements (derivative cyclic voltammetry) revealed that the actual decomposition involved a chemical follow-up reaction from the $1e^-$ oxidized, formally Rh(IV), radical intermediate $[Rh^{IV}(Cp^*)Me_2(PPh_3)]^{\bullet+}$. Obtained kinetic parameters were indicative of a *first-order* process with minor solvent–electrolyte effects on the decomposition rate [$k = 96\,s^{-1}$, activation parameters $\Delta H = 51\,kJ\,mol^{-1}$, $\Delta S^\ddagger = -26\,J.K^{-1}.mol^{-1}$]. These data were considered indicative for a direct concerted reductive elimination of ethane from the 17 VE species $[Rh^{IV}(Cp^*)Me_2(PPh_3)]^{\bullet+}$ to give the 15 VE species $[Rh^{II}(Cp^*)(PPh_3)]^{\bullet+}$. The latter is further oxidized to yield $[Rh^{III}(Cp^*)(PPh_3)(NCMe)_2]^{2+}$ (see Fig. 32). A stepwise mechanism, involving elimination of a

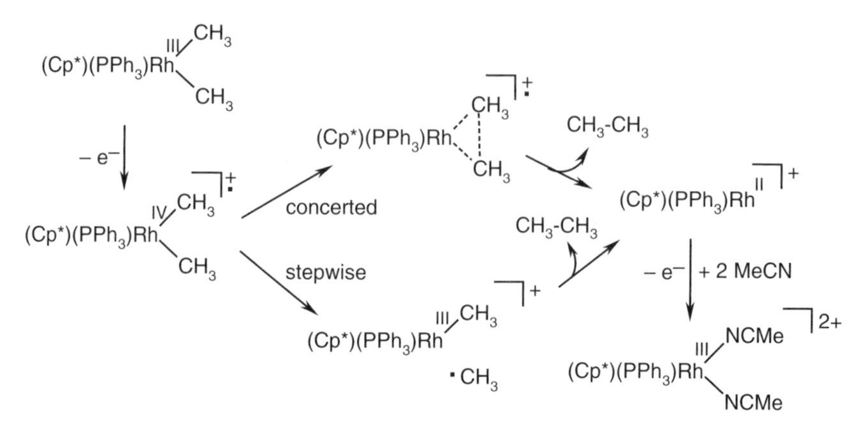


Figure 32. Reductive elimination of ethane triggered by oxidation of $[(Cp^*)Rh^{III}(PPh_3)Me_2]$ in MeCN.

methyl radical (i.e., homolytic M—C bond splitting, as described below for $[Pt(R)_2(L)_2]$ and $[Pd(R)_2(L)_2]$), which subsequently abstracts a methyl group from $[Rh^{III}(Cp^*)Me(PPh_3)]^+$ within the solvent cage was considered unlikely. In that case, substantial formation of methane by hydrogen abstraction from the solvent should have been observed. Furthermore, the negative activation entropy indicated an ordered transition state, consistent with the proposed concerted mechanism.

In $CH_2Cl_2$, besides ethane also some methane was observed (~ 6:1 ratio). It was suggested that in this solvent a minor part of the reaction could thus proceed via the homolysis pathway to give the methyl radical. Nevertheless, the kinetic data suggested that also in this solvent the most important reaction is still a direct concerted reductive elimination of ethane. The formation of mainly methyl-containing species $[Rh^{III}(Cp^*)Me(PPh_3)(X)]$ (the unknown X can be replaced by MeCN) after intramolecular elimination of ethane from $[Rh^{IV}(Cp^*)Me_2(PPh_3)]^{\bullet+}$ in dichloromethane is remarkable. This was explained by a methyl group transfer from $[Rh^{III}(Cp^*)Me_2(PPh_3)]$ or $[Rh^{IV}(Cp^*)Me_2(PPh_3)]^{\bullet+}$ to the reductive elimination product $[Rh^{II}(Cp^*)(PPh_3)]^{\bullet+}$, (see Fig. 33).

Oxidation of $[Rh^{III}(Cp^*)Me_2(dmso)]$ in DMSO (dimethyl sulfoxide) also resulted in reductive elimination of ethane from the $1e^-$ oxidized species

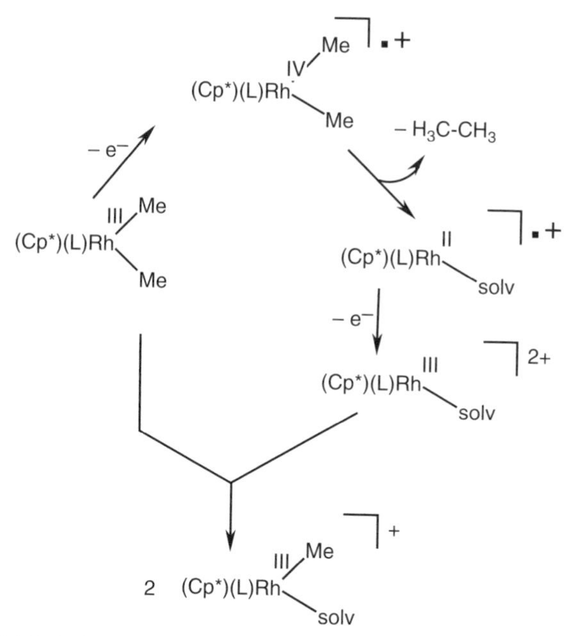


Figure 33. Oxidatively induced reductive elimination of ethane from $[Rh^{IV}(Cp^*)Me_2(L)]$ species followed by a comproportionation reaction between $[Rh^{III}(Cp^*)Me_2(L)]$ and $[Rh^{III}(Cp^*)(solv)(L)]$ to form $[Rh(Cp^*)Me(L)(solv)]$.

$[Rh^{IV}(Cp^*)Me_2(PPh_3)]^{\bullet+}$. The reaction yielded exclusively ethane and $[Rh^{III}(Cp^*)Me(dmso)_2]^+$. Formation of the latter was explained by a comproportionation reaction between $[Rh^{III}(Cp^*)Me_2(dmso)]$ and $[Rh^{III}(Cp^*)(dmso)_2]^{2+}$ (Fig. 33) (112).

Quite similar results were obtained upon oxidation of the triazacyclononane complex $[Rh^{III}Me_2(tacn)]$ (109). This led to almost exclusive formation of ethane (only ~1% $CH_4$). The organometallic products formed in this reaction proved to be $[Rh^{III}Me_2(tacn)(NCMe)]^+$ (tacn = 1,47-trimethyl-1,47-triazocyclononane) and $[Rh^{III}Me(tacn)(NCMe)_2]^{2+}$ in a constant ratio of 3:1, irrespective of the reaction conditions. The kinetics of the reaction ($k = 6\,s^{-1}$, activation parameters $\Delta H = 57\,kJ\,mol^{-1}$ and $\Delta S = -35\,J\,(K\,mol^{-1})$, negligible solvent effects) are consistent with a concerted intramolecular reductive elimination of ethane. To explain the constant 3:1 ratio of the organometallic products and the observed methyl scrambling between *unconsumed* $[Rh^{III}Me_2(tacn)]$ and $[Rh^{III}(CD_3)_3(tacn)]$ under oxidative conditions (no scrambling in absence of an oxidant), the authors proposed formation of a nonsymmetrical binuclear intermediate from the oxidized species $[Rh^{III}(CH_3)_3(tacn)]^{\bullet+}$ and its primary reductive elimination product $[Rh^{III}Me(tacn)(NCMe)]^{\bullet+}$. Attack of MeCN to to either of the two nonequivalent Rh sites then yields the observed products (Fig. 34).

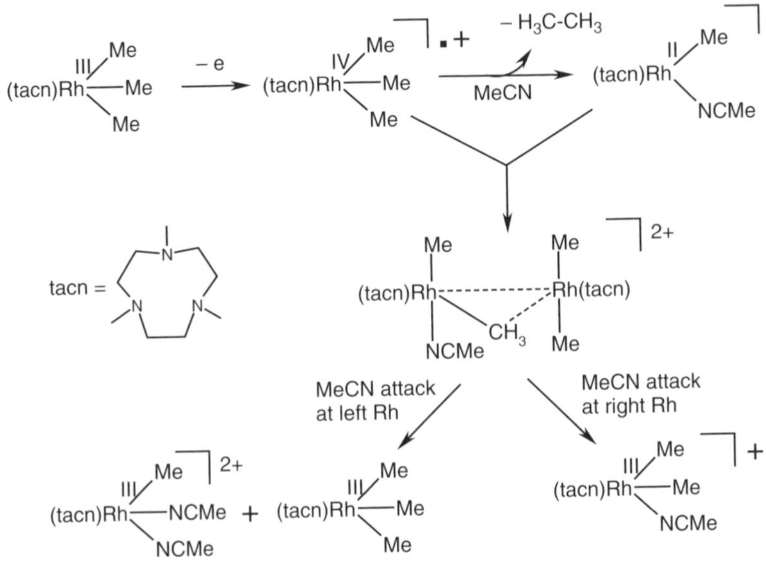

Figure 34. Binuclear reactions between $[RhMe_3(tacn)]^{\bullet+}$ and its reductive elimination product $[RhMe(NCMe)(tacn)]^{\bullet+}$ following oxidation of $[RhMe_3(tacn)]$.

The same authors also reported the redox chemistry of the analogous iridium compounds $[Ir^{III}(Cp^*)Me_2(PPh_3)]$ and $[Ir^{III}(Cp^*)(H)Me(PPh_3)]$ (110). For $[Ir^{III}(Cp^*)(H)Me(PPh_3)]$, oxidation resulted in a rapid reductive elimination of methane. The authors suggested that the overall $2e^-$ oxidation proceeds via a direct concerted reductive elimination of methane from the 17 VE species $[Ir^{IV}(Cp^*)(H)Me(PPh_3)]^{\bullet+}$ to yield $[Ir^{II}(Cp^*)(PPh_3)]^{\bullet+}$, which is then further oxidized in MeCN to yield $[Ir^{III}(Cp^*)(PPh_3)(NCMe)_2]^{2+}$. This behavior is similar to the reductive elimination of ethane from $[Rh^{IV}(Cp^*)Me_2(PPh_3)]^{\bullet+}$. Interestingly, the exact Ir analogue $[Ir^{III}(Cp^*)Me_2(PPh_3)]$ behaves differently. Both the chemical outcome of the reaction and the mechanistic details are remarkably different. Oxidation of $[Ir^{III}(Cp^*)Me_2(PPh_3)]$ is chemically reversible on the cyclic voltammetry time scale, and the cation radical $[Ir^{IV}(Cp^*)Me_2(PPh_3)]^{\bullet+}$ proved quite persistent at $-20°C$. Chemical oxidation of $[Ir^{III}(Cp^*)Me_2(PPh_3)]$ at RT with only 1 equiv of a ferrocenium derivative resulted in a slow reaction. The reaction mixture revealed mainly methane, and $[Ir^{III}(Cp^*)Me(PPh_3)(NCMe)]^+$ as the dominant organometallic product. Interestingly, the oxidation of $[Ir^{III}(Cp^*)Me_2(PPh_3)]$ did not produce $[Ir^{III}(Cp^*)(PPh_3)(NCMe)_2]^{2+}$, which on the basis of the chemistry of the rhodium analogue $[Rh^{III}(Cp^*)Me_2(PPh_3)]$ would be the expected product from reductive elimination of ethane from $[Ir^{IV}(Cp^*)Me_2(PPh_3)]^{\bullet+}$. Labeling experiments revealed that D had been incorporated into the Cp* methyl groups of the product. No evidence for D incorporation into the Ir—Me groups or into the phenyl rings of $PPh_3$ was found. The major reaction producing methane was thus suggested to involve Cp* C—H activation at some stage, but the exact mechanism remained unclear.

Around the same time Diversi and co-workers reported similar studies concerning the oxidation of $[Ir^{III}(Cp^*)Me_2(PR_3)]$ (R = Me, Ph) (111). In contrast to the studies of Tilset and co-workers, they used nonpolar and poorly coordinating aromatic solvents. Under these circumstances, catalytic amounts (4–6 mol %) of an oxidant (ferrocenium or $Ag^+$) proved sufficient to induce rapid and quantitative methane elimination from $[Ir^{III}(Cp^*)Me_2(PR_3)]$ at room temperature. Rapid follow-up reactions with C—H bonds of various aromatic solvents (Ar—H) then yielded $[Ir^{III}(Cp^*)Me(Ar)(PR_3)]$. Similar reactions in absence of an oxidant required severe conditions (110°C) for prolonged times (2 weeks), thus demonstrating efficient electron-transfer catalysis in the activation of aromatic C—H bonds (Fig. 35).

A detailed analysis (113) demonstrated that oxidation of $[Ir^{III}(Cp^*)$-$Me_2(PPh_3)]$ initially yields the orthometalated product $[Ir^{III}(Cp^*)Me(\kappa^2\text{-}P,C\text{-}\{-C_5H_4(Ph)_2P-\})]$, which could be prepared as a pure compound in $CH_2Cl_2$. In the presence of aromatic solvents and a catalytic amount of an oxidant, the latter converts to $[Ir^{III}(Cp^*)Me(Ar)(PPh_3)]$. The outcome of the reactions is independent of the use of either $Ag^+$ or $Fc^+$ as the chemical redox catalyst. A small, but

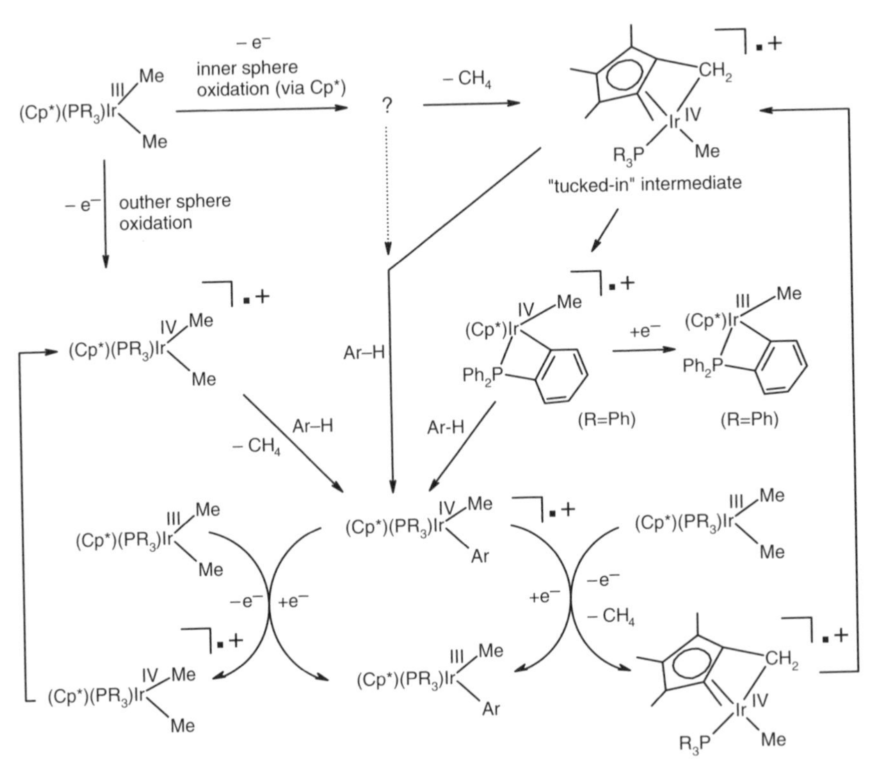


Figure 35. Electron-transfer catalysis in aromatic C–H activation by iridium complexes.

significant kinetic isotope effect for activation of benzene ($k_H/k_D \sim 2.2$), was observed under electron-transfer conditions.

Electrochemical oxidation of $[Ir^{III}(Cp^*)Me_2(PR_3)]$ revealed a different intermediate than observed upon chemical oxidation with $Ag^+$. The electrochemical oxidation reveals rhombic EPR spectra with $g$ values characteristic for a metal-centered radical (Table XI). This species does not convert to orthometalated $[Ir^{III}(Cp^*)Me(\kappa^2\text{-}P,C\text{-}\{-C_5H_4(Ph)_2P-\})]$ as was observed in the chemical oxidation. On the other hand, chemical oxidation with $Ag^+$ initially yields metal-centered radicals of a different type (for EPR data see Table XI). Upon raising the temperature >253 K, vigorous evolution of methane and the rise of new (isotropic) EPR spectra were observed. These spectra were characteristic for mainly (Cp*-type) ligand centered radicals. Spectral simulation revealed hyperfine coupling with iridium ($A_{Ir} = 2.7\,G \sim 7.6\,MHz$), two sets of methyl groups ($A_H = 14.6\,G \sim 41\,MHz$ and $A_H = 2.7\,G \sim 7.6\,MHz$), and a methylene group ($A_H = 1.35\,G \sim 3.8\,MHz$). The $g$ values were not reported, but must be close to $g_e$. The spectra were assigned to a follow-up intermediate proposed to

TABLE XI
The EPR Data Reported for Cationic $[Ir^{IV}(Cp^*)(R')(R'')(PR_3)]^{\bullet+}$ Intermediates Obtained by Electrochemical and Chemical Oxidation of $[Ir^{IV}(Cp^*)(R')_2(PR_3)]^a$

| $PR_3$ | R', R" | Oxidant–/ Reductant | $g_1$ | $g_2$ | $g_3$ | $A_1$ | $A_2$ | $A_3$ | Reference |
|---|---|---|---|---|---|---|---|---|---|
| $PPh_3$ | Me, Me | Electrode | 2.387 | 2.315 | 1.846 | nr | nr | nr | 113 |
| $PPh_3$ | Me, Me | $Ag^{+a}$ | 2.172 | 2.172 | 1.812 | nr | nr | nr | 113 |
| $PPh_3$ | Me, Me | $NO^+$ | 2.067 | 2.033 | 1.879 | 67.9 | 87.6 | 68.6 | 114 |
| $PPh_2Me$ | Me, Me | $NO^+$ | 2.011 | 1.975 | 1.830 | 53.7 | 84.9 | 61.5 | 114 |
| $PPhMe_2$ | Me, Me | $NO^+$ | 2.009 | 1.977 | 1.830 | 56.5 | 81.9 | 63.4 | 114 |
| $PMe_3$ | Me, Me | $NO^+$ | 2.013 | 1.979 | 1.831 | 50.9 | 84.3 | 57.5 | 114 |
| $PMe_3$ | Me, Me | $Ag^{+b}$ | (*c*) | (*c*) | (*c*) | nr | nr | nr | 113 |
| $PPh_3$ | $-CH_2SiMe_3$, $-CH_2SiMe_3$ | $Ag^+$ | 2.191 | 2.191 | 1.963 | nr | nr | nr | 113 |
| $PPh_3$ | $-CH_2SiMe_2CH_2-$ | $Ag^+$ | 1.964 | 1.964 | 2.210 | nr | nr | nr | 113 |

[a]Hyperfine couplings in megahertz.
[b]Converts to an organic radical 253 K.
[c]Two species. The EPR spectrum is shown in Ref. (113), but the *g* values were not reported.

have the "tucked-in" structure depicted in Fig. 35, consistent with the observed H/D scrambling at the Cp* methyl groups under these conditions.

On the basis of these observations, the electron-transfer catalysis in the activation of aromatic C–H bonds was proposed to proceed via two possible pathways. Outher-sphere oxidation, which yields $[Ir^{IV}(Cp^*)Me_2(PR_3)]^{\bullet+}$ and inner-sphere oxidation via the Cp* ligand yielding another type of iridium(IV) radical cation that easily converts to a so-called tucked-in intermediate (Fig. 35). Both intermediates were claimed to be capable of subsequent activation of aromatic C–H bonds. Apparently, the redox and chemical noninnocence of the Cp* ligand plays a crucial role in at least part of the observations.

The $[Ir^{III}(Cp^*)Me_2(PR_3)]$ compounds later also proved to catalyze the coupling of $Me_2PhSiH$ to $Me_2PhSiSiMe_2Ph$ in the presence of $1e^-$ oxidants, via very similar mechanisms (115).

To shed some light on the role of coordinating reagents on the different outcome upon oxidation of $[Ir^{III}(Cp^*)Me_2(PR_3)]$, the Diversi group studied the oxidation of $[Ir^{III}(Cp^*)Me_2(PR_3)]$ with ferrocenium salts in the presence of pyridine (116). Under these conditions, methane and $[Ir^{III}(Cp^*)Me(PR_3)(Py)]$ are produced and the reaction requires one equivalent of the oxidant, in good agreement with the reactions in MeCN reported by Tilset and co-workers (110). Again labeling experiments indicated that hydrogen abstraction from a Cp* methyl group provides the hydrogen required to eliminate methane. The tucked-in intermediate then abstracts a H/D atom from the solvent to restore the $\eta^5$-$C_5Me_5$ ligand (Fig. 36).

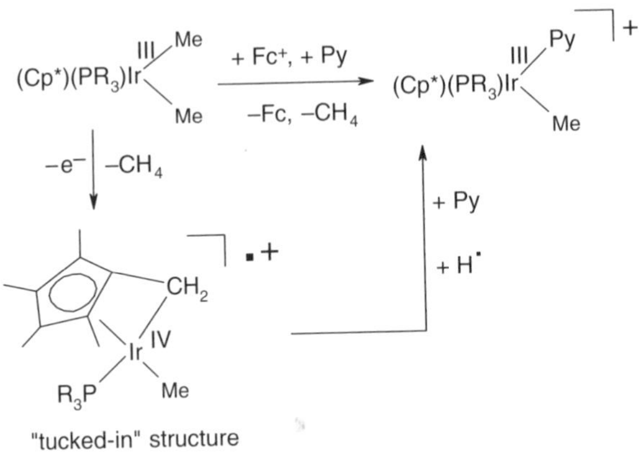

Figure 36. Oxidation of $[Ir^{III}(Cp^*)Me_2(PR_3)]$ in the presence of pyridine.

Studies by the Tilset group, in which they replaced $L = PR_3$ by L = dmso, proved that both the outcome and the mechanistic details of oxidatively induced reductive elimination reactions at $[Ir^{III}(Cp^*)Me_2(L)]$ are remarkably sensitive to the nature of the ligand L (112). Oxidation of $[Ir^{III}(Cp^*)Me_2(dmso)]$ yields mainly methane, but via a different mechanism as found for $[Ir^{III}(Cp^*)$-$Me_2(PR_3)]$ complexes. For $[Ir^{III}(Cp^*)Me_2(PR_3)]$ the Cp* methyl groups proved to be source of hydrogen atoms required to eliminate $CH_4$ from Ir–Me fragments (Fig. 36). However, for $[Ir^{III}(Cp^*)Me_2(dmso)]$ H/D scrambling at the Cp* methyl groups was not observed at all. In this case, oxidation in the presence of $D_2O$ lead to clear-cut formation of MeD, thus indicating that water can also be a source of hydrogen atoms. Since $[Ir^{III}(Cp^*)Me_2(dmso)]$ did not directly react with $D_2O$, a protonolysis of (a solvent adduct of) the $[Ir^{IV}(Cp^*)Me_2(dmso)]^{\bullet+}$ intermediate was proposed.

Besides methane, also some ethene ($CH_2{=}CH_2$) was formed upon $1e^-$ oxidation. At voltammetric conditions, where the local concentration of electrode generated $[Ir^{IV}(Cp^*)Me_2(dmso)]^{\bullet+}$ is high, formation of ethene occurs in much higher yields. Under these conditions, kinetic measurements (derivative cyclic voltammetry), revealed a second-order reaction in [Ir]. Most likely, this involves a reaction between two $[Ir^{IV}(Cp^*)Me_2(dmso)]^{\bullet+}$ species generated at the electrode [$k = 3 \times 10^4\,M^{-1}\,s^{-1}$, activation parameters $\Delta H^{=} + 10\,kJ\,mol^{-1}$ and $\Delta S = -190\,J.K^{-1}.mol^{-1}$]. Furthermore, a kinetic isotope effect of $k_H/k_D \sim 2$ was observed upon comparing the rates of $[Ir^{III}(Cp^*)Me_3(dmso)]$ and $[Ir(Cp^*)(CD_3)_2(dmso)]$. These data have led the authors to propose a hydrogen atom-transfer reaction from an Ir– fragment of $[Ir^{IV}(Cp^*)Me_2(dmso)]^{\bullet+}$ to the iridium center of another $[Ir^{IV}(Cp^*)Me_2(dmso)]^{\bullet+}$ species.

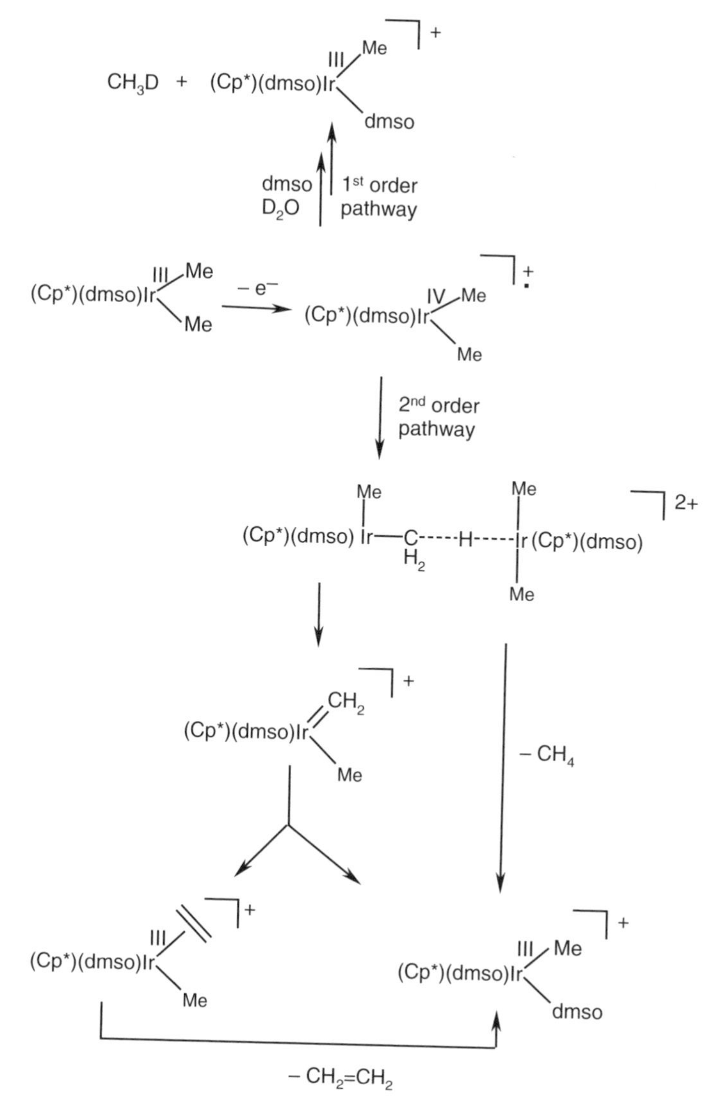


Figure 37. Oxidation of $[Ir^{III}(Cp^{*})Me_2(dmso)]$ in dmso revealing: (*a*) water to be the source of hydrogen atoms for methane elimination, and (*b*) a second-order pathway for formation of $CH_2{=}CH_2$ (and $CH_4$) involving hydrogen abstraction from Ir–Me by another Ir.

Carbene coupling involving such obtained iridium–carbene species thus explains the observed formation of ethene (see Fig. 37), in analogy with ethene formation upon oxidation of $[Re^{II}(Cp)Me(NO)(PPh_3)]$ (117).

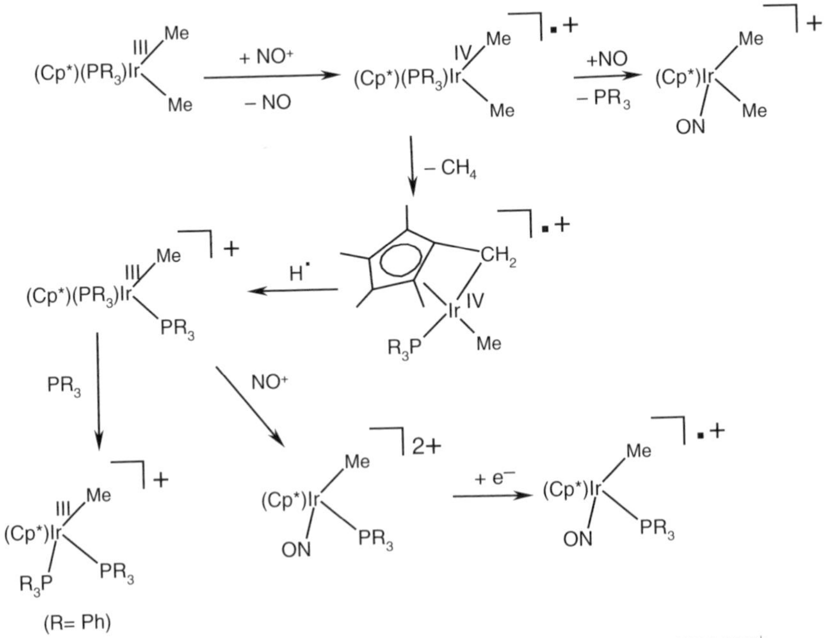

Figure 38. Redox chemistry of $[Ir^{III}(Cp^*)Me_2(PR_3)]$ in the precence of $NO/NO^+$.

Chemical oxidation of $[Ir^{III}(Cp^*)Me_2(PR_3)]$ species with the coordinating oxidant $NO^+$ was also investigated (Fig. 38). Depending on the nature of the $PR_3$ ligand, this either led to displacement of the phosphine or reductive elimination of methane. Again formation of a tucked-in intermediate that subsequently abstracts a hydrogen atom from the solvent was proposed (based on the observed H/D scrambling of the Cp* methyl groups). The resulting unsaturated $[Ir^{III}(Cp^*)Me(PR_3)]$ can either bind $NO^+$ or $PR_3$, again depending on the nature of the $PR_3$ ligand (114).

The use of $NO^+$ as an oxidant proved convenient to study the paramagnetic cationic intermediates with EPR. Good resolution EPR spectra were reported, which were assigned to monocationic $[Ir^{IV}(Cp^*)Me_2(PPh_3)]^{\bullet+}$ species. Quite remarkably, these spectra do not at all resemble those obtained upon oxidation with $Ag^+$, neither do they resemble the EPR spectra upon electrochemical oxidation (Table XI). Apparently, the use of different oxidants also leads to formation of different isomers of the cationic $[Ir^{IV}(Cp^*)Me_2(PPh_3)]^{\bullet+}$ radicals (or their solvent adducts?).

The obtained diamagnetic $[Ir(Cp^*)Me(PR_3)(NO)]^{2+}$ products also proved redox active, and could be reversibly reduced to the radicals $[Ir(Cp^*)Me(PR_3)(NO)]^{\bullet+}$ (Fig. 38). The EPR spectra of the persistent radicals are shown in Table XII, and revealed only iridium hyperfine coupling (no

TABLE XII
The EPR Data Reported for Cationic $[Ir(Cp^*)Me(PR_3)(NO)]^{\bullet+}$ Species Obtained by Chemical Reduction of Their Dicationic Precursors $[Ir(Cp^*)Me(PR_3)(NO)]^{2+a}$

| $PR_3$ | $g_1$ | $g_2$ | $g_3$ | $A_1$ | $A_2$ | $A_3$ | Reference |
|---|---|---|---|---|---|---|---|
| $PPh_2Me$ | 2.009 | 1.975 | 1.819 | 51.7 | 95.4 | 41.8 | 114 |
| $PPhMe_2$ | 2.014 | 1.978 | 1.814 | 44.6 | 93.3 | 65.1 | 114 |
| $PMe_3$ | 1.949 | 1.916 | 1.762 | 45.7 | 90.2 | 62.4 | 114 |

[a] Hyperfine couplings in megahertz.

hyperfine coupling to N of NO). Apparently, these species are predominantly metal-centered radicals, and should formally be regarded as Ir(II) species with an $NO^+$ ligand or Ir(IV) with an $NO^-$ ligand.

### 2. *Palladium and Platinum Systems*

One-electron oxidation of $[Pd^{II}(R)_2(dmpe)]$ (R = Me or $CH_2SiMe_3$) in MeCN results in homolytic cleavage of the Pd—C bond with formation of alkyl radicals and $[Pd^{II}(R)(dmpe)(NCMe)]^4$ (118). The radical intermediate $[Pd^{III}(R)_2(dmpe)]^{\bullet+}$ was not observed. The homolytic bond splitting of the $Pd^{III}$—R bond to yield Pd(II) and $R^{\bullet}$ radicals represents an oxidatively induced reductive elimination.

Two-electron oxidation of *cis*-$[Pt^{II}(R)_2(L)_2]$ (R = Me, Et; L = $PPh_3$, $PMe_2Ph$) with $IrCl_6^{2-}$ in MeCN results in formation of *cis*-$[Pt^{II}(R)(X)(L)_2]^{n+}$ and $[Pt^{IV}(R)_2(X)_2(L)_2]^{n+}$ species (X = $Cl^-$, MeCN) (119). The product distribution depends on the nature of both R and L. Formation of the Pt(II) species *cis*-$[Pt^{II}(R)(X)(L)_2]^{n+}$ is accompanied by formation of R—Cl and R—H. Trapping of alkyl radicals formed in this reaction with *t*-BuNO or $O_2$ is indicative for homolytic bond splitting of the Pt—C bond of the $1e^-$ oxidized species *cis*-$[Pt^{III}(R)_2(L)_2]^{\bullet+}$ (not detected). Alkyl radical elimination apparently competes with redox oxidation of *cis*-$[Pt^{III}(R)_2(L)_2]^{\bullet+}$, which accounts for the formation of $[Pt^{IV}(R)_2(X)_2(L)_2]^{n+}$.

Oxidation of the platinacyclobutanes $[Pt^{II}(\kappa^2\text{-}C,C\text{-}CH_2CH_2CH_2\text{—})(L)_2]$ (L = $PPh_3$, $(L)_2$ = bpy) did not result in elimination of organic products, but the fate of the oxidized Pt complex was not clear (120).

A number of reports describe electrochemically irreversible oxidations of other $[Pt^{II}(R)_2(L)_2]$ species (R = hydrocarbyl, $L_2$ = bidentate N-donor ligands (e.g., see Table IV). The irreversible oxidation waves have been suggested to be the result of a rapid attack on, or Pt—C homolysis of, the resulting Pt(III) ($d^7$) species (121). In most cases, the nature of the obtained products is not clear.

Oxidation of $[Pt^{IV}(Me)_4(\alpha\text{-diimine})]$ was investigated and compared to oxidation of $[Pt^{II}(Me)_2(\alpha\text{-diimine})]$ (Fig. 39) (122–124). The HOMO consists

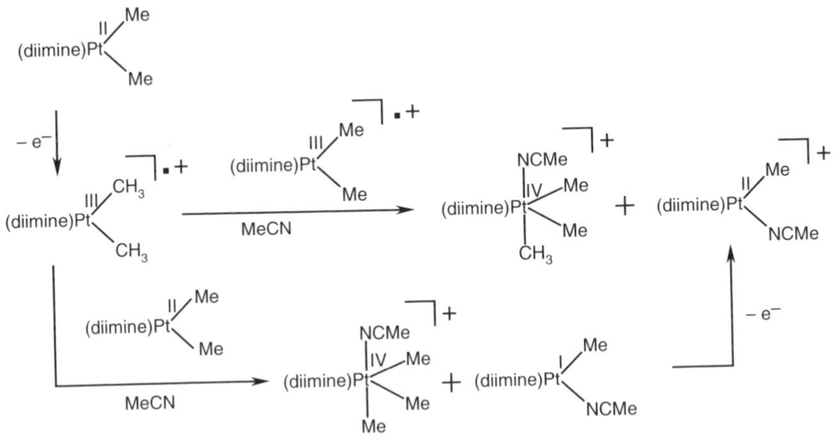

Figure 39. Bimolecular methyl transfer following oxidation of $[Pt^{II}Me_2(\alpha\text{-diimine})]$.

of a metal-centered *d* orbital in case of the $Pt^{II}$ complexes and for $Pt^{IV}$ it is a Pt—$C_{ax}$ σ orbital with some contribution of a platinum $p_x$ orbital (123–125). Both the $Pt^{II}Me_2$ and the $Pt^{IV}Me_4$ complexes show irreversible oxidation waves in their cyclic voltammograms at 0.4 V versus $Fc/Fc^+$. This irreversible behavior was assumed to be the result of the coordinatively unsaturated and axially nonprotected $Pt^{II}Me_2$ complexes, which likely react with nucleophiles. Based on their calculated HOMOs, the $PtMe_4$ species were proposed to undergo a Pt—$C_{ax}$ bond cleavage after oxidation. Isosbestic points in the UV–vis spectra indicate that the follow up reactions are selective. However, this was not further investigated (124).

Chemical and electrochemical oxidation of these $[Pt^{II}Me_2(\alpha\text{-diimine})]$ species in MeCN was later studied in more detail in the Tilset group (38). Irreversible electrochemical waves were again observed, and bulk electrolysis revealed consumption of $\sim 1.1\text{–}1.6\,F\,mol^{-1}$ indicative of a $1e^-$ oxidation process. (Electro)chemical bulk oxidation leads to only marginal formation of methane or ethane, and almost quantitative formation of the species $[Pt^{IV}Me_3(\alpha\text{-diimine})(NCMe)]^+$ and $[Pt^{II}Me(\alpha\text{-diimine})(NCMe)]^+$ in a 1:1 ratio. It was proposed that the short-lived $[Pt^{III}Me_2(\alpha\text{-diimine})]^{\bullet +}$ decomposes via a bimolecular methyl transfer from one platinum to another. This explains the product distribution and the lack of products derived from free alkyl radicals. This could either involve a reaction between two $[Pt^{III}Me_2(\alpha\text{-diimine})]^{\bullet +}$ intermediates or a reaction of $[Pt^{III}Me_2(\alpha\text{-diimine})]^{\bullet +}$ with the starting material $[Pt^{II}Me_2(\alpha\text{-diimine})]$.

The photoreactivity and photophysical properties of $[Pt^{IV}Me_4(\alpha\text{-diimine})]$ proved to be very different from those of $[Pt^{II}Me_2(\alpha\text{-diimine})]$. The $Pt^{II}Me_2$

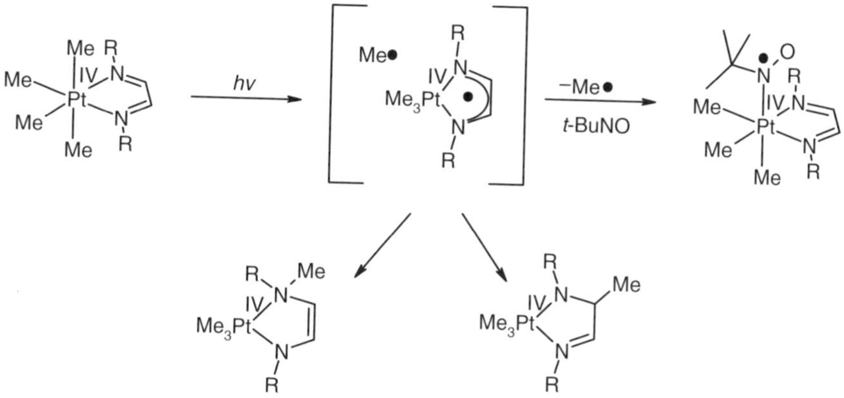

Figure 40. Photoreactivity of [$Pt^{IV}Me_4$(α-diimine)]. Irradiation of [$Pt^{IV}Me_4$(α-diimine)] results in Pt—C bond homolysis. Radical recombination yields alkylated ligands.

species are luminescent, whereas the $Pt^{IV}Me_4$ complexes proved to be photoreactive (Fig. 40) (123, 124).

Photolysis of [$Pt^{IV}Me_4$(α-diimine)] species was proposed to proceed via an intramolecular electron transfer from a filled M—$C_{ax}$ σ-bonding orbital to the empty π* level of the α-diimine ligand. The half-filled M—$C_{ax}$ σ-bonding orbital renders the axial Pt—C bond weak, resulting in a homolytic bond cleavage. Since only weak EPR signals centered ~ $g = 2.00$ were observed, the spin-trap reagent 2-methyl-2-nitrosopropane (*t*-Bu—NO) was used. At 240 K hyperfine coupling with a N atom (44 MHz) with $^{195}Pt$ satellites (112 MHz) was observed. The *g*-value of 2.0060 is clearly influenced by spin–orbit coupling (platinum influence). Both the *g* value and a $^{195}Pt$ coupling suggest that the spin trap has intercepted the organometallic product of the photopromoted homolytic Pt—C bond splitting reaction (123). Apparently, the spin trap is less effective to capture the methyl radical. The reactions in Fig. 40 actually represent a photopromoted intramolecular electron transfer, followed by an oxidatively induced reductive elimination.

Whereas all hydrocarbyl [$Pt^{II}(R)_2(L)_2$] species (R = hydrocarbyl, $L_2$ = bidentate N-donor ligands) reveal irreversible oxidation waves in cyclic voltammetry, mesityl [$Pt^{II}mes_2(L)_2$] complexes reveal reversible oxidation waves. In these cases, oxidation to stable, paramagnetic, but EPR silent [$Pt^{III}mes_2(L_2)]^+$ species is possible.

The oxidation occurs at relatively low potentials due to the electron-donating mesityl groups. Apparently, axial shielding of the coordinatively unsaturated metal center by the bulky mesityl groups stabilizes the rare mononuclear $Pt^{III}$ center toward dimerization or other aggregation, but more importantly also toward attack of nucleophilic solvent molecules or electrolyte ions at the very reactive electrophilic metal center of the paramagnetic cations $[Pt^{III}mes_2(L)]^+$ (34, 35). The absence of typical UV–vis bands supports the assignment of a metal-centered radical. The oxidation products $[Pt^{III}mes_2(L)]^+$ do not show any EPR signals, even upon cooling to 3 K, presumably due to fast relaxation processes.

Whereas one-electron oxidation of mononuclear mesityl–$Pt^{II}$ complexes has led to formation of a persistent paramagnetic Pt(III) species, electrochemical oxidation of the dinuclear $[Pt^{II}mes_2(bpym)Pt^{II}mes_2]$ proceeds via a $2e^-$ process yielding a diamagnetic $Pt^{III}$—$Pt^{III}$ species (125). In the case of the bptz and bpip ligands, a very small, but detectable, splitting between the $Pt^{II}$—$Pt^{II}$/$Pt^{II}$—$Pt^{III}$ and $Pt^{II}$—$Pt^{III}$/$Pt^{III}$—$Pt^{III}$ redox couples was observed with cyclic voltammetry (126) [bpym = 2,2'-bipyrimidine; bptz = 3,6-bis(2-pyridyl)-1,2,4,5-tetrazine; bpip = 2,5-bis(1-plenyliminoethyl)pyrazine]. In contrast to $d^5/d^6$ mixed-valent species, there is no stability of the $d^7/d^8$ mixed-valence state (127). Apparently, there is only little or no communication between both platinum centers via the ligand π system.

### C. Reactions of Paramagnetic Porphyrinato Radicals $[(por)M^{II}]$ (M = Rh, Ir, where por = porphyrin)

The reactivity of porphyrinato Rh(II) and Ir(II) complexes has been intensively investigated. Earlier work in this field (up to 1995) is concisely summarized in the review of DeWitt in a general way without specific emphasis on their organometallic chemistry (2). This chapter tries to give a detailed and complete overview of their metalloradical behavior in organometallic chemistry, including newer literature up to 2006.

Although $[M^{II}(por)]$ (M = Rh, Ir) species are obviously not organometallic in nature themselves, they reveal an interesting reactivity pattern toward a variety of substrates (olefins, $CO/H_2$, isocyanides, C—H bonds) with formation of M—C bonds.

Generally, $[M^{II}(por)]$ species (Fig. 41) are prepared by thermally or photochemically induced homolytic M—M, M—H, or M—C bond splitting of their corresponding M—M, $M^{III}$—Me or $M^{III}$—H precursors (2). Occasionally, (electro)chemical reduction or oxidation has been applied as well (128–130).

In all cases, the mononuclear square-planar $d^7$ complexes are very reactive. As a result, the least hindered complexes containing the porphyrins OEP (octaethylporphyrinato), TPP tetraphenylporphyrinato, TTP (tetratolylporphyrinato), and TXP [tetrakis(3,5-dimethylphenyl)porphyrinato] easily dimerize (Fig. 42) to form diamagnetic metal–metal bonded dinuclear species (in which a description as $M^I$—$M^{III}$ is indistinguishable from a antiferromagnetically coupled $M^{II}$—$M^{II}$ system). The Rh—Rh bond is, however, rather weak, and allows thermal homolytic Rh—Rh bond splitting to study the reactivity of the

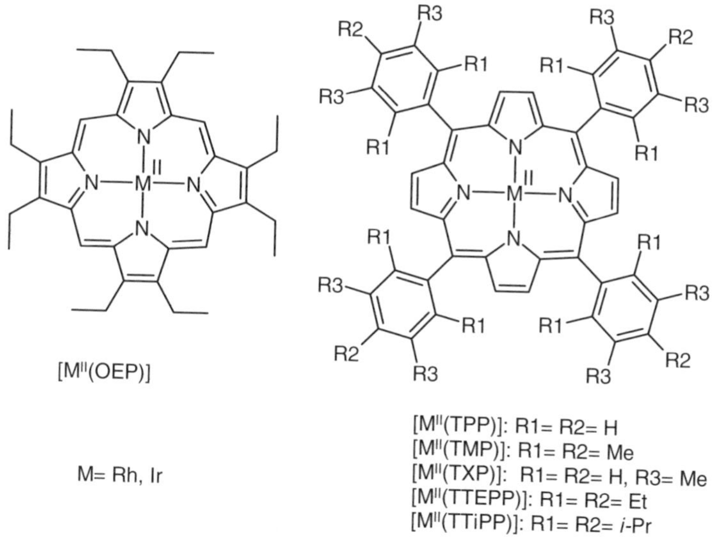

$[M^{II}(OEP)]$

M= Rh, Ir

$[M^{II}(TPP)]$: R1= R2= H
$[M^{II}(TMP)]$: R1= R2= Me
$[M^{II}(TXP)]$: R1= R2= H, R3= Me
$[M^{II}(TTEPP)]$: R1= R2= Et
$[M^{II}(TTiPP)]$: R1= R2= *i*-Pr

Figure 41. Ligand coding in Rh(II) and Ir(II) porphyrin complexes.

mononuclear species. To promote dissociation of the dinuclear Rh—Rh species in mononuclear square-planar $d^7$ species, porphyrin ligands have been supplied with bulky groups at their periphery. The steric bulk of the porphyrin dianion decreases in the order TTiPP > TTEPP > TMP > TXP > TTP > TPP > OEP (see Fig. 45). More bulky porphyrins result in weaker Rh—Rh bonds (Rh—Rh bond dissociation enthalpies: OEP: $16.5\,\text{kcal mol}^{-1}$; TXP: $\sim 12\,\text{kcal mol}^{-1}$; TMP: $\sim 0\text{kcal mol}^{-1}$) and consequently more reactive systems (131–133).

The complexes $[Rh^{II}(TMP)]$, $[Rh^{II}(TTEPP)]$, $[Rh^{II}(TTiPP)]$, $[Ir^{II}(TTEPP)]$, and $[Ir^{II}(TTiPP)]$ are sufficiently bulky to completely prevent dimerization (TMP = tetramesitylporphyrinato; TTEPP = tetra(2,4,6-triethylphenyl)porphyrinato TTiPP = tetra(2,4,6-triisopropyl- phenyl)porphyrinato). The EPR spectroscopy, $^1$H NMR paramagnetic shifts and line-broadening studies have proven useful to study the structure and reactivity of these paramagnetic complexes (and their adducts with ethene and CO, see below) (133–135). The EPR parameters of the rhodium complexes $[Rh^{II}(TMP)]$ and $[Rh^{II}(TTiPP)]$ are

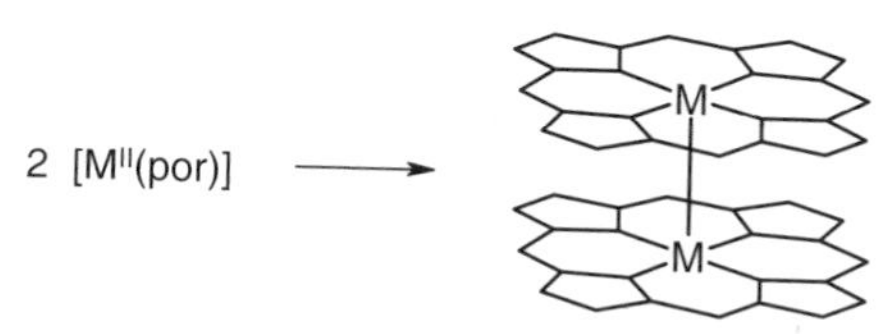


Figure 42. Formation of dinuclear M—M bonded species from $[M^{II}(por)]$ species (M = Rh, Ir).

consistent with a prevalent $d_{z^2}$ SOMO associated with a $(d_{xy})^2(d_{xz,yz})^4(d_{z^2})^1$ ground state. Remarkably, the iridium complex [$Ir^{II}$(TTiPP)] does not reveal an observable EPR spectrum, which was explained by the presence of one or more excited states with energies close to the ground state, causing rapid electron spin relaxation (33). This was further confirmed by the thermal behavior of the pyrrole chemical shifts. While a linear dependence of the paramagnetic shift of [$Rh^{II}$(TTiPP)] against 1/*T* (inverse of the temperature) indicates a simple Curie paramagnetic behavior associated with a single contributing state $[d_{z^2})^1]$, the curvature of the plot for [$Ir^{II}$(TTiPP)] suggests that several states are thermally populated. The upfield chemical shifts of the pyrrole hydrogens ($\delta_{pyr} = -20.9$ ppm) compared to the downfield chemical shift of [$Rh^{II}$(TTiPP)] ($\delta_{pyr} = +17.5$ ppm) was taken as evidence for the spin population of the porphyrin π orbitals in the case of iridium. This can be explained by assuming an $(d_{xy})^2(d_{z^2})^2(d_{xz,yz})^3$ ground state, in contrast to the $(d_{xy})^2(d_{xz,yz})^4(d_{z^2})^1$ ground state observed for the rhodium analogue.

**Reactivity of [$M^{II}$(por)] Species.** The mononuclear square-planar $d^7$ [$M^{II}$(por)]$^{\bullet}$ (M = Rh, Ir) complexes (either as stable entities for the TMP, TTEPP, or TTiPP complexes, or as transient intermediates for the other complexes) behave very much like organic radicals. For example, they tend to dimerize, either directly via metal–metal bonds (only for TXP, TTP, TPP, or OEP) or via reducible ligands like CO or olefins (with net $2e^-$ ligand reduction). The [(por)$M^{II}$] complexes also show a remarkable reactivity toward a variety of otherwise rather inert substrates. Activation under mild conditions of $H_2$, benzylic, and allylic C—H bonds, and even methane has been reported.

### 1. *Reactions With CO*

Under a CO atmosphere, [$Rh^{II}$(por)] complexes tend to dimerize with formation of carbonyl bridges. As far as we know, reactions of the iridium analogues toward CO have not been reported.

The least hindered OEP complexes reveal an equilibrium mixture containing [$Rh^{II}$(OEP)]$_2$ (and its CO adduct), a predominant monocarbonyl bridged species [$Rh^{III}$(OEP)($\mu_2$-CO)$Rh^{III}$(OEP)] (formulated as a metalloketone) and an oxalyl bridged species [$Rh^{III}$(OEP)($\mu_2$-C(O)C(O)—)$Rh^{III}$(OEP)] in minor amounts (Fig. 43) (136, 137).

Although the above reactions could (partly) involve direct CO insertions into the M—M or M—C bonds, a similar reaction for the sterically more hindered and thus mononuclear complex [$Rh^{II}$(TMP)] yields an EPR observable mononuclear CO adduct [$Rh^{II}$(TMP) (CO)] (<1%), in equilibrium with the oxalyl bridged species [$Rh^{III}$(TMP)($\mu_2$-C(O)C(O)—)$Rh^{III}$(TMP)] (>99%, 0.3 atm. CO, T < 298 K).

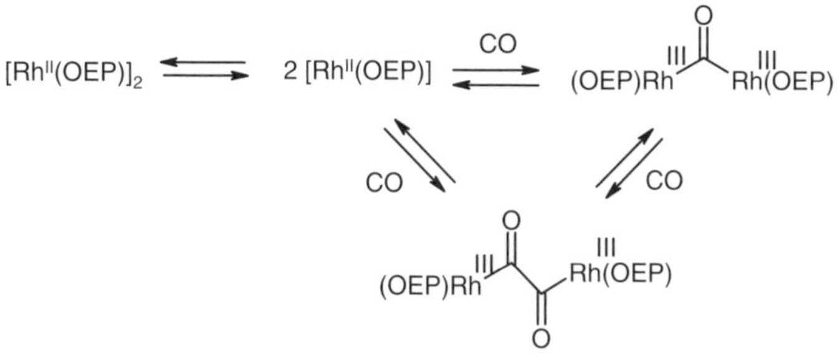

Figure 43. Single and double CO insertion between two $[Rh^{II}(OEP)]$ complexes.

These data are indicative for a mechanism involving C–C coupling of two Rh–C(O)• fragments (Fig. 44) (138, 139). Similar results were obtained for $[Rh^{II}(TXP)]$ reacting with CO, yielding exclusively $[Rh^{III}(TXP)$ ($\mu_2$-C(O)C(O)–)$Rh^{III}(TXP)]$. The intermediate $[Rh^{II}(TXP)(CO)]$ species was not detected in this case (139, 140).

The more hindered $[Rh^{II}(TTiPP)(CO)]$ analogue proved stable with respect to dimerization. The EPR parameters of the transient $[Rh^{II}(TMP)(^{13}CO)]$ ($g_1 = 2.176, g_2 = 2.147, g_3 = 1.995$) and stable $[Rh^{II}(TTiPP)(^{13}CO)]$ ($g_1 = 2.167, g_2 = 2.138, g_3 = 2.000$) in both cases revealed a rhombic $g$-tensor, relatively small Rh hyperfine couplings (along $g_3 \sim 65$–$67$ MHz) and relatively

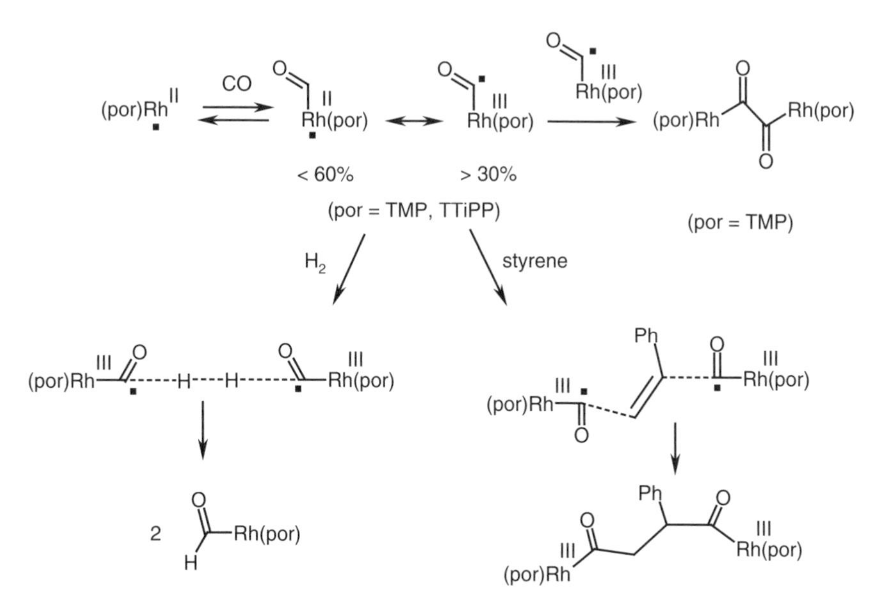


Figure 44. Bent $Rh^{II}CO$ complexes that react like acyl radicals.

large $^{13}C$ hyperfine couplings (>300 MHz along all three principal axes). These data indicate the presence of a bent M—CO unit with a spin density distribution intermediate between that of a metal and a carbonyl carbon centered radical (134, 138, 139). The spin density was estimated to be located for > 30% at CO and <60% at Rh (Fig. 44).

The acyl radical contribution to the electronic structure of $[Rh^{II}(por)(CO)]$ species also allows hydrogen radical abstraction reactions from (OEP)Rh—H bonds (only for the least hindered OEP complexes) (141), and $H_2$ or $HSn(Bu)_3$ (with formation of $[Sn(Bu)_3]_2$ ) to form the formyl species $[Rh^{III}(por)(C(O)H)]$ (139). Binuclear coupling of two Rh—C(O)• radicals to styrene yielding $[Rh^{III}(por)(\mu\text{-}C(O)CH_2CH(Ph)C(O)\text{—})Rh^{III}(por)]$ has also been reported (Fig. 44) (139). The failure of the more hindered complex $[Rh^{II}(TMP)]$ to abstract a hydrogen atom from $[Rh^{III}(por)(H)]$ to produce a formyl complex (in contrast to observations for OEP complexes), was ascribed to the steric bulk of TMP. The transition state for hydrogen atom transfer to the carbonyl center is sterically inaccessible for the TMP complex. The reaction with $H_2$ was proposed to proceed via a tri-molecular transition state involving simultaneous action of two $[Rh^{II}(por)(CO)]$ species breaking the H—H bond in a concerted reaction (Fig. 44), very similar to the transition state proposed for reactions of $[Rh^{II}(por)]$ toward C—H bonds (see Section IV.C.6 on C—H activation). The observation that (TMP)Rh—H species fail to react with CO to form formyl species $[Rh^{III}(TMP)(C(O)H)]$ indicates that the H—H bond breaking involves the metalloformyl radical units $Rh\text{—}CO^{\bullet}$, and not the metalloradical $[Rh^{II}(por)]$ species.

This was confirmed by a number of experiments using a tethered binuclear $[Rh^{II}(por)]$ analogue with steric properties comparable to $[Rh^{II}(TMP)]$ (Fig. 45) (142). Whereas hydride species obtained by reaction with $H_2$ failed to react subsequently with CO over a period of weeks, the sequential reaction in the reversed order (i.e., first CO, and then $H_2$) gave the bis(formyl) species as the only species observed. Reactions with water yielded formyl-hydride species, presumably via an intermediate formyl-carboxylic acid complex that loses $CO_2$.

Additional evidence for a concerted reaction of two metalloformyl radical units $Rh\text{—}CO^{\bullet}$ comes from studies of reactions with ethanol, resulting in predominant formation of the formyl-ethoxy ester. Reaction of ethanol with $Rh\text{—}CO^{\bullet}$ units from two separate molecules would yield the nonobserved symmetrical products $[(H(O)C)Rh^{III}(por)\text{-tether-}(por)Rh^{III}(C(O)H)]$ and $[(OEt(O)C)Rh^{III}(por)\text{-tether-}(por)Rh^{III}\text{-}(C(O)OEt)]$. In addition, nonconcerted reaction of ethanol with a monocarbonyl complex $[Rh^{II}(por)\text{-tether-}(por)Rh^{II}(CO)]$ would yield nonobserved species containing Rh—H units. Taken together, it seems most likely that the formation of $Rh^{III}\text{—}C(O)H$ and $Rh^{III}\text{—}C(O)Y$ units proceeds a concerted cleavage of the H—Y bonds via a trimolecular transition state, as shown in Fig. 44.

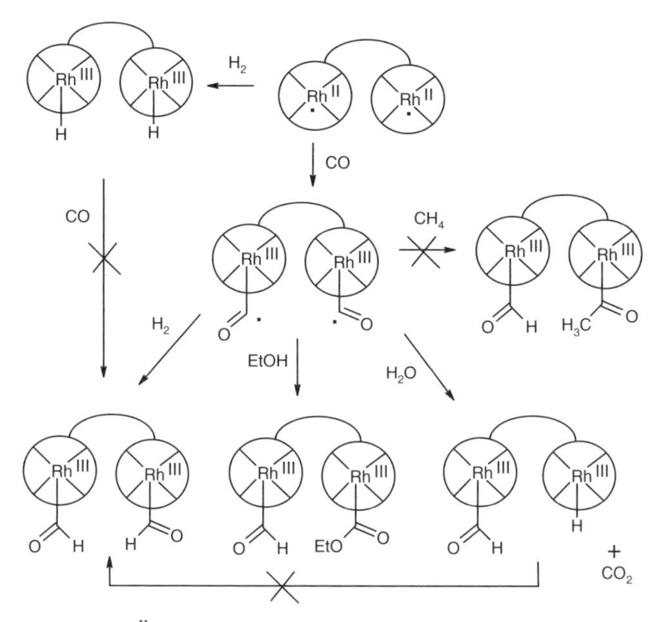


Figure 45. Bent $Rh^{II}CO$ complexes that react like acyl radicals in tethered complexes.

### 2. *Radicaloid Insertions into (por)Rh–H Bonds*

The above results might seem conflicting with reported insertion reactions of CO, olefins, aldehydes, and isocyanides into the Rh–H bonds of the less hindered complexes [$Rh^{III}$(OEP)(H)] and [$Rh^{III}$(TPP)(H)] (141, 143–150). Since [$Rh^{III}$(por)(H)] species lack any cis vacant sites, these insertion reactions are quite remarkable and do not likely proceed via a common (migratory) insertion requiring coordination of the CO, CNR, or olefinic substrates cis to the Rh–H bond. In fact, the insertion reactions are reported to proceed via radical mechanisms. The hydride species are in equilibrium with the [$Rh^{II}$(por)]$_2$ dimeric species (Fig. 46) through bimolecular reductive elimination of $H_2$ (a process not accessible for the more hindered TMP or TTiPP complexes) (151). From thereon, thermal homolytic bond splitting of the Rh–Rh bond and binding of CO (or aldehydes–isonitrils–olefins) is likely to yield a Rh–C(O)• acyl-like species, which can subsequently react with $H_2$ or [$Rh^{III}$(por)(H)] to abstract a hydrogen atom to give [$Rh^{III}$(por)(C(O)H)]. Similar mechanisms were envisaged to yield [$Rh^{III}$(por)(–CH=NR))] / [$Rh^{III}$(por)(–CHROH)] / [$Rh^{III}$(por)(–$CH_2CH_2R$)] in the case of isocyanides–aldehydes–olefins. Reaction of the iridium analogue [$Ir^{III}$(OEP)(H)] with CO does not lead to CO insertion, and only adduct formation was observed (148).

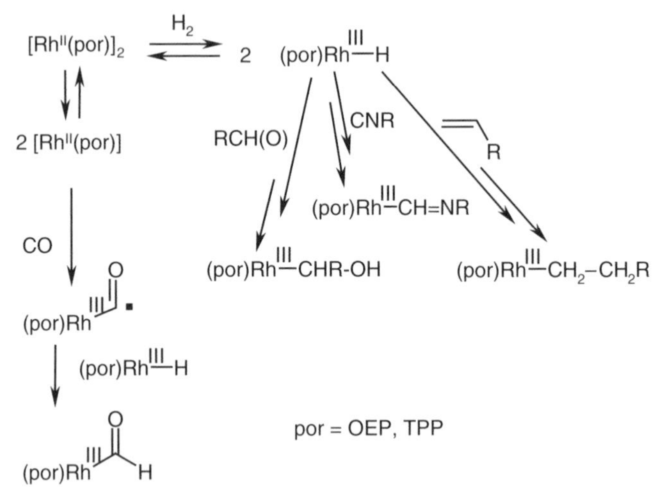


Figure 46. Net insertion of CO, aldehydes, isocyanides, and olefins into Rh–H bonds of $[Rh^{III}(TPP)(H)]$ and $[Rh^{III}(OEP)(H)]$ complexes involving radical pathways.

The CO insertion into the Rh–C bond of $[Rh^{III}(TPP)Me]$ does not occur thermally (<350 K), but only occurs after photochemical homolytic bond splitting of the Rh–C bond, thus yielding $[Rh^{III}(TPP)(C(O)Me)]$ (148). Thermally induced homolytic bond splitting of Rh–C bonds of $[Rh^{III}(OEP)(R)]$ with weaker binding R groups is possible, but requires quite high temperatures and prolongued reaction times [R = CH(Ar)$Et_3$: 120 h at 373 K; R = $CH_2CH_2CH_2$ Ar: 24 h at 443 K, which in the absence of air leads to formation of olefins and alkanes and in the presence of air to aldehydes and ketones] (152). Taking advantage of photochemically induced homolytic bond splitting of Rh–C bonds, $[Rh^{II}(OEP)]_2$ complexes have also been applied as catalysts in the photochemical production of formaldehyde (RT) and methanol (353 K) from CO and $H_2$ (153).

Radicaloid insertions of olefins into the Rh–H bond of $[Rh^{III}(TPP)(H)]$ has been used to obtain $Rh^{III}$–$CH_2$–(alkyl)–Nu*H* species (Nu*H* = OH, NH) using olefins functionalized with end-on –OH and –NH functionalities. Under basic conditions, intramolecular $S_N2$–type attack of the Nu at the α-carbon atom of $Rh^{III}$–$CH_2$–(alkyl)–Nu yields $[Rh^{I}(TPP)]^-$ and cyclic organic products (–$CH_2$–(alkyl)–Nu–) (see Fig. 47). Protonation of $[Rh^{I}(TPP)]^-$ then allows regeneration of $[Rh^{III}(TPP)(H)]$. The combination of these reactions constitutes a new method for selective intramolecular anti-Markovnikov hydrofunctionalization of olefins with O–H and N–H functionalities (150). In this way, three- and five-membered ring compounds (epoxides, furan derivatives, pyrrolidine derivatives) were readily obtained. Formation of four- or six-membered rings

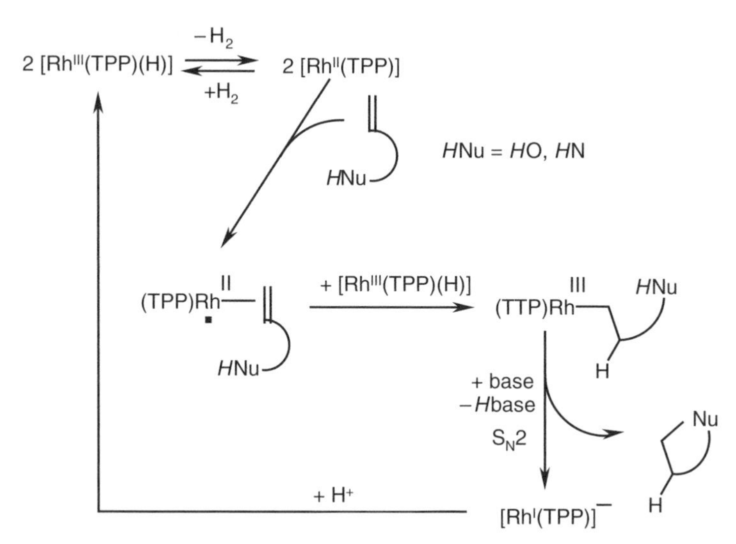


Figure 47. Selective intramolecular anti-Markovnikov hydrofunctionalization of olefins with O–H and N–H functionalities via radical insertion of olefins in the Rh–H bond.

proved to be impossible, and lead to regeneration of the starting olefins instead (via a net β-hydrogen elimination).

The (TPP)Rh system efficiently mediates all steps, but the system is not capable of mediating the anti-Markovnikov hydro-functionalization reactions in a one-pot catalytic reaction. So far, the reaction conditions required for the different reaction steps proved incompatible.

### 3. *Reactions With Alkenes and Alkynes*

A variety of olefins and acetylenes has been reported to react with $[Rh^{II}(OEP)]$ to form binuclear alkyl bridged $[Rh^{III}(OEP)(\mu\text{-}CH_2CHR\text{–})Rh^{III}(OEP)]$ complexes. The reaction of $[Rh^{II}(OEP)]$ with styrene was claimed to proceed via a radical chain mechanism (141). In the absence of $[Rh^{III}(OEP)(H)]$, the initially formed $[Rh^{II}(OEP)(CH_2{=}CHPh)]$ intermediate reacts with another $[Rh^{II}(OEP)]_2$ complex abstracting a $[Rh^{II}(por)]$ radical, thus yielding the alkyl bridged species $[Rh^{III}(OEP)(\mu_2\text{-}CH_2CHPh\text{–})Rh^{III}(OEP)]$. In the presence of $[Rh^{III}(OEP)(H)]$, the intermediate $[Rh^{II}(OEP)(CH_2{=}CHPh)]$ abstracts a hydrogen atom from $[Rh^{III}(OEP)(H)]$ to form a mononuclear alkyl complex $[Rh^{III}(OEP)(CH_2CH_2Ph)]$ (Fig. 48). Other olefins like acrylates (154) have also been reported to give similar alkyl bridged species $[Rh^{III}(OEP)(\mu_2\text{-}CH_2CHR\text{–})Rh^{III}(OEP)]$, and acetylenes give the corresponding $\mu_2\text{-}CH{=}CR\text{–}$ bridged species $[Rh^{III}(OEP)(\mu_2\text{-}CH{=}CHR\text{–})Rh^{III}(OEP)]$ (155).

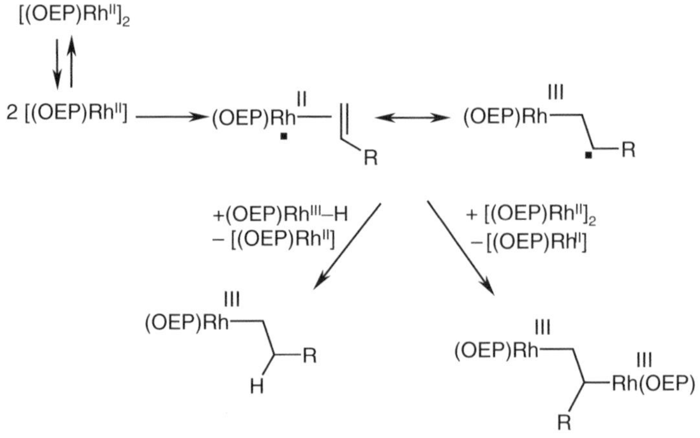

[(OEP)Rh$^{II}$]$_2$

2 [(OEP)Rh$^{II}$] ⟶ (OEP)Rh$^{II}$—‖—R ⟷ (OEP)Rh$^{III}$—CH$_2$—ĊH—R

+(OEP)Rh$^{III}$–H
− [(OEP)Rh$^{II}$]

+ [(OEP)Rh$^{II}$]$_2$
−[(OEP)Rh$^{II}$]

(OEP)Rh$^{III}$—CH$_2$—CH(H)—R

(OEP)Rh$^{III}$—CH$_2$—CH(R)—Rh$^{III}$(OEP)

Figure 48. Reactivity of [Rh$^{II}$(por)] and [Rh$^{III}$(por)(H)]) toward olefins without allylic hydrogen atoms.

Olefins containing allylic hydrogen atoms (e.g., allylbenzene) (R = CH$_2$Ph), allylcyanide (R$_1$ = CH$_2$CN), and 1-hexene (R$_1$ = CH$_2$C$_3$H$_8$) are reported to form σ-allyl complexes [Rh$^{III}$(OEP)(CH$_2$CH=CHR)] in ~ 50% yield (Fig. 49) (155). A more detailed study concerning the behavior of [Rh$^{II}$(OEP)] toward propene was reported (156). Propene initially forms the alkyl bridged species [Rh$^{III}$(OEP)(μ$_2$-CH$_2$CHMe–)Rh$^{III}$(OEP)]. This species reveals an interesting 1,2-exchange of the two metal sites, reflecting the lability of the Rh–C bond of the Rh–CHMe fragment. The species is not stable and converts over the course of some days to a 1:1 mixture of the σ-allyl species [Rh$^{III}$(OEP)

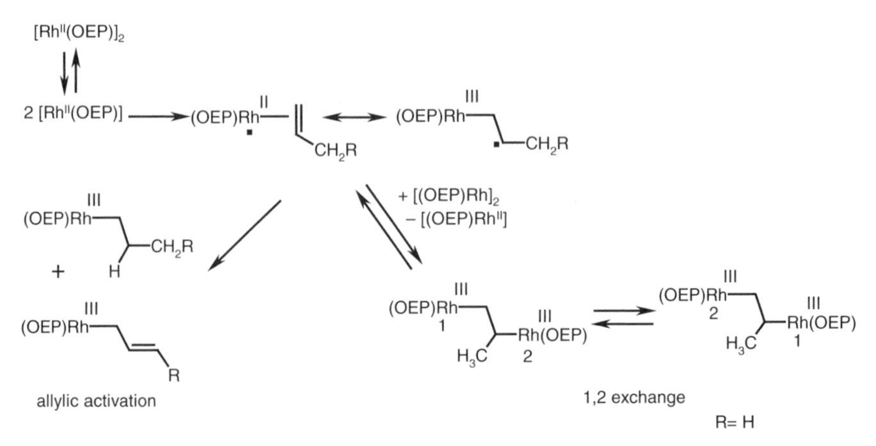


Figure 49. Reactivity of [Rh$^{II}$(por)] species toward olefins containing allylic hydrogen atoms.

$(CH_2CH{=}CH_2)$] and the propyl species [$Rh^{III}$(OEP)($CH_2CH_2CH_2$)], which likely involves a net transfer of an allylic hydrogen atom from [$Rh^{II}$(OEP)(propene)] to another [$Rh^{II}$(OEP)(propene)] species (obtained from [$Rh^{III}$(OEP)($\mu_2$-$CH_2CH$(Me)–) $Rh^{III}$(OEP)] by Rh–C homolysis) (see Fig. 49).

A comparable allylic activation of methyl methacrylate [$CH_2{=}C$(Me)(COOMe),MMA] was reported to give a 1:1 mixture of [$Rh^{III}$(OEP)($CH_2C$(COOMe)$=CH_2$)] and [$Rh^{III}$(OEP)($CH_2CH$(Me)(COOMe))]. Reaction of the sterically more hindered [$Rh^{II}$(TMP)] with MMA yields a 1:1 mixture of [$Rh^{III}$(TMP)($CH_2C$(COOMe)–$CH_2$)] and [$Rh^{III}$(TMP)(H)]. The reactions were proposed to proceed via abstraction of an allylic hydrogen atom from the metalloradical alkene complex [$Rh^{II}$(por)($CH_2{=}C$(Me)(COOMe)] by the metalloradical [$Rh^{II}$(por)] to yield the $\sigma$-allyl complex [$Rh^{III}$(por)($CH_2C$(COOMe)$=CH_2$)] and the hydride species [$Rh^{III}$(por)(H)] directly. As described above, the OEP complexes allow insertion reactions into the Rh–H bond, which for MMA yields [$Rh^{III}$(OEP)($CH_2CH$(Me)(COOMe))]. Olefin insertion into the Rh–H bond of the TMP complexes does not occur, presumably for steric reasons (see above).

More hindered [$Rh^{II}$(por)] complexes with por = TTiPP, TTEPP, and TMP provide further insight in the behavior of the [$Rh^{II}$(por)(olefin)] intermediates. Like [$Rh^{II}$(OEP)], [$Rh^{II}$(TMP)] with ethene still rapidly converts to the diamagnetic ethylene bridged species [$Rh^{III}$(TMP)($\mu_2$-$C_2H_4$)$Rh^{III}$(TMP)] involving a bimolecular M–C coupling reaction (139). Increasing the steric bulk by using the complex [$Rh^{II}$(TTEPP)], however, prevents direct M–C coupling, and bimolecular C–C coupling via two rhodium–ethene radicals yields the butylene bridged species [$Rh^{III}$(TTEPP)($\mu_2$–$CH_2CH_2$–$CH_2CH_2$–$Rh^{III}$(TTEPP)], (see Fig. 50) (157).

No paramagnetic intermediates have been observed with EPR spectroscopy for either [$Rh^{II}$(TMP)] or for [$Rh^{II}$(TTEPP)]. However, a further increase of the steric bulk by using [$Rh^{II}$(TTiPP)] gives rise to a reasonably stable paramagnetic ethene adduct [$Rh^{II}$(TTiPP)($CH_2{=}CH_2$)]$^\bullet$, revealing a rhombic EPR spectrum at 90 K ($g$ values: $g_1 = 2.323$, $g_2 = 2.222$ and $g_3 = 1.982$). A substantial spin population at ethene (~0.29) was derived from the $^{13}C$ hyperfine couplings of [$Rh^{II}$(TTiPP)($^{13}CH_2{=}^{13}CH_2$)] ($A_{zz}$: 38 MHz, $A_{xx} = A_{yy} \sim 5$ MHz) (134, 157). The spin population at rhodium was estimated at ~0.67. The equivalence of the two $^{13}C$ atoms in the EPR spectrum reveals a symmetrically bound ethene $\pi$ complex, and the delocalization of the unpaired electron to the ethene fragment (30%) occurs almost entirely via the $C_{2p}$ orbitals. At temperatures >200 K, [$Rh^{II}$(TTiPP)(ethene)] easily loses the ethene fragment to form [$Rh^{II}$(TTiPP)] (135).

Reactions of the iridium complexes [$Ir^{II}$(TMP)] and [$Ir^{II}$(TTEPP)] with ethene proceed analogously to the above described rhodium analogues (Fig. 50) (33). Like its rhodium analogue, reaction of [$Ir^{II}$(TTiPP)] with ethene

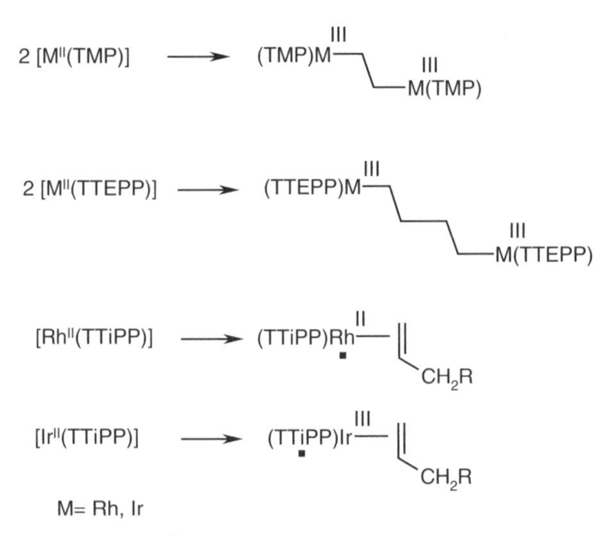


Figure 50. The reactivity of $[M^{II}(por)(olefin)]$ species changes from M–C to C–C coupling to no coupling at all upon increasing the steric bulk.

does not result in M—C or C—C coupling, and formation of the ethene adduct is witnessed by NMR, UV–Vis ($\lambda_{max}$ 444 and 730 nm), and EPR spectroscopy [$<g>= 1.987$ (290 K); $g_{\parallel} = 1.96$, $g_{\perp} = 1.998$ (90 K)]. However, these spectroscopic data are indicative of a donor induced intramolecular electron transfer from the $Ir^{II}$ center to the porphyrin ligand $\pi^*$ orbital to yield a $Ir^{III}(por^{\bullet\ 3-})$ species. This contrasts the results obtained by ethene binding to $[Rh^{II}(TTiPP)]$, where the unpaired electron remains in the $d_{z^2}$ orbital.

Selective binding of ethene and butadiene between the two rhodium sites of a tethered bimolecular bis(por)Rh species with steric demands comparable to TMP, results in selective intramolecular coupling (Fig. 51) (142). This gives further support to the idea that formation of alkyl bridged species from olefins and $[M^{II}(por)]$ requires a concerted action of two metalloradical sites (157).

The reactions of $[Rh^{II}(por)]$ with olefins reveal that the $[Rh^{II}(por)(olefin)]$ intermediates have a substantial spin density at the olefin fragment. This prompted the Wayland group to investigate attempts to initiate radical polymerization of acrylates with these species (154). Although photopromoted radical initiation (involving homolytic Rh—C bond splitting) was achieved, direct radical initiation with $[Rh^{II}(OEP)]$ or $[Rh^{II}(TMP)]$ was not observed.

Based on an estimated Rh—$CH_2R$ bond dissociation enthalpy (BDE) of $\sim$50 kcal mol$^{-1}$, the energy associated with olefin binding to $[Rh^{II}(por)]$ estimated at 0–8 kcal mol$^{-1}$ (a much weaker affinity for olefins than for CO) (135, 154), and the energy required to convert ethene into its biradical $\bullet CH_2$—$CH_2\bullet$ (64 kcal mol$^{-1}$), formation of an intermediate $[Rh^{III}(por)$

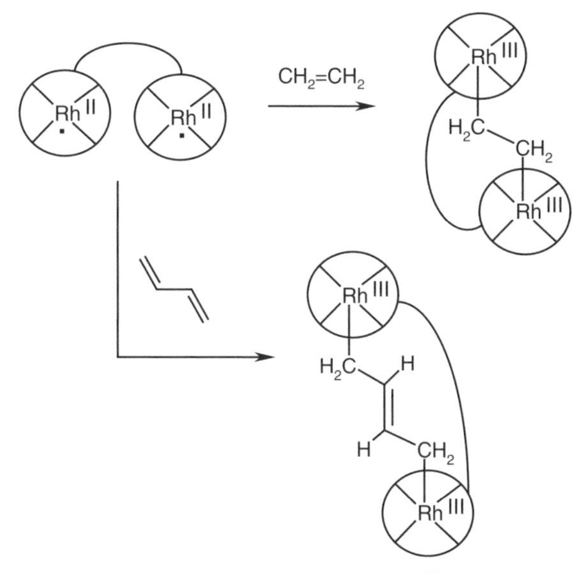


Figure 51. Reaction of olefins with tethered [$Rh^{II}$(por)] complexes.

($CH_2$—$CH_2$•)] species with 100% spin population at the β-carbon atom requires ~14 kcal $mol^{-1}$. Relaxation of this species to the more stable π-complex [$Rh^{II}$(por)($CH_2$=$CH_2$)] should be exothermic by ~14–22 kcal $mol^{-1}$ (154). Based on these arguments (and the above EPR results), the authors argued that [$Rh^{II}$(por)(alkene)] complexes are not authentic alkyl radicals, and therefore cannot initiate alkene polymerization reactions (Fig. 52).

These arguments also imply that formation of bent $M^{II}$-olefin species could be rather unfavorable. On the other hand, coupling of two (por)$M^{II}$(olefin) species with C—C bond formation to yield $M^{III}$—$CH_2CH_2$—$CH_2CH_2$—$M^{III}$ cannot be envisaged without a species having at least some [(por) $Rh^{III}$($CH_2$—$CH_2$•)] contribution to its electronic structure. Furthermore, substituents known to stabilize radicals should favor the [$Rh^{III}$(por)($CH_2$CHR•)] contribution [e.g., R = Ph, C(O)R, CN, alkyls in Fig. 52].

In addition, coordinating ligands might facilitate formation of the bent [$Rh^{III}$(por)($CH_2CH_2$•)] structure. Recent DFT investigations concerning

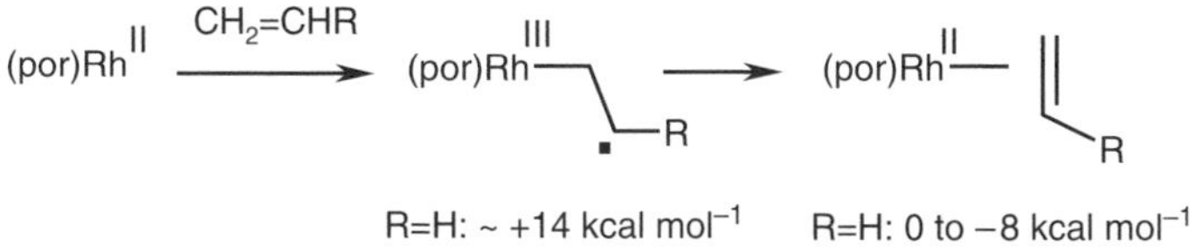

(por)$Rh^{II}$ —($CH_2$=CHR)→ (por)$Rh^{III}$—$CH_2$–ĊHR → (por)$Rh^{II}$—($CH_2$=CHR)

R=H: ~ +14 kcal $mol^{-1}$    R=H: 0 to −8 kcal $mol^{-1}$

Figure 52. The [$Rh^{II}$(por)(olefins)]; metal or carbon centered radicals?

$[Ir^{II}(Me_3tpa)(ethene)]^{2+}$ have indicated that MeCN coordination to the metal can be sufficient to overcome the unfavorable bending of the olefin, thus yielding an iridium–olefin species with an electronic structure intermediate between a slipped olefin $\bullet Ir^{II}{-}CH_2{=}CH_2$ and an alkyl radical $Ir^{III}{-}CH_2{-}CH_2\bullet$ description (see Section IV.D). In addition to the energy gain from MeCN coordination, delocalization of the unpaired spin to iridium renders formation of such species less unfavorable in this case.

Assuming formation of $Ir^{III}{-}CH_2{-}CH_2\bullet$ with 100% spin population at the β-carbon atom (which is not the case), we estimate a stronger $Ir^{III}{-}CH_2R$ BDE ($\sim$59 kcal $mol^{-1}$) compared to $Rh^{III}{-}CH_2R$ (BDE: $\sim$50 kcal $mol^{-1}$) from the relative DFT energies of $[Ir^{II}(Me_3tpa)(MeCN)(ethene)]^{2+}$ (0), $[Ir^{II}(Me_3tpa)(MeCN)]^{2+}$ ($-4.8$ kcal $mol^{-1}$) and the energy required to convert ethane into its biradical $\bullet CH_2{-}CH_2\bullet$ (64 kcal $mol^{-1}$). This might provide a further contributing factor to stabilization of the bent olefin structure in $[Ir(Me_3tpa)(MeCN)(CH_2CH_2)]^{2+}$.

### 4. *[M^II(por)(olefin)] Intermediates in Reactions of [M^II(por)(alkyls)] With Nitroxyl Radicals*

The $[M^{III}(por)(alkyl)]$ species reveal a very remarkable reactivity toward (TEMPO) and related nitroxyl radicals (M = Rh, Ir, por = TTP, TMP and some less common OMP, BTPP, and BOCP porphyrinato ligands shown in Fig. 53).

The $[M^{III}(por)(alkyl)]$ complexes lacking β-hydrogen atoms do not react, but (por)Rh-alkyl species with β-hydrogen atoms eliminate olefin fragments upon

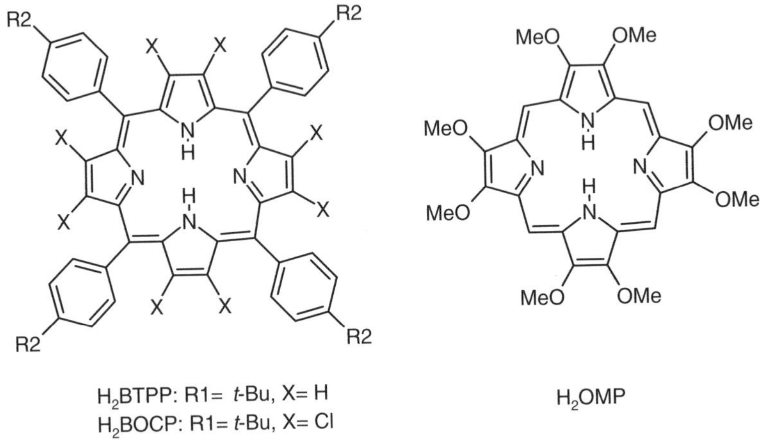

$H_2$BTPP: R1= *t*-Bu, X= H
$H_2$BOCP: R1= *t*-Bu, X= Cl

$H_2$OMP

Figure 53. Ligand coding of the pohyrins BTPP, BOCP, and OMP.

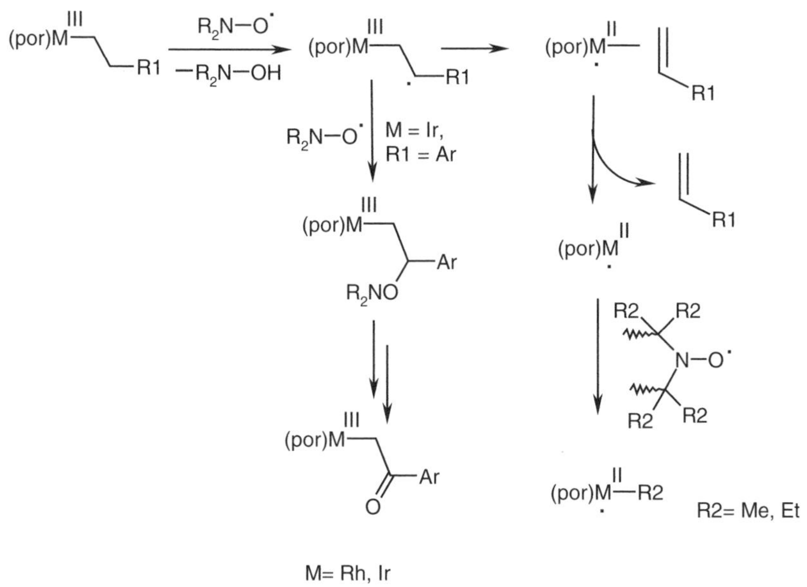

Figure 54. Reaction of TEMPO and related nitroxyl radicals with $[M^{III}(por)(alkyl)]$ species ($M = Rh$, Ir) via $[M^{II}(por)(olefin)]$ intermediates.

heating in the presence of TEMPO (158, 159). For $R = -CH_2-CH_2-Ph$, formation of styrene was confirmed with NMR (Fig. 54).

Abstraction of β-hydrogen atoms from alkyl species $[M^{III}(por)(-CH_2CH_2R)]$ yields alkyl radical species $[M^{III}(por)(-CH_2CHR\bullet)]$, which are in fact the same activated β-alkyl radicals resulting from bending the M—C—C bond in the olefin complexes $[M^{II}(por)(CH_2{=}CHR)]$. These $[M^{III}(por)(-CH_2CHR\bullet)]$ species should thus easily relax to the more stable olefin complexes $[M^{II}(por)(CH_2{=}CHR]$. Since $[Rh^{II}(por)]$ species have only a weak affinity for olefins (135), $[M^{II}(por)(CH_2{=}CHR]$ should readily lose the olefin (Fig. 54). The resulting 15 VE $[Rh^{II}(por)]$ species subsequently react with TEMPO to form $[Rh^{III}(por)Me]$ via a C—C bond activation reaction with net abstraction of a methyl radical from TEMPO. Formation of $[Rh^{II}(TMP)(R)]$ species by a net abstraction of an alkyl radical R• from a nitroxyl radical $(R'R_2C)_2NO\bullet$ has also been observed for isolated $[Rh^{II}(TMP)]$ species in the presence of nitroxyl radicals (see Fig. 54) (160). While the mechanism of this reaction is unknown (as is the fate of the nitroxyl radicals) it is highly likely that the Rh(II) radical is the reacting species rather than Rh(I) or Rh(III) species. Both $[Rh^{I}(TMP)]$ and $[Rh^{III}(TMP)](ClO_4)$ were independently shown to be nonreactive towards the nitroxyl radicals. So far, these systems provide the only examples of transition metal complexes capable of intermolecular activation of an unstrained C—C bond.

For the analogous iridium complexes [$Ir^{III}$(TTP)(R)], similar results were obtained upon reaction with TEMPO. Complexes lacking β-hydrogen atoms did not react, and nonaromatic alkyl complexes eliminate olefins allowing formation of [$Ir^{III}$(TTP)Me] via C—C activation of TEMPO by [$Ir^{II}$(TTP)]. Arylethyl complexes [$Ir^{III}$(TTP)(—$CH_2CH_2$Ar)], however, reveal different products, which further support the proposed radical abstraction mechanism (161). Under a $N_2$ atmosphere, TEMPO oxidizes the [$Ir^{III}$(TTP)(—$CH_2CH_2$Ar)] species selectively at the beta position to yield [$Ir^{III}$(TTP)(—$CH_2$C(O)Ar)] (Fig. 54). The mechanism proposed for this reaction involves abstraction of a β-hydrogen atom from [$Ir^{III}$(TTP)(—$CH_2CH_2$Ar)] by TEMPO to give [$Ir^{III}$(TTP)(—$CH_2$CH•Ar)] and TEMPO*H*. Apparently, due to stabilization of the beta-carbon radical by the aromatic group and a substantially stronger Ir—C bond compared to a Rh—C bond, this species is sufficiently long lived to react with a second TEMPO moiety. Coupling of the [$Ir^{III}$(TTP)(—$CH_2$CH•Ar)] radical with the TEMPO radical was proposed to yield a [$Ir^{III}$(TTP)(—$CH_2$CHAr—O—$NR_2$] intermediate. The latter could then decompose to the observed [$Ir^{III}$(TTP)(—$CH_2$C(O)Ar)] species, either directly via a hydrogen shift (with elimination of $R_2$NH) or by abstraction of the remaining β-hydrogen by an additional TEMPO moiety. This mechanism resembles the proposed mechanism of [($Me_3$tpa)$Ir^{II}$(ethene)] species with $O_2$ via β-oxidation of intermediate $Ir^{III}$—$CH_2CH_2$• carbon radicals (162).

Reaction of TEMPO with [$Ir^{III}$(TTP)(—$CH_2CH_2CH_2$Ar)] species, despite the fact that these species also contain reactive benzylic protons, only yields [(TTP)$Ir^{III}$Me]. This illustrates the importance of both the iridium center at the beta position and the arene at the alpha position to stabilize the carbon-centered radical intermediates upon reaction of TEMPO with [$Ir^{III}$(TTP)(—$CH_2$ $CH_2$Ar)].

### *5. C—X, C—N, and C—O Activation Reactions*

The reaction of electrochemically generated [$Rh^{II}$(TPP)] species with olefins and alkyl-halogenides has been investigated (129). These species were generated by reduction of the cationic [$Rh^{III}(TPP)(L)_2]^+$ precursors (L = dimethylamine). The behavior of these transient species toward olefins is rather different from the reported reactivity of all other [$Rh^{II}$(por)] species (described above). Instead of the expected oxidative addition of two (TPP)$Rh^{II}$ fragments to the carbon–carbon double bonds to form binuclear [$Rh^{III}$(TPP)($\mu_2$-$CH_2$—CHR—)$Rh^{III}$(TPP)] species, or allylic activation to form [$Rh^{III}$(TPP) ($CH_2$CH=CHR)] species, as described for OEP complexes (Figs. 48 and 49), C—C bond cleavage at the position adjacent to the double bond was observed in this case (Fig. 55). The reactivity toward alkynes proceeds similarly.

The authors reported a featureless EPR spectrum ($g = 2.00$, line width 17 G) of an intermediate in the presence of 1-pentyne, which was ascribed to the $\pi$

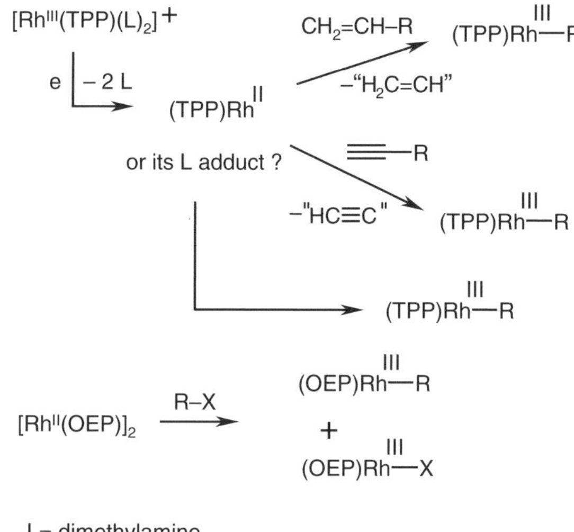


Figure 55. Reactivity of olefins, alkynes, and organic halides toward electrochemically generated $[Rh^{II}(TPP)]$ (or its dimethylamine adduct?).

adduct. However, the EPR data are very different from those reported for olefin adducts of $[Rh^{II}(TTiPP)]$ (see above). Perhaps a more likely explanation would be formation of six-coordinate $[Rh^{III}(TPP^{\bullet\,3-})(L)(olefin)]$ species as intermediates resulting from the incomplete dissociation of the dimethylamine ligands (L) after reduction of $[Rh^{III}(TPP)(L)_2]^+$. Nitrogen donors like pyridine have been reported to induce intramolecular electron transfer from the $Rh^{II}$ center to the porphyrin ligand $\pi^*$ to form $[Rh^{III}(por^{\bullet\,3-})]$ species (134), which might explain the EPR results and the different reactivity of these species compared to their OEP analogues. Alternatively, the use of the more polar THF solvent compared to benzene might give an explanation for the observed differences.

Similar experiments in the presence of organic halides lead to radical-based C—X bond splitting to obtain alkyl complexes. No formation of $[Rh^{III}(TPP)(X)]$ by halide abstraction was observed in these experiments (128, 130). This contrasts formation of both $[Rh^{III}(TPP)(X)]$ and $[Rh^{III}(TPP)(R)]$ upon reaction of organic halides (R—X) with $[Rh^{II}(OEP)]$ (Fig. 55) (155).

Further carbon abstraction reactions from alkylisocyanides (147) and $P(OMe)_3$ (163) by $[Rh^{II}(OEP)]$ and $[Rh^{II}(TMP)]$ demonstrate the radical nature of these species (Fig. 56). Arylisocyanides do not undergo C—C cleavage, and an EPR spectrum of the 2,6-dimethylphenyl isocyanide adduct of $[Rh^{II}(TMP)]$ was reported [$g_{\parallel} = 1.995$, $g_{\perp} = 2.158$, $A_{\parallel}^{Rh} = 60\,MHz$, $A_{\parallel}^{N} = 28\,MHz$ (90 K)], consistent with a primarily rhodium-based radical with the unpaired electron occupying a $d_{z^2}$ based MO (Fig. 56).

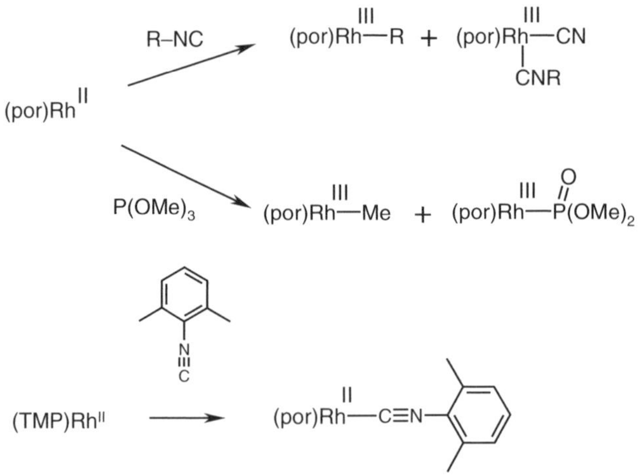

Figure 56. Radical-type reactivity of $[Rh^{II}(por)]$ complexes toward isocyanides and $P(OMe)_3$.

### 6. *The C—H Bond Activation Reactions*

The metalloradical $[Rh^{II}(por)]$ complexes are capable of breaking C—H bonds of hydrocarbon fragments in a controlled way. The $R_3C$—H bonds are broken between two metalloradical $[Rh^{II}(por)]$ fragments to produce $[Rh^{III}(por)(CR_3)]$ and $[Rh^{III}(por)(H)]$.

The complexes containing the least-hindered porphyrins OEP are dinuclear, and these metal–metal bound $[Rh^{II}(OEP)]_2$ and $[Ir^{II}(OEP)]_2$ species are only capable of breaking activated C—H bonds. These involve allylic C—H bonds of olefins (see Section IV.C.3), benzylic C—H bonds of toluene and related alkyl aromatics (152, 164), benzyl alcohols, and α-CH bonds of aldehydes and ketones (165), (see Fig. 57). Reactions with alkyl aromatics and benzyl alcohols invariably results in C—H activation at the benzylic position. The aromatic C—H bonds do not react. For the benzyl alcohols, the initial $M^{III}$—CR(OH)Ar products from benzylic C—H activation subsequently decompose under formation of organic carbonyl species ArC(O)R and $[M^{III}(OEP)(H)]$ via a net β-hydrogen elimination. Ketones and aldehydes preferentially react at the alkyl group adjacent to the carbonyl position to yield β-formyl complexes $M^{III}$—$CR_2C(O)H$ and $M^{III}$—$CR_2C(O)R$. For the aldehydes, this is quite remarkable because the aldehydic (O)C—H bond (BDE $\sim$ 86 kcal mol$^{-1}$) is substantially weaker than an alkyl C—H bond (BDE $\sim$ 100 kcal mol$^{-1}$). Therefore, the reaction with aldehydes and ketones was proposed to proceed via their enol tautomers

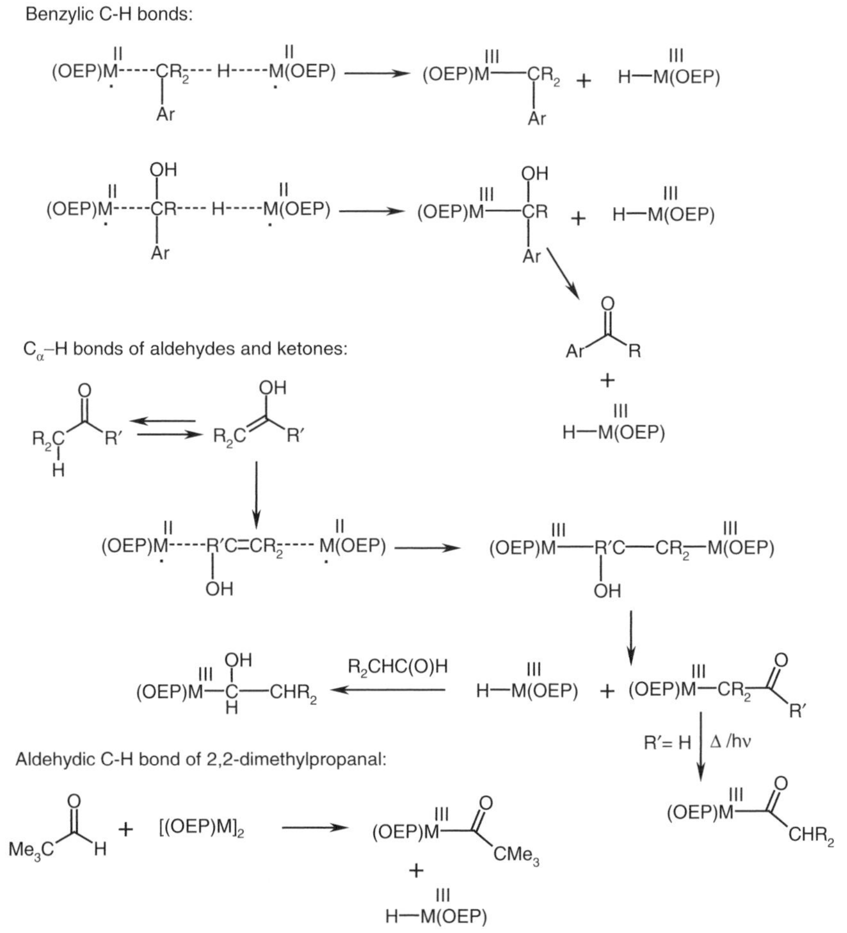

Benzylic C-H bonds:

$C_\alpha$–H bonds of aldehydes and ketones:

Aldehydic C-H bond of 2,2-dimethylpropanal:

Figure 57. Breaking activated benzylic and $C_\alpha$–H bonds of aldehydes and ketones between two $[M^{II}(OEP)]$ metalloradicals.

($R_2C{=}C(OH)R$) (165). Similar to the reactivity with other olefins (see Section IV.C.3), the enol tautomer $R_2C{=}C(OH)R$ can react with two $[M^{II}(OEP)]$ species to form a bridged $[M^{III}(OEP)(\mu_2\text{-}CR_2C(OH)R\text{–})M^{III}(OEP)]$ intermediate. A net β-hydrogen elimination then yields the observed β-formyl species $[M^{III}(OEP)(CR_2C(O)R)]$ and $[M^{III}(OEP)(H)]$. The $[M^{III}(OEP)(H)]$ species then react further to yield $[M(OEP)]_2$ or $[M^{III}(OEP)(CH(OH)R)]$ species by dinuclear $H_2$ elimination or a net aldehyde insertion, as described above in Section IV.C.2.

Breaking stronger C–H bonds, like those of methane, ethane, or methanol, proved difficult with $[Rh^{II}(OEP)]_2$ complexes. Reversible formation of only

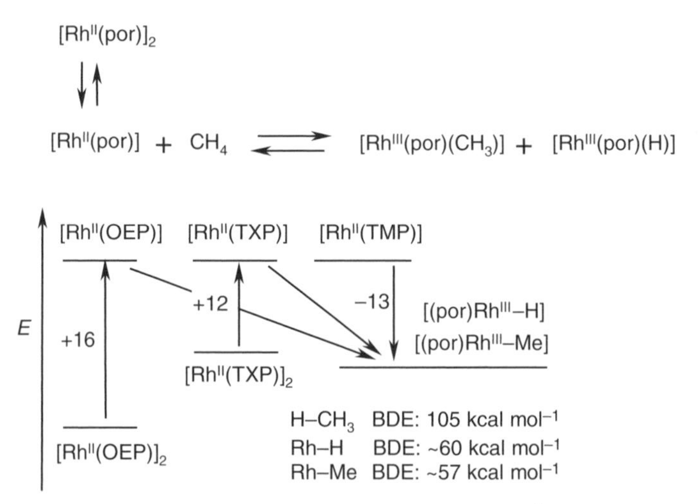


Figure 58. Reversible oxidative addition of methane between two $[Rh^{II}(por)]$ metalloradicals and the approximate relative energy of the OEP, TXP, and TMP species.

small quantities of $[Rh^{III}(OEP)Me]$ and $[Rh^{III}(OEP)(H)]$ was observed in the reaction with methane, but their fraction was insufficient for quantitative kinetic or thermodynamic studies.

The reaction with the more hindered complexes $[Rh^{II}(TXP)]_2$ and $[Rh^{II}(TMP)]$ toward $CH_4$ is more favorable (166). Both complexes react reversibly with methane to form $[(por)Rh^{III}—Me]$ and $[(por)Rh^{III}—H]$ complexes (Fig. 58). While for $[Rh^{II}(TXP)]_2$, the equilibrium is still in favor of $[Rh^{II}(TXP)]_2$ ($\Delta H^0 \sim 0\,\mathrm{kcal\,mol^{-1}}; \Delta S^0 \sim -7\,\mathrm{cal\,(K.mol^{-1})}$), for $[Rh^{II}(TMP)]$ the equilibrium lies to the right in favor of $[Rh^{III}(TMP)Me]$ and $[Rh^{III}(TMP)(H)]$ $[\Delta H^0 \sim -13\,\mathrm{kcal\,mol^{-1}}; \Delta S^0 \sim -19\,\mathrm{cal\,(K.mol^{-1})}]$. The thermodynamic differences are dominated by the Rh—Rh BDE required to convert the $[Rh^{II}(OEP)]_2$ ($\sim 16\,\mathrm{kcal\,mol^{-1}}$) and $[Rh^{II}(TXP)]_2$ ($\sim 12\,\mathrm{kcal\,mol^{-1}}$) species into their monomeric species. While the mononuclear species $[Rh^{II}(TMP)]$ can react directly, $[Rh^{II}(OEP)]_2$ and $[Rh^{II}(TXP)]_2$ must dissociate before they can react with methane (Fig. 58). Introducing further steric requirements beyond those of TMP (as in the porphyrins TTEPP or TTiPP) is expected to result in weaker Rh—C bonds, thereby disfavoring the C—H activation process (167).

The C—H activation processes were proposed to proceed via a trimolecular pathway, involving a linear four-centered concerted breaking of the C—H bond between two Rh(II) metalloradicals [see Fig. 59 (*a*)]. In principle, this reaction could also proceed via two subsequent separate steps: hydrogen-atom abstraction from the hydrocarbon $R_3C—H$ bond, thus generating a $Rh^{III}—H$ species and a carbon-centered radical $\bullet CR_3$, which is rapidly followed by capture of the thus (*a*) $\dot{Rh}^{II}\text{-----}H\text{------}CR_3\text{--}\dot{Rh}^{II} \longrightarrow Rh^{III}\text{—}H + CR_3\text{—}Rh^{III}$

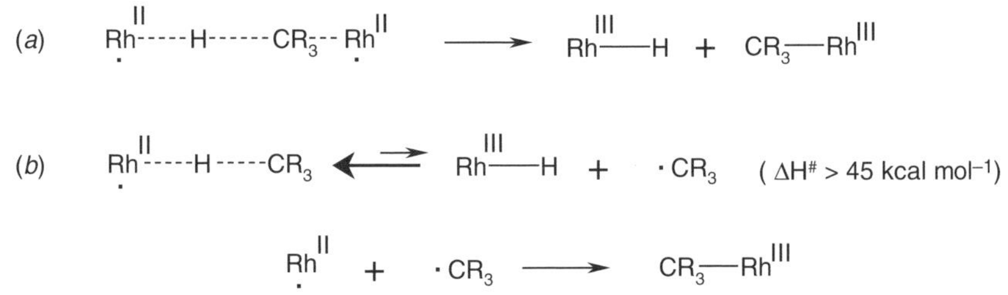

(*b*) $\dot{Rh}^{II}\text{-----}H\text{-----}CR_3 \rightleftarrows Rh^{III}\text{—}H + \cdot CR_3$ $(\Delta H^{\#} > 45 \text{ kcal mol}^{-1})$

$\dot{Rh}^{II} + \cdot CR_3 \longrightarrow CR_3\text{—}Rh^{III}$

Figure 59. Concerted four-centered pathway proposed for C–H activation by $[Rh^{II}(por)]$ radicals.

generated carbon-centered radical $\bullet CR_3$ by another Rh(II) radical to give $Rh^{III}$-$CR_3$ [see Fig. 59 (*b*)]. However, the relatively weak Rh–H bond ($\sim 60\,\text{kcal mol}^{-1}$) compared to the C–H bond ($105\,\text{kcal mol}^{-1}$) makes this route prohibitively endothermic at room temperature.

The activation parameters (TMP : $\Delta H \sim +7\,\text{kcal mol}^{-1}$; $\Delta S \sim -39$ cal. $K^{-1}$.$\text{mol}^{-1}$; TXP : $\Delta H \sim +17\,\text{kcal mol}^{-1}$; $\Delta S \sim -25\,\text{cal.K}^{-1}.\text{mol}^{-1}$, the rate law (rate $= k[(por)Rh\bullet]^2[CH_4]$) and large kinetic isotope effects ($k_H/k_D > 8$) derived from kinetic studies of the reactions in Fig. 58 (166, 167), are consistent with the proposed concerted four-centered pathway with a near linear transition state in Fig. 59(*a*). Similar kinetic studies suggest that toluene and methane C–H bond cleavage reactions (and also H–H bond reactions of $H_2$) proceed via related mechanisms (166, 168, 169).

It seems probable that the reactions proceed via initially formed precomplexes of the type $[Rh^{II}(por)(HCR_3)]$ by binding of $HCR_3$ to $[Rh^{II}(por)]$. Without precomplex formation, the proposed trimolecular transition state would have a very limited reaction probability according to standard collision theory. So far, such a precomplex $[(por)Rh^{II}(HCR_3)]$ has not actually been observed yet.

To enhance the kinetics of the C–H activation process, Wayland and co-workers prepared homodinuclear complexes consisting of two (porphyrinato)-Rh(II) complexes tethered by different spacers (170–172). This significantly accelerates the C–H bonds activation process by diminishing the unfavorable entropy contribution to the activation barrier [Fig. 59(*a*)]. Within one dinuclear complex, one Rh-site receives a hydride ligand and the other Rh-site receives an alkyl ligand (Fig. 60).

Reactions of $CH_4$ or MeOH with $Rh^{II}\bullet$ units in different molecules would yield the symmetrically derivatized alkyl and hydride products $[(XCH_2\text{—})Rh^{III}$ (por)-tether-(por)$Rh^{III}(\text{—}CH_2X)]$ (X = H, OH) and $[(H)Rh^{III}$(por)-tether-(por)-$Rh^{III}(H)]$. Fast and selective formation of the unsymmetrical $[(XCH_2\text{—})\ Rh^{III}$ (por)-tether-(por)$Rh^{III}(H)]$ species (Fig. 60), which convert more slowly to symmetrical species via different pathways, provides additional evidence for the proposed concerted pathway in Fig. 59(*a*).

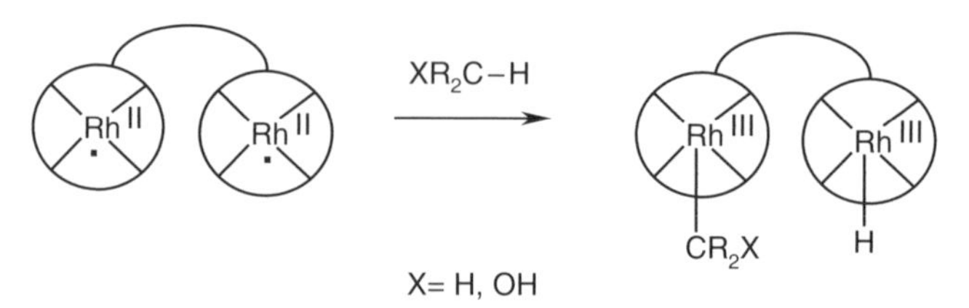


Figure 60. Homodinuclear tethered [$Rh^{II}$(por)] complexes for C—H activation.

Activation of ethane with these tethered dinuclear species was observed to give exclusive C—H activation, despite the fact that C—C bond cleavage would be thermodynamically more favorable (170). For steric reasons, C—H activation of methane is kinetically favored over C—H activation of ethane or methanol substrates in these systems (second-order rate constants in the order $H_2 > CH_4 >$ MeOH > Et > MePh). Activation parameters and kinetic isotope effects derived from kinetic studies of C—H activation processes with these tethered complexes are consistent with the previous conclusions derived from reactions with [$Rh^{II}$(TMP)].

Recent investigations aimed at taking advantage of C—H activation processes by [$Rh^{II}$(por)] species in synthetic transformations, aiming for subsequent functionalization of the resulting alkyl species. The C—H activation of methane with monomeric [$Rh^{II}$($F_{28}$TPP)] species has been reported, in which $F_{28}$TPP is an electron-deficient fluorinated analogue of the TPP porphyrin that stabilizes the Rh(I) oxidation state significantly. As a result, thus formed Rh—$CH_3$ fragments are now more sensitive toward nucleophilic attack ($S_N2$) by $PPh_3$ to form $[Me_3—PPh_3]^+$ and [$Rh^I$($F_{28}$TPP)]$^-$ (173). Protonation of the latter to $Rh^{III}$—H species, and binuclear reductive elimination of $H_2$ regenerates the monomeric [$Rh^{II}$($F_{28}$TPP)] species. So far, this system is quite slow, and does not catalyze one-pot reactions yet, but efficient catalytic systems for selective C—H functionalization might well result from future studies of related systems.

## 7. *Chemistry of Similar Nonporphyrinato Planar M(II) Metalloradicals*

Radical-type chemistry of square-planar Rh(II) species has been mainly studied for species containing porphyrinato ligands, but similar chemistry for the dianionic planar coordinating $N_4$- and $N_2O_2$-ligands TMTAA (174), OETAP (175), DBPB (176), and TTBS (177, 178). (Fig. 61) has been reported, including formation of $M^{III}$($\mu_2$-$CH_2$CHR—)$M^{III}$ bridged species in reactions with olefins, C—N bond cleavage in reactions with isocyanides, O—C cleavage in reactions with $P(OMe)_3$, and formation of M—C(O)H formyl species in reactions with $CO/H_2$. [TMTAA = benzotetromethylazo[*1a*]anulene dianion; OETAP = octaethyltetraazoporphyrin dianion; $H_2$DBPB = 4,5-dimethyl-1,2

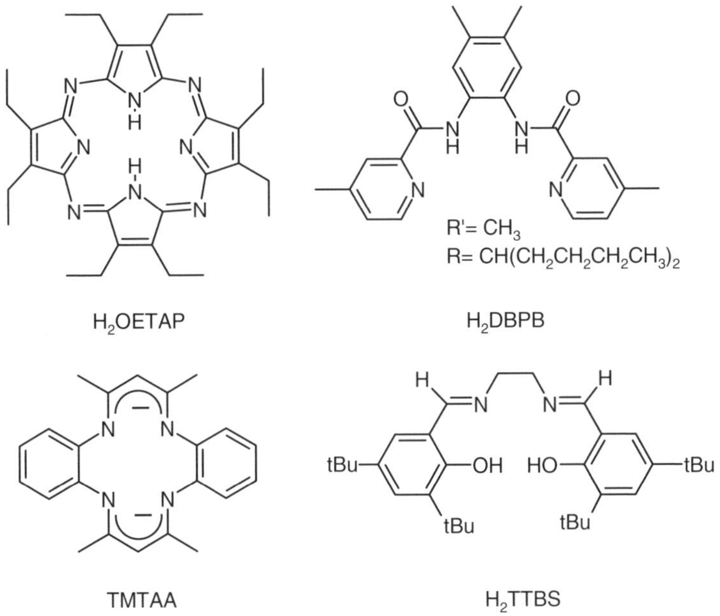

Figure 61. Alternative ligands used in the organometallic chemistry of planar $Rh^{II}$ metalloradicals.

bis-(Ca-(1-butylpentyl))pyridine-2-carboxamido)benzene; TTBS = [*N*,*N*'-ethylene bis(3,5-di-*tert*-butyl salicyladimine dianion]. These reactions do not proceed in all cases as selective as with porphyrinato ligands, presumably due to the fact that some of these ligands are more flexible. Also, in the case of the DBPB and TTBS ligands, group transfer to the negatively charged phenolate oxygen and amido nitrogen donors has been observed. Nevertheless, these examples reveal that the intriguing radical-type chemistry of neutral square-planar M(II) complexes is not at all restricted to the use of porphyrinato ligands.

## D. Reactivity of Five-Coordinate $[M^{II}(N\text{-ligand})(\text{olefin})]^{2+}$ Complexes (M = Rh, Ir)

A series of five-coordinate, dicationic Rh(II) and Ir(II) olefin complexes stabilized by N-donor ligands has been investigated in detail. Species of the type $[Ir^{II}(N_4\text{-ligand})(\text{ethene})]^{2+}$ (162, 179, 180) $[Rh^{II}(N_3\text{-ligand})(\text{nbd})]^{2+}$ (181) (nbd = norbornadiene) and $[M^{II}(N_3\text{-ligand})(\text{cod})]^{2+}$ (182, 183) have been obtained from their M(I) precursors by one-electron oxidation using either $[Fc]^+$ or $Ag^+$ as an oxidant (Fig. 62).

Cyclic voltammograms of these complexes revealed reversible waves for the M(I)/M(II) couple in all cases, except for some of the less bulky cyclooctadiene

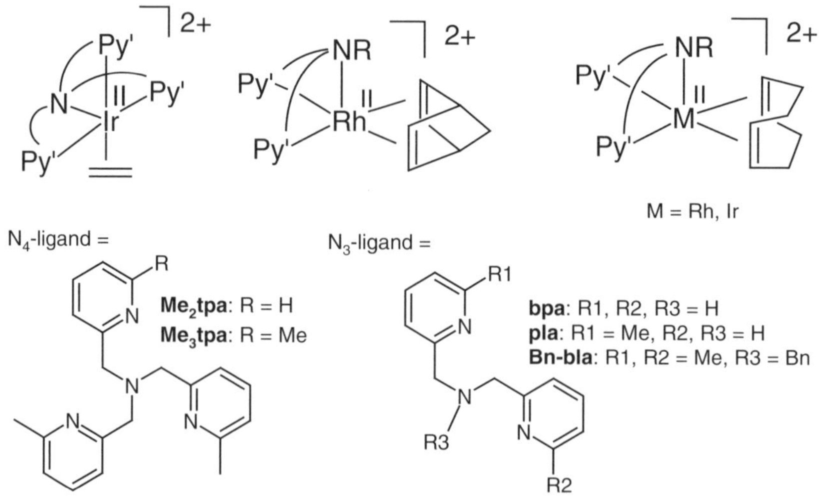

Figure 62. Structure of $[\mathrm{Ir^{II}}(N_4\text{-ligand})(\mathrm{ethene})]^{2+}$, $[\mathrm{Rh^{II}}(N_3\text{-ligand})(\mathrm{nbd})]^{2+}$, and $[\mathrm{M^{II}}(N_3\text{-ligand})(\mathrm{cod})]^{2+}$ complexes.

complexes. The M(II)/M(III) couple was not observed at potentials below the oxidation potential of the solvent. This indicates that direct electron-transfer (ET) disproportionation of the dicationic M(II) species to monocationic M(I) and tricationic M(III) species is substantially endergonic for all these species ($>$22–30 kcal $mol^{-1}$ in both acetone and $CH_2Cl_2$). Electron-transfer disproportionation of M(II) to M(I) and tricationic M(III) is therefore unlikely. However, in the presence of chloride anions, irreversible oxidation waves are observed. It seems that $Cl^-$ coordination triggers the ET disproportionation, most likely because formation of dicationic $[\mathrm{M^{III}(Cl)}]$ products is less unfavorable than formation of tricationic M(III) species. The $Cl^-$ ion triggered disproportionation of $[\mathrm{Rh^{II}(pla)(nbd)}]^{2+}$ in the presence of water or alcohols, led to formation of $[\mathrm{Rh^{I}(pla)(nbd)}]^{+}$ and $[\{\mathrm{Rh^{III}(Cl)(nbd{-}OR)(pla)}\}_2]^{2+}$ (Fig. 63). The latter species likely result from a Wacker-type attack of water–alcohols to the $[\mathrm{Rh^{III}(Cl)(pla)(nbd)}]^{2+}$ intermediate. Heating of $[\{\mathrm{Rh^{III}(Cl)(nbd{-}OH)(pla)}\}_2]^{2+}$ in the presence of an excess of the oxidant $FeCl_3$ led to catalytic formation of norbornenone, albeit with only low total turnover numbers (181).

The X-ray structures of $[\mathrm{Ir^{II}(Me_3tpa)(ethene)}]^{2+}$ (162, 179) and $[\mathrm{Rh^{II}(Bn{-}bla)(cod)}]^{2+}$(183) as well as their M(I) precursors have been reported. The M(I) species adopt a trigonal-bipyramidal geometry. Upon oxidation the geometry changes to a square pyramid with the amine at the apical position and the olefin and the pyridines coordinating in the basal plane. The pyridine-methyl groups shield the vacant position trans to the amine (Fig. 64). On going from

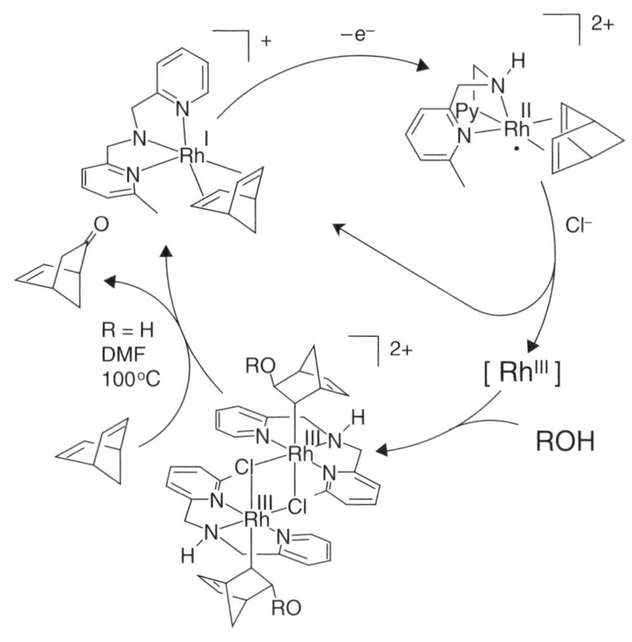


Figure 63. Catalytic formation of norbornenone via chloride triggered disproportionation of $[Rh^{II}(nbd)(pla)]^{2+}$.

M(I) to M(II), the M → olefin π back-bonding becomes weaker, resulting in longer M—C and shorter C=C bonds. The M–N interactions become stronger. The structures elucidated by X-ray diffraction are consistent with DFT calculations and EPR data.

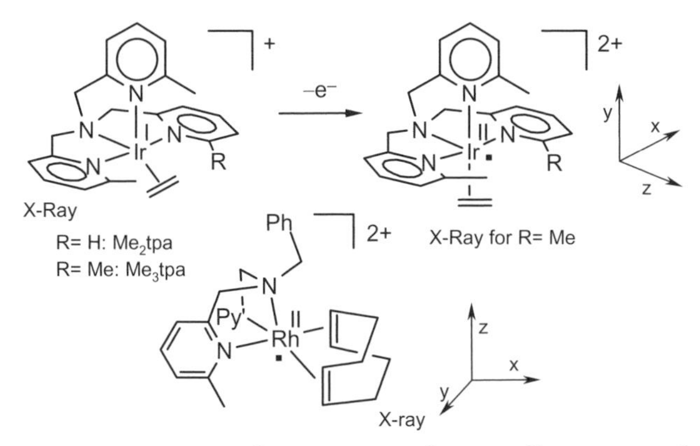


Figure 64. Sterically protected $[Ir^{II}(ethene\ (Me_ntpa)]^{2+}$ and $[Rh^{II}(cod)(Bn–bla)]^{2+}$ species.

Similar to other square-pyramidal structures in which the unpaired electron resides in the $d_{z^2}$ orbital, $[Rh^{II}(N_3\text{-ligand})(nbd)]^{2+}$ and $[Rh^{II}(N_3\text{-ligand})(cod)]^{2+}$ show EPR spectra with $g_x$, $g_y > 2$ and $g_z \sim 2$ (see also Fig. 4), with $g_z$ revealing a strong (super)hyperfine coupling with both the metal and the apical $N_{amine}$ donor (Table XI) (162, 181, 183). For all analogous Ir compounds (Table XIII) the $g_z$ value is notably lower than $g_e$, pointing to some mixing of the $d_{z^2}$ orbital with a $d_{x^2-y^2}$ orbital ($d_{xy}$ for the cod complexes) in the SOMO. For most of the complexes, the hyperfine couplings in the $g_x$ and $g_y$ directions are not resolved, except for $[Ir^{II}(N_4\text{-ligand})(ethene)]^{2+}$ revealing iridium hyperfine couplings along $g_x$. The equal spacing of the four line pattern is, however, substantially distorted by the strong Ir quadrupole interactions, causing this signal to appear

TABLE XIII
The EPR Values for $[Ir^{II}(N_4\text{-ligand})(ethene)]^{2+}$, $[Rh^{II}(N_3\text{-ligand})(nbd)]^{2+}$, and $[M^{II}(N_3\text{-ligand})(cod)]^{2+}$ (M = Rh, Ir) [a]

| Complex | $g_x$<br>$g_y$<br>$g_z$ | $A_x(^{103}Rh/^{193}Ir)$<br>$A_y(^{103}Rh/^{193}Ir)$<br>$A_z(^{103}Rh/^{193}Ir)$ | $A_x(^{14}N_{amine})$<br>$A_y(^{14}N_{amine})$<br>$A_z(^{14}N_{amine})$ | NQI (η) | References |
|---|---|---|---|---|---|
| $[(Me_2tpa)Ir^{II}(ethene)]^{2+}$ | 2.52 | 47 | | | 162 |
| | 2.27 | | | | |
| | 1.98 | 43 | 18 | | |
| $[(Me_3tpa)Ir^{II}(ethene)]^{2+}$ | 2.54 | 45 | | −10 | 162 |
| | 2.27 | | | −16 | |
| | 1.98 | 46 | 17 | 26 | |
| $[(pla)Rh^{II}(nbd)]^{2+}$ | 2.24 | | | | 181 |
| | 2.18 | | | | |
| | 2.01 | 57 | 60 | | |
| $[(bpa)Rh^{II}(cod)]^{2+}$ | 2.23 | | | | 182, 183 |
| | 2.20 | | | | |
| | 2.01 | 61 | 61 | | |
| $[(bpa)Ir^{II}(cod)]^{2+}$ | 2.45 | | | | 183 |
| | 2.36 | | | | |
| | 1.95 | 130 | 60 | | |
| $[(pla)Rh^{II}(cod)]^{2+}$ | 2.22 | | | | 183 |
| | 2.20 | | | | |
| | 2.01 | 11 | 35 | | |
| $[(pla)Ir^{II}(cod)]^{2+}$ | 2.43 | | | −54 | 183 |
| | 2.34 | | | −54 | |
| | 1.95 | 112 | 60 | +108 | |
| $[(Bn\text{–}bla)Rh^{II}(cod)]^{2+}$ | 2.25 | | 46 | | 183 |
| | 2.20 | | 42 | | |
| | 1.99 | 60 | 64.4 | | |
| $[(Bn\text{–}bla)Ir^{II}(cod)]^{2+}$ | 2.50 | | | −60 | 183 |
| | 2.36 | | | −60 | |
| | 1.92 | 108 | 60 | +120 | |

[a]Hyperfine couplings in megahertz.

as a triplet. In addition, the nuclear quadrupole interactions also cause the occurance of forbidden $\Delta M_I = 2$ transitions as weak satellite signals along this direction (162). The EPR spectra of $[Ir^{II}(N_3\text{-ligand})(cod)]^{2+}$ species, as well as $[Ir^{II}(HBPz_3)(CO)_2]$ (46) also show features caused by strong nuclear quadrupole interactions.

The EPR parameters calculated with DFT were in satisfactory agreement with the experimentally derived parameters. Experimental ENDOR measurements, as well as DFT spin density plots, reveal that the spin density is substantially delocalized over the $N_{amine}$ donor (15–18%) and the metal (73–78%) (183). Equally large spin densities have been observed for amines coordinated to $[Rh^{II}(por)]$ species (134). Delocalization of the unpaired electron over the metal and the amine might contribute to the relative stability of these species.

Although the $[M^{II}(N_3\text{-ligand})(cod)]^{2+}$ species can be isolated as pure compounds, they decompose upon standing in solution. The $[M^{II}(N_3\text{-ligand})(cod)]^{2+}$ complexes invariably decompose to 1:1 mixtures of $[M^{III}(N_3\text{-ligand})(cyclooctadienyl)]^{2+}$ and species that could be described as protonated $M^I(cod)$ species. For the latter protonated $M^I(cod)$ species, the position of the proton depends on the nature of the metal. The proton ends-up at the metal for iridium, forming $[Ir^{III}(H)(N_3\text{-ligand})(cod)]^{2+}$, and at the $N_3$-ligand Py moiety for rhodium, forming $[Rh^I(N_3\text{-ligand } H)(cod)]^{2+}$ (Fig. 65) (182, 183). The reaction rates decrease upon increasing the steric bulk of the $N_3$-ligand.

The product ratio can be explained by a hydrogen atom transfer from one $M^{II}(cod)$ species to another $M^{II}(cod)$ species. An ET disproportionation reaction between two $M^{II}(cod)$ species to give $M^I(cod)$ and $M^{III}(cod)$ followed by a proton transfer from an allylic position of $M^{III}(cod)$ to $M^I(cod)$ was ruled out (182, 183). Electrochemical data suggest that the solvent is easier to oxidize

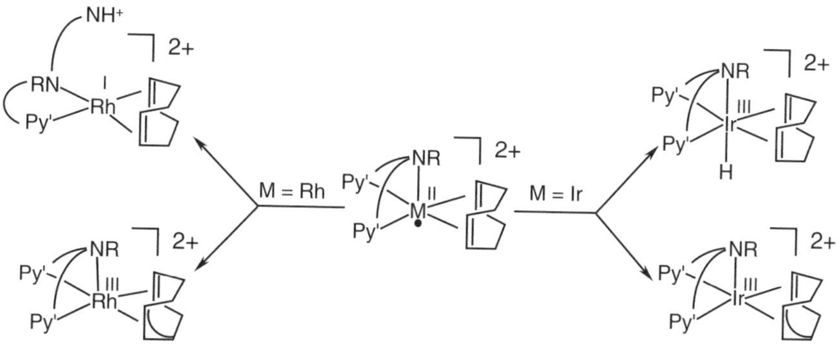

Figure 65. Selective conversion of the $[M^{II}(cod)(N_3\text{-ligand})]^{2+}$ complexes to allyl species $[M^{III}(cyclooctadienyl)(N_3\text{-ligand})]^{2+}$ (M = Rh, Ir) and equimolar amounts of $[Rh^I(cod)(N_3\text{-ligandH})]^{2+}$ (for M = Rh) or $[Ir^{III}(H)(cod)(N_3\text{-ligand})]^{2+}$ (for M = Ir).

than the $M^{II}$(cod) species. The $\Delta G^{act}$ values for hypothetical initial ET steps, as derived from electrochemical data (>30 kcal $mol^{-1}$ depending on the species) are higher than the value obtained from kinetic studies of $[Ir^{II}(Bn\text{—}bla)(cod)]^{2+}$ ($\sim$ 20 kcal $mol^{-1}$). All other species studied react much faster, some of which even decompose slowly at −78°C. Clearly, the electrochemical data do not support an ET disproportionation mechanism. Furthermore, very similar $[M^{II}(N_3\text{-ligand})(nbd)]^{2+}$ species are stable with respect to ET disproportionation (in the absence of chloride). Another mechanism must be responsible for the observed decompositions. A direct hydrogen-atom abstraction (at least in case of the more hindered Bn—bla ligands) of an allylic hydrogen atom of $M^{II}$(cod) by another M(II) species seems unlikely, because the metal is rather shielded by the $N_3$-ligand PyMe bulk. The kinetic rate expression $v = k_{obs}\,[M^{II}]^2$ in acetone, with $k_{obs} = k[H^+][S]$ and [S] being the concentration of additional coordinating reagents (MeCN), is in agreement with a solvent assisted dissociation of one of the pyridine donors (183). Solvent coordination results in formation of more open, reactive species. Protonation of the noncoordinating pyridine increases the concentration of this species, and thus $[H^+]$ appears in the kinetic rate expression. Both the rate expression and the kinetic activation parameters ($\Delta H^{\ddagger} = 12 \pm 2$ kcal $mol^{-1}$, $\Delta S^{\ddagger} = -27 \pm 10$ cal $K^{-1}$ $mol^{-1}$ and $\Delta G_{289K} = 20 \pm 5$ kcal $mol^{-1}$) are in agreement with a subsequent bimolecular hydrogen atom transfer from an allylic $M^{II}$(cod) position to another $M^{II}$ species (Fig. 66). This mechanism is quite similar to the bimolecular C—H activation reactions observed for $[Rh^{II}(por)]$ species described by Wayland et al. (166) (see Section IV.C.6).

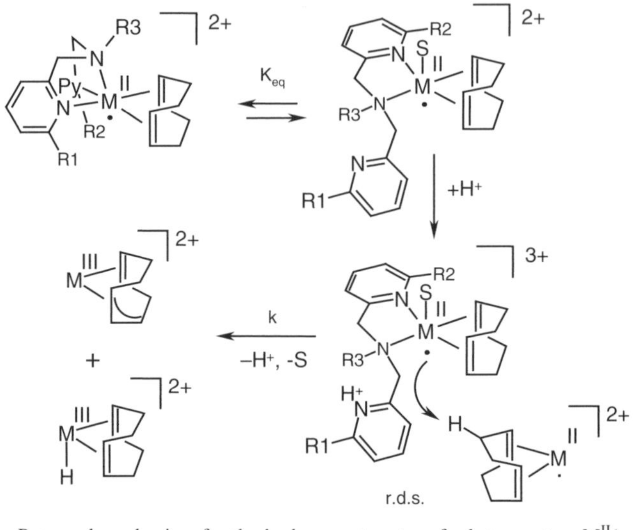

Figure 66. Proposed mechanism for the hydrogen atom transfer between two $M^{II}$(cod) species.

Treatment of $[Ir^{II}(Me_3tpa)(ethene)]^{2+}$ and $[M^{II}(dpa)(cod)]^{2+}$ with dioxygen leads to formation of new EPR spectra (179, 182–184). A substantial decrease in the *g* anisotropy indicates a shift of the spin density from the metal to dioxygen. The observed spectra are consistent with formation of the superoxide complexes $[Ir^{III}(O_2\bullet)(Me_3tpa)(ethene)]^{2+}$ and $[M^{III}(O_2\bullet)(dpa)(cod)]^{2+}$. Formation of superoxide complexes from $[M^{II}(N_3\text{-ligand})(cod)]^{2+}$ and $O_2$ appears to be reversible. Increasing the *N*-ligand steric bulk decreases the oxygen affinities. Reaction of $[Rh^{II}(pla)(cod)]^{2+}$ with $O_2$ only leads to small concentrations of $[Ir^{III}(O_2\bullet)(pla)(cod)]^{2+}$ where pla = (2-pyradylmethyl). The $[Rh^{II}$(R—bla)(cod)$]^{2+}$ (R = H, Bn) and $[Rh^{II}(pla)(nbd)]^{2+}$ complexes proved to be air stable as a result of shielding by the bulky Py—Me groups. No C—O bond formation reactions were observed for any of the $[M^{III}(O_2\bullet)(N_3\text{-ligand})(cod)]^{2+}$ complexes. These species decompose to the same mixture of compounds as observed for $[M^{II}(N_3\text{-ligand})(cod)]^{2+}$ in the absence of $O_2$ (see above) (182). $[Ir^{II}(Me_2tpa)(ethene)]^{2+}$ reacts rapidly with $O_2$, but not selectively (all solvents).

The fate of the more hindered analogue $[Ir^{III}(O_2\bullet)(Me_3tpa)(ethene)]^{2+}$ depends on the solvent. The reaction in acetone leads to yet unidentified products. The reaction in MeCN takes a different course (as described hereafter). To shed some light on these reactions, $[(Me_3tpa)Ir^{III}(O_2\bullet)(ethene)]^{2+}$ was trapped with DMPO (DMPO = 5,5-diserlyl-2-pyrrolidine-1-oxide) (184). This led to dissociation of ethene and formation of a C—O bond between the superoxide and DMPO (Fig. 67). The peroxide bridged species likely rearranges to a species with an acetal group, thus explaining the absence of proton hyperfine couplings in EPR. The unusual large N hyperfine coupling ($A^N = 74$ MHz) compared to trapped organic radicals ($A^N = 42$ MHz) indicates that the unpaired electron resides mainly at the nitrogen part of the nitroxyl group (185). Therefore, DMPO appears to be coordinated to the iridium center. This seems to be confirmed by the unusually large *g* anisotropy for a DMPO trapped radical. At room temperature, this species converts to a diamagnetic species with a coordinated oxygenated DMPO fragment (Fig. 67). This likely involves hydrogen abstraction from the solvent and elimination of $H_2O$ (Fig. 68). Remarkably, such obtained solvent radicals were trapped by Ir(II) species, rather than the spin-trap DMPO. Clear formation of $[Ir^{III}(CH_2CN)(Me_3tpa)]^{2+}$ in acetonitrile and $[Ir^{III}(CH_2COMe)(Me_3tpa)]^{2+}$ in acetone has been detected with ESI–MS (ESI–MS = electron-spray ionisation mass spectrometry).

Although the PyMe groups shield the vacant site of $[Ir^{II}(Me_3tpa)(ethene)]^{2+}$ from attack by bulky reagents, acetonitrile is small enough to coordinate to iridium trans to the amine (Fig. 68). The DFT calculations of the MeCN adduct indicate that the ethene fragment of this species coordinates in a slipped way, leading to an iridium–alkyl radical-type species (162). Whereas the α-carbon of the slipped olefin fragment seems to be tetrahedral and $sp^3$ hybridized, the

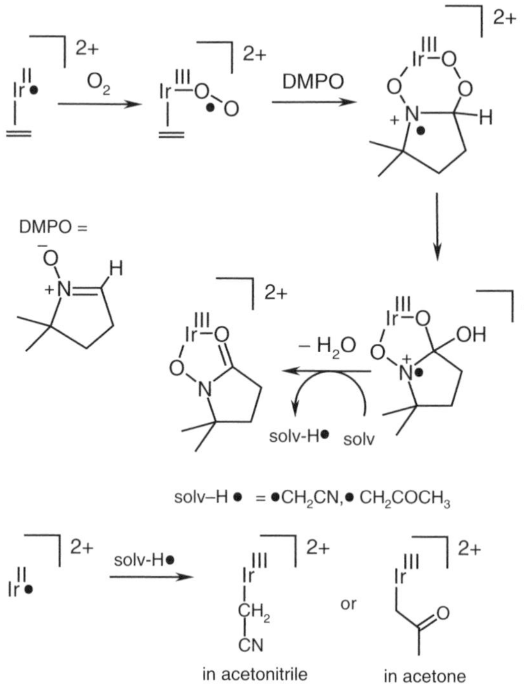

Figure 67. Proposed mechanism for the oxygenation of DMPO at Ir(II).

β-carbon atom is planar and *sp*$^2$ hydridized. A spin-density plot (Fig. 68) reveals that this species is intermediate between the slipped-olefin description $Ir^{II}$—$CH_2$—$CH_2^+$ and the ethyl radical description $Ir^{III}$—$CH_2$—$CH_2$•; the latter prevails.

Dissociation of ethene from [$Ir^{III}$(—$CH_2CH_2$•)($Me_3$tpa)(NCMe)]$^{2+}$ leads to the reactive metalloradical [$Ir^{II}$($Me_3$tpa)(NCMe)]$^{2+}$. This species contains only N-donor ligands and has its unpaired electron located in an orbital that is not sterically shielded by the ligand bulk. Although there is no spectroscopic evidence for formation of [$Ir^{III}$($CH_2CH_2$•)($Me_3$tpa)(NCMe)]$^{2+}$ or [$Ir^{II}$($Me_3$tpa)(NCMe)]$^{2+}$, their existence can be rationalized from the reaction products of the $Ir^{II}$(ethene) species in acetonitrile. In the presence of the stable radical TEMPO, radical coupling of the metalloradical [$Ir^{II}$($Me_3$tpa)(NCMe)]$^{2+}$ with TEMPO takes place, leading to the formation of [$Ir^{III}$($TEMPO^-$) ($Me_3$tpa)(NCMe)]$^{2+}$. Assisted by hydrogen-bond formation between the heteroatoms of the coordinated ligands and water, the nitrile is quickly hydrolyzed to an amide in the presence of water, leading to [$Ir^{III}$(NHCOMe)(TEMPO*H*) ($Me_3$tpa)]$^{2+}$. In the absence of TEMPO, [$Ir^{II}$($Me_3$tpa)(NCMe)]$^{2+}$ recombines with [$Ir^{II}$($Me_3$tpa)($CH_2CH_2$)]$^{2+}$, or perhaps more likely [$Ir^{III}$($CH_2CH_2$•)($Me_3$tpa) (MeCN)]$^{2+}$ to

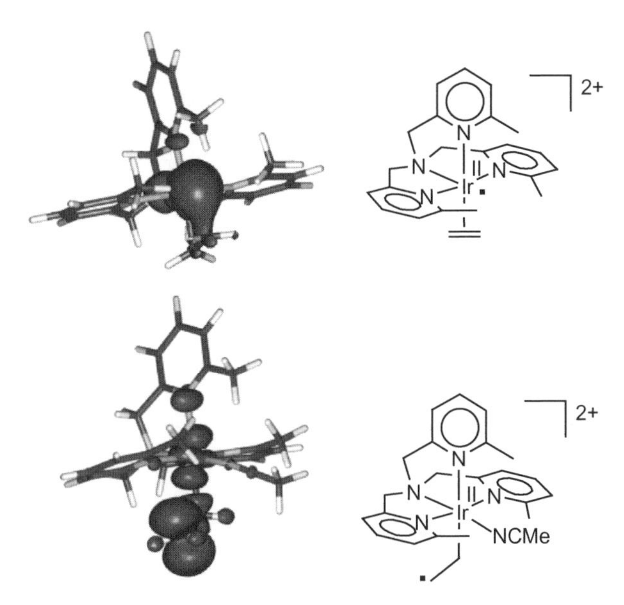


Figure 68. Spin density plots of $[Ir^{II}(CH_2{=}CH_2)(Me_3tpa)]^{2+}$ (above) and $[Ir^{III}(-CH_2-CH_2\bullet)(Me_3tpa)(NCMe)]^{2+}$ (below).

form an ethylene bridged species (Fig. 69) (162, 180). The increased reactivity on going from $Me_3$tpa to the less bulky $Me_2$tpa complex in formation of the ethylene bridged species, is in good agreement with the proposed associative addition of acetonitrile to $[Ir^{II}(Me_n tpa)(ethene)]^{2+}$ being the first step of the dimerization process. In contrast to observations concerning reactions of $[M^{II}(por)]$ species with olefins (see Section IV.C.3), no butylene bridged species were formed. Apparently, the concentration of $[Ir^{III}(CH_2CH_2\bullet)(N_4\text{-ligand})(MeCN)]^{2+}$ is too low for carbon–carbon coupling reactions to take place, and metal–carbon coupling with $[Ir^{II}(N_4\text{-ligand})\ (NCMe)]^{2+}$ species prevails.

In the presence of dioxygen, the $[Ir^{III}(CH_2CH_2\bullet)(N_4\text{-ligand})(MeCN)]^{2+}$ species are prone to C—O bond formation. Quite likely, the triplet biradical dioxygen reacts directly with the β-carbon centered radical to give $Ir^{III}-CH_2CH_2OO\bullet$ species. The latter alkylperoxo radical species then abstracts a hydrogen atom from acetonitrile, leading to a hydroperoxo intermediate $Ir^{III}-CH_2CH_2OOH$, which eliminates water to give the observed $Ir^{III}-CH_2COH$ COH species (Fig. 70) (184). Also, in this case the remaining solvent radical is trapped by the Ir(II) starting material. An alternative mechanism via $[Ir^{III}(O_2\bullet)(N_4\text{-ligand})(ethene)]^{2+}$ species cannot be ruled out completely.

The formation of the similar products $[Ir^{III}(CH_2COH)(Me_3tpa)(H_2O)]^{2+}$, $[\{Ir^{III}(Me_3tpa)(NCMe)\}_2(\mu_2\text{-}CH_2CH_2-)]^{4+}$, $[\{Ir^{III}(Me_3tpa)(NCPh)\}_2(\mu_2\text{-}CH_2$

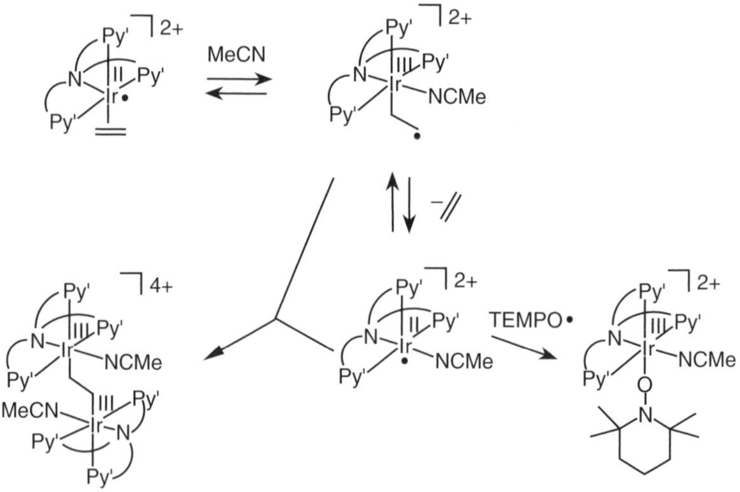

Figure 69. Formation of $Ir^{III}$–$CH_2CH_2$–$Ir^{III}$ and $Ir^{III}$–TEMPO via $Ir^{III}(CH_2CH_2\bullet)(NCMe)$ and $Ir^{II}(NCMe)$ species.

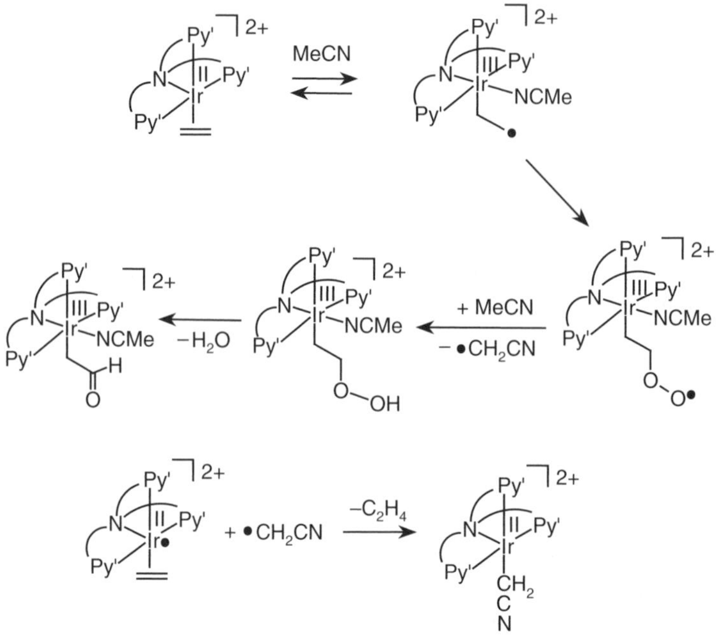

Figure 70. Proposed mechanism for the monooxygenation of $[Ir^{II}(ethene)(Me_3tpa)]^{2+}$ in acetonitrile.

$CH_2—)]^{4+}$, and $[\{Ir^{III}(Cl)(Me_3tpa)\}_2(\mu_2\text{-}CH_2CH_2—)]^{2+}$ indicate that this chemistry is not restricted to acetonitrile and expands to other donors ligands (e.g., other nitriles, $H_2O$, and chloride) (162, 184).

The $[Ir^{II}(Me_3tpa)$ ethene)$]^{2+}$ complex also reacts with nitrogen monoxide, leading to the formation of $[Ir(Me_3tpa)(NO)]^{2+}$. Nitrogen monoxide binds to the iridium center in a linear fashion, as indicated by the X-ray structure determination and an absorption frequency of $\nu = 1800\ cm^{-1}$ in the IR spectrum. The $[Ir(Me_3tpa)(NO^+)]^{2+}$ complex can also be prepared by treatment of $[Ir^I(Me_3tpa)(ethene)]^+$ with $[NO]^+[PF_6]^-$.

## E. Metal–Carbon Bond Formations at Rh(II) Relevant in Catalytic Carbene-Transfer Reactions

Bergman and co-workers (186, 187) reported on some metal–carbon bond-formation reactions starting from a number of $[Rh^{II}(Cl)_2$bis(oxozaline)] complexes. Heating these complexes results in formation of the hydride species $[Rh^{III}(Cl)_2(H)$ bis(oxozaline)] and the cyclometalated complex $[Rh^{III}(\eta^3$-bis(oxozalinato)$(Cl)_2]$ (Fig. 71) (187). The first step is believed to involve an oxidative addition of an aromatic C—H bond of the ligand to give a Rh(IV) intermediate. A subsequent intermolecular hydrogen atom transfer was proposed to yield the two observed diamagnetic products. Slow precipitation of rhodium metal prevented a detailed kinetic study.

Treatment of $[Rh^{II}(Cl)_2$bis(oxozaline)] with AgOTf yields $[Rh^{II}(Cl)(Tf)$bis(oxozaline)], which can be used as an olefin cyclopropanation catalyst. Although the catalytic precursor does not contain a Rh—C bond, the carbene-transfer reaction is highly likely to involve organometallic intermediates. Catalyst

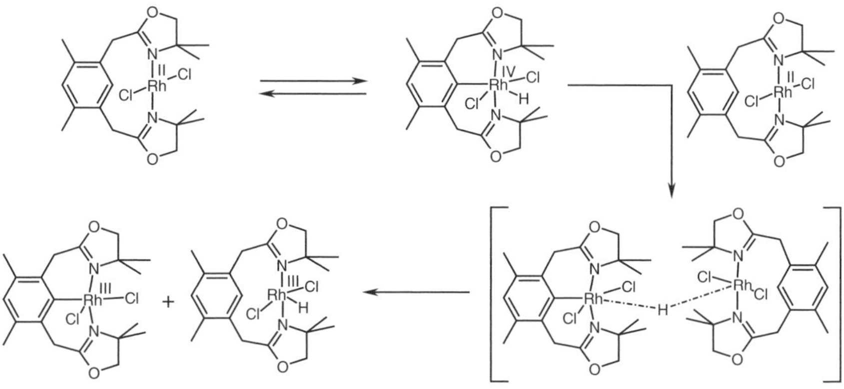

Figure 71. Disproportionation of $[Rh^{II}(Cl)_2$bis(oxozaline)] via intramolecular C—H activation and binuclear hydrogen transfer.

activity can be enhanced by addition of $NaBAr_F$, which yields $[Rh^{II}bis(oxozaline)(Cl)(CH_2Cl_2)]^+$ (Tf = triflate, $BAr_F$ = tetrakis[3,5-(trifluoromethyl)phenyl)borate] (Fig. 71) (188). The active species is believed to be a low-valent 15 VE species with two bis(oxozaline) nitrogens in mutual trans positions and a chloro ligand occupying a third position. Two sites of the rhodium center are sterically hindered by the bis(oxozaline) methyl groups and the bridging arene. Trans to the chloro ligand, a labile triflate anion, or dichloromethane ligand is bound allowing easy formation of a vacant site. In the presence of diazoacetates, dimerization to fumarates and maleates takes place, indicative of carbene-transfer reactions (188). Also selective cyclopropanation (Fig. 72) of olefins and formation of aziridines from imines is catalyzed by these complexes. Dinuclear carboxylate bridged Rh(II)—Rh(II) species have been extensively studied as catalysts for carbene-transfer reactions (189). It has always been assumed that the Rh(II)—Rh(II) species stay intact during catalysis, but no synergistic effects could be proven. The elegant study by Bergman clearly reveals that mononuclear Rh(II) species can also be active catalysts for carbene-transfer reactions. One might therefore consider open-shell species to play an active role in other carbene-transfer reactions as well.

Whereas carbene dimerization to (maleates and) fumarates occurs as an unwanted side reaction lowering the yields of the above carbene-transfer reactions, fumarates can also stabilize Rh(II) species (190). Electrochemical two-electron oxidation of the hydride-olefin species $[Rh^I(H)\{CH(COOMe)=CH(COOMe)\}(P_3\text{-ligand})]$ results in an oxidatively induced migratory insertion of the coordinated fumarate into the Rh—H bond, yielding $[Rh^{III}(-CH(COOMe)CH_2COOMe)(P_3\text{-ligand})]^{2+}$ shown in Fig. 73 [$P_3$-ligand = $MeC(CH_2PPh_2)_3$]. Cyclic voltammetry investigations revealed an electrochemically irreversible $2e^{-1}$ oxidation. Electrochemical $1e^{-1}$ reduction of the oxidized species is, however, reversible on the cyclic voltammetry time scale.

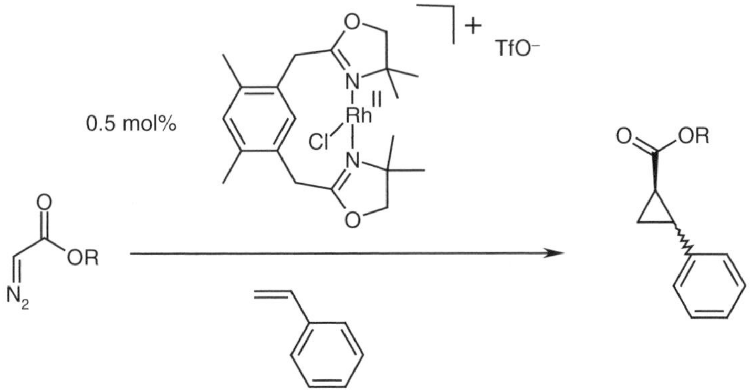

Figure 72. Cyclopropanation of styrene mediated by a mononuclear $[Rh^{II}(Cl)_2(benbox)]^+$.

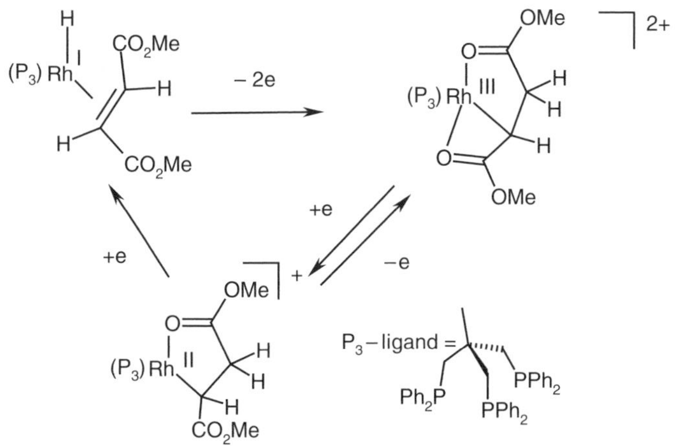

Figure 73. Redox behavior of $[Rh^{II}\{CH(COOMe)CH_2(COOMe)\}(triphos)]^+$.

Apparently, one-electron reduction of the Rh(III) species does not cause deinsertion/β-hydrogen elimination in the Rh(II) state, as observed upon further reduction to Rh(I) (irreversible $1e^{-1}$ reduction wave at a 0.56 V lower potential). The Rh(II) species (Fig. 73) was characterized by means of EPR spectroscopy. The obtained *g* values in frozen solution and at room temperature indicate a square-pyramidal geometry with one P atom at the apical position (Table XIV) (190). Quite similar EPR spectra and geometries were observed for the related complexes $[Rh^{II}(CN)(PP_3)]^+$, $[Rh^{II}(CN(NP_3))]^+$, and $[Rh^{II}(C{\equiv}CPh)(NP_3)]^+$ ($PP_3 = P(CH_2CH_2PPh_2)_3$, $NP_3 = N(CH_2CH_2PPh_2)_3$) (191–193) (see Fig. 74 and Table XII.

The $[Rh^{II}(CN)(PP_3)]^+$ complex was prepared electrochemically and consists of a mixture of two isomeric square-pyramidal species at room temperature, with one of the isomers having the $PP_3$ bridgehead phosphine at the apical position and the other isomer having one of the terminal phosphines of the $PP_3$ ligand at the apical position (Fig. 74). These species decompose via disproportionation to a mixture of $[Rh^{III}(CN)_2(PP_3)]^+$ and $[Rh^{I}(PP_3)]^+$ (191). Similar results were obtained for $[Rh^{II}(CN)(NP_3)]^+$. The $[Rh^{II}(C{\equiv}CPh)(NP_3)]^+$ complex was prepared by protonation of $[Rh^{I}(C{\equiv}CPh)(PP_3)]^+$, followed by thermal decomposition of the resulting vinylidene species $[Rh^{III}(C{=}CHPh)(PP_3)]^+$ in refluxing THF with net release of dihydrogen (Fig. 74). Alternatively, $[Rh^{II}(C{\equiv}CR)(NP_3)]^+$ and $[Rh^{II}(C{\equiv}CR)(PP_3)]^+$ can be obtained directly by one electron oxidation of their Rh(I) precursors $[Rh^{I}(C{\equiv}CPh)(NP_3)]$ and $[Rh^{I}(C{\equiv}CR)(PP_3)]$ (192, 193). Like the cyanide complexes, the $[Rh^{II}(C{\equiv}CR)(PP_3)]^+$ species also consist of a mixture of two isomers.

The complex $[M^{I}\{\eta^3\text{-}S(C_6H_4)CH{=}CH_2\}(P_3\text{-ligand})]$ undergoes two subsequent one-electron oxidations to $[M^{II}\{\eta^3\text{-}S(C_6H_4)CH{=}CH_2\}(P_3\text{-ligand})]^+$ and

TABLE XIV

The EPR Values of the $(P_3)Rh^{II}$, $(PP_3)Rh^{II}$, and $(NP_3)Rh^{II}$ Species.[a]

| | $g$ (Frozen Solution) | A (Frozen Solution) | $g$ Value (RT) | Hyperfine Couplings (RT) | Reference |
|---|---|---|---|---|---|
| $[Rh^{II}\{CH(COOMe)CH_2(COOMe)\}(P_3)]^+$ | $g_\perp = 2.108$<br>$g_\parallel = 2.002$ | $A_\perp = 670$<br>$A_\parallel = 762$ | $g = 2.070$ | $A_{P\text{-apical}} = 681$ | 190 |
| $[Rh^{II}(C{\equiv}CPh)(NP_3)]^+$ | $g_\perp = 2.082$<br>$g_\parallel = 2.007$ | $A_\perp = 586$<br>$A_\parallel = 702$ | $g = 2.062$ | $A_{P\text{-apical}} = 641$ | 192, 193 |
| $[Rh^{II}(C{\equiv}CCO_2Et)(NP_3)]^+$ | $g_\perp = 2.085$<br>$g_\parallel = 2.002$ | $A_\perp = 627$ | $g = 2.063$ | $A_{P\text{-apical}} = 632$ | 193 |
| $[Rh^{II}(C{\equiv}CPh)(PP_3)]^{+b}$ | $g_\perp = 2.086$<br>$g_\parallel = 1.996$ | $A_\perp = 524$<br>$A_\parallel = 679$ | a: $g = 2.057$<br>b: $g = 2.057$ | a: $A_{P\text{-apical}} = 547$<br>b: $A_{P\text{-apical}} = 602$ | 193 |
| $[Rh^{II}(C{\equiv}CCO_2Et)(PP_3)]^{+b}$ | $g_\perp = 2.086$<br>$g_\parallel = 1.995$ | $A_\perp = 537$<br>$A_\parallel = 679$ | a: $g = 2.055$<br>b: $g = 2.053$ | a: $A_{P\text{-apical}} = 547$<br>b: $A_{P\text{-apical}} = 601$ | 193 |
| $[Rh^{II}(C{\equiv}CCOH)(PP_3)]^+$ | $g_\perp = 2.086$<br>$g_\parallel = 1.999$ | $A_\perp = 540$<br>$A_\parallel = 671$ | a: $g = 2.056$<br>b: $g = 2.054$ | a: $A_{P\text{-apical}} = 547$<br>b: $A_{P\text{-apical}} = 601$ | 193 |
| $[Rh^{II}(CN)(PP_3)]^+$ | $g_\perp = 2.090$<br>$g_\parallel = 2.004$ | $A_\perp = 550$<br>$A_\parallel = 679$ | a: $g = 2.061$<br>b: $g = 2.066$ | a: $A_{P\text{-apical}} = 635$<br>b: $A_{P\text{-apical}} = 593$ | 191 |
| $[Rh^{II}(CN)(NP_3)]^+$ | $g_\perp = 2.102$<br>$g_\parallel = 2.002$ | $A_\perp = 597$<br>$A_\parallel = 787$ | $g = 2.066$ | $A_{P\text{-apical}} = 651$ | 191 |
| $[Rh^{II}\{\eta^3\text{-}S(C_6H_4)CH{=}CH_2\}(P_3)]^+$ | $g_\perp = 2.080$<br>$g_\parallel = 2.018$ | $A_\perp = 422$<br>$A_\parallel = 494$ | $g = 2.058$ | $A_{P\text{-apical}} = 446$ | 194 |
| $[Rh^{II}\{\eta^4\text{-}S(C_6H_4)CHMe\}(P_3)]$ | $g_{11} = 2.091$<br>$g_{22} = 2.008$<br>$g_{33} = 1.997$ | $A_{11} < 85$<br>$A_{22} < 85$<br>$A_{33} < 85$ | $g = 2.033$ | $A = <85$ | 194 |
| $[Rh^{II}\{\eta^4\text{-}S(C_6H_4){-}CH(CH_2CPh_3)\}(P_3)]$ | $g_{11} = 2.094$<br>$g_{22} = 2.008$<br>$g_{33} = 2.003$ | $A_{11} < 85$<br>$A_{22} < 85$<br>$A_{33} < 85$ | $g = 2.032$ | $A = <70$ | 194 |
| $[Ir^{II}\{\eta^4\text{-}S(C_6H_4){-}CHMe\}(P_3)]$ | $g_{11} = 2.051$<br>$g_{22} = 2.003$<br>$g_{33} = 2.003$ | $A_{11} < 85$<br>$A_{22} < 70$<br>$A_{33} < 70$ | $g = 2.047$ | $A = <10$ | 194 |
| $[Ir^{II}(\eta^2\text{-}C,S\text{-}C_8H_6S)(P_3)]$ (two isomers *a* and *b*) | *a:*<br>$g_\perp = 2.116$<br>$g_\parallel = 1.928$<br>*b:*<br>$g_\perp = 2.121$<br>$g_\parallel = 1.908$ | *a:*<br>$A_\perp = 583$<br>$A_\parallel = 623$<br>*b:*<br>$A_\perp = 591$<br>$A_\parallel = 612$ | *a:*<br>$g = 2.050$<br>*b:*<br>$g = 2.043$ | *a:*<br>$A_{P\text{-apical}} = 654$<br>*b:*<br>$A_{P\text{-apical}} = 663$ | 194 |
| $[Ir^{II}(DTBC)(P_3)]^c$ | $g_\perp = 2.062$<br>$g_\parallel = 1.960$ | $A_\perp = 687$<br>$A_\parallel = 850$ | $g = 2.026$ | $A_{P\text{-apical}} = 737$ | 195 |

[a]Hyperfine couplings in megahertz.

[b]In frozen solutions, only one species is observed, but in solution a mixture of two isomeric species (a and b) is present.

[c]DTBC = 3,5-di-*tert*-butylcatecholase.

$[M^{III}\{\eta^3\text{-}S(C_6H_4)CH{=}CH_2\}(P_3\text{-ligand})]^{2+}$, respectively, both processes being complicated by chemical follow-up reactions (M = Rh, Ir; $P_3$-ligand = $MeC(CH_2PPH_2)_3$) (194). $[M^I\{\eta^3\text{-}S(C_6H_4)CH{=}CH_2\}(P_3\text{-ligand})]$ reveals substantial π-back-bonding to the olefin, allowing a metalla(III)cyclopropane

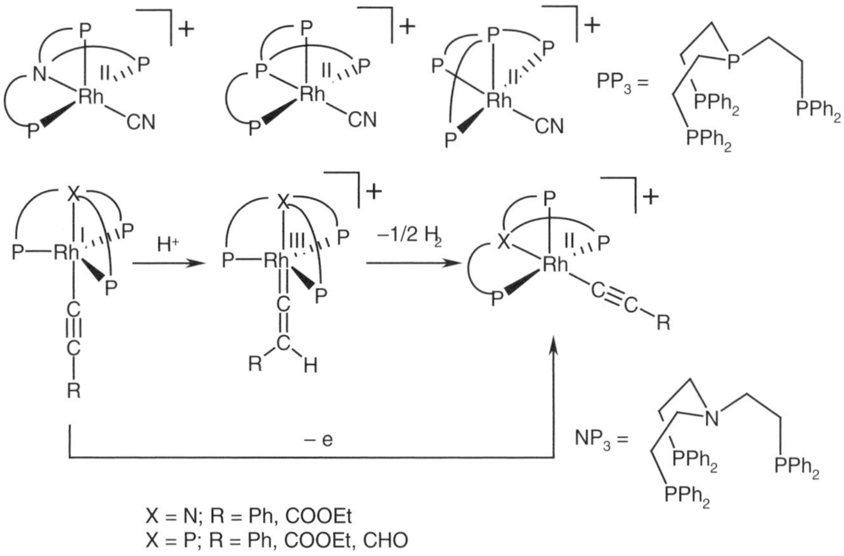

X = N; R = Ph, COOEt
X = P; R = Ph, COOEt, CHO

Figure 74. Structures of $Rh^{II}(CN)$ complexes and synthesis of $Rh^{II}(C{\equiv}CR)$ species.

description as indicated by X-ray diffraction and NMR measurements. The metal-to-olefin π-back-bonding in $[M^{II}\{\eta^3\text{-}S(C_6H_4)CH{=}CH_2\}(P_3\text{-ligand})]^+$ and $[M^{III}\{\eta^3\text{-}S(C_6H_4)CH{=}CH_2\}(P_3\text{-ligand})]^{2+}$ is substantially reduced. As a result, EPR spectra of $[Rh^{II}\{\eta^3\text{-}S(C_6H_4)CH{=}CH_2\}(P_3\text{-ligand})]^+$ reveal large hyperfine coupling constants with one phosphorus atom. This is indicative of a $d^7$ metal ion in an octahedral geometry with the unpaired electron in the $d_{z^2}$ orbital (Table XII). For the iridium analogue, no EPR data could be obtained, presumably due to the high reactivity of the species.

In solution, the $[M^{II}\{\eta^3\text{-}S(C_6H_4)CH{=}CH_2\}(P_3\text{-ligand})]^+$ species abstract a hydrogen atom (presumably from the solvent) and convert to diamagnetic $[M^{III}\{\eta^4\text{-}S(C_6H_4)CH(Me)\}\ (P_3\text{-ligand})]^+$ species. In these species, besides the anionic C and S donors, an aromatic double-bond also coordinates to the metal. The initially formed anti conformer rapidly converts to the thermodynamically more stable syn conformation (the one depicted in Fig. 75). One-electron reduction of the syn conformer $[M^{III}\{\eta^4\text{-}S(C_6H_4)CH(Me)\}(P_3\text{-ligand})]^+$ yields the neutral, paramagnetic $[M^{II}\{\eta^4\text{-}S(C_6H_4)CH(Me)\}(P_3\text{-ligand})]$ species. For the rhodium species, a rhombic EPR spectrum (no resolved hyperfine couplings) was observed, suggestive of a trigonal-bipyramidal species. The *formal* Rh(II) oxidation state of this species (for some reason assigned in this chapter as $Rh^0$) might be misleading, regarding the potential redox noninnocence of the dianionic $\eta^4\text{-}S(C_6H_4)CH(Me)$ ligand (Fig. 75). The radical was assigned as a predominantly metal-centered radical based on the (albeit rather small) deviations of $g_{iso}$ from $g_e$. This might be correct, but additional spectroscopic

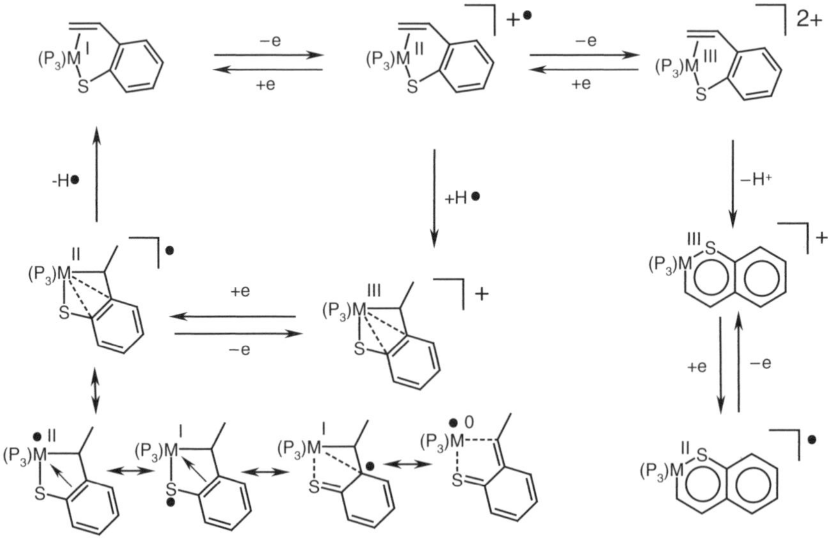

Figure 75. Redox induced conversion pathways in rhodium and iridium supported by benzo[*b*]thiophene complexes.

measurements and/or DFT calculations would be required to leave this assignment beyond any doubt. Quite confusing, the authors assigned the EPR spectrum of the iridium analogue to a low-spin $d^7$ species [which would correspond to Ir(II) not Ir(0)]. It thus seems most safe to leave the assignment of the electronic structure of these species as a subject requiring additional future investigations.

The paramagnetic, *formally* M(II) species $[M^{II}\{\eta^4\text{-}S(C_6H_4)CH(Me)\}(P_3\text{-ligand})]$ are not stable in solution at higher temperatures, resulting in a net loss of a hydrogen atom to regenerate the M(I) starting material $[M^{I}\{\eta^3\text{-}S(C_6H_4)CH{=}CH_2\}(P_3\text{-ligand})]$ (194). This reaction implies a homolytic C—H bond-cleavage reaction at the aliphatic β-carbon position. Experimental proof for this pathway was obtained. The authors prepared $[Rh^{II}\{\eta^4\text{-}S(C_6H_4)CH(CH_2CPh_3)\}(P_3\text{-ligand})]$ with EPR parameters very similar to $[Rh^{II}\{\eta^4\text{-}S(C_6H_4)CH(Me)\}(P_3\text{-ligand})]$. This species decomposes via a homolytic C—C bond cleavage with net loss of a trityl radical, thus generating the same $[Rh^{I}\{\eta^3\text{-}S(C_6H_4)CH{=}CH_2\}(P_3\text{-ligand})]$ species (Fig. 76). The greater stability of the trityl radical as compared to a hydrogen atom seems to be the driving force causing C—C bond cleavage to prevail over C—H bond cleavage in this case. The fate of the trityl and hydrogen-atom radicals has not been investigated.

The two-electron oxidized species $[M^{III}\{\eta^3\text{-}S(C_6H_4)CH{=}CH_2\}(P_3\text{-ligand})]^{2+}$ quickly converts to the metalla-thia-aromatic species $[M^{III}$

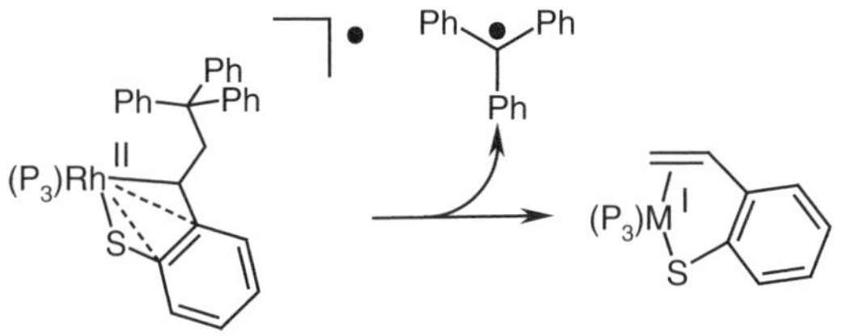

Figure 76. The C–C bond cleavage with net elimination of a trityl radical from $[(P_3)Rh^{II}\{\eta^3\text{-}S(C_6H_4)CH(CH_2CPh_3)\}]$.

$(\eta^2\text{-}C,S\text{-}C_8H_6S)(P_3\text{-ligand})]^+$ upon release of a proton. This species, in turn, was electrochemically reduced to $[M^{II}(\eta^2\text{-}C,S\text{-}C_8H_6S)(P_3\text{-ligand})]$. Again, the *formal* M(II) oxidation state of these species might be misleading due to the expected redox noninnocence of the ligand. The EPR spectrum of $[Ir^{II}(\eta^2\text{-}C,S\text{-}C_8H_6S)(P_3\text{-ligand})]$ could only be satisfactory simulated assuming the presence of two isomeric species (194). The spectra of both species are consistent with a distorted square-pyramidal geometry. The structure of these species seems to be similar to that of the related Ir(II) complex $[Ir^{II}(DTBC)(P_3\text{-ligand})]$ (DTBC = 3,5-di-*tert*-butylcatecholate), revealing comparable EPR parameters (195). The presence of two isomeric species was explained by assuming different Jahn–Teller distortions related to the asymmetry of the ligand. All paramagnetic species described in this section have a short lifetime of ~10–25 s.

## F. On the Involvement of Paramagnetic Metal Complexes in Catalysis

Despite the diverse chemistry of the paramagnetic species presented in this chapter, the number of reports claiming their involvement in catalysis is so far quite limited. Paramagnetic Rh(II) species have been suggested to play an important role as intermediates mediating chemical transformations of diamagnetic $Rh^{III}$-allyl species in reactions with $CCl_3Br$ (196). The Rh(II) intermediates, or $Rh^{III}$–superoxo complexes, have also been reported to mediate the catalytic isomerization of 1-nonene to 2-nonene (in the presence of oxygen) (197). In hydrogenation of ketones and olefins with polymer-bound rhodium catalysts, the observation of Rh(II) species *after the reaction* was taken as an indication for the possible involvement of these species under catalytic conditions (198). Furthermore, iridium-catalyzed carbonylation of alcohols to carboxylic acids has been claimed to involve Ir(II) species (199).

The experimental evidence for the involvement of paramagnetic catalysts in the above examples is not very strong. Stronger evidence for the involvement of paramagnetic intermediates in catalytic reactions is present in more recent examples concerning cyclopropanation of olefins with ethyldiazoacetate mediated by well-defined Rh(II) species (Section IV.E) (188), and Wacker-type oxidation of norbornadiene to norbornenone (Section IV.D) (181). These examples demonstrate that organometallic radicals of Rh and Ir might actually play an important role in future (or even existing?) catalytic reactions, and one might start to think about taking advantage of the rich chemistry displayed by these type of complexes.

## V. CONCLUSIONS

Although examples of paramagnetic organometallic (OM) complexes of Rh and Ir are not abundant, this chapter shows that they can neither be considered a rarity any longer. Paramagnetic OM species of Pd and Pt, on the other hand, are still quite rare.

While many of the species discussed in this chapter are metal-centered radicals, it is clear that formation of ligand-centered radicals by oxidation of redox noninnocent radicals can also play an important role. This includes oxidation of the known redox noninnocent dithiole-thionate-, indole-, and catechol-based ligands, as well as the rather unexpected formation of aminyl radicals upon oxidation of $Rh^{I}(NR_2)$ amido species.

*Rh(II) and Ir(II) Species.* Quite a large number of mononuclear $d^7$ Rh(II) and Ir(II) species have been isolated. In fact, formation of these species by either oxidation of the M(I) precursors (Table XV) or reduction of the M(III) precursors (Table XVI) occurs at remarkably accessible redox potentials. This suggests that such species could well play an important role in catalytic reactions mediated by these metals. Regarding their high reactivity, formation of Rh(II) and Ir(II) species will easily lead to catalyst deactivation. The paramagnetic $d^7$ M(II) species could even be active intermediates, although proven examples of catalytically active M(II) species are (still) quite limited.

Persistent, nonreactive examples are all protected by steric bulk. All other examples are highly reactive species. The radical-type reactivity of such OM Rh(II) and Ir(II) species is quite comparable to that of organic radicals. Metal centered reactivity is dominated by radicaloid C—X, C—C, and C—H bond cleavage reactions, radical couplings with other (organic and metal centered) radicals, electron-transfer disproportionation, and formation of superoxides upon reaction with $O_2$. For Rh(II) and Ir(II) species containg carbon monoxide-, olefinic ligands- or cyclopentadienyl-type ligands, the unpaired electron is partly located at these $\pi$-accepting ligands. The EPR data and DFT calculations confirm the redox

TABLE XV

Selected M(I)/M(II) Redox Couples Relevant to Formation of M(II) species (M = Rh, Ir)[a]

| Compound | $E_{1/2}$ (V) | $E_{peak}$ (Irreversible) | Solvent | References |
|---|---|---|---|---|
| $[Rh^I(C_6Cl_5)_2(cod)]^-/[Rh^{II}(C_6Cl_5)_2(cod)]$ | −0.524 | | $CH_2Cl_2$ | 29 |
| *cis*-$[Rh^I(C_6Cl_5)_2\{P(OPh)_3\}_2]^-$/ *cis*-$[Rh^{II}(C_6Cl_5)_2\{P(OPh)_3\}_2]$ | −0.529 | | $CH_2Cl_2$ | 29 |
| $[Rh^I(C_6Cl_5)_2(CO)_2]^-/ [Rh^{II}(C_6Cl_5)_2(CO)_2]$ | | −0.457 | $CH_2Cl_2$ | 29 |
| $[Rh^I(C_6Cl_5)_2(CO)(PPh_3)]^-/ [Rh^{II}(C_6Cl_5)_2(CO)(PPh_3)]$ | −0.507 | | $CH_2Cl_2$ | 29 |
| *trans*-$[Rh^I(C_6Cl_5)_2\{P(OPh)_3\}_2]^-$/ *trans*-$[Rh^{II}(C_6Cl_5)_2\{P(OPh)_3\}_2]$ | −0.549 | | $CH_2Cl_2$ | 29 |
| $[Rh^I(C_6Cl_5)_2(PPh_3)_2]^-/ [Rh^{II}(C_6Cl_5)_2(PPh_3)_2]$ | −0.579 | | $CH_2Cl_2$ | 29 |
| $[Rh^I(C_6Cl_5)_2(dpe)]^-/ [Rh^{II}(C_6Cl_5)_2(dpe)]$ | −0.568 | | $CH_2Cl_2$ | 29 |
| $[Ir^I(C_6Cl_5)_2(cod)]^-/ [Ir^{II}(C_6Cl_5)_2(cod)]$ | −0.515 | | $CH_2Cl_2$ | 30 |
| $[Ir^I(Me_3tpa)(ethene)]^+/ [Ir^{II}(Me_3tpa)(ethene)]^{2+}$ | −0.368 | −0.289 | Acetone $CH_2Cl_2$ | 180, 162 |
| $[Ir^I(Me_2tpa)(ethene)]^+/ [Ir^{II}(Me_2tpa)(ethene)]^{2+}$ | −0.249 | −0.173 | Acetone $CH_2Cl_2$ | 180, 162 |
| $[Ir^I(Metpa)(ethene)]^+/ [Ir^{II}(Metpa)(ethene)]^{2+}$ | −0.122 | −0.043 | Acetone $CH_2Cl_2$ | 180, 162 |
| $[Ir^I(tpa)(ethene)]^+/ [Ir^{II}(tpa)(ethene)]^{2+}$ | −0.144 | −0.010 | Acetone $CH_2Cl_2$ | 180, 162 |
| $[Rh^I(dpa)(cod)]^+/ [Rh^{II}(dpa)(cod)]^{2+}$ | 0.065 | 0.119 | Acetone $CH_2Cl_2$ | 182 |
| $[Rh^I(pla)(nbd)]^+/ [Rh^{II}(pla)(nbd)]^{2+}$ | 0.113 | 0.192 | Acetone $CH_2Cl_2$ | 181 |
| $[Rh^I(CO)_2\{(HB(pz')_3\}]$/ $[Rh^{II}(CO)_2\{(HB(pz')_3\}]^+$ | 0.15 V | | $CH_2Cl_2$ | 43 |
| $[Rh^I(CO)\{P(OPh)_3\}\{(HB(pz')_3\}]$/ $[Rh^{II}(CO)\{P(OPh)_3\}\{(HB(pz')_3\}]^+$ | −0.17 | | $CH_2Cl_2$ | 43, 44 |
| $[Rh^I(CO)(PPh_3)\{(HB(pz')_3\}]$/ $[Rh^{II}(CO)(PPh_3)\{(HB(pz')_3\}]^+$ | −0.25 | | $CH_2Cl_2$ | 44 |
| $[Rh^I(\eta^3\text{-TMPP})_2]^+/[Rh^{II}(\eta^3\text{-TMPP})_2]^{2+}$ | −1.397 | | $CH_2Cl_2$ | 14 |
| $[Rh^I(\eta^3\text{-TMPP})_2(CN\textit{t}\text{-Bu})_2]^+$/ $[Rh^{II}(\eta^3\text{-TMPP})_2(CN\textit{t}\text{-Bu})_2]^{2+}$ | −0.787 | | $CH_2Cl_2$ | 16 |
| $[Ir^I(H)(CO)(PPh_3)_3]$/ $[Ir^{II}(H)(CO)(PPh_3)_3]^+$ | −0.47 | | $CH_2Cl_2$ | 11,12 |
| $[Rh^I(\eta^1:\eta^6:\eta^1$-1,4-bis {4-(diphenylphosphino)butyl}-2, 3,5,6-tetramethylbenzene)]$^+$ $Rh^I/Rh^{II}$ | 0.41 | | $CH_2Cl_2$ | 21 |
| $[Rh^I(\eta^1:\eta^6:\eta^1$-1,4-bis {4-(diphenylphosphino)propoxy}-2, 3,5,6-tetramethylbenzene)]$^+$ $Rh^I/Rh^{II}$ | 0.528 | | $CH_2Cl_2$ | 21 |
| $[Rh^I(\eta^1:\eta^6:\eta^1$-1,4-bis {4-(diphenylphosphino)ethoxy}-2,3 ,5,6-tetramethylbenzene)]$^+$ $Rh^I/Rh^{II}$ | 0.56 | | $CH_2Cl_2$ | 21 |
| $[Rh^I(CN)(NP_3)]$/ $[Rh^{II}(CN)(NP_3)]^+$ | −0.707 | | $CH_2Cl_2$ | 191 |
| $[Rh^I(CN)(PP_3)]$/ $[Rh^{II}(CN)(PP_3)]^+$ | −0.647 | | $CH_2Cl_2$ | 191 |
| $[Rh^I(C{\equiv}CPh)(NP_3)]$/ $[Rh^{II}(C{\equiv}CPh)(NP_3)]^+$ | −0.562 | | $CH_2Cl_2$ | 191, 193 |
| $[Rh^I(C{\equiv}CPh)(PP_3)]$/ $[Rh^{II}(C{\equiv}CPh)(PP_3)]^+$ | −0.546 | | $CH_2Cl_2$ | 191, 193 |
| $[Rh^I(C{\equiv}CCO_2Et)(NP_3)]$/ $[Rh^{II}(C{\equiv}CCO_2Et)(NP_3)]^+$ | −0.543 | | $CH_2Cl_2$ | 193 |
| $[Rh^I(C{\equiv}CCO_2Et)(PP_3)]$/ $[Rh^{II}(C{\equiv}CCO_2Et)(PP_3)]^+$ | −0.538 | | $CH_2Cl_2$ | 193 |
| $[Rh^I(C{\equiv}CCHO)(PP_3)]$/ $[Rh^{II}(C{\equiv}CCHO)(PP_3)]^+$ | −0.534 | | $CH_2Cl_2$ | 193 |
| $[Rh^I(Cp^*)(abpy)]$/ $[Rh^{II}(Cp^*)(Cl)(abpy^{2-})]^-$ | −2.04 | | MeCN | 96 |
| $[Ir^I(Cp^*)(abpy)]$/ $[Ir^{II}(Cp^*)(Cl)(abpy^{2-})]^-$ | −2.38 | | MeCN | 96 |
| $[Rh^I(Cp)(cot)]$/ $[Rh^{II}(Cp)(cot)]^+$ | | +0.15 | $CH_2Cl_2$ | 97 |
| $[Rh^I(CpPh_5)(1,5\text{-cod})]$/ $[Rh^{II}(CpPh_5)(1,5\text{-cod})]$ | +0.09 | | $CH_2Cl_2$ | 101 |
| $[Rh^I(CpPh_5)(1,3\text{-cod})]$/ $[Rh^{II}(CpPh_5)(1,3\text{-cod})]$ | −0.01 | | $CH_2Cl_2$ | 101 |

[a] Redox potentials versus $Fc/Fc^+$.

TABLE XVI
Selected M(III)/M(II) Redox Couples Relevant to Formation of Paramagnetic M(II) Species (M = Rh, Ir)[a]

| Compound | $E_{1/2}$ | $E_{peak}$ (irreversible) | Solvent | References |
|---|---|---|---|---|
| $[Rh^{III}(\eta^3\text{-TMPP})_2]^{3+}$/ $[Rh^{II}(\eta^3\text{-TMPP})_2]^{2+}$ | −0.287 | | $CH_2Cl_2$ | 14 |
| $[Rh^{III}(CN)(NP_3)]^{2+}$/ $[Rh^{II}(CN)(NP_3)]^{+}$ | −0.327 | | $CH_2Cl_2$ | 191 |
| $[Rh^{III}(CN)(PP_3)]^{2+}$/ $[Rh^{II}(CN)(PP_3)]^{+}$ | −0.177 | | $CH_2Cl_2$ | 191 |
| $[Rh^{III}(C{\equiv}CPh)(NP_3)]^{2+}$/ $[Rh^{II}(C{\equiv}CPh)(NP_3)]^{+}$ | −0.497 | | $CH_2Cl_2$ | 191, 193 |
| $[Rh^{III}(C{\equiv}CPh)(PP_3)]^{2+}$/ $[Rh^{II}(C{\equiv}CPh)(PP_3)]^{+}$ | −0.207 | | $CH_2Cl_2$ | 191, 193 |
| $[Rh^{III}(C{\equiv}CCO_2Et)(NP_3)]^{2+}$/ $[Rh^{II}(C{\equiv}CCO_2Et)(NP_3)]^{+}$ | −0.367 | | $CH_2Cl_2$ | 193 |
| $[Rh^{III}(C{\equiv}CCO_2Et)(PP_3)]^{2+}$/ $[Rh^{II}(C{\equiv}CCO_2Et)(PP_3)]^{+}$ | −0.237 | | $CH_2Cl_2$ | 193 |
| $[Rh^{III}(C{\equiv}CCHO)(PP_3)]^{2+}$/ $[Rh^{II}(C{\equiv}CCHO)(PP_3)]^{+}$ | −0.137 | | $CH_2Cl_2$ | 193 |
| $[Rh^{III}(Cp)_2]^{+}$/ $[Rh^{II}(Cp)_2]$ | −1.81 | | MeCN | 80 |
| $[Rh^{III}(Cp^*)_2]^{+}$/ $[Rh^{II}(Cp^*)_2]$ | −2.22 | | THF | 83 |
| $[Rh^{III}(Cp)(Cp^*)]^{+}$/ $[Rh^{II}(Cp)(Cp^*)]$ | −1.91 | | THF | 83 |
| $[Rh^{III}(\eta^5\text{-}C_9H_7)(Cp^*)]^{+}$/ $[Rh^{II}(\eta^5\text{-}C_9H_7)(Cp^*)]$ | −1.54 | | THF | 83 |
| $[Rh^{III}(CpPh_4)_2]^{+}$/ $[Rh^{II}(CpPh_4)_2]$ | −1.44 | | DMF | 102 |
| $[Ir^{III}(Cp^*)_2]^{+}$/ $[Ir^{II}(Cp^*)_2]$ | −2.67 | | THF | 83 |
| $[Ir^{III}(Cp)(Cp^*)]^{+}$/ $[Ir^{II}(Cp)(Cp^*)]$ | −2.43 | | THF | 83 |
| $[Ir^{III}(\eta^5\text{-}C_9H_7)(Cp^*)]^{+}$/ $[Ir^{II}(\eta^5\text{-}C_9H_7)(Cp^*)]$ | −1.90 | | THF | 83 |

[a]Redox potentials versus $Fc/Fc^+$.

noninnocence of these ligands, which is also reflected in the observation of a variety of metal–ligand and ligand–ligand coupling reactions.

*Pd(III) and Pt(III) Species*. Only a limited number of persistent $d^7$ OM Pd(III) and PtIII species is known to date. These involve some sterically protected $Pt^{III}(\sigma\text{-Aryl})_n$ complexes, a couple of Pd(III)-carbolides and $[Pd(CN)_4(Cl)_2]^{3-}$ in KCl/NaCl host lattices. All $Pd^{III}$—R and $Pt^{III}$—R alkyl complexes are highly instable, and tend to decompose via $M^{III}$—C bond homolysis pathways to form diamagnetic M(II) species and carbon-centered organic radicals R•.

*Rh(IV) and Ir(IV) Species*. High valent $d^5$ OM $Rh^{IV}$ and $Ir^{IV}$ species are also very rare. Only 3–4 isolated and persistent Ir(IV) examples have been reported. All other OM $d^5$ Rh(IV), and Ir(IV) species have only been detected as very short-lived intermediates in oxidatively induced reductive elimination reactions concerning $M^{III}(R)_n$ alkyl and $M^{III}(H)(R)_n$ alkyl hydride species. Oxidation of these species to form M(IV) intermediates often leads to concerted reductive eliminations of organic C—C or C—H coupled products, but more complicated radical-type reactions have also been observed. For Cp*-type complexes, the cyclopentadienyl ligand often participates in the reaction sequence as a chemically noninnocent ligand, facilitating C—H elimination, with the hydrogen atom of the R—H product stemming from the Cp* ligand.

## ACKNOWLEDGMENTS

This work was supported by the Netherlands Organisation for Scientific Research (NWO-CW), the Radboud University Nijmegen, the University of Amsterdam, and the ETH Zürich.

## ABBREVIATIONS

| | |
|---|---|
| abpy | 2,2′-Azobipyridyl |
| Acacen$^{2-}$ | *N,N′*-Ethylenebis(acetylacetoneiminate) |
| ax/eq | Axial / equatorial |
| BDE | Bond dissociation enthalpy |
| Bn—bla | *N*-Benzyl-*N,N*-bis(6-methyl-2-pyridylmethyl)amine |
| bpa | *N,N*-Bis(2-pyridylmethyl)amine |
| bpip | 2,5-Bis(1-phenyliminoethyl)pyrazine |
| bpm | 4,4′-Bipyrimidine |
| bptz | 3,6-Bis(2-pyridyl)-1,2,4,5-tetrazine |
| bpy | 2,2′-Bipyridine |
| bpym | 2,2′-Bipyrimidine |
| bpz | 2,2′-Bipyrazine |
| cod | 1,5-Cyclooctadiene |
| cot | 1,3,5,7-Cyclooctatetraene |
| Cp | Cyclopentadienyl |
| Cp* | Pentamethylcyclopentadienyl |
| Cy | Cyclohexyl |
| 3,6-DBCat$^{-2}$ | 3,6-Di-*tert*-butylcatecholate dianion |
| DFT | Density functional theory |
| DMSO | Dimethyl sulfoxide (solvent) |
| dmso | Dimethyl sulfoxide (ligand) |
| dmphen | Dimethylphenantroline |
| dppe | 1,2-Bis(diphenylphosphino)ethane |

| | |
|---|---|
| dppm | Bis(diphenylphosphino)methane |
| dppz | Dipyro(3.2-*a*:2′,3′*c*)phenazine |
| DTBC | 3,5-Di-*tert*-butylcatecholate |
| dmphen | Dimethylphenantroline |
| dppz | Dipyro(3.2-*a*:2′,3′*c*)phenazine |
| DMPO | 5,5-Dimethyl-2-pyrrolidine-1-oxide |
| EPR | Electron paramagnetic resonance |
| EC mechanism | Electrochemical step followed by a chemical reaction step |
| ENDOR | Electron nuclear double resonance |
| ESI–MS | Electron-spray ionization mass spectrometry |
| ET | Electron transfer |
| $Fc/Fc^+$ | Ferrocene/ferrocenium |
| *fac* / *mer* | Facial/meridional |
| $HBPz'_3$ | Hydro-tris-pyrazolylborate derivative |
| $HCPz'_3$ | Tris-pyrazolylmethyl derivative |
| HFP | 1,1,1,3,3,3-Hexaflouropropan-2-ol |
| $^1H$ NMR | Proton NMR |
| *H*tbu-iepp | 3-[*N*-2-Pyridylmethyl-*N*-2-hydroxy-3,5-di(*tert*-butyl)benzylamine]-ethylindole *(H* denotes a dissociable proton) |
| HOMO | Highest occupied MO |
| HOMO-1 | Occupied MO closest in energy to the HOMO/SOMO |
| *H*p-iepp | 3-[*N*-2-Pyridylmethyl-*N*-4-hydroxybenzylamine] ethylindole |
| p-iepp-c | piepp bound to a metal through a carbon atom |
| IR | Infrared |
| LUMO | Lowest unoccupied MO |
| mes | Mesityl |
| $Me_2tpa$ | *N,N,N*-(2-Pyridylmethyl)-bis(6-methyl-2-pyridyl-methyl)amine |
| $Me_3tpa$ | *N,N,N*-Tris(6-methyl)-2-pyridylmethyl)amine |
| MMA | Methyl methacrylate |
| MO | Molecular orbital |
| napy | 5,7-Dimethyl-1,8-naphtyridine-2-onate |
| nbd | Norbornadiene |
| NMR | Nuclear magnetic resonance |
| OEP | Octaethylporhyrin |
| OLED | Organic light emitting diode |
| OM | Organometallic |
| pd | Pentane-2,4-dionate |

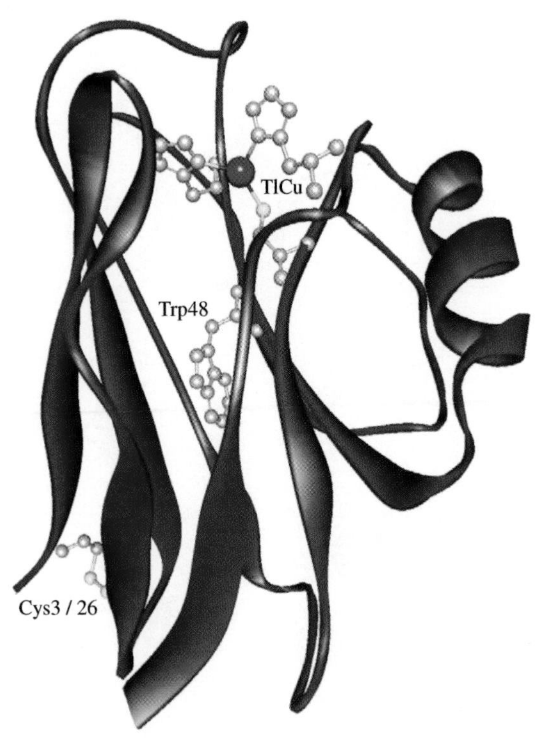


Figure 1 (Chapter 1). Three-dimensional structure of *Pseudomonas aeruginosa* azurin (16). In addition to the protein backbone, the side chains of three copper ligating residues, His46, His117, and Cys112 are shown near the top together with the disulfide bridge (bottom) and Trp48 (center). Coordinates were taken from the Protein Data Bank (PDB), code 4AZU.

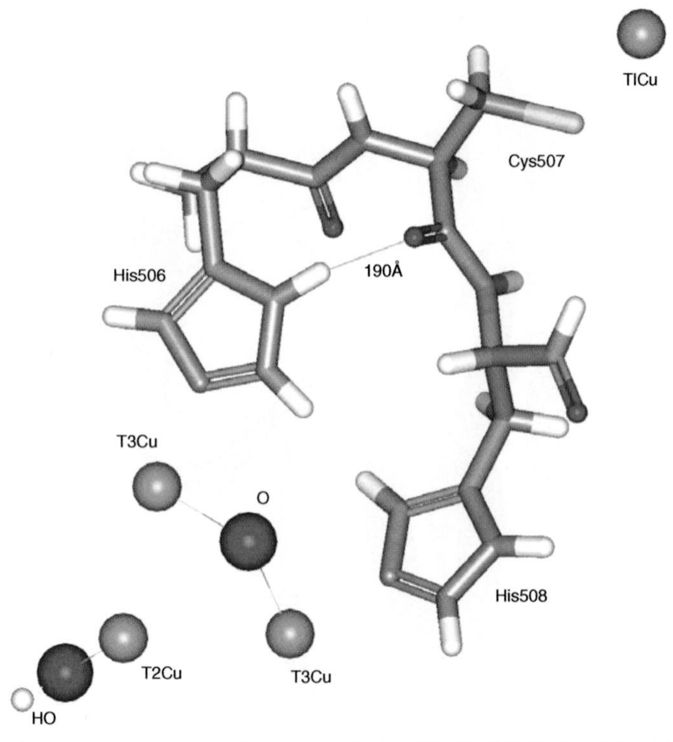


Figure 9 (Chapter 1). Electron-transfer pathways in AO. The Cys507 binds to T1, while the two neighboring residues, His506 and His508 bind to the two T3 copper ions. This provides a very short electronic coupling between Cys507 and either His506 or His508, both consisting of nine covalent bonds. An alternative pathway may be envisaged through the carbonyl oxygen atom of Cys507 and $N_\delta$ of His506. Calculations, showing that all three pathways have essentially the same electronic coupling (48), were based on the Beratan and Onuchic model (6, 7). Coordinates were taken from the PDB, code 1AOZ.

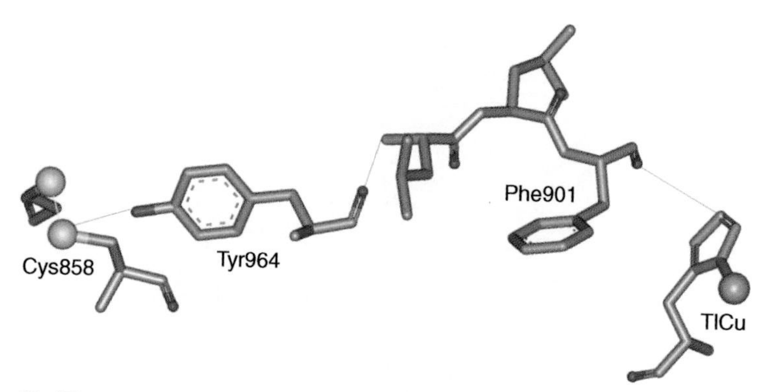


Figure 13 (Chapter 1). Result of ET pathway calculations in human Ceruloplasmin of the connection between the Cys858–Cys881 disulfide group in domain 5 and T1($Cu^{II}$) of domain 6 (72). The pathway consists of 20 covalent bonds, one hydrogen bond, and two van der Waals contacts. The distance between the T1 copper ion and $S_\gamma$ of Cys858 is 2.59 nm. Calculations were based on the Beratan and Onuchic model (6, 7). Coordinates were taken from the PDB, code 1KCW.

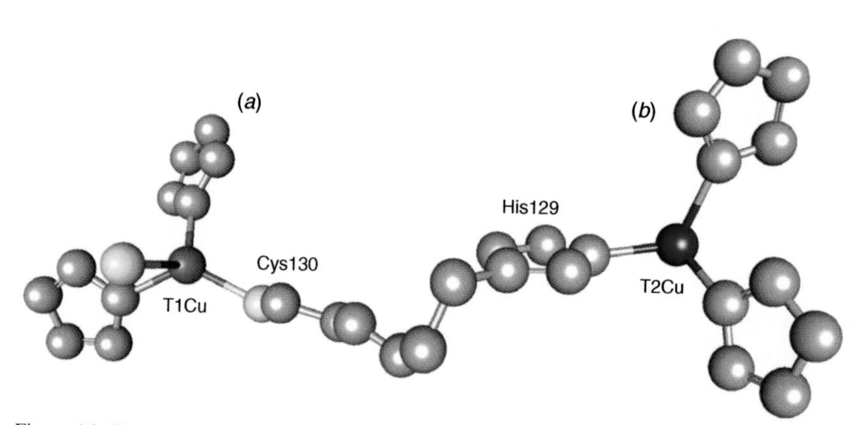


Figure 16 (Chapter 1). Calculated ET pathway between T1Cu (*a*) and T2Cu (*b*) in CuNiR from *A. cycloclastes* (AcNiR). The short path consists of the T1 ligand, Cys130 and the neighboring His129, which coordinates to T2Cu. The connection consists of 11 covalent bonds, and the distance between the copper centers is 1.26 nm. Calculations were based on the Beratan and Onuchic model (6, 7). Coordinates were taken from the PDB, code 1NDT.

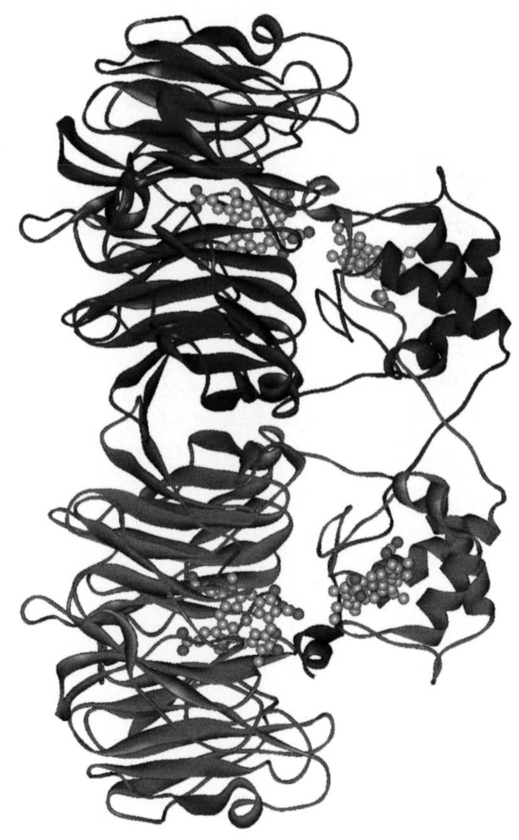

Figure 17 (Chapter 1). The 3D structure of *P. aeruginosa* $cd_1$NiR. The two subunits containing the heme-*c* centers are seen on the right-hand side while the heme-$d_1$ centers are buried in the subunits shown to the left in the figure. The coordinates were taken from the PDB, code 1NNE.

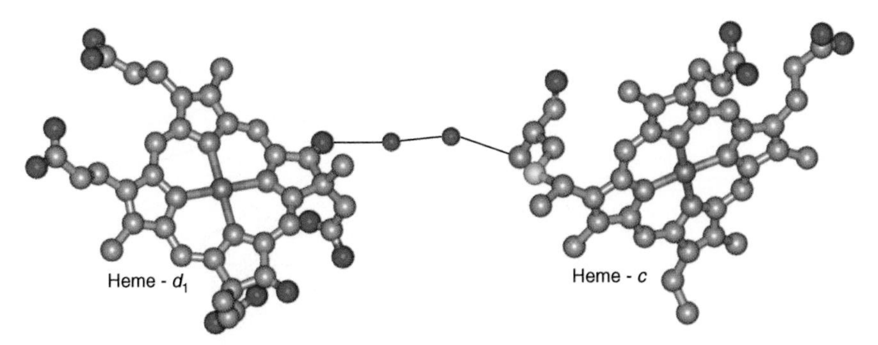


Figure 19 (Chapter 1). Calculated ET pathway between heme-$c$ and heme-$d_1$ in *P. aeruginosa* $cd_1$NiR. The pathway includes Cys50, covalently connected to heme-$c$, a van der Waals contact to a water molecule, and two hydrogen bonds, the latter of which connects to a carbonyl oxygen on the $d_1$ heme ring. The distance between the iron atoms is 2.0 nm. Calculations were based on the Beratan and Onuchic model (6, 7). Coordinates were taken from the PDB, code 1NNE.

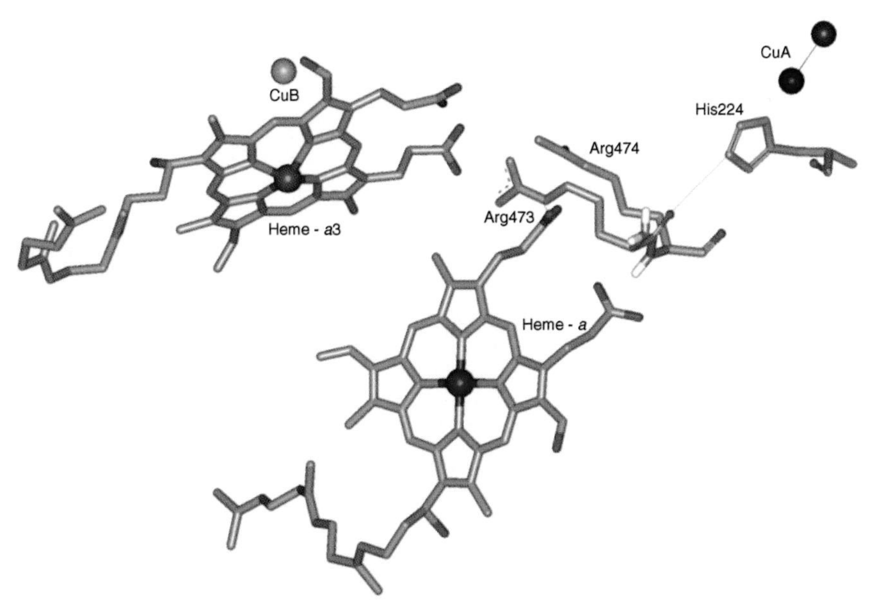


Figure 24 (Chapter 1). Electron-transfer pathway between $Cu_A$ and heme-$a$ in *P. denitrificans* COX. The path consists of 14 covalent bonds and two hydrogen bonds. The direct distance between to two metal ion centers is 2.0 nm. The binuclear heme-$a_3$/$Cu_B$ site is also shown. Calculations were based on the Beratan and Onuchic model (6, 7). Coordinates were taken from the PDB, code 1QLE.

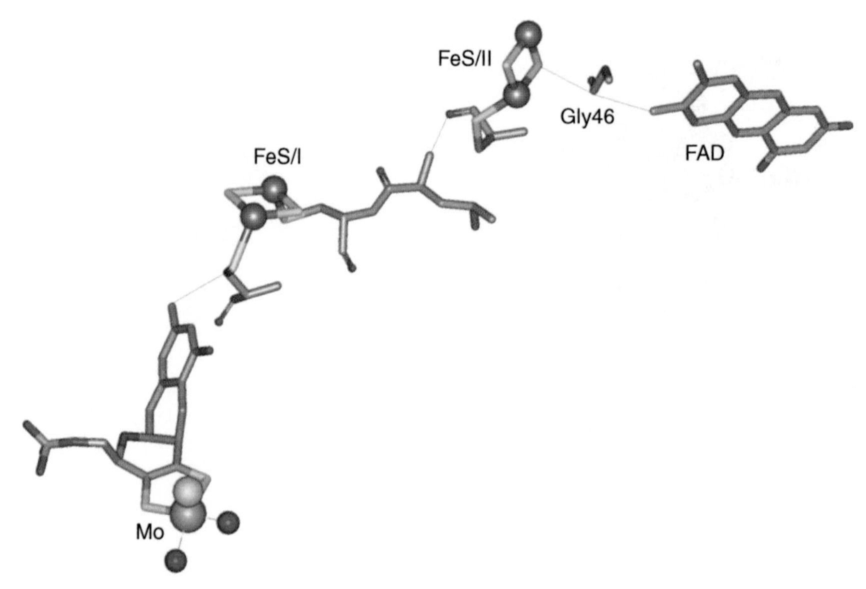


Figure 26 (Chapter 1). Electron-transfer pathway from the molybdenum–pterin center to FAD in xanthine oxidase. Besides the two mercapto groups, Mo is shown coordinated to inorganic sulfur and two oxygen atoms. The total distance between Mo and FAD is 2.9 nm. The redox centers are coupled through a series of covalent bonds, three short van der Waals contacts, and a single hydrogen bond. Calculations were based on the Beratan and Onuchic model (6, 7). Coordinates were taken from the PDB, code 1FIQ.

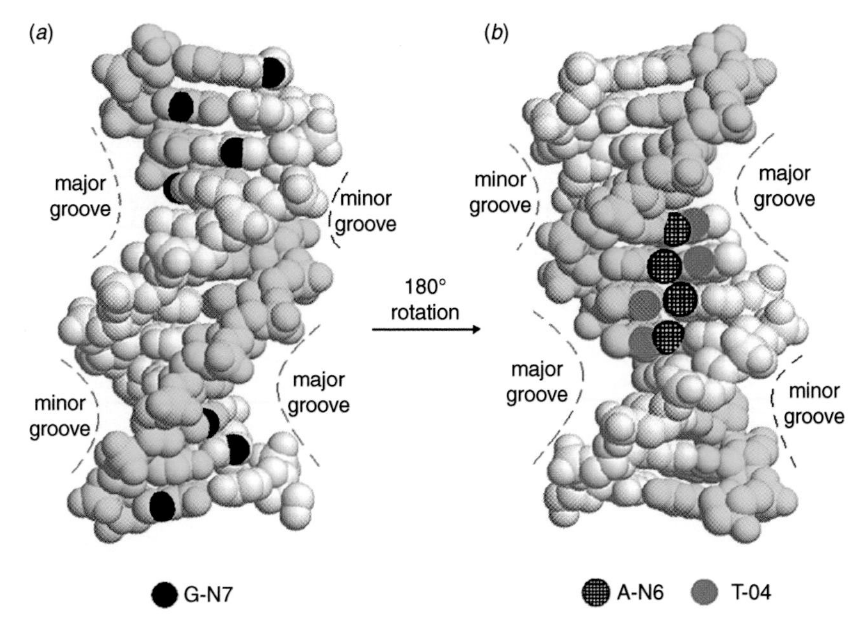


Figure 6 (Chapter 8). The G-N7 (*a*) and A-N6 and T-O4 sites (*b*) in a typical B-DNA duplex (Dickerson–Drew dodecamer d(5′-CGCGAATTCGCG-3′)$_2$, (PDB ID: 1BNA). The duplex in the right view has been rotated 180° with respect to the left view.

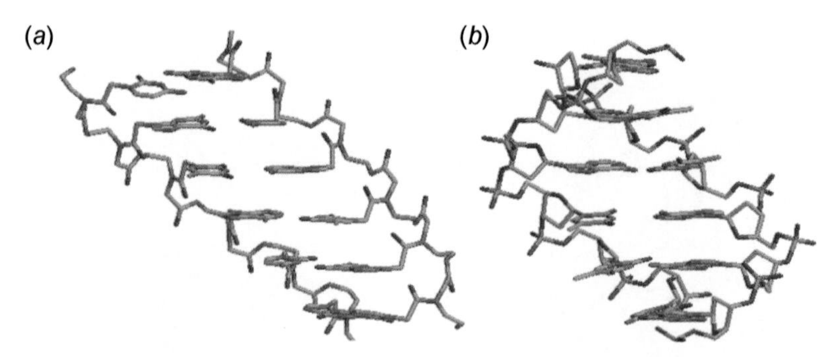


Figure 8 (Chapter 8). (*a*) The 6-bp PNA duplex based on the $H_2N$-CGTACG-H strand (PDB: 1PUP), and (*b*) the 6-bp central fragment $d(GAATTC)_2$ of the Dickerson-Drew DNA dodecamer (PDB: 1BNA).

| | |
|---|---|
| phen | 1,10-Phenantroline |
| pla | *N,N*-(2-Pyridylmethyl)(6-methyl-2-pyridylmethyl) amine |
| por | Porphyrin |
| Py | Pyridine |
| Pz | Pyrazolyl |
| SOMO | Singly occupied MO |
| tacn | 1,4,7-Trimethyl-1,4,7-triazacyclononane |
| tap | 1,4,7,10-Tetraazaphenanthrene |
| tbu-iepp-c | tbu-iepp Bound to a metal through a carbon atom |
| TEMPO | 2,2,6,6-Tetramethylpiperidine-1-oxyl |
| Tf | Triflate |
| thf | Tetrahydrofuran (ligand) |
| tht | Tetrahydrothiophene |
| TMP | *meso*-Tetramesitylporhyrin |
| TMPP | Tris(2,4,6-trimethoxyphenyl)phosphine |
| tol | Tolyl |
| TPP | *meso*-Tetraphenylporhyrin |
| TXP | *meso*-Tetraxylylporhyrin |
| TTiPP | *meso*-Tetrakis(2,4,6-tri-isopropylphenyl)porhyrin |
| TTEPP | *meso*-Tetrakis(2,4,6-tri-ethylphenyl)porhyrin |
| trop | 5-H-dibenzo [a,d] cycloheptene-5-yl |
| VE | Valence electron |
| Q / $SQ^-$ / $cat^{2-}$ | Quinone / semiquinone anion / catecholate dianion |
| $\mu_{eff}(\mu_B)$ | Effective magnetic moment (in Bohr magnetons) |

## REFERENCES


1. K. K. Pandey, *Coord. Chem. Rev.*, 121, 1 (1991).
2. D. G. DeWitt, *Coord. Chem. Rev.*, 147, 209 (1996).
3. C. Sporer, H. Heise, K. Wurst, D. Ruiz-Molina, H. Kopacka, P. Jaitner, F. Köhler, J. J. Novoa, and J. Veciana, *Chem. Eur. J.*, 10, 1355 (2004).
4. B. L. Tumanskii, R. G. Gasanov, M. V. Tsikalova, A. V. Usatov, E. V. Martynova, and Yu. N. Novikov, *Russ. Chem. Bull.*, 53, 2051 (2004).
5. B. L. Tummanskii, V. V. Bashilov, O. G. Kalina, and V. I. Sokolov, *J. Organomet. Chem., 599*, 28 (2000).
6. B. A. Goodman and J. B. Raynor, *Advances in Inorganic Chemistry and Radiochemistry*, H. J. Emeléus and A. G. Sharpe Ed., *13*, 136 (1970).
7. S. Zecchin, G. Zotti, and G. Pilloni, *J. Organomet. Chem.*, 294, 379 (1985).
8. M. Kubota, M. K. Chan, D. C. Boyd, and K. R. Mann, *Inorg. Chem.*, 26, 3261 (1987).

9. G. Pilloni, G. Schiavon, G. Zotti, and S. Zecchin, *J. Organomet. Chem.*, 134, 305 (1977).
10. S. Valcher, G. Pilloni, and M. Martelli, *J. Electroanal. Chem.*, *42*, App. 5 (1973).
11. D. Menglet, A. M. Bond, K. Coutinho, R. S. Dickson, G. G. Lazarev, S. A. Olsen, and J. R. Pilbrow, *J. Am. Chem. Soc.*, 120, 2086 (1998).
12. A. M. Bond, D. G. Humprey, D. Menglet, G. G. Lazarev, R. S. Dickson, and T. Vu, *Inorg. Chim. Acta*, 300, 565 (2000).
13. K. R. Dunbar, S. C. Haefner, and L. E. Pence, *J. Am. Chem. Soc.*, 111, 5504 (1989).
14. S. C. Haefner, K. R. Dunbar, and C. Bender, *J. Am. Chem. Soc.*, 113, 9540 (1991).
15. K. R. Dunbar, S. C. Haefner, and P. N. Swepston, *J. Chem. Soc., Chem. Commun.*, 460 (1991).
16. K. R. Dunbar and S. C. Haefner, *Organometallics*, 11, 1431 (1992).
17. A. Aràneo, F. Morazzoni, and T. Napoletano, *J. Chem. Soc., Dalton Trans.*, 2039 (1975).
18. G. Mercati and F. Morazonni, *J. Chem. Soc., Dalton Trans.*, 569 (1979).
19. E. T. Singewald, C. S. Slone, C. L. Stern, C. A. Mirkin, G. P. A. Yap, L. M. Liable-Sands, and A. L. Rheingold, *J. Am. Chem. Soc.*, 119, 3048 (1997).
20. F. M. Dixon, J. R. Farrell, P. E. Doan, A. Williamson, D. A. Weinberger, C. A. Mirkin, C. Stern, C. D. Incarvito, L. M. Liable-Sands, L. N. Zakharov, and A. L. Rheingold, *Organometallics*, 21(15), 3091 (2002).
21. F. M. Dixon, M. S. Masar III, P. E. Doan, J. R. Farrell, F. P. Arnold Jr., C. A. Mirkin, C. D. Incarvito, L. N. Zakharov, and A. L. Rheingold, *Inorg. Chem.*, 42, 3245 (2003).
22. E. T. Singewald, C. A. Mirkin, A. D. Levy, and C. L. Stern, *Angew. Chem., Int. Ed. Engl.*, 33, 2473 (1994).
23. R. Usón, J. Forniés, M. Tomás, B. Menjón, K. Sünkel, and R. Bau, *J. Chem. Soc., Chem. Commun.*, 751 (1984).
24. R. Usón, J. Forniés, M. Tomás, B. Menjón, R. Bau, K. Sünkel, and E. Kuwabara, *Organometallics*, *5*, 1576 (1986).
25. M. P. García, M. V. Jiménez, A. Cuesta, C. Siurana, L. A. Oro, F. J. Lahoz, J. A. López, and M. P. Catalán, A. Tiripicchio, M. Lanfranchi, *Organometallics*, *16*, 1026 (1997).
26. A. A. Danopouolos, G. Wilkinson, B. Hussain-bates, and M. B. Hursthouse, *J. Chem. Soc., Dalton. Trans.* 3165 (1992).
27. R. S. Hay-Motherwell, S. U. Koschmieder, G. Wilkinson, B. Hussain-bates, and M. B. Hursthouse, *J. Chem. Soc., Dalton Trans.*, 2821 (1991).
28. M. P. García, M. V. Jiménez, L. A. Oro, F. J. Lahoz, M. C. Tiripicchio, and A. Tiripicchio, *Organometallics*, 12, 4660 (1993).
29. M. P. García, M. V. Jiménez, L. A. Oro, F. J. Lahoz, J. M. Casas, and P. J. Alonso, *Organometallics*, *12*, 3257 (1993).
30. M. P. García, M. V. Jiménez, L. A. Oro, F. J. Lahoz, and P. J. Alonso, *Angew. Chem., Int. Ed. Engl.*, 31, 1527 (1992).
31. M. P. García, M. V. Jiménez, F. J. Lahoz, J. A. López, and L. A. Oro, *J. Chem. Soc., Dalton Trans.*, 4211 (1998).
32. P. J. Alonso, R. Alcalá, R. Usón, and J. Forniés, *J. Phys, Chem. Solids*, *52*, 975 (1991).
33. H. Zhai, A. Bunn, and B. B. Wayland, *Chem. Commun.*, 1294 (2001).
34. A. Klein and W. Kaim, *Organometallics*, *14*, 1176 (1995).
35. A. Klein, H.-D. Hausen, and W. Kaim, *J. Organomet. Chem.*, 440, 207 (1992).
36. A. Klein, E. J. L. McInnes, and W. Kaim, *J. Chem. Soc., Dalton Trans.* 2371 (2002).
37. L. F. Warren and M. F. Hawthorne, *J. Am. Chem. Soc.*, 4823 (1968).

38. L. Johansson, O. B. Ryan, C. Rømming, and M. Tilset *Organometallics*, *17*, 3957 (1998).
39. (a) R. S. Hay-Motherwell, G. Wilkinson, B. Husssain-Bates, and M. B. Hursthouse, *Polyhedron*, 10/12, 1457 (1991). (b) R. S. Hay-Motherwell, G. Wilkinson, B. Husssain-Bates, and M. B. Hursthouse, *J. Chem. Soc., Dalton Trans.* 3477 (1991).
40. Z. Lu, C.-H. Jun, S. R. de Gala, M. P. Sigalas, O. Eisenstein, and R. H. Crabtree, *Organometallics*, 14, 1168 (1995).
41. A. S. Ionkin and W. Marshall, *Organometallics*, 23, 6031 (2004).
42. N. G. Connelly, D. J. H. Emslie, B. Metz, A. G. Orpen, and M. J. Quayle, *Chem. Commun.* 2289 (1996).
43. N. G. Connelly, D. J. H. Emslie, W. E. Geiger, W. D. Hayward, E. B. Linehan, A. G. Orpen, M. J. Quayle, and P. H. Rieger, *J. Chem. Soc., Dalton Trans.*, 670 (2001).
44. W. E. Geiger, N. C. Ohrenberg, B. Yeomans, N. G. Connelly, and D. J. H. Emslie, *J. Am. Chem. Soc.*, *125*, 8680 (2003).
45. C. J. Adams, N. G. Connelly, D. J. H. Emslie, O. D. Hayward, T. Manson, A. G. Orpen, and P. H. Rieger, *Dalton Trans.*, 2835 (2003).
46. N. G. Connelly, D. J. H. Emslie, P. Klangsinsirikul, and P. H. Rieger, *J. Phys. Chem. A*, *106*, 12214 (2002).
47. R. P. A. Muniz, N. V. Vugman, and J. Danon, *J. Chem. Phys.*, *54*(3), 1284 (1971).
48. R. P. A. Muniz, N. V. Vugman, and J. Danon, *J. Chem. Phys.*, *57*(3), 1297 (1972).
49. N. V. Vugman and W. O. Franco, *J. Chem. Phys.*, 78(4), 2099 (1983).
50. N. V. Vugman and W. O. Franco, *Phys. Lett. A*, *155*(8), 516 (1991).
51. N. V. Vugman, A. O. Caride, and J. Danon, *J. Chem. Phys.*, *59*(8), 4418 (1973).
52. N. V. Vugman, M. L. Netto Grillo, and V. K. Kain, *Chem. Phys. Lett.*, *188*(5/6), 419 (1992).
53. N. V. Vugman, M. F. Elia, and R. P. A. Muniz, *Mol. Phys.*, *30*(6), 1813 (1975).
54. L. F. Warren and M. F. Hawthorne, *J. Am. Chem. Soc.*, *92*, 1157 (1970).
55. S. A. Jasper, J. C. Huffman, and L. J. Todd, *Inorg. Chem.*, *37*, 6060 (1998).
56. K. Krogmann, *Angew. Chem., Int. Ed. Engl.*, *8*, 35 (1969).
57. N. G. Connelly, A. C. Loyns, M. A. Ciriano, M. J. Fernandez, L. A. Oro, and B. E. Villarroya, *J. Chem. Soc., Dalton Trans.*, 689 (1989).
58. B. E. Villarroya, L. A. Oro, F. J. Lahoz, A. J. Edwards, M. A. Ciriano, P. J. Alonso, A. Tiripicchio, and M. T. Camellini, *Inorg. Chim. Acta*, *250*, 241 (1996).
59. M. A. Casado, J. J. Pérez-Torrente, J. A. López, M. A. Ciriano, P. J. Alonso, F. J. Lahoz, and L. A. Oro, *Inorg. Chem.*, *40*, 4785 (2001).
60. D. C. Boyd, N. G. Connelly, G. G. Herbosa, M. G. Hill, K. R. Mann, C. Mealli, A. G. Orpen, K. E. Richardson, and P. H. Rieger, *Inorg. Chem.*, *33*, 960 (1994) and references cited therein.
61. N. Kanematsu, M. Ebihara, and T. Kawamura, *Inorg. Chim. Acta*, *323*, 96 (2001).
62. J. M. Poblet, and M. Bernard, *Inorg. Chem.*, *27*, 2935 (1988).
63. K. M. Kadish, T. D. Phan, L. Giribabu, E. Van Caemelbecke, and J. L. Bear, *Inorg. Chem.*, *42*, 8663 (2003).
64. N. G. Connelly, O. D. Hayward, P. Klangsinsirikul, A. G. Orpen, and P. H. Rieger, *Chem. Commun.* 963 (2000).
65. C. J. Adams, R. A. Baber, N. G. Connelly, P. Harding, O. D. Hayward, M. Kandiah, and A. G. Orpen, *Dalton Trans.*, 1749 (2006).

66. (a) T. Büttner, J. Geier, G. Frison, J. Harmer, C. Calle, A. Schweiger, H. Schönberg, and H. Grützmacher, *Science*, *307*, 235 (2005). (b) P. Maire, M. Königsmann, A. Sreekanth, J. Harmer, A. Schweiger, and H. Grützmacher, *J. Am. Chem. Soc.*, 128, 6578 (2006).
67. P. Maire, A. Sreekanth, T. Büttner, J. Harmer, I. Gromov, H. Rüegger, F. Breher, A. Schweiger, and H. Grützmacher, *Angew. Chem. Int. Ed. Engl.*, *45*, 3265 (2006).
68. C. W. Lange and C. G. Pierpont, *J. Am. Chem. Soc.*, *114*, 6582 (1992).
69. C. G. Pierpont and A. S. Attia, *Coll. Czech. Chem. Commun.*, *66*(1), 33 (2001).
70. C. G. Pierpont and C. W. Lange, *Prog. Inorg. Chem.*, *41*, 331 (1994).
71. D. N. Hendrickson and C. G. Pierpont, *Top. Curr. Chem. 234*, 63 (2004).
72. D. Herebian, P. Ghosh, H. Chun, E. Bothe, T. Weyhermuller, and K. Wieghardt, *Eur. J. Inorg. Chem.* 1957 (2002).
73. H. Chun, C. N. Verani, P. Chaudhuri, E. Bothe, E. Bill, T. Weyhermuller, and K. Wieghardt, *Inorg. Chem.*, *40,* 4157 (2001).
74. P. Chaudhuri, C. N. Verani, E. Bill, E. Bothe, T. Weyhermuller, and K. Wieghardt, *J. Am. Chem. Soc.*, *123,* 2213 (2001).
75. C. W. Lange, M. Foldeaki, V. I. Nevodchikov, V. K. Cherkasov, G. A. Abakumov, and C. G. Pierpont, *J. Am. Chem. Soc.*, *114*, 4220 (1992).
76. T. Motoyama, Y. Shimazaki, T. Yajima, Y. Nakabayashi, Y. Naruta, and O. Yamauchi, *J. Am. Chem. Soc.*, *126*, 7378 (2004).
77. K. Kawabata, M. Nakano, H. Tamura, and G. Matsubayashi, *J. Organomet. Chem.*, *689*, 405 (2004).
78. K. Kawabata, M. Nakano, H. Tamura, and G. Matsubayashi, *Inorg. Chim. Acta.*, *357*, 4373 (2004).
79. E. O. Fischer, and H. Wawersik, *J. Organomet. Chem.*, *5*, 559 (1966).
80. N. El Murr, J. E. Sheats, W. E. Geiger, Jr., and J. D. L. Holloway, *Inorg. Chem.*, *18*(6), 1443 (1979).
81. O. V. Gusev, L. I. Denisovich, M. G. Peterleitner, A. Z. Rubezhov, N. A. Ustynyuk, and P. M. Maitlis, *J. Organomet. Chem.*, *452*, 219 (1993).
82. O. V. Gusev, S. Sergeev, I. M. Saez, and P. M. Maitlis, *Organometallics*, *13*, 2059 (1994).
83. O. V. Gusev, M. G. Peterleitner, M. A. Ievlev, A. M. Kal'sin, P. V. Petrovskii, L. I. Denisovich, and N. A. Ustynyuk, *J. Organomet. Chem.*, *531*, 95 (1997).
84. H. J. Keller and H. Wawersik, *J. Organomet. Chem.*, *8*, 185 (1967).
85. E. O. Fischer and H. H. Lindner, *J. Organomet. Chem.*, *1*, 307 (1964).
86. H. El Amouri, G. Gruselle, J. Vaisserman, M. J. McGlinchey, and G. Jaouen, *J. Organomet. Chem.*, *485*, 79 (1995).
87. R. J. McKinney, *J. Chem. Soc., Chem. Commun.*, 603 (1980).
88. N. G. Connelly, A. R. Lucy, J. D. Payne, A. R. M. Galas, and W. E. Geiger, *J. Chem. Soc., Dalton Trans*, 1879 (1983).
89. M. J. Freeman, A. G. Orpen, N. G. Conneley, I. Manners, and S. J. Raven, *J. Chem. Soc., Dalton Trans*, 2283 (1985).
90. E. Fonseca, W. E. Geiger, T. E. Bitterwolf, and A. L. Rheingold, *Organometallics*, *7*, 567 (1988).
91. M. Carano, F. Cicogna, I. D'Ambra, B. Gaddi, G. Ingrosso, M. Maracicco, D. Paolucci, F. Paolucci, C. Pinzini, and S. Roffia, *Organometetallics.*, *21*, 5583 (2002).
92. R. E. Dessy, R. B. King, and M. Waldrop, *J. Am. Chem. Soc.*, *88* (3), 471 (1966).

93. R. E. Dessy, R. B. King, and M. Waldrop, *J. Am. Chem. Soc.*, *88* (22), 5112 (1966).

94. R. E. Dessy, R. Kornmann, C. Smit, and R. Haytor, *J. Am. Chem. Soc.*, *90*(8), 2001 (1968).

95. W. Kaim, *Coord. Chem. Rev.*, 219, 463 (2001).

96. W. Kaim, R. Reinhardt, S. Greulich, and J. Fiedler, *Organometallics*, *22*, 2240 (2003).

97. L. Brammer, N. G. Connelly, J. Edwin, W. E. Geiger, A. G. Orpen, and J. B. Sheridan, *Organometallics*, *7*, 1259 (1988).

98. O. V. Gusev, L. N. Morozova, T. A. Peganova, M. G. Peterleitner, S. M. Peregudova, L. I. Denisovich, P. V. Petroskii, Y. F. Oprunenko, and N. A. Ustynyuk, *J. Organomet. Chem.*, *493*, 181 (1995).

99. F. Morandini, G. Pilloni, G. Consiglio, A. Sironi, and M. Moret, *Organometallics*, *12*, 3495 (1993).

100. M. I. Bruce, P. A. Humphrey, M. L. Williams, B. W. Skelton, and A. H. White, *Austr. J. Chem.*, *42*, 1847 (1989).

101. (a) M. J. Shaw, W. E. Geiger, J. Hyde, and C. White, *Organometallics.*, *17*, 5486 (1998). (b) M. J. Shaw, J. Hyde, C. White, and W. E. Geiger, *Organometallics.* *23*, 2205 (2004).

102. J. E. Collins, M. P. Catellani, A. L. Rheingold, E. J. Miller, W. E. Geiger, A. L. Rieger, and P. H. Rieger, *Organometallics*, *14*, 1232 (1995).

103. B. de Bruin, unpublished results.

104. S.-I. Ohnishi and I. Nitta, *J. Chem. Pys.*, *39*, 2848 (1963).

105. G. R. Liebling and H. M. McConnel, *J. Chem. Pys.*, *42*(11), 3931 (1965).

106. J. Scott, S. Gambarotta, I. Korobkov, Q. Knijnenburg, B. de Bruin, and P. H. M. Budzelaar *J. Am. Chem. Soc.*, *127*, 17204 (2005).

107. N. G. Connelly and S. J. Raven, *J. Chem. Soc., Dalton Trans*, *8*, 1613 (1986).

108. A. Pedersen and M. Tilset, *Organometallics*, *12*, 56 (1993).

109. E. Fooladi and M. Tilset, *Inorg. Chem.*, *36*, 6021 (1997).

110. A. Pedersen and M. Tilset, *Organometallics*, *13*, 4887 (1994).

111. P. Diversi, S. Iacoponi, G. Ingrosso, F. Laschini, A. Lucherini, and P. Zanello, *J. Chem. Soc. Dalton Trans.*, 351 (1993).

112. E. Fooladi, T. Graham, M. L. Turner, B. Dalhus, P. M. Maitlis, and M. Tilset, *J. Chem. Soc., Dalton Trans*, 975 (2002).

113. P. Diversi, S. Iacoponi, G. Ingrosso, F. Laschini, A. Lucherini, C. Pinzino, G. Ucello-Baretta, and P. Zanello, *Organometallics.*, *14*, 3275 (1995).

114. P. Diversi, F. Fabrizi de Biani, G. Ingrosso, F. Laschini, A. Lucherini, C. Pinzino, and P. Zanello, *J. Organomet. Chem.*, *584*, 73 (1999).

115. P. Diversi, F. Marchetti, V. Ermini, and S. Matteoni, *J. Organomet. Chem.*, *593–594*, 154 (2000).

116. P. Diversi, V. Ermini, G. Ingrosso, A. Lucherini, C. Pinzino, and F. Simoncini, *J. Organomet. Chem.*, *555*, 135 (1998).

117. M. Tilset, G. S. Bodner, D.ZR. Senn, J. A. Gladysz, and V. D. Parker, *J. Am. Chem. Soc.*, *109*, 7551 (1987).

118. A. L. Seligson and W. C. Trogler, *J. Am. Chem. Soc.*, *114*, 7085 (1992).

119. J. Y. Chen and J. K. Kochi, *J. Am. Chem. Soc.*, *99*, 1450 (1977).

120. R. J. Kringler, J. C. Huffman, and J. K. Kochi, *J. Am. Chem. Soc.*, *104*, 2147 (1982).

121. C. Vogler, B. Schwederski, A. Klein, and W. Kaim, *J. Organomet. Chem.*, *436*, 367 (1992).

122. S. Hasenzahl, H.-D. Hausen, and W. Kaim, *Chem. Eur. J.*, 1, 95 (1995).

123. A. Klein, S. Hasenzahl, and W. Kaim, *J. Chem. Soc., Perkin Trans. 2*, 2573 (1997).

124. W. Kaim, A. Klein, S. Hasenzahl, H. Stoll, S. Záliš, and J. Fiedler, *Organometallics*, *17*, 237 (1998).

125. A. Klein, W. Kaim, F. M. Hornung, J. Fiedler, and S. Záliš, *Inorg. Chim. Acta*, *264*, 269 (1997).

126. A. Klein, S. Hasenzahl, W. Kaim, and J. Fielder, *Organometallics*, *17*, 3532 (1998).

127. (a) W. Kaim and S. Kohlmann, *Inorg Chem.*, *26*, 68 (1987). (b) S. D. Ernst and W. Kaim, *Inorg Chem.*, *28*, 1520 (1989). (c) W. Kaim and S. Kohlmann, *Inorg. Chem.*, *29*, 2909 (1990). (d) W. Kaim and V. Kasack, *Inorg. Chem.*, *29*, 4696 (1990). (e) M. Hunzicker and A. Ludi, *J. Am. Chem. Soc.*, *99*, 7370 (1977). (b) R. R. Ruminski and J. D. Petersen, *Inorg. Chem.*, *21*, 3706 (1982).

128. J. E. Anderson, C.-L. Yao, and K. M. Kadish, *Inorg. Chem.*, *25*, 718 (1986).

129. J. E. Anderson, C.-L. Yao, and K. M. Kadish, *Organometallics*, *6*, 706 (1987).

130. J. E. Anderson, C.-L. Yao, and K. M. Kadish, *J. Am. Chem. Soc.*, *109*, 1106 (1987).

131. B. B. Wayland, *Polyhedron*, *16/17*, 1545 (1988).

132. B. B. Wayland, S. Ba, and A. E. Sherry, *J. Am. Chem. Soc.*, *119*, 5305 (1991).

133. B. B. Wayland, V. LK. Coffin, and M. D. Farnos, *Inorg. Chem.*, *27*, 2745 (1988).

134. B. B. Wayland, A. E. Sherry, and A. G. Bunn, *J. Am. Chem. Soc.*, *115*, 7675 (1993).

135. L. Basickes, A. W. Bunn, and B. B. Wayland, *Can. J. Chem.*, *79*, 854 (2001).

136. B. B. Wayland, B. A. Woods, and V. L. Coffin, *Organometallics*, *5*, 1059 (1986).

137. V. L. Coffin, W. Brennen, and B. B. Wayland, *J. Am. Chem. Soc.*, *110*, 6063 (1988).

138. A. E. Sherry and B. B. Wayland, *J. Am. Chem. Soc.*, *111*, 5010 (1989).

139. B. B. Wayland, A. E. Sherry, G. Pozmik, and A. G. Bunn, *J. Am. Chem. Soc.*, *114*, 1673 (1992).

140. B. B. Wayland, A. E. Sherry, and V. L. Coffin, *J. Chem. Soc., Chem. Commun*, *10*, 662 (1989).

141. R. S. Paonessa, N. C. Thomas, and J. Halpern, *J. Am. Chem. Soc.*, *107*, 4333 (1985).

142. X.-X. Zhang, G. F. Parks, and B. B. Wayland, *J. Am. Chem. Soc.*, *119*, 7938 (1997).

143. B. B. Wayland, B. A. Woods, and R. Pierce, *J. Am. Chem. Soc.*, *104*, 302 (1982).

144. B. B. Wayland and B. A. Woods, *J. Chem. Soc., Chem. Commun*, *14*, 700 (1981).

145. B. B. Wayland, B. A. Woods, and V. M. Minda, *J. Chem. Soc., Chem. Commun*, *11*, 634 (1982).

146. B. B. Wayland, A. Duttaahmed, and B. A. Woods, *J. Chem. Soc., Chem. Commun*, 142 (1983).

147. G. Pozmik, P. J. Carrol, and B. B. Wayland, *Organometallics*, *12*, 3410 (1993).

148. M. D. Farnos, B. A. Woods, and B. B. Wayland, *J. Am. Chem. Soc.*, *108*, 3659 (1986).

149. B. B. Wayland, S. L. Van Voorhees, and C. Wilker, *Inorg. Chem.*, *25*, 4039 (1986).

150. M. S. Sanford and J. T. Groves, *Angew. Chem. Int. Ed. Engl.*, *43*, 588 (2004).

151. B. B. Wayland and K. J. Del Rossi, *J. Organomet. Chem.*, *276*(1), C27 (1984).

152. K. J. Del Rossi and B. B. Wayland, *J. Am. Chem. Soc.*, *107*, 7941 (1985).

153. H. W. Bosch and B. B. Wayland, *J. Chem. Soc., Chem. Commun*, 900 (1986).

154. B. B. Wayland, G. Poszmik, and M. Fryd, *Organometallics*, *11*, 3534 (1992).

155. H Ogoshi, J. Setsunu, and Z. Yoshida, *J. Am. Chem. Soc.*, *99*(11), 3869 (1977).

156. B. B. Wayland, Y. Feng, and S. Ba, *Organometallics*, *8*, 1438 (1989).

157. A. G. Bunn and B. B. Wayland, *J. Am. Chem. Soc.*, *114*, 6917 (1992).

158. M. Feng and K. S. Chan, *J. Organomet. Chem.*, *584*, 235 (1999).

159. K. W. Mak, S. K. Yeung, and K. S. Chan, *Organometallics*, *21*, 2362 (2002).

160. M. K. Tse and K. S. Chan, *J. Chem. Soc., Dalton Trans*, 510 (2001).

161. S. K. Yeung and K. S. Chan, *Organometallics*, *24*, 6426 (2005).
162. D. G. H. Hetterscheid, J. Kaiser, E. Reijerse, T. P. J. Peters, S. Thewissen, A. N. J. Blok, J. M. M. Smits, R de Gelder, and B. de Bruin, *J. Am. Chem. Soc.*, *127*, 1895 (2005).
163. B. B. Wayland and B. A. Woods, *J. Chem. Soc., Chem. Commun.*, 475 (1981).
164. K. J. Del Rossi and B. B. Wayland, *J. Chem. Soc., Chem. Comun.*, 1653 (1986).
165. K. J. Del Rossi, X.-X. Zhang, and B. B. Wayland, *J. Organomet. Chem.*, *504*, 47 (1995).
166. B. B. Wayland, S. Ba, and A. E. Sherry, *J. Am. Chem. Soc.*, *113*, 5305 (1991).
167. A. E. Sherry and B. B. Wayland, *J. Am. Chem. Soc.*, *112*, 1259 (1990).
168. B. B. Wayland, S. Ba, and A. E. Sherry, *Inorg. Chem.*, *31*, 148 (1992).
169. X-X- Zhang and B. B. Wayland, *Inorg. Chem.*, *39*, 5318 (2000).
170. X-X- Zhang and B. B. Wayland, *J. Am Chem. Soc.*, *116*, 7897 (1994).
171. W. H. Cui, X. P. Zhang, and B. B. Wayland, *J. Am. Chem. Soc.*, *125*, 4994 (2003).
172. W. H. Cui and B. B. Wayland, *J. Am. Chem. Soc.*, *126*, 8266 (2004).
173. A. P. Nelson and S. G. DiMagno, *J. Am. Chem. Soc.*, *122*, 8569 (2000).
174. S. L. van Voorhees and B. B. Wayland, *Organometallics*, *6*, 204 (1987).
175. Y. Ni, J. P. Fitzgerald, P.Carrol, and B. B. Wayland, *Inorg. Chem.*, *33*, 2029 (1994).
176. M. Wei and B. B. Wayland, *Organometallics*, *15*, 4681 (1996).
177. A. G. Bunn, M. Wei, and B. B. Wayland, *Organometallics*, *13*, 3390 (1994).
178. A. G. Bunn, Y. Ni, M. Wei, and B. B. Wayland, *Inorg. Chem.*, *39*, 5576 (2000).
179. B. de Bruin, T. P. J. Peters, S. Thewissen, A. N. J. Blok, J. B. M. Wilting, R. de Gelder, J. M. M. Smits, and A. W. Gal, *Angew. Chem., Int. Ed. Engl.*, *41*, 2135 (2002).
180. B. de Bruin, S. Thewissen, T.-W. Yuen, T. P. J. Peters, J. M. M. Smits, and A. W. Gal, *Organometallics*, *21*, 4312 (2002).
181. D. G. H. Hetterscheid, J. M. M. Smits, and B. de Bruin, *Organometallics*, *23*, 4236 (2004).
182. D. G.. H. Hetterscheid, B. de Bruin, J. M. M. Smits, and A. W. Gal, *Organometallics*, *22*, 3022 (2003).
183. D. G. H. Hetterscheid, M. Klop, R. J. N. A. M. Kicken, J. M. M. Smits, E. J. Reijerse, and B. de Bruin, *Chem. Eur. J.*, *13*, 3386 (2007).
184. D. G. H. Hetterscheid, M. Bens, and B. de Bruin, *Dalton Trans.*, 979 (2005).
185. For examples see the NIEST spin-trap database, which can be found at the url: https://dir-apps.niehs.nih.gov/stdb/index.cfm. Supporting references are provided therein.
186. M. Gerisch, J. R. Krumper, R. G. Bergman, and T. D. Tilley, *J. Am. Chem. Soc.*, *123*, 5818 (2001).
187. M. Gerisch, J. R. Krumper, R. G. Bergman, and T. D. Tilley, *Organometallics*, *22*, 47 (2003).
188. J. R. Krumper, M. Gerisch, J. M. Suh, R. G. Bergman, and T. D. Tilley, *J. Org. Chem.*, *68*, 9705 (2003).
189. See, for example, H. M. L. Davies and R. E. J. Beckwith, *Chem. Rev.*, *103*, 2861 (2003) and references cited therein.
190. C. Bianchini, F. Laschi, A. Meli, M. Peruzzini, P. Zanello, and P. Frediani, *Organometallics*, *7*, 2575 (1988).
191. C. Bianchini, F. Laschi, M. F. Ottaviani, M. Peruzzini, P. Zanello, and F. Zanobini, *Organometallics*, *8*, 893 (1989).
192. C. Bianchini, F. Laschi, M. F. Ottavini, M. Peruzzini, and P. Zanello, *Organometallics*, *7*, 1660 (1988).

193. C. Bianchini, A. Meli, M. Peruzzini, A. Vacca, F. Laschi, P. Zanello, and F. M. Ottaviani, *Organometallics*, *9*, 360 (1990).

194. C. Bianchini, V. Herrera, M. V. Jiménez, F. Laschi, A. Meli, R. Sánchez-Delgado, F. Vizza, and P. Zanello, *Organometallics*, *14*, 4390 (1995).

195. P. Barbaro, C. Bianchini, K. Linn, C. Mealli, A. Meli, and F. Vizza, *Inorg. Chim. Acta*, 198, 31 (1992).

196. A. E. Crease, B. D. Gupta, M. D. Johnson, and S. Moorhouse, *J. Chem. Soc. Dalton, Trans.*, 1821 (1978).

197. L. D. Tyutchenkova, V. G. Vinogradova, and Z. K. Maizus, *Izv. Akad. Nauk, SSSR, Ser. Khim.*, *4*, 773 (1978).

198. G. Strukul, M. Bonivento, M. Graziani, E. Cerna, and N. Palladino, *Inorg. Chim. Acta*, *12*, 15 (1975).

199. S. Padhye, R. Yerande, R. P. Patil, A. A. Kelkar, and R. V. Chaudhari, *Inorg. Chim. Acta*, *156*, 23 (1989).

## CHAPTER 6

# Unique Metal–Diyne, –Enyne, and –Enediyne Complexes: Part of the Remarkably Diverse World of Metal–Alkyne Chemistry


**SIBAPRASAD BHATTACHARYYA, SANGITA and JEFFREY M. ZALESKI**

*Department of Chemistry*
*Indiana University*
*Bloomington, IN 47405*


CONTENTS




*Progress in Inorganic Chemistry, Vol. 55* Edited by Kenneth D. Karlin

# I. INTRODUCTION

## A. Background

The application of metal–alkyne σ ($\eta^1$) and π($\eta^2$) linkages have enabled a truly remarkable array of novel chemical properties to emerge over the last 30 years. These include the development of unique inorganic–organometallic architectures (1, 2), alkyne transformations (3–6) and metathesis reactions (7–9), specific and stereoselective diyne and enyne ring-closing strategies for organic chemistry (10, 11), as well as metal-containing polymeric constructs that exhibit electrical–photoconductivity or long wavelength luminescence (12–17). Many of these diverse geometric and electronic properties can be directly traced to the direct spatial and energetic overlap of the metal and alkyne π-orbital frameworks, as well as the rigid-rod nature of the conjugated alkyne backbone. The genesis of synthetic routes to simple metal–alkynes has made possible the preparation of metal–polyynes (18) that perform unimolecular electron transfer, form metallomesogens (19), and structures possessing nonlinear optical properties (19–24). More recently, the preparation of σ and π metal–enediyne complexes has emerged as a metal-mediated route to traditional organic Bergman cyclization (25–27). Once again, the confluence

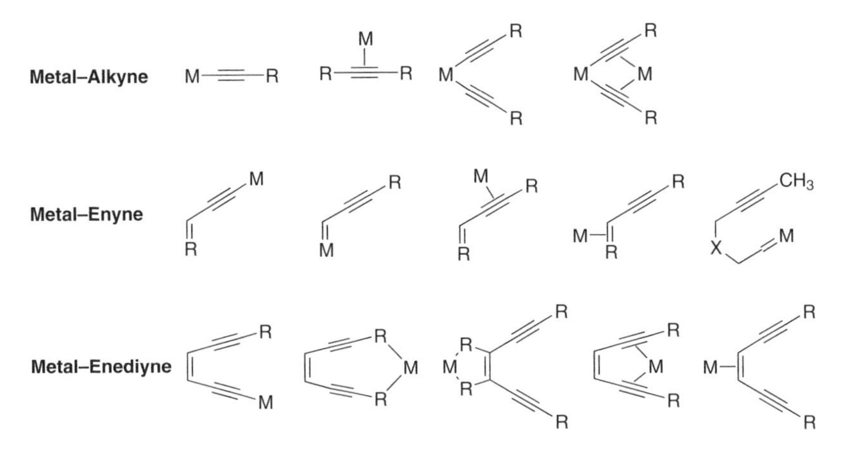


Figure 1. General σ and π metal–alkyne interactions.

of metal and alkyne geometric and electronic structure is the enabling feature that translates into new chemical properties and reactivities, further evolving the already diverse world of metal–alkyne chemistry.

## B. Chapter Scope

The field of metal–alkyne chemistry has exploded since the early 1980s. The discovery that metals enhance the potential of these already electronically rich ligands to perform remarkable synthetic transformations, or exhibit unique physical properties deriving from a marriage of metal and ligand geometric or electronic structures, has fueled the excitement. Due to the vastness of the metal–alkyne field, we are forced to be selective in our examination of the area. The goal of this chapter is therefore to specifically highlight distinct molecular constructs within the metal–alkyne domain, with special focus on σ and π metal cis-diyne, enyne, and enediyne complexes (Fig. 1) that produce intriguing geometric architectures, or perform key chemical transformations important to the broad field of chemistry.

# II. TRANSITION METAL ACETYLIDES

## A. Syntheses and Structures of Metal–Alkyne and cis-Diyne σ Complexes

Metal–alkyne σ complexes (2) and poly-yne analogues (28) are generally prepared by one of the following strategies: (a) reaction of alkali metal alkynylating reagents (M = Li, Na, Mg) with metal halides; (b) Cu(I) catalyzed alkyne-transfer reactions; (c) metathesis of metal halides with trimethylstannyl alkyne-transfer reagents; (d) direct dehydrohalogenation of activated alkynes with metal halides; (e) oxidative addition of terminal diynes, and (f) small molecule ($N_2$, $H_2$, $CH_4$) elimination. These approaches are specific to the system of interest, and thus synthetic methodologies as a function of periodicity have been carefully documented (1, 2). Examples of each of these are given below.

(a) Alkali metal alkyne-transfer metathesis reactions with metal halides (29–31) is a versatile method to prepare mono-, cis-di-, or polyacetylide products as a function of stoichiometry (*m* and *x*).

$$L_nMX \quad + \quad xM'\text{-}(C{\equiv}CR)_m \longrightarrow L_nM\{(C{\equiv}CR)_m\}_x$$

(X = Cl, Br, I) (M′ = Li, Na, Mg) (R = H, alkyl, trialkylsilyl, aryl)

Alkyne anion salts can be readily prepared by reaction of a terminal alkyne with 1 equiv of BuLi (32), while protected alkynes and diynes are accessible from the doubly protected $Si(Me)_3$ alkyne with MeLi·LiBr (29, 32). The acidity of the acetylenic hydrogen ($pK_a \sim 25$) permits the former reaction to tolerate basic protecting and functional groups (R = H, alkyl, trialkylsily, aryl), however, the sensitivity of the alkyne anion salt to degradation or polymerization can lead to low product yields (29, 31). Additionally, alkynyl Grignard reagents (29, 33, 34) can be prepared by treatment of the formal alkyne with EtMgBr to yield alkynyl–MgBr species. Reaction of $MgBr_2$ with alkynyl–Li salts is also an effective route to such Grignard reagents (29). More recently, calcium, strontium, and barium triphenylsilylethynyl derivatives of the form [M([18]crown-6)($C{\equiv}CSiPh_3)_2$] (M = Ca, Sr, Ba) have been prepared via treatment of alkaline earth amides [$M(N(SiMe_3)_2)_2(thf)_2$] (M = Ca, Sr, Ba; thf = tetrahydrofuran, ligand), with $Ph_3SiC{\equiv}CH$ in the presence of [18]crown-6 in THF (advent) (35).

As examples, reaction of $Ru_2Cl_4Ar_2$ with $MC{\equiv}CR$ (M = Na, Li) in the presence of simple ligands (e.g., phenanthroline or $PMe_3$) generates the corresponding mono- or bis(alkynyl) ruthenium arene species (Scheme 1) (36). Similarly, the stepwise replacement of chloride in $TiCl_2Cp_2SiR_2$ using the 1,3-diyne $LiC{\equiv}CC{\equiv}CR$ (R = Et, $SiMe_3$, Fc, or Rc) (37) or reaction of ZrCl[2-N($SiMe_3$)-4-$MeC_5H_4N]_3$ leads to generation of the mono- or bis(diyne) $\sigma$ complexes (38) (Fig. 2) via nucleophilic attack by the alkyne anion. Also, Stang and co-workers (39) used this route to prepare chiral $d^8$ tetranuclear macrocycles based on cis-diynyl

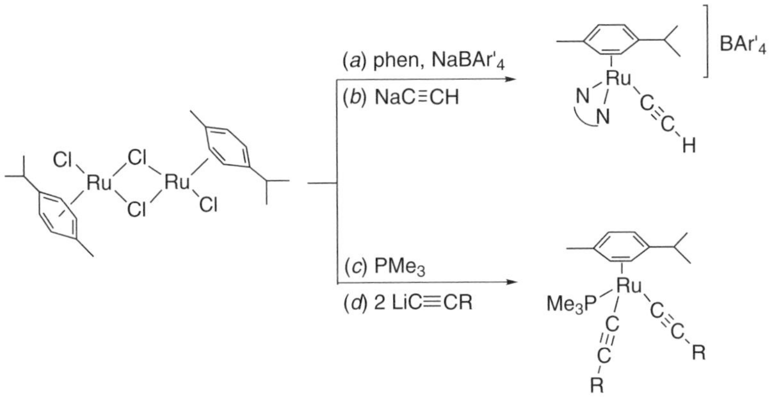

Scheme 1. Synthesis of $Ru^{II}$ σ-alkyne and diyne constructs by alkali metal metathesis.

coordination as host–guest inclusion complexes (Scheme 2). Finally, Ren and co-workers (40, 41) used this synthetic approach to prepare a series of Ru-capped metallaynes as intriguing prototypes for organometallic wires (Scheme 3). Finally, a peripheryl, but related polymeric synthon of the form $[AgC{\equiv}CR]_x$, has been shown to act as an alkynylating agent in reactions with platinum substrates, displacing weakly coordinating ligands and stabilizing the resulting complex by $\eta^2$-silver acetylide and weak Pt—Ag bonding interactions (42–45).

(b) The Cu(I)-catalyzed dehydrohalogenation reaction was originally developed in the 1970s (M = Pd, Pt) (46, 47) as a direct route to σ-alkyne complexes, bypassing the need for generation of an unstable, alkali metal–alkyne species. It is now one of the most versatile methods for preparing metal alkyne σ complexes.

$$L_nMX + xH(C{\equiv}CR)_m \xrightarrow[NR'_3]{CuI} \left[Cu\{(C{\equiv}CR)_m\}_x\right] \longrightarrow L_nM\{(C{\equiv}CR)_m\}_x + NR'_3HX$$

(X = Cl, Br, I)

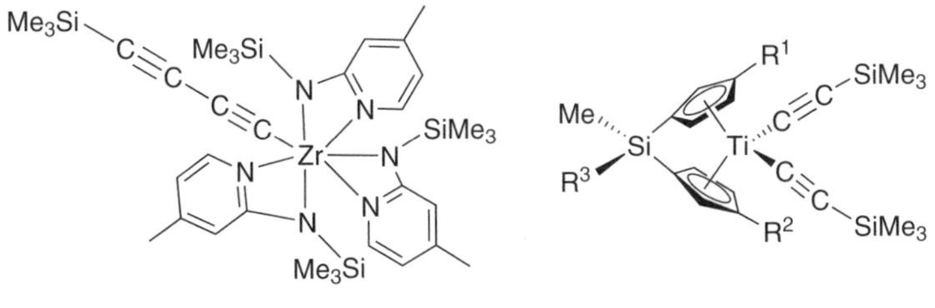

Figure 2. Early transition metal σ-alkyne complexes formed via alkali metal metathesis.

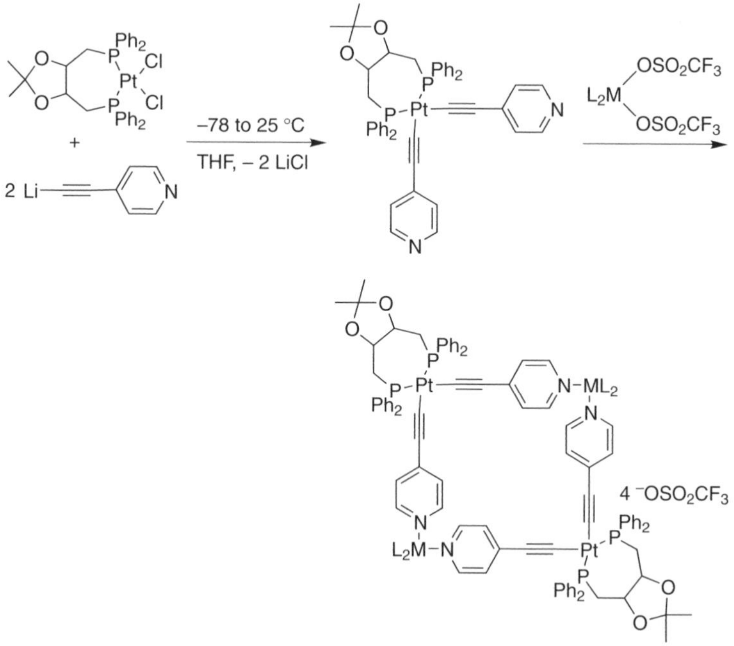

Scheme 2. Preparation of chiral $d^8$ macrocycles for host–guest inclusion chemistry. L = *(R)*- or *(S)*-2,2′-bis(diphenyphosphino)-1,1′-binaphthyl.

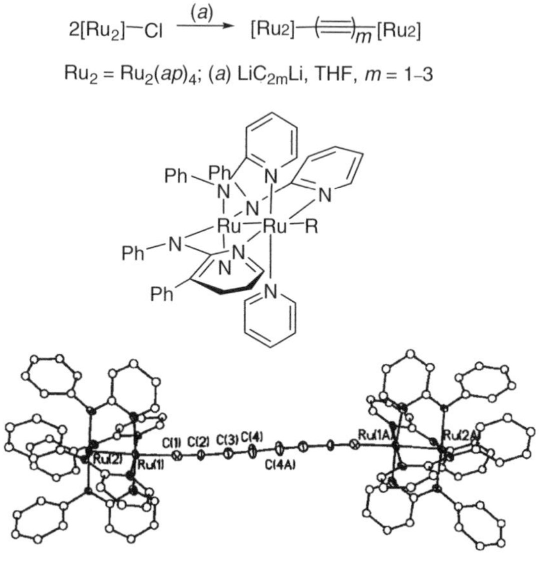

$$2[Ru_2]{-}Cl \xrightarrow{(a)} [Ru2]{-}(\equiv)_m{-}[Ru2]$$

$Ru_2 = Ru_2(ap)_4$; (*a*) $LiC_{2m}Li$, THF, $m = 1$–3



Scheme 3. General strategy for preparation of Ru-capped metallaynes as organometallic wires, and the X-ray structure of a tetraruthenium tetrayne prototype. Z-Anilonopyridinate is ap.

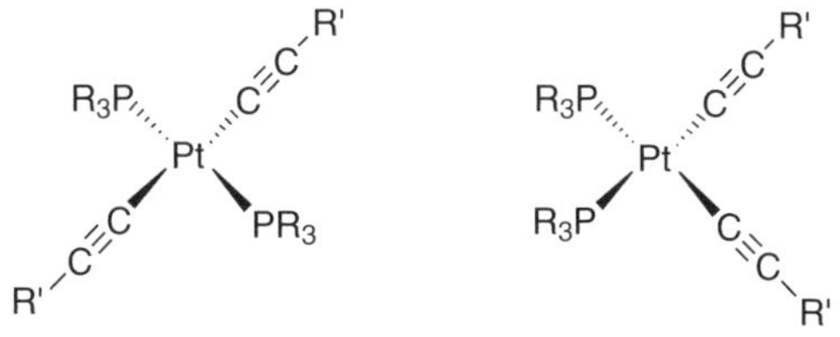

Figure 3. The Pt(II) diyne σ-complexes prepared by Cu(I)-catalyzed dehydrohalogenation.

Reaction of the desired alkyne with a metal halide in an amine solvent in the presence of catalytic amounts of CuI leads to the generation of mono- or bis(acetylide) compounds in both the trans or less stable cis conformations (48) (Fig. 3). The general synthetic mechanism has been studied in detail for M = Pd (49), and is believed to involve formation of a Cu–alkyne species that undergoes reversible migration to the metal halide center. Alkyne ligand transfer between Pd and Cu ultimately leads to formation of cis and trans products, the latter being more favored due to labilizing–stabilizing ability (i.e., trans-influence) of the R group that affects molecular orbital energetics–bonding in ground and transition states (Scheme 4). In the presence of a phenyl ligand on Pd, an aryl–alkyne coupling product is also formed by reductive elimination, which is the basis for the widely used Sonogashira coupling reaction (50).

The formation of molecular photochemical devices (51), as well as chiral metallacyclophanes for asymmetric catalysis (52) from fundamentally similar *cis*-Pt(acetylide)$_2$ constructs reveals their chemical diversity and utility as basic molecular building blocks (Fig. 4). The same synthetic approach has been extended to the preparation of metal–diynyl complexes from across the periodic table (53). Rigid-rod σ-alkynyl polymeric species

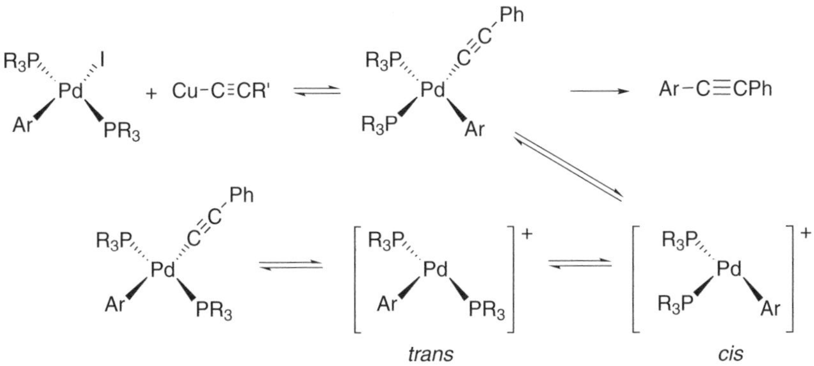

Scheme 4. Example of Cu(I)-catalyzed dehydrohalogenation as a route to metal σ-alkynes.

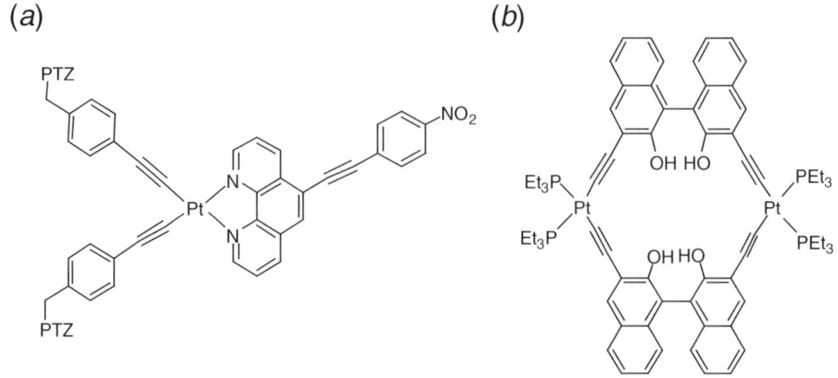

Figure 4. Molecular photochemical device prototype (*a*) and chiral metallacyclophane for asymmetric catalysis (*b*) based on bis(acetylide)Pt(II) compounds.

(48, 54) with $M_w > 10^6$ ($M_w$ = weight-average molecular weight) are also obtained from direct dehydrohalogenation (55, 56), as well as alkyne ligand transfer (57), and $O_2$-dependent coupling (58) strategies using cuprous halide and divalent, group 10 (VIII) metal halides.

(c) Synthesis of the versatile group 14 (IV A) alkyne-transfer reagents (59–61) (e.g., $Me_3Sn{-}C{\equiv}CR$ from reaction of the alkyne with BuLi and $Me_3SnCl$, and subsequent metathesis with transition metal chlorides yields the desired metal alkyne species with concomitant formation of the alkyl tin halide (62).

$$L_nMCl + Me_3Sn{-}(C{\equiv}CR)_m \longrightarrow L_nM{-}(C{\equiv}CR)_m + Me_3SnCl$$

A wide range of simple alkyne complexes, as well as polyynyl and polymeric species have been prepared from this route as a function of stoichiometry (12, 63–68). Molecular weights as high as 210,000 amu for *trans*-$PtCl_2(PBu_3)_2$ and as low as 20,000 amu for M = Ni, Pd have been observed. Other examples include reaction of $Rh(PMe_3)_4Cl$ with 1 equiv of $Me_3Sn(C{\equiv}CPh)$, which generates the mono-σ-acetylide compound (64, 67 ,68), as well as conversion of $Ru(dppm)Cl_2$ to the monodiyne or *cis*-σ-diacetylide species (via $LiC{\equiv}CR$) (69–71). As shown in Scheme 5, tetraalkylsilyl reagents are widely used as alkyne protecting groups in metal–alkyne reactions, in alkyne precursor preparations, and to generate polyyne compounds (72–74). The silyl group can be readily protiodesilylated using KOH/MeOH to generate the terminal alkyne if desired.

(d) σ-Acetylide complexes can also be prepared by reaction of a terminal alkyne with a $d^6$ metal chloride precursor (M = Ru, Os; L = dppm), which

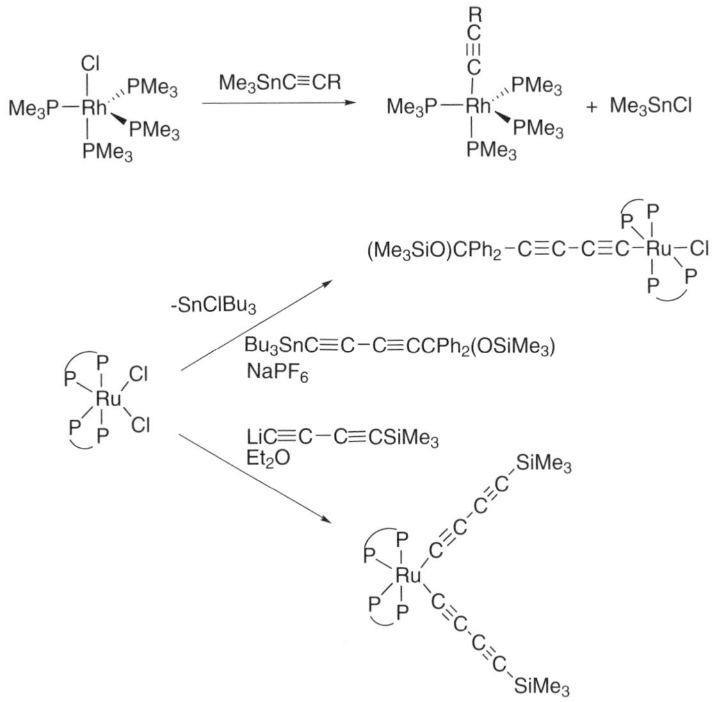

$$\mathrm{Me_3P{-}RhCl(PMe_3)_3} \xrightarrow{\mathrm{Me_3SnC{\equiv}CR}} \mathrm{Me_3P{-}Rh(C{\equiv}CR)(PMe_3)_3} + \mathrm{Me_3SnCl}$$

$$\mathrm{RuCl_2(P\text{-}P)_2} \xrightarrow[\mathrm{NaPF_6}]{\mathrm{Bu_3SnC{\equiv}C{-}C{\equiv}CCPh_2(OSiMe_3)},\ -\mathrm{SnClBu_3}} \mathrm{(Me_3SiO)CPh_2{-}C{\equiv}C{-}C{\equiv}C{-}Ru(P\text{-}P)_2{-}Cl}$$

$$\mathrm{RuCl_2(P\text{-}P)_2} \xrightarrow[\mathrm{Et_2O}]{\mathrm{LiC{\equiv}C{-}C{\equiv}CSiMe_3}} \mathrm{(P\text{-}P)_2Ru(C{\equiv}C{-}C{\equiv}C{-}SiMe_3)_2}$$

Scheme 5. Illustration of group 14 (IVA) alkylstannyl and alkylsilyl compounds as alkyne-transfer reagents and alkynyl protecting groups.

generates the isolable vinylidene intermediate as a salt. Subsequent deprotonation forms the stable metal–alkyne product (75–80).

$$\mathrm{L}_n\mathrm{MCl} + 2\mathrm{HC{\equiv}CR} \longrightarrow \left[\mathrm{L}_n\mathrm{M{=}C{=}C}\begin{smallmatrix}\mathrm{R}\\ \mathrm{H}\end{smallmatrix}\right]^+ \xrightarrow{\text{base}} \mathrm{L}_n\mathrm{M{-}C{\equiv}CR}$$

Complexes with M = Fe can also be prepared by direct deprotonation of the terminal alkyne without isolation of the vinylidene species (80) (Scheme 6). Moreover, for diyne ligands, a series of bimetallic mixed-valence constructs have been prepared that show Class II mixed-valence delocalization (80). Similarly, reaction of $RuCl_2(PR_3)(\eta\text{-}C_6Me_6)$ (where $PR_3 = PMe_3$, $PMe_2Ph$, $PMePh_2$) with a nonnucleophilic base gives the Ru–alkyne complex once again via the vinylidene intermediate (81–83).

(e) Terminal alkynes and diynes can oxidatively add to low-valent, unsaturated metal centers to yield a σ-metal–alkyne or vinylidene species (84, 85).

$$\mathrm{L}_n\mathrm{MCl} \xrightarrow{\mathrm{HC{\equiv}CR}} \mathrm{L}_n\mathrm{HClM{-}C{\equiv}CR} \longrightarrow \mathrm{L}_n\mathrm{ClM{=}C{=}C}\begin{smallmatrix}\mathrm{R}\\ \mathrm{H}\end{smallmatrix}$$

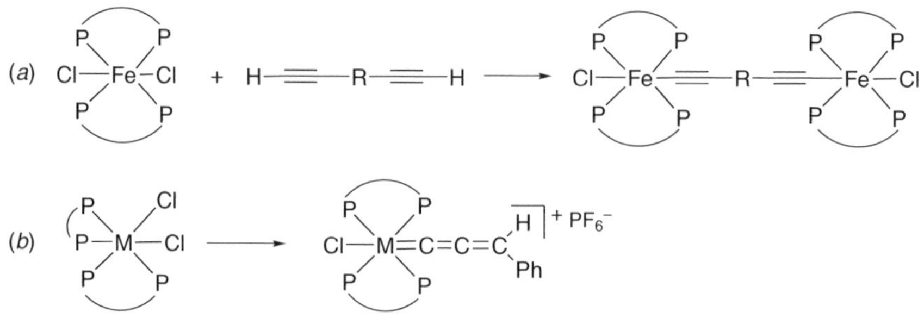

Scheme 6. Routes to $d^6$ metal acetylides via direct deprotonation or isolation of vinylidene precursors and subsequent deprotonation.

Reactions of this type are especially operative with the $RhCl(Pi\text{-}Pr_3)_2$ synthon (Scheme 7). Generation of the corresponding M(I) vinylidene species (M = Rh, Ir) have also been established, for example, by direct reaction of $RhCl(PPi\text{-}Pr_3)_2$ or *trans*-$IrCl(C_8H_{14})(PPi\text{-}Pr_3)_2$ with an alkylvinylchloride in the presence of sodium metal (Scheme 8) (86).

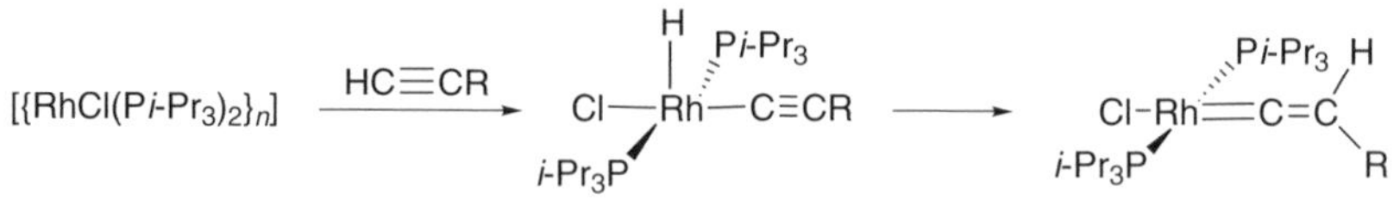

Scheme 7. The Rh–vinylidene species derived from oxidative addition of an alkyne followed by hydride transfer within the metal–acetylide precursor.

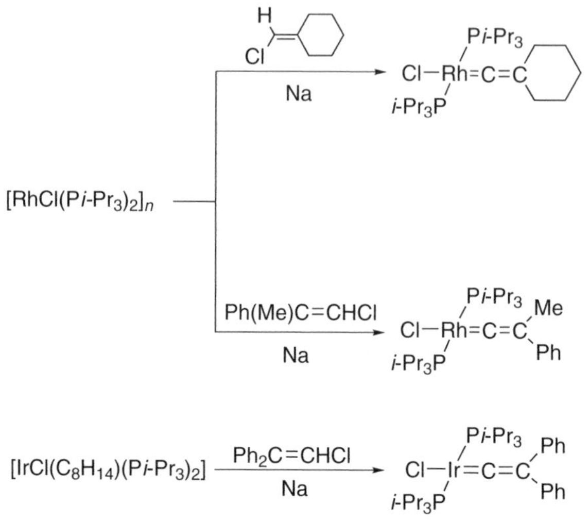

Scheme 8. Oxidative addition strategies to Rh and Ir acetylides.

The proposed mechanism involves formation of a vinyl radical that attacks the M(I) reagent, leading to a transient M(II) species that reacts with a second equivalent of radical, affording the M(I) vinyldene and isobutene.

Similarly, Rh and Ir σ-alkynes can also be prepared by generation of the triflate salt of the alkynyliodonium cation $RC{\equiv}CI^{+}Ph$, which serves as an electrophile for a nucleophilic, monovalent metal center (87, 88).

$$L_nMCl + [RC{\equiv}C\text{-}I^{+}Ph][^{-}OSO_2CF_3] \longrightarrow L_nM(C{\equiv}CR)(OSO_2CF_3) + PhI$$

$$L_nMCl + [PhI^{+}C{\equiv}C\text{-}I^{+}Ph][(^{-}OSO_2CF_3)_2] \longrightarrow [(ClL_nM{-}C{\equiv}C{-}ML_nCl)]\,[(OSO_2CF_3)_2] + 2PhI$$

The corresponding M(III) derivatives are isolated in good yields with either coordinated or noncoordinating triflate anions and concomitant release of phenyliodide.

(f) Small molecule elimination ($H_2$, $N_2$, $CH_4$) has been used as an effective means to generate coordinative unsaturation thereby driving metal–alkyne complexation reactions to produce vinylidene and metal–alkyne σ complexes. These routes employ the metal–small molecule complex as the starting material that is susceptible to either entropic loss to free a coordination position for alkyne binding by simple ligand substitution ($N_2$) (89), or reductive elimination of small molecules [$H_2$ (90); $CH_4$, Rh (91, 92), Co (93)] and oxidative addition of the alkyne.

$$L_nMN_2 + RC{\equiv}CH \xrightarrow[-N_2]{} L_nM(RC{\equiv}CH) \longrightarrow L_nM(C{\equiv}CR)(H)$$

$$L_nMH + RC{\equiv}CH \xrightarrow[-H_2]{} [L_nM{-}C{\equiv}CR]^{+} \xrightarrow{H^{+}} [L_nM{-}C{=}C(R)(H)]^{+}$$

Oxidative addition of a terminal alkyne to the $L_4MCH_3$ species generates the $L_4MH(C{\equiv}CR)$ complex and methane as a byproduct (91, 92). Basic workup regenerates the neutral alkyne species.

$$L_nMCH_3 + HC{\equiv}C{-}X{-}C{\equiv}CH \longrightarrow L_nM{-}C{\equiv}C{-}X{-}C{\equiv}CH + CH_4$$

Methane byproducts are also generated by bimolecular reaction of $Rh(NP_3)H$ compounds with $CH_3^{+}$ leaving a coordinatively unsaturated four-coordinate complex that oxidatively adds alkyne (94).

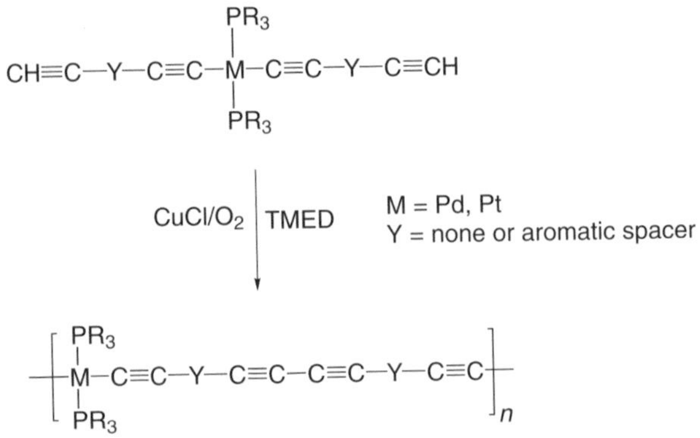

$$\mathrm{CH{\equiv}C{-}Y{-}C{\equiv}C{-}M(PR_3)_2{-}C{\equiv}C{-}Y{-}C{\equiv}CH} \xrightarrow[\text{TMED}]{\text{CuCl/O}_2} \left[\mathrm{{-}M(PR_3)_2{-}C{\equiv}C{-}Y{-}C{\equiv}C{-}C{\equiv}C{-}Y{-}C{\equiv}C{-}}\right]_n$$

M = Pd, Pt
Y = none or aromatic spacer

Scheme 9. Homocoupling of metal alkynes by Hay's coupling reaction (TMED=$N, N, N', N'$-tetramethylethylene).

## B. Thermal Reactivities of Metal–Acetylides

The scope of reactivity of metal–acetylide constructs is extremely broad and divergent, and as such, only select examples of these reactions will be presented here.

1. Polymerization of metal–acetylide complexes is a common reaction that leads to organometallic polymers by oxidative coupling and alkynyl–ligand exchange. In the former case, a catalytic amount of a cuprous halide and $O_2$ (Hay's coupling reaction) (95) is used to couple species with two terminal alkynyl units (Scheme 9) (58). This is a useful preparative route that leads to a high degree of polymerization. The method is also particularly convenient as there is no stoichiometric restriction on a reactant bearing identical functional groups. The alkynyl–ligand exchange method is a copper-catalyzed process taking place in amine solvents that has been used successfully to produce Ni-containing polymers (Scheme 10) (57). There are several examples (17, 77, 96–101) of formation

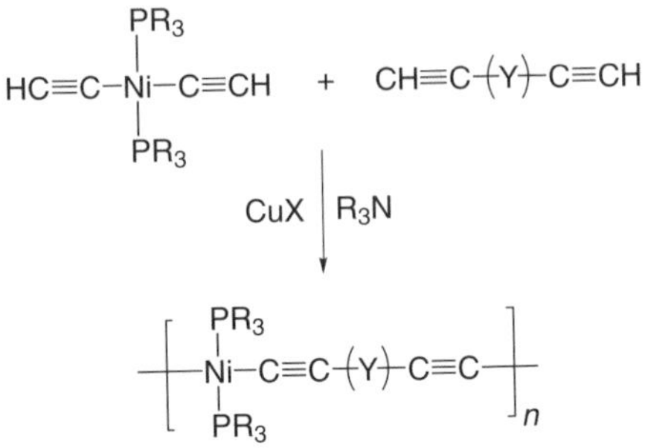

$$\mathrm{HC{\equiv}C{-}Ni(PR_3)_2{-}C{\equiv}CH} + \mathrm{CH{\equiv}C{-}(Y){-}C{\equiv}CH} \xrightarrow[\text{R}_3\text{N}]{\text{CuX}} \left[\mathrm{{-}Ni(PR_3)_2{-}C{\equiv}C{-}(Y){-}C{\equiv}C{-}}\right]_n$$

Scheme 10. Copper-catalyzed alkynyl ligand exchange.

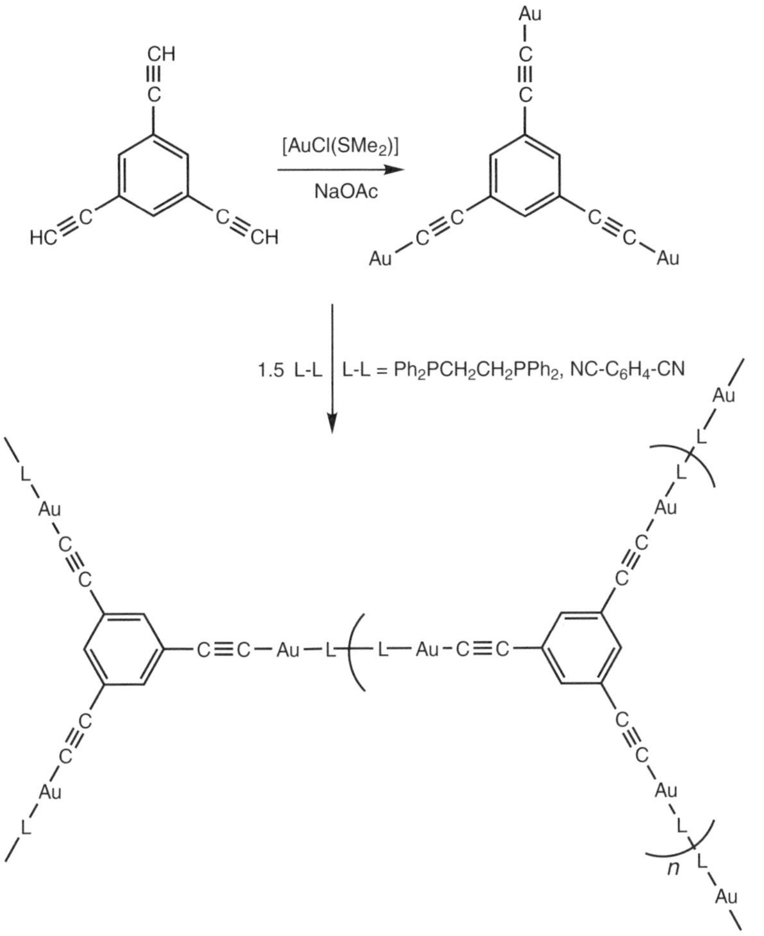

Scheme 11. Formation of alkynyl gold dendrimeric constructs.

of organometallic oligomer, coordination polymers, and dendrimeric species (Scheme 11) using metal–acetylide precursors and bridging ligands to create a multimetallic alkynyl scaffold.

2. Polynuclear assemblies can also be readily formed due to the ability of the alkyne ligands to form strong σ-bonds with metals, as well as their propensity for π coordination ($\eta^2$-fashion) (102) with electrophilic metal centers like Cu(I), Ag(I), and Au(I). Fornies and co-workers (44, 45, 103, 104) prepared interesting examples of di- and polynuclear alkynyl platinum complexes (Scheme 12). The $[PtCu_4(C{\equiv}CPh)_8]$ (105–107)

$[NBu_4]_2[Pt(C{\equiv}CPh)_4] + 2[Cu(MeCN)_4][PF_6] \longrightarrow [Pt_2Cu_4(C{\equiv}CPh)_8]_2 + [Pt_2Cu_4(C{\equiv}CPh)_8]_3$

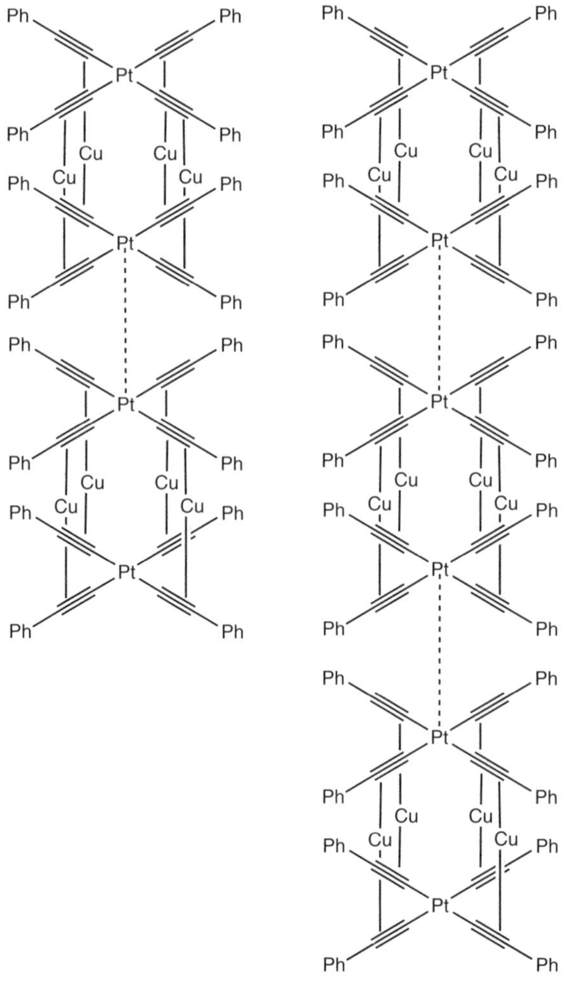

Scheme 12. Dimeric and trimeric aggregates of hexanuclear complex $[Pt_2Cu_4(C{\equiv}CPh)_8]$.

clusters reveal strong σ coordination to Pt(II) and additional metal–π interactions of Cu(I). The individual molecular units then aggregate in the solid state via weak Pt•••Pt interactions to form extended polynuclear assemblies.

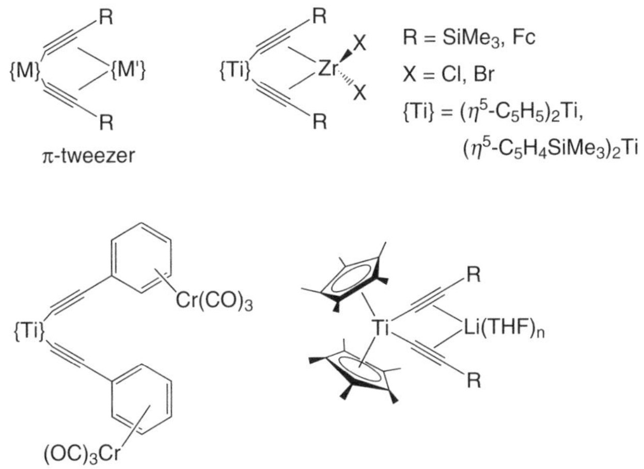

$R = SiMe_3$, Fc
$X = Cl$, Br
$\{Ti\} = (\eta^5\text{-}C_5H_5)_2Ti$, $(\eta^5\text{-}C_5H_4SiMe_3)_2Ti$

Figure 5. Example of π-tweezer organometallic compounds.

3. Organometallic σ-alkynyl compounds can react with stoichiometric metal salts to form σ-alkynyl–π-bonded "tweezer" systems for a range of transition metals (Fig. 5) (108–110). Like the metal–acetylides, these tweezer compounds can form unusual polynuclear assemblies that are both homo- and heteromultimetallic and show properties reflecting their σ/π-coordinated nature. Interest in the π-bound coordination chemistry and organometallic transformations using catalysts of these forms has led to a wealth of reactivity that has been comprehensively reviewed elsewhere (111, 112) (Fig. 6). In the case of Pt(0) complexes, some interesting single-bond cleavage reactions with early transition metal tweezer units have been reported and are reviewed along with the general chemistry of titanocenes and zirconocene complexes of diynes and polyynes (9). Here Rosenthal et al. (9, 113) reported the reaction of σ-disubstituted titanocene and zirconocene butadiyne complexes in which the σ/π-dinuclear complex results in the presence of a second equivalent of metallocene (Scheme 13).
4. Alkyne cross-coupling reactions over the last 25 years have become one of the most valuable assets in the synthetic chemist's toolbox. The now famous Sonogashira coupling (50, 114) of terminal alkynes with aryl or vinyl halides is readily achieved with a palladium catalyst, a copper(I) cocatalyst, and amine base. In the catalytic cycle (Scheme 14a), copper– and palladium–alkyne complexes are the key intermediates that lead to coupling of R and R′ units via the alkyne. Analogously, the Stille coupling (115) reaction performs alkyne coupling without base by generation of a trimethylstannyl alkyne (116–119) in place of the Cu(I) cocatalyst.

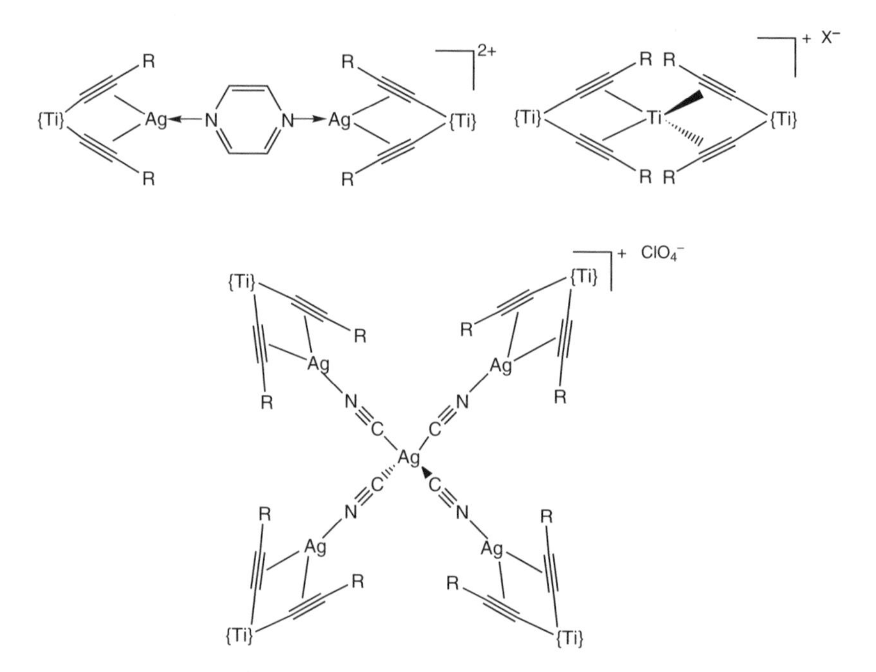


Figure 6. Multimetallic π-tweezer assemblies.

A somewhat different but mechanistically related reaction is the [2 + 3] cycloaddition of a functionalized alkyne or nitrile to an azide to form a disubstituted triazole (120) or tetrazole ring (121, 122), linking the respective functionalities irreversibly (Scheme 14b). This "click chemistry" was used by Sharpless and co-workers (120) in 2001 as a tool to probe biochemical catalysis and substrate activation. The ease of the Cu(I)-catalyzed reaction has created a true explosion (120–160) of simple coupling–functionalization chemistry of all types of biochemical components (sugars, DNA, proteins, enzymes, substrates, inhibitors) (131, 135, 136, 139, 142, 155, 157–160), polymers (126, 134, 140, 147, 154),

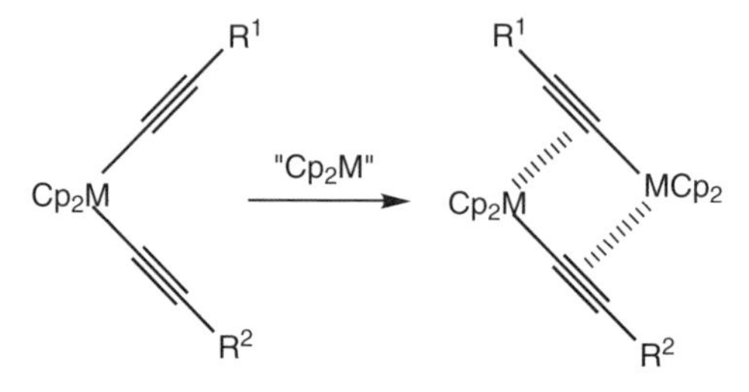


Scheme 13. The $\eta^1$ and $\eta^2$ alkynyl bonding modes of dinuclear "metallocene" complexes.

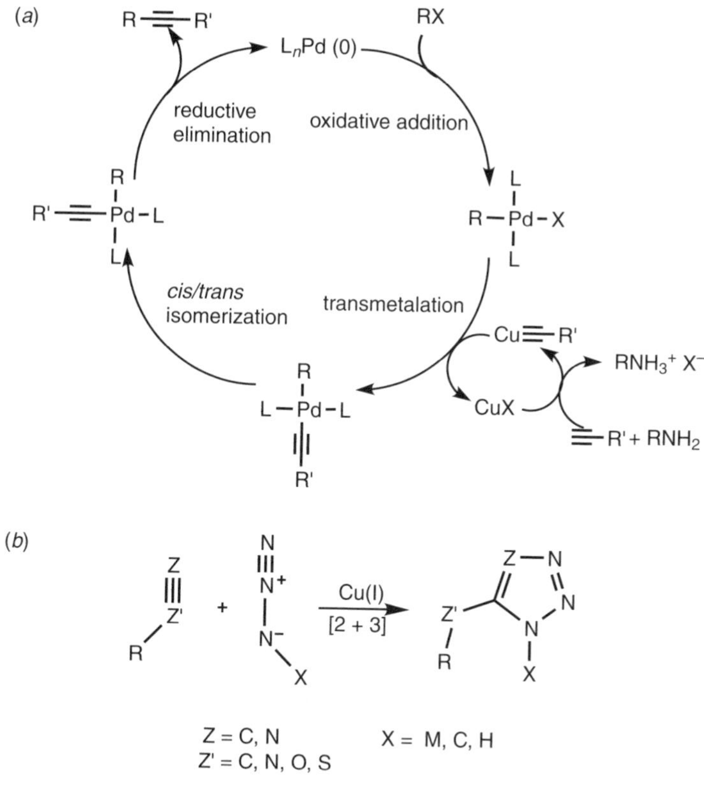


Scheme 14. (*a*) Sonogashira cross-coupling reaction mechanism. (*b*) "Click" chemistry; [2 + 3] cycloaddition of an azide to an alkyne or nitrile to form 1,5-triazole- or tetrazole-linked R and X. The analogous substituted 2,5-tetrazole or 1,4-triazoles are also part of the reaction scope.

nanoparticles–supramolecules (147–149, 156), and liquid crystals (128, 138) via the "click" approach. Clearly, simple coupling stategies (e.g., click chemistry) will continue to be widely utilized and make complex functionalization very accessible.

5. Alkynyl-transfer reactions from metal–acetylides are uncommon, but not unprecedented. Among the metal–acetylides, mercury alkynyl complexes are somewhat unique because of their ability to undergo ligand exchange and transfer alkyne to metal halide complexes. Due to their inherent instability, studies of mercury alkynyl complexes are sparse compared to their isoelectronic [Au(I), Ag(I)] counterparts. As an example, the reaction of dialkynyl mercury species with *cis*-$[PtCl_2(CO)L]$ has been thoroughly investigated, and shown to occur in a stereospecific fashion (161) (Scheme 15). Moreover, an unususal alkynyl-transfer reaction is observed between $[Pt(dppm)_2Cl_2]$ and a series of $[Hg(C{\equiv}CR)_2]$ species. The reaction occurs nearly quantitatively,

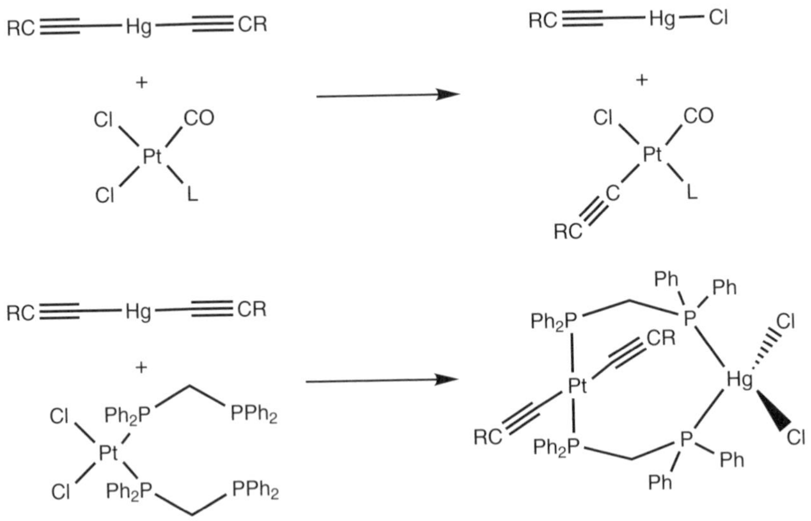

Scheme 15. Alkynyl-transfer reactions involving Pt and Hg.

and involves the formation of an eight-membered ring containing both platinum and mercury centers (162) (Scheme 15).

6. Metathesis is one of the most fundamental and important reactions in chemistry as evidenced by Schrock, Grubbs, and Chauvin receiving the 2005 Nobel Prize for development of the alkene metathesis method in organic synthesis. By comparison, alkyne methathesis is in its infancy and is only now being recognized as a synthetically useful approach to the preparation of organic compounds and specifically, various natural products (163). Much like the changing partners dance used to illustrate the mechanism of alkene metathesis, in alkyne metathesis, alkylidyne components of disubstituted alkynes are scrambled statistically to give all possible alkynes (Scheme 16) (164, 165). Schrock and co-workers (166, 167)

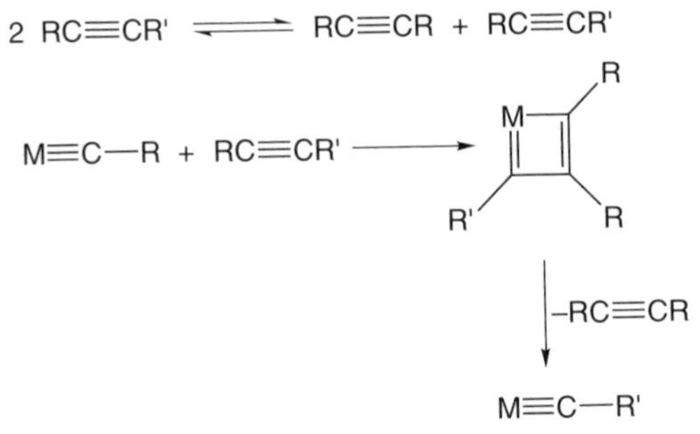

$$2\ RC{\equiv}CR' \rightleftharpoons RC{\equiv}CR + RC{\equiv}CR'$$

Scheme 16. Mechanism of alkyne metathesis.

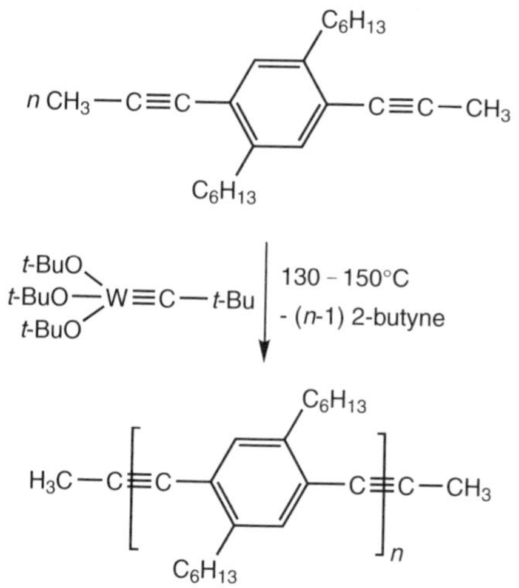

Scheme 17. Formation of poly(arylenethynylene)s via a tungsten–carbyne metathesis catalyst.

showed that well-defined molybdenum– or tungsten–carbyne complexes are active in metathetical reactions with alkynes much like those deomonstrated in alkene methathesis. In 1997, Weiss et al. (168) reported the first use of Schrock's high oxidation state tungsten–carbyne species for the formation of a range of poly(aryleneethynylene)s (Scheme 17) and showed that the method was competitive with Pd-catalyzed strategies for alkyne coupling.

In an unusual photochemical example, Rosenthal and co-workers (169) showed that photocatalytic C–C single-bond metathesis could be operative for butadiyne in the presence of excess titanocene. If butadiynes *t*-BuC≡C–C≡C*t*-Bu and $Me_3SiC{\equiv}C{-}C{\equiv}CSiMe_3$ (1:1) are treated with an excess of $Cp_2Ti$ reagent and the mixture is then irradiated, the unsymmetrically substituted diyne $Me_3SiC{\equiv}C{-}C{\equiv}C$*t*-Bu is afforded after oxidative workup, in addition to the symmetrically substituted starting diynes (Scheme 18). Dimeric

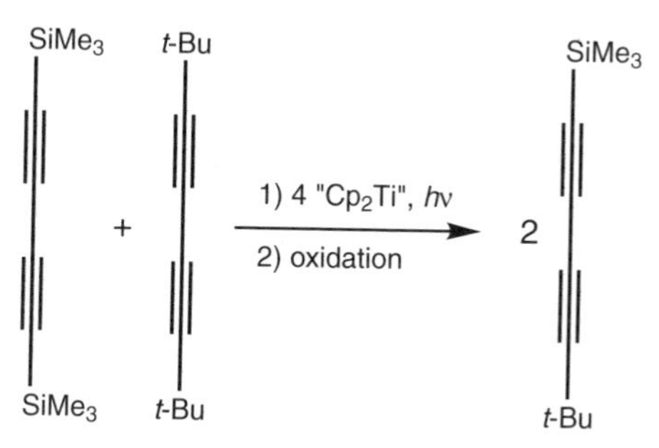


Scheme 18. Titanocene-mediated photocatalytic butadiyne metathesis reaction.

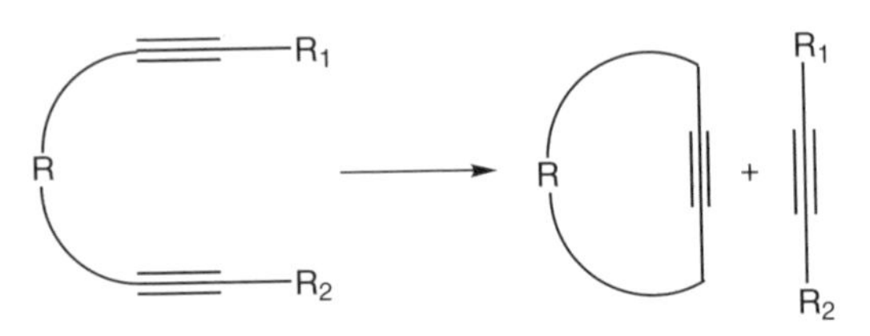


Scheme 19. Ring-closing alkyne metathesis reaction.

$Cp_2Ti$-acetylides of the form shown in Scheme 13 have been proposed as plausible intermediates in this metathesis reaction.

Ring-closing alkyne metathesis (RCAM) is also an important alkyne metathesis reaction that produces cyclic alkynes via ring closing of acyclic diynes (Scheme 19). While early reports of alkyne metathesis focused on dimerization or cross-metathesis of simple acetylene derivatives and speciality polymers, ongoing developments in the field revealed the considerable potential of the methodology for efficient syntheses of functionalized macrocycles by ring closing alkyne metathesis (170, 171). Subsequent semireduction of the cyclic alkynes yields a stereospecific methodology for the syntheses of (*Z*)-alkenes in various macrocyclic natural products (170, 171). Recent review articles by Bunz (172) and Furstner and Davies (163) illustrate the utility of Mo/W-catalyzed alkyne metathesis reactions for construction of a range of chemical architectures (173–177), some of which are shown in Scheme 20.

Recently Moore and co-worker further demonstrated the utility of molybdenum–alkylidyne-mediated alkyne metathesis in the development of shape persistent arylene–ethynylene macrocycles (178, 179). The approach utilizes a precipitation-driven metathesis that drives the reaction very efficiently producing ethynylene macrocycles on the multigram scale at room temperature (Scheme 21).

## C. Physical Properties and Electronic Considerations

The chemistry of metal alkynyls has attracted enormous attention, in particular, with the emerging interest in their potential applications in the field of materials science (2, 14, 16, 180–197). The alkynyl group, with its linear geometry, structural rigidity, extended π-electron delocalization, and ability to interact with metal centers via $p\pi$–$d\pi$ overlap, has allowed its use as a building block for the generation of carbon-rich, metal-containing constructs that may possess potential applications in nonlinear optical materials, molecular wires (2, 16, 186, 194–196) and organometallic oligomeric (193) and polymeric materials (14, 57, 182). Many of these materials may have potential in molecular electronics as luminescent materials with unique properties (e.g., optical nonlinearity, electrical conductivity, and liquid crystallinity) (181, 184, 185, 187, 190–192, 197).

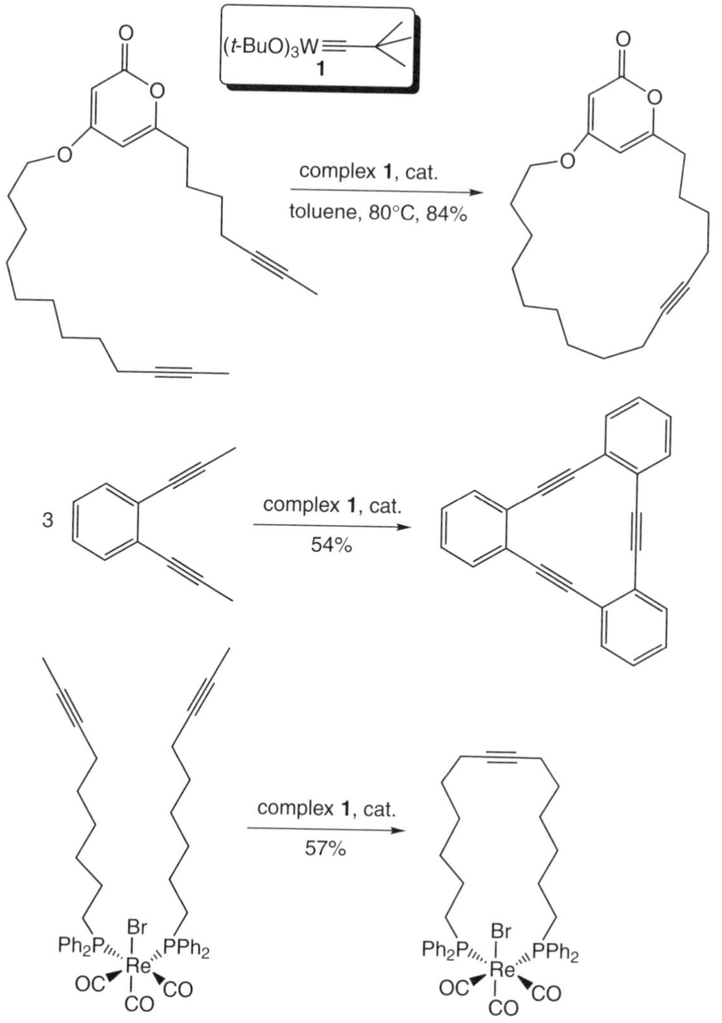

Scheme 20. Examples of ring-closing alkyne metathesis.

The luminescent polynuclear metal complexes may also find utility as chemosensors and in signaling applications.

1. The photophysical properties of metal–alkyne materials are rich due to strong metal–ligand orbital overlap and unsaturated $\pi$ system (16). Photoactive materials can be readily constructed from copper(I) (198–200), rhenium(I) (101, 186, 201–203) platinum(II) alkynyls (13, 203–219), as well as various gold(I) derivatives (210, 220–226).

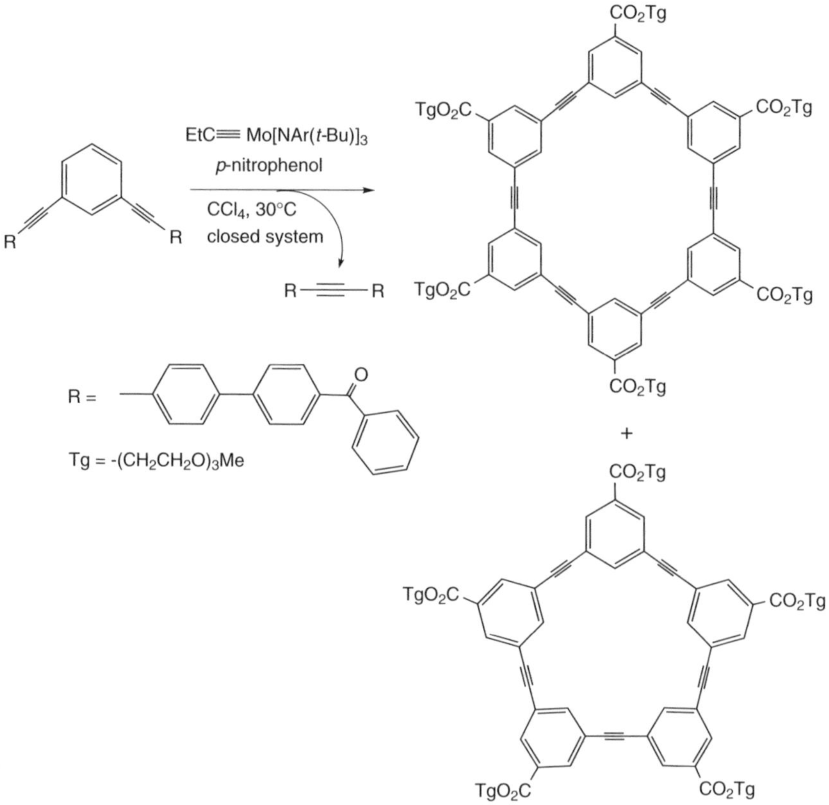
EtC≡ Mo[NAr(t-Bu)]3
p-nitrophenol
CCl4, 30°C
closed system
R—≡—R
R
R
R =
O
Tg = -(CH2CH2O)3Me
CO2Tg
TgO2C
CO2Tg
TgO2C
CO2Tg
CO2Tg
+
CO2Tg
TgO2C
CO2Tg
TgO2C
CO2Tg

Scheme 21. Formation of arylene ethynylene macrocycles by alkyne metathesis.

The trinuclear copper(I) acetylide complexes $[Cu_3(\mu\text{-dppm})_3(\mu_3\text{-}\eta^1\text{-}C{\equiv}CC{-}Ph)_2]^+$ and $[Cu_3(\mu\text{-dppm})_3(\mu_3\text{-}\eta^1{-}CC{\equiv}Ph)]^{2+}$ exhibit intense and long-lived luminescence in the solid state ($\tau \sim 14\ \mu s$ at 298 K) and solution ($\tau \sim 6.5\ \mu s$ in acetonitrile) upon photoexcitation (227–229). Complexes containing electron-rich acetylides show a progression to lower energy of their emission maximum as a function of donor ability of the acetylide ligand. For example, the emission maxima of $[Cu_3(\mu\text{-dppm})_3(\mu_3\text{-}\eta^1\text{-}CC{\equiv}R)]^{2+}$ in acetone solution occur in the order: $R = C_6H_4{-}OMe\text{-}4$ (483 nm) $\sim$ Ph (499 nm) $>$ $C_6H_4{-}NH_2{-}4$ (504, 564 nm) $>$ *t*-Bu (640 nm) $>$ *n*-$C_6H_{13}$ (650 nm). Based on these data, the origin of the emission has been proposed to involve substantial $^3$LMCT [acetylide $\rightarrow$ $Cu_3$] character ($^3$LMCT = triplet ligand-to-metal charge transfer).

The tetranuclear copper(I) acetylide complex exhibits strong and long-lived greenish-yellow emission upon photoexcitation. The solid sample emits at 509 nm ($\tau_o = 9.8$ μs) and 551 nm at 298 and 77 K, respectively. In solution (acetone), the emission occurs at $\lambda_{max} = 562$ nm with an appreciable lifetime ($\tau_o = 16$ μs) and quantum yield ($\Phi = 0.22$). The emission has been proposed to originate from an excited state with considerable $^3$LMCT [$(C{\equiv}C)^{2-}$ $\rightarrow$ $Cu_4$] character. As expected based on the ease of Cu(I) oxidation, these luminescent polynuclear copper(I) acetylide complexes also possess rich photoredox chemistry (227–229) as evidenced by strong bimolecular quenching by electron acceptors (e.g., pyridinium ion), indicative of a highly reducing excited state ($E^\circ \sim -1.77$ eV).

Two triangular copper(I) arrays bridged by a diethynylbenzene unit give the luminescent, hexanuclear complex shown in Fig. 7 (199). Excitation of this species in the solid state and in fluid solutions results in long-lived and intense luminescence. The copper(I) monoynyl derivatives exhibit strong phosphorescence between 416 and 697 nm, with the emission energy dependent on the electron-donating ability of the alkynyl ligand. The origin of the emission has been proposed to involve substantial LMCT [alkynyl $\rightarrow$ $Cu_3$] character, with contributions from the metal-centered (MC) $3d^94s^1$ configuration. Therefore, the lowest lying emissive state could best be considered as an admixture of $^3$LMCT and MC $3d^94s^1$ components modulated by the degree of copper–copper interaction. In cases of more extended π conjugation of the alkynyl, mixing of ligand-centred $\pi$–$\pi^*$ into the emissive excited state has also been proposed. The luminescence properties of the Ag(I) counterparts of the form [$Ag_3(\mu$-P—P$)_3(\mu_3$-$\eta^1$-C≡C—R$)]^{2+}$ {P—P = dppm, R = Ph, $p$-$C_6H_4$—$OCH_3$, $p$-$C_6H_4$—$NO_2$; P—P = $(Ph_2P)_2N$—$CH_2CH_2CH_3$, R = Ph and [$Ag_3(\mu$-dppm$)_3((\mu_3$-$\eta^1$-C≡C—$p$-$C_6H_4$—$NO_2)_2]^+$ have also been investigated

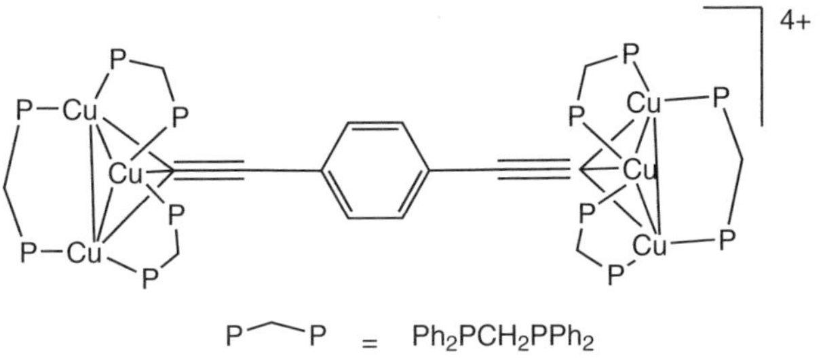

P⌒P = $Ph_2PCH_2PPh_2$

Figure 7. Diethynylbenzene-bridged trigangular {$[Cu(I)]_3$}$_2$ luminescent arrays.

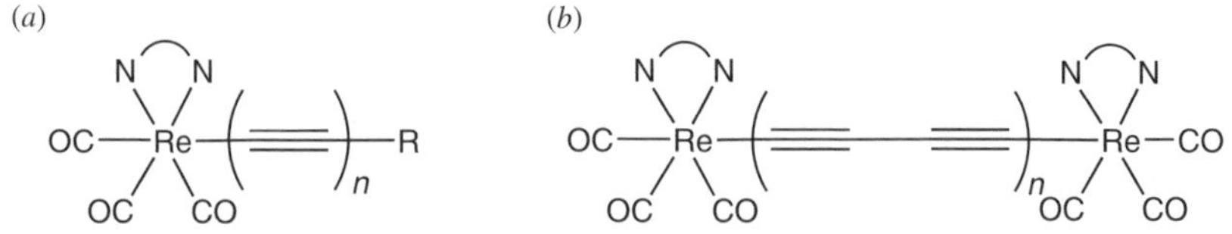

(*a*) (*b*)

N⌒N = bpy, *t*-$Bu_2$bpy, $Me_2$bpy, phen

$n = 1$

R = *t*-Bu, $SiMe_3$, Ph, $C_6H_4$-OMe-4

$C_6H_4$-$C_2H_5$-4, $C_6H_4$-Ph-4, $C_6H_4$N-4

$nC_{10}H_{21}$, H, Ph

$n = 2$

R = Ph, H, $SiMe_3$

Figure 8. Phosphorescent Re(I) alkynyl compounds.

(216). The complexes are isostructural and isoelectronic to the trinuclear copper(I) acetylide derivatives.

Rhenium(I) alkynyls involve the coordination of the alkynyl unit to the rhenium(I) center in a terminal σ-bonding mode to form monomeric [Fig. 8(*a*)] or dimeric [Fig. 8(*b*)] and strongly luminescent rhenium(I) dinuclear rhenium(I) alkynyls (101). In contrast to their copper alkynyl analogues, the rhenium(I) alkynyls with polypyridyl ancillary ligands exhibit intense phosphorescence typical of the metal-to-polypyridyl CT excited-state properties, perturbed by mixing of $\pi$ (C≡C) $\rightarrow \pi^*$(NN) ligand-to-ligand charge transfer (LLCT) character into the excited state. In general, the better the electron–donor ability of the alkynyl ligand, the lower is the $^3$MLCT [$d\pi$ (Re) $\rightarrow \pi^*$(NN)] emission energy.

Heterobimetallic and polymetallic structures featuring bridging acetylides can also be prepared from the respective Cu(I), [or Ag(I)], and Re(I) substructures described above. Excitation of both the penta- (224, 230) and decanuclear (203) mixed-metal alkynyl complexes in the solid state and in solution at $\lambda > 400$ nm results in strong red luminescence (680–700 nm) characteristic of predominantly [$d\pi$(Re) $\rightarrow \pi^*$(NN)] MLCT excitation mixed with $\pi$(C≡C) $\rightarrow \pi^*$(NN) LLCT perturbed by the $M_3$ or $M_6$ cores of the $d^{10}$ components. The MLCT state energy in the absorption spectrum for M = Ag(I) is slightly blue-shifted relative to M = Cu(I) due to the poorer donating ability of the $Ag_3$ versus $Cu_3$ core, which causes the Re(I) center of the $[Re(CO)_3(NN)]^+$ fragment to be less electron rich.

In 1989, the first luminescent platinum(II)–alkynyl construct, the dinuclear platinum(II) phenylethynylidene complex [$Pt_2$(μ-C=CHPh)(C≡CPh)$(PEt_3)_3$Cl], was shown to exhibit red-orange luminescence at

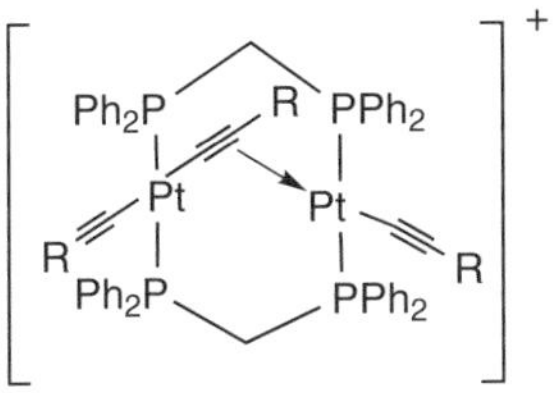

Figure 9. Dinuclear Pt(II)-alkynyl A-frame structures exhibiting strong luminescence.

77 K (207), opening the door for examination of many other platinum(II) alkynyl complexes that exhibit similar luminescence behavior both in the solid state and solution (215, 219). As a result, considerable interest in the luminescence properties of simple mononuclear Pt(II)–alkynyl complexes of the form *trans*-$[Pt(C{\equiv}CR)_2(PEt_3)_4]$ has revealed intense, structured emission at 77 K derived from Pt(II)-to-$\pi^*$(C≡C) MLCT excitation (212). Analogous *trans*-$[Pt(C{\equiv}CR)_2(dppm\text{-}P)_2]$ complexes also show intense luminescence. Their electronic absorption spectra in solution display low-energy bands between 400 and 450 nm that are red-shifted with respect to the related mononuclear complexes. The A-frame structured dinuclear $[Pt_2(\mu\text{-dppm})_2(\mu\text{-}C{\equiv}CR)(C{\equiv}CR)_2]^+$ complexes (Fig. 9) exhibit long-lived intense luminescence both in the solid-state and in solution at room temperature, with broad emission bands in the 570–650-nm region (215, 219). The maximum emission wavelength inversely correlates with the trend in aromaticity of the R group, supporting the role of the $\pi^*$(C≡CR) orbital in the frontier electronic structure. Increase in the conjugation of the alkynyl ligand can also change the nature of the emission from a predominantly $^3$MLCT[$d\pi$(M) → $\pi^*$(C≡CR)] origin to one that has increasing intraligand (IL) character, that is a mixed IL[$\pi \rightarrow \pi^*$(C≡CR)] / $^3$MLCT[$d\pi$ (M) → $\pi^*$(C≡CR)] state with predominantly IL character.

Similar electronic structures and optical properties translate to branched alkynyl complexes of the form shown in Fig. 10. The 77 K solid state and EtOH–MeOH glass of these complexes show intense green-to-yellow luminescence with microsecond lifetimes upon photoexcitation (231). Branched, Pt(II)–alkynyl complexes with polyaromatic ligands also show analogous properties, but with red-shifted emission in the range of 600–800 nm. Similar to the electronic absorption, the red shift in the emission energy with alkynyl substitution is suggestive of an emission origin of mixed IL[$\pi \rightarrow \pi^*$(C≡CR)]/MLCT[$d\pi$ (M) → $\pi^*$(C≡CR)] triplet state with predominantly IL character.

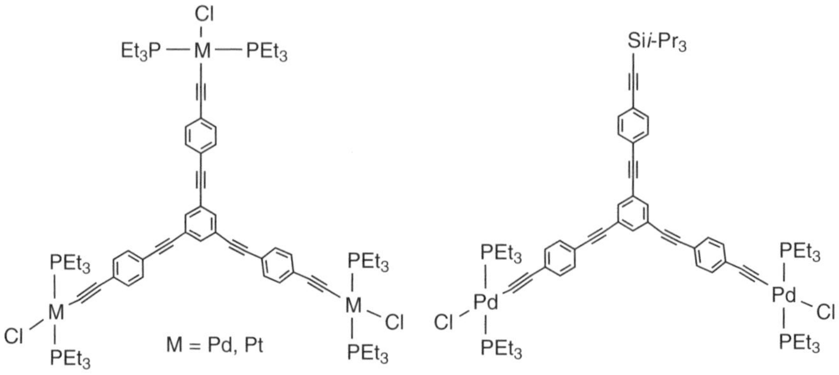

Figure 10. Branched Pd(II)–, Pt(II)–alkynyl complexes that exhibit considerable intraligand charge-transfer luminescence character.

The role of the alkynyl orbitals in the photophysical properties made possible the extension of small molecule luminescence to metal-containing, acetylenic polymers (232). To this end, Raithby and co-workers (13) reported a blue-luminescent [Pt(P*n*-$Bu_3$)$_2$]-based polymer with acetylide ligands containing low-lying $\pi^*$ orbitals in the main chain. The emission was assigned to derive from a $^3$IL CT excited state of the bridging acetylides. Ligation of oligopyridines, which also feature low-lying $\pi^*$ orbitals, allows harnessing of the luminescence of Pt(II)-$\sigma$-alkynyl core for energy-transfer applications.

In 1994, the first luminescent Pt(II)–acetylide complex with aromatic diimine ligands was reported (204). The title compound, [Pt(phen)(C$\equiv$CPh)$_2$] (phen = 1,10-phenanthroline), exhibits intense $^3$[5*d*(Pt) $\rightarrow$ $\pi^*$(phen)] MLCT emission in fluid solution at room temperature. Synthetic challenges and low solubilities of these complexes have been overcome in order to probe the excited-state properties of a range of systems of the general form [Pt($\alpha$-diimine)(C$\equiv$CAr)$_2$] each with systematic modification of both the diimine and acetylide functionalities (206, 208, 209, 211, 213, 214). The prominent MLCT absorption and high quantum yield for photoluminescence has helped these constructs ascend to become promising materials in the organic light-emitting diode (OLEDs) technology (233), as well as key components for photoinduced charge-separation architectures (51).

Much like the emission properties of the Re(I)—Cu(I) or Re(I)—Ag(I) multinuclear bimetallic systems described above, the luminescence of the hexanuclear bimetallic acetylide complexes containing Pt(II) of the form [$Pt_2Cu_4(C_6F_5)_4$(C$\equiv$C*t*-Bu)$_4$(acetone)$_2$] (45) have attracted considerable interest for their potential applications in nonlinear optics and luminescence applications. These are highly luminescent materials even at room temperature, due in part to the presence of short metal–metal contacts. Complexes with elongated metal–metal distances tend to be nonemissive, suggesting a role for the metal–metal electronic manifold in the transition. The emission by these complexes is quite temperature dependent, exhibiting a $\sim$3000-cm$^{-1}$ shift to lower energy of the emission band at 77 K. This behavior has many precedents in polynuclear Cu(I) compounds (234). It is interesting to note that at room temperature the solid (KBr) emission and excitation spectra are similar to those observed in fluid solution ($\lambda_{exc}$ 424 nm, $\lambda_{em}$ 620 nm), indicating no significant structural rearrangements upon dissolution.

The platinum–copper compound $[PtCu_2(C_6F_5)_2(C{\equiv}CPh)_2]_x$ is also luminescent in the solid state. Excitation at 465 nm generates a broad emission band ($\lambda_{em} = 642$ nm), with a shoulder at lower energies. Upon excitation to lower energies ($\lambda_{exc}$ 537 nm), the low-energy shoulder is strongly enhanced and becomes dominant, exhibiting a maximum at $\lambda_{em} = 787$- with 635-nm excitation. The atypical photophysics suggests a dual excited-state emission process.

The luminescence characteristics of $d^{10}$ gold(I) complexes are well known as their emission properties derive from the classic $d\sigma^* \rightarrow p\sigma$ electronic transition (235–239) and as a consequence, the metal–metal distance is reduced in the excited state. The luminescence properties of Au(I)–alkynyl complexes were first reported in 1993 for the dimeric phosphine compound $[\{Au(C{\equiv}CPh)\}_2(\mu{-}dppe)]$ (dppe = diphenylphosphino-ethane) (240). The X-ray crystal structure reveals a fairly long intramolecular Au•••Au separation of 3.153(2) Å, indicating a weak metal–metal contact (240). The compound exhibits ligand-centered emission at $\lambda_{em} = 420$ nm in dichloromethane solution at 298 K, and luminescence at 550 nm at 298 K in the solid state. The higher energy emission band is consistent with a ligand-centered excited state while the long wavelength luminescence is proposed to derive from the same type of metal antibonding–metal bonding transition ($d\delta^* \rightarrow p\sigma$) as in the normal $d^8$–$d^8$ and $d^{10}$–$d^{10}$ cases (235–239). Compounds supporting bridging alkynyl ligands with extended aromaticity (e.g., 1,4-diethynylbenzene) begin to show characteristics of the low-lying $\pi^*$ orbital of the alkyne, in much the same manner as the Cu(I)– or Pt(II)–alkynyl systems do. Here, the metal-based excited-state manifold reduces the alkyne $\pi^*$ orbital in the excited state leading to a MMLCT configuration (Au–Au $d\sigma^* \rightarrow \pi^*$ (C≡C) (226).

Gold(I)–alkynyls have also been employed as $\eta^2$-metalloligands in the construction of mixed-metal gold(I)–copper(I) and silver(I) complexes of the form $[\eta^2\text{-}(R_3P)AuC{\equiv}CC({=}CH_2)Me_2Cu]PF_6$ (R = Ph, $p$-Tol) and

$[(dppf)Au_2(\eta^2\text{-}C{\equiv}CC(=CH_2)Me_2M)]X$ (M = Cu, X = $PF_6$; M = Ag, X = OTf) (224). The heterometallic nature of these constructs was found to show ion-induced luminescence switching behavior. The simple monomeric phosphine compounds $[(R_3P)AuC{\equiv}CC(=CH_2)Me]$ (R = Ph, *p*-Tol) show luminescence at $\lambda_{em} = 454$ and 463 nm in the solid state, which have been assigned as deriving from metal-modulated $^3[\pi \rightarrow \pi^*(C{\equiv}C)]$ IL and $^3[\sigma(Au{-}P) \rightarrow \pi^*(C{\equiv}C)]$ MLCT states. Upon $\pi$ coordination to Cu(I), the $\pi^*(C{\equiv}C)$ orbital energy is lowered in energy due to the bonding interaction with the metal $d\pi$ orbital, giving rise to longer wavelength emission between 600 and 665 nm. As an exemption to this general trend, the complex $[(dppf)Au_2(\eta^2\text{-}C{\equiv}CC(=CH_2)Me_2)]$, is nonemissive in the solid state due to efficient quenching by the ferrocenyl unit of the ancillary dppf ligand. Gold(I) alkynylcalix[4]crown-6 complexes (241) also show rich luminescence properties upon excitation with $\lambda_{exc} = 370$ nm at both 77 K and room temperature that are red-shifted due to short Au•••Au distances in the solid state. Such intramolecular Au•••Au interactions give rise to a narrowing of the $d\sigma^*/p\sigma$ highest occupied molecular orbital (HOMO)–lowest unoccupied molecular orbital (LUMO) energy gap, which is influenced by the Au•••Au interactions and mixing with metal-perturbed intraligand $\pi \rightarrow \pi^*$ (C≡C) states.

In contrast to the related Au(I) (224, 234, 241–256), and the isoelectronic Pt(II) compounds (16, 250, 257, 258) that show excellent luminescence properties, room temperature luminescent Au(III) compounds are rare with only a limited number of examples known (259–261). This same trend holds for alkynyl complexes; Au(I) (240–242, 246, 249) and Pt(II) species (16, 262, 263) are widely known and possess rich photoluminescence properties, while the number Au(III) examples is quite limited (257, 259, 264, 265). The primary reason for this is the low-energy ligand-field states that are capable of quenching the emissive excited state (266). Additionally, the ease in metal center reduction to Au(I) makes these complexes susceptible to participation in photoinduced electron transfer as a secondary quenching pathway to luminescence. One of the advantages of coupling strong $\sigma$-donating alkynyl ligands to Au(III) is that the strong bonding interaction raises the energy of the ligand-field states, at the same time decreasing the electrophilicity of the Au(III) center, which enhances the overall luminescence of the compound (259, 264). For example, cyclometalated Au(III)–alkynyl compounds **1–5** (Fig. 11) display intense luminescence between 470 and 610 nm in solution at room temperature upon excitation with $\lambda_{exc} \geq 360$ nm. The luminescence is proposed to originate from a metal-perturbed IL $^3\pi\pi^*$ state of the tridentate CNC ligand. Overall, the enhanced luminescence character of

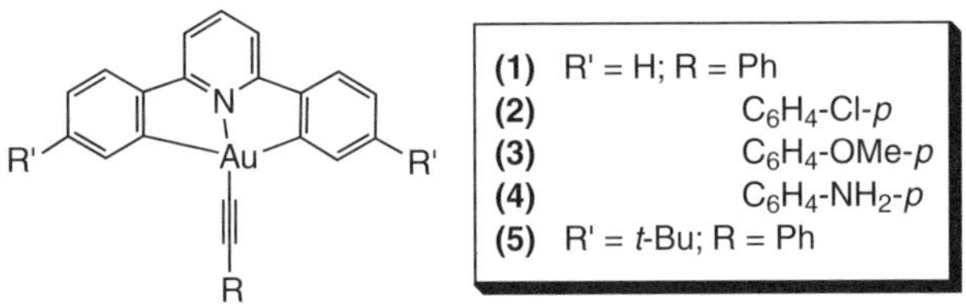

(1) R' = H; R = Ph
(2) $C_6H_4$-Cl-*p*
(3) $C_6H_4$-OMe-*p*
(4) $C_6H_4$-$NH_2$-*p*
(5) R' = *t*-Bu; R = Ph

Figure 11. Cyclometalated, luminescent Au(III)-alkynyl compounds.

these complexes support the idea that strong σ-donating alkynyl ligands defeat luminescence deactivation pathways.

Taking advantage of the thermal stability and neutral charge of these cyclometalated Au(III)–alkynyl complexes, fabrication of organic OLEDs can be realized by vacuum deposition. Compounds of the form [Au(CNC)(C≡C–R)] [HCNCH = 2,6-diphenylpyridine, R = Ph and *p*-$C_6H_4N(Ph_2)$, serve as a new class of electrophosphorescent material, and can play the role of both emitter and dopant for OLEDs to generate electroluminescence (EL) with high brightness and efficiency (264). The color of the EL is tunable by variation in the applied direct current (dc) voltage, as well as the dopant concentration. A maximum external quantum efficiency of 5.5%, corresponding to a current efficiency of 17.6 $cdA^{-1}$ and luminance power efficiency of 14.5 $lmW^{-1}$, has been obtained in one of the multilayer OLEDs.

2. Square-planar platinum(II) polypyridine complexes are well known to exhibit rich polymorphism in the solid state (267–269). The differences in their electronic absorption spectra in the solid state is dependent on the extent of metal–metal interactions and π–π stacking of the polypyridyl ligands. Unusual polymorphic characteristics are also observed in specific Pt(II)–terpyridine–alkynyl complexes (tpy = terpyridine). For example, the butadiynyl derivative [Pt(tpy)(C≡CC≡CH)]OTf exists in two forms, a dark green species and a red material, both of which have been structurally characterized and shown to exhibit different crystal packing arrangements (269). The dark green form exists as a linear chain with the platinum atoms equally spaced, and short intermolecular Pt•••Pt contacts of 3.39 Å. The red material, however, is comprised of a dimeric structure with alternating Pt•••Pt distances of 3.39 and 3.65 Å (Fig. 12). In contrast, the [Pt(*t*-$Bu_3$tpy)(C≡CC≡CH)]OTf derivative exists as discrete monomers without Pt•••Pt contacts due to the bulky 4,4′,4″-tri-*tert*-butyl-2,2′:6′,2″-terpyridine ligand. Dissolution of either the green or red forms of [Pt(tpy)(C≡CC≡CH)]OTf in acetonitrile or acetone generates a yellow solution that turns blue ($\lambda_{max}$ = 615 nm) as the media becomes more

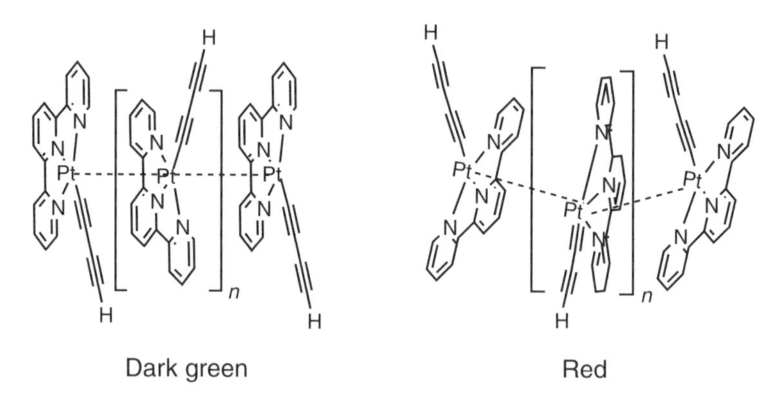


Figure 12. Solid-state polymorphism of Pt(II)-alkynyl complexes.

nonpolar. Concomittant enhancement of the luminescence in the near-infrared (IR) region ($\lambda_{em} = 785$ nm) is also observed. These unusual optical properties are proposed to derive from aggregation induced by poor solvation of the monomeric complex. Aggregate formation is likely facilitated by the propensity to form metal–metal interactions and $\pi$–$\pi$ stacking (270). Thus the 615-nm band in the electronic absorption spectrum has been assigned to a metal-to-metal charge transfer (MMLCT) transition, and the 785-nm phosphorescence to a state of $^3$MMLCT character. In contrast, the related [Pt($t$-Bu$_3$tpy)(C≡CC≡CH)]OTf complex with bulky tri-*tert*-butyl-2,2′:6′,2″-terpyridine ligand exhibits no detectable solvent-dependent color changes, and as such, the aggregation phenomena giving rise to the unusual optical properties of these complexes is inhibited by the steric bulk of the ligand.

3. Beyond molecular characteristics, metal–polyynes also possess interesting macroscopic properties (e.g., formation of lyotropic liquid crystalline materials or metallomesogens). Analogous to rigid-rod polyamides and polyisocyanates (271), Pt and Pd polymers form nematic liquid crystalline mesophases in simple organic solvents. (54, 192, 272) The main chain of these polymers orients parallel or perpendicular (Fig. 13) to the applied magnetic field of a simple nuclear magnetic resonance (NMR) magnet (56, 273). Those polymers containing diethynyl aromatic bridges align parallel to the field while those with only dialkynyl units orient perpendicular to the external field due to the negative magnetic anisotropy of the alkyne units (273).
4. Organic (21, 274–278) and organometallic (21, 273–276, 278) alkyne constructs prominently exhibit nonlinear optical (NLO) behavior and can rank among the most efficient quadratic NLO molecules. When light

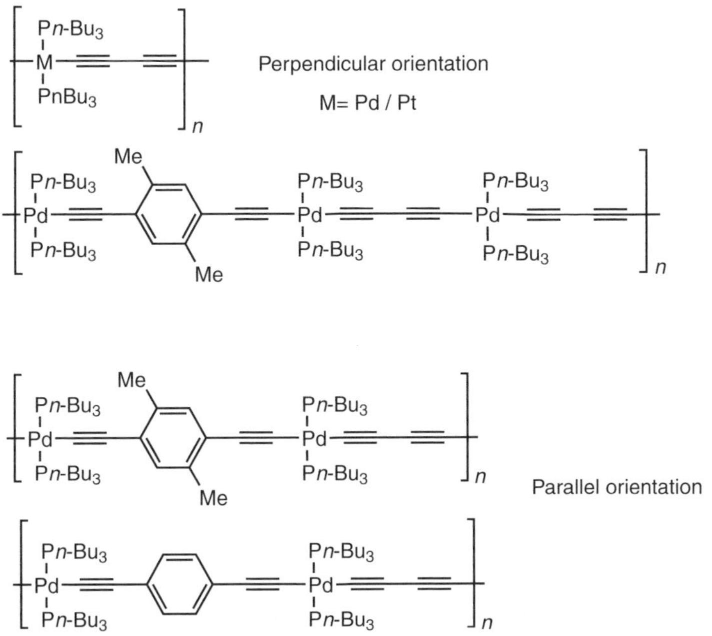

Figure 13. The $d^8$ [Pt(II), Pd(II)]-polyyne metallomesogens that align parallel or perpendicular to an applied magnetic field.

interacts with materials possessing NLO properties, the electromagnetic wave of the incident radiation is perturbed producing phase, frequency, polarization, amplitude, and/or trajectory shifts of the transmitted radiation via harmonic generation, frequency mixing, and optical parametric oscillation. Optical transformations of this type have considerable applications to optical data storage and communication, as well as advanced image processing. Many organic molecules containing conjugated $\pi$ systems with unsymmetrical charge distribution (e.g., donor–acceptor substituted azo dyes, Schiff bases, and stilbenes) possess considerable NLO responses (184, 279–292) derived mainly from electronic nonlinearities. Nonlinearities can be enhanced by either increasing the conjugation length (improving delocalization) or increasing the strength of donor and/or acceptor groups (improving electron asymmetry). Such organic NLO materials are often advantageous for their higher optical damage threshold than inorganic crystals, ease of synthesis and fabrication, structural diversity and facile casting into thin films for which electric field poling can introduce the asymmetry needed for the appearance of second-order NLO effects. However, organic NLO materials have several disadvantages

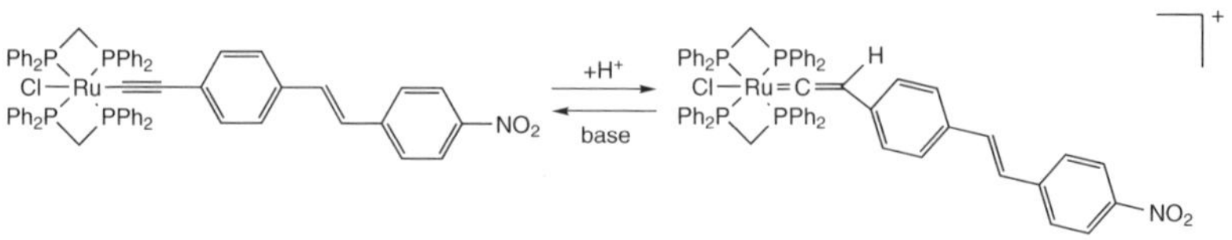

Figure 14. Protonation–deprotonation switching mechanism of quadratic NLO response.

including modest optical transparency at the high energy edge of the visible spectrum, and low thermal stability which can lead to facile relaxation to random orientations.

The NLO properties of organometallic and coordination complexes are also rich (21, 184, 279–296). Metal–alkyne complexes were first reported ~1960 (297) and have recently attracted significant interest because of their potential in materials applications (2, 298). Studies of these types (299) have resulted in the development of structure–NLO response relationships for quadratic optical nonlinearities (β-value), which increase with valence electron count and ease of oxidation of metal. The amplitude is also tunable by ancillary ligand modification and substitution. Select small alkynyl complexes have been shown to exhibit β values at 1064 nm $> 2600 \times 10^{-30}$ esu (299).

Metal–alkynyl complexes have also found favor in NLO switching applications (300). A range of switching methodologies have been proposed including photoisomerization (299), protonation–deprotonation (301), and oxidation–reduction (300, 302). For example, as much as a fivefold increase in the β values (corrected and uncorrected for resonance enhancement) is observed upon deprotonating the Ru–vinylidene complex shown in Fig. 14. Analogously, the branched, Ru–alkynyes in Fig. 15 reveal third-order (cubic) nonlinear optical enhancements of almost 7× upon three, one-electron oxidations of the ruthenium centers. These NLO effects are frequently enhanced for Ru– and Ni–alkynyl compounds due to appreciable two-photon absorption cross-section in the visible region. The optical transparency of Au–alkynyl compounds relieves the resonance condition and allows for a true evaluation of intrinsic off-resonance hyperpolarizabilities.

The efficiencies for these metal alkynyl compounds are dependent on the nature of the bridging alkynyl ligand where $C_6H_4 < C_6H_4C_6H_4 < C_6H_4C{\equiv}CC_6H_4 < (E)$-$C_6H_4CH{=}CHC_6H_4$. This trend can be rationalized by considering the π-bridge lengthening, torsion effects at the phenyl–phenyl linkage (e.g., biphenyl compound), and the orbital energy mismatch of *p* orbitals (*sp*-hybridized acetylenic carbons versus $sp^2$ hybridized phenyl carbons for the diphenylacetylene compound) (303). Substitution on the alkynyl ligand also contributes significantly to Quadratic nonlinearities in the order β(H) < β (CHO) < β($NO_2$), and confirming that the

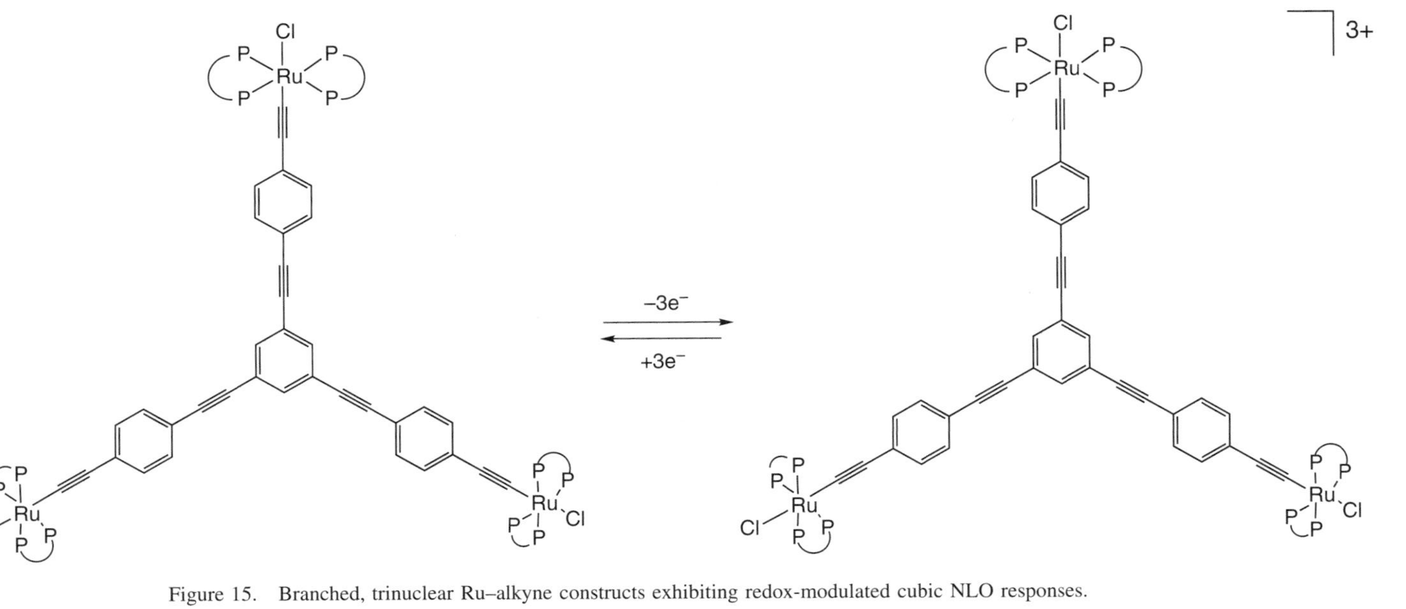


Figure 15. Branched, trinuclear Ru–alkyne constructs exhibiting redox-modulated cubic NLO responses.

acceptor group is more efficient when conjugated with the metal alkyne. Positional substitutions also play a role in nonlinear behavior as β(3-CHO) < β(4-CHO) (301). Bridge geometry affects quadratic nonlinearities as β [(*Z*)-isomer] < β[(*E*)-isomer], which can be explained from a combination of the greater dipole moment and more intense optical transition for the latter (303).

For the metal–alkyne fragment, nonlinearity also increases upon increasing valence electron count [14 valence electron (triphenylphosphine) gold alkynyl compounds < 18 valence electron (cyclopentadienyl) (triphenylphosphine)nickel, and (cyclopentadienyl) bis(triphenylphosphine) ruthenium alkynyl compounds] and increasing ease of oxidation (less easily oxidizable (cyclopentadienyl)(triphenylphosphine)nickel alkynyl complexes < more easily oxidizable (cyclopentadienyl)bis(triphenylphosphine)ruthenium alkynyl complexes).

## III. METAL–ALKYNE π-COMPLEXATION

### A. The $\eta^2$ Bonding to Metals

Transition metal–alkyne π-bonding interactions are similar to their alkene counterparts in that both can be considered as donation from the alkyne π-to-metal *d* orbitals, with concomitant back-donation from the metal *d*-to-alkyne π* orbitals leading to bent alkyne structures (Fig. 16) (304, 305) Commonly, the donor interactions are primarily σ type while the back-donation occurs mainly

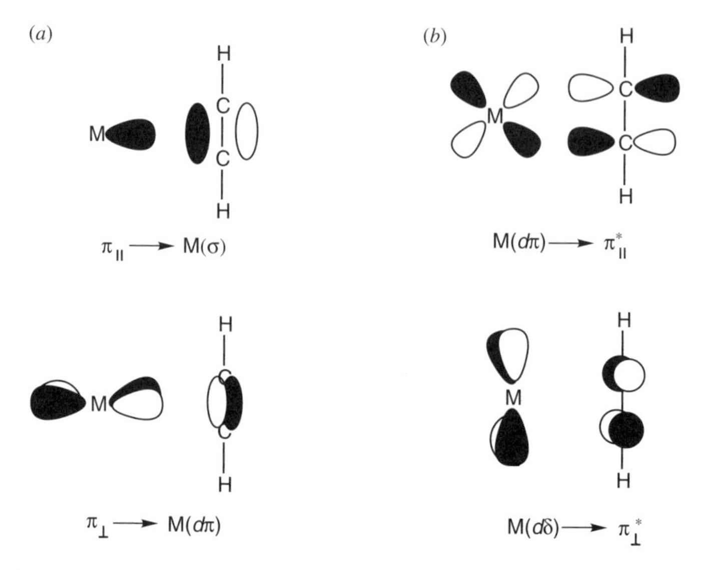


Figure 16. Metal–alkyne orbital bonding interactions. Alkyne ligand σ and π donation to metal *d*σ and *d*π orbitals (*a*), and corresponding metal-to-π* back-donation (*b*).

through orbitals of π symmetry. The primary difference between metal–alkene and –alkyne complexes is that in addition to the in-plane $\pi_{\parallel}$ MO, a second π orbital that lies orthogonal to the $MC_2$ plane ($\pi_{\perp}$) (306) may participate in metal–alkyne $\eta^2$ bonding (307, 308). Depending on the symmetry of the complex, both in-plane and out-of-plane π orbitals of the alkyne may participate in M–alkyne bonding (309), thus alkyne ligands may be viewed as either two- or four-electron donors. Early computational models of transition metal–acetylene complexes proposed that early elements ($Sc^+ - V^+$ and $Y^+ - Nb^+$) form three-membered metallacyclic compounds, where the metal ion inserts into the in-plane $\pi_{\parallel}$ bond of acetylene leading to a significant increase of C–C distance compared to free acetylene, while late transition metals form π complexes with alkynes that can be considered less covalent in nature (310, 311).

## B. Metal–Diyne π-Complexation

Simple, stable homoleptic bis(alkyne) transition metal π complexes are of interest as starting materials for organometallic reactions, as well as model compounds for understanding the interactions between alkynes and metal surfaces. Although a wide range of π metal–alkyne compounds are known (102), most of the more unusual examples derive from the metals of groups 6 (VI B), 10 (VIII), and 11 (I B). Thus, due to the vastness of this area, our discussion will be mainly restricted these metal–bis(alkyne) constructs.

1. One of the early group 6 (VI B) metal–alkyne examples, W(CO)(3-hexyne)$_3$, was prepared by Tate and Augl in the early 1960s according to (312, 313).

$$W(CO)_3(MeCHN)_3 + 3CEt{\equiv}CEt \xrightarrow{\Delta} W(CO)(EtC{\equiv}CEt)_3$$

This first report of the correctly assigned pseudotetrahedral tungsten derivatives of this complex (Fig. 17), suggested that the alkyne was doubly

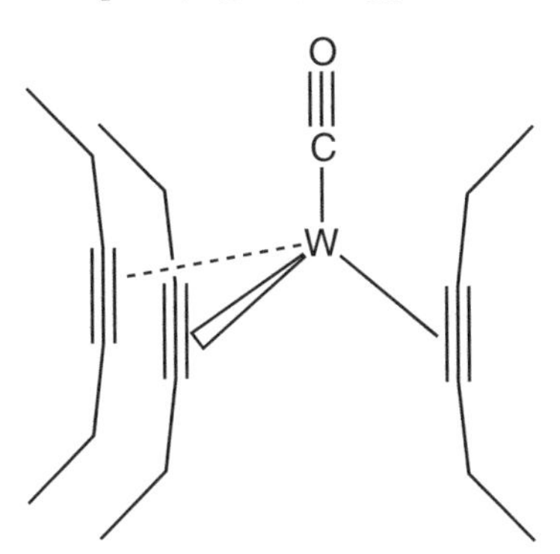


Figure 17. Pseudotetrahedral W(CO)(EtC≡CEt)$_3$ geometry.

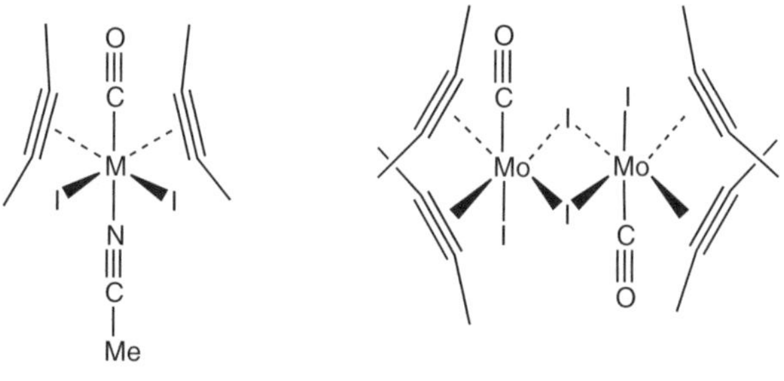

M = Mo, W

Figure 18. The Mo(II) and W(II) bis(alkyne) monomeric and dimeric π complexes.

π bonded to the tungsten atom. The strong π-acid character of the alkyne ligands was evident in the high monocarbonyl stretching frequency ($2050 \pm 20\,cm^{-1}$) observed for these formal $d^6$ monomers. A symmetry-based molecular orbital description of the unusual four coordinate $C_{3v}$ $W(CO)(RC{\equiv}CR)_3$ series of molecules was presented by King in 1968 (306). According to this description, the electron count adheres to the effective atomic number rule: $W(0)\ (6) + 3\pi_{\parallel}\ (6) + 2\pi_{\perp}\ (4) + 1\sigma\ (2) = 18$ electrons, and hence these are stable compounds.

Group 6 (VIB) $d^4$ alkyne π-complexes illustrate the features of four-electron donor alkyne chemistry. Early work in the syntheses, structures, and reactivities of $d^4$ alkyne complexes of Mo and W (308) showed that the reaction of the complexes $[MI_2(CO)_3(NCMe)_2]$ (M = Mo or W) with 2 equiv of the alkyne ligands $RC_2R$ in $CH_2Cl_2$ at room temperature affords high yields of new bis(alkyne) complexes $[MI_2(CO)(NCMe)(\eta^2\text{-}MeC_2Me)_2]$ (Fig. 18) (314, 315). Initially, the monomeric molybdenum complex could not be isolated as it rapidly dimerizes with loss of acetonitrile to generate the iodo-bridged $[Mo(\mu\text{-}I)I(CO)(\eta^2\text{-}MeC_2Me)_2]_2$ species, but monomeric bis(alkyne) complexes of Mo were later obtained and crystallographically characterized (316). The analogous tungsten compound is inert to dimerization, even under reflux in $CHCl_3$ for 24 h (315), but upon treatment with nucleophiles [e.g., $PPh_3$, dithiocarbamates, and 4,4′-bipyridine(bpy)] (314, 317, 318), it can form the corresponding bis(alkyne) complexes (Fig. 19). Analogously, 1,4-diphenylbutadiyne complexes $[\{W(\mu\text{-}I)I(CO)(NCMe)(\eta^2\text{-}PhC_2C_2Ph)\}_2]$ and $[WI_2(CO)(NCMe)(\eta^2\text{-}PhC_2C_2Ph)_2]$ have been prepared by the reaction of $[WI_2(CO)_3(NCMe)_2]$ with 1 or 2 equiv of 1,4-diphenylbutadiyne, respectively (319). Equimolar quantities of $[WI_2(CO)_3(NCMe)_2]$ and $[WI_2(CO)(NCMe)(\eta^2\text{-}PhC_2C_2Ph)_2]$ react to give the bimetallic 1,4-diphenylbutadiyne-bridged complex (Fig. 20) (319). Trinuclear tungsten–alkyne scaffolds with the $W^{II}{}_2W^{IV}\text{-}(\mu_3\text{-}O)$ core (Fig. 21) can be

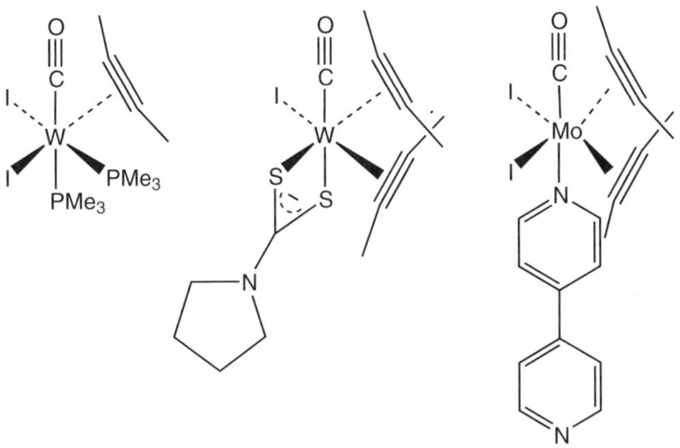

Figure 19. Stable mono- and bis(alkyne) complexes of group 6 (VIB).

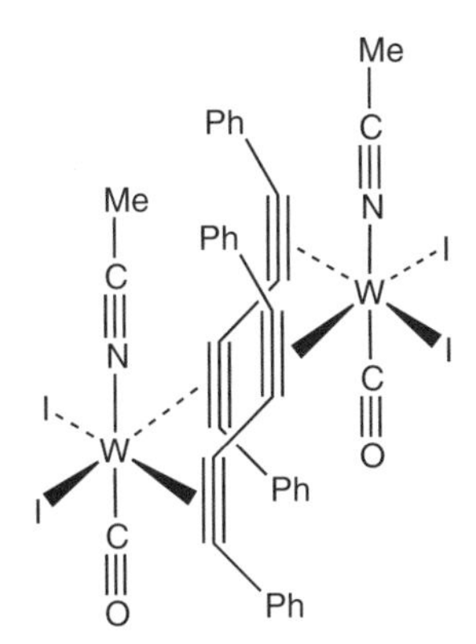


Figure 20. Illustration of the 1,4-diphenylbutadiyne-bridged tetraalkynyl $W_2$(II,II) dimer.

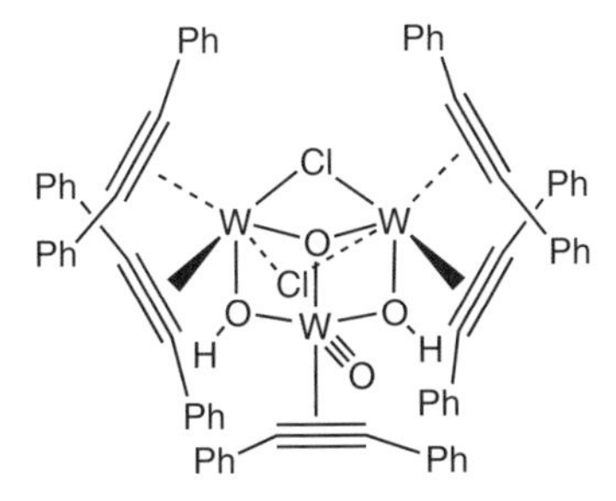


Figure 21. Structure of $[W_3(\mu_3\text{-O})(\mu\text{-OH})_2(\mu\text{-Cl})_2(O)(\eta^2\text{-PhC}{\equiv}\text{CPh})_5]$.

accessed by the reaction of $[WCl_2(CO)_2(\eta^2\text{-PhC}\equiv\text{CPh})_2]$ and $[WCl(SnCl_3)(CO)\ (\eta^2\text{-PhC}\equiv\text{CPh})_2]$ in moist air (320). Similar $\pi$-complexation strategies also extend to bioorganometallic chemistry (321). In this approach, (322) a peptide bearing two alkyne groups (308) coordinates each alkyne $\eta^2$ to the W(II) center to form either a metallacyclic ring or dipeptide system (Scheme 22) (323).

Examples of $d^6$ metal (Mo and W) bis(alkyne) complexes are comparatively rare. Such octahedral $d^6$ bis(alkyne) species deserve particular attention in light of the presence of electron–electron repulsion between the filled $d\pi$ orbitals of the metal center and $\pi_\perp$ orbitals of the alkyne ligands. Nevertheless, $d^6$ metal–alkyne complexes can be formed with electronically poor alkyne ligands, where electron-withdrawing functionality on the alkyne making it weakly $\pi$ donating. In this case, the $\pi$ bond is weak in comparison to the $d^4$ alkyne $\pi$ complexes. As an example, the $d^6$ octahedral bis(alkyne) complex $[W(CO)_2(dppe)(DMAC)_2]$ (DMAC = dimethylacetylenedicarboxylate) was first prepared by Templeton and co-workers (324). Analogously, Wang and co-workers (325, 326) synthesized similar bis(alkyne) complexes of the type $[W(CO)_2(NN)(DMAC)_2]$ and $[Mo(CO)_2(NN)(DMAC)_2]$, where NN is a simple bidentate neutral nitrogen-donor ligand. Most of these complexes containing asymmetric alkyne ligands show interesting conformational isomers due to rotation of alkyne ligands (Scheme 23).

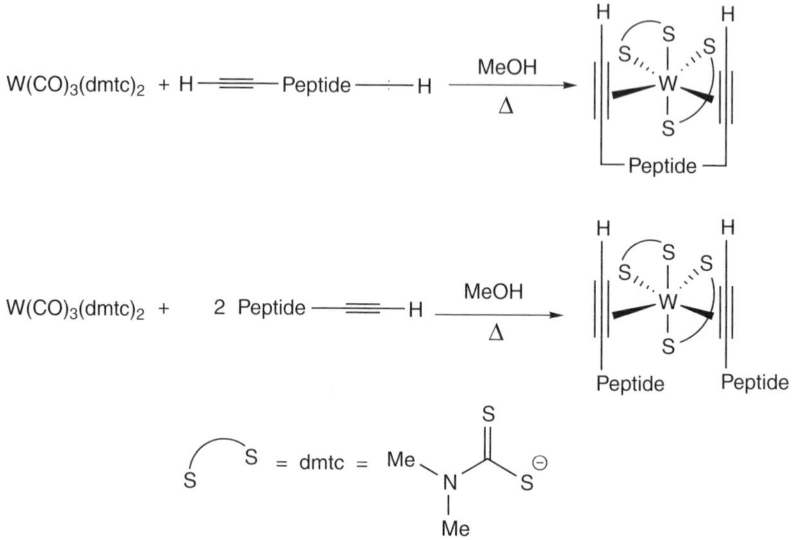

Scheme 22. Metallcyclic and dipeptide bis(alkyne) compounds of W(II), where dmtc = dimethyldithiocarbamate.

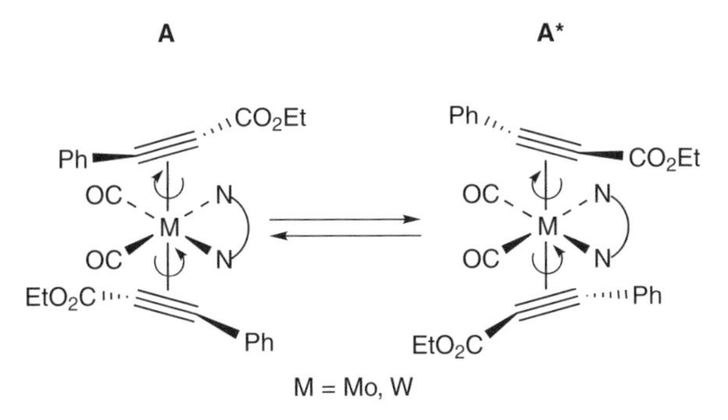


Scheme 23. Alkyne rotation in $d^6$ bis(alkyne) complexes of Mo, and W by disrotatory mechanism. The astrisk(*) denotes enantiomeric structure.

As a first-row transition element, chromium does not typically form stable π complexes with alkynes, and therefore a modest number of stable Cr–bis(alkyne) complexes are known. In 1975 Connor and Hudson (327) synthesized low valent, Cr(0)–alkynyl thioether complexes of the form $[Cr(CO)_2(MeSC{\equiv}CSMe)_2]$. This photosensitive compound is proposed to rapidly convert to *cis*-$[Cr(CO)_4(MeSC{\equiv}CSMe)_3]$ and the three alkyne ligands rapidly cyclotrimerize to produce hexakis(methylthio)benzene chelated to the chromium center (Fig. 22). Analogously, in 1982 the stable chromium bis(alkyne) complex $[Cr(Me_3SiC{\equiv}CSiMe_3)_2(CO)_2]$ was synthesized by reacting a pentacarbonyl-carbene complex of chromium with excess bis(trimethylsilyl)acetylene (328). When this species was treated with excess diphenylacetylene, ligand substitution leads to the tris(diphenylacetylene) compound, which subsequently promotes dimerization of two acetylenes to form tetraphenylcyclobutadiene that binds to the metal in the $\eta^4$ fashion (Scheme 24) (329). These early cyclodimerization and -trimerization strategies revealed the utility of metal–alkyne chemistry for organic transformations.

2. Although alkyne π complexes of group 10 (VIII) and 11 (IB) metals have been known for many years, recent work in this field has revealed that the

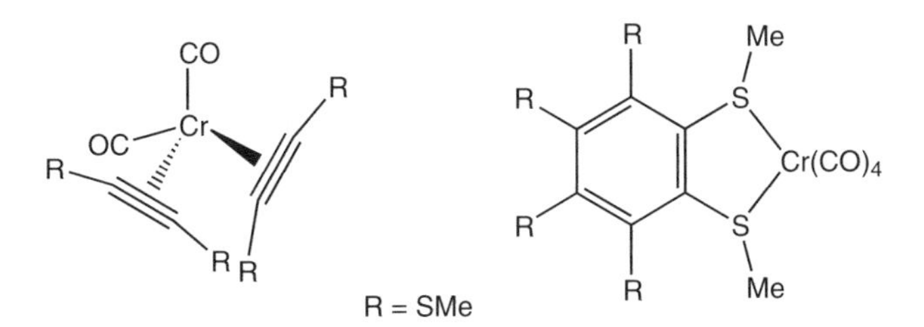


Figure 22. Chromium(0) bis(alkyne) compound (*a*) and cyclotrimerization product (*b*).

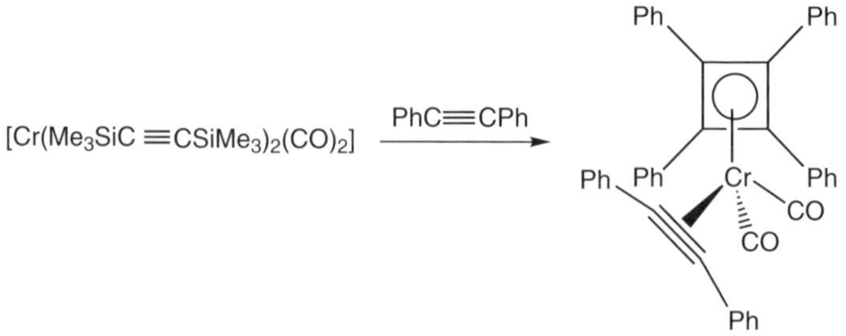

$[Cr(Me_3SiC{\equiv}CSiMe_3)_2(CO)_2] \xrightarrow{PhC{\equiv}CPh}$

Ph Ph Ph Ph Cr Ph CO CO Ph

Scheme 24. Reaction of Cr–bis(alkyne) with excess diphenylacetylene to form a coordinated tetraphenylcyclobutadiene product.

chemistry of these systems remains rich and is continuously evolving. The interest in heterometallic and polymetallic compounds led to the use of metal–alkyne σ compounds (cf, Section II) as starting materials to achieve larger constructs by subsequent metal π complexation to the alkyne units. For example, Fornies and co-workers (44, 45, 104, 330) used the $[Pt(C{\equiv}CR)_4]^{2-}$ unit to prepare a wide range of multimetallic assemblies. Lang and co-workers (111, 331–334) also used $\{[Ti](C{\equiv}CR)_2\}$ σ compounds {where $[Ti] = (\eta^2\text{-}C_5H_4SiMe_3)_2Ti$} and subsequent π complexation to form *π-tweezer* systems, such as those shown in Fig. 5–6.

Interesting, Au(I) catenanes have been synthesized and isolated as yellow single crystals by Mingos et al. (335, 336) using labile $[Au(NH_3)_2]^+$ as the starting material. The X-ray structure of the product features two interlocking six-membered rings of gold atoms, and therefore the compound is formulated as $[\{Au(C_2t\text{-}Bu)\}_6]_2$. Adoption of this novel structure may be attributed to a combination of the three coordination modes shown in Scheme 25. Analogously, Mingos and co-workers also developed corresponding Ag(I) polymetallic assemblies in which noncovalent Ag•••Ag interactions are observed. This unique cationic, cage species $[Ag_{14}(C{\equiv}Ct\text{-}Bu)_{12}Cl]^+$ (335, 337) is synthesized by $CHCl_3$ treatment of polymeric $[Ag(C{\equiv}Ct\text{-}Bu)]_n$. The X-ray crystallographic characterization of this cage reveals that the Ag atoms are arranged in a near-regular rhombic dodecahedron with edge-to-edge Ag•••Ag distances in the 2.953(2)–2.986(2)-Å range (Fig. 23).

Kovacs and Frenking (338) recently reported a theoretical study of bis($\eta^2$-alkyne) complexes of Cu(I), Ag(I), and Au(I), and suggest that isolation of Cu(I) complexes with two simple, nonbridging alkyne ligands is extremely challenging due to the relatively weaker π-donating ability of Cu(I) relative to Ag(I) and Au(I) (335, 337). By using bulky alkyne ligands like cycloheptynes, Behrens and co-workers (339) successfully synthesized and X-ray crystallographically characterized the three-coordinate Cu(I)–bis(alkyne) complex

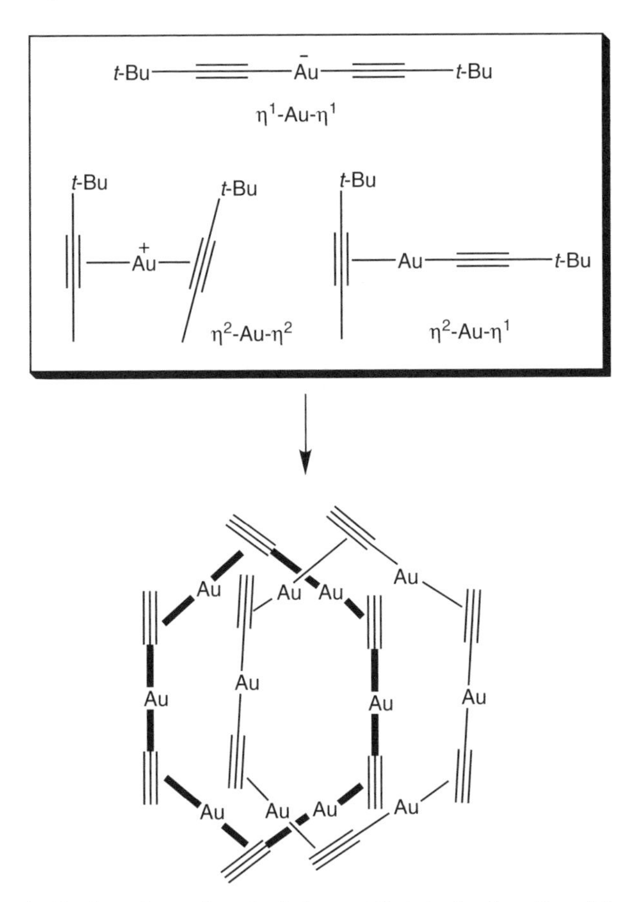


Scheme 25. Au(I)–alkyne interactions (top) that contribute to the formation of the polymetallic Au–catenate $[\{Au(C{\equiv}Ct\text{-}Bu)_6\}_2]$ (bottom).

$[CuX(cycloheptyne)_2]$ (Fig. 24), illustrating that such compounds are accessible under well-controlled coordination conditions. Organometallic, substituted alkynes [e.g., $(OC)_5ReC{\equiv}CR$, where $R = Re(CO)_5$, $SiMe_3$] can be used as alkyne π-donor ligands to also afford group 11 (IB) bis(alkyne) complexes. Beck and co-workers (340) synthesized a series of $[M\{(OC)_5ReC{\equiv}CR\}_2]$ complexes of Cu(I), Ag(I), and Au(I) (Fig. 25).

Finally, in 1994, only platinum systems of the type bis(alkyne)M(0) were known (341, 342), partly because attempts to isolate the corresponding Ni(0) analogue had been unsuccessful (343). However, cyclotriyne ligands have been shown to react $(cod)_2Ni$, where cod=1,5-cyclooctadiene, to give Ni(0)–tris(alkyne) macrocyclic π complexes (344). In 1996, Walther et al. (345, 346)

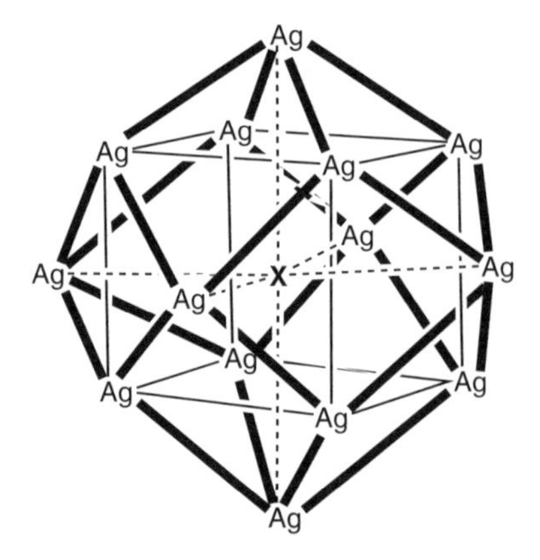


Figure 23. Rhombohedral core of the silver cage $[Ag_{14}(C{\equiv}C\textit{t}\text{-}Bu)_{12}Cl][OH]$ (X = OH).

successfully synthesized the first crystalline bis(alkyne)–Ni(0) complex, (2,5-dimethyl-hex-3-yne-2,5-diol)$_2$Ni, (Fig. 26), which was structurally characterized by single-crystal X-ray diffraction (346). These constructs once again show the importance of metal–alkyne π bonding and at the same time, the potentially reversible nature of the metal–ligand π interaction that has important implications in catalysis.

## C. Ligand Transformations and Reactivity

Section III.B illustrated the complex and unusual structures that can be generated from the different binding modes of metal–alkyne constructs. Clearly, metal–alkyne π-complex chemistry is rich and in some ways has a distinct

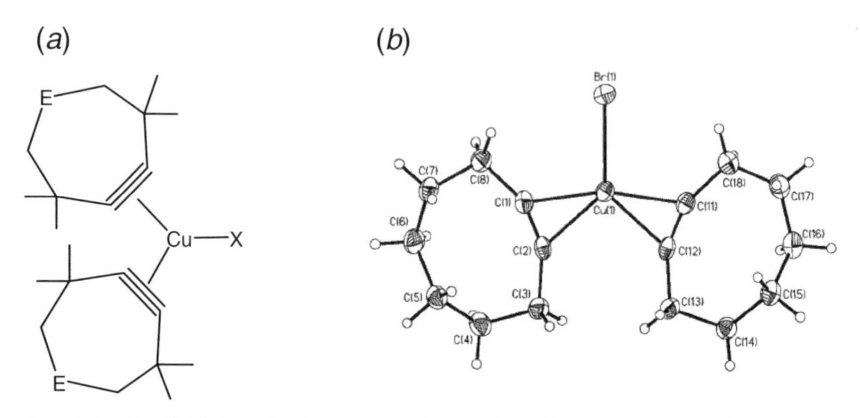


Figure 24. The Cu(I)–bis(alkyne) constructs (*a*) and the molecular structure of [CuBr(cyclooctyne)$_2$] (*b*).

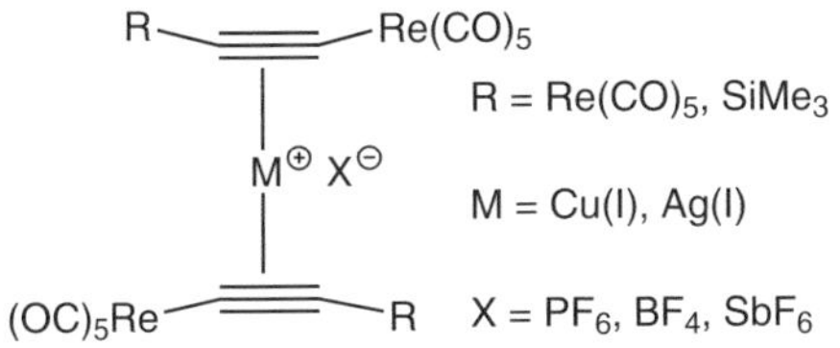

R—≡—Re(CO)5
M⊕ X⊖
(OC)5Re—≡—R

R = $Re(CO)_5$, $SiMe_3$

M = Cu(I), Ag(I)

X = $PF_6$, $BF_4$, $SbF_6$

Figure 25. The Cu(I)–and Ag(I)–bis(alkyne) complexes with mono- and dirhenoethyne ligands.

advantage over σ-alkyne coordination; that is the ability to serve as a π acid upon binding (i.e., oxidative addition) and exhibit reversible or competitive dissociation by reductive elimination. Many of structures shown in Section III.B are stable, and hence they have been crystallographically characterized. In contrast, this section will highlight examples of metal–alkyne π complexation that rapidly leads to ligand transformation, and thus the intermediate π species is not immediately isolable. In order to illustrate specific reactivity, a few enyne and enediyne compounds are discussed prior to their Sections IV and V, respectively.

1. One of the simplest and most widely applicable metal–alkyne π-mediated transformations is the intramolecular cycloisomerization reaction (347–353). As an example, Murai and co-workers (347, 350) reported the cycloisomerization of ω-aryl-1-alkynes in which the aromatic rings act as nucleophiles toward the π-coordinated, transition metal bound alkyne without invoking metallation of the arene moiety. In this process, $PtCl_2$ or select Ru(II) complexes catalyzed the cycloisomerization of different aryl alkynes to yield dihydronaphthalenes or dihydrobenzocycloheptenes (Scheme 26). Different cyclization modes are observed depending upon the length of the aryl–alkyne tethers. Exo–dig cyclization is preferred when the alkyne and arene are separated by three or four carbon atoms, while the endo–dig mode is observed in substrates with only two carbon atoms in the tether.

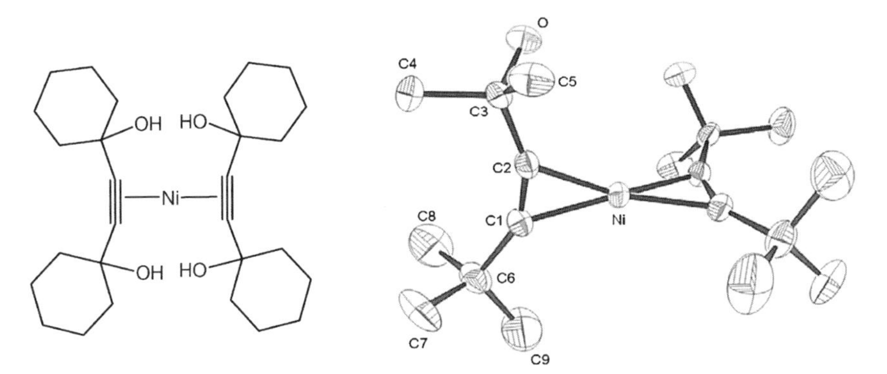


Figure 26. The Ni(0)–bis(alkyne) compound and a representative X-ray structure.

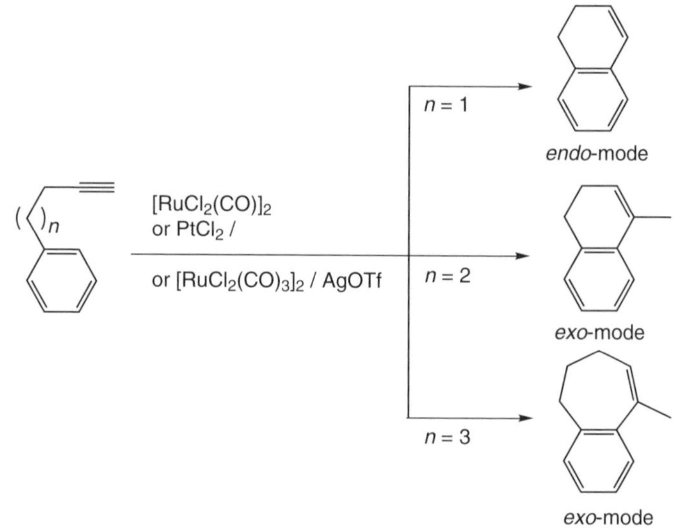

Scheme 26. Exo–endo–dig cyclization modes involving nucleophilic attack of the arene upon the metal–alkyne π complex.

Addition of an electrophilic metal salt [e.g., $PtCl_2$ (348), $AuCl_3$ (349), $GaCl_3$ (350), or $InCl_3$ (351)] or metal complex to a biphenyl derivative bearing an alkyne unit at one of its ortho positions leads to an equilibrium between the alkyne and the corresponding $\eta^2$ complex. Nucleophilic attack by adjacent aromatic ring leads to C–C bond formation with concomitant release of the catalyst (Scheme 27). The 6-endo cyclization

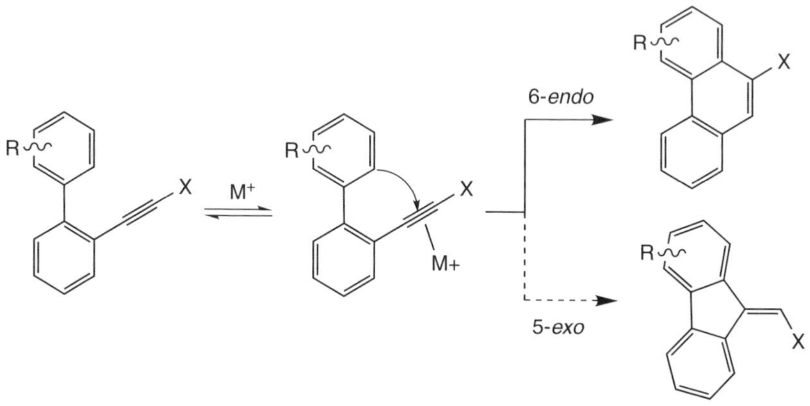

Scheme 27. Metal-catalyzed cycloisomerization of ortho-alkynylated biphenyl derivatives feature 6-endo- and 5-exo–dig cyclization pathways.

mode is preferred in most cases; only alkynes bearing an electron-withdrawing group are prone to cyclization by a 5-exo–dig pathway to generate the 9-alkylidene fluorene derivatives. This reaction strategy is readily applicable to the preparation of heterocyclic systems (e.g., benzoindoles, benzocarbazoles, naphthothiophenes), as well as bridgehead nitrogen heterocycles (e.g., pyrrolo[1,2-*a*]quinolines).

The more complex [2 + 2 + 2] cycloisomerization reaction of acetylene units is also catalyzed by transition metal–alkyne π complexation and can be readily utilized for the synthesis of a variety of polysubstituted benzene derivatives in a straightforward manner (10, 352, 353). Recently, this methodology has been applied to the cyclization of 15-membered, nitrogen-containing di- and triacetylenic macrocycles. Upon coordination with Pd(0) to the triacetylenic macrocycle at ambient temperature, the π-coordinated Pd(0) complex results. Subsequent refluxing of this species in toluene promotes cycloisomerization to the hexasubstituted arene (354) (Scheme 28). The Rh(I) [e.g., $RhCl(CO)(PPh_3)_2$] complex also catalyzes these same transformations in high (>80%) yields.

Cyclopentadienones are also good synthons for polycyclic compounds (355–359) and polymers (360, 361) and are prepared generally by alkyne–alkyne–CO coupling (362, 363). However, most preparations of cyclopentadienones are stepwise processes involving catalysts such as $CpCo(PPh_3)_2$ (364) and $RhCl(PPh_3)_3$ (365) that mediate alkyne–alkyne coupling to give metallacyclopentadienes. The products are then converted into the desired cyclopentadienones by insertion of CO and elimination of catalyst. Catalysts, such as $CpCo(CO)_2$ (366, 367) and $Fe(CO)_5$ (368–370), have been shown to mediate carbonylative alkyne–alkyne coupling, but the cyclopentadienones are obtained as $\eta^4$ metal complexes and oxidative demetallation is necessary to obtain uncomplexed cyclopentadienones. The recent development of catalytic transition metal-mediated carbonylative alkyne–alkyne coupling serves as a direct and effective synthesis of cyclopentadienones (371). The IrCl(CO)diphos catalyst promotes carbonylative alkyne–alkyne coupling to form cyclopentadienones in high isolated yields (Scheme 29). Binding of a diyne to Ir(I) releases CO, leading to the corresponding π complex, which cyclometallates and subsequently inserts CO into one of the Ir—C bonds. A second equivalent of CO releases and regenerates the catalyst. The transformation is also tolerant of a number of functional groups.

2. Carbon–carbon bond coupling reactions that link alkynyl ligands are potentially important reactions for the preparation of alkyne oligomers (372) and nonlinear optical materials, as well as carbocyclic and heterocyclic compounds (373, 374). As a metal-mediated strategy to these

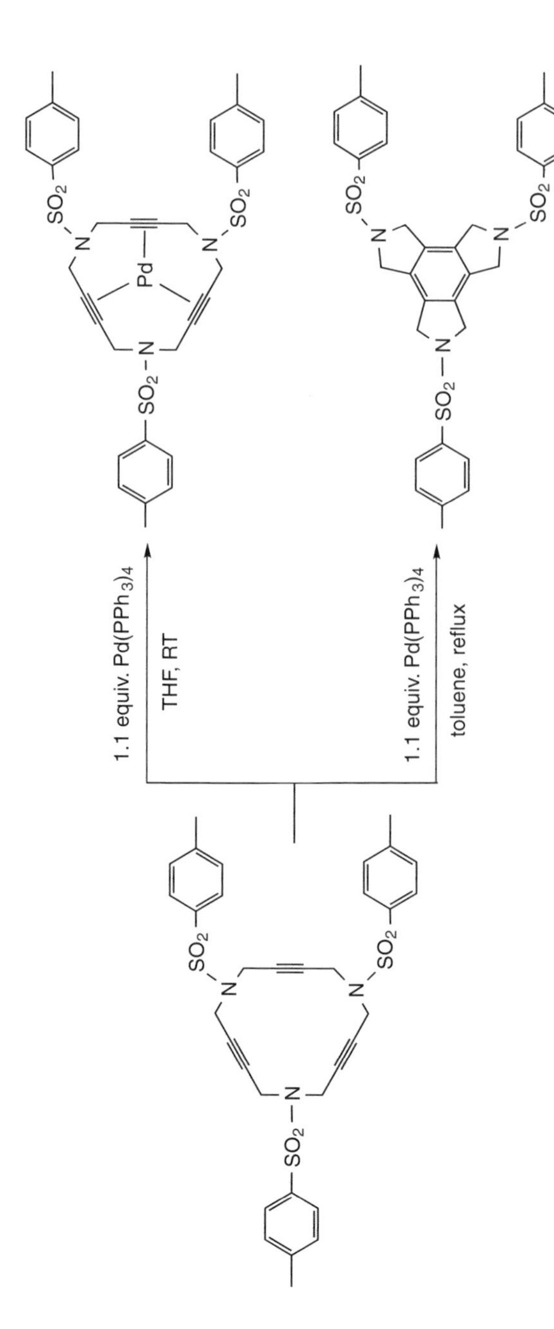


Scheme 28. The Pd(0)-promoted cycloisomerization of a macrocyclic triyne. (Room temperature = RTS equivalent = equiv.)

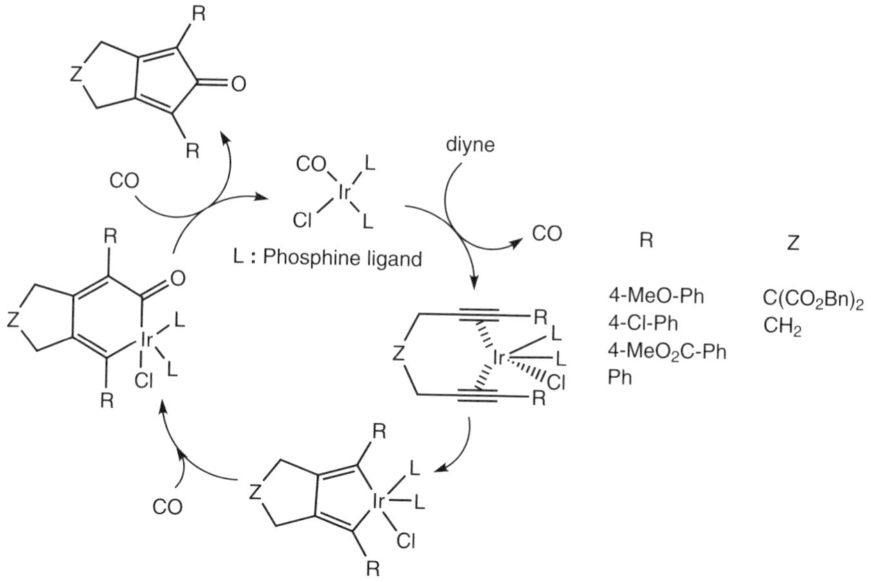

Scheme 29. The Ir(I)-mediated carbonylative alkyne–alkyne coupling for entry into substituted cyclopentadienone synthons.

couplings, acyclic bis(acetylide)Pt(II) complexes can be utilized to cross-couple alkynes (375). Specifically, reaction of *cis*-Pt(C≡CPh)$_2$(dppe) with metal carbonyls in refluxing toluene leads to formation of unusual metallacycles (Scheme 30) of the form [$Mn_2$Pt(PhCCCCPh)$(CO)_6$(dppe)] and [$PtRu_3$(PhCCCCPh)$(CO)_{10}$(dppe)] (376). In these constructs, the original alkyne ligands from the bis(acetylide)Pt(II) starting material are cross-coupled in a head-to-head C—C coupling orientation. The mechanism involves formation of an $\eta^2$ $\pi$ complex of Pt(II) with the coupled 1,3-diyne upon treatment with the metal carbonyl. The coupling reaction presumably also occurs via $\pi$ complexation of the metal carbonyl.

Within this theme, cyclic cis-enediynes can be readily prepared by a two-step one-pot procedure (377). The first step is the copper-catalyzed addition of $\alpha,\omega$-diynes to manganese carbyne complexes to give an intermediate bis(alkynylcarbene) species that rearranges to an enediyne unit below room temperature. In the second step, the manganese catalyst, which Magnus has been used effectively as an alkyne protecting–stabilizing group in natural product syntheses (378–383), is released from the enediyne by photolysis, copper-catalyzed air oxidation, or stoichiometric Cu(II) oxidation. These new procedures have been applied to a variety of five-, six-, and seven-membered ring cyclic enediynes containing a range of ether, ester, and ketone functional groups (Scheme 31).

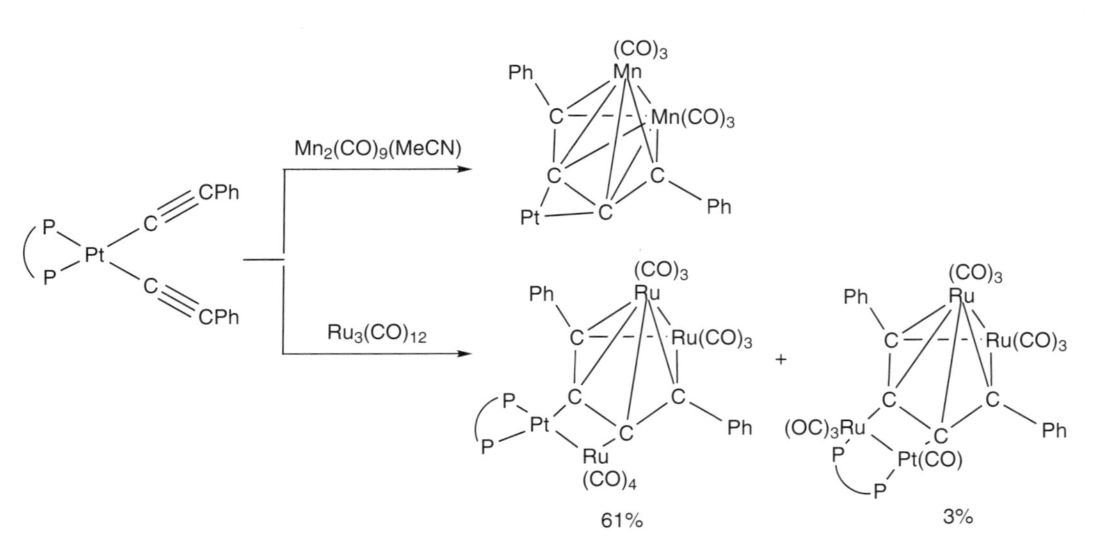


Scheme 30. Heterobimetallic carbon–carbon coupling reactions that yield metallacyclic, head-to-head 1,3-diyne products.

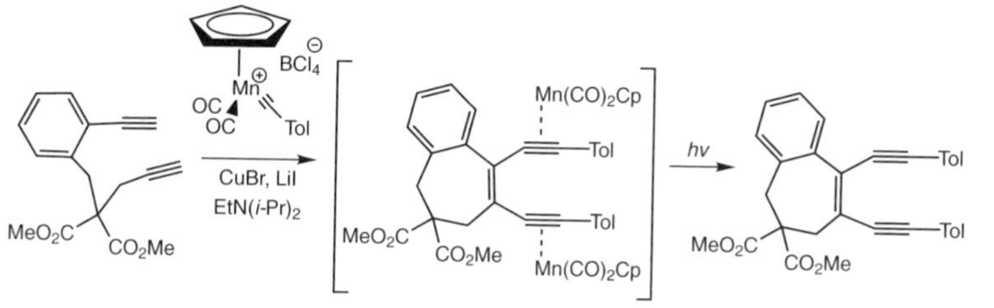

Scheme 31. *cis*-Enediyne synthesis via a bis(alkynylcarbene) intermediate. (Tol = Tolyl.)

3. Metal-catalyzed hydroarylation of alkynes catalyzed by electrophilic transition metal complexes has received much attention as a valuable synthetic alternative to the Heck and cross-coupling processes for the synthesis of alkenyl arenes (384). Metal trifluoromethanesulfonates (metal triflates) [$M(OTf)_n$; M = Sc, Zr, In] catalyze the hydroarylation of alkynes via π complexation to give 1,1-diarylalkenes in very good yields (Scheme 32) (385). The reaction likely proceeds by a Friedel–Crafts mechanism via the alkenyl cation intermediate where the aryl starting material also serves as the solvent.

Hydroarylation can also be mediated by Au(I) and Au(III) (Scheme 33) (384). In the case of aryl substituted alkynes, the Au(III) π complex undergoes electrophilic aromatic substitution with the electron-rich arene to give alkenyl–Au(III) complex, which is immediately protonated by the $H^+$ generated upon C—C bond formation. For the Au(I)-catalyzed hydroarylation, the cationic gold complex π coordinates the alkyne, with subsequent nucleophilic attack by the arene from the opposite face leading to an alkenyl–gold complex, which is protonated to the desired products. The nature of the reaction causes the regioselectivity of this reaction to be sensitive to electronic rather than steric factors.

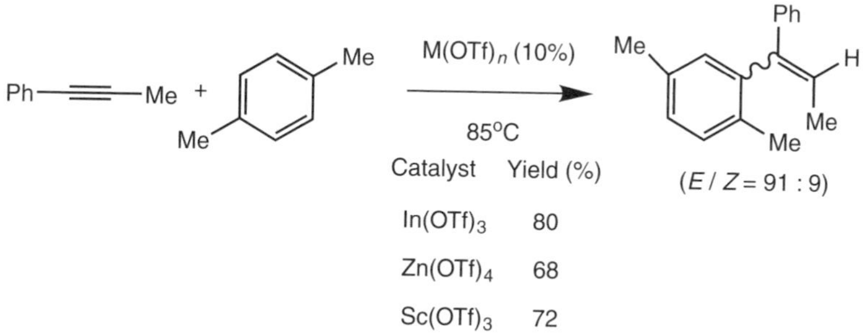

| Catalyst | Yield (%) |
|---|---|
| $In(OTf)_3$ | 80 |
| $Zn(OTf)_4$ | 68 |
| $Sc(OTf)_3$ | 72 |

Scheme 32. Metal–triflate (OTf) hydroarylation of alkynes.

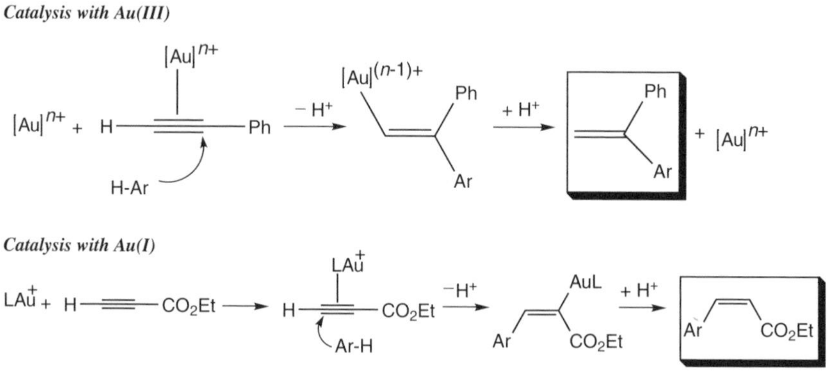

*Catalysis with Au(III)*

*Catalysis with Au(I)*

Scheme 33. Catalytic hydroarylation of alkynes by gold complexes.

4. Finally, Katz and subsequently Rooney (386–388) showed that $(CO)_5W[C(OMe)Ph]$ catalyzes the polymerization of alkynes via carbene–alkyne and purported metallacyclobutene intermediates (Scheme 34). Later computational work by Hofmann (388b) called into question the formation of the latter species, suggesting that the reaction proceeds directly via the vinyl–carbene. Formation of the alkyne–carbene intermediate by low-temperature photolysis of $(CO)_5W[C–(OMe)Ph]$ in the presence of alkynes leads to polymerization upon warming. The use of a pre-formed carbene complex is not required since active polymerization catalysts can be formed by photolysis of $W(CO)_5$ in the presence of terminal alkynes in hydrocarbon solutions. A key step in catalyst generation is rearrangement of a coordinated alkyne to a vinylidene ligand (389).

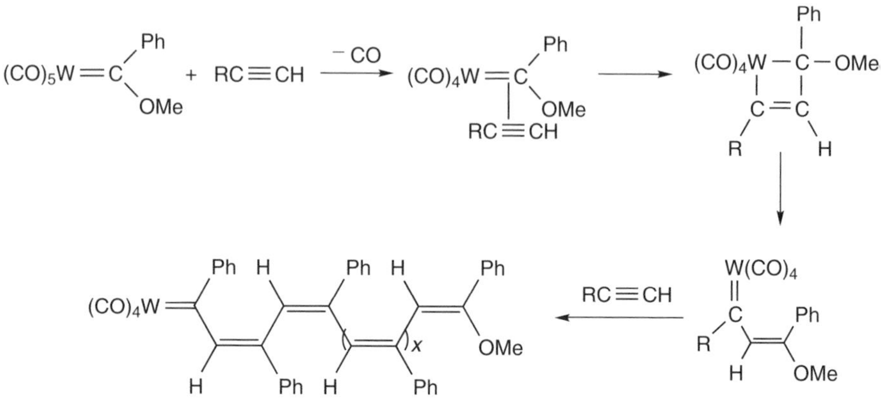

Scheme 34. Carbene–alkyne polymerization mechanism to generate polyacetylene derivatives.

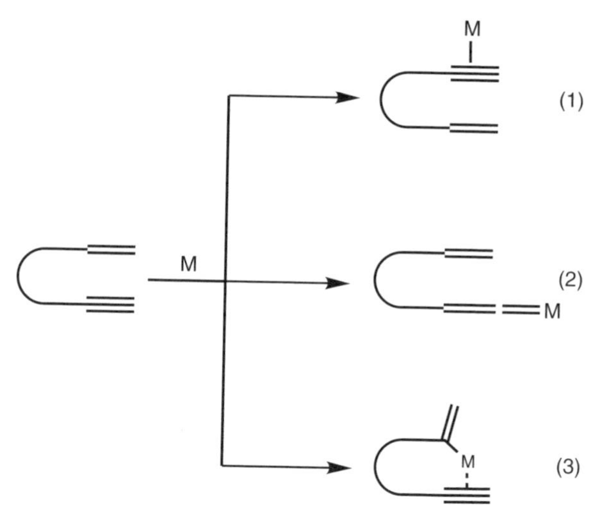


Scheme 35. Modes of metal–enyne bonding. (1) $\pi$ complex; (2) $\sigma$ carbenoid construct; (3) both $\sigma$, $\pi$ interactions.

# IV. ENYNE CONSTRUCTS AND METAL-MEDIATED REACTIVITY

## A. Transition Metal–Enyne Structure

Conjugated enynes possess several of the same transition metal binding modes exhibited by their diyne counterparts, including $\pi$-coordination, $\sigma$-carbenoid formation, and $\sigma/\pi$ metal complexation (Scheme 35). In some cases, these constructs are very stable and can be isolated, but they are frequently transient structures in which the metal ion acts as a catalyst for enyne cycloisomerisation and ligand transformation reactions.

Transition metals mainly form $\pi$ complexes with enynes. Nearly all the metal-mediated enyne cycloisomerization reactions occur via this structural pathway (11). In some cases, it has been shown (Scheme 36) that transition

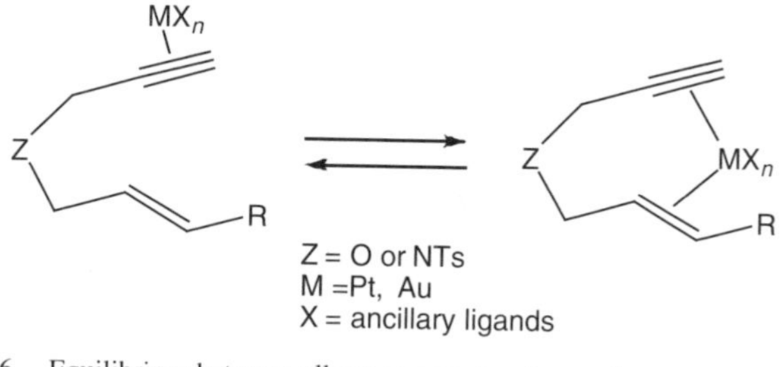

$MX_n$ Z R ⇌ Z $MX_n$ R

Z = O or NTs
M =Pt, Au
X = ancillary ligands

Scheme 36. Equilibrium between alkyne $\pi$ complexation and enyne $\pi$ complexation.

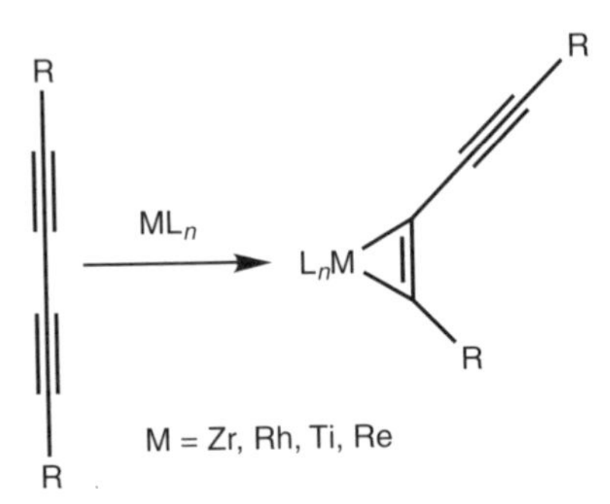


Scheme 37. Metal-1,3-diyne π interaction leading to $\eta^2$-enyne structure.

metals can form π bonds with both the alkyne and -enefragments of an enyne (390, 391). The π-bonding interaction of transition metals with 1,3-diynes, which have been discussed in earlier sections (1), resembles a coordinated enyne, despite the diyne ligand origin (Scheme 37).

The σ-carbenoid enyne is a reactive intermediate usually generated during the metal ion-induced enyne cycloisomerization (Scheme 38). Ohe and coworkers (392–395) showed that it is possible to isolate this intermediate as a stable species. However, they have also demonstrated that in the presence of base, the transformation is catalytic at reflux temperature. A series of transition metals (e.g., Ru, Ir, Rh, Au, Pt, Pd, Ti, Co) have been used for this type of transformation (11, 390, 396, 397). Casey et al. (398, 399) synthesized alkynylcarbenes of the form M=C(OR)C≡CR′ (Fig. 27), which have rich chemistry, especially as starting materials to synthesize a large array of target products. Transition metals can also form unusual carbenoid enyne constructs (1) of the form M=C=CH–C≡CR with 1,3-diynes (Fig. 27). Finally, several new types of organometallic enyne compounds having unique bonding modes

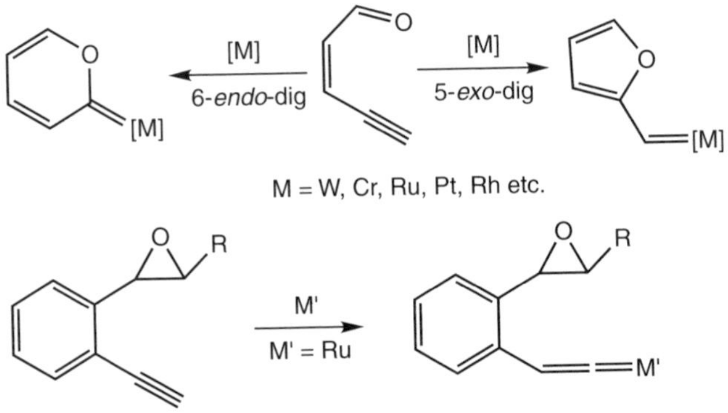

Scheme 38. Metal–enyne carbenoid intermediates.

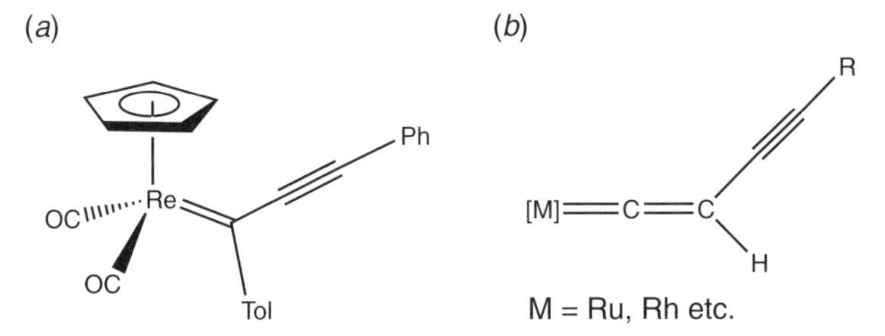


Figure 27. Metal-σ-carbenoid structures: (*a*) alkynyl–carbene; (*b*) alkynyl alkyne–carbenoid.

(Fig. 28) have been nicely described in recent articles (1, 399, 400). Nearly all of these compounds have been generated from reactions of 1,3-diynes with suitable transition metals via metal-assisted rearrangements.

## B. Intramolecular Ring-Closing Enyne Metathesis

Beyond structure, metal-mediated enyne reactivity and bond reorganization (metathesis) has a rich history, beginning with its serendipitous discovery ~20 years ago. Excellent reviews have been compiled highlighting different aspects of the advances in these cyclization reactions (11, 363, 401–408). By definition, enyne metathesis is a bond reorganization of an alkene and an alkyne to produce a 1,3-diene (403, 406 ,407) (Scheme 39). The enyne metathesis reaction is completely atom economical (409) (i.e., no olefin-containing byproduct is released during the process) and is sometimes called an alkylidene migration reaction because the alkylidene part migrates from the alkene to the alkyne carbons. The synthetic value of this reaction is enhanced by the fact that, in addition to being a means to an end in itself, the 1,3-diene systems thus formed

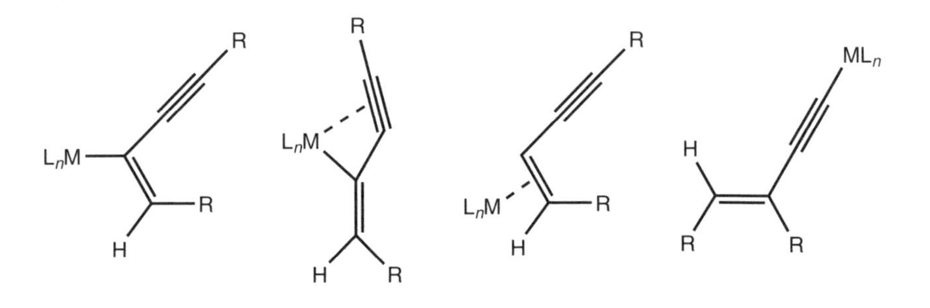


Figure 28. Different types of σ- and/or π-bonded enyne motifs.

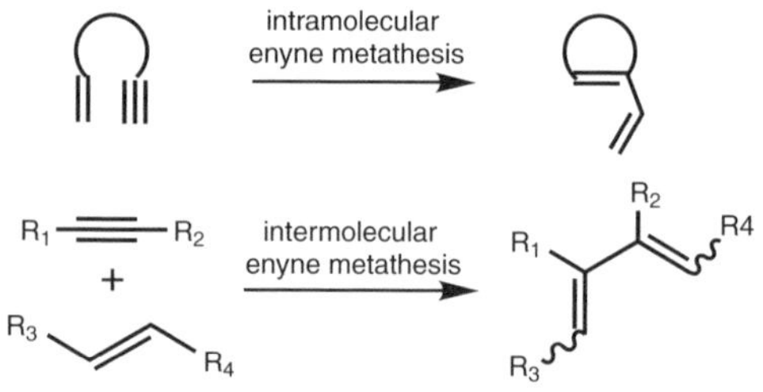

Scheme 39. Intra and intermolecular enyne metathesis.

are themselves versatile synthetic reagents that can undergo further selective transformations (e.g., cycloaddition reactions). The metathesis reaction can be catalyzed by metal carbenes or by low-valent transition metals (e.g., chromium, ruthenium, iridium, palladium, platinum, gold, and their salt complexes) (401, 410–420).

Synthetic applications of enyne metathesis can be separated into three main areas: intramolecular (ring-closing) enyne metathesis; intermolecular (cross) enyne metathesis, and tandem enyne metathesis. The intramolecular (ring-closing) enyne-metathesis reaction is a useful method for the synthesis of macrocyclic ring systems. Intermolecular (cross-metathesis) reactions have been employed much less frequently owing to the perceived difficulties in realizing acceptable levels of selectivity. Tandem enyne metathesis (ring-closing enyne metathesis followed by a second ring-closing diene metathesis) transformations are very powerful for the construction of bicyclic ring systems from acyclic starting materials. The latter three topics will be discussed in Sections IV B–IV.D.

Intramolecular enyne metathesis reactions are commonly called enyne ring-closing or cycloisomerization reactions because a ring is produced by the metathesis. This reaction can be further subdivided into three categories. The first is metal carbene-mediated enyne metathesis. The metal carbene reacts first with the -yne unit and the resulting carbene complex further reacts with the –ene functionality to give intermediate **A** (Scheme 40), which subsequently produces 1,3-dienes together with the metal carbene catalyst (pathway 1). The second category involves cycloisomerization through metallacyclic intermediates. The oxidative metallacycloaddition to enynes forms the metallacycle B, which undergoes subsequent $\beta$-hydride elimination to give a mixture of the 1,3- and 1,4-dienes (pathway 2). The third category involves the $\pi$ complexation of an electrophilic transition metal onto the alkyne, leading to the formation of the polarized $\eta^1$-alkyne complex **C**, which contains a positive charge at the $\beta$-position (pathway 3). Potential delocalization of a positive charge produces isomers **D, E, F**, and **G**, while intermediate **H** produces the 1,3-diene through a bond-rearrangement process.

The catalytic enyne metathesis reaction was first discovered by Katz and Sivavec (421) in 1985 in which they reported the first examples of this process in

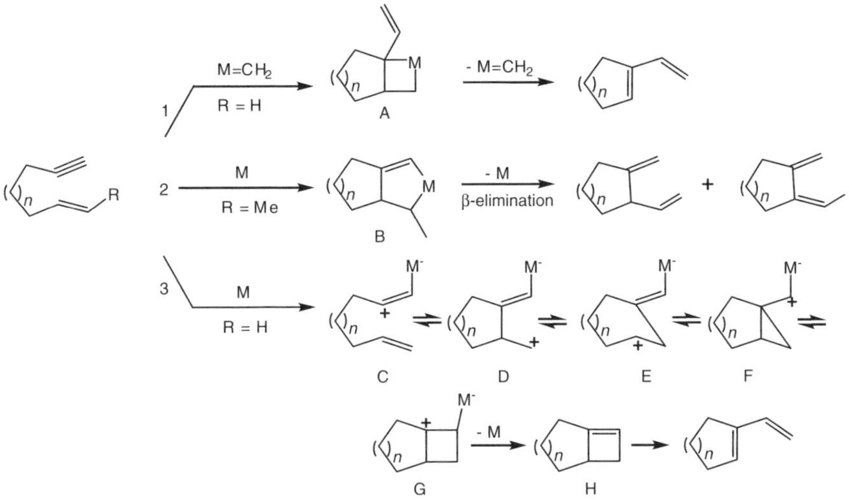

Scheme 40. Generalized cycloisomerization reactivity of 1,*n*-enynes (478).

the presence of substoichiometric amounts of tungsten Fischer carbene complexes [Scheme 41(*a*)]. Subsequently, Hoye, Mori, and co-workers (419, 422 ,423) reported similar reactions with molybdenum and chromium Fischer carbene complexes [Scheme 41(*b*)]. As extensions of earlier studies (424, 425), Fischer carbene complexes of chromium were shown to catalyze the enyne metathesis of 1,6-enynes in which the alkene terminus supports an alkoxy group (419, 420). However, Grubbs' catalyst (426) was later found to be more broadly applicable (427, 428).

Shortly after the discovery of enyne metathesis, Trost began developing cycloisomerization reactions of enynes using Pd(II) and Pt(II) metallacyclic catalysts (429–433), which are mechanistically divergent from the metal–carbene reactions. The first of these metal catalyzed cycloisomerization reactions of 1,6-enynes appeared in 1985 (434). The reaction mechanism is proposed to involve initial enyne $\pi$ complexation of the metal catalyst, which in this case is a cyclometalated Pd(II) cyclopentadiene, followed by oxidative cyclometalation of the enyne to form a tetradentate, putative Pd(IV) intermediate [Scheme 42(*a*)]. Subsequent reductive elimination of the cyclometalated catalyst releases a cyclobutene that rings opens to the 1,3-diene product. Although this scheme represents the fundamental mechanism for enyne metathesis and is useful in the synthesis of complex 1,3-cyclic dienes [Scheme 42(*b*)], variations in the reaction pathway due to selective $\pi$ complexation or alternative cyclobutene reactivity (e.g., isomerization, $\beta$-hydride elimination, path 2, Scheme 40) leads to variability in the reaction products. Strong evidence for intermediacy of cyclobutene species derives from the stereospecificity of the reaction. Alkene

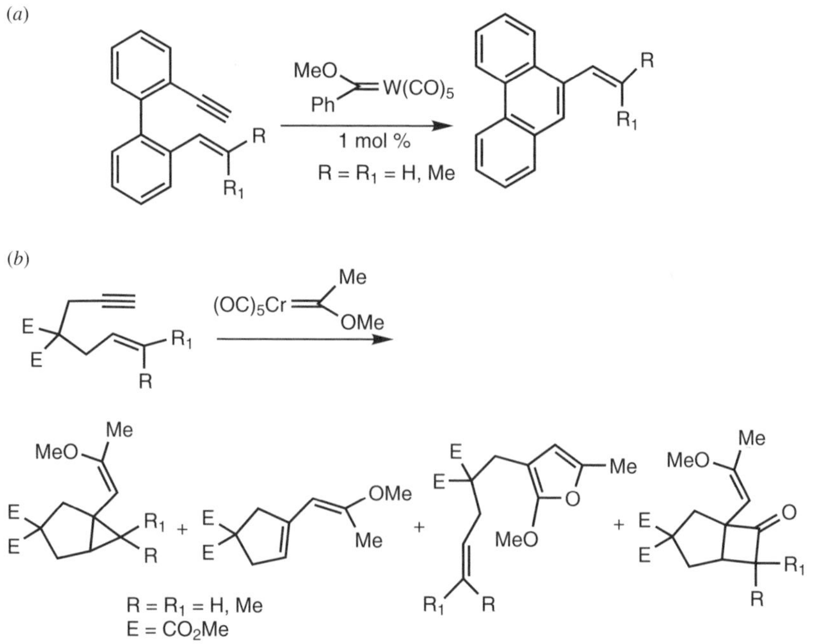

Scheme 41. Examples of (*a*) tungsten and (*b*) chromium Fischer carbene-mediated enyne cycloisomerization reactions.

substitution can also play a role in the nature of the reaction pathway, where the traditional palladacycle and/or alternative $\eta^2$-$\pi$-activation mechanisms can be operative [Scheme 42(*c*) (431).

Following the seminal contributions by Trost, the first enyne metathesis reaction utilizing a ruthenium carbene complex was reported in 1994 by Kinoshita and Mori (428) and described the formation of five-, six-, and seven-membered nitrogen-containing heterocycles by intramolecular ring-closing enyne metathesis [Scheme 43(*a*)]. Much of the mechanistic insight into the pathways for enyne metathesis by these metal–carbenes derives from the concurrent development of bicycle-forming tandem enyne cyclization (cf. Section IV.D) (427), and the extensive work on olefin metathesis by Grubbs and coworkers (435–438). Although the mechanism for the tandem reaction is thought to involve formation of a vinyl carbene intermediate via an alkyne-derived cyclobutene that spontaneously ring opens, the intramolecular reaction mechanism is not as clear and may involve variable modes of enyne $\pi$ complexation. Two mechanistic hypotheses have been proposed [Scheme 43(*b*)]; the yne-then-ene pathway involves immediate formation of a metallacyclobutene species that

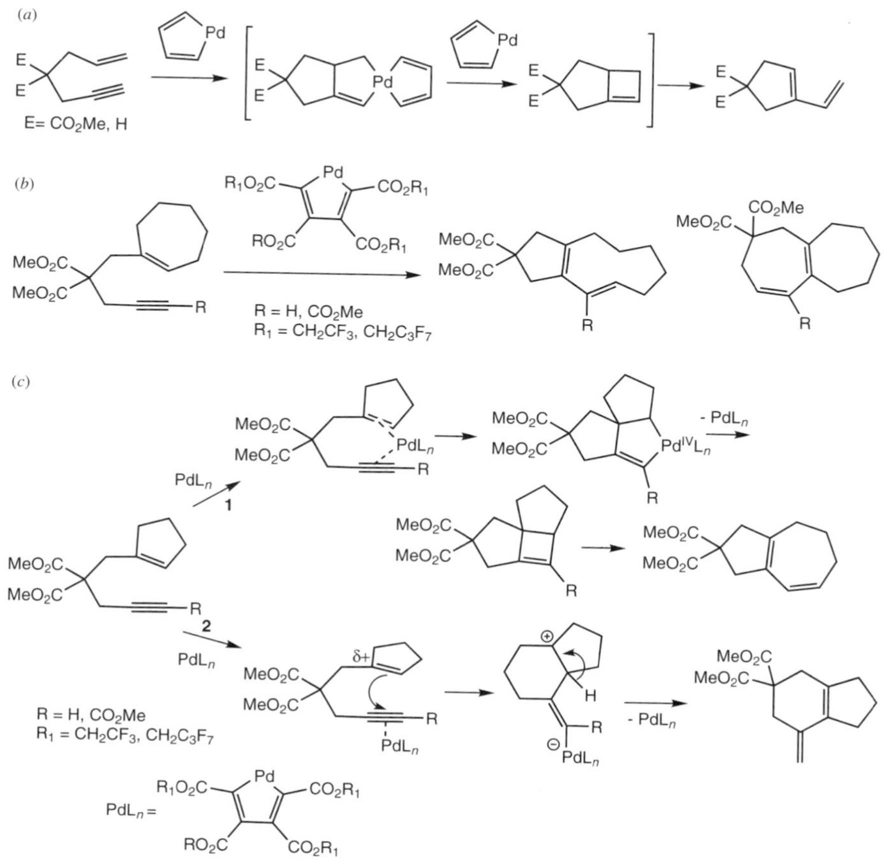

Scheme 42. (*a*) General scheme for Trost palladacycle-catalyzed enyne metathesis reaction. (*b*) the Pd(IV) mediated construction of extended ring structures. (*c*) Dual reaction pathways leading the traditional enyne metathesis product (upper) versus the alternative $\eta^2$-mediated product (lower).

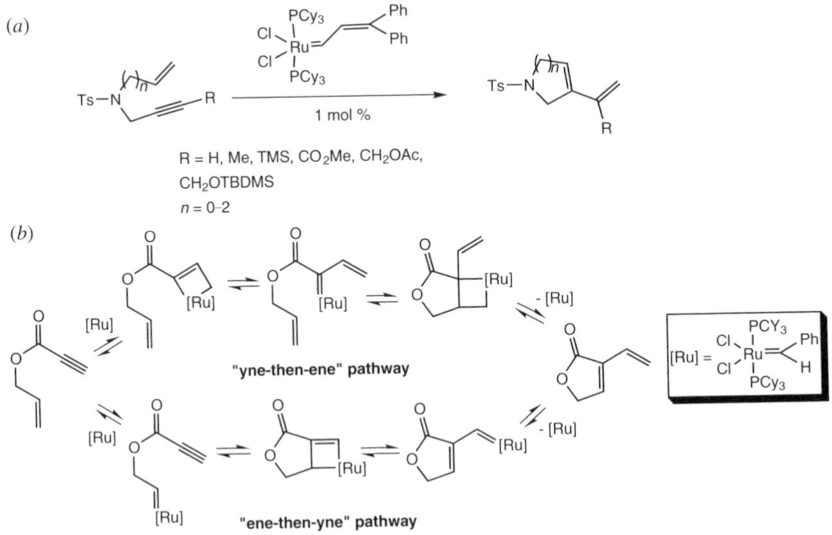

(*a*)

PCy3 Cl Ru Cl PCy3 Ph Ph

Ts—N R → Ts—N R

1 mol %

R = H, Me, TMS, $CO_2Me$, $CH_2OAc$, $CH_2OTBDMS$

$n = 0\text{–}2$

(*b*)

O O [Ru] [Ru] [Ru] [Ru] - [Ru]

"yne-then-ene" pathway

[Ru] = Cl Ru PCY3 Ph Cl PCy3 H

[Ru] [Ru] [Ru] [Ru] - [Ru]

"ene-then-yne" pathway

Scheme 43. (*a*) Enyne metathesis using first generation Grubbs' Ru–carbene catalyst. (*b*) Mechanistic pathways yne-then-ene (upper) and ene-then-yne (lower). (TBDMS = *tert*-butyldimethylsilyloxy.)

ring-closes via a vinyl carbene to generate the 1,3-diene, while the 'ene-then-yne' pathway initially forms the terminal Ru–alkylidene that ring-closes via the metallacyclobutene to form the vinyl carbene and final product by methylene transfer (439). As an extension of their previous contribution, in 1996 Mori and co-workers (428a) reported the first application of a ring-closing enyne metathesis reaction in a total synthesis, namely, that of the tricyclic alkaloid (−)-stemoamide in 9% overall yield (Scheme 44). Within this theme, natural products have more recently been constructed using Grubbs' catalyst to drive the ring-closing

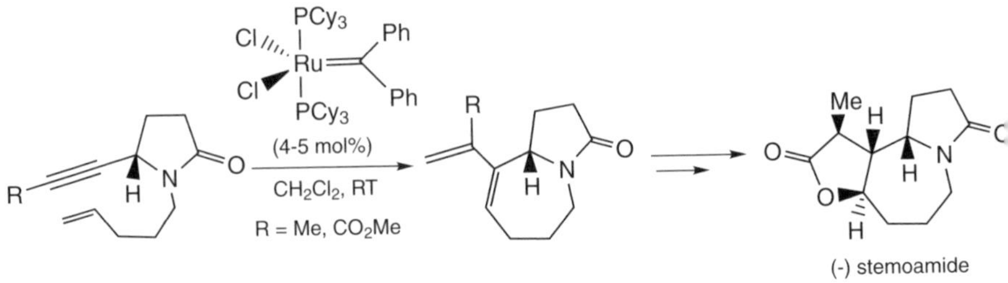

PCy3 Cl Ru Cl PCy3 Ph Ph

R H N O (4-5 mol%) $CH_2Cl_2$, RT R = Me, $CO_2Me$ → R H N O → O Me H H N O H O H

(-) stemoamide

Scheme 44. Ring-closing enyne metathesis using Grubbs' catalyst for the synthesis of (−)-stemoamide.

metathesis of aromatic enynes to produce 1,3-dienes that subsequently undergo a Diels–Alder reaction to give the desired product (440, 441).

The early period of enyne metathesis also saw a growth in non-carbene-based catalysts for 1,6-enyne metathesis and cycloisomerisation with skeletal rearrangement. Murai reported the complex $[RuCl_2(CO)_3]_2$ as a very efficient catalyst for 1,6- and 1,7-enyne cycloisomerization to afford vinyl cyclopentenes (and vinyl cyclohexenes) with high stereoselectivity (418). Murai and co-workers (410, 442) also reported examples of the cyclorearrangement of 1,7-enynes catalyzed by metal halides (e.g., $PtCl_2$ and $GaCl_3$) leading to the formation of six-membered rings. Also note that halide complexes of Rh, Re, Ir, Pt, and Au were active for these same transformations. Recently, the use of simple platinum salts and complexes have emerged as remarkable catalysts for molecular reorganizations during enyne cycloisomerizations. The first $PtCl_4$-catalyzed enyne cycloisomerization was reported by Blum et al. (443) and showed that allyl propargyl ethers are cycloisomerized to 3-oxabicylo[4.1.0]hept-4-enes in moderate yields. Subsequently, Murai and co-workers (410), Fürstner et al. (444), and Echavarren and co-workers (445) reported extensively on the use of $PtCl_2$ to catalyze formation of vinyl cyclopentenes and vinyl methylene pentanes from 1,6-enynes. Oi et al. (446) found that although addition of phosphine ligands inhibited catalysis, the activity could be regained if the halide (Cl) was abstracted. Pérez-Castells and co-workers reported the formation of a naphthalene derivative as the side product of the intramolecular Pauson–Khand reaction of aromatic enynes in the presence of a stoichiometric amount of $Co_2(CO)_8$ (447). Intramolecular enyne metathesis has also been applied to ring-fused carbohydrate synthesis (448–453). Van Boom and co-workers (450, 451) initiated this approach by using the ring-closure reaction carbohydrate-derived enynes to form spiroacetals.

The synthetic utility of the metathesis reaction has fostered considerably more effort in that arena than probing the fundamental parameters of the reaction. To this end, Hansen and Lee (144) systematically probed the role of ring size in ring-closing enyne metathesis. For these types of tartrate-based macrocyclic systems, the alkylidene can add in either of two regiochemical addition modes, leading to either an exocyclic diene (exo mode, via an exocyclic vinylcarbene) or an endocyclic diene (endo mode, via an endocyclic vinyl carbene) (Scheme 45). The methodology has proven effective for the preparation of 10–12-membered rings, albeit by variations in the reaction pathway as dictated by ring size and stability of the bicyclic Ru–cyclobutene intermediate.

With this foundation and historical perspective of transition metal-promoted intramolecular enyne metathesis reactions of 1,*n*-enynes in place, the remainder of this section will describe specific metal–catalyst systems with examples of their utilities for cycloisomerization reactions involving various substrates.

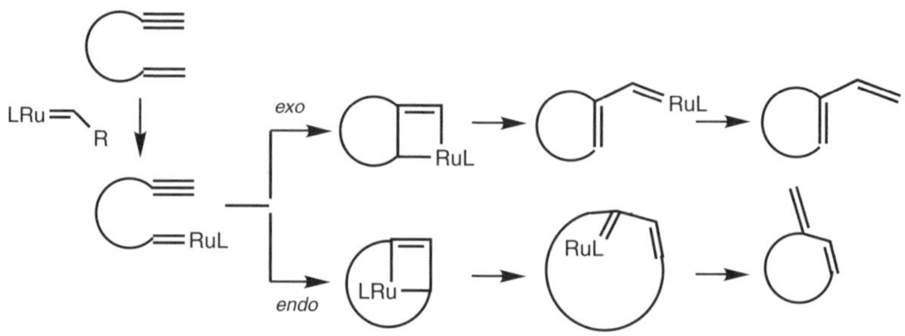

Scheme 45. Ring-closing enyne metathesis to form 10–12-membered macrocycles.

1. With the backdrop of the pioneering work on palladium–platinum-catalyzed intramolecular enyne metathesis by Trost et al. (429–433), simple platinum salts were later developed by Blum et al. (443) and Murai and co-workers (410) (e.g., $PtCl_4$ and $PtCl_2$, respectively) to cyclize 1,6-enynes and select 1,7-enynes. Murai observed that skeletal reorganizations of enynes lead to two products (**I, II**; Scheme 46). The first (**I**) is the expected product of a metathesis reaction, while the second (**II**) involves an unusual carbon–carbon bond formation and is the major product with unsubstituted substrates, or when electron-withdrawing groups are present. In 1998, Fürstner et al. (444) suggested that the mechanism for Pt(II)-catalyzed cycloisomerization of 1,6-enynes involves formation of a carbocation as the reactive intermediate. Attack by $PtCl_2$ at the alkene would render a cation that can be delocalized and represented by several structures (Scheme 47). Subsequent cyclization leads to several potential cationic resonance forms that lead to respective products upon hydride shift and ring opening or catalyst elimination. Echavarren and co-workers used milder nucleophilic agents (e.g., allylstannanes and allylsilanes) to attack the carbocationic intermediate (454, 455). In this case, no traditional 1,3-diene metathesis products are formed and the carbocyclic products reach high yield (Scheme 48). The reaction is proposed to occur

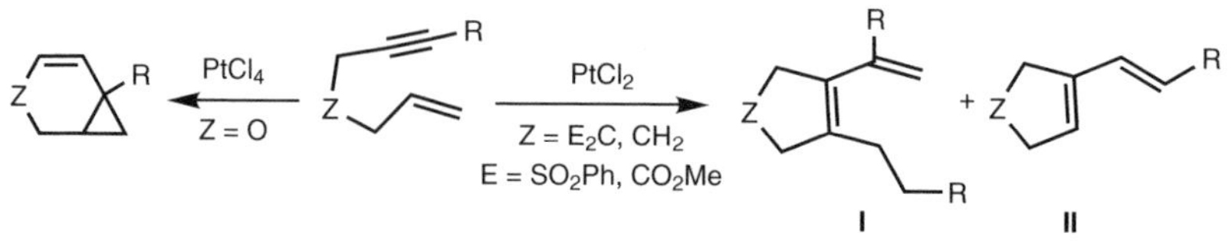

Scheme 46. Intramolecular Pt(II)–Pt(IV) enyne metathesis featuring substitution-dependent skeletal rearrangement.

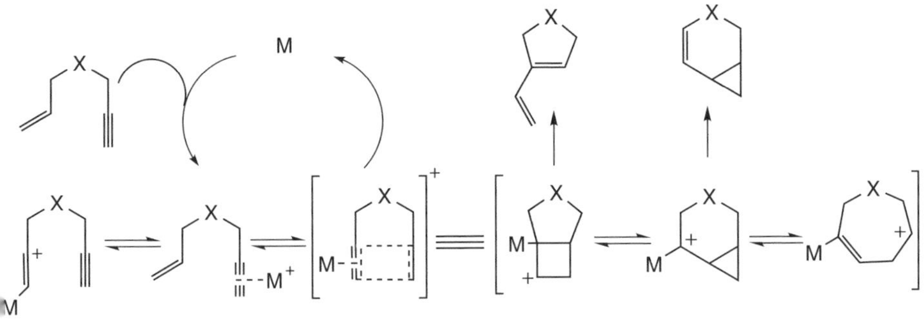

Scheme 47. Postulated mechanism for Lewis acid (e.g., $PtCl_2$) catalyzed enyne metathesis via a carbocation intermediate.

by coordination of the metal to the alkyne, followed by the anti attack of the allyl, and proceeds with anti stereoselectivity. In select cases, an endo cyclization product is detected (Scheme 49). The alkoxycyclization process competes favorably with cycloisomerization unless the triple bond is disubstituted; under these conditions a mixture of products is obtained. Formation of cyclopropanes is only observed when a heteroatom is placed in the tether (445, 456). Within this mechanistic family, transannular $PtCl_2$-catalyzed cyclization of enynes opened a new route to cyclopropanic tricyclic systems (457).

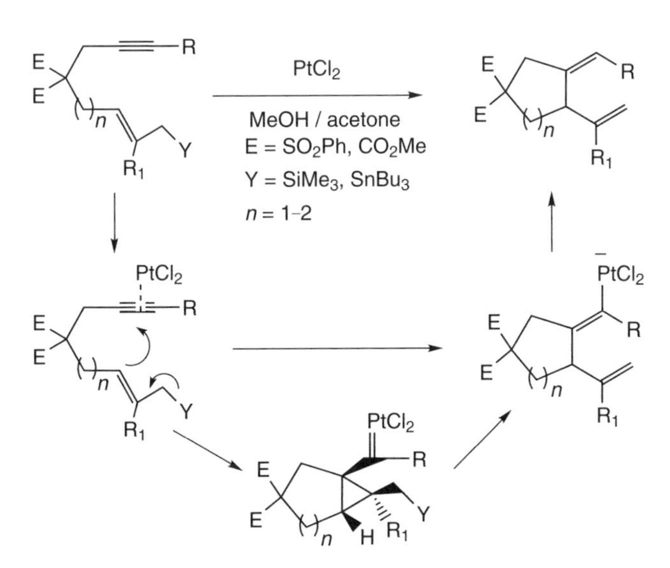


Scheme 48. Carbocyclic product formation during Pt(II)–enyne cycloisomerization.

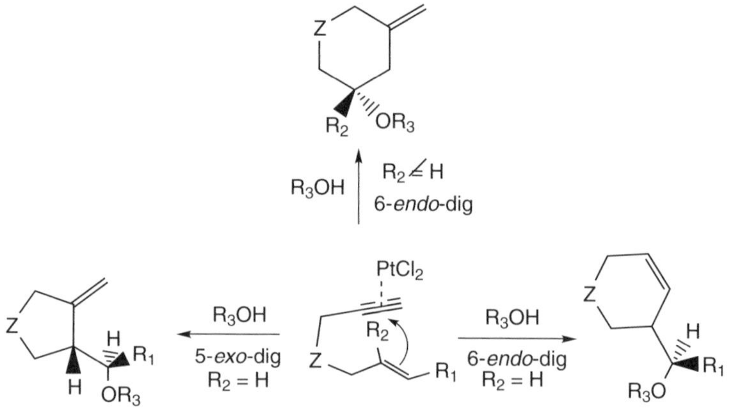

Scheme 49. The Pt(II)-catalyzed endo/exo alkoxycyclization.

Simple Pd(II) salts also catalyze the ring closing of 1,6-enynes leading exclusively to 1,3-dienes when there is no allylic substituent bearing hydrogen (Scheme 50). This Pd-catalyzed enyne intramolecular cyclization reaction represents a new process for the synthesis of stereodefined α-arylmethylene-γ-butyrolactones, lactams, multifunctional THFs pyrolidines, and cyclopentanes. Successful formation of five-membered products via a Pd(0)-catalyzed pathway is another example of an unusual alkenylpalladium species-forming process involving a π-allylpalladium and an alkyne [Scheme 51(*a,b*)]. Oxidative addition of Pd(0)-to-allyl halide or carbonate generates a π-allylpalladium intermediate **II**. Insertion of the alkyne to the Pd–allyl bond affords **III**. The transient vinylpalladium complex **III** can be subsequently trapped via transmetalation with aryl boronic acids and reductive elimination gives cyclized products **V** with regeneration of the Pd(0) catalyst in accordance with established organopalladium chemistry (458, 459) [Scheme 51(*b*)].

2. Non-carbenoid, ruthenium-catalyzed intramolecular enyne metathesis shares a mechanistic resemblance to the Pd-mediated reactions above.

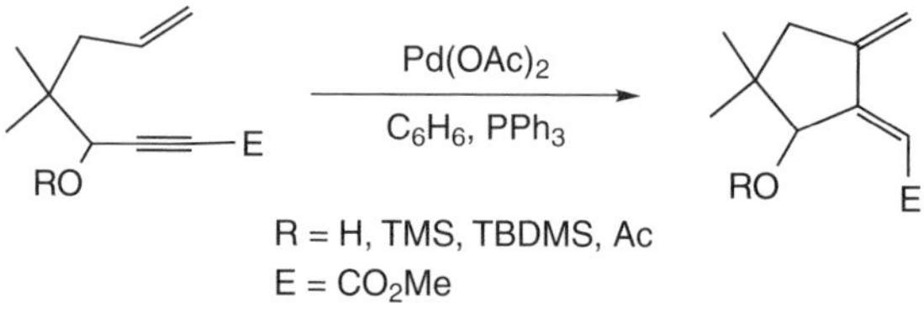

Scheme 50. Formation of 1,3-dienes by Pd(II)–enyne cycloisomerization.

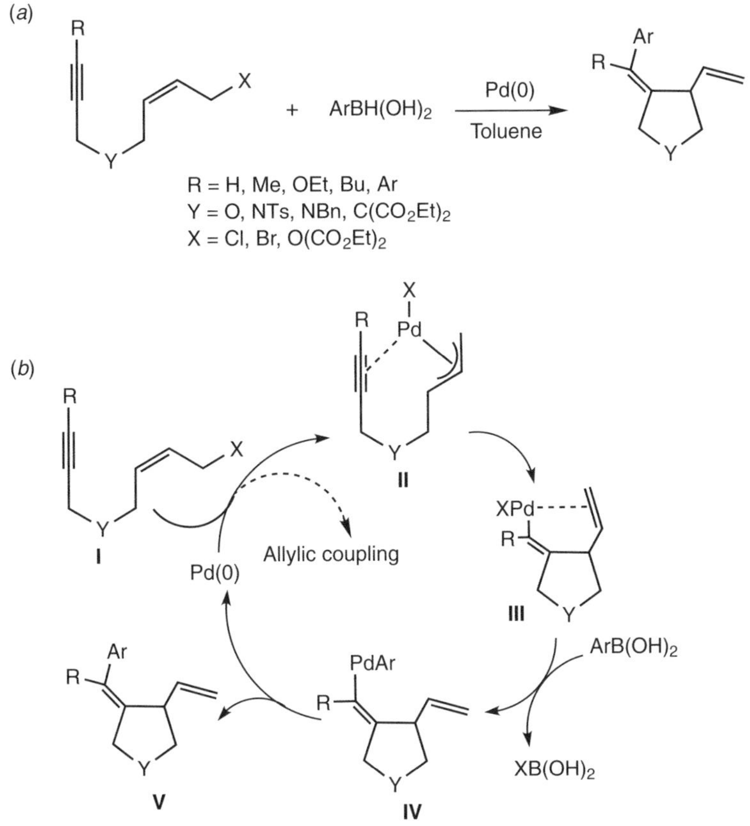

Scheme 51. (*a*) Enyne metathesis to form five-membered products using Pd(0). (*b*) Mechanism illustrating the formation of an unusual alkenylpalladium intermediate via a π-allylpalladium species.

Here catalysis involves the formation of a ruthenium vinylidene, an anti-Markovnikov addition of water (368), and cyclization of an acylmetal species onto the alkene. Although cyclization may occur via hydroacylation (Scheme 52, path A) (460–462) or the Michael addition reaction (Scheme 52, path B) (463, 464), the requirement for an electron-withdrawing substituent on the alkene and the absence of aldehyde formation suggest path B to be the more likely mechanism (465, 466). Trost discovered that the use of the cationic ruthenium catalyst $CpRu(MeCN)_3^+PF_6^-$ is tolerant of 1,2-di- and trisubstituted alkenes and promotes cyclization of 1,6- and 1,7-enynes to five- and six-membered ring products (467). In a number of examples, the ruthenium reaction is complementary to the Pd-catalyzed cyclization described above, selectively forming the 1,4-diene over the traditional 1,

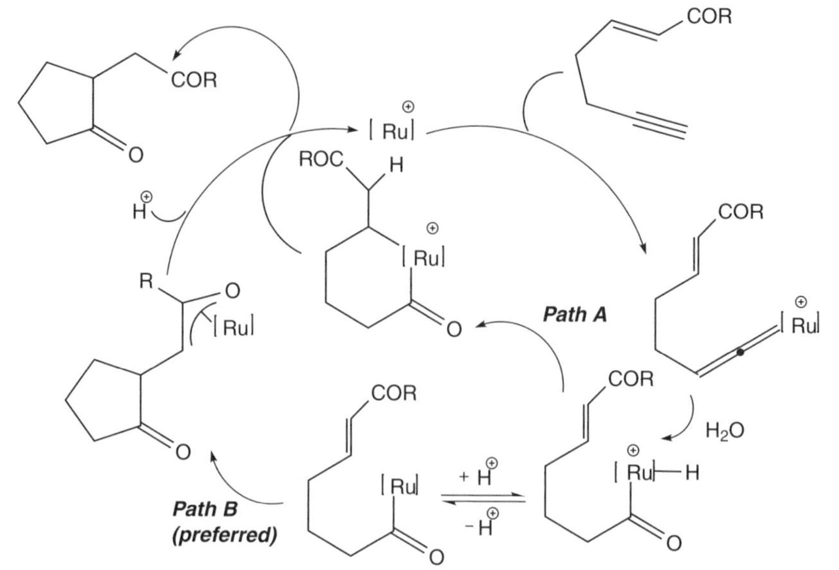

Scheme 52. Mechanistic pathways for non-carbenoid, Ru-catalyzed enyne metathesis.

3-diene (11) (Scheme 53). Trost also observed the formation of seven-membered rings from the cyclization of alkenes and alkynoates [Scheme 54(*a*)] (468). It has been postulated that cyclization of these cycloheptenes is initiated by activation of the allylic C—H bond to form a π-allylruthenium intermediate [**I**, Scheme 54(*b*)]. A 7-exo-dig carboruthenation of the alkynoate produces the ruthenium-hyrdido enolate **II**, while subsequent β-elimination gives the corresponding cycloheptenes and regenerates the cationic ruthenium catalyst.

3. Zero-valent, nickel-mediated intramolecular enyne metathesis is also an effective method for the synthesis of functionalized ring structures. Oxidative cycloaddition of enynes having an electron-withdrawing group on the alkenyl group generates nickelacyclopentene intermediates that

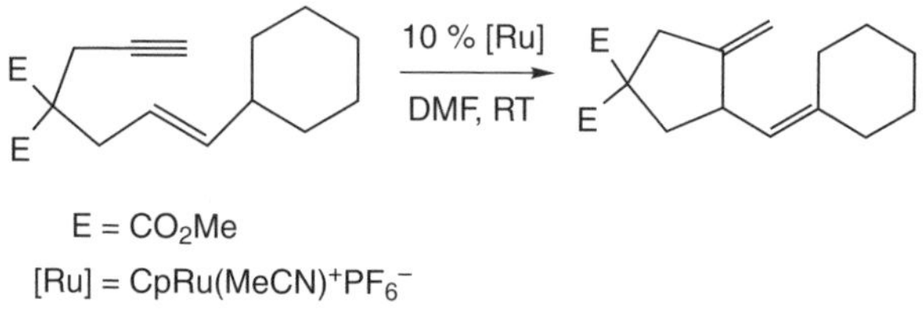

Scheme 53. Cationic $CpRu(MeCN)_3^+$ catalyzed, five-membered ring-forming 1,6-enyne cyclization to selectively yield 1,4-diene products.

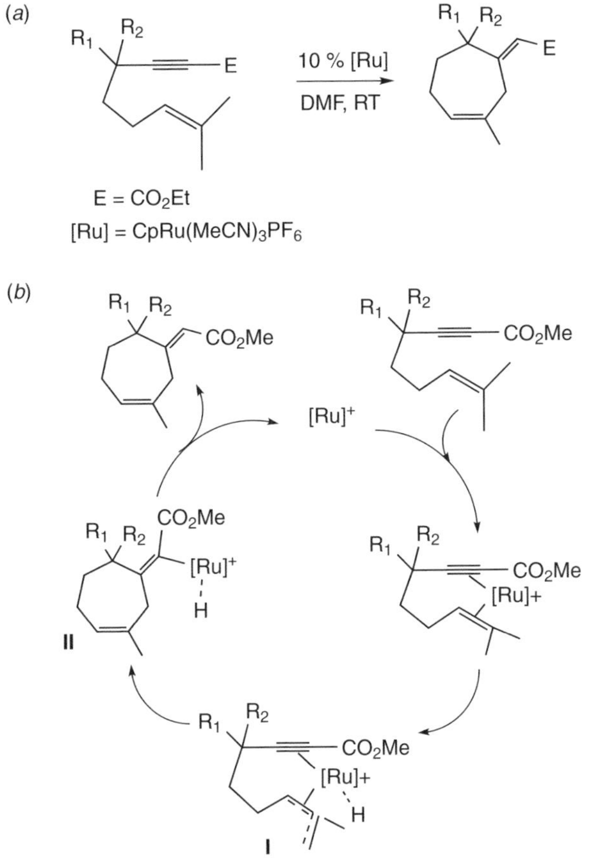

Scheme 54. (*a*) Reaction conditions and (*b*) mechanism for formation of 1,4-cycloheptenes by cationic ruthenium complex-mediated cycloisomerization. (DMF = Dimethylformamide.)

regioselectively add $CO_2$ at the C$sp^3$—Ni bond, giving cyclized carboxylation products in good yields (469). The utility of this novel method was demonstrated by application to the synthesis of various carboxylic acid derivatives of five- to-six-membered ring backbones (Scheme 55). Nickel(0)-mediated intramolecular cyclization of enynes has also been effective for the one-step synthesis of polycyclic hydrocarbons bearing functional groups (470) [Scheme 56(*a*)]. The mechanism involves stoichiometric π complexation of Ni(0) followed by reaction with isocyanide to form the expected bicyclic iminocyclopentene species via five- and six-membered nickel metallacycle intermediates (470) [Scheme 56(*b*)].

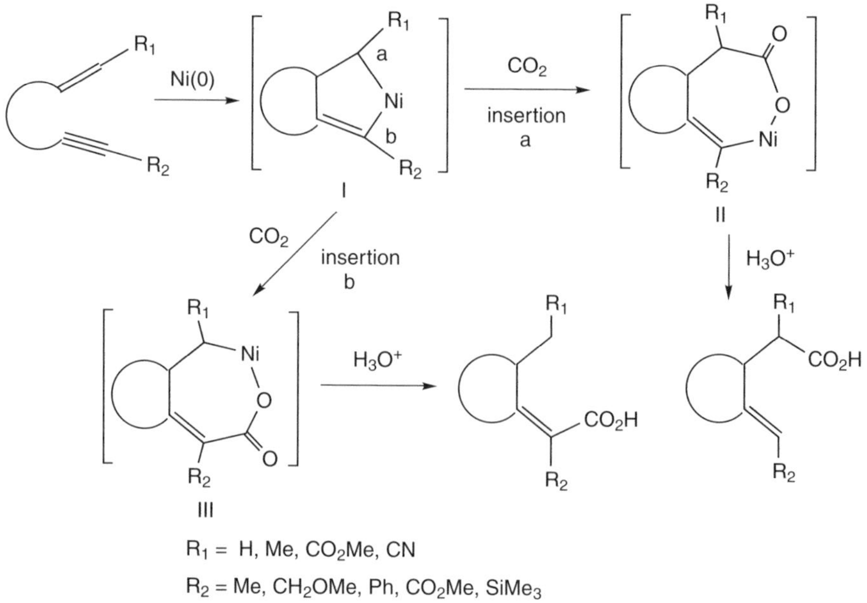

$R_1$ Ni(0) $R_2$ $R_1$ a Ni b $R_2$ I $CO_2$ insertion a $R_1$ O O Ni $R_2$ II

$CO_2$ insertion b $H_3O^+$

$R_1$ Ni O O $R_2$ III $H_3O^+$ $R_1$ $CO_2H$ $R_2$ $R_1$ $CO_2H$ $R_2$

$R_1$ = H, Me, $CO_2Me$, CN

$R_2$ = Me, $CH_2OMe$, Ph, $CO_2Me$, $SiMe_3$

Scheme 55. Formation of cyclic carboxylic acid derivatives by enyne cycloisomerization and $CO_2$ insertion at Ni(0).

4. The Rh-catalyzed cycloisomerization of 1,6-enynes that occurs via rhodium vinylidene-mediated intermediates was first described in 1988 by Grigg et al. (471). Cyclization of 1,6-enynes (**I**, Scheme 57) with $[RhCl(cod)]_2/P(p\text{-}FC_6H_4)_3$, where cod=1,5-lyclooctadiene as the catalyst generates a metal–vinylidene that undergoes $[2 + 2]$ cycloaddition and

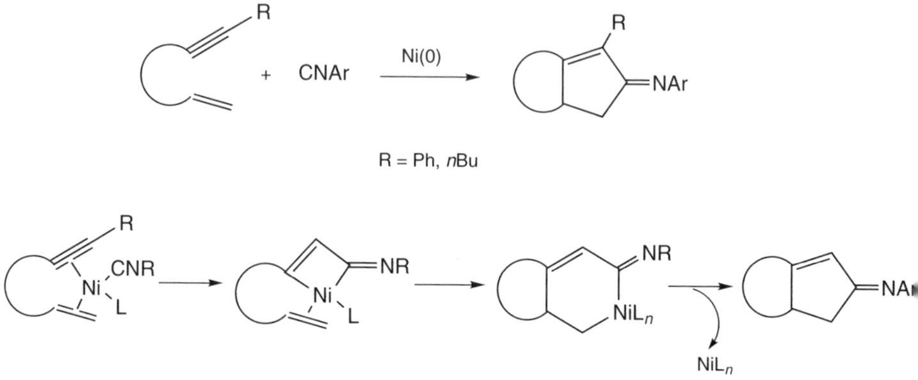

R + CNAr Ni(0) R =NAr

R = Ph, *n*Bu

R CNR Ni L =NR Ni L NR $NiL_n$ =NAr $NiL_n$

Scheme 56. (*a*) Reaction and (*b*) mechanism for formation of Ni(0)-mediated iminocyclopentenes.

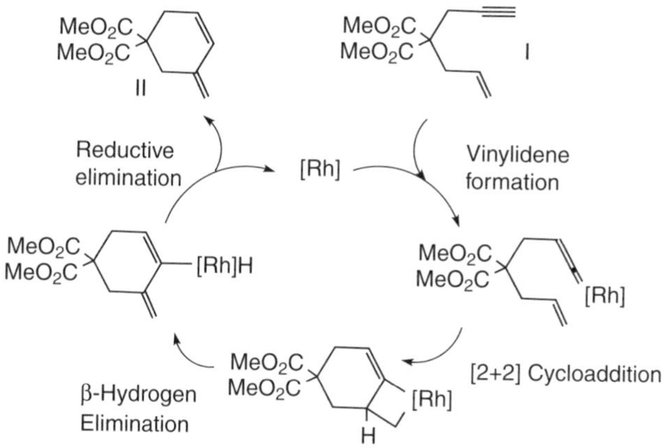

Scheme 57. Cyclization of 1,6-enynes by $[RhCl(Cod)]_2/P(p\text{-}FC_6H_4)_3$.

subsequent β-hydride elimination leading to cyclic diene isomer **II** at 25°C in modest yields (∼5%) (472, 473). When the reaction is performed at 85°C, diene **II** is produced in >80% yield with exo-to-endo selectivity of the new alkene formed as a consequence of the cyclization. Analogous cyclization reactions have been observed for $RhCl(PPh_3)_3$ and cationic Au(I) catalyst systems, wherein hydridorhodium and aurocarbene intermediates, respectively, are proposed (471, 474).

5. Electrophilic Au(I) complexes or their halide AuX analogoues typically cyclize enynes (**I**, Scheme 58) (475) by a 5-exo–dig pathway to give a variety of cycloisomerization and addition derivatives. The mechanism is proposed to involve formation of a cyclopropyl gold–carbene intermediate

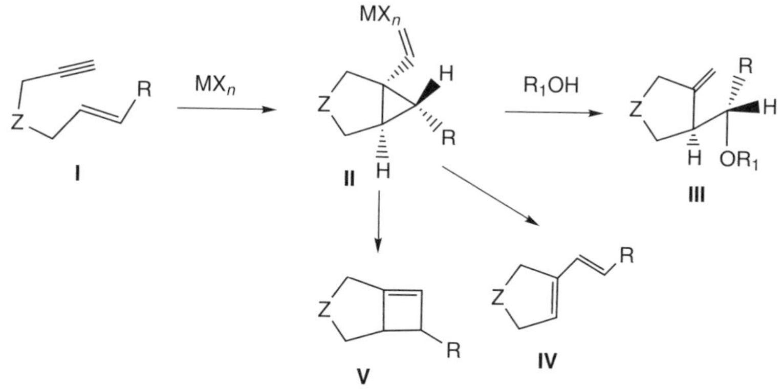

Scheme 58. Cycloisomerization or alkoxycyclization reactions catalyzed by Au(I) halides.

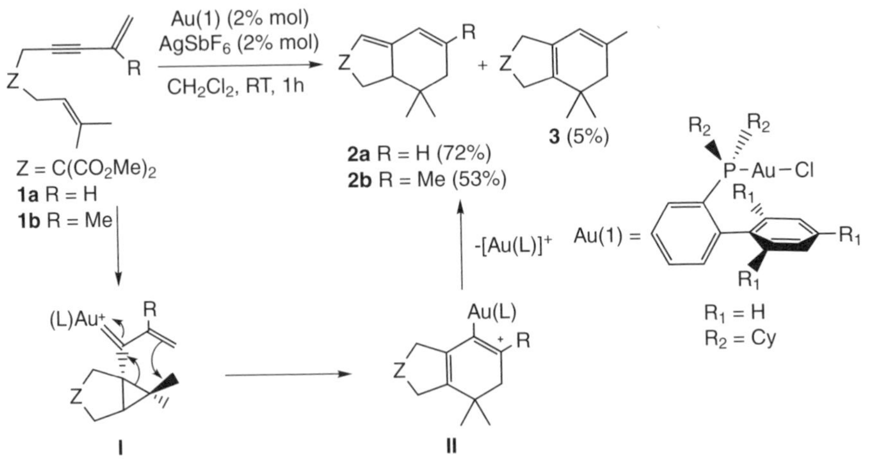

Scheme 59. 1,8-Dien-3-yne substrate cycloisomerization and ring expansion via a cyclopropyl gold–carbene intermediate.

(**II**, Scheme 58) (11, 391, 403, 405, 445, 476, 477). Subsequent reaction with nucleophiles $R_1OH$ (alcohols or water) gives the nucleophilic addition products of type **III**, (11, 391, 403, 405, 445, 476, 477), whereas in the absence of nucleophiles, 1,3-dienes of the form **IV** are generated by skeletal rearrangement or, less commonly, cyclobutenes (Scheme 58, **V**) can be obtained (11, 391, 403, 405, 410, 413, 478–480). For some of these transformations, $[Au(PPh_3)]X$ complexes proved to be the most effective catalysts (474, 481–484). However, enynes containing substituted alkynes or those supporting an aryl group do not undergo cycloisomerization or alkoxycyclization reactions (474). More complex 1,8-dien-3-yne substrates (e.g., **1a,** Scheme 59) cyclize by a 5-exo–dig pathway to give six-membered ring hydrindane products with Au(I) as a catalyst (Scheme 59) (483, 485). Similarly, substrates with R = Me lead to a mixture of regioisomeric dienes (e.g., **2b** and **3**, Scheme 59). The reaction presumably proceeds through the cyclopropyl gold–carbene intermediate **I**, which undergoes ring expansion to form allyl cation **II**. This is followed by loss of a proton and demetalation to give dienes **2a**, **2b**, and **3** (483) (Scheme 59).

6. Cyclopentadienyl compounds of the form $CpCo(CO)_2$ are also known to convert 1,*n*-enynes into the tranditional 1,3-diene cycloisomerization products (486, 487). Remarkably, the metathesis reaction is independent of tether length between both unsaturated functional groups, and leads to five-membered ring carbocycles exclusively (Scheme 60).

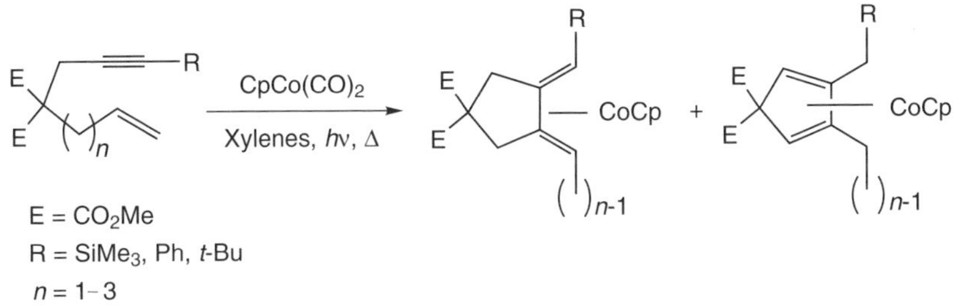

E = $CO_2Me$
R = $SiMe_3$, Ph, *t*-Bu
*n* = 1–3

Scheme 60. Traditional 1,3-diene cycloisomerization products upon metathesis by $CpCo(CO)_2$.

7. Titanium(II)-mediated intramolecular enyne cycloisomerization is particularly effective as a new approach to cyclic siloxanes (Scheme 61) (488). Similarly, silyl-, stannyl-, or sulfur-containing functional groups (e.g., sulfides, sulfoxides, and sulfones) are important electronic and protecting groups in organic synthesis. Enynes bearing these functionalities can be readily cyclized, like their traditional carbon counterparts, by stoichiometric addition of specific group 4(IVB) transition metals, in addition to the more classic late metals (489–492). Reaction of the Ti(IV) alkoxide $Ti(O\textit{i}\text{-}Pr)_4$ with the Grignard *i*-PrMgCl reductant (489, 491) leads to transient formation of a Ti(II) species of the form ($\eta^2$-alkyne)$Ti(O\textit{i}\text{-}Pr)_2$ that converts to the titanacycle, which is protonated–deuterated to yield the cyclic, demetalated product (493) (Scheme 62).
8. Iridium-catalyzed intramolecular 1,*n*-enyne metathesis has been studied as a unique tool for the synthesis of various types of cyclic compounds. Reactions of this type depend on both the structure of substrates and the nature of catalyst systems used (411). Recently, the cycloisomerization of various 1,6-enynes have been shown to be catalyzed by $[Ir(cod)Cl]_2$/dppf (494). These reactions are highly stereoselective, and generate the (*Z*)-isomer preferentially over the (*E*)-isomer (Scheme 63). The proposed mechanism (Scheme 64) involves oxidative cyclization of the enyne at Ir(I) to give the trivalent iridacyclopentene. The intermediate undergoes β-hydride elimination to give the irida-1,3-diene, which experiences steric repulsion between the metal fragment and the cis substituent on the

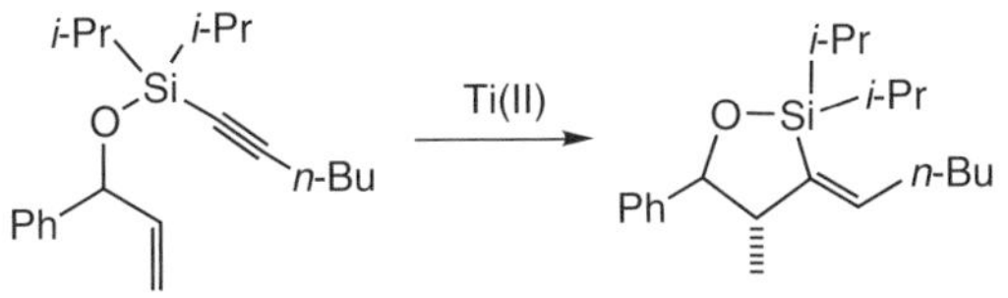

Scheme 61. The Ti(II)–enyne cycloisomerization as a route to cyclic siloxane.

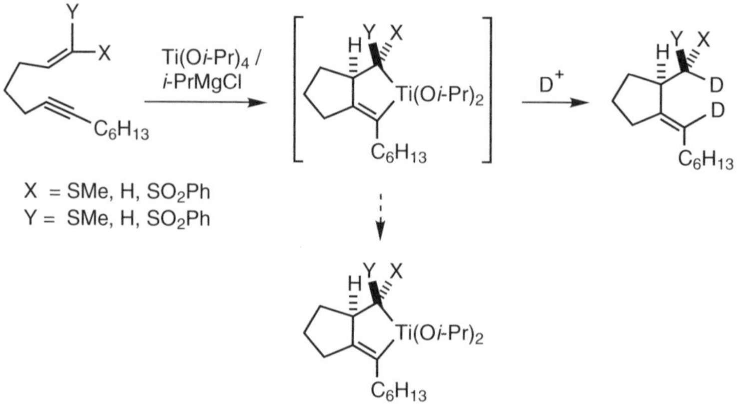

X = SMe, H, $SO_2Ph$
Y = SMe, H, $SO_2Ph$

Scheme 62. Enyne cyclization of sulfur-containing substrates by alkyoxy–titanium species.

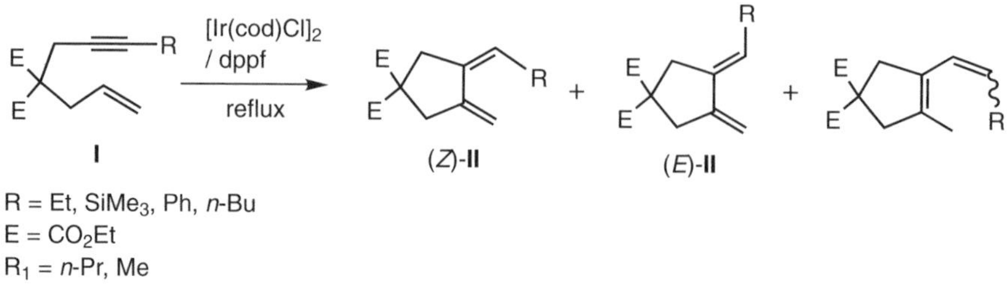

R = Et, $SiMe_3$, Ph, *n*-Bu
E = $CO_2Et$
$R_1$ = *n*-Pr, Me

Scheme 63. Stereoselective enyne metathesis by iridium to form the desired 1,3-diene.

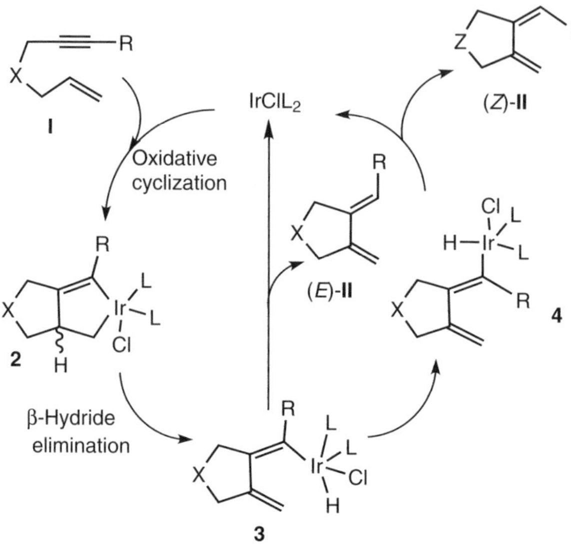

Scheme 64. Mechanism for oxidative enyne cycloisomerization at Ir(I).

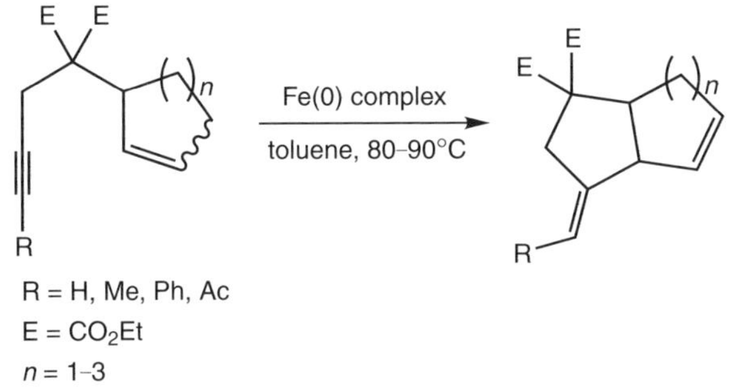

Scheme 65. Iron(0) salt $[CpFe(C_2H_4)_2][Li(tmeda)]$ generation of 1,4-dienes by enyne cycloisomerization.

alkene. Steric repulsion is relieved via formation of a zwitterionic carbene species and isomerization to give (*Z*)-**II**. Competitive reductive elimination generates the (*E*)-**II** isomer, and thus, a mixture of (*E*)- and (*Z*)-**II** is obtained.

9. Very recently, Fürstner et al. (495) reported the cyclization of acyclic 1,6-enynes catalyzed by the 18-electron half-sandwich iron(0) salt $[CpFe(C_2H_4)_2][Li(tmeda)]$ (Scheme 65). The 1,4-diene products are tolerant of alkene ring size ($n = 1–3$) and substitution at the alkyne.
10. Murai and co-workers (442) used $GaCl_3$ to promote 1,6-enyne cyclization producing 1-vinylcycloalkenes in >75% yield (442) (Scheme 66). The $GaCl_3$-catalyzed route offers several advantages: First, enynes bearing two substituents at the olefin undergo efficient bond rearrangement, and second, enynes that are monosubstituted at the terminal olefinic carbon undergo a stereospecific skeletal reorganization with respect to the geometry of the olefin. The reaction mechanism is proposed to involve production of a cyclobutene intermediate that subsequently undergoes conrotatory ring opening, thereby explaining retention of the olefin geometry. The electrophilic addition of $GaCl_3$ to the alkyne gives the zwitterionic vinyl–gallium species, (350) which is stabilized by olefin ring

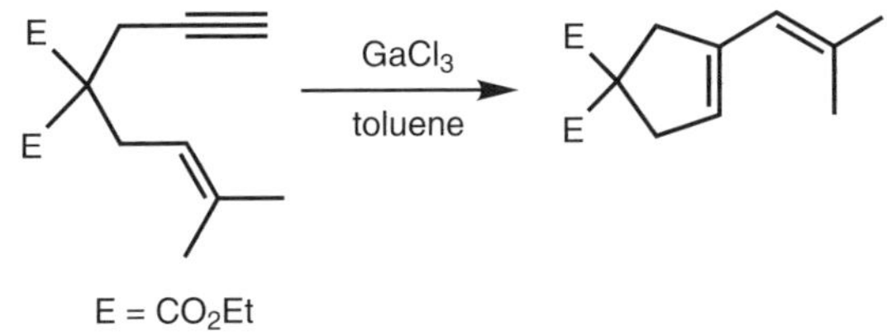

Scheme 66. The Ga(III)-mediated 1,6-cycloisomerization to form the 1,3-diene.

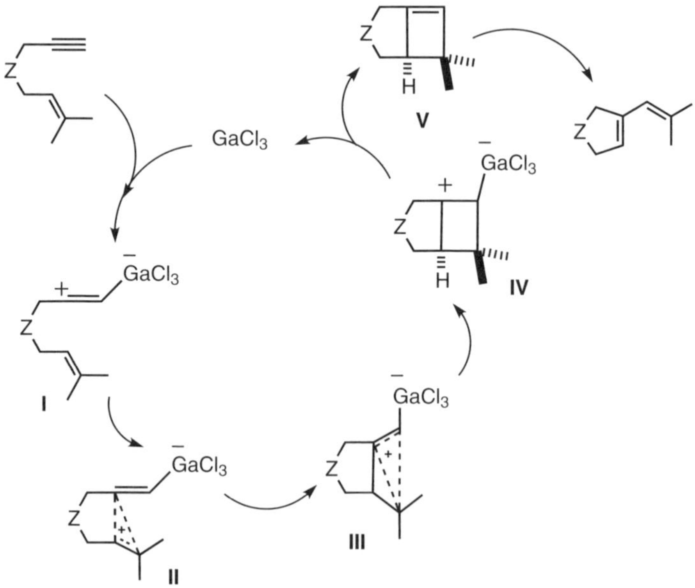

Scheme 67. Proposed mechanism for the Ga(III)-catalyzed 1,6-enyne metathesis.

closure (445) (Scheme 67). The dotted triangle and positive charge represent a cationic three-center two-electron bond. After formation of the cyclopentane ring, the newly formed carbocation is stabilized by the Ga-substituted olefin. The formation of the second C–C bond results in the production of the cyclobutane ring, which is followed by elimination of $GaCl_3$ to yield the cyclobutene intermediate. Subsequent ring opening affords the 1-vinylcyclopentene (412).

## C. Intermolecular Enyne Cross-Metathesis

In contrast to the reliable, high-yielding, and selective intramolecular 1,*n*-enyne metathesis reaction, intermolecular enyne metathesis (enyne cross-metathesis) has seen less use in the synthesis of complex molecules due to limited selectivity, despite its potential in fragment-coupling processes (404). The most common use of intermolecular enyne metathesis employs ethylene as the alkene component, providing a particularly convenient method for the production of 2,3-disubstituted butadiene systems (or 2-substituted butadienes in the case of terminal alkynes), which are important and synthetically useful structural motifs (Scheme 68). This protocol was introduced by Mori, who subsequently applied it to an expedient synthesis of the complex product anolignan (Scheme 69) (428, 496). The cross-metathesis of the internal alkyne was induced with the

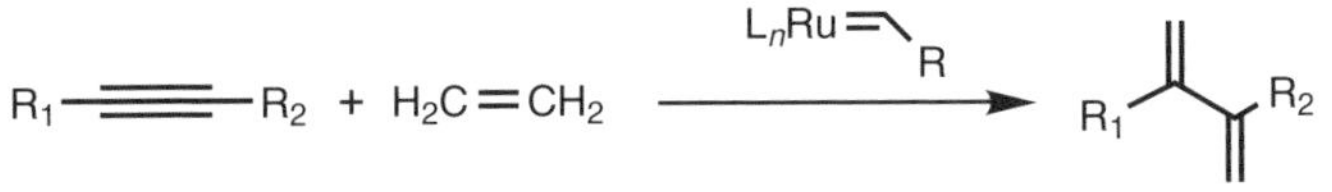

Scheme 68. Use of Ru–carbene catalyst for enyne cross-metathesis to generate 2,3-butadienes.

required stereochemistry by treatment with Grubbs' catalyst (10 mol%) in toluene at 80°C under ethylene at atmospheric pressure in 86% yield. The first enyne-cross-metathesis reactions of substituted alkenes to afford acyclic 1,3-disubstituted butadiene systems were reported by the Blechert and co-workers in 1997 (497).

Cross-enyne metathesis has also been applied to solid-phase synthesis by immobilizing either the alkyne or the alkene (498, 499). By combination with a Diels–Alder reaction, this process has been used to create a combinatorial library of various six-membered ring structures (500). Within this theme, intermolecular enyne metathesis is the integral synthetic method used to produce novel amino acid derivatives (501, 502). To this end, the aromatic ring of highly substituted phenylalanines was synthesized through enyne cross-metathesis and cycloaddition.

Transition metal catalyzed intermolecular carbocyclization has been used in the construction of six-membered ring systems (10, 352) and provides a powerful approach to the construction of complex polycyclic systems. The rhodium-catalyzed intermolecular [2 + 2 + 2] carbocyclization of heteroatom-tethered 1,6-enynes with symmetrical and unsymmetrical alkynes affords the corresponding bicyclohexadienes in a highly efficient and regioselective

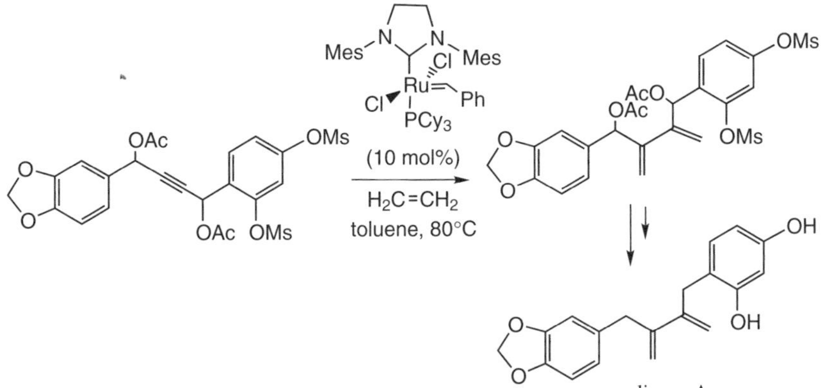

Scheme 69. Second-generation Grubbs' catalyst applied to intermolecular enyne cross-metathesis for the preparation of anolignan. (Mes = Mesityl; Cy = cyclohexyl; Ms = mesylate.)

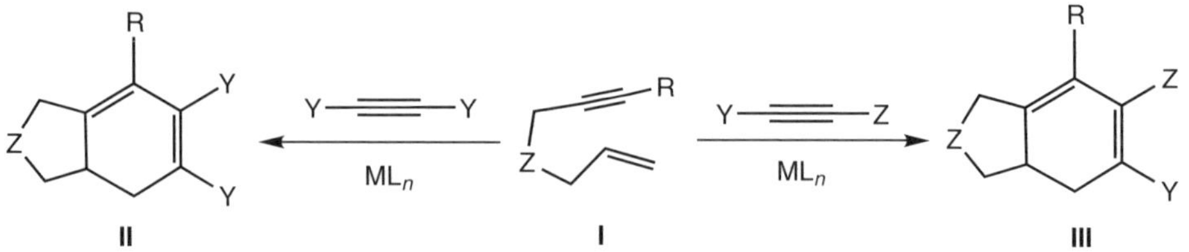

Scheme 70. The Rh-catalyzed [2 + 2 + 2] carbocyclization of 1,6 enynes with symmetric and unsymmetrically substituted alkynes.

manner, respectively (503, 504) (Scheme 70). Mechanistically, the terminal acetylenic C–H bond of the tethered enyne undergoes oxidative insertion to afford hydrido-rhodium σ-complex (ii, Scheme 71). Coordination of the free alkyne followed by hydrometalation is proposed to generate the σ/π complex, en route to the key metallacyclopentadiene (505, 506). Intramolecular migratory insertion across the tethered alkene, followed by a reductive elimination of metallacycle then forms the [2 + 2 + 2] carbocyclization adduct (**vi**, Scheme 71). The Rh(I) methodology also serves as as a new route to the synthesis of chiral compounds possessing a quaternary carbon stereocenter. The asymmetric carbon center is generated by the oxidative coupling of enynes, followed by carbonyl insertion and reductive elimination of the metal catalyst [Scheme 72(*a*)]. The enantioselective coupling of an enyne possessing a

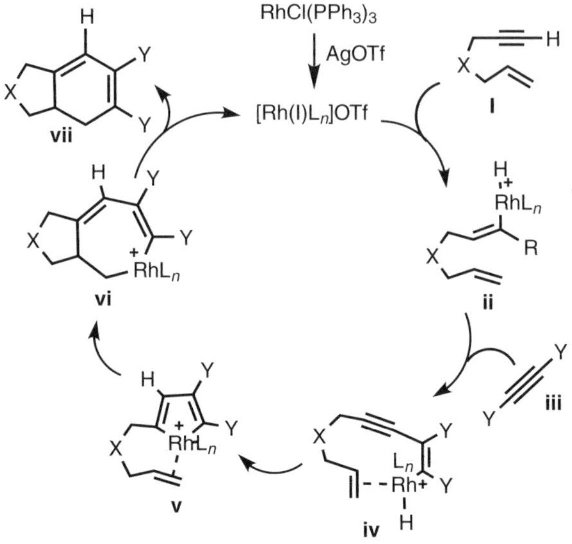

Scheme 71. The Rh-catalyzed cross-enyne metathesis for the preparation of bicyclic 1,3-dienes.

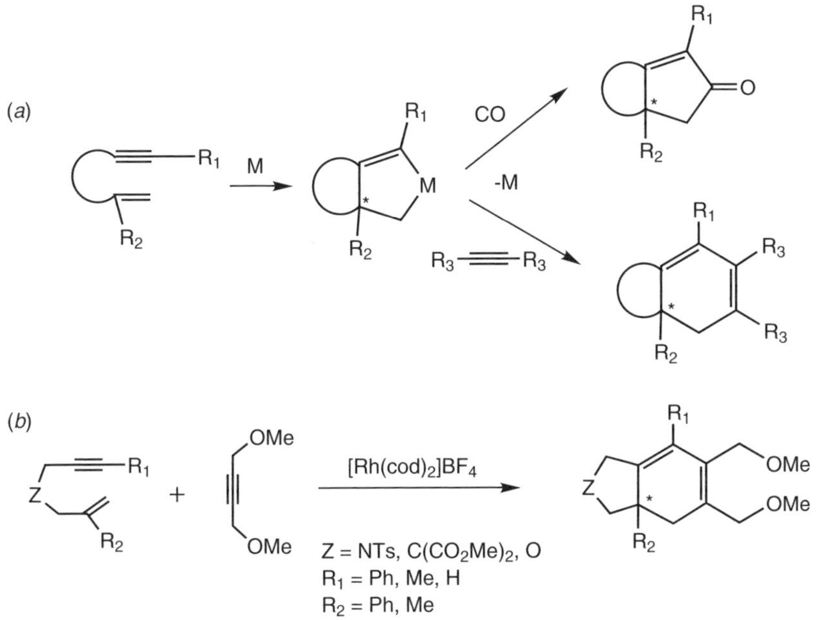

Scheme 72. (*a*) Illustration and (*b*) example of enantioselective enyne-cross metathesis using Rh(I) as a catalyst.

disubstituted olefin with alkyne insertion could provide a chiral bicyclic 1,3-diene with a quaternary carbon stereocenter (371) [Scheme 72(*b*)].

Inter- and intramolecular palladium-catalyzed [4 + 2] homocycloaddition reactions (i.e., benzannulation reaction) of conjugated enynes and bis-(enynes) have been reported by Yamamoto and co-workers (10, 507–509). Reaction of two-substituted conjugated enynes in the presence of $Pd(PPh_3)_4$ (2 mol%) smoothly undergoes [4 + 2] intermolecular homocycloaddition to afford 1,4-disubstituted benzenes in good yields (10) (Scheme 73).

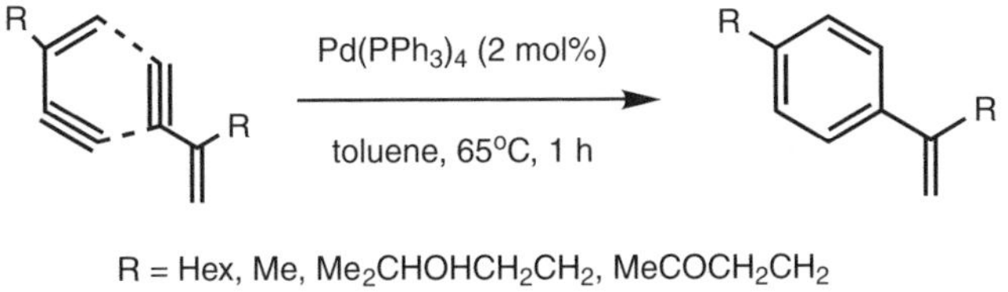

Scheme 73. Formation of disubstituted benzenes by [4 + 2] enyne cross-metathesis at Pd(0).

## D. Tandem Enyne Metathesis

An initial ring-closing enyne metathesis followed by second ring-closing metathesis, is known as a tandem or cascade metathesis. Metathetical transformations of this type are very powerful for the construction of bicyclic ring systems from acyclic starting materials. The first examples of tandem enyne metathesis of dienynes to form carbobicycles was reported by Grubbs in 1994, where the product from the initial enyne metathesis reaction undergoes ring-closing diene metathesis with another alkene in the substrate to produce a fused bicycle (427). The reaction mechanism likely involves initial alkene metathesis and then ring-closing metathesis onto the hindered internal alkyne (Scheme 74). The resulting vinyl carbene participates in ring-closing alkene metathesis to form the bicycle. In this way, the [4.3.0] and [5.4.0] bicyclic rings can be formed in very good yields (427, 510). Within this general theme, Blechert and co-workers employed tandem diyne–alkene cross-metathesis to produce conjugated trienes (511). The first-generation Grubbs' carbene promotes the reaction between a threefold excess of alkene and terminal diynes (Scheme 75). In 2003, Liu and co-workers (512) reported formation of dioxabicyclic ring systems using tandem enyne metathesis. Also, Lee et al. (513) developed a three-component tandem metathesis/Diels–Alder reaction. The initial ring-closing enyne metathesis is performed using the second generation Grubbs' catalyst and produces a cyclic diene that undergoes cross-metathesis and a subsequent cycloaddition to form functionalized tricyclic systems (Scheme 76).

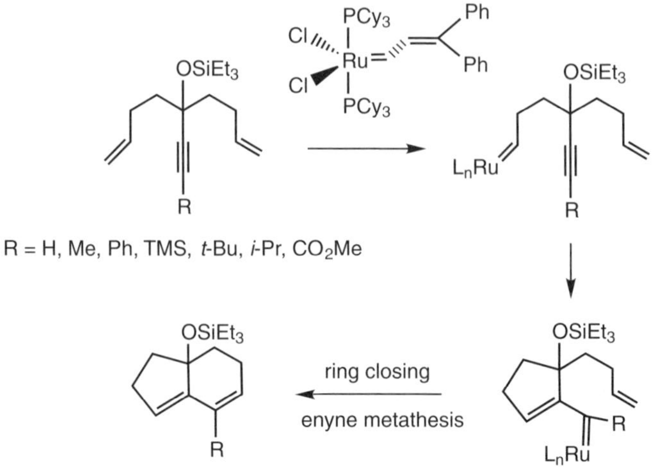

R = H, Me, Ph, TMS, *t*-Bu, *i*-Pr, $CO_2Me$

Scheme 74. Tandem ring-closing enyne metathesis using first generation Grubbs' catalyst to form bicyclic ring systems.

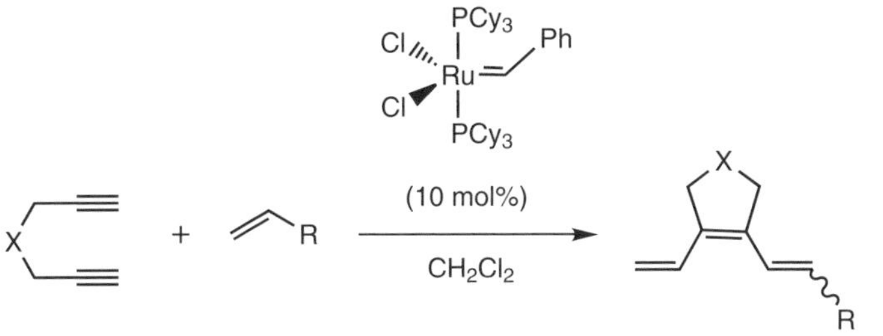

X = NTs, NAc, O, C(COO)$_2$Et
R = $CH_2SiMe_3$, $(CH_2)_3OCPh_3$, $CH_2OAc$, $CH_2Ph$, $CH_2OTBDMS$

Scheme 75. Tandem diyne-ene cross-metathesis to yield cyclic trienes via first generation Grubbs' catalyst.

## V. METALLOENEDIYNE FRAMEWORKS: METAL-MEDIATED BERGMAN CYCLIZATION

### A. Synthesis and Bergman Cyclization of Organometallic σ/π Enediynes

From the preceding sections describing advances in metal–diyne and enyne coordination and reactivity, it is evident that metal–enediyne (metalloenediyne) complexes might be expected to form readily and to lead to cycloisomerization, a reaction established in the organic literature three decades ago by Masamune (514) and Bergman and their co-workers (515, 516). In fact, organometallic reagents have been employed for both activation and stabilization of enediyne motifs by both σ and π complexation mechanisms. Recent reviews highlighting both the role of metals in Bergman cyclization with respect to developments in organic chemistry, as well as the emerging focus on metalloenediyne complexes

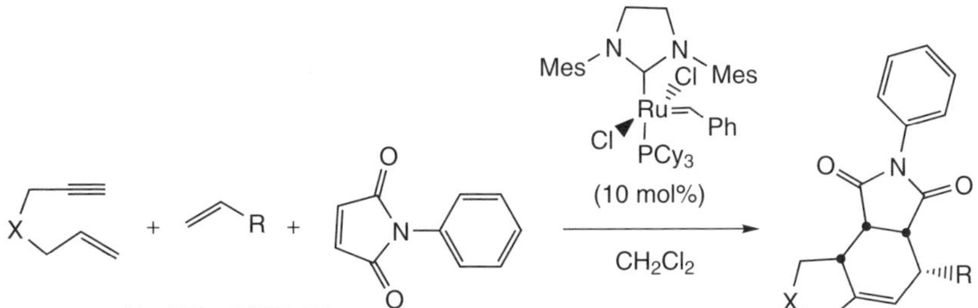

X = NTs, $C(CO_2Et)_2$
R = Bu, $(CH_2)_2Br$, Bn, Ph

Scheme 76. A three-component tandem metathesis/Diels–Alder reaction promoted by second generation Grubbs' catalyst to form functionalized tricycles.

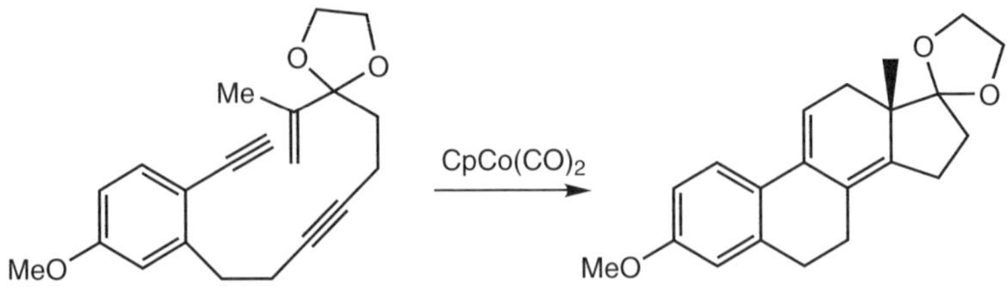

Scheme 77. Intramolecluar, metal-mediated [2 + 2 + 2] cycloaddition of an enediyne to form polycycles of medical interest.

in organometallic and inorganic chemistry have been recently published (25–27) and serve as supplementary reference material. This section focuses on only metal-mediated cyclization strategies.

1. Metals have been long recognized as an effective tool in organometallic chemistry for promoting basic reactivity, stability, and syntheses of general diyne–ene motifs. Vollhardt and co-workers (517) were the first to show that metal complexes, such as $\eta^5$-$CpCo(CO)_2$, could bind cyclobutadienylene–enediynes and promote isomerization of the enediyne framework. Moreover, $\pi$-complexation of synthetic enediyne precursors by $\eta^5$-$CpCo(CO)_2$ can also promote [2 + 2 + 2] cycloaddition reactions as a strategy for the total synthesis of polycycles of medicinal interest (Scheme 77) (518, 519). Similarly, Magnus et al. employed $\eta^2$-$Co_2CO_6$-alkylidyne complex formation to stabilize enediyne precursors (**I, II**) during the synthesis of bicyclic natural product analogues (Scheme 78)

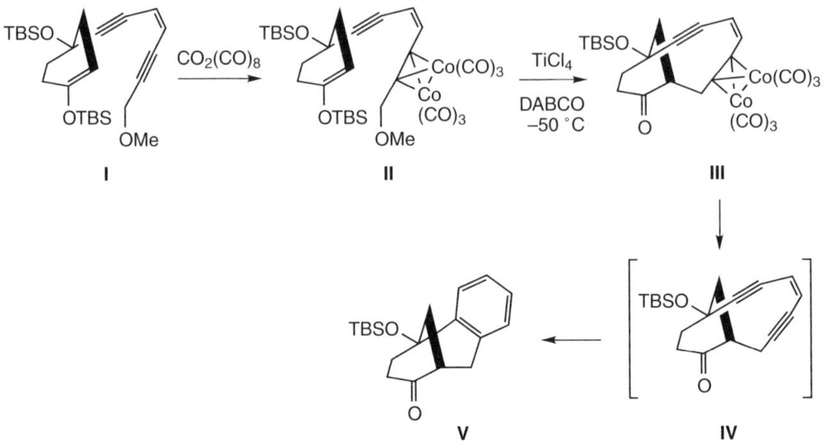

Scheme 78. The $\eta^2$-$Co_2CO_6$-alkylidyne stabilization of **II** toward Bermgan cyclization. (TBS = TBDMS = *tert*-trelhyldemethylsiyl.)

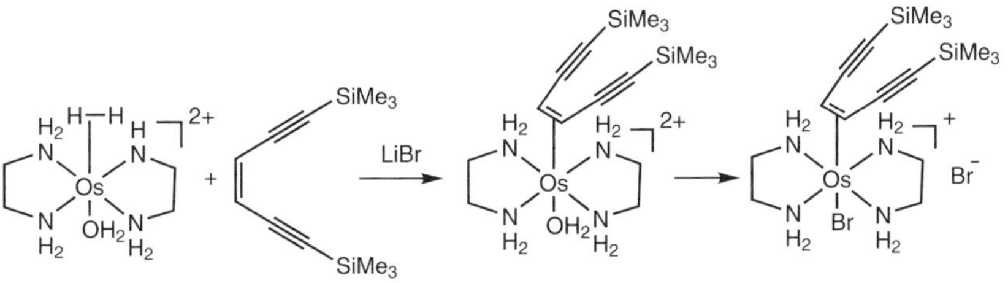

Scheme 79. Synthesis of a water soluble, Os(II) metal–ene–diyne complex.

(378–383, 520, 521). Subsequent ring closure (**III**) and oxidative decomplexation of the $\eta^2$-$Co_2CO_6$ cluster frees the strained enediyne unit (**IV**), leading to facile Bergman cyclization (**V**).

In addition to metal-yne $\pi$ complexation, Pu and co-workers (522) synthesized and crystallographically characterized a water soluble, metal-ene $\pi$-complex by reacting $[Os(en)_2H_2H_2O]^{2+}$ with a TMS-substituted enediyne (TMS=Tetramethylethylene). The X-ray structure shows $\eta^2$-enediyne complexation to osmium (Scheme 79), however, no metal-promoted Bergman cyclization reactivity is observed, and only enediyne dissociation occurs upon oxidation or photolysis of the metal complex.

Additional developments in the synthesis of enediyne motifs, beyond the now well-established Pd-catalyzed routes, have recently been reported. Casey et al. (377, 399) demonstrated that *cis*-enediyne complexes can be readily obtained via dimerization of $Cp(CO)_2Re$–alkynylcarbene complexes at 100°C (cf. Fig. 27), as well as from the addition of $\alpha,\omega$-diynes to manganese carbyne complexes that rearrange to enediynes below room temperature. In parallel, Cummins and co-workers have demonstrated that sequential reductive coupling of Mo(IV) acetylides and alkyne metathesis can also be used as a novel route to both (*E*)- and (*Z*)-enediyne constructs (523).

Despite the early work of Vollhardt and Magnus, the potential of organometallic enediyne complex formation for inhibiting or activating diradical formation was not fully appreciated until several years later. In 1995, Finn and co-workers (524) realized that although cycloaromatization required heating, allene-ene-ynes often undergo $\sigma,\pi$-diradical ring closure more rapidly by the Myers–Saito type mechanism. Using the well-established syntheses of metal–alkylidenes from terminal alkynes, rapid cycloaromatization of a benzannulated enediyne was shown to proceed via a $RuCp(PMe_3)_3$ vinylidene-ene-yne species (Scheme 80). Analogously, rhodium(I)-catalyzed reactions of acyclic enediynes via the Myers–Saito (525–528) cycloaromatization mechanism have been reported to proceed in good yields. The mechanism involves cyclization of the *in situ* formed rhodium–vinylidene–allene–enyne species

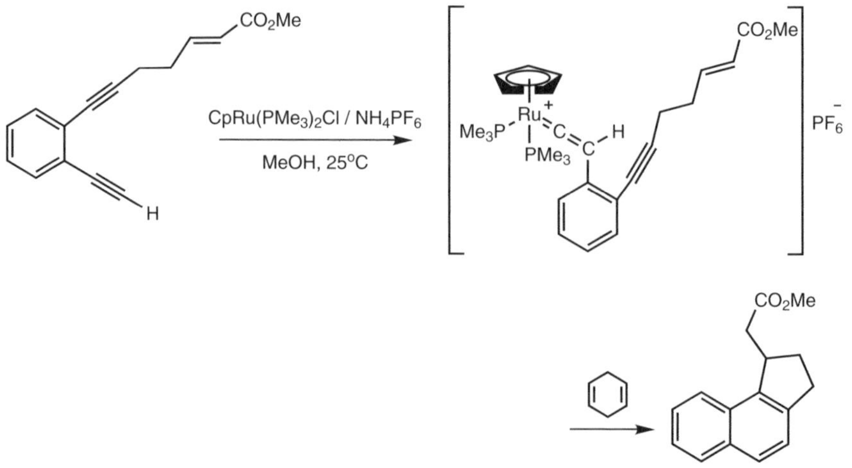

CO2Me
H
CpRu(PMe3)2Cl / NH4PF6
MeOH, 25°C
Ru+
Me3P
PMe3
C
H
CO2Me
PF6-
CO2Me


Scheme 80. Cyclization of asymmetric enediynes via a Ru–allene–enyne intermediate.

which proceeds to the seven- or eight-membered rhodacycle intermediate prior to release of the product from catalyst (529, 530) (Scheme 81).

In addition to the metal–alkylidene σ-complex activation, O'Connor and coworkers (531–533) elegantly demonstrated that simple enediynes will undergo

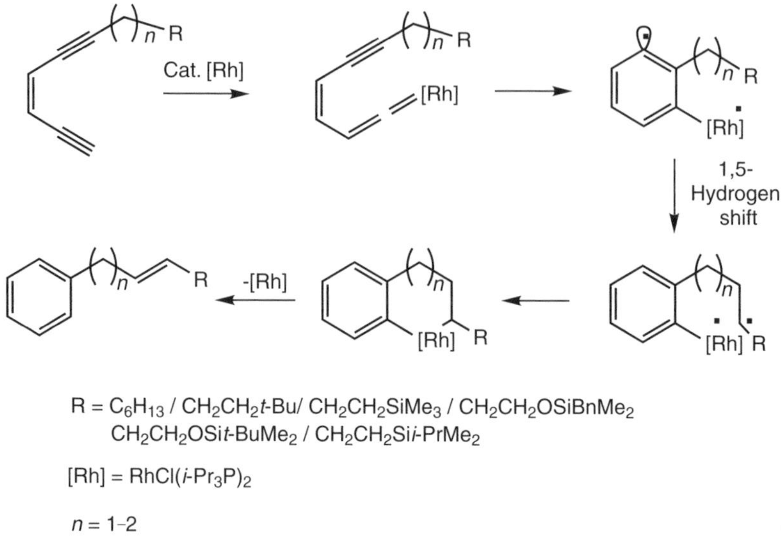

R
n
Cat. [Rh]
[Rh]
1,5-
Hydrogen
shift
-[Rh]
R = C6H13 / CH2CH2t-Bu/ CH2CH2SiMe3 / CH2CH2OSiBnMe2
CH2CH2OSit-BuMe2 / CH2CH2Sii-PrMe2
[Rh] = RhCl(i-Pr3P)2
n = 1-2


Scheme 81. The Rh(I)-catalyzed Myers–Saito enediyne cyclization via a metal–vinylidene–allene–enyne intermediate.

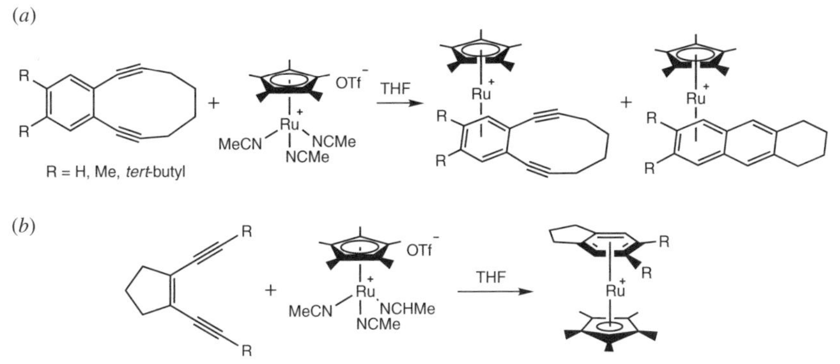

(*a*)

R = H, Me, *tert*-butyl

(*b*)

Scheme 82. Bergman cyclization of (*a*) cyclic and (*b*) acyclic enediynes by Ru–alkyne π complexation.

direct Bergman cyclization upon π-complexation by $[\eta^5\text{-}(C_5Me_5)Ru(MeCN)_3]OTf$. Reaction of cyclic enediyne **I** with results in either ruthenium binding to the arene and a stable π complex, or ruthenium–alkyne complexation and facile Bergman cyclization [Scheme 82(*a*)] (532). Similarly, the same ruthenium coreagent is also capable of cyclizing acyclic enediynes in 64–88% yields at ambient temperature [Scheme 82(*b*)]. The facile thermal reactivity of these systems engenders an interest in evolving the reaction into a catalytic cycle. The main obstacle to catalysis here is release of the cyclized product from the metal. To this end, O'Connor and co-workers (534) further developed the cycloaromatization reaction with iron analogue $[(\eta^5\text{-}C_5Me_5)Fe(MeCN)_3]PF_6$ and have cleverly employed photochemical liberation of the arene product from the iron center, a process that is significantly more facile than in the case of ruthenium. The involvement of photoexcitation in the reaction scheme provides a novel method for cycloaromatization of trans-enediynes as well, simply by first promoting photochemical trans-to-cis isomerization. These constructs are normally inert toward direct thermal and photochemical cyclization, however, in the presence of acetone-$d_6$ containing 1,4-cyclohexadiene, acyclic enediynes undergo a rapid (30-min) reaction with $[(\eta^5\text{-}C_5Me_5)Fe(MeCN)_3]PF_6$ to give in cyclized product 80–92% yield depending upon substituents. The catalytic cycle in (Scheme 83) represents the first transition metal catalyzed Bergman cycloaromatization reaction. These developments are significant in that they may lead to the potential for a suite of iron–arene complexes as air-stable reagents for the room temperature cycloaromatization of enediynes.

Activation of acyclic enediynes by ruthenium-catalyzed nucleophilic aromatization has also been reported (396). The importance of this cyclization is indicated by the regioselectivity of nucleophilic addition, which occurs only at

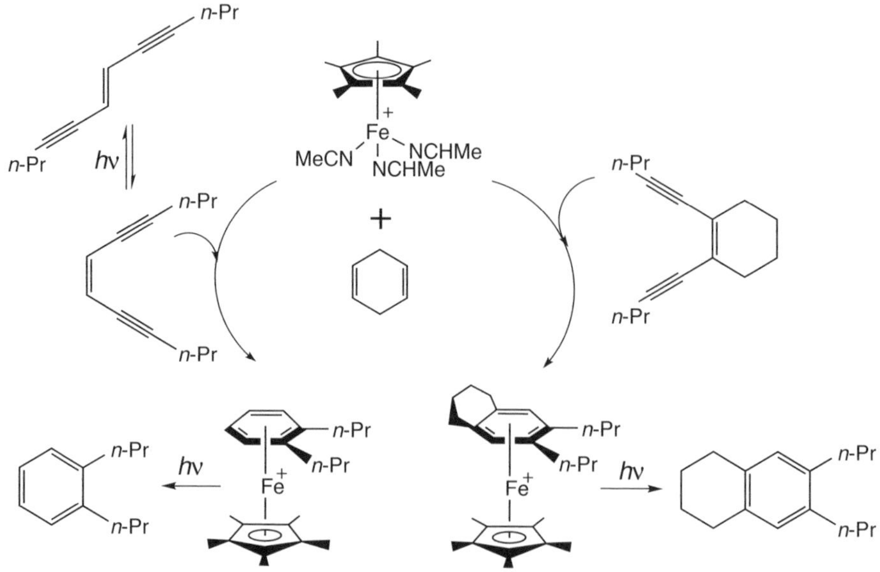

Scheme 83. Reaction scheme for photopromoted, catalytic acyclic enediyne cyclization.

the more electron-rich alkyne carbon. These cyclization reactions are highly regioselective and are compatible with a wide range of nucleophiles including water, alcohols, aniline, pyrrole, ethyl acetylacetonate, and dimethyl malonate, which reflects nucleophilic C–R (R = O, N, C) bond formation (Scheme 84) (396). The method is very useful because it provides easy access to functionalized aromatic compounds from readily available unfunctionalized enediynes. The mechanism of cycloaromatization is proposed to involve ruthenium-$\pi$-alkyne intermediates (396), analogous to those reported by O'Connor. Unfortunately, this catalytic cyclization failed to work for hydrohalogenation (NuH = HCl, HBr, HI) to give desired 1-halonaphthalenes, which are important building blocks in many synthetic applications.

Continuing in this theme, Liu and co-workers (535) demonstrated Pt-catalyzed hydrohalogenation of 1,2-bis(ethynyl)benzenes that efficiently yield 1-halonaphthalene products [Scheme 85(*a*)]. These catalytic reactions were performed via treatment of enediynes with aqueous HX (2 equiv) and $PtCl_2$ (5–10 mol%) in hot 3-pentanone (100°C, 2–10 h). Under these conditions, cyclization proceeds via platinum-catalyzed 6$\pi$-electrocyclization of 1,2-bis(1′-haloethenyl)benzene intermediates, which is the key step in the haloaromatization of 1,2-bis(ethynyl)benzenes. This reaction is mechanistically distinct from the authors' previous demonstration of nucleophilic, ruthenium-catalyzed aromatization of enediynes (396) [Scheme 85(*b*)]. The proposed mechanism identifies the origin of the regioselective halogenation of 1,2-bis(ethynyl)

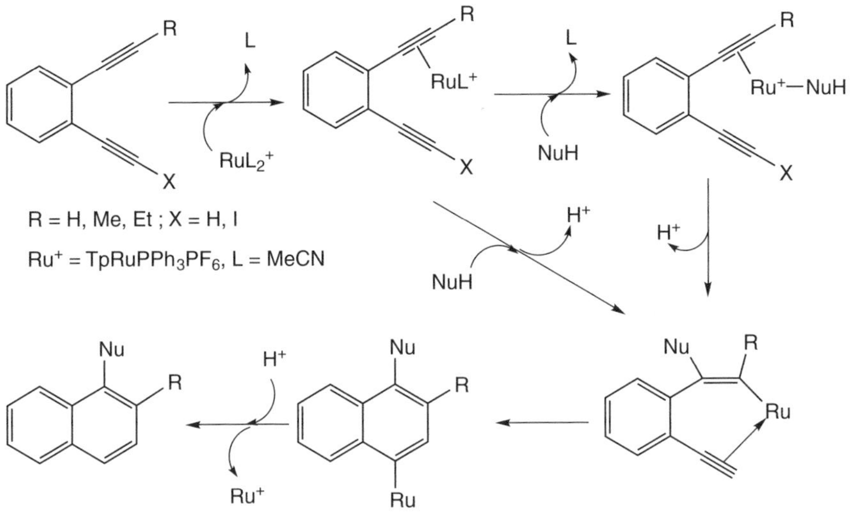

R = H, Me, Et ; X = H, I

$Ru^+ = TpRuPPh_3PF_6$, L = MeCN

Scheme 84. Regioselective nucleophilic enediyne cycloaromatization by Ru-catalysts.

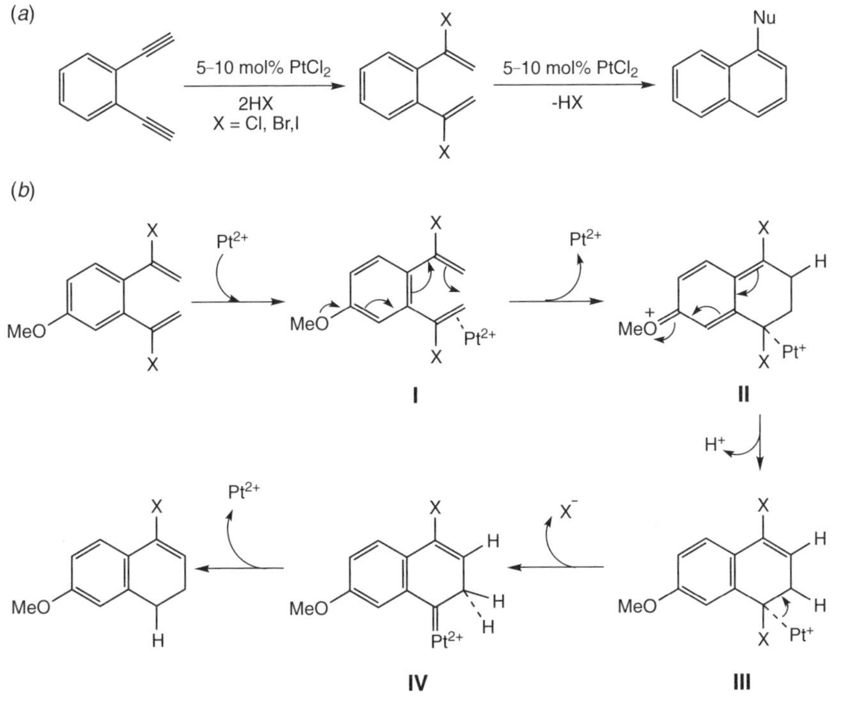

Scheme 85. (a) Overall reaction and (b) mechanism of Pt-catalyzed, hydrohalogenation-promoted enediyne cycloaromatization.

benzene bearing a methoxy group. In this case, $PtCl_2$ serves as a precursor for generation of a reactive platinum species that coordinates to the terminal vinyl group as the methoxy activates the olefin to generate the platinated cyclohexyl species **II** [Scheme 85(*b*)]. Subsequent deprotonation generates the aromatic intermediates **III** (536, 537), which coverts to the Pt–alkylidene **IV**, finally eliminating $Pt^{2+}$ to release the 1-halonaphthalene product through a 1,2-hydrogen shift (538–540) with regeneration of catalyst. This mechanism resembles that of the Pd(II)-catalyzed Cope rearrangement of acyclic 1,5-dienes (541).

Other, less traditional metals have recently been shown to promote Bergman cyclization of enediynes. In 1991, Tsuchiya and co-workers (542) explored the synthesis of benzotellurepines by the hydrotelluration of an enediyne. While the expected benzotellurepine was formed in good yield, the authors reported that naphthalene (i.e., the cycloaromatized product) is a major decomposition product of this reaction. It was not until very recently that Anthony and co-workers (543) reinvestigated tellurium-mediated cycloaromatization of acyclic enediynes bearing bulky TMS functionality at the alkyne termini. Such cyclization reactions are notoriously difficult and thermally unfavorable due to the steric clash of the TMS protecting groups in the product. Moreover, the stoichiometric $Na_2Te$ reagent is also a less robust reagent than desired. Anthony and co-workers (543) showed that desilylation/cycloaromatization of the silylated arenediyne occurs in good yield at 40°C, using stoichiometric Te(0) with *in situ* generation of the telluride (Scheme 86). The requirement for base promotes desilyation which drives the reaction to Bermgan product, a process accelerated in some cases by ultrasonic mixing.

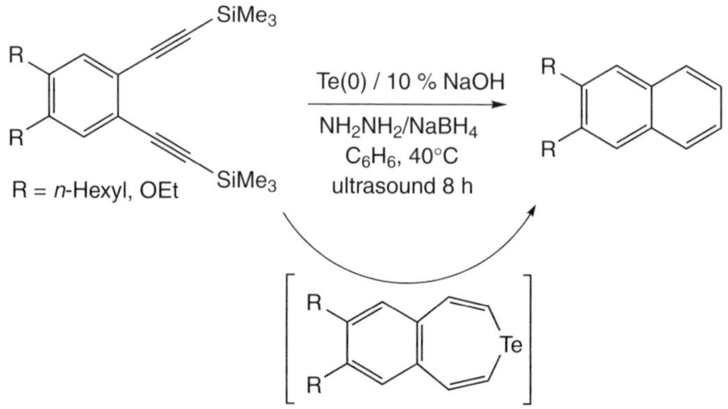

Scheme 86. Tellurium-mediated, desilylative cycloaromatization of benzannulated enediynes.

## B. Metalloenediyne Coordination Complexes and Their Thermal Reactivities

### *1. Geometric Factors Influencing Bergman Cyclization*

Coordination chemistry approaches to enediyne cyclization have been shown to be extremely dependent on the conformation of the bound enediyne ligand, which is strongly influenced by the geometry of the resulting transition metal coordination complex (25–27). The first proposals suggesting metal coordination could influence the conformation of the bound enediyne ligand were reported in 1995–1996 by Buchwald and co-workers (544) and König et al. (545, 546). Here the 1,2-bis(diphenylphosphinoethynyl)benzene ligand, which is stable in the solid state to highly elevated temperatures [Differential Scanning Caloumetry (DSC): 243°C], undergoes spontaneous Bergman cyclization in solution at ambient temperature upon binding $Pd^{2+}$ and $Pt^{2+}$ as the dichloride salts. In the solid state, these complexes are stable enough to interrogate, exhibiting DSC-determined cyclization temperatures of 61°C ($Pd^{2+}$) and 81°C ($Pt^{2+}$) for the neat powders (547). Analogously, the substituted bipyridyl–enediyne adopts a nonplanar ground structure due to free rotation about the C—C bond adjoining the rings. Complexation with $Hg^{2+}$ draws the chelating nitrogens into a planar geometry, thereby forcing the enediyne into a more planar and active conformation for Bergman cyclization. The resulting complex cyclizes in the solid state ~100°C lower in temperature (145°C) than the free ligand (237°C) (Scheme 87). Basak demonstrated analogous results with very stable, benzannulated en-based enediyne ligands upon Cu(II) and Ni(II) complexation (548, 549) (Fig. 29). Despite the apparent clarity of mechanism, the Buchwald study reported one negative result; the $Hg^{2+}$ compound did not

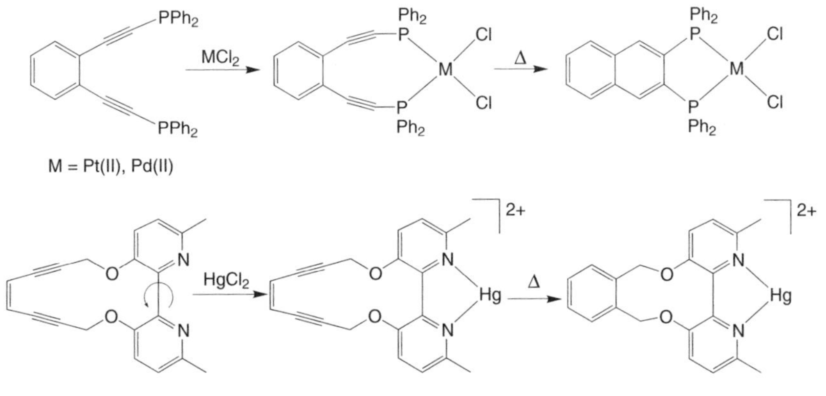

Scheme 87. Structural control of enediyne conformation using metal coordination.

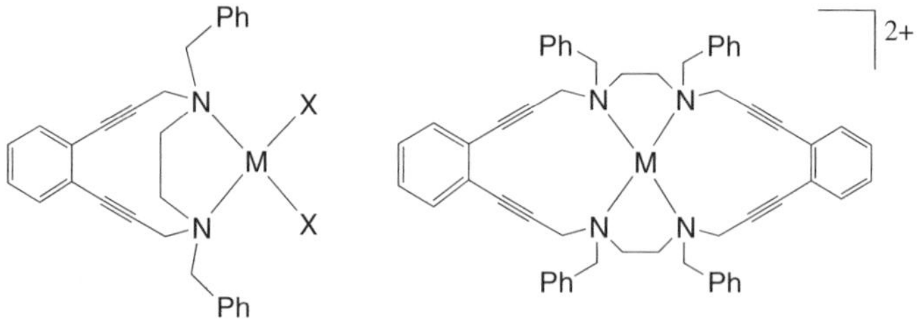

Figure 29. Ethylenediamine-based enediyne ligands that bind Cu(II) and Ni(II).

undergo Bergman cyclization due to an enhanced alkyne termini separation ($\geq$3.4 Å) as determined by MMX calculations.

The notion that metal geometry could be responsible for the disparate reactivities of the tetrahedral $d^{10}$ $Hg^{2+}$ versus square-planar $d^8$ $Pd^{2+}$ and $Pt^{2+}$ compounds was presented and demonstrated in four subsequent studies; two involving the 1,2-bis(diphenylphosphinoethynyl)benzene enediyne ligand with various $d^{10}$ metals (550, 551), and two describing a previously unreported bis(pyridyl) enediyne where the bond between the two rings (e.g., Scheme 86) was removed to allow for enhanced conformational flexibility and response to changes in metal geometry (552, 553). Formation of Pd(0) in the presence of 1,2-bis(diphenylphosphinoethynyl)benzene affords the tetrahedral, $d^{10}$ bis(enediyne) compound (Scheme 88). Similarly, reaction of the same ligand with Cu(I) or Ag(I) produces the isostructural $d^{10}$ analogues, all of which exhibit solid-state cyclization temperatures $>$ 220°C (551). Moreover, the thermal reactivities [Pd(0) (223°C) $\sim$ Cu(I) 228°C $<$ Ag(I) 269°C] loosely correlate with the alkyne termini separations, where Pd(0) (3.47 Å) $\sim$ Cu(I) 3.42 Å $<$ Ag(I) 3.61 Å, primarily due to the increased M–P bond lengths (2.3 vs. 2.5 Å) in the latter case.

The bis(pyridyl) enediyne analogues also show a pronounced reactivity based on the geometry of the metal. For the more flexible bis(pyridyl-oxy) enediynes (Scheme 89), the response to changes in metal geometry [tetrahedral Cu(I) vs. tetragonal Cu(II)] is $\sim$90°C, whereas their smaller, rigid analogous reflect the same differences in thermal reactivity ($\sim$170°C) that the 1,2-bis(diphenylphosphinoethynyl)benzene ligand exhibits when complexed to $d^{10}$ versus $d^8$ metals. This same geometry–ligand flexibility trend is also observed in macrocyclic and tetradentate metalloenediynes (Scheme 90) (554, 555). These enediyne ligands are prepared by sequential coupling of alkyne fragments to the desired enyne precursor allowing systematic variation of the size of the macrocycle–ligand and the nature–steric bulk of two of the metal-binding R groups (554). Subsequent complexation with Cu(II) or Zn(II) generates macrocyclic complexes that exhibit cyclization temperatures by DSC, which are sensitive to the structure of metal site. The more rigid macrocyles containing Cu(II) or Zn(II) tend toward

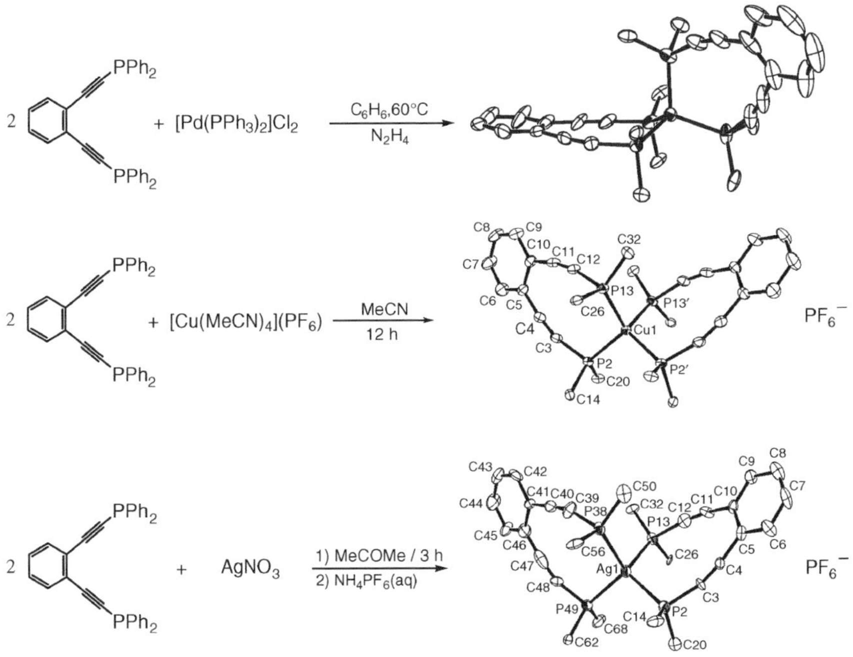

Scheme 88. Synthesis and X-ray structures of tetrahedral $d^{10}$ metalloenediynes.

more planar–tetragonal structures based on correlation to previous studies (552, 553) and undergo reaction between 120 and 140°C in the solid state. As the size of the macrocle increases, the Zn(II) derivative adopts a more distorted geometry, increasing the cyclization temperature to ~160°C, while the Cu(II) structure remains planar–tetragonal. The situation is geneally analogous for the tetradentate macrocycles with the exception that the increased steric bulk and rigidity of the two non-ethylenediamine R groups now causes the Zn(II) complexes to dimerize, stabilizing the DSC-evaluated cyclization temperatures (Fig. 30) to near 300°C (554).

The ability of this general suite of ligands to potentially build bridging structures, especially the rigid 3-pyridyl and quinoline R-groups constructs, coupled with the rigid 2-pyridyl trans-binding ligand reported by Thummel and co-workers (556), has contributed to the development of dimeric (557–559) and higher order (560–567) metal-containing architectures (Fig. 31). Extended chain coordination polymers (556, 563 ,568) [e.g., Fig. 31(*a*)] can also be formed over the length of the crystalline lattice. The bridging nature of the structures inhibits the Bergman reaction, at the same time allowing the ligand to effectively template complex multimetal assemblies for bio- and materials-related inorganic chemistry.

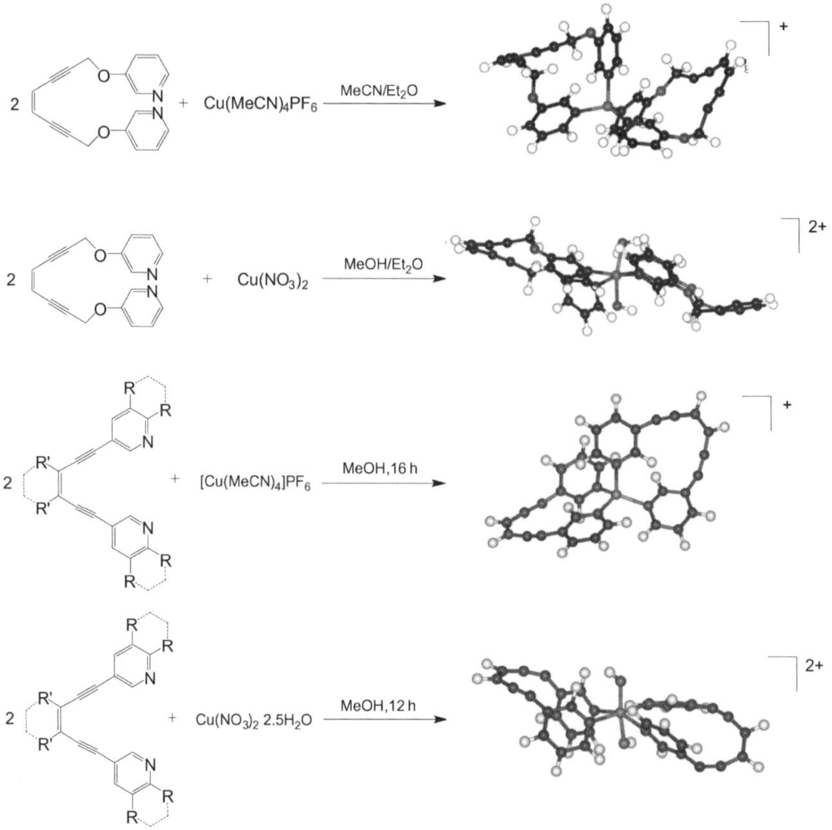

Scheme 89. Preparation of flexible and rigid bis(pyridyl) enediynes of Cu(I) and Cu(II).

This area is likely to continue to grow as new enediyne ligands are created for the original purpose of probing the cyclization reaction.

In light of the potential therapeutic value of enediynes due to their natural product origin and pronounced antitumor properties (27), interest in controlled, ambient or physiological temperature thermal activation has been a primary interest in the metalloenediyne area. Achievement of this goal requires balance between thermal stability of the structure and the ability to activate diradical formation. Based on the structure–activity correlation from the above ligand set, flexible pyridyl–quinoline designs bearing small metal-binding imine functionalities near the alkyne termini (Scheme 91) have been developed and reported to cyclize at or near ambient temperature upon addition of simple, innocuous metal salts (e.g., $MgCl_2$, $CaCl_2$, and $ZnCl_2$) (569, 570).

Despite the implication that trans metal geometries and commensurate ligand conformations are inherently stable, trans-to-cis conformational changes, have

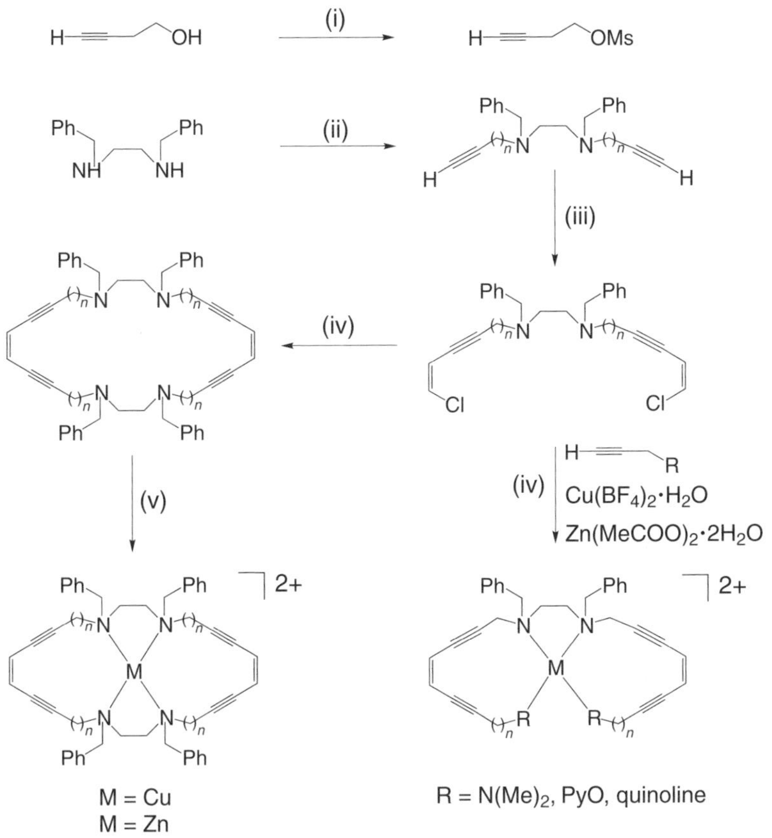

M = Cu
M = Zn

R = $N(Me)_2$, PyO, quinoline

(i) TEA (triethylamine), MsCl, methylene chloride, 20 °C; (ii) ROMs or propargyl bromide, $K_2CO_3$, DMF; (iii) *cis*-1,2-dichloroethylene, Pd(0), CuI, *n*-$BuNH_2$, benzene, 40–50°C; (iv) Pd(0), CuI, *n*-$BuNH_2$, benzene, 40–50°C; (v) $Cu(BF_4)_2 \cdot H_2O$, MeCN, 40°C / $Zn(MeCOO)_2 \cdot 2H_2O$, MeCN, 40°C

Scheme 90. Synthesis of macrocyclic and tetradentate metalloenediynes of Cu(II) and Zn(II).

been used to activate stable metalloenediyne starting materials (Scheme 92) toward Bergman cyclization (571, 572). Here, heating in solution to 75°C drives the Bergman reaction and leads to isolation of the fully cyclized ligand product via removal of the metal with excess pyridine or EDTA. Reconsititution of the metal in stoichiometric amounts leads to formation of the *cis*-metalloenediyne product, which is the expected structure based on the now structurally restricted ligand backbone. Mechanistically, this reaction is proposed to proceed via dissociation of the weakly coordinated thioether linkage (M–S ~ 2.3 Å) and rotation about the metal–nitrogen bond that relieves the constraint of the trans

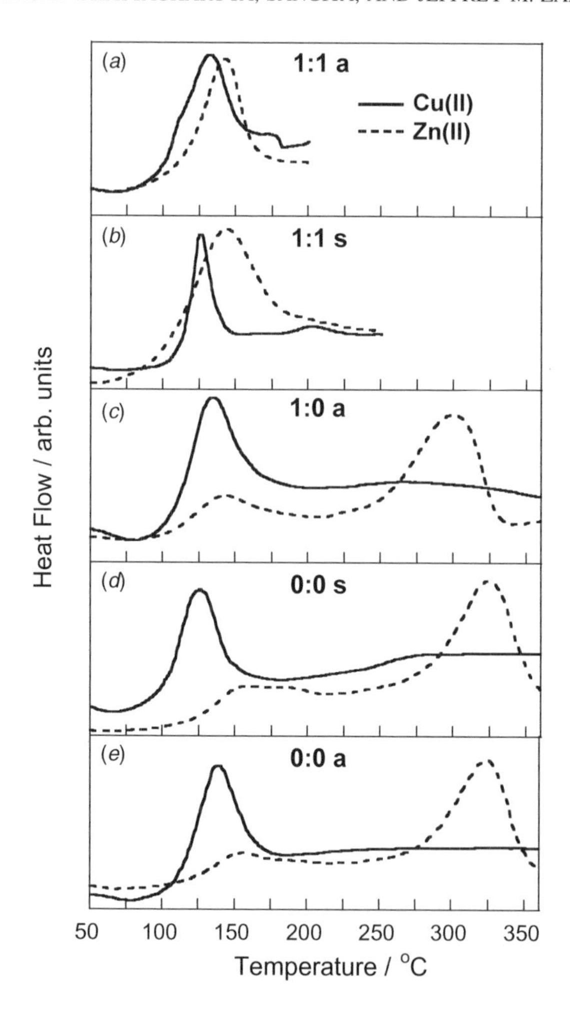


Figure 30. The DSC traces of mononuclear Cu(II) (*a–e*, solid trace) and Zn(II) (*a–b*, dotted) metallenediynes along with their dinuclear Zn(II) counterparts (*c–e*, dotted). The code in each spectrum denotes the number of additional carbons in the arms at the alkyne termini on each side of the molecule and **s** or **a** represents symmetric or asymmetric R group (*a*) 3-hydroxypyridine:dimethylamine; (*b*) 3-hydroxypyridine (*c*) 3-hydroxypyridine:quinoline; (*d*) quinoline; (*e*) 3-pyridine:quinoline) substitution.

structure and permits formation of a transient cisoid geometry that is downhill to product (Scheme 93) (571). This reaction can also be induced by addition of 1 equiv of a weak base (e.g., pyridine), which promotes displacement of the thioether by ligand substitution leading to cyclization, under these conditions at ambient temperature.

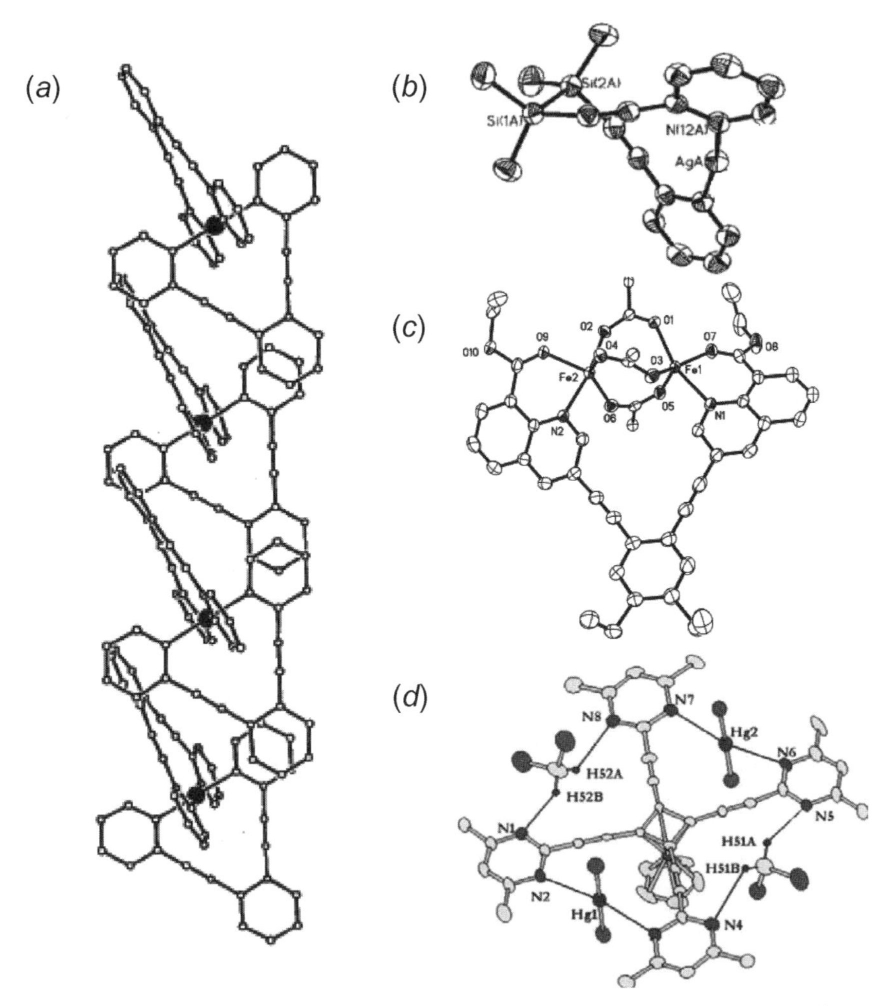


Figure 31. Employment of rigid, (*a–c*) bis(pyridyl)- and (*d*) bis(pyrimidyl)-derived enediyne scaffolds for construction of multinuclear metal–enediyne architectures. (*Note*: in *b*) the Si–Si single bond precludes categorization of this structure as a true "enediyne" framework, but the obvious structural analogy makes it worthy of mention in this context.)

## 2. *Electronic Structure Contributions to Thermal Bergman Cyclization*

Despite the now well-established geometric contributions to Bergman cyclization over the past 10 years, many fewer examples and much less is known about potential electronic contributions to the metal-mediated cyclization event. Although it is clear that the organometallic $\sigma$–$\pi$ complexation mechanisms

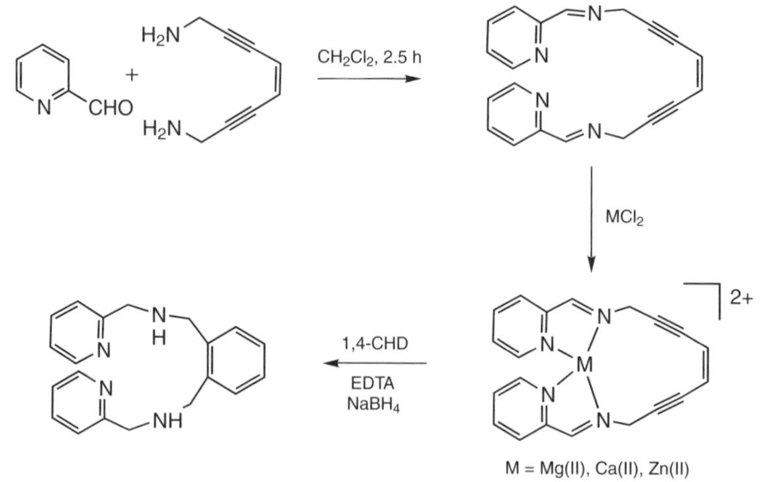

$CH_2Cl_2$, 2.5 h

$MCl_2$

1,4-CHD
EDTA
$NaBH_4$

M = Mg(II), Ca(II), Zn(II)

Scheme 91. Preparation and Bergman cyclization of tetradentate enediynes by addition of simple divalent metal salts at ambient temperature. (Ethylenediaminetetraaceticacid = EDTA.)

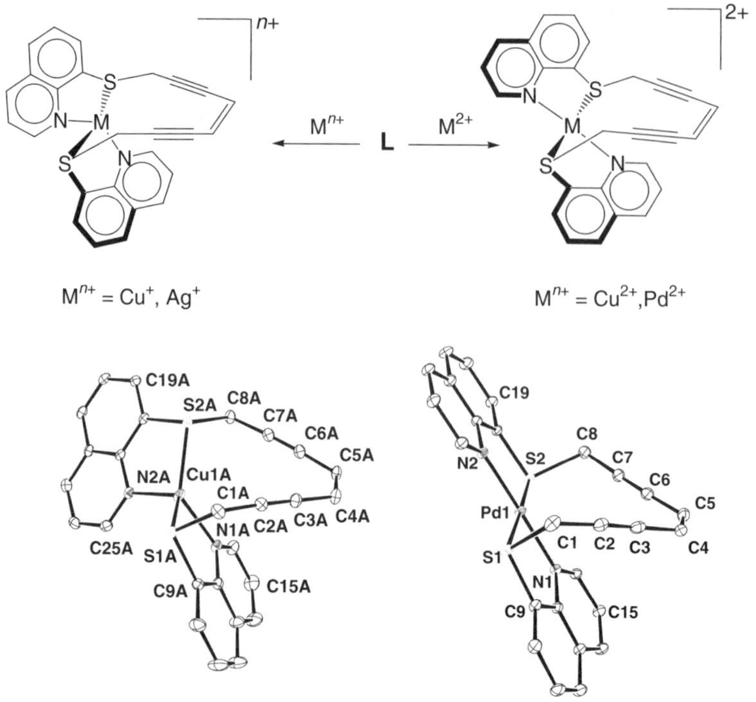

$M^{n+}$ L $M^{2+}$

$M^{n+} = Cu^+, Ag^+$

$M^{n+} = Cu^{2+}, Pd^{2+}$

Scheme 92. Synthesis and X-ray structures of *trans*-Cu(I)- and Pd(II)-mercaptoquinoline enediynes.

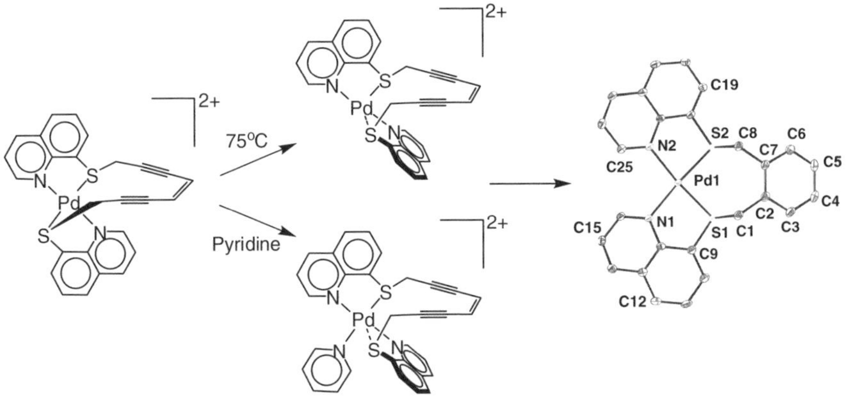

Scheme 93. trans-to-cis Geometric switch to Bergman cyclization.

contain an electronic component to their cyclization reactivity, this contribution has yet to be explored and systematically evaluated in detail for these systems. However, some information is available on the electronic structural contributions to cyclization of metalloenediyne coordination complexes (573, 574). Isolation of the electronic contribution from the normally dominant geometric effect has been achieved by using a rigid metal-chelating functionality that allows the enediyne motif to adopt a conformation independent of the geometry at the metal center. The first example is based on a tertiary en scaffold, where the additional enediyne linkage bridges the nitrogens supporting ancillary phenyl or methyl groups (26, 573). The X-ray structures of Cu(II), Zn(II), and Pd(II) analogues (Fig. 32) reveal that the tetragonal–planar geometries of the $d^9$ and $d^8$ systems, respectively, offset the ancillary halogen ligands in a plane perpendicular to the enediyne ligand. In contrast, the tetrahedral Zn(II) derivatives show a weak electronic interaction between the halogen lone pair and the developing C2•••C7 bond that tracks with the metal–halogen bond length (26, 573). The enhanced electronic repulsion due to the shorter M–L bond length assists in increasing the overall barrier to Bergman product.

The second example of an electronic contribution to metalloenediyne Bergman cyclization establishes the premise that much like for electron withdrawing effects of halogen substitution at the alkyne termini positions (575), introduction of high-valent metal in the same vicinity can promote a reduction in the thermal barrier to cyclization by LMCT polarization (574). Once again, elimination of the geometric contributions to Bergman cyclization by restricting the ligand conformation and response to metal geometry is required. This is achieved by preparation of the enedithiolate chelate with the enediyne motif appended by two adjacent thioether linkages (Scheme 94). The double bond of the enedithiolate isolates the geometric contributions of the metal center from the conformation

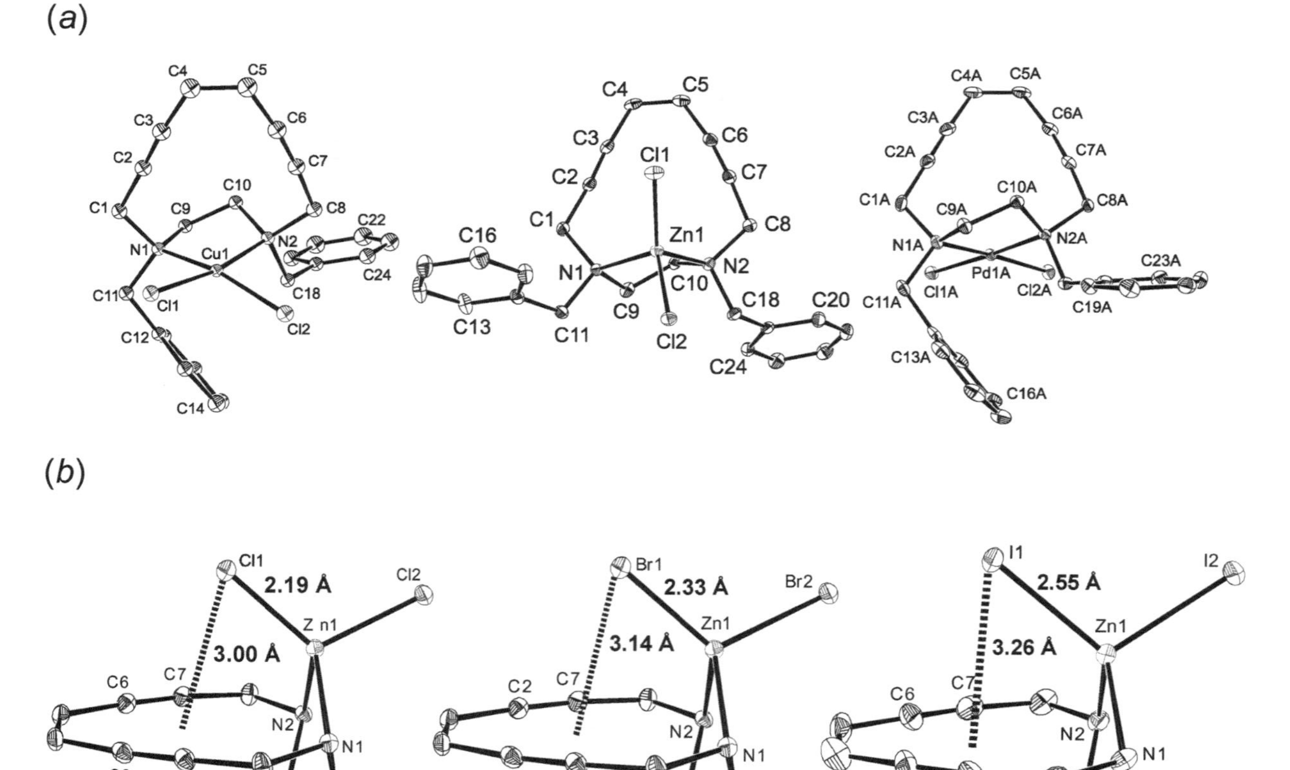


Figure 32. (*a*) Front and (*b*) side view of the X-ray structures of restricted $MCl_2$ metalloenediynes.

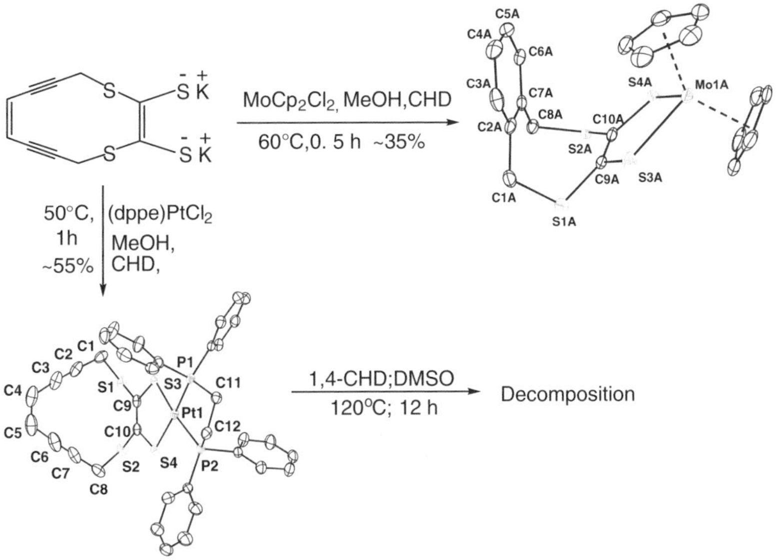

$MoCp_2Cl_2$, MeOH,CHD
60°C,0. 5 h ~35%

50°C, 1h ~55%
$(dppe)PtCl_2$ MeOH, CHD,

1,4-CHD;DMSO
120°C; 12 h
Decomposition

Scheme 94. Electronic activation of Bergman cyclization by LMCT polarization. (DMSO = Dimethyl sulfoxide.)

of the enediyne backbone, and at the same time, providing a good donor ligand set for LMCT polarization. Remarkably, complexation of the ligand salt with $MoCp_2Cl_2$ generates the fully cyclized product within 30 min at ambient temperature, while reaction with the softer $(dppe)PtCl_2$ motif leads only to stable complex formation and a considerably higher activation energy. Systematic evaluation of the binding–activation event by arylation and benzannulation (Scheme 95) reveals that indeed metal complex binding to the thiolates is responsible for the unusual reactivity. Direct evidence for the differential barrier heights was determined from solution kinetics, while detailed mechanistic insight was obtained from density functional theory (DFT) calculations of the potential surface to the diradical (574). The computations show that for the benzylated ligand, the electron–electron repulsion in the transition state is greater than for the Mo(IV) metalloenediyne complex (Fig. 33) due to the energetic removal of the sulfur lone-pair–diradical interaction. This is also reflected in the differential charge across the atoms of the enediyne motif upon progression to the transition state; the Mo(IV) complex shows a less polar, and thus lower energy electronic structure (Fig. 34).

## C. Photochemical Bergman Cyclization of Metalloenediynes

Much like the sparse demonstrations of electronically activated metalloenediyne Bergman cyclization, only two examples of photochemically induced

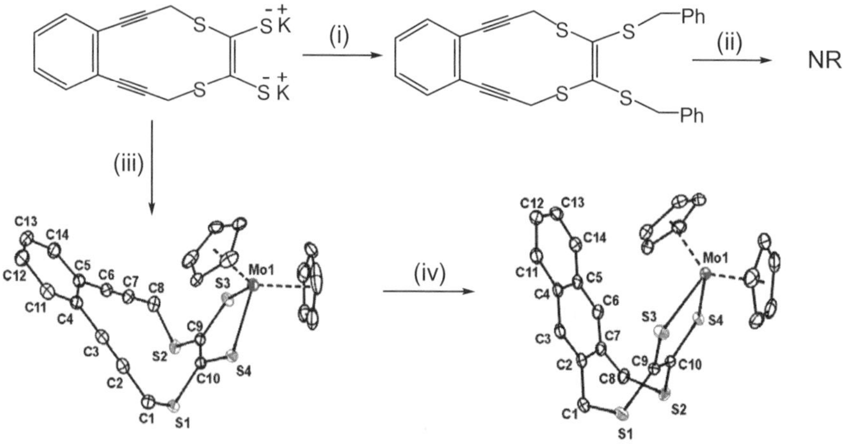

(i) $PhCH_2Br$, 25°C, 80%; (ii)DMSO, CHD, 180°C, 24h; (iii) MeOH, $MoCp_2Cl_2$, 60oC,0.5 h, 42%; (iv) DMSO, CHD, 120°C, 5h,15%.

Scheme 95. Ligand stabilization by arylation (top) and benzannulation (bottom).

Bergman cyclization (576) of simple metalloenediynes have been reported (577, 578). The first of these involves direct copper–pyridine charge-transfer photoexcitation (cf. Scheme 89); Cu(I) → pyridyloxy and pyridyloxy → Cu(II) excitation at $\lambda \geq 395$ nm leads to electronically initiated cyclization of both Cu(I) and Cu(II) in ~50% isolated yield within 12 h (577). Previous work suggested that population of the $^3\pi\pi^*$ state, which is localized on the double bond, would generate the trans or (*E*)-enediyne. Since the Zn(II)–metalloenediyne complex does not absorb in the excitation window, sensitization using acetophenone was used to exclusively populate the excited $^3\pi\pi^*$ state by energy transfer (Scheme 96). The photoproduct formed is indeed the trans enediyne, leading to questions concerning the origin of the photocyclization mechanism. Chemical R-group substitution conbined with electrochemical evaluation of the electronic structure and time-dependent density functional theory (TDDFT) jointly revealed the energetic ordering of the frontier molecular orbitals involved in the electronic structure and the orbital parentage of the electronic transitions (Fig. 35) (579). Within this framework, photochemical Bergman cyclization is envisioned to occur by population of an excited state localized on the pyridine due to direct metal–ligand orbital overlap, with population transfer to the other, lower lying $\pi$ system of the enediyne. Population of the enediyne $\pi$ system can also occur by a hyperconjugative interaction via the intervening ether linkage, thereby making the electronic transitions essentially mulitconfigurational (579). The electronic photoactivation mechanism is operative for *in vitro* DNA-cleaving activity of both calf thymus and double-stranded, 25mer DNA

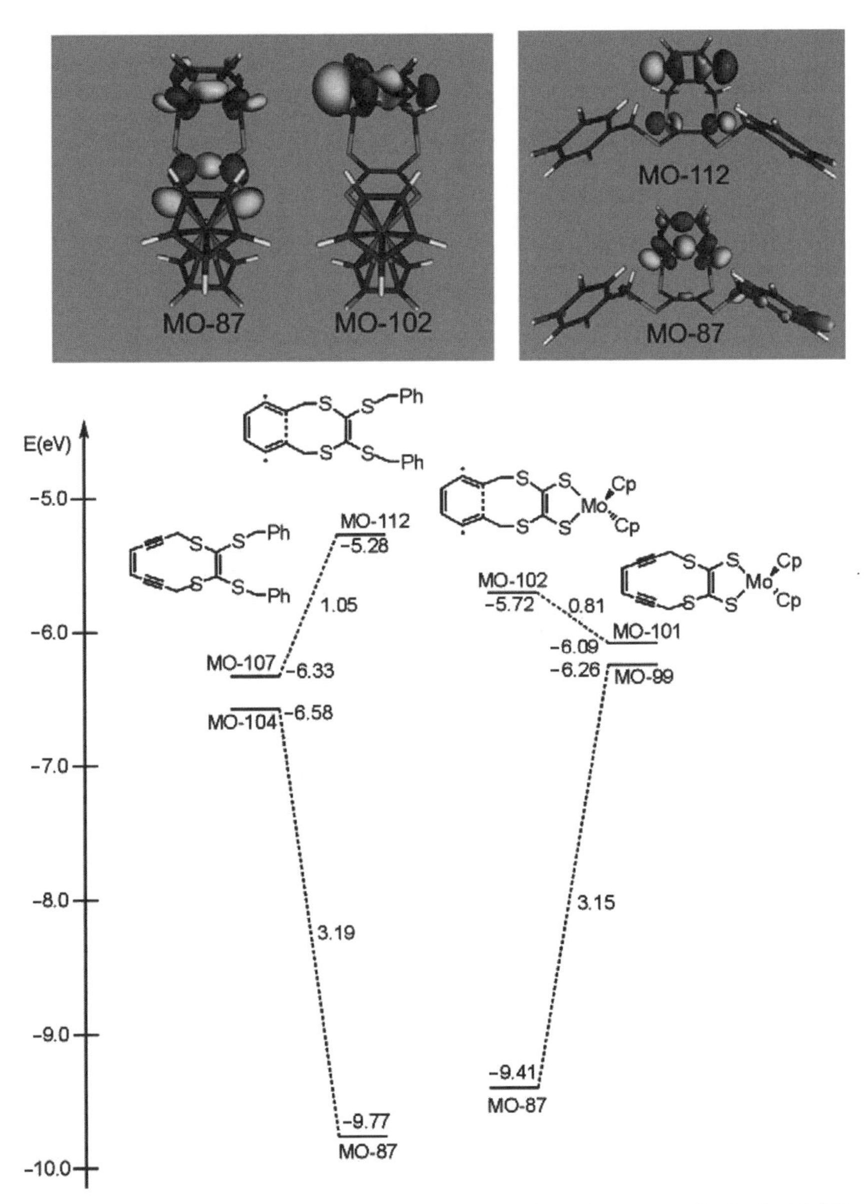


Figure 33. Molecular orbital description of the reduced barrier to Bergman cyclization by Mo(IV) complexation.

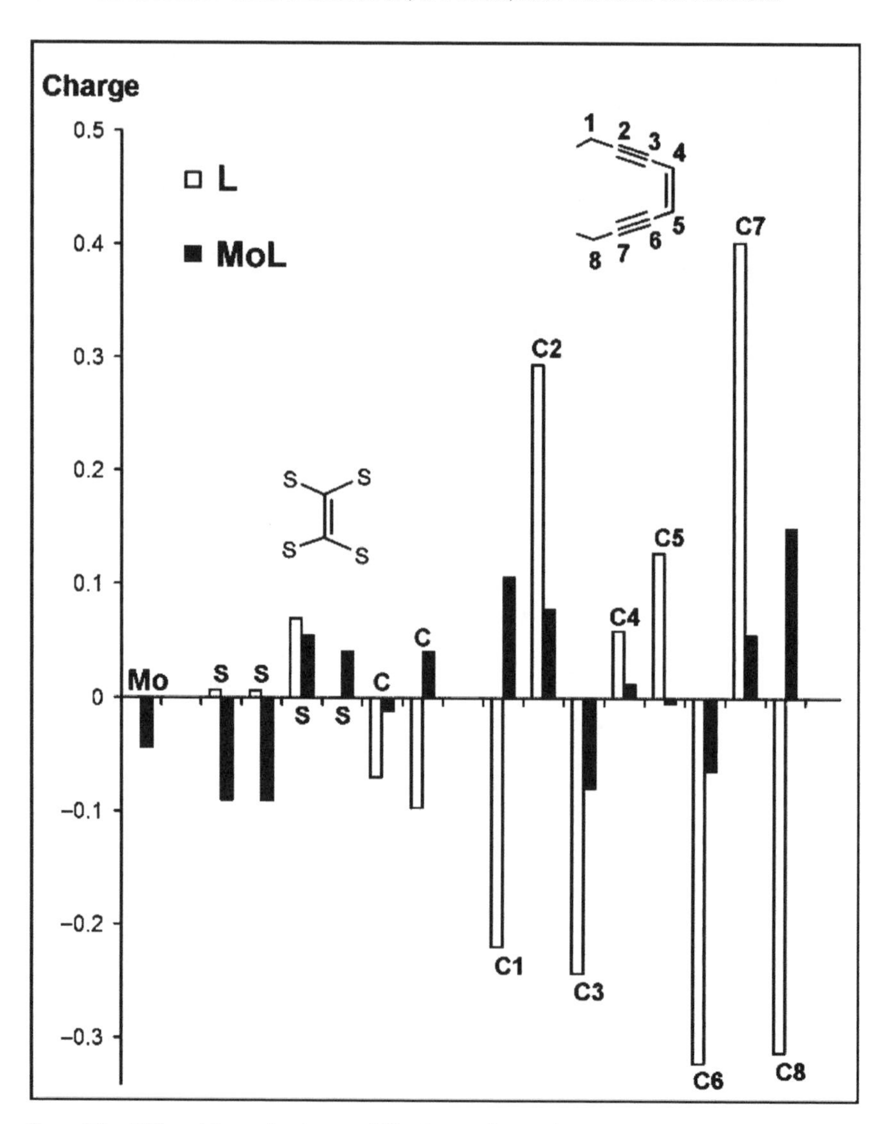


Figure 34. Differential atomic charge within the enediyne unit upon progression from ground to transition states for the Mo(IV) metalloenediyne complex (MoL) and corresponding benzylated ligand (L).

sequences, which have been shown to degrade by hydrogen-atom abstraction at the C-4′ position (577). Binding constants to DNA for these copper metalloenediynes and a [Ru(phen)$_2$(dppz-enediyne)] analogue (Fig. 36) (580) have been determined to be on the order of $10^4$ $M^{-1}$ for (R = H, Fig. 37), values that are typical for such inorganic complexes.

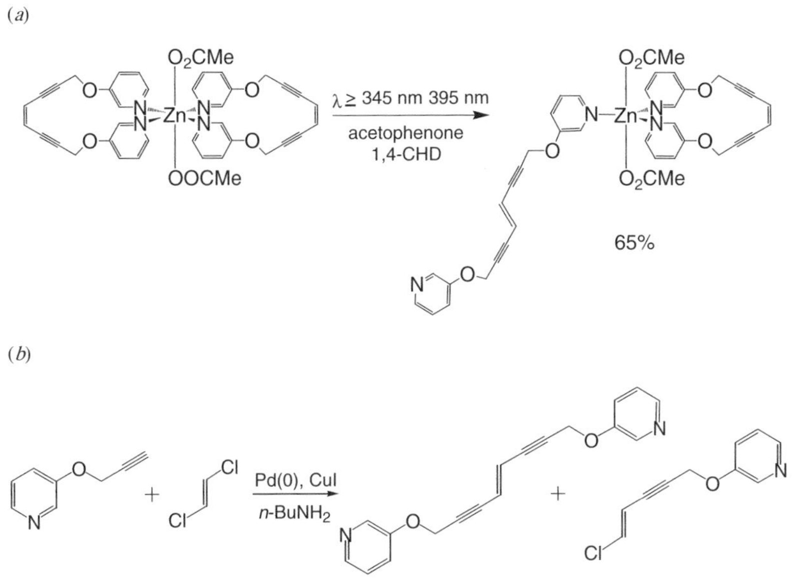

(*a*)

(*b*)

Scheme 96. (*a*) Photosensitization of the ${}^3\pi\pi^*$ state of the Zn(II) pyridyl-oxy metalloenediyne to generate the trans enediyne product. (*b*) Synthesis of the authentic trans enediyne ligand.

In a slightly different approach to photochemical Bergman cyclization by CT excitation, the development of a long-wavelength, enediyne LMCT transition was used to thermally heat the metalloenediyne complex by nonradiative decay. The choice of V(V) coupled with a catechol–enediyne motif (Scheme 97) generates a strong LMCT transition ($\varepsilon \sim 8000\ M^{-1}\text{cm}^{-1}$) at 1032 nm that is used to photochemically ($\lambda = 785$ or 1064 nm) generate sufficient thermal energy to create Bergman and polymerized product (578). The photoreaction is monitored by Raman spectroscopy that shows luminescence from the photoproducts that builds during the timecourse of the photolysis. Product analysis confirms both the presence of the cyclized product and extended polymer chains from 5000–274,000 amu (578).

## D. Porphyrin–Enediyne Conjugates

To date, photoelectronic Bergman cyclization of organic chromophores has been generally restricted to ultraviolet (UV) excitation due to the lack of an extended chromophore that absorbs strongly in the visible spectral region (for examples see, 581–583). The recent syntheses of dialkynylporphyrins (i.e., porphyrin–enediyne motifs) permitted the extension of typical Bergman cyclization studies to these large conjugated structures with and without central metal ion

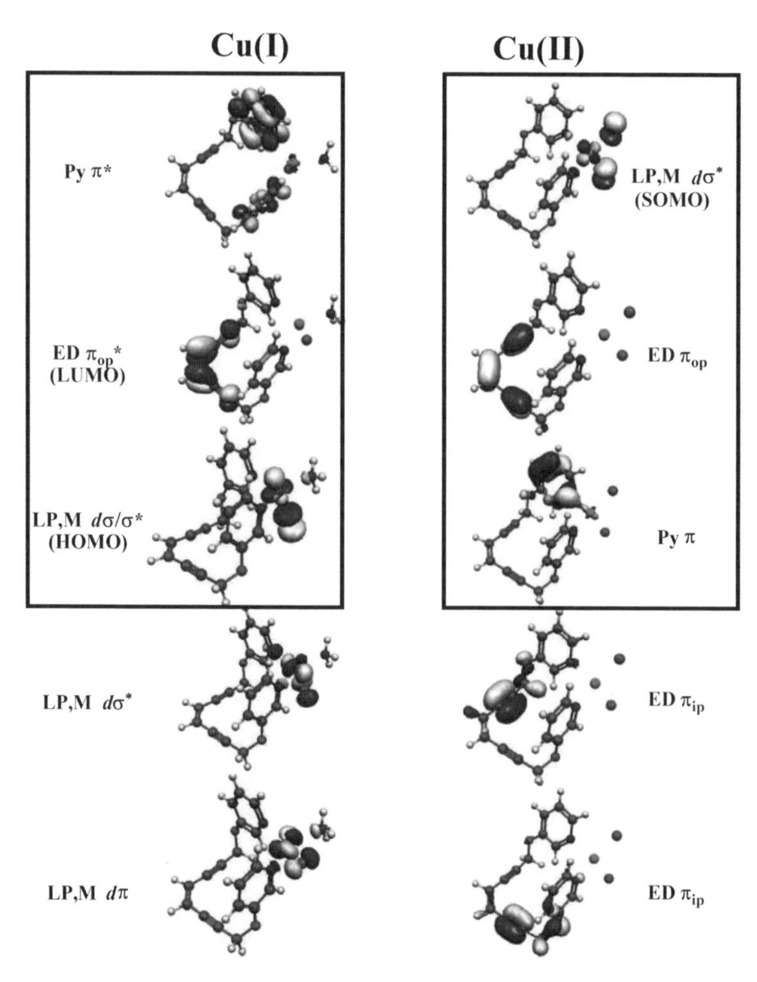


Figure 35. Frontier MOs (obtained from TDDFT) involved in the photoelectronic Bergman cyclization of copper metalloenediynes. The metal frontier orbitals (yellow) are highlighted along with the pyridine π and π* sets for a respective copper oxidation states.

(116–119, 584). Smith and co-workers (116) were the first to report the preparation of the target compounds using Stille coupling of the two (trimethystannyl)phenylacetylene units to the β-dibromotetraphenylporphyrin. Subsequent Bergman cyclization of the resulting conjugate generates the planar, aromatic piceno unit, which derives from electrophilic, 1,4-phenyl diradical addition to the adjacent phenyl rings of the TPP backbone. The cyclization temperatures for these reaction are very high (solid: 290°C; solution: 190°C). Substitution of the bulkyl phenyl R group (Scheme 98) with R = H or R = Br markedly lowers

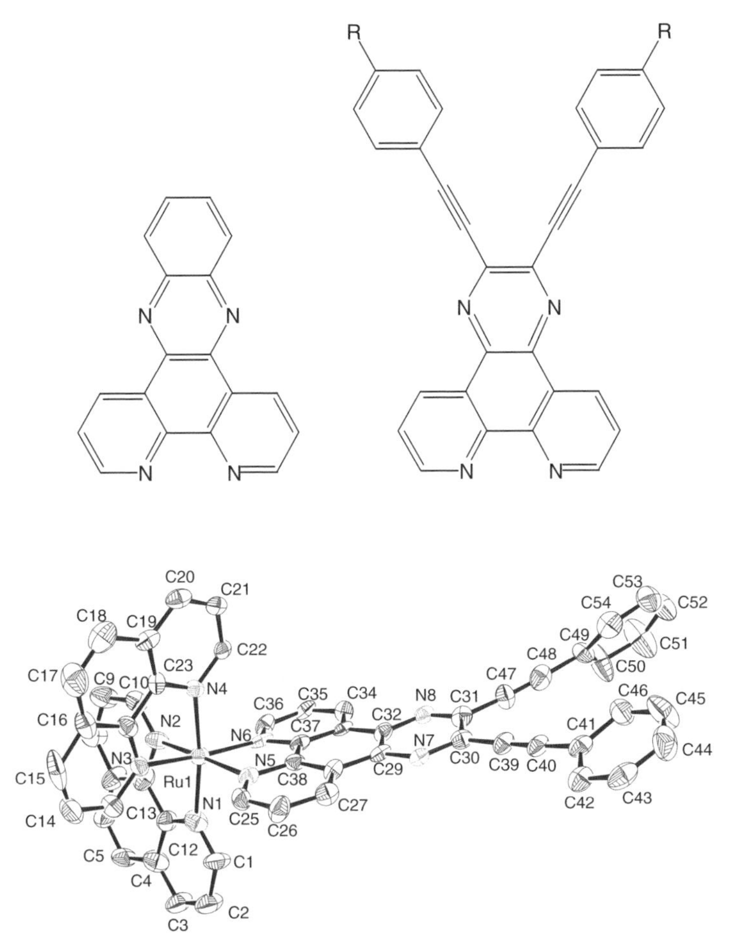


Figure 36. Structure of "light-switch" metalloenediynes (R = H, *t*-Bu).

these temperatures [solid (R = H): 170°C; solid (R = Br): 134°C] relative to R = Ph, TMS due to reduced steric clash in the product and the electron-withdrawing nature of the halogen (Fig. 38) (117). For each of these thermal cyclization reactions, the reaction coordinate consists of the traditional Bergman cyclization step, as well as a second barrier to the picenoporphyrin product that depicts the rearomatization process by loss of $H_2$ (116, 118). This latter step can be markedly influenced by addition of quinone as an oxidant, converting thermally activated Bergman reactions into ambient temperature processes (118).

Incorporation of the enediyne motif into a large chromophores, such as porphyrins (116–119, 584) or their sister phthalocyanines (585–596) has led to the exposure of several intriguing electronic properties of these constructs.

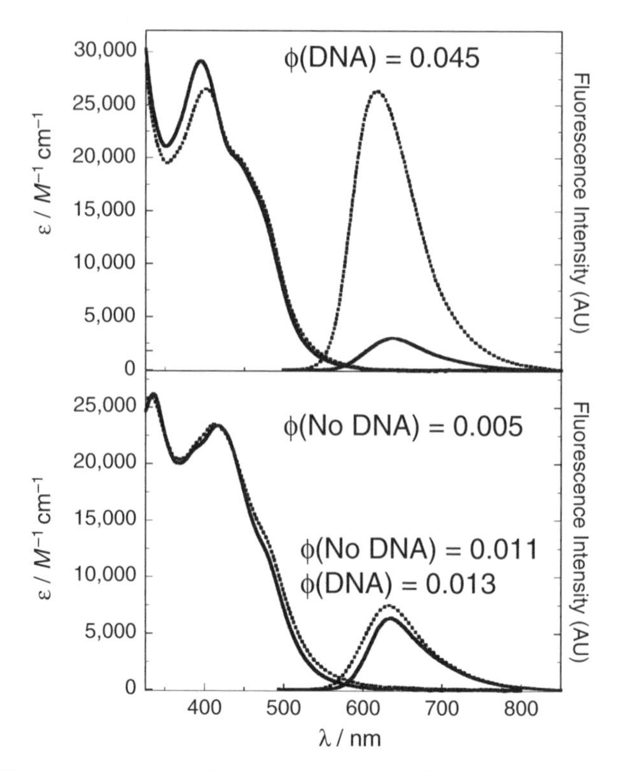


Figure 37. Electronic absorption and luminescence spectra of light switch metalloenediyne. Upper plot: R = H; Lower plot: R = *t*-Bu. The luminescence quantum yields $\phi$ reveal that the R = H derivative binds to calf thymus DNA with a constant of $\sim 10^4\ M^{-1}$, while binding is inhibited for R = *t*-Bu.

Conjugation of the alkyne units into the porphyrin or phthalocyanine backbone electronic structure leads to absorptivity enhancements and large red-shifts in their electronic spectra that have been sought for nonlinear optics, dyes, photonic materials, and photodynamic therapy types of applications.

In the context of enediyne reactivity, the electronic structure properties and their relationship to the cyclization chemistry of only the porphyrins have been probed in detail to date. Electronic absorption measurements and Raman analysis of the octa(phenylacetylene) derivatives (Scheme 99) show a correlated redshift in the absorption profile of ~13 nm/alkyne (Fig. 39), as well as a concomitant decrease in the frequency of the porphyrin marker bands between 1300 and 1600 $cm^{-1}$ as a function of increasing number of alkynes and increasing planarity of the macrocycle (Fig. 40) (584). These compounds also exhibit structure-dependent Bergman cyclization temperatures, but due to the steric crowding and large substituents at the alkyne termini positions,

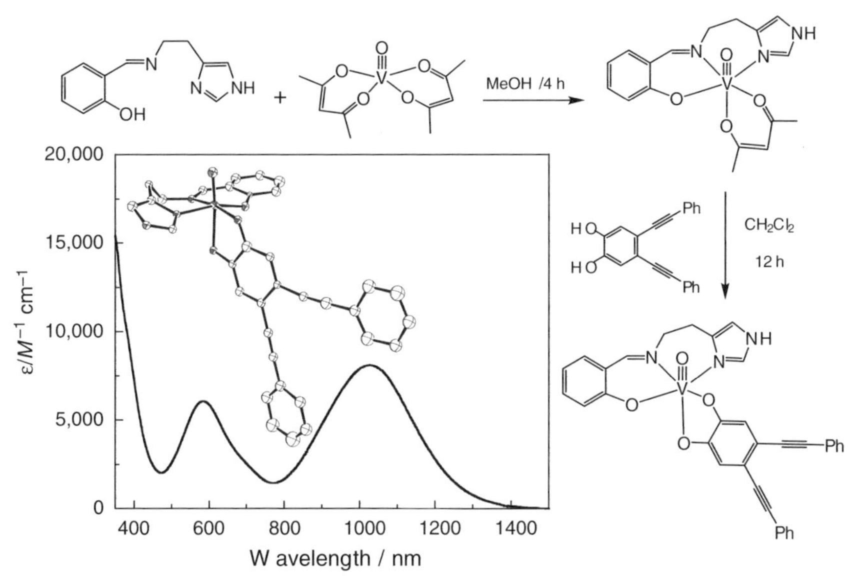


Scheme 97. Preparation, X-ray structure and electronic spectrum of V(V)-catechol-enediyne that undergoes photothermal Bergman cyclization with $\lambda = 785$ or 1064 nm.

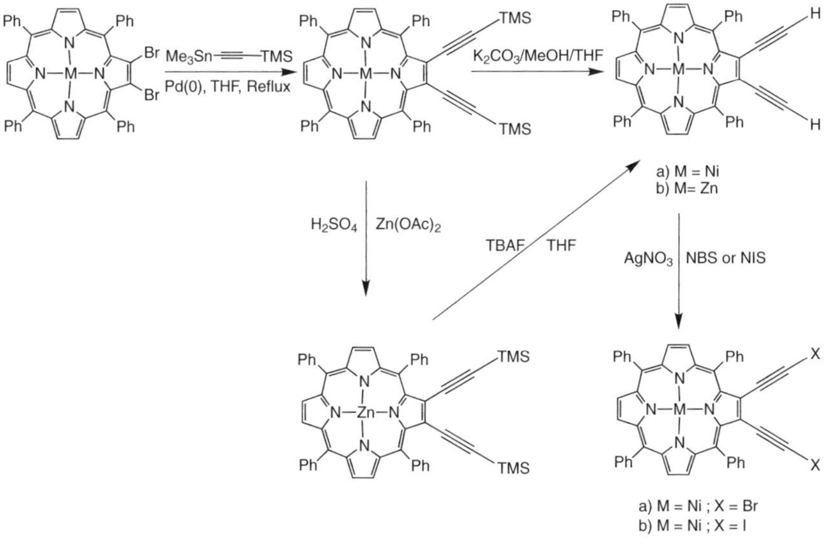

Scheme 98. Synthetic strategy for preparation of substituted dialkynylporphyrins. (TMS = Te-Tetramethysilane; NBS = *N*-bromosuccinimide NIS = *N*-iodosuccinimide.)

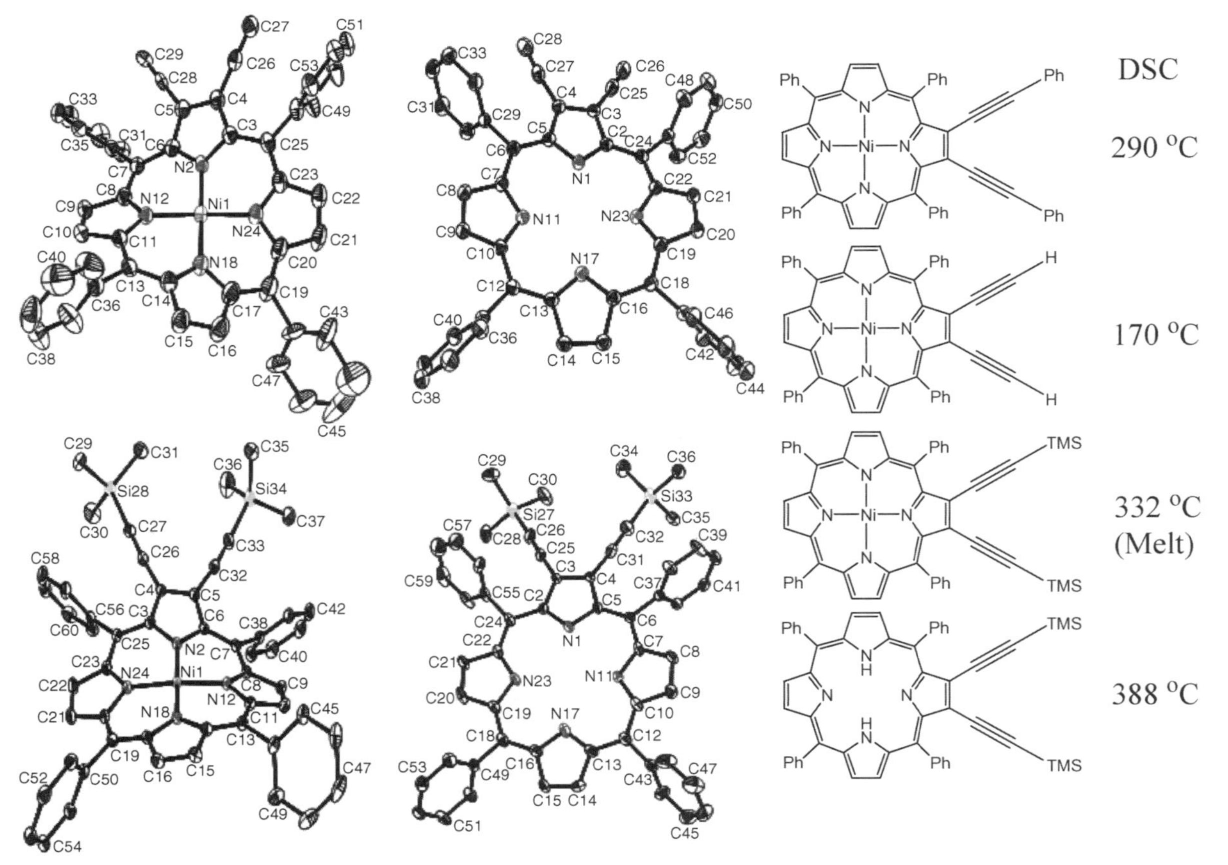


Figure 38. X-ray structures and DSC temperatures for alkyne-substituted dialkynylporphyrins.

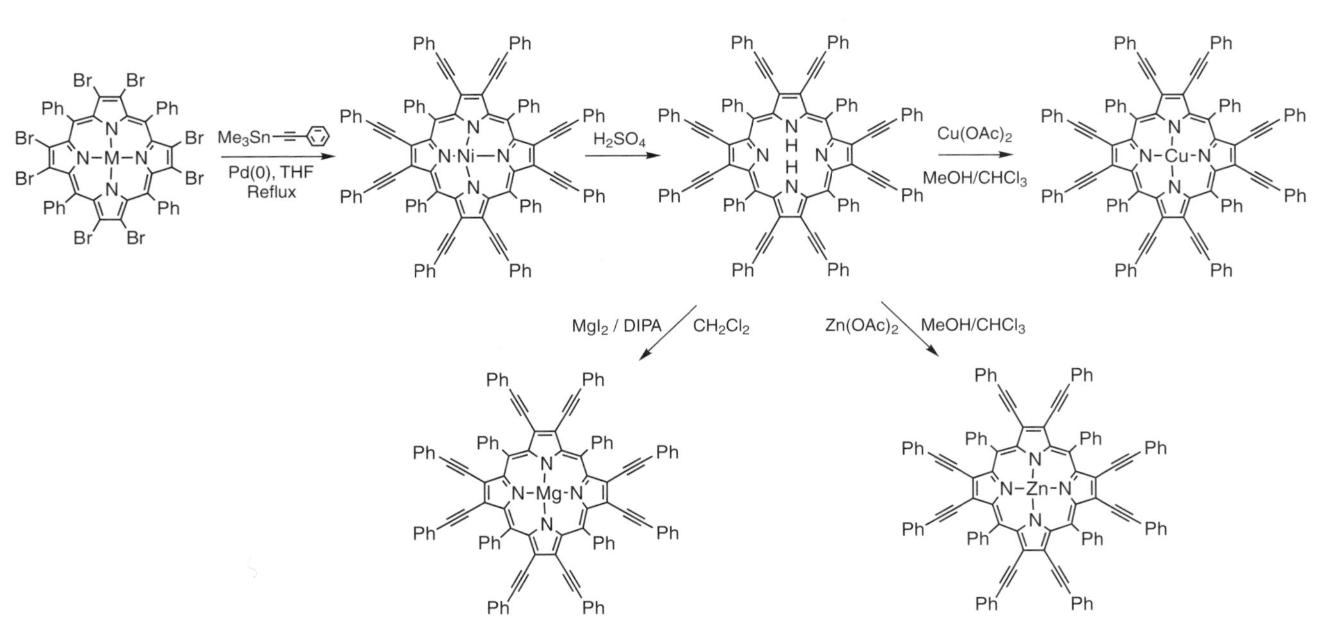


Scheme 99. Synthesis of octaalkynyl porphyrins. (DIPA = diisopropylamine.)

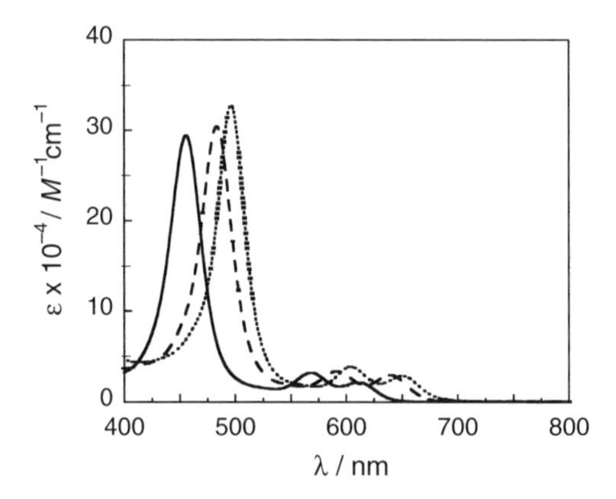


Figure 39. Electronic spectra of alkynyltetraphenylporphyrins showing a red shift as a function of increasing β-alkyne substitution ($n = 5$–8).

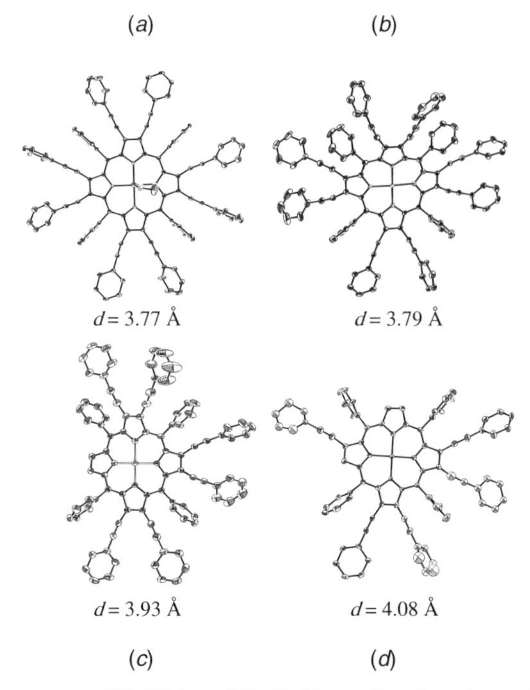


Figure 40. X-ray stuctures of Mg(II) (a) and Cu(II) (b) octaalkynyl porphyrins (c) along with their Ni(II) hexa- and pentaalkynyl derivatives (d). The *d*-value represents the average alkyne termini separation.

240–290°C is required for reaction (584). To this end, photochemical excitation of free base derivative yields only meso-reduced products, consistent with the lowered redox potential due to alkyne conjugation and the high thermal barrier to Bergman product.

The high cyclization temperatures of the octaalkynyl compounds, and their propensity for reductive photochemistry suggested that the dialkynyl compounds with simple acetylenes (R = H) may be more likely to generate Bergman photoproduct in light of their markedly lower cyclization temperatures and reduced steric hindrance. Indeed, excitation of the diacetylene derivative (cf. Fig. 38) with $\lambda \geq 395$ nm leads to photocyclization to yield the same picenoporphyrin product as the thermal reaction along the ground-state potential surface (Fig. 41) (119). Analogous to the thermal reaction, the temperature at which the photoreaction occurs (R = H; 10°C) is also dependent on the steric bulk of the R group at the alkyne termini position. Overall, the photoreaction temperature is generally lower than the ground-state thermal reaction, indicating

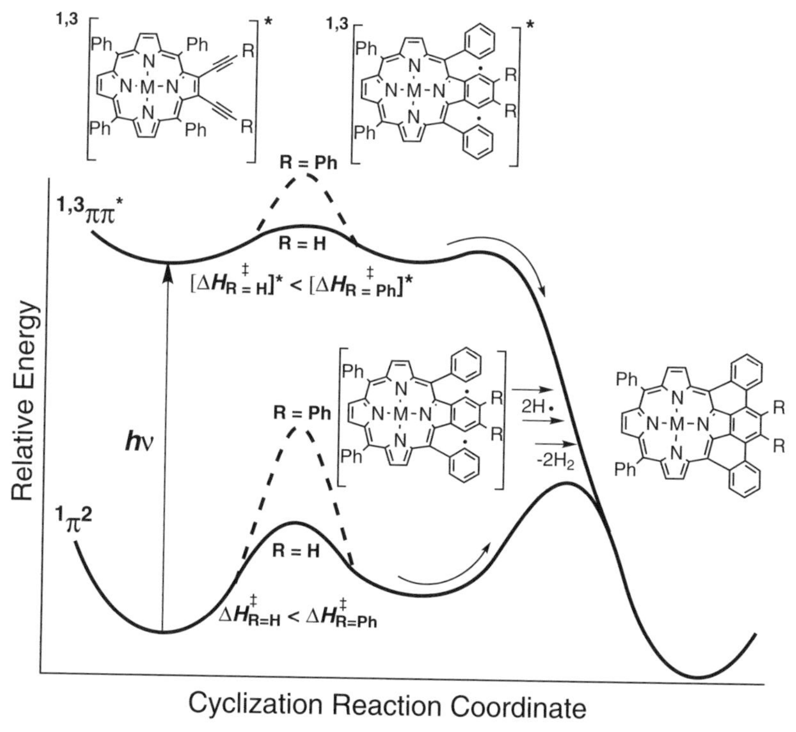


Figure 41. Reaction profile for R-dependent thermal and photocyclization reactions.

that similar, but not identical reaction intermediates are formed along the excited-state surface. The lack of an obvious quinone dependence to the reaction supports this assertion. Clearly, more work is required in the photochemistry of this reaction to fully appreciate the nature of these excited state intermediates and pathways.

Beyond the initial impetus for study of the Bergman photoreaction, the ability to perform radical-mediated macrocycle periphery modification is an intriguing and important emerging area that could have significant impact in porphyrin-related fields. To this end, employment of stoichiometric or catalytic metal systems, such as ruthenium or platinium (397, 532–535, 597, 598) to cyclize enediynes, and in this case porphrin–enediyne conjugates, leads to the potential to generate asymmetric porphyrinoids. Zaleski and co-workers (599, 600) utilized $PtCl_2$ in stoichiometric and catalytic amounts (~0.1 equiv) to activate both di- and tetralkynylporphyrins to their phenanthroporphyrin derivatives in good yields (55–65%) (Scheme 100). The origin of the phenanthroporphyrin (vs. the picenoporphyrin) derives from the proposed Pt–carbene species (535, 598) formed at the alkyne carbon, which inhibits substitution of the traditional 1,4-diradical across the meso-aromatic rings at one of the two positions. Such strategies involving radical-promoted periphery modifications will assuredly continue to develop new porphyrinoid constructs with novel structural and electronic properties.

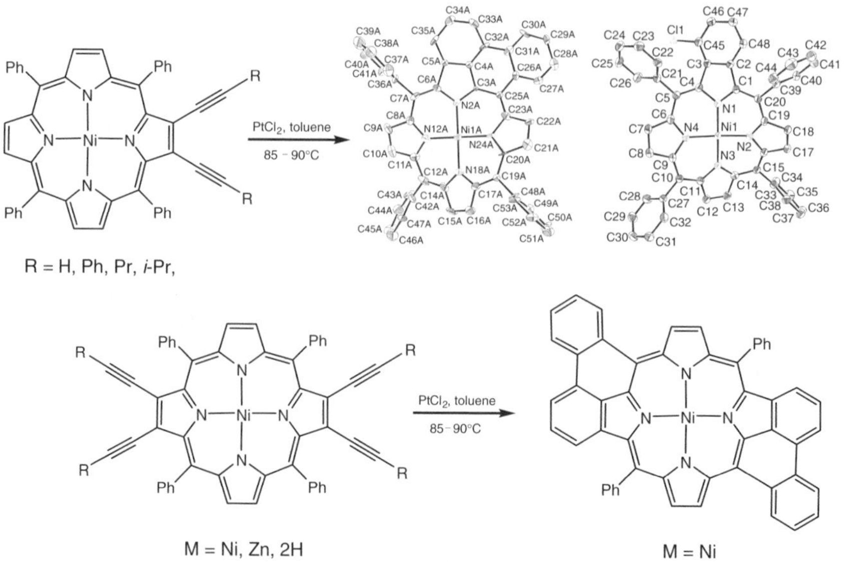

R = H, Ph, Pr, *i*-Pr,

M = Ni, Zn, 2H

M = Ni

Scheme 100. The $PtCl_2$-catalyzed cyclization of porphyrin–enediynes to the phenanthroporphyrin product.

## VI. CONCLUSIONS

It is clear from their widely diverse geometric and electronic structures that metal–alkyne chemistry is truly a rich subdiscipline with broad impact in the field of chemistry. Moreover, the remarkable metal-mediated chemical transformations of diyne and enyne substrates are critical to the ability to build complex organic frameworks in a concise and reproducible manner. Just beyond these well-established areas of metal–alkyne chemistry lie reactions of enediynes with metal salts or complexes. This is an emerging and rapidly developing area that promises to be an important mechanistic route to Bergman cyclization and the DNA-damaging 1,4-phenyl diradical via thermal or photochemical excitation. Enediyne transformations of this type also serve as building blocks to unusual organic frameworks. The ability to make the enediyne motif structurally rigid also makes them well suited as ligand scaffolds for complex multinuclear assemblies of materials and biological interest. To summarize, continued advances in metal–alkyne structures, properties, and reaction chemistry, will allow these motifs to evolve into a key branch of future chemical applications.

## ACKNOWLEDGMENTS

The authors would like to thank David Dye for assistance during the preparation of this manuscript. JMZ would also like to extend deepest gratitude to all of the graduate students and postdoctoral fellows that contributed to aspects of the science discussed in Section V. Without their tireless efforts and important scientific developments, this portion of the chapter would not have been possible. Additionally, funding from the NIH, NSF, and The American Cancer Society, as well as Indiana University is gratefully acknowledged.

## ABBREVIATIONS

| | |
|---|---|
| ap | 2-Anilinopyridinate |
| Bn | Benzyl |
| cod | 1,5-Cyclooctadiene |
| Cp | Cyclopentadienyl |
| Cy | Cyclohexyl |
| DABCO | 1,4-Diazobicyclo[2.2.2]octane |
| DMAC | Dimethylacetamide |
| Dmtc | Dimethyldimethylcarbomate |
| dppe | Diphenyl phosphinoethane |
| dppf | 1,1′-Bis(diphenylphosphino)ferrrrocene |
| dppm | Diphenylphosphinomethane |

| | |
|---|---|
| DSC | Differential scanningcalorimetry |
| EDTA | Ethylenediaminetetraacetic acid |
| HOMO | Highest occupied molecular orbital |
| IR | Infrared |
| LLCT | Ligand-to-ligand charge transfer |
| LMCT | Ligand-to-metal charge transfer |
| LUMO | Lowest unoccupied molecular orbital |
| Mes | 2,4,6-Trimethylphenyl (or mesityl) |
| ML | Metal centered |
| MMLRT | Metal-to-metal charge transfer |
| Ms | Mesylate (or methanesulfonyl) |
| $M_w$ | Molecular weight |
| NBS | *N*-Bromosuccinimide |
| NCO | Nonlinear optics |
| NIS | *N*-lodosuccinimide |
| NMR | Nuclear magnetic resonance |
| OLED | Organic light emitting diode |
| OTf | Trifluoromethanesulfonate (or triflate) |
| phen | 1,10-Phenanthroline |
| RCAM | Ring-closing alkene–alkyne metathesis |
| PTZ | Phenothiozine |
| RT | Room temperature |
| TBS = TBDMS | *tert*-Butyldimethylsilyl |
| TDDFT | Time-dependent density functional theory |
| TEA | Triethanolamine |
| Tg | Triglyme |
| thf | Tetrahydrofuran (ligand) |
| THF | Tetrahydrofuran (solvent) |
| TMED | Tetramethylethylenediamine |
| tol | Tolyl |
| TMS | Trimethylsilyl |
| p-Tol | *p*-Toluine |
| tpy | Terpyridine |
| UV | Ultraviolet |

## REFERENCES


1. P. J. Low and M. I. Bruce, *Adv. Organomet. Chem.*, *48*, 71 (2001).
2. N. J. Long and C. K. Williams, *Angew. Chem., Int. Ed. Engl.*, *42*, 2586 (2003).
3. H. Werner, *Angew. Chem. Int. Ed. Engl.*, *29*, 1077 (1990).

4. M. I. Bruce, *Chem. Rev., 91*, 197 (1991).
5. M. I. Bruce, *Chem. Rev.*, *98*, 2797 (1998).
6. D. Touchard and P. H. Dixneuf, *Coord. Chem. Rev.*, *178–180*, 409 (1998).
7. H. Lang and M. Weinmann, *Synlett*, 1 (1996).
8. R. Choukroun and P. Cassoux, *Acc. Chem. Res.*, *32*, 494 (1999).
9. U. Rosenthal, P.-M. Pellny, F. G. Kirchbauer, and V. V. Burlakov, *Acc. Chem. Res.*, *33*, 119 (2000).
10. S. Saito and Y. Yamamoto, *Chem. Rev.*, *100*, 2901 (2000).
11. C. Aubert, O. Buisine, and M. Malacria, *Chem. Rev.*, *102*, 813 (2002).
12. J. Lewis, M. S. Khan, A. K. Kakkar, B. F. G. Johnson, T. B. Marder, H. B. Fyfe, F. Wittmann, R. H. Friend, and A. E. Dray, *J. Organomet. Chem.*, *425*, 165 (1992).
13. M. Younus, A. Kohler, S. Cron, N. Chawdhury, M. R. A. Al-Mandhary, M. S. Khan, J. Lewis, N. J. Long, R. H. Friend, and P. R. Raithby, *Angew. Chem., Int. Ed. Engl.*, *37*, 3036 (1998).
14. N. Chawdhury, A. Koehler, R. H. Friend, M. Younus, N. J. Long, P. R. Raithby, and J. Lewis, *Macromolecules*, *31*, 722 (1998).
15. V. W.-W. Yam, K. Kam-Wing Lo, and K. Man-Chung Wong, *J. Organomet. Chem.*, *578*, 3 (1999).
16. V. W.-W. Yam, *Acc. Chem. Res.*, *35*, 555 (2002).
17. M. J. Irwin, J. J. Vittal, and R. J. Puddephatt, *Organometallics*, *16*, 3541 (1997).
18. R. P. Kingsborough and T. M. Swager, *Prog. Inorg. Chem.*, *48*, 123 (1999).
19. D. W. Bruce and D. O'Hare, *Inorganic Materials, 2nd ed.*, John Wiley & Sons, Inc., New York, 1997.
20. R. J. Lagow, J. J. Kampa, H.-C. Wei, S. L. Battle, J. W. Genge, D. A. Laude, C. J. Harper, R. Bau, I. F. Haw, and E. Munson., *Science*, *267*, 362 (1995).
21. N. J. Long, *Angew. Chem., Int. Ed. Engl.*, *34*, 21 (1995).
22. S. Barlow and D. O'Hare, *Chem. Rev.*, *97*, 637 (1997).
23. I. R. Whittall, A. M. McDonagh, M. G. Humphrey, and M. Samoc, *Adv. Organomet. Chem.*, *42*, 291 (1998).
24. I. R. Whittall, A. M. McDonagh, M. G. Humphrey, and M. Samoc, *Adv. Organomet. Chem.*, *43*, 349 (1998).
25. A. Basak, S. Mandal, and S. S. Bag, *Chem. Rev.*, *103*, 4077 (2003).
26. S. Bhattacharyya and J. M. Zaleski, *Curr. Top. Med. Chem.*, *4*, 1637 (2004).
27. D. S. Rawat and J. M. Zaleski, *Synlett*, 393 (2004).
28. S. Szafert and J. A. Gladysz, *Chem. Rev.*, *103*, 4175 (2003).
29. J. L. Brefort, R. J. P. Corriu, P. Gerbier, C. Guerin, B. J. L. Henner, A. Jean, T. Kuhlmann, F. Garnier, and A. Yassar, *Organometallics*, *11*, 2500 (1992).
30. M. L. H. Green, *Organometallic Compounds*, Methuen, London, 1968.
31. M. F. Shostakovskii and A. V. Bogdanova, *The Chemistry of Diacetylenes*, John Wiley & Sons, Inc., New York, 1974.
32. A. B. Holmes, C. L. D. Jennings-White, A. H. Schulthess, B. Akinde, and D. R. M. Walton, *Chem. Commun.*, 840 (1979).
33. M. I. Bruce, R. Clark, J. Howard, and P. Woodward, *J. Organomet. Chem.*, *42*, C107 (1972).
34. O. M. Abu Salah and M. I. Bruce, *Dalton Trans.*, 2302 (1974).
35. D. C. Green, U. Englich, and K. Ruhlandt-Senge, *Angew. Chem., Int. Ed. Engl.*, *38*, 354 (1999).

36. H. Le Bozec, K. Ouzzine, and P. H. Dixneuf, *Organometallics*, *10*, 2768 (1991).
37. H. Lang, S. Blau, H. Pritzkow, and L. Zsolnai, *Organometallics*, *14*, 1850 (1995).
38. M. Oberthur, G. Hillebrand, P. Arndt, and R. Kempe, *Chem. Ber./Rec.*, *130*, 789 (1997).
39. C. Mueller, J. A. Whiteford, and P. J. Stang, *J. Am. Chem. Soc.*, *120*, 9827 (1998).
40. G.-L. Xu, C. G. Jablonski, and T. Ren, *J. Organomet. Chem.*, *683*, 388 (2003).
41. G.-L. Xu, G. Zou, Y.-H. Ni, M. C. DeRosa, R. J. Crutchley, and T. Ren, *J. Am. Chem. Soc.*, *125*, 10057 (2003).
42. P. Espinet, J. Fornies, F. Martinez, M. Tomas, E. Lalinde, M. T. Moreno, A. Ruiz, and A. J. Welch, *Dalton Trans.*, 791 (1990).
43. J. Fornies, M. A. Gomez-Saso, F. Martinez, E. Lalinde, M. T. Moreno, and A. J. Welch, *New J. Chem.*, *16*, 483 (1992).
44. I. Ara, J. Fornies, J. Gomez, E. Lalinde, and M. T. Moreno, *Organometallics*, *19*, 3137 (2000).
45. I. Ara, J. R. Berenguer, E. Eguizabal, J. Fornies, J. Gomez, and E. Lalinde, *J. Organomet. Chem.*, *670*, 221 (2003).
46. K. Sonogashira, T. Yatake, Y. Tohda, S. Takahashi, and N. Hagihara, *Chem. Commun.*, 291 (1977).
47. K. Sonogashira, S. Takahashi, and N. Hagihara, *Macromolecules*, *10*, 879 (1977).
48. K. Sonogashira, Y. Fujikura, T. Yatake, N. Toyoshima, S. Takahashi, and N. Hagihara, *J. Organomet. Chem.*, *145*, 101 (1978).
49. K. Osakada, R. Sakata, and T. Yamamoto, *Organometallics*, *16*, 5354 (1997).
50. K. Sonogashira, Y. Tohda, and N. Hagihara, *Tetrahedron Lett.*, 4467 (1975).
51. J. E. McGarrah, Y.-J. Kim, M. Hissler, and R. Eisenberg, *Inorg. Chem.*, *40*, 4510 (2001).
52. J. Hua and W. Lin, *Org. Lett.*, *6*, 861 (2004).
53. M. I. Bruce, B. C. Hall, P. J. Low, M. E. Smith, B. W. Skelton, and A. H. White, *Inorg. Chim. Acta*, *300–302*, 633 (2000).
54. N. Hagihara, K. Sonogashira, and S. Takahashi, *Adv. Pol. Sci.*, *41*, 149 (1981).
55. K. Sonogashira, S. Kataoka, S. Takahashi, and N. Hagihara, *J. Organomet. Chem.*, *160*, 319 (1978).
56. S. Takahashi, H. Morimoto, E. Murata, S. Kataoka, K. Sonogashira, and N. Hagihara, *J. Polym. Sci., Polym. Chem.*, *20*, 565 (1982).
57. K. Sonogashira, K. Ohga, S. Takahashi, and N. Hagihara, *J. Organomet. Chem.*, *188*, 237 (1980).
58. S. Takahashi, E. Murata, K. Sonogashira, and N. Hagihara, *J. Polym. Sci., Polym. Chem.*, *18*, 661 (1980).
59. P. J. Stang and F. Diederich, Eds., *Modern Acetylene Chemistry*, John Wiley & Sons, Inc., VCH, Weinheim, 1995.
60. V. Farina, V. Krishnamurthy, and W. J. Scott, *The Stille Reaction*, 1997.
61. T. N. Mitchell, *Metal-Catalyzed Cross-Coupling Reactions*, 167 (1998).
62. M. E. Wright, *Macromolecules*, *22*, 3256 (1989).
63. B. Cetinkaya, M. F. Lappert, J. McMeeking, and D. E. Palmer, *Dalton Trans.*, 1202 (1973).
64. S. J. Davies, B. F. G. Johnson, M. S. Khan, and J. Lewis, *Chem. Commun.*, 187 (1991).
65. B. F. G. Johnson, A. K. Kakkar, M. S. Khan, J. Lewis, A. E. Dray, R. H. Friend, and F. Wittmann, *J. Mat. Chem.*, *1*, 485 (1991).
66. B. F. G. Johnson, A. K. Kakkar, M. S. Khan, and J. Lewis, *J. Organomet. Chem.*, *409*, C12 (1991).

67. M. S. Khan, S. J. Davies, A. K. Kakkar, D. Schwartz, B. Lin, F. G. Johnson, and J. Lewis, *J. Organomet. Chem.*, *424*, 87 (1992).
68. M. S. Khan, N. A. Pasha, A. K. Kakkar, P. R. Raithby, J. Lewis, K. Fuhrmann, and R. H. Friend, *J. Mat. Chem.*, *2*, 759 (1992).
69. N. Pirio, D. Touchard, P. H. Dixneuf, M. Fettouhi, and L. Ouahab, *Angew. Chem., Int. Ed. Engl.*, *31*, 651 (1992).
70. D. Touchard, N. Pirio, L. Toupet, M. Fettouhi, L. Ouahab, and P. H. Dixneuf, *Organometallics*, *14*, 5263 (1995).
71. L. Dahlenburg, A. Weiss, M. Bock, and A. Zahl, *J. Organomet. Chem.*, *541*, 465 (1997).
72. R. Eastmond and D. R. M. Walton, *Tetrahedron*, *28*, 4591 (1972).
73. R. Eastmond, T. R. Johnson, and D. R. M. Walton, *Tetrahedron*, *28*, 4601 (1972).
74. R. Eastmond, T. R. Johnson, and D. R. M. Walton, *J. Organomet. Chem.*, *50*, 87 (1973).
75. P. Haquette, N. Pirio, D. Touchard, L. Toupet, and P. H. Dixneuf, *Chem. Commun.*, 163 (1993).
76. D. Touchard, C. Morice, V. Cadierno, P. Haquette, L. Toupet, and P. H. Dixneuf, *Chem. Commun.*, 859 (1994).
77. O. Lavastre, M. Even, P. H. Dixneuf, A. Pacreau, and J.-P. Vairon, *Organometallics*, *15*, 1530 (1996).
78. O. Lavastre, J. Plass, P. Bachmann, S. Guesmi, C. Moinet, and P. H. Dixneuf, *Organometallics*, *16*, 184 (1997).
79. D. Touchard, P. Haquette, S. Guesmi, L. Le Pichon, A. Daridor, L. Toupet, and P. H. Dixneuf, *Organometallics*, *16*, 3640 (1997).
80. M. C. B. Colbert, J. Lewis, N. J. Long, P. R. Raithby, M. Younus, A. J. P. White, D. J. Williams, N. N. Payne, L. Yellowlees, D. Beljonne, N. Chawdhury, and R. H. Friend, *Organometallics*, *17*, 3034 (1998).
81. A. Romero, D. Peron, and P. H. Dixneuf, *Chem. Commun.*, 1410 (1990).
82. D. Peron, A. Romero, and P. H. Dixneuf, *Gazz. Chim. Ital.*, *124*, 497 (1994).
83. D. Peron, A. Romero, and P. H. Dixneuf, *Organometallics*, *14*, 3319 (1995).
84. T. Rappert, O. Nuernberg, and H. Werner, *Organometallics*, *12*, 1359 (1993).
85. H. Werner, O. Gevert, P. Steinert, and J. Wolf, *Organometallics*, *14*, 1786 (1995).
86. J. Wolf, R. W. Lass, M. Manger, and H. Werner, *Organometallics*, *14*, 2649 (1995).
87. P. J. Stang and C. M. Crittell, *Organometallics*, *9*, 3191 (1990).
88. P. J. Stang and R. Tykwinski, *J. Am. Chem. Soc.*, *114*, 4411 (1992).
89. C. Bianchini, A. Meli, M. Peruzzini, F. Zanobini, and P. Zanello, *Organometallics*, *9*, 241 (1990).
90. C. Bianchini, F. Laschi, F. Ottaviani, M. Peruzzini, and P. Zanello, *Organometallics*, *7*, 1660 (1988).
91. P. Chow, D. Zargarian, N. J. Taylor, and T. B. Marder, *Chem. Commun.*, 1545 (1989).
92. H. B. Fyfe, M. Miekuz, D. Zargarian, N. J. Taylor, and T. B. Marder, *Chem. Commun.*, 188 (1991).
93. H. F. Klein and H. H. Karsch, *Chem. Ber.*, *108*, 944 (1975).
94. C. Bianchini, D. Masi, A. Meli, M. Peruzzini, J. A. Ramirez, A. Vacca, and F. Zanobini, *Organometallics*, *8*, 2179 (1989).
95. A. S. Hay, *J. Org. Chem.*, *27*, 3320 (1962).
96. R. Dembinski, T. Lis, S. Szafert, C. L. Mayne, T. Bartik, and J. A. Gladysz, *J. Organomet. Chem.*, *578*, 229 (1999).

97. M. J. Irwin, L. Manojlovic-Muir, K. W. Muir, R. J. Puddephatt, and D. S. Yufit, *Chem. Commun.*, 219 (1997).
98. G. Jia, R. J. Puddephatt, J. D. Scott, and J. J. Vittal, *Organometallics*, *12*, 3565 (1993).
99. C. P. McArdle, M. C. Jennings, J. J. Vittal, and R. J. Puddephatt, *Eur. J. Chem.*, *7*, 3572 (2001).
100. V. W.-W. Yam, S. W.-K. Choi, and K.-K. Cheung, *Organometallics*, *15*, 1734 (1996).
101. V. W.-W. Yam, V. C.-Y. Lau, and K.-K. Cheung, *Organometallics*, *15*, 1740 (1996).
102. H. Lang, R. Packheiser, and B. Walfort, *Organometallics*, *25*, 1836 (2006).
103. J. Fornies and E. Lalinde, *Dalton Trans.*, 2587 (1996).
104. J. Fornies, J. Gomez, E. Lalinde, and M. T. Moreno, *Inorg. Chem.*, *40*, 5415 (2001).
105. J. P. H. Charmant, J. Fornies, J. Gomez, E. Lalinde, R. I. Merino, M. T. Moreno, and A. G. Orpen, *Organometallics*, *18*, 3353 (1999).
106. V. Wing-Wah Yam, K.-L. Yu, and K.-K. Cheung, *Dalton Trans.*, 2913 (1999).
107. V. W.-W. Yam, C.-K. Hui, S.-Y. Yu, and N. Zhu, *Inorg. Chem.*, *43*, 812 (2004).
108. H. Lang, N. Mansilla, and G. Rheinwald, *Organometallics*, *20*, 1592 (2001).
109. S. Kocher and H. Lang, *J. Organomet. Chem.*, *641*, 62 (2002).
110. F. G. Kirchbauer, P.-M. Pellny, H. Sun, V. V. Burlakov, P. Arndt, W. Baumann, A. Spannenberg, and U. Rosenthal, *Organometallics*, *20*, 5289 (2001).
111. H. Lang and T. Stein, *J. Organomet. Chem.*, *641*, 41 (2002).
112. H. Lang, D. S. A. George, and G. Rheinwald, *Coord. Chem. Rev.*, *206–207*, 101 (2000).
113. U. Rosenthal, A. Ohff, W. Baumann, R. Kempe, A. Tillack, and V. V. Burlakov, *Organometallics*, *13*, 2903 (1994).
114. K. Sonogashira, *Handbook of Organopalladium Chemistry for Organic Synthesis*, Vol. 1, John Wiley & Sons, Inc., 2002, pp. 493–529.
115. D. Milstein and J. K. Stille, *J. Am. Chem. Soc.*, *100*, 3636 (1978).
116. H. Aihara, L. Jaquinod, D. J. Nurco, and K. M. Smith, *Angew. Chem. Int. Ed. Engl.*, *40*, 3439 (2001).
117. M. Nath, J. C. Huffman, and J. M. Zaleski, *J. Am. Chem. Soc.*, *125*, 11484 (2003).
118. M. Nath, J. C. Huffman, and J. M. Zaleski, *Chem. Commun.*, 858 (2003).
119. M. Nath, M. Pink, and J. M. Zaleski, *J. Am. Chem. Soc.*, *127*, 478 (2005).
120. H. C. Kolb, M. G. Finn, and K. B. Sharpless, *Angew. Chem., Int. Ed. Engl.*, *40*, 2004 (2001).
121. Z. P. Demko and K. B. Sharpless, *Angew. Chem., Int. Ed. Engl.*, *41*, 2113 (2002).
122. Z. P. Demko and K. B. Sharpless, *Angew. Chem., Int. Ed. Engl.*, *41*, 2110 (2002).
123. T.-D. Kim, J. Luo, Y. Tian, J.-W. Ka, N. M. Tucker, M. Haller, J.-W. Kang, and A. K. Y. Jen, *Macromolecules*, *39*, 1676 (2006).
124. T. Hasegawa, M. Umeda, M. Numata, T. Fujisawa, S. Haraguchi, K. Sakurai, and S. Shinkai, *Chem. Lett.*, *35*, 82 (2006).
125. P. Rodriguez Loaiza, S. Loeber, H. Huebner, and P. Gmeiner, *J. Comb. Chem.*, *8*, 252 (2006).
126. R. Luxenhofer and R. Jordan, *Macromolecules*, *39*, 3509 (2006).
127. F. S. Hassane, B. Frisch, and F. Schuber, *Bioconj. Chem.*, *17*, 849 (2006).
128. D. D. Diaz, K. Rajagopal, E. Strable, J. Schneider, and M. G. Finn, *J. Am. Chem. Soc.*, *128*, 6056 (2006).
129. A. H. Yap and S. M. Weinreb, *Tetrahedron Lett.*, *47*, 3035 (2006).
130. V. Ladmiral, G. Mantovani, G. J. Clarkson, S. Cauet, J. L. Irwin, and D. M. Haddleton, *J. Am. Chem. Soc.*, *128*, 4823 (2006).

131. M. Whiting, J. Muldoon, Y.-C. Lin, S. M. Silverman, W. Lindstrom, A. J. Olson, H. C. Kolb, M. G. Finn, K. B. Sharpless, J. H. Elder, and V. V. Fokin, *Angew. Chem., Int. Ed. Engl.*, *45*, 1435 (2006).
132. V. D. Bock, R. Perciaccante, T. P. Jansen, H. Hiemstra, and J. H. Van Maarseveen, *Org. Lett.*, *8*, 919 (2006).
133. D. A. Ossipov and J. Hilborn, *Macromolecules*, *39*, 1709 (2006).
134. D. J. V. C. van Steenis, O. R. P. David, G. P. F. van Strijdonck, J. H. van Maarseveen, and J. N. H. Reek, *Chem. Commun.*, 4333 (2005).
135. V. P. Mocharla, B. Colasson, L. V. Lee, S. Roeper, K. B. Sharpless, C.-H. Wong, and H. C. Kolb, *Angew. Chem., Int. Ed. Engl.*, *44*, 116 (2004).
136. K. Kacprzak, M. Migas, A. Plutecka, U. Rychlewska, and J. Gawronski, *Heterocycles*, *65*, 1931 (2005).
137. Q. Wang, S. Chittaboina, and H. N. Barnhill, *Lett. Org. Chem.*, *2*, 293 (2005).
138. H. Gallardo, F. Ely, A. Bortoluzzi, and G. Conte, *Liq. Cryst.*, *32*, 667 (2005).
139. R. Franke, C. Doll, and J. Eichler, *Tetrahedron Lett.*, *46*, 4479 (2005).
140. B. C. Englert, S. Bakbak, and U. H. F. Bunz, *Macromolecules*, *38*, 5868 (2005).
141. L. D. Pachon, J. H. van Maarseveen, and G. Rothenberg, *Adv. Synth. Catal.*, *347*, 811 (2005).
142. R. L. Weller and S. R. Rajski, *Org. Lett.*, *7*, 2141 (2005).
143. A. Krasinski, Z. Radic, R. Manetsch, J. Raushel, P. Taylor, K. B. Sharpless, and H. C. Kolb, *J. Am. Chem. Soc.*, *127*, 6686 (2005).
144. M. Malkoch, K. Schleicher, E. Drockenmuller, C. J. Hawker, T. P. Russell, P. Wu, and V. V. Fokin, *Macromolecules*, *38*, 3663 (2005).
145. F. Pagliai, T. Pirali, E. Del Grosso, R. Di Brisco, G. C. Tron, G. Sorba, and A. A. Genazzani, *J. Med. Chem.*, *49*, 467 (2006).
146. S. Hotha and S. Kashyap, *J. Org. Chem.*, *71*, 364 (2006).
147. P. Wu, M. Malkoch, J. N. Hunt, R. Vestberg, E. Kaltgrad, M. G. Finn, V. V. Fokin, K. B. Sharpless, and C. J. Hawker, *Chem. Commun.*, 5775 (2005).
148. A. Trabolsi, M. Elhabiri, M. Urbani, J. L. Delgado de la Cruz, F. Ajamaa, N. Solladie, A.-M. Albrecht-Gary, and J.-F. Nierengarten, *Chem. Commun.*, 5736 (2005).
149. R. K. O'Reilly, M. J. Joralemon, K. L. Wooley, and C. J. Hawker, *Chem. Mater.*, *17*, 5976 (2005).
150. H. Gao, G. Louche, B. S. Sumerlin, N. Jahed, P. Golas, and K. Matyjaszewski, *Macromolecules*, *38*, 8979 (2005).
151. M. Malkoch, R. J. Thibault, E. Drockenmuller, M. Messerschmidt, B. Voit, T. P. Russell, and C. J. Hawker, *J. Am. Chem. Soc.*, *127*, 14942 (2005).
152. N. V. Tsarevsky, K. V. Bernaerts, B. Dufour, F. E. Du Prez, and K. Matyjaszewski, *Macromolecules*, *37*, 9308 (2004).
153. D. B. Ramachary and C. F. Barbas, III, *Eur. J. Chem.*, *10*, 5323 (2004).
154. B. Helms, J. L. Mynar, C. J. Hawker, and J. M. J. Frechet, *J. Am. Chem. Soc.*, *126*, 15020 (2004).
155. R. Manetsch, A. Krasinski, Z. Radic, J. Raushel, P. Taylor, K. B. Sharpless, and H. C. Kolb, *J. Am. Chem. Soc.*, *126*, 12809 (2004).
156. P. Wu, A. K. Feldman, A. K. Nugent, C. J. Hawker, A. Scheel, B. Voit, J. Pyun, J. M. J. Frechet, K. B. Sharpless, and V. V. Fokin, *Angew. Chem., Int. Ed. Engl.*, *43*, 3928 (2004).
157. A. E. Speers and B. F. Cravatt, *Chem. Biol.*, *11*, 535 (2004).
158. L. V. Lee, M. L. Mitchell, S.-J. Huang, V. V. Fokin, K. B. Sharpless, and C.-H. Wong, *J. Am. Chem. Soc.*, *125*, 9588 (2003).

159. T. S. Seo, Z. Li, H. Ruparel, and J. Ju, *J. Org. Chem.*, *68*, 609 (2003).
160. W. G. Lewis, L. G. Green, F. Grynszpan, Z. Radic, P. R. Carlier, P. Taylor, M. G. Finn, and K. B. Sharpless, *Angew. Chem., Int. Ed. Engl.*, *41*, 1053 (2002).
161. R. J. Cross and M. F. Davidson, *Dalton Trans.*, 1987 (1986).
162. C. R. Langrick, D. M. McEwan, P. G. Pringle, and B. L. Shaw, *Dalton Trans.*, 2487 (1983).
163. A. Fürstner and P. W. Davies, *Chem. Commun.*, 2307 (2005).
164. R. R. Schrock, *Dalton Trans.*, 2541 (2001).
165. T. J. Katz and J. McGinnis, *J. Am. Chem. Soc.*, *97*, 1592 (1975).
166. S. A. Krouse and R. R. Schrock, *Macromolecules*, *22*, 2569 (1989).
167. R. R. Schrock, *Chem. Commun.*, 2773 (2005).
168. K. Weiss, A. Michel, E.-M. Auth, U. H. F. Bunz, T. Mangel, and K. Muellen, *Angew. Chem., Int. Ed. Engl.*, *36*, 506 (1997).
169. S. Pulst, F. G. Kirchbauer, B. Heller, W. Baumann, and U. Rosenthal, *Angew. Chem., Int. Ed. Engl.*, *37*, 1925 (1998).
170. A. Fuerstner, O. Guth, A. Rumbo, and G. Seidel, *J. Am. Chem. Soc.*, *121*, 11108 (1999).
171. A. Fuerstner, K. Radkowski, J. Grabowski, C. Wirtz, and R. Mynott, *J. Org. Chem.*, *65*, 8758 (2000).
172. U. H. F. Bunz, *Chem. Rev.*, *100*, 1605 (2000).
173. D. Song, G. Blond, and A. Furstner, *Tetrahedron*, *59*, 6899 (2003).
174. M. Ijsselstijn, B. Aguilera, G. A. van der Marel, J. H. van Boom, F. L. van Delft, H. E. Schoemaker, H. S. Overkleeft, F. P. J. T. Rutjes, and M. Overhand, *Tetrahedron Lett.*, *45*, 4379 (2004).
175. G. Brizius, N. G. Pschirer, W. Steffen, K. Stitzer, H.-C. zur Loye, and U. H. F. Bunz, *J. Am. Chem. Soc.*, *122*, 12435 (2000).
176. E. B. Bauer, S. Szafert, F. Hampel, and J. A. Gladysz, *Organometallics*, *22*, 2184 (2003).
177. E. B. Bauer, F. Hampel, and J. A. Gladysz, *Adv. Syn. Catal.*, *346*, 812 (2004).
178. W. Zhang and J. S. Moore, *J. Am. Chem. Soc.*, *126*, 12796 (2004).
179. W. Zhang and J. S. Moore, *J. Am. Chem. Soc.*, *127*, 11863 (2005).
180. J. Manna, K.D. John and M.D. Hopkins, *Adv. Organomet. Chem*, *38*, 79 (1998).
181. T. Kaharu, H. Matsubara, and S. Takahashi, *J. Mat. Chem.*, *1*, 145 (1991).
182. H.B. Fyfe, M. Mlekuz, G. Stringer, N.J. Taylor and T.B. Marder, *Inorganic and Organometallic Polymers with Special Properties.*, Kluwer Academic Publishers, Dordrecht, The Netherlands, 1992.
183. D.W. Bruce and D. O'Hare, *Inorganic Materials*, John Wiley & Sons, Inc., London, 1992.
184. P.N. Prasad and D.J. Williams, *Introduction to Nonlinear Optical Effects in Molecules and Polymers.*, Wiley-Interscience, New York, 1991.
185. T. L. Schull, J. G. Kushmerick, C. H. Patterson, C. George, M. H. Moore, S. K. Pollack, and R. Shashidhar, *J. Am. Chem. Soc.*, *125*, 3202 (2003).
186. V. W.-W. Yam, *Chem. Commun.*, 789 (2001).
187. M. H. Chisholm, *Angew. Chem.*, *103*, 690 (1991).
188. H. Lang, *Angew. Chem., Int. Ed. Engl*, *33*, 547 (1994).
189. U. H. F. Bunz, *Angew. Chem., Int. Ed. Engl*, *35*, 969 (1996).
190. S. K. Hurst, M. P. Cifuentes, A. M. McDonagh, M. G. Humphrey, M. Samoc, B. Luther-Davies, I. Asselberghs, and A. Persoons, *J. Organomet. Chem.*, *642*, 259 (2002).

191. S. K. Hurst, M. G. Humphrey, J. P. Morrall, M. P. Cifuentes, M. Samoc, B. Luther-Davies, G. A. Heath, and A. C. Willis, *J. Organomet. Chem.*, *670*, 56 (2003).
192. A. Abe, N. Kimura, and S. Tabata, *Macromolecules*, *24*, 6238 (1991).
193. A. La Groia, A. Ricci, M. Bassetti, D. Masi, C. Bianchini, and C. Lo Sterzo, *J. Organomet. Chem.*, *683*, 406 (2003).
194. M. I. Bruce, *Coord. Chem. Rev.*, *166*, 91 (1997).
195. S. Rigaut, J. Perruchon, L. Le Pichon, D. Touchard, and P. H. Dixneuf, *J. Organomet. Chem.*, *670*, 37 (2003).
196. R. Dembinski, T. Bartik, B. Bartik, M. Jaeger, and J. A. Gladysz, *J. Am. Chem. Soc.*, *122*, 810 (2000).
197. M. Mayor, C. Von Hanisch, H. B. Weber, J. Reichert, and D. Beckmann, *Angew. Chem., Int. Ed. Engl.*, *41*, 1183 (2002).
198. V. W.-W. Yam, W. K.-M. Fung, and K.-K. Cheung, *Organometallics*, *17*, 3293 (1998).
199. V. W.-W. Yam, W. K.-M. Fung, and K.-K. Cheung, *Chem. Commun.*, 963 (1997).
200. V. W.-W. Yam, W.-K. Lee, K.-K. Cheung, B. Crystall, and D. Phillips, *J. Chem. Soc., Dalton Trans.: Inorg. Chem.*, 3283 (1996).
201. V. W.-W. Yam, S. H.-F. Chong, and K.-K. Cheung, *Chem. Commun.*, 2121 (1998).
202. V. W.-W. Yam, S. H.-F. Chong, C.-C. Ko, and K.-K. Cheung, *Organometallics*, *19*, 5092 (2000).
203. V. W.-W. Yam, W.-Y. Lo, and N. Zhu, *Chem. Commun.*, 2446 (2003).
204. C. W. C. Chan, Luk Ki; Che, Chi Ming, *Coord. Chem. Rev.*, *132*, 87 (1994).
205. C. L. Choi, Y. F. Cheng, C. Yip, D. L. Phillips, and V. W.-W. Yam, *Organometallics*, *19*, 3192 (2000).
206. W. B. Connick, D. Geiger, and R. Eisenberg, *Inorg. Chem.*, *38*, 3264 (1999).
207. E. A. B. Ed. Baralt, J. N. Demas, P. Galen Lenhert, C. M. Lukehart, Andrew T. McPhail, Donald R. McPhail, James B. Myers, LouAnn Sacksteder, and William R. True, *Organometallics*, *8*, 2417 (1989).
208. S. Fernandez, J. Fornies, B. Gil, J. Gomez, and E. Lalinde, *Dalton Trans.*, 822 (2003).
209. M. Hissler, W. B. Connick, D. K. Geiger, J. E. McGarrah, D. Lipa, R. J. Lachicotte, and R. Eisenberg, *Inorg. Chem.*, *39*, 447 (2000).
210. T. E. Mueller, S. W.-K. Choi, D. M. P. Mingos, D. Murphy, D. J. Williams, and V. W.-W. Yam, *J. Organomet. Chem.*, *484*, 209 (1994).
211. I. E. Pomestchenko, C. R. Luman, M. Hissler, R. Ziessel, and F. N. Castellano, *Inorg. Chem.*, *42*, 1394 (2003).
212. L. Sacksteder, E. Baralt, B. A. DeGraff, C. M. Lukehart, and J. N. Demas, *Inorg. Chem.*, *30*, 2468 (1991).
213. T. J. Wadas, R. J. Lachicotte, and R. Eisenberg, *Inorg. Chem.*, *42*, 3772 (2003).
214. C. E. Whittle, J. A. Weinstein, M. W. George, and K. S. Schanze, *Inorg. Chem.*, *40*, 4053 (2001).
215. V. W.-W. Yam, L. P. Chan, and T. F. Lai, *Organometallics*, *12*, 2197 (1993).
216. V. W.-W. Yam, W. K.-M. Fung, and K.-K. Cheung, *Organometallics*, *16*, 2032 (1997).
217. V. W.-W. Yam, C.-K. Hui, K. M.-C. Wong, N. Zhu, and K.-K. Cheung, *Organometallics*, *21*, 4326 (2002).
218. V. W.-W. Yam, C.-H. Tao, L. Zhang, K. M.-C. Wong, and K.-K. Cheung, *Organometallics*, *20*, 453 (2001).
219. V. W.-W. Yam, P. K.-Y. Yeung, L.-P. Chan, W.-M. Kwok, D. L. Phillips, K.-L. Yu, R. W.-K. Wong, H. Yan, and Q.-J. Meng, *Organometallics*, *17*, 2590 (1998).

220. C.-M. Che, H.-Y. Chao, V. M. Miskowski, Y. Li, and K.-K. Cheung, *J. Am. Chem. Soc.*, *123*, 4985 (2001).
221. K.-K. C. Xiao Hong, Chun-Xiao Guo, and Chi-Ming Che, *Dalton Trans.*, *13*, 1867 (1994).
222. W. Lu, N. Zhu, and C.-M. Che, *J. Organomet. Chem.*, *670*, 11 (2003).
223. S. W.-K. C. Vivian W.-W. Yam, and K.-K. Cheung, *Organometallics*, *15*, 1734 (1996).
224. V. W.-W. Yam, K.-L. Cheung, E. C.-C. Cheng, N. Zhu, and K.-K. Cheung, *Dalton Trans.*, 1830 (2003).
225. V. W.-W. Yam, S. W.-K. Choi, and K.-K. Cheung, *Dalton Trans.*, *16*, 3411 (1996).
226. V. W.-W. Yam and S. W.-K. Choi, *Dalton Trans.*, *22*, 4227 (1996).
227. V. W. W. Yam, W. K. Lee, and T. F. Lai, *Organometallics*, *12*, 2383 (1993).
228. V. W.-W. Yam, W. K.-M. Fung, and K.-K. Cheung, *J. Cluster Sci.*, *10*, 37 (1999).
229. V. W.-W. Yam, W. K.-M. Fung, and M.-T. Wong, *Organometallics*, *16*, 1772 (1997).
230. V. W.-W. Yam, W. K.-M. Fung, K. M.-C. Wong, V. C.-Y. Lau, and K.-K. Cheung, *Chem. Commun.*, 777 (1998).
231. C.-H. Tao, N. Zhu, and V. W.-W. Yam, *Eur. J. Chem.*, *11*, 1647 (2005).
232. Y. Fujikura, K. Sonogashira, and N. Hagihara, *Chem. Lett.*, 1067 (1975).
233. S.-C. Chan, M. C. W. Chan, Y. Wang, C.-M. Che, K.-K. Cheung, and N. Zhu, *Eur. J. Chem.*, *7*, 4180 (2001).
234. P. C. Ford, E. Cariati, and J. Bourassa, *Chem. Rev.*, *99*, 3625 (1999).
235. D. M. Roundhill, H. B. Gray, and C. M. Che, *Acc. Chem. Res.*, *22*, 55 (1989).
236. P. D. Harvey, R. F. Dallinger, W. H. Woodruff, and H. B. Gray, *Inorg. Chem.*, *28*, 3057 (1989).
237. S. F. Rice, V. M. Miskowski, and H. B. Gray, *Inorg. Chem.*, *27*, 4704 (1988).
238. P. D. Harvey and H. B. Gray, *J. Am. Chem. Soc.*, *110*, 2145 (1988).
239. A. E. Stiegman, S. F. Rice, H. B. Gray, and V. M. Miskowski, *Inorg. Chem.*, *26*, 1112 (1987).
240. D. Li, X. Hong, C.-M. Che, W.-C. Lo, and S.-M. Peng, *Dalton Trans.*, *19*, 2929 (1993).
241. S.-K. Yip, E. C.-C. Cheng, L.-H. Yuan, N. Zhu, and V. W.-W. Yam, *Angew. Chem., Int. Ed. Engl.*, *43*, 4954 (2004).
242. H.-Y. Chao, W. Lu, Y. Li, M. C. W. Chan, C.-M. Che, K.-K. Cheung, and N. Zhu, *J. Am. Chem. Soc.*, *124*, 14696 (2002).
243. K.-L. Cheung, S.-K. Yip, and V. W.-W. Yam, *J. Organomet. Chem.*, *689*, 4451 (2004).
244. P. C. Ford and A. Vogler, *Acc. Chem. Res.*, *26*, 220 (1993).
245. W.-F. Fu, K.-C. Chan, K.-K. Cheung, and C.-M. Che, *Eur. J. Chem.*, *7*, 4656 (2001).
246. W. J. Hunks, M.-A. MacDonald, M. C. Jennings, and R. J. Puddephatt, *Organometallics*, *19*, 5063 (2000).
247. A. Kishimura, T. Yamashita, and T. Aida, *J. Am. Chem. Soc.*, *127*, 179 (2005).
248. Y.-A. Lee and R. Eisenberg, *J. Am. Chem. Soc.*, *125*, 7778 (2003).
249. W. Lu, N. Zhu, and C.-M. Che, *J. Am. Chem. Soc.*, *125*, 16081 (2003).
250. G. M. M. Agostina Cinellu, M. V. Pinna, S. Stoccoro, A. Zucca, and M. Manassero, *Dalton Trans.*, *16*, 2823 (1999).
251. A. Johnson and R. J. Puddephatt, *Dalton Trans.*, *14*, 1384 (1977).
252. M. A. Rawashdeh-Omary, M. A. Omary, J. P. Fackler, Jr., R. Galassi, B. R. Pietroni, and A. Burini, *J. Am. Chem. Soc.*, *123*, 9689 (2001).
253. R. L. White-Morris, M. M. Olmstead, F. Jiang, D. S. Tinti, and A. L. Balch, *J. Am. Chem. Soc.*, *124*, 2327 (2002).

254. V. W.-W. Yam, E. C.-C. Cheng, and K.-K. Cheung, *Angew. Chem., Int. Ed. Engl.*, *38*, 197 (1999).
255. V. W.-W. Yam, E. C.-C. Cheng, and Z.-Y. Zhou, *Angew. Chem., Int. Ed. Engl.*, *39*, 1683 (2000).
256. V. W.-W. Yam and K. K.-W. Lo, *Chem. Soc. Rev.*, *28*, 323 (1999).
257. L. A. Mendez, J. Jimenez, E. Cerrada, F. Mohr, and M. Laguna, *J. Am. Chem. Soc.*, *127*, 852 (2005).
258. D. R. McMillin and J. J. Moore, *Coord. Chem. Rev.*, *229*, 113 (2002).
259. V. W.-W. Yam, K. M.-C. Wong, L.-L. Hung, and N. Zhu, *Angew. Chem., Int. Ed. Engl.*, *44*, 3107 (2005).
260. V. W.-W. Yam, S. W.-K. Choi, T.-F. Lai, and W.-K. Lee, *Dalton Trans.*, *6*, 1001 (1993).
261. C.-W. Chan, W.-T. Wong, and C.-M. Che, *Inorg. Chem.*, *33*, 1266 (1994).
262. V. W.-W. Yam, K. M.-C. Wong, and N. Zhu, *Angew. Chem., Int. Ed. Engl.*, *42*, 1400 (2003).
263. W. Lu, B.-X. Mi, M. C. W. Chan, Z. Hui, C.-M. Che, N. Zhu, and S.-T. Lee, *J. Am. Chem. Soc.*, *126*, 4958 (2004).
264. K. M.-C. Wong, X. Zhu, L.-L. Hung, N. Zhu, V. W.-W. Yam, and H.-S. Kwok, *Chem. Commun.*, 2906 (2005).
265. M. Bardaji, A. Laguna, J. Vicente, and P. G. Jones, *Inorg. Chem.*, *40*, 2675 (2001).
266. A. Vogler and H. Kunkely, *Coord. Chem. Rev.*, *219*, 489 (2001).
267. J. A. Bailey, M. G. Hill, R. E. Marsh, V. M. Miskowski, W. P. Schaefer, and H. B. Gray, *Inorg. Chem.*, *34*, 4591 (1995).
268. R. Buchner, C. T. Cunningham, J. S. Field, R. J. Haines, D. R. McMillin, and G. C. Summerton, *Dalton Trans.*, 711 (1999).
269. V. W.-W. Yam, K. M.-C. Wong, and N. Zhu, *J. Am. Chem. Soc.*, *124*, 6506 (2002).
270. C. Yu, K. M.-C. Wong, K. H.-Y. Chan, and V. W.-W. Yam, *Angew. Chem., Int. Ed. Engl.*, *44*, 791 (2005).
271. S. M. Aharoni, *Macromolecules*, *12*, 94 (1979).
272. S. Takahashi, E. Murata, M. Kariya, K. Sonogashira, and N. Hagihara, *Macromolecules*, *12*, 1016 (1979).
273. S. Takahashi, Y. Takai, H. Morimoto, and K. Sonogashira, *Chem. Commun.*, 3 (1984).
274. S. Di Bella, *Chem. Soc. Rev.*, *30*, 355 (2001).
275. I.R. Whittall, A.M. McDonagh, M.G. Humphrey, and M. Samoc, *Adv. Organomet. Chem.*, *42*, 291 (1998).
276. I.R. Whittall, A.M. McDonagh, M.G. Humphrey, and M. Samoc, *Adv. Organomet. Chem*, *43*, 349 (1999).
277. S. R. Marder, *Inorganic Materials*, John Wiley & Sons, Inc., Chichester, 1992.
278. T. Verbiest, S. Houbrechts, M. Kauranen, K. Clays, and A. Persoons, *J. Mat. Chem.*, *7*, 2175 (1997).
279. A. J. Heeger, J. Orenstein, and D. Ulrich, *Nonlinear Optical Properties of Polymers*, Materials Research Society, Pittsburgh, 1988.
280. D. S. Chemla, *Nonlinear Optical Properties of Organic Molecules and Crystals I*, Academic Press, Orlando, FL, 1987.
281. D. S. Chemla, *Nonlinear Optical Properties of Organic Molecules and Crystals II*, Academic Press, Orlando, FL, 1987.
282. G.J. Ashwell and D. Bloor, *Organic Materials for Non-linear Optics III*, Royal Society of Chemistry, Cambridge, 1993.

283. R. A. Hann, *Organic Materials for Non-Linear Optics*, Royal Society of Chemistry, London, 1989.

284. F. K. J. Messier, P. Prasad, and D. Ulrich, *Nonlinear Optical Effects in Organic Polymers*, Kluwer Academic Publishers, Dordrecht, The Netherlands, 1989.

285. F. K. J. Messier and P. Prasad, *Organic Molecules for Nonlinear Optics and Photonics*, Kluwer Academic Publishers, Dordrecht, The Netherlands, 1991.

286. B. Kirtman and B. Champagne, *Inter. Rev. Phys. Chem.*, *16*, 389 (1997).

287. T. Kobayashi, *Nonlinear Optics of Organics and Semiconductors*, Springer-Verlag, Berlin, 1989.

288. M. H. Lyons, *Materials for Non-Linear and Electro-Optics*, Institute of Physics, Bristol, 1989.

289. E. R.A. Hann, *Organic Materials for Non-linear Optics II*, Royal Society of Chemistry, London, 1991.

290. S.R. Marder, J.E. Sohn, and G.D. Stucky, *Materials for Nonlinear Optics, Chemical Perspectives*, American Chemical Society, Washington, DC, 1991.

291. D. J. Williams, *Nonlinear Optical Properties of Organic and Polymeric Materials*, American Chemical Society, Washington, DC, 1983.

292. D. J. Williams, *Angew. Chem. Int. Edengl.*, *23*, 690 (1984).

293. M. A. R. David R. Kanis, and Tobin J. Marks, *Chem. Rev*, *94*, 195 (1994).

294. H. S. Nalwa, *Appl. Organomet. Chem*, *5*, 1 (1991).

295. H. S. Nalwa, *Adv. Mater.*, *5*, 341 (1993).

296. H. S. Nalwa, T. Watanabe, and S. Miyata, *Adv. Mater.*, *7*, 754 (1995).

297. R. Nast, *Angew. Chem. Int. Edengl.*, *72*, 26 (1960).

298. M. G. Humphrey, *J. Organomet. Chem.*, *670*, 1 (2003).

299. M. P. Cifuentes and M. G. Humphrey, *J. Organomet. Chem.*, *689*, 3968 (2004).

300. B. J. Coe, *Eur. J. Chem.*, *5*, 2464 (1999).

301. S. K. Hurst, M. P. Cifuentes, J. P. L. Morrall, N. T. Lucas, I. R. Whittall, M. G. Humphrey, I. Asselberghs, A. Persoons, M. Samoc, B. Luther-Davies, and A. C. Willis, *Organometallics*, *20*, 4664 (2001).

302. T. Weyland, I. Ledoux, S. Brasselet, J. Zyss, and C. Lapinte, *Organometallics*, *19*, 5235 (2000).

303. I. R. Whittall, M. G. Humphrey, S. Houbrechts, A. Persoons, and D. C. R. Hockless, *Organometallics*, *15*, 5738 (1996).

304. J. Chatt and L. A. Duncanson, *J. Chem. Soc.*, 2939 (1953).

305. M. J. S. Dewar, *Bull. Soc. Chim. Fr.*, C71 (1951).

306. R. B. King, *Inorg. Chem.*, *7*, 1044 (1968).

307. G. Frenking and N. Froehlich, *Chem. Rev.*, *100*, 717 (2000).

308. J. L. Templeton, *Adv. Organomet. Chem.*, *29*, 1 (1989).

309. M. Iyoda, Y. Kuwatani, M. Oda, K. Tatsumi, and A. Nakamura, *Angew. Chem.*, *30*, 1670 (1991).

310. M. Sodupe, C. W. Bauschlicher, Jr., S. R. Langhoff, and H. Partridge, *J. Phys. Chem.*, *96*, 2118 (1992).

311. M. Sodupe and C. W. Bauschlicher, Jr., *J. Phys. Chem.*, *95*, 8640 (1991).

312. D. P. Tate and J. M. Augl, *J. Am. Chem. Soc.*, *85*, 2174 (1963).

313. D. P. Tate, J. M. Augl, W. M. Ritchey, B. L. Ross, and J. G. Grasselli, *J. Am. Chem. Soc.*, *86*, 3261 (1964).

314. E. M. Armstrong, P. K. Baker, M. E. Harman, and M. B. Hursthouse, *Dalton Trans.*, 295 (1989).

315. E. M. Armstrong, P. K. Baker, and M. G. B. Drew, *Organometallics*, *7*, 319 (1988).
316. N. G. Aimeloglou, P. K. Baker, M. M. Meehan, and M. G. B. Drew, *Polyhedron*, *17*, 3455 (1998).
317. E. M. Armstrong, P. K. Baker, K. R. Flower, and M. G. B. Drew, *Dalton Trans.*, 2535 (1990).
318. P. K. Baker and P. L. Veale, *Trans. Met. Chem.*, *28*, 418 (2003).
319. T. Ajayi-Obe, E. M. Armstrong, P. K. Baker, and S. Prakash, *J. Organomet. Chem.*, *468*, 165 (1994).
320. T. Szymanska-Buzar and T. Glowiak, *J. Organomet. Chem.*, *523*, 63 (1996).
321. K. Severin, R. Bergs, and W. Beck, *Angew. Chem., Int. Ed. Engl.*, *37*, 1635 (1998).
322. T. P. Curran, A. L. Grant, R. A. Lucht, J. C. Carter, and J. Affonso, *Org. Lett.*, *4*, 2917 (2002).
323. T. P. Curran, R. S. H. Yoon, and B. R. Volk, *J. Organomet. Chem.*, *689*, 4837 (2004).
324. K. R. Birdwhistell, T. L. Tonker, and J. L. Templeton, *J. Am. Chem. Soc.*, *109*, 1401 (1987).
325. T. Y. Hsiao, P. L. Kuo, C. H. Lai, C. H. Cheng, C. Y. Cheng, and S. L. Wang, *Organometallics*, *12*, 1094 (1993).
326. C. H. Lai, C. H. Cheng, C. Y. Cheng, and S. L. Wang, *J. Organomet. Chem.*, *458*, 147 (1993).
327. J. A. Connor and G. A. Hudson, *J. Organomet. Chem.*, *97*, C43 (1975).
328. K. H. Doetz and J. Muehlemeier, *Angew. Chem.*, *94*, 936 (1982).
329. H. H. Wenk, M. Winkler, and W. Sander, *Angew. Chem. Int. Ed. Engl.*, *42*, 502 (2003).
330. J. Fornies, E. Lalinde, A. Martin, and M. T. Moreno, *J. Organomet. Chem.*, *490*, 179 (1995).
331. M. Al-Anber, B. Walfort, T. Stein, and H. Lang, *Inorg. Chim. Acta*, *357*, 1675 (2004).
332. H. Lang, A. Del Villar, B. Walfort, and G. Rheinwald, *J. Organomet. Chem.*, *682*, 155 (2003).
333. H. Lang and A. del Villar, *J. Organomet. Chem.*, *670*, 45 (2003).
334. H. Lang, E. Meichel, T. Stein, C. Weber, J. Kralik, G. Rheinwald, and H. Pritzkow, *J. Organomet. Chem.*, *664*, 150 (2002).
335. D. M. P. Mingos, R. Vilar, and D. Rais, *J. Organomet. Chem.*, *641*, 126 (2002).
336. D. M. P. Mingos, J. Yau, S. Menzer, and D. J. Williams, *Angew. Chem., Int. Ed. Engl.*, *34*, 1894 (1995).
337. D. Rais, J. Yau, M. P. Mingos, R. Vilar, A. J. P. White, and D. J. Williams, *Angew. Chem., Int. Ed. Engl.*, *40*, 3464 (2001).
338. A. Kovacs and G. Frenking, *Organometallics*, *18*, 887 (1999).
339. G. Groeger, U. Behrens, and F. Olbrich, *Organometallics*, *19*, 3354 (2000).
340. S. Mihan, K. Sunkel, and W. Beck, *Eur. J. Chem.*, *5*, 745 (1999).
341. N. M. Boag, M. Green, J. A. K. Howard, F. G. A. Stone, and H. Wadepohl, *Dalton Trans.*, 862 (1981).
342. N. M. Boag, M. Green, D. M. Grove, J. A. K. Howard, J. L. Spencer, and F. G. A. Stone, *Dalton Trans.*, 2170 (1980).
343. G. A. Ozin, D. F. McIntosh, W. J. Power, and R. P. Messmer, *Inorg. Chem.*, *20*, 1782 (1981).
344. L. Guo, J. D. Bradshaw, C. A. Tessier, and W. J. Youngs, *Organometallics*, *14*, 586 (1995).
345. D. Walther, T. Klettke, A. Schmidt, H. Goerls, and W. Imhof, *Organometallics*, *15*, 2314 (1996).
346. D. Walther, A. Schmidt, T. Klettke, W. Imhof, and H. Goerls, *Angew. Chem.*, *106*, 1421 (1994).
347. N. Chatani, H. Inove, T. Ikeda, and S. Murai, *J. Org. Chem.*, *65*, 4913 (2000).
348. A. Fürstner and V. Mamane, *J. Org. Chem.*, *67*, 6264 (2002).
349. J. H. Teles, S. Brode, M. Chabanas, *Angew. Chem., Int. Ed. Engl.*, *110*, 1475 (1999).
350. H. Inoue, N. Chatani, and S. Murai, *J. Org. Chem.*, *67*, 1414 (2002).

351. A. Fürstner and V. Mamane, *Chem. Commun.*, 2112 (2003).
352. M. Lautens, W. Klute, and W. Tam, *Chem. Rev.*, *96*, 49 (1996).
353. N. E. Schore, *Chem. Rev.*, *88*, 1081 (1988).
354. A. Torrent, I. Gonzalez, A. Pla-Quintana, A. Roglans, M. Moreno-Manas, T. Parella, and J. Benet-Buchholz, *J. Org. Chem.*, *70*, 2033 (2005).
355. T. Jikyo, M. Eto, and K. Harano, *J. Chem. Soc., Perkin Trans. 1: Org. Bio-Org. Chem.*, 3463 (1998).
356. T. Jikyo, M. Eto, and K. Harano, *Tetrahedron*, *55*, 6051 (1999).
357. Y. Yoshiro and M. Masumura, *Tetrahedron Lett.*, *20*, 1765 (1979).
358. J. D. Rainier and J. E. Imbriglio, *Org. Lett.*, *1*, 2037 (1999).
359. Y. Yamashita, Y. Miyauchi, and M. Masumura, *Chem. Lett.*, 489 (1983).
360. S.-H. Jung, S. J. Park, and H.-N. Cho, *Poly. Preprints (Ame. Chem. Soc., Div. Poly. Chem.)*, *45*, 885 (2004).
361. U. Kumar and T. X. Neenan, *Macromolecules*, *28*, 124 (1995).
362. H.-W. Fruehauf, *Chem. Rev.*, *97*, 523 (1997).
363. I. Ojima, M. Tzamarioudaki, Z. Li, and R. J. Donovan, *Chem. Rev.*, *96*, 635 (1996).
364. H. Yamazaki and N. Hagihara, *J. Organomet. Chem*, *7*, 21 (1967).
365. E. Müller, *Synthesis*, 761 (1974).
366. E. R. F. Gesing, J.-P. Tane; Vollhardt, K. P. C., *Angew. Chem., Int. Ed. Engl*, *19*, 1023 (1980).
367. R. L. V. Halterman, K. P. C. , *Organometallics*, *7*, 883 (1988).
368. H.-J. Knoelker, E. Baum, and J. Heber, *Tetrahedron Lett.*, *36*, 7647 (1995).
369. A. J. Pearson and A. Perosa, *Organometallics*, *14*, 5178 (1995).
370. A. J. Pearson and X. Yao, *Synlett*, 1281 (1997).
371. T. Shibata, N. Toshida, M. Yamasaki, S. Maekawa, and K. Takagi, *Tetrahedron*, *61*, 9974 (2005).
372. S. Yamazaki and Z. Taira, *J. Organomet. Chem.*, *578*, 61 (1999).
373. R. Choukroun, B. Donnadieu, J.-S. Zhao, P. Cassoux, C. Lepetit, and B. Silvi, *Organometallics*, *19*, 1901 (2000).
374. U. Rosenthal, *Angew. Chem., Int. Ed. Engl.*, *43*, 3882 (2004).
375. S. Yamazaki, Z. Taira, T. Yonemura, and A. J. Deeming, *Organometallics*, *24*, 20 (2005).
376. S. Yamazaki, Z. Taira, T. Yonemura, A. J. Deeming, and A. Nakao, *Chem. Lett.*, 1174 (2002).
377. C. P. Casey, T. L. Dzwiniel, S. Kraft, and I. A. Guzei, *Organometallics*, *22*, 3915 (2003).
378. P. Magnus and F. Bennett, *Tetrahedron Lett.*, *30*, 3637 (1989).
379. P. Magnus, P. Carter, J. Elliott, R. Lewis, J. Harling, T. Pitterna, W. E. Bauta, and S. Fortt, *J. Am. Chem. Soc.*, *114*, 2544 (1992).
380. P. Magnus and P. A. Carter, *J. Am. Chem. Soc.*, *110*, 1626 (1988).
381. P. Magnus, R. Lewis, and F. Bennett, *J. Am. Chem. Soc.*, *114*, 2560 (1992).
382. P. Magnus and R. T. Lewis, *Tetrahedron Lett.*, *30*, 1905 (1989).
383. P. Magnus, R. T. Lewis, and J. C. Huffman, *J. Am. Chem. Soc.*, *110*, 6921 (1988).
384. C. Nevado and A. M. Echavarren, *Synthesis*, 167 (2005).
385. T. Tsuchimoto, T. Maeda, E. Shirakawa, and Y. Kawakami, *Chem. Commun.*, 1573 (2000).
386. H. H. Thoi, K.J. Ivin, and J.J. Rooney, *Faraday Trans. I*, *78*, 2227 (1982).
387. T. J. Katz and S.J. Lee, *J. Am. Chem. Soc.*, *102*, 422 (1980).

388. (a) T. J. Katz, T. H. Ho, N. Y. Shih, Y. C. Ying, and V. I. W. Stuart, *J. Am. Chem. Soc.*, *106*, 2659 (1984). (b) P. Hofmann and M. Haemmerle, *Angew. Chem., Int. Ed. Engl.*, *28*, 908 (1989); P. Hofmann, M. Haemmerle, and G. Unfried, *New J. Chem.*, *15*, 769 (1991).
389. S. J. Landon, P. M. Shulman, and G. L. Geoffroy, *J. Am. Chem. Soc.*, *107*, 6739 (1985).
390. C. Bruneau, *Angew. Chem., Int. Ed. Engl.*, *44*, 2328 (2005).
391. A. M. Echavarren and C. Nevado, *Chem. Soc. Rev.*, *33*, 431 (2004).
392. K. Miki, S. Uemura, and K. Ohe, *Chem. Lett.*, *34*, 1068 (2005).
393. K. Miki, T. Yokoi, F. Nishino, K. Ohe, and S. Uemura, *J. Organomet. Chem.*, *645*, 228 (2002).
394. K. Ohe, M.-a. Kojima, K. Yonehara, and S. Uemura, *Angew. Chem., Int. Ed. Engl.*, *35*, 1823 (1996).
395. K. Ohe, K. Miki, T. Yokoi, F. Nishino, and S. Uemura, *Organometallics*, *19*, 5525 (2000).
396. S. Datta, A. Odedra, and R.-S. Liu, *J. Am. Chem. Soc.*, *127*, 11606 (2005).
397. R. J. Madhushaw, C.-Y. Lo, C.-W. Hwang, M.-D. Su, H.-C. Shen, S. Pal, I. R. Shaikh, and R.-S. Liu, *J. Am. Chem. Soc.*, *126*, 15560 (2004).
398. C. P. Casey, S. Kraft, and M. Kavana, *Organometallics*, *20*, 3795 (2001).
399. C. P. Casey, S. Kraft, and D. R. Powell, *J. Am. Chem. Soc.*, *124*, 2584 (2002).
400. C. Bruneau and P. H. Dixneuf, *Acc. Chem. Res.*, *32*, 311 (1999).
401. L. Anorbe, G. Dominguez, and J. Perez-Castells, *Eur. J. Chem.*, *10*, 4938 (2004).
402. A. J. Fletcher and S. D. R. Christie, *Perkin 1*, 1657 (2000).
403. S. T. Diver and A. J. Giessert, *Chem. Rev.*, *104*, 1317 (2004).
404. S. T. Diver and A. J. Giessert, *Synthesis*, 466 (2004).
405. G. C. Lloyd-Jones, *Org. Biomol. Chem.*, *1*, 215 (2003).
406. K. C. Nicolaou, P. G. Bulger, and D. Sarlah, *Angew. Chem., Int. Ed. Engl.*, *44*, 4490 (2005).
407. C. S. Poulsen and R. Madsen, *Synthesis*, 1 (2003).
408. B. M. Trost and M. J. Krische, *Synlett*, 1 (1998).
409. B. M. Trost, *Acc. Chem. Res.*, *35*, 695 (2002).
410. N. Chatani, N. Furukawa, H. Sakurai, and S. Murai, *Organometallics*, *15*, 901 (1996).
411. N. Chatani, H. Inoue, T. Morimoto, T. Muto, and S. Murai, *J. Org. Chem.*, *66*, 4433 (2001).
412. A. Fürstner, F. Stelzer, and H. Szillat, *J. Am. Chem. Soc.*, *123*, 11863 (2001).
413. A. Fürstner, H. Szillat, and F. Stelzer, *J. Am. Chem. Soc.*, *122*, 6785 (2000).
414. E. C. Hansen and D. Lee, *J. Am. Chem. Soc.*, *125*, 9582 (2003).
415. E. C. Hansen and D. Lee, *J. Am. Chem. Soc.*, *126*, 15074 (2004).
416. E. C. Hansen and D. Lee, *Org. Lett.*, *6*, 2035 (2004).
417. S. Mix and S. Blechert, *Org. Lett.*, *7*, 2015 (2005).
418. N. Chatani, T. Morimoto, T. Muto, and S. Murai, *J. Am. Chem. Soc.*, *116*, 6049 (1994).
419. S. Watanuki and M. Mori, *Organometallics*, *14*, 5054 (1995).
420. S. Watanuki, N. Ochifuji, and M. Mori, *Organometallics*, *13*, 4129 (1994).
421. T. J. Katz and T. M. Sivavec, *J. Am. Chem. Soc.*, *107*, 737 (1985).
422. P. F. Korkowski, T. R. Hoye, and D. B. Rydberg, *J. Am. Chem. Soc.*, *110*, 2676 (1988).
423. T. R. Hoye and J.A. Suriano, *Organometallics*, *11*, 2044 (1992).
424. S. Watanuki and M. Mori, *Heterocycles*, *35*, 679 (1993).
425. M. Mori and S. Watanuki, *Chem. Commun.*, *15*, 1082 (1992).

426. R. H. Grubbs, S. J. Miller, and G.C. Fu, *Acc. Chem. Res.*, *28*, 446 (1995).

427. S.-H. Kim, N. Bowden, and R. H. Grubbs, *J. Am. Chem. Soc.*, *116*, 10801 (1994).

428. (a) A. Kinoshita and M. Mori, *Synlett*, 1020 (1994). (b) A. Kinoshita and M. Mori, *J. Org. Chem.*, *61*, 8356 (1996).

429. B. M. Trost and A. S. K. Hashmi, *Angew. Chem. Int. Ed. Engl.*, *32*, 1085 (1993).

430. B. M. Trost and G. J. Tanoury, *J. Am. Chem. Soc.*, *110*, 1636 (1988).

431. B. M. Trost and M. K. Trost, *Tetrahedron Lett.*, *32*, 3647 (1991).

432. B. M. Trost and M. K. Trost, *J. Am. Chem. Soc.*, *113*, 1850 (1991).

433. B. M. Trost, M. Yanai, and K. Hoogsteen, *J. Am. Chem. Soc.*, *115*, 5294 (1993).

434. B. M. Trost and M. Lautens, *J. Am. Chem. Soc.*, *107*, 1781 (1985).

435. M. S. Sanford, J. A. Love, and R. H. Grubbs, *J. Am. Chem. Soc.*, *123*, 6543 (2001).

436. M. S. Sanford, M. Ulman, and R. H. Grubbs, *J. Am. Chem. Soc.*, *123*, 749 (2001).

437. T. M. Trnka and R. H. Grubbs, *Acc. Chem. Res.*, *34*, 18 (2001).

438. T. M. Trnka, J. P. Morgan, M. S. Sanford, T. E. Wilhelm, M. Scholl, T.-L. Choi, S. Ding, M. W. Day, and R. H. Grubbs, *J. Am. Chem. Soc.*, *125*, 2546 (2003).

439. T. R. Hoye, S. M. Donaldson, and T. J. Vos, *Org. Lett.*, *1*, 277 (1999).

440. M. Rosillo, L. Casarrubios, G. Dominguez, and J. Pérez-Castells, *Tetrahedron Lett.*, *42*, 7029 (2001).

441. M. Rosillo, G. Dominguez, L. Casarrubios, U. Amador, and J. Pérez-Castells, *J. Org. Chem.*, *69*, 2084 (2004).

442. N. Chatani, H. Inoue, T. Kotsuma, and S. Murai, *J. Am. Chem. Soc.*, *124*, 10294 (2002).

443. J. Blum, H. Beer-Kraft, and Y. Badrieh, *J. Org. Chem.*, *60*, 5567 (1995).

444. A. Fürstner, H. Szillat, B. Gabor, and R. Mynott, *J. Am. Chem. Soc.*, *120*, 8305 (1998).

445. M. Mendez, M. P. Munoz, C. Nevado, D. J. Cardenas, and A. M. Echavarren, *J. Am. Chem. Soc.*, *123*, 10511 (2001).

446. S. Oi, I. Tsukamoto, S. Miyano, Y. Inoue, *Organometallics*, *20*, 3704 (2001).

447. J. Blanco-Urgoiti, L. Casarrubios, G. Dominguez, and J. Pérez-Castells, *Tetrahedron Lett.*, *42*, 3315 (2001).

448. F.-D. Boyer, I. Hanna, and L. Ricard, *Org. Lett.*, *3*, 3095 (2001).

449. J. S. Clark and O. Bamelin, *Angew. Chem., Int. Ed. Engl.*, *39*, 372 (2000).

450. M. A. Leeuwenburgh, C. C. M. Appeldoorn, P. A. V. Van Hooft, H. S. Overkleeft, G. A. Van der Marel, and J. H. Van Boom, *Eur. J. Org. Chem.*, 873 (2000).

451. M. A. Leeuwenburgh, C. Kulker, H. I. Duynstee, H. S. Overkleeft, G. A. Van der Marel, and J. H. Van Boom, *Tetrahedron*, *55*, 8253 (1999).

452. J. Marco-Contelles, N. Arroyo, and J. Ruiz-Caro, *Synlett*, 652 (2001).

453. C. S. Poulsen and R. Madsen, *J. Org. Chem.*, *67*, 4441 (2002).

454. C. Fernandez-Rivas, M. Mendez, and A. M. Echavarren, *J. Am. Chem. Soc.*, *122*, 1221 (2000).

455. M. Mendez and A. M. Echavarren, *Eur. J. Org. Chem.*, 15 (2002).

456. M. P. Munoz, M. Mendez, C. Nevado, D. J. Cardenas, and A. M. Echavarren, *Synthesis*, 2898 (2003).

457. C. Blaszykowski, Y. Harrak, M.-H. Goncalves, J.-M. Cloarec, A.-L. Dhimane, L. Fensterbank, and M. Malacria, *Org. Lett.*, *6*, 3771 (2004).

458. E.-I. Negishi, *Handbook of Organopalladium Chemistry for Organic Synthesis*, Vol. 1, John Wiley & Sons, Inc., Hoboken, NJ, 2002.

459. J. Tsuji, *Palladium Reagents and Catalysts: Innovations in Organic Synthesis*, John Wiley & Sons, Inc., Chichester, UK, 1995.
460. B. Bosnich, *Acc. Chem. Res.*, *31*, 667 (1998).
461. C.-H. Jun, J.-B. Hong, and D.-Y. Lee, *Synlett*, 1 (1999).
462. T. Kondo, Y. Tsuji and Y. Watanabe, *Tetrahedron Lett.*, *28*, 6229 (1987).
463. D. Seyferth and R. C. Hui, *J. Am. Chem. Soc*, *107*, 4551 (1985).
464. M. P. Cooke and R. M. Parlman, *J. Am. Chem. Soc.*, *99*, 5222 (1977).
465. M. A. Esteruelas, A. V. Gomez, F. J. Lahoz, A. M. Lopez, E. Onate, and L. A. Oro, *Organometallics*, *15*, 3423 (1996).
466. H. Kim and C. Lee, *J. Am. Chem. Soc.*, *127*, 10180 (2005).
467. B. M. Trost and F. D. Toste, *J. Am. Chem. Soc.*, *122*, 714 (2000).
468. B. M. Trost and F. D. Toste, *J. Am. Chem. Soc.*, *121*, 9728 (1999).
469. M. Takimoto, T. Mizuno, Y. Sato, and M. Mori, *Tetrahedron Lett.*, *46*, 5173 (2005).
470. K. Tamao, K. Kobayashi, and Y. Ito, *Synlett*, 539 (1992).
471. R. Grigg, P. Stevenson, and T. Worakun, *Tetrahedron*, *44*, 4967 (1988).
472. T. Ohmura, Y. Yamamoto, and N. Miyaura, *J. Am. Chem. Soc.*, *122*, 4990 (2000).
473. B. M. Trost and Y. H. Rhee, *J. Am. Chem. Soc.*, *125*, 7482 (2003).
474. C. Nieto-Oberhuber, M. P. Munoz, E. Bunuel, C. Nevado, D. J. Cardenas, and A. M. Echavarren, *Angew. Chem., Int. Ed. Engl.*, *43*, 2402 (2004).
475. S. Ma, S. Yu, and Z. Gu, *Angew. Chem., Int. Ed. Engl.*, *45*, 200 (2006).
476. M. P. Munoz, J. Adrio, J. C. Carretero, and A. M. Echavarren, *Organometallics*, *24*, 1293 (2005).
477. C. Nevado, D. J. Cardenas, and A. M. Echavarren, *Eur. J. Chem.*, *9*, 2627 (2003).
478. G. B. Bajracharya, I. Nakamura, and Y. Yamamoto, *J. Org. Chem.*, *70*, 892 (2005).
479. Y. Miyanohana, H. Inoue, and N. Chatani, *J. Org. Chem.*, *69*, 8541 (2004).
480. B. P. Peppers and S. T. Diver, *J. Am. Chem. Soc.*, *126*, 9524 (2004).
481. M. R. Luzung, J. P. Markham, and F. D. Toste, *J. Am. Chem. Soc.*, *126*, 10858 (2004).
482. V. Mamane, T. Gress, H. Krause, and A. Fuerstner, *J. Am. Chem. Soc.*, *126*, 8654 (2004).
483. C. Nieto-Oberhuber, S. Lopez, and A. M. Echavarren, *J. Am. Chem. Soc.*, *127*, 6178 (2005).
484. L. Zhang and S. A. Kozmin, *J. Am. Chem. Soc.*, *126*, 11806 (2004).
485. N. Mezailles, L. Ricard, and F. Gagosz, *Org. Lett.*, *7*, 4133 (2005).
486. O. Buisine, C. Aubert, and M. Malacria, *Eur. J. Chem.*, *7*, 3517 (2001).
487. D. Llerena, C. Aubert, and M. Malacria, *Tetrahedron Lett.*, *37*, 7353 (1996).
488. G. W. O'Neil and A. J. Phillips, *Tetrahedron Lett.*, *45*, 4253 (2004).
489. O. G. Kulinkovich and A. de Meijere, *Chem. Rev.*, *100*, 2789 (2000).
490. M. Mori, T. Hirose, H. Wakamatsu, N. Imakuni, and Y. Sato, *Organometallics*, *20*, 1907 (2001).
491. F. Sato, H. Urabe, and S. Okamoto, *Synlett*, 753 (2000).
492. E.-I. Negishi and T. Takahashi, *Acc. Chem. Res.*, *27*, 124 (1994).
493. M. Narita, H. Urabe, and F. Sato, *Angew. Chem., Int. Ed. Engl.*, *41*, 3671 (2002).
494. S. Kezuka, T. Okado, E. Niou, and R. Takeuchi, *Org. Lett.*, *7*, 1711 (2005).
495. A. Fürstner, R. Martin, and K. Majima, *J. Am. Chem. Soc.*, *127*, 12236 (2005).
496. M. Mori, K. Tonogaki, and N. Nishiguchi, *J. Org. Chem.*, *67*, 224 (2002).
497. R. Stragies, M. Schuster, and S. Blechert, *Angew. Chem. Int. Ed. Engl.*, *36*, 2518 (1997).

498. S. C. Schuerer and S. Blechert, *Synlett*, 166 (1998).
499. M. Schuster and S. Blechert, *Tetrahedron Lett.*, *39*, 2295 (1998).
500. S. C. Schurer and S. Blechert, *Synlett*, 1879 (1999).
501. S. Kotha, S. Halder, and E. Brahmachary, *Tetrahedron*, *58*, 9203 (2002).
502. S. Kotha, S. Halder, E. Brahmachary, and T. Ganesh, *Synlett*, 853 (2000).
503. P. A. Evans, K. W. Lai, and J. R. Sawyer, *J. Am. Chem. Soc.*, *127*, 12466 (2005).
504. P. A. Evans, J. R. Sawyer, K. W. Lai, and J. C. Huffman, *Chem. Commun.*, 3971 (2005).
505. K. Tanaka and G. C. Fu, *J. Am. Chem. Soc.*, *123*, 11492 (2001).
506. K. Tanaka and G. C. Fu, *Angew. Chem., Int. Ed. Engl.*, *41*, 1607 (2002).
507. V. Gevorgyan, A. Takeda, and Y. Yamamoto, *J. Am. Chem. Soc.*, *119*, 11313 (1997).
508. V. Gevorgyan and Y. Yamamoto, *J. Organomet. Chem.*, *576*, 232 (1999).
509. D. Weibel, V. Gevorgyan, and Y. Yamamoto, *J. Org. Chem.*, *63*, 1217 (1998).
510. Y. Hayashi, M. Osawa, K. Kobayashi, and Y. Wakatsuki, *Chem. Commun.*, 1617 (1996).
511. R. Stragies, M. Schuster, and S. Blechert, *Chem. Commun.*, 237 (1999).
512. C.-J. Wu, R. J. Madhushaw, and R.-S. Liu, *J. Org. Chem.*, *68*, 7889 (2003).
513. H.-Y. Lee, H. Y. Kim, H. Tae, B. G. Kim, and J. Lee, *Org. Lett.*, *5*, 3439 (2003).
514. N. Darby, C. U. Kim, J. A. Salaun, K. W. Shelton, S. Takada, and S. Masamune, *Chem. Commun.* (1971).
515. R. G. Bergman, *Acc. Chem. Res.*, *6*, 25 (1973).
516. R. R. Jones and R. G. Bergman, *J. Am. Chem. Soc.*, *94*, 660 (1972).
517. J. R. Fritch and K. P. C. Vollhardt, *J. Am. Chem. Soc.*, *100*, 3643 (1978).
518. E. D. Sternberg and K. P. C. Vollhardt, *J. Org. Chem.*, *49*, 1564 (1984).
519. E. D. Sternberg and K. P. C. Vollhardt, *J. Org. Chem.*, *49*, 1574 (1984).
520. P. Magnus, *Tetrahedron*, *50*, 1397 (1994).
521. P. Magnus and T. Pitterna, *Chem. Commun.*, 541 (1991).
522. T. Hasegawa and L. Pu, *J. Organomet. Chem.*, *527*, 287 (1997).
523. J. M. Blackwell, J. S. Figueroa, F. H. Stephens, and C. C. Cummins, *Organometallics*, *22*, 3351 (2003).
524. Y. Wang and M. G. Finn, *J. Am. Chem. Soc.*, *117*, 8045 (1995).
525. R. Nagata, H. Yamanaka, E. Murahashi, and I. Saito, *Tetrahedron Lett.*, *31*, 2907 (1990).
526. R. Nagata, H. Yamanaka, E. Okazaki, and I. Saito, *Tetrahedron Lett.*, *30*, 4995 (1989).
527. A. G. Myers, E. Y. Kuo, and N. S. Finney, *J. Am. Chem. Soc.*, *111*, 8057 (1989).
528. A. G. Myers and P. S. Dragovich, *J. Am. Chem. Soc.*, *111*, 9130 (1989).
529. K. Ohe, M-a. Kojima, K. Yonehara, and S. Uemura *Angew. Chem. Int. Ed. Engl.*, *35*, 1823 (1996).
530. T. Manabe, S.-i. Yanagi, K. Ohe, and S. Uemura, *Organometallics*, *17*, 2942 (1998).
531. K. K. Baldridge, B. T. Donovan-Merkert, J. M. O'Connor, L. I. Lee, A. Closson, D. Fandrick, T. Tran, K. D. Bunker, M. Fouzi, and P. Gantzel, *Org. Biomol. Chem.*, *1*, 763 (2003).
532. J. M. O'Connor, S. J. Friese, and M. Tichenor, *J. Am. Chem. Soc.*, *124*, 3506 (2002).
533. J. M. O'Connor, L. I. Lee, P. Gantzel, A. L. Rheingold, and K.-C. Lam, *J. Am. Chem. Soc.*, *122*, 12057 (2000).
534. J. M. O'Connor, S. J. Friese, and B. L. Rodgers, *J. Am. Chem. Soc.*, *127*, 16342 (2005).
535. C.-Y. Lo, M. P. Kumar, H.-K. Chang, S.-F. Lush, and R.-S. Liu, *J. Org. Chem.*, *70*, 10482 (2005).

536. G. Bastian, R. Royer, and R. Cavier, *Eur. J. Med. Chem.*, *18*, 365 (1983).
537. F. Kaluza and G. Perold, *Chem. Ber.*, *88*, 597 (1955).
538. C. Roger, G. S. Bodner, W. G. Hatton, and J. A. Gladysz, *Organometallics*, *10*, 3266 (1991).
539. H. Kusama, J. Takaya, and N. Iwasawa, *J. Am. Chem. Soc.*, *124*, 11592 (2002).
540. R. S. Bly, G. S. Silverman, and R. K. Bly, *J. Am. Chem. Soc.*, *110*, 7730 (1988).
541. L. E. Overman and F. M. Knoll, *J. Am. Chem. Soc.*, *102*, 865 (1980).
542. H. Sashida, H. Kurahashi, and T. Tsuchiya, *Chem. Commun.*, 802 (1991).
543. C. A. Landis, M. M. Payne, D. L. Eaton, and J. E. Anthony, *J. Am. Chem. Soc.*, *126*, 1338 (2004).
544. B. P. Warner, S. P. Millar, R. D. Broene, and S. L. Buchwald, *Science*, *269*, 814 (1995).
545. B. König, H. Hollnagel, B. Ahrens, and P. G. Jones, *Angew. Chem., Int. Ed. Engl.*, *34*, 2538 (1995).
546. B. König, W. Pitsch, and I. Thondorf, *J. Org. Chem.*, *61*, 4258 (1996).
547. Typically, cyclization temperatures evaluated by DSC are ∼60–90°C higher in temperature than their corrsponding values in solution.
548. A. Basak and K. R. Rudra, *Tetrahedron Lett.*, *41*, 7231 (2000).
549. A. Basak and J. C. Shain, *Tetrahedron Lett.*, *39*, 3029 (1998).
550. N. L. Coalter, T. E. Concolino, W. E. Streib, C. G. Hughes, A. L. Rheingold, and J. M. Zaleski, *J. Am. Chem. Soc.*, *122*, 3112 (2000).
551. E. W. Schmitt, J. C. Huffman, and J. M. Zaleski, *Chem. Commun.*, 167 (2001).
552. P. J. Benites, D. S. Rawat, and J. M. Zaleski, *J. Am. Chem. Soc.*, *122*, 7208 (2000).
553. D. S. Rawat, P. J. Benites, C. D. Incarvito, A. L. Rheingold, and J. M. Zaleski, *Inorg. Chem.*, *40*, 1846 (2001).
554. T. Chandra, R. A. Allred, B. J. Kraft, L. M. Berreau, and J. M. Zaleski, *Inorg. Chem.*, *43*, 411 (2004).
555. T. Chandra, M. Pink, and J. M. Zaleski, *Inorg. Chem.*, *40*, 5878 (2001).
556. Y.-Z. Hu, C. Chamchoumis, J. S. Grebowicz, and R. P. Thummel, *Inorg. Chem.*, *41*, 2296 (2002).
557. J. Kuzelka, J. R. Farrell, and S. J. Lippard, *Inorg. Chem.*, *42*, 8652 (2003).
558. J. J. Kodanko, D. Xu, D. Song, and S. J. Lippard, *J. Am. Chem. Soc.*, *127*, 16004 (2005).
559. J. J. Kodanko, A. J. Morys, and S. J. Lippard, *Org. Lett.*, *7*, 4585 (2005).
560. N. Schultheiss, C. L. Barnes, and E. Bosch, *Synth. Commun.*, *34*, 1499 (2004).
561. P. Sengupta, H. Zhang, and D. Y. Son, *Inorg. Chem.*, *43*, 1828 (2004).
562. M. R. A. Al-Mandhary and P. J. Steel, *Aust. J. Chem.*, *55*, 705 (2002).
563. N. Schultheiss, C. L. Barnes, and E. Bosch, *Cryst. Growth Des.*, *3*, 573 (2003).
564. T. Kawano, J. Kuwana, and I. Ueda, *Bull. Chem. Soc. Jpn.*, *76*, 789 (2003).
565. S. Shotwell, H. L. Ricks, J. G. M. Morton, M. Laskoski, J. Fiscus, M. D. Smith, K. D. Shimizu, H.-C. zur Loye, and U. H. F. Bunz, *J. Organomet. Chem.*, *671*, 43 (2003).
566. E. Bosch and C. L. Barnes, *J. Coord. Chem.*, *56*, 329 (2003).
567. T. Kawano, J. Kuwana, C.-X. Du, and I. Ueda, *Inorg. Chem.*, *41*, 4078 (2002).
568. T. Chandra, J. C. Huffman, and J. M. Zaleski, *Inorg. Chem. Commun.*, *4*, 434 (2001).
569. D. S. Rawat and J. M. Zaleski, *J. Am. Chem. Soc.*, *123*, 9675 (2001).
570. Sangita and J. M. Zaleski, *J. Am. Chem. Soc*, submitted (2007).
571. S. Bhattacharyya, D. F. Dye, M. Pink, and J. M. Zaleski, *Chem. Commun.*, 5295 (2005).
572. S. Bhattacharyya, M. Pink, J. C. Huffman, and J. M. Zaleski, *Polyhedron*, *25*, 550 (2006).

573. S. Bhattacharyya, A. E. Clark, M. Pink, and J. M. Zaleski, *Chem. Commun.*, 1156 (2003).
574. S. Bhattacharyya, M. Pink, M.-H. Baik, and J. M. Zaleski, *Angew. Chem., Int. Ed. Engl.*, *44*, 592 (2005).
575. M. Prall, A. Wittkopp, A. A. Fokin, and P. R. Schreiner, *J. Comput. Chem.*, *22*, 1605 (2001).
576. A. E. Clark, E. R. Davidson, and J. M. Zaleski, *J. Am. Chem. Soc.*, *123*, 2650 (2001).
577. P. J. Benites, R. C. Holmberg, D. S. Rawat, B. J. Kraft, L. J. Klein, D. G. Peters, H. H. Thorp, and J. M. Zaleski, *J. Am. Chem. Soc.*, *125*, 6434 (2003).
578. B. J. Kraft, N. L. Coalter, M. Nath, A. E. Clark, A. R. Siedle, J. C. Huffman, and J. M. Zaleski, *Inorg. Chem.*, *42*, 1663 (2003).
579. A. E. Clark, E. R. Davidson, and J. M. Zaleski, *Chem. Commun.*, 2876 (2003).
580. B. J. Kraft, C. G. Hughes, B. Robinson, M. Pink, and J. M. Zaleski, *Inorg. Chem.*, submitted (2006).
581. I. V. Alabugin and S. V. Kovalenko, *J. Am. Chem. Soc.*, *124*, 9052 (2002).
582. A. Evenzahav and N. J. Turro, *J. Am. Chem. Soc.*, *120*, 1835 (1998).
583. R. L. Funk, E. R. R. Young, R. M. Williams, M. F. Flanagan, and T. L. Cecil, *J. Am. Chem. Soc.*, *118*, 3291 (1996).
584. T. Chandra, B. J. Kraft, J. C. Huffman, and J. M. Zaleski, *Inorg. Chem.*, *42*, 5158 (2003).
585. R. Li, X. Zhang, P. Zhu, D. K. P. Ng, N. Kobayashi, and J. Jiang, *Inorg. Chem.*, *45*, 2327 (2006).
586. N. Kobayashi and C. C. Leznoff, *J. Porphyrins Phthalocyanines*, *8*, 1015 (2004).
587. A. Kalkan, A. Koca, and Z. A. Bayir, *Polyhedron*, *23*, 3155 (2004).
588. E. M. Garcia-Frutos, S. M. O'Flaherty, E. M. Maya, G. De La Torre, W. Blau, P. Vazquez, and T. Torres, *J. Mat. Chem.*, *13*, 749 (2003).
589. C. C. Leznoff and B. Suchozak, *Can. J. Chem.*, *79*, 878 (2001).
590. R. Faust, *Eur. J. Org. Chem.*, 2797 (2001).
591. R. Faust and F. Mitzel, *Perkin 1*, 4526 (2000).
592. E. M. Maya, C. Garcia, E. M. Garcia-Frutos, P. Vazquez, and T. Torres, *J. Org. Chem.*, *65*, 2733 (2000).
593. C. C. Leznoff, Z. Li, H. Isago, A. M. D'Ascanio, and D. S. Terekhov, *J. Porphyrins Phthalocyanines*, *3*, 406 (1999).
594. H. Isago, D. S. Terekhov, and C. C. Leznoff, *J. Porphyrins Phthalocyanines*, *1*, 135 (1997).
595. H. Isago, C. C. Leznoff, M. F. Ryan, R. A. Metcalfe, R. Davids, and A. B. P. Lever, *Bull. Chem. Soc. Jpn.*, *71*, 1039 (1998).
596. D. S. Terekhov, K. J. M. Nolan, C. R. McArthur, and C. C. Leznoff, *J. Org. Chem.*, *61*, 3034 (1996).
597. A. Odedra, C.-J. Wu, T. B. Pratap, C.-W. Huang, Y.-F. Ran, and R.-S. Liu, *J. Am. Chem. Soc.*, *127*, 3406 (2005).
598. B. P. Taduri, Y.-F. Ran, C.-W. Huang, and R.-S. Liu, *Org. Lett.*, *8*, 883 (2006).
599. L. K. Boerner, M. Nath, M. Pink, and J. M. Zaleski, *Inorg. Chem.*, submitted (2007).
600. M. Nath, L. K. Boerner, M. Pink, and J. M. Zaleski, *Chem. Commun.*, submitted (2007).

CHAPTER 7

# Oxygen Activation Chemistry of Pacman and Hangman Porphyrin Architectures Based on Xanthene and Dibenzofuran Spacers

**JOEL ROSENTHAL and DANIEL G. NOCERA**

*Department of Chemistry, 6-335*
*Massachusetts Institute of Technology*
*Cambridge, MA 02139*

CONTENTS



*Progress in Inorganic Chemistry, Vol. 55* Edited by Kenneth D. Karlin

## I. INTRODUCTION

The activation and functionalization of small-molecule substrates at biological and chemical active sites typically proceed by bond-making and bond-breaking processes that encompass multiple oxidation–reduction events. The catalysis may be kinetically constrained when the multielectron bond rearrangements are driven by uncoupled, single-electron steps. In such instances, intermediates can form that require energetically unfavorable steps to complete the overall process. Natural systems often resolve the kinetic challenge of multielectron catalysis by utilizing two or more metals or proximate redox-active cofactors to deliver multiple oxidizing or reducing equivalents to small-molecule substrates in cooperative fashion (1). Inspired by biology, designers of synthetic catalysts have explored bimetallic cooperativity on the tenet two metal centers working in concert might better promote chemical transformations along multielectron pathways by avoiding nonproductive and uncontrollable one-electron–radical side reactions during small molecule activation. Against this backdrop of biomimetic catalyst design, Chang and Collman first prepared and investigated the chemistry of cofacial porphyrins. These bimetallic architectures comprise two porphyrins linked in a face-to-face arrangement, with the two general connectivites shown in Fig. 1(a): (a) porphyrins linked by two or more flexible strapping units (FTF = face to face) (2–27) and (b) by a single rigid pillar (28–53). As has been documented in several reviews (54–58), these two classes of cofacial bisporphyrins are distinguished by their ability to activate small molecules by more than one electron. The singly bridged Pacman systems have proven to be the most successful in this regard due to their impaired ability to adopt conformations in which there is significant lateral slippage between porphyrin rings, thus allowing for efficient multielectron small-molecule activation with only the need for minor structural reorganization of juxtaposed subunits (2).

(*a*) **Original Pacman Assemblies**

$_n(H_2C)$ $(CH_2)_n$ NH HN O

FTF DPA DPB

(*b*) **Second Generation Pacman Assemblies**

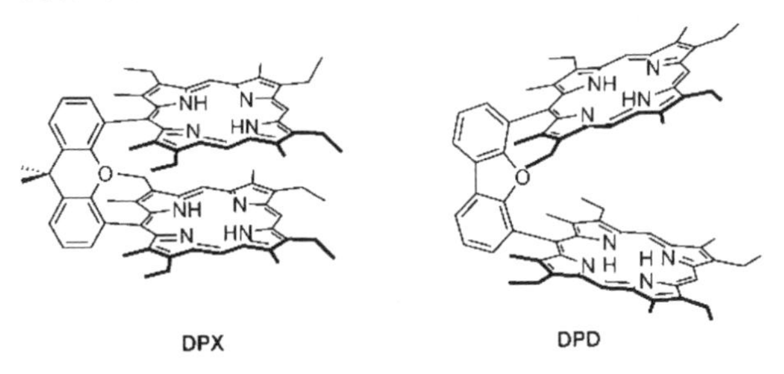


(*c*) **Hangman Porphyrin Assemblies**

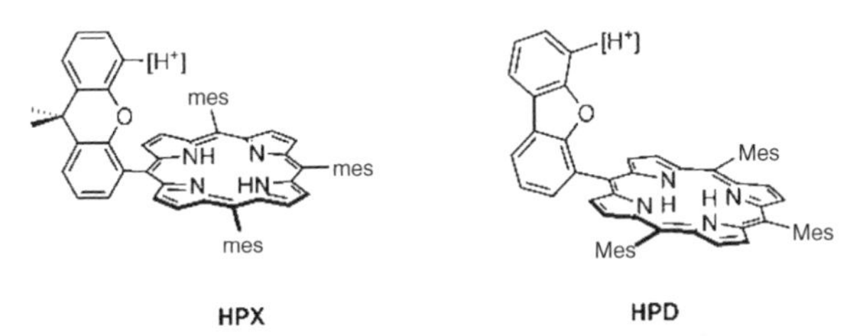


Figure 1. (*a*) Original strapped (FTF) and pillared Pacman bisporphyrin architectures ($n = 1$–3). (*b*) More synthetically accessible pillared Pacman frameworks. (*c*) Hangman porphyrin architecture with acid–base functionality ($[H^+]$ = general proton-transfer functionality). (Mesityl = mes).

Over the past 30 years, the singly bridged "Pacman" systems, which have been most successful in this regard, have featured two spacers, anthracene (DPA = 1,8-bis[5-(2,8,13,17-tetraethyl -3,7,12,18-tetramethylporphyrinyl)]anthracene) and biphenylene (DPB = 1,8- bis[5-(2,8,13,17-tetraethyl -3,7,12,18-tetramethylporphyrinyl)]biphenylene) shown in Fig. 1(*a*). These types of cofacial bisporphyrins containing either cobalt or iron are electrocatalysts for the reduction of dioxygen by two and four electrons (59–64). Ruthenium and osmium cofacial bisporphyrin complexes reduce protons to dihydrogen (65), and dinitrogen by four and six electrons to generate hydrazine and ammonia (50, 51, 66), respectively. In addition, various architectures containing Rh metal centers are able to activate H—H and C—H bonds (53, 67, 68). The exceptional multielectron reactivity displayed by the pillared DPA and DPB bisporphyrins compared to their strapped FTF counterparts has led to the suggestion that cofacial bisporphyrin architectures have a vertical Pacman flexibility that allows the binding pocket to accommodate reaction intermediates structurally during catalysis (54). Nevertheless, the DPA and DPB systems of Fig. 1(*a*) differ in vertical pocket size by only ~1 Å (43), offering a limited range of conformational flexibility for examining structure–reactivity relationships.

With interest in examining the structural limits of vertical flexibility within the Pacman motif, we sought to develop methods for the facile assembly of new cofacial bisporphyrins that exhibit variable pocket sizes with minimal lateral displacements. To this end, Pacman bisporphyrins based on the DPX (DPX = diporphyrin xanthene) and DPD (DPD = 4,6- bis[5-(2,8,13,17-tetraethyl- 3,7,12,18-tetramethylporphyrinyl)] dibenzofuran) architectures shown in Fig. 1(*b*) were targeted. The requisite dialdehyde pillars needed for cofacial assembly, obtained through arduous multistep routes for DPB and DPA (44, 46), are much more easily prepared for the DPX and DPD spacers. The DPX and DPD series of cofacial bisporphyrins display an active multielectron catalysis that is predicated on a proton-coupled electron-transfer (PCET) reactivity (69–73). This reactivity can be preserved upon the removal of one porphyrin subunit from the DPX or DPD pillar, as long as it is replaced by an acid–base functionality appropriately positioned over the remaining porphyrin platform, as shown in Fig. 1(*c*). Much like the Pacman porphyrin constructs, these "Hangman" porphyrins display an active multielectron activation chemistry of small-molecule substrates. This chapter reviews the structural and reaction chemistry of Pacman and Hangman porphyrin complexes, with particular emphasis placed on oxygen activation and oxidation catalysis.

## II. SYNTHESIS AND PHYSICAL PROPERTIES OF PACMAN PORPHYRIN COMPLEXES

### A. General Synthesis

A large vertical cleft size is spanned by DPX and DPD cofacial bisporphyrins. The 4,5-substitution on xanthene affords a DPX cleft with lateral and vertical preorganization comparable to preexisting DPA and DPB systems (Fig. 1). The introduction of porphyrinic substituents at the 4 and 6 positions of dibenzofuran imparts a significantly larger cleft angle on the cofacial bisporphyrin motif to produce a DPD pocket with lateral preorganization and greatly enhanced vertical flexibility. In this section, the ligand syntheses, coordination chemistry, and structural analysis of DPX and DPD cofacial bisporphyrins are described. The results and data collected within this section provide a firm foundation for the discussion of detailed studies of Pacman catalytic applications in subsequent sections.

Installation of xanthene and dibenzofuran spacers into pillared cofacial bisporphyrin architectures is adapted from the original work of Rebek and co-workers (74) and Cram and co-workers (75) involving supramolecular cleft design. The xanthene-bridged cofacial bisporphyrin $H_4$(DPX), as well as the dibenzofuran-bridged homologue $H_4$(DPD) have been prepared as outlined in Schemes 1 and 2. This three-branch approach (Scheme 1) borrows from methods originally developed by Chang et al. (44, 46) and Collman et al. (49) for the preparation of pillared bisporphyrins bridged by anthracene and biphenylene. This convergent route involves the coupling of rigid dicarboxaldehyde bridges with the appropriate α-free pyrrole ethyl esters and dipyrrylmethane dialdehydes to form the cofacial bisporphyrin (Scheme 2).

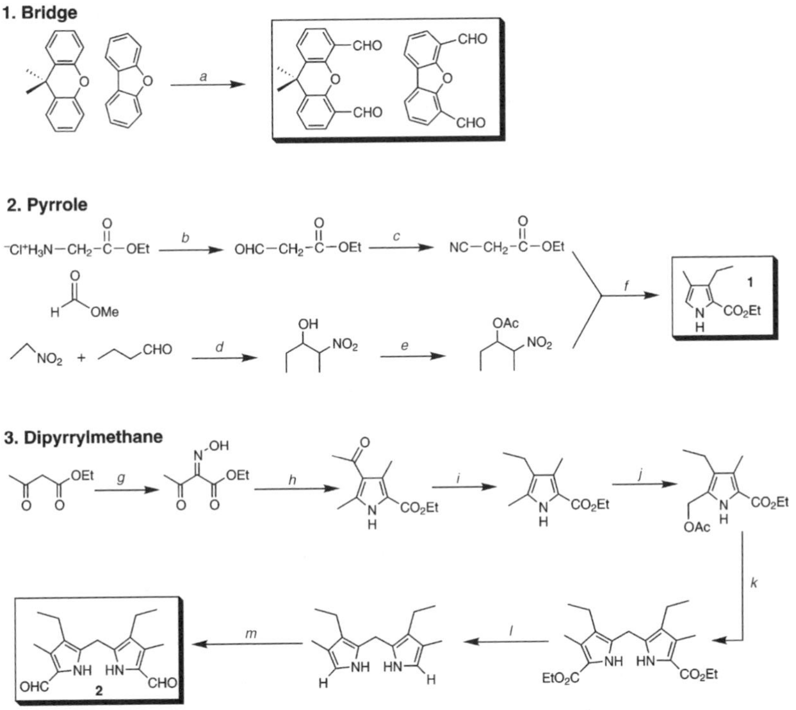

Scheme 1. (*a*) i. *n*-BuLi, TMEDA (TMEDA = *N,N,N′,N′*-tetraethylenediamine); ii. DMF (DMF = dimethylformamide); iii. $H_2O$ (*b*) $Et_3N$, reflux; (*c*) $POCl_3$, DMF; (*d*) KF, IPA (IPA = isopropyl alcohol), <40°C; (*e*) $Ac_2O$, <60°C; (*f*) THF/IPA, DBU (DBU = 1,8-diazobicyclo[5.4.0]undec-7-ene), <30°C; (*g*) $NaNO_2$, $H_2O$, <180°C; (*h*) Zn, 2,4-pentanedione, AcOH, NaOAc, <70°C; (i) $NaBH_4$, $BF_3$•$OEt_2$, 0°C; (*j*) $Pb(OAc)_4$, AcOH, 100°C; (*k*) AcOH, $H_2O$, reflux; (*l*) NaOH, $NH_2NH_2$•$H_2O$, EtOH/$H_2O$, reflux; (*m*) i. $POCl_3$, DMF, ii. $Na_2CO_3$.

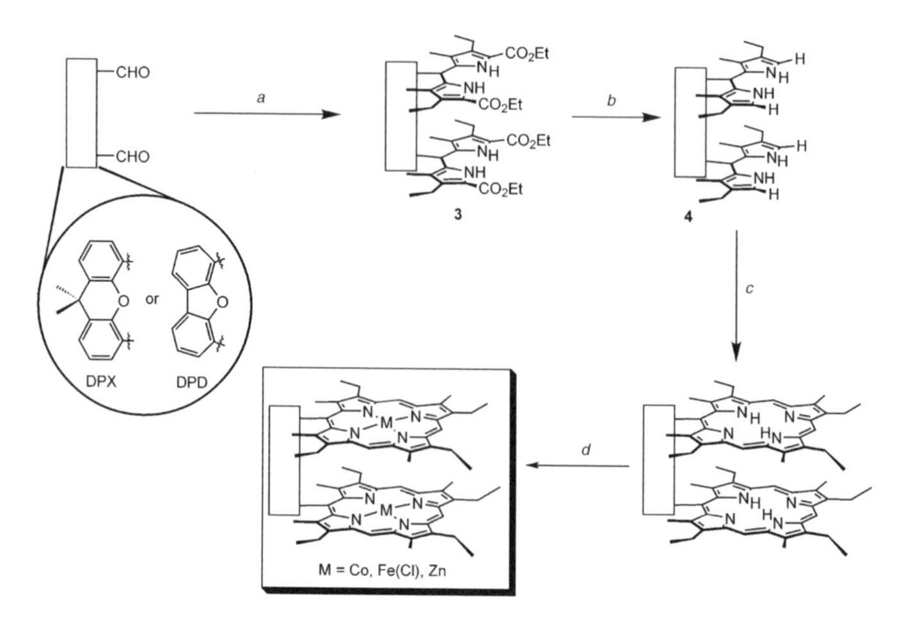


Scheme 2. (*a*) **2**, EtOH, HCl, reflux; (*b*) NaOH, ethylene glycol, reflux; (*c*) i. **2**, PTSA (PTSA = *p*-toluenesulfonic acid), MeOH; ii. *o*-chloranil; (*d*) $MX_2$.

In Scheme 1, the first synthetic branch consists of the regioselective dilithiation of 9,9-dimethylxanthene or dibenzofuran in the presence of dry DMF followed by hydrolysis of the intermediate imidate salt. Both the xanthene and dibenzofuran dialdehyde bridges are afforded in high yield via this facile one-pot reaction offering a significant synthetic improvement over the DPA and DPB systems, which require five-to-eight synthetic steps to deliver the dialdehyde pillars. The second branch of Scheme 1 entails the α-free pyrrole ethyl ester (**1**) synthesis via a five-step procedure using a Barton–Zard strategy (76, 77), while the third branch delineates the preparation of the appropriate dipyrrylmethane dialdehyde (**2**) by a typical seven step synthesis (78).

The three branch coupling shown in Scheme 2 allows for Pacman construction. Reaction of either of the rigid dialdehyde spacers with 4 equiv of **1** in boiling ethanol affords the corresponding ester-protected bisdipyrrylmethane derivatives (**3**). Subsequent saponification of the α-ethyl esters of **3** with sodium hydroxide in ethylene glycol generates the α-free tetrapyrrole **4**. Cyclization of **4** with dipyrrylmethane dialdehyde (**2**) in the presence of a catalytic amount of *p*-toluenesulfonic acid (PTSA) followed by oxidation with *o*-chloranil gives the corresponding bisporphyrins $H_4$(DPX) and $H_4$(DPD). The yield for the final coupling step is comparable to that for the optimized procedure employed by

Collman et al. for the synthesis of $H_4$(DPB) (49). The free-base cofacial bisporphyrins $H_4$(DPD) and $H_4$(DPX) are obtained in 15 and 16 steps, respectively, in an overall yield of 3%. For comparison, the related anthracene- and biphenylene-bridged bisporphyrins $H_4$(DPA) and $H_4$(DPB) are synthesized in 20 and 23 steps, respectively.

The cofacial disposition of the porphyrin rings in $H_4$(DPX) is reflected by the compound's $^1$H NMR spectrum. The further upfield shift of the internal NH-pyrrolic resonances in the proton nuclear magnetic resonance ($^1$H NMR) spectrum of $H_4$(DPX) (−6.42 and −6.80 ppm) as compared to the monomer subunit (−3.10 and −3.29 ppm) is characteristic of a cofacial arrangement; enhanced shielding is caused by ring current interactions between the closely separated porphyrin macrocycles. Sharp NMR signals observed for the meso and methyl protons of $H_4$(DPX) provide support for a nonslipped, stacked conformation of two adjoined porphyrin rings. In contrast, the internal NH-pyrrolic resonances of $H_4$(DPD) are only slightly upshifted (−3.85, −3.91 ppm) as compared to monomeric porphyrins, providing evidence for the greater distance between porphyrin planes of the DPD homologue (see below).

Homobimetallic complexes of $H_4$(DPX) and $H_4$(DPD) are easily prepared by direct reaction of the free-base Pacman assemblies with the appropriate metal salts (Fig. 2) (79). The coordination chemistry of both $H_4$(DPX) and $H_4$(DPD)

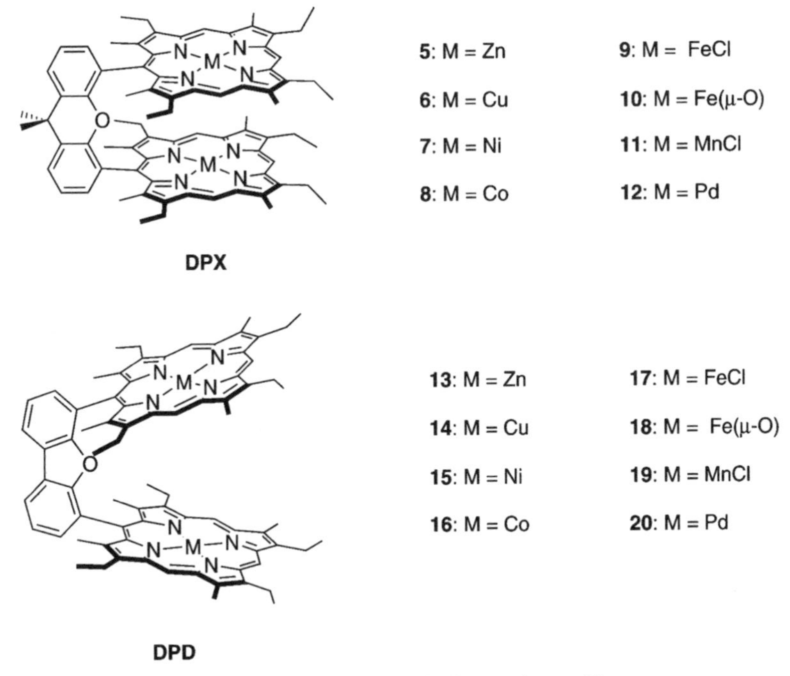


Figure 2. Cofacial Pacman porphyrin complexes of interest.

has been examined most extensively for first-row transition metals. For example, treatment of $H_4$(DPX) with $Zn(OAc)_2 \bullet 2H_2O$ in methanol–chloroform mixtures affords the corresponding binuclear zinc(II) complex $Zn_2$(DPX) (**5**) in excellent yield. Figure 2 presents the various homobimetallic DPX and DPD complexes that will be discussed throughout the chapter.

## B. Structural Characterization

A systematic structural study of the DPX Pacman framework has been provided by a homologous series of dinuclear zinc, copper, and nickel derivatives. The molecular structures of the homobimetallic zinc(II), copper(II), and nickel(II) complexes of the DPX construct (**5**–**7**) are shown in Fig. 3. Trends in bond lengths and angles of macrocyclic core structures and side chains agree well with those observed in related systems including $H_4$(DPA), $H_4$(DPB), and 1,2-bis[5-(2,3,7,8-12,13,17,18-octaethylporphyrinato)]-*cis*-ethene porphyrins (80–83).

The three side-on views of the molecular structures for **5**–**7**, shown in Fig. 4, confirm the ability of the xanthene bridge to hold two porphyrin rings in a cofacial arrangement, akin to what is observed for the analogous anthracene and biphenylene systems. Pertinent geometrical parameters are summarized in Table I. Scheidt and Lee's semiquantitative scheme for the pairwise overlap of the $\pi$ systems of spatially oriented porphyrin monomers within the crystalline lattice (84) provides a useful framework to analyze the structures of the xanthene bisporphyrins described here. The most important geometric features are the lateral shift between the metal centers and the mean separation of the macrocycle planes, which are defined in Table I. Authentic $\pi$–$\pi$ interactions between aromatic macrocycles (as opposed to crystal packing effects) are signified by small lateral shifts ($< 4$ Å) (84).

The data of Table I are unique insofar as lateral shifts are collected for a homologous series of cofacial bisporphyrins anchored by the same pillar. Owing to the inflexibility of the xanthene spacer, the lateral shifts of **5**–**7**, defined by the methine–methine separation perpendicular to the bridge, are small and similar to one another (Fig. 4). Of the three structures, **7** exhibits the least $\pi$ overlap, though the observed lateral shift is only moderately larger than that observed for related Ni(II) cofacial bisporphyrins (i.e., lateral shift of $Ni_2$(DPA) is 2.40 Å). The splayed structure of **7** suggests that the association of the nickel subunits is smaller than that of zinc and copper congeners. This finding is consistent with NMR investigations, which also show the same trend for the aggregation of monomeric, asymmetric metalloporphyrins along this metal series. The interplanar mean plane separations for **5**–**7** do not vary widely, ranging from 3.4 to 3.6 Å.

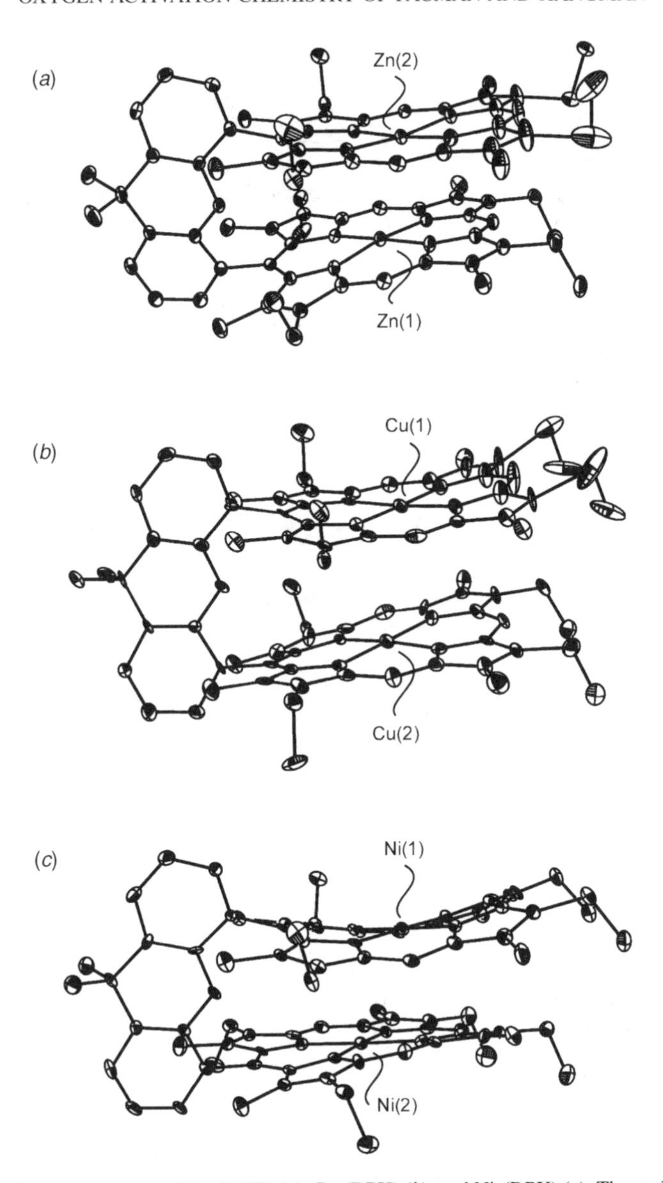


Figure 3. Crystal structures of $Zn_2$(DPX) (*a*), $Cu_2$(DPX) (*b*), and $Ni_2$(DPX) (*c*). Thermal ellipsoids are drawn at the 25% probability level. Hydrogen atoms and solvent molecules within the lattices have been omitted for clarity.

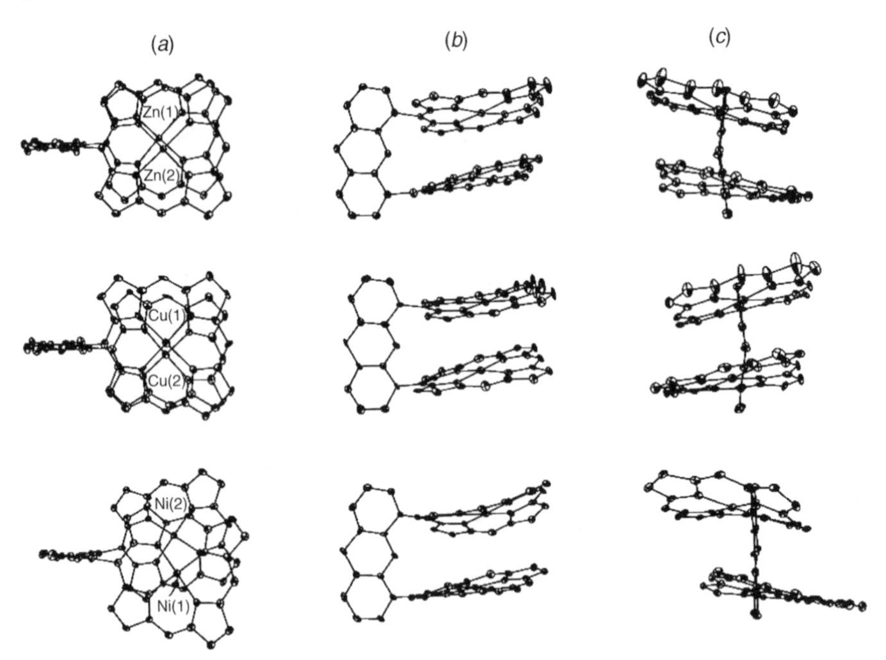


Figure 4. Comparative views, from top to bottom, of the crystal structures of **5**–**7**: (*a*) top view, perpendicular to porphyrin planes, (*b*) side view, perpendicular to bridge plane, (*c*) side view, parallel to bridge plane. Side groups and hydrogen atoms have been omitted for clarity.

TABLE I
Crystallographically Derived Intradimer Geometrical Features for $Zn_2$(DPX) (**5**), $Cu_2$(DPX) (**6**), and $Ni_2$(DPX) (**7**).[a]

| Metrics | **5** | **6** | **7** |
|---|---|---|---|
| M—M distance (Å) | 3.708 | 3.910 | 4.689 |
| M—$N_4$ displacement (Å) | 0.0918 | 0.0290 | 0.0061 |
| Ct—Ct distance (Å) | 3.863 | 3.978 | 4.698 |
| MPS (Å) | 3.417 | 3.611 | 3.666 |
| Interplanar angle (deg) | 4.4 | 2.3 | 1.9 |
| Slip angle (deg) | 27.8 | 24.8 | 38.7 |
| Lateral shift (Å) | 1.802 | 1.668 | 2.937 |

[a] Metrics were derived as follows. Macrocyclic centers (Ct) were calculated as the centers of the four nitrogen planes (4-N plane) for each macrocycle. Interplanar angles were measured as the angle between the 4-N least-squares planes. Plane separations were measured as the perpendicular distance from one macrocycle's 4-N least-square plane to the center of the other macrocycle; mean-plane separations (MPS) were the average of the two plane separations. Slip angles ($\alpha$) were calculated as the average angle between the vector connecting the two metal centers and the unit vectors normal to the two macrocyclic 4-N least-squares planes ($\alpha = \alpha_1 + \alpha_2/2$). The lateral shift was defined as $[\sin(\alpha) \times (\text{Ct–Ct distance})]$.

A direct structural comparison between DPX and DPD homologues is afforded by analysis of the molecular structures of the free-base Pacman architectures $H_4$(DPX) and $H_4$(DPD) along with the corresponding bispalladium(II) complexes $Pd_2$(DTX) (**12**) and $Pd_2$(DTD) (**20**), as shown in Fig. 5 (85). Trends in bond lengths and angles of the macrocyclic core and side chains agree well with those observed in related cofacial bisporphyrins. Conformational analysis of the two macrocycles of $H_4$(DPX) reveals inequivalent ring systems. Ring 1 exhibits a slightly ruffled $S_4$ conformation, with a mean deviation from planarity of 0.2885 Å. In contrast, ring 2 displays a saddle conformation with a mean deviation from planarity of 0.3548 Å. A similar ring inequivalency has been observed in a number of pillared bisporphyrin systems, and it is ascribed to a localized macrocycle distortion induced by the steric crowding of a large meso substituent flanked by proximate alkyl groups. A notable feature concerning the structure of $H_4$(DPX) is the severe butterfly-fold of the xanthene backbone along its center O(1)–C(47) axis. This bent distortion is best described in terms of an intraplanar angle through the center axis-fold. For $H_4$(DPX), the butterfly angle is 36.1°. In contrast, metal derivatives of DPX typically display a flattened

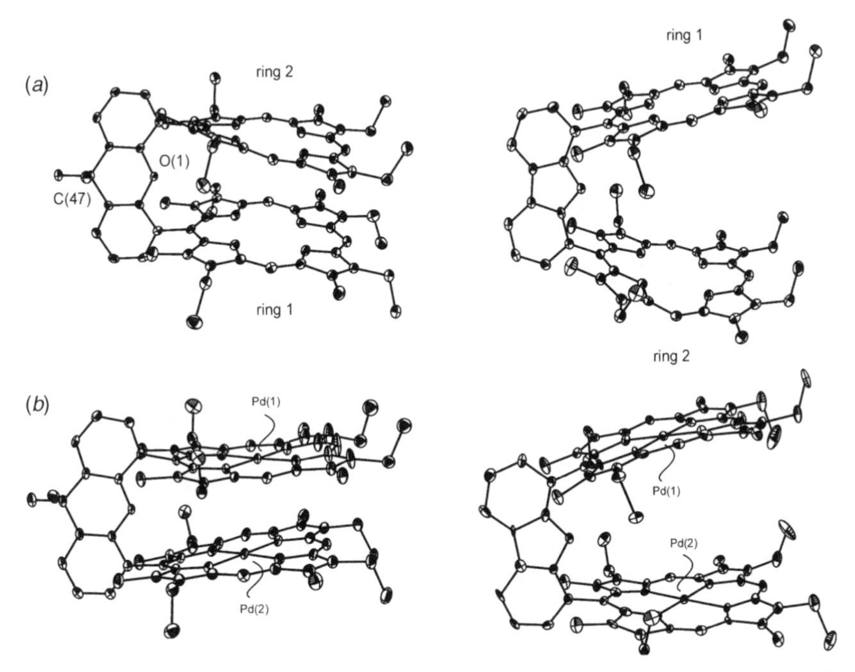


Figure 5. Crystal structure of $H_4$(DPX) and $H_4$(DPD) (*a*) and $Pd_2$(DPX) (**12**) and $Pd_2$(DPD) (**20**) (*b*). Thermal ellipsoids are drawn at the 25% probability level. Hydrogen atoms and solvent molecules within the lattices have been omitted for clarity.

spacer with butterfly angles in the range of 6–15°, with smaller metal cations causing a greater distortion of the bridge from its naturally bent fold. The two ring systems of $H_4$(DPD) are also structurally inequivalent in the solid state. However, unlike $H_4$(DPX), both porphyrin subunits of the DPD homologue display a ruffled $S_4$ conformation. Ring 1 of the Pacman assembly is nearly planar, with a mean deviation of just 0.0553 Å for the macrocyclic atoms from the porphyrin mean plane. The meso carbons are alternately displaced from the mean porphyrin plane, ranging from 0.0758 Å above to 0.0822 Å below the 24-atom macrocyclic unit. Ring 2 has a more pronounced ruffled structure. The mean deviation of 0.1639 Å from the porphyrin plane is greater than that observed for ring 1, and the meso carbons are displaced from 0.3121 Å below to 0.3938 Å above the mean porphyrin plane.

The Pd(II) ions of $Pd_2$(DPX) (**12**) are situated in an approximately $N_{pyrrole}$ square, as the N—Pd—N bond angles are $90 \pm 2.2°$. The average Pd—N bond length is 2.017 Å. The two macrocycles of **12** are structurally inequivalent in the solid state. The ring with Pd(1) is nearly planar, with a mean deviation of 0.0925 Å from the 24-atom macrocyclic plane. The porphyrin displays a ruffled $S_4$ conformation with meso carbons displaced from 0.1838 to 0.1982 Å above the mean plane. The macrocycle containing Pd(2) is also nearly planar, with a mean deviation of 0.0870 Å from the 24-atom plane; however, it exhibits a slight saddle conformation, with β carbons ranging from 0.2198 Å below to 0.1686 Å above the mean ring plane. The square geometry for the Pd(II) core of $Pd_2$(DPD) (**20**) is confirmed by the average N—Pd—N bond angles of $90 \pm 2.5°$. The relatively large deviation of Pd—N bond lengths from the 2.022 Å average [1.926(10) to 2.099(8) Å] are caused by aggregation and packing forces. For the porphyrin ring containing Pd(1), the average Pd—N bond lengths are markedly shorter (1.984 Å) than for the core with Pd(2) (2.059 Å). As observed for the analogous bis(palladium)(II) DPX complex **12**, the two-ring systems of **20** are structurally inequivalent in the solid state. The macrocycle with Pd(1) exhibits a pronounced dome conformation with a mean deviation of 0.2802 Å for the macrocyclic atoms from the porphyrin mean plane. The meso carbons directly connected to and trans to the spacer are slightly displaced from the porphyrin mean plane (−0.0509; and −0.1054 Å, respectively), while the β carbons span up to 0.3986 Å above the mean plane. The ring with Pd(2) displays a saddle geometry, with a mean deviation of 0.2946 Å from the 24-atom porphyrin mean plane. The meso carbons are alternately displaced from the mean porphyrin plane, ranging from 0.1706 Å above to 0.3514 Å below the 24-atom macrocycle unit.

Three mutually perpendicular views of the molecular structures of $H_4$(DPX) and $H_4$(DPD) are shown in Fig. 6 and confirm the ability of the xanthene and dibenzofuran bridges to confine the two porphyrin rings within a face-to-face arrangement while producing molecular clefts with a wide range of vertical

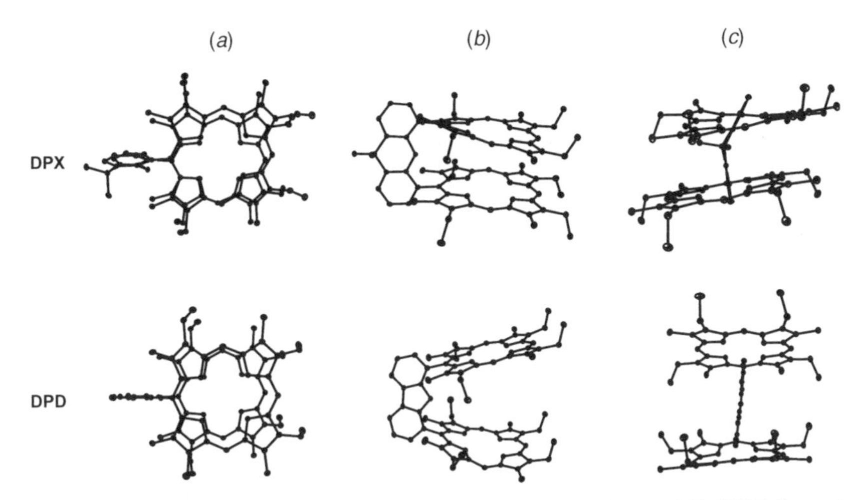


Figure 6. Comparative views of the crystal structures of $H_4$(DPX) (top) and $H_4$(DPD) (bottom): (*a*) top view; (*b*) side view, perpendicular to the bridge plane; (*c*) side view, parallel to the bridge plane. Hydrogen atoms omitted for clarity.

pocket sizes. Pertinent geometrical intradimer features for DPX and DPD complexes characterized by X-ray crystallography are summarized in Table II. Definitions for geometric measurements are given in Table II and Fig. 7.

A number of structural features can be used to tune the Pacman metal–metal distances, not the least of which is porphyrin axial ligation. This fact is especially evident for DPD, where extremely compressed or splayed geometries can be achieved by the addition of axial ligands. With external ligation, the metal–metal distances of DPD can range from 3.504 Å for $Fe_2O$(DPD) (**17**), which clamps its Pacman bite around a bridging oxo ligand, to 7.775 Å for $Zn_2$(DPD) (**13**), which allows the DPD platform to attain its natural splayed conformation (see below). More subtle structural changes are available through metal substitution, which can influence the π overlap between porphyrin subunits. Within the DPX series, $Ni_2$(DPX) (**17**) exhibits the least π overlap, giving the largest torsional twist (~20°) and metal–metal separations (~ 4.6 Å) in the absence of any external axial ligands.

The collected data illustrate the significant effect of replacing a single six-membered ether ring (DPX) for the corresponding five-membered congener (DPD). For example, in the absence of axial ligands, the center-to-center distances vary from 3.8 to 4.7 Å for the DPX platforms to 7.6–8.2 Å for the DPD compounds. The vertical range of almost 4 Å observed for DPX and DPD is in stark contrast to DPA and DPB, which differ by only ~1 Å in vertical proportion. Even more notable is the vast array of interplanar angles between the two macrocyclic subunits. The ring-parallel DPX complexes exhibit interplanar

TABLE II
Crystallographically Derived Intradimer Geometrical Features for DPX and DPD Compounds.[a]

| Compound | M–M (Å) | Ct–Ct (Å) | MPS (Å) | Interplanar Angle (deg) | Torsional Twist (deg) | a–b Distance (Å) | c–d Distance (Å) |
|---|---|---|---|---|---|---|---|
| $H_4$(DPX) | NA[b] | 4.002 | 3.609 | 4.7 | 14.3 | 4.355 | 4.324 |
| $Zn_2$(DPX) (**5**) | 3.708 | 3.863 | 3.417 | 4.4 | 7.9 | 4.619 | 4.272 |
| $Cu_2$(DPX) (**6**) | 3.910 | 3.978 | 3.611 | 2.3 | 7.4 | 4.594 | 4.321 |
| $Ni_2$(DPX) (**7**) | 4.689 | 4.698 | 3.666 | 1.9 | 22.2 | 4.657 | 4.466 |
| $Co_2$(DPX) (**8**) | 4.582 | 4.630 | 3.519 | 2.5 | 21.1 | 4.591 | 4.403 |
| $Pd_2$(DPX) (**12**) | 3.970 | 4.002 | 4.058 | 3.9 | 14.6 | 4.613 | 4.318 |
| $H_4$(DPD) | NA[b] | 8.220 | 8.220 | 23.0 | 1.9 | 4.853 | 5.654 |
| $Zn_2$(DPD) (**13**) | 7.775 | 7.587 | 7.356 | 29.6 | 1.2 | 4.800 | 5.577 |
| $Co_2$(DPD) (**16**) | 8.624 | 8.874 | 8.794 | 56.5 | 9.3 | 4.826 | 5.730 |
| $Fe_2O$(DPD) (**17**) | 3.504 | 4.611 | 4.871 | 21.1 | 1.9 | 4.763 | 5.082 |
| $Pd_2$(DPD) (**20**) | 6.809 | 6.781 | 6.541 | 11.0 | 3.5 | 4.823 | 5.570 |

[a] Metrics were derived as follows. Macrocyclic centers (Ct) were calculated as the centers of the four nitrogen planes (4-N plane) for each macrocycle. Interplanar angles were measured as the angle between the 4-N least-squares planes. Plane separations were measured as the perpendicular distance from one macrocycle's 4-N least-square plane to the center of the other macrocycle; mean-plane separations (MPS) were the average of the two plane separations. Torsional twists were measured as the angle between the two meso carbon to spacer bonds. The a–b and c–d distances were obtained according to the convention shown in Fig. 7. Data is reproduced from Ref. (86).

[b] Not available = NA.

angles of $<5^\circ$, while their ring-splayed DPD congeners display angles in the range of 20–25°. For comparison, the corresponding values for unligated DPA and DPB complexes lie between 2° and 7°. Lastly, the cofacial geometry of DPX and DPD is evident from their small torsional angles, which are all $<25^\circ$. The DPD complexes are especially notable in this regard, with torsion angles of $<10^\circ$ in all cases.

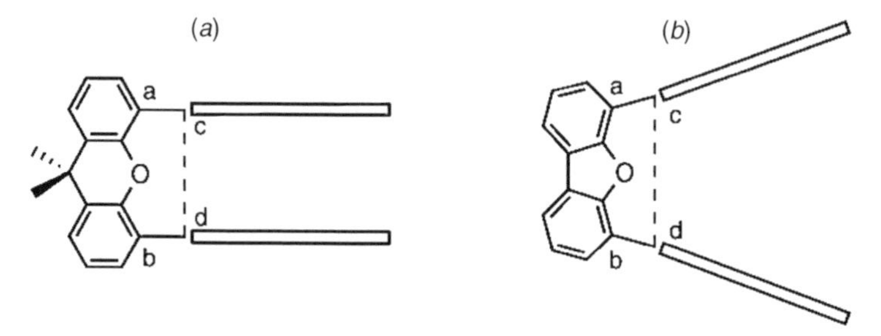


Figure 7. Illustrated distances a–b and c–d for the cofacial bisporphyrin systems DPX (*a*) and DPD (*b*). Table II sub-caption defines the methods by which the crystallographically derived geometric features were measured.

## C. Magnetic Properties

The electron paramagnetic resonance (EPR) spectra of homobimetallic copper(II) derivatives $Cu_2$(DPX) (**6**) and $Cu_2$(DPD) (**14**) in frozen solution provide a useful complement to the crystallographic studies presented in Section II.B. Prior studies have demonstrated the validity of such methods for structure determination of related cofacial bisporphyrins (87). The interspin geometry of the two spatially separated paramagnetic Cu(II) centers can be obtained by examining their triplet spectrum, and the interspin distance, $r$, can be determined from the ratio of the intensity of the half-field transitions to the intensity of the allowed transitions (88). The dipolar splitting of both the copper parallel and perpendicular lines in **6** and **14** are clearly resolved with no evidence of aggregation. Simulation gives $r$ values of 4.2 and 8.0 Å for **6** and **14**, respectively. The separations between the copper parallel lines [two-dimensional (2D) values] give Cu—Cu distances that agree with those obtained from the dipolar splitting simulations.

The spectra of the half-field transitions of **6** and **14** provide another estimate of metal–metal separation. The ratio of the intensity of the half-field transitions to the intensity of the allowed transitions were measured to be $4.8 \times 10^{-3}$ for **6** and $8.7 \times 10^{-5}$ for **14**, giving interspin distances of 4.0 and 7.8 Å, respectively. The data in Table III are generally consistent with the crystallographic results of Table II, with EPR providing a systematically higher value for metal-metal and interplanar separations owing presumably to the absence of crystal packing effects and $\pi$–$\pi$ stacking in frozen solution (84).

## D. Photophysical Properties

The absorption and emission properties of $H_4$(DPX) and $H_4$(DPD) and their metal complexes are consistent with the structural geometries observed in the

TABLE III
Comparison of Geometric Intradimer Parameters Derived from EPR and X-ray Crystallography for Dicopper(II) Cofacial Bisporphyrins

| Compound | $r_{exp}$ (Å) [a] | $r_{sim}$ (Å) [b] | M—M (Å) [c] | Interplanar (Å) [d] | MPS (Å) [c] |
|---|---|---|---|---|---|
| $Cu_2$(DPX) (**6**) | 4.00 | 4.20 | 3.90 | 4.00 | 3.60 |
| $Cu_2$(DPD) (**14**) | 7.80 | 8.00 | | 7.30 | |
| $Cu_2$(DPA) | 4.90 | 4.90 | 4.57 | 4.60 | 3.90 |
| $Cu_2$(DPB) | 4.14 | 4.13 | 3.81 | 3.90 | 3.50 |

[a] Determined from the ratio of the intensities of the half-field to allowed EPR transitions.
[b] Estimated from dipolar splitting simulations.
[c] From X-ray crystal structures.
[d] Interplanar distances determined from EPR simulation.

TABLE IV
The ultraviolet (UV)-Visible (vis) Absorption Data for DPX and DPD Complexes in Dichloromethane at 298 K, $\lambda_{abs}$/nm ($\varepsilon \times 1000\, M^{-1}\, cm^{-1}$).[a]

| Compound | Soret (B) | | | Q region | |
|---|---|---|---|---|---|
| $H_4$(DPX) | 382 (202) | 509 (11.6) | 541 (6.4) | 578 (7) | 631 (2.9) |
| $Zn_2$(DPX) (**5**) | 389 (290) | 541 (14.3) | | 576 (13.2) | |
| $Cu_2$(DPX) (**6**) | 387 (265) | 534 (16.7) | | 571 (19.8) | |
| $Ni_2$(DPX) (**7**) | 388 (305) | 526 (13.6) | | 564 (23) | |
| $H_4$(DPD) | 397 (276) | 501 (25) | 536 (13.5) | 571 (11.8) | 625 (5.2) |
| $Zn_2$(DPD) (**13**) | 400 (512) | 534 (30.6) | | 571 (29.6) | |
| $Cu_2$(DPD) (**14**) | 397 (400) | 528 (17.6) | | 565 (23.4) | |
| $Ni_2$(DPD) (**15**) | 395 (411) | 521 (20.6) | | 560 (39.4) | |

[a] Extinction coefficients ($\varepsilon$) are reported per molecule.
The $\varepsilon$ values are in parentheses.

solid state and in solution. Electronic absorption and emission data for $H_4$(DPX) and $H_4$(DPD) and their homobimetallic zinc, copper, and nickel complexes **5–7** and **13–15** are collected in Tables IV and V, respectively. Soret (B) and Q absorption bands arising from the standard Gouterman four-orbital model for porphyrin spectra (89, 90) undergo varying degrees of perturbation upon cofacially disposing two porphyrin rings using xanthene or dibenzofuran.

The spectra of the biszinc (II) derivatives $Zn_2$(DPX) (**5**) and $Zn_2$(DPD) (**13**), shown in Fig. 8, are exemplary of each series. Figure 8(*a*) compares the electronic absorption spectra of $Zn_2$(DPX) (**5**), $Zn_2$(DPD) (**13**), and their monomer analogue Zn(Etio) (Etio = Etioporphyrin-I) in dichloromethane. Characteristic of strongly interacting porphyrin subunits (10, 91), the B band of $Zn_2$(DPX) (**5**) is blue-shifted ($\lambda_{abs} = 389$ nm, $\varepsilon = 290{,}000\, M^{-1}cm^{-1}$) and broadened relative to that of both $Zn_2$(DPD) (**13**) ($\lambda_{abs} = 400$ nm, $\varepsilon = 512{,}000$

TABLE V
Emission Data and Singlet Excited State Parameters for DPX and DPD Complexes in Dichloromethane Solution at 298 K

| Compound | $\lambda_{em}$ (nm) [a] | $\Phi_{fl}$ [b] | $\tau_{fl}$ (ns) [c] |
|---|---|---|---|
| $H_4$(DPX) | 637, 671, 703 | $0.0298 \pm 0.009$ | $10.55 \pm 0.01$ |
| $Zn_2$(DPX) (**5**) | 591, 640 | $0.0084 \pm 0.001$ | $1.35 \pm 0.01$ |
| $H_4$(DPD) | 628, 658, 694 | $0.0870 \pm 0.010$ | $10.70 \pm 0.10$ |
| $Zn_2$(DPD) (**13**) | 578, 631 | $0.0541 \pm 0.001$ | $1.50 \pm 0.02$ |
| $H_2$(Etio) | 623, 649, 674, 689 | $0.1200 \pm 0.008$ | $11.50 \pm 0.10$ |
| Zn(Etio) | 572, 625 | $0.0298 \pm 0.009$ | $1.45 \pm 0.01$ |

[a] The parameter $\lambda_{em}$ is the corrected emission fluorescence energy maximum.
[b] The parameter $\Phi_{fl}$ is the quantum yield for emission fluorescence.
[c] The parameter $\tau_{fl}$ is the observed fluorescence lifetime.

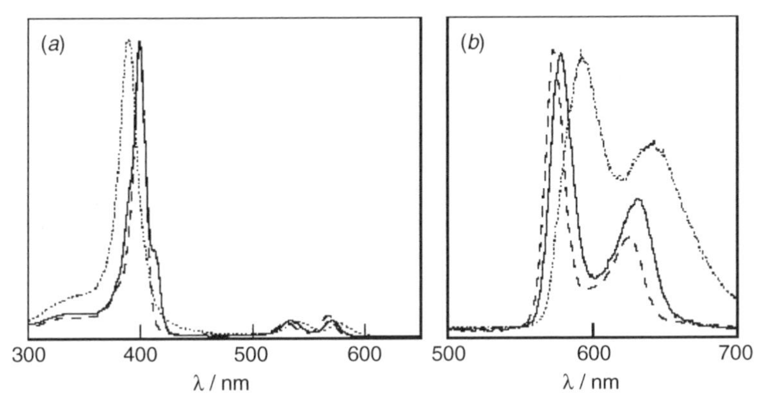


Figure 8. (*a*) Absorption and (*b*) emission spectra of $Zn_2$(DPX) (**16**) (dotted line), $Zn_2$(DPD) (**28**) (solid line), and Zn(Etio) (dashed line) in dichloromethane solution at room temperature. Spectra are normalized in intensities.

$M^{-1}cm^{-1}$) and the monomer ($\lambda_{abs} = 400$ nm, $\varepsilon = 200{,}000\ M^{-1}cm^{-1}$), while the Q(1,0) ($\lambda_{abs} = 541$ nm, $\varepsilon = 14{,}300\ M^{-1}cm^{-1}$) and Q(0,0) ($\lambda_{abs} = 576$ nm, $\varepsilon = 13{,}200\ M^{-1}cm^{-1}$) maxima are shifted to the red [$\lambda_{abs} = 534$ and 571 nm for **13**, 529 and 567 nm for Zn(Etio)]. In contrast, the absorption characteristics of $Zn_2$(DPD) (**13**) more closely approximate that of the porphyrin monomer. The observed spectral differences are consistent with the enlarged distance between the two porphyrin chromophores imparted by the dibenzofuran backbone of DPD. The increased separation decreases $\pi$–$\pi$ overlap and exciton coupling between the porphyrin subunits. Similar trends are observed for DPA and DPB, where the exciton coupling in the latter series is slightly greater due to its more compressed cofacial structure (92).

The absorption spectra for compounds $H_4$(DPX) and $H_4$(DPD) show a phyllo-type splitting pattern that is typical for mono meso-substituted free base porphyrins (93, 94), with the intensity of the Q bands displaying the following pattern: $Q_y$ (1,0) > $Q_y$ (0,0) > $Q_x$ (1,0) > $Q_x$ (0,0). The spectra of the biscopper(II) and bisnickel(II) derivatives are of the hypso type, due to the overlap between the filled $d_{xz}$ and $d_{yz}$ orbitals and the empty porphyrin $\pi^*$ levels. The B band maximum in the absorption spectrum of $Cu_2$(DPX) (**6**) appears at 387 nm ($\varepsilon = 265{,}000\ M^{-1}\ cm^{-1}$), with the Q(1,0) and Q(0,0) transitions occurring at 534 ($\varepsilon = 16{,}700\ M^{-1}cm^{-1}$) and 571 nm ($\varepsilon = 19{,}800\ M^{-1}\ cm^{-1}$), respectively, and the B band maximum in the absorption spectrum of $Cu_2$(DPD) (**14**) appears at 397 nm ($\varepsilon = 400{,}000\ M^{-1}\ cm^{-1}$), with the Q(1,0) and Q(0,0) transitions occurring at 528 ($\varepsilon = 17{,}600\ M^{-1}cm^{-1}$) and 565 nm ($\varepsilon = 23{,}400\ M^{-1}\ cm^{-1}$), respectively. The Soret bands in the absorption spectra of $Ni_2$(DPX) (**7**) and $Ni_2$(DPD) (**15**) are centered at 388 nm ($\varepsilon = 250{,}000\ M^{-1}\ cm^{-1}$) and 395 nm ($\varepsilon = 411{,}000\ M^{-1}$

$cm^{-1}$), respectively. The Q bands of **7** and **15** are shifted to higher energy with respect to their biscopper(II) counterparts, appearing at 526 ($\varepsilon = 13{,}600\ M^{-1}\ cm^{-1}$) and 564 nm ($\varepsilon = 23{,}000\ M^{-1}\ cm^{-1}$) for **7** and 521 ($\varepsilon = 20{,}600\ M^{-1}\ cm^{-1}$) and 560 nm ($\varepsilon = 39{,}400\ M^{-1}\ cm^{-1}$) for **15**. The relative energies of the Q bands of the DPX and DPD derivatives follow the trend biszinc(II) > biscopper (II) > bisnickel(II) due to the decreased perturbation of the porphyrin $\pi$–$\pi^*$ transition by the transition metals with lower *d*-electron counts. This trend is expected from Gouterman's four-orbital model analysis (89, 90).

Both $Zn_2$(DPX) (**5**) and $Zn_2$(DPD) (**13**) produce strong fluorescence typical of Q(0,0) and Q(1,0) excitation. As with the Q-band absorption profile, a correspondent red shift of these emission bands relative to monomer is observed for **5** ($\lambda_{em} = 591$ and 640 nm for **5**, $\lambda_{em} = 572$ and 625 nm for monomer), while **13** ($\lambda_{em} = 578$ and 631 nm) is similar to the monomer [Fig. 8(*b*)]. The free-base porphyrins $H_4$(DPX) and $H_4$(DPD) also exhibit fluorescence that follow the trend of their biszinc(II) congeners. The fluorescence maxima for free base porphyrin $H_4$(DPX), centered at $\lambda_{em} = 637$, 671, and 703 nm, are red-shifted in comparison to $H_2$(Etio-I) ($\lambda_{em} = 623$, 649, 674, and 689 nm). The phyllo-type splitting pattern [$\lambda_{abs}$ ($Q_y(1,0)$)] = 509 nm, $\varepsilon = 10{,}100\ M^{-1}cm^{-1}$; $\lambda_{abs}$ [$Q_y(0,0)$] = 543 nm, $\varepsilon = 5800\ M^{-1}cm^{-1}$; $\lambda_{abs}$ [$Q_x(1,0)$] = 579 nm, $\varepsilon = 6200\ M^{-1}\ cm^{-1}$; $\lambda_{abs}$ [$Q_x(0,0)$] = 631 nm, $\varepsilon = 2400\ M^{-1}cm^{-1}$), which is characteristic of mono meso-substituted porphyrins, though observed in the absorption profile of $H_4$(DPX), is not apparent in the fluorescence spectrum. $H_4$(DPD) also displays a strong fluorescence with maxima at $\lambda_{em} = 628$, 658, and 694 nm; these values are comparable to the monomer.

Time-resolved spectroscopic measurements are also in agreement with the notion that the compressed, parallel arrangement of the porphyrin chromophores of DPX compared to the splayed wedge geometry of the DPD porphyrins leads to greater exciton coupling in the former. The singlet excited-state properties of the free base and biszinc(II) derivatives are in accord with the steady-state electronic absorption and emission measurements; data are collected in Table V. For example, $Zn_2$(DPX) (**5**) has a shortened singlet excited-state lifetime ($\tau_{fl} = 1.35 \pm 0.01$ ns) and reduced fluorescence quantum yield ($\Phi_{fl} = 0.0084 \pm 0.001$) compared to $Zn_2$(DPD) (**13**) ($\tau_{fl} = 1.50 \pm 0.02$ ns, $\Phi_{fl} = 0.0541 \pm 0.001$). The free base porphyrins also follow the same trend, with $H_4$(DPX) ($\tau_{fl} = 10.55 \pm 0.05$ ns and $\Phi_{fl} = 0.0298 \pm 0.009$) having a shorter singlet excited-state lifetime and reduced fluorescence quantum yield compared to $H_4$(DPD) ($\tau_{fl} = 10.70 \pm 0.10$ ns, $\Phi_{fl} = 0.0870 \pm 0.001$).

The situation is more complex for the triplet excited states for the DPX and DPD frameworks. The dynamics for these excited states has been examined for the bispalladium(II) complexes $Pd_2$(DPX) (**12**) and $Pd_2$(DPD) (**20**) (95). For the bispalladium(II) Pacman systems, the heavy metal cores promote efficient intersystem crossing from singlet-to-triplet excited states by spin–orbit coupling

TABLE VI
Emission Data and Triplet Excited State Parameters for DPX and DPD Complexes in Cyclohexane Solution at 298 K

| Compound | $\lambda_{em}$ (nm) [a] | $\Phi_p$ [b] | $\tau_p$ (μs) [c] |
|---|---|---|---|
| $Pd_2$(DPX) (**12**) | 678 | $35.2(3) \times 10^{-3}$ | 262(12) |
| $Pd_2$(DPD) (**20**) | 673 | $0.993(108) \times 10^{-3}$ | 0.52(1) |
| Pd(Etio) | 668 | $67.3(48) \times 10^{-3}$ | 203(7) |
| Pd(PhEtio) | 672 | $0.598(49) \times 10^{-3}$ | 0.46(1), 0.16(1) |

[a] The parameter $\lambda_{em}$ is the corrected emission phosphorescence energy maximum.
[b] The parameter $\Phi_p$ is the quantum yield for emission phosphorescence.
[c] The parameter $\tau_p$ is the observed phosphorescence lifetime.

(94). Emission data are collected in Table VI. In contrast to trends observed for the singlet excited states of the free base and biszinc(II) DPX and DPD complexes, the phosphorescence quantum yield [$\Phi_p = 1.07(54) \times 10^{-3}$] and lifetime ($\tau_p = 0.53$ μs) of $Pd_2$(DPD) (**20**) are significantly attenuated as compared to its condensed $Pd_2$(DPX) (**12**) congener ($\Phi_p = 35.2(3) \times 10^{-3}$ and $\tau_p = 262$ μs). Electronic absorption and emission spectroscopy of $Pd_2$(DPX) (**12**) and $Pd_2$(DPD) (**20**) establish that the porphyrin rings of the DPX cofacial analogue are in closer proximity than the porphyrin rings of the DPD analogue in solution (Table VI). These results suggest that the enhanced triplet excited-state decay of the DPD system in comparison to DPX is not a consequence of simple interplanar interactions, such as exciton coupling, but rather arises from the increased conformational flexibility of the splayed Pacman congener. The DPD architecture enables the porphyrin subunits to rotate about the aryl ring of the wedge-shaped dibenzofuran pillar. Such rotations are sterically hindered for the case of xanthene. In support of this contention, the photophysical properties of the monomer compounds Pd(Etio) and Pd(PhEtio) [PhEtio = 5-(4′-bromophenyl)-2,8,13,17-tetraethyl-3,7,12,18-tetramethylporphyrin] were assessed and found to be similar to $Pd_2$(DPX) (**12**) and $Pd_2$(DPD) (**20**), respectively. In addition, nonlocal density functional calculations on structurally similar Pd(PhOMP) confirm the possibility of ring rotation about the C(meso)–C(aryl) bond, and show that such torsional motions are accompanied by non-planar distortions of the porphyrin ring with a decrease in the $T_1$–$S_0$ energy gap (Fig. 9). The triplet state of bispalladium(II) Pacman complexes with varying cleft sizes and orientations can undergo energy transfer (96). The excited-state reactivity has been exploited for the development of $O_2$ sensors with detection limits $<1$ ppm (97).

The multielectron reactivity of cofacial bisporphyrins complexes in both the ground and excited state is reliant on the ability of the Pacman cleft to accommodate reaction intermediates during catalysis. Crystallographic analysis confirms the ability of the DPX and DPD platforms to provide cofacial pockets

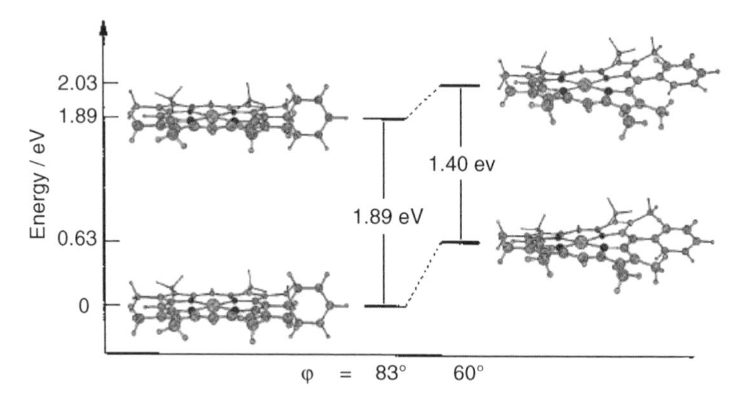


Figure 9. Energy level diagram showing the relative energies of both singlet and triplet states of Pd(PhOMP) on going from the planar structure to the distorted structure. The DFT-optimized structures of planar and distorted Pd(PhOMP) in both singlet and triplet states are also shown; φ is the torsional angle between the porphyrin plane and the meso-phenyl substituent. The value of the $T_1$–$S_0$ energy gap upon nonplanar distortion is depicted between the highest occupied molecular orbital (HOMO) and lowest unoccupied molecular orbital (LUMO) levels.

with disparate vertical cleft dimensions that can be tuned by axial ligation (i.e., substrate binding) and/or metal substitution. Especially noteworthy, is the ability of the DPD system to span a vertical range of metal–metal distances of >4 Å (see Section V.B). The extent to which these structural differences impact the binding of small molecules is the subject Section III.

## III. SUPRAMOLECULAR INVESTIGATION OF THE PACMAN EFFECT

The propensity of Pacman constructs to promote multielectron catalysis in the ground state provided impetus for investigating the excited-state chemistry associated with the Pacman effect. To this end, host–guest chemistry of the DPD platform was developed by using stable adducts that were amenable to structural and spectroscopic scrutiny. Sanders' approach (22–26, 98–109) of using multifunctional zinc(II) porphyrin hosts with neutral nitrogen donors to form Lewis acid–base inclusion complexes provided an attractive starting point for this strategy. For our purposes, pyrimidine-based guests were selected for investigation. Although few metal-based receptors are known to encapsulate pyrimidine and its derivatives (110, 111), these heterocycles provide an excellent geometric match for the wedge-shaped DPD framework.

The complexation of 2-aminopyrimidine with **13** has been probed by a combination of $^1$H NMR, UV–vis spectroscopy, and crystallographic studies. As

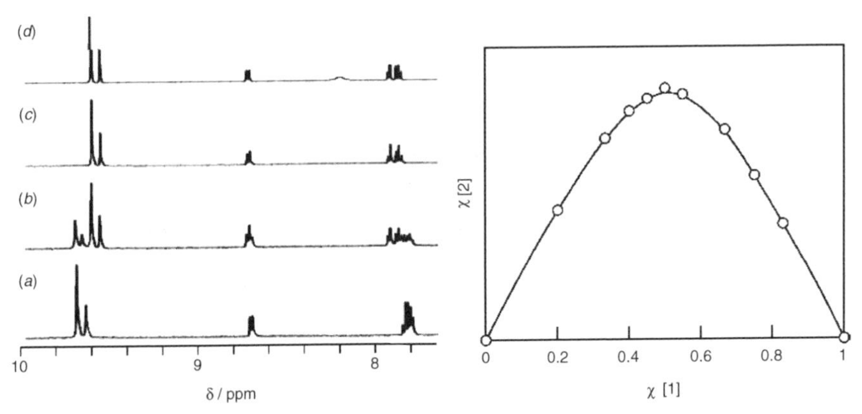


Figure 10. (*a*) Selected $^{1}H$ NMR spectra of **13** containing (1) 0, (2) 0.67, (3) 1, and (4) 3 equiv of 2-aminopyrimidine in $CD_2Cl_2$ solution at 25°C. The spectral range captures the signals for the meso protons of free host **13** and **13**·(2-aminopyrimidine), which are found in the range of 9–10 ppm. (*b*) Job plot establishing the 1:1 stoichiometry for binding of 2-aminopyrimidine inside the cleft of host **13**. χ[**1**] and χ[**2**] are the mole fractions of host and host–guest complexes, respectively.

highlighted by the NMR spectra of Fig. 10, the presence of up to 1 equiv of the pyrimidine guest results in the formation of a single host–guest complex. The upfield shift of the aromatic pyrimidine resonances (9.58 and 9.63 ppm) is consistent with coordination of the bidentate pyrimidine within the splayed cleft of **13**. A Job plot (112) of the $^{1}H$ NMR titration data (Fig. 10) shows that the complex is optimally formed at equimolar concentrations of porphyrin host and pyrimidine guest (i.e., a 0.5-mol fraction). The formation of a stable 1:1 complex of **13** and 2-aminopyrimidine was also observed in the solid state. The structure of the host–guest complex **13•**(2-aminopyrimidine) (Fig. 11) confirms the encapsulation of 2-aminopyrimidine inside the Pacman cavity. The framework of **13** undergoes a structural alteration upon ligation of the pyrimidine guest to produce **13•**(2-aminopyrimidine). The binding of the pyrimidine guest triggers a substantial compression along the Pacman vertical axis, affording a complex with a reduced Zn—Zn distance of 6.684 Å (Zn—Zn = 7.775 Å in free host **13**).

The binding between **13** and 2-aminopyrimidine is also evident from UV–vis absorption spectra. The addition of 2-aminopyrimidine to dichloromethane solutions of **13** at room temperature results in the red shifts for the Pacman Soret and Q-band regions as shown in Fig. 12. Such shifts are indicative of axial coordination of a nitrogen donor to a zinc(II) porphyrin. The association constant of $K_a = 9.6 \times 10^7\ M^{-1}$ is among the highest observed for the binding of nitrogen heterocycles to zinc(II) porphyrin-based hosts. The binding constant

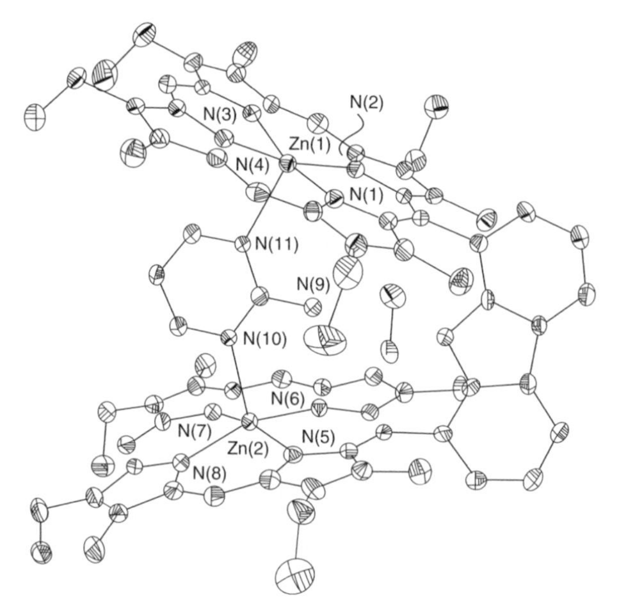


Figure 11. Crystal structure of **13** • (2-aminopyrimidine). Thermal ellipsoids are drawn at the 25% probability level. Hydrogen atoms have been omitted for clarity.

decreases significantly as the donor properties of the pyrimidine are attenuated with the introduction of electron-withdrawing groups on the aromatic pyrimidine ring. As shown in Table VII, $K_a$ decreases by five orders of magnitude upon the replacement of the amino group by bromide. The large $K_a$ of

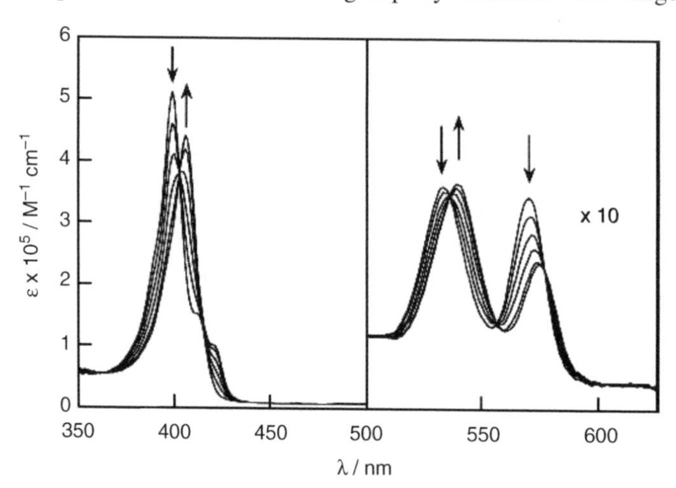


Figure 12. Absorption spectra of **13** in the presence of 0, 0.2, 0.4, 0.6, 0.8, and 1.0 equiv of 2-aminopyrimidine at 25°C.

TABLE VII
Association Constants for Binding of Various Pyrimidines to $Zn_2$(DPD) (**13**) in Dichloromethane at 25°C

| Guest | $K_a$ ($M^{-1}$) |
|---|---|
| 2-Aminopyrimidine | $9.6(7) \times 10^7$ |
| Pyrimidine | $4.0(2) \times 10^4$ |
| 2-Chloropyrimidine | $6.2(1) \times 10^2$ |
| 2-Bromopyrimidine | $7.0(1) \times 10^2$ |

**13**•(2-aminopyrimidine) may also reflect a hydrogen-bonding contribution of the substrate within the cleft. Thermodynamic parameters of $\Delta H = -63(6)$ kJ mol$^{-1}$ and $\Delta S = -61(12)$ J mol$^{-1}$ K$^{-1}$ were obtained from titrations of **13** with 2-aminopyridine at various temperatures. The Pacman effect is evident in the smaller entropic penalty associated with formation of **13**•(2-aminopyrimidine) as compared to the binding of ditopic ligands to doubly strapped dizinc(II) porphyrin FTF dimers ($S \sim -100$ to $-130$ J mol$^{-1}$ K$^{-1}$). In the case of the former, the Pacman flexing occurs along a constrained vertical dimension, whereas the latter involves motion along both lateral and vertical dimensions.

The Pacman effect is not only confined to ground-state properties of cofacial porphyrins but is manifested in excited-state properties as well. The transient absorption spectra of the zinc porphyrins decay on the microsecond time scale; the susceptibility of these transient species to quenching by $O_2$ confirms the triplet character of such excited states. The time-resolved absorption decays for **13**, **13**•(2-aminopyrimidine), **5**, Zn(OEP) and Zn(PhEtio) are shown in Fig. 13, and the triplet lifetimes for each species are collected in Table VIII. The general trend in the lifetimes of the zinc porphyrin monomers and Pacman complexes is reminiscent of those for the respective palladium congeners (Section II.D). The monoexponential excited-state decays of **13** and **5** occur on markedly disparate time scales, with Zn(OEP) exhibiting a long-lived excited state that is commensurate with that of the DPX system. By comparison, the shortened triplet lifetimes observed for **13** and Zn(PhEtio) are a consequence of nonplanar distortions of the porphyrin macrocycle owing torsional motion about the C(meso)–C(aryl) bridge bonds. Such distortions ultimately lead to enhancements in nonradiative decay pathways. This motion is suppressed for the DPX homologue. A compressed porphyrin–porphyrin axis hinders rotation of the macrocyclic subunits with respect to the aryl bridge owing to the steric congestion imparted by the proximate cofacial porphyrin. The substantial increase in the triplet excited-state lifetime for the host–guest complex **13**•(2-aminopyrimidine) over **13** confirms that substrate binding within the Pacman cleft serves to suppress the torsional undulations that promote nonradiative decay in the excited states of the splayed Pacman architecture.

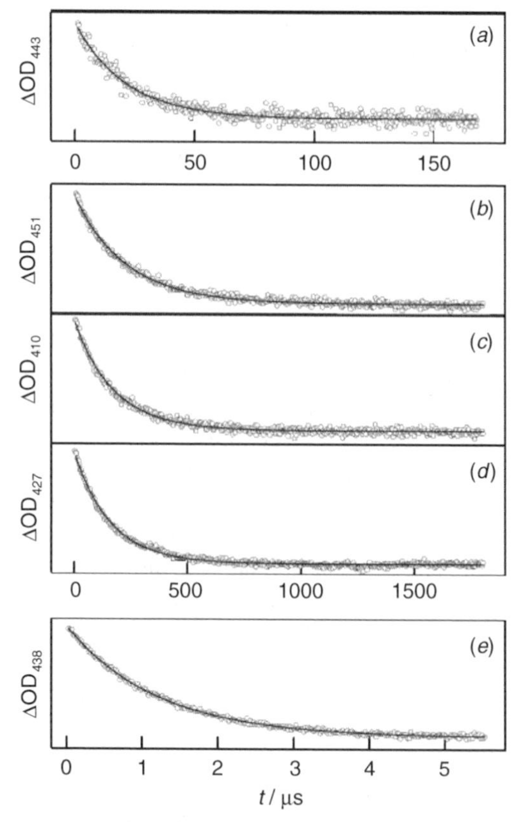


Figure 13. Transient absorption decays monitored at ΔOD maximum of (*a*) $Zn_2$(DPD) (**13**), (*b*) **13•**(2-aminopyrimidine), (*c*) $Zn_2$(DPX) (**5**), (*d*) Zn(OEP), and (*e*) Zn(PhEtio) following excitation by a 3-ns pulse at 532 nm.

The Pacman effect allows the cofacial bisporphyrin motif to create a suitable microenvironment for mediating dynamic host–guest chemistry for catalysis in ground and excited states. In particular, the DPD architecture has exceptional vertical flexibility that is accompanied by a minimized entropic penalty. An

TABLE VIII
Transient Absorption Maxima and Decay Lifetimes of Zinc Porphyrins

| | λ(nm) | τ (μs) |
|---|---|---|
| $Zn_2$(DPD)(**13**) | 443 | 21.4(12) |
| 13•(2-Aminopyrimidine) | 451 | 227(8) |
| $Zn_2$(DPX) (5) | 410 | 169(9) |
| Zn(OEP) | 427 | 161(5) |
| Zn(PhEtio) | 438, 463 | 1.19(1), 0.0773(15) |

accessible pocket that can collapse about a guest with minimal reorganization energy should lower the transition state energy for substrate activation. Accordingly, the Pacman architecture provides an attractive platform for investigations of proton-coupled, multielectron activation of small molecules, particularly, those involving oxygen.

## IV. CATALYTIC DIOXYGEN REDUCTION BY DICOBALT(II) PACMAN ARCHITECTURES

In Nature, multimetallic active sites control the delivery of electrons and protons in the Ca/$Mn_4$ cluster of the oxygen-evolving complex (OEC) of Photosystem II for O—O bond-making and in the dinuclear iron-hemencopper assembly in cytochrome *c* oxidase (C*c*O) for O—O bond breaking. As emphasized by Babcock and co-workers (113), the PCET chemistry of $O_2$ bond-forming chemistry at the OEC active site in PS II has remarkable chemical, mechanistic, and structural similarities with the $O_2$ bond-breaking chemistry at cytochrome *c* oxidase. Against this backdrop of microscopic reversibility, we initiated reactivity studies of the thermodynamically favorable process of O—O bond breaking needed for oxygen reduction to water prior to attacking the more challenging, thermodynamically uphill O—O bond making chemistry needed for water oxidation. To this end, the O—O bond-breaking chemistry of Pacman porphyrins was investigated. Seminal studies of Collman and Chang had already established that Pacman porphyrins containing late transition metals (e.g., Co). were effective electrocatalysts for the reduction of oxygen to water (54, 55, 63, 64). The studies described below provide new mechanistic insights for the O—O bond-breaking process, especially as it pertains to PCET reactivity.

Molecular oxygen is typically reduced along the two pathways shown by Fig. 14 (114, 115). Complete $O_2$ reduction ($4e^- + 4H^+$, horizontal pathway)

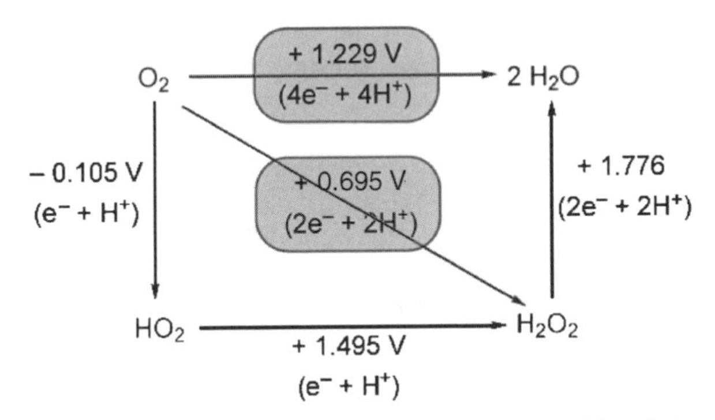


Figure 14. Standard reduction potentials ($E^\circ$) for oxygen in acidic solution (113).

gives 2 equiv of water, whereas partial reduction ($2e^- + 2H^+$, diagonal pathway) yields $H_2O_2$ (116). The two ($\Delta G° = -0.695$ V) and four-electron ($\Delta G° = -1.229$ V) reductions of $O_2$ are both energetically downhill, but more than one-half of a volt of energy is squandered in the former reaction. Accordingly, biological processes coupled to $O_2$ reduction (i.e., respiration) are highly selective for the complete $4e^- + 4H^+$ pathway in order to maximize the energy available for adenosine triphosphate (ATP) synthesis. The pursuit of structural and functional models for $O_2$ activation and its complete reduction has emphasized bimetallic reaction centers positioned within well-defined rigid pockets (117–121). Dicobalt(II) cofacial bisporphyrins are among the few molecular catalysts capable of mediating the selective four-electron, four-proton reduction of oxygen to water. As discussed by Collman et al. (2, 54) and Taube (122) these systems fulfill the two general requirements for an effective four-electron, four-proton oxidation-reduction catalyst: (a) they bypass one- and two-electron redox transformations through cooperative bimetallic reactivity, and (b) they hinder the protonation and release of the two-electron peroxo-type intermediates. The substantial body of work on cofacial bisporphyrins as $O_2$-reduction catalysts has clarified the structural attributes of an effective architecture, including the restriction of macrocyclic subunits to a face-to-face arrangement with minimal lateral displacements, while at the same time allowing sufficient vertical flexibility to bind and activate the $O_2$ substrate. The dicobalt(II) derivatives of both DPA and DPB (45, 46, 49) have been shown to efficiently electrocatalyze the complete four-electron reduction of oxygen to water (as opposed to the two-electron pathway involving peroxide). Given that the intermetallic distances between the DPA and DPB derivatives generally only differ by ~1 Å (see below) it was unclear as to the effect that Pacman cleft size and oxygen orientation has on the efficiency of the complete $4e^- + 4H^+$ oxygen reduction process. To address this issue, the $O_2$ reduction chemistry of the four dicobalt(II) Pacman complexes shown in Fig. 15 was probed by a combination of electrocatalytic and homogeneous reduction experiments.

Initial experiments carried out for the dicobalt(II) DPX (**8**) and DPD (**16**) architectures revealed that both Pacman designs are effective electrocatalysts for

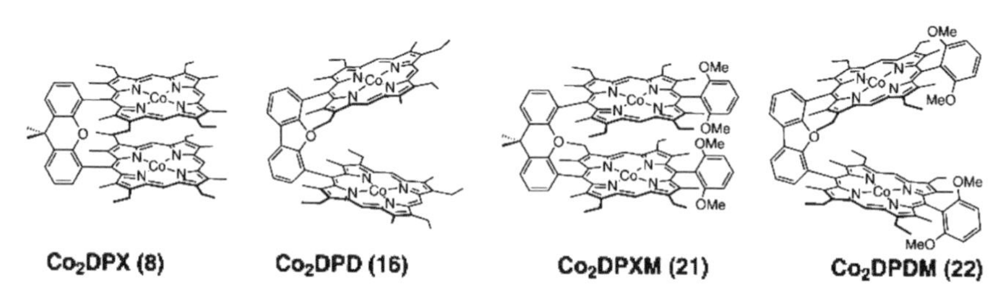


Figure 15. Cobalt Pacman porphyrins for $O_2$ reduction.

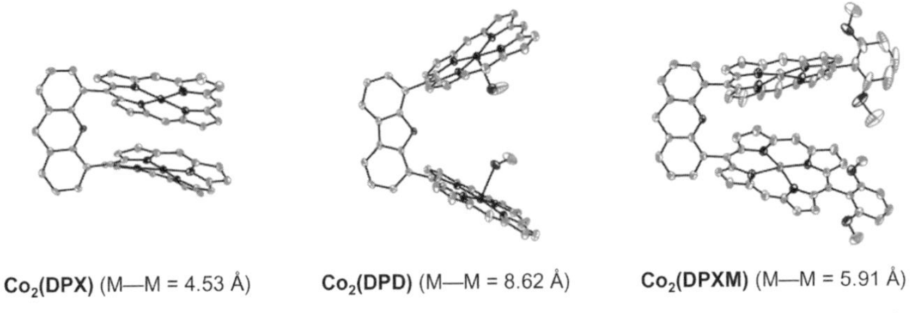


Figure 16. Crystal structures of dicobalt(II) Pacman porphyrins exhibit large variabilities in metal–metal distances.

the complete $4e^- + 4H^+$ reduction of $O_2$ to $H_2O$, over the more commonly observed pathway to produce $H_2O_2$, despite a roughly 4 Å difference in their metal–metal distal separation (Fig. 16) (123). The $Co_2$(DPX) (**8**) complex reduces $O_2$ to $H_2O$ at a potential of +0.38 V versus Ag/AgCl, with 72% selectivity and $Co_2$(DPD) (**16**) catalyzes $O_2$ reduction at +0.37 V, with direct production of $H_2O$ at 80% selectivity (black bars, Fig. 17 and Table IX).

The similar efficacy of **8** and **16** to produce water suggests that the substantial Pacman flexibility of the molecular clefts allows the molecular-binding pocket to structurally accommodate the various reaction intermediates during multi-electron catalysis. Subsequent experiments in which the pocket sizes of the

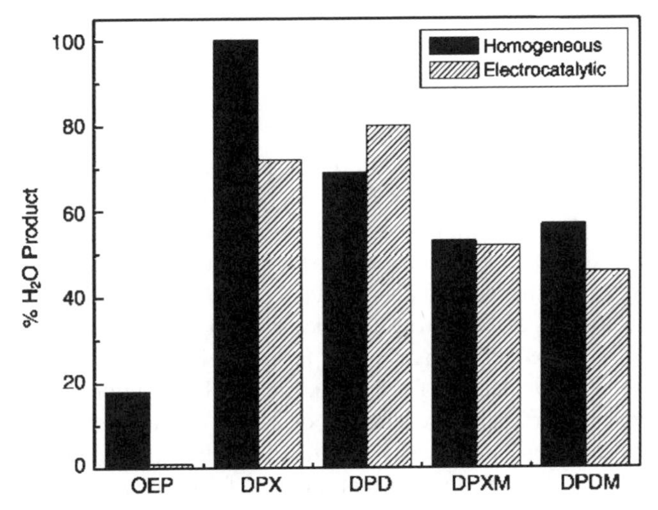


Figure 17. Efficiency of $H_2O$ production from $O_2$ by dicobalt(II) Pacman systems under homogeneous (black) and electrocatalytic (striped) conditions.

TABLE IX
Electrochemistry and $O_2$ Reducing Ability of Dicobalt(II) Pacman Porphyrins

| Compound | $E_{ox}$/V[a] | $E_{disk}$[a] | % $H_2O$ Product Electrocatalytic | % $H_2O$ Product Homogeneous |
|---|---|---|---|---|
| $Co_2$(DPX) (**8**) | +0.28, +0.17 | +0.38 | 72 | 100 |
| $Co_2$(DPD) (**16**) | +0.33 | +0.37 | 80 | 69 |
| $Co_2$(DPXM) (**21**) | +0.31, +0.14 | +0.24 | 52 | 57 |
| $Co_2$(DPDM) (**22**) | +0.33 | +0.25 | 46 | 53 |

[a]All electrochemical measurements referenced to Ag/AgCl.

Pacman motif was further tuned by incorporation of sterically demanding aryl groups, which were installed trans to the spacer following Scheme 2, led to a surprising result, however. Both $Co_2$(DPXM) (**21**) (DPXM = diporphyrin xanthene methoxyaryl, and $Co_2$(DPDM) (**22**) (DPDM = diporphyrin dibenzofuran methoxyaryl) display markedly lower $H_2O/H_2O_2$ product ratios than their parent DPX and DPD counterparts while exhibiting similar selectivities compared to each other. As shown in Figure 17, $Co_2$(DPXM) (**21**) catalyzes $O_2$ reduction to $H_2O$ at +0.24 V, and 52% efficiency, whereas $Co_2$(DPDM) (**22**) catalyzes $O_2$ reduction to $H_2O$ at +0.25 V and 46% efficiency (Table IX). The installation of a trans-aryl group into the Pacman motif clearly results in decreased selectivity for the direct reduction of $O_2$ to $H_2O$, with a comparable effect for both xanthene- and dibenzofuran-bridged scaffolds. Given that the structural flexibility and redox behavior of the cofacial bisporphyrin systems are largely unperturbed by trans-aryl substitution, more subtle electronic and protonation effects were considered as the cause for this change in catalytic selectivity (124).

To probe these issues, $O_2$ reduction experiments were carried out under homogeneous conditions for **8**, **16**, **21**, and **22**, using ferrocene as the chemical reductants and proton sources of differing strengths in order to optimize the $O_2$ reduction efficiency for water production (125). These types of experiments have been developed by Fukuzumi and co-workers (126, 127). Such homogeneous $O_2$ reduction experiments are powerful because they allow for conversions to be investigated stoichiometrically, and hence allow for greater mechanistic insight as compared to studies of $O_2$ reduction at an electrode surface. As the $O_2$ reduction takes place, ferrocene is oxidized and converted to ferrocenium. This oxidation of the ferrocene can be monitored by absorption spectroscopy. Accordingly, the number of electron equivalents that are consumed in this reaction can be easily monitored. For the complete four-equivalent reduction of $O_2$ to water, the concentration of ferrocenium at the end of the reaction is equal to four times the initial concentration of $O_2$ in solution. Conversely, for the two-equivalent reduction of $O_2$ to $H_2O_2$, the concentration of ferrocenium at the end of the reaction is only two times the initial concentration of $O_2$ in solution. Catalytic systems that produce both $H_2O_2$ and $H_2O$ upon $O_2$ reduction, will form ferrocenium at concentrations between the these two extremes. In this manner, the efficacy of $O_2$ reduction by the various catalysts of Fig. 15 has been assesed.

As is shown in Fig. 17, the results of the homogeneous reduction corroborate those obtained from electrocatalytic reductions. The disparate efficiencies for $H_2O$ production from $O_2$ between the dicobalt(II) methoxyaryl substituted Pacman architectures and the parent DPD and DPX complexes imply that the selective reduction of $O_2$ to $H_2O$ versus $H_2O_2$ steps beyond the idea of simple redox cooperativity between proximate metal centers. Noting that $O_2$ reduction requires both proton and electron equivalents, we sought to interrogate the role of proton delivery in determining $O_2$ reduction pathways catalyzed by bimetallic systems. The DFT calculations of the $O_2$ adducts of dicobalt cofacial systems reveal that the nature of the HOMOs is reversed for Pacman porphyrins with and without the methoxyaryl functionality. Figure 18 shows the results for the oxygen adducts of the cobalt DPX/DPXM congeners. Significant electron density resides on $O_2$ included within the DPX pocket. Proton transfer to $O_2$ can therefore occur, leading to efficient O—O bond breakage and the production of water. Conversely, the HOMO of the DPXM analogue shows no electron density at the bound $O_2$. We have confirmed this difference in the basicity of the oxygen adducts by spectroscopic and p$K_a$ studies.

Taken together, these results have led to the mechanistic model shown in Fig. 19. The cycle emphasizes the role of PCET activation for O—O bond breaking. Our results show that it is the basicity (p$K_a \sim 12.5$ in PhCN) of the superoxo complex (shown in the box of Fig. 19) that is the key determinant of the selectivity for $O_2$ reduction. Targeted proton transfer to the $[Co_2O_2]$

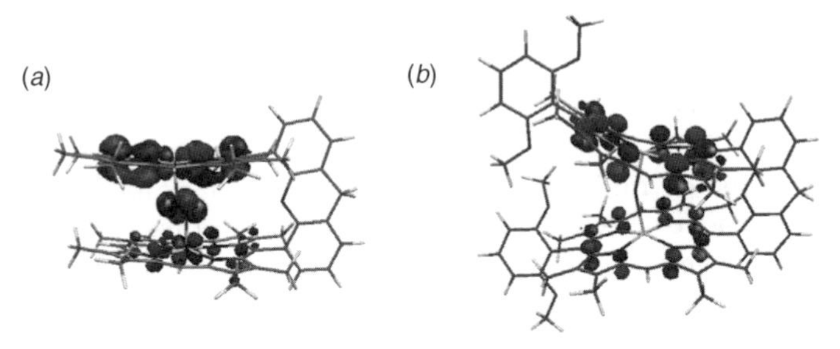


Figure 18. The HOMO of the superoxide complexes of (*a*) $[Co_2(DPX)(O_2)]^+$ and (*b*) $[Co_2(DPXM)(O_2)]^+$. The $Co_2$DPX complex has significant electron density at the bound oxygen and consequently is able to accept a proton to drive O—O bond breaking by PCET and water is obtained. This is not the case for $Co_2$DPXM, which cannot be protonated, thus leading to peroxide as the oxygen reduction product.

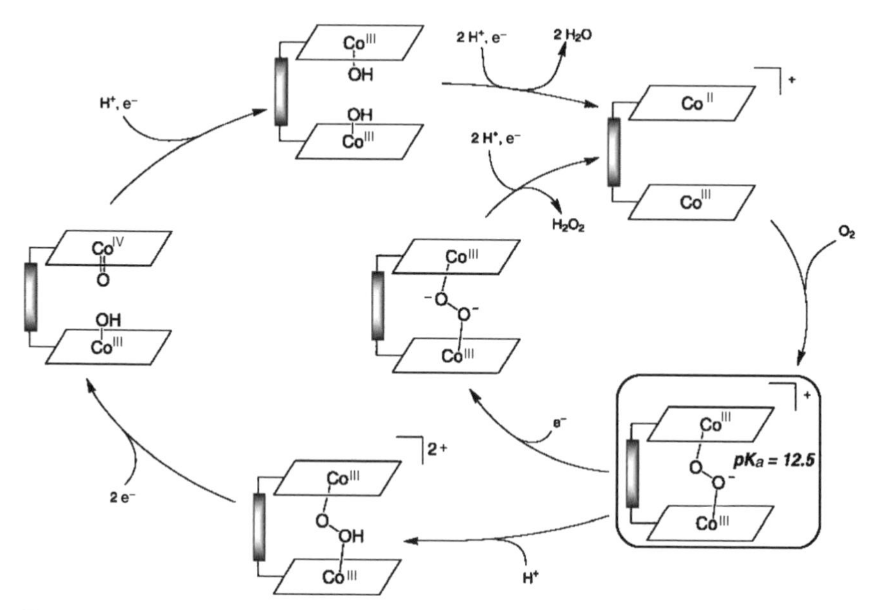


Figure 19. Cycle for $O_2$ reduction by cofacial bis(porphyrins). Protonation of the superoxo intermediate (highlighted by box) is the key to efficient reduction of $O_2$ to $H_2O$.

superoxo core of $[Co_2(bisporphyrin)(O_2)]^+$ triggers a two-electron transfer. The net result is a three-electron transfer process that leads to O—O bond cleavage to produce an oxo-hydroxy species (outer cycle of Fig. 19). Subsequent reduction by an additional electron produces the fully reduced water product. This result concurs with the chemistry of C*c*O. Spectroscopic and kinetics studies of partially and fully reduced enzymes establish that the O—O bond is broken upon the addition of three electrons and one proton to produce the oxo-hydroxy product (128–131). In the case of the enzyme, the peroxo species is stable to dissociation, and hence the third electron adds to peroxide. Conversely, it is believed that superoxide of the cofacial porphyrins is more susceptible to release. Hence, if the superoxo is not sufficiently basic, one-electron reduction of the $[Co_2(bisporphyrin)(O_2)]^+$ ensues in the absence of a proton and peroxide is produced (inner cycle of Fig. 19). Thus protonation of the superoxo permits the system to bypass peroxo-type intermediates and in doing so drive the necessary three-electron equivalents that results in O—O bond cleavage.

The overall mechanism in Fig. 19 is satisfying on several counts. First, as mentioned above, a proper stoichiometry of one proton and three electrons needed for O—O bond cleavage is satisfied. Second, it clearly identifies the importance of the superoxo as the "resting state" for catalysis, a fact long known for the $O_2$ reduction chemistry of Pacman porphyrins (54). Finally, the mechanistic cycle clarifies some perplexing observations of synthetic constructs. Several porphyrin templates bearing a distal metal-binding cap exhibit comparable selectivities for the four-proton, four-electron pathway *with* or *without* a second redox-active metal ion bound in the distal cap (121, 132–134) In addition, cofacial Pacman porphyrins containing two cobalt centers (45, 123), one cobalt and one Lewis acidic metal (39), or one cobalt and one metal-free subunit (63, 64) are all effective catalysts for the direct production of $H_2O$ from $O_2$. The mechanism of Fig. 19 suggests that the second functional site, whether that be another porphyrin, metal complex or metal-free coordination sphere, is to adjust the $pK_a$ of the oxygen adduct. Since a cooperative redox activity is not required from two cofacial porphyrins, the second site may be redox inactive or completely absent. Kadish et al. (135) have shown that similar selectivities for the $4e^-$ reduction of $O_2$ to $H_2O$ can be accomplished using dicobalt(III)-biscorrole Pacman (136) and porphyrin–corrole Pacman assemblies (137). Moreover, the $4e^-$ pathway is accessible for porphyrin-corrole Pacman systems in which the porphyrin macrocycle is unmetalated (138). These results mirror those observed for FTF bisporphyrin systems containing one free-base porphyrin and one cobalt porphyrin (64). The picture, which is starting to emerge indicate that selective $O_2$ reduction to yield water, may be able to be carried out using a diverse set metal-based redox platforms juxtaposed by functionalities capable of functioning as efficient proton shuttles (i.e., the Hangman construct). Additional studies along these lines are emphasized in Section VI.

## V. PACMAN PORPHYRIN COMPLEXES AS OXIDATION CATALYSTS

### A. $H_2O_2$ Disproportionation by Dimanganese(III) Chloride Pacman Architectures

When earlier transition metals are used in the cofacial DPX and DPD porphyrin motifs, oxidation reactions are promoted by a metal oxo intermediate. As for reductive catalysis, the oxidative catalytic reactions of the Pacman porphyrins are derived from PCET. The catalase reaction ($H_2O_2$ disproportionation to give 0.5 $O_2$ and $H_2O$) is exemplary of this PCET reactivity as the reaction includes both O—O bond making and bond-breaking catalysis coupled to proton transport.

The TON for $O_2$ production by dimanganese(III) chloride complexes of both DPX (**11**) and DPD (**19**) are graphically represented in Fig. 20. Both systems are poor catalase mimics, displaying turnover numbers (TONs) for $O_2$ evolution that are $<40$ (139). Given that the etioporphyrin-type substitution patterns of the parent DPX and DPD systems are generally unstable to oxidizing conditions,

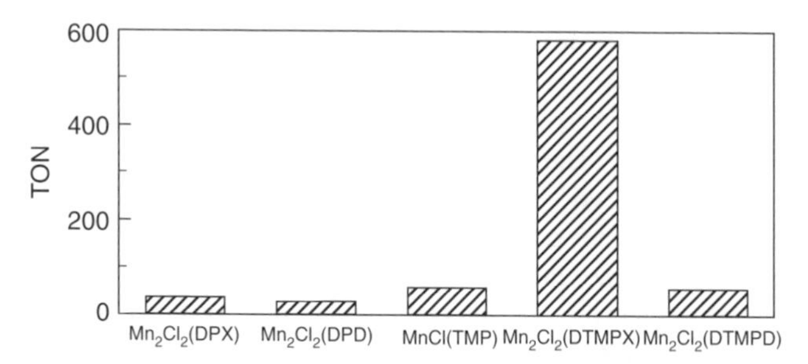


Figure 20. Total turnover numbers recorded for $H_2O_2$ disproportionation to give 0.5 $O_2$ and $H_2O$ for a set of dimanganese(III) Pacman systems and a monomeric control.

owing to the unprotected meso positions on the porphyrin rings, it was reasoned that the construction of meso-substituted bisporphyrin Pacman derivatives might display higher catalase (and more generally, oxidative) efficiencies. Scheme 3 presents a modular and general synthetic route for the facile preparation of meso-tetraaryl cofacial bisporphyrins affixed to xanthene (DTMPX = ditrimesitylpor-itylporphyrin xanthene) or dibenzofuran (DTMPD = ditrimesitylporphyrin dibenzofuran) scaffolds using Suzuki cross-coupling methods (140–154). The mesityl (mes) substituents incorporated into the Pacman motif provide steric bulk, as well as oxidative stability for high-valent metal oxo species (155, 156).

Oxygen evolution experiments of the resulting $Mn_2Cl_2$(DTMPX) (**23**) and $Mn_2Cl_2$(DTMPD) (**24**) derivatives obtained by the methods of Scheme 3 were related to MnCl(TMP) monomer (TMP = tetramesityl porphyrin) in a manner similar to that for the DPX and DPD complexes (**11**) and (**19**) above. Complex **23**, containing the xanthene bridge with *meso*-tetraarylporphyrin subunits, is the most active (579 ± 30 TON), catalyzing $H_2O_2$ disproportionation process by 10-fold more than other compounds examined. Notably, the analogous *meso*-tetraaryl-substituted DPD platform (**24**) showed no enhanced reactivity over that of the monomeric MnCl(TMP) (58 ± 5 TON) control catalyst. These results clearly indicate the importance of both the appropriate substitution pattern about the porphyrin periphery, as well as the control that can be imparted on some catalytic processes by Pacman cleft orientation.

## B. Photooxidation Chemistry of Diiron(III) μ-Oxo Pacman Architectures

The synthesis of diiron(III) Pacman porphyrin constructs can be prepared in a manner similar to that shown in Schemes 2 and 3. Initial synthetic targets were centered on diiron(III) hydroxide bisporphyrin Pacman systems for applications involving electrochemical or photochemical water oxidation via a scheme that is the reverse of those implicated for $O_2$ reduction by dicobalt(II) bisporphyrin

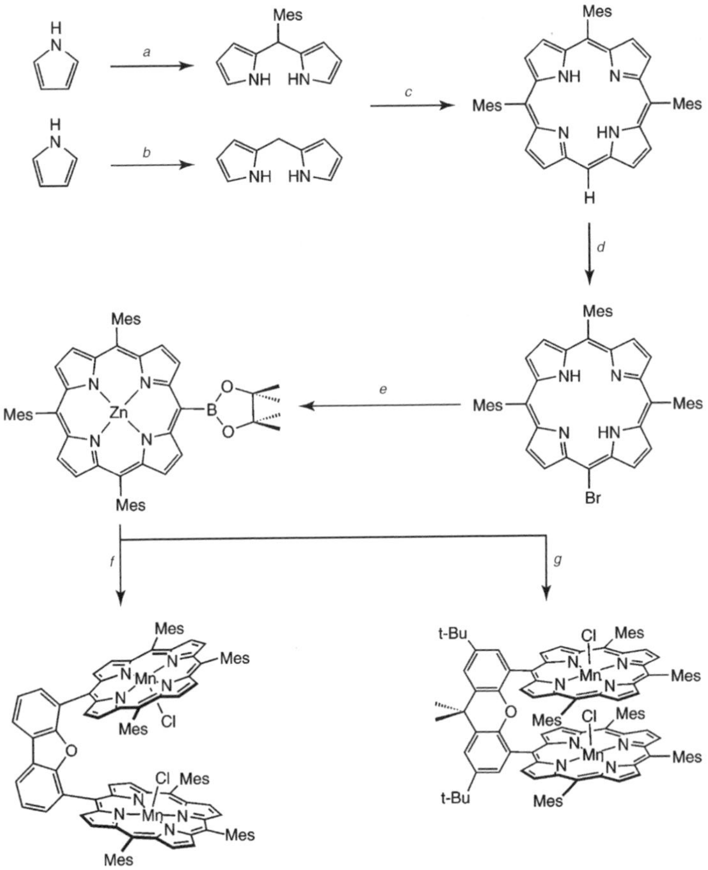

Scheme 3. (*a*) The $BF_3 \cdot OEt_2$, mesitaldehyde; (*b*) $BF_3 \cdot OEt_2$, $CH_2(OMe_3)_2$; (*c*) i. $BF_3 \cdot OEt_2$, mesitaldehyde; ii. DDQ; (*d*) NBS; (*e*) i. $Zn(OAc)_2 \cdot 2H_2O$; ii. pinacol borane, $Pd(PPh_3)_2Cl_2$, $Et_3N$ (*f*) i. 4,6-dibromodibenzofuran, $Pd(PPh_3)_4$, $K_3PO_4$; ii. 6 *N* HCl; iii. $Mn(OAc)_2 \cdot 4H_2O$, DMF; iv. aq NaCl, HCl; (*g*) i. 4,5-dibromo-2,7-di-*tert*-butyl-9,9-dimethylxanthene, $Pd(PPh_3)_4$, $Ba(OH)_2 \cdot 8H_2O$; ii. 6 *N* HCl; iii. $Mn(OAc)_2 \cdot 4H_2O$, DMF; iv. aq NaCl, HCl; actual position of axial chloride atoms for the dimanganese(III) Pacman assemblies is unknown.

Pacman architectures (Section IV). This proposed oxidative pathway, which is shown in Fig. 21, however, can be circumvented by the formation of the highly stable diiron(III)-μ-oxo bisporphyrin Pacman via loss of $H_2O$. One motivation for the preparation of the DPD framework was the hope that installation of the splayed, rigid spacer would hinder intramolecular μ-oxo formation due to the torsional strain associated with clamping down the Pacman bite angle. However, as Fig. 22 shows, the dibenzofuran to be a surprising flexible unit. Somewhat

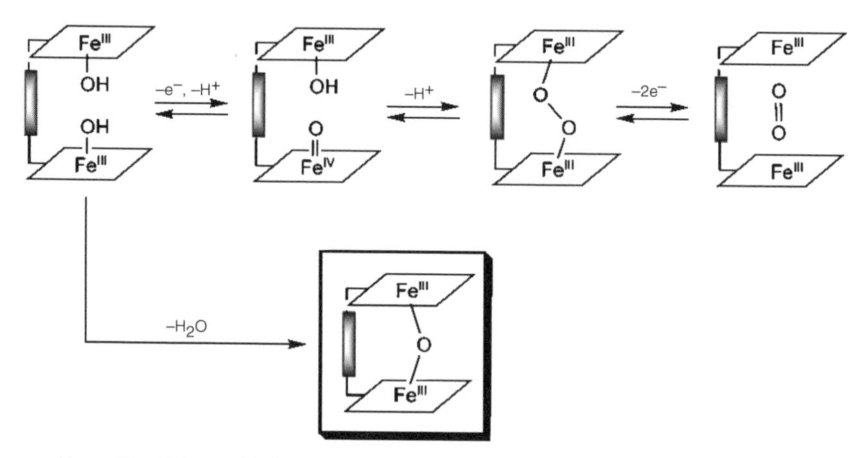


Figure 21. Water oxidation by diiron(III) bisporphyrins is thwarted by μ-oxo formation.

remarkably, metathesis of the diiron(III) chloride dibenzofuran Pacman construct (**17**) with NaOH results in the isolation of diiron(III)-μ-oxo bisporphyrin Pacman derivative (**18**). The structural characterization of DPD architecture **18** provided the first direct experimental support for drastic vertical conformational change in a single cofacial platform. A comparison between $Zn_2$(DPD) (**5**) and $Fe_2O$(DPD) (**18**), shown in Fig. 22, demonstrates the unprecedented ability of the DPD framework to open and close its binding pocket by a vertical distance of $> 4$ Å in the presence of exogenous ligands, thus structurally confirming the "Pacman effect" for pillared cofacial porphyrin systems.

Unlike most unbridged diiron(III)-μ-oxo bisporphyrins, in which the rings are arranged in a staggered conformation in order to maximize π interactions (84) and minimize steric repulsions, the rigid dibenzofuran spacer restricts the porphyrin cores from rotation with respect to each other. This restriction results

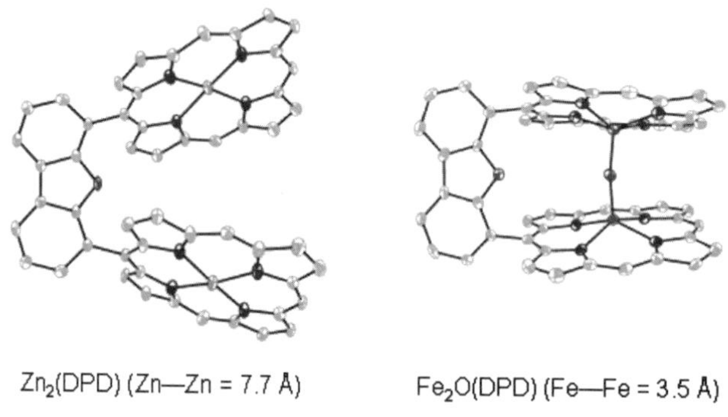


Figure 22. Structural comparison of $Zn_2$(DPD) (**5**) and $Fe_2O$(DPD) (**18**).

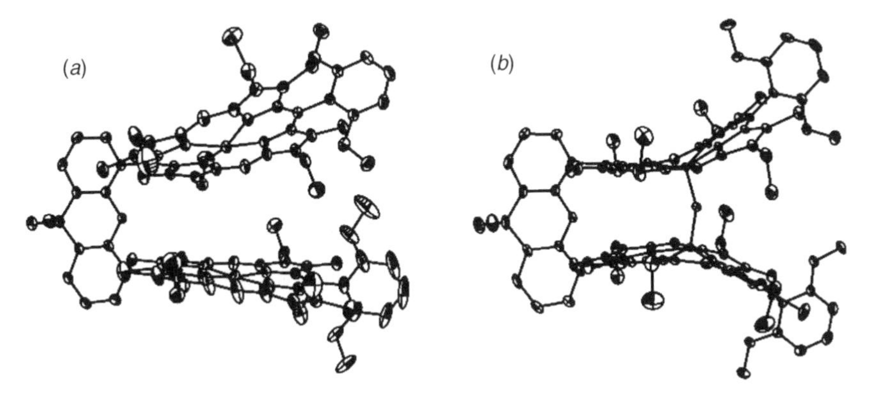


Figure 23. Solid state structures of (*a*) $Zn_2$(DPDM) and $Fe_2O$(DPDM) (**25**).

in an eclipsed conformation with an average torsion angle of a mere 1.65° between the two porphyrin macrocycles. Perhaps more striking is the bent Fe—O—Fe angle of 158.7° in the structure of **18**. The Fe—O—Fe unit of most diiron(III)-μ-oxo bisporphyrins is virtually linear with angles ranging from 172 to 179° to minimize the nonbonded interactions between the porphyrinic cores (157, 158). For comparison, the smallest Fe—O—Fe angle observed prior to that of **18** was 161.1° in a urea-linked bisporphyrin, which arose due to hydrogen bonding of the central oxygen to two solvent water molecules (159).

Installation of sterically demanding groups trans to the rigid spacer are also ineffective in preventing μ-oxo formation for diiron(III) Pacman systems. The complex $Fe_2O$(DPXM) (**25**) is readily prepared by iron insertion into $H_4$(DPXM) followed by basic workup. The solid-state structure obtained for **25** (Fig. 23) clearly shows the presence of an intramolecular μ-oxo bridge, which results in a severely distorted Pacman cleft, as compared to the dizinc(II) congener. The ability of the DPXM Pacman to clamp down in the diiron(III)-μ-oxo complex (**25**) results in a reduced interplanar angle of roughly 25° and shortened metal–metal distance of 3.489 Å compared to the relaxed $Zn_2$(DPXM) compound (interplanar angle = 32.9°, Zn–Zn = 5.913 Å) (160). Reminiscent of the unsubstituted $Fe_2O$(DPD) structure of Fig. 22, only a small torsional twist between the porphyrin subunits is observed for methoxyaryl congener **25**. The severely bent Fe—O—Fe angle of 155.2° exceeds that observed in **18** and indeed is the smallest recorded for a diiron(III)-μ-oxo bisporphyrin complex (159, 161). Additionally, the structure of **25** is distinguished by a the severe bending of porphyrin rings; an intraporphyrin angle of 150.6° attests to the thermodynamic inclination for diiron(III)-μ-oxo formation.

The extreme stability of diiron(III)-μ-oxo bisporphyrin would seem to pose a significant hurdle to carrying out the oxidation chemistry of Fig. 21, and other small molecule activation chemistry within the Pacman cleft. However, an active

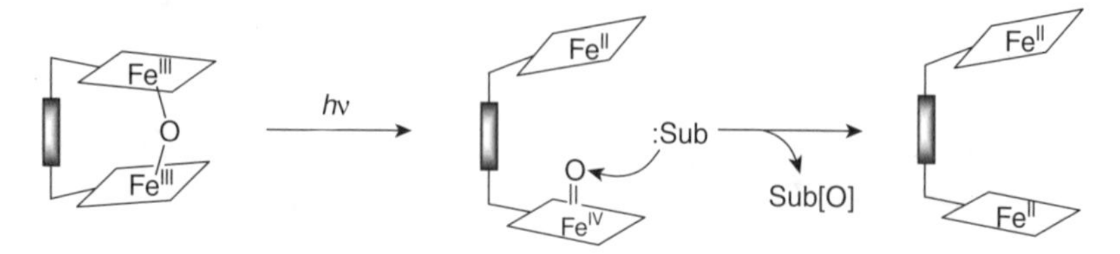


Figure 24. Generalized scheme for OAT via photochemical metal-oxo generation.

oxidation intermediate, cofacial $(PFe^{II})(PFe^{IV}{=}O)$, may be delivered by using light excitation to cleave the thermally inert $Fe^{III}$–O–$Fe^{III}$ bond via the internal disproportionation shown in Fig. 24 (162). The active catalyst, which is the ferryl $(PFe^{IV}{=}O)$ complement of the pair, is capable of oxygenating electron-rich organic oxygen atom acceptors, such as phosphines (86) and sulfides (163) to generate an oxidized substrate and two equiv of $PFe^{II}$ (Fig. 24). Reaction of two ferrous porphyrin subunits with $O_2$ re-forms the diiron(III)-μ-oxo assembly via a Balch-type mechanism (164, 165) for reentry into the oxidative process. In this way, a photocycle for organic substrate oxidation can be constructed using molecular oxygen as the terminal oxidant and oxygen atom source. A general depiction of this cycle appears in Fig. 25. The mechanism for such OAT reactions

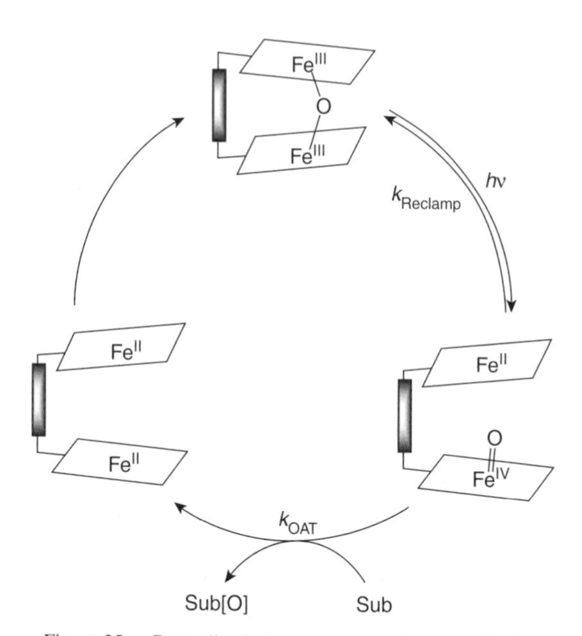


Figure 25. Generalized photocycle for substrate oxidation.

TABLE X
Product Quantum Yields ($\phi_p$) for the Photoinduced Oxidation of Phosphites by $Fe_2O$(DPX) (**10**) and $Fe_2O$(DPD) (**18**)

| | | $\Phi_p$ | |
|---|---|---|---|
| Phosphites | Cone angle (°) | (DPX)$Fe_2O$ (**10**) | (DPD)$Fe_2O$ (**18**) |
| $P(OMe)_3$ | 107 | $9.0(1) \times 10^{-8}$ | $7.4(7) \times 10^{-4}$ |
| $P(OEt)_3$ | 109 | $5.0(1) \times 10^{-8}$ | $3.2(4) \times 10^{-4}$ |
| P(O–*i*-Pr)$_3$ | 130 | No reaction | $2.2(2) \times 10^{-5}$ |
| $P(OTMS)_3$ | >172 | No reaction | $2.3(2) \times 10^{-6}$ |

is believed to proceed under stereoelectronic control, which is preferred for the side-on attack of substrate on the electrophilic M=O subunit (166–171). To this end, the oxidative photocatalysis builds on the chemistry of unbridged $PFe^{III}$–O–$Fe^{III}P$ constructs (172–177) with the enhancement stemming from the fact that the Pacman assembled from a single rigid pillar sterically directs substrate to the electronically preferred side-on approach to the metal–oxo bond.

Comparative quantum yield and TA experiments have shown that $Fe_2O$(DPD) (**18**) exhibits a superior photoreactivity (~10,000-fold) as compared to $Fe_2O$(DPX) (**10**) in which substrate attack is impaired by the smaller Pacman cleft. Additionally, the DPX Pacman congener is far more sensitive to the size (cone angle) of the attacking substrate (162) as shown by Table X. Investigation of the mechanistic details of the oxidative photocycle by time-resolved spectroscopy revealed that the rates of net OAT for spring-loaded DPD Pacman platforms are four orders of magnitude greater than those of relaxed platforms (Table XI).

Transient absorption measurements show that the spring action of the bisporphyrin cleft does little to impede reclamping to form the μ-oxo species, but rather is important to opening the cofacial cleft to allow substrate access to the photogenerated ferryl oxidant (Fig. 26). This photochemical and reactivity data support the contention that the DPD framework is structurally spring loaded

TABLE XI
Lifetimes, $k_{reclamp}$, $\Phi_p$, and Calculated $k_{OAT}$ for a Series of Diiron(III)-μ-Oxo Bis(porphyrins)

| | | | $P(OMe)_3$ Oxidation | |
|---|---|---|---|---|
| Compound | $\tau_o$ (ns) | $k_{reclamp}$ ($10^8$ s$^{-1}$) | $k_{OAT}$ ($M^{-1}$ s$^{-1}$) | $\Phi_p$ |
| $Fe_2O$(DPD) (**18**) | 1.36 | 7.4 | $1.1 \times 10^7$ | $7.4 \times 10^{-4}$ |
| $Fe_2O$(DPX) (**10**) | 1.26 | 7.9 | $1.4 \times 10^3$ | $9.0 \times 10^{-8}$ |
| $Fe_2O$(DPXM) (**25**) | 1.27 | 7.9 | $9 \times 10^4$ | $5.5 \times 10^{-6}$ |
| $Fe_2O(Etio)_2$ | 0.97 | 10.3 | $1.3 \times 10^7$ | $6.5 \times 10^{-4}$ |

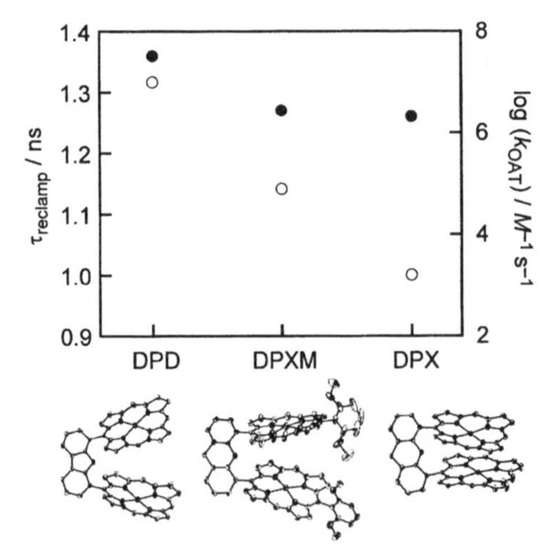


Figure 26. Values of $k_{\text{reclamp}}$ (•) and $k_{\text{OAT}}$ (○) for cofacial bisporphyrins.

(162, 163), resulting from the ability of the cofacial porphyrin to open and close its binding pocket by a vertical distance of over 4 Å (see above).

Although (bis)etioporphyrin Pacman architectures are capable of oxidizing electron-rich organic substrates (i.e., phosphines and sulfides), photooxidation of more thermodynamically and kinetically challenging substrates, such as olefins and alkanes, does not proceed readily due to the relatively low oxidation potential of the parent etioporphyrin ferryl subunit (178). The redox properties of porphyrins can be made more oxidizing by modulating the substituents on the porphyrin periphery. As has been shown previously, the oxidizing power of metalloporphyrins can be greatly enhanced by introducing electron-withdrawing groups onto the porphyrin macrocycle (179, 180). In an effort to expand the photooxidation chemistry of Fig. 25 to a broader range of potential organic substrates, a spring-loaded bisporphyrin DPD architecture appended with *meso*-pentafluorophenyl porphyrins, $Fe_2O$(DPDF) (**26**), was prepared as shown in Scheme 4 (181). Fluorinated Pacman **26** is a superior catalyst for sulfide oxidation, with TONs approaching 10,000, as compared to a TON of roughly

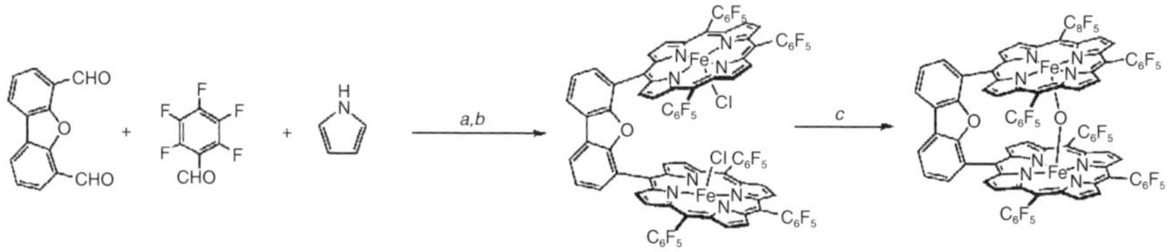

Scheme 4. (*a*) i. $BF_3 \cdot OEt_2$; ii. DDQ; (*b*) i. $FeBr_2$; ii. HCl; (*c*) NaOH, $H_2O$.

TABLE XII
Summary of Kinetic Data for Reaction of Various Hydrocarbons With $Fe_2O$ (DPDF) (**26**).[a]

| Substrate | $BDE_{C-H}$ (kcal $mol^{-1}$) | IE (eV) | $\Phi_P$ | $k_{ox}$ ($M^{-1}s^{-1}$) |
|---|---|---|---|---|
| Fluorene | 80 | 7.91 | $1.52 \times 10^{-2}$ | $1.36 \times 10^{7}$ |
| Diphenylmethane | 84.5 | 8.73 | $2.76 \times 10^{-3}$ | $2.41 \times 10^{6}$ |
| Cumene | 84.8 | 8.8 | $1.99 \times 10^{-3}$ | $1.74 \times 10^{6}$ |
| Toluene | 90 | 8.83 | $1.51 \times 10^{-3}$ | $1.32 \times 10^{6}$ |
| Toluene-$d^8$ | 90 | 8.83 | $9.79 \times 10^{-4}$ | $8.53 \times 10^{5}$ |

[a] Table reproduced from Ref. 183. All ionization energy (IE) values obtained from National Institute of Standards and Technology (NIST) (http://webbook.nist.gov/chemistry/); BDE values were obtained from Ref. 185.

600 for etio homologue **18** (181). Moreover, the electron deficiency of the fluorinated porphyrin results in a red–shift of the absorption spectrum, allowing for photooxidation chemistry to be trigger using visible light ($\lambda_{exc} = 425$–$460$ nm). This result is in contrast with etioporphyrin DPD Pacman **18**, which requires light in the near-UV region ($\lambda_{exc} = 360$–$380$ nm) (182).

Pacman **26** is also capable of catalytic olefin oxidation using $O_2$ as the terminal oxidant and oxygen atom source. The photoreaction is efficient and clean, as photocatalysis can be carried out with turnover numbers exceeding 1000 for styrene oxidation (181). More recently, it has been found that hydrocarbons can also be oxidized under similar photooxidation conditions (183). Product appearance quantum yields ($\Phi_p$) are listed in Table XII for the photoreactions of **26** with hydrocarbon substrates. The correlation between hydrocarbon IE and $\log(k_{ox})$ was taken as evidence of a mechanism in which $H^{\bullet}$ abstraction occurs via an asynchronous PCET to the ferryl operates for this system (71, 184). As shown in Fig. 27, the corresponding relation employing the

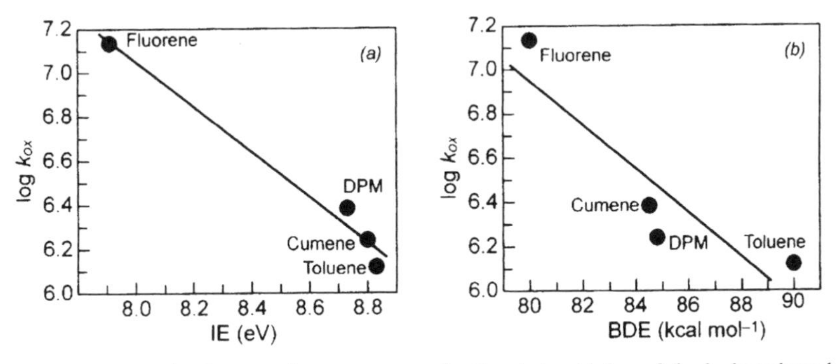


Figure 27. Correlation between the rate constant for the photooxidation of the hydrocarbons by $Fe_2O$(DPDF) (**26**) and (*a*) the ionization energy of the substrate and (*b*) the C–H bond dissociation energy.

substrate C–H bond dissociation energy ($BDE_{C-H}$) (185) is not nearly as strongly correlated (186). The contention that an asynchronous PCET reaction is the crucial step in the hydroxylation chemistry is further bolstered by the small kinetic isotope effect (KIE) for hydrogen atom abstraction from toluene-$d^8$ as compared to that recorded for the protio substrate ($k_H/k_D = 1.55$) (70, 181). The isotope effect observed for the Pacman $PFe^{IV}{=}O$ oxidant is significantly smaller than those observed for the oxidation of ethylbenzene by nonheme $Fe^{IV}{=}O$ species, in which highly synchronous hydrogen tunneling mechanisms have been suggested to be involved in the C–H bond activation process (187).

Hydrocarbon photooxidation can also be carried out catalytically with $Fe_2O$(DPDF) (**26**) in the presence of $O_2$. Figure 28 shows the substrate TONs and various oxidation products formed under catalytic conditions. The trend in the TONs roughly parallels the substrate reactivity (lower IE correlates with higher TON). Competitive oxidation experiments with labeled substrates indicate that autooxidation chemistry does not contribute to the overall oxidation process, giving rise to the proposed mechanism shown in Fig. 28. The $Fe_2O$(DPDF) Pacman photocatalyst is distinguished from other C–H oxidants because it (1) catalytically oxidizes C–H bonds; (2) with oxygen as the terminal oxidant and oxygen atom source; and (3) without the need for a coreductant, allowing for oxidative cycles to be created in which only $O_2$ and substrate are consumed.

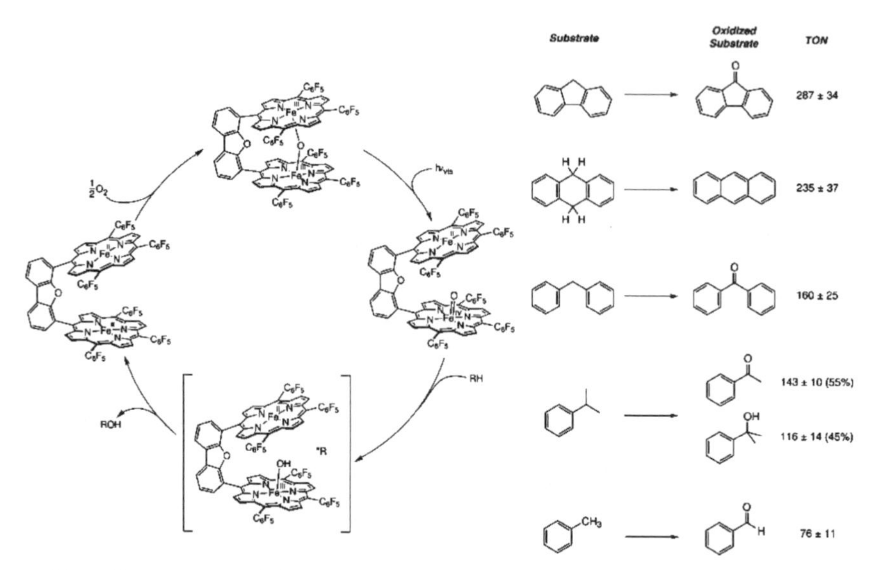


Figure 28. Photocatalytic hydrocarbon oxidation using visible light and molecular oxygen.

# VI. HANGMAN PORPHYRIN CONSTRUCTS FOR OXYGEN ACTIVATION

The importance of PCET to the efficacy of oxygen activation is highlighted by the $O_2$ reduction chemistry of the Pacman porphyrins. As described in Section IV, the proton affinity of the substrate is affected by the electronic properties of the capping porphyrin. The acid–base properties of substrates bound to the porphyrin platform may more precisely be controlled when one porphyrin redox subunit of the Pacman construct is replaced with a protonic functionality. The framework resulting from "hanging" a hydrogen-bond functionality from the xanthene has been designated Hangman porphyrins (HPX) (188). These proton-redox shuttle hybrids were prepared by following methods developed for the construction of symmetric Pacman assemblies (Scheme 5). These methods were subsequently extended to enable the facile preparation of the Hangman superstructure via more modular Suzuki cross-coupling methodologies (138).

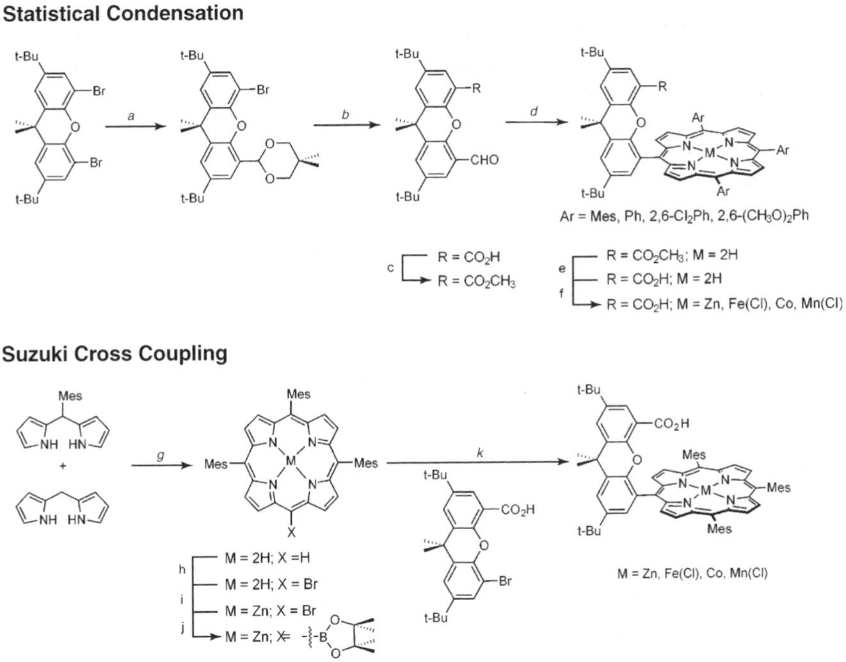

**Statistical Condensation**

Ar = Mes, Ph, 2,6-$Cl_2$Ph, 2,6-$(CH_3O)_2$Ph

c: R = $CO_2H$ → R = $CO_2CH_3$

e, f: R = $CO_2CH_3$; M = 2H → R = $CO_2H$; M = 2H → R = $CO_2H$; M = Zn, Fe(Cl), Co, Mn(Cl)

**Suzuki Cross Coupling**

h, i, j: M = 2H; X = H → M = 2H; X = Br → M = Zn; X = Br → M = Zn; X = pinacol boronate

M = Zn, Fe(Cl), Co, Mn(Cl)

Scheme 5. (*a*) i. PhLi, TMEDA; ii. DMF; iii. $H_2O$ ; iv. neopentyl glycol, benzenesulfonic acid, toluene, reflux; (*b*) i. PhLi, TMEDA; ii. $CO_2$; iii. $H_3O^+$; (*c*) $Me_3OH$, $H_2SO_4$; (d) i. Ar–CHO, pyrrole, $BF_3{\bullet}OEt_2$; ii. DDQ; (*e*) NaOH, $H_2O$, THF; (*f*) $MX_2$, DMF, reflux; (*g*) i. $BF_3{\bullet}OEt_2$; ii. DDQ; (*h*) *N*-bromosuccinimide (NBS); (i) $Zn(OAc)_2$, DMF, reflux; (*j*) pinacol borane, $Pd(PPh_3)_2Cl_2$, $Et_3N$, 1,2-dichloroethane, 90°C;(*k*) i. $Pd(PPh_3)_4$, $Na_2CO_3$, DMF, reflux; ii. 6 *N* HCl, $CHCl_3$; iii. $MX_2$, DMF, reflux.

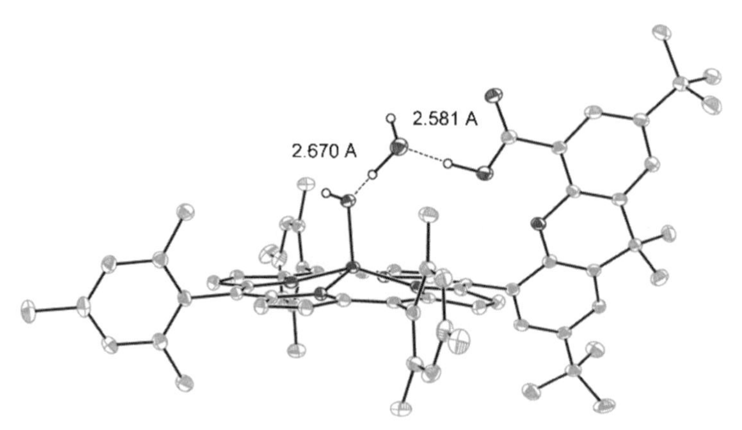


Figure 29. The molecular structure of the heme water channel model, $Fe^{III}(OH)(HPX{-}CO_2H)$.

The Hangman construct allows for precise control over the functional nature of the hydrogen-bonding group in terms of proton-donating ability and arrangement in relation to the metalloporphyrin redox site. The carboxylic acid derivative has led to an especially fruitful study of PCET. The structure of the iron complex, shown in Fig. 29 (189), represents the first structurally characterized monomeric iron(III) hydroxide porphyrin. Even more striking is the water molecule bound between the xanthene carboxylic acid and the hydroxide ligand. This is the first model of a redox-active site displaying a structurally well-defined proton-transfer network. This unique structural motif captures the essence of the monooxygenases in which structured water is proposed to finely tune heme electronic structure and redox potential, as well as providing a possible proton-relay during multielectron catalysis. Notably, spectroscopic data for $Fe^{III}OH(HPX)$ indicate that the water molecule remains bound in solution as well as in the solid state, and that this binding is chemically reversible (189). The measured binding energy of water is 5.8 $kcal\,mol^{-1}$ (190). The HPX platform thus provides a geometrically matched microcavity for examination of PCET-mediated $O \cdots O$ bond forming and breaking reactions by juxtaposing two oxygen atoms between proton and electron-transfer sites.

The Hangman construct represents a simplified model of metalloenzymes with engineered distal sites (191–193). These hybrid architectures capture control of both proton and electron transfer, but with greatly reduced complexity since secondary and tertiary protein structure are not required to impose the structural proximity of the "hanging" proton network and redox center. The pendant acid group of the Hangman porphyrin plays the same role as the water channels of natural monooxygenases such as cytochrome P450 shown in Fig. 30(*a*) (194). Protons are managed by the water channel, which is directed

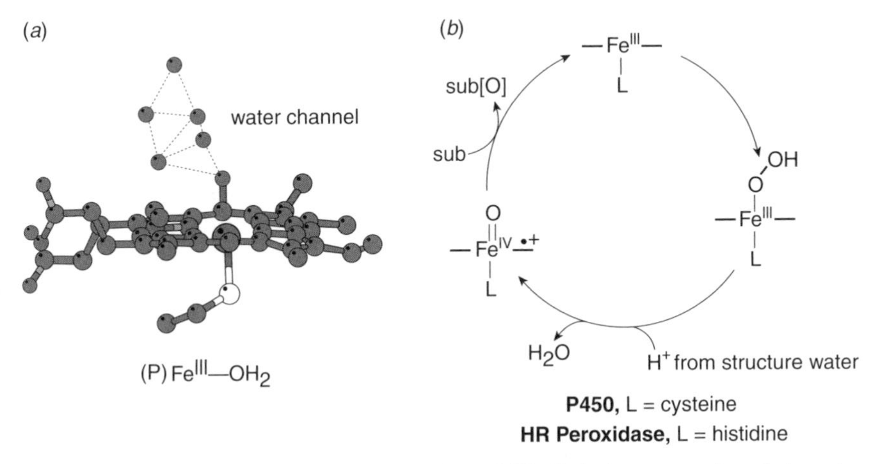


Figure 30. (*a*) High-resolution structure of cytochrome P450, displaying a water channel above the heme. (*b*) Peroxo shunt mechanism of monooxygenases producing compound **I** ((P$^{\bullet+}$)Fe$^{IV}$=O), which oxidizes substrates by their nucleophilic attack on the electrophilic oxo of the (P$^{\bullet+}$)Fe$^{IV}$=O core.

along the coordination axis of oxygen activation. The highly activated ferryl oxygen of the redox cofactor, (P$^+$)Fe$^{IV}$=O (compound I-type intermediate in which an Fe$^{IV}$=O is ligated by a porphyrin cation radical, P$^+$), is produced by proton transfer from the water channel to a ferric peroxy intermediate, as shown in Fig. 30(*b*). Formation of the high-valent metal–oxo fragment is thus accomplished by coupling proton transfer to an internal 2e$^-$ redox event. The Hangman porphyrins capture this monooxygenase activity precisely. Stopped-flow kinetics for the reaction of (HPX)Fe$^{III}$(OH) and *m*-chloroperoxybenzoic acid (*m*-CPBA) (195) show that proton transfer from the acid–base hanging group in (HPX)Fe$^{III}$ peroxide complexes yields (HP$^+$X)Fe$^{IV}$=O. The kinetics for this process are reproduced in Fig. 31. Substitution of perbenzoate for hydroxide first affords a ferric acylperoxo species in seconds. The absorption spectrum of the acyl peroxide ($\lambda_{max}$ = 416 and 506 nm) smoothly disappears with the concomitant growth of the absorption spectrum for (HP$^{\bullet+}$X)Fe$^{IV}$=O ($\lambda_{max}$ = 678 nm) (196). No intermediates are observed in the process, thereby indicating exclusive O—O bond heterolysis to generate the highly oxidizing metal–oxo intermediate.

The metal–oxo, in much the same way as it does it biology, promotes a prolific catalytic activity. The oxo reacts with peroxide in a dismutation reaction similar to that observed for the bis(manganese) Pacman complexes of Section V to produce oxygen and water. The O—O bond activation was evaluated for a set of HPX and HPD compounds with varying hydrogen-bonding and proton-transfer abilities (197). From a family of eight compounds shown in Fig. 32,

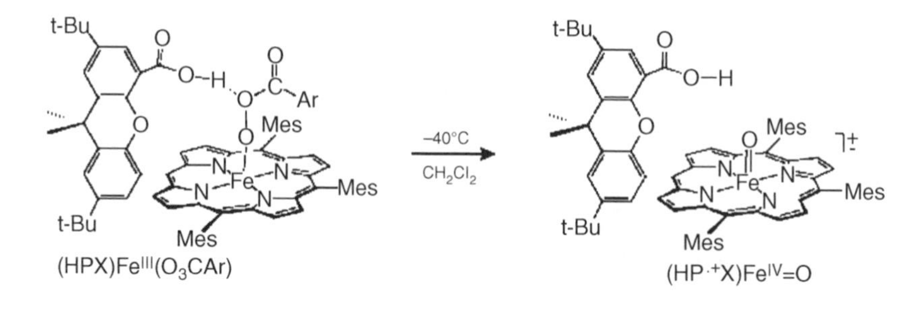


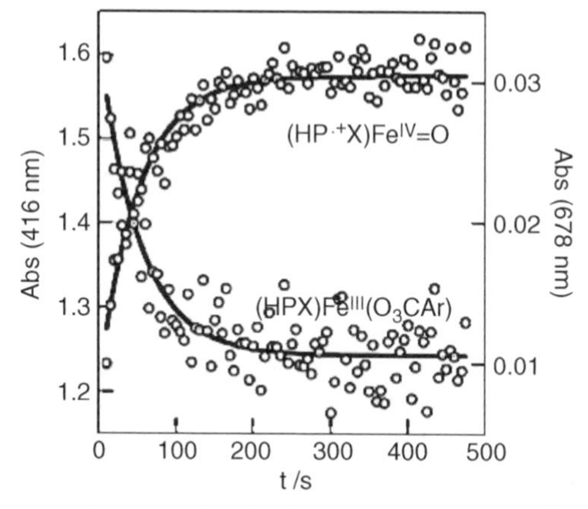


Figure 31. Selected stopped-flow data for the disappearance of (HPX)Fe$^{III}$ acyl peroxide ($\lambda_{max}$ ) 416 nm and the concomitant appearance of (HP$^{\bullet+}$X)Fe$^{IV}$=O ($\lambda_{max}$) 678 nm. Global analysis of the full spectral window (400–700 nm) for the disappearance and appearance traces using a first-order kinetic model gives $k_{obs} = (1.9 \pm 0.1) \times 10^{-2}\ s^{-1}$ for O–O bond heterolysis.

(p$K_a$ values ranging from $< 2$–25), we have observed that the most effective disproportionation catalysts possess hanging groups with p$K_a < 5$ ($R = CO_2H$ and $P(O)(OH)_2$]. Control experiments using a simple redox-only porphyrin analogue with similar steric and electronic properties [Fe(Cl)TMP] and a hanging group with no acid–base functionality (hanging group = $CO_2Me$) show negligible reactivity as compared to **28**, establishing that a single well-positioned proton-transfer site engenders greatly enhanced oxygen activation chemistry. The proposed catalytic pathway for the O–O reactivity of the most active xanthene-bridged HPX–COOH platform, shown in Fig. 33, bears many similarities with the reaction cycle of monooxygenases (Fig. 30). Reaction of $H_2O_2$ with the porphyrin in the presence of an axial ligand results in a metal hydroperoxide complex (239) similar to that observed in the stopped-flow experiments (Fig. 31). Directed proton transfer to the bound $HO_2^-$ species

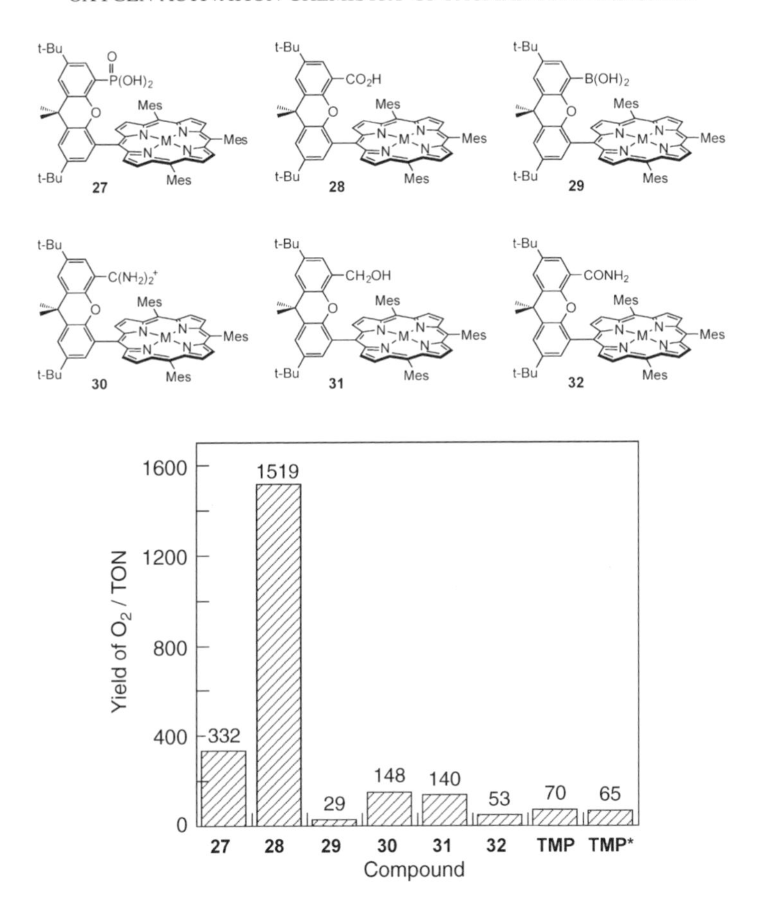


Figure 32. Turnover numbers for oxygen production by $H_2O_2$ dismutation catalyzed by iron Hangman complexes **27**–**32** and FeCl(TMP). The TMP denotes FeCl(TMP) and TMP* denotes FeCl(TMP) with 1 equiv benzoic acid.

promotes heterolytic O—O bond cleavage to produce the active $Fe^{IV}{=}O$ porphyrin cation radical [or $Mn^{V}{=}O$ for peroxide bound to the Mn(III) HPX catalyst].

The electrophilic oxo of the Hangman platform is also susceptible to nucleophilic attack by the two-electron bond of olefins. The reaction parallels the peroxide shunt cycle (198–200) of cytochrome P450 and peroxidase enzymes while building on the results of the observed biomimetic catalase activity. The common olefins styrene and *cis*-cyclooctene were chosen as substrates for epoxidation by the manganese HPX and HPD derivatives MnCl(HPX—$CO_2$H) (**33**), The MnCl(HPX—$CO_2$Me) complex (**34**) and MnCl(HPD—$CO_2$H) (**35**) with MnCl(TMP) as the standard baseline compound.

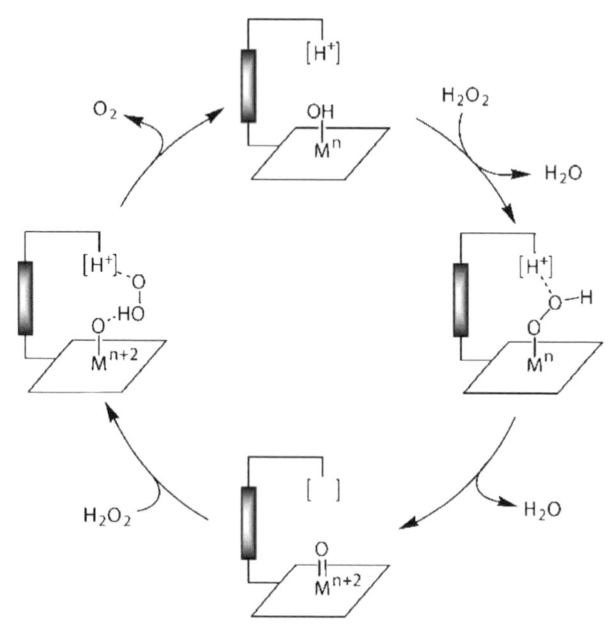


Figure 33. Proposed pathway for proton-assisted metal–oxo formation and substrate oxidation by HPX Hangman porphyrins.

Table XIII lists the product yields for olefin epoxidation with $H_2O_2$ catalyzed by the manganese complexes at 0.2 mol% catalyst loading.

Epoxide products are formed in >95% selectivity with little to no allylic oxidation side products, indicating that the involvement of radical species, such as HO• and HOO•, are unlikely (166, 201). The results for epoxidation reactivity of the manganese derivatives follow the same general trend observed

TABLE XIII
Epoxidation of Olefins Catalyzed by Manganese Hangman Complexes.[a]

| Catalyst | Substrate | |
|---|---|---|
| | Styrene (% Yield) | *cis*-Cyclooctene (% Yield) |
| MnCl(HPX–$CO_2$H) (**33**) | $70 \pm 2$ | $92 \pm 6$ |
| MnCl(HPX–$CO_2$Me) (**34**) | $38 \pm 2$ | $42 \pm 5$ |
| MnCl(HPD–$CO_2$H) (**35**) | $4 \pm 1$ | $5 \pm 1$ |
| MnCl(TMP) | $4 \pm 2$ | $5 \pm 1$ |
| MnCl(TMP) + 1 equiv PhCOOH | $6 \pm 1$ | $9 \pm 2$ |

[a]Yields were determined by analysis of the epoxide products using GC–MS.

for the iron-mediated catalase reactions. Complex **33** bearing a protic carboxylic acid group on a xanthene platform is the most active, affording styrene and *cis*-cyclooctene oxides in 70% ($350 \pm 10$ TON) and 92% ($460 \pm 30$ TON), respectively. These values are almost twice as high as those obtained for the ester derivative **34** ($191 \pm 8$ TON for styrene, $208 \pm 22$ TON for *cis*-cyclooctene). Under these conditions, both HPD complex **35** and MnCl(TMP) yield significantly less epoxide product upon reaction with either styrene or *cis*-cyclooctene (<10%, <50 TON). As observed for the catalase reactions, negligible olefin epoxidation is detected in the absence of catalyst or in the presence of catalyst without axial ligand. Furthermore, the external addition of 1 equiv of either benzoic acid or methyl benzoate to MnCl(TMP) fails to yield an increase in epoxide product. Lastly, we also examined the stereochemistry of the epoxidation of *cis*-stilbene by **33** (202–205). The high cis-epoxide/trans-epoxide ratio of $6 \pm 1$ obtained is consistent with a predominant oxomanganese(V) porphyrin oxidizing agent (206–209).

## VII. OUTLOOK AND FUTURE DIRECTIONS

### A. O—O Bond Formation at Hangman Platforms

The microscopic reverse of O—O bond scission, in which water is oxidized in a $4e^-/4H^+$ PCET process to generate $O_2$, is a reaction of great consequence to the generation of carbon-neutral energy (182, 195, 210–212). The mechanistic parallels of the $H_2O \leftrightarrow O_2$ interconversion, as first enunciated by Babcock and co-workers (213), are reflected by the two natural energy conversion systems shown in Fig. 34. The photoinduced oxidation of water to oxygen by Photosystem II (PSII) is the critical solar conversion process that is life's energy source, which is harnessed in the reduction of oxygen to water by cytochrome *c* oxidase (C*c*O), which serves as Nature's fuel cell. Both O—O bond formation and bond cleavage proceed through a common hydroxo–metal oxo intermediate.

Against this backdrop of bioenergy conversion, the scheme shown in Fig. 35 may be developed. Whereas the Pacman porphyrins containing late transition metals, such as cobalt (Section IV), can drive $O_2$ reduction (to the right), the cofacial Pacman motif poses a problem when porphyrins contain earlier transition metals, such as Fe and Mn, which are needed for water oxidation and O—O bond coupling (to the left). The thermodynamic predisposition for M—(μ-O)—M (μ-oxo) formation of the type shown in Figs. 22 and 23 poses a significant impediment to water assembly (see Fig. 21). On this count, the Hangman porphyrins are superior. The hanging acid–base group permits the secondary coordination sphere of the porphyrin to be controlled for water

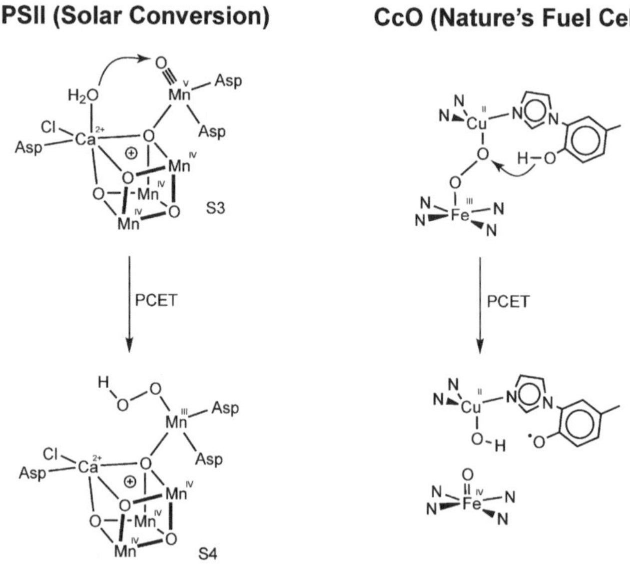

**PSII (Solar Conversion)** **CcO (Nature's Fuel Cell)**

PCET PCET

S3 S4

Figure 34. The PCET driven processes for O–O bond formation and cleavage in Nature.

assembly. In the absence of a second metal, μ-oxo formation is circumvented. Moreover, the electrophilic oxo required for water oxidation in Fig. 35 is supported on the Hangman platform (see Fig. 31) and the oxo exhibits reactivity that is essential for water activation. Specifically, the electrophilic oxo of $(HP^{\bullet+}X)Fe^{IV}{=}O$ is attacked by the two-electron bond of olefins (187, 214). To effect water splitting, the olefinic nucleophile needs to be replaced by hydroxide. Practically, $OH^-$ is thermodynamically more difficult to oxidize than olefins, and hence the electrophilicity of the oxo of $(HP^{\bullet+}X)Fe^{IV}{=}O$ needs to be maximized. Moreover, the proton inventory of the current HPX platform, is incompatible for the formation and stabilization of nucleophilic hydroxide from a suspended water molecule. The carboxylic acid hanging group of the Hangman architecture needs to be replaced with a Brønstead–Lowry base to generate the required hydroxide nucleophile.

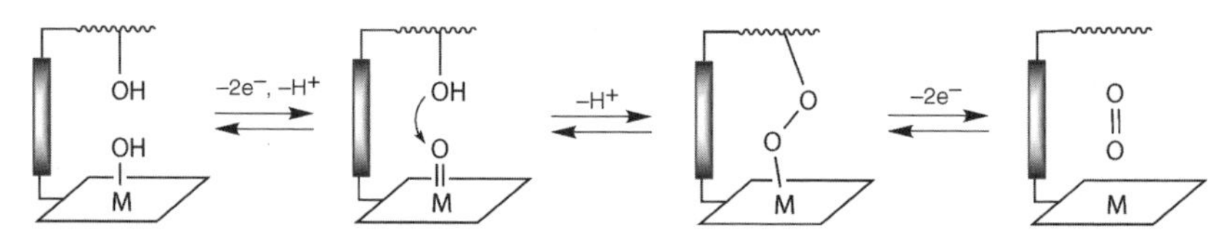


Figure 35. The PCET mechanism for the reversible interconverstion of $H_2O$ and $O_2$.

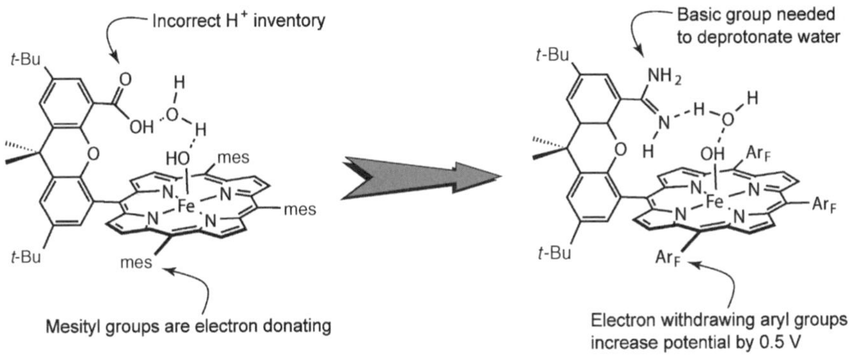

Figure 36. General strategy for adopting the Hangman porphyrin construct for (O–O bond coupling.

Figure 36 shows a reconstructed Hangman platform for the water-splitting reaction. In order to boost the oxidation potential of the porphyrin subunit, electron-withdrawing groups have been incorporated onto the periphery of the porphyrin macrocycle. Introduction of ancillary fluorinated phenyl groups onto the meso positions of the porphyrin framework increases the oxidizing power of the porphyrin macrocycle by >0.4 V (215, 216). The fluorination of the β-positions of the porphyrin ring may further augment the electron-withdrawing properties of the redox platform (217). We have also developed a successful methodology for the incorporation of a wide variety of functional acid–base groups including *proton-accepting* amidine and pyridine functionalities. These hanging groups have the same bond number connectivity as carboxylate, so it is geometrically matched to bind a water molecule in the Hangman cleft and is basic enough to generate the hydroxide nucleophile *in situ*. We are currently evaluating the capacity of the Hangman platform shown in Fig. 36 to promote O–O bond coupling by electrochemical and photochemical means.

## B. Other Hangman Platforms

Significant challenges are posed by the lengthy and tedious preparation of porphyrin platforms and their intractability to modular modifications. Initial inroads in this area have been made with the development of Hangman salophens (Fig. 37). The R groups of new Hangman salophen xanthene (HSX) ligands can be altered to easily tune the properties of the redox platform. A wide variety of HSX derivatives with distinct electronic properties have been prepared in a modular fashion and display disparate catalytic activities with respect to oxygen activation (218). For M = Mn and R = OMe, direct spectroscopic evidence for the production of Mn(V)–oxo upon O–O bond heterolysis

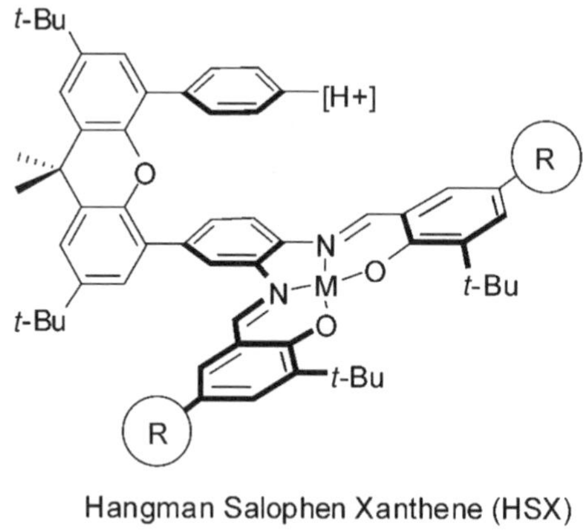

Hangman Salophen Xanthene (HSX)

Figure 37. Modular nature of Hangman salophen allows for facile synthetic alteration of the redox and acid–base properties.

at the Hangman salophen platforms has been provided by stopped flow experiments. The use of *m*-CPBA as the oxygen atom source allows the catalase cycle to be arrested and intermediates at the Hangman salophen platform detected. Proton transfer to the aryl peroxide coordinated to the Mn(III) center furnishes the high-valent Mn(V) oxo intermediate, though elusive (219, 220), is often cited as the active catalyst in model and enzymatic systems (221). Unlike the Hangman porphyrin systems, oxo production at the salophen is not rate-determining and for this reason the acid-to-metal distance is less crucial for the first step in of a catalase cycle. Instead, subsequent oxidation of peroxide by the Mn(V) oxo appears to be rate-determining, a contention that is supported by recent theoretical investigations of Mn salen [salen = bis(salicylidene)ethyne)ethylenediamine] catalase activity (222, 223). The catalase cycle deduced from computational studies is shown in Fig. 38. The calculations establish that the end-on binding of the $H_2O_2$ is the most productive approach for substrate oxidation by the Mn(V) oxo center. Hydrogen bonding to the oxygens of the salen platform leads to temporary catalyst deactivation by forming kinetically stable intermediates, sending the reaction off-cycle (Fig. 38, **A**). The hanging acid–base group of the Hangman construct can keep the catalyst on cycle by promoting the end-on assembly of $H_2O_2$ via a hydrogen-bonding network (Fig. 38, **B**). In promoting the end-on association of $H_2O_2$ in the Hangman pocket, the hanging group averts temporary catalyst deactivation arising from geometrically disadvantageous binding modes of $H_2O_2$.

The chiral framework that has been the centerpiece of the development of metal salens as asymmetric oxidation catalysts (224, 225) cannot be installed in the HSX platform since the acid functionalized xanthene is bound to a salophen

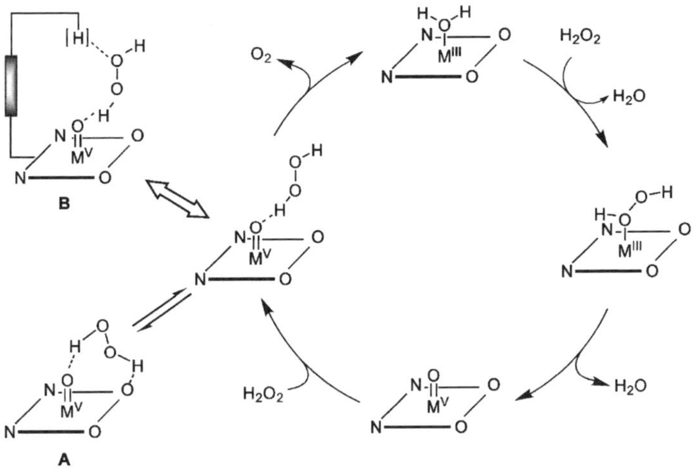

Figure 38. Cycle for catalase activity of Hangman salophen complexes.

ligand through the phenylenediamine bridge. To address this design issue, the Hangman salen xanthane (HSX*) and $H_{ph}SX^*$ ligands shown in Fig. 39 have been designed (226). The rigid hydrogen-bond functionalities are poised over a salen platform that incorporates a chiral cyclohexanediimine bridge. In order to accommodate the aliphatic bridge, the attachment point of the functionalized xanthene spacer to the phenolic arms of the ligand has been modified. The incorporation of two xanthene scaffolds forms a convergent and symmetric molecular cleft of two different sizes formed from a hydrogen-bond group directly attached to the xanthene and via a phenylene spacer. The HSX* and $H_{ph}SX^*$ platforms exhibit both catalase and epoxidation activity. Comparison of the reactivity profiles and overall TON for $H_2O_2$ dismutation between Hangman salens with pendant acid and methyl ester groups establishes the importance of the hydrogen-bonding group in facilitating catalysis. Computational studies suggest that only one carboxylic acid is needed to promote the favored end-on

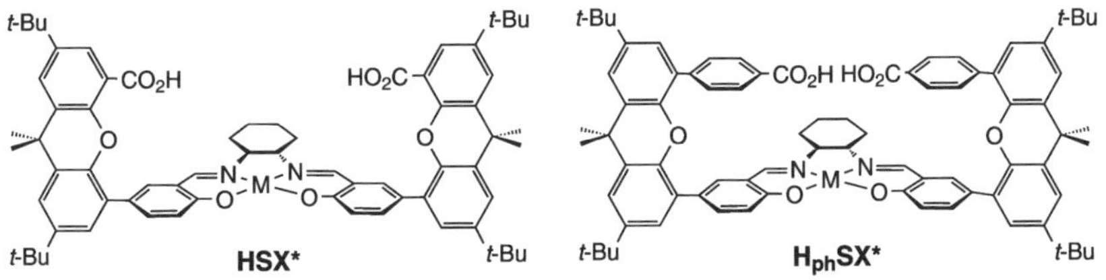

Figure 39. Hangman salens with convergent hydrogen-bonding clefts.

hydrogen peroxide assembly with the metal oxo catalytic intermediate. When the functional group is extended with a phenylene spacer, the two carboxylic acids are sufficiently close to form a $-(COOH)_2-$ dimer, which reduces the propensity of the hanging group to promote substrate assembly; accordingly, lower catalytic activity is observed. When the $H_2O_2$ substrate is replaced by prochiral substrates, such as 1,2-dihydronapthalene, asymmetric epoxidation is observed with 23% enantiomeric excess (ee), demonstrating communication between the chiral backbone and substrate despite the bulky xanthene functionalizations situated at the 5 and 5′ position of the macrocycle. Although this ee is inferior to the best salen epoxidation catalysts, it is on par with the "introductory" *ee*s observed for first generation catalysts. More specifically, the ee is consistent with other salen macrocycles which are unfunctionalized in the 3 and 3′ positions (224, 227, 228). The addition of bulky functional groups in the 3 and 3′ positions is believed to enhance the selectivity in epoxidation reactions by directing olefin approach over the chiral inducing cyclohexyl bridge by sterically blocking alternative substrate orientations (227). With this in mind, increases in *ee* will be sought by introducing steric substituents at crucial positions on the salen macrocycle so as to promote substrate interaction with the chiral cyclohexyldiamine backbone of the salen macrocycle.

## VIII. CONCLUSIONS

The xanthene and dibenzylfuran pillars are effective scaffolds for the construction of cofacial architectures. Herein, the incorporation of porphyrins and salens into the Pacman design are described. The benefit afforded by the xanthene and dibenzofuran scaffolds for preorganizing metal coordination complexes (229–235) and redox-active organic chromophores such as perylenediimide (PDI) (236–238) has since been realized and exploited. A general feature of the work described here and these subsequent efforts is the facility with which the dimensions of the cleft size may be varied for preorganization of redox cofactors. We have shown that this property of dimensional control may be augmented by using the scaffold to tailor the secondary coordination sphere of a redox-active center for PCET reactivity. The ability to bring these two features together by use of the xanthene and dibenzofuran scaffolds offers a powerful new tool for the activation of small molecule substrates.

## ACKNOWLEDGMENTS

J.R. thanks the Fannie and John Hertz Foundation for a predoctoral fellowship. The work on PCET and the activation of small molecules by PCET has been supported by grants from the NIH (GM47274), DOE (DE-FG02-05ER15745) and the NSF (CHE-0132680).

## ABBREVIATIONS

| | |
|---|---|
| ATP | Adenosine triphosphate |
| C*c*O | Cytochrome *c* oxidase |
| Ct | Center |
| BDE | Bond dissociation energy |
| DBU | 1,8-Diazabicyclo[5.4.0]undec-7-ene |
| DDQ | 2,3-Dicyano-5,6-dichloroparabenzoquinone |
| DMF | Dimethylformamide |
| DPA | 1,8-Bis[5-(2,8,13,17-tetraethyl-3,7,12,18-tetramethylporphyrinyl)]anthracene |
| DPB | 1,8-Bis[5-(2,8,13,17-tetraethyl-3,7,12,18-tetramethylporphyrinyl)]biphenylene |
| DPD | 4,6-Bis[5-(2,8,13,17-tetraethyl-3,7,12,18-tetramethylporphyrinyl)]dibenzofuran |
| DPDF | 4,6-Bis[5-(10,15,20-tripentafluorophenylporphy rinyl)]dibenzofuran |
| DPDM | 4,6-Bis[5-(2,8,13,17-tetraethyl-3,7,12,18-tetramethyl-15-(2,6-dimethoxyphenyl)porphyrinyl)] dibenzofuran |
| DPX | 4,5-Bis[5-(2,8,13,17-tetraethyl-3,7,12,18-tetramethylporphyrinyl)]-9,9-dimethylxanthene |
| DPXM | 4,5-Bis[5-(2,8,13,17-tetraethyl-3,7,12,18-tetramethyl-15-(2,6-dimethoxyphenyl)porphyrinyl)]-9,9-dimethylxanthene |
| DTMPD | 4,6-Bis[5-(10,15,20-trimesitylporphyrinyl)]dibenzofuran |
| DTMPX | 4,5-Bis[5-(10,15,20-trimesitylporphyrinyl)]-9,9-dimethylxanthene |
| ee | Enantiomeric excess |
| EPR | Electron paramagnetic resonance |
| Etio | Etioporphyrin I |
| EtOH | Ethanol |
| FTF | Face to face |
| KIE | Kinetic isotope effect |
| $^{1}$H NMR | Proton nuclear magnetic resonance |
| HPD | Hangman porphyrin dibenzofuran |
| HOMO | Highest occupied molecular orbital |

| | |
|---|---|
| HPX | Hangman porphyrin xanthene |
| HSX | Hangman salophen xanthene |
| HSX* | Hangman salen xanthene |
| IE | Ionization energy |
| IPA | Isopropyl alcohol |
| LUMO | Lowest unoccupied molecular orbital |
| mes | mesetyl |
| *m*-CPBA | *m*-Chloroperbenzoic acid |
| MPS | Mean-plane separations |
| NBS | *N*-Bromosuccinimide |
| NIST | National Institute of Standards and Technology |
| NMR | Nuclear magnetic resonance |
| OAT | Oxygen atom transfer |
| OD | Optical density |
| OEC | Oxygen evolving complex |
| OEP | Octaethylporphyrin |
| P | Porphyrin |
| PCET | Proton-coupled electron transfer |
| PDI | Perylenediimide |
| PhEtio | 5-(4-Bromophenyl)-2,8,13,17-tetraethyl-3,7,12,18-tetramethylporphyrin |
| PhOMP | 5-phenyl-2,3,7,8,12,13,17,18-octamethylporphyrin |
| PSII | Photosystem II |
| PTSA | *p*-Toluenesulfonic acid |
| salen | Bis(saliecylidene)ethylenediamine |
| salophen | Bis(salicyclidene)diaminobenzene |
| TA | Transient obsorption |
| THF | Tetrahydrofuran |
| TMEDA | *N*,*N*,*N*′,*N*′-Tetramethylethylenediamine |
| TMP | 5,10,15,20-Tetramesitylporphyrin |
| TON | Turnover number |
| UV | Ultraviolet |
| vis | Visible |

## REFERENCES

1. S. J. Lippard and J. M. Berg, *Principles of Bioinorganic Chemistry*; University Science Books: Mill Valley, CA, 1994.
2. J. P. Collman, C. M. Elliot, T. R. Halbert, and B. S. Tovrog, *Proc. Natl. Acad. Sci. U.S.A.*, *74*, 18 (1977).
3. J. P. Collman, A. O. Chong, G. B. Jameson, R. T. Oakley, E. Rose, E. R. Schmittou, and J. A. Ibers, *J. Am. Chem. Soc. 103*, 516 (1981).

4. J. P. Collman, F. C. Anson, C. E. Barnes, C. S. Bencosme, T. Geiger, E. R. Evitt, R. P. Kreh, K. Meier, and R. B. Pettman, *J. Am. Chem. Soc. 105*, 2694 (1983).
5. J. P. Collman, C. S. Bencosme, C. E. Barnes, and B. D. Miller, *J. Am. Chem. Soc. 105*, 2704 (1983).
6. J. P. Collman, C. S. Bencosme, R. R. Durand, Jr., R. P. Kreh, and F. C. Anson, *J. Am. Chem. Soc. 105*, 2699 (1983).
7. K. Kim, J. P. Collman, and J. A. Ibers, *J. Am. Chem. Soc. 110*, 4242 (1988).
8. C. K. Chang, M. S. Kuo, and C. B. Wang, *J. Heterocyl. Chem. 14*, 943 (1977).
9. C. K. Chang, *J. Am. Chem. Soc. 99*, 2819 (1977).
10. C. K. Chang, *J. Heterocyl. Chem. 14*, 1285 (1977).
11. C. K. Chang, *J. Chem. Soc. Chem. Commun.* 800 (1977).
12. N. E. Kagan, D. Mauzerall, and R. B. Merrifield, *J. Am. Chem. Soc. 99*, 5484 (1977).
13. H. Ogoshi, H. Sugimoto, and Z. Yoshida, *Tetrahedron Lett.* 169 (1977).
14. J. B. Paine III, D. Dolphin, and M. Gouterman, *Can. J. Chem. 56*, 1712 (1978).
15. K. Ichimura, *Chem. Lett.* 641 (1977).
16. B. C. Bookser and T. C. Bruice, *J. Am. Chem. Soc. 113*, 4208 (1991).
17. R. Karaman, O. Almarsson, and T. C. Bruice, *J. Org. Chem. 57*, 1555 (1992).
18. R. Karaman, O. Almarsson, A. Blasko, and T. C. Bruice, *J. Org. Chem. 57*, 2169 (1992).
19. R. Karaman, S. W. Jeon, O. Almarsson, and T. C. Bruice, *J. Am. Chem. Soc. 114*, 4899 (1992).
20. R. Karaman, A. Blasko, O. Almarsson, R. Arasasingham, and T. C. Bruice, *J. Am. Chem. Soc. 114*, 4889 (1992).
21. R. Karaman and T. C. Bruice, *J. Org. Chem. 56*, 3470 (1991).
22. S. Anderson, H. L. Anderson, and J. K. M. Sanders, *J. Chem. Soc. Perkin Trans. 1*, 2255 (1995).
23. S. Anderson, H. L. Anderson, and J. K. M. Sanders, *Acc. Chem. Res. 26*, 469 (1993).
24. H. L. Anderson, C. A. Hunter, M. N. Meah, and J. K. M. Sanders, *J. Am. Chem. Soc. 112*, 5780 (1990).
25. C. A. Hunter and J. K. M. Sanders, *J. Am. Chem. Soc. 112*, 5525 (1990).
26. C. A. Hunter, M. N. Meah, and J. K. M. Sanders, *J. Am. Chem. Soc. 112*, 5773 (1990).
27. J. A. Cowan and J. K. M. Sanders, *J. Chem. Soc. Chem. Commun.* 1213 (1985).
28. A. Harriman, and J. P. Sauvage, *Chem. Soc. Rev. 25*, 41 (1996).
29. Y. Naruta, M. Sasayama, and K. Maruyama, *Chem. Lett.* 1267 (1992).
30. Y. Naruta and M. Sasayama, *J. Chem. Soc. Chem. Commun.* 2667 (1994).
31. Y. Naruta, N. Sawada, and M. Tadokoro, *Chem. Lett.* 1713 (1994).
32. Y. Naruta and M. Sasayama, *Chem. Lett.* 2411 (1994).
33. Y. Naruta, M. Sasayama, and T. Sasaki, *Angew. Chem. Int. Ed. Engl. 33*, 1839 (1994).
34. Y. Naruta, M. Sasayama, and K. Ichihara, *J. Mol. Cat. A 117*, 115 (1997).
35. Y. Naruta and K. Maruyama, *J. Am. Chem. Soc. 113*, 3595 (1991).
36. A. Brun, A. Harriman, V. Heitz, and J. P. Sauvage, *J. Am. Chem. Soc. 113*, 8657 (1991).
37. S. Chardon-Noblat, J. P. Sauvage, and P. Mathis, *Angew. Chem. Int. Ed. Engl. 28*, 593 (1989).
38. K. Ichihara and Y. Naruta, *Chem. Lett.* 185 (1998).
39. R. Guilard, S. Brandes, C. Tardieux, A. Tabard, M. L'Her, C. Miry, P. Gouerac, Y. Knop, and J. P. Collman, *J. Am. Chem. Soc. 117*, 11721 (1995).
40. R. Guilard, S. Brandes, A. Tabard, N. Bouhmaida, C. Lecomte, P. Richard, and J. M. Latour, *J. Am. Chem. Soc. 116*, 10202 (1994).

41. R. Guilard, M. A. Lopez, A. Tabard, P. Richard, C. Lecomte, S. Brandes, J. E. Hutchison, and J. P. Collman, *J. Am. Chem. Soc. 114*, 9877 (1992).
42. L. M. Proniewicz, J. Odo, J. Goral, C. K. Chang, and K. Nakamoto, *J. Am. Chem. Soc. 111*, 2105 (1989).
43. J. P. Fillers, K. G. Ravichandran, I. Abdalmuhdi, A. Tulinsky, and C. K. Chang, *J. Am. Chem. Soc. 108*, 417 (1986).
44. C. K. Chang and I. Abdalmuhdi, *J. Org. Chem. 48*, 5388 (1983).
45. C. K. Chang, H. Y. Liu, and I. Abdalmuhdi, *J. Am. Chem. Soc. 106*, 2725 (1984).
46. C. K. Chang and I. Abdalmuhdi, *Angew. Chem. Int. Ed. Engl. 23*, 164 (1984).
47. J. P. Collman, K. Kim, and J. M. Garner, *J. Chem. Soc. Chem. Commun.* 1711 (1986).
48. J. P. Collman, K. Kim, and C. R. Leidner, *Inorg. Chem. 26*, 1152 (1987).
49. J. P. Collman, J. E. Hutchison, M. A. Lopez, A. Tabard, R. Guilard, W. K. Seok, J. A. Ibers, and M. L'Her, *J. Am. Chem. Soc. 114*, 9869 (1992).
50. J. P. Collman, J. E. Hutchison, M. A. Lopez, and R. Guilard, *J. Am. Chem. Soc. 114*, 8066 (1992).
51. J. P. Collman, J. E. Hutchison, M. S. Ennis, M. A. Lopez, and R. Guilard, *J. Am. Chem. Soc. 114*, 8074 (1992).
52. J. P. Collman, Y. Ha, P. S. Wagenknecht, M. A. Lopez, and R. Guilard, *J. Am. Chem. Soc. 115*, 9080 (1993).
53. J. P. Collman, Y. Y. Ha, R. Guilard, and M. A. Lopez, *Inorg. Chem. 32*, 1788 (1993).
54. J. P. Collman, P. S. Wagenknecht, and J. E. Hutchison, *Angew. Chem. Int. Ed. Engl. 33*, 1537 (1994).
55. J. P. Collman, *Angew Chem. 106*, 1620 (1994).
56. J. P. Collman, R. Boulatov, C. J. Sunderland, and Fu, *Chem. Rev.* 104, 561 (2004).
57. J. P. Collman, R. Boulatov, and C. J. Sunderland, in *The Porphyrin Handbook*, Vol. 11, K. M. Kadish, K. M. Smith, R. Guilard, Eds., Academic Press, San Diego, CA, 2003, pp. 1–49.
58. C. K. Chang, *Adv. Chem. Ser. 173*, 162 (1979).
59. Y. Le Mest, M. L'Her, J. Courtot-Coupez, J. P. Collman, E. R. Evitt, and C. S. Bencosme, *J. Chem. Soc. Chem. Commun.* 1286 (1983).
60. Y. Le Mest and M. L'Her, *J. Chem. Soc. Chem. Commun.* 1441 (1995).
61. Y. Le Mest, M. L'Her, and J. Y. Saillard, *Inorg. Chim. Acta 248*, 181 (1996).
62. Y. Le Mest, C. Inisan, A. Laouenan, M. L'Her, J. Talarmain, M. El Khalifa, and J. Y. Saillard, *J. Am. Chem. Soc. 119*, 6905 (1997).
63. H.-Y. Lui, I. Abdalmuhdi, C. K. Chang, and F. C. Anson, *J. Phys. Chem. 89*, 665 (1985).
64. C.-L. Ni, I. Abdalmuhdi, C. K. Chang, and F. C. Anson, *J. Phys. Chem. 91*, 1158 (1987).
65. J. P. Collman, J. E. Hutchison, P. S. Wagenknecht, N. S. Lewis, M. A. Lopez, and R. Guilard, *J. Am. Chem. Soc. 112*, 8207 (1990).
66. J. P. Collman, J. E. Hutchison, M. A. Lopez, R. Guilard, and R. A. Reed, *J. Am. Chem. Soc. 113*, 2794 (1991).
67. W. Cui and B. B. Wayland, *J. Am. Chem. Soc. 126*, 8266 (2004).
68. W. Cui and B. B. Wayland, *J. Porphyrins Phthalocyanines 8*, 103 (2004).
69. C. J. Chang, J. D. K. Brown, M. C. Y. Chang, E. A. Baker, and D. G. Nocera, in *Electron Transfer in Chemistry*, Vol. 3.2.4, V. Balzani, Ed., Wiley-VCH, Weinheim, Germany, 2001, pp. 409–461.
70. R. I. Cukier and D. G. Nocera, *Annu. Rev. Phys. Chem. 49*, 337 (1998).

71. J. M. Hodgkiss, J. Rosenthal, and D. G. Nocera, in *Handbook of Hydrogen Transfer. Physical and Chemical Aspects of Hydrogen Transfer*, Vol. 1, J. T. Hynes, R. L. Schowen, and H. H. Limbach, Eds., Wiley-VCH, Germany, 2006.
72. J. Stubbe, D. G. Nocera, C. S. Yee, and M. C. Y. Chang, *Chem. Rev. 103*, 2167 (2003).
73. C. J. Chang, M. C. Y. Chang, N. H. Damrauer, and D. G. Nocera, *Biochem. Biophys. Acta 1655*, 13 (2004).
74. J. S. Nowick, P. Ballester, F. Ebmeyer, and J. Rebek Jr., *J. Am. Chem. Soc. 112*, 8902 (1990).
75. E. B. Schwartz, C. B. Knobler, and D. J. Cram, *J. Am. Chem. Soc. 114*, 10775 (1992).
76. D. H. R. Barton and S. Z. Zard, *J. Chem. Soc. Chem. Commun.* 1098 (1985).
77. J. L. Sessler, A. Mozaffari, and M. R. Johnson, *Org. Synth. 70*, 68 (1992).
78. R. Young and C. K. Chang, *J. Am. Chem. Soc. 107*, 898 (1985).
79. J. W. Buchler, in *Porphyrins and Metalloporphyrins*; 2nd ed., K. M. Smith, Ed., Elsevier Scientific, Oxford, 1975, pp. 157–232.
80. M. O. Senge, K. R. Gerzevske, M. G. H. Vincente, T. P. Forsyth, and K. M. Smith, *Angew. Chem., Int. Ed. Engl. 32*, 750 (1993).
81. D. P. Arnold, V. V. Borovkov, and G. V. Ponomarev, *Chem. Lett. 6*, 485 (1996).
82. S. I. Shishporenok and V. S. Chirvonyi, *J. App. Spec. 6*, 968 (2001).
83. S. I. Shishporenok and V. S. Chirvony, *Optics Spectrosc. 92*, 877 (2002).
84. W. R. Scheidt and Y. J. Lee, *Struct. Bonding (Berlin) 64*, 1 (1987)
85. F. Bolze, C. P. Gros, M. Drouin, E. Espinosa, P. D. Harvey, and R. Guilard, *J. Organomet. Chem. 643–644*, 89 (2002).
86. C. J. Chang, E. A. Baker, B. J. Pistorio, Y. Deng, Z.-H. Loh, S. E. Miller, S. D. Carpenter, and D. G. Nocera, *Inorg. Chem. 41*, 3102 (2002).
87. S. S. Eaton, G. R. Eaton, and C. K. Chang, *J. Am. Chem. Soc. 107*, 3177 (1985).
88. S. S. Eaton, K. M. More, B. M. Sawant, and G. R. Eaton, *J. Am. Chem. Soc. 105*, 6560 (1983).
89. M. Gouterman, *J. Mol. Spectrosc. 6*, 138 (1961).
90. M. Gouterman, D. Holten, and E. Lieberman, *Chem. Phys. 25*, 139 (1977).
91. J. T. Fletcher and M. J. Therien, *Inorg. Chem. 41*, 331 (2002).
92. Y. Le Mest, M. L'Her, N. H. Hendricks, K. Kim, and J. P. Collman, *Inorg. Chem. 31*, 835 (1992).
93. K. M. Kadish, G. Royal, E. V. Caemelbecke, and L. Gueletti, *The Porphyrin Handbook*, Vol. 9, K. M. Kadish, K. M. Smith, R. Guilard, Eds., Academic Press, San Diego, CA, 2000, pp. 1–220.
94. K. Kalyanasundaram, *Photochemistry of Polypyridine and Porphyrin Complexes*; Academic Press, London, 1992.
95. Z.-H. Loh, S. E. Miller, C. J. Chang, S. D. Carpenter, and D. G. Nocera, *J. Phys. Chem. A. 106*, 11700 (2002).
96. S. Faure, C. Stern, E. Espinosa, J. Douville, R. Guilard, and P. D. Harvey, *Chem. Eur. J. 11*, 3469 (2005).
97. S. Faure, C. Stern, R. Guilard, and P. D. Harvey, *Inorg. Chem. 44*, 9232 (2005).
98. J. K. M. Sanders, *Pure Appl. Chem. 72*, 2265 (2000).
99. M. Nakash and J. K. M. Sanders, *J. Chem. Soc., Perkin Trans. 2,* 2189 (2001).
100. M. Nakash and J. K. M. Sanders, *J. Org. Chem. 65*, 7266 (2000)
101. M. Nakash, Z. Clyde-Watson, N. Feeder, J. E. Davies, S. J. Teat, and J. K. M. Sanders, *J. Am. Chem. Soc. 122*, 5286 (2000).
102. Y. Uemori, A. Nakatsubo, H. Imai, S. Nakagawa, and E. Kyuno, *Inorg. Chem. 31*, 5164 (1992).

103. M. J. Crossley and P. Thordarson, *Angew. Chem., Int. Ed. Engl. 41*, 1709 (2002).

104. X. Huang, B. Borhan, B. H. Rickman, K. Nakanishi, and N. Berova, *Chem. Eur. J. 6*, 216 (2000).

105. J. Brettar, J.-P. Gisselbrecht, M. Gross, and N. Solladie, *Chem. Commun.* 733 (2001).

106. V. V. Borovkov, J. M. Lintuluoto, and Y. Inoue, *J. Am. Chem. Soc. 123*, 2979 (2001).

107. C. A. Hunter, M. N. Meah, and J. K. M. Sanders, *J. Chem. Soc., Chem. Commun.* 692 (1988).

108. C. A. Hunter, M. N. Meah, and J. K. M. Sanders, *J. Chem. Soc., Chem. Commun.* 694 (1988).

109. C. A. Hunter, P. Leighton, and J. K. M. Sanders, *J. Chem. Soc., Perkin Trans. 1*, 547 (1989).

110. D. E. Richardson and H. Taube, *J. Am. Chem. Soc. 105*, 40 (1983).

111. H. E. Katz, *J. Org. Chem. 54*, 2179 (1984).

112. K. A. Connors, *Binding Constants*; John Wiley& Sons, enc. New York, 1987.

113. C. W. Hoganson, M. A. Pressler, D. A. Proshlyakov, and G. T. Babcock, *Biochim. Biophys. Acta 1365*, 170 (1998).

114. G. Milazzo and S. Caroli, *Tables of Standard Electrode Potentials*, Wiley-Interscience, New York, 1978, p. 229.

115. N. N. Greenshaw and A. Earnshaw, *Chemistry of the Elements*; Reed Education and Professional Publishing: Oxford, 2001.

116. F. A. Cotton and G. Wilkinson, *Advanced Inorganic Chemistry*, 5th ed., Wiley-Interscience, New York, 1988.

117. J. P. Collman, M. Rapta, M. Bröring, L. Raptova, R. Schwenninger, B. Boitrel, L. Fu, and M. L'Her, *J. Am. Chem. Soc. 121*, 1387 (1999).

118. J. P. Collman, L. Fu, P. C. Herrmann, Z. Wang, M. Rapta, M. Broring, R. Schwenninger, and B. Boitrel, *Angew. Chem. Int. Ed. Engl. 37*, 3397 (1998).

119. J. O. Baeg and R. H. Holm, *Chem. Commun.* 571 (1998).

120. H. C. Liang, M. Dahan, and K. D. Karlin, *Curr. Opin. Chem. Biol. 3*, 168 (1999).

121. D. Ricard, B. Andrioletti, M. L'Her, and B. Boitrel, *Chem. Commun.* 1523 (1999).

122. H. Taube, *Prog. Inorg. Chem. 34*, 607 (1986).

123. C. J. Chang, Y. Deng, C. Shi, C. K. Chang, F. C. Anson, and D. G. Nocera, *Chem. Commun.* 1355 (2000).

124. C. J. Chang, Z.-H. Loh, C. Shi, F. C. Anson, and D. G. Nocera, *J. Am. Chem. Soc. 126*, 10013 (2004).

125. J. Rosenthal and D. G. Nocera, *manuscript in preparation*.

126. S. Fukuzumi, K. Okamoto, C. P. Gros, and R. Guilard, *J. Am. Chem. Soc. 126*, 10441 (2004).

127. S. Fukuzumi, K. Okamoto, Y. Tokuda, C. P. Gros, and R. Guilard, *J. Am. Chem. Soc. 126*, 17059 (2004).

128. Ö. Hansson, B. Karlsson, R. Aasa, T. Vänngård, B. G. Malmström, *EMBO J. 1*, 1295 (1982).

129. D. F. Blair, S. N. Witt, and S. I. Chan, *J. Am. Chem. Soc. 107*, 7389 (1985).

130. J. E. Morgan, M. I. Verkhovsky, G. Palmer, and M. Wikström, *Biochemistry 40*, 6882 (2001).

131. M. R. A. Blomberg, P. E. M. Siegbahn, G. T. Babcock, and M. Wikström. *J. Am. Chem. Soc. 122*, 12848 (2000).

132. R. Boulatov, J. P. Collman, I. M. Shiryaeva, and C. J. Sunderland, *J. Am. Chem. Soc. 124*, 11923 (2002).

133. D. Ricard, A. Didier, M. L'Her, and B. Boitrel, *ChemBioChem 2*, 144 (2001).

134. D. Ricard, M. L'Her, P. Richard, and B. Boitrel, *Chem. Eur. J. 7*, 3291 (2001).

135. K. M. Kadish, L. Fremond, Z. Ou, J. Shao, C. Shi, F. C. Anson, F. Burdet, C. P. Gros, J. M. Barbe, and R. Guilard, *J. Am. Chem. Soc. 127*, 5625 (2005).
136. R. Guilard, F. Burdet, J. M. Barbe, C. P. Gros, E. Espinosa, J. Shao, Z. Ou, R. Zhan, and K. M. Kadish, *Inorg. Chem. 44*, 3972 (2005).
137. K. M. Kadish, Z. Ou, J. Shao, C. P. Gros, J. M. Barbe, F. Jerome, F. Bolze, F. Burdet, and R. Guilard, *Inorg. Chem. 41*, 3990 (2002).
138. K. M. Kadish, J. Shao, Z. Ou, L. Fremond, R. Zhan, R. Burdet, J. M. Barbe, C. P. Gros, and R. Guilard, *Inorg. Chem. 44*, 6744 (2005).
139. L. L. Chng, C. J. Chang, and D. G. Nocera, *J. Org. Chem. 68*, 4075 (2003).
140. B. Vaz, R. Alvarez, M. Nieto, A. I. Paniello, and A. R. de Lera, *Tetrahedron Lett. 42*, 7409 (2001).
141. W. M. Sharman and J. E. Van Lier, *J. Porphyrins Phthalocyanines 4*, 441 (2000).
142. C. M. Muzzi, C. J. Medforth, L. Voss, M. Cancilla, C. Lebrilla, J.-G. Ma, J. A. Shelnutt, and K. M. Smith, *Tetrahedron Lett. 40*, 6159 (1999).
143. C.-J. Liu, W.-Y. Yu, S.-M. Peng, T. C. W. Mak, and C.-M. Che, *J. Chem. Soc. Dalton Trans.* 1805 (1998).
144. M. K. Tse, Z.-Y. Zhou, T. C. W. Mak, and K. S. Chan, *Tetrahedron 56*, 7779 (2000).
145. X. Zhou and K. S. Chan, *J. Org. Chem. 63*, 99 (1998).
146. X. Zhou, M. K. Tse, T. S. M. Wan, and K. S. Chan, *J. Org. Chem. 61*, 3590 (1996).
147. K. S. Chan, X. Zhou, M. T. Au, and C. Y. Tam, *Tetrahedron 51*, 3129 (1995).
148. Y. Deng, C. K. Chang, and D. G. Nocera, *Angew. Chem. Int. Ed. Engl. 39*, 1066 (2000).
149. P. M. Iovine, M. A. Kellet, N. P. Redmore, and M. J. Therien, *J. Am. Chem. Soc. 122*, 8717 (2000).
150. A. G. Hyslop, M. A. Kellett, P. M. Iovine, and M. J. Therien, *J. Am. Chem. Soc. 120*, 12676 (1998).
151. C. K. Chang and N. Bag, *J. Org. Chem. 60*, 7030 (1995).
152. R. S. Loewe, K.-Y. Tomizaki, W. J. Youngblood, Z. Bo, and J. S. Lindsey, *J. Mater. Chem. 12*, 3438 (2002).
153. L. Yu and J. S. Lindsey, *Tetrahedron 57*, 9285 (2001).
154. B. Shi and R. W. Boyle *J. Chem. Soc., Perkin Trans. 1*, 1397 (2002).
155. J. T. Groves, R. C. Haushalter, M. Nakamura, T. E. Nemo, and B. J. Evans, *J. Am. Chem. Soc. 103*, 2884 (1981).
156. R.-J. Cheng, L. Latos-Grazynski, and A. L. Balch, *Inorg. Chem. 21*, 2412 (1982).
157. O. P. Anderson, C. K. Schauer, and W. S. Caughey, *Am. Cryst. Assoc. 10*, 23 (1982).
158. A. B. Hoffman, D. M. Collins, V. W. Day, E. B. Fleischer, T. S. Srivastava, and J. L. Hoard, *J. Am. Chem. Soc. 94*, 3620 (1972).
159. J. T. Landrum, D. Grimmett, K. J. Haller, W. R. Scheidt, and C. A. Reed, *J. Am. Chem. Soc. 103*, 2640 (1981).
160. C. J. Chang, Y. Deng, G.-H. Lee, S.-M. Peng, C.-Y. Yeh, and D. G. Nocera, *Inorg. Chem. 41*, 3008 (2002).
161. Y. Deng, C. J. Chang, and D. G. Nocera, *J. Am. Chem. Soc. 122*, 410 (2000).
162. J. M. Hodgkiss, C. J. Chang, B. J. Pistorio, and D. G. Nocera, *Inorg. Chem. 42*, 8270 (2003).
163. B. J. Pistorio, C. J. Chang, and D. G. Nocera, *J. Am. Chem. Soc. 124*, 7884 (2004).
164. D.-H. Chin, J. Del Gaudio, G. N. La Mar, and A. L. Balch, *J. Am. Chem. Soc. 99*, 5486 (1977).

165. D.-H. Chin, G. N. La Mar, and A. L. Balch, *J. Am. Chem. Soc. 102*, 4344 (1980).
166. J. T. Groves and Y.-Z. Han, in *Cytochrome P-450: Structure, Mechanism and Biochemistry*; 2nd Ed., P. R. Ortiz de Montellano, Ed., Plenum, New York, 1995, pp. 3–48.
167. J. T. Groves, *J. Chem. Ed. 62*, 928 (1985).
168. J. T. Groves, Y. Han, and D. van Engen, *Chem. Commun.* 436 (1990).
169. J. T. Groves, and R. S. Myers, *J. Am. Chem. Soc. 107*, 5791 (1983).
170. A. S. Veige, L. M. Slaughter, P. T. Wolczanski, N. Matsunaga, S. A. Decker, and T. R. Cundari, *J. Am. Chem. Soc. 123*, 6419 (2001).
171. E. N. Jacobsen, *Acc. Chem. Res. 33*, 421 (2000).
172. M. W. Peterson, D. S. Rivers, and R. M. Richman, *J. Am. Chem. Soc. 107*, 2907 (1985).
173. M. W. Peterson and R. M. Richman, *Inorg. Chem. 24*, 722 (1985).
174. L. Weber, R. Hommel, J. Behling, G. Haufe, and H. Hennig, *J. Am. Chem. Soc. 116*, 2400 (1994).
175. L. Weber, G. Haufe, D. Rehorek, and H. Hennig, *J. Mol. Cat. 60*, 267 (1990).
176. D. N. Hendrickson, M. G. Kinnaird, and K. S. Suslick, *J. Am. Chem. Soc. 109*, 1243 (1987).
177. K. S. Suslick and R. A. Watson, *New J. Chem.16*, 633 (1992).
178. K. M. Kadish, K. M. Smith, and R. Guilard, *The Porphyrin Handbook*, Vol. 9, Academic Press, San Diego, CA, 2000.
179. S. Neya and N. Funasaki, *J. Heterocyclic Chem. 34*, 689 (1997).
180. E. K. Woller and S. G. DiMagno, *J. Org. Chem.* 62, 1588 (1997).
181. J. Rosenthal, B. J. Pistorio, L. L. Chng, and D. G. Nocera, *J. Org. Chem. 70*, 1885 (2005).
182. J. Rosenthal, J. Bachman, J. L. Dempsey, A. J. Esswein, T. G. Gray, J. M. Hodgkiss, D. R. Manke, T. D. Luckett, B. J. Pistorio, A. S. Veige, and D. G. Nocera, *Coord. Chem. Rev. 249*, 1316 (2005).
183. J. Rosenthal, T. D. Luckett, J. M. Hodgkiss, and D. G. Nocera, *J. Am. Chem. Soc. 128*, 6546 (2006).
184. D. C. Rosenfeld, D. S. Kuiper, E. B. Lobkovsky, and P. T. Wolczanski, *Polyhedron 25*, 251 (2006).
185. R. Y. Luo, *Handbook of Bond Dissociation Energies in Organic Compounds*, CRC Press: Boca Raton, FL, 2003.
186. J. M. Mayer, *Annu. Rev. Phys. Chem. 55*, 363 (2004).
187. J. Kaizer, E. J. Klinker, N. Y. Oh, J.-U. Rohde, W. J. Song, A. Stubna, J. Kim, E. Münck, W. Nam, and L. Que, Jr., *J. Am. Chem. Soc. 126*, 472 (2004).
188. C. J. Chang, C. Y. Yeh, and D. G. Nocera, *J. Org. Chem. 67*, 1403 (2002).
189. C. Y. Yeh, C. J. Chang, and D. G. Nocera, *J. Am. Chem. Soc. 123*, 1513 (2001).
190. C. J. Chang, L. L. Chng, and D. G. Nocera, *J. Am. Chem. Soc. 125*, 1866 (2003).
191. S. I. Ozaki, M. P. Roach, T. Matsui, and Y. Watanabe, *Acc. Chem. Res. 34*, 818 (2001).
192. A. R. Dunn, I. J. Dmochowski, A. M. Bilwes, H. B. Gray, and B. R. Crane, *Proc. Natl. Acad. Sci. U.S.A. 98*, 12420 (2001).
193. I. Hamachi, S. Tsukiji, S. Shinkai, and S. Oishi, *J. Am. Chem. Soc. 121*, 5500 (1999).
194. T. L. Poulos, B. C. Finzel, and A. J. Howard, *Biochemistry 25*, 5314 (1986).
195. J. L. Dempsey, A. J. Esswein, D. R. Manke, J. Rosenthal, J. D. Soper, and D. G. Nocera, *Inorg. Chem. 44*, 6879 (2005).
196. J. T. Groves and Y. Watanabe, *J. Am. Chem. Soc. 110*, 8443 (1988).
197. L. L. Chng, C. J. Chang, and D. G. Nocera, *Org. Lett. 5*, 2421 (2003).

198. P. R. Ortiz de Montellano, *Cytochrome P450: Structure, Mechanism and Biochemistry*; 2nd ed., Plenum: New York, 1995.
199. M. Sono, M. P. Roach, E. D. Coulter, and J. H. Dawson, *Chem. Rev. 96*, 2841 (1996).
200. T. L. Poulos, *Curr. Opin. Struct. Biol. 5*, 767 (1995).
201. R. A. Sheldon, *Metalloporphyrins in Catalytic Oxidations*; Marcel Dekker, New York, 1994.
202. J. T. Groves and M. K. Stern, *J. Am. Chem. Soc. 110*, 8628 (1988).
203. O. Bortolini and B. Meunier, *J. Chem. Soc. Perkin Trans 2* 1967 (1984).
204. P. Battioni, J. P. Renaud, J. F. Bartoli, M. Reina-Artiles, M. Fort, and D. Mansuy, *J. Am. Chem. Soc. 110*, 8462 (1988).
205. A. J. Castellino and T. C. Bruice, *J. Am. Chem. Soc. 110*, 158 (1988).
206. N. Jin, J. L. Bourassa, S. C. Tizio, and J. T. Groves, *Angew. Chem. Int. Ed. Engl. 39*, 3849 (2000).
207. N. Jin and J. T. Groves, *J. Am. Chem. Soc. 121*, 2923 (1999).
208. W. Nam, I. Kim, M. H. Lim, H. J. Choi, J. S. Lee, and H. G. Jang, *Chem. Eur. J. 8*, 2067 (2002).
209. Gross has characterized oxomanganese(V) corroles: Z. Gross, G. Golubkov, and L. Simkovich, *Angew. Chem. Int. Ed. Engl. 39*, 4045 (2000).
210. N. S. Lewis and D. G. Nocera, *Proc. Natl. Acad. Sci. U.S.A. 103*(43) 15729 (2006).
211. R. Eisenberg and D. G. Nocera, *Inorg. Chem. 44*, 6799 (2005).
212. D. G. Nocera, *Daedalus 135*(4), 112 (2006).
213. C. W. Hoganson, M. A. Pressler, D. A. Proshlyakov, and G. T. Babcock, *Biochim. Biophys. Acta 1365*, 170 (1998).
214. J. T. Groves, A. Gross, and M. K. Stern, *Inorg. Chem. 33*, 5065 (1994).
215. S. Neya and N. Funasaki, *J. Heterocyclic Chem. 34*, 689 (1997).
216. E. K. Woller and S. G. DiMagno, *J. Org. Chem. 62*, 1588 (1997).
217. K. M. Kadish, E. V. Caemelbecke and G. Royal, in *The Porphyrin Handbook*, Vol. 8, K. M. Kadish, K. M. Smith, and R. Guilard, Eds., San Diego: CA, 2000.
218. S.-Y. Liu and D. G. Nocera, *J. Am. Chem. Soc. 127*, 5278 (2005).
219. M. J. Sabater, M. Alvaro, H. Garcia, E. Palomares, and J. C. Scaiano, *J. Am. Chem. Soc. 123*, 7074 (2001).
220. K. Srinivasan, P. Michaud, and J. K. Kochi, *J. Am. Chem. Soc. 108*, 2309 (1986).
221. E. M. McGarrigle and D. G. Gilheany, *Chem. Rev. 105*, 1563 (2005).
222. Y. G. Abashkin and S. K. Burt, *Inorg. Chem. 44*, 1425 (2005).
223. Y. G. Abashkin and S. K. Burt, *J. Phys. Chem. B 108*, 2708 (2004).
224. W. Zhang, J. L. Loebach, D. R. Wilson, and E. N. Jacobsen, *J. Am. Chem. Soc. 112*, 2081 (1990).
225. R. Irie, K. Noda, Y. Ito, N. Matsumoto, and T. Katsuki, *Tetrahedron Lett. 31*, 7345 (1990).
226. J. Y. Yang, J. Bachmann, and D. G. Nocera, *J. Org. Chem. 71*(23), 8706 (2006).
227. E. N. Jacobsen and M. H. Wu, in *Comprehensive Asymmetric Catalysis*, E. N. Jacobsen, A. Pfaltz, and H. Yamamoto, Eds., Springer: New York, 1999; pp. 649–677.
228. G.-Q. Lin, Y.-M. Li, and A. S. C. Chan, in *Principles and Applications of Asymmetric Synthesis*; Wiley-Interscience, New York, 2001, pp. 237–241.
229. M. J. McNevin and J. R. Hagadorn, *Inorg. Chem. 43*, 8547 (2004).
230. M. L. Hlavinka, M. J. McNevin, R. Shoemaker, and J. R. Hagadorn, *Inorg. Chem. 45*, 1815 (2006).
231. M. L. Hlavinka and J. R. Hagadorn, *Organometallics* 24, 5335 (2005).

232. B. Clare, N. Sarker, R. Shoemaker, and J. R. Hagadorn, *Inorg. Chem. 43*, 1159 (2004).
233. J. R. Hagadorn and M. J. McNevin, *Organometallics 22*, 609 (2003).
234. J. R. Hagadorn, M. J. McNevin, G. Wiedenfeld, and R. Shoemaker, *Organometallics 22*, 4818 (2003).
235. M. L. Hlavinka, Hagadorn J. R. *Chem. Commun.* 2686 (2003).
236. A. S. Lukas, P. J. Bushard, and M. R. Wasielewski, *J. Phys. Chem. A 106*, 2074 (2002).
237. M. J. Tauber, R. F. Kelley, J. M. Giaimo, B. Rybtchinski, and M. R. Wasielewski, *J. Am. Chem. Soc. 128*, 1782 (2006).
238. J. M. Giaimo, A. V. Gusev, and M. R. Wasielewski, *J. Am. Chem. Soc. 124*, 8531 (2002).
239. J. D. Soper, S. V. Kryatov, E. V. Rybak-Akimova, and D. G. Nocera, *J. Am. Chem. Soc. 129*, 5069 (2007).

# CHAPTER 8

# Metal-Containing Nucleic Acid Structures Based on Synergetic Hydrogen and Coordination Bonding

**WEI HE, RAPHAEL M. FRANZINI, and CATALINA ACHIM**

*Department of Chemistry*
*Carnegie Mellon University*
*Pittsburgh, PA 15213*

CONTENTS



*Progress in Inorganic Chemistry, Vol. 55* Edited by Kenneth D. Karlin

# I. INTRODUCTION

Interactions of metal ions with nucleic acids have been investigated since 1924, when Hammarsten recognized that the polyanionic character of what was later to be called deoxyribonucleic acid (DNA) requires neutralization by metal cations (1). The systematic investigation of metal binding to DNA and how this binding depends on the DNA sequence, the ionic strength and concentration of DNA solutions, as well as the temperature began in the 1950s and 1960s. A historic perspective of these early research efforts has been presented in a recent review by Lippert (2). It was determined early on that there are two modes of metal ion binding to DNA. One of them is nonspecific and primarily due to electrostatic and steric interactions between the polyanionic DNA and metal cations and causes no changes in the hydration sphere of the metal ions. The other metal-binding mode is site specific and involves direct metal coordination to binding sites in DNA (e.g., phosphates or nucleobases). The latter binding mode occurs, for example, between the nucleobases and soft metal ions (e.g., $Cu^{2+}$, $Hg^{2+}$, and $Ag^{+}$) and can interfere with the nucleic acid duplex formation (3, 4).

The indirect and direct interactions between nucleic acids and metals are relevant for biology, therapeutics, and materials science, which explains the extensive research efforts dedicated to their study. For example, in biology, metal homeostasis in prokaryotes is achieved using metalloregulatory proteins, which activate, repress, or derepress the transcription of operons that encode proteins involved in metal transport and storage (5). In cancer chemotherapeutics, one of the most potent and thoroughly studied drugs is cisplatin, a simple metal complex that exerts its activity by interaction with DNA (6–8). The partially aquated forms of this complex bind to the N7 atoms of the purine bases

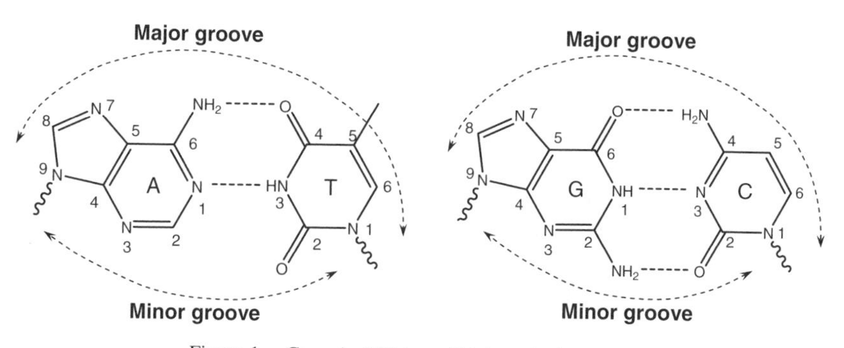


Figure 1. Canonical Watson–Crick nucleobase pairs.

of the DNA duplex (Fig. 1), thus forming adducts that contain 1,2-intrastrand cross-links as major products. These cross-links change the DNA structure by bending the major groove and exposing the minor groove (9), thus facilitating the binding to DNA of several proteins, including some involved in DNA repair and transcription (6). The DNA has been used also as a synthon for biomimetic nanotechnology because of its unique ability to store and transfer information through Watson–Crick base pairing (Fig. 1). In the past 20 years, this property has been extensively exploited starting with the pioneering contributions from the laboratory of N.C. Seeman, which generated a wealth of DNA-based nanostructures (10–12). This DNA nanotechnology entails connecting motifs derived from branched-DNA molecules that have intermolecular sticky ends, which leads to predictable affinity of components and structure of the assembled system.

Transition metal ion incorporation in hybrid inorganic–nucleic acid structures adds another dimension to the field of DNA nanotechnology because metal ions, in particular those with unpaired *d* electrons, have a broad range of electronic and magnetic properties. In most applications in material science, DNA is used as a scaffold for uniform binding of metal ions either to nucleobases or to ligands attached to nucleobases. For example, Braun et al. (13) induced the formation of a conductive silver wire on a DNA scaffold by treatment with an $Ag^+$ solution, which leads to both nonspecific substitution of the $Na^+$ counterions by $Ag^+$, as well as $Ag^+$ binding to nucleobases. This was followed by the reduction of $Ag^+$ to metallic silver using hydroquinone and subsequent uniform deposition of Ag on the Ag clusters initially formed on DNA. Sequence specificity has been achieved in the Ag deposition by using RecA proteins (14). These proteins mediate homologous recombination of a single-stranded DNA (ssDNA) probe onto a double-stranded DNA (dsDNA) substrate based on sequence homology between the probe and substrate (14). The ssDNA probe that hybridized onto an aldehyde-derivatized DNA duplex

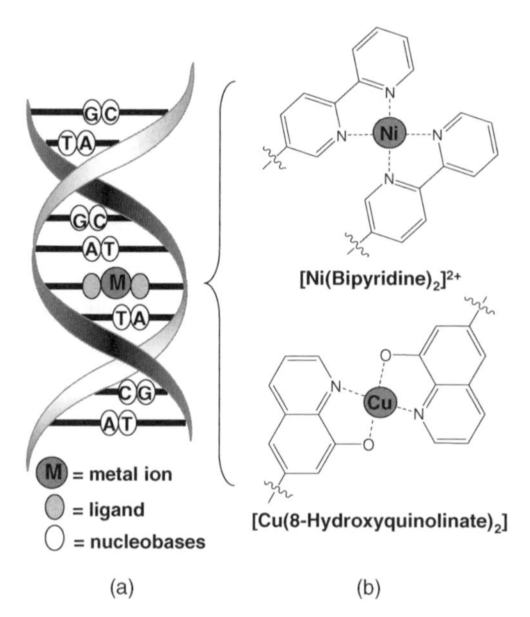


Figure 2. Metal-containing, ligand-modified nucleic acid duplex (*a*) and examples of metal–ligand alternative base pairs (*b*).

"protected" the aldehyde groups on the duplex. Consequently, $Ag^+$ ions were reduced on the dsDNA substrate only at the positions where there was no ssDNA. The Ag clusters generated by this process catalyzed the subsequent deposition of Au onto DNA. More recently, the binding of RecA itself to the DNA substrate was used to protect the DNA from being derivatized with an aldehyde and thus from being the site of Ag cluster growth when treated with $Ag^+$ (15). Both procedures achieved the same final goal, namely, the Ag deposition at predefined positions on the DNA substrate.

The ability to localize the metal ions at specific positions in the nucleic acid structures can be achieved by modification of the nucleic acids with ligands that have a higher affinity for metal ions than the natural nucleobases or the sugar-phosphate backbone (Fig. 2). The choice of ligands is based on their metal-binding constants and the structure of the metal–ligand complexes they form. This strategy has been used frequently in recent years to create nucleic acid duplexes in which metal–ligand complexes function either as alternative nucleobase pairs or as metal-dependent connectors between nucleic acid structures. This chapter describes the current state of this emerging field of supramolecular chemistry that encompasses several aspects of coordination and nucleic acid chemistry.

To date, several strategies have been used to create systems based on synergetic formation of Watson–Crick base pairs (bp) and of coordination

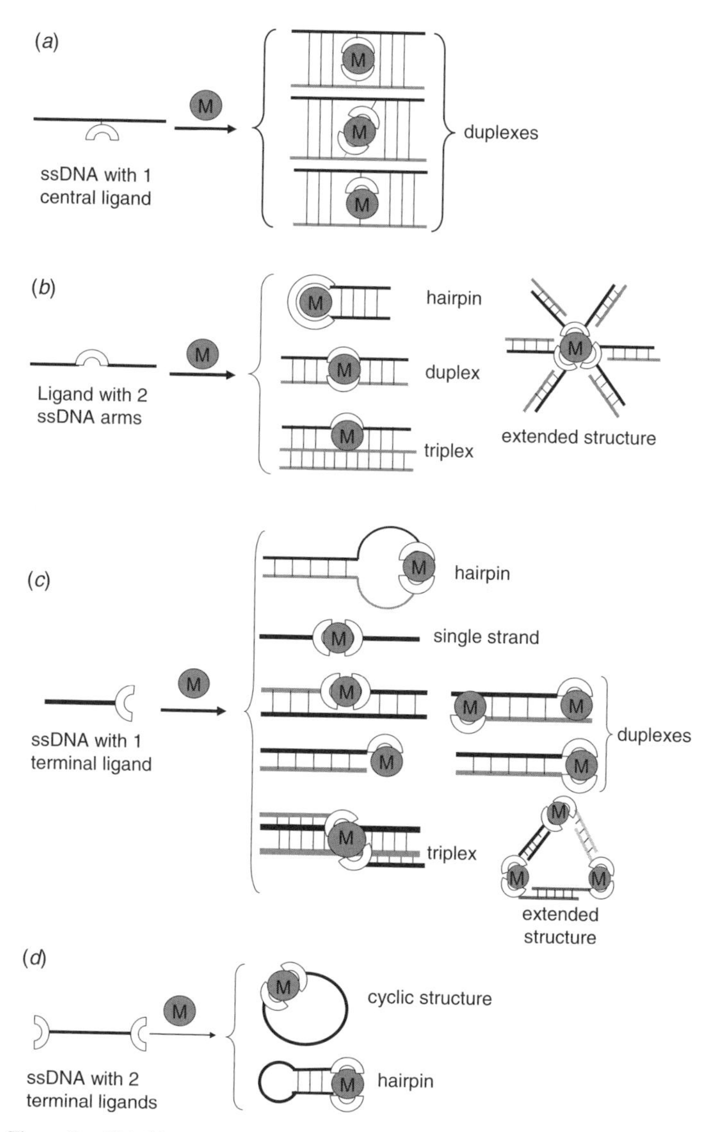


Figure 3. Hybrid structures based on coordination complexes and nucleic acids.

bonds (Fig. 3). Ligand incorporation in the center of nucleic acid oligomers [Fig. 3(*a*)] was used to create metal-containing, ligand-modified duplexes (Section IV). In this approach, the coordination complex formed between the metal ion and the ligands plays the role of an artificial base pair and will be referred to as a metal–ligand alternative base pair or metal-mediated base pair.

Ligands have also been used as connectors of ss nucleic acids [Fig. 3(*b*)]. Terminal modification of nucleic acid oligomers with ligands was used to create a broad range of structures, including metal-containing, single strands, hairpins, duplexes, and triplexes (Section V) [Fig. 3(*c* and *d*)]. Within these hybrid inorganic–nucleic acid structures, there are synergetic relationships between (a) metal coordination and oligonucleotide hybridization, and (b) the properties of the metal complexes and those of the nucleic acid structures that contain the metal complexes. For example, upon hybridization of ligand-modified oligonucleotides to form a duplex that contains ligands in complementary positions, the binding constant for a metal ion to the ligands brought in close proximity of each other by the duplex is generally larger than that to each independent ligand (16, 17). Conversely, ligand-modified oligonucleotides that may form two distinct structures [e.g., a duplex or a hairpin, Fig. 3(*b*)], preferably will adopt the structure in which the metal ion achieves its most common coordination number and geometry (18).

Besides natural DNA, several synthetic DNA analogues have been subject to ligand modification, namely, peptide nucleic acid (PNA), locked nucleic acid (LNA), and glycol nucleic acid (GNA) (Fig. 4). After introducing these types of nucleic acid analogues (Section II), the coordination properties of DNA to late 3*d* transition metal ions and to $Hg^{2+}$ (Section III), which have been the ions most often incorporated in nonmodified nucleic acid duplexes, are briefly presented. The 3*d* metal ions usually coordinate to guanine, which acts as a monodentate ligand coordinated through the N7 position. The $Hg^{2+}$ forms interstrand cross-links that involve thymine. The use of ligands that have higher binding constants than the nucleobases allows the synthetic chemist to create high affinity metal-binding sites within DNA duplexes. In separate sections, the studies aimed at the construction of (a) metal-containing, ligand-modified, nucleic acid duplexes from oligomers that contain ligand modifications in the middle of their sequence (Section IV), and (b) metal-containing nucleic acid structures from oligomers that contain ligands at their terminus (Section V) are described.

## II. OVERVIEW OF NUCLEIC ACIDS: DNA, PNA, LNA, AND GNA

Deoxyribonucleic acid and several of its synthetic analogues that are capable of forming duplexes based on Watson–Crick base pairing have been used to generate hybrid metal–nucleic acid structures by incorporation of metal–ligand alternative base pairs in the nucleic acid. This strategy for metal incorporation has not been applied to ribonucleic acid (RNA), although this molecule can be chemically synthesized and, implicitly, modified with ligands, which is likely because of the diverse set of structures that it can adopt and its chemical instability.

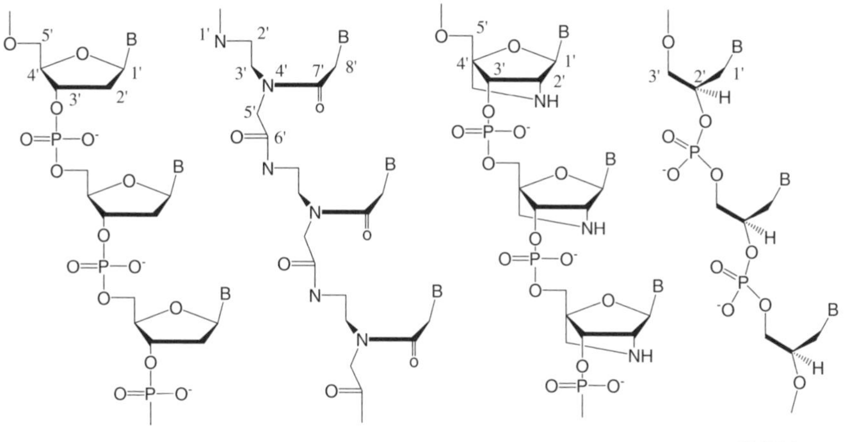

DNA PNA Amino-LNA (*R*)-GNA

Figure 4. The chemical structure of DNA and of DNA analogues with different backbones.

In contrast to RNA, DNA occurs mostly as a double helix, formed by Watson–Crick hybridization of two DNA oligomers. These oligomers consist of nucleotides, which contain a sugar, a phosphate, and a heterocyclic nucleobase (Fig. 5), linked by phosphodiester bonds. The major interactions that can stabilize DNA duplexes are (a) hydrogen bonding between complementary bases; (b) stacking interactions of adjacent bases, which include hydrophobic and van der Waals interactions; and (c) steric interactions (19). The three major DNA conformers, A-, B-, and Z-forms, have characteristic ranges of helical parameters (Table I) (20–22). Within each of these categories, the exact DNA structure depends significantly on the sequence and the environmental conditions (e.g., DNA concentration, solvent, and ionic strength). For canonical

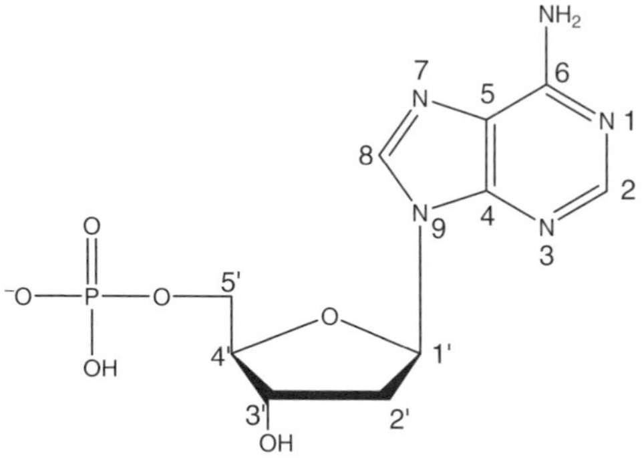

Figure 5. Deoxyadenylic acid.

TABLE I
Mean Value of Parameters for Regular A-, B-, Z-DNA (RNA) (20–22) and PNA (23, 24) Duplexes[a]

| | DNA | | | PNA |
|---|---|---|---|---|
| **Structural Type** | **A-Form** | **B-Form** | **Z-Form** | **P-Form** |
| **Example** | ds DNA•RNA ds RNA | *in vivo* DNA | GC-rich sequences | Homo-duplex of PNA |
| **Helix sense** | Right handed | Right handed | Left handed | Both right (r) and left (l) handed |
| **Major groove** | Narrow and deep | Wide and deep | Wide and shallow | Wide and deep |
| **Minor groove** | Wide and shallow | Narrow | Narrow and deep | Narrow and shallow |
| **Repeating unit** (bp) | 1 | 1 | 2 | 1 |
| **Diameter** (Å) | 24 | 24 | 20 | 28 |
| **X-displacement relative to long axis of helix (Å)** | −4.5 | 0.23 | 3.3 | 8.3 |
| **Rise per bp along axis** (Å) | 2.56 | 3.4 | 3.8 | 3.2–3.4 |
| **No. base pair/helical turn** | 11 | 10 | 12 | 18–20 |
| **Pitch per turn of helix** (Å) | 28 | 33.2 | 45.6 | 58 |
| **bp inclination** (°) | 12 | 2.4 | 7.1 | 0.3−26.7(r),−8.9(l) |
| **Tip** (°) | 11.0 | 0 | 0.7 | – |
| **Buckle** (°) | 2.4 | 0.2 | 3.1 | – |
| **Propeller twist** (°) | 8.3 | 11.0 | 2.6 | 19(r), −18(l) |
| **Roll** (°) | 6.3 | 0.6 | −0.4 | – |
| **Twist** (°) | 33 | 36 | −60/2 | 19.1–19.8 |
| **Base tilt** (°) | 20 | 0 | −1.7 | 0.1 |

[a]Definitions of helical and base pair parameters in DNA and PNA helices.

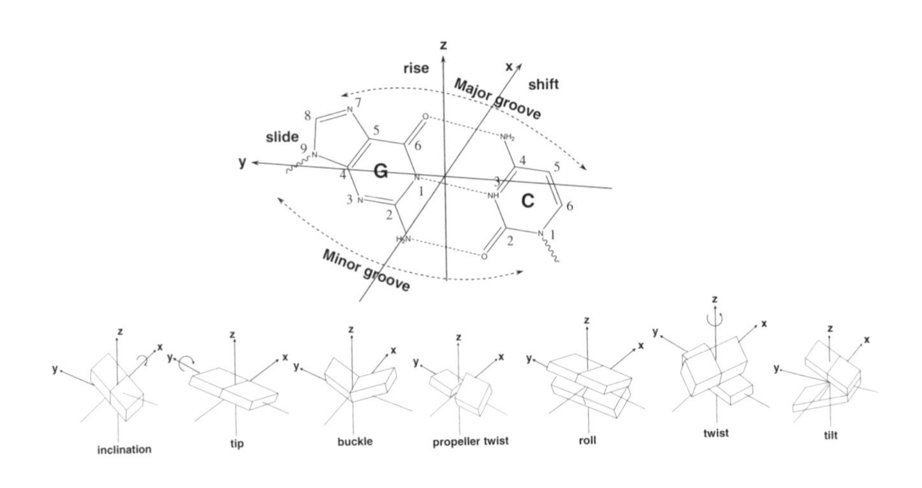

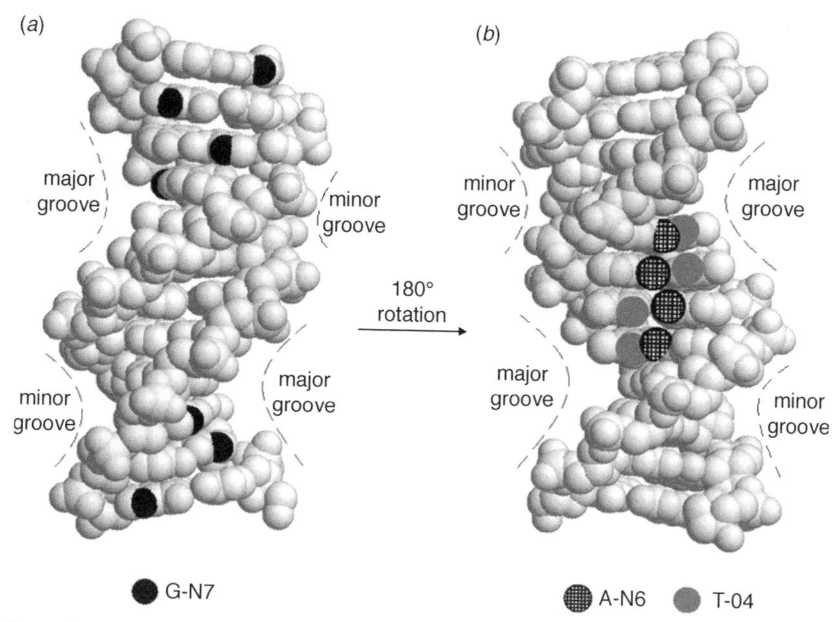


Figure 6. The G-N7 (*a*) and A-N6 and T-O4 sites (*b*) in a typical B-DNA duplex (Dickerson–Drew dodecamer d(5′-CGCGAATTCGCG-3′)$_2$, (PDB ID: 1BNA). The duplex in the right view has been rotated 180° with respect to the left view. (See color insert.)

Watson–Crick base pairs, the G-N7 and A-N7 atoms and the A-N6/T-O4 atom pairs are oriented toward the DNA major groove (Fig. 1 and 6), which is wide and deep for B-DNA, and wide and shallow for Z-DNA (Table I). The exposure of the N7 atoms of the purine bases and of the bidentate (N, O) site of the AT base pairs makes them good candidates for metal coordination (see Section III). The highest accessibility to metal ions is provided by Z-DNA, which has a wide and shallow major groove (Fig. 7 and Table I). The 2.6–3.8 Å rise between nucleobase pairs is only slightly larger than metal–ligand distances, and thus it is possible that a metal ion incorporated in a metal–ligand alternative base pair coordinates donor atoms from adjacent nucleobase pairs without major distortions of the nucleic acid duplex. This distance is also comparable to intermetal distances in metal clusters with single-atom bridges, for example, oxo or hydroxo, thus boding well for the formation of arrays of bridged metal ions within the duplex.

The last 15 years have brought about the development of several synthetic oligonucleotide analogues, which can bind to DNA and RNA, and thus have potential for antisense and antigene applications. These analogues contain the A, G, T, C bases and can form duplexes or triplexes with DNA or on their own by Watson–Crick base pairing (Fig. 4). Their backbone is different from that of

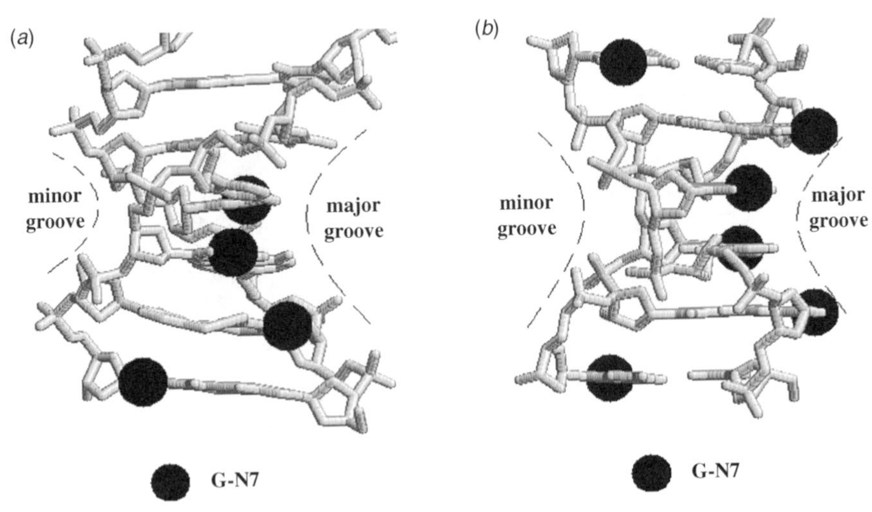


Figure 7. The G-N7 sites in B-DNA (*a*) expanded view of a fragment (underlined) from Dickerson–Drew dodecamer d(5′-CGCGAATTCGCG-3′)$_2$, (PDB: 1BNA) and Z-DNA (*b*) hexamer d(5′-CGCGCG-3′)$_2$, PDB ID: 1WOE) duplexes.

DNA, which makes them chemically and biochemically more stable than DNA, and thus more appropriate for nanotechnological and certain biological applications.

PNA is a synthetic DNA analogue, which was developed in 1991 (25). The PNA pseudo-peptide backbone is based on aminoethylglycine (Fig. 4), which is neutral. This explains why the stability of PNA•DNA duplexes is higher than that of DNA•DNA duplexes, and why its solubility in water is lower than that of DNA or RNA. Amino acids with groups that are positively charged at neutral pH (e.g., lysine and arginine) have been introduced into PNA oligomers to improve their solubility, enabling the preparation of millimolar solutions of such oligomers. The synthesis of PNA monomers is relatively simple (26). The PNA monomers containing the A, G, C, and T bases are commercially available and numerous PNA monomers with different pseudo-peptide backbones or with alternative nucleobases have been prepared (27). For example, alanyl PNA is a peptide-based nucleic acid structure with an alanyl-derived backbone having the nucleobases attached to the β-carbon (28). The PNA oligomers are prepared by modified solid-phase peptide synthesis using either Boc or Fmoc protection strategies (29). These methods made possible the attachment of ligands or of metal complexes that contain carboxylic acid groups to the $NH_2$ end of the PNA oligomers. Several of these metal complexes can be used as infrared (IR) [e.g., (benzene)chromiumtricarbonyl)] or electrochemistry probes (e.g., ferrocene or cobaltocene) (30–33).

The duplexes PNA•PNA and PNA•DNA are formed by Watson–Crick base pairing and their stability is higher than that of DNA•DNA duplexes. Single point mismatches have a larger destabilization effect on the thermal stability of PNA•PNA duplexes than on PNA•DNA or DNA•DNA duplexes and thus make PNA particularly suitable for single-point mutation discrimination. The large effect of the mismatches suggests that the effect of metal–complex incorporation in nucleic acid duplexes will also be stronger for PNA than for DNA.

In contrast to the chiral sugar-phosphate backbone of DNA, the aminoethylglycine backbone of PNA does not contain stereocenters. A preferred helical handedness can be induced in PNA duplexes by the attachment of a D- or L-amino acid at the C-terminus of the duplex (34). The X-ray structures of a 6 bp PNA duplex and of a partly self-complementary ssPNA dodecamer that forms a 4 bp duplex part in the crystal show that the PNA•PNA duplexes adopt a distinct p-form helix, which can be either right- or left-handed (Fig. 8) (23, 24). The p-helix has a very large helical pitch and a wider diameter than that of the DNA or RNA helices (Fig. 8 and Table I).

Oxy-LNA is an RNA analogue developed in 1998, in which the ribose group of the sugar-phosphate backbone contains a methylene bridge between the 2′-oxygen atom of the ribose and the 4′-carbon atom (35–37). Several C-glycoside analogues of LNA have been prepared including an amino-LNA that has a 2′-NH group instead of the 2′-oxygen (Fig. 4). The bridge preorganizes the LNA monomer into a locked C3′-*endo* conformation similar to that observed in A-type DNA, and thus reduces the entropic penalty for the formation of duplexes containing LNA. As a consequence, LNA•LNA and hetero-LNA•RNA and LNA•DNA duplexes are more stable than DNA•DNA or DNA•RNA ones. Currently, there is no solution structure for a fully modified LNA•LNA duplex,

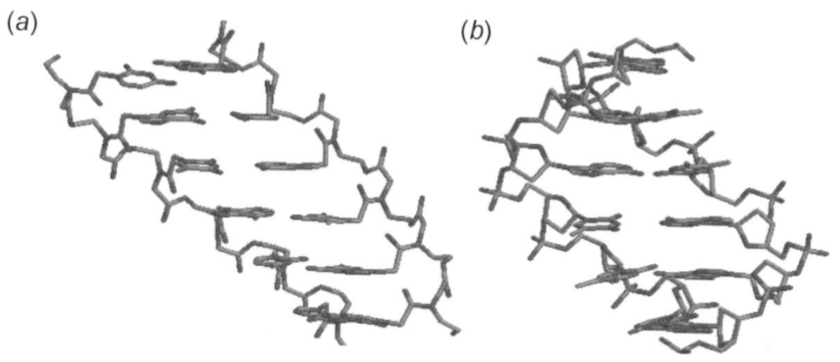


Figure 8. (*a*) The 6-bp PNA duplex based on the $H_2N$-CGTACG-H strand (PDB: 1PUP), and (*b*) the 6-bp central fragment $d(GAATTC)_2$ of the Dickerson-Drew DNA dodecamer (PDB: 1BNA). (See color insert.)

but NMR structures of partially or completely modified LNA•RNA and LNA•DNA hybrids show that the LNA nucleotides induce a preference for an A-type, right-handed duplex structure because of their C3′-*endo* conformation and because they steer the sugar conformations of neighboring nucleotides into similar C3′-*endo* conformations (38).

Glycol nucleic acid (GNA) is a simplified DNA analogue that has an acyclic three-carbon propylene glycol phosphodiester backbone (Fig. 4) (39). Variable temperature ultraviolet (UV) spectroscopy was used to demonstrate that complementary 18-mer GNA oligomers that contain T and A bases form antiparallel, helical duplexes based on Watson–Crick base pairing, which are more stable than the analogous DNA or RNA duplexes. No NMR or X-ray structure of GNA or hybrid GNA–DNA duplexes is yet available. Both GNA and LNA oligomers can be synthesized by standard solid-phase DNA synthetic methods, because the backbone of both molecules is formed by phosphodiester bonds.

To make possible the metal ion incorporation at specific locations in the DNA, PNA, LNA, or GNA duplexes, ligands have been included in oligonucleotides of these nucleic acids, typically in complementary positions. All ligands used to date contain aromatic rings, and can participate in $\pi$–$\pi$ stacking interactions with adjacent bases, but generally, they cannot form hydrogen bonds to bridge the two complementary oligonucleotides forming the duplex and may differ in size from nucleobases. Consequently, in all studies published so far, the stability of the ligand-modified duplexes was similar or lower than that of the duplexes that have AT or GC base pairs instead of the pairs of ligands.

## III. INTERACTION OF NONMODIFIED DNA WITH METAL IONS

### A. Binding of 3*d* Transition Metal Ions to DNA Duplexes

Early in the twentieth century, it was discovered that DNA is a strong acid that exists as a polyanion at neutral pH and requires cations for charge balance (1). Metal cation binding to DNA by electrostatic interactions is nonspecific and does not cause a change in the coordination sphere of the metal. The metal ions can be mobile around the nucleic acid or localized as a consequence of the accumulation of negative charge in certain areas of the secondary structure of the nucleic acid. These interactions are predominant for alkali and alkali earth metals, which act most frequently as counterions to DNA and are available in the millimolar concentration range within cells. The phosphate groups and the nucleobases in the DNA and RNA can act as ligands to metal ions. In this case, the binding is specific. In the case of both specific and nonspecific metal binding, the water molecules coordinated to the metal ions can act as hydrogen-bond donors and can mediate indirect interactions between the metal and phosphate groups or nucleophilic sites of the nucleobases.

The coordination properties of the nucleobases have been reviewed by Houlton (40) and by Lippert (2). In a recent review, Lippert discussed the influence of the metal coordination on the p$K_a$ of the nucleobases (41), which correlates with their coordination properties. While the coordination properties of nucleobases, nucleosides, and nucleotides have been extensively studied and reviewed, the number of articles dedicated to the coordination properties of nucleic acids is significantly smaller. DeRose et al. (42) recently published a systematic review of the site-specific interactions between both main group and transition metal ions with a broad range of nucleic acids from 10 bp DNA duplexes to 300–400 nucleotide RNA molecules as well as with some nucleobases, nucleosides, and nucleotides. They focused on results obtained primarily from X-ray crystallographic studies. Egli also presented information on the metal ion coordination to DNA in reviews dedicated to X-ray studies of nucleic acids (43, 44). Sletten and Frøystein (45) reviewed NMR studies of the interaction between nucleic acids and several late transition metal ions and $Zn^{2+}$. Binding of metal complexes to DNA by $\pi$ interactions has been reviewed by Dupureur and Barton (46).

To put in perspective the binding of metal ions to ligand-modified nucleic acid duplexes, we summarize the conclusions of these reviews on metal binding to DNA duplexes and emphasize those conclusions that are relevant for the influence of metal coordination on the secondary structure of DNA. These studies have been the prologue of today's efforts to use DNA or its synthetic analogues as scaffold for transition metal ions.

Eichhorn et al. (3) made the earliest phenomenological observations on the transition metal ion coordination to nucleic acids by noting that the preference of these cations for phosphate over nucleobase coordination decreases in the order $Co^{2+} > Ni^{2+} > Mn^{2+} > Zn^{2+} > Cd^{2+} > Cu^{2+} > Ag^{+} > Hg^{2+}$. Research conducted in the last four decades revealed that the factors that affect metal binding to DNA include (a) the nucleophilicity of the phosphates and nucleobases and their hard and soft Lewis base character, respectively; (b) the accessibility of these groups to metal ions; (c) the molecular electrostatic potential of DNA; (d) the ability of the groups neighboring the metal coordination site to form hydrogen bonds with the water molecules coordinated to the metal ion; and, in the case of binding of multiple metal ions to the same DNA duplex, (e) the changes in the secondary structure of the DNA and in the nucleophilicity of the phosphate and nucleobases induced by the first metal ions that bind to DNA and that influence subsequent binding of metal ions to

DNA. The number of DNA duplexes with different sequences for which complexes with transition metal ions have been structurally characterized is relatively small and the conclusions summarized above about the relative importance of the factors that affect the metal binding have begun to emerge only recently.

In the case of purines, the N1 and N7 atoms have the highest affinity for metal ions but, in DNA duplexes N1 is involved in hydrogen bonding with complementary bases (Fig. 1) (47, 48). As a consequence, the G-N7 and A-N7 positions, which are most accessible in the major groove of DNA (Fig. 7), are the most common sites to which transition metal ions bind, with the G-N7 being preferred over the A-N7 site (42), which is in agreement with the lower basicity of A-N7 (52). Based on density functional theory (DFT) calculations, the higher reactivity toward G than A of mono or diaqua $Pt^{2+}$ complexes generated by partial aquation of cisplatin has been attributed to (a) stronger hydrogen bonding and electrostatic interactions with guanine, which stabilize the trigonal-bipyramidal transition state of the reaction between the metal complex and the base; and (b) stronger electronic interactions between the platinum complex and the guanine compared to adenine (49). Numerous studies have show that the metal binding to DNA is strongly related with the sequence and structure of the DNA. For example, the higher accessibility to metal ions of the G-N7 sites in Z-DNA than in B-DNA contributes to the fact that $Ni^{2+}$ coordination to G-N7 favors the transition from B- to Z-DNA for double-stranded poly(dG-dC) (50, 51) (Fig. 7). On the other hand, binding of $Ni^{2+}$ to poly(dA-dT) DNA is nonspecific despite the presence of adenines (53).

The NMR solution studies carried out by the Sletten group examined the sequence dependence of the metal binding to DNA duplexes (Entries 1–4, Table II) and proposed a trend for the sequence selectivity of $Zn^{2+}$, $Mn^{2+}$, and $Co^{2+}$ binding to G bases in DNA, which is $5'$-*G*G > *G*A > *G*T ≫ *G*C (the italicized letter indicates the nucleobase to which the metal ion is coordinated) (54, 55, 63, 64). This order can also be invoked in explaining the binding of $Zn^{2+}$ to the self-complementary DNA dodecamer $d(A_1T_2G_3G_4G_5T_6A_7C_8C_9C_{10}A_{11}T_{12})_2$ investigated by Marzilli and co-workers by NMR spectroscopy (Entry 5, Table II) (56); the metal ion binds preferentially to G bases and the binding affinity decreases in the order $G_4 > G_3 > G_5$. If the neighboring bases are included in this sequence of binding affinity, the series becomes $5'$-$G_4G_5$ >-$G_3G_4$ >-$G_5T_6$. The difference between $5'$-$G_3G_4$ and $5'$-$G_5T_6$ could be explained using Sletten's arguments based on sequence, and the one between 5-$G_4G_5$ and $5'$-$G_3G_4$ could be attributed to the decrease in the size of the molecular electrostatic potential of DNA from the center to the end of the duplex (65).

More recently, X-ray structures of 6–10 bp DNA duplexes crystallized in the presence of $Zn^{2+}$, $Co^{2+}$, or $Ni^{2+}$ have been obtained at 2.9–3.1 Å resolution (Entries 6–9, Table II) (57–60). These structures showed that the coordination sphere of the metal ions is constituted by one or two guanines coordinated

TABLE II
Complexes of Selected 3*d* Transition Metal Ions With Nucleic Acid Duplexes

| | Sequence(5′—>3′ If Not Otherwise Specified) | Metal | Binding Sites | Reference |
|---|---|---|---|---|
| 1 | $d(G_1G_2T_3A_4C_5C_6G_7G_8T_9A_{10}C_{11}C_{12})_2$ | $Zn^{2+}$, $Mn^{2+}$ | $G_7$-N7 > $G_1$-N7 | 54 |
| 2 | $d(G_1G_2C_3G_4C_5C_6)_2$ | $Zn^{2+}$, $Mn^{2+}$ | $G_1$-N7 | 54 |
| 3 | $d(G_1G_2T_3A_4T_5A_6T_7A_8T_9A_{10}C_{11}C_{12}G_{13}G_{14}T_{15}A_{16}T_{17}A_{18}T_{19}A_{20}T_{21}A_{22}C_{23}C_{24})_2$ | $Zn^{2+}$, $Mn^{2+}$ | $G_7$-N7 | 54 |
| 4 | $d(C_1G_2C_3G_4A_5A_6T_7T_8C_9G_{10}C_{11}G_{12})_2$ | $Zn^{2+}$ | $G_4$-N7;5′-*GG*> = *GA*>*GT* ≫ *GC* | 55 |
| 5 | $d(A_1T_2G_3G_4G_5T_6A_7C_8C_9C_{10}A_{11}T_{12})_2$ | $Zn^{2+}$ | $G_4$-N7 > $G_3$-N7 > $G_5$-N7 | 56 |
| 6 | $d(C_1G_2T_3A_4T_5A_6T_7A_8C_9G_{10})_2$ | $Ni^{2+}$ | $G_2$-N7 and $G_{10}$-N7 | 57 |
| 7 | $d(C_1G_2T_3G_4T_5A_6C_7A_8C_9G_{10})_2$ | $Ni^{2+}$ | $G_2$-N7 and $G_{10}$-N7 | 58 |
| 8 | $d(G_1G_2C_3G_4C_5C_6)_2$ | $Ni^{2+}$, $Co^{2+}$, $Zn^{2+}$ | $G_1$-N7 | 59 |
| 9 | MeCO-$(Arg)_4$-NH-p-$d(C_1G_2C_3A_4A_5T_6T_7G_8C_9G_{10})_2$ | $Zn^{2+}$ | Phosphate of $C_1$,$G_{10}$;$G_2$-N7,$G_{10}$-N7 | 60 |
| 10 | $d(C_1G_2C_3G_4C_5G_6)_2$ | $Cu^{2+}$ | all G-N7 | 61 |
| 11 | $d(G_1A_2A_3T_4T_5C_6G_7)_2$ | $Ni^{2+}$ | $G_7$-N7 | 62 |

through N7, by phosphate groups, which are situated in the close proximity of the coordinating guanines, and by water molecules, which may form hydrogen bonds with nearby phosphates or nucleobases. The observation that the metal ions coordinate exclusively to the terminal guanines of the duplex was attributed to the higher accessibility of these nucleobases, and gave support to the suggestion made earlier that transition metal ions cannot coordinate to the internal G-N7 atoms in B-DNA (59, 60, 66). However, in a more recent article, Sletten and co-workers (54) pointed out that the analysis of the DNA sequence for which the X-ray structures have been determined shows that the coordination of metal ions to the terminal Gs in these DNA duplexes could also be explained by the rule of sequence selectivity proposed based on their NMR solution studies. Another 1.2 Å crystal structure of a 6-bp DNA duplex, where all the G bases were in a 5′-*GC* context, determined using a crystal soaked in $CuCl_2$, showed that $Cu^{2+}$ ions can bind to all six Gs, with the metal site coordinated to the terminal $G_6$ having the largest occupancy (Entry 10, Table II) (61). This result suggests that the accessibility of coordination sites is very important for metal binding.

These studies indicate that the sequence of duplexes and the accessibility of bases are the dominant factors that determine the binding pattern of the labile metal ions to DNA, and that the molecular electrostatic potential further affects the metal binding. Of importance for this chapter is that in all of the reported solution and solid-state structures, the 3*d* transition metal ions coordinate to the N7 of a G nucleobase and complete their octahedral coordination sphere with water molecules or with phosphate groups from adjacent nucleotides within the duplexes. Thus they do not form direct interstrand bridges.

### B. M-DNA

Lee et al. (67) reported in 1993 that at high pH, complexes termed M-DNA form between the nucleobase pairs within calf thymus (CT), bacterial or synthetic DNA, and the divalent metal ions, $Zn^{2+}$, $Ni^{2+}$, or $Co^{2+}$. The chemical nature of M-DNA has been investigated using methods that probe changes in DNA properties, but do not directly probe the coordination of the metal ions. Titration of a $d(T\text{-}G)_{15}\bullet d(C\text{-}A)_{15}$ duplex with $Zn^{2+}$ at pH 8.5 monitored by NMR spectroscopy showed a decrease in the intensity of the imino peaks with the increasing concentration of $Zn^{2+}$ and the complete disappearance of these peaks when the $Zn^{2+}$ concentration became almost equal to that of the DNA base pairs (67). The transformation induced by $Ni^{2+}$ of B- to M-DNA for CT DNA in solutions with 1.1 mM base pair concentrations led to the release of 1.1 equiv of protons per base pair (68). Based on these results, Lee and co-workers (69) suggested that in the transformation of B- to M-DNA, the imino protons of the GC and AT base pairs are released and that the divalent metal ions bridge the nucleobases in each base pair of M-DNA (Fig. 9).

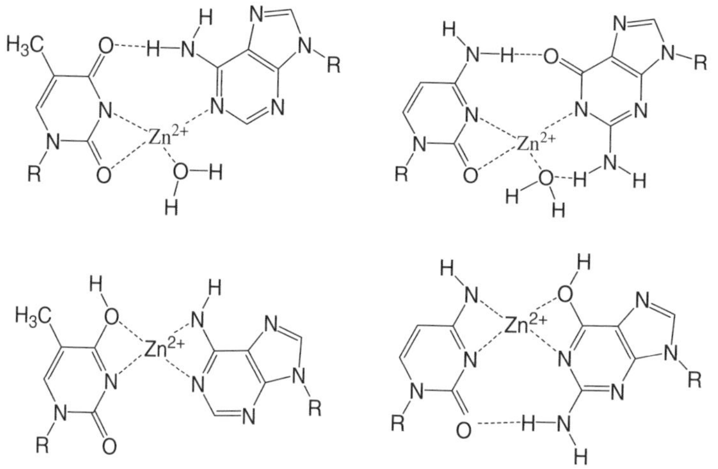

**A-Zn²⁺-T** **G-Zn²⁺-C**

Figure 9. Proposed coordination motifs for $Zn^{2+}$ in M-DNA (adapted from Reference 68).

In most studies published after the initial reports on the existence of M-DNA (67, 68), the formation of M-DNA for a variety of DNA duplexes (Table III) was demonstrated using an ethidium fluorescence assay (70, 71, 73–77). This assay is based on the fact that ethidium bromide intercalates in B-DNA and consequently fluoresces, but it cannot intercalate in M-DNA because of electrostatic repulsion between ethidium and the metal ions within the duplex.

A conformation change in the B-DNA to M-DNA transformation was inferred based on surface plasmon resonance studies (70). For the B- to M-DNA transformation to take place, it was necessary that a threshold concentration of metal ions be present in solution. For example, the transformation of B- to M-DNA in a solution with a DNA base pair concentration of ~1.1 mM took place after the concentration of $Ni^{2+}$ reached a threshold concentration of 0.7 m*M* (68). Surface plasmon resonance experiments on biotinylated DNA immobilized on a streptavidin-containing matrix confirmed that conversion of B-DNA to M-DNA by $Ni^{2+}$ requires a threshold $Ni^{2+}$ concentration (70). A concentration of 0.25 m*M* of $Zn^{2+}$ or $Co^{2+}$ was required for a 50% conversion of B-DNA to M-DNA at a DNA basepair concentration of 15 μ*M* in solutions with a pH of 8.5 (71). The threshold concentration depended on (a) the DNA's base composition, sequence, and concentration, (b) the metal ion, and (c) the solution pH (68, 69, 71, 72). The reason why a minimum concentration of metal ion is required before the B- to M-DNA transition takes place is not clear. The effect was attributed to metal ions binding to the DNA backbone prior to binding

TABLE III
Sequences and Metal Ions That Have Been Studied With Respect to Formation of M-DNA

| Entry | DNA Sequence | Metal | References |
|---|---|---|---|
| 1 | CT DNA *Escherichia coli* DNA<br>Bacterial DNAs<br>$d(TG)_{15} \cdot d(AC)_{15}$<br>poly[d(AT)], poly[d(GC)],<br>poly[d(TTC)] · poly[d(GAA)]<br>poly[d(TCC)] · poly[d(GGA)]<br>poly(dT) · poly(dA) | $Ni^{2+}$<br>$Co^{2+}$<br>$Zn^{2+}$ | 67, 73 |
| 2 | R-5′-GTCACGATGGCCCAGTAGTT-3′<br>3′-CAGTGCTACCGGGTCATCAA-5′-X<br>R = 5-Carboxyfluorescein or disulfide linkage<br>X = Rhodamine or none | $Ni^{2+}$<br>$Co^{2+}$<br>$Zn^{2+}$ | 68,74,75 |
| 3 | CT DNA<br>Phage λ-DNA<br>poly(dG)·poly(dC), poly[d(GC)], poly[d(GGCC)]<br>poly(dA)·poly(dT), poly[d(AT)]<br>poly[d(AU)], poly[r(AU)] | $Ni^{2+}$<br>$Co^{2+}$<br>$Zn^{2+}$ | 70 – 73 |
| 4 | 5′–GTGGCTAACTACGCATTCCACGACCAAATG-3′<br>3′-CACCGATTGATGCGTAAGGTGCTGGTTTAC-Bt-5′<br><br>5′–Bt-d(C)$_{30}$ (ss)[a]<br>5′–Bt-d(A)$_{30}$ (ss)[a]<br>5′–Bt-d(T)$_{30}$ (ss)[a] | $Ni^{2+}$<br>$Zn^{2+}$<br>$Cd^{2+}$ | 70 |

[a]Bt = Biotinylated

to the nucleobases (68), or alternatively it could also be a consequence of the cooperative character of the metal binding to the base pairs.

Lee and co-workers (69) reported that fluorescence lifetime measurements on fluoresceine–rhodamine labeled Zn-DNA duplexes of varying lengths were indicative of long-range electron transfer by electron hopping between the metal ions within Zn-DNA. As a first step toward building electronic devices based on M-DNA, electron transfer was studied in a three-arm junction built from DNA and transformed into M-DNA by addition of $Zn^{2+}$ at pH 8.5 (77). Fluorescence quenching studies led to the conclusion that the oxidation state of an electron acceptor situated on one of the arms influenced the electron transfer between a donor and acceptor each situated on either of the other two arms of the junction, an effect resembling the behavior of a transistor. Electrochemical measurements on self-assembled monolayers of thiol-modified DNA duplexes on gold electrodes indicated a significant increase in the electron-transfer rate upon addition of $Zn^{2+}$ to the DNA monolayers at high pH (73, 74). For example, the rate constant for electron transfer through dsDNA monolayers after addition of $Zn^{2+}$ at pH 8.6 was $1.2 \times 10^{-4}\ cm\ s^{-1}$ while at the same pH, but in the absence of $Zn^{2+}$, electron transfer through the dsDNA monolayers was not detectable (74). This change in electron-transfer properties of the dsDNA was attributed to conversion of B- to M-DNA, but a recent scanning electrochemical microscopy study showed that the changes in the surface charge of DNA monolayers that take place upon metal binding to DNA have a significant effect on the electron-transfer rate (75), which must be considered in the analysis and interpretation of the earlier electrochemistry results. Finally, metallic conduction was reported for M-DNA prepared by addition of $Zn^{2+}$ to phage λ-DNA bundles placed between gold electrodes (76). This property was assigned to the existence of a band of Zn *d* states situated close to the Fermi level of the electrodes. Several subsequent papers presented alternative interpretations of this result or reported non-metallic conduction of Zn-DNA (78).

### C. Binding of $Hg^{2+}$ Ions to DNA Duplexes

The kinetically inert $Hg^{2+}$ ion, which is a soft Lewis acid and prefers linear coordination, binds to DNA duplexes by forming interstrand bridges that involve thymine (Table IV). Two models have been proposed for these bridges. One model involves coordination of $Hg^{2+}$ to thymine pairs formed by chain slippage within DNA to form T-Hg-T bridges [Figs. 10(*a*) and 11, entry 2 of Table IV] (80,82, 83), and the other model involves coordination to the AT base pairs existent in DNA to form A-Hg-T bridges (Fig. 10*b* and *c* and entry 1 of Table IV) (79, 84). The former model does not exclude initial binding of $Hg^{2+}$ to the AT base pairs followed by structural reorganization to form T-Hg-T cross-links (Fig. 11).

TABLE IV
Complexes of $Hg^{2+}$ With Nucleic Acid Duplexes

| | DNA Duplex | Proposed Metal-Binding Mode | $T_m$(°C) | Reference |
|---|---|---|---|---|
| 1 | 5′-CGCGAATTCGCG-3′<br>3′-GCGCTTAAGCGC-5′ | AN6-Hg-$TO_4$ | | 79 |
| 2 | Poly (dA-dT) | T-N3-Hg-T-N3 | | 80 |
| 3 | 5′-AAAAAAAAAA**T**AAAAAAAAAA-3′<br>3′-TTTTTTTTTT**T**TTTTTTTTTT-5′ | T-N3-Hg-T-N3 | 37 (no $Hg^{2+}$)<br>47 (4 equiv $Hg^{2+}$/T•T) | 81 |
| | Control:<br>5′-AAAAAAAAAAAAAAAAAAAAA-3′<br>3′-TTTTTTTTTTTTTTTTTTTTT-5′ | | 44 | |
| 4 | 5′-GTGACCA**T**AGCAGTG-3′<br>3′-CACTGGT**T**TCGTCAC-5′ | T-N3-Hg-T-N3 | 47 (no $Hg^{2+}$)<br>56 (4 equiv $Hg^{2+}$/T•T) | 81 |
| 5 | 5′-GTGACCA**TT**TGCAGTG-3′<br>3′-CACTGGT**TT**ACGTCAC-5′ | T-N3-Hg-T-N3 | 45 (no $Hg^{2+}$)<br>58 (4 equiv $Hg^{2+}$/T•T) | 81 |
| 6 | 5′-GTGACCA**TTT**TGCAGTG-3′<br>3′-CACTGGT**TTT**ACGTCAC-5′ | T-N3-Hg-T-N3 | 43 (no $Hg^{2+}$)<br>61 (4 equiv $Hg^{2+}$/T•T) | 81 |

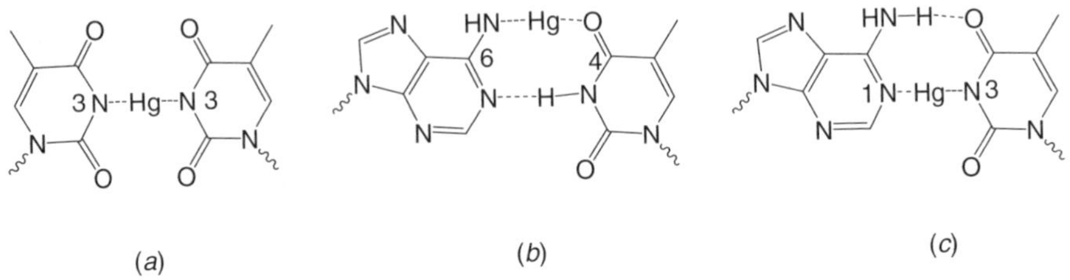

(a) (b) (c)

Figure 10. Proposed coordination modes of $Hg^{2+}$ to DNA: (*a*) TN3-Hg-TN3; (*b*) AN6-Hg-TO4; (*c*) AN1-Hg-TN3.

Evidence for the model in which $Hg^{2+}$ binds to AT base pairs came from NMR spectroscopy. The $Hg^{2+}$ addition to the $d(C_1G_2C_3G_4A_5A_6T_7T_8C_9G_{10}C_{11}G_{12})_2$ duplex (Entry 1 of Table IV) led to significant changes in the the AT region of the NMR spectra, with the resonances of the T-N3 protons disappearing after addition of 5 equiv $Hg^{2+}$/duplex, but not in the GC region (79, 84). Based on these results, it was concluded that $Hg^{2+}$ binds to the four central AT base pairs, creating a bulge in the middle of the duplex. The coordination mode in which $Hg^{2+}$ bridges the A-N6 and T-O4 [Fig. 10(*b*)] is more likely than the one involving A-N1 and T-N3 [Fig. 10(*c*)], because in the first case $Hg^{2+}$ binding does not interfere with hydrogen bonding between AN1 and T-N3 [Fig. 10(*b* and *c*)] (45). The T-Hg-T coordination, which is part of the chain slippage model (Fig. 11), was supported by the crystal structure of a 2:1 complex formed between 1-methylthymine and $Hg^{2+}$ (85). The model was also supported by the NMR studies of Young et al. (80), who showed that addition of 0.5 equiv $Hg^{2+}$/AT base pair to poly(dA-dT) leads to the complete disappearance of the N3 proton signals from the one dimensional (1D) $^1H$ NMR spectrum of the DNA duplex. Recent experiments indicated that $Hg^{2+}$

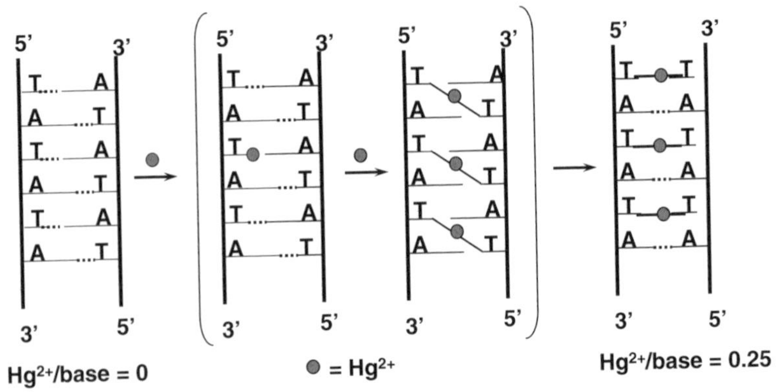


Figure 11. Proposed chain slippage model for $Hg^{2+}$ binding (adapted from Reference 80).

binds preferentially to TT rather than to AT base pairs (18). For example, addition of 1 equiv of $Hg^{2+}$ to the partially self-complementary oligonucleotide $A_1T_2G_3G_4G_5T_6T_7C_8C_9C_{10}A_{11}T_{12}$ triggered the formation of a duplex containing two central T-Hg-T cross-links (18) even if an A-Hg-T coordination mode for $Hg^{2+}$ was also possible. Only addition of a second equivalent of $Hg^{2+}$ led to changes in the NMR spectrum that were indicative of $Hg^{2+}$ binding to the terminal AT base pairs, in a manner similar to that described by Sletten and co-workers (79, 84). The stability of T-Hg-T complexes has been utilized to introduce mercury-mediated base pairs at specific positions of a duplex (see Section IV.B).

Beside $Hg^{2+}$, other late transition metal ions (e.g., $Ag^+$ and $Au^{3+}$) can coordinate to DNA nucleobases to form interstrand cross-links. Several DNA coordination modes have been suggested for $Ag^+$ ions, including coordination to the purine N1 and the pyrimidine N3 (86). At $Ag^+$/base pair ratios $<0.5$, the metal ion coordinates to the N7 position of purines, in particular to guanine, while at a $Ag^+$/bp ratio of 0.5, $Ag^+$ forms cross-links between the two strands of the duplex by binding to two nucleobases (89). The $Au^{3+}$ ion was also shown to coordinate to solvent-exposed GC base pairs in RNA duplexes to form a G-Au-C bridge that resembles T-Hg-T (88).

## IV. METAL-CONTAINING LIGAND-MODIFIED NUCLEIC ACID DUPLEXES

An interesting approach to incorporate transition metal ions in nucleic acid duplexes is the formation of metal-mediated-base pairs (Fig. 2) (89, 90). This approach was first described by Tanaka and Shionoya (91) and it involves the substitution of nucleobase pairs with ligands that have high affinity for metal ions. Ligands juxtaposed at complementary positions within a nucleic acid duplex form a high affinity binding site for metal ions. The DNA monomers in which the nucleobase was replaced with a ligand were termed by Weizman and Tor (92, 93) ligandosides.

This novel class of metal-DNA hybrid structures is of high interest for several reasons. On one hand, it adds a novel dimension to the growing field of artificial base pairs, which were previously based either on alternative hydrogen-bonding patterns or on hydrophobic interactions (94, 95). The stronger coordinative bonds when compared to that of hydrogen bonds bode for the formation of alternative base pairs whose stability surpasses that of the natural Watson–Crick base pairs and which may be of interest for applications in biotechnology. The extension of the genetic code by using metal-ligand alternative base pairs is also a possibility (96), although this goal can be accomplished in a biological system only if the metal-containing alternative base pairs can be recognized and extended by polymerases. For the inorganic chemist, the interest in such structures is linked to the possibility of using the nucleic acid duplex as a scaffold in which metal ions are incorporated in a rational and well-defined manner. The possibility of bridging the gap between extended metal clusters, which have been produced by materials science, and polynuclear metal complexes, which have been synthesized by supramolecular inorganic chemistry, makes the efforts aimed at using nucleic acids as scaffold for transition metal ions worthwhile.

## A. Design Strategy

Several ligands have been used in the design of metal-containing nucleic acid duplexes (Chart 1 and Table V). The ligands were chosen such that when situated in complementary positions within nucleic acid duplexes, they form square-planar complexes and participate in stacking interactions with the adjacent base pairs. Most of the ligands fulfil these conditions by having 1–3 metal-binding sites that are part of an aromatic system. To ensure that the metal ions are incorporated selectively at the positions where the duplex has been modified, the affinity of the ligands for the metal ions must be higher than that of the natural nucleobases, in particular that of the G bases, which bind in a monodentate fashion through G-N7. Preferably, the metal ions should have the ability to form either square-planar complexes or octahedral complexes in which the ligands occupy equatorial positions and the axial positions may be occupied by donor atoms from the adjacent nucleobases. An important condition that artificial base pairs must fulfil to be useful for the extension of the genetic code

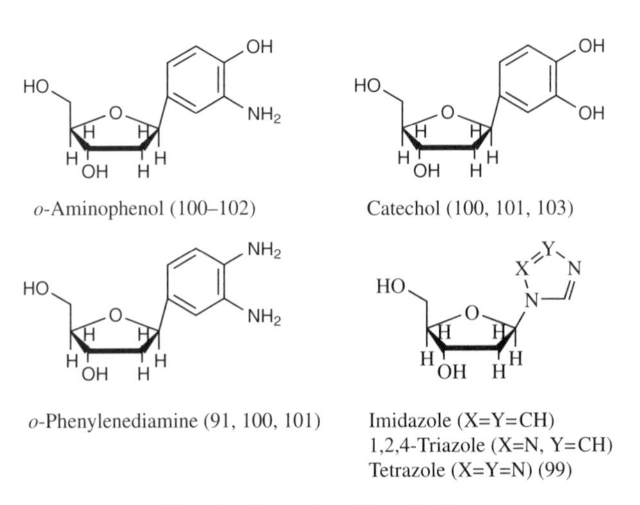


**Chart 1.** Examples of ligandosides.

is to be orthogonal to natural base pairs. Therefore, the metal ion should form complexes only with the extraneous ligands and not mixed-ligand complexes with the natural bases. Incorporation of multiple metal ions in adjacent positions within the duplexes is likely to depend on the overall charge of the metal complex and of the nucleic acid. In DNA duplexes, the phosphodiester backbone can act as an intrinsic counteranion for positively charged metal complexes incorporated in the duplex, and thus can mitigate electrostatic repulsion between adjacent metal–ligand complexes.

## B. Alternative Metal-Ligand Base Pairs Based on Nucleobase Ligands

The studies in which ligands extraneous to the nucleic acid duplexes have been employed to create high-affinity metal-binding sites have been preceded by position-specific, metal ion binding that exploited the demonstrated ability of guanine to act as a monodentate ligand to late 3*d* transition metal ions and of thymines to coordinate $Hg^{2+}$ (see Section III). For example, G overhangs have been introduced at each end of DNA duplexes with the goal of forming G-$M^{2+}$-G complexes, which would contribute to the organization of the duplexes in crystals (Entry 11, Table II) (62). The intended G-$Ni^{2+}$-G complexes did not form when the DNA duplexes were crystallized in the presence of $Ni^{2+}$, which is likely due to the fact that the interactions between the guanine and $Ni^{2+}$ are comparable in strength to other intermolecular forces that govern the crystal formation. Instead, adjacent duplexes were bridged by $[Ni(G_7)(H_2O)_5]^{2+}$ complexes, with the G ligand from one DNA duplex and the coordinated water molecules forming hydrogen bonds to phosphates in the adjacent DNA duplex.

In contrast, the strong binding of $Hg^{2+}$ to T has been used successfully to direct the metal ion binding to T•T mismatches introduced in DNA duplexes (Entries 3–6 in Table IV) (81, 97). Ultraviolet titrations of these duplexes showed stoichiometric binding of 1 $Hg^{2+}$ per T•T pair, confirming that $Hg^{2+}$ binds specifically to these pairs. 1:1 complexes between $Hg^{2+}$ and the T•T-containing DNA duplexes have been also evidenced by electrospray ionization mass spectrometry (ESI MS) (81). Further support for the binding of $Hg^{2+}$ to T-N3 came from NMR titrations of a DNA duplex with two central T•T mismatches (81, 97), which showed that addition of 2 equiv of $Hg^{2+}$ led to the almost complete disappearance of the thymine N3 protons. When considered together with the similarity between the changes in UV spectra observed for mismatch T•T DNA (81) and for d(AT) DNA upon addition of $Hg^{2+}$ (86), these results support the chain slippage model for $Hg^{2+}$ binding to DNA (see Section III.C). The DNA duplex formed by the combination of hydrogen bonding and Hg-T coordination was more stable than the corresponding one based exclusively on Watson–Crick base pairing (Entry 3, Table IV) (81). The effect of multiple T-Hg-T complexes in the DNA duplex was cooperative, as evidenced by the increase in the duplex stabilization with the number of adjacent T-Hg-T alternative base pairs (T•T, $T_m$: 47→56°C; 2T•T, $T_m$: 45→58°C; 3T•T, $T_m$: 43→61°C, Entries 4–6, Table IV) (81).

The affinity of T for $Hg^{2+}$ has been also used to control the DNA structure adopted by sequences that could form either hairpins or duplexes. Kuklenyik and Marzilli (18) designed DNA oligonucleotides that contained stretches of several thymines between two complementary ends. This sequence design allows the oligonucleotide to form either a duplex with a $T_n$ bulge or a hairpin with a $T_n$ loop [Fig. 12(*a*)]. Binding of $Hg^{2+}$ to oligonucleotides containing four consecutive Ts led to intrastrand $T_5$-Hg-$T_8$ cross-links and stabilization of a hairpin [Fig. 12(*b*)]. Even in the absence of $Hg^{2+}$, these oligonucleotides adopted a hairpin structure in which the $T_5$ and $T_8$ bases formed a wobble base pair and were thus prepositioned for binding to a metal ion. The oligonucleotides that contained two or three consecutive T bases formed hairpins in the absence of $Hg^{2+}$, but formed duplexes with central T-Hg-T cross-links in the presence of $Hg^{2+}$ [Fig. 12(*c* and *d*)]. The metal-induced transformation of hairpins into duplexes is likely due to the fact that T-Hg-T intrastrand cross-links pose a high steric constraint on hairpins with short stretches of Ts.

The $Pt^{2+}$ ion, which has high affinity for the G-N7 sites, has been used to direct the stabilization of specific DNA structures by formation of interstrand G-$Pt^{2+}$-G cross-links (98). The reaction of a palindromic, antiparallel DNA duplex with trans-$[Pt(NH_3)_2(H_2O)Cl]^+$, followed by melting and rehybridization led to a DNA duplex that contained a terminal G-$[Pt(NH_3)_2]^{2+}$-G interstrand complex, which is kinetically inert. Analysis of the circular dichroism (CD) spectrum, the thermodynamic parameters for melting determined from UV melting curves, and the nuclear Overhouser effect spectra (NOESY) of the Pt-containing DNA duplex indicated that the duplex was parallel stranded and contained Hoogsteen base pairs. Watson–Crick base pairing of a pyrimidine-rich strand onto this duplex led to the formation of a triplex. Notably, melting of the Watson–Crick base pairs of the triplex preceded melting of the Hoogsteen pairs. This melting order is opposite to that observed for triplexes that contain exclusively natural nucleobases and is determined by the coexistence of the Hoogsteen and coordination bonds within the duplex part of the triplex.

### C. Metal Complexes With Ligandoside Monomers

To date, ligand incorporation in duplexes of DNA or synthetic DNA analogues has been achieved by using monomers in which the nucleobase is substituted with a ligand that has strong metal-binding affinity. Several β-C ligandosides that have been synthesized, but have not been incorporated

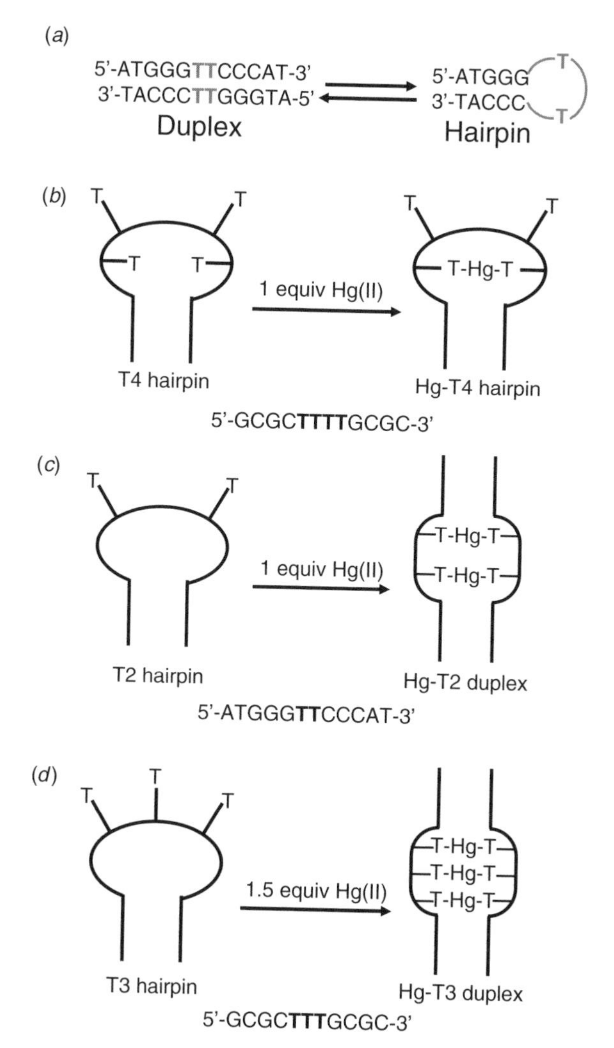


Figure 12. (*a*) Duplex-hairpin equilibrium for a partly self-complementary oligonucliotide; (*b-d*) influence of $Hg^{2+}$ binding on the secondary structure of DNA (adapted from reference 18).

yet in nucleic acid duplexes, contained either the monodentate imidazole, 1,2,4-triazole, or tetrazole ligands (99), or the bidentate *o*-phenylenediamine (91, 100), *o*-aminophenol (100–102), or catechol ligands (100, 101, 103, 104) (Chart 1). The sugar moieties of the *p*-toluoyl-protected 1,2,4-triazole and tetrazole ligandosides adopted a C3′ *endo* conformation in crystals, which is in contrast to the situation usually found in deoxyribonucleosides. This conformation is present in A-type DNA, RNA, and LNA and it would be interesting to determine the structure and stability of DNA and LNA duplexes that contain these ligandosides.

The phenylenediamine- or aminophenol-containing ligandosides formed 2:1 ligand/metal complexes with divalent metal ions (e.g., $Pd^{2+}$, $Cd^{2+}$, or $Zn^{2+}$) (100, 104) and the catechol ligandoside formed a bis(1,2-benzenediolato)borate anion upon reaction with trimethyl borate in dimethyl sulfoxide (DMSO) (103). The latter anion has a tetrahedral geometry (105) and, thus it is likely to have unfavorable steric interactions with the adjacent base pairs in a DNA duplex. The basicity of the ligandosides containing imidazole, 1,2,4-triazole, and tetrazole decreases in this order. The metal-binding affinity parallels this trend and, as a consequence, the imidazole ligandoside formed 2:1 complexes with both $Ag^{+}$ and $Hg^{2+}$, but the tetrazole did not form complexes with these metal ions (99). Based on DFT calculations, the two imidazole ligands have a coplanar arrangement in the $Hg^{2+}$ complex (99). This ligand arrangement favors good stacking interactions between the complex and the neighboring bases. It would be interesting to examine the possibility that the chelate effect due to the positioning of two ligands in close proximity of each other within a duplex could affect the $\beta_{ML2}$ binding constant to the extent that a $Hg^{2+}$–tetrazole complex could form within a nucleic acid duplex.

Of the ligandosides discussed in this section, only the aminophenol-containing one in its benzyl-protected form was incorporated in two complementary DNA strands (Entry 1, Table V). The thermal stability of the DNA duplex formed from these two strands was slightly lower than that of a duplex with an AT base pair, both in the absence and presence of $Cu^{2+}$. This result is not surprising, as the benzyl protection group renders the ligand significantly bulkier than the A or T bases and reduces the binding affinity of the ligand for metal ions (104).

### D. Duplexes That Contain One Alternative Metal–Ligand Base Pair

Incorporation of a pair of ligands in complementary positions in the middle of DNA duplexes creates a metal-binding site of $[n + m]$ type, where $n$ and $m$ are the number of donor atoms through which the two ligands coordinate to the metal ion. Metal binding sites of the $[n + n]$ type are simpler to create because the same ligand-containing phosphoramidite can be used to incorporate the $n$-dentate ligand in each of the two complementary strands of the duplex. In this section, we describe the nucleic acid duplexes that contain metal-mediated alternative base pairs reported to date according to their $[n + m]$ coordination scheme. In this chapter abbreviations for ligands (e.g., bpy) will appear in light face type, while these for ligands attached to a nucleic acid backbone are in boldface (e.g., **bpy**).

TABLE V
Cartoon Representation of Alternative Metal–Ligand or Bis(ligand) Base Pairs Within Duplexes

| Entry | Base Pair Structure<br>Ligand Name (Abbr.) | Sequence | $T_m$ (°C) | Conc. (μ*M*) | References |
|---|---|---|---|---|---|
| 1 | *o*-Bn-2-Aminophenol (**2AP**) | 5′-dA$_{10}$XdA$_{10}$-3′<br>3′-dT$_{10}$XdT$_{10}$-5′<br>X = 2**AP** | 42.5<br>42.5 (1 equiv $Cu^{2+}$) | 1.19 | 104 |
| | | Control: A-T | 46.5 (A-T) | | |
| 2 | Pyridine (**py**) | 5′-CACATTAXTGTTGTA-3′<br>3′-GTGTAATXACAACAT-5′<br>X = **py** | 25.4 (1 equiv $Ag^{+}$) | 1 | 106 |
| | | Control: A-T, C-G | 39.4 (A-T)<br>41.1 (C-G) | | |
| 3 | Pyridine (**py**) | 5′-dT$_{10}$XdT$_{10}$-3′<br>3′-dA$_{10}$XdA$_{10}$-5′<br>X = **py** | 34.2<br>38.0 (1 equiv $Ag^{+}$)<br>41.0 (3 equiv $Ag^{+}$)<br>33.7 (1 equiv $Cu^{2+}$)<br>34.3 (1 equiv $Ni^{2+}$)<br>34.0 (1 equiv $Pd^{2+}$)<br>33.7 (1 equiv $Hg^{2+}$) | 1.2 | 17 |
| | | Control: T-A | 46.5 (T-A) | | |

(*continued*)

TABLE V
*(Continued)*

| Entry | Base Pair Structure Ligand Name (Abbr.) | Sequence | $T_m$ (°C) | Conc. (μ*M*) | References |
|---|---|---|---|---|---|
| 4 | DNA, N, N, N, N, DNA<br>2′2-Bipyridine (**bpy**) | 5′-GATGACXGCTAGCTAGGAC-3′<br>3′-CTACTGXCGATCGATCCTG-5′<br>X = **bpy** | 67.4 | 1.2 | 107 |
| | | Control: A-T, C-G | 64.0 (A-T)<br>66.0 (C-G) | | |
| | | 5′-GATGACX$_n$GCTAG-3′<br>3′-CTACTGX$_n$CGATC-5′<br>X = **bpy** | 48.0 ($n = 1$)<br>42.9 ($n = 2$)<br>40.7 ($n = 3$)<br>39.4 ($n = 4$)<br>37.5 ($n = 5$)<br>35.0 ($n = 6$) | 1.2 | 108, 109 |
| | | Control: $n = 0$ | 45.0 ($n = 0$) | | |
| 5 | DNA, N, $M^{2+}$, N, N, N, DNA<br>5-Methylenebipyridine (**bpy**) | 5′-AGTCGXCGACT-3′<br>3′-TCAGCXGCTGA-5′<br>X = **bpy** | 56.5<br>64 (1 equiv $Cu^{2+}$) | 1 | 92 |
| | | Control (10 mer):<br>5′-AGTCGCGACT-3′<br>3′-TCAGCGCTGA-5′ | 56.5 (10 mer) | | |
| 6 | PNA, N, N, $Ni^{+2}$, N, N, PNA | H-GTAGXTCACT-Lys-$NH_2$<br>$NH_2$-Lys-CATCXAGTGA-H<br>X = **Mebpy** | 48<br>59 (1 equiv $Ni^{2+}$) | 5 | 110, 111 |
| | | Control: A-T | 66.5 (A-T) | | |

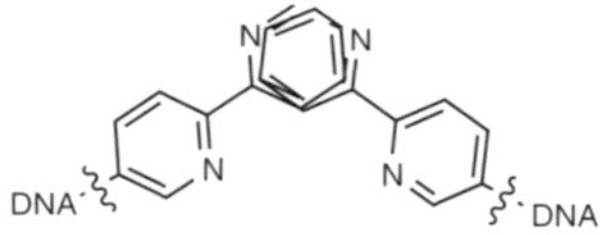

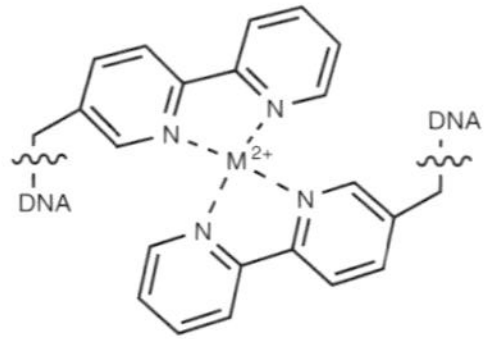

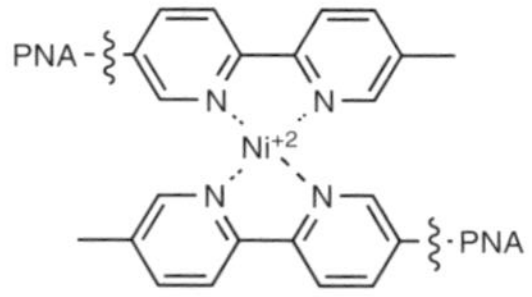

| | | | | | |
|---|---|---|---|---|---|
| 7 | 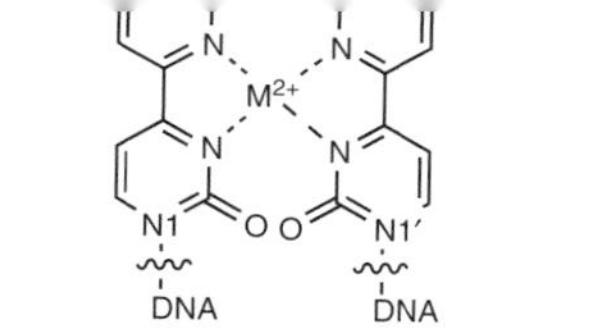 4-(2′-Pyridyl)-pyrimidinone (**Pyr^P^**) | 5′-CTTTCTXTCCCT-3′<br>3′-GAAAGAXAGGGA-5′<br>X = **Pyr^P^** | 24.7<br>41.2 (4 equiv $Ni^{2+}$)<br>29.9 (4 equiv $Co^{2+}$)<br>26.8 (4 equiv $Cu^{2+}$)<br>24.1 (4 equiv $Zn^{2+}$)<br>24.0 (4 equiv $Fe^{2+}$)<br>23.9 (4 equiv $Mn^{2+}$) | 2.5 | 112 |
| | | Control : T-A, C-G | 36.8 (T-A)<br>40.2 (C-G) | | |
| 8 | 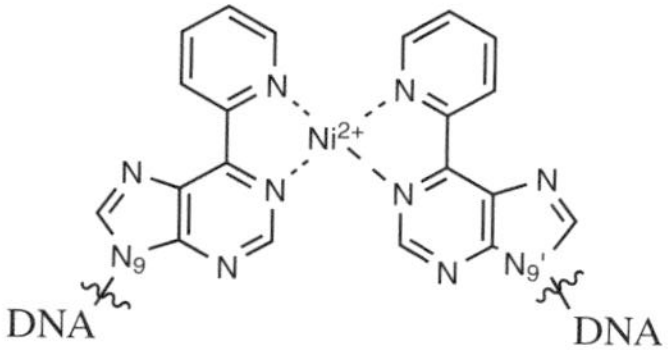 6-(2′-Pyridyl)-purine (**Pur^P^**) | 5′-CTTTCTXTCCCT- 3′<br>3′-GAAAGAXAGGGA-5′<br>X = **Pur^P^** | 28.5<br>46.1 (2 equiv $Ni^{2+}$)<br>38.8 (2 equiv $Co^{2+}$)<br>31.4 (2 equiv $Cu^{2+}$)<br>30.8 (2 equiv $Zn^{2+}$)<br>30.5 (2 equiv $Ag^{+}$)<br>28.8 (2 equiv $Fe^{2+}$)<br>29.2 (2 equiv $Mn^{2+}$)<br>29.1 (2 equiv $Eu^{3+}$)<br>27.3 (2 equiv $Pd^{2+}$) | 2.5 | 113 |
| | | Control: T-A, C-G | 36.8 (T-A)<br>40.1 (C-G) | | |
| | | 5′-CTTTCTXXXTCCCT-3′<br>3′-GAAAGAXXXAGGGA-5′ | 20.6<br>64.3 (1.3 equiv $Ni^{2+}$) | | |
| 9 | 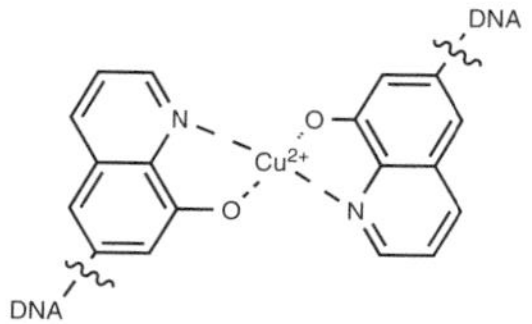 8-Hydroxyquinoline (**Q**) | 5′-CACATTAQTGTTGTA-3′<br>3′-GTGTAATQACAACAT-5′ | 36.1<br>65 (1 equiv $Cu^{2+}$) | 2.0 | 114 |
| | | Control: A-T, G-C<br>Q-N (N = A, T, C, G) | 41.3 (A-T)<br>44.6 (G-C)<br>34.7 (Q-A)<br><30 (Q-T)<br>32.5 (Q-C)<br>35.4 (Q-G) | | |

*(continued)*

TABLE V
(*Continued*)

| Entry | Base Pair Structure<br>Ligand Name (Abbr.) | Sequence | $T_m$ (°C) | Conc. (μ*M*) | References |
|---|---|---|---|---|---|
| 10 | 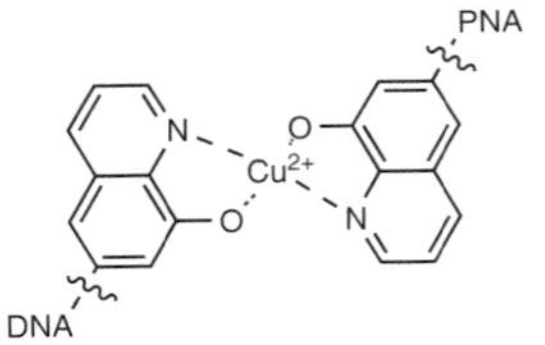 8-Hydroxyquinoline (**Q**) | H-GTAGQTCACT-Lys-$NH_2$<br>$H_2N$-Lys-CATCQAGTGA-H | 46<br>>79 (1 equiv $Cu^{2+}$) | 5 | 115 |
| | | H-GTAGQTCACT-Lys-$NH_2$<br>$H_2N$-Lys-TCACTQGATG-H | 72 (1 equiv $Cu^{2+}$) | | |
| | | H-AGTGAQCTAC-Lys-$NH_2$<br>$H_2N$-Lys-CATCQAGTGA-H | >79 (1 equiv $Cu^{2+}$) | | |
| | | H-TGAGQTCACT-Lys-$NH_2$<br>$H_2N$-Lys-TCACTQGAGT-H | 77 (1 equiv $Cu^{2+}$) | | |
| | | H-TGACQTGACT-Lys-$NH_2$<br>$H_2N$-Lys-TCAGTQCAGT-H | 66 (1 equiv $Cu^{2+}$) | | |
| | | H-ACACQACACA-Lys-$NH_2$<br>$H_2N$-Lys-ACACAQCACA-H | n.d. (1 equiv $Cu^{2+}$) | | |
| | | Control: A-T | 67 (A-T) | | |
| 11 | 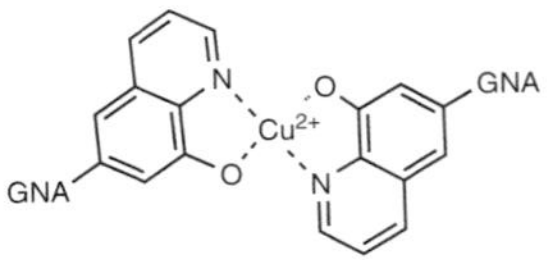 8-Hydroxyquinoline (**Q**) | 5′-CACATTAQTGTTGTA-3′<br>3′-GTGTAATQACAACAT-5′ | n.d. w/o $Cu^{2+}$<br>70.5 (1 equiv $Cu^{2+}$) | 2.0 | 114 |
| | | Control: A-T, G-C | 41.3 (A-T)<br>44.6 (G-C) | | |
| 12 | 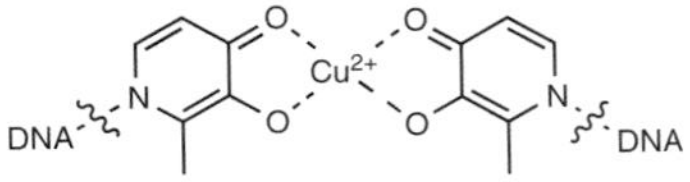 Hydroxypyridone (**H**) | 5′-CACATTAHTGTTGTA-3′<br>3′-GTGTAATHACAACAT-5′ | 37.0<br>50.1 (1 equiv $Cu^{2+}$) | 2.0 | 116 |
| | | Control: A-T | 44.2 (A-T) | | |
| | | 5′-G(H)$_n$C-3′<br>3′-C(H)$_n$G-5′ | | 2.0 | 117 |

| | | | | | |
|---|---|---|---|---|---|
| 13 | DNA–salen–DNA ($M^{2+}$ complex)<br>*N,N′*-Bis(salicylidene)ethylendiamine (**salen**) | 5′-CACATTAXTGTTGTA-3′<br>3′-GTGTAATXACAACAT-5′<br>X = **salen** | 40.6<br>82.4 (1 equiv $Cu^{2+}$, 33 equiv en)<br>54.9 (1.3 equiv $Cu^{2+}$)<br>52.3 (1.3 equiv $Cu^{2+}$, 67 equiv $MeNH_2$)<br>68.8 (1 equiv $Mn^{2+}$, 33 equiv en)<br>40.7 (2 equiv $Mn^{2+}$)<br>48.8 (133 equiv $Zn^{2+}$, 33 equiv en)<br>36.5 (133 equiv $Ni^{2+}$, 33 equiv en) | 3.0 | 118 |
| | | Control: A-T | 50.1 (A-T) | | |
| 14 | **Dipic · py**<br>Pyridine-2,6-dicarboxylate (**Dipic**) | 5′-CACATTAYTGTTGTA-3′<br>3′-GTGTAATXACAACAT-5′<br>X = **Dipic**, Y = **py** | n.d. w/o $Cu^{2+}$<br>38.6 (1 equiv $Cu^{2+}$)<br>39.5 (2 equiv $Cu^{2+}$)<br>40.1 (5 equiv $Cu^{2+}$) | 2 | 96,120 |
| | | Control: A-T | 41.1 (A-T) | | |
| | | 5′-CACATTXYTGTTGTA-3′<br>3′-GTGTAAYXACAACAT-5′ | 38.5 (15 equiv $Cu^{2+}$) | | |
| | | 5′-CACATPYXXYGTTGTA-3′<br>3′-GTGTAXYYXCAACAT-5′ | 39.6 (15 equiv $Cu^{2+}$) | | |

*(continued)*

TABLE V
(*Continued*)

| Entry | Base Pair Structure Ligand Name (Abbr.) | Sequence | $T_m$ (°C) | Conc. (μ*M*) | References |
|---|---|---|---|---|---|
| 15 | 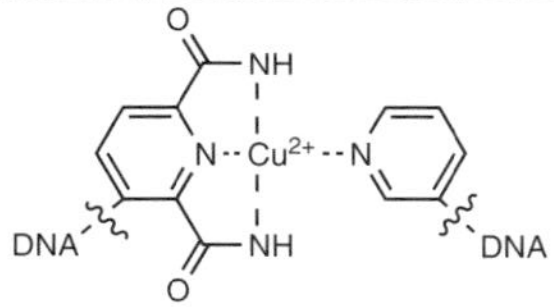 **Dipam · py**<br>Pyridine-2,6-dicarboxamide (**Dipam**) | 5′-CACATTAXTGTTGTA-3′<br>3′-GTGTAATYACAACAT-5′<br>X = **Dipam**, Y = **py** | 28.0<br>43.0 (15 equiv $Cu^{2+}$) | 1 | 120 |
| | | 5′-CACATTXYTGTTGTA-3′<br>3′-GTGTAAYXACAACAT-5′ | 43.4 (15 equiv $Cu^{2+}$) | | |
| | | 5′-CACATYXXYGTTGTA-3′<br>3′-GTGTAXYYXCAACAT-5′ | 47.3 (15 equiv $Cu^{2+}$) | | |
| 16 | 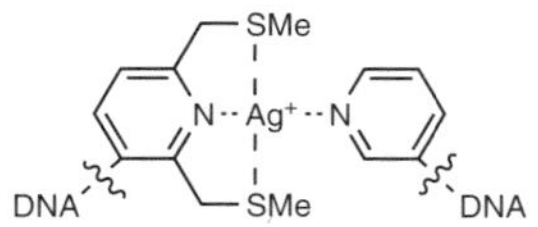 **SPy · py**<br>2,6-Bis-(ethylthiomethyl) pyridine (**SPy**) | 5′-CACATTAXTGTTGTA-3′<br>3′-GTGTAATYACAACAT-5′<br>X = **SPy**; Y = **py** | 23.5<br>35.0 (1 equiv $Ag^{+}$) | 1 | 106 |
| | | Control: A-T, C-G | 39.1 (A-T)<br>39.6 (C-G) | | |
| | | 5′-CACAXTACTGXTGTA-3′<br>3′-GTGTYATGACYACAT-5′ | 37.0 (2 equiv $Ag^{+}$) | | |
| | | 5′-CACXTTAYTGTXGTA-3′<br>3′-GTGYAATXACAYCAT-5′ | 25.6 (3 equiv $Ag^{+}$) | | |
| 17 | 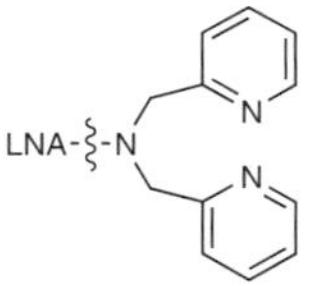 *N,N*-bis(2-pyridylmethyl)-amine (**2PA**) | 5′-GTGAXATGC-3′<br>3′-CACTATACG-5′<br>(LNA duplex)<br>X = **2PA** | 34<br>40 (1 equiv $Ni^{2+}$)<br>36 (1 equiv $Cu^{2+}$)<br>38 (1 equiv $Zn^{2+}$) | 1 | 121 |
| | | 5′-GTGAXAXGC-3′<br>3′-CACTATACG-5′ | 38<br>51 (1 equiv $Ni^{2+}$) | | |

| | | | | | |
|---|---|---|---|---|---|
| | | 5′-GTGAXATGC-3′<br>3′-CACXATACG-5′ | 34<br>53/24 (1/10 equiv $Ni^{2+}$)<br>40/20 (1/10 equiv $Cu^{2+}$)<br>36/31 (1/10 equiv $Zn^{2+}$) | | |
| | | 5′-GTGAXATGC-3′<br>3′-CACTAXACG-5′ | 39<br>47/50 (1/10 equiv $Ni^{2+}$)<br>42/39 (1/10 equiv $Cu^{2+}$)<br>46/47 (1/10 equiv $Zn^{2+}$) | | |
| | | 5′-GTGAXAXGC-3′<br>3′-CACTXTXCG-5′ | 47<br>48/20 (1/10 equiv $Ni^{2+}$)<br>56/<10 (1/10 equiv $Cu^{2+}$)<br>50/37 (1/10 equiv $Zn^{2+}$) | | |
| | | Control : 5′-GTGATATGC-3′<br>3′-CACTATACG-5′ | 28 ( w/ or w/o M) | | |
| 18 | 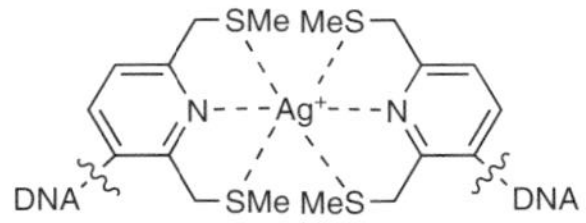 SPy · SPy | 5′-CACATTAXTGTTGTA-3′<br>3′-GTGTAATXACAACAT-5′<br>X = **SPy** | 42.5 (1 equiv $Ag^{+}$) | 1 | 106 |
| | | Control: A-T, C-G | 39.1 (A-T)<br>39.6 (C-G) | | |
| | | 5′-CACAXTACTGXTGTA-3′<br>3′-GTGTXATGACXACAT-5′ | 39.1 (2 equiv $Ag^{+}$) | | |
| | | 5′-CACXTTAXTGTXGTA-3′<br>3′-GTGXAATXACAXCAT-5′ | 44.6 (3 equiv $Ag^{+}$) | | |

[a]n.d. = not detected

[b]All $T_m$ values have been measured in phosphate buffer unless otherwise noted. If present during $T_m$ measurement, transition metal ions are specified in parentheses. Ligands are shown including the linkers used to connect them to the nucleic acid backbone (e.g., the C1' of DNA and C8' of PNA) (see Fig. 4 for atom numbering information). Concentration is of duplex.

### *1. [1+1] Coordination*

The $Ag^+$ ions can form linear, two-coordinate complexes with monodentate ligands, including pyridine (py). The binding constants of py to $Ag^+$ are low, and an $[Ag(py)_2]^+$ complex does not form in solutions containing micromolar concentrations of $Ag^+$ and py (122). Nevertheless, at this concentration range, a $[Ag(\mathbf{py})_2]^+$ complex formed within DNA duplexes that contained a pair of **py** in complementary positions, which can be attributed to a duplex-induced, supramolecular chelate effect (Entry 2, Table V) (17, 106). The coordination of 1 equiv of $Ag^+$ to the pair of **py** was confirmed by $^1$H NMR titrations, which showed that the $Ag^+$ binding exclusively affected the protons of the **py** and not those of the natural nucleobases. The $Ag^+$ ion coordination to **py** had a stabilization effect on the modified DNA duplexes that depended on the sequence and length of the DNA duplex. For example, $Ag^+$ had no effect on the stability of a 15 bp, **py**-modified DNA duplex (106), but increased the stability of a 21 bp, **py**-modified DNA duplex (17) (Entries 2 and 3, Table V). The stabilization effect of $Ag^+$, on the latter duplex increased with the $Ag^+$ concentration up to 3 equiv of $Ag^+$. Linear [1 + 1] complexes of azole and purine ligands have been proposed as potential alternative base pairs, but ligandosides with these ligands have not been synthesized yet (Chart 1) (123, 124).

The $Ag^+$ can also form three-coordinate trigonal complexes. This property was used to create an $[Ag(\mathbf{py})_3]^+$ complex within a homopurine–pyrimidine DNA triplex in which each DNA strand had a central **py** ligand (17). Specific binding of $Ag^+$ to the **py**-containing triplex was inferred based on the observation that the melting temperature for the triplex–duplex transition increased in the presence of $Ag^+$, while $Ag^+$ lowered the melting temperature for a triplex without **py** modification. This is one of the very few examples of metal-containing nucleic acid triplex. Other metal-containing triplexes currently reported contain a platinum or ruthenium ion, which did not play a determinant role in the triplex formation (125, 126) or were formed by template synthesis of a metal-containing single strand on a DNA duplex template (see Section III.2).

### *2. [2+2] Coordination*

Several groups have developed metal-mediated alternative base pairs with [2 + 2] coordination by using the bidentate ligands bipyridine, 8-hydroxyquinoline, hydroxypyridone, or salicylaldehyde. *N, N′*-Bis(salicylidene)ethylenediamine (salen), a four-dentate ligand that coordinates through two nitrogen and two oxygen atoms, is also included in this section because of the similarity between its coordination complexes and those with bis(salicyladehyde) coordination.

**a. Bipyridine and Related Aromatic, Bis(imine) Ligands** 2,2′-Bipyridine (bpy) has been used most extensively to incorporate metal ions in nucleic acids because of its high affinity for a variety of transition metal ions (127) and its ability to π-stack efficiently with adjacent base pairs. For incorporation in ssDNA, bpy has been connected through its 5 position, either directly (107) or through a methylene linker (92), to the 1′ position of the 2′-deoxy-D-ribose to obtain a β-isomeric C-ligandoside (Entries 4 and 5, Table V). The methylene linker increases the conformational flexibility of **bpy** and it can increase the entropic penalty for the formation of the **bpy**-modified duplex. This inference is supported by experiments with **bpy**-modified DNAs. For example, in experiments by Brotschi et al. (107), a 19 bp duplex with a central **bpy•bpy** pair in which the two **bpy** were directly attached to the 2′-deoxyribose was as stable as a duplex with a GC pair in place of the **bpy•bpy** pair (Entry 4, Table V), although the two **bpys** cannot form hydrogen bonds and are not complementary in shape. Weizman and Tor (92), incorporated a central **bpy•bpy** pair in a 10 bp DNA duplex to create an 11 bp DNA duplex, where the bpys were attached to the DNA backbone through a methylene linker. The 11 bp, **bpy**-containing duplex was as stable as the nonmodified, 10 bp DNA duplex, although in general, extension of a DNA duplex by a base pair leads to an increase in the duplex stability. This result suggests that the **bpy•bpy** pair connected through the methylene linkers to the duplex backbone has a lower stabilization effect than a Watson–Crick base pair, and thus than a **bpy•bpy** pair directly attached to the DNA backbone. Nevertheless it is possible that the effect of **bpy** substitution depends also on the sequence and/or the length of the DNA duplex, because these parameters were different in the experiments by Weizman and by Brotschi.

5′-Methyl-bpy (Mebpy) and bpy have been introduced in PNA duplexes (Entry 6, Table V) (16, 110, 111). The methyl group did not affect the stability of the PNA double helix (16). A 10 bp PNA duplex with a central pair of **bpys** was significantly less stable than a PNA duplex with the same sequence, but a central AT base pair. The destabilization effect of **bpy** on the PNA duplexes was larger than the effect exerted by the same ligand in DNA duplexes. This finding correlates with the superior mismatch discrimination of PNA over DNA.

An increase in the number of adjacent **bpy** pairs introduced in DNA or PNA duplexes led to a systematic decrease of the duplex stability, as did the pairing of **bpy** against either a natural base or an abasic site (16, 108, 109). Brotschi proposed that incorporation of consecutive **bpy** pairs in adjacent positions in DNA creates a zipper-like interstrand stacking motif with a stretched phosphate backbone and all **bpys** having their distal rings involved in π-stacking interactions (see schematic representation in Entry 4, Table V) (109).

Studies of PNA and DNA duplexes that contained a **bpy** opposite a natural nucleobase showed that metal ion coordination has either no effect or causes a decrease in these duplexes' thermal stability (110, 128). This result suggests that the metal ion does not form an alternative base pair with mixed **bpy**- nucleobase coordination. The decrease in the melting temperature may be due to coordination of the metal ions to **bpy** only, which causes a distortion of the duplex and/or a loss of stacking interactions.

In contrast, when a pair of **bpys** was placed in complementary positions in DNA or PNA duplexes, addition of certain metal ions led consistently to duplex stabilization with respect to the metal-free, ligand-containing duplex. The $Cu^{2+}$ ion coordinated two **bpys** situated in a DNA duplex and the $Cu^{2+}$-**bpy** bonds stabilized the DNA duplex (92). As a consequence, addition of 1 equiv of $Cu^{2+}$ to an 11 bp DNA duplex with a central pair of **bpys** led to (a) a significant increase in the melting temperature of the duplex; (b) an increase in the slope of the thermal denaturation curve indicative of increased cooperativity; and (c) the appearance of a new absorption band at ~305 nm, which is characteristic of $\pi$–$\pi^*$ transitions of the coordinated bpy (Entry 5, Table V) (92). Addition of 1 equiv of $Ni^{2+}$ to 10 bp PNA duplexes with a central pair of **Mebpys** partially restored the loss in thermal stability caused by substitution of an AT base pair with the **bpy** ligands (110) (Entry 6, Table V). The UV titrations monitored at 320 nm, the wavelength where the $Ni^{2+}$-coordinated **bpy** has a $\pi$–$\pi^*$ band, demonstrated that one $Ni^{2+}$ ion binds to two **Mebpy** ligands (110, 111). Addition of $Fe^{2+}$, $Zn^{2+}$, $Ag^{+}$, $Pd^{2+}$, or $Pt^{2+}$ did not significantly affect the thermal stability of the duplex. The position in the PNA duplex of the ligand substitution moderated the effect of the ligand substitution and of the metal binding to the ligands on the duplex thermal stability (16). This position dependence occurs because the nucleobase pairs are subject to fraying when situated close to the end of the duplex, but not when situated in the middle of the duplex. Consequently, substitutions of the natural base pairs with alternative base pairs that are less stable have smaller impact at the end of the duplex than in the middle of the duplex, and substitutions with more stable alternative base pairs have a stronger impact close to the end of the duplex than in the middle of the duplex. Circular dichroism spectra of the centrally **bpy**-modified PNA duplexes, in the absence or presence of $Ni^{2+}$, showed a pattern characteristic for a left-handed PNA helix (110). This handedness is induced by an L-Lysine situated at the C-terminus of the PNA strands that form the duplex. Two additional spectral features at 290–320 nm appeared in the CD spectrum upon $Ni^{2+}$ binding, suggesting that the duplex exerts a chiral induction effect onto the prochiral $[Ni(\textbf{Mebpy})_2]^{2+}$ complex.

Two ligand-containing nucleosides that have the same cis-diimine metal-binding site as **bpy** have been synthesized and used in the synthesis of ligand-modified DNA duplexes (112, 113). 4-(2′-Pyridyl)-pyrimidinone (**Pyr$^{P}$**) (Entry 7, Table V) and 6-(2′-pyridyl)-purine (**Pur$^{P}$**) (Entry 8, Table V) are formally obtained by substitution of the 4-amino group of cytosine and of the 6-amino group of adenine with Py, respectively. Incorporation of a pair of **Pyr$^{P}$** or **Pur$^{P}$**

ligands in the middle of a 12 bp DNA duplex reduced the duplex stability below that of duplexes containing an AT or a GC base pair in the same position. The decrease in stability for duplexes containing **Pyr$^P$** ligands was larger than that for **Pur$^P$**-modified DNA duplexes, which could be a result of better π-stacking for the larger **Pur$^P$** ligands.

The increase in the stability of ligand-modified duplexes due to metal-binding was the same for **Pyr$^P$** or **Pur$^P$** despite the fact that the N1–N1′ (Entry 7, Table V) distance is significantly shorter in $[Ni(\mathbf{Pyr^P})_2]^{2+}$ (4.9 Å) than the N9–N9′ (Entry 8, Table V) distance in $[Ni(\mathbf{Pur^P})_2]^{2+}$ (9.54 Å), which in turn is similar to the N9–N1′ distance (Fig. 1) in a natural base pair (9.1–9.6 Å). The formation of the $[Ni(\mathbf{Pyr^P})_2]^{2+}$-containing DNA duplex despite this structural dichotomy was attributed to (a) stacking interactions between the ligands and the bases, and (b) the coordinative bonds between $Ni^{2+}$ and **Pyr$^P$**, which may be strong enough to overcome the steric effect of the small complex size.

The metal induced stabilization of **Pyr$^P$**- or **Pur$^P$**-modified duplexes decreased according to the sequence $Ni^{2+} > Co^{2+} > Cu^{2+} > Fe^{2+} \sim Zn^{2+} \sim Mn^{2+}$, which was similar to that observed for a 10 bp PNA duplex that contained a **bpy** pair in the penultimate position of the duplex (16). This trend correlates with that of the increasing binding constants between the $M^{2+}$ ions and bpy, with the exception of $Cu^{2+}$, which based on the binding constant would be expected to induce an increase in $T_m$ larger than that induced by $Co^{2+}$.

The stabilization conferred to the DNA or PNA duplexes by the $[ML_2]$ complexes with **bpy**, **Pur$^P$**, or **Pyr$^P$** ligands is at most equal to that of a GC base pair, despite the fact that the complexes contain coordination bonds, which are stronger than hydrogen bonds. This discrepancy may be due to the fact that in $[ML_2]$ complexes, the two aromatic cis-diimine ligands cannot be strictly coplanar, because of the steric clash between the 6,6′-protons of the carbon atoms adjacent to the imine nitrogens (129). Consequently, these complexes undergo either a tetrahedral or a bow-step distortion (Fig. 13). In general, π-stacking interactions between the $[M(bpy)_2]$ complexes in the solid state favor the latter distortion. The DFT calculations for the $[Ni(\mathbf{Pur^P})_2]^{2+}$ predict a bow-step square-planar geometry, which may be further reinforced by stacking interactions between the **Pur$^P$** ligands and adjacent nucleobases (113).

**b. 8-Hydroxyquinoline** 8-Hydroxyquinoline (Q) has been used for the efficient incorporation of metal ions into DNA, PNA, and GNA duplexes (Entries 9–11 in Table V). In this section, we understand by GNA duplex, a DNA duplex that contains one base pair with a C3 glycol backbone (Fig. 4). The phenolic group and the aromatic nitrogen of Q are in an optimal 1,4-relationship for coordinating metal ions and as a consequence, the affinity of Q for metal ions is very high. This ligand is particularly amenable for creating alternative base pairs because it can form neutral $[MQ_2]$ complexes with divalent metal ions, which have a rigorously

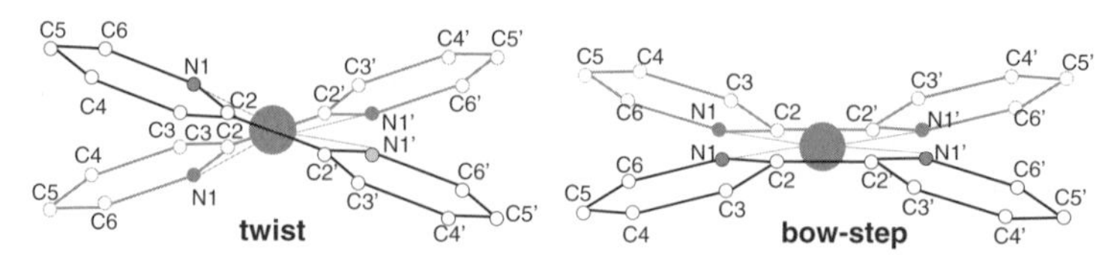


Figure 13. Schematic representations of typical distortions in $[M(bpy)_2]^{2+}$ complexes (adapted from Reference 129).

square-planar geometry. The fact that the $[MQ_2]$ complexes with divalent metal ions are neutral bodes well for the incorporation of multiple, adjacent [M**Q**$_2$] complexes within a nucleic acid duplex. Furthermore, the extended hydrophobic and aromatic surface of the ligand is ideal for $\pi$ stacking.

Substitution of an AT base pair with a pair of **Q** ligands in the middle of DNA, PNA, or GNA duplexes led to a significant or complete loss of stability for the duplexes, which suggests that the two ligands are not hydrogen bonded at pH 7 (Entries 9–11, in Table V) (114, 115). Addition of $Cu^{2+}$ increased the thermal stability of **Q**-DNA, **Q**-PNA, and **Q**-GNA duplexes by >29°C. Interestingly, in the presence of $Cu^{2+}$, the melting curve for **Q**-containing PNA duplexes was similar to that measured for **Q**-containing GNA duplexes and close to that for **Q**-containing DNA duplexes. This similarity suggests that the strong coordinative bonds in the [M**Q**$_2$] alternative base pair are the dominant factor for the stability of the duplexes. In the presence of $Cu^{2+}$, the thermal stability of DNA or GNA duplexes that contain a natural base across from **Q** was significantly lower than that of duplexes containing [Cu**Q**$_2$] (114). This indicates that the [Cu**Q**$_2$] alternative base pair is highly specific, irrespective of the nucleic acid backbone.

In the presence of $Cu^{2+}$, partly self-complementary PNA single strands that contained a **Q** ligand formed duplexes with melting temperatures close to those of corresponding [Cu**Q**$_2$]-containing PNA duplexes formed from fully complementary oligomers (Entry 10, Table V) (115). The high duplex stability is likely due to the fact that [Cu**Q**$_2$] complexes bridge the nucleic acid strands at all temperatures, reducing the loss of translational entropy associated with the duplex formation. Indeed, titrations with $Cu^{2+}$ of PNA duplexes that contained two **Q** ligands in complementary positions showed that [Cu**Q**$_2$] complexes form at a $Cu^{2+}$:**Q**-PNA oligomers ratio of 1:2, irrespective of the temperature (25–95°C) (115). As a result, the gain in enthalpy from $\pi$ stacking of several, but not all base pairs, is a sufficient driving force for duplex formation, even in the absence of Watson Crick complementarity between several base pairs in the duplex.

The $Cu^{2+}$-containing, **Q**-modified GNA or PNA duplex adopted a structure similar to that of duplexes made exclusively from Watson–Crick base pairs, as indicated by CD spectroscopy (36, 112). In the case of **Q**-PNA, only the fully complementary duplexes showed a CD spectrum in the absence or in the presence of $Cu^{2+}$, suggesting that structural differences due to the presence of mismatches prevent the transmission of the chiral induction effect of the L-lysines situated at the C-terminus of the PNA strands. Nevertheless, the Electron paramagnetic resonance (EPR) spectra of the **Q**-PNA duplexes formed from partly complementary strands were similar to each other and to the spectrum of the fully complementary **Q**-PNA duplex (115). They also confirmed the coordination of two **Q** ligands to $Cu^{2+}$. The similarity of the EPR spectra suggests that the environment created by the duplexes around the [Cu**Q**$_2$] complexes is conserved irrespective of the degree of complementarity of the two strands that form the duplex.

**c. Hydroxypyridone** Hydroxypyridone (H) can form neutral, square-planar [$MH_2$] complexes with divalent metal ions (130–132). Tanaka et al. (116) synthesized an **H**-containing phosphoramidite and used it to incorporate H ligands into DNA oligomers (Entry 12, Table V) (116). Titration of the corresponding hydroxypyridone ligandoside with $Cu^{2+}$ monitored by UV spectroscopy (116) showed the concomitant disappearance of the free hydroxypyridone-based $\pi$–$\pi^*$ transitions and appearance of $\pi$–$\pi^*$ transitions characteristic for the deprotonated ligand (130). Spectral changes at both wavelengths ceased at a $Cu^{2+}$/**H** ratio of 1:2, confirming the formation of a [Cu**H**$_2$] complex, which was also supported by mass spectrometry (MS) (116). As observed for other ligand-modified duplexes, replacement of an AT base pair in the middle of a DNA duplex with a pair of **H** ligands decreased the stability of the DNA duplex. Binding of $Cu^{2+}$ to the **H** ligands led to a stabilization of the duplex with respect to both the ligand-modified duplex and the duplex that contained an AT base pair instead of the pair of **H** ligands.

**d. Salen** Schiff bases are versatile ligands and have a high affinity for metal ions. The salen bis(salicylidene)ethylenediamine version is widely used for its catalytic properties. The formation of salen from two salicylaldehyde and one ethylenediamine (en) is a reversible process and coordination of a metal ion is generally required as the driving force for the reaction to occur in an aqueous environment. The two aromatic rings of salen are coplanar in complexes of several metal ions, and thus well positioned for $\pi$–$\pi$ interactions with adjacent nucleobase pairs. Conversely, the preorganization of the two salicylaldehyde (**sal**) ligands inside the duplex provides a driving force for the formation of **salen** in dilute samples (140). Indeed, room temperature CD spectra of an unmetalated **sal**-modified duplex and of duplexes that contained a [Cu(**sal**)$_2$] or a [Mn(**sal**)$_2$] complex, showed that they have a regular B-DNA structure (118), and thus that there is steric compatibility between the pairs of ligands or the complexes, respectively, and the duplex.

An acetal-protected salicylaldehyde phosphoramidite with the ligand connected directly to the $C_1$’ position of the deoxyribose was used in the synthesis of complementary DNA strands, each having one ligand in the middle of the sequence (Entry 13, Table V) (118). The DNA duplex that had a central pair of sal ligands was less stable than a corresponding duplex with an AT base pair, which is in agreement with the observations made for all DNA duplexes in which a pair of ligands substitutes a nucleobase pair. At pH 9.0, the stability of the **sal**-containing DNA duplex increased in the presence of $Cu^{2+}$ or of $Cu^{2+}$ and methylamine, a gain that was attributed to the formation of an interstrand, square-planar [Cu(**sal**)$_2$] or $Cu^{2+}$–bis(hydroxo-imine) complex, respectively.

Comparison of the stabilization effect exerted on the **sal**-containing duplex by either $Cu^{2+}$ ($\Delta T_m = 15°C$) or by en ($\Delta T_m = 5.6°C$) with the effect of the combination of $Cu^{2+}$ and en ($\Delta T_m > 40°C$), which leads to a [Cu(**salen**)] complex, indicated a cooperative stabilization effect exerted by the metal coordination and ethylene cross-link (118). The stabilization of the duplex upon formation of [M(**salen**)] complexes depended importantly on the metal ion. A stoichiometric amount of $Mn^{2+}$ produced a large duplex stabilization, similar to that of $Cu^{2+}$, but $Zn^{2+}$ and $Ni^{2+}$ caused a relatively small stabilization and only if added in excess to duplex solutions containing en (118). The effect of the pair of **sal** ligands, [Cu(**sal**)$_2$], **salen**, or [Cu(**salen**)] on the stability of the duplex depended on the point of attachment of the **sal** ligand to the DNA backbone (133). The largest stabilization was observed for a DNA duplex that included the [Cu(**salen**)] complex attached to the sugar backbone in such a way that the distance between the two $C_1$ atoms that link the complex to the two strands is similar to the corresponding distance for an AT base pair. Also, the melting temperature of the DNA duplexes that contained [Cu(**salen**)] or [Mn(**salen**)]$^+$ complexes decreased with the number of Watson–Crick base pairs formed within the duplex (133).

### 3. *[3+1] Coordination*

The first metal–ligand alternative base pair to be incorporated in DNA duplexes was of [3 + 1] type and was reported by Meggers et al. in 2000 (96). This pair was based on tridentate pyridine-2,6-dicarboxylate (**Dipic**) and monodentate pyridine (**py**) ligands, which were introduced in complementary positions in the middle of a DNA duplex (Entry 14, Table V). The same group reported the incorporation in DNA duplexes of three other tridentate ligands that bear structural similarity to **Dipic**, namely, pyridine-2,6-dicarboxamide (**Dipam**) (Entry 15, Table V) (120), pyridine-2,6-(N-methyl-)dicarboxamide (**Me-Dipam**) (120), and 2,6-bis(ethylthiomethyl)pyridine (**SPy**) (Entry 16, Table V) (106). In the absence of a metal ion, incorporation in DNA duplexes of any of the four tridentate ligands opposite a pyridine had a destabilization effect similar to that of a mismatch or completely prevented the formation of a duplex.

Carboxamide ions, of which Dipam is an example, are the conjugate bases of amides and coordinate to metal ions through the amide nitrogen. For example, in neutral solutions of picolinamide and $Cu^{2+}$ or $Ni^{2+}$, a complex forms in which the ligand is deprotonated and coordinated to the metal ion through the pyridine and amide nitrogen (134, 135). The methyl carboxamide group in Me-Dipam has a higher $pK_a$ than the carboxamide group in Dipam. Thus the protonation equilibrium at neutral pH is shifted toward the neutral form of the ligand, and reduces the ligand's metal binding affinity. This reduction is further accentuated by the steric effect of the methyl group. These considerations explain the relative stability of DNA duplexes containing one of the three tridentate ligands across **py** in the presence of $Cu^{2+}$, which decreases in the order [M(**Dipam**)-**py**] > [M(**Dipic**)**py**] > [M(**Me-Dipam**)**py**] (Entries 14 and 15, Table V). In the presence of $Cu^{2+}$, the stability of the duplexes that contained a central **Dipic•py** or a **Dipam•py** pair of ligands was similar or higher than that of duplexes with all-natural base pairs (96, 120).

Interestingly, to achieve stabilization of the modified duplex, a minimum amount of 15 equiv of $Cu^{2+}$ per duplex was required for the **Dipam•py**-modified double helix, although for the apparently less stable **Dipic•py** duplex, 1 equiv of $Cu^{2+}$ was sufficient. Possible explanations for this observation could be that the binding constant of $Cu^{2+}$ to **Dipic•py** is higher than that to **Dipam•py**, and that either (a) the binding constant does not correlate directly with the stabilization of the duplex, or (b) small steric (or electronic) factors related to the presence of the proton on the carboxamide of **Dipam** have a subtle influence on the duplex stability.

In the presence of $Cu^{2+}$, a surprisingly high melting temperature was measured for a duplex containing **Dipic** opposite to an adenine (96). This duplex had a melting temperature comparable to that of the duplex with a **Dipic**-$Cu^{2+}$-**py** pair, and significantly higher than that for duplexes that contained an AG or CC mismatch. It would be interesting to investigate the EPR spectrum of this duplex to determine if the metal ion forms a mixed-ligand complex with the **Dipic•**A pair. Also, the influence of $Cu^{2+}$ on the stability of a duplex with a **Dipic** ligand could be even stronger when **Dipic** would be paired to a purine that lacks the sterically demanding amino group at the N6 position. EPR spectroscopy would be particularly useful in answering this question. This method has been already used to demonstrate quantitative binding of $Cu^{2+}$ to DNA duplexes that contained a central **Dipic•py** pair of ligands (96). The EPR spectrum was rhombic, which is characteristic of a complex with square-planar geometry. The $g$ and hyperfine $A_{Cu}$ values for $Cu^{2+}$ and the superhyperfine $A_N$ values determined by simulation of the spectrum were close to those measured for the synthetic [Cu(Dipic)(py)] complex (136). The square-planar geometry of the [3 + 1] $Cu^{2+}$ complex was further confirmed by a crystal structure of a related DNA duplex, which contained two isolated **Dipic**-$Cu^{2+}$-**py** alternative base pairs (119) (see Section IV.E). The crystal structure has shown additional axial coordination by neighboring nucleotides to $Cu^{2+}$.

By virtue of its coordination through sulfur, the SPy ligand is a soft Lewis base and has a high affinity for soft metal ions. Addition of 1 equiv of $Ag^{+}$ to a duplex that contained an **SPy•py** pair led to an increase of the duplex stability compared to that of the metal-free, ligand-modified duplex (Entry 16, Table V). Although this ligand pair was designed specifically to bind a soft metal ion, addition of other soft Lewis acids (e.g., $Pd^{2+}$, $Pt^{2+}$, or $Au^{3+}$) had no effect on the duplex stability. The selectivity for one metal ion has been observed also, for example, for DNA duplexes that contained a **Dipic•py** pair of ligands. While $Cu^{2+}$ coordination increased the **Dipic•py** duplex stability, addition of $Ce^{3+}$, $Mn^{2+}$, $Fe^{2+}$, $Co^{2+}$, $Ni^{2+}$, $Zn^{2+}$, $Pd^{2+}$, or $Pt^{2+}$ had no effect. It would be interesting to determine if some of these metal ions coordinate to one or both ligands, which is likely, even if they do not affect the melting temperature of the duplex.

Bis(2-pyridylmethyl)amine (2PA) is a neutral ligand that can coordinate to transition metal ions in a tridentate fashion (Entry 17, Table V). This ligand has been introduced in ssLNA and ssPNA, which each formed duplexes with ssDNA (121, 137). Duplexes formed in the absence of metal ions between an ssLNA that contained a **2PA** ligand and an ssDNA had a thermal stability higher than that of nonmodified duplexes (121). This effect is opposite to that observed for ligand-modified DNA duplexes, which are typically less stable than their nonmodified counterparts, but is similar to the effect previously observed for LNA•DNA duplexes containing N-functionalized 2′-amino–LNAs (138).

When one or several isolated **2PA** ligands were introduced in LNA•DNA duplexes, addition of 1 equiv of $Ni^{2+}$, $Zn^{2+}$, or $Cu^{2+}$ per **2PA** ligand significantly increased the duplex stability with respect to that of the metal-free duplexes (121). A similar increase in stability was observed upon addition of 1 equiv of $Ni^{2+}$ or $Cu^{2+}$ to **2PA**-modified PNA•DNA duplexes (137). The effect of $Ni^{2+}$ was larger than that of $Cu^{2+}$ for both types of duplexes. In the case of $Zn^{2+}$, 2 equiv of metal ions were necessary to measure reversible melting transitions for PNA•DNA duplexes, which was attributed to the low-binding constant of 2PA for $Zn^{2+}$ (137).

It is conceivable that an interstrand complex can be created by coordination of a metal ion to the **2PA** situated on the PNA or LNA strand and to a nucleobase or phosphodiester group from the opposite, non-modified DNA strand in the nucleic acid duplex. Molecular modeling of LNA•DNA duplexes showed that a [3 + 1] **2PA**–$M^{2+}$–$PO_4R_2^{-}$ coordination mode can be realized by coordination of a **2PA** ligand from one strand and a phosphodiester from either the same strand or from the opposite strand of the LNA•DNA duplex (121). This model was supported by data obtained for **2PA**-modified PNA•DNA duplexes showing that (a) the stability of a $Ni^{2+}$-containing, **2PA**-modified PNA•DNA duplex did not depend on the duplex sequence, and (b) the effect of $Ni^{2+}$ on the duplex stability depended on the ionic strength of the solution (137). In most cases, addition of >1 equiv of metal ion per pair of adjacent **2PA** ligands destabilized the duplex, suggesting that steric and electrostatic repulsion occurs when each of the **2PA** ligands binds a metal ion.

In experiments using PNA strands that contained a ligand and a naphthalene diimide group (NADI) in adjacent positions (139,140), the $Zn^{2+}$-dependent binding of the **2PA**-modified PNA probe to a DNA target and the NADI intercalation into the PNA•DNA duplex were cooperative (140) (Fig. 14). The effect of the metal ion increased with the decreasing size of the PNA•DNA duplex and was independent of the sequence of the overhang in the DNA strand. Studies of the effect of mismatch base pairs situated close to the end of the PNA•DNA duplexes on the thermal stability of the duplex and molecular modeling indicated that both NADI and the adjacent $Zn^{2+}$–**2PA** complex intercalate between the terminal and penultimate base pairs of the duplex, a process driven by the electrostatic attraction between the positively charged $Zn^{2+}$–**2PA** complex and the negatively-charged DNA backbone. This system, which incorporates both a PNA and a ligand with high affinity for metal ions, represents a step toward *in vivo*, sequence-specific, nucleic acid probes that could be activated by metal ions present in high concentrations in specific cells.

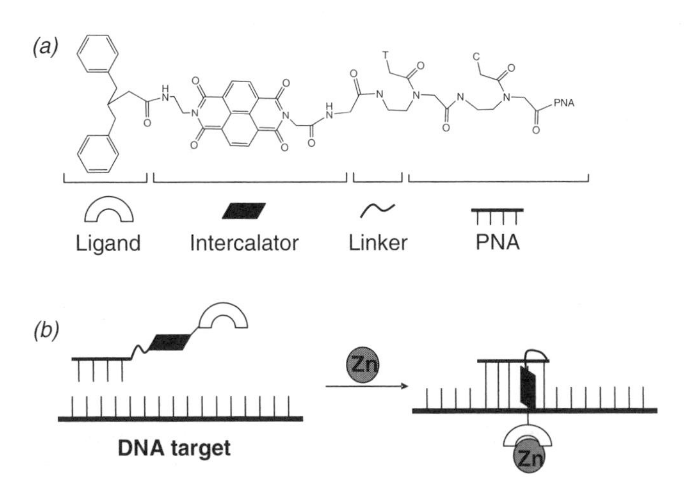


Figure 14. (a) PNA probe; (b) $Zn^{2+}$-dependent interaction between PNA probe and ssDNA (adapted from Reference 140).

### 4. *Other Proposed Coordination Modes*

Observation of an increase in the thermal stability of ligand-modified, nucleic acid duplexes in the presence of a metal ion is indicative of metal coordination to the ligands to form a nucleic acid duplex stabilized by $\pi$-stacking and hydrogen bonds between nucleobases and by coordinative bonds between the metal and the ligands from both strands. Studies of the thermal stability of duplexes do not provide information on the actual metal coordination, which may involve not only ligands extraneous to the nucleic acid, but also nucleobase or phosphate groups. The coordination modes presented in this section have been proposed based only on variable temperature UV spectroscopy, and thus are yet to be confirmed using characterization methods that provide direct information on the coordination of the metal ion.

Six coordination is possible for $Ag^+$ in a complex with a 15 bp DNA duplex that contained two tridentate **SPy** ligands in complementary positions (Entry 18, Table V) (106). The stability of the DNA duplex with a central **SPy**-$Ag^+$-**SPy** pair surpassed that of the duplex with a **SPy**-$Ag^+$-**py** pair (Entries 16 and 18, Table V) although six coordination is less common for $Ag^+$ than four coordination. The two tridentate **SPy** ligands provide a total of six potential coordination sites, and theoretically allow octahedral coordination of $Ag^+$, but it is possible that in the **SPy**-$Ag^+$-**SPy** complex, $Ag^+$ coordinates fewer than the ligands' six donor atoms.

Two adjacent pairs of histidines have been introduced in alanyl, alanyl/homoalanyl, homoalanyl, and alanyl/norvalyl PNA duplexes, which have a duplex width of 14, 15, 16.5, and 18 Å, respectively (141). A hypothesis was set forth, but not proven, that the four histidines coordinate one metal ion in the narrower duplexes and a dimer of metal ions bridged by water ligands in the broader duplexes. The $Zn^{2+}$ ion, but not $Cu^{2+}$, stabilized the modified duplexes. The EPR spectroscopy and extended X-ray absorption fine structure (EXAFS) were used to study solutions of a single-stranded, alanine–norvaline, 8-mer PNA that had two central histidines in the presence of either 1 or 0.5 equivalents of $Cu^{2+}$, but did not provide enough information to elucidate the $Cu^{2+}$ coordination.

## E. Duplexes That Contain Multiple Metal-Binding Sites

Application of the metal–ligand incorporation in nucleic acid duplexes either to expand the genetic code or to create extended, metal-containing nanostructures requires that several metal complexes are incorporated into one nucleic acid duplex. This is a difficult task because the presence of multiple ligands situated in close proximity of each other within the same oligomer makes possible the formation of different structures for the nucleic acid, (e.g., hairpin or duplex), and of metal complexes with different coordination numbers and geometries. As a consequence, the characterization of these duplexes is more complicated than that of duplexes containing a single metal-binding site.

The first DNA duplex in which more than one metal–ligand alternative base pair was incorporated was the palindromic d(5′-$C_1G_2C_3G_4$**Dipic**$_5A_6T_7$ **py**$_8C_9G_{10}C_{11}G_{12}$-3′)$_2$, which included two isolated [Cu(**Dipic**)(**py**)] complexes (119). The crystal structure of this duplex clearly showed that the incorporation of multiple, isolated, metal–ligand alternative base pairs is compatible with the formation of a DNA duplex. More recently, two isolated [Cu(**salen**)] complexes have been included in a 13 bp DNA duplex (142), as determined by UV titrations and electron ionization mass spectrometry (ESI MS). The melting temperature of the duplex increased with the increasing concentration of $Cu^{2+}$ and was 75°C in the presence of two equivalents of $Cu^{2+}$ per duplex.

In the **Dipic•py** modified DNA duplex, the $Cu^{2+}$ ions were coordinated by the **Dipic** and **py** ligands in a square-planar arrangement, with two additional donor atoms from adjacent base pairs, namely, the O4′ of the $T_7$ nucleotide and the O6 of $G_4$, weakly coordinated in axial positions (119). The metal complexes influenced the structure adopted by the ligand-modified duplex. In the crystal, the metal-containing DNA duplex adopted a *Z* conformation and CD spectra showed that this conformation is also adopted by the duplex in solution (120). The preference for this conformation was likely due to the fact that (a) in Z-DNA, $Cu^{2+}$ can coordinate two axial ligands from neighboring base pairs, and/ or (b) the nucleic acid part of the metal-containing duplex had an alternating purine-pyrimidine (APP) sequence, which is typical for Z-DNA duplexes. The effect of the metal complex on the structure of the modified DNA duplex depended on the DNA sequence. For example, the d(5′-$C_1G_2C_3G_4A_5$**Dipic**$_6$-**py**$_7T_8C_9G_{10}C_{11}G_{12}$-3′) duplex that contained two **Dipic•py**-binding sites, but did not have an APP sequence adopted a B-conformation in solution in the presence of $Cu^{2+}$ ions (119).

Several adjacent [Cu**H**$_2$] metal–ligand alternative base pairs also have been incorporated in DNA duplexes (Entry 12, Table V) (117). Duplexes containing one-to-five [Cu**H**$_2$] complexes flanked on each side by a GC base pair have been prepared. The 1:2 Cu/**H** stoichiometry of the complexes was confirmed by UV titrations and by ESI MS. Circular dichroism spectra revealed the existence in solution of a right-handed, helical duplex similar to B-DNA. A CD feature at 324 nm was attributed to the coordinated hydroxypyridone and was indicative of a chiral induction effect exerted by the duplex on the [Cu**H**$_2$] complexes. The EPR spectra showed that the spins of adjacent $Cu^{2+}$ ions were coupled ferromagnetically, which is a consequence of spatial proximity and parallel alignment of the metal complexes within the duplex. The $Cu^{2+}$—$Cu^{2+}$ distance was estimated to be ~3.7 Å based on the magnitude of the dipolar coupling between the adjacent $Cu^{2+}$ ions. When a **py** ligand and several **H** ligands were incorporated in the same oligonucleotide, that is, d(5′-**GHpyHC**-3′) and d(5′-**GHHpyHHC**-3′), it was possible to create a heterometallic duplex that included both $Cu^{2+}$ and $Hg^{2+}$ coordinated to **H** and **py** ligands, respectively (143). The same strategy of creating heterometallic arrays of metal ions arranged in an order encoded in the sequence of ligands was successfully applied to duplexes that included up to 10 [Cu(**salen**)] and T-Hg-T complexes between two short stretches of GC nucleobase pairs (143). The stoichiometry of these duplexes was determined usign CD and UV titrations and ESI MS.

Quantum mechanical calculations have been performed on models for duplexes that contain multiple, adjacent [$\mathbf{ML}_n$] alternative base pairs in an effort to evaluate their potential for charge transfer (144, 145). The systems were modeled as stacks of several T-Hg-T pairs situated between terminal GC pairs (145), as stacks of [Cu$\mathbf{H}_2$] complexes, or as infinite wires based on the periodic repetition of the [Cu$\mathbf{H}_2$] stacks (144). The DNA backbone was not included in these models. The study of the model that contained 2 or 3 T-Hg-T base pairs led to the conclusion that in the case of hole transfer betweeen the terminal GC pairs by a superexchange mechanism, the metal-containing pairs mediate donor–acceptor coupling that is more efficient than that mediated by AT base pairs (145). The metal ions are not responsible for the superior coupling as their contribution to the highest occupied molecular orbital (HOMO) is small, but may have an important role if electron transfer would be mediated by the T-Hg-T pairs because the metal orbitals are important contributors to lowest unoccupied molecular orbital (LUMO). The theoretical study of duplexes formed from [Cu$\mathbf{H}_2$] was based on spin polarized DFT (144) and it determined that, in the ferromagnetic state of the array of $Cu^{2+}$ ions, the highest occupied energy levels are discrete and a conduction band is not formed. It would be interesting to synthesize and study arrays of other metal ions (e.g., $Ni^{2+}$ or $Cu^{2+}$), bridged by small ligands, which may be more efficient in mediating charge transfer. In [$M_3(dpa)_4Cl_2$] where dpa = anion of di(2-pyridyl)amine, complexes in which the metal ions are situated at 2.2–2.6 Å (146), the latter two metal ions have shown delocalized metal–metal bonding stronger than that observed for $Cu^{2+}$.

Recent studies of bipyridine-modified PNA duplexes have shown that several positively charged and nonsquare-planar complexes also can be incorporated in duplexes (16). The UV–vis and EPR spectroscopic studies showed that the 10 bp PNA duplexes that contained two or three adjacent pairs of **bpy** ligands contained one or two octahedral [M(**bpy**)$_3$]$^{2+}$ complexes (M = Ni, Cu), respectively (16). The melting curves of the **bpy**-modified PNA duplexes in the presence of metal ions showed low hypochromicity, suggesting that the octahedral complexes interfere with the stacking of the base pairs. Addition of 3 equiv of $Ni^{2+}$ to the PNA duplexes that contained three pairs of **bpy** did not lead to conversion of the two tris-**bpy** complexes into three bis-**bpy** complexes, which is likely due to the electrostatic repulsion.

Several other reports explored DNA duplexes that contain multiple ligand pairs, but their focus was only on the effect of the metal ions on the duplexes' melting temperature. Switzer et al. (113) examined the effect of $Ni^{2+}$ on 15 bp DNA duplexes containing three adjacent pairs of **Pur**$^{P}$ ligands (Entry 8, Table V). They reported that addition of 4 equiv of $Ni^{2+}$ relative to the duplex increased the thermal stability of the duplexes drastically. The fact that a melting temperature could be determined for the **Pur**$^{\mathbf{P}}$-modified duplexes in the presence of $Ni^{2+}$ suggests that the stacking of the bp is not affected by the three metal complexes. This is in contrast to the behavior of the **bpy**-modified PNA duplexes in the presence of the same metal ion (16). The difference in behavior of the **Pur**$^{\mathbf{P}}$–DNA and **bpy**–PNA duplexes may be due to (a) the fact that the larger **Pur**$^{\mathbf{P}}$ ligand could form bis-ligand, but not tris-ligand, complexes; (b) the difference in duplex length (i.e., 15 bp DNA duplex versus 10 bp PNA duplex); and (c) the fact that the DNA backbone acts as a counteranion for the complex, thus reducing the electrostatic repulsion between the adjacent, positively-charged $[Ni\mathbf{L}_n]^{2+}$ complexes, while the PNA backbone is neutral.

Zimmermann et al. (120) reported the incorporation of two or four [3 + 1]-type **Dipam•py** or **Dipic•py** ligand pairs in adjacent positions in the center of DNA duplexes (Entries 14 and 15, Table V), and measured the $T_m$ for these duplexes in the presence of 15 equiv of $Cu^{2+}$ per pair of ligands (120). A systematic, but relatively small, increase in the duplex stability was observed when the number of metal–ligand alternative base pairs was increased from 1 to 4. Incorporation of two or three **SPy•py** or **SPy•SPy** pairs at isolated positions in the same 15 bp duplex (Entries 16 and 18, Table V) led to nonsystematic changes in the duplex stability in the presence of stoichiometric amounts of $Ag^{+}$ ions (106). For example, a duplex that contained a single **SPy•SPy** pair was more stable than a duplex with two **SPy•SPy** pairs, but less stable than a duplex containing three **SPy•SPy** pairs. This observation is indicative of sequence- and structure-dependent effects on the metal incorporation.

Several metal ions have been coordinated to DNA and PNA dimers (DNA–**bpy**$_2$) and trimers (PNA–**bpy**$_3$), respectively, which consisted exclusively of **bpy** ligands and did not contain natural nucleobases (Chart 2) (93, 147). UV titrations and ESI MS showed that DNA–**bpy**$_2$ formed an intramolecular [Cu(DNA–**bpy**$_2$)] complex with $Cu^{2+}$. In contrast, $Ag^{+}$ coordination led to a double-stranded [$Ag_2$(DNA–**bpy**$_2$)$_2$] structure that contained two [Ag(**bpy**)$_2$]$^{+}$ complexes. Interestingly, in the presence of $Pd^{2+}$, the same DNA–**bpy**$_2$ ligand formed a mixture of intrastrand, mononuclear and interstrand, binuclear complexes (93).

Gilmartin et al. (147) monitored metal-to-ligand charge transfer (MLCT) transitions and used MS to characterize the complexes formed between PNA–**bpy**$_3$ and $Fe^{2+}$ or $Cu^{2+}$. Titrations of the PNA with $Fe^{2+}$ and $Cu^{2+}$ showed inflection points in the absorbance change at PNA–**bpy**$_3$/M ratios of 1:1 and

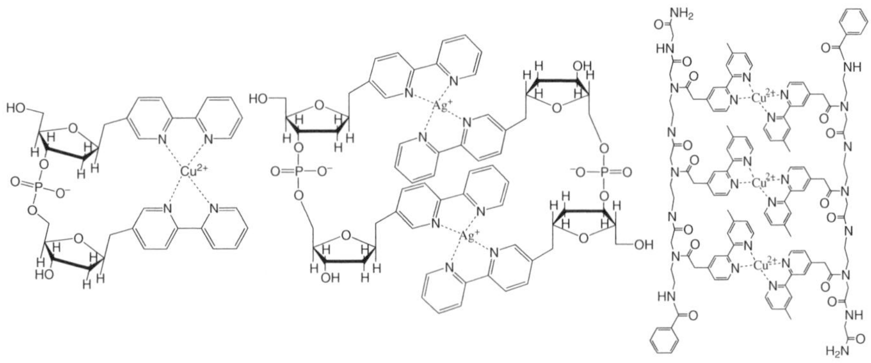

**Chart 2.** Metal complexes with DNA-**bpy**$_2$ and PNA-**bpy**$_3$ (93, 147).

~1:1.6, respectively, which correspond to **bpy**/M ratios of 3:1 and ~2:1. High-resolution mass spectrometry for the $Fe^{2+}$ complex showed the formation of $[Fe_2(PNA–\mathbf{bpy}_3)_2 + 4H]^{4+}$ ions. Low-resolution was used to argue for the formation of a $\{[Cu_3(PNA–\mathbf{bpy}_3)_2] + 2Na\}^{2+}$ species. Based on these results and molecular modeling, it was concluded that double-stranded structures bridged by two octahedral $[Fe(\mathbf{bpy})_3]^{2+}$ or by three four-coordinate $[Cu(\mathbf{bpy})_2]^{2+}$ complexes are formed. The EPR spectrum of the $Cu^{2+}$ complex showed at $g = 2$ a narrow, possibly isotropic signal and another broad signal, which were superimposed on a very broad feature that peaked at $g > 4$. The complexity of the spectrum was attributed to (a) heterogeneity in the sample arising from the presence of structural isomers, and (b) weak magnetic coupling between the $Cu^{2+}$ sites within the same duplex.

The research aimed at the specific metal ion incorporation into ligand-modified nucleic duplexes has confirmed several general design principles. The ligands used in this strategy need to have a high affinity for transition metal ions and to possess an extended aromatic surface that allows stacking with adjacent nucleobases. Metal complexes of the [2 + 2] or [3 + 1] type have been widely used for the successful formation of metal-containing alternative base pairs. Beside these rather obvious considerations, the studies described above have revealed that the effect of metal base pairs on the properties of nucleic acid duplexes is not straightforward to predict. The higher strength of coordinative bonds compared to that of hydrogen bonds does not imply that the stability of duplexes that contain both types of bonds is always higher than that of corresponding duplexes that are based exclusively on Watson–Crick base pairing. The effect of metal-containing base pairs on duplexes also depends on the geometry and charge of the metal complex and is influenced by the nucleic acid backbone and structure, and by the duplex sequence. Duplexes containing one metal–ligand alternative base pair have been investigated more extensively than the incorporation of multiple metal complexes in duplexes. Furthermore, the analysis of duplexes that contained multiple metal ions revealed a direct electronic contact between adjacent metal sites within a duplex only in one case. Recent results on **bpy**-modified PNA structures showed that in duplexes with several adjacent ligand pairs, the metal ions can adopt different coordination modes (16). Therefore, investigations of such duplexes need to address the stoichiometry of metal ion coordination besides the thermal stability of the duplex. Finally, molecular modeling, X-ray and NMR structural studies, and investigations of the electronic structure of the metal-containing duplexes are scarce and would be useful for both the interpretation of the properties of the duplexes, as well as the rational design of metal-containing duplexes with specific physical or chemical properties.

## V. HYBRID STRUCTURES BASED ON OLIGONUCLEOTIDES THAT CONTAIN TERMINAL LIGANDS

The rich coordination chemistry of transition metal ions has been used not only to create metal–ligand complexes that play the role of alternative nucleobase pairs within nucleic acid duplexes, but also to influence the secondary structure adopted by the nucleic acid, for example, hairpin, duplex, or triplex, and to create connectors for such nucleic acid structures. In this context, oligonucleotides that contain terminal ligands can lead to structures distinct from those accessible by using centrally-modified oligonucleotides, such as cyclic structures or hairpins (Fig. 3).

Coordination of metal ions to ligands attached to each end of one oligonucleotide can generate circular DNA structures [Fig. 15(*a*)]. If the two ends of the oligonucleotide are complementary, metal-binding to the terminal ligands leads to the formation of a hairpin [Fig. 15(*b* and *c*)] in which the metal complex plays a role equivalent to that played by the Watson–Crick base pairs in the stem of the hairpin. A metal complex also can be part of the loop of a hairpin if either one of its ligands is itself a linker for two (partly) complementary oligonucleotides [Fig. 15 (*d*–*f*)] or two of its ligands are each attached to the 3′ and 5′ ends of two oligonucleotides [Fig.15 (*g* and *h*)].

### A. Circular Structures

Terpyridine (tpy) forms [M(tpy)$_2$] complexes, which are not good candidates for metal-mediated base pairs within a nucleic acid duplex because of their octahedral geometry, in which the two tpy ligands are situated in reciprocally perpendicular planes. Therefore, they have been used instead to connect the ends of terminally modified oligonucleotides. The **tpy**-modified ssDNA, which did not

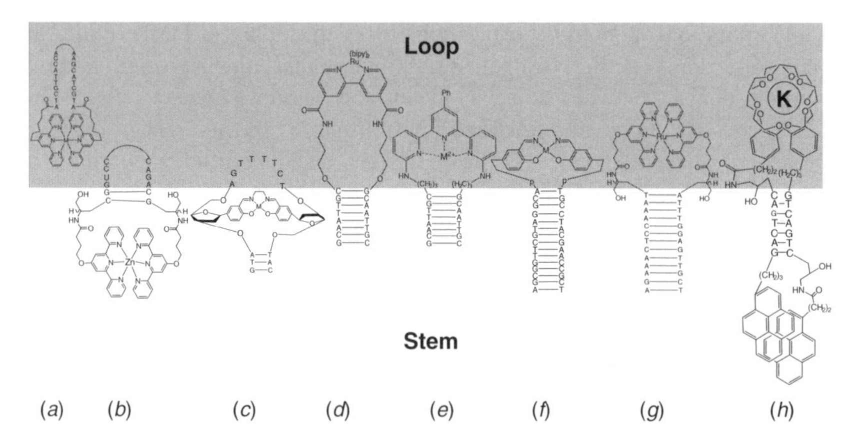


Figure 15. Metal-containing circular (*a*) or hairpin nucleic acids (*b–h*). For each hairpin, the loop is within the gray part and the stem is within the white part of the figure.

have any self-complementary parts, but had **tpy** ligands at the 3′- and 5′-ends, formed a stable circular structure upon binding of 1 equiv of $Fe^{2+}$, $Zn^{2+}$, or $Cu^{2+}$ [Fig. 15(*a*)] (148, 149). This property was surmised from the fact that the mobility of the metal-containing, **tpy**–DNA complexes in gel electrophoresis was slightly higher than that of the **tpy**–DNA in the absence of the metal ions (148). The 1:2 $M^{2+}$/**tpy** stoichiometry of complexes formed between the metal ions and the **tpy**-modified DNA was demonstrated by titrations, which monitored the absorbance of visible MLCT and UV $\pi$–$\pi^*$ ligand-based transitions, by ESI MS, and by EPR spectroscopy in the case of $Cu^{2+}$.

The loop sequence of the **tpy**-modified DNA strand was designed such that the strand could hybridize to a molecular beacon [Fig. 16(*a*)] only when in its extended, linear form [Fig. 16(*b*)]. Upon hybridization to the **tpy**-modified DNA, the molecular beacon switches from its closed, nonfluorescent form to its open, fluorescent form [Fig. 16(*b*)]. Both $Fe^{2+}$ and $Zn^{2+}$ form very stable $[M(tpy)_2]^{2+}$ complexes and consequently, the circular structure adopted by the **tpy**-modified DNA strands in the presence of 1 equiv of these metal ions represented a steric barrier for its hybridization to the molecular beacon. The metal ion acted as an allosteric "off" switch for DNA hybridization (148). This study demonstrated that the incorporation of metal-binding ligands in nucleic acid strands can be used to create metal-dependent allosteric switches for the Watson–Crick interactions between the nucleic acid strands, a mechanism that may find applications in creating artificial systems with regulatory function. In contrast, in the presence of $Cu^{2+}$ the DNA hybridization to the molecular beacon was energetically more favorable than the formation of circular structures from the **tpy**-containing DNA strands. Therefore, hybridization of a complementary ssDNA to the

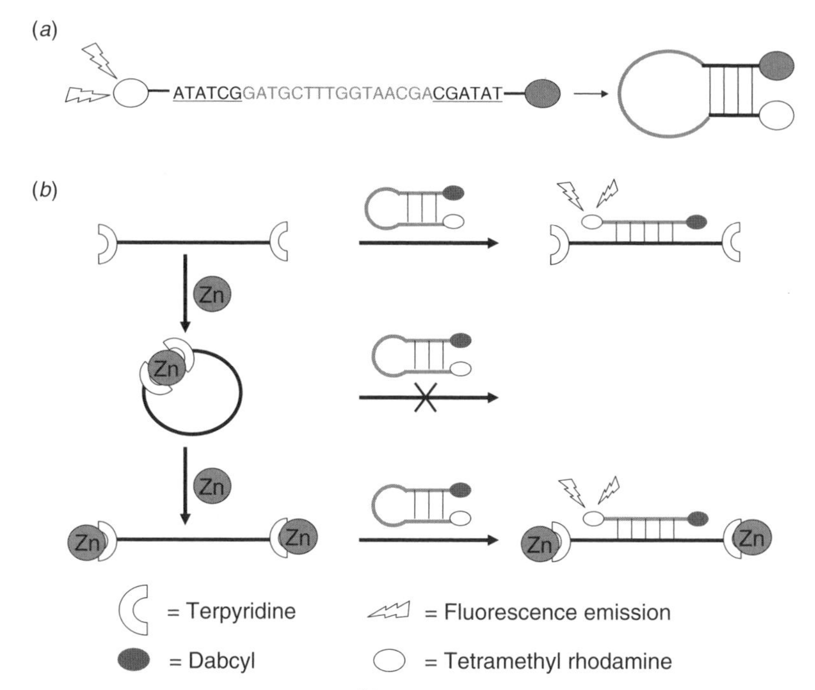


Figure 16. (*a*) Molecular beacon; (*b*) $Zn^{2+}$-dependent switch of the hybridization between the molecular beacon and the **tpy**-modified ssDNA. (4-[[4-(Dimethylamine)phenyl]azo]-benzoic acid = dabcyl) (adapted from Reference 148).

$Cu^{2+}$-containing, **tpy**-modified circular DNA triggered the conversion of the circular DNA to linear DNA and the release of $Cu^{2+}$ (149a). Once released, the metal ion could coordinate phenanthroline (phen), a precatalyst present in solution. The Cu-phen complex catalyzed the oxidative synthesis of a fluorescent dye, thus amplifying the allosteric signal determined by the DNA hybridization process. The same strategy was applied to detect ssDNA by enzymatic amplification (149b). In this case, $Zn^{2+}$ released from a $Zn^{2+}$-**tpy** DNA complex as a consequence of sequence-specific interactions between the complex and a complementary ssDNA was captured by apo-carbonic anhydrase, to form the holoenzyme, which catalyzes the $CO_2$ conversion to $HCO_3^-$ (149b).

The transformation between linear and circular **tpy**–DNA structures was dependent on the $M^{2+}$/**tpy** stoichiometry and the relative stability of $[M(tpy)]^{2+}$, $[M(tpy)_2]^{2+}$, and the metal complexes with other chelating agents. For example, the linear-to-circular transformation of **tpy**-modified DNA induced by 1 equiv of $Zn^{2+}$ could be reversed by a second equivalent of $Zn^{2+}$, which transformed the $[Zn(\mathbf{tpy})_2]^{2+}$ complex into two terminal $[Zn(\mathbf{tpy})]^{2+}$ complexes [Fig. 16(*b*)] (148). Also, addition of ethylenediaminetetracetic acid (EDTA), which has a higher affinity for $Zn^{2+}$ than tpy, had the same effect of opening the circular DNA. In contrast, addition of EDTA had no effect on circular, $[Fe(\mathbf{tpy})_2]^{2+}$-DNA because the $[Fe(tpy)_2]^{2+}$ is both kinetically inert and thermodynamically stable (148).

## B. Hairpin-Like Structures

Replacement of terminal base pairs with metal–ligand alternative base pairs can exert a significantly larger stabilization of double helical nucleic acid structures than similar substitutions made in the middle of nucleic acid duplexes because the terminal nucleobase pairs are subject to fraying while the central base pairs are not (16). This property can be exploited to form short duplexes or hairpins that otherwise would not be stable. The approach has been used to create $Zn^{2+}$-containing, 2′-O-Me-RNA and $Cu^{2+}$-containing DNA hairpins from short, **tpy**-modified 2′-O-Me-RNA and **sal**-modified DNA strands, respectively, that each had complementary 3′- and 5′-end sequences [Fig. 15(*b* and *c*)] (118, 150). In both systems, the metal complex reinforced the stem of the hairpin. The $[Zn(\mathbf{tpy})_2]^{2+}$ complex contributed to the thermal stability of the RNA hairpin, which was higher than that of a hairpin with the same sequence, but which had an AU base pair instead of the $[Zn(\mathbf{tpy})_2]^{2+}$ complex (150). In the presence of en and $Cu^{2+}$, the **sal**-containing ssDNA formed a hairpin with an intramolecular [Cu(**salen**)] complex. The thermal stability of this $Cu^{2+}$-containing hairpin was significantly higher than that of the hairpin formed exclusively by Watson–Crick base pairing of the two complementary end stretches of nucleobases existent in the **sal**-containing DNA strand [Fig. 15(*c*)] (118).

While the **tpy**–RNA and **sal**–DNA hairpins described above contained the metal complex in their stem, it is also possible that a metal complex be part of a hairpin's loop [Fig. 15(*d*–*h*)]. In the simplest example, a ligand connects two complementary DNA strands and plays the role of the hairpin loop. The ligand also makes possible site-specific metal-binding to the hairpin loop. The metal-binding influences the hairpin formation, stability, and structure. Both **bpy** and **tpy** ligands have been used to create this type of DNA hairpin [Fig. 15(*d* and *e*)] (151, 152). The **bpy** that covalently linked two complementary DNA strands was one of the three ligands of a $[Ru(\mathbf{bpy})(bpy)_2]^{2+}$ complex, which also contained two unmodified bpys [Fig. 15(*d*)] (152). Molecular modeling indicated that it was possible for the two $[Ru(\mathbf{bpy})](bpy)_2]^{2+}$-linked DNA strands to hybridize and form a hairpin stem. The independence of the melting temperature of the Ru-containing DNA structure on the DNA concentration confirmed that the melting was intramolecular and thus typical of a hairpin. The hairpin based on the **tpy**–loop provided a rare example of anticooperativity between metal binding and nucleic acid hybridization, with the metal binding reducing the stability of the hairpin [Fig. 15(*e*)]. Coordination of transition metal ions to **tpy** reduced the hairpin's stability in the order $Co^{2+} \sim Ni^{2+} < Zn^{2+} < Cu^{2+} < Pd^{2+}$ (151). A computational analysis identified a metal-induced change in the **tpy** conformation, which hindered the stacking between **tpy** and the adjacent base pair, as a possible cause for the hairpin destabilization. The effect of metal ions on the hairpin was reversible and chelation of the metal ions by EDTA restored the hairpin's thermal stability.

Metal ion coordination to ligands situated at the 3′- and 5′-ends of two complementary DNA strands creates a metal complex, which as a whole can play the role of a hairpin loop [Fig. 15(*f–h*)]. **Sal**, **tpy**, and benzo-15-crown-5-ether (**benzo-15C5**) have been used as terminal ligands and have connected the DNA strands through [M(**salen**)] ($M = Mn^{2+}$ or $Ni^{2+}$) (153), $[M(\mathbf{tpy})_2]^{2+}$ ($M = Zn^{2+}$ or $Fe^{2+}$) (150), and $[K(\mathbf{benzo\text{-}15C5})_2]^+$ (154) coordination complexes, respectively. These ligands were attached to the DNA through flexible aliphatic or ethylene glycol linkers to alleviate steric interactions between the metal complexes and adjacent nucleobase pairs. The hybridization of the two ssDNA connected by each coordination complex was demonstrated by variable temperature UV spectroscopy. The general structure of these metal-containing systems is formally similar to that of a hairpin, with the stem being the DNA duplex formed by Watson–Crick base pairing and a loop, which is the metal complex that connects the two DNA strands of the duplex. Nevertheless, if the metal ion is not coordinated to the ligands at high temperature, the assemblies are more accurately described as duplexes and will be referred herein as metal-bridged duplexes. Experimentally, this distinction can be made by high-temperature UV titrations to verify the metal coordination or by measuring the concentration dependence of the melting temperature of the nucleic acid structures.

The duplex with strands bridged by a [M(**salen**)] ($M = Ni^{2+}$ and $Mn^{2+}$) complex had higher thermal stability than a hairpin with identical stem sequence, but with a three nucleotide loop [Fig. 15(f)] (153). The melting temperature of a **tpy**-modified duplex measured in the presence of $Zn^{2+}$ was comparable to that of a duplex formed from the same two DNA strands covalently linked by a hexaethylene glycol bridge, but it augmented with the increasing concentration of the two DNA strands and metal ion, indicating that the formation of the $[Zn(\mathbf{tpy})_2]^{2+}$-bridged DNA duplex is not an intramolecular process [Fig. 15(g)] (150). Two scenarios that could explain this observation are that (a) the metal ion is not coordinated to **tpy** ligands at high temperature, or (b) the metal ion links at all temperatures two ssDNA into extended DNA strands, and the extended strands hybridize into a duplex that contains two central $[Zn(\mathbf{tpy})_2]^{2+}$ complexes.

The $K^+$ triggered the hybridization of two complementary DNA strands that each had a terminal **benzo-15C5** ligand [Fig. 15(*e*)] (154). This process was evidenced by a change in the fluorescence of pyrene fluorophores situated at the ssDNA termini opposite to those bearing the **benzo-15C5** ligands. The change in fluorescence occurred when the two pyrenes were brought in close proximity of each other by the duplex formation, because the excimer and monomer pyrene emitt at different wavelengths (155). The $T_m$ of the duplex formed from the **benzo-15C5**-modified DNA strands was higher by ~10°C in the presence of excess $K^+$. The duplex stabilization in the presence of $K^+$ was due to both the formation of the [K(**benzo-15C5**)$_2$]$^+$ complex and the nonspecific shielding of the negative charge of the DNA single strands in solutions with increasing ionic strength. The design of this DNA conjugate is appropriate for creating a fluorescence $K^+$ sensor, but in practice the system had low selectivity relative to $Na^+$.

### C. DNA-Templated Synthesis of Metal-Containing, ssDNA

The DNA-template strategies have been utilized extensively to direct chemical reactions between reagents attached to short nucleic acid strands. The DNA template preorganizes shorter ssDNA to which reactive groups are appended such that these groups are brought in close proximity of each other (156). Noncovalent interactions between the oligonucleotides situated on the DNA template have been shown to stabilize the ternary DNA complex (157, 158). The same method can be used to direct the synthesis of a metal-containing, ssDNA from two shorter, ligand-modified DNA strands that are preorganized on a complementary ss or ds DNA template (Fig. 17). In this case, the metal coordination to ligands attached at the 3′- or 5′-end of the two short DNA strands can also stabilize the ternary nucleic acid structure. This approach for cooperative oligonucleotide association can be used to sense nucleic acid strands because the nucleic acid used as template is not modified, and the two strands that associate on the template are short, making their binding sequence selective.

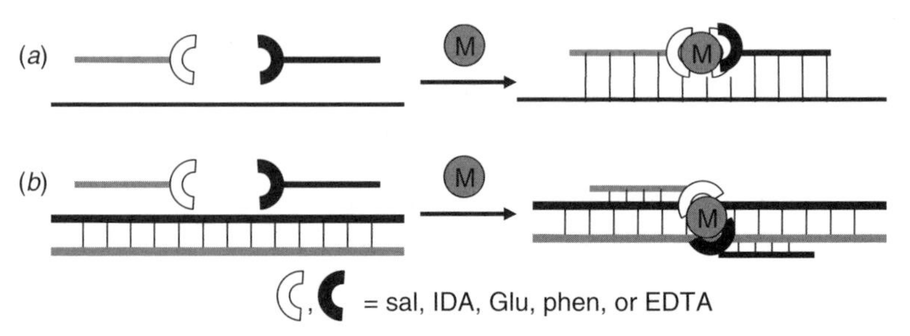


Figure 17. DNA templated synthesis of metal-containing ssDNA based on a ssDNA template (*a*) or a dsDNA template (*b*) (adapted from References 161 and 162).

Salicylaldehyde and iminodiacetate (**IDA**) have been used as terminal ligands in ssDNAs that were hybridized onto a ssDNA or RNA template [Fig. 17(*a*)] (159–161). The use of RNA as template has the advantage that it can be selectively digested by RNase H, thus "releasing" the metal-containing ssDNA conjugate formed on the template (159).

In the absence of metal ions, two short DNA strands containing **IDA** ligands hybridized to adjacent sites on a DNA template and formed a helix with a slightly lower melting temperature than that of the helix formed with nonmodified DNA strands (161). The destabilization was attributed to charge repulsion between the negatively charged, IDA ligands situated in close proximity of each other. The melting temperature of the ternary complex formed after incubation with excess $Gd^{3+}$ ions was significantly higher than that of the complex formed in the absence of $Gd^{3+}$, indicating that $Gd^{3+}$ coordinates to **IDA** and bridges the two short DNA strands.

After incubation with $Mn^{2+}$ or $Ni^{2+}$ and a diamine, a covalently linked interstrand [M(**salen**)] complex was formed in relatively high yield between two strands modified with **sal** ligands and arranged onto a ssDNA template (159, 160). The duplexes formed between the [M(**salen**)]-ligated DNA strand and the DNA template adopted a B-DNA conformation and were only slightly less stable than duplexes that had two nucleobases instead of the metal complex. The templated synthesis of **salen** took place in the absence of the metal ion too, although with a significantly lower yield. Notably, the DNA template was essential for the formation of the [M(**salen**)] complex. The linker between the **sal** ligand and each ssDNA ensured the flexibility necessary to accommodate the metal complex in the modified DNA duplex and influenced the yield for the formation of the **salen**–metal complex. Furthermore, the linker's metal-binding or hydrogen-bonding abilities have been shown to influence the preorganization of the ligand and the metal ions that participate in the formation of the [M(**salen**)] complex.

Ihara and co-workers (162, 163) demonstrated metal-driven, cooperative formation of triplexes by using a DNA duplex as template for the association of two short DNA strands that each contained one terminal **IDA** or glutamate (**Glu**) ligand [Fig. 17(*b*)]. The sequence of the ligand-modified short DNA strands was pyrimidine rich and the dsDNA template had a central palindromic sequence flanked by two purine-rich tracks (162, 163). Both $Lu^{3+}$ and $Cu^{2+}$ have been used as allosteric effectors to form bis-ligand complexes with the two short, DNA oligomers modified with either **IDA** or **Glu**. Formation of the DNA triplex in solutions containing the DNA duplex template, 2 equiv of ligand-containing short DNA strands, and 1 equiv of metal ions was confirmed by CD spectroscopy. Given the low-binding constant between $Cu^{2+}$ and **Glu**, a metal complex with the two short strands would not form in the absence of the DNA duplex template. The metal coordination increased the melting temperature of the triplex-to-duplex transition for both the **IDA** and **Glu** systems in the presence of $Lu^{3+}$ and $Cu^{2+}$, respectively. When the DNA duplex template contained only one purine track and thus could bind only one of the two short DNA strands, metal ions did not have a significant effect on the melting of the triplex. This result confirmed the hypothesis that the enhancement of the metal-binding affinity of the ligand-containing DNA strands was due to cooperative dimerization onto the template.

A similar DNA-templated strategy was applied to form mixed-ligand **edta–phen** complexes, where edta (ethylenediaminetetracetate) is a ligand, with $Tb^{3+}$ under conditions in which, according to the stability constants, only **edta** was expected to coordinate to $Tb^{3+}$ [Fig. 17(*a*)] (164). The ligands have been attached to the 3′- and 5′-ends of two short ssDNA strands that were hybridized on a ssDNA template. The system formed between the DNA template strand and the short, ligand-modified strands showed two melting transitions that were attributed to the melting of each of the two short strands. The transition attributed to the **edta**-containing ssDNA was sensitive to the presence of the metal ion, only if the **phen**-modified ssDNA was also present. This evidence suggested that the metal ion coordinates both the **edta** and the **phen** ligands. Arguments for the formation of the mixed ligand complex were also formulated based on the results of time-resolved emission spectroscopy. The template-dependent interaction between the lanthanide ion and the **phen** and **edta** ligands translated into strong luminescence and could be a valuable signal for detection of ss oligonucleotides.

### D. Extended, Hybrid Inorganic–Nucleic Acid Nanostructures

The ability of DNA to form a broad range of structural motifs through Watson–Crick base pairing has been exploited to create two-dimensional (2D) and three-dimensional (3D) nanostructures, which contain a combination of DNA hairpins, duplexes, or triplexes, and of multiple-arm DNA junctions (165–167). As metal ions have relatively rigid and diverse coordination geometries, they also can be used to construct inorganic junctions between DNA strands that are covalently linked to ligands. The self-assembly of the hybrid inorganic–nucleic acid structures that contain metal ions is based on a set of assembly rules, which includes both the Watson–Crick base pairing and the coordination properties of the metal ion, and consequently leads to a diverse set of topologies. These nanostructures have a broad range of potential applications, including the arrangement of transition metal ions at distances that make possible electron transfer or magnetic interactions between the metal ions, the construction of periodic structures of metal ions that can be used in the crystallization of biological molecules, and the synthesis of catalytic systems sensitive to external factors, such as temperature or redox-active small molecules.

The number of DNA arms connected to an inorganic junction is determined by the number of ssDNA strands attached to each ligand, the denticity of the ligand, and the coordination number of the metal ion (Fig. 18). An example of a

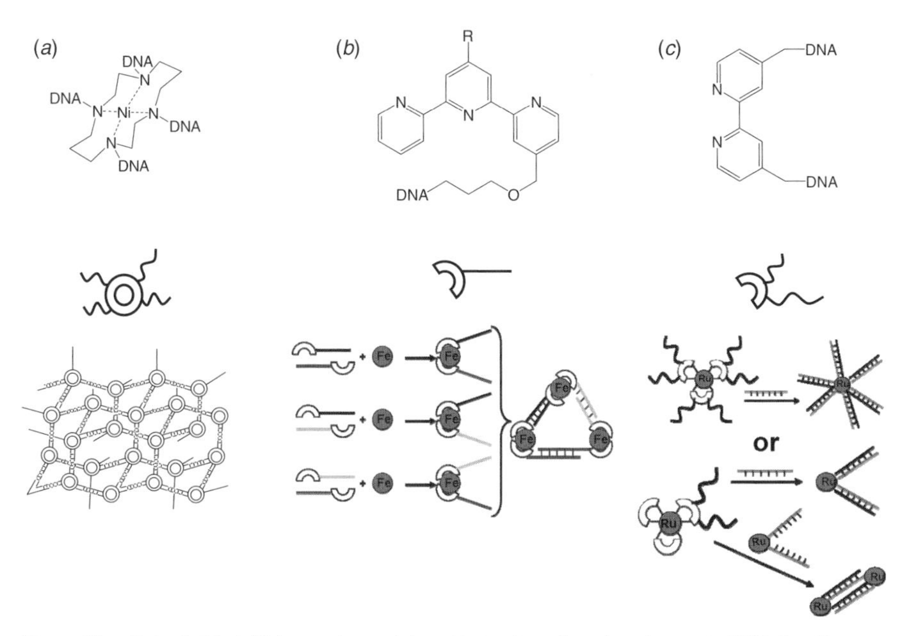


Figure 18. Extended hybrid inorganic–nucleic acid structures based on inorganic DNA junctions (adapted from References 168,169, and 172).

DNA junction whose connectivity was independent of the metal coordination was based on 1,4,8,11-tetraazacyclotetradecane (cyclam), which had a DNA single strand attached to each of the four secondary nitrogen atoms [Fig. 18(*a*)]. $Ni^{2+}$ coordination to **cyclam** was used only to increase the rigidity of the junction (168). Metal binding to **tpy**-modified ssDNA led to octahedral $[Fe(\mathbf{tpy})_2]^{2+}$ complexes (169) or $[Ru(\mathbf{tpy})_2]^{2+}$ complexes (170), which functioned as two-arm DNA junctions [Fig. 18(*b*)]. $Fe^{2+}$ binding to DNA strands that had one terminal **bpy** ligand formed three-arm DNA junctions (171). The $[Ru(\mathbf{bpy})_3]^{2+}$-based DNA junctions have been created from **bpy** ligands, which had two ssDNA arms attached to the 4 and 4′ positions [Fig. 18(*c*)]. The same type of ligand has been used to synthesize $[Ru(\mathbf{bpy})(bpy)_2]^{2+}$ complexes in which only one of the **bpy** ligands had two DNA arms (172). Both the DNA-modified **bpy** ligand and the $Ru^{2+}$ complex that contained one modified ligand functioned as two-arm DNA junctions (172). The complex in which all three **bpy** ligands carried DNA strands acted as a six-arm DNA junction [Fig. 18(c)] (173). The number of DNA arms of each metal-based DNA junction was confirmed by stepwise hybridization with complementary ssDNA monitored by nondenaturing gel electrophoresis (168, 173).

Typically, the ligands are attached to ssDNA through a flexible linker, such as an aliphatic carbon chain, to minimize the steric and/or electronic effect of the metal complex on the hybridization properties of the ssDNA arms. For example, a ssDNA linked to **bpy** hybridized to a complementary ssDNA if the linker was an aliphatic chain of nine carbon atoms but not if the chain had only two carbon atoms (173). Titrations with complementary DNA strands of a four-arm, [Ni(**cyclam**)]$^{2+}$-based DNA junction, which contained relatively long hexamethylene linkers, showed that the four arms can hybridize simultaneously (168). Variable-temperature UV spectroscopy studies indicated that the hybridization ability of the DNA strands was not affected by their attachment to the nonmetalated or metalated **cyclam** ligand (168). The linker also modulates the effect of coordination geometry of the metal ion on the overall structure. Despite the fact that the [Ni(**cyclam**)]$^{2+}$ complex adopts a square-planar geometry, the four DNA arms attached to **cyclam** through hexamethylene linkers are likely to adopt a tetrahedral orientation that minimizes the electrostatic repulsion between the DNA arms (168). Support for this hypothesis came from nondenaturing gel electrophoresis experiments. These experiments showed a migration anomaly for the DNA junction with one-to-four hybridized DNA arms, with the apparent length of the complex being larger than expected based on the increase in sequence length and in mass upon hybridization of each arm. The increase in the extent of the anomaly with the number of hybridized arms suggested an increase in the cross-section of the junction and a star-like shape, such as that of a tetrahedral arrangement of the four DNA duplexes connected to the [Ni(**cyclam**)]$^{2+}$ complex.

Extended structures can be rationally built by appropriate combination of the metal-based DNA junctions with nucleic acid entities that have a specific degree of complementarity. Hybridization of two four-arm, [Ni(**cyclam**)]$^{2+}$-based DNA junctions that had complementary DNA arms showed results consistent with the formation of high-order, infinite structures [Fig. 18(*a*)] (168). Mixing of complementary three-arm, [Fe(**bpy**)$_3$]$^{2+}$-based DNA junctions led to mesoscopic structures (171). The combination of three two-arm, [Fe(**tpy**)$_2$]$^{2+}$-based DNA junctions that had arms intermolecularly, pairwise complementary led to DNA triangles with distinct DNA duplexes as edges and [Fe(**tpy**)$_2$]$^{2+}$ vertices [Fig. 18(*b*)] (169). Hybridization by slow cooling of 1:1 mixtures of two-arm DNA junctions based on **bpy**–Ru$^{2+}$ or **tpy**–Ru$^{2+}$ complexes that had intramolecularly identical but intermolecularly complementary DNA arms led to infinite, linear DNA polymer formation (170, 172). In contrast, room temperature hybridization of the same two-arm DNA junctions based on **bpy**–Ru$^{2+}$ led to the formation of a mixture of structures, the majority of which were dimeric and cyclic [Fig. 18(*c*)] (172).

The ssDNA oligonucleotides have been also used to "clamp" by hybridization organic, 2D modules in a predetermined assembly in which reactive groups of different organic modules are situated in close proximity of each other, and can be efficiently "glued" together by subsequent ractions. For example, ethynylbenzene molecules [Fig. 19(*a* and *b*)] have been linked to DNA oligonucleotides, whose

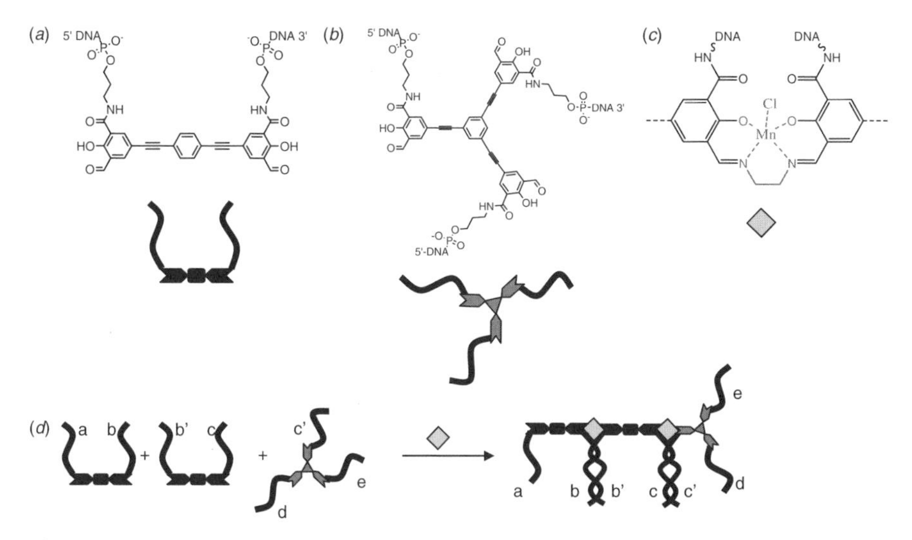


Figure 19. Extended hybrid inorganic–nucleic acid structures (*d*) based on organic (*a* and *b*) and inorganic (*c*) DNA junctions (adapted from Reference 174).

sequences were designed such that hybridization leads to a predetermined topology of ethynylbenzene and **sal** ligands (174). Upon DNA hydridization, **sal** were situated in adjacent positions and reacted with $M^{n+}$ and en to form covalent [M(**salen**)] cross-links [Fig. 19 (*c* and *d*)] (175). Attempts to enzymatically digest the oligonucleotide strands that encoded the structure to afford programmed oligo(phenylethylene)-metallo-salen complexes resulted in incomplete removal of the oligonucleotides. To circumvent this problem, redox-sensitive disulfide groups were used as linkers between the ligands and nucleic acid strands (176) and the redox-inert $Al^{3+}$ was coordinated to **salen**. Treatment with a reducing agent of the $[Al(\mathbf{salen})]^{+}$ disulfide-bridged, DNA structures led indeed to the isolation of the metallo-salen-oligo(phenylethylene) backbone.

## VI. CONCLUSIONS

This chapter presented the results of research aimed at the combined use of transition metal coordination and nucleic acid hybridization to construct hybrid inorganic–nucleic acid supramolecular structures. The interaction of alkali metal ions with DNA and RNA is nonspecific, and has been long recognized to be important for the charge neutralization and the nucleic acid secondary structure. In contrast, the binding of transition metal ions to DNA duplexes is site specific, but it is still relatively weak compared to their binding to many polydentate ligands used in coordination chemistry. This property makes possible the site-specific incorporation of transition metal ions in nucleic acid structures that are modified with ligands. If complementary DNA oligonucleotides are modified with ligands situated in complementary positions, DNA duplexes that contain artificial metal–ligand base pairs are formed upon hybridization. The metal–ligand bonds play a role similar to that of the hydrogen bonds in Watson–Crick base pairs for molecular recognition and duplex stabilization. The two types of interactions determine the properties of the metal-containing duplex. Terminal ligand modification of ssDNA is the basis for the synthesis of metal-based DNA junctions between ss- or ds-DNA, which can be used to create 2D and 3D structures by self-assembly. The structure and properties of the metal-containing supermolecules depend on the coordination geometry of the metal complexes that act as rigid junctions, on the flexibility of the linkers that connect the ligands to the nucleic acids, and on the stability of the DNA duplexes. It can be easily envisioned that the range of structures generated by a strategy that combines metal coordination and nucleic acid hybridization is broad in terms of both topology and function.

In the synthesis of metal-containing nucleic acid structures, DNA and its synthetic analogues have been preferred to RNA because they are chemically and biochemically more stable and structurally simpler. These properties make them appropriate for use in nanotechnology applications in which the integrity and stability of the nanostructures is important. The discovery of well-defined, stable DNA junctions was essential for the development of structural DNA nanotechnology as a research field in the last 25 years. Metal complexes have the potential to form such junctions too and will diversify the structural motifs available for DNA-based nanoassemblies. Furthermore, given the electronic structure and spectroscopic and chemical properties of transition metal ions, these ions can be used as intrinsic reporters of the structure or as functional elements. For example, metal ions situated at specific locations in a periodic DNA lattice could direct the organization of proteins for crystallization experiments. To date, DNA nanostructures have been made functional by exploiting structural changes induced in DNA by small molecules, DNA single strands, or DNA-binding proteins to create nanomechanical devices. Distinct from these applications, the catalytic properties of the transition metal ions integrated in DNA nanostructures together with the template effect exerted by the nucleic acid part of the structure could be used to catalyze reactions of DNA-attached chemical substrates. Finally, as metal coordination is orthogonal to hydrogen bonding, the former interactions could be used to expand the genetic code, provided that methods for *in vivo* incorporation of ligands into DNA are developed.

Achievement of the potential of hybrid metal ion–nucleic acid structures depends on the ability of researchers to synthesize ligand-containing, nucleic acid oligomers and to characterize the complexes formed by these oligomers with transition metal ions. The use of synthetic and characterization methods specific to the fields of coordination and nucleic acid chemistry creates a concomitant opportunity and challenge. For example, the spectroscopic methods used to investigate the stoichiometry and coordination geometry of the metal complexes attached to nucleic acids may require concentrations exceeding those typically used or practically achievable for nucleic acid solutions. Also, the new structural motifs brought about by the stereochemistry of the metal complexes create isomers that are close in energy and need to be isolated and characterized. The effort necessary to find solutions for these problems will be compensated for by the properties of the hybrid transition metal ion–nucleic acid structures, which are promising.

## ACKNOWLEDGMENTS

This work was supported by NSF Grant No. CHE-0347140 and by the Camille and Henry Dreyfus Foundation.

## ABBREVIATIONS

| | |
|---|---|
| 2AP | 2-Aminophenol |
| 1D | One dimensional |
| 2D | Two dimensional |
| 3D | Three dimensional |
| 2PA | Bis(2-pyridylmethyl)amine |
| APP | Alternating purine-pyrimidine |
| benzo-15C5 | Benzo-15-crown-5-ether |
| Bn | Benzyl |
| Bt | Biotinylated |
| bpy | 2, 2′-Bipyridine |
| CD | Circular dichroism |
| CT | Calf thymus |
| Cyclam | 1,4,8,11-Tetraazacyclotetradecane |
| Dabcyl | 4-[[4-(dimethylamino)phenyl]azo]-benzoic acid |
| DFT | Density function theory |
| Dipic | Pyridine-2,6-dicarboxylate |

| | |
|---|---|
| Dipam | Pyridine-2.6-dicarboxamide |
| DMSO | Dimethyl sulfoxide |
| DNA | Deoxyribonucleic acid |
| dpa | Anion of di(2-pyridyl)amine |
| ds | Double stranded |
| en | Ethylenediamine |
| EDTA | Ethylenediaminetetraacetic acid |
| edta | Ethylenediaminetetraacetato (ligand) |
| ESI MS | Electrospray ionization mass spectrometry |
| EPR | Electron paramagnetic resonance |
| EXAFS | Extended X-ray absorption fine structure |
| equiv | Equivalent |
| Glu | Glutamate |
| GNA | Glycol nucleic acid |
| H | Hydroxypyridone |
| His | Histidine |
| $^1$H NMR | Proton nuclear magnetic resonance |
| HOMO | Highest occupied molecular orbital |
| IDA | Iminodiacetate |
| L | Ligand |
| LNA | Locked nucleic acid |
| LUMO | Lowest unoccupied molecular orbital |
| Mebpy | Methyl bipyridine |
| MLCT | Metal-to-ligand charge transfer |
| MS | Mass spectrometry |
| NADI | Naphthalene diimide |
| NMR | Nuclear magnetic resonance |
| NOESY | Nuclear Overhauser effect spectroscopy |
| phen | Phenantroline |
| PNA | Peptide nucleic acid |
| $Pur^P$ | 6-(2′-Pyridyl)-purine |
| Py | Pyridine |
| $Pyr^P$ | 4-(2′-Pyridyl)-pyrimidine |
| Q | 8-Hydroxyquinoline |
| sal | Salicylaldehyde |
| salen | *N,N′*-Bis(salicylidene)ethylendiame |
| SPy | 2,6-Bis(ethylthiomethyl)pyridine |
| ss | Single stranded |
| tpy | Terpyridine |
| UV | Ultraviolet |
| Vis | Visible |

## REFERENCES


1. E. Hammarsten, *Biochem. Z.*, *144*, 383 (1924).
2. B. Lippert, *Coord. Chem. Rev.*, *200–202*, 487 (2000).
3. G. L. Eichhorn, J. M. Rifkind, and Y. A. Shin, *Adv. Chem. Ser.*, *162*, 304 (1977).
4. Y. A. Shin and G. L. Eichhorn, *Biopolymers*, *16*, 225 (1977).
5. M. A. Pennella and D. P. Giedroc, *BioMetals*, *18*, 413 (2005).
6. D. Wang and S. J. Lippard, *Nat. Rev. Drug Disc.*, *4*, 307 (2005).
7. K. R. Barnes and S. J. Lippard, *Met. Ions Biol. Syst.*, *42*, 143 (2004).
8. C. X. Zhang and S. J. Lippard, *Curr. Opin. Chem. Biol.*, *7*, 481 (2003).
9. E. R. Jamieson and S. J. Lippard, *Chem. Rev.*, *99*, 2467 (1999).
10. N. C. Seeman, *Methods Mol. Biol. (Totowa, NJ, U. S.)*, *303*, 143 (2005).
11. N. C. Seeman, *Chem. Biol.*, *10*, 1151 (2003).
12. N. C. Seeman, *Biochemistry*, *42*, 7259 (2003).
13. E. Braun, Y. Eichen, U. Sivan, and G. Ben-Yoseph, *Nature*, *391*, 775 (1998).
14. K. Keren, M. Krueger, R. Gilad, G. Ben-Yoseph, U. Sivan, and E. Braun, *Science*, *297*, 72 (2002).
15. K. Keren, R. S. Berman, and E. Braun, *Nano Lett.*, *4*, 323 (2004).
16. R. M. Franzini, R. M. Watson, G. K. Patra, R. M. Breece, D. L. Tierney, M. P. Hendrich, and C. Achim, *Inorg. Chem.*, *45*, 9798 (2006).
17. K. Tanaka, Y. Yamada, and M. Shionoya, *J. Am. Chem. Soc.*, *124*, 8802 (2002).
18. Z. Kuklenyik and L. G. Marzilli, *Inorg. Chem.*, *35*, 5654 (1996).
19. E. T. Kool, *Ann. Rev. Bioph. Biom. Structure*, *30*, 1 (2001).
20. G. M. Blackburn, M. J. Gait, ed., *Nucleic Acids in Chemistry and Biology, 2nd ed.*, Oxford Univ. Press, Oxford, UK, 1996.
21. S. Neidle, ed., *Oxford Handbook of Nucleic Acid Structure*, Oxford Univ. Press, Oxford UK, 1999.
22. R. E. Dickerson, H. R. Drew, B. N. Conner, R. M. Wing, A. V. Fratini, and M. L. Kopka, *Science*, *216*, 475 (1982).
23. B. Petersson, B. B. Nielsen, H. Rasmussen, I. K. Larsen, M. Gajhede, P. E. Nielsen, and J. S. Kastrup, *J. Am. Chem. Soc.*, *127*, 1424 (2005).
24. H. Rasmussen, J. S. Kastrup, J. N. Nielsen, J. M. Nielsen, and P. E. Nielsen, *Nat. Struct. Biol.*, *4*, 98 (1997).
25. P. E. Nielsen, M. Egholm, R. H. Berg, and O. Buchardt, *Science*, *254*, 1497 (1991).
26. F. Beck, *Methods Mol. Bio.*, *208*, 29 (2002).
27. F. Beck and P. E. Nielsen, *Art. DNA*, 91 (2003).
28. U. Diederichsen, *Angew. Chem., Int. Ed. Eng.*, *35*, 445 (1996).
29. P. E. Nielsen, Ed., *Peptide Nucleic Acids: Protocols and Applications, 2nd ed.,* Horizon Bioscience, Wymondham, UK, 2004.
30. A. Hess and N. Metzler-Nolte, *Chem. Commun.*, 885 (1999).
31. J. C. Verheijen, G. A. van der Marel, J. H. van Boom, and N. Metzler-Nolte, *Bioconjugate Chem.*, *11*, 741 (2000).
32. A. Maurer, H.-B. Kraatz, and N. Metzler-Nolte, *Eur. J. Inorg. Chem.*, 3207 (2005).
33. R. Hamzavi, T. Happ, K. Weitershaus, and N. Metzler-Nolte, *J. Organomet. Chem.*, *689*, 4745 (2004).

34. P. Wittung, M. Eriksson, R. Lyng, P. E. Nielsen, and B. Norden, *J. Am. Chem. Soc.*, *117*, 10167 (1995).
35. S. Obika, D. Nanbu, Y. Hari, J.-I. Andoh, K.-I. Morio, T. Doi, and T. Imanishi, *Tetrahedron Lett.*, *39*, 5401 (1998).
36. S. K. Singh, P. Nielsen, A. A. Koshkin, and J. Wengel, *Chem. Commun.*, 455 (1998).
37. J. Wengel, M. Petersen, M. Frieden, and T. Koch, *Lett. Pept. Sci.*, *10*, 237 (2004).
38. K. E. Nielsen, J. Rasmussen, R. Kumar, J. Wengel, J. P. Jacobsen, and M. Petersen, *Bioconjugate Chem.*, *15*, 449 (2004).
39. L. Zhang, A. Peritz, and E. Meggers, *J. Am. Chem. Soc.*, *127*, 4174 (2005).
40. A. Houlton, *Adv. Inorg. Chem.*, *53*, 87 (2002).
41. B. Lippert, *Prog. Inorg. Chem.*, *54*, 385 (2005).
42. V. J. De Rose, S. Burns, N. K. Kim, and M. Vogt, *Comp. Coord. Chem. II*, *8*, 787 (2004).
43. M. Egli, *Chem. Biol.*, *9*, 277 (2002).
44. M. Egli, *Curr. Opin. Chem. Biol.*, *8*, 580 (2004).
45. E. Sletten and N. A. Froeystein, *Met. Ions Biol. Syst.*, *32*, 397 (1996).
46. C. M. Dupureur and J. K. Barton, *Comp. Supramol. Chem.*, *5*, 295 (1996).
47. R. B. Martin, *Met. Ions Biol. Syst.*, *32*, 61 (1996).
48. R. B. Martin, *Acc. Chem. Res.*, *18*, 32 (1985).
49. M.-H. Baik, R. A. Friesner, and S. J. Lippard, *J. Am. Chem. Soc.*, *125*, 14082 (2003).
50. T. Schoenknecht and H. Diebler, *J. Inorg. Biochem.*, *50*, 283 (1993).
51. M. Gueron, J. P. Demaret, and M. Filoche, *Biophys. J.*, *78*, 1070 (2000).
52. H. Sigel, N. A. Corfu, L. N. Ji, and R. B. Martin, *Comments Inorg. Chem.*, *13*, 35 (1992).
53. J. C. Sitko, E. M. Mateescu, and H. G. Hansma, *Biophys. J.*, *84*, 419, (2003).
54. J. Vinje and E. Sletten, *Chem.—Eur. J.*, *12*, 676 (2006).
55. N. A. Froeystein, J. T. Davis, B. R. Reid, and E. Sletten, *Acta Chem. Scand.*, *47*, 649 (1993).
56. X. Jia, G. Zon, and L. G. Marzilli, *Inorg. Chem.*, *30*, 228 (1991).
57. N. G. A. Abrescia, L. Malinina, L. G. Fernandez, T. Huynh-Dinh, S. Neidle, and J. A. Subirana, *Nucleic Acids Res.*, *27*, 1593 (1999).
58. N. G. A. Abrescia, H.-D. Tam, and J. A. Subirana, *J. Biol. Inorg. Chem.*, *7*, 195 (2002).
59. S. L. Labiuk, T. J. Delbaere Louis, and J. S. Lee, *J. Biol. Inorg. Chem.*, *8*, 715 (2003).
60. M. Soler-Lopez, L. Malinina, V. Tereshko, V. Zarytova, and J. A. Subirana, *J. Biol. Inorg. Chem.*, *7*, 533 (2002).
61. T. F. Kagawa, B. H. Geierstanger, A. H. J. Wang, and P. S. Ho, *J. Biol. Chem.*, *266*, 20175 (1991).
62. N. Valls, I. Uson, C. Gouyette, and J. A. Subirana, *J. Am. Chem. Soc.*, *126*, 7812 (2004).
63. E. Moldrheim, B. Andersen, N. A. Froystein, and E. Sletten, *Inorg. Chim. Acta*, *273*, 41 (1998).
64. J. Vinje, J. A. Parkinson, P. J. Sadler, T. Brown, and E. Sletten, *Chem. A Eur. J.*, *9*, 1620 (2003).
65. A. Pullman, B. Pullman, and R. Lavery, *Theochem*, *10*, 85 (1983).
66. Y.-G. Gao, M. Sriram, and A. H. J. Wang, *Nucleic Acids Res.*, *21*, 4093 (1993).
67. J. S. Lee, L. J. P. Latimer, and R. S. Reid, *Biochem. Cell Biol.*, *71*, 162 (1993).
68. P. Aich, S. L. Labiuk, L. W. Tari, L. J. T. Delbaere, W. J. Roesler, K. J. Falk, R. P. Steer, and J. S. Lee, *J. Mol. Biol.*, *294*, 477 (1999).
69. S. D. Wettig, D. O. Wood, P. Aich, and J. S. Lee, *J. Inorg. Biochem.*, *99*, 2093 (2005).
70. D. O. Wood and J. S. Lee, *J. Inorg. Biochem.*, *99*, 566 (2005).

71. D. O. Wood, M. J. Dinsmore, G. A. Bare, and J. S. Lee, *Nucleic Acids Res.*, *30*, 2244 (2002).
72. P. Aich, R. J. S. Skinner, S. D. Wettig, R. P. Steer, and J. S. Lee, *J. Biomol. Struct. Dyn.*, *20*, 93 (2002).
73. S. D. Wettig, D. O. Wood, and J. S. Lee, *J. Inorg. Biochem.*, *94*, 94 (2003).
74. C.-Z. Li, Y.-T. Long, H.-B. Kraatz, and J. S. Lee, *J. Phys. Chem. B*, *107*, 2291 (2003).
75. B. Liu, A. J. Bard, C.-Z. Li, and H.-B. Kraatz, *J. Phys. Chem. B*, *109*, 5193 (2005).
76. A. Rakitin, P. Aich, C. Papadopoulos, Y. Kobzar, A. S. Vedeneev, J. S. Lee, and J. M. Xu, *Phys. Rev. Lett.*, *86*, 3670 (2001).
77. S. D. Wettig, G. A. Bare, R. J. S. Skinner, and J. S. Lee, *Nano Lett.*, *3*, 617 (2003).
78. (a) M. Fuents-Cabrera, B. G. Sumpter, J. E. Sponer, J. Sponer, L. Petit, J.C. Wells, *J. Phys. Chem. B*, *111*, 870, (2007) and refs. therein; (b) E. M. Conwell, *Topics Curr. Chem.*, *237*, 73 (2004).
79. N. A. Froeystein and E. Sletten, *J. Am. Chem. Soc.*, *116*, 3240 (1994).
80. P. R. Young, U. S. Nandi, and N. R. Kallenbach, *Biochemistry*, *21*, 62 (1982).
81. Y. Miyake, H. Togashi, M. Tashiro, H. Yamaguchi, S. Oda, M. Kudo, Y. Tanaka, Y. Kondo, R. Sawa, T. Fujimoto, T. Machinami, and A. Ono, *J. Am. Chem. Soc.*, *128*, 2172 (2006).
82. S. Katz and V. Santilli, *Biochim. Biophys. Acta*, *55*, 621 (1962).
83. S. Katz, *Biochim. Biophys. Acta*, *68*, 240 (1963).
84. S. Steinkopf, W. Nerdal, A. Kolstad, and E. Sletten, *Acta Chem. Scand.*, *50*, 775 (1996).
85. L. D. Kosturko, C. Folzer, and R. F. Stewart, *Biochemistry*, *13*, 3949 (1974).
86. R. M. Izatt, J. J. Christensen, and J. H. Rytting, *Chem. Rev.*, *71*, 439 (1971).
87. H. Arakawa, J. F. Neault, and H. A. Tajmir-Riahi, *Biophys. J.*, *81*, 1580 (2001).
88. E. Ennifar, P. Walter, and P. Dumas, *Nucleic Acids Res.*, *31*, 2671 (2003).
89. M. Shionoya and K. Tanaka, *Curr. Opin. Chem. Biol.*, *8*, 592 (2004).
90. H.-A. Wagenknecht, *Angew. Chem., Int. Ed. Engl.*, *42*, 3204 (2003).
91. K. Tanaka and M. Shionoya, *J. Org. Chem.*, *64*, 5002 (1999).
92. H. Weizman and Y. Tor, *J. Am. Chem. Soc.*, *123*, 3375 (2001).
93. H. Weizman and Y. Tor, *Chem. Commun.*, 453 (2001).
94. A. A. Henry and F. E. Romesberg, *Curr. Opin. Chem. Biol.*, *7*, 727 (2003).
95. E. T. Kool, *Acc. Chem. Res.*, *35*, 936 (2002).
96. E. Meggers, P. L. Holland, W. B. Tolman, F. E. Romesberg, and P. G. Schultz, *J. Am. Chem. Soc.*, *122*, 10714 (2000).
97. Y. Tanaka, H. Yamaguchi, S. Oda, Y. Kondo, M. Nomura, C. Kojima, and A. Ono, *Nucleos. Nucleot. Nucl. Acids*, *25*, 613 (2006).
98. J. Müller, M. Drumm, M. Boudvillain, M. Leng, E. Sletten, and B. Lippert, *J. Biol. Inorg. Chem.*, *5*, 603 (2000).
99. J. Müller, D. Boehme, P. Lax, M. M. Cerda, and M. Roitzsch, *Chem.—Eur. J.*, *11*, 6246 (2005).
100. M. Shionoya and K. Tanaka, *Bull. Chem. Soc. Jpn.*, *73*, 1945 (2000).
101. K. Tanaka, M. Tasaka, H. Cao, and M. Shionoya, *Eur. J. Pharm. Sci.*, *13*, 77 (2001).
102. M. Tasaka, K. Tanaka, M. Shiro, and M. Shionoya, *Supramol. Chem.*, *13*, 671 (2001).
103. H. Cao, K. Tanaka, and M. Shionoya, *Chem. Pharmaceu. Bull.*, *48*, 1745 (2000).
104. K. Tanaka, M. Tasaka, H. Cao, and M. Shionoya, *Supramol. Chem.*, *14*, 255 (2002).
105. W. Clegg, A. J. Scott, F. J. Lawlor, N. C. Norman, T. B. Marder, C. Dai, and P. Nguyen, *Acta Crystallogr., Sec. C: Crystal Structure Commun.*, *C54*, 1875 (1998).

106. N. Zimmermann, E. Meggers, and P. G. Schultz, *J. Am. Chem. Soc.*, *124*, 13684 (2002).
107. C. Brotschi, A. Haberli, and C. J. Leumann, *Angew. Chem., Int. Ed. Engl.*, *40*, 3012 (2001).
108. C. Brotschi and C. J. Leumann, *Angew. Chem., Int. Ed. Engl.*, *42*, 1655 (2003).
109. C. Brotschi, G. Mathis, and C. J. Leumann, *Chem.—Eur. J.*, *11*, 1911 (2005).
110. D.-L. Popescu, T. J. Parolin, and C. Achim, *J. Am. Chem. Soc.*, *125*, 6354 (2003).
111. R. Franzini, R. M. Watson, D.-L. Popescu, G. K. Patra, and C. Achim, *Polymer Preprints*, *45*, 337 (2004).
112. C. Switzer and D. Shin, *Chem. Commun.*, 1342 (2005).
113. C. Switzer, S. Sinha, P. H. Kim, and B. D. Heuberger, *Angew. Chem., Int. Ed. Engl.*, *44*, 1529 (2005).
114. L. Zhang and E. Meggers, *J. Am. Chem. Soc.*, *127*, 74 (2005).
115. R. M. Watson, Y. Skorik, G. K. Patra, and C. Achim, *J. Am. Chem. Soc.*, *127*, 14628 (2005).
116. K. Tanaka, A. Tengeiji, T. Kato, N. Toyama, M. Shiro, and M. Shionoya, *J. Am. Chem. Soc.*, *124*, 12494 (2002).
117. K. Tanaka, A. Tengeiji, T. Kato, N. Toyama, and M. Shionoya, *Science*, *299*, 1212 (2003).
118. G. H. Clever, K. Polborn, and T. Carell, *Angew. Chem., Int. Ed. Engl.*, *44*, 7204 (2005).
119. S. Atwell, E. Meggers, G. Spraggon, and P. G. Schultz, *J. Am. Chem. Soc.*, *123*, 12364 (2001).
120. N. Zimmermann, E. Meggers, and P. G. Schultz, *Bioorg. Chem.*, *32*, 13 (2004).
121. B. R. Babu, P. J. Hrdlicka, C. J. McKenzie, and J. Wengel, *Chem. Commun.*, 1705 (2005).
122. R. M. Smith and A. E. Martell, *Critical Stab. Const. II*, Plemum Press: New York and London, (1975).
123. J. Müller, E. Gil Bardaji, and F. A. Polonius, *Acta Crystallogr., Sec. E: Structure Rep. Online*, *E62*, 223 (2006).
124. J. Müller, F.-A. Polonius, and M. Roitzsch, *Inorg. Chim. Acta*, *358*, 1225 (2005).
125. S. K. Sharma and L. W. McLaughlin, *J. Inorg. Biochem.*, *98*, 1570 (2004).
126. K. Wiederholt and L. W. McLaughlin, *Nucleic Acids Res.*, *27*, 2487 (1999).
127. E. C. Constable, *Adv. Inorg. Chem.*, *34*, 1 (1989).
128. C. Brotschi and C. J. Leumann, *Nucleos. Nucleot. Nucl. Acids*, *22*, 1195 (2003).
129. B. Milani, A. Anzilutti, L. Vicentini, A. S. Santi, E. Zangrando, S. Geremia, and G. Mestroni, *Organometallics*, *16*, 5064 (1997) and references therein.
130. A. El-Jammal, P. L. Howell, M. A. Turner, N. Li, and D. M. Templeton, *J. Med. Chem.*, *37*, 461 (1994).
131. S. I. Ahmed, J. Burgess, J. Fawcett, S. A. Parsons, D. R. Russell, and S. H. Laurie, *Polyhedron*, *19*, 129 (2000).
132. W. P. Griffith and S. I. Mostafa, *Polyhedron*, *11*, 2997 (1992).
133. G. H. Clever, Y. Soelk, W. Spahl, and T. Carell, *Chem.-Eur. J 12*, 8708 (2006).
134. H. L. Conley, Jr., and R. B. Martin, *J. Phys. Chem.*, *69*, 2914 (1965).
135. Y. Nawata, H. Iwasaki, and Y. Saito, *Bull. Chem. Soc. Jpn.*, *40*, 515 (1967).
136. R. P. Bonomo, V. Cucinotta, and F. Riggi, *J. Mol. Struct.*, *69*, 295 (1980).
137. A. Mokhir, R. Stiebing, and R. Krämer, *Bioorg. Med. Chem. Lett.*, *13*, 1399 (2003).
138. P. J. Hrdlicka, B. R. Babu, M. D. Sorensen, and J. Wengel, *Chem. Commun.*, 1478 (2004).
139. A. A. Mokhir and R. Krämer, *Bioconjugate Chem.*, *14*, 877 (2003).
140. A. Mokhir, R. Krämer, and H. Wolf, *J. Am. Chem. Soc.*, *126*, 6208 (2004).

141. A. Küsel, J. Zhang, M. A. Gil, A. C. Stueckl, W. Meyer-Klaucke, F. Meyer, and U. Diederichsen, *Eur. J. Inorg. Chem.*, 4317 (2005).
142. G. H. Elever and T. Earell, *Angew. Chem., Int. Ed. Engl.*, *46*, 250 (2007).
143. K. Tanaka, G. H. Elever, Y. Takezaura, Y. Yamada, C. Kaul, M. Shionoya, and T. Carell, *Nat. Nanotech.*, *1*, 190 (2006).
144. H. Y. Zhang, A. Calzolari, and R. Di Felice, *J. Phys. Chem. B*, *109*, 15345 (2005).
145. A. A. Voityuk, *J. Phys. Chem. B*, *110*, 21010 (2006).
146. M. Benard, J. F. Berry, F. A. Cotton, C. Gaudin, X. Lopez, C. A. Murillo, and M.-M. Rohmer, *Inorg. Chem.*, *45*, 3932 (2006).
147. B. P. Gilmartin, K. Ohr, R. L. McLaughlin, R. Koerner, and M. E. Williams, *J. Am. Chem. Soc.*, *127*, 9546 (2005).
148. M. Göritz and R. Krämer, *J. Am. Chem. Soc.*, *127*, 18016 (2005).
149. (a) N. Graf, M. Göritz, and R. Krämer, *Angew. Chem., Int. Ed. Engl.*, *45*, 4013 (2006); (b) N. Graf and R. Krämer, *Chem. Comm.*, 4375 (2006).
150. L. Zapata, K. Bathany, J.-M. Schmitter, and S. Moreau, *Eur. J. Org. Chem.*, 1022 (2003).
151. G. Bianke and R. Haener, *ChemBioChem*, *5*, 1063 (2004).
152. F. D. Lewis, S. A. Helvoigt, and R. L. Letsinger, *Chem. Commun.*, 327 (1999).
153. J. L. Czlapinski and T. L. Sheppard, *ChemBioChem*, *5*, 127 (2004).
154. K. Fujimoto, Y. Muto, and M. Inouye, *Chem. Commun.*, 4780 (2005).
155. P. L. Paris, J. Langenhan, and E. T. Kool, *Nucleic Acids Res.*, *26*, 3789 (1998).
156. X. Li and D. R. Liu, *Angew. Chem., Int. Ed. Engl.*, *43*, 4848 (2004).
157. M. D. Distefano, J. A. Shin, and P. B. Dervan, *J. Am. Chem. Soc.*, *113*, 5901 (1991).
158. M. D. Distefano and P. B. Dervan, *J. Am. Chem. Soc.*, *114*, 11006 (1992).
159. J. L. Czlapinski and T. L. Sheppard, *J. Am. Chem. Soc.*, *123*, 8618 (2001).
160. J. L. Czlapinski and T. L. Sheppard, *Bioconjugate Chem.*, *16*, 169 (2005).
161. I. Horsey, Y. Krishnan-Ghosh, and S. Balasubramanian, *Chem. Commun.*, 1950 (2002).
162. S. Sueda, T. Ihara, and M. Takagi, *Chem. Lett.*, 1085 (1997).
163. T. Ihara, Y. Takeda, and A. Jyo, *J. Am. Chem. Soc.*, *123*, 1772 (2001).
164. Y. Kitamura, T. Ihara, Y. Tsujimura, M. Tazaki, and A. Jyo, *Chem. Lett.*, *34*, 1606 (2005).
165. U. Feldkamp and C. M. Niemeyer, *Angew. Chem., Int. Ed. Engl.*, *45*, 1856 (2006).
166. N. C. Seeman, *Trends Biochem. Sci.*, *30*, 119 (2005).
167. N. C. Seeman and P. S. Lukeman, *Rep. Prog. Phy.*, *68*, 237 (2005).
168. K. M. Stewart and L. W. McLaughlin, *J. Am. Chem. Soc.*, *126*, 2050 (2004).
169. J. S. Choi, C. W. Kang, K. Jung, J. W. Yang, Y.-G. Kim, and H. Han, *J. Am. Chem. Soc.*, *126*, 8606 (2004).
170. K. M. Stewart and L. W. McLaughlin, *Chem. Commun.*, 2934 (2003).
171. S. Takeneka, Y. Funatu, and H. Kondo, *Chem. Lett.*, *10*, 891 (1996).
172. D. Mitra, N. Di Cesare, and H. F. Sleiman, *Angew. Chem., Int. Ed. Engl.*, *43*, 5804 (2004).
173. K. M. Stewart, J. Rojo, and L. W. McLaughlin, *Angew. Chem., Int. Ed. Engl.*, *43*, 5808 (2004).
174. K. V. Gothelf, A. Thomsen, M. Nielsen, E. Clo, and R. S. Brown, *J. Am. Chem. Soc.*, *126*, 1044 (2004).
175. K. V. Gothelf and R. S. Brown, *Chem.—Eur. J.*, *11*, 1062 (2005).
176. R. S. Brown, M. Nielsen, and K. V. Gothelf, *Chem. Commun.*, 1464 (2004).

CHAPTER 9

# Bispidine Coordination Chemistry

**PETER COMBA, MARION KERSCHER, and WOLFGANG SCHIEK**

*Universität Heidelberg*
*Anorganisch-Chemisches Institut*
*Heidelberg, Germany, 69120.*

CONTENTS



*Progress in Inorganic Chemistry, Vol. 55* Edited by Kenneth D. Karlin

# I. INTRODUCTION

The bispidine backbone (bispidine = 3,7-diazabicyclo[3.3.1]nonane) **1** is a simplified derivative of the natural product sparteine, **2** (see Chart 1; shown is the structure of α-isosparteine). Sparteine, a polycyclic alkaloid, occurs in a wide range of plants and has a variety of medical and chemical applications. It has been investigated extensively, for example, as a cardiac antiarrhythmia (1, 2), an ion channel blocker (3, 4), and an uterotonic drug (5). It was also used as a chiral base (6) or ligand (7) for asymmetric synthesis and kinetic optical resolution (8).

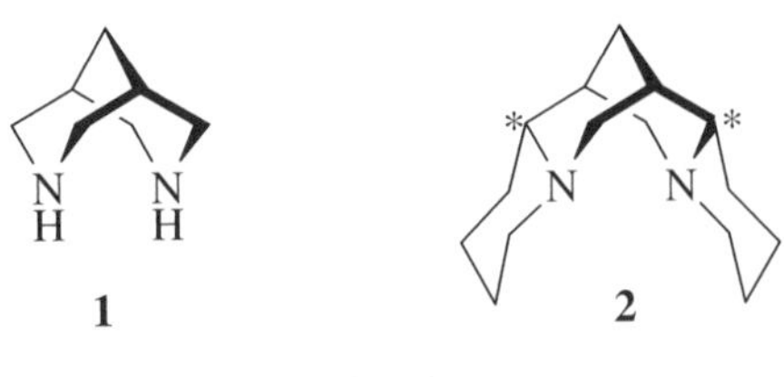


Chart 1

Sparteine is, due to its rigidity and ability to efficiently shield the coordination site, an interesting bidentate ligand. The coordination chemistry of sparteine is not extensively explored, but recent reports have described the synthesis, spectroscopic properties and structures of a row of its transition metal complexes (9–15). One possible reason for the poorly studied coordination chemistry of sparteine is the fact that differently substituted and enantiomerically pure derivatives of sparteine require some effort and are expensive to synthesize (16, 17), which limits their application to a few small-scale experiments. Bispidine-type ligands in contrast are readily obtained, substituted in 2-, 3-, 4-, 6-, 7- and / or 8-position with a wide range of donor sets, which opens the door for multidentate and oligonucleating ligands with a variety of possible donor sets and varying structural forms, without significant loss of the enormous diazaadamantane-based rigidity.

The first report of bispidine-type molecules dates back to 1930, when Mannich reported the synthesis of compounds **3–5** (see Scheme 1)(18). Due to their antiarrhythmic (19–21) and analgetic (22–24) activity, compounds **1, 6–8** (see Charts 1, 2) and derivatives have attracted much attention, and much of this work has been reviewed and patented comprehensively (25–29).

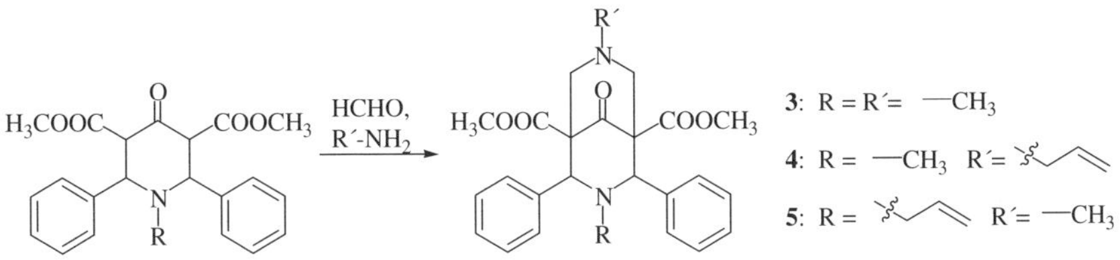

$H_3COOC$, $COOCH_3$, HCHO, $R'$-$NH_2$ →

3: R = R′ = —$CH_3$

4: R = —$CH_3$ R′ = allyl

5: R = allyl R′ = —$CH_3$

Scheme 1

First transition metal coordination compounds with bidentate bispidine ligands were described in 1957 (30). The initial report with metal complexes of tetradentate bispidine ligands dates back to 1969 (31). Following these early reports, there have been a number of studies on the complexation properties of several bipidine derivatives (32–35). However, extensive, broad, and thorough studies of the bispidine coordination chemistry began only <10 years ago. These studies will be reviewed here. They include structural and theoretical work, spectroscopy, electron-transfer studies, metal ion selective complexation, and applications in biomimetic chemistry, catalysis, and molecular magnetism.

In general, the chair–chair (cc) conformation of bispidines is preferred over the chair–boat (cb) and boat–boat (bb) isomers (see Section II.B and Chart 9). Therefore, bidentate bispidine ligands are very rigid and highly preorganized, and were expected to lead to very stable complexes with a pronounced selectivity with respect to the size (and electronic requirements) of the metal ion (36–38). Bispidines may also be regarded as sterically reinforced and, therefore, highly preorganized derivatives of daco, **9** (daco = 1,5-diazacyclooctane, see Chart 3). In view of the versatility of daco and the well-developed coordination chemistry of that diamine (39), the bispidine coordination chemistry was expected to be of general interest. Indeed, with respect to many fundamental principles in coordination chemistry (e.g., preorganization, rigidity and elasticity, complementarity, and entasis), bispidine coordination chemistry provides excellent examples for the discussion, understanding, and further development of these principles.

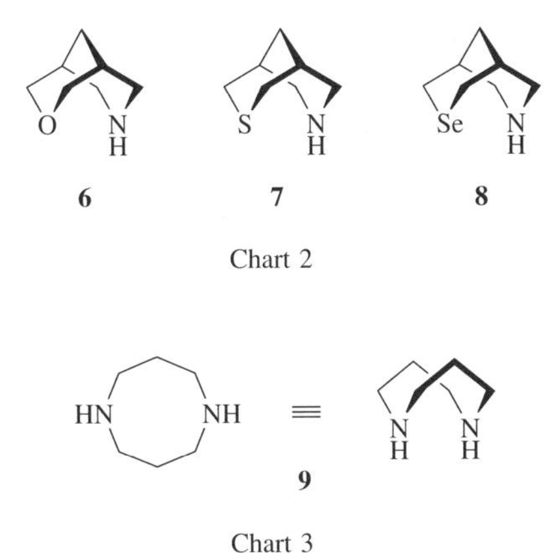


Chart 2

Chart 3

The plasticity of bonds and angles around metal ions is generally larger than that of organic molecules (40). Therefore, reinforced ligands with restricted conformational flexibility may enforce specific, not necessarily the most stable, coordination geometries (38, 40–42). This approach is important for the energization of coordination compounds, which is used to induce specific reactivities and is known as the entatic state principle. This principle was developed from observations of metalloproteins (41, 43–48). A careful analysis of structures, reactivities, and thermodynamic properties of metalloproteins and low molecular weight coordination compounds (38, 40–42, 49–51), indicates that ligand preorganization, steric and electronic complementarity, ligand-enforced coordination geometries, and entasis are important general principles in bioinorganic and classical coordination chemistry, specifically in catalysis (41, 42).

There is a growing interest in the quickly developing field of coordination chemistry of bispidine ligands over the last decade. The basis is a set of many new ligands, close to 100 experimentally determined structures of bispidine complexes, the thorough analysis of the structures and electronics of bispidine coordination compounds, and their stabilities and reactivities. This provides us with a data set, which may give us some impression of the wealth of coordination compounds in general and leads us to a thorough understanding of what is needed to design metal complexes and to develop new principles and theories in coordination chemistry.

## II. BISPIDINE LIGANDS

### A. Synthesis

#### *1. Mannich Reaction*

The most common and versatile route to bispidines is a variant of the Mannich condensation (see Scheme 1). The reaction of a component with C—H acidic hydrogen atoms, an aldehyde, and a primary amine or ammonia in the ratio 1:2:1 leads to a piperidone, which in some cases can be further condensed with an aldehyde (usually formaldehyde) and a primary amine to yield a bispidone. For symmetric bispidine derivatives, the preparation can be conducted in one step with the aldehyde, the C—H acidic component, and the amine in a 4:1:2 ratio.

To a large extent the substitution pattern of the amine is variable (24, 52–54). There are more restrictions for the aldehyde component. For the piperidone synthesis, most five- or six-ring aromatic or heteroaromatic aldehydes, as well as their ring-substituted derivatives, can be used (53, 55–57), but so far only a few aliphatic mono- and dialdehydes (e.g., formaldehyde, acetaldehyde, isobutyr-, glutar-, and succinaldehyde) have found their way into the bispidine preparation.

The pool of C—H acidic components is restricted to compounds that allow a double Mannich reaction. Most common are esters of acetonedicarbonic acid, especially dimethyl acetonedicarboxylate, which is the general substrate for the synthesis of 2,3,4,7-substituted bispidine-type ligands, or diphenylketone, which gives access to N-substituted bispidine derivatives. Others have also been used (34, 58–60). The variety of amines and aldehydes opens the door for a wide range of rigid multidentate ligands with a variety of donor atoms and, therefore, to the rational ligand design and to complexes with tunable properties. Examples of bispidones with different donor sets are shown in Tables I and II (see Charts 4 and 5, respectively, for the types of bispidine ligands listed in these tables).

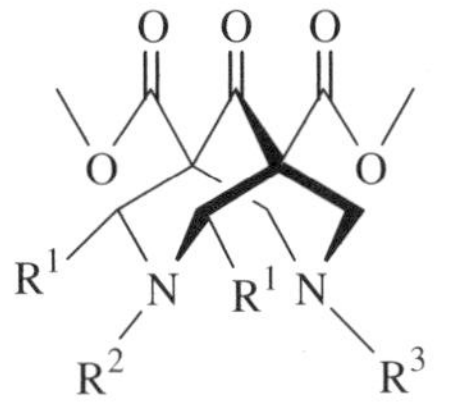

Chart 4

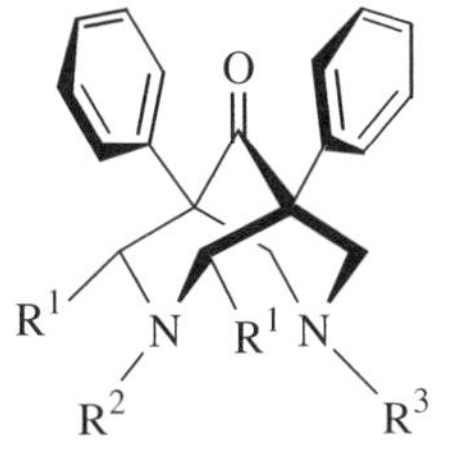

Chart 5

In addition and with reference to earlier work, which describes the synthesis of a bicyclic piperidone, derived from treating diethyl acetonedicarboxylate, an aldehyde and γ-aminobutyraldehyde (79), it is possible to synthesize the bispidine derivative **57** with an unsymmetric substitution pattern (see Chart 6)(80). This finding may be the basis for a whole new series of chiral bispidine ligands.

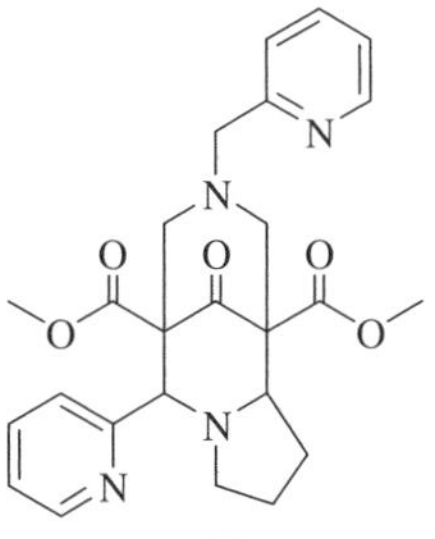

**57**

Chart 6

TABLE I
Bi-, Tetra-, Penta-, and Hexadentate 1,5-Dicarboxylic Acid Ester Substituted Bispidine Ligands[a]

| No. | $R^1$ | $R^2$ | $R^3$ | References |
|---|---|---|---|---|
| **10** | H | $CH_3$ | $CH_3$ | 61 |
| **11** | $CH_3$ | $CH_3$ | $CH_3$ | 61, 62 |
| **12** | H | phenyl | phenyl | 59 |
| **13** | 2-pyridyl | H | $CH_3$ | 52 |
| **14** | 2-pyridyl | $CH_3$ | $CH_3$ | 63 |
| **15** | 2-pyridyl | $CH_3$ | $CH_2CH_2N(CH_3)_2$ | 64 |
| **16** | 2-pyridyl | $CH_3$ | $CH_2CH_2N(C_2H_5)_2$ | 64 |
| **17** | 2-pyridyl | $CH_3$ | $CH_2CH_2N(CH(CH_3)_2)_2$ | 64 |
| **18** | 6-Br-2-pyridyl | $CH_3$ | $CH_3$ | 65 |
| **19** | 5-Br-2-pyridyl | $CH_3$ | $CH_3$ | 65 |
| **20** | 5-$OCH_3$-2-pyridyl | $CH_3$ | $CH_3$ | 65 |
| **21** | 5-$CH_3$-2-pyridyl | $CH_3$ | $CH_3$ | 65 |
| **22** | 2-pyridyl | $CH_3$ | $CH_2CH_2OH$ | 66 |
| **23** | 2-pyridyl | $CH_3$ | 2-hydroxybenzyl | 66 |
| **24** | 2-pyridyl | $CH_3$ | 2-pyridylmethyl | 66 |

TABLE I
*(Continued)*

| No. | $R^1$ | $R^2$ | $R^3$ | References |
|---|---|---|---|---|
| **25** | 2-pyridyl | $CH_3$ | 1-(pyridin-2-yl)ethyl | 67 |
| **26** | 2-pyridyl | $CH_3$ | quinolin-2-ylmethyl | 65 |
| **27** | 6-bromo-2-pyridyl (Br) | $CH_3$ | pyridin-2-ylmethyl | 65 |
| **28** | 2-pyridyl | pyridin-2-ylmethyl | $CH_3$ | 66 |
| **29** | 2-pyridyl | pyridin-2-ylmethyl | ethyl | 68 |
| **30** | 2-pyridyl | pyridin-2-ylmethyl | isopropyl | 68 |
| **31** | 2-pyridyl | pyridin-2-ylmethyl | isobutyl | 68 |
| **32** | 2-pyridyl | pyridin-2-ylmethyl | pyridin-2-ylmethyl | 66 |
| **33** | 2-pyridyl | 2-(pyridin-2-yl)ethyl | 2-(pyridin-2-yl)ethyl | 69 |
| **34** | 6-methyl-2-pyridyl | H | $CH_3$ | 52 |
| **35** | 6-methyl-2-pyridyl | $CH_3$ | $CH_3$ | 70 |
| **36** | 6-methyl-2-pyridyl | $CH_3$ | pyridin-2-ylmethyl | 71 |
| **37** | 6-methyl-2-pyridyl | $CH_3$ | (6-methylpyridin-2-yl)methyl | 71 |
| **38** | 2-quinolyl | $CH_3$ | $CH_3$ | 72 |
| **39** | 2-quinolyl | $CH_3$ | pyridin-2-ylmethyl | 71 |
| **40** | 2-quinolyl | $CH_3$ | (6-methylpyridin-2-yl)methyl | 71 |

TABLE I
(*Continued*)

| No. | $R^1$ | $R^2$ | $R^3$ | References |
|---|---|---|---|---|
| **41** | 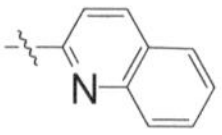 | $CH_3$ | 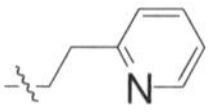 | 71 |
| **42** | 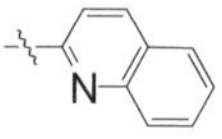 | $CH_3$ | 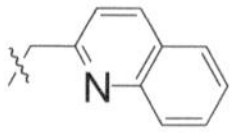 | 65 |
| **43** | 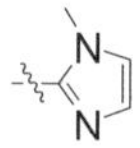 | $CH_3$ | $CH_3$ | 71 |
| **44** | 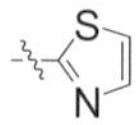 | $CH_3$ | $CH_3$ | 73 |
| **45** | 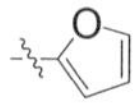 | $CH_3$ | $CH_3$ | 73 |
| **46** | 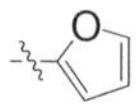 | 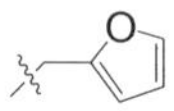 | 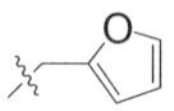 | 73 |
| **47** | 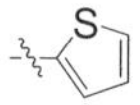 | $CH_3$ | $CH_3$ | 73 |
| **48** | 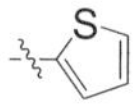 | 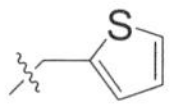 | 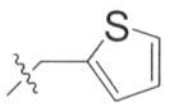 | 73 |

[a]See Chart 4.

Another interesting feature for the coordination chemistry of bispidine-type ligands is that diamines may be used in their synthesis. This finding gives us access to ditopic ligand systems (see Chart 7) (24, 81, 82).

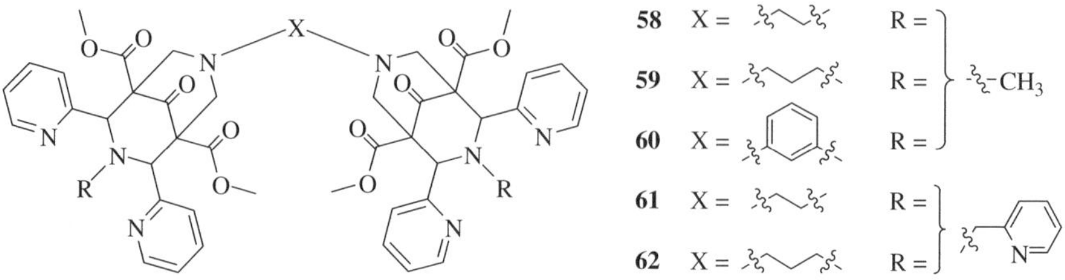

**58** X = R = $CH_3$
**59** X = R = $CH_3$
**60** X = R = $CH_3$
**61** X = R =
**62** X = R =

Chart 7

The mechanism of the Mannich reaction was for a long time subject to discussion (83–86). It is most reasonably described by an aldehyde and an amine, which form an imminium ion in a first step, which then attacks the nucleophilic center of the C—H acidic component. With primary amines, the resulting secondary amine reacts further to yield a tertiary amine.

TABLE II
Bi- and Tetradentate 1,5-Diphenyl-Substituted Bispidine Ligands[a]

| No. | $R^1$ | $R^2$ | $R^3$ | References |
|---|---|---|---|---|
| **49** | $-H$ | $-CH_3$ | $-CH_3$ | 74 |
| **50** | $-H$ | 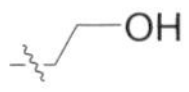 $-CH_2CH_2OH$ | 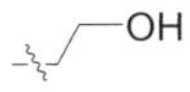 $-CH_2CH_2OH$ | 74 |
| **51** | $-H$ | 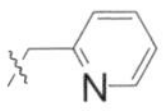 $-CH_2$-(2-pyridyl) | 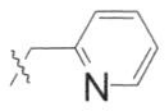 $-CH_2$-(2-pyridyl) | 34 |
| **52** | $-H$ | 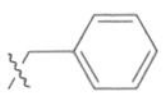 $-CH_2C_6H_5$ | 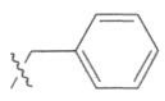 $-CH_2C_6H_5$ | 59 |
| **53** | $-H$ |  $-CH_2CH=CH_2$ |  $-CH_2CH=CH_2$ | 75 |
| **54** | $-H$ |  $-CH_2CH_2C{\equiv}N$ |  $-CH_2CH_2C{\equiv}N$ | 76 |
| **55** | $-H$ | 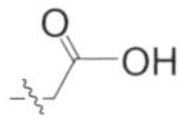 $-CH_2C(O)OH$ | 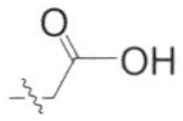 $-CH_2C(O)OH$ | 77 |
| **56** | $-H$ | 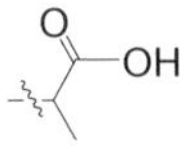 $-CH(CH_3)C(O)OH$ | 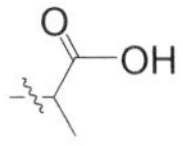 $-CH(CH_3)C(O)OH$ | 78 |

[a]See Chart 5.

Unfortunately, no general conventions for the conditions in Mannich reactions to yield piperidones or bispidones can be given. One of the major problems is a competitive aldol reaction (23, 87), which needs to be suppressed by a careful choice of the reaction parameters. These individually depend on the type of components used in the specific reaction. Only a few general rules for the working conditions for the Mannich bispidone synthesis can be given. Higher temperatures are usually applied, when the synthesis is conducted in one step. When conducted in two steps, the first usually is carried out at 0°C, and the second step by refluxing the reaction mixture. Generally, the reaction temperature should not exceed 150°C, because many Mannich bases degenerate at higher temperatures. The reaction solution of the Mannich condensation needs to be rather concentrated. Possible solvents are methanol, ethanol, isopropyl alcohol, butanol, and acetic acid (i.e., polar protic solvents). However, for other Mannich condensations, dioxane, tetrahydrofuran (THF), nitrobenzene, nitromethane, and toluene have also been used as solvent.

Formaldehyde can be used as an aqueous solution, as paraformaldehyde or, if needed and useful for the desired products, condensed with ammonia (e.g., in the form of hexamethylenetetramine). In the form of condensates, acetic acid is necessary to make the aldehyde available. In alcoholic solutions, an excess of formaldehyde is required since some of it is consumed by the formation of acetals. Piperidones are unstable in dilute mineral acids. Therefore, hydrochloric acid is usually avoided. Aldehydes that are used to form the 2,4-substituents are usually introduced in the sequence, which leads to the first six-membered ring. The variation of temperature in the second step may lead to different isomers of the bispidine ligands (see below).

Control of the pH is also of importance for the Mannich-based bispidine synthesis. Formation of an aldol product competes with the Mannich condensation in the basic pH region. It is for this reason that, in some cases, the reaction is sensitive to the order in which the reactants are added to the reaction mixture. It is possible to add the aldehyde and amine components one after another to a solution of the CH-acidic compound, but sometimes the aldol reaction can be disfavored by changing the order. This allows the imminium ion to be formed in advance. The precursors of **45** and **46** have been prepared by this method. In some cases, it has been useful to use a protonated amine component as the acetate salt (e.g., **49** or precursors for **44**, **47**, and **48**), as the chloride salt (e.g., **11**) or to carry out the reaction in acetic acid. Aromatic amines (e.g., aniline) give rise to para-substituted aromatic amines if the solution is not approximately neutral. In a very elegant procedure, a condensate of formaldehyde and aniline, which is the trimeric methyleneaniline, was prepared separately, and treated in the Mannich reaction with dimethyl acetonedicarboxylate and formaldehyde to yield the 3,7-diphenylbispidone (**12**) (58).

### 2. *Ring Fission of Diazaadamantanes*

The Mannich reaction with dibenzeneketone, paraformaldehyde, and ammonium acetate in a 1:4:2 ratio yields 1,5-diphenylbispidine-9-one (74). There was a controversy about the reaction pathway until it was shown that the initial product was 5,7-diphenyl-1,3-diazaadamantane-6-one (88). Following this observation the synthesis of various 1,3-diazaadamantane-6-ones with different substituents in the 5,7-position was reported (88–90), and these compounds found their way into the preparation of bispidine derivatives.

The aminal component of the 1,3-diazaadamantane derivatives can be cleaved under acidic conditions to yield N,N′-disubstituted bispidinine derivatives (see Scheme 2). Depending on the ratio of the reactants, N-monosubstituted bispidine derivatives can be obtained as well. Various acids (e.g., sulfuryl chloride, thionyl chloride, phosgene, tosyl chloride, benzoyl chloride or nitrous acid), as well as the anhydrides or chlorides of carbonic or sulfonic acids, have been used as cleaving reagents (88, 89, 91–94).

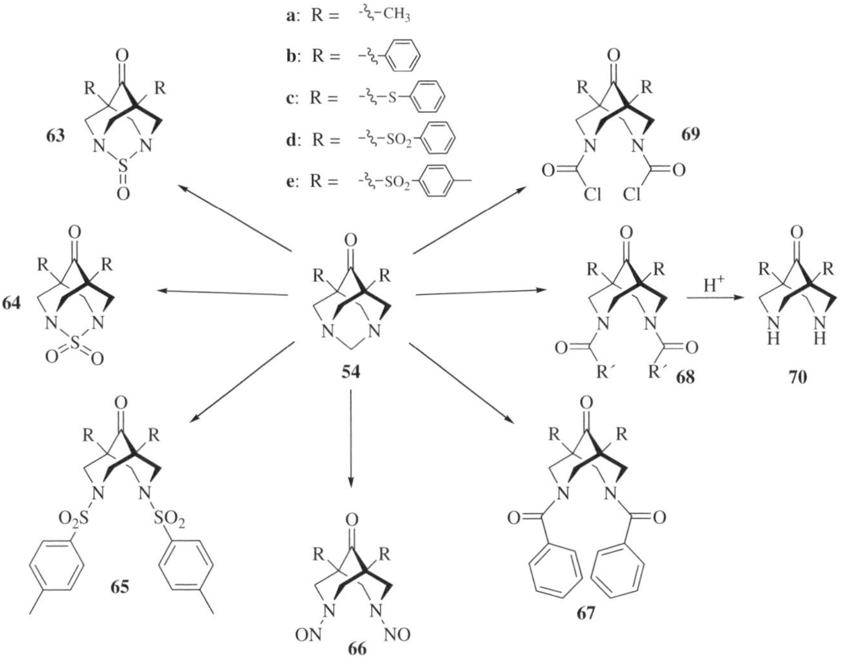

Scheme 2

From the 3,7-diacetyl compounds **68a** and **b** (with $R' = CH_3$) the bispidinine dervatives **70a** and **b** are obtained by heating in mineral acid. The resulting secondary amines open the door to introducing a variety of substituents, which are difficult to obtain with the Mannich approach to bispidine ligands.

The bis-α-haloamides, resulting from the reaction of 1,3-diazaadamantanes with α-halo derivatives of acetic anhydrid or chloracetic acid (95, 96), are useful agents for cyclization reactions (35, 97–99), following known cyclization methods (100), to yield the macrocyclic compounds **71–74** (see Chart 8). Other interesting macrocyclic bispidine compounds with crown ether fragments were also described (101).

### *3. Intramolecular Cyclization*

3,5-Disubstituted pyridines are an important group of substrates for the bispidine synthesis. Diethylpyridin-3,5-dicarboxylate can be converted to *N*-tosylpiperidine-3,5-dicarboxamide, which generates 2,4-dioxo-7-tosylbispidine upon heating in methylnaphthalene. Reduction with lithium aluminium hydride ($LiAlH_4$) yields the bispidine **1** under simultaneous elimination of the *p*-tosyl

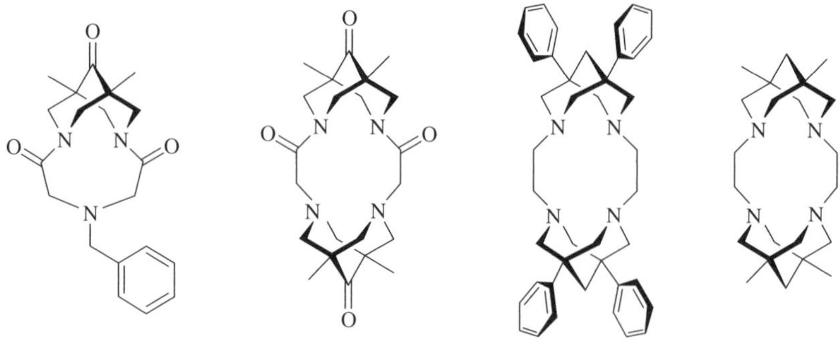

71 72 73 74

Chart 8

residue (see Scheme 3) (30, 102). 3,5-Bis(bromomethyl-piperidine), obtained from pyridine-3,5-dicarboxylate was also converted to the bispidine **1** by condenzation with ammonia (Scheme 4) (103). Catalytic hydrogenation of 5-cyano-nicotinic acid ethylester leads to the bispidin-2-one, which is readily reduced to the bispidine **1**. Under severe conditions, the dicyano derivative can be hydrogenated and cyclized in one step (Scheme 5) (104).

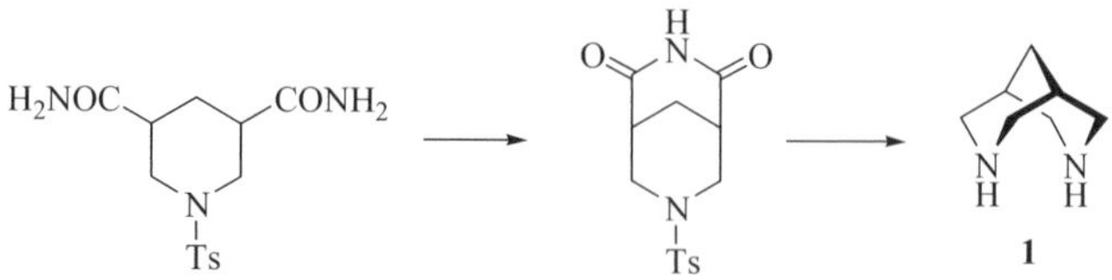

Scheme 3

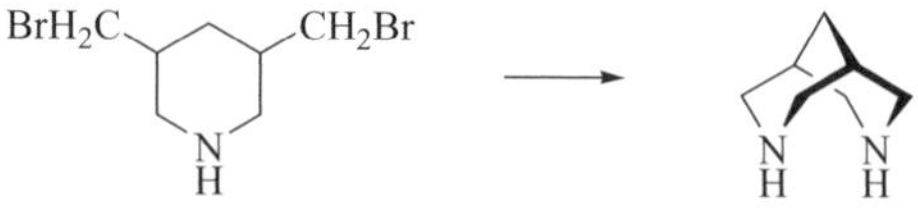

Scheme 4

Scheme 5

Additional strategies, which deviate from the synthetic pathways described above, include intramolecular cyclization reactions to obtain *N,N′*-disubstituted bispidine ligands. The cyclization may be started with *N,N*-bis

(2-ethoxycarbonylethyl)-*N*-(2-pyridylmethyl)amine to form the 1-(2-pyridylmethyl)-piperidin-4-one, which can be treated with formaldehyde and picolylamine in a Mannich reaction to the *N*,*N*′-bis(2-pyridylmethyl)-bispidone **75**, followed by the reduction of the keto function to yield the bis(picolyl) substituted bispidine **76** (Scheme 6) (105).

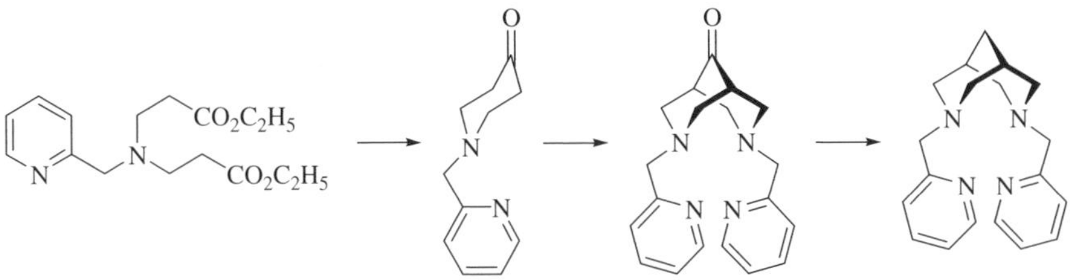

Scheme 6

Double cyclization in one step is also possible with the 1,5-diiodo-2,4-bis(iodomethyl)-pentane precursor. Following that route, the treatment of the precursor with aniline leads to the *N*,*N*′-diphenyl substituted bispidine **77** (106), whereas the reaction of the tetraiodo compound with 1-phenylethylamine yields the bispidine **78** (Scheme 7) (107).

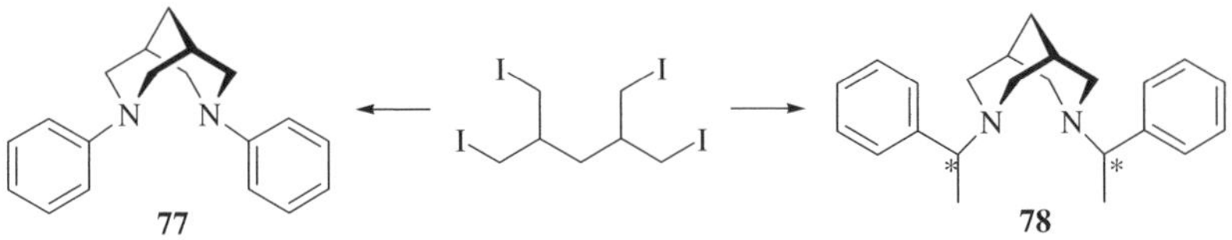

Scheme 7

### 4. *Reduction of Bispidones*

Due to the synthetic requirement for sufficiently C–H acidic starting materials most of the bispidine compounds described above have a keto group in the 9-position (i.e., most of the ligands discussed in this chapter are bispidine-9-ones or bispidones). It is exactly this functional group that is resposible for the sensitivity of these compounds with respect to retro-Mannich reactions. This type of cleavage of bispidones can be suppressed by the simple reduction of the keto group. Depending on the type of substituents and the substitution pattern on the bispidone backbone, there exist different reduction pathways.

In the case of highly symmetric bispidones, such as compounds **49–56** (Table II), as well as the comparable 1,5-dialkyl- or 1,5-unsubstituted derivatives, the Wolff–Kishner reaction with hydrazine hydrate (32, 88, 108–112) or tosyl hydrazide (113, 114) leads to the corresponding bispidines, while reduction of these compounds with $LiAlH_4$ (58, 88) or $NaBH_4$ (115) yields bispidoles (Scheme 8).

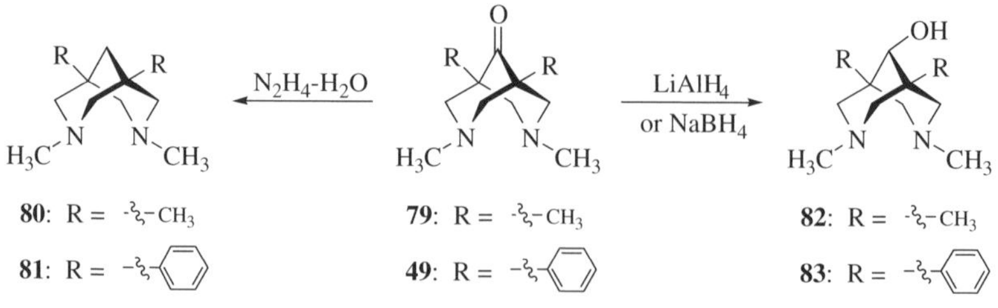

Scheme 8

With 2,4-substituted, bispidones of lower symmetry and analoguous 1,5-dicarboxylic acid esters (e.g., compounds **10–48** in Table I) the reduction under Wolff–Kishner conditions fails, but reaction with complex hydrides leads to the corresponding bispidoles. The treatment with $LiAlH_4$ yields a 1:1 mixture of the epimeric tris alcohols, whereas reduction with $NaBH_4$ at ambient temperature in various solvents leads to the epimeric mono-alcohols in different ratios up to epimerically pure compounds (see Scheme 9) (116, 117), which can be further reduced to the tris alcohols with $LiAlH_4$. For example, the reduction of **14** with $NaBH_4$ in dry methanol yields 65% *syn-* (**84b**) and 35% *anti*-product (**84a**), while the reaction in a mixture of dioxane–water leads exclusively to the *anti*-configuration of alcohol **84a** (117).

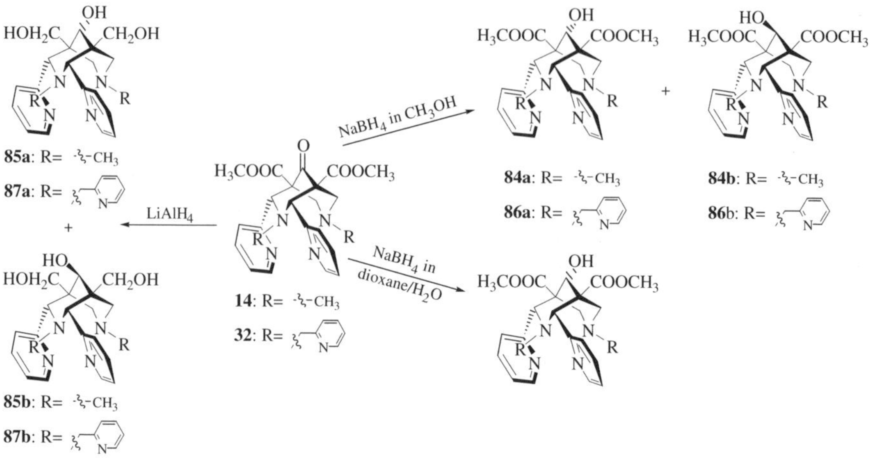

Scheme 9

In contrast to these methods the amide precursors for the macrocyclic ligands **73** and **74** (e.g., **72**) are reduced with diisobutylaluminum hydride (DIBAL-H), followed by a KF workup (35, 99).

## B. Stereochemistry

The stereochemical variation of the bispidine backbone includes the following structural forms:

1. Conformational transformation of the two six-membered rings of the bicycle, where each six-membered ring can either adopt a chair or a boat form.
2. Epimerization of the substituents in the 2- and 4-position leading to exo/exo or endo/exo (or endo/endo) isomers.
3. Atropisomerism caused by the hindered rotation of large substituents in the 2- and 4-position.

The conformation of derivatives of bicyclo[3.3.1]nonane has been the subject of many studies, based on proton nuclear magnetic resonance ($^{1}$H-NMR), $^{13}$C-NMR, infrared (IR) and Raman spectroscopy, dipole measurements, X-ray crystallography, complexation experiments and various types of computational studies. Most of this work has been reviewed in detail (26, 118, 119), and here we only report a summary of the general aspects.

For the unsubstituted diazabicyclo[3.3.1]nonane carbohydrate backbone there are three stable strain energy minimized conformations; cc, cb, and bb (see Chart 9). While all three conformations are free from angular strain, each is to some extent destabilized by nonbonded interactions.

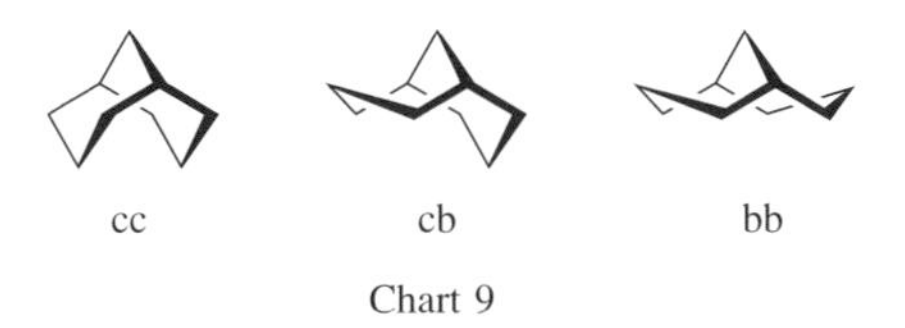


Chart 9

The predominant conformation of the mother compound bicyclo[3.3.1]nonane in the gas phase is the double chair (cc), as detected by electron diffraction studies (120–123). The most important destabilization in the cc conformer of this structure is van der Waals repulsion of the endo-3- and endo-7- hydrogen atoms. In the twin–chair conformation with ideal tetrahedral angles, the C3–C7 separation would be 2.5 Å and the separation of the corresponding endo-hydrogen atoms would have the physically impossible value of 0.75 Å (124). Repulsion induces an elongation of the C3$\cdots$C7 distance to $\sim$ 3.1 Å, which translates to an endo-hydrogen distance of $\sim$2.2 Å, accompanied by a flattening of the two "wings" of the double chair and increased angular strain (118, 119).

The 3,7-diazaderivative, that is, the bispidin **1** in the twin–chair conformation, does not have any endo-hydrogen atoms and, therefore, there is no such steric repulsion. The two secondary amine nitrogen atoms are separated by the short distance of ~3 Å, and this leads to other types of unfavorable interactions. First, there is steric repulsion (van der Waals interaction). Second, the dipole–dipole interaction of the two hetero atoms leads to a destabilization. Third, there is a repulsive orbital interaction of the two lone pairs, called the hockey stick effect (see Chart 10) (125–128). Diazaadamantane derivatives have the nitrogen lone pairs pointing outside the cavity and, therefore, have no lone-pair repulsion and a nitrogen···nitrogen separation of ~2.5 Å (60, 129). When the hockey stick interaction is turned on by cleaving the methylene bridge between the two nitrogen atoms of diazaadamantane, the N3···N7 distance increases to ~3 Å, as in the 3,7-diphenylbispidin **77** (Chart 11) or in α-isosparteine **2** (130, 131). Nevertheless, the double chair is the lowest energy conformation in most of the 3,7-disubstituted bispidine derivatives, although, for bicyclononane systems with or without heteroatoms, it was shown that bulky 3- and 7-substituents may lead to a preference for the cb conformation (132).

N N

Chart 10

Me N N Me

**88**

Ph N N Ph

**77**

Chart 11

By $^1$H-NMR spectroscopy and dipole measurements, the 3,7-dimethyl-3,7-diazabicyclo[3.3.1]nonane (dimethylbispidine) **88** was shown to adopt a cc conformation with flattened wings (108). This was later confirmed by $^{13}$C-NMR spectroscopy and semiempirical molecular orbital (MO) calculations (EH, CNDO/2: the double-chair form was found to be more stable than the cb isomer by ~83.8 kJ mol$^{-1}$, and the bb form is even more destablized (133). This preference for cc is not due to solvent or packing effects since **88** has the same conformational preference in the gas phase, as shown by a computational study, which involved a scan of the potential energy surface (PES) and the calculated ionization potentials (MINDO/3, MNDO, and ab initio calculations at the STO-3G level). Therefore, the experimental results are best interpreted by a flattened double-chair conformation with both methyl substituents in exo orientation (134). The 3,7-diphenylbispidine (**77**) also adopts the cc conformation, as shown by X-ray diffraction in the solid and by $^{1}$H-NMR in solution (106, 130).

Due to the removal of some flagpole–flagpole repulsion when a carbonyl group is introduced at C9, the cb conformation must become more stable. This indeed is the case, as shown with the *N*-methyl-*N*′-benzyl derivative **90** (see Chart 12, Table III), which has a cb conformation in the crystalline form. This might, however, be due to packing effects, and cc conformations are adopted in most cases in solution (135). For example, the twinchair conformation was found for molecules **89–92** in $(CD_3)_2SO$ and $CDCl_3$ on the basis of Raman, IR, and $^{1}$H-/$^{13}$C-NMR spectroscopies (136, 137). In addition, a series of 3-alkyl-7-methyl-3.7-diazabicyclo[3.3.1]nonan-9-ones (**93–95**) have been studied by NMR-spectroscopy in $CDCl_3$ solution, and these have been shown to adopt cc conformations with the N-substituents in equatorial orientation. An increased distortion of the six-membered rings was observed with increasing size of the pendant groups (138).

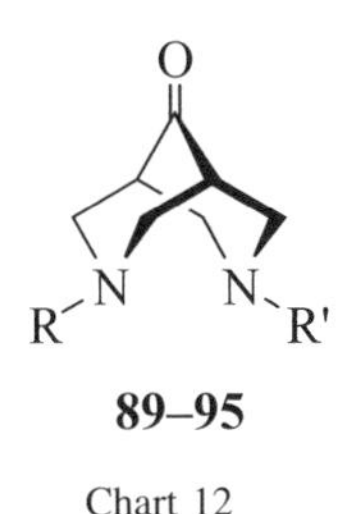


**89–95**

Chart 12

The 9-carbonyl group has a significant influence on the first protonation constant by about four orders of magnitude toward lower values. This has been related to electronic interactions between the carbonyl group and the nitrogen donors through $\sigma$ bonds (59, 139). The same conclusion emerged from 1,3-diazaadamantane derivatives on the basis of photoelectron spectroscopy and semiempirical MO calculations (140).

Reduction of the C9-carbonyl groups to an alcohol changes the conformational behavior drastically. The cb conformation is favored in nonpolar solvents for some 9-hydroxy derivatives, as determined by NMR, IR, and Raman spectroscopies. The stabilization of the cb conformation is due to an intramolecular OH–N interaction (see Scheme 10) (136). In very dilute solutions in

Table III
*N,N′*-Disubstituted 3,7-Diazabicyclo[3.3.1]nonane-9-one Derivatives[a]

| No. | R | R′ |
|---|---|---|
| **89** | $-CH_3$ | $-CH_3$ |
| **90** | $-CH_3$ | 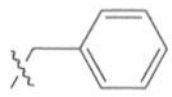 |
| **91** | $-CH_3$ | 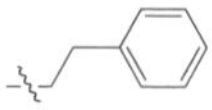 |
| **92** | 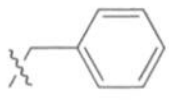 | 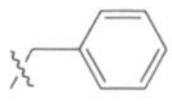 |
| **93** | $-CH_3$ | |
| **94** | $-CH_3$ | 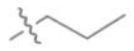 |
| **95** | $-CH_3$ | 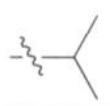 |

[a]See Chart 12.

$CDCl_3$, $CCl_4$, and $CS_2$ only a small amount of free OH groups related to the cc conformer could be detected. However, the spectra in $(CD_3)_2SO$ indicated, that in more polar solvents there is a competitive solute–solvent interaction, which leads to a switch from the cb to the cc form. X-ray data of 3,7-dimethyl-3,7-diazabicyclo[3.3.1]nonan-9-ol*$2H_2O$*EtOH have shown that both nitrogen atoms take part in hydrogen bonds, either with water or ethanol, and the molecule adopts a cc conformation. With IR spectroscopy of a water-free crystal, the cb conformation was assigned to the same molecule, and this is stabilized by an intramolecular N···H—O bond (136).

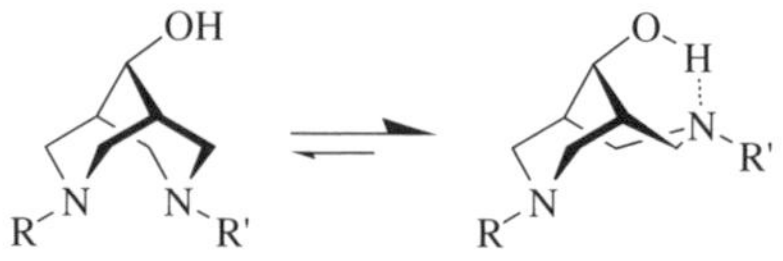

Scheme 10

There are numerous X-ray studies of monoprotonated bispidine derivatives (141–144). The N3···N7 distances decrease to 2.66 Å, due to relief of the lone-pair repulsion. Only three published X-ray data sets of diprotonated molecules with the bispidine backbone in the cc conformation exist (145–147). On the basis of PM3 calculations, it has been shown for several bispidinines that the singly protonated forms prefer the cc conformation, and these have low activation barriers for proton exchange between the two nitrogen centers. The diprotonated bispidones are, independent on their conformation, unfavorable, and this is based on heats of formation of $\sim$1000 kJ mol$^{-1}$ (148). Methylation instead of protonation of a bispidine amine nitrogen atom makes the stabilization by an intramolecular N···H—N bond impossible. Therefore, for substituted 3,7-diazabicyclo [3.3.1]nonan-9-oles, a switch to the cb conformation is observed, and this is due to the steric demand of the endo-methyl group, as analyzed by nuclear over hauser effect (NOE) experiments (116).

Substituents in the 1- and 5-positions are not expected to significantly change the conformational behavior of bispidines. A minor effect is that these substituents decrease the flattening of the wings, as observed in bicyclo[3.3.1]-nonane (149), to some extent. This should, therefore, increase the N3–N7 interactions and make the cb conformation for 1,5-substituted bispidones more favorable. Indeed, in the group of bispidones derived from dibenzylaceton, the conformational equilibrium depends on the N-pendent substituents. Compounds with pyramidal nitrogen atoms (e.g., **49**, **50–54**, **65b** or their diisopropyl-substituted analogues), tend to adopt a cb form in the solid. This is attributed to the phenyl groups, which prevent a flattening of the wings to some extent, because they do not allow a significant increase of the angles around C1 and C5, and therefore lead to a severe N···N repulsion (76, 93, 94, 150–153). In solution, the measurement of dipole moments has led to the assumption that these compounds undergo a fast interconversion between the degenerate cb and bc conformations (154). This is confirmed by variable temperature $^{1}$H- and $^{13}$C NMR experiments (94, 152, 154). The small activation barrier for the chair-to-boat interconversion is also of importance for the coordination of bispidine-type ligands to metal ions (see Section III).

Transformation of the amine-to-amide nitrogens leads to a trigonalization. This decreases the endo–endo repulsion and leads to a preference of the double-chair form, as, for example, observed in molecules **66b** and **68b** (94, 155). For this type of bispidine compounds, different isomers, due to a parallel (pa) or antiparallel (ap) arrangement of the amide carbonyl groups, exist (see Chart 13), and, depending on the size of the energy barrier, both isomers can be observed (156).

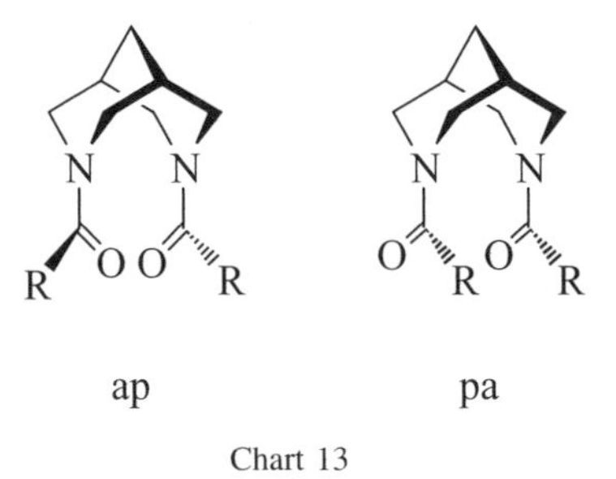


Chart 13

For the 1,5-dimethyl-3,7-di-*tert*-butyl-3,7-diazybicyclo[3.3.1]nonan-9-one (**96**) the cb conformation is observed in the solid state (see Chart 14). However, it is not clear whether this conformation is a result of the bulky N-substituents or an effect of the hindered flattening of the wings due to the 1,5-methyl substituents (157). Interestingly, the 1,5-dinitro derivative **97,** which has no sterically demanding substituents, shows a cb conformation in the solid state and an interconversion between the degenerate cb and bc forms in solution. The activation barrier was found by low temperature $^{13}$C-NMR experiments to be $\sim$27.2 kJ mol$^{-1}$ (158). Direct conversion of **97** yields the 1,5-diamino derivative **98**, from which the 1,5-dibromo- (**99**) and the 1,5-dihydroxo-derivatives (**100**) are obtained (see Scheme 11). These all adopt the twin–chair conformation in the solid state (144).

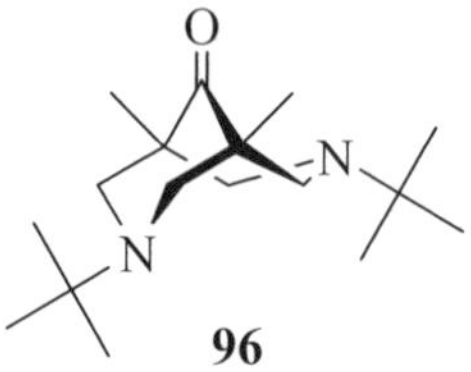

**96**

Chart 14

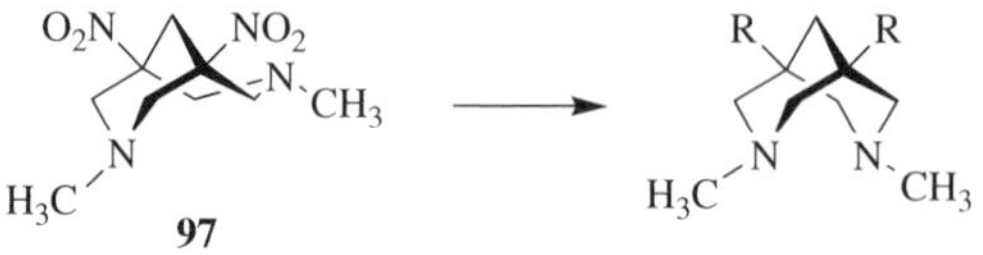

**97**

**98**: R = $NH_2$

**99**: R = OH

**100**: R = Br

Scheme 11

The conformational behavior is significantly changed by exo-2-substituents. Compounds **101–103** (see Chart 15) exist in the cb conformation with the substituted ring in the boat form. This is the result of the preference of the pendant group to take an equatorial position and thus omitting steric interaction (159). Endo-2-substituted bispidines have not yet been reported.

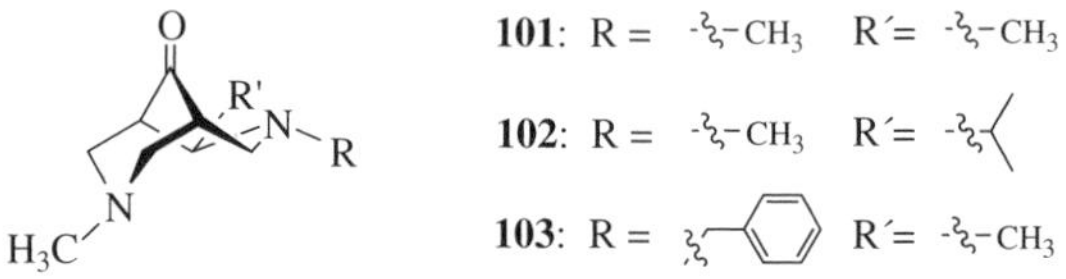

**101**: R = $-CH_3$ R′= $-CH_3$

**102**: R = $-CH_3$ R′= isopropyl

**103**: R = phenyl R′= $-CH_3$

Chart 15

Substituents in 2,4-position generaly prefer equatorial orientations in a chair conformation of the six-membered ring. Bis-axial orientation of these substituents would lead to strong repulsion, but in some examples an axial–equatorial conformation is observed, as will be disscussed below. The X-ray analysis of the 2,4-diaryl-3,7-dimethyl-1,5-dimethoxycarbonylbispidin-9-one (**3**) reveals a flattened chair–chair conformation with the aryl substituents, as expected, in the equatorial position. From comparison of the IR-, $^{1}$H-, and $^{13}$C NMR spectra with those of the analoguous *p*-methoxyphenyl- and *p*-chlorophenyl-substituted derivatives, it follows that these all adopt a double-chair conformation (56). The same behavior was also found for the bispidones **14** and **35** (70) **46–48** and **104** (see Chart 16) (160). The PM3 calculations for various bispidones with aromatic $R^1$ substituents have shown that the cc conformation is slightly more stable than the cb form, and this is much more stable than the bb conformation in all cases (161).

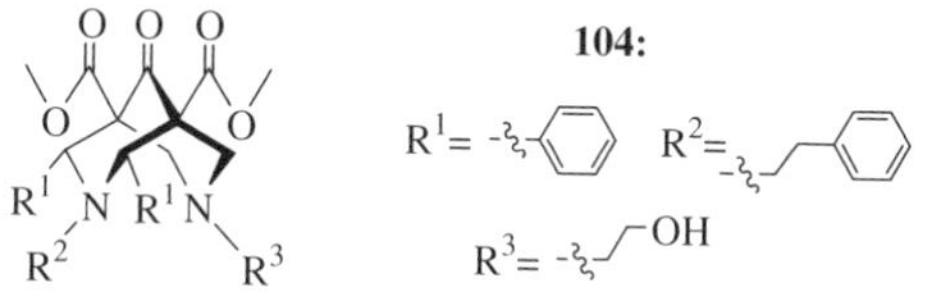

Chart 16

The dependence of the conformation on the size of the N-substituents was studied with several N7-substituted 2,4-*o*-chlorphenyl-3-methyl bispidines in solution (see Scheme 12) (55). Equilibration of the compounds in hot ethanol or chloroform always leads to a mixture of the cc and the cb conformers. The ratio was analyzed by NMR and showed a systematic dependence on the size of the N7 substituent. With increasing steric demand of the N7 substituent, the amount of the cb conformer increases up to 34% in the case of the adamantyl derivative. The higher substituted of the two six-membered rings never changes conformation and has both aromatic rings in the equatorial position. There is no report on other substituted bispidines and, therefore, this effect cannot be generalized.

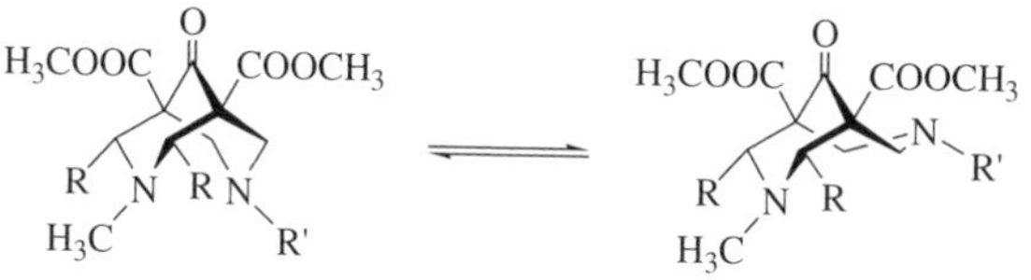

Scheme 12

There is one report on a 2,4,6,8-substituted bispidine. Here, the structure is the result of a minimization of close contacts of all substituents. In analogy to the systems reported above, all aryl substituents prefer equatorial positions, and this is possible in a cb conformation with two exo- and two endo-phenyl groups (60, 162, 163).

For the most stable double-chair conformation, which is the geometry that coordinates to metal ions, substituents in the C2,C4-positions can have exo–exo, exo–endo, or endo–endo orientation (Chart 17). Since these substituents prefer equatorial positions the exo–exo configuration is the most unfavorable for the cc conformation and has not been observed so far. Also, the exo–endo configuration is energetical less favorable than the endo–endo form in the double-chair conformation and consequently is rarely reported. It was shown that the transformation from exo–endo to endo–endo is possible in some cases by refluxing the bispidine compound in protic (usually alcoholic) solution. The transformation mechanism is characterized by the breaking and closing of the covalent C1–C2 or C4–C5 bonds (see Scheme 13). This corresponds to a retro-Mannich reaction, where in the ring-opened form rotation around the N3–C2 or the N3–C4-bond is possible (22, 161).

exo–exo exo–endo endo–endo

Chart 17

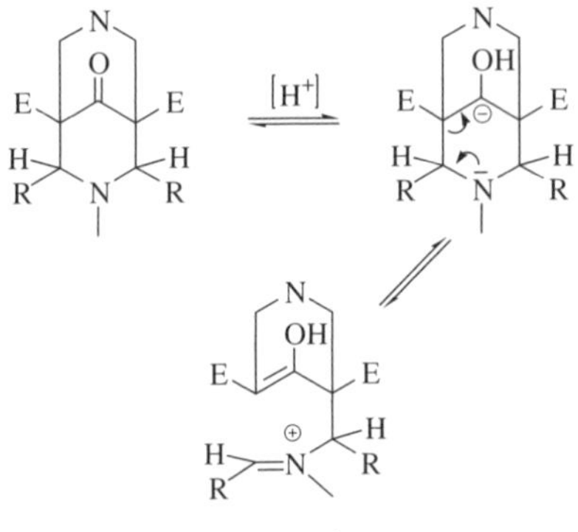

N O E E H H R N R [H+] N OH E E H H R N R N OH E E H H R N R

Scheme 13

This leads to the conclusion that the formation of either epimeric structure can be controled by the conditions under which the reaction is carried out, and for most reactions it turns out that the endo–endo epimer is the thermodynamically controled product (see Chart 18). For *p*-nitrophenyl substituented bispidones, the free activation enthalphy for the isomerization, deduced from NMR experiments, was calculated to be $\sim$105 kJ mol$^{-1}$. For phenyl- or *m*-nitrophenyl-substituted bispidinones, the endo–endo isomer is $\sim$8.4 kJ mol$^{-1}$

more stable than the exo–endo isomer (PM3-calculations), and therefore preferred thermodynamically. For the 2-quinolyl-substituted derivative **38**, calculations (PM3) of the heat of formation indicate that the endo–endo and exo–endo configurations are similar in energy. This is in agreement with the experimental observation that in some cases only the exo–endo form is isolated (161). In other examples, the endo–endo isomer, which is that needed for coordination chemistry, is obtained as a pure product (71).

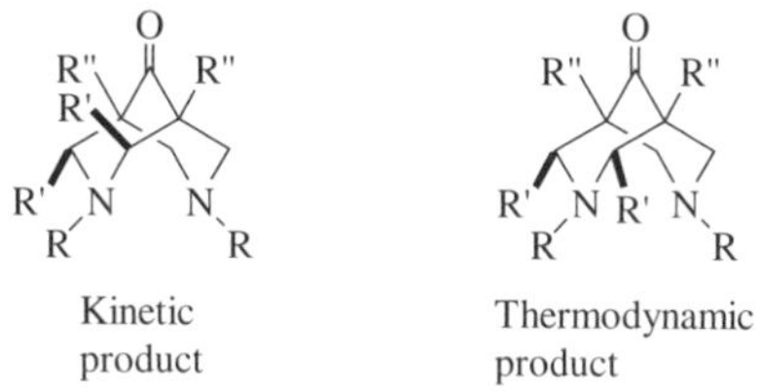

Kinetic product

Thermodynamic product

Chart 18

An interesting conversion from the endo–endo to the exo–endo configuration is observed when the 2,4-di-2-pyridyl-substituted bispidinone **14** is dissolved in hot ethanol and hydrochloric or perchloric acid are added. The $^{1}$H-NMR spectra show a loss of symmetry of the compound, which can be explained by the formation of the hemiaminal cyclization product **105** (see Chart 19) (23, 164). This unusual product also yielded crystals for X-ray analysis (147).

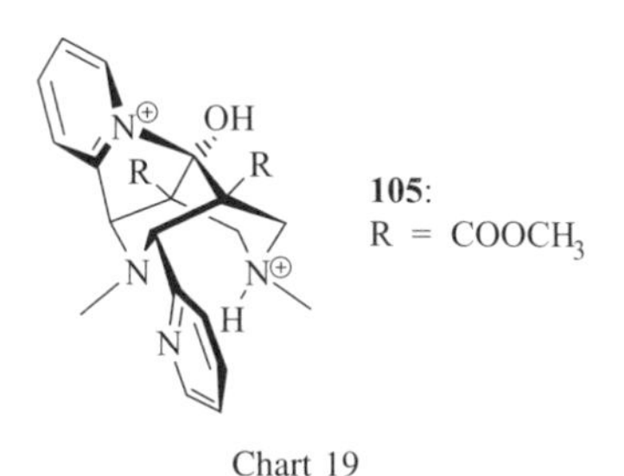


Chart 19

Aromatic substituents at C2 and C4 are to some extent hindered in rotation. This leads to a third type of isomerism in bispidine chemistry (i.e., atropisomerism). The activation energy for the rotation around the C2/C4–aryl bond for bispidones with various meta-substituted phenyl groups was determined by various NMR methods and found to be 70–75 kJ mol$^{-1}$ (23). For a rotation of 180°, which is usually necessary for the coordination of bispidine ligands to metal ions (e.g., **14**, see Scheme 14), two energy barriers have to be overcome. The higher is the result of an interaction of the ortho-disposed proton of the aromatic ring with the proton or the alkyl substituent at the N3 amine nitrogen atom. The second, smaller barrier, is due to an interaction of the ortho protons of the aromatic substituent with the equatorially oriented hydrogen atoms at C6 and C8.

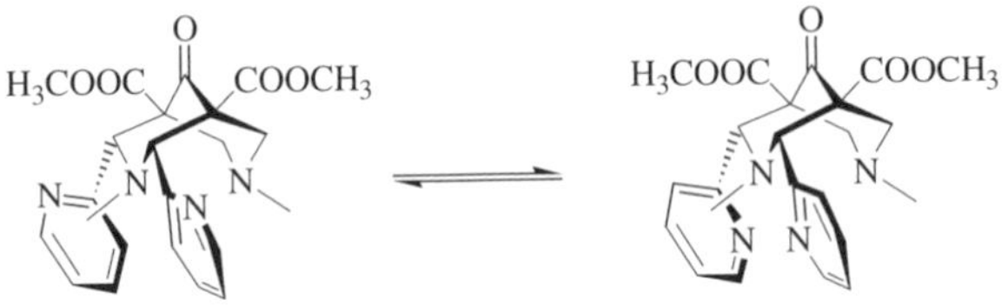

Scheme 14

Nuclear magnetic resonance studies and PM3 calculations have shown that 2,4-di-4-quinolyl substituted bispidinones in cc conformation exist in two different endo–endo geometries, which differ in the arrangement of the 4-quinolyl rings (161). One configurational isomer has a symmetric orientation of the two aromatic rings, the other has one of the two groups rotated by 180°. The two isomers differ in their heats of formation by ~20 kJ mol$^{-1}$. As in the case of 2-naphthyl-substituted derivatives, these compounds suffer from stereochemical overcrowding, and thus form stable rotamers.

Obviously, in 2,4,6,8-substituted bispidine molecules there is a large steric hindrance for the rotation of the substituents. Cooperative effects in the restricted rotation of the aryl groups in 2,4,6,8-tetraphenyl-3,7-diazabicyclo[3.3.1]nonanes have been described (165). The rotation of the 2,4-phenyl substituents seems to be correlated with that of the 6,8-phenyl groups. This emerges from large negative activation entropies for the rotation.

## III. BISPIDINE COORDINATION CHEMISTRY

### A. Complex Synthesis

The synthesis of transition metal bispidine complexes generally follows known procedures for the preparative chemistry of the corresponding metal ions with neutral polydentate amine–pyridine ligands. Due to the highly preorganized diazaadamantane-derived ligand backbone of the bidentate chelate and the low-energy barriers for rearrangement of the pendant donor groups, there generally are relatively high complex stabilities and fast complex formation kinetics. Note that the stabilities are dependent on the metal ion, reduced by a lack of complementarity (42) due to a less than optimal relative orientation of the ligand lone pairs and the relevant metal *d* orbitals (misfit, see Section III.D.1).

Due to the relatively large bispidine cavities, the redox potentials are generally rather high (positive) and the stabilization of higher oxidation states can be difficult. For example, the bispidine–cobalt(II) complexes usually are air stable and oxidation only occurs with strong oxidants (e.g., $H_2O_2$; see Section III.D.3) (166, 167). However, due to the high complementarity for copper(II) (see, however, Section III.D.1) the corresponding copper(I) complexes are, as expected, highly air sensitive (70, 81, 168, 169). Oxidation of the iron(II) complexes to hydroperoxo- and peroxo-iron(III) and to oxo-iron(IV) complexes is generally easily accomplished (170–172), however, other simple iron(III) complexes (e.g., aqua, chloro, hydroxo) have not yet been isolated or spectroscopically characterized. This might also be due to the rigid ligand structures and relatively large cavities, which are responsible for subtle spin equilibria at the various oxidation states (66, 171, 173–175).

Rather difficult is the complexation with the macrocyclic bis(bispidine) ligand **74** (see Chart 8 above), and the reported yield of the copper(II) complex is $<10\%$ (99). This result has been improved, but there are still problems which are related to two features of this type of ligands. (1.) Their rigidity prevents a folding that is necessary for the usual step-wise coordination of a macrocyclic ligand (176–180). Therefore, the formation rate is slow and the corresponding activation energy is large. (2.) The relatively tight cycle with the four amine lone pairs ideally pointing toward the center (42) leads to stabilization of the protonated form, and these ligands therefore are efficient proton sponges with $pK_a$ values $\sim 25$ (181, 182). These ligands were found to abstract protons from chloroform (181) but, more recently, proton exchange was found to be possible in aqueous solution and various solvent mixtures (182).

## B. Molecular Structures of Bispidine Complexes

### 1. *General Aspects*

Selected structural data of 99 first-row transition metal bispidine complexes are listed in Table IV, most are available in the CSD, 28 of the structures have not yet been published. Complexes of alkali metal ions, heavier transition metal ions, and lanthanoids have also been described, but not structurally characterized (69). Chart 20 defines the structural parameters [note that the two torsional angles (tor1 and tor2, C1/5-C2/4-Car-Npy2/1) are given without sign; the plane twist parameter is the angle between the planes defined by the donors py1 and py2, which usually are pyridine groups]. $X_E$ and $X_A$ are in-plane equatorial (trans to N3) and axial (trans to N7) coligands. The type in Table IV refers to the structural type as defined in Chart 20 and the corresponding examples given in Fig. 1, where plots of selected structures listed in Table IV are presented. Hydrogen atoms are omitted in these and other plots in the following sections.

The most impressive parameter to show the rigidity of the bispidine ligands is the N3···N7 distance, which is 2.9 Å on average. The shortest distances are observed for tetrahedral copper(II) complexes of the bidentate ligand (**49**) and

TABLE IV
Selected Structural Parameters of Bispidine–Metal Complexes[a]

| M | L | $X_1$, $X_2$/ $X_E$, $X_A$ | Type | Reference | M-N3 | M-N7 | M-D1/ M-py1 | M-D2/ M-py2 | M-$X_1$/ M-$X_E$ | M-$X_2$/ M-$X_A$ | N3-N7 | tor 1 | tor 2 | Plane Twist |
|---|---|---|---|---|---|---|---|---|---|---|---|---|---|---|
| ***Bidentate Ligands*** | | | | | | | | | | | | | | |
| Cu(II) | **49** | Cl, Cl | 1 | 183 | 2.004 | 1.965 | | | 2.182 | 2.262 | 2.714 | | | |
| Cu(II) | **49** | Br, Br | 1 | 184 | 1.992 | 1.975 | | | 2.311 | 2.395 | 2.765 | | | |
| Cu(II) | **100** | Cl, Cl | 1[b] | 185 | 2.116 | 2.074 | | | 2.285 | 2.306 | 2.847 | | | |
| Cu(II) | **52** | Cl, Cl | 1 | 186 | 2.013 | 2.042 | | | 2.251 | 2.193 | 2.836 | | | |
| Cu(II) | **53** | Cl, Cl | 1 | 187 | 2.021 | 2.003 | | | 2.211 | 2.242 | 2.798 | | | |
| | | | | | 2.010 | 2.022 | | | 2.221 | 2.229 | 2.808 | | | |
| Cu(II) | **53** | Br, Br | 1 | 188 | 2.023 | 1.999 | | | 2.337 | 2.344 | 2.827 | | | |
| Cu(II) | **54** | Cl, Cl | 1 | 188 | 1.979 | 2.015 | | | 2.205 | 2.231 | 2.836 | | | |
| | | | | | 1.989 | 2.004 | | | 2.225 | 2.241 | 2.814 | | | |
| Cu(II) | **80** | Cl, Cl | 1 | 111 | 1.984 | 1.984 | | | 2.248 | 2.248 | 2.784 | | | |
| ***3,7–Disubstituted Tetradentate Ligands*** | | | | | | | | | | | | | | |
| Cu(II) | **56** | $H_2O$, [Cu(**52**)($H_2O$)] | 2[b] | 141 | 2.031 | 1.988 | 1.920 | 1.963 | | 2.198 | 2.819 | | | 11.56 |
| | | | | | 2.030 | 1.991 | 1.949 | 1.923 | | 2.230 | 2.841 | | | 16.82 |
| Cu(II) | **76** | | 2 | 105 | 2.002 | 2.002 | 1.976 | 1.976 | | | 2.822 | | | 47.96 |
| ***2,4-Disubstituted Tetradentate Ligands*** | | | | | | | | | | | | | | |
| Cr(III) | **14**[c] | Cl, Cl | 3 | 189 | 2.095 | 2.222 | 2.049 | 2.060 | 2.287 | 2.304 | 2.868 | 89.25 | 83.57 | 26.35 |
| Mn(II) | **14** | Cl, Cl | 3 | 190 | 2.388 | 2.530 | 2.230 | 2.255 | 2.404 | 2.462 | 2.962 | 90.48 | 84.69 | 34.72 |
| Mn(II) | **14**[c] | $OSO_2CF_3$, $OSO_2CF_3$ | 3 | 189 | 2.303 | 2.351 | 2.233 | 2.211 | 2.112 | 2.157 | 2.959 | 85.62 | 88.56 | 30.58 |
| Mn(II) | **14** | $O_2NO$, $ONO_2$ | 3[d] | 189 | 2.348 | 2.356 | 2.304 | 2.295 | 2.284 | 2.166 | 2.938 | 80.34 | 87.07 | 39.19 |
| | | | | | | | | | 2.408 | | | | | |
| Mn(II) | **14**[c] | $O_2CCH_3$, $ONO_2$ | 3[d] | 189 | 2.373 | 2.388 | 2.299 | 2.276 | 2.255 | 2.192 | 2.944 | 87.53 | 92.56 | 21.33 |
| | | | | | | | | | 2.425 | | | | | |
| Fe(II) | **14** | SCN, SCN | 3 | 66 | 2.242 | 2.373 | 2.170 | 2.176 | 2.038 | 2.117 | 2.922 | 87.75 | 83.82 | 33.39 |

| | | | | | | | | | | | | | | |
|---|---|---|---|---|---|---|---|---|---|---|---|---|---|---|
| Fe(II) | **14** | $OCOCH_3$, $OCOCH_3$ | 3 | 66 | 2.274 | 2.455 | 2.184 | 2.179 | 1.971 | 2.122 | 2.909 | 87.47 | 87.08 | 24.99 |
| Fe(II) | **14** | $OCOCH_3$, $OSO_2CF_3$ | 3 | 66 | 2.271 | 2.362 | 2.188 | 2.187 | 1.960 | 2.228 | 2.931 | 84.14 | 86.65 | 26.04 |
| Co(II) | **14**[c] | $H_2O$, $H_2O$ | 3 | 189 | 2.160 | 2.222 | 2.122 | 2.091 | 2.042 | 2.200 | 2.915 | 86.89 | 87.34 | 26.23 |
| Co(II) | **14**[c] | $H_2O$, $H_2O$ | 3 | 189 | 2.149 | 2.213 | 2.143 | 2.124 | 2.065 | 2.140 | 2.903 | 80.55 | 86.81 | 30.88 |
| Co(II) | **14**[c] | $O_2NO$ | 3 | 166 | 2.139 | 2.134 | 2.085 | 2.093 | 2.066 | 2.281 | 2.882 | 84.93 | 88.26 | 32.79 |
| Co(II) | **14**[c] | $O_2CCH_3$ | 3 | 189 | 2.147 | 2.169 | 2.111 | 2.097 | 2.061 | 2.235 | 2.904 | 87.47 | 85.39 | 25.38 |
| Co(II) | **14**[c] | $H_2O$, $H_2O$ | 3 | 167 | 2.115 | 2.132 | 2.081 | 2.089 | 2.070 | 2.118 | 2.832 | 86.43 | 86.00 | 27.57 |
| Co(III) | **14**[c] | $O_2CO$ | 3 | 167 | 1.941 | 2.027 | 1.934 | 1.913 | 1.905 | 1.916 | 2.783 | 83.35 | 84.60 | 18.40 |
| | | | | | 1.932 | 2.036 | 1.924 | 1.928 | 1.912 | 1.915 | 2.789 | 83.66 | 83.53 | 27.06 |
| Co(III) | **14**[c] | Cl, Cl | 3 | 167 | 1.965 | 2.018 | 1.918 | 1.957 | 2.254 | 2.246 | 2.755 | 85.08 | 84.18 | 26.92 |
| Co(III) | **14**[c] | Cl, $H_2O$ | 3 | 167 | 1.956 | 1.977 | 1.945 | 1.927 | 2.214 | 1.966 | 2.717 | 81.68 | 84.28 | 25.13 |
| Co(III) | **14**[c] | OCHCH2, $H_2O$ | 3 | 167 | 1.951 | 1.978 | 1.937 | 1.930 | 1.910 | 1.955 | 2.731 | 82.38 | 84.25 | 27.06 |
| Cu(I) | **14** | $NCCH_3$ | 3 | 70 | 2.203 | 2.160 | 2.066 | (3.118) | 1.873 | *f* | 2.914 | 82.73 | 64.61 | 51.82 |
| Cu(I) | **14** | $NCCH_3$ | 3 | 70 | 2.243 | 2.158 | 2.251 | 2.380 | 1.893 | *f* | 2.947 | 81.71 | 83.57 | 41.68 |
| | | | | | 2.292 | 2.186 | 2.169 | 2.247 | 1.936 | *f* | 2.973 | 85.47 | 83.61 | 34.85 |
| Cu(II) | **14** | $NCCH_3$, $OSO_2CF_3$ | 3 | 73 | 2.022 | 2.355 | 1.992 | 1.998 | 1.980 | 2.608 | 2.922 | 89.57 | 93.81 | 23.22 |
| Cu(II) | **14** | Cl | 3 | 191 | 2.049 | 2.351 | 1.999 | 2.010 | 2.234 | *f* | 2.909 | 93.19 | 86.38 | 24.84 |
| Cu(II) | **14**[c] | Cl | 3 | 81 | 2.042 | 2.273 | 2.020 | 2.024 | 2.232 | *f* | 2.921 | 90.68 | 86.41 | 30.03 |
| Cu(II) | **14**[c] | $O_2NO$ | 3 | 189 | 1.999 | 2.280 | 1.999 | 1.999 | 1.976 | 2.613 | 2.904 | 92.41 | 90.57 | 16.14 |
| Cu(II) | **14**[c] | $^{-}O$, $^{-}O$, Cl, Cl, Cl, Cl | 3 | 192 | 2.042 | 2.433 | 2.031 | 2.009 | 1.909 | 2.456 | 2.920 | 87.75 | 93.43 | 24.66 |
| Zn(II) | **14**[c] | Cl | 3 | 189 | 2.191 | 2.103 | 2.092 | 2.124 | 2.261 | *f* | 2.942 | 92.50 | 82.27 | 24.57 |
| | | | | | 2.206 | 2.110 | 2.114 | 2.095 | 2.261 | *f* | 2.939 | 90.38 | 88.28 | 26.61 |
| Zn(II) | **14**[c] | $O_2NO$ | 3 | 189 | 2.160 | 2.130 | 2.112 | 2.106 | 2.087 | 2.370 | 2.951 | 82.80 | 87.02 | 34.60 |
| Cu(II) | **19**[c] | $NCCH_3$, $FBF_3$ | 3 | 65 | 2.010 | 2.257 | 2.022 | 2.020 | 1.957 | 2.726 | 2.921 | 88.87 | 87.28 | 31.68 |
| Cu(II) | **21**[c] | $NCCH_3$ | 3 | 65 | 2.007 | 2.245 | 1.991 | 2.007 | 1.966 | *f* | 2.903 | 89.12 | 88.97 | 25.92 |
| Mn(II) | **35** | Cl, Cl | 3 | 193 | 2.334 | 2.449 | 2.356 | 2.365 | 2.413 | 2.486 | 2.955 | 86.76 | 85.25 | 32.13 |
| Cu(I) | **35** | $NCCH_3$ | 3 | 73 | 2.203 | 2.184 | 2.096 | (2.897) | 1.900 | *f* | 2.925 | 89.67 | 80.28 | 30.59 |
| Cu(II) | **35** | $NCCH_3$, $FBF_3$ | 3 | 70 | 2.005 | 2.376 | 2.052 | 2.075 | 1.951 | 2.911 | 2.934 | 98.73 | 95.18 | 20.55 |
| Cu(II) | **35**[c] | $O_2NO$ | 3 | 194 | 1.976 | 2.092 | 2.259 | 2.347 | 1.964 | 2.283 | 2.853 | 82.02 | 81.14 | 36.78 |
| Cu(II) | **35**[c] | *e*, $H_2O$ | 3 | 195 | 2.121 | 1.987 | 2.011 | 2.032 | *f* | 1.883 | 2.901 | 81.06 | 77.89 | 16.70 |

*(continued)*

TABLE IV
(*Continued*)

| M | L | $X_1$, $X_2$/ $X_E$, $X_A$ | Type | Reference | M-N3 | M-N7 | M-D1/ M-py1 | M-D2/ M-py2 | M-$X_1$/ M-$X_E$ | M-$X_2$/ M-$X_A$ | N3-N7 | tor 1 | tor 2 | Plane Twist |
|---|---|---|---|---|---|---|---|---|---|---|---|---|---|---|
| Cu(II) | **35**[c] | *e*, Cl | 3 | 70 | 2.147 | 2.120 | 2.061 | 2.064 | *f* | 2.221 | 2.930 | 77.96 | 77.18 | 34.62 |
| Cu(II) | **38**[c] | $H_2O$, $H_2O$ | 3 | 195 | 1.987 | 2.330 | 2.041 | 2.041 | 1.967 | 2.838 | 2.897 | 99.57 | 103.07 | 9.72 |
| Cu(II) | **38**[c] | F | 3 | 71 | 2.006 | 2.253 | 2.087 | 2.093 | 1.836 | *f* | 2.902 | 93.18 | 96.01 | 19.77 |
| Cu(II) | **38**[c] | *e*, Cl | 3 | 71 | 2.143 | 2.135 | 2.017 | 2.033 | *f* | 2.255 | 2.913 | 80.43 | 79.83 | 29.31 |
| | | | | | 2.149 | 2.131 | 2.004 | 2.019 | *f* | 2.265 | 2.922 | 82.06 | 78.88 | 20.29 |
| Cu(II) | **43**[c] | Cl | 3 | 168 | 2.115 | 2.316 | 1.967 | 1.971 | 2.228 | *f* | 2.917 | 85.71 | 86.37 | 28.71 |
| 2 Mn(II) | **58**[c] | $CH_3COCH_2COCH_3$ | 3 | 189 | 2.299 | 2.437 | 2.230 | 2.212 | 2.065 | 2.119 | 2.953 | 86.69 | 83.76 | 19.82 |
| 2 Mn(II) | **58**[c] | $2ONO_2$, $2ONO_2$ | 3 | 189 | 2.307 | 2.406 | 2.310 | 2.291 | 2.217 | 2.178 | 2.979 | 82.25 | 80.12 | 46.37 |
| 2 Mn(II) | **58**[c] | $2OCOCH_3$, $2H_2O$ | 3 | 189 | 2.329 | 2.450 | 2.267 | 2.295 | 2.132 | 2.153 | 2.981 | 87.77 | 88.33 | 24.40 |
| 2 Cu(I) | **58** | $2NCCH_3$ | 3 | 196 | 2.010 | 2.336 | 1.988 | 2.010 | 1.986 | *f* | 2.934 | 92.82 | 86.55 | 23.83 |
| 2 Cu(I) | **58**[h] | $2NCCH_3$ | 3 | 196 | 2.206 | 2.192 | | 2.014 | 1.864 | *f* | 2.926 | *f* | 87.12 | exo–endo |
| 2 Cu(I) | **59** | $2NCCH_3$ | 3 | 196 | 2.205 | 2.193 | 2.222 | 2.498 | 1.897 | *f* | 2.945 | 84.86 | 80.24 | 42.31 |
| | | | | | 2.235 | 2.178 | 2.338 | 2.254 | 1.925 | *f* | 2.952 | 82.06 | 83.92 | 45.00 |
| 2 Cu(II) | **59**[c] | 2Cl | 3 | 196 | 2.036 | 2.356 | 1.992 | 2.016 | 2.230 | *f* | 2.916 | 97.10 | 93.46 | 15.02 |
| | | | | | 2.029 | 2.307 | 2.018 | 2.033 | 2.221 | *f* | 2.920 | 91.27 | 90.63 | 29.76 |
| 2 Cu(II) | **59** | $H_2O$, $H_2O$ $OSO_2CF_3$, $H_2O$ | 3 | 66 | 2.194 | 2.359 | 2.195 | 2.156 | 2.062 | 2.121 | 2.935 | 85.15 | 90.28 | 22.63 |
| | | | | | 2.201 | 2.352 | 2.180 | 2.175 | 2.070 | 2.129 | 2.936 | 83.75 | 89.48 | 28.38 |
| 2 Cu(II) | **59**[c] | $^{-}O$, $^{-}O$, Cl, Cl, Cl, Cl | 3 | 192 | 2.027 | 2.360 | 1.985 | 2.004 | 1.898 | *f* | 2.944 | 90.39 | 92.11 | 18.24 |
| | | | | | 2.029 | 2.374 | 2.020 | 2.021 | 1.892 | *f* | *f* | 94.25 | 91.10 | 19.58 |
| Mn(II) | **85b** | Cl, Cl | 3 | 190 | 2.343 | 2.446 | 2.241 | 2.217 | 2.395 | 2.523 | 2.979 | 81.71 | 86.56 | 36.50 |
| | | | | | 2.345 | 2.437 | 2.232 | 2.232 | 2.434 | 2.530 | 2.984 | 89.27 | 81.62 | 38.95 |
| ***2,3,4-Trisubstituted Pentadentate Ligands*** | | | | | | | | | | | | | | |
| Fe(II) | **28**[c] | $OSO_3$ | 4 | 66 | 2.231 | 2.261 | 2.211 | 2.177 | 1.968 | 2.186 | 2.896 | 79.59 | 84.84 | 40.12 |
| Fe(II) | **28**[c] | $OFe^{III}Cl_3$ | 4[b] | 66 | 2.205 | 2.218 | 2.165 | 2.121 | 1.779 | 2.147 | 2.894 | 79.74 | 89.15 | 39.07 |

| | | | | | | | | | | | | | |
|---|---|---|---|---|---|---|---|---|---|---|---|---|---|
| Co(III) | **28**[c] | OH | 4 | 167 | 1.930 | 2.046 | 1.929 | 1.947 | 1.876 | 1.931 | 2.803 | 83.75 | 82.64 | 32.58 |
| Cu(II) | **28** | $NCCH_3$ | 4 | 82 | 2.105 | 2.106 | 2.260 | 2.353 | 2.015 | 2.027 | 2.884 | 80.51 | 74.84 | 35.13 |
| Cu(II) | **28**[c] | $H_2O$ | 4 | 82 | 2.056 | 2.072 | 2.519 | 2.315 | 1.983 | 2.033 | 2.847 | 68.94 | 79.50 | 48.21 |
| | | | | | 2.058 | 2.082 | 2.472 | 2.330 | 1.972 | 2.038 | 2.847 | 70.46 | 79.69 | 50.70 |
| Cu(II) | **28**[c] | Cl | 4 | 168 | 2.070 | 2.479 | 2.012 | 1.987 | 2.255 | 2.544 | 2.931 | 90.83 | 96.20 | 12.10 |
| Cu(II) | **29** | $NCCH_3$ | 4 | 197 | 2.065 | 2.476 | 2.000 | 2.001 | 2.021 | 2.281 | 2.978 | 91.28 | 89.94 | 20.36 |
| Cu(II) | **29**[c] | $NCCH_3$ | 4 | 197 | 2.049 | 2.422 | 2.010 | 1.995 | 1.994 | 2.526 | 2.937 | 89.21 | 91.97 | 18.43 |
| Cu(II) | **30** | $NCCH_3$ | 4 | 197 | 2.055 | 2.603 | 2.003 | 2.000 | 2.013 | 2.291 | 3.015 | 92.70 | 93.54 | 16.77 |
| Cu(II) | **30**[c] | $H_2O$ | 4 | 197 | 2.033 | 2.585 | 1.980 | 2.004 | 1.995 | 2.340 | 3.006 | 94.56 | 91.87 | 12.59 |
| Cu(II) | **31**[c] | $H_2O$ | 4 | 197 | 2.031 | 2.465 | 1.992 | 1.983 | 1.958 | 2.499 | 2.931 | 89.82 | 94.00 | 11.10 |
| 2 Cu(II) | **61** | 2 $NCCH_3$ | 4 | 82 | 2.029 | 2.560 | 2.021 | 1.976 | 1.972 | 2.372 | 3.002 | 85.58 | 101.72 | 18.17 |
| 2 Cu(II) | **61**[c] | 2 $NCCH_3$ | 4 | 82 | 2.052 | 2.429 | 2.026 | 1.992 | 2.033 | 2.331 | 2.938 | 86.43 | 91.80 | 26.26 |
| ***2,4,7-Trisubstituted Pentadentate Ligands*** | | | | | | | | | | | | | | |
| Fe(II) | **22**[c] | Cl | 5 | 66 | 2.195 | 2.300 | 2.196 | 2.174 | 2.123 | 2.361 | 2.874 | 86.50 | 81.05 | 31.57 |
| Cu(I) | **22** | $NCCH_3$ | 5 | 189 | 2.212 | 2.208 | 2.290 | 2.304 | 1.914[i] | *f* | 2.933 | 81.85 | 87.55 | 38.15 |
| | | | | | 2.239 | 2.216 | 2.217 | 2.387 | 1.919[i] | *f* | 2.955 | 92.12 | 83.03 | 33.82 |
| Cu(II) | **22**[c] | $OHCH_3$ | 5 | 73 | 2.016 | 2.336 | 2.003 | 2.000 | 2.003 | 2.406 | 2.932 | 91.10 | 90.09 | 25.47 |
| Cu(II) | **22**[c] | Cl | 5 | 189 | 2.029 | 2.357 | 2.004 | 2.009 | 2.224[i] | *f* | 2.940 | 93.35 | 94.66 | 21.77 |
| Cu(II) | **23**[c] | | 5 | 193 | 2.033 | 2.233 | 1.997 | 2.033 | 1.895 | *f* | 2.912 | 93.88 | 87.82 | 20.59 |
| V(IV) | **24**[c] | O | 5 | 198 | 2.117 | 2.338 | 2.103 | 2.112 | 2.094 | 1.599 | 2.871 | 83.20 | 81.58 | 30.71 |
| V(IV) | **24**[c] | O | 5 | 198 | 2.106 | 2.359 | 2.098 | 2.097 | 2.075 | 1.586 | 2.869 | 83.97 | 84.51 | 25.38 |
| | | | | | 2.098 | 2.364 | 2.097 | 2.109 | 2.078 | 1.589 | 2.836 | 86.99 | 84.31 | 26.13 |
| V(V) | **24**[c] | $O_2$, O | $5^d$ | 198 | 2.191 | 2.556 | 2.114 | 2.114 | 1.839[i] | 1.603 | 2.906 | 94.65 | 89.77 | 24.74 |
| | | | | | | | | | 1.808 | | | | | |
| Mn(II) | **24** | Cl | 5 | 193 | 2.283 | 2.415 | 2.271 | 2.262 | 2.191 | 2.391 | 2.943 | 85.40 | 86.83 | 29.49 |
| Fe(II) | **24**[c] | Cl | 5 | 66 | 2.194 | 2.362 | 2.182 | 2.143 | 2.134 | 2.417 | 2.879 | 85.73 | 90.78 | 24.63 |
| Fe(II) | **24**[c] | OSO3 | 5 | 66 | 2.181 | 2.375 | 2.143 | 2.192 | 2.129 | 2.015 | 2.880 | 86.91 | 84.77 | 20.84 |
| Co(III) | **24**[c] | $H_2O$ | 5 | 167 | 1.977 | 2.022 | 1.969 | 1.929 | 1.941 | 1.962 | 2.797 | 86.06 | 83.46 | 26.97 |
| Cu(I) | **24** | | 5 | 73 | 2.275 | 2.238 | 1.981 | 2.958 | 1.948 | *f* | 2.922 | 88.70 | 83.68 | 41.02 |
| Cu(II) | **24**[c] | $OClO_3$ | 5 | 195 | 1.995 | 2.278 | 1.997 | 1.978 | 1.960 | 2.696 | 2.940 | 87.37 | 91.07 | 26.24 |
| Cu(II) | **24**[c] | Cl | 5 | 168 | 2.036 | 2.368 | 2.028 | 2.028 | 2.029 | 2.717 | 2.915 | 93.13 | 89.21 | 23.47 |
| Cu(II) | **39** | Cl | 5 | 71 | 2.089 | 2.114 | 2.344 | 2.609 | 2.019 | 2.305 | 2.853 | 78.26 | 69.92 | 50.81 |

(continued)

TABLE IV
(*Continued*)

| M | L | $X_1$, $X_2$/ $X_E$, $X_A$ | Type | Reference | M-N3 | M-N7 | M-D1/ M-py1 | M-D2/ M-py2 | M-$X_1$/ M-$X_E$ | M-$X_2$/ M-$X_A$ | N3-N7 | tor 1 | tor 2 | Plane Twist |
|---|---|---|---|---|---|---|---|---|---|---|---|---|---|---|
| Cu(II) | **39**[e] | $OHCH_3$ | 5 | 71 | 2.041 | 2.093 | 2.376 | 2.522 | 2.000 | 2.057 | 2.849 | 80.90 | 73.01 | 46.04 |
| Cu(II) | **40** | Cl | 5 | 71 | 2.159 | 2.079 | 2.686 | 2.320 | 2.159 | 2.264 | 2.836 | 65.15 | 76.55 | 58.39 |
| Cu(II) | **41**[c] | | 5 | 71 | 1.953 | 2.095 | 2.311 | 2.269 | 1.927 | [f] | 2.834 | 90.33 | 79.18 | 38.47 |
| Cu(II) | **41**[c] | $FBF_3$ | 5 | 71 | 2.000 | 2.168 | 2.235 | 2.263 | 1.984 | 2.811 | 2.876 | 79.12 | 91.56 | 38.36 |
| Cu(II) | **42**[c] | $NCCH_3$ | 5 | 65 | 2.122 | 2.015 | 2.288 | 2.929 | 2.108 | 1.988 | 2.818 | 75.16 | 73.13 | 31.43 |
| ***2,3,4,7-Tetrasubstituted Hexadentate Ligands*** | | | | | | | | | | | | | | |
| Fe(II) | **32**[c] | $OSO_3$ | 6 | 66 | 2.214 | 2.273 | 2.172 | 2.206 | 1.957[i] | 2.195 | 2.883 | 87.86 | 78.06 | 39.96 |
| Ni(II) | **32**[c] | | 6 | 69 | 2.074 | 2.156 | 2.153 | 2.103 | 2.078 | 2.081 | 2.880 | 83.49 | 83.98 | 34.97 |
| Cu(II) | **32**[c] | | 6 | 69 | 2.093 | 2.037 | 2.608 | 2.012 | 2.031 | 2.009 | 2.830 | 80.05 | 81.66 | 41.34 |
| | | | | | 2.087 | 2.045 | 2.280 | 2.573 | 2.028 | 2.012 | 2.841 | 77.59 | 80.07 | 39.95 |
| Zn(II) | **32**[c] | | 6 | 69 | 2.166 | 2.234 | 2.285 | 2.170 | 2.069 | 2.112 | 2.905 | 84.74 | 83.47 | 36.61 |
| | | | | | 2.138 | 2.224 | 2.147 | 2.172 | 2.132 | 2.223 | 2.911 | 80.95 | 87.39 | 32.10 |
| | | | | | 2.110 | 2.253 | 2.126 | 2.157 | 2.132 | 2.251 | 2.880 | 88.38 | 80.93 | 31.87 |
| Cu(II) | **33**[c] | | 6 | 69 | 2.016 | 2.195 | 2.024 | 2.006 | 1.966 | [f] | 2.903 | 88.72 | 89.53 | 23.89 |
| Li(I) | **87b** | | 6 | | 2.120 | 2.237 | 2.313 | 2.273 | 2.090 | 2.190 | 2.890 | 80.46 | 82.16 | 36.87 |

[a]See Chart 20 and Fig. 1.
[b]Dinuclear complex.
[c]Hydrated at C9.
[d]Seven-coordinated complex.
[e]Demethylated at N7.
[f]Not available.
[g]Hydrolyzed ester groups at C1 and C5.
[h]exo–endo isomer of the ligand.
[i]Coligand $X_E$ coordinated to the metal center instead of a bispidine pendant donor group.

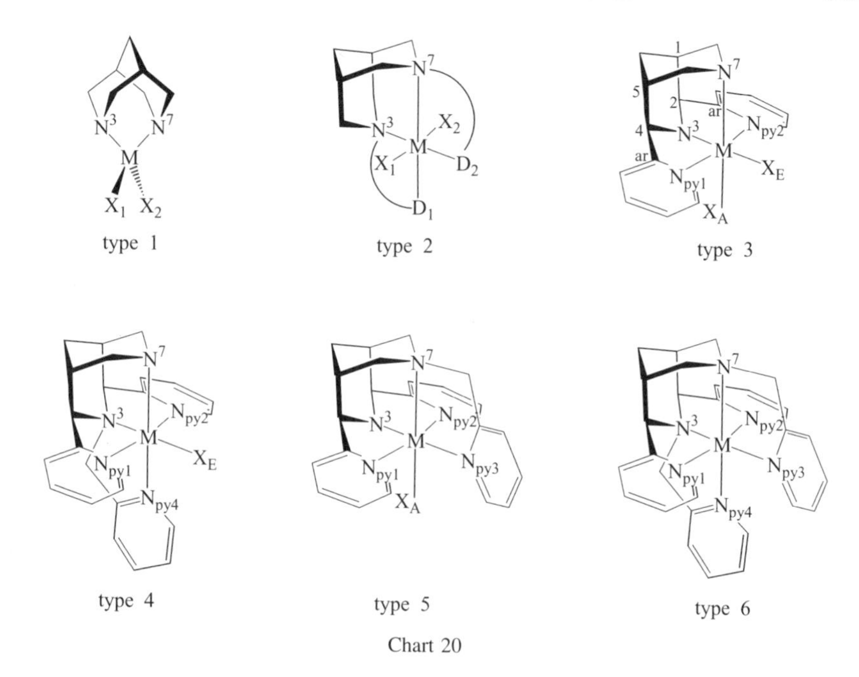


Chart 20

for the cobalt(III) complexes, specifically with the N7-demethylated ligand **14c**, where N3···N7 is 2.71 Å. The longest distance, which is observed for ligands with sterically demanding substituents at N7, is 3.0 Å. Therefore, it appears that the ligand cavity may be tuned to some extent by the steric demand of the N3 and N7 substituents. This finding is of some importance for the various types of isomerism discussed below (Section III.D.1).

With the bidentate and the N, N′-disubstituted tetradentate ligands (**49**, **52–56**, **76**), there is no ligand-enforced asymmetry with respect to the metal–N3,N7 distances. All other ligands have additional 2,4-based donor groups. Depending on the electronic and geometric properties of these donors, there is a pronounced difference in the metal ion to N3 and N7 distances and bond strengths. In many cases (but not exclusively), there is a long distance to N7 and a shorter one to N3. This situation has been studied and discussed in detail for the tetradentate ligand **14** and its derivatives (189), for the hexadentate ligands (69) and for the Jahn–Teller active copper(II) systems (71, 82, 194, 199) (see Section III.D). Interestingly, this asymmetry usually has a strong influence on the distances and bond strengths of the in-plane and axial coligands $X_E$ and $X_A$, where $X_A$ often (but not exclusively) has a weaker and longer bond than $X_E$ (see Section III.D.1). This design principle is important in bispidine coordination chemistry with respect to thermodynamic properties (Section III.D.2) and reactivities (Section III.D.3).

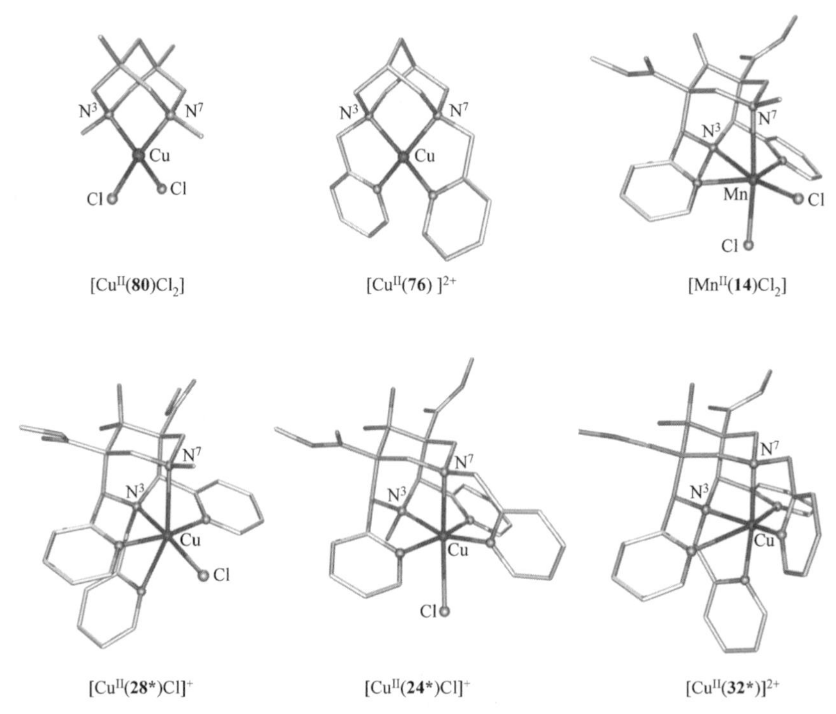


Figure 1. Plots of the molecular cations of examples of the six types of bispidine complexes (hydrogen atoms omitted; see Table IV for details and references).

Many of the coordinated ligands have a hydrated ketone at C9 [$R_2C{=}O$ vs. $R_2C(OH)_2$]. This finding generally is observed when the complex synthesis is not performed in strictly anhydrous conditions (69, 82, 189). From p$K_a$ values, photoelectron spectroscopy, and MO calculations of diazaadamantane derivatives, it was found that ketones lead to a reduction of the nucleophilicities of the aza-groups. This result was attributed to through-bond interactions and inductive effects (140). and is supported by a number of examples in Table IV (e.g., ligands **14, 14*; 35, 35*; 28,** and **28***, where the ligands with an arc are hydrated at C9).

There is a very large variation of bond distances of the metal centers to the bispidine donors. The shortest (and longest) distances to N3 are 1.93 Å (2.39 Å), those to N7 are 1.98 Å (2.60 Å), and those to the 2,4-substituted donors are 1.91 Å (2.39 Å). It emerges that the very rigid ligands allow a coordination geometry of considerable elasticity (flat potential energy surface), which leads to interesting types of isomerism and possibilities to tune the properties (see Section III.C). The individual bond distances are generally in the expected ranges for the corresponding metal ions (200), but usually at the longer limit of the corresponding average distances (69). This finding is also confirmed by force-field calculations (42, 69, 189, 201), which indicate that there is basically no strain induced to the ligand upon coordination to metal ions, except for small metal centers (short metal–donor distances, see Section III.D.1). The result is that bispidine ligands quite generally have a preference for large metal ions, which leads to interesting stabilities and selectivities (see Section III.D.2) (69, 201).

Earlier force-field calculations (105) are at variance with these results and interpretations, but the structural data in Table IV and the experimentally determined complex stabilities (Table IX in Section III.D.2) support these predictions. Why this is so however is an interesting question. For the macrocyclic ligand **74**, there are, in contrast to complexes of the other bispidine ligands, smaller than average metal–donor distances (99). It appears that in this example the lone pairs are at an ideal orientation with respect to the metal center (42). Incidentally, and in contrast to the other ligands, the macrocycle **74** is a proton sponge (also see Table IX). One general observation is that metal-ion-based electronic preferences are much less pronounced than ligand-centered electronic and steric effects, that is, the angular distortion due to elongated metal–donor bonds does not cost much energy (40). A possible reason for the larger destabilization due to short metal–donor bonds is a build-up of van der Waals repulsion, when the 2,4-substituents of the bispidine backbone and/or the coligands are getting close to the bispidine bicycle (also see Section III.C).

### 2. *Novel Bispidine-Derived Chromophores*

The bispidine coordination chemistry described in this chapter, specifically the properties reported in Section III.D, are largely based on coordination geometries enforced by very few ligands, primarily the tetradentate ligand **14** and the two isomeric pentadentate ligands **24** and **28**, and structurally analoguous derivatives. Some of the properties reported in Section III.D are unique. These and other properties are directly related to the structural features discussed above. Some of the ligands and complexes reported in this section and their structures presented in Fig. 2 are based on experiments, others are too tedious to prepare in large enough quantity for explorative coordination chemistry, some are based on the design of ligands and studies in progress, others are pure fiction and perhaps will never be made.

The spectroscopic properties and structure of $[Cu(\mathbf{74})(OH_2)_n]^{2+}$ $n = 0$, 2 [Fig. 2(*f*) with $n = 0$], based on an MM-AOM analysis (202, 203) have been reported (99). The computed structure of the ligand part of the molecule agrees well with the crystal-structural analysis of the metal-free ligand, which is available as the free base and in the diprotonated form (99, 181, 182). There are two isomers with respect to the conformation of the ethylene bridges. That in Fig. 2(*f*) is the more stable structure of the complex, those in Fig. 3 are the

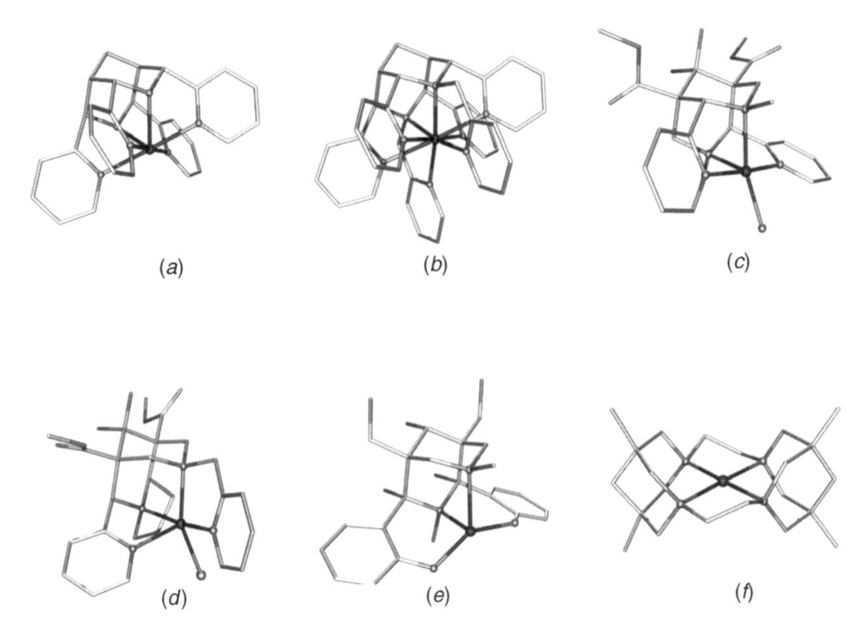


Figure 2. Plots of the molecular cations of complexes with novel bispidine-type ligands: (*a*) 2,4,6,8-tetrakispyridine-3,7-diazabicyclo[3.3.1]nonane, Co(III), computed; (*b*) 2,3,4,6,7,8-hexakispyridine-3,7-diazabicyclo[3.3.1]nonane, Co(III), computed; (*c*) dimethyl-2,6-bispyridine-3,7-diazabicyclo[3.3.1]nonane-9-diol-1,5-dicarboxylate, X-ray, Cu(II), Cl; (*d*) **57**, X-ray, Cu(II), Cl; (*e*) 2,4-bisphenol-3,7-diazabicyclo[3.3.1]nonane, X-ray, Cu(II); (*f*) **74**, Cu(II), computed.

slightly more stable conformations of the ligand. The fact that the ligand structure does not change much upon coordination confirms the rigidity and high degree of preorganization of these macrocyclic ligands. Structural and computational studies indicate that, due to lone-pair repulsion in the metal-free ligand, this is even more strained and distorted than the coordinated macrocycle and leads to an increase of the complex stabilities (42). Recently, crystals of the copper(II) complex of **74** were obtained. Due to severe disorder related to the five-memebered chelate rings, so far we were unfortunately not able to solve the structure. However, a preliminary analysis allows to confirm the very short Cu—N bonds of ~1.98 Å and the correspondingly strong ligand field (99).

The synthesis of the 2,4-bisphenol derivative of the standard tetradentate ligand **14** is quite tedious and so far has not been optimized (204). Apart from the fact that the noninnocent phenolate donors may lead to interesting ligand systems and metal complexes, it is the six-membered chelate rings involving the "in-plane" phenolate donors that lead to structural properties quite different from those of the other bispidine ligands [shown in Fig. 2(*e*) is a plot of a preliminary X-ray molecular structure of one of the possible conformers (meso form) of a copper(II) complex].

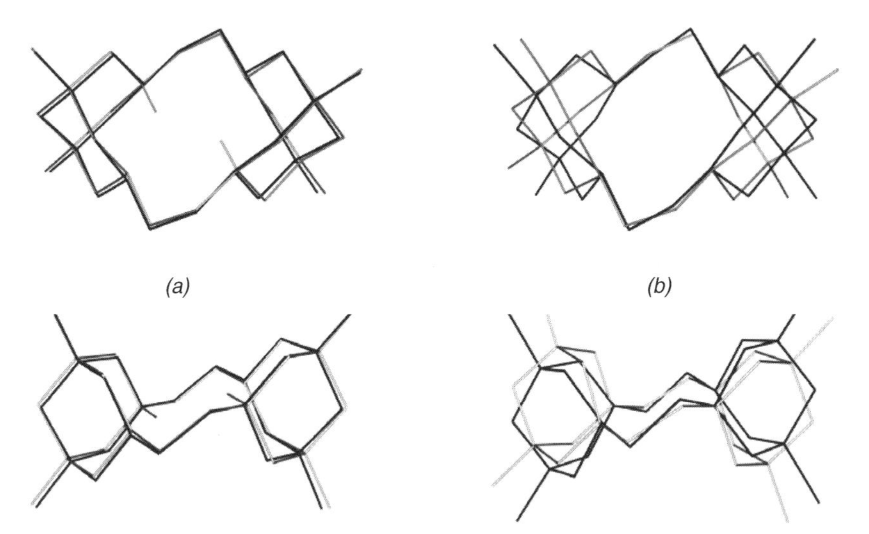


Figure 3. Overlay plots of crystallographically determined structures of (*a*) the diprotonated (light) and uprotonated (dark) metal-free ligands **74**, and of (*b*) the unprotonated metal-free ligand **74** (light) and the ligand part of the corresponding copper(II) complex (dark, disordered structure, see text).

An interesting possible and stable isomer of the 2,4-bispyridine substituted ligand **14** is the corresponding 2,6-bispyridine–bispidone [shown in Fig. 2(*c*) is a plot of the X-ray molecular structure of its copper(II) complex] (205). There are two major differences between this structure and complexes of **14**. (1.) In structure **2*c***, the two amine donors N3 and N7 are symmetry related, and therefore are identical in contrast to ligands and their complexes derived from **14**. While much of the interesting properties of bispidine complexes are based on this difference, it certainly is of interest to develop the coordination chemistry of the isomeric tetradentate ligand. (2.) While **14** and its complexes have $C_s$ symmetry the ligand and complex of the isomer shown in Fig. 2(*c*) are $C_2$ symmetric, and therefore chiral. If catalytically active (such as complexes based on **14**), complexes derived from that shown in Fig. 2(*c*) should be interesting for enantioselective reactions.

The piperidone precursor of **57** (see Chart 6) is a possible starting material for the synthesis of 2,6-substituted chiral, tetradentate bispidine ligands. The copper(II) complex of **57** is chiral [see Fig. 2(*d*) for a plot of the experimental structure] and has an interesting set of bond distances. Quite unexpectedly on the basis of the chelate ring sizes, the 2-pyridine substituent of the bispidine backbone is the axial donor (2.25 Å), while N3 (2.08 Å), N7 (2.05 Å), the N7-pendant pyridine (2.01 Å) and the chloride (2.28 Å) are the in-plane donors (80).

The other two complexes in Fig. 2 are based on the 2,4,6,8-tetrakispyridine bispidine backbone, with a hexadentate [Fig. 2(*a*)] and an octadentate ligand

[Fig. 2(*b*)]. These are expected to be stable organic molecules. However, based on the problems of stereocontrol (see Section II.B) it is unlikely that this chemistry will soon start to develop. The computed structures indicate that the hexadentate ligand will enforce a trigonal-prismatic coordination geometry [Fig. 2(*a*)], and the octadentate ligand has a preference for square-prismatic structures [Fig. 2(*b*)].

### C. Ligand Rigidity and Elasticity of the Coordination Sphere

The rigidity of bispidine ligands has been analyzed on the basis of cavity size calculations with molecular mechanics [and density functional theory (DFT) — in order to check the accuracy and reliability of the force-field calculations] and a comparison of the computed and corresponding experimental structures and their analysis, based on the computed strain energy curves (42, 69, 189, 201).

An efficient measure for the rigidity of the bispidine ligands is the N3···N7 distance, which is very similar for metal-free and coordinated ligands and is practically constant in all the complexes listed in Table IV (see Section III.B.1). Cavity shapes, sizes, and plasticities of asymmetrical ligands are best computed by a stepwise variation of the constrained sum of all distances of the donors to an interactionless metal ion. These curves are computed such that the sum of all bond distances (six in a pseudo-octahedral complex) is constrained to a specific value, but there is flexibility with respect to each individual bond (i.e., the position of the metal ion within the cavity is flexible) (206). The minimum of the resulting curves is related to the optimum shape and size of the chromophore (metal–donor distances), the steepness is a measure for the rigidity of the coordination geometry, and structures along the potential energy curve should show the experimental modes of distortion. Computed cavity size curves for the various types of bispidine ligands have been produced with MOMEC (69, 189, 207–209) and are presented in Fig. 4 (also included in the plots are cavity size curves with a variant of the method described above) (69, 189). Note that these potential energy curves and the concomitant interpretations do not include the energetics of the metal–donor bonding (preference of the metal ion for specific distances, angular geometries, distortion modes). These are the basis for the "electronic complementarity" and are disucussed in Section III.D.1.

From these cavity size curves of the relevant bispidine-type ligands presented in Fig. 4 it emerges that there is basically no loss of ligand-based steric energy for large metal ions [$\lesssim 10\,\mathrm{kJ\,mol^{-1}}$ from the optimum metal ion size at $\sim(\mathrm{M{-}N})_{av}$ ~2.2 Å to $(\mathrm{M{-}N})_{av}$ ~2.6 Å]. This is due to the fact that the rigid ligands are open on one face (opposite to the adamantanoide cap; the obvious exception is the macrocyclic ligand **74**), and there is little constraint to fix the metal ion at a particular position in the ligand cavity (also see the discussion on bonding and complementarity below). For averaged metal–donor distances <2.0–2.1 Å, there is a considerable build-up of ligand-based strain. The reason

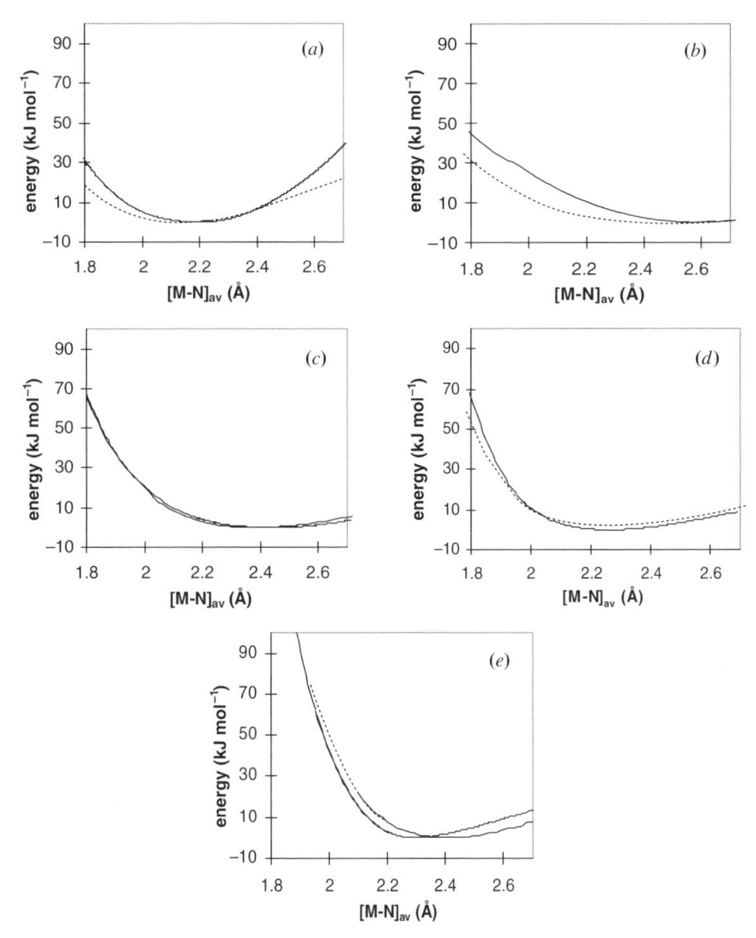


Figure 4. Hole size curves of bispidine ligands (molecular mechanics, no metal center-dependent energy terms included, see text; broken lines are with sum constraints, solid lines are with an approach with individual, asymmetric variations of all six M—N bonds) of (*a*) **76**, (*b*) **14**, (*c*) **28**, (*d*) **24**, (*e*) **32** (69).

for this is that a reduction of the sum of bond distances is only possible when ligand-based angles and distances are distorted. This leads to a dramatic increase of the ligand-based steric energy. In addition, a compression of the cavity leads to a build-up of intraligand van der Waals repulsion (see Section III.B.1). The conclusion from the empirical force-field calculations is that all bispidine ligands are very rigid they do not change their shape along the entire curves, and this is in agreement with the observations from experimental structures (see Fig. 8 in Section III.D.1). However, there is a significant elasticity of the coordination geometries and, therefore, the ligands are not size- and shape-selective (69, 189, 201). This finding is in good agreement with experimental observations (see stability constants, Section III.D.2), but it is at variance with earlier predictions and interpretations based on another type of molecular mechanics based approach with the bispidine-based ligand **76** (32, 105).

## D. Properties of Bispidine Complexes

### *1. Bonding, Isomerism, and Dynamics*

A number of interesting modes of isomerism have been observed with bispidine complexes where isomerism is not always used in a puristic sense, and structural variation probably is a more correct term (42, 70, 71, 82, 189, 194, 199). Some or most of these observations with bispidine complexes may be related in some way to distortional or bond-stretch isomerism and probably are not among the least appropriate examples in this area (210–212). On a more pragmatic basis, these observations are all related to the fact that, quite generally, bispidine complexes have an extremely flat potential energy surface (also see Section III.C). To a large extent this is the result of the very rigid ligand framework (42). The fact that the ligands are rigid and the coordination geometries are elastic is not a contradiction, but the latter is born out of the former and both together are the prerequisite for much of the interesting, and in some examples, unusual reactivity emerging from bispidine coordination chemistry. These ligand–geometry-enforced features are basically steric in nature, but enhanced by subtle electronic effects (also partially derived from ligand enforced distortions). This leads in the cis-octahedral bispidine complex geometry, to a clear electronic differentiation of the in-plane (equatiorial $X_E$, in the plane of the two pyridine donors, trans to N3) and the axial ($X_A$, perpendicular to the two pyridine donors, trans to N7) sites in complexes of **14** and corresponding derivatives (see Fig. 5, also see Section III.D).

Many of the copper(II) bispidine complexes have a square-pyramidal coordination geometry with an elongated Cu—N7 bond, and the substrate coordinated trans to N3 and in-plane with the two pyridine groups ($X_E$). The stability of this, for copper(II) unusual structure, enforced by the (tetradentate) bispidine ligands, has been qualitatively assumed to be due to a high complementarity (42, 81) and is the reason for the relatively high complex stabilities (Section III.D.2), the strong bonds to substrates (Section III.D.2, III.D.3), and the interesting biomimetic copper chemistry (Section III.D.3; see, however, the discussion of other structural forms and of the bonding properties below) (81, 169, 192, 196, 213).

In a preliminary DFT study, a simplified model for a $[Cu(\mathbf{14})(Cl)]^+$ complex with two simple imines and two amines (Chart 21) was used to determine the

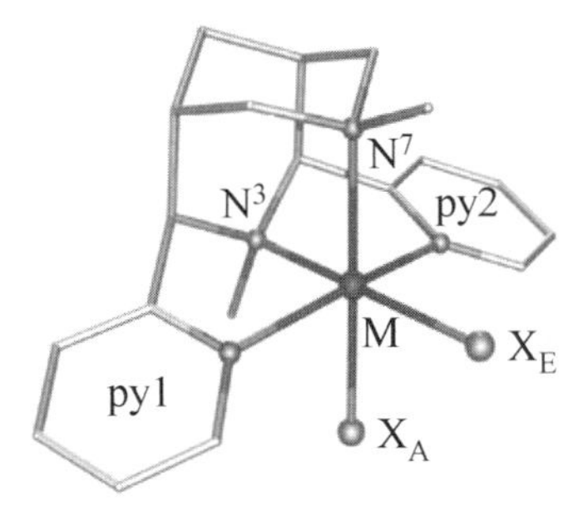


Figure 5. Plot of the structure of a bispidine complex showing the electronically and sterically different positions of a trans to N3 ($X_E$) and a trans to N7 ($X_A$) coordinated substrate.

loss of energy when the bispidine structure is enforced to copper(II) (214). Three of the possible isomers of $[Cu(imine)_2(amine)_2Cl]^{2+}$ and an overlay plot of the refined structures of the bispidine complex and that of the simplified model are given in Fig. 6, together with the important energies. From these model calculations it appears that, as expected, the square-pyramidal structure with an apical $Cl^-$ ion [the "normal" Cu(II) structure] is the most stable. The bispidine structure, with $Cl^-$ in plane with the two imines (pyridines) is destabilized (energized, "entatic state") (41), but only by $\sim$10 kJ mol$^{-1}$. The structure with two perpendicular imine groups in-plane with an amine (N7) and the $Cl^-$ substrate (as observed with **35** with methylated pyridine groups) is destabilized by an additional $\sim$20 kJ mol$^{-1}$. The Cu—Cl bond is strongest for the normal bispidine structure, predicted to be weaker by $\sim$10 kJ mol$^{-1}$ for the structure with the methylated ligand and much weaker ($\sim$85 kJ mol$^{-1}$) for the normal Cu(II) structure with an apical $Cl^-$ (214). These predicted energies are qualitatively in excellent agreement with experimental data (note that partial quenching of the pseudo-Jahn–Teller elongation in the structure with the methylated ligand **35** by the tight five-membered chelates involving the pyridine donors need to be added), that is, with the complex stabilities, the formation constants of the chloro complexes and the relevant redox potentials (199).

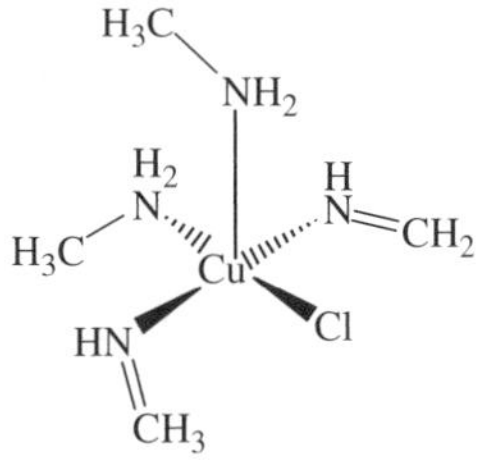

Chart 21

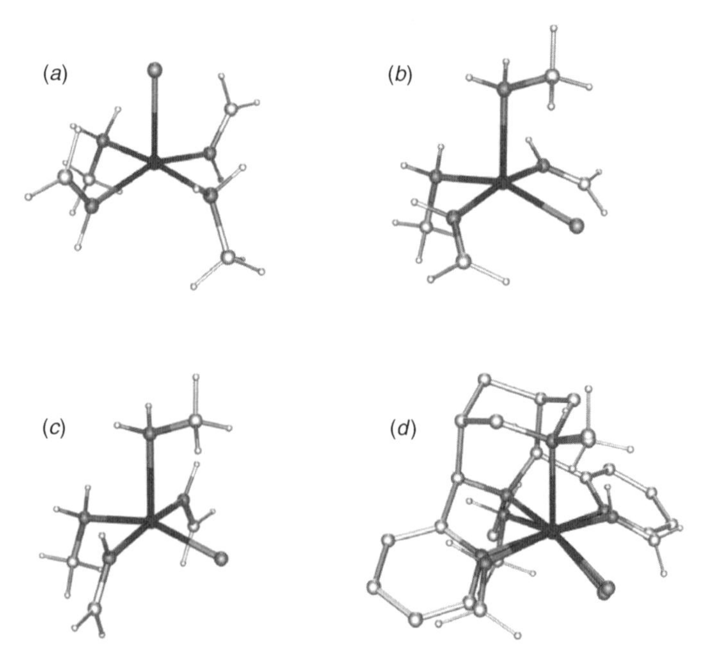


Figure 6. Computed structures (DFT) of three of the possible isomers of Cu(II) complexes with a simplified bispidine model-chromophore (see Chart 21) (214). (*a*) One of the "normal" Cu(II) isomers (axial $Cl^-$ $E_{tot} = -11\,kJ\,mol^{-1}$, $E_{Cu-Cl} = 84\,kJ\,mol^{-1}$, Cu—Cl = 2.52 Å; (*b*) bispidine isomer ($Cl^-$ trans to N3), $E_{tot} = 0\,kJ\,mol^{-1}$, $E_{Cu-Cl} = 0\,kJ\,mol^{-1}$, Cu—Cl = 2.29 Å; (*c*) methylbispidine isomer (Cl-trans to N7), $E_{tot} = +18\,kJ\,mol^{-1}$. $E_{Cu-Cl} = 12\,kJ\,mol^{-1}$, Cu—Cl = 2.25 2.25 Å; (*d*) overlay plot of the isomer shown in (*b*) with the computed $[Cu(\mathbf{14})Cl]^+$ structure.

A more thorough DFT- and ligand-field density functional theory (LFDFT)-based study of the bonding situation is currently in progress. From the *d*-orbital splitting diagram, it emerges that π-bonding effects are, at least for copper(II), of minor importance. A subtle compromise between π donation by the substrate and π backbonding from the pyridines might contribute to some of the stabilization of π-donor substrates. The most interesting observation from these preliminary data, however, is (see Fig. 7) that a distortion of the Cu—N7 bond (misdirected valence) leads to a weakening of the Cu—N7 and a concomitant strengthening of the Cu—Cl interaction (215). This clearly is a feature that is not only operational in Jahn–Teller active electronic configurations, such as that of copper(II), and is a valuable explanation for the fact that many bispidine complexes with other metal centers have similar (but less pronounced) structural features (see Section III.B and III.D.2). Obviously, this is also of importance for the reactivity of the complexes, specifically for the extensively studied iron systems (see Section III.D.3).

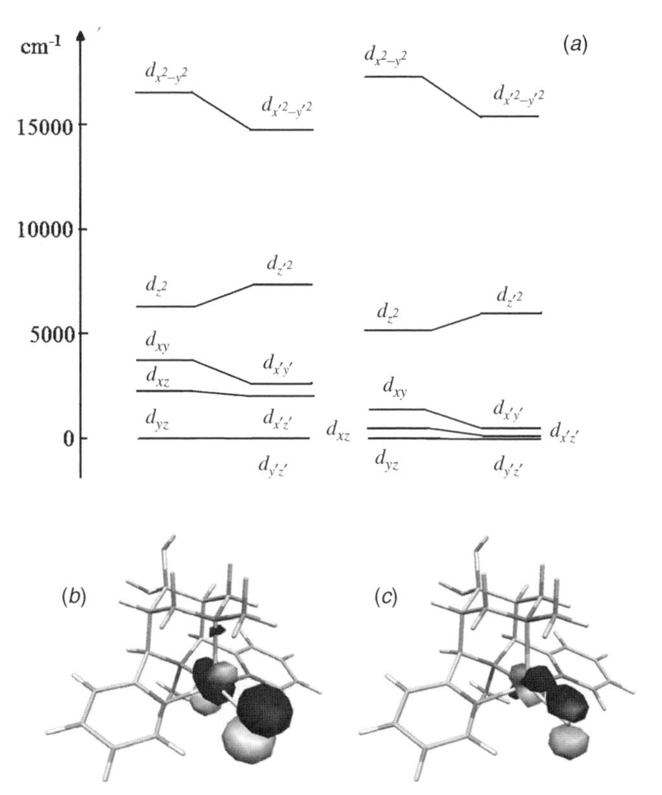


Figure 7. (*a*) The *d*-orbital scheme of square-pyramidal bispidine–copper(II)-coligand complexes (ligand **14***; ester groups at C1 and C5 are replaced by hydrogen atoms); from left to right: $Cl^-$ trans to N3, $Cl^-$ trans to N7, $NH_3$ trans to N3, $NH_3$ trans to N7; (*b*) computed structure of $[Cu(\mathbf{14}^*)Cl]^+$ with the $d_{z^2}$-type orbital; (*c*) computed structure of $[Cu(\mathbf{14}^*)Cl]^+$ with the $d_{xz}$-type orbital (215).

The structural analysis of transition metal complexes of the tetradentate ligand **14** reveals that, in terms of the chromophore, there are two structural types (isomers), one with M-N3 < M-N7 (as argued above to be the electronically more stable geometry), and one with M-N3 > M-N7 (42, 189, 199). Geometries with M-N3 > M-N7 are, as expected, primarily observed with electronically innocent metal ions [Zn(II), Cu(I), Mn(II)] or with the α-methylated ligand **35**. This is another example for the flat potential energy surface, which directs the structure to one of the (in this case) two shallow minima, as a consequence of a subtle competition between steric and electronic effects. This is visualized in Fig. 8(*a*), which is an overlay plot of the ligand portion of 40 complexes with **14** (and **35**, where the pyridine methyl substituents have been deleted). Also shown in Fig. 8 are plots of the average ligand structure with the metal ion included [Cu(I), Mn(II) for the two isomers; Cu(II) is included as the third metal ion, which has,

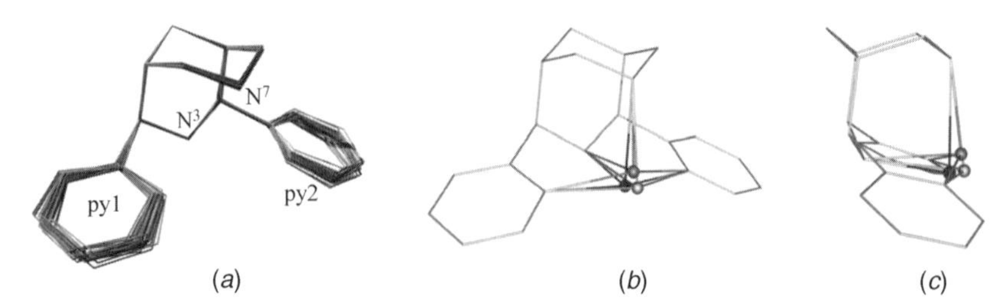


Figure 8. (*a*) Overlay plot of the ligand portion of 40 X-ray structures of complexes with ligand **14**; (*b*) and (*c*) average structure of (*a*) with three metal centers [Cu(II): dark, Cu(I): middle, Mn(II): light] included in their crystallographically determined sites (189).

due to the Jahn–Teller vibration, a similar structure to the complex with Mn(II), but a more pronounced distortion]. A hole size curve of **14** and two nearly degenerate curves of the two isomers of the cobalt(II) complexes (axial coligands deleted) are shown in Fig. 9 (also see Section III.C) (189, 199). The ligand-only curve shows that metal coordination does not lead to significant strain induced by the metal ion to the ligand, and the two curves with the metal–bispidine donor bonds included, indicate that the two isomers are nearly degenerate, that is, the potential energy surface is very flat with two shallow minima. This is also shown experimentally: the two isomeric forms of $[Co(\mathbf{14})(X)(Y)]^{n+}$ (although with different coligands X and Y) have been isolated and structurally characterized: X, Y = $NO_3^-$: Co—N3,7 = 2.14, 2.13 Å(166); X, Y = $(OH_2)_2$: Co—N3,7 = 2.15, 2.22 Å(189).

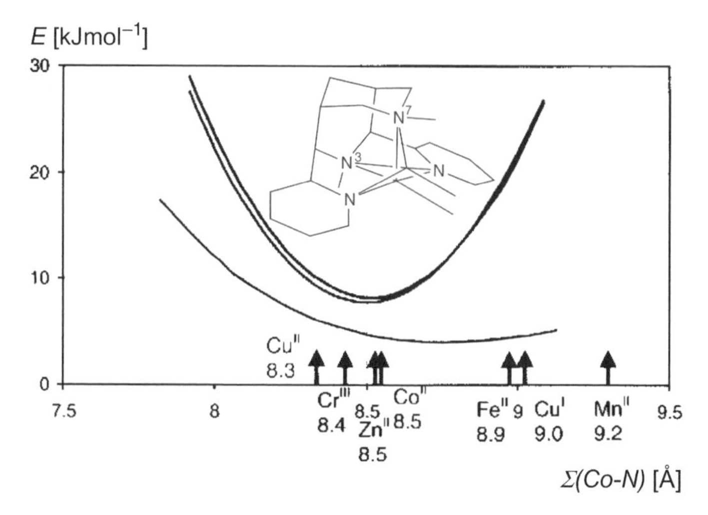


Figure 9. Hole size curve of ligand **14** (lowest energy curve, no metal–donor interaction terms included), and hole size curves of the two isomers of $[Co(\mathbf{14})]^{2+}$ with the cobalt–bispidine interaction terms included. The arrows are averaged sums of observed M—L bond distances (189).

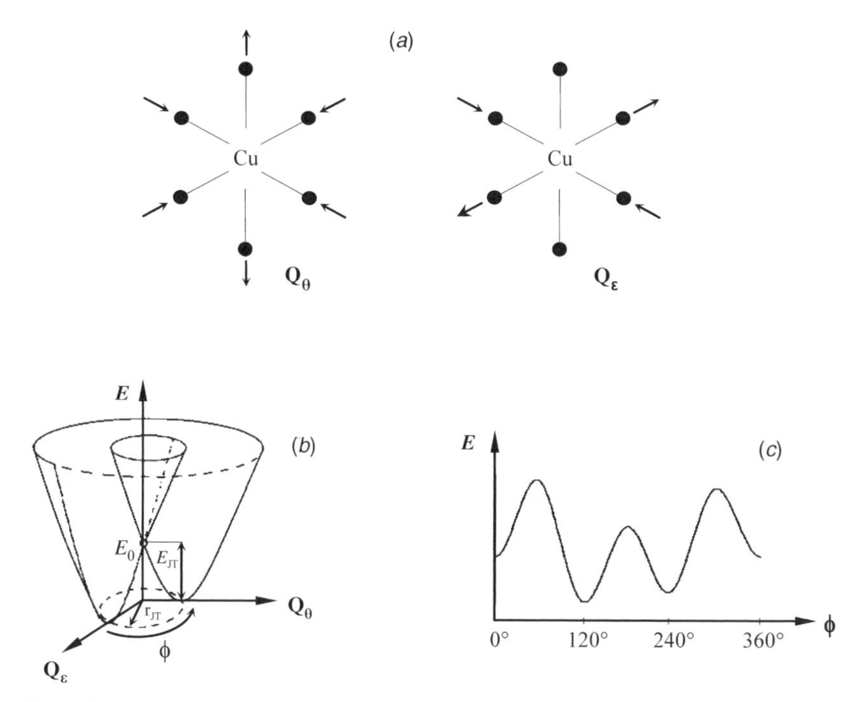


Figure 10. (*a*) The $Q_\theta$ and $Q_\varepsilon$ components of the $e_g$ vibrational mode, (*b*) the Mexican hat potential energy surface for copper(II) in an octahedral ligand field, and (*c*) a cross-section through the warped rim [dotted line in (*b*)] of the Mexican hat potential (217).

The copper(II) coordination chemistry is dominated by multivarious coordination geometries and structural plasticity, due to electronic effects based on Jahn–Teller distortions. Geometries derived from tetragonal symmetry are usually observed with hexacoordinate complexes, and for copper(II) compounds with six identical ligands, this is due to the Jahn–Teller lability of the degenerate $^2D$ ground state (216, 217). Analoguous distortions are also observed in copper(II) complexes of asymmetric multidentate ligands that cannot lead to degenerate ground states. The corresponding Jahn–Teller vibrational modes are visualized with the Mexican hat potential energy surface shown in Fig. 10(*b*) (217, 218). A circular cross-section through the warped rim [dotted line in Fig. 10(*b*)] describes all linear combinations of the $Q_\theta$ and $Q_\varepsilon$ modes [Fig. 10(*a*)] and has three minima. These minima correspond to the tetragonal elongations along each Cartesian coordinate axes, and the three saddle points are the corresponding compressed geometries [Fig. 10(*c*)] (217, 218). Generally, the energy barriers along the rim are relatively small (219).

With minima of similar energy and low barriers, one expects fluxionality (dynamic Jahn–Teller effects) (217, 218, 220), and stable Jahn–Teller isomers are only expected with at least two structures of similar stability and separated by a relatively large energy barrier. With the exception of the thoroughly studied Tutton salts, where the structural variation is based on lattice effects (221), and a thorough study on a system with dynamic Jahn–Teller distortions (222), Jahn–Teller isomers have not been observed until recently (71, 82, 194, 199).

The first examples reported were the copper(II) complexes of **28** and **61**, both with $NCCH_3$ as the sixth ligand, and crystallized as the triflate salt (82). Both were available in the 9-ketone and the hydrated forms. These complexes have identical chromophores and only differ in the substituent at N7 [see Fig. 11 (*a* and *d*)]. Interestingly, the two structures are drastically different. The dinuclear complex $[Cu_2(\mathbf{61})(NCCH_3)_2]^{4+}$ [Fig. 11(*d*)] is elongated along N7—Cu—py3. The complex $[Cu(\mathbf{28})(NCCH_3)]^{2+}$ [Fig. 11(*a*)] has short Cu—N7 and Cu—py3 bonds and an elongated py1—Cu—py2 axis [see Fig. 11(*e*), Table V] (82). A range of ligands with N7-substituents of various steric demand are available, and the corresponding crystal structures of the copper(II) complexes are assembled in Fig. 11 and Table V (68). From variable temperature solution and single-crystal electronic spectroscopy, as well as temperature-dependent X-ray crystallography, it was concluded that with the thoroughly studied pair of complexes $[Cu(\mathbf{28})(NCCH_3)]^{2+}$ and $[Cu_2(\mathbf{61})(NCCH_3)_2]^{4+}$ there is no isomerization in the temperature range $313\,K > T > 10\,K$. The color change with temperature was found to be due to the usual vibrational expansion and a large anhamonicity of the potential energy surface (82).

Also presented in Fig. 11 and Table V are three copper(II) complexes of ligand **24** (168, 223). Interestingly, the trinuclear hexacyanoferrate-based complex with $[Cu_2(\mathbf{24})NC]^{+}$-caps has an elongated py1—Cu—py2 axis, and a short bond from $Cu^{2+}$ to N7 and the cyano bridge, which leads to very interesting magnetic properties (see Section III.D.4). All other complexes of **24** have an elongated N7—Cu(II) substrate axis. Long metal-N7 bonds are also found in the iron(II) complexes, where no Jahn–Teller lability is expected, and probably also in the corresponding iron(III) and iron(IV) complexes (see Section III.D.3). This result is of importance for the catalytic activities. The fact that the copper(II) complexes with the pentadentate quinoline-based ligands **39–42** all have an elongated quinoline–copper(II)–quinoline axis (also listed in Table V) indicates that the various minima are close in energy. Unusual and of specific interest are the two structures of ligand **41** [see Table VI, Fig. 11(*g*)] (71). These have elongated quinoline–copper(II)–quinoline axes and short bonds to N3, N7, and the pyridine. However, the fourth in-plane ligand (trans to N7) is missing or at a very long distance ($Cu—FBF_3 = 2.81$ Å). A thorough investigation of the fundamental reasons for the various types of distortion is under way (215, 224). It is not unexpected that an empirical force-field model, which includes a ligand-field electronic term (ligand-field molecular mechanics, LFMM) (225) allows to predict the observed structural types of all bispidine–copper(II)

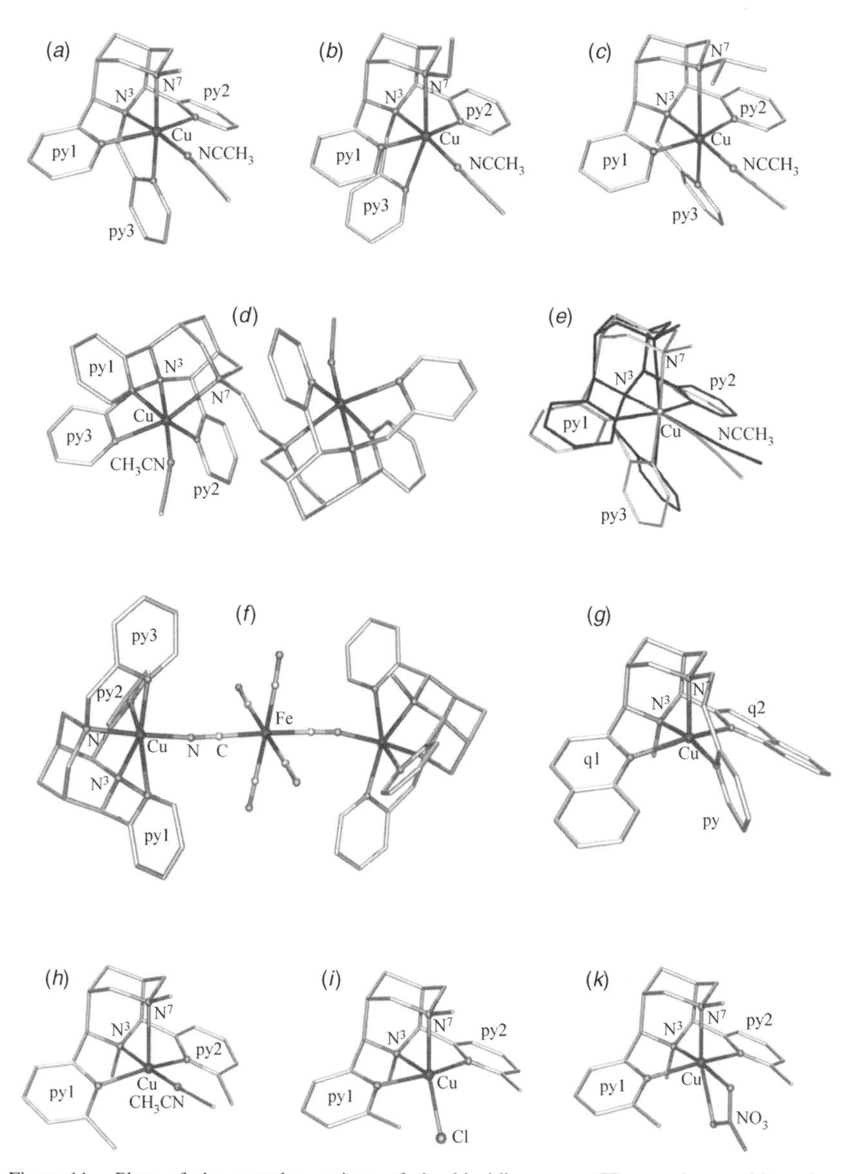


Figure 11. Plots of the complex cations of the bispidine–copper(II) complexes with various directions of the Jahn–Teller elongation (given in brackets), determined by X-ray crystallography (hydrogen atoms and substituents at C1, C5, and C9 omitted; see Table VI for details and references); (*a*) $[Cu(\mathbf{28}(NCCH_3)]^{2+}$ (py1–Cu–py2), (*b*), $[Cu(\mathbf{29})(NCCH_3)]^{2+}$ (N7–Cu–py3), (*c*) $[Cu(\mathbf{30})(NCCH_3)]^{2+}$ (N7–Cu–py3), (*d*) $[Cu(\mathbf{61})(NCCH_3)]^{2+}$ (N7–Cu–py3), (*e*) overlay plot of the chromophores of (*a*) and (*d*), (*f*) $\{Fe(CN)_6[Cu(\mathbf{24})]_2\}^+$ (py1–Cu–py2), (*g*) $[Cu(\mathbf{41})]^{2+}$ (q1–Cu–q2), (*h*) $[Cu(\mathbf{35})(NCCH_3)]^{2+}$ (Cu–N7), (*i*) $[Cu(\mathbf{35})(Cl)]^+$ (Cu–N3), (*k*) $[Cu(\mathbf{35})(NO_3)]^+$ (py1–Cu–py2).

TABLE V
Selected Structural Parameters of Bispidine–Copper(II) Complexes[a]

| L | $X_E$,$X_A$ | Reference | M-N3 | M-N7 | M-py1 | M-py2 | M-$X_E$ | M-$X_A$ |
|---|---|---|---|---|---|---|---|---|
| **14** | $NCCH_3$, $OSO_2CF_3$ | 73 | 2.022 | ***2.355*** | 1.992 | 1.998 | 1.980 | ***2.608*** |
| **14** | Cl | 191 | 2.049 | ***2.351*** | 1.999 | 2.010 | 2.234 | [b] |
| **14**[c] | Cl | 81 | 2.042 | ***2.273*** | 2.020 | 2.024 | 2.232 | [b] |
| **14**[c] | $O_2NO$ | 189 | 1.999 | ***2.280*** | 1.999 | 1.999 | 1.976 | ***2.613*** |
| **14**[c] | Cl, $^-O$, Cl, $^-O$, Cl, Cl | 192 | 2.042 | ***2.433*** | 2.031 | 2.009 | 1.909 | ***2.456*** |
| **19**[c] | $NCCH_3$, $FBF_3$ | 65 | 2.010 | ***2.257*** | 2.022 | 2.020 | 1.957 | ***2.726*** |
| **21**[c] | $NCCH_3$ | 65 | 2.007 | ***2.245*** | 1.991 | 2.007 | 1.966 | [b] |
| **22**[c] | $OHCH_3$ | 73 | 2.016 | ***2.336*** | 2.003 | 2.000 | 2.003 | ***2.406*** |
| **22**[c] | Cl | 189 | 2.029 | ***2.357*** | 2.004 | 2.009 | 2.224[d] | [b] |
| **23**[c] |  | 193 | 2.033 | ***2.233*** | 1.997 | 2.033 | 1.895 | [b] |
| **24**[c] | $OClO_3$ | 195 | 1.995 | ***2.278*** | 1.997 | 1.978 | 1.960 | ***2.696*** |
| **24**[c] | $[FeCN_6]^{3-}$ | 223 | 2.048 | 2.104 | ***2.556*** | ***2.371*** | 2.056 | 1.956 |
| **24**[c] | Cl | 168 | 2.036 | ***2.368*** | 2.028 | 2.028 | 2.029 | ***2.717*** |
| **28** | $NCCH_3$ | 82 | 2.105 | 2.106 | ***2.260*** | ***2.353*** | 2.015 | 2.027 |
|  |  |  | 2.056 | 2.072 | ***2.519*** | ***2.315*** | 1.983 | 2.033 |
| **28**[c] | $H_2O$ | 82 | 2.058 | 2.082 | ***2.472*** | ***2.330*** | 1.972 | 2.038 |
| **28**[c] | Cl | 168 | 2.070 | ***2.479*** | 2.012 | 1.987 | 2.255 | ***2.544*** |
| **29** | $NCCH_3$ | 197 | 2.065 | ***2.476*** | 2.000 | 2.001 | 2.021 | ***2.281*** |
| **29**[c] | $NCCH_3$ | 197 | 2.049 | ***2.422*** | 2.010 | 1.995 | 1.994 | ***2.526*** |
| **30** | $NCCH_3$ | 197 | 2.055 | ***2.603*** | 2.003 | 2.000 | 2.013 | ***2.291*** |
| **30**[c] | $H_2O$ | 197 | 2.033 | ***2.585*** | 1.980 | 2.004 | 1.995 | ***2.340*** |
| **31**[c] | $H_2O$ | 197 | 2.031 | ***2.465*** | 1.992 | 1.983 | 1.958 | ***2.499*** |
| **32**[c] |  | 69 | 2.093 | 2.037 | ***2.608*** | 2.012 | 2.031 | 2.009 |
|  |  |  | 2.087 | 2.045 | ***2.280*** | ***2.573*** | 2.028 | 2.012 |
| **33**[c] |  | 69 | 2.016 | 2.195 | 2.024 | 2.006 | 1.966 | [b] |
| **35** | $NCCH_3$, $FBF_3$ | 70 | 2.005 | ***2.376*** | 2.052 | 2.075 | 1.951 | ***2.911*** |
| **35**[c] | $H_2O$ | 195 | ***2.121*** | 1.987 | 2.011 | 2.032 | [b] | 1.883 |
| **35**[c] | $NO_3^-$ | 194 | 1.976 | 2.092 | ***2.259*** | ***2.347*** | 1.964 | 2.283 |
|  |  |  | 1.987 | 2.032 | ***2.377*** | ***2.376*** | 1.982 | 2.139 |
| **35**[c] | Cl | 70 | ***2.147*** | 2.120 | 2.061 | 2.064 | [b] | 2.221 |
| **38**[c] | $H_2O$, $H_2O$ | 195 | 1.987 | ***2.330*** | 2.041 | 2.041 | 1.967 | ***2.838*** |
| **38**[c] | F | 71 | 2.006 | ***2.253*** | 2.087 | 2.093 | 1.836 | [b] |
| **38**[c] | Cl | 71 | ***2.143*** | 2.135 | 2.017 | 2.033 | 2.255 | [b] |
|  |  |  | ***2.149*** | 2.131 | 2.004 | 2.019 | [b] | 2.265 |
| **39** | Cl | 71 | 2.089 | 2.114 | ***2.344*** | ***2.609*** | 2.019 | 2.305 |
| **39**[c] | $OHCH_3$ | 71 | 2.041 | 2.093 | ***2.376*** | ***2.522*** | 2.000 | 2.057 |
| **40** | Cl | 71 | 2.159 | 2.079 | ***2.686*** | ***2.320*** | 2.159 | 2.264 |
| **41**[c] |  | 71 | 1.953 | 2.095 | ***2.311*** | ***2.269*** | 1.927 | [b] |
| **41**[c] | $FBF_3$ | 71 | 2.000 | 2.168 | ***2.235*** | ***2.263*** | 1.984 | 2.811 |
| **42**[c] | $NCCH_3$ | 65 | 2.122 | 2.015 | ***2.288*** | ***2.929*** | 2.108 | 1.988 |
| **43**[c] | Cl | 168 | 2.115 | ***2.316*** | 1.967 | 1.971 | 2.228 | [b] |

TABLE V
(*Continued*)

| L | $X_E$,$X_A$ | Reference | M-N3 | M-N7 | M-py1 | M-py2 | M-$X_E$ | M-$X_A$ |
|---|---|---|---|---|---|---|---|---|
| **59**[c] | 2 Cl | 196 | 2.036 | ***2.356*** | 1.992 | 2.016 | 2.230 | [b] |
| | | | 2.029 | ***2.307*** | 2.018 | 2.033 | 2.221 | [b] |
| **59**[c] | $H_2O$, $H_2O$ | 66 | 2.194 | ***2.359*** | 2.195 | 2.156 | 2.062 | 2.121 |
| | $OSO_2CF_3$, $H_2O$ | | 2.194 | ***2.352*** | 2.180 | 2.175 | 2.070 | 2.129 |
| **59**[c] | 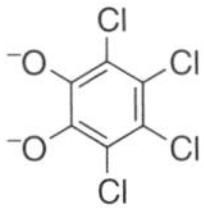 Cl, $^-$O, Cl, $^-$O, Cl, Cl | 192 | 2.027 | ***2.360*** | .985 | 2.004 | 1.898 | [b] |
| | | | 2.029 | ***2.374*** | 2.020 | 2.021 | 1.892 | [b] |
| **61** | 2 $NCCH_3$ | 82 | 2.029 | ***2.560*** | 2.021 | 1.976 | 1.972 | ***2.372*** |
| **61**[c] | 2 $NCCH_3$ | 82 | 2.052 | ***2.429*** | 2.026 | 1.992 | 2.033 | ***2.331*** |

[a]Elongated bonds in bold and italics. See Chart 20 and Table IV for more details.
[b]Not available.
[c]Hydrated at C9.
[d]Coligand $X_E$ coordinated to the metal center instead of a bispidine pendant donor group.

structures reported here (Table VI) (224). However, this does not necessarily help us to understand the fundamental reasons behind the observed structural features (215). Of interest is that conventional force-field calculations also allow us to predict the structures accurately (see Section III.D.2) (201), which suggests that the direction of the elongation is a ligand-based property, and that the electronic effects are merely an additional perturbation. This finding also explains why Jahn–Teller–innocent complexes [e.g., Zn(II)] have similar structures.

Another interesting set of three structures is that involving ligand **35**. Depending on the size and denticity (steric and electronic properties) of the coligand X, all three possible Jahn–Teller isomers of $[Cu(\mathbf{35})(X)]^{n+}$ are trapped and structurally as well as spectroscopically characterized (Fig. 11(*h*, *i*, and *k*), Table VI) (194, 199). Of specific interest here, in terms of applications, is that with X=$Cl^-$ or $OH_2$, the complex with ligand **35** has the coligand coordinated trans to N7 with an elongated Cu—N3 bond, while the analoguous complex with ligand **14** has the coligands trans to N3 with a long Cu—N7 bond. This is due to steric effects related to methylation of the pyridine donors and, due to the bispidine ligand geometry, leads to partially quenched Jahn–Teller distortions, which are of importance for the tuning of the redox potentials (see Section III.D.2) and reactivities (see Section III.D.3).

Probably the most appropriate example of distortional isomerism is that observed with copper(I) complexes of the tetradentate ligands **14** and **35**. Shown in Fig. 12(*a*) and (*b*) are plots of the X-ray crystal structures of two forms of $[Cu(\mathbf{14})(NCCH_3)]^+$ [the five-coordinate form (*b*) is red, the four-coordinate

TABLE VI
Experimental Structural Parameters of Isomers of the Copper(I) Complexes of Various Bispidine Ligands[a]

| | $[Cu(\mathbf{14})NCCH_3]^+$ | | $[Cu(\mathbf{35})NCCH_3]^+$ | $[Cu(\mathbf{24})]^+$ | $[Cu(\mathbf{22})NCCH_3]^+$ | |
|---|---|---|---|---|---|---|
| | **(a)** | **(b)** | **(c)** | **(d)** | **(e)** | |
| **Bond Length (Å)** | | | | | **(1)** | **(2)** |
| Cu–N3 | 2.20 | 2.24 | 2.20 | 2.28 | 2.21 | 2.24 |
| Cu–N7 | 2.16 | 2.16 | 2.18 | 2.24 | 2.21 | 2.22 |
| $Cu–N_{py1}$ | (3.12) | 2.38 | (2.90) | (2.96) | 2.30 | 2.39 |
| $Cu–N_{py2}$ | 2.07 | 2.25 | 2.10 | 1.98 | 2.29 | 2.22 |
| $Cu–N_{py3}$ | | | | 1.95 | | |
| $Cu–NCCH_3$ | 1.87 | 1.98 | 1.90 | | 1.91 | 1.92 |
| **Distances (Å)** | | | | | | |
| N3–N7 | 2.91 | 2.95 | 2.92 | 2.92 | 2.93 | 2.95 |
| $N_{py1}–N_{py2}$ | 4.95 | 4.41 | 4.75 | 4.63 | 4.40 | 4.37 |
| **Bond Angles (°)** | | | | | | |
| N3–Cu–N7 | 83.77 | 84.07 | 83.62 | 80.70 | 83.13 | 83.11 |
| $N3–Cu–N_{py1}$ | 65.74 | 72.73 | 66.94 | 63.38 | 73.81 | 73.11 |
| $N3–Cu–N_{py2}$ | 79.58 | 75.11 | 79.35 | 78.89 | 75.37 | 74.22 |
| $N3–Cu–N_{py3}$ | | | | 136.63 | | |
| $N3–Cu–NCCH_3$ | 134.18 | 149.33 | 150.09 | | 152.73 | 153.13 |
| $N_{py1}–Cu–N_{py2}$ | 144.31 | 144.71 | 143.60 | 138.22 | 146.22 | 143.21 |
| **Torsions (°)** | | | | | | |
| $C1–C2–C_{A1}–N_{py1}$ | 64.6 | 83.5 | 80.28 | 83.68 | 87.52 | 82.95 |
| $C5–C4–CA2–N_{py2}$ | 82.7 | 85.5 | 89.67 | 88.70 | 81.84 | 92.10 |

[a](See Fig. 12, (*a*)–(*e*) in Fig. 12 and Table VII refer to the same structures) (226).

from (*a*) is yellow] (70). Also shown in Fig. 12 are the experimental structure of the four-coordinate form with the methylated ligand **35** and the structures with ligands **22** and **24**. Selected structural parameters are given in Table VI. An interesting detail is that, while with **14** the noncoordinated pyridine donor is rotated by ~20° with respect to the position in the bonded form [Fig. 12(*a*)], in Fig. 12(*c*) with the methylated pyridine group of ligand **35** it is not, suggesting that it is still partially bonded (2.9 Å vs. 2.1–2.2 Å) (70, 226). This suggests that the rotational barrier is low and the orientation of the dangling pyridine group primarily depends on crystal lattice effects. Altogether, this is a further indication for a flat energy surface with various shallow minima. Note again that this is the result of a very rigid ligand, where the only flexibility is the rotation of the pyridine rings (also see Fig. 8).

It is not unexpected that this fluxionality is also observed in NMR spectroscopy. Interestingly, the behavior of the complexes with the two ligands **14** and **35** is quite different. While in the complex with the unsubstituted ligand **14** the

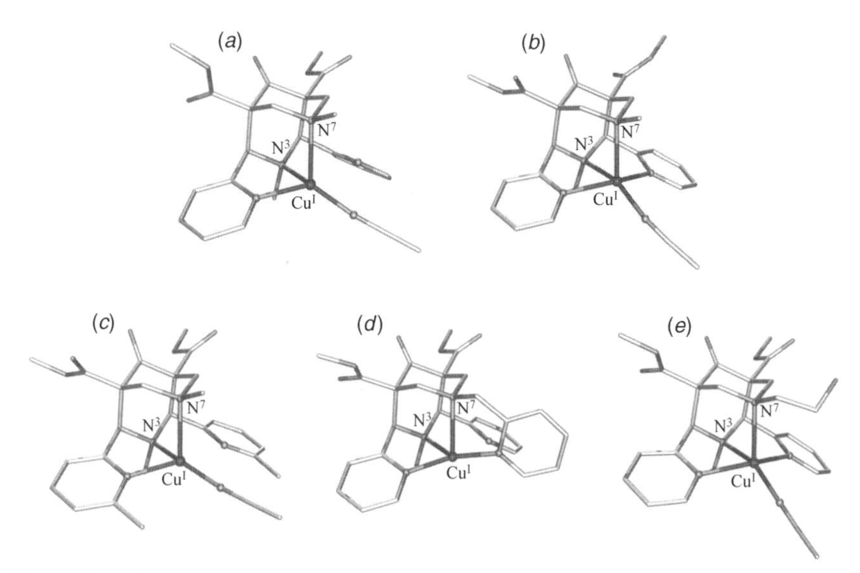


Figure 12. Plots of the complex cations of various isomers of bispidine–copper(I) complexes, determined by X-ray crystallography (hydrogen atoms omitted; the structural data and references are given in Table VII; (*a*) $[Cu(\mathbf{14})(NCCH_3)]^+$, (*b*) $[Cu(\mathbf{14})(NCCH_3)]^+$, (*c*) $[Cu(\mathbf{35}(NCCH_3)]^+$, (*d*) $[Cu(\mathbf{24})]^+$, (*e*) $[Cu(\mathbf{22})(NCCH_3)]^+$ (226).

spectra suggest a symmetrical five-coordinate structure (mirror plane), even at low temperature and with line broadening at room temperature [population of the two enantiomeric four-coordinate species, as shown in Fig. 13(*a*)], the ambient temperature spectrum of the complex with the methylated ligand **35** [Fig. 13(*b*)] indicates, that the four-coordinate species is more stable in this system. There is a second form of this complex, however, which is assigned to a highly symmetrical four-coordinate complex with both pyridine donors coordinated [see Fig. 13(*c*)] (226).

This behavior is nicely simulated by DFT calculations (226). The computed structures are in acceptable agreement with experimental data (see Table VII), and the energies indicate that for **14** the five-coordinate structure is slightly more stable while the four-coordinate structure is more stable for the methylated ligand **35** (here, the five-coordinate structure could only be located as a transition state). Interestingly, for this ligand there also is an optimized four-coordinate minimum structure with both pyridine groups coordinated, but no $CH_3CN$, and this is only $5\,kJ\,mol^{-1}$ less stable than the corresponding unsymmetrical four-coordinate structure with a coordinated $CH_3CN$. For **14**, this structural type is $\sim 15\,kJ\,mol^{-1}$ less stable than the optimum structure. This is exactly what is observed by NMR spectroscopy.

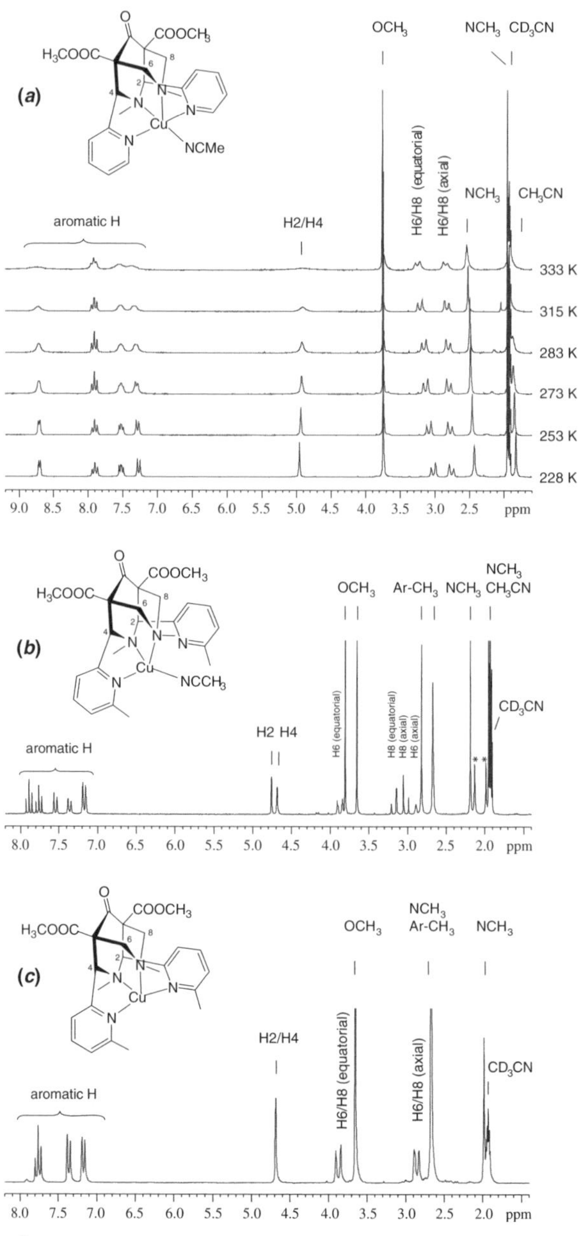


Figure 13. The $^1H$ NMR spectra (200 MHz, $CD_3CN$) of the isomeric copper(I) complexes of the tetradentate bispidine ligands **14** and **35**; (*a*) temperature dependent spectra of $[Cu(\mathbf{14})(NCCH_3)]^+$; (*b*) ambient temperature spectrum of the unsymmetrical four-coordinate form of $[Cu(\mathbf{35})(NCCH_3)]^+$; (*c*) ambient temperature spectrum of $[Cu(\mathbf{35})]^+$ (226).

TABLE VII
Selected Calculated and Experimental Geometric Parameters for the $[Cu(L)CH_3CN]^+$ Complexes, for L = **14**, **35**.[a]

| Ligand | 14 | | | | 35 | | | |
|---|---|---|---|---|---|---|---|---|
| Parameters | 5-Coord. | ***5-coord.*** | 4-Coord. | ***4-Coord.*** | 5-Coord.[b] | 4-Coord. | ***4-Coord.*** | 4-Coord.[c] |
| Cu-N3 | 2.418 | ***2.292*** | 2.228 | ***2.203*** | 2.166 | 2.223 | ***2.203*** | 2.217 |
| Cu-N7 | 2.288 | ***2.188*** | 2.234 | ***2.160*** | 2.203 | 2.224 | ***2.184*** | 2.200 |
| Cu-Npy1 | 2.249 | ***2.169*** | 2.225 | ***2.066*** | 2.632 | 2.186 | ***2.096*** | 2.019 |
| Cu-Npy2 | 2.249 | ***2.247*** | 2.745 | ***3.118*** | 2.632 | 3.188 | ***2.897*** | 2.019 |
| Cu-$NCCH_3$ | 1.993 | ***1.936*** | 1.935 | ***1.873*** | 1.918 | 1.920 | ***1.900*** | [d] |
| Angles(°) | | | | | | | | |
| N7-Cu-N3 | 82.46 | ***83.15*** | 84.58 | ***83.77*** | 86.56 | 84.91 | ***83.63*** | 86.59 |
| N3-Cu-Nac | 160.71 | ***154.95*** | 146.85 | ***134.16*** | 147.05 | 139.64 | ***150.09*** | [d] |
| Torsion angles (°) | | | | | | | | |
| N3-C-C-Npy1 | 37.26 | ***38.39*** | 39.69 | ***41.53*** | 46.58 | 41.18 | ***35.69*** | 37.29 |
| N3-C-C-Npy2 | −37.26 | ***−39.94*** | −44.54 | ***−61.15*** | 46.58 | −55.60 | ***−45.44*** | 37.29 |
| Relative energies ($kJ\ mol^{-1}$) | | | | | | | | |
| Solvated by $CH_3CN$ | 0.00 | [d] | 1.98 | [d] | 3.08 | 0.00 | [d] | 5.05 |

[a]Experimental values are given in bold and italics.
[b]The structure is a transition state and not a minimum.
[c]The four-coordinate complex without $NCCH_3$ (226).
[d]Not available.

TABLE VIII
Reduction Potentials of Bispidine–Transition-Metal Couples

| Couple | Ligand | Coligand | Solvent | $E_{1/2}$(mV) | (vs.) | Reference |
|---|---|---|---|---|---|---|
| Co(III/II) | **14** | $CH_3CN$ | $CH_3CN$ | 653 | ($Ag/AgNO_3$) | 167 |
| Co(III/II) | **14** | $NO_3^-$ | $CH_3CN$ | 372 | ($fc/fc^+$) $Ag/AgNO_3$ | 69 |
| Co(III/II) | **14**[a] | $CH_3CN$ | $CH_3CN$ | 250 | ($Ag/AgNO_3$) | 167 |
| Co(III/II) | **32** | | $CH_3CN$ | 224 | ($fc/fc^+$) $Ag/AgNO_3$ | 69 |
| Co(III/II) | **33** | $CH_3CN$ | $CH_3CN$ | 496 | ($fc/fc^+$) $Ag/AgNO_3$ | 69 |
| Cu(II/I) | **14** | $CH_3CN$ | $CH_3CN$ | −417 | ($Ag/AgNO_3$) | 168 |
| Cu(II/I) | **18** | $CH_3CN$ | $CH_3CN$ | 53 | ($Ag/AgNO_3$) | 65 |
| Cu(III/I) | **19** | $CH_3CN$ | $CH_3CN$ | −272 | ($Ag/AgNO_3$) | 65 |
| Cu(II/I) | **20** | $CH_3CN$ | $CH_3CN$ | −387 | ($Ag/AgNO_3$) | 65 |
| Cu(II/I) | **21** | $CH_3CN$ | $CH_3CN$ | −410 | ($Ag/AgNO_3$) | 65 |
| Cu(II/I) | **24** | $CH_3CN$ | $CH_3CN$ | −603 | ($Ag/AgNO_3$) | 168 |
| Cu(II/I) | **26** | $CH_3CN$ | $CH_3CN$ | −552 | ($Ag/AgNO_3$) | 65 |
| Cu(II/I) | **27** | $CH_3CN$ | $CH_3CN$ | −389 | ($Ag/AgNO_3$) | 65 |
| Cu(II/I) | **28** | $CH_3CN$ | $CH_3CN$ | −489 | ($Ag/AgNO_3$) | 168 |
| Cu(II/I) | **28**[b] | $CH_3CN$ | $CH_3CN$ | −460 | ($Ag/AgNO_3$) | 68 |
| Cu(II/I) | **29**[b] | $CH_3CN$ | $CH_3CN$ | −400 | ($Ag/AgNO_3$) | 68 |
| Cu(II/I) | **30** | $CH_3CN$ | $CH_3CN$ | −369 | ($Ag/AgNO_3$) | 68 |
| Cu(II/I) | **30**[b] | $CH_3CN$ | $CH_3CN$ | −343 | ($Ag/AgNO_3$) | 68 |
| Cu(II/I) | **31** | $CH_3CN$ | $CH_3CN$ | −462 | ($Ag/AgNO_3$) | 68 |
| Cu(II/I) | **32** | | $CH_3CN$ | −573 | ($Ag/AgNO_3$) | 69 |
| Cu(II/I) | **33** | $CH_3CN$ | $CH_3CN$ | −478 | ($fc/fc^+$) $Ag/AgNO_3$ | 69 |
| Cu(II/I) | **35** | $CH_3CN$ | $CH_3CN$ | −98 | ($Ag/AgNO_3$) | 168 |
| Cu(II/I) | **36** | $CH_3CN$ | $CH_3CN$ | −450 | ($Ag/AgNO_3$) | 71 |
| Cu(II/I) | **36** | $Cl^-$ | $CH_3CN$ | −691 | ($Ag/AgNO_3$) | 71 |
| Cu(II/I) | **37** | $CH_3CN$ | $CH_3CN$ | −94 | ($Ag/AgNO_3$) | 71 |
| Cu(II/I) | **38** | $CH_3CN$ | $CH_3CN$ | −74 | ($Ag/AgNO_3$) | 71 |
| Cu(II/I) | **38**[c] | $CH_3CN$ | $CH_3CN$ | 17 | ($Ag/AgNO_3$) | 65 |
| Cu(II/I) | **39** | $CH_3CN$ | $CH_3CN$ | −383 | ($Ag/AgNO_3$) | 71 |
| Cu(II/I) | **40** | $CH_3CN$ | $CH_3CN$ | −69 | ($Ag/AgNO_3$) | 71 |
| Cu(II/I) | **41** | $CH_3CN$ | $CH_3CN$ | −78 | ($Ag/AgNO_3$) | 71 |
| Cu(II/I) | **42** | $CH_3CN$ | $CH_3CN$ | −35 | ($Ag/AgNO_3$) | 65 |
| Cu(II/I) | **43** | $CH_3CN$ | $CH_3CN$ | −440 | ($Ag/AgNO_3$) | 168 |
| Cu(II/I) | **58** | $CH_3CN$ | $CH_3CN$ | −345 | ($Ag/AgNO_3$) | 213 |
| Cu(II/I) | **59** | $CH_3CN$ | $CH_3CN$ | −334 | ($Ag/AgNO_3$) | 213 |
| Cu(II/I) | **61** | $CH_3CN$ | $CH_3CN$ | −385 | ($Ag/AgNO_3$) | 213 |
| Cu(II/I) | **62** | $CH_3CN$ | $CH_3CN$ | 436 | ($Ag/AgNO_3$) | 213 |
| Fe(III/II) | **23** | $Cl^-$ | $CH_3CN$ | −264 | ($Ag/AgNO_3$) | 66 |
| Fe(III/II) | **24** | $SO_4^{2-}$ | $CH_3OH$ | 53 | ($Ag/AgNO_3$) | 66 |
| Fe(III/II) | **24** | $Cl^-$ | $CH_3OH$ | 156 | ($Ag/AgNO_3$) | 66 |
| Fe(III/II) | **28** | $SO_4^{2-}$ | $CH_3OH$ | 54 | ($Ag/AgNO_3$) | 66 |
| Fe(III/II) | **28** | $Cl^-$ | $CH_3OH$ | 143 | ($Ag/AgNO_3$) | 66 |
| Fe(III/II) | **32** | | $CH_3CN$ | 661 | ($Ag/AgNO_3$) | 66 |

[a]Demethylated at N7.
[b]9-one ligands, all the others are hydrolyzed at C9.
[c]4-Cl substituted quinoline groups.

### 2. *Thermodynamics*

Due to the relatively short time in which this field has started to be developed (166), and the exciting observations in the areas of structures, bonding and reactivity, complexation and ligand exchange rates have not yet been studied, with the evaluation of electron-transfer kinetics we are just at the beginning (see Section III.D.3) (169), and data on complex stabilities and redox potentials are incomplete (69, 201). However, the interpretation of some trends is possible, specifically in relation to the structural data and some theoretical studies (see Sections III.C and III.D.3). The available redox potentials are given in Table VIII and complex stabilities are assembled in Table IX.

The most extensive series of complex stabilities is that of the copper complexes. As generally observed, the stabilities of the copper(I) complexes are rather insensitive to the ligand (227). Therefore, a linear correlation between the copper(II) stability constants and the potential of the Cu(II/I) couple with the expected slope of $-59\,\text{mV}$ per log unit is observed. The comparably low copper(II) complex stability with the methylated tetradentate ligand **35** [7 log units difference compared to that with ligand **14**] is well understood and primarily due to partial quenching of the Jahn–Teller effect (see Section III.D.1)) (199). More difficult to appreciate are the relative stabilities of the copper(II) complexes with the two pentadentate and the hexadentate ligands: the stability of the complex with the pentadentate ligand **24** is ~2 log units larger than that with the

TABLE IX
Complex Stabilities of Bispidine Complexes ($T = 25.0°C$, $H_2O$, $\mu = 0.1\ M$ (KCl)[a]

| Protonation Constants of the Ligands | | | | | | |
|---|---|---|---|---|---|---|
| | **14**[b] | **35**[b] | **24**[b] | **28**[b] | **32**[b] | **87b**[b] |
| $pK_{a1}(LH^+)$ | 1.78(12) | | 1.86(10) | 2.50(4) | | 1.75(12) |
| $pK_{a2}(LH_2^{2+})$ | 2.36(13) | 2.98(9) | 3.95(10) | 5.21(4) | 4.72(13) | 5.65(9) |
| $pK_{a3}(LH_3^{3+})$ | 9.10(10) | 9.13(7) | 7.44(8) | 8.89(3) | 6.68(8) | 8.93(5) |
| **Complex Stabilities (ML)** | | | | | | |
| Co(II) | 5.46(5) | [c] | 6.23(5) | 13.69(5) | 7.30(6) | 10.60(4) |
| Ni(II) | [c] | 7.50(9) | 6.10(8) | 9.54(6) | 5.02(7) | 7.20(10) |
| Cu(II) | 16.56(5) | 9.60(7) | 18.31(12) | 15.66(3) | 16.28(10) | 17.70(7) |
| Cu(I) | 5.61(32)[d] | 5.48(32)[a] | 5.69(22)[a] | 6.29(43)[d] | 4.97(52)[d] | [c] |
| Zn(II) | 11.37(1) | [c] | 8.28(5) | 13.57(4) | 9.18(5) | 12.52(5) |
| Li(I) | 2.70(9) | 3.90(10) | [c] | 3.65(9) | 3.70(8) | ~ 3 |

[a]Refs. 69, 201.
[b]Values in parentheses are standard deviations.
[c]Not available.
[d]Acetonitrile.

tetradentate ligand **14**, however, the stability with the hexadentate ligand **32** is similar to that with the tetradentate ligand **14** and that with the other pentadentate ligand **28** (isomer of **24**) is even ~1 log unit smaller.

Qualitatively, the relative stabilities of the five copper(II) complexes (which are correlated to their redox potentials; as mentioned above) may be explained with a high complementary of the tetradentate ligand **14** and also of the pentadentate ligand **24** for Cu(II) (see also, Section III.D.1). The isomeric pentadentate ligand **28** is believed to suffer from strain induced by its coordination to copper(II), and this is supported by the two close-to-degenerate minima of the two Jahn–Teller isomers (201). An interesting question is, why the order of complex stabilites with other metal ions without Jahn–Teller active ground states also does not follow the naively expected pattern and why, in addition, it is different from the order of stabilites observed for Cu(II) (see Table IX). Here, the analysis of the experimental structures combined with a molecular mechanics based analysis of the corresponding copper(II) and zinc(II) structures yields a semiquantitative interpretation (201). Interestingly, the zinc(II) complexes have similar, albeit smaller distortions than the copper(II) compounds. It follows that, for copper(II), the Jahn–Teller isomers might be structurally enforced by the ligands. The electronic (Jahn–Teller) effects then are merely an additional perturbation (201). The combination and partial cancellation of structural and electronic effects (including the misdirected valences, see Section III.D.1) leads to stability constants that do not follow the usual Irving–Williams behavior (69, 201). This might be the basis for the possibility of designing, preparing, and studing novel metal ion selective ligands, specifically for cobalt(II).

### 3. *Reactivity*

**a. Copper Catalyzed Aziridination.** Aziridines, the nitrogen analogues of epoxides, are found in natural products, where some are known to have cytotoxic properties, and in organic synthesis aziridines are attractive intermediates (228–231). An extensively studied and, depending on the catalyst, efficient and highly stereoselective preparative method is the copper-catalyzed synthesis of aziridines (see Scheme 15) (232–238). The most frequently used nitrene source in these reactions is [*N*-(*p*-toluenesulfonyl) imino]phenyliodinane (PhINTs), but others have also been used (239–241). Interestingly, copper(I) as well as copper(II) complexes have been found to be catalytically active, and copper(II)-catalyzed reactions are of particular interest since the catalysts are air stable. Among the classical copper(II) coordination compounds that have been reported to efficiently catalyze olefin aziridination (168, 242–244), the bispidine complexes are of particular interest because the wide variability of the ligands leads to a set of catalysts, where the redox potential and the shape of the cavity for the substrate coordination may be efficiently tuned, and which also may include chiral catalysts for enantioselective reactions. Of specific interest in terms of the electronics is the pair of tetradentate ligands **14** and **35** with 6-H- or 6-Me-pyridine donor groups because α-methylation is known to lead to destabilization of the copper(II) complex (see Section III.D.1), and therefore leads to an ~300 mV higher Cu(II/I) redox potential (see Table VIII) (168, 199). This finding is of interest because a correlation of the activity with the redox potential and supporting computational work might lead to important mechanistic information. There still is some dispute whether the copper complexes are Lewis acid catalysts or whether electron transfer is involved and, in the latter proposal, whether copper(I)–nitrene or copper(III)–imido complexes are the catalytically active species (236, 245–248).

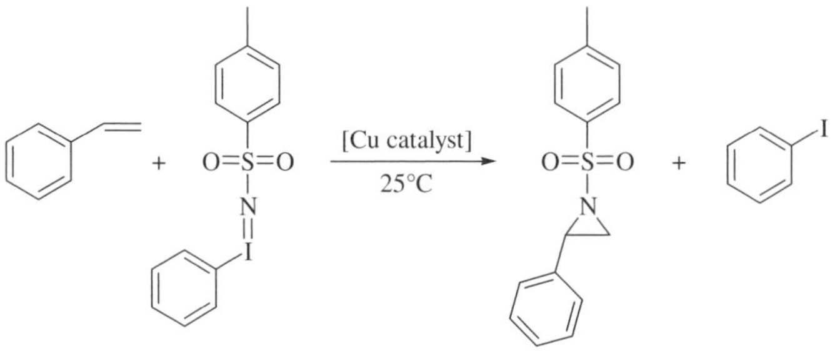

Scheme 15

So far there is only one report on bispidine–copper catalyzed aziridination reactions (168). This reaction includes the ligands **14**, **24**, **28**, **35**, **43** (see Table X).

TABLE X

Results of the Catalytic Aziridination of Styrene in $CH_3CN$[a]

| Catalyst | [PhINTs]/mol | [catalyst]/mol% | Yield/% | ton |
|---|---|---|---|---|
| $[Cu^{II}($**14**$)](BF_4)_2$ | 0.4 | 5 | 41 | $9 \pm 1$ |
| $[Cu^{II}($**35**$)](BF_4)_2$ | 0.4 | 5 | 94 | $19 \pm 1$ |
| $[Cu^{II}($**35**$)](BF_4)_2$ | 0.6 | 3.5 | 67 | $19 \pm 1$ |
| $[Cu^{I}($**35**$)](BF_4)$ | 0.9 | 1.7 | 80 | $47 \pm 2$ |
| $[Cu^{II}($**43**$)](BF_4)_2$ | 0.4 | 5 | 29 | $6 \pm 0.5$ |
| $[Cu^{II}($**28**$)](BF_4)_2$ | 0.4 | 5 | 0 | 0 |
| $[Cu^{I}($**28**$)](BF_4)$ | 0.4 | 1.7 | 7 | $7.5 \pm 1$ |
| $[Cu^{II}($**24**$)](BF_4)_2$ | 0.4 | 5 | 0 | 0 |
| $[Cu^{I}($**24**$)](BF_4)$ | 0.4 | 1.7 | 6 | $5.5 \pm 1$ |
| $[Cu^{II}($**38**$)](BF_4)_2$[b] | 0.4 | 5 | | 17 |
| $[Cu^{II}($**36**$)](BF_4)_2$[b] | 0.4 | 5 | | 3 |

[a]See Scheme 15. Ref. 168.

[b]Ref. (65).

From the observations with these systems it follows that (1.) the copper(I) complexes generally are more efficient than the copper(II) precatalysts; (2.) the efficiency is correlated to the Cu(II/I) redox potential; (3.) pentadentate ligands seem in general to lead to less efficient catalysts than the tetradentate bispidine ligands. A reason for the latter observation might be that coordination of the substrate (nitrene source, PhINTs) is partially blocked with the pentadentate ligands, which might lead to inhibition of catalysis. Therefore, steric effects might also, at least partially, be responsible for the efficiency of the catalyst based on ligand **35**.

Also included in Table X are preliminary unpublished data of the copper(II) complexes of **36** and **38**, which are part of a series of new ligands with substituted pyridine and quinoline donors (**18–21**, **26**, **27**, **36–42**), which have modified donor sets and interesting steric and redox properties (see Table VIII).(65) These seem to confirm that the catalytic efficiency is correlated to ligands, which destabilize the copper(II) oxidation state.

The most intriguing feature of the bispidine copper(II) coordination chemistry is that three strikingly different structural forms (the three Jahn–Teller isomers, see above) are available (71, 82, 194, 199). Although the landscape around these three local minima has been studied to some extent, and some possibilities to stabilize–destabilize specific structures have been identified (199), unambiguous rules have not yet been defined. The striking structural difference between $[Cu(\mathbf{14})Cl]^+$ and $[Cu(\mathbf{35})Cl]^+$, and the difference in redox potential ($\Delta E^\circ = 300$ mV) and aziridination efficiency (see above) (168) indicate that a thorough understanding of these systems in a wider context might lead to complexes and catalysts with the possibility to switch between various structures and properties (redox potential, catalytic efficiency) with a variation of the temperature or other external parameters.

**b. Biomimetic Copper Chemistry.** The major role of copper in biological systems is related to redox chemistry and includes electron-transfer mediators (blue-copper proteins), dioxygen transport (hemocyanin), oxidation catalysts (e.g., catechol oxidase), and dioxygen activation enzymes (e.g., tyrosinase) (249–262). The copper proteins have been classified in type-1, -2, and -3 centers; more recently this has been extended due to novel structures with distinct electronic properties found in Nature and low molecular weight model systems (263).

The copper coordination chemistry of bispidine ligands has been studied extensively, and this has been particularly rewarding (70, 71, 81, 82, 168, 169, 192, 194, 196, 199, 201, 213, 214). The main reasons are that (1.) the bispidine backbone is complementary with respect to copper(II) and, therefore, complex stabilities may be relatively high, comparable to those with macrocyclic ligands (69, 201), and that modifications of the ligand backbone can be used to tune the stabilities and redox potentials (199); (2.) due to the ligand rigidity, the copper(II) and copper(I) geometries are very similar; the corresponding similarity in blue copper proteins has in a simplistic view been taken as a main reason for fast electron self-exchange and a basis for the entatic state paradigm (41, 169); (3.) bispidine ligands enforce square-pyramidal or cis-octahedral coordination geometries. In one of the three possible Jahn–Teller isomers a tertiary amine is the axial and the substrate (coligand) is a strongly bound in-plane ligand, leading to exceptionally stable copper(II)–substrate complexes (169, 214). Bispidine ligand modifications can be used to tune these interactions toward weaker bonding (71, 214, 215).

While the very rigid and, with respect to the type of donor groups and their geometric disposition (two trans-disposed pyridine donors and two cis-oriented tertiary amines), enforced and inflexible geometry precludes an accurate structural and spectroscopic modeling of copper proteins, it was especially feature (3.) that lead to the isolation and characterization of novel model complexes with hemocyanine- and catechol oxidase activities properties (81, 192, 196, 213). In the latter case, it was possible to isolate and structurally characterize complexes with coordinated catechol model substrates with structural features, which have been proposed to be of relevance in the enzyme catalysis cycle, but have not been observed before in low molecular weight complexes (192, 213).

1. Electron Transfer. An important feature of transition metal ion bispidine complexes is the elastic coordination geometry (see Sections III.C and III.D.1) (42, 189). For copper(II), this leads to close-lying minima for various isomeric chromophores along Jahn–Teller-active vibrational coordinates (82, 194) and for copper(I) this manifests itself in dynamic processes, which involve isomers with changes in coordination number (70, 199, 226). In both these and other cases, this is a result of a flat potential energy surface with various shallow minima and (in many cases) relatively low-energy barriers for isomer interconversion; that is, in both oxidation states a manifold of geometries are easily accessible. For the copper(II/I) electron self-exchange it follows that the inner-sphere reorganization energy $\Delta G^*_{\text{in}}$ is small, but this does not mean that electron transfer is as fast as, for example, in blue copper proteins. Electron-transfer kinetics have been studied for the $[\text{Cu}(\mathbf{14})(\text{solvent})]^{2+/1+}$ couple (169).

   The computed value of $\Delta G^*_{\text{in}}$, based on three different molecular mechanics based methods and on preliminary DFT calculations is only $10–20\,\text{kJ mol}^{-1}$ (see Fig. 14), and the calculated outer-sphere reorganization energy (reorganization of the solvent sheath) is of about the same order of magnitude (169). Based on similar studies with hexamine cobalt(III/II) couples and on experimental data with the bispidine–copper(II/I) couple the entropy term $[T\Delta S^*\ (T = 298\ \text{K})]$ is $\sim 20\,\text{kJ mol}^{-1}$. The resulting electron self-exchange rate $k^{\text{calc}}(298\ \text{K}) = 1.5\,M^{-1}\text{s}^{-1}$ is in acceptable agreement

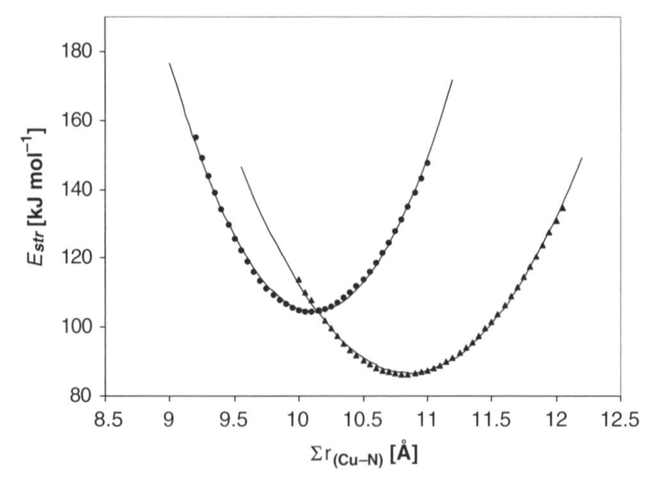


Figure 14. Computed potential energy curves for $[Cu(\mathbf{14})(NH_3)]^{2+}$ (●) and $[Cu(\mathbf{14})(NH_3)]^{+}$ (▲) (169).

with the experimentally observed value of $k^{obs}(298\ K) = 15 \pm 11\ M^{-1}s^{-1}$ (169). With other bispidine ligands similar experimental rates are observed (264). A number of possible reasons have been discussed for the observation that small inner-sphere reorganization energies do not necessarily lead to fast electron self-exchange rates (169, 265). This is in contrast to the often used, but naïve interpretation of the fast electron self-exchange in blue copper proteins, and is another interesting argument in the long dispute on entatic states (41).

2. Oxygenation of Copper(I). Reactions between copper proteins with mono- and coupled oligonuclear active sites and dioxygen involve mono-, di-, and trivalent copper and are of relevance, for example, for dioxygen transport (hemocyanin) and superoxide dispropotionation (superoxide dismutase, SOD), and copper enzymes in general function as oxidases (e.g., catechol oxidase), monooxygenases (e.g., tyrosinase), and dioxygenases (e.g., quercetin 2,3-dioxygenase) (261, 262). Various structural forms of copper–dioxygen adducts have been characterized by X-ray crystallography, studied spectroscopically, with computational methods and in terms of their formation and decay kinetics and mechanisms (266–272). Among the most relevant forms in biological systems, model studies and catalytically active complexes are 1:1 and 2:1 complexes with end-on or side-on superoxo-, peroxo-, and dioxo adducts with copper(II) or copper(III) centers (Scheme 16) (262).

The type-3 copper proteins hemocyanin, tyrosinase, and catechol oxidase with dicopper active sites and three histidine imidazole donors per copper

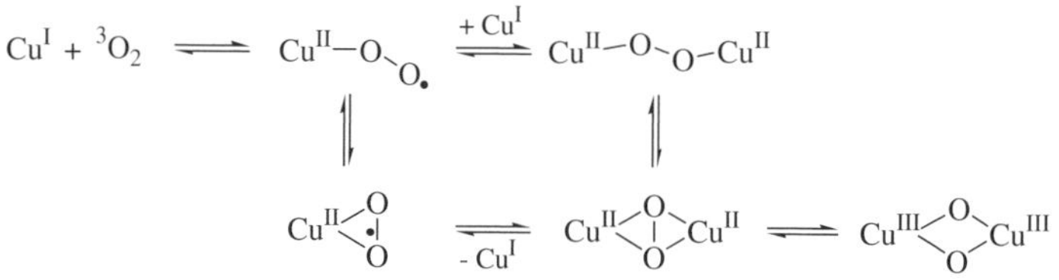

$$Cu^{I} + {}^{3}O_2 \rightleftharpoons Cu^{II}{-}O{-}O^{\bullet} \underset{}{\overset{+\,Cu^{I}}{\rightleftharpoons}} Cu^{II}{-}O{-}O{-}Cu^{II}$$

$$Cu^{II}(O{\cdot}O) \underset{-\,Cu^{I}}{\rightleftharpoons} Cu^{II}(\mu{-}O{-}O)Cu^{II} \rightleftharpoons Cu^{III}(\mu{-}O)_2Cu^{III}$$

Scheme 16

center have very similar structures, but a strikingly different function (273–278). Bispidine ligands have been used successfully to stabilize dicopper peroxo complexes and copper, as well as dicopper catecholate compounds as structural models for the catecholase enzyme. The stabilization of these intermediates is the result of the unusual coordination geometry and bonding with an in-plane coordinated substrate, enforced by the rigid bispidine ligand to the copper(II) center, as discussed above. As expected, the copper–oxygen adducts with the methylated ligand (**35**) are much less stable (quenched Jahn–Teller effect, see above) and indeed could not be observed so far (196). Also, due to repulsion of the in-plane pyridine donors of the two copper sites, copper–copper distances of less than ~4.0 Å are sterically unfavorable. Therefore, the biologically relevant and catalytically active side-on peroxo- and dioxo forms with the typical diamond core (average Cu–Cu distances of 3.5 and 2.8 Å, respectively) are not expected to be observed with the 2,4-substituted bispidine ligands.

Oxygen binding to bispidine–dicopper complexes is not reversible, and the peroxo complexes decay to stable copper(II) products. With half-lifes at room temperature of up to ~1 h, the end-on μ-peroxo bispidine–dicopper(II) complexes are among the most stable species of that kind (81, 196), but so far there are no experimental structural data available. However, the computed structure, based on a force field tuned and validated with well-characterized systems (279), is a valuable model (81) and has also been supported by DFT calculations (201). The computed structure is shown in Fig. 15 and spectroscopic data are given in Table XI. The ligand- to-metal change transfer (LMCT) transition and the vibrational spectra are very similar within the group of bispidine-derived dicopper peroxo complexes and similar to corresponding complexes with other ligand systems. This has not necessarily been expected, since in contrast to the square-pyramidal bispidine complexes with the in-plane-coordinated peroxo group, other typical peroxo complexes are trigonal bipyramidal with axially coordinated peroxo bridges (280). Interestingly, a qualitative analysis leads to the conclusion that, based on the thorough spectroscopic/MO analysis of the only structurally characterized μ-peroxo–dicopper(II) complex (281) and a similar assignment of the LMCT transitions, the very similar spectroscopic

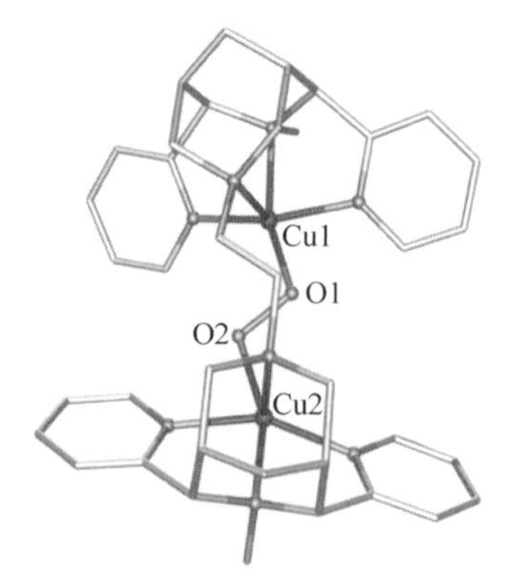


Figure 15. Computed structure (molecular mechanics) of $[Cu_2(\mathbf{58})(O_2)]^+$ (81).

data are expected (262). While this seems to need a confirmation, based on a more elaborate analysis, it is interesting that an independent DFT-based analysis of the *d* orbitals also comes to the conclusion that the bispidine-enforced geometry leads to a *d*-orbital arrangement that is related to that which emerges from a trigonal-bipyramidal ligand field (see above) (215).

Time-resolved spectroscopy (stopped-flow ultraviolet–visible (UV-vis) spectroscopy at -90°C, proprionitrile or acetonitrile, $[O_2] \gg$ [complex]) has been used to characterize intermediates and evaluate the mechanism of the peroxo complex formation (see Fig. 16) (196). Based on the similarity of the spectral features with known superoxo copper(II) and peroxo–dicopper(II) complexes (262, 268, 281) the mechanism shown in Scheme 17 was proposed, and the spectra of the superoxo copper(II) and peroxo–dicopper(II) complexes were determined (see Table XI). For steric reasons and in

TABLE XI
Charge Transfer (nm) and Raman ($\nu_{o-o}$, $cm^{-1}$) Bands of Peroxo Dicopper(II)–Bispidine Complexes.[a]

| Complex | CT | CTsh | $\nu_{o-o}$ |
|---|---|---|---|
| $[Cu_2(\mathbf{14})_2O_2]^{2+}$ | 504 | 630 | 840 |
| $[Cu_2(\mathbf{58})O_2]^{2+}$ | 488 | 649 | 824 |
| $[Cu_2(\mathbf{59})O_2]^{2+}$ | 499 | 650 | 837 |
| $[Cu_2(\mathbf{60})O_2]^{2+}$ | 503 | 650 | 847 |
| $[Cu_2(\mathbf{61})O_2]^{2+}$ | 516 | 640 | [b] |
| $[Cu_2(\mathbf{62})O_2]^{2+}$ | 526 | [b] | [b] |
| $[Cu_2(L)_2O_2]^{2+c}$ | 525 | 590 | 832 |

[a]Ref. 196.
[b]Not available.
[c]Dinucleating tpa-type ligand (282); tpa = tris(2-pyridylmethyl)amine.

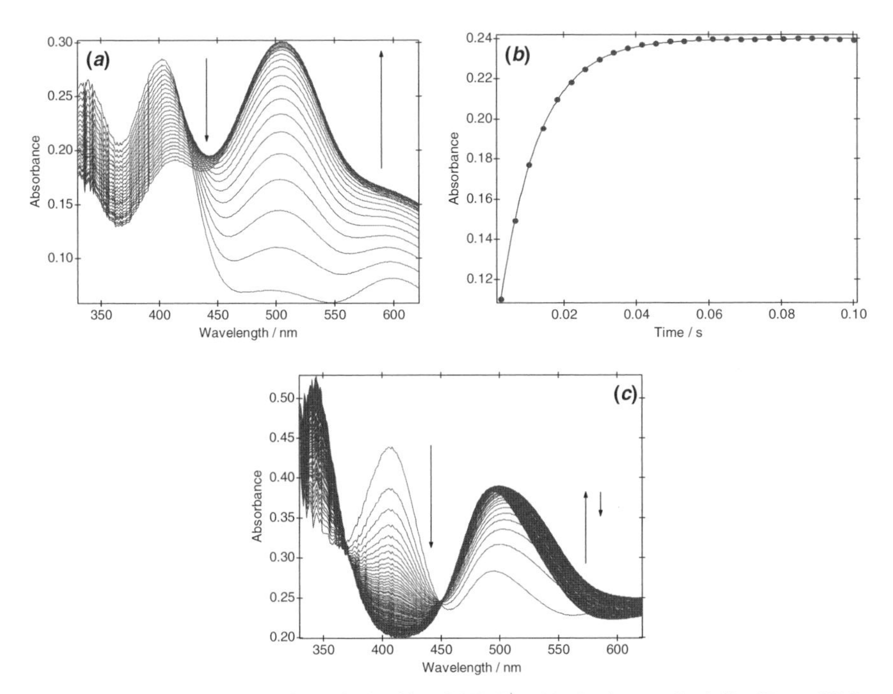


Figure 16. (*a*) and (*b*) Reaction of $[Cu(\mathbf{14})(NCCH_3)]^+$ with $O_2$ in proprionitrile ($T = -90°C$, [complex] = 0.2 m*M*, $[O_2] = 4.4$ m*M*; (*a*) $\Delta t = 0.27$s, total *t* is 7s; (*b*) absorbance at 404 nm), and (*c*) reaction of $[Cu_2(\mathbf{58})(NCCH_3)_2]^+$ with $O_2$ in acetone ($T = -60°C$, [complex] = 0.2 m*M*, $[O_2] = 5.1$ m*M*; $\Delta t = 1.2$s, total *t* is 100 s (196).

comparison with published data, the mononuclear superoxo complexes are believed to have monodentate end-on coordinated superoxide (283, 284). The mononuclear $[Cu(\mathbf{14})(NCCH_3)]^+$ complex reacts with $O_2$ yielding a dinuclear peroxo complex of similar stability to those with tris(2-pyridylmethyl)amine (tpa), but the methylated ligand (**35**) which, in the copper(II) complex enforces axial coordination of the peroxide, is so unstable that no oxygen-based intermediate was detected during the oxidation reaction. Preorganization of the dinuclear complex with the bridged bis (bispidines) **58**–**62** leads to a dramatically enhanced stability (electronic stabilization), and the ethyl-bridged ligands **58** and **61** lead to the most stable peroxo complexes, as predicted by molecular mechanics (minimization of enthalpic destabilization) (279).

$$[LCu^I\text{-}NCCH_3]^+ + O_2 \rightleftharpoons [LCu^{II}\text{-}O_2]^+ + CH_3CN$$

$$[LCu^{II}\text{-}O_2]^+ + [LCu^I\text{-}NCCH_3]^+ \rightleftharpoons [LCu^{II}\text{-}O_2\text{-}Cu^{II}L]^{2+} + CH_3CN$$

Scheme 17

3. Catechol Oxidase Activity. Catechol oxidase has, as hemocyanin and tyrosinase, a type-3 dicopper core with three histidines coordinated to each of the copper centers. Hemocyanin is an oxygen carrier without phenolase (ortho-hydroxylation) and diphenolase (catechol oxidase) activity, although enzyme modification can induce significant catalytic activity (285). Tyrosinase has both phenolase and diphenolase activity, and monophenolase activity has also been demonstrated for catecholase (286, 287). Based on biochemical, low molecular weight experimental modeling, spectroscopic, structural, and computational work, a number of proposals for the catechol oxidase catalytic cycle were suggested (see Scheme 18). Main variations in the different proposals are (1.) the stoichiometry of the reaction, that is two [catalytic cycle (*a*)] or one [catalytic cycle (*b*)] catechol to quinone transformation steps per cycle; (2.) the nature of the catechol to dicopper center binding, that is a monodentate (oxy-m) or a bridged (oxy-b) intermediate; and (3.) the nature of the intermediates and transition states, in particular for electron and proton transfer (288).

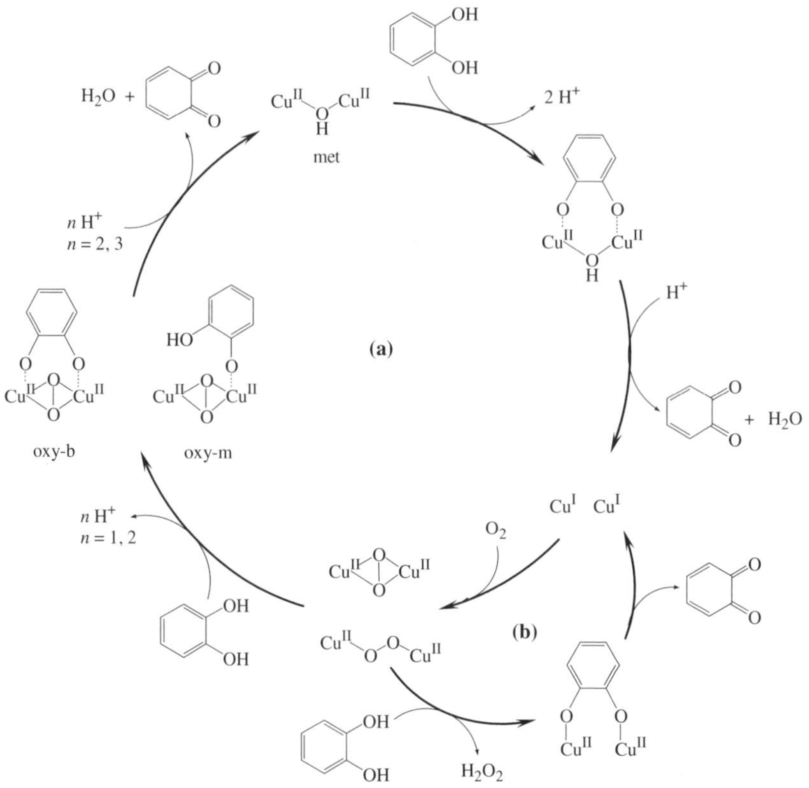

Scheme 18

Protein crystal structures of three forms of catechol oxidase have been determined (277). These include the reduced Cu(I)Cu(I) state with a bridging water and a Cu–Cu distance of 4.4 Å (deoxy), the oxidized Cu(II)Cu(II) form with a hydroxo bridge and a Cu–Cu distance of 2.9 Å (met), and an intermediate Cu(II)Cu(II) form with a bound catechol model substrate (the inhibitor phenylthiourea), coordinated as a monodentate ligand to one of the copper centers with a Cu–Cu distance of 4.2 Å (277). A number of low molecular weight model systems have been described, and relevant experimental structural data have been obtained for the dicopper(I) (289–292), the μ-peroxo dicopper(II) (280), the μ-$\eta^2$:$\eta^2$-peroxo dicopper(II) (269), the bis-μ-oxo-dicopper(III) (293), the μ-hydroxo-dicopper(II) (289), and various dicopper(II) catecholate complexes with bridging catecholate (192, 294), monodentate and various forms of chelating catecholate bound to one copper center (these generally have the deactivated tetrachlorocatechol (tcc) coordinated as the substrate) (192). Among the most relevant high-resolution structures as models for catecholase are the first structure with a dicopper(II) center bridged with a catecholate (294), the recently published structure of the hydroxo-bridged-dicopper(II) met form (289), and the series of structures, which emerged from copper(II)–bispidine chemistry and includes a structure with a bridging catecholate, one with a monodentate catecholate (the first structure of that kind and a structural model for an intermediate in one of the catalytic cycles described above) and structures with two types of chelating catecholates (192), see Fig. 17 and Table XII.

Spectroscopic and electrochemical data of the copper–bispidine–catechol systems are given in Table XIII, which also includes data of the catalytic activity [as in other investigations the kinetic data have been obtained with 3,5-di-*tert*-butylcatechol (dtbc)]. The spectrophotometric titration data indicate that there are two catecholate binding modes in the bispidine complexes. Based on the stoichiometry, the structural data and DFT calculations, the 530 and the 450 nm charge-transfer transitions are attributed to the bridging and chelating coordination modes of catechol, respectively (213). Based on this assignment, DFT calculations and the observed reactivities, it appears that for the copper bispidine complexes a bridging catecholate is needed for its activation toward oxidation to the quinoline product. Interestingly, the dicopper complex of a recently studied macrocyclic ligand has very similar reactivities, but the catecholase catalysis seems to follow a different pathway (289).

Various kinetic experiments and product analyses indicate that the catechol oxidation catalyzed by the bispidine–copper(II) complexes proceeds via the mechanism shown in Scheme 19. There is one quinone product per cycle and dioxygen is reduced to hydrogen peroxide. Interestingly, all individual steps ($1 \rightarrow 2 \rightarrow 3 \rightarrow 1$) are relatively fast

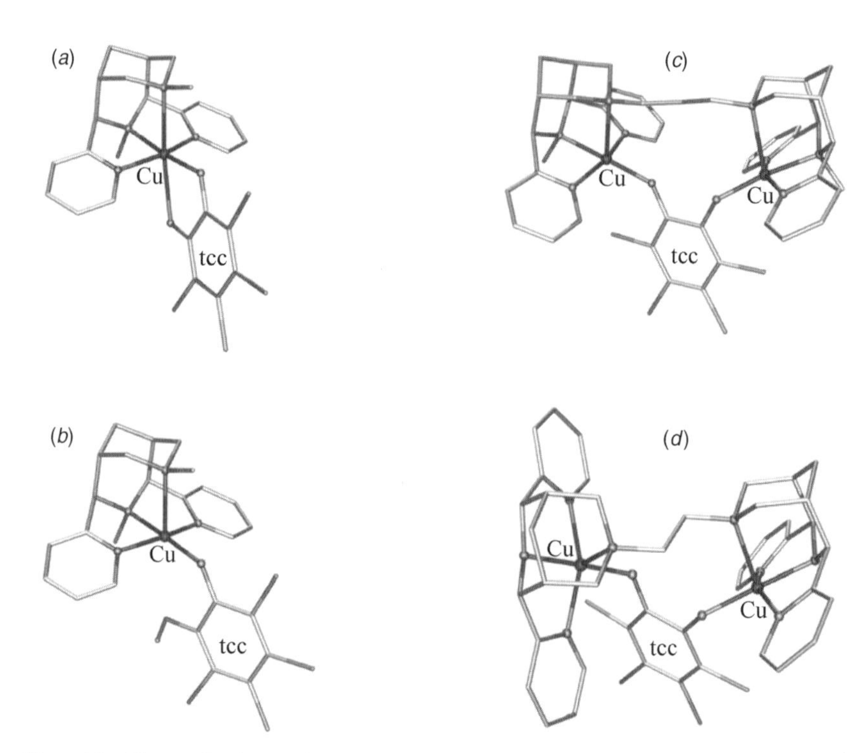


Figure 17. Plots of the molecular cations of (*a*) [Cu(**14**)(tcc)], (*b*) [Cu(**14**)(tcc)$^+$, (*c*) [Cu$_2$(**59**)(tcc)]$^{2+}$ (all X-ray); (*d*) [Cu$_2$(**58**)(tcc)]$^{2+}$ (DFT) (192, 213).

(~two orders of magnitude faster than $k_{cat}$). The reason proposed for the slow and inefficient catalytic process is that the catecholate-bridged dicopper(II) active state (3) is in equilibrium with an inactive dicopper(II) intermediate with two chelating catecholates, that is, the substrate inhibits the reaction if present at high concentration (213).

It is of interest that, based on conclusive experimental data, different types of complexes oxidize catechole by strikingly different reaction mechanisms (see Scheme 18) (213, 289). Therefore, this is an excellent example to show that catalysts that are similar in terms of activity might differ considerably in terms of the reaction mechanism. More importantly: structural and/or spectroscopic models of metalloenzymes are not necessarily useful biomimetic model complexes, even if they catalyze the reaction.

**c. Iron Oxidation Chemistry.** The iron bispidine chemistry has developed to be a particularly fruitful subject. The first preparative studies have produced a variety of stable mono- and dinuclear iron(II) complexes with tetra-, penta-, and hexadentate bispidine ligands with amine, pyridine, alcoholate, and phenolate

TABLE XII

Structural Parameters of Copper(II)–Bispidine Complexes Relevant for Catecholase Modeling (213)
[(*a*)–(*d*) are X-ray structures; (*e*)–(*j*) are computed structures]

| Parameter[a] | (a)[b] | (b)[c] | (c)[d] | (d)[e] | (e)[f] | (f)[g] | (g)[h] | (h)[i] | (i)[j] |
|---|---|---|---|---|---|---|---|---|---|
| Cu–Cu′ | 7.04 | 5.96 | 7.92 | 7.01 | 5.47 | 5.46 | 6.20 | 5.67 | 9.10 |
| Cu-N3 | 2.03/2.03 | 2.03 | 2.07 | 2.08/2.06 | 2.10/2.10 | 2.07/2.59 | 2.17/2.06 | 2.16/2.21 | 2.16/2.12 |
| Cu-N7 | 2.36/2.30 | 2.36 | 2.61 | 2.54/2.46 | 2.44/2.56 | 2.77/2.07 | 3.41/2.11 | 2.24/2.36 | 3.55/2.68 |
| Cu-py1 | 2.01/2.03 | 2 | 2.00 | 2.03/2.02 | 2.06/2.05 | 2.03/1.99 | 1.93/2.68 | 2.05/1.99 | 1.92/1.97 |
| Cu-py2 | 1.99/2.0 | 1.99 | 1.98 | 2.27/2.25 | 2.05/2.04 | 1.92/1.96 | 1.94/2.37 | 2.11/2.11 | 1.96/2.04 |
| Cu-py3 | l | l | 2.52 | 2.63/2.39 | l | l | 2.04/l | l | 2.03/l |
| Cu-O1/C1 | 2.23/2.22 | 1.9 | 2.25 | 2.27/2.25 | 1.95/1.93 | 1.90/2.00 | l/1.94 | 1.93/1.97 | l/1.95 |
| Cu-O2 | l | l | l | l | l | l | l/2.83 | l | –/2.43 |
| N3-Cu-N7 | 82.8/84.4 | 83.9 | 77.1 | 81.2/79.7 | 72.4/82.7 | 79.7/87.5 | 69.7/91.4 | 88.4/86.9 | 66.9/79.6 |
| N3-Cu-py1 | 81.4/81.0 | 81.5 | 83.4 | 81.9/81.7 | 81.1/81.6 | 80.5/74.8 | 85.9/74.0 | 79.6/79.2 | 86.0/83.1 |
| N3-Cu-py2 | 81.2/80.2 | 82.4 | 78.7 | 82.2/81.4 | 80.0/80.7 | 80.1/74.2 | 87.6/77.8 | 77.7/76.9 | 86.6/80.8 |
| N3-Cu-py3 | l | l | 81.8 | l | l | l | 86.0/– | l | 87.1/l |
| N3-Cu-O1/Cl | 169/168 | 178 | 179 | 176/171 | 173/166 | 172/167 | l/157 | 160/151 | l/179 |
| N7-Cu-py1 | 98.4/97.0 | 95.6 | 86.1 | 88.9/92.3 | 97.6/94.3 | 85.6/101 | 74.8/95.3 | 98.9/99.4 | 69.7/87.9 |
| py1-Cu-py2 | 159/157 | 162 | 165 | 164/163 | 157/161 | 156/138 | 148/149 | 149/150 | 144/164 |

[a] Distances in angstroms (Å), angles in degrees.

[b] $[Cu_2(\mathbf{59})Cl_2]$.

[c] $[Cu_2(\mathbf{59})(tcc)]$ (bridged).

[d] $[Cu_2(\mathbf{61})(Cl_2)]$.

[e] $[Cu_2(\mathbf{62})(Cl_2)]$.

[f] $[Cu_2(\mathbf{58})(tcc)]$ (bridged).

[g] $[Cu_2(\mathbf{61})(tcc)]$ (bridged).

[h] $[Cu_2(\mathbf{61})(tcc)]$ (bidentate to one $Cu^{2+}$).

[i] $[Cu_2(\mathbf{62})(tcc)]$ (bridged).

[j] $[Cu_2(\mathbf{62})(tcc)]$ (bidentate to one $Cu^{2+}$)(201).

[k] Not available.

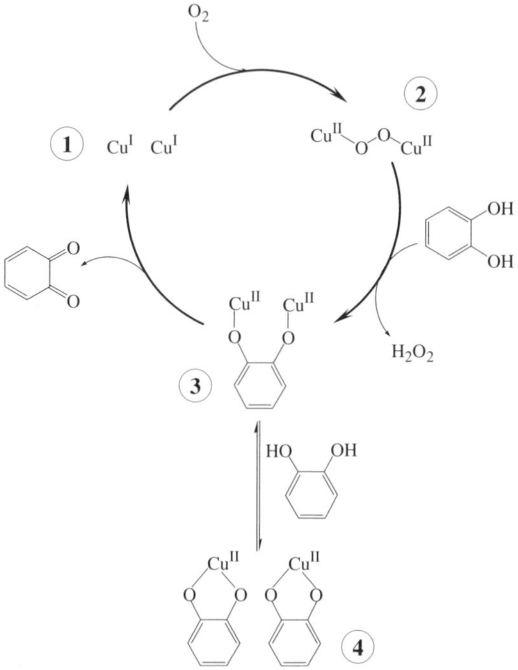

Scheme 19

TABLE XIII

Experimental Data of the Bispidine–Copper Catecholase Model systems: Electronic Spectra ($CH_3OH$), Reduction Potentials (MeCN vs. $Ag/AgNO_3$), Bispidine–copper(II)/tcc stability constants, and Michaelis–Menten Catalysis Rates for the dtbc to dtbq Transformation (213)

| Complex | Cu:tcc | $\lambda_{max}$ (nm) | $\varepsilon_{max}$ ($L\ mol^{-1}\ cm^{-1}$)[a] | $E_{1/2}$ (mV) | $\log K_1$ 2:1 | $\log K_2$ 2:2 | $k_{cat}$ ($10^{-3}\ s^{-1}$) |
|---|---|---|---|---|---|---|---|
| $[Cu^{II}(\mathbf{14})]^{2+}$ | 1:1 | 450 | 510 | −417 | [b] | 5.17[a] | 0.0 |
| $[Cu_2^{II}(\mathbf{58})]^{4+}$ | 2:1 | 529 | 2631 | −345 | 6.90 | 3.06 | 5.96 |
| | 2:2 | 453 | 577 | | | | |
| $[Cu_2^{II}(\mathbf{59})]^{4+}$ | 2:1 | 534 | 1541 | −334 | 4.53 | 3.64 | 0.78 |
| | 2:2 | 456 | 685 | | | | |
| $[Cu_2^{II}(\mathbf{61})]^{4+}$ | 2:1 | 545 | 1312 | −385 | 3.36 | 4.12 | 0.04 |
| | 2:2 | 468 | 651 | | | | |
| $[Cu_2^{II}(\mathbf{62})]^{4+}$ | 2:1 | 532 | 1640 | −436 | 5.09 | 2.01 | 17.22 |

[a] Per dicopper(II) complex [except for the Cu(**14**) catalyst].
[b] Not available.

donors (66). As expected, from the preference of bispidine ligands for large metal ions, these have comparably low ligand fields, and therefore all are preferentially high spin and close to the spin-crossover limit at room temperature. They also have rather positive redox potentials (see Table IX).

Of particular interest is the oxidation of the ferrous to the corresponding ferric and ferryl complexes and their application as catalysts for the oxidation of saturated and unsaturated hydrocarbon compounds. Mononuclear non-heme iron complexes have been suggested as valuable models for biological oxygen activation, bioinspired oxidation catalysis, and for technical processes (295–302). A number of ligand systems, many of them with chromophores similar to those of the bispidine ligands, have been used to study the mechanisms of the iron(II) to iron(III) and iron(IV) reactions and of substrate oxidation involving kinetic studies, as well as structural, spectroscopic, and theoretical work (295–310). Some of the mechanistic studies have been influenced by recent experimental and theoretical developments in the area of heme iron oxidation processes (311, 312). and a comparison of the electronic structures and reactivities of heme and non-heme $Fe^{IV}{=}O$ species is therefore of particular interest (313). While there is good structural, spectroscopic, and computational evidence for stable non-heme $S = 1$ $Fe^{IV}{=}O$ species (306–308) and for their activity as oxidants, evidence for $S = 2$ $Fe^{IV}{=}O$ [except for the biochemical TaudD system (305)] or $Fe^{V}{=}O$ (299, 309, 310) is more circumstantial and disputable.

The bispidine ligand-based studies involve primarily the tetradentate ligand **14** and the two isomeric pentadentate ligands **24** and **28**. The tetradentate ligand-based complexes have two sites for substrate (peroxide, oxide, hydroxide) coordination, one with a strong (equatorial) and one with a weaker bond (axial). In the pentadentate ligand-based complexes, one each is blocked by a third pyridine donor (see Fig. 18). This leads to interesting differences of the two isomeric pentacoordinate systems in terms of the stabilities and catalytic activities of the high-valent metastable intermediates. The studies so far include mechanistic investigations of the Fe(II) oxidation with $H_2O_2$, *tert*-butylhydroperoxide, or iodosylbenzene under aerobic and unaerobic conditions, and the

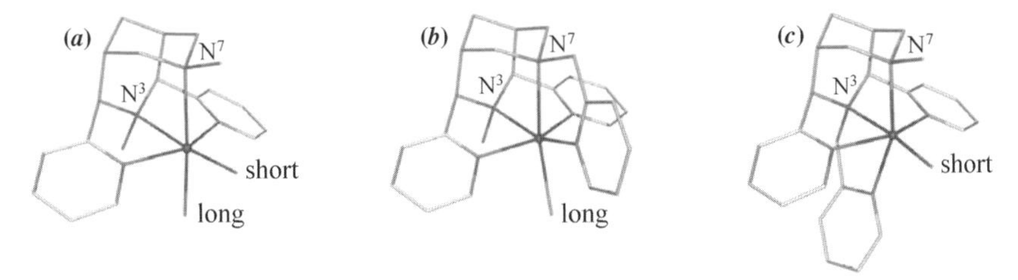


Figure 18. Plots of the structures of bispidine–metal complexes with the tetradentate ligand **14** (*a*) with two sterically and electronically different sites for substrate coordination, and with the isomeric pentadentate ligands **24** (*b*) and **28** (*c*), which have one of those sites blocked by a pendant pyridine.

spectroscopic and quantum mechanical characterization of the high-valent species involved (170–175), as well as alkene oxidation studies with cyclooctene as substrate (314–316) and potential applications as bleaching catalysts (64, 317–319).

Reaction of the pentadentate bispidine-based systems with $H_2O_2$ in methanol at low-temperature yields a purple solution of the low-spin iron(III) end-on hydroperoxo complexes; that of the isomer with an in-plane coordination of the hydroperoxide is, as expected, considerably more stable than that with an axially coordinated substrate. Deprotonation to the end-on high-spin peroxo complex (seven-coordinate) is reversible for the more stable isomer, with a p$K$ value of 4.5. These complexes have been fully characterized spectroscopically and by DFT calculations (170). The corresponding ferryl complexes are prepared *in situ* by addition of excess iodosylbenzene to the iron(II) complexes, and these were also characterized spectroscopically and assigned an $S = 1$ ground state (172, 314). Reaction of $[Fe^{II}(\mathbf{28})(OH_2)]^{2+}$ in aqueous solution ($2 < pH < 6$) with excess $H_2O_2$ directly produces $[Fe^{IV}{=}O(\mathbf{28})]^{2+}$ (see Fig. 19) (172). The difference in reactivity might be due to subtle solvent-dependent changes of the redox potential and/or spin states, but a thorough interpretation has to await further data. Preliminary kinetic data with various bispidine complexes suggest that, depending on the reaction conditions, Fe(III) species result from the decay of directly formed $Fe{=}O^{2+}$ rather than being a precursor of the ferryl compounds. The direct formation of ferryl complexes from iron(II) and $H_2O_2$ and their unprecedented stability in acidic aqueous solution, is of relevance for

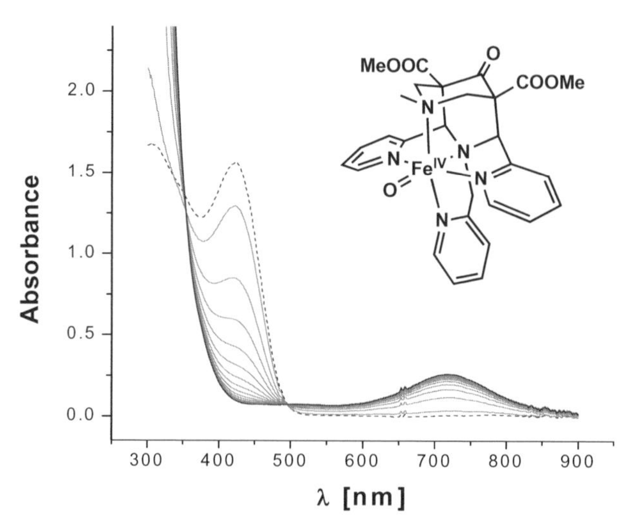


Figure 19. The reaction of $[Fe^{II}(\mathbf{28})(OH_2)]^{2+}$ (1 m$M$, dashed line) with 10 eq $H_2O_2$ at 0°C in an aqueous citrate buffer (pH 5.4), to form $[Fe^{IV}{=}(O)(\mathbf{28})]^{2+}$ (dark solid line). (172).

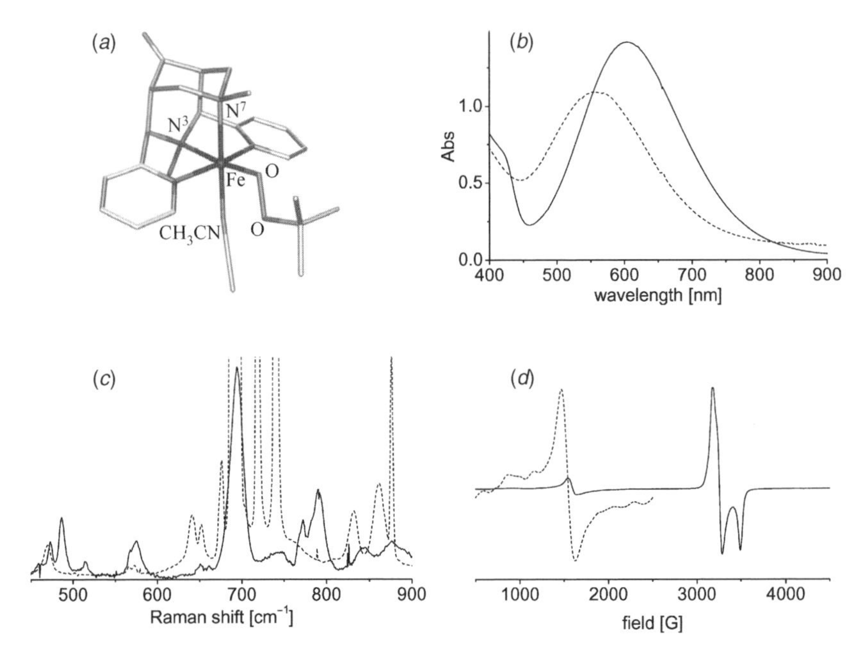


Figure 20. Structural and spectroscopic properties of the spin-crossover complex $[Fe(\mathbf{14})(OO\mathit{t}\text{-}Bu)(X)]^{2+}$ (X = solvent or no ligand). (*a*) The DFT- optimized structure, (*b*) electronic spectra (233 K), (*c*) resonance Raman spectra (77 K), (*d*) X-band electron paramagnetic resonance (EPR) spectra (4 K); full lines: $CH_3CN$, dotted lines: $CH_2Cl_2$ (171).

Fenton chemistry (320–322) and indicates that ferryl species (at least when supported by bispidine ligands) rather than $^{\bullet}OH$ radicals may be involved in substrate oxidation. For iron(II) aqua ions, this has also been proposed on the basis of DFT calculations (323), and a similar mechanism is confirmed by DFT calculations with the bispidine complexes (174).

No high-valent iron complexes have been trapped and spectroscopically characterized with the iron complex based on the tetradentate ligand **14**, although DFT calculations indicate that ferryl complexes may be formed (173). Also, $[Fe(II)(\mathbf{14})X_2]^{n+}$ (X = solvent)/$H_2O_2$ systems catalyze the oxidation of cyclooctene, and this suggests that high-valent iron species are present (315). The reaction of $[Fe(II)(\mathbf{14})(X)_2]^{n+}$ with *tert*-butyl hydroperoxide indeed yields an iron(III) alkylperoxo complex, which, in dependence of the solvent has high-spin ($CH_2Cl_2$) or low-spin ($CH_3CN$) electronic configuration (see Fig. 20) (171). This type of alkylperoxo–iron(III) complex is known to decay heterolytically to iron(III) and alkylperoxide (high-spin form) or homolytically to a ferryl complex and an O-based radical (low-spin form; see Scheme 20) (324). No well-resolved spectroscopic data of the ferryl decay product are available so far, and this might be related to its instability and the DFT-predicted $S = 2$ configuration (173).

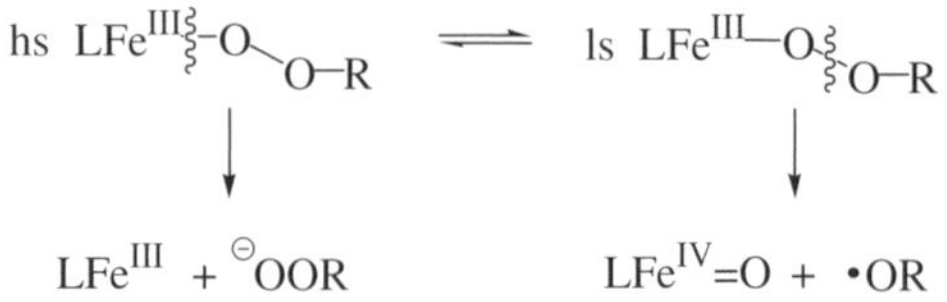

$$\text{hs LFe}^{\text{III}}\text{–O–O–R} \rightleftharpoons \text{ls LFe}^{\text{III}}\text{–O–O–R}$$

$$\text{LFe}^{\text{III}} + {}^{\ominus}\text{OOR} \qquad \text{LFe}^{\text{IV}}\text{=O} + \bullet\text{OR}$$

Scheme 20

In the iron(II)-catalyzed oxidation of olefins with hydrogen peroxide to epoxide and diol products, the two complexes [Fe(II)(**24**)X]$^{n+}$ and [Fe(II)(**14**)$X_2$]$^{n+}$ are among the most efficient catalysts; the isomer of the former complex with ligand **28** leads to more stable intermediates and has a similar reactivity but, as expected, on a much longer time scale (314, 315). Selected experimental data are given in Table XIV. Interesting features are that the yields and product distributions are strikingly different under aerobic and unaerobic conditions, and also that they are strongly temperature dependent (314). This has not been reported for other systems, and it is not clear whether or not this has been thoroughly evaluated (303). It emerges that the tetra- and pentadentate-ligand-based catalysts operate by very different mechanisms. No pentadentate ligand systems with comparable reactivity have been reported, and these generally produce epoxide as the only product.

Interesting observations with the pentadentate ligand-based catalyst are

1. In aerobic reactions, epoxide is the only product. Under unaerobic conditions only ~20% of the epoxide is formed and about the same amount of diol product is also obtained, which is a mixture of cis and trans diol. Labeling indicates that in ~80% of the epoxide in the aerobic reaction the oxygen atom arises from dioxygen.
2. The stoichiometric reaction with *in situ* produced $Fe^{IV}{=}O$ produces only epoxide, with a larger yield under aerobic conditions, where dioxygen from air appears in the epoxide product.
3. The yield in methanol as solvent is much reduced, in aerobic and unaerobic reactions it is the same, and there is no diol product formed.

From 2, it was concluded that the ferryl complex is the catalytically active species. Observation 1 suggested that 80% of the epoxide product in the aerobic reaction is derived from a carbon-based radical, which is quenched by $O_2$ (autoxidation), and this is known to produce epoxide in reactions with cyclooctene (325). Methanol (observation 3) is known to quench radicals. The fact that the diols formed are a mixture of cis and trans products (observation 1; this is very unusual in iron-catalyzed olefin oxidations) suggested that the diol results from the capture of $^{\bullet}OH$ radicals by the putative carbon-based radical.

This is supported by observation 2, that is no diols are formed in the absence of the $H_2O_2$ oxidant. The proposed mechanism (Scheme 21) (314) is in agreement with all these observations, and it is supported by DFT calculations (Fig. 21) (315, 316).

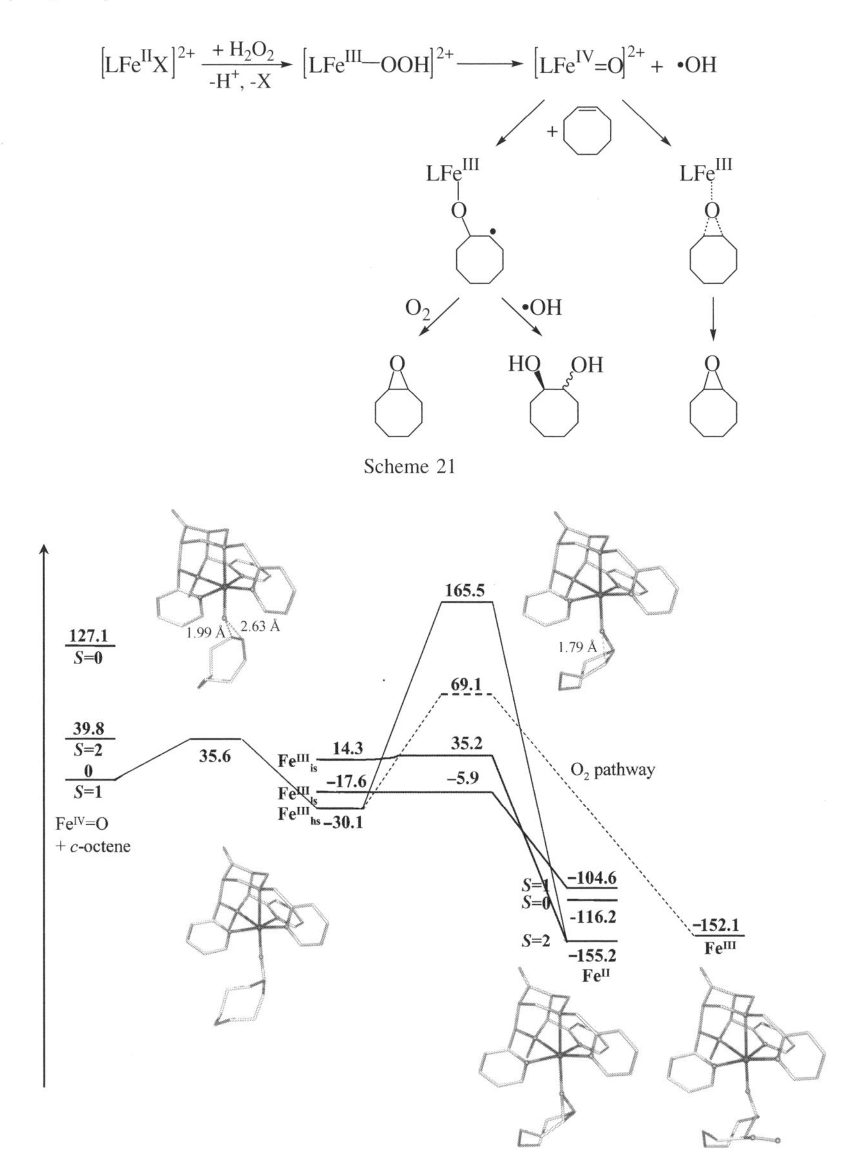


Figure 21. The reaction profile (DFT) of the oxygenation of cyclooctene by $[(\mathbf{24})Fe{=}O]^{2+}$ (see Scheme 21) (316).

Important observations with the tetradentate ligand-based catalyst are

1. The diol product selectively has a cis configuration and both hydroxyl oxygen atoms arise from $H_2O_2$. Also, the yield of diol is independent of the atmosphere, (i.e., ~1.5 ton).
2. The yield of the epoxide product in the aerobic reaction (2.2 ton) is about twice that of the unaerobic reaction (1.0 ton), and the labeling suggests that slightly more than one-half of the label has its origin from dioxygen from air, the rest arises from $H_2O_2$.

From these observations, one may conclude that the diol product arises directly from an iron-based oxidant, rather than from a carbon-based radical intermediate that captures $^{\bullet}$OH or other oxygen-based radicals as in the mechanism proposed for the pentadentate ligand based catalyst. The formation of epoxide, however, is proposed to occur in a similar pathway to that described above, although the Fe—O—C—C-type intermediate then must have a shorter lifetime. Based on this interpretation, which admittedly is not the only possible one, one would conclude that no $^{\bullet}$OH are present, and this is in agreement with the DFT-based suggestion that there is a direct formation of the iron(IV) catalyst from its iron(II) precursor and $H_2O_2$, a proposal that is in agreement with the fact that no iron(III) intermediates are observed (see Scheme 22, Fig. 22).

The reported experimental data, DFT-derived structural electronic and energetic information, and mechanistic interpretations, lead to a consistent picture, which is in line with observations with other ligand systems, although there is some disagreement in terms of possible interpretations. However, there are a number of open questions (e.g., regarding the spin state of the ferryl species, which for both the tetra- and pentadentate ligand-based systems is predicted by DFT to be close to the spin cross-over limit) (175). Therefore, some of the current interpretations might be revised in the future.

**d. Other Metal Ions**. The bispidine coordination chemistry of a number of other transition metal ions has been studied in some detail (166, 167, 189, 190, 198). Of specific interest and discussed here are the redox chemistry of vanadium, in combination with some structural and spectroscopic studies (198), and the cobalt(II)–bispidine hydrogen peroxide chemistry (167).

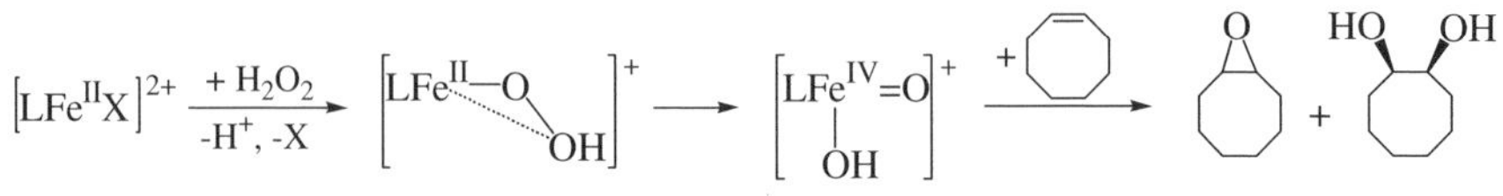

$$[LFe^{II}X]^{2+} \xrightarrow[-H^+,\ -X]{+\ H_2O_2} [LFe^{II}\text{—}O\text{—}OH]^+ \longrightarrow [LFe^{IV}{=}O(OH)]^+ \xrightarrow{+\ \text{cyclooctene}} \text{epoxide} + \text{cis-diol}$$

Scheme 22

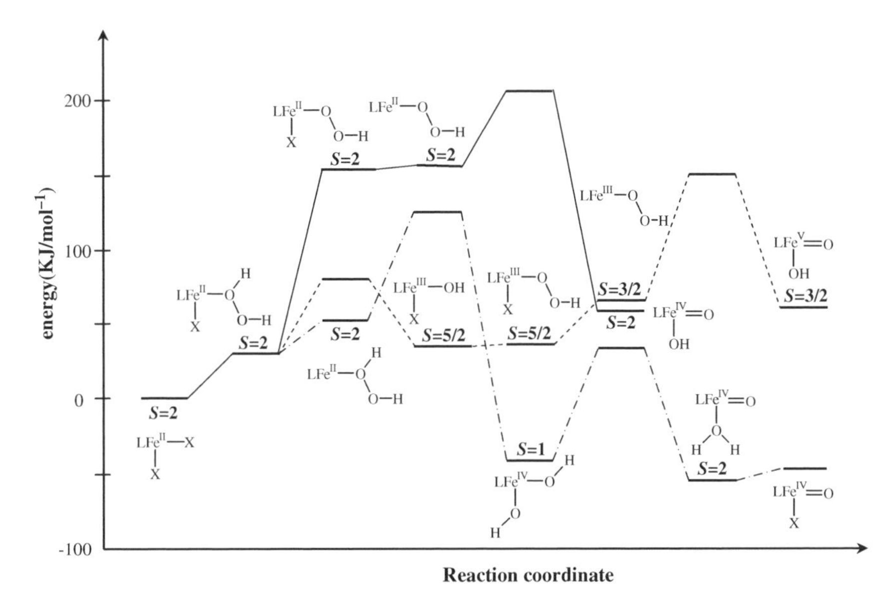


Figure 22. The three main reaction pathways for the formation of high-valent iron oxo species from $[Fe(\mathbf{14})(HOOH)]^{2+}$, entropy and solvation not included (173).

1. *Vanadium*. Vanadyl complexes of the three ligands **14**, **24**, and **28** were prepared from vanadyl sulfate and the metal-free ligands. They have the expected spectroscopic properties, and this is confirmed by the X-ray crystal structure of the vanadyl complex of **24**, which is shown in Fig. 23(*a*) and has metal–donor bond distances in the expected range. In terms of reactivity, these complexes are rather unspectacular. No haloperoxidase reactivity was observed, and oxidation of alkanes and alkenes is rather sluggish (138). This is not unexpected since with the pentadentate ligands there is no free site at the vanadium(IV) center for metal–substrate interaction. The same is true for the vanadium(V) oxo–peroxo complexes with tetradentate coordination of the pentadentate ligand.

With an excess of $H_2O_2$, $[V^{IV}{=}O(L)(OH_2)]^{2+}$ can be oxidized in a clean pseudo-first-order reaction with a half-life of ~15 min. For the complex with the pentadentate ligand **24** a, single-crystal X-ray structure of the oxo–peroxo complex was obtained at 100 K (see Fig. 23). The complex was also isolated (but not crystallized) as a nitrate and as the perchlorate salt with a $^{18}O$ labeled peroxide. An oxo–peroxo vanadium(V) complex with **23** (deprotonated phenolate) was also isolated and characterized. Spectroscopically (IR, NMR) these compounds are as similar as expected, and the series allowed for a meaningful analysis of the V—O and

TABLE XIV

Catalytic Efficiency and Selectivity of the Bispidine–Iron Catalyzed Oxidation of Olefins With $H_2O_2$[a]

| | tpa[b] | 14 | | | 24 | | | | 28 | N4py[c] |
|---|---|---|---|---|---|---|---|---|---|---|
| | Under $O_2$ | Under $O_2$ | Under Ar | Under $O_2$ | Under Ar | In $CH_3OH$ Under $O_2$ | $Fe^{IV}=O$ Under $O_2$ | $Fe^{IV}=O$ Under Ar | Under $O_2$ | Under $O_2$ |
| Diol; epoxide | 4.0; 3.4 | 1.6; 2.2 | 1.5; 1.0 | 0.0; 5.0 | 1.0; 1.0 | 0.0; 1.9 | 0.0; 1.0 | 0.0; 0.4 | 0.1; 0.8 | 0.0; 0.6 |
| Epoxide, % $^{18}O_2$ [$H_2\,^{18}O_2$/$H_2\,^{18}O$/$^{18}O_2$ | 90/9/1 | 45/0/55 | 100/0/- | 15/~0/85 | 82/18/- | [d] | 45/0/55 | 60/40/- | [d] | [d] |
| cis-Diol % $^{18}O$ from | | | | | | | | | | |
| $H_2\,^{18}O_2$ [no O/1O/2O] | 0/97/3 | 0/0/100 | 0/0/100 | No diol | 0/23/77 | No diol | No diol | No diol | No diol | No diol |
| $H_2\,^{18}O$ [no O/1O/2O] | 13/86/1 | 92/8/0 | 93/7/0 | | 0/30/0 | | | | | |
| trans-Diol % $^{18}O$ from | | | | | | | | | | |
| $H_2\,^{18}O_2$ [no O/1O/2O] | No trans | No trans | No trans | No diol | 0/56/44 | No diol | No diol | No diol | No diol | No diol |
| $H_2\,^{18}O$ [no O/1O/2O] | | | | | 0/50/0 | | | | | |

[a]See Refs. 314 and 315; yields given in ton, ton $_{max}$ = 10

[b]tpa = Tris(2-pyridylmethyl)amine

[c]N4py = Bis(2-pyridylmethyl)bis(2-pyridyl)methylamine.

[d]Not available.

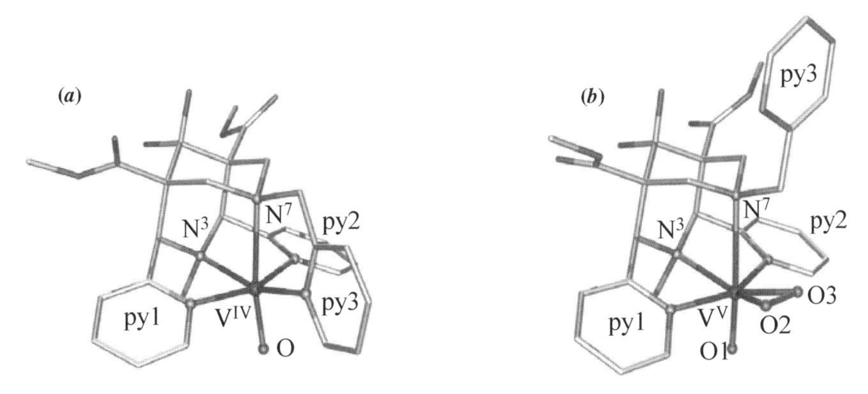


Figure 23. Plots of the structures (X-ray) of the complex cations of (*a*) $[V(\mathbf{24}^*)(O)]^{2+}$ and (*b*) $[V(\mathbf{24}^*)(O)(O_2)]^+$ (198).

O—O vibrations. This was of importance because the original analysis of the peroxo O—O distance [1.281(5) Å] suggested that this complex might be described as a vanadium(IV) oxo–superoxo complex (326). Based on the IR spectra of the two salts and labeling studies, the V=O and the O—O stretching modes are found at 962 and 890 (932) $cm^{-1}$, respectively (198). These are typical for peroxo complexes (O—O $\lesssim 1000\ cm^{-1}$; superoxo $\gtrsim 1100\ cm^{-1}$), and the linear relation between the O—O distance and vibrational frequency of peroxo- and superoxo compounds suggests that the O—O distance in $[V(O)(O_2)(\mathbf{24})]^+$ is underestimated by ~0.1 Å (326).

Close inspection of the crystal structure (see Fig. 24) reveals much larger atomic displacement parameter (ADP) values for O2 and O3 in comparison with the other donor atoms. In addition, there is a systematic

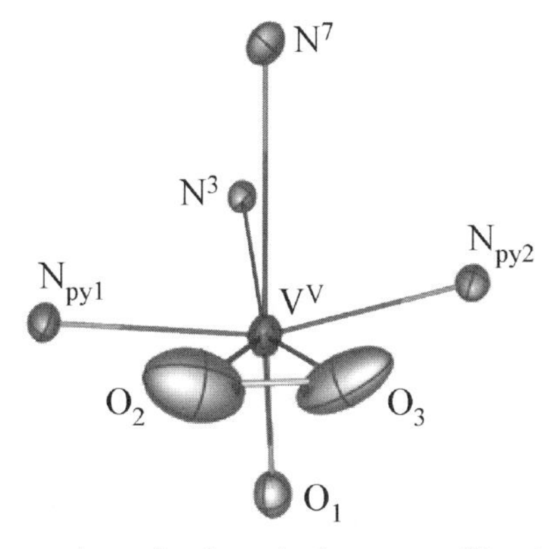


Figure 24. Plot of the experimentally determined structure (X-ray) of $[V(\mathbf{24}^*)(O)(O_2)]^+$ to visualize the disordered peroxo ligand (198).

deviation from spherical shapes of the thermal envelopes for O2 and O3. Normally, such effects can be attributed to libration of nonrigid groups (327). which leads to an apparently shorter bond, as seen in the structure discussed here. However, the fact that the structure was obtained at low temperature, where librational effects are expected to be small, and that modeling of libration was not successful suggests that the apparent O—O bond compression and the observed thermal envelopes might be due to static disorder effects.

2. *Cobalt.* Cobalt(II) bispidine complexes in general are air stable (69, 166, 167, 189). This finding is unusual in cobalt–amine chemistry, where the cobalt(II)–$O_2$ ($H_2O_2$) coordination chemistry is dominated by the formation of μ-peroxo and μ-superoxo dicobalt(III) complexes and their hydrolysis to mononuclear cobalt(III) compounds (328); in addition, there are few reports on mononuclear cobalt(II) and cobalt(III) hydroperoxo and peroxo complexes (328–334). The exceptionally high redox potentials, in line with the low ligand fields (see Table VIII) are the result of the rigid bispidine ligands with a preference for relatively large metal ions (189, 199). A number of X-ray crystal structures of cobalt(II) complexes have been published, and these are discussed above in relation to the elasticity of the coordination geometry and isomerism of the chromophore (189).

Cobalt(III) complexes of bispidine ligands were prepared from the cobalt(III) complex $Na_3[Co(OCO_2)_3]$ (335) and by oxidation of the cobalt(II) precursor with $H_2O_2$ (167). Both lead to unexpected and interesting products. From X-ray crystal structures, it emerges that the product of **14**, based on the synthesis with the tris-carbonato-cobalt(III) method, has hydrolyzed ester groups (see Fig. 25) (167). This coordinated ligand is of particular interest for further preparative chemistry and the synthesis of bispidine derivatives with functional side chains (see below). Even more interesting is the observation of an oxidative *N*-dealkylation in the reaction of various bispidine–cobalt(II) complexes with $H_2O_2$ (Scheme 23, Fig. 25) (167).

Oxidative N-dealklylation is an important reaction in enzymatic drug metalbolism (336), and metal-based electron transfer as well as C-based radical mechanisms that have been proposed as relevant pathways (337). A number of low molecular weight iron- and cobalt-based biomimetic models have been described (259, 336). The oxidation of cobalt(II)–bispidine complexes is a rather sluggish reaction with pseudo-first-order rate constants of ~(1–10) $10^{-3} s^{-1}$. The N3 dealkylation was observed with the two tetradentate ligands **14** and **35** with N3 methyl or benzyl substituents. The oxidation with 3 equiv of $H_2O_2$ leads to ~40% N-dealkylated cobalt(III) complex; with ≥ 20 equiv of $H_2O_2$, the conversion is essentially quantitative. The product $[Co(III)(\mathbf{14}^{c,d})(X)_2]^{n+}$ ($\mathbf{14}^{c,d}$) is the N7-dealkylated ligand

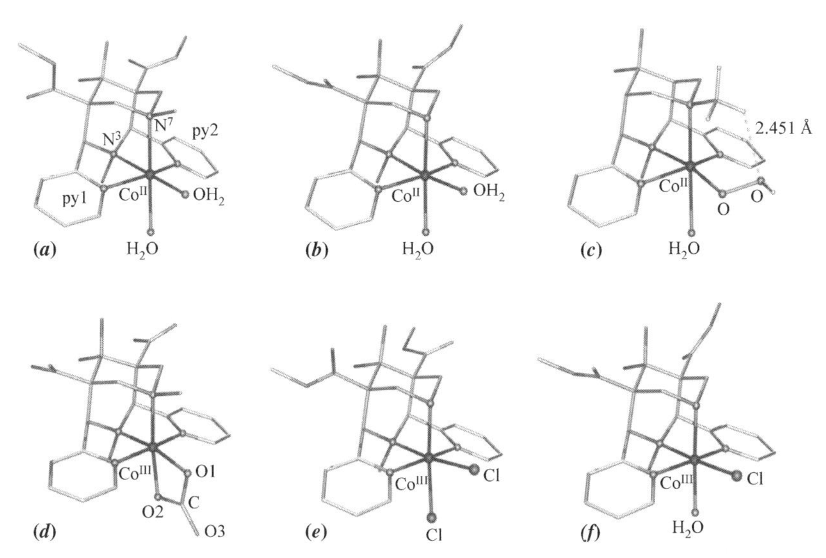


Figure 25. Plots of the complex cations of the X-ray structures of (*a*) [Co(**14***)$(OH_2)_2$]$^+$; (*b*) [Co(**14**$^*$)$(OH_2)_2$]$^{2+}$; (*c*) [Co(**14***)(OOH)($OH_2$)]$^+$, computed (DFT); (*d* [Co(**14**$^*$)($CO_3$)]$^+$; (*e*)[Co(**14**$^*$)(Cl)(Cl)]$^+$; (*f*) [Co(**14**$^*$(Cl)($OH_2$)]$^{2+}$ (hydrogen atoms omitted; see Table IV for the nomenclature of the modified ligands) (167).

(see Fig. 25(*e*,*f*) and Table IV; X = solvent, anion) was characterized by X-ray crystallography, NMR spectroscopy, and electrospray ionization–mass spectrometry (ESI–MS). The organic products of the *N*-dealkylation, formaldehyde in the case with **14** or benzaldehyde with ligand **35**, were quantitatively analyzed as the corresponding hydrazones (338). The N-dealkylation only occurs in an aerobic atmosphere, suggesting that a carbon-based radical mechanism, which involves autoxidation, is involved. The fact that the reaction only occurs in MeCN as solvent probably is due to a large shift of the redox potential (up to ~400 mV). Based on these observations and supported by $H_2{}^{18}O_2$ labeling studies and preliminary DFT calculations, the mechanism shown in Scheme 23 was proposed.

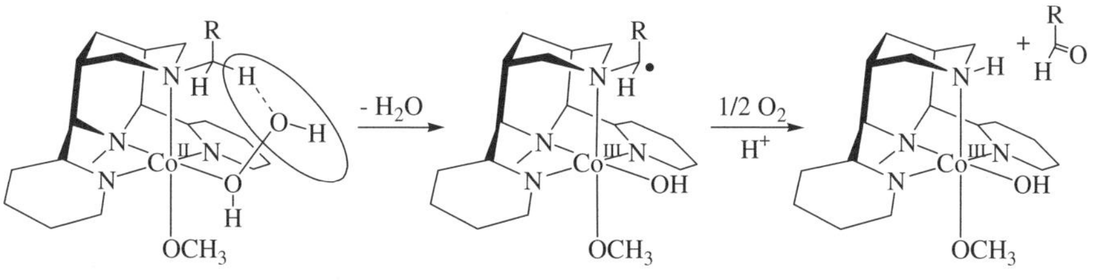

Scheme 23

### 4. *New Developments*

Preliminary experiments and ideas based on published work with other ligand systems exist, but are too incomplete to be reported here in detail. These include complexes and their properties with novel bispidine ligands (see Section III.B.2), complexes with manganese, chromium, as well as second- and third-row transition metal ions (e.g., ruthenium); oxidation reactions with dioxygen alone or with other peroxides (e.g., *tert*-butyl-peroxide); the stabilization and spectroscopic characterization of mononuclear superoxo, peroxo, and oxo complexes; other catalytic processes (e.g., the iron-catalyzed aziridination), enantioselective reactions with chiral bispidine ligands; and the iron oxidation chemistry continues to produce novel and exciting results.

An interesting possible further extension is the functionalization of bispidine ligands with hydrophobic groups, for example, for metal ion selective extractions (69, 339). biopolymers for nuclear medicinal applications (340), solids for heterogeneous catalysis and sensors, or additional coordination sites for the synthesis of heterodinuclear complexes with applications in biomimetic chemistry, catalysis, and as luminescence sensors. There is a variety of possible sites for ligand modification. Of particular interest is the C9 position, which has been selectively and stereospecifically reduced to an alcohol (190), and the two hydrolyzed C1,C5 ester groups (167).

For some time, we have been involved in the area of molecular magnetism. These studies were based on melamine-derived mono-, bis-, and tris-macrocyclic copper(II) and nickel(II) complexes and their host–guest compounds with cyanometalates (341–345). Magnetic exchange through the cyano bridge was shown to lead to multivarious effects and based on a thorough interpretation, there is the possibility to tune the magnetic properties of heterooligonuclear complexes (345, 346). A thorough theoretical study indicates that $\sigma$, as well as $\pi$-MO effects, are of importance, and these can be varied through the geometry of the cyano bridges and the electronic ground states of the interacting metal centers (346). The relatively well-understood steric and electronic properties of bispidine coordination compounds, in particular the interesting variation of substrate (CN) binding to the trans-to-N3 and trans-to-N7 site, as well as the variation of the binding strength through substitution of the bispidine backbone, makes bispidine complexes interesting building blocks for the synthesis of heterooligonuclear cyanometalates. A number of such complexes have been prepared, characterized and studied structurally, spectroscopically, and magnetically (347). One relevant structure is presented in Fig. 11(*f*).

## IV. CONCLUSIONS

Bispidines have been around for ~70 years (18), and for ~30 years they have been used in pharmaceutical studies (19). The first transition metal coordination compounds were reported ~50 years ago (30). A few groups have started to show some interest in or study various specific aspects of bispidine coordination chemistry (32–34, 156, 348). However, it was <10 years ago that bispidine ligands really started to attract the attention of coordination chemists (166). One of the main reasons for the versatility and broad applicability of bispidine ligands is their fairly easy synthesis and the rather broad variation of coordination geometries and donor sets for mono- and dinuclear complexes (Sections II and III.2). Important features of bispidine ligands and their complexes are (1.) a highly preorganized and, for many transition metal ions, complementary ligand structure that leads in general to comparably stable complexes and fast complexation reactions; (2.) an enforced cis-octahedral (square-pyramidal) coordination geometry with two cis-disposed free coordination sites for substrate binding with sterically and electronically very different binding characteristics; and (3.) a very rigid ligand structure and, nevertheless, a high elasticity of the coordination geometry.

Structures and structure correlations have been studied in detail (see Sections III.B and III.C) (42, 69, 71, 189, 194, 199), and specific features of the bonding of bispidine ligands to transition metal ions and of substrates to metal–bispidine fragments have been evalutated (see Section III.D.1) (214, 215, 349). It is especially the enforced cis-octahedral geometry and the elasticity of the coordination sphere (flat potential energy surfaces with various shallow minima), together with the interesting bonding properties of substrates that have lead to a series of interesting observations and discoveries in various fields of coordination chemistry. Among these are

1. The first examples of structurally characterized isomers of penta- and hexacoordinate copper(II) chromophores along Jahn–Teller active vibrations (82).
2. The spectroscopic characterization and mechanistic evaluation of the formation and decay of one of the most stable μ-peroxo–dicopper(II) complexes (81, 196).
3. The structural characterization of a series of copper–catecholate complexes, including model structures for catechol oxidase, which have not been known before, and the investigation of a catechol oxidase mechanism (192, 213).
4. The study of novel features of iron-catalyzed oxidation processes of relevance for enzyme-catalyzed and technical processes (172, 315, 316).

Many of these new developments are related to the very specific properties of bispidine ligands, others might also have been observed with other ligand systems. New bispidine ligands (Section II) whose coordination chemistry has not yet been developed in detail, novel bispidine chromophores (Section

III.B.2), and new areas in coordination chemistry, where bispidine complexes are just starting to produce the first results (Section III.D.4) will make sure that the fascination emerging from bispidine complexes will continue.

## ACKNOWLEDGMENTS

Our contributions in this area were generously supported by the University of Heidelberg, the German Science Foundation (DFG, SFB 623, Graduate College 850. SPP 1118, SPP 1137, Co 188/17-3), industrial cooperations (Unilever Vlaardingen) and support of guest scientists by the Alexander von Humboldt Society and the German Academic Exchange Program (DAAD). We are grateful for this invaluable support, the many scientific contributions of our co-workers and collaboration partners, whose names appear in the references, and for the help of Marlies von Schoenebeck-Schilli and Karin Stelzer in the preparation of this chapter.

## ABBREVIATIONS

| | |
|---|---|
| ADP | Atomic displacement parameter |
| ap | Antiparallel |
| bb | Boat–boat |
| cb | Chair–boat |
| cc | Chair–chair |
| CNDO | Complete neglect of differential overlap |
| daco | 1,5-Diazacyclooctane |
| DFT | Density functional theory |
| DIBAL-H | Diisobutylaluminum hydride |
| dtbc | 3,5-Di-*tert*-butylcatechol |
| EH | Expended Hückal |
| EPR | Electronic paramagnetic resonance |
| ESI–MS | Electrospray ionization–mass spectrometry |
| $^1H^1$NMR | Proton nuclear magnetic resonance |
| IR | Infrared |
| LFDT | Ligand-field density functional theory |
| LFMM | Ligand-field molecular mechanics |
| $LiAlH_4$ | Lithium aluminium hydroxide |
| LMCT | Ligand-to-metal change transfer |
| MINDO | Modified intermediate neglect of differential overlap |
| MNDO | Modified neglect of differential overlap |
| MO | Molecular orbital |
| NMR | Nuclear magnetic resonance |

| | |
|---|---|
| NOE | Nuclear overhauser effect |
| pa | Parallel |
| PES | Potential energy surface |
| SOD | Superoxide dismutase |
| STO-3G | Slater-type orbital, three gaussians |
| tcc | Tetrachlorocatechol |
| tpa | Tris(2-pyridylmethyl)amine |
| UV | ultraviolet |
| viz | Visible |

## REFERENCES


1. M. K. Pugsley, D. A. Saint, E. Hayes, K. D. Berlin, and M. J. A. Walker, *Eur. J. Pharmacol.*, *294*, 319 (1995).
2. P. W. Thies, *Pharm. Z.*, *15*, 172 (1986).
3. S. Körper, M. Wink, and R. H. A. Fink, *FEBS Lett.*, *436*, 251 (1998).
4. C. L. Schauf, C. A. Colton, J. S. Colton, and F. A. Davis, *J. Pharmacol. Exp. Ther.*, *197*, 414 (1976).
5. H. G. Goetz, *Pharm. Praxis*, *9*, 265 (1967).
6. D. Hoppe and G. Christoph, *The Chemistry of Organolithium Compounds*, Z. Rappoport and I. Marek, Eds., Wiley, Chichester, 2004.
7. O. Chuzel and O. Riant, *Top. Organomet. Chem.*, *15*, 59 (2005).
8. E. Vedejs and M. Jure, *Angew. Chem.*, *117*, 4040 (2005).
9. S. C. Bart, E. J. Hawrelak, A. K. Schmisseur, E. Lobkovsky, and P. J. Chirik, *Organometallics*, *23*, 237 (2004).
10. B. Jasiewicz, E. Sikorska, I. V. Khmelinskii, B. Warzajtis, U. Rychlewska, W. Boczon, and M. Sikorski, *J. Mol. Struc.*, *707*, 89 (2004).
11. Y.-M. Lee, S. K. Kang, G. Chung, Y.-K. Kim, S.-Y. Won, and S.-N. Choi, *J. Coord. Chem.*, *56*, 635 (2003).
12. C. Lorber, R. Choukron, J.-P. Costes, and B. Donnadieu, *C. R. Chim.*, *5*, 251 (2002).
13. Y. Funahashi, K. Nakaya, S. Hirota, and O. Yamauchi, *Chem. Lett.*, 1172 (2000).
14. E. Boschmann, L. M. Weinstock, and M. Carmack, *Inorg. Chem.*, *13*, 1297 (1974).
15. M.-C. Hung, M.-C. Tsai, G.-H. Lee, and W.-F. Liaw, *Inorg. Chem.*, *45*, 6041 (2006).
16. R. Iyengar and V. Gracias, *Chemtracts-Org. Chem.*, *17*, 92 (2004).
17. F. Binnig, *Arzneimittelforschung*, *24*, 752 (1974).
18. C. Mannich and P. Mohs, *Chem. Ber.*, *B63*, 608 (1930).
19. P. C. Ruenitz and C. M. Mokler, *J. Med. Chem.*, *20*, 1668 (1977).
20. B. R. Bailey, III, K. D. Berlin, E. M. Holt, B. J. Scherlag, R. Lazzara, J. Brachmann, D. Van der Helm, D. R. Powell, N. S. Pantaleo, and P. C. Ruenitz, *J. Med. Chem.*, *27*, 758 (1984).

21. M. D. Thompson, G. S. Smith, K. D. Berlin, E. M. Holt, B. J. Scherlag, D. van der Helm, S. W. Muchmore, and K. A. Fidelis, *J. Med. Chem.*, *30*, 780 (1987).
22. A. Samhammer, U. Holzgrabe, and R. Haller, *Arch. Pharm.*, *322*, 551 (1989).
23. U. Holzgrabe and E. Ericyas, *Arch. Pharm. (Weinheim)*, *325*, 657 (1992).
24. U. Kuhl, W. Englberger, M. Haurand, and U. Holzgrabe, *Arch. Pharm. Pharm. Med. Chem.*, *333*, 226 (2000).
25. G. Lesma, A. Sacchetti, A. Silvani, and B. Danieli, *New Methods for the Asymmetric Synthesis of Nitrogen Containing Heterocycles*, J. L. Vicario et al., Eds., Research Signpost, Trivandrum, India, 2005.
26. R. Jeyaraman and S. Avila, *Chem. Rev.*, *81*, 149 (1981).
27. F. Binnig and M. Raschack, Ges. Offer. DE2744248 (1979).
28. K. Andersson, A. Bjoere, M. Bjoersne, F. Ponten, G. Strandlund, P. Svensson, and L. Tottie, PCT Int. Appl. WO 2002004446 (2002).
29. W. Cautreels, C. Steinborn, M. Straub, K. Beckmann, and J. W. C. M. Jansen, PCT Int. Appl. WO 200502690 (2005).
30. H. Stetter and R. Merten, *Chem. Ber.*, *90*, 868 (1957).
31. R. Haller, *Arch. Pharm.*, *302*, 113 (1969).
32. G. D. Hosken and R. D. Hancock, *J. Chem. Soc., Chem. Commun.*, 1363 (1994).
33. S. Z. Vatsadze, N. V. Zyk, R. D. Rakhimov, K. P. Butin, and N. S. Zefirov, *Russ. Chem. Bull.*, *44*, 440 (1995).
34. D. S. C. Black, G. B. Deacon, and M. Rose, *Tetrahedron*, *51*, 2055 (1995).
35. Y. Miyahara, K. Goto, and T. Inazu, *Chem. Lett.*, 620 (2000).
36. S. P. Artz and D. J. Cram, *J. Am. Chem. Soc.*, *106*, 2160 (1984).
37. R. D. Hancock, *J. Chem. Educ.*, *69*, 615 (1992).
38. P. Comba, *Coord. Chem. Rev.*, *185*, 81 (1999).
39. W. K. Musker, *Coord. Chem. Rev.*, *117*, 133 (1992).
40. P. Comba, *Coord. Chem. Rev.*, *182*, 343 (1999).
41. P. Comba, *Coord. Chem. Rev.*, *200–202*, 217 (2000).
42. P. Comba and W. Schiek, *Coord. Chem. Rev.*, *238–239*, 21 (2003).
43. B. L. Vallé and R. J. P. Williams, *Biochemistry*, *59*, 498 (1968).
44. H. Eyring, R. Lumry, and J. D. Spikes and W. D. G. Mc Elroy, Eds., John Hopkins Press, Baltimore, 1954.
45. R. Lumry and H. Eyring, *J. Phys. Chem.*, *58*, 110 (1954).
46. R. J. P. Williams, *Inorg. Chim. Acta*, *5*, 137 (1971).
47. H. B. Gray and B. G. Malmström, *Biochemistry*, *28*, 7499 (1989).
48. B. G. Malmström, *Eur. J. Biochem.*, *223*, 711 (1994).
49. R. J. P. Williams and J. J. R. Frausto da Silva, *The Natural Selection of the Chemical Elements*, Clarendon Press, Oxford, New York, 1996.
50. J. J. R. Frausto da Silva and R. J. P. Williams, *The Biological Chemistry of the Elements*, Clarendon Press, Oxford, New York, 1991.
51. R. J. P. Williams and J. J. R. Frausto da Silva, *Bringing Chemistry to Life–From Matter to Man*, Oxford University Press, Oxford, 1999.
52. R. Haller, *Arzneimittelforschung*, *15*, 1327 (1965).
53. R. Caujolle and A. Lattes, *C. R. Acad. Sci., Ser. C*, *288*, 217 (1979).

54. A. Cambareri, D. P. Zlotos, U. Holzgrabe, W. Englberger, and M. Haurand, *J. Heterocyclic Chem.*, *39*, 789 (2002).
55. R. Caujolle, P. Castera, and A. Lattes, *Bull. Soc. Chim. Fr.*, *9–10*, 413 (1984).
56. M. J. Fernandez, J. M. Casares, E. Galvez, P. Gómez-Sal, R. Torres, and P. Ruiz, *J. Heterocyc. Chem.*, *29*, 1797 (1992).
57. M. Gdaniec, M. Pham, and T. Polonski, *J. Org. Chem.*, *62*, 5619 (1997).
58. S. Chiavarelli, F. Töffler, and D. Misiti, *Ann. Ist. Super. Sanità*, *4*, 157 (1968).
59. A. Gogoll, H. Grennberg, and A. Axen, *Organomet.*, *16*, 1167 (1997).
60. H. Quast, B. Müller, E.-M. Peters, K. Peters, and H. G. Von Schnering, *Chem. Ber.*, *115*, 3631 (1982).
61. H. Hennig and W. Pesch, *Arch. Pharm.*, *307*, 569 (1974).
62. C. Mannich and F. Veit, *Chem. Ber.*, *68*, 506 (1935).
63. R. Haller and H. Unholzer, *Arch. Pharm.*, *305*, 855 (1972).
64. P. Comba, J. H. Koek, A. Lienke, M. Merz, and L. Tsymbal, *PCT Int. Appl. WO 2003104234 (2003).*
65. P. Comba, C. Lopez de Laorden, and M. Zajaczkowski, to be published.
66. H. Börzel, P. Comba, K. S. Hagen, M. Merz, Y. D. Lampeka, A. Lienke, G. Linti, H. Pritzkow, and L. V. Tsymbal, *Inorg. Chim. Acta*, *337*, 407 (2002).
67. P. Comba and S. Wiesner, to be published.
68. A. Benz, P. Comba, M. Kerscher, B. Seibold, M. Zajaczkowski, and H. Wadepohl, to be published.
69. C. Bleiholder, H. Börzel, P. Comba, R. Ferrari, A. Heydt, M. Kerscher, S. Kuwata, G. Laurenczy, G. A. Lawrance, A. Lienke, B. Martin, M. Merz, B. Nuber, and H. Pritzkow, *Inorg. Chem.*, *44*, 8145 (2005).
70. H. Börzel, P. Comba, K. S. Hagen, C. Katsichtis, and H. Pritzkow, *Chem. Eur. J.*, *6*, 914 (2000).
71. P. Comba, C. Lopez de Laorden, and H. Pritzkow, *Helv. Chim. Acta*, *88*, 647 (2005).
72. T. Siener, A. Cambareri, U. Kuhl, W. Engelberger, M. Haurand, B. Kögel, and U. Holzgrabe, *J. Med. Chem.*, *43*, 3746 (2000).
73. P. Comba and M. Kerscher, to be published.
74. Z.-Y. Kyi and W. Wilson, *J. Org. Chem.*, *7*, 1706 (1951).
75. S. Chiavarelli and G. Settimj, *Gazz. Chim. Ital.*, *88*, 1246 (1958).
76. S. Z. Vatsadze, S. E. Sosonyuk, N. V. Zyk, K. A. Potekhin, O. I. Levina, Y. T. Stuchkov, and N. S. Zefirov, *Dokl. Akad. Nauk SSSR*, *341*, 201 (1995).
77. H. Stetter and K. Dieminger, *Chem. Ber.*, *92*, 2658 (1959).
78. S. Chiavarelli, G. P. Valsecchi, F. Toffler, and L. Gramiccioni, *Boll. Chim. Farm.*, *106*, 301 (1967).
79. F. Lions and A. M. Willison, *J. Proc. R. Soc. New South Wales*, *73*, 240 (1940).
80. P. Comba, M. Merz, H. Pritzkow, and S. Wiesner, to be published.
81. H. Börzel, P. Comba, C. Katsichtis, W. Kiefer, A. Lienke, V. Nagel, and H. Pritzkow, *Chem. Eur. J.*, *5*, 1716 (1999).
82. P. Comba, A. Hauser, M. Kerscher, and H. Pritzkow, *Angew. Chem. Int. Ed., Engl.*, *42*, 4536 (2003).
83. S. V. Lieberman and E. C. Wagner, *J. Org. Chem.*, *14*, 1001 (1949).
84. T. F. Cummings and R. J. Shelton, *25*, 419 (1960).
85. M. Tramontini and L. Angiolini, *Tetrahedron*, *46*, 1791 (1990).

86. M. Arend, B. Westerman, and N. Risch, *Angew. Chem. Int. Ed. Engl.*, *38*, 1044 (1998).
87. W. Hänsel and R. Haller, *Arch. Pharm.*, *303*, 334 (1970).
88. H. Stetter, J. Schäfer, and K. Dieminger, *Chem. Ber.*, *91*, 598 (1958).
89. H. Stetter, K. Dieminger, and E. Rauscher, *Chem. Ber.*, *92*, 2057 (1959).
90. A. I. Kuznetsov, E. B. Basargin, A. S. Moskovkin, M. K. Ba, I. V. Miroshnichenko, M. Y. Botnikov, and B. V. Unkovskii, *Chem. Hetero. Comp.*, *12*, 1679 (1985).
91. S. Chiavarelli and G. Settimj, *Gazz. Chim. Ital.*, *88*, 1234 (1958).
92. D. Misiti and S. Chiavarelli, *Gazz. Chim. Ital.*, *96*, 1696 (1966).
93. O. I. Levina, K. A. Potekhin, E. N. Kurkutova, Y. T. Struchkov, V. A. Palyulin, and N. S. Zefirov, *Cryst. Struct. Commun.*, *11*, 1909 (1982).
94. P. H. McCabe, N. J. Milne, and G. A. Sim, *J. Chem. Soc., Chem. Commun.*, 625 (1985).
95. S. Chiavarelli and G. Settimj, *Gazz. Chim. Ital.*, *88*, 1253 (1958).
96. T. E. Agadzhanyan and G. L. Arutyunyan, *Arm. Khim. Zh.*, *34*, 963 (1981).
97. T. E. Agadzhanyan and G. L. Arutyunyan, *Arm. Khim. Zh.*, *36*, 730 (1983).
98. G. G. Minasyan, T. E. Agadzhanyan, and G. G. Adamyan, *Chem. Hetero. Comp.*, 106 (1994).
99. P. Comba, H. Pritzkow, and W. Schiek, *Angew. Chem., Int. Ed. 40,* 2465 (2001).
100. J. S. Bradshaw, K. E. Krakowiak, and R. M. Izatt, *Aza-Crown Macrocycles, The Chemistry of Heterocyclic Compounds*, Vol. 51, John Wiley & Sons, Inc., New York, Chichester Brisbane, Toronto, Singapore, 1993.
101. D. S. C. Black, M. A. Horsham, and M. Rose, *Tetrahedron*, *51*, 4819 (1995).
102. H. Stetter and H. Hennig, *Chem. Ber.*, *88*, 789 (1955).
103. F. Galinovsky and H. Langer, *Mh. Chem.*, *86*, 449 (1955).
104. F. Bohlmann, N. Ottawa, and R. Keller, *Liebigs Ann. Chem.*, *587*, 162 (1954).
105. G. D. Hosken, C. C. Allan, J. C. A. Boeyens, and R. D. Hancock, *J. Chem. Soc., Dalton Trans.*, 3705 (1995).
106. N. S. Zefirov and S. V. Rogozina, *Tetrahedron*, *30*, 2345 (1974).
107. A. Gogoll, C. Johansson, A. Axen, and H. Grennberg, *Chem. Eur. J.*, *7*, 396 (2001).
108. J. E. Douglass and T. B. Ratliff, *J. Org. Chem.*, *33*, 355 (1968).
109. E. E. Smissman and P. C. Ruenitz, *J. Org. Chem.*, *41*, 1593 (1976).
110. S. A. Zisman, K. D. Berlin, and B. J. Scherlag, *Org. Prep. Proc. Int.*, *22*, 255 (1990).
111. V. A. Palyulin, S. V. Emets, K. A. Potekhin, A. E. Lysov, Y. G. Sumskaya, and N. S. Zefirov, *Dokl. Akad. Nauk SSSR*, *381*, 789 (2001).
112. D. M. Hodgson, T. J. Buxton, I. D. Cameron, E. Gras, and E. H. M. Kirton, *Org. Biomol. Chem.*, *1*, 4293 (2003).
113. O. Huttenloch, E. Laxman, and H. Waldmann, *Chem. Commun.*, 673 (2002).
114. G. Lesma, B. Danieli, D. Passarella, A. Sacchetti, and A. Silvani, *Tetrahedron: Asym.*, *14*, 2453 (2003).
115. A. Gogoll, H. Grennberg, and A. Axen, *Magn. Res. Chem.*, *35*, 13 (1997).
116. R. Haller and H. Unholzer, *Arch. Pharm.*, *304*, 654 (1971).
117. A. Samhammer, U. Holzgrabe, and R. Haller, *Arch. Pharm. (Weinheim)*, *322*, 545 (1989).
118. N. S. Zefirov and V. A. Palyulin, *Top. Stereochem.*, *20*, 171 (1991).
119. N. S. Zefirov, *Russ. Chem. Rev.*, *44*, 196 (1975).
120. V. S. Mastryukov, E. L. Osina, O. V. Dorofeeva, M. V. Popik, L. V. Vilkov, and N. A. Belikova, *J. Mol. Struct.*, *52*, 211 (1979).

121. V. S. Mastryukov, M. V. Popik, O. V. Dorofeeva, A. V. Golubinskii, L. V. Vilkov, N. A. Belikova, and N. L. Allinger, *Tetrahedron Lett.*, 4339 (1979).

122. V. S. Mastryukov, M. V. Popik, O. V. Dorofeeva, A. V. Golubinskii, L. V. Vilkov, N. A. Belikova, and N. L. Allinger, *J. Am. Chem. Soc.*, 1333 (1981).

123. E. L. Osina, V. S. Mastryukov, L. V. Vilkov, and N. A. Belikova, *J. Chem. Soc., Chem. Commun.*, 12 (1976).

124. G. A. Sim, *Tetrahedron*, *39*, 1181 (1983).

125. N. S. Zefirov, *Russ. J. Org. Chem*, *6*, 1768 (1970).

126. N. S. Zefirov and N. M. Shektman, *Russ. Chem. Rev.*, *40*, 315 (1971).

127. N. S. Zefirov, V. S. Blagoveshchenskiy, I. Kazimirchic, and N. S. Surova, *Tetrahedron*, *27*, 3111 (1971).

128. E. L. Eliel and S. A. Evans, *J. Am. Chem. Soc.*, *94*, 8587 (1972).

129. H. Küppers, A. Samhammer, and R. Haller, *Acta Cryst., Sect. C*, *43*, 1974 (1987).

130. O. I. Levina, K. A. Potekhin, E. N. Kurkutova, J. T. Struchkov, V. A. Palyulin, and N. S. Zefirov, *Dokl. Akad. Nauk SSSR*, *277*, 367 (1984).

131. M. Przybylska and W. H. Barnes, *Acta Cryst., Sect. B*, *6*, 377 (1953).

132. S. K. Bhattacharjee and K. K. Chako, *Tetrahedron*, *35*, 1999 (1979).

133. M. R. Chakrabarty, R. L. Ellis, and J. L. Roberts, *J. Org. Chem.*, *35*, 541 (1970).

134. P. Livant, K. A. Roberts, M. D. Eggers, and S. D. Worley, *Tetrahedron*, *37*, 1853 (1981).

135. P. Smith-Verdier, F. Florencio, and S. Garcia-Blanco, *Acta Cryst., Sect. C*, *39*, 101 (1983).

136. E. Galvez, M. S. Arias, J. Bellanato, J. V. Garcia-Ramos, F. Florencio, P. Smith-Verdier, and S. Garcia-Blanco, *J. Mol. Struct.*, *127*, 185 (1985).

137. T. Briekwicki and M. Wiewiorowski, *Bull. Pol. Acad. Sci. Chem.*, *34*, 205 (1986).

138. M. S. Arias, E. Galvez, J. C. Del Castillo, J. J. Vaquero, and J. Chicharro, *J. Mol. Struct.*, *156*, 239 (1987).

139. G. Settimj, M. R. Del Giudice, S. D'Angelo, and L. Di Simone, *Ann. Chim. (Rome)*, *64*, 281 (1974).

140. R. Gleiter, M. Kobayashi, and J. Kuthan, *Tetrahedron*, *32*, 2775 (1976).

141. A. N. Chekhlov, *J. Struct. Chem.*, *41*, 116 (2000).

142. M. J. Fernández, R. M. Huertas, E. Galvez, J. Server-Carrió, M. Martinez-Ripoll, and J. Bellanato, *J. Mol. Struct.*, *355*, 229 (1995).

143. M. J. Fernández, R. M. Huertas, E. Gálvez, A. Orjales, A. Berisa, L. Labeaga, A. G. Garcia, G. Uceda, J. Server-Carrió, and M. Martinez-Ripoll, *J. Mol. Struct.*, *372*, 203 (1995).

144. V. A. Palyulin, O. M. Grek, S. V. Emets, K. A. Potekhin, A. E. Lysov, and N. S. Zefirov, *Dokl. Akad. Nauk SSSR*, *370*, 53 (2000).

145. A. V. Goncharov, K. A. Potekhin, Y. T. Struchkov, A. M. Svetlanova, S. V. Chenadanova, V. A. Palyulin, and N. S. Zefirov, *Dokl. Akad. Nauk SSSR*, *323*, 285 (1992).

146. M. J. Fernandez, M. S. Toledano, E. Gálvez, E. Matesanz, and M. MartÀnez-Ripoll, *J. Heterocyc. Chem.*, *29*, 723 (1992).

147. U. Kuhl, A. Cambareri, C. Sauber, F. Sörgel, R. Hartmann, H. Euler, A. Kirfel, and U. Holzgrabe, *J. Chem. Soc., Perkin Trans. 2*, 2083 (1999).

148. W. Brandt, S. Drosihn, M. Haurand, U. Holzgrabe, and C. Nachtsheim, *Arch. Pharm. Pharm. Med. Chem.*, *329*, 311 (1996).

149. G. A. Sim, *Acta Cryst. Sect. B*, *35*, 2455 (1979).

150. O. I. Levina, K. A. Potekhin, V. G. Rau, J. T. Struchkov, V. A. Palyulin, and N. S. Zefirov, *Cryst. Struct. Commun*, *11*, 1073 (1982).

151. M. Z. Buranbaev, Y. T. Gladii, T. T. Omarov, A. S. Gubasheva, V. A. Palyulin, and N. S. Zefirov, *Zh. Strukt. Khim.*, *31*, 189 (1990).

152. S. Z. Vatsadze, D. P. Krut'ko, N. V. Zyk, N. S. Zefirov, A. V. Churakov, and J. A. Howard, *Mendeleev Commun.*, 103 (1999).

153. N. S. Zefirov, V. A. Palyulin, S. V. Starovoitova, K. A. Potekhin, and J. T. Struchkov, *Dokl. Akad. Nauk SSSR*, *347*, 637 (1996).

154. Y. Takeuchi, P. Scheiber, and K. Takada, *J. Chem. Soc., Chem. Commun*, 403 (1980).

155. O. I. Levina, K. A. Potekhin, E. N. Kurkutova, Y. T. Struchkov, V. A. Palyulin, I. I. Baskin, and N. S. Zefirov, *Dokl. Akad. Nauk SSSR*, *281*, 1367 (1985).

156. V. A. Palyulin, S. V. Emets, V. A. Chertkov, C. Kasper, and H.-J. Schneider, *Eur. J. Org. Chem.*, 3479 (1999).

157. K. A. Potekhin, O. I. Levina, Y. T. Struchkov, A. M. Svetlanova, R. S. Idrisova, V. A. Palyulin, and N. S. Zefirov, *Mendeleev Commun.*, 87 (1991).

158. N. S. Zefirov, V. A. Palyulin, G. A. Efimov, O. A. Subbotin, O. I. Levina, K. A. Potekhin, and Y. T. Struchkov, *Dokl. Akad. Nauk SSSR*, *320*, 1392 (1991).

159. P. C. Ruenitz and E. E. Smissman, *J. Org. Chem.*, *42*, 937 (1977).

160. R. Caujolle, A. Lattes, J. Jaud, and J. Galy, *Acta Cryst., Sect. B*, *37*, 1699 (1981).

161. T. Siener, U. Holzgrabe, S. Drosihn, and W. Brandt, *J. Chem. Soc., Perkin Trans. 2*, 1827 (1999).

162. W. Quast and B. Müller, *Chem. Ber.*, *113*, 2959 (1980).

163. L. M. Jackman, T. S. Dunne, B. Müller, and H. Quast, *Chem. Ber.*, *115*, 2872 (1982).

164. U. Holzgrabe and W. Brandt, *J. Med. Chem.*, *46*, 1383 (2003).

165. L. M. Jackman, T. S. Dunne, J. L. Roberts, B. Müller, and H. Quast, *J. Mol. Struct.*, *126*, 433 (1985).

166. P. Comba, B. Nuber, and A. Ramlow, *J. Chem. Soc., Dalton Trans.*, 347 (1997).

167. P. Comba, S. Kuwata, M. Tarnai, and H. Wadepohl, *J. Chem. Soc., Chem. Commun.*, 2074 (2006).

168. P. Comba, M. Merz, and H. Pritzkow, *Eur. J. Inorg. Chem.*, 1711 (2003).

169. P. Comba, M. Kerscher, and A. Roodt, *Eur. J. Inorg. Chem.*, *23*, 4640 (2004).

170. M. R. Bukowski, P. Comba, C. Limberg, M. Merz, L. Que, Jr., and T. Wistuba, *Angew. Chem. Int. Ed.*, *43*, 1283 (2004).

171. J. Bautz, P. Comba, and L. Que Jr., *Inorg. Chem.*, *45*, 7077 (2006).

172. J. Bautz, M. Bukowski, M. Kerscher, A. Stubna, P. Comba, A. Lienke, E. Münck, and L. Que, Jr, *Angew. Chem., Int. Ed. 45*, 5681 (2006).

173. P. Comba, G. Rajaraman, and H. Rohwer, *Inorg. Chem.*, (2007) in press.

174. A. E. Anastasi, A. Lienke, P. Comba, H. Rohwer, and J. E. McGrady, *Eur. J. Inorg. Chem.*, *65* (2007).

175. A. Anastasi, P. Comba, J. McGrady, A. Lienke, and H. Rohwer, submitted for publication.

176. C. T. Lin, D. B. Rorabacher, G. R. Cayley, and D. W. Margerum, *Inorg. Chem.*, *14*, 919 (1975).

177. J. A. Drumhiller, F. Montavon, J. M. Lehn, and R. W. Taylor, *Inorg. Chem.*, *25*, 3751 (1986).

178. L. L. J. Diaddario, L. A. Ochrymowycz, and D. B. Rorabacher, *Inorg. Chem.*, *31*, 2347 (1992).

179. D. Aronne, C. W. Ochrymowycz, and D. B. Rorabacher, *Inorg. Chem.*, *34*, 1844 (1995).

180. C. Csiki, K. M. Norenberg, C. M. Shoemaker, and M. Zimmer, *Molecular Modeling and Dynamics of Bioinorganic Systems*, L. Banci and P. Comba, Eds., Kluwer, Dordrecht, The Netherlands, 1997.
181. Y. Miyahara, K. Goto, and T. Inazu, *Tetrahedron Lett.*, *42*, 3097 (2001).
182. P. Comba, M. Kerscher, L. Kleditzsch, W. Schiek, and H. Wadepohl, to be published.
183. O. I. Levina, K. A. Potekḥin, E. N. Kurkutova, Y. T. Struchkov, O. N. Zefirova, V. A. Palyulin, and N. S. Zefirov, *Dokl. Akad. Nauk SSSR*, *289*, 876 (1986).
184. S. Z. Vatsadze, N. V. Zyk, A. V. Churakov, and L. G. Kuzmina, *Chem. Hetero. Comp.*, 1266 (2000).
185. V. A. Palyulin, S. V. Emets, K. A. Potekhin, A. E. Lysov, Y. G. Sumskaya, and N. S. Zefirov, *Dokl. Akad. Nauk SSSR*, *374*, 634 (2000).
186. S. V. Emets, N. I. Kurto, V. A. Palyulin, N. S. Zefirov, K. A. Potekhin, and A. E. Lysov, *Bull. Moscow Univ., Chem.*, *42*, 390 (2001).
187. S. Z. Vatsadze, S. E. Sosonyuk, N. V. Zyk, K. A. Potekhin, O. I. Levina, J. T. Struchkov, and N. S. Zefirov, *Khim. Getero. Soedin.*, 770 (1996).
188. S. Z. Vatsadze, V. K. Beĺsky, S. E. Sosonyuk, and N. S. Zefirov, *Chem. Hetero. Comp.*, 356 (1997).
189. P. Comba, M. Kerscher, M. Merz, V. Müller, H. Pritzkow, R. Remenyi, W. Schiek, and Y. Xiong, *Chem. Eur. J.*, *8*, 5750 (2002).
190. P. Comba, B. Kanellakopulos, C. Katsichtis, A. Lienke, H. Pritzkow, and F. Rominger, *J. Chem. Soc., Dalton Trans.*, 3997 (1998).
191. H. Börzel, P. Comba, K. S. Hagen, C. Katsichtis, and H. Pritzkow, to be published.
192. H. Börzel, P. Comba, and H. Pritzkow, *J. Chem. Soc., Chem. Commun.*, 97 (2001).
193. P. Comba, M. Tarnai, and H. Wadepohl, to be published.
194. P. Comba, B. Martin, A. Prikhod'ko, H. Pritzkow, and H. Rohwer, *C. R. Chim.*, *6*, 1506 (2005).
195. P. Comba, B. Seibold, and H. Wadepohl, to be published.
196. H. Börzel, P. Comba, K. S. Hagen, M. Kerscher, H. Pritzkow, M. Schatz, S. Schindler, and O. Walter, *Inorg. Chem.*, *41*, 5440 (2002).
197. A. Benz and P. Comba, to be published.
198. P. Comba, S. Kuwata, G. Linti, M. Tarnai, and H. Wadepohl, *Eur. J. Inorg. Chem.*, 657 (2007).
199. P. Comba and M. Kerscher, *Cryst. Eng.*, *6*, 197 (2004).
200. A. G. Orpen, L. Bramner, F. H. Allen, O. Kennard, D. G. Watson, and R. Taylor, *J. Chem. Soc., Dalton Trans.*, 1 (1989).
201. K. Born, P. Comba, R. Ferrari, S. Kuwata, G. A. Lawrance and H. Wadepohl, *Inorg. Chem. 46*, 458 (2007).
202. P. Comba, *Inorg. Chem.*, *33*, 4577 (1994).
203. P. Comba, T. W. Hambley, M. A. Hitchman, and H. Stratemeier, *Inorg. Chem.*, *34*, 3903 (1995).
204. P. Comba, A. Lienke, M. Merz, and M. Tarnai, to be published.
205. P. Comba, A. Prik'hodko, and H. Wadepohl, to be published.
206. P. Comba, N. Okon, and R. Remenyi, *J. Comput. Chem.*, *20*, 781 (1999).
207. P. Comba, T. W. Hambley, G. Lauer, and N. Okon, *MOMEC97, A Molecular Modeling Package for Inorganic Compounds*, Heidelberg, 1997, www.comba-group.uni-hd.de.
208. J. E. Bol, C. Buning, P. Comba, J. Reedijk, and M. Ströhle, *J. Comput. Chem.*, *19*, 512 (1998).
209. P. Comba, A. Daubinet, B. Martin, H. J. Pietzsch, and H. Stephan, *J. Organomet. Chem.*, *691*, 2495 (2006).

210. W.-D. Stohrer and R. Hoffmann, *J. Am. Chem. Soc.*, *74*, 779 (1972).
211. G. Parkin, *Chem. Rev.*, *93*, 887 (1993).
212. M. M. Rohmer and M. Bénard, *Chem. Soc. Rev.*, *30*, 340 (2001).
213. K. Born, P. Comba, A. Daubinet, A. Fuchs, and H. Wadepohl, *J. Biol. Inorg. Chem.*, **12**, 36 (2007).
214. P. Comba and A. Lienke, *Inorg. Chem.*, *40*, 5206 (2001).
215. M. Atanasov and P. Comba, to be published.
216. H. A. Jahn and E. Teller, *Proc. R. Soc.* (1937).
217. B. N. Figgis and M. A. Hitchman, *Ligand Field Theory and its Applications*, Wiely-VCH, Weinheim, New York, 2000.
218. I. B. Bersuker, *Chem. Rev.*, *101*, 1067 (2001).
219. M. A. Hitchman, Y. V. Yablokov, V. E. Petrashen, M. A. Augustyniak-Jablokov, H. Stratemeier, M. J. Riley, K. Lukaszewicz, P. E. Tomaszewski, and A. Pietraszko, *Inorg. Chem.*, *41*, 229 (2002).
220. T. Astley, P. J. Ellis, H. C. Freeman, M. A. Hitchman, F. R. Keene, and E. R. T. Tiekink, *J. Chem. Soc, Dalton Trans.*, 595 (1995).
221. C. J. Simmons, M. A. Hitchman, H. Stratemeier, and A. J. Schultz, *J Am Chem Soc*, *115*, 11304 (1993).
222. C. Simmons, *New J. Chem.*, *17*, 77 (1993).
223. M. Atanasov, C. Busche, P. Comba, and H. Wadepohl, to be published.
224. P. Comba and R. J. Deeth, to be published.
225. V. J. Burton, R. J. Deeth, C. M. Kemp, and P. J. Gilbert, *J. Am. Chem. Soc.*, *117*, 8407 (1995).
226. K. Born, P. Comba, M. Kerscher, and H. Rohwer, to be published.
227. E. A. Ambundo, M.-V. Deydier, A. J. Grall, N. Aguera-Vega, L. T. Dressel, T. H. Cooper, M. J. Heeg, L. A. Ochrymowycz, and D. B. Rorabacher, *Inorg. Chem.*, *38*, 4233 (1999).
228. D. Tanner, *Angew. Chem., Int. Ed. Engl.*, *35*, 599 (1994).
229. H. C. Kolb, M. G. Finn, and K. B. Sharpless, *Angew. Chem. Int. Ed. Engl.*, *40*, 2004 (2001).
230. S. Shizeki, M. Ohtsuka, K. Irinoda, K. Kukita, K. Nagaoka, and T. J. Nakashina, *Antibiot*, *40*, 60 (1987).
231. R. S. Coleman and J.-S. Kong, *J. Am. Chem. Soc.*, *120*, 3538 (1998).
232. D. A. Evans, M. M. Faul, M. T. Bilodeau, B. A. Anderson, and D. M. Barnes, *J. Am. Chem. Soc.*, *115*, 5328 (1993).
233. D. A. Evans, M. M. Faul, and M. T. Bilodeau, *J. Am. Chem. Soc.*, *116*, 2742 (1994).
234. D. A. Evans and J. A. Ellman, *J. Am. Chem. Soc.*, *11*, 1063 (1991).
235. D. A. Evans, M. M. Faul, and M. T. Bilodean, *J. Org. Chem.*, *56*, 6744 (1991).
236. E. N. Jacobsen, *Compr. Asymmetric Catal. I–III*, 1999.
237. Z. Li, R. W. Quan, and E. N. Jacobsen, *J. Am. Chem. Soc.*, *117*, 5889 (1995).
238. Z. Li, K. R. Conser, and E. N. Jacobsen, *J. Am. Chem. Soc.*, *115*, 5326 (1993).
239. T. P. Albone, P. S. Aujnla, P. C. Taylor, S. Challenger, and A. M. Derrick, *J. Org. Chem.*, *63*, 9569 (1998).
240. P. Dauban and R. H. Dodd, *J. Org. Chem.*, *64*, 5304 (1999).
241. A. V. Gontcharov, H. Liu, and K. B. Sharpless, *Org. Lett.*, *1*, 783 (1999).
242. J. A. Halfen, J. M. Uhan, D. C. Fox, M. P. Mehn, and L. Que, Jr., *Inorg. Chem.*, *39*, 4913 (2000).

243. J. A. Halfen, D. C. Fox, M. P. Mehn, and L. Que, *Inorg. Chem.*, *40*, 5060 (2001).
244. J. A. Halfen, J. K. Hallman, J. A. Schultz, and J. P. Emerson, *Organometallics*, *18*, 5435 (1999).
245. B. D. Brandes and E. N. Jacobsen, *Tetrahedron Lett.*, *36*, 5123 (1995).
246. M. M. Diaz-Requejo, P. J. Perez, M. Brookhart, and J. L. Templeton, *Organometallics*, *16*, 4399 (1997).
247. P. Brandt, M. J. Södergren, P. G. Andersson, and P.-O. Norrby, *J. Am. Chem. Soc.*, *122*, 8013 (2000).
248. K. M. Gillespie, E. J. Crust, R. J. Deeth, and P. Scott, *J. Chem. Soc., Chem. Commun.*, 785 (2001).
249. W. Kaim and B. Schwederski, *Bioanorgan. Chem.*, B.G. Teuber, Stuttgart, 1995.
250. S. J. Lippard and J. M. Berg, *Principles of Bioinorganic Chemistry*, University Science Books, Mill Valley, CA, 1994.
251. J. Reedijk, *Bioinorganic Catalysis*, Marcel Dekker Inc., New York, Basel, 1993.
252. J. Reedijk and E. Bouwman, *Bioinorganic Catalysis*, 2nd ed., Marcel Dekker, New York, 1999.
253. K. D. Karlin and Z. Tyeklar, *Bioinorganic Chemistry of Copper*, Chapman & Hall, London, New York, 1993.
254. K. D. Karlin, Z. Tyeklar, and A. D. Zuberbühler, in *Bioinorganic Catalysis*, J. Reedijk, Ed., Marcel Dekker Inc., New York, 1993.
255. K. D. Karlin and A. D. Zuberbühler, in *Bioinorganic Catalysis*, J. Reedijk and E. Bouwman, Eds., Marcel Dekker, New York, Basel, 1999.
256. K. D. Karlin and J. Zubieta, *Copper Coordination Chemistry: Biochemical & Inorganic Perspectives*, Adenine Press, Guilderland, New York, 1982.
257. S. Schindler, *Eur. J. Inorg. Chem.*, 2311 (2000).
258. W. B. Tolman, R. L. Rardin, and S. J. Lippard, *J. Am. Chem. Soc.*, *111*, 4532 (1989).
259. J. Klinmann, *Chem. Rev.*, *96*, 2541 (1996).
260. E. I. Solomon, U. M. Sundaram, and T. E. Machonkin, *Chem. Rev.*, *96*, 2563 (1996).
261. E. A. Lewis and W. B. Tolman, *Chem. Rev.*, *104*, 1047 (2004).
262. L. M. Mirica, X. Ottenwaelder, and T. D. P. Stack, *Chem. Rev.*, *104*, 1013 (2004).
263. E. I. Solomon, R. K. Szilagyi, S. D. George, and L. Basumallick, *Chem. Rev.*, *104*, 419 (2004).
264. K. Born, P. Comba, M. Kerscher, and B. Seibold, to be published.
265. C. Buning, G. W. Canters, P. Comba, C. Dennison, L. Jeuken, M. Melter, and J. Sanders-Loehr, *J. Am. Chem. Soc.*, *122*, 204 (2000).
266. N. Kitajima and Y. Moro-oka, *Chem. Rev.*, *94*, 737 (1994).
267. A. G. Blackman and W. B. Tolman, in *Structure and Bonding*, B. Meunier, Ed., Springer, Berlin, 2000.
268. M.-A. Kopf and K. D. Karlin, in *Biomimetic Oxidations Catalyzed by Transition Metal Complexes*, E. B. Meunier, Ed., ICP London, 2000.
269. Kitajima, *J. Am. Chem. Soc.*, *111*, 8975 (1989).
270. N. Kitajima, T. Koda, S. Hashimoto, T. Kitagawa, and Y. Moro-oka, *J. Am. Chem.Soc.*, *113*, 5664 (1991).
271. T. D. Westmoreland, D. E. Wilcox, M. J. Baldwin, W. B. Mims, and E. I. Solomon, *J. Am. Chem. Soc.*, *111*, 6106 (1989).
272. M. J. Henson, J. Mukherjee, D. E. Root, T. D. P. Stack, and E. I. Solomon, *J. Am. Chem. Soc.*, *121*, 10332 (1999).

273. K. A. Magnus, H. Ton-That, and J. E. Carpenter, *Chem. Rev.*, *94*, 727 (1994).

274. B. Hazes, K. A. Magnus, C. Bonaventura, J. Bonaventura, Z. Danter, K. H. Kalk, and W. G. J. Hol, *Protein Sci.*, *2*, 597 (1993).

275. A. Volbeda and W. G. J. Hol, *J. Mol. Biol.*, *209*, 249 (1989).

276. K. A. Magnus, B. Hazes, H. Ton-That, C. Bonaventura, L. Bonaventura, Z. Danter, K. H. Kalk, and W. G. J. Hol, *Proteins: Struct. Funct. Genef.*, *19*, 302 (1994).

277. T. Klabunde, C. Eicken, J. C. Saccettini, and B. Krebs, *Nat. Struct. Biol.*, *5*, 1084 (1998).

278. Y. Matoba, T. Dumagai, A. Yamamoto, H. Yoshitsu, and M. Sugiyama, *J. Biol. Chem.*, *281*, 8981 (2006).

279. P. Comba, P. Hilfenhaus, and K. D. Karlin, *Inorg. Chem.*, *36*, 2309 (1997).

280. R. R. Jacobson, Z. Tyeklar, A. Farooq, K. D. Karlin, S. Liu, and J. Zubieta, *J. Am. Chem. Soc.*, *110*, 3690 (1988).

281. M. J. Baldwin, P. K. Ross, J. E. Pate, Z. Tyeklar, K. D. Karlin, and E. I. Solomon, *J. Am. Chem. Soc.*, *113*, 8671 (1991).

282. D.-H. Lee, N. Wei, N. N. Murthy, Z. Tyeklár, K. D. Karlin, S. Kaderli, B. Jung, and A. D. Zuberbühler, *J. Am. Chem. Soc.*, *117*, 12498 (1995).

283. M. Schatz, V. Raab, S. Foxon, G. Brehm, S. Schneider, M. Reiher, M. C. Holthausen, J. Sundermeyer, and S. Schindler, *Angew. Chem., Int. Ed. Engl.*, *43(33)*, 4360 (2004).

284. C. Würtele, S. Schindler, E. Gaontchenova, K. Harms, J.Sundermeyer, and M. C. Holthausen, *Angew. Chem. Int. Ed. Engl.*, **45**, 3867 (2006).

285. H. Decker and T. Rimke, *J. Biol. Chem.*, *273*, 25889 (1998).

286. J. C. Espin, M. Morales, R. Varon, J. Tudela, and F. Garcia-Canovas, *Phytochem.*, *44*, 17 (1997).

287. J. C. Espin, M. Ocha, J. Tudela, and F. Garcia-Canovas, *Phytochem.*, *45*, 667 (1997).

288. P. E. M. Siegbahn, *J. Biol. Inorg. Chem.*, *9*, 577 (2004).

289. I. A. Koval, C. Belle, K. Selmeczi, C. Philouze, E. Saint-Aman, A. M. Schuitema, P. Gamez, J.-L. Pierre, and J. Reedijk, *J. Biol. Inorg. Chem.*, *10*, 739 (2005).

290. P. Comba, T. W. Hambley, P. Hilfenhaus, and D. T. Richens, *J. Chem. Soc., Dalton Trans.*, 533 (1996).

291. D. Utz, F. W. Heinemann, F. Hampel, D. T. Richens, and S. Schindler, *Inor. Chem.*, *42*, 1430 (2003).

292. D. Ma, M. Allmendinger, U. Thewalt, A. Lentz, M. Klinga, and B. Rieger, *Eur. J. Inorg. Chem.*, 2857 (2002).

293. J. A. Halfen, S. Mahapatra, E. C. Wilkinson, S. Kaderli, V. G. Young, L. Que Jr., A. D. Zuberbühler, and W. B. Tolman, *Science*, *271*, 1397 (1996).

294. K. D. Karlin, Y. Gultneh, T. Nicholson, and J. Zubieta, *Inorg. Chem.*, *24*, 3725 (1985).

295. B. Meunier, in *Biomimetic Oxidations Catalyzed by Transition Metal Complexes*, B. Meunier, Ed., Imperial College Press, London, 2000.

296. B. Meunier, in *Structure & Bonding*, E. B. Meunier, Ed., Springer, Berlin, 2000.

297. L. Que, Jr., and R. Y. N. Ho, *Chem. Rev.*, *96*, 2607 (1996).

298. E. I. Solomon, T. C. Brunold, M. I. Davis, J. N. Kensley, S.-K. Lee, N. Lehnert, F. Neese, A. J. Skulan, Y.-S. Yang, and J. Zhon, *Chem. Rev.*, *100*, 235 (2000).

299. M. Costas, M. P. Mehn, M. P. Jensen, and L. Que, Jr., *Chem. Rev.*, *104*, 939 (2004).

300. T. J. Collins, *Acc. Chem. Res.*, *35*, 782 (2002).

301. C. Limberg, *Angew. Chem.*, *115*, 6112 (2003).

302. R. Hage and A. Lienke, *Angew. Chem.*, *118*, 212 (2006).
303. K. Chen, M. Costas, and L. Que, Jr., *J. Chem. Soc., Dalton Trans.*, 672 (2002).
304. N. Lehnert, F. Neese, R. Y. N. Ho, L. Que, Jr., and E. I. Solomon, *J. Am. Chem. Soc.*, *124*, 10810 (2002).
305. J. C. Price, E. W. Barr, B. Tirupati, M. Bollinger, Jr., and C. Krebs, *Biochemistry*, *42*, 7497 (2003).
306. C. A. Grapperhaus, B. Mienert, E. Bill, T. Weyhermüller, and K. Wieghardt, *Inorg. Chem.*, *39*, 5306 (2000).
307. J.-U. Rohde, J.-H. In, M. H. Lim, W. W. Brennessel, M. R. Bukowski, A. Stubna, E. Münck, W. Nam, and L. Que, Jr., *Science*, *299*, 1037 (2003).
308. A. Decker, J. U. Rohde, L. Que, Jr., and E. I. Solomon, *J. Am. Chem. Soc.*, *126*, 5378 (2004).
309. A. Ghosh, E. Tangen, H. Ryeng, and P. R. Taylor, *Eur. J. Inorg. Chem.*, 4555 (2004).
310. D. Quinonero, K. Morokuma, D. G. Musaev, K. Morokuma, R. Mas-Balleste, and J. Que, L., *J. Am. Chem. Soc.*, *126*, 6548 (2005).
311. S. Shaik, S. P. deVisser, F. Ogliaro, H. Schwarz, and D. Schröder, *Curr. Opinion Chem. Biol.*, *6*, 556 (2002).
312. B. Meunier, S. P. de Visser, and S. Shaik, *Chem. Rev.*, *104*, 3947 (2004).
313. A. Decker and E. I. Solomon, *Angew. Chem.*, *117*, 2292 (2005).
314. M. R. Bukowski, P. Comba, A. Lienke, C. Limberg, C. Lopez de Laorden, R. Mas-Balleste, M. Merz, and L. Que, Jr., *Angew. Chem.*, *118*, 3524 (2006).
315. J. Bautz, P. Comba, C. Lopez de Laorden, M. Menzel, and G. Rajaraman, *Angew. Chem.*, submitted for publication.
316. P. Comba and G. Rajaraman, *Inorg. Chem.*, submitted for publication.
317. H. Börzel, P. Comba, R. Hage, M. Kerscher, A. Lienke, and M. Merz, PCT Inst. Appl. WO 2002048301 (2002).
318. R. Hage, J. Lienke, P. Veeran. Petersen. PCT Int. Appl. WO 2005042532 (2005).
319. R. Hage, J. Lienke, P. Veeran. Petersen. PCT Int. Appl. WO 2005049778 (2005).
320. H. J. H. Fenton, *J. Chem. Soc.*, *65*, 899 (1894).
321. W. C. Bray and M. H. Gorin, *J. Am. Chem. Soc.*, *554*, 2124 (1932).
322. F. Haber and J. Weiss, *Proc. R. Soc. (London)*, *147*, 332 (1934).
323. F. Buda, B. Ensing, M. C. M. Gribnau, and E. J. Baerends, *Chem. Eur. J.*, *7*, 2775 (2001).
324. N. Lehnert, R. Y. N. Ho, L. Que, Jr., and E. I. Solomon, *J. Am. Chem. Soc.*, *123*, 12802 (2001).
325. D. E. van Sickle, F. R. Mayo, and R. M. Arluck, *J. Am. Chem. Soc.*, *87*, 4824 (1965).
326. C. J. Cramer, W. B. Tolman, K. H. Theopold, and A. L. Rheingold, *Proc. Natl. Acad. Sci. USA*, *100*, 3635 (2003).
327. J. D. Dunitz, E. F. Maverick, and K. N. Trueblood, *Angew. Chem.*, *100*, 910 (1988).
328. S. Fallab and P. R. Mitchell, *Adv. Inorg. Bioinorg. Mech.*, *3*, 311 (1984).
329. S. A. Mirza, B. Bocquet, C. Robyr, S. Thani, and A. F. Williams, *Inorg. Chem.*, *35*, 1332 (1996).
330. B. Bosnich, W. G. Jackson, S. T. D. Lo, and J. W. McLaren, *Inorg. Chem.*, *13*, 2605 (1974).
331. W. L. I. Johnson and J. W. Geldard, *Inorg. Chem.*, *17*, 1675 (1978).
332. B. Bosnich, H. Boucher, and C. Marshall, *Inor. Chem.*, *15*, 634 (1976).
333. S. Hikichi, M. Akita, and Y. Moro-oka, *Coord. Chem. Rev.*, *198*, 61 (2000).
334. A. F. M. Mokhlesur Rahman, W. G. Jackson, and A. C. Willig, *Inorg. Chem.*, *43*, 7558 (2004).

335. M. Shibata, *Modern Syntheses of Cobalt(III) Complexes*, Vol. 110, Springer, Berlin Heidelberg New York, 1983.

336. M. Sono, M. P. Roach, E. D. Coulter, and J. H. Dawson, *Chem. Rev.*, *96*, 2841 (1996).

337. J. T. Groves and D. V. Adhyam, *J.Am. Chem. Soc.*, *106*, 2177 (1984).

338. N. Deno and R. E. Fruit, *J. Am. Chem. Soc.*, *90*, 3502 (1968).

339. K. Yoshizuka, K. Inoue, K. Ohto, K. Gloe, H. Stephan, T. Rambusch, and P. Comba, Solvent Extraction for the 21st Century, Proceedings of ISEC 'gg, Barcelona 1999, 2001, p. 687.

340. P. Comba, S. Juran, and H. Stephan, *work in progress.*

341. P. Comba, Y. D. Lampeka, A. Y. Nazarenko, A. I. Prikhod'ko, and H. Pritzkow, *Eur. J. Inorg. Chem.*, 1464 (2002).

342. P. Comba, Y. D. Lampeka, L. Lötzbeyer, and A. I. Prikhod'ko, *Eur. J. Inorg. Chem.*, 34 (2003).

343. M. Atanasov, P. Comba, Y. D. Lampeka, G. Linti, T. Malcherek, R. Miletich, A. I. Prikhod'ko, and H. Pritzkow, *Chem. Eur. J.*, *12*, 737 (2006).

344. P. Comba, Y. D. Lampeka, A. Prik'hodko, and G. Rajaraman, *Inorg. Chem.*, *45*, 3632 (2006).

345. M. Atanasov, P. Comba, S. Förster, G. Linti, T. Malcherek, R. Miletich, A. Prikhod'ko, and H. Wadepohl, *Inorg. Chem.*, **45**, 7722 (2006).

346. M. Atanasov, P. Comba, and C. Daul, *J. Phys. Chem. (A)*, **110**, 13332 (2007).

347. M. Atanasov, C. Busche, P. Comba, B. Martin, and H. Wadepohl, to be published.

348. G. A. Lawrance, *work in progress.*

349. M. Atanasov, P. Comba, B. Martin, V. Müller, G. Rajaraman, H. Rohwer, and S. Wunderlich, *J. Comp. Chem.*, *27*, 1263 (2006).

# Subject Index



*Progress in Inorganic Chemistry, Vol. 55* Edited by Kenneth D. Karlin

# Cumulative Index, Volumes 1–55



*Progress in Inorganic Chemistry, Vol. 55* Edited by Kenneth D. Karlin